Laser Spectroscopy 2

Wolfgang Demtröder

Laser Spectroscopy 2

Experimental Techniques

5th Edition

Prof. Dr. Wolfgang Demtröder
University of Kaiserslautern
Kaiserslautern, Germany

ISBN 978-3-662-51728-4 ISBN 978-3-662-44641-6 (eBook)
DOI 10.1007/978-3-662-44641-6
Springer Heidelberg New York Dordrecht London

Printed on acid-free paper

Springer is part of Springer Science+Business Media (www.springer.com)

Preface to the Fifth Edition

The second volume of "Laser Spectroscopy" covers the different experimental techniques, necessary for the sensitive detection of small concentrations of atoms or molecules, for Doppler-free spectroscopy, laser-Raman-spectroscopy, double-resonance techniques, multi-photon spectroscopy, coherent spectroscopy and time-resolved spectroscopy. In these fields the progress of the development of new techniques and improved experimental equipment is remarkable. Many new ideas have enabled spectroscopists to tackle problems which could not be solved before. Examples are the direct measurements of absolute frequencies and phases of optical waves with frequency combs, or time resolution within the attosecond range based on higher harmonics of visible femtosecond lasers. The development of femtosecond non-collinear optical parametric amplifiers (NOPA) has considerably improved time-resolved measurements of fast dynamical processes in excited molecules and has been essential for detailed investigations of important processes, such as the visual process in the retina of the eye or the photosynthesis in chlorophyl molecules.

In particular, the applications of laser spectroscopy in chemistry, biology medicine and for the solution of technical problems have made rapid progress. This is illustrated by several examples in the last chapter.

In this new edition some recent developments are discussed, as for instance the application of optical combs to precision molecular spectroscopy and its extension into the far UV region, to metrology, to astronomy and as frequency normal for the global positioning system. The progress in the controllable generation of high harmonics with frequencies up into the X-ray region and with a time resolution in the attosecond time scale are outlined and also new techniques of cooling and trapping of atoms or molecules in various realizations of optical traps are presented.

For several sections of Vol. 2 some basic knowledge of spectroscopic techniques or instrumentation is necessary. Therefore, cross links to Vol. 1 are given, where the fundamentals of laser spectroscopy are discussed.

At the end of each chapter some problems are given, which should help the students to check their understanding of the subject treated in the corresponding chapter. The solutions are given at the end of the book.

The author would like to thank Dr. Th. Schneider from Springer-Verlag for his patience and encouragement and Mrs. St. Hohensee from le-tex, who has taken care of the layout and printing. The author is grateful to many people who have contributed to the improvement of this new edition by pointing to errors and possible revisions of the former edition. He thanks many colleagues for their permission to use figures from their work.

He will appreciate any future cooperation of readers for improving this textbook. Any mail with questions or suggestions for corrections will be answered as soon as possible.

Kaiserslautern, Germany
November 2014

Wolfgang Demtröder

Preface to the Fourth Edition

Nearly 50 years after the realization of the first laser in 1960, laser spectroscopy is still a very intense field of research which has expanded with remarkable progress into many areas of science, medicine and technology, and has provided an ever-increasing number of applications. The importance of laser spectroscopy and its appreciation by many people is, for instance, proved by the fact that over the last ten years three Nobel Prizes have been awarded to nine scientists in the field of laser spectroscopy and quantum optics.

This positive development is partly based on new experimental techniques, such as improvements of existing lasers and the invention of new laser types, the realization of optical parametric oscillators and amplifiers in the femtosecond range, the generation of attosecond pulses, the revolution in the measurements of absolute optical frequencies and phases of optical waves using the optical frequency comb, or the different methods developed for the generation of Bose–Einstein condensates of atoms and molecules and the demonstration of atom lasers as a particle equivalent to photon lasers.

These technical developments have stimulated numerous applications in chemistry, biology, medicine, atmospheric research, materials science, metrology, optical communication networks, and many other industrial areas.

In order to cover at least some of these new developments, a single volume would need too many pages. Therefore the author has decided to split the book into two parts. The first part contains the foundations of laser spectroscopy, i.e., the basic physics of spectroscopy, optical instruments and techniques. It furthermore provides a short introduction to the physics of lasers, and discusses the role of optical resonators and techniques for realizing tunable narrowband lasers, the working horses of laser spectroscopy. It gives a survey on the different types of tunable lasers and represents essentially the updated and enlarged edition of the first six chapters of the third edition. In order to improve its value as a textbook for students, the number of problems has been increased and their solutions are given at the end of Vol. 1. The second volume discusses the different techniques of laser spectroscopy. Compared to the third edition, it adds many new developments and tries to bring the reader up to speed on the present state of laser spectroscopy.

The author wishes to thank all of the people who have contributed to this new edition. There is Dr. Th. Schneider at Springer-Verlag, who has always supported the author and has shown patience when deadlines were not kept. Claudia Rau from LE-TeX has taken care of the layout, and many colleagues have given their permission to use figures from their research. Several readers have sent me their comments on errors or possible improvements. I thank them very much.

The author hopes that this new edition will find a similar friendly approval to the former editions and that it will enhance interest in the fascinating field of laser spectroscopy. He would appreciate any suggestions for improvement or hints about possible errors, and he will try to answer every question as soon as possible.

Kaiserslautern, Germany
February 2008

Wolfgang Demtröder

Preface to the Third Edition

Laser Spectroscopy continues to develop and expand rapidly. Many new ideas and recent realizations of new techniques based on old ideas have contributed to the progress in this field since the last edition of this textbook appeared. In order to keep up with these developments it was therefore necessary to include at least some of these new techniques in the third edition.

There are, firstly, the improvement of frequency-doubling techniques in external cavities, the realization of more reliable cw-parametric oscillators with large output power, and the development of tunable narrow-band UV sources, which have expanded the possible applications of coherent light sources in molecular spectroscopy. Furthermore, new sensitive detection techniques for the analysis of small molecular concentrations or for the measurement of weak transitions, such as overtone transitions in molecules, could be realized. Examples are Cavity Ringdown Spectroscopy, which allows the measurement of absolute absorption coefficients with great sensitivity or specific modulation techniques that push the minimum detectable absorption coefficient down to 10^{-14} cm^{-1}!

The most impressive progress has been achieved in the development of tunable femtosecond and subfemtosecond lasers, which can be amplified to achieve sufficiently high output powers for the generation of high harmonics with wavelengths down into the X-ray region and with pulsewidths in the attosecond range. Controlled pulse shaping by liquid crystal arrays allows coherent control of atomic and molecular excitations and in some favorable cases chemical reactions can already be influenced and controlled using these shaped pulses.

In the field of metrology a big step forward was the use of frequency combs from cw mode-locked femtosecond lasers. It is now possible to directly compare the microwave frequency of the cesium clock with optical frequencies, and it turns out that the stability and the absolute accuracy of frequency measurements in the optical range using frequency-stabilized lasers greatly surpasses that of the cesium clock. Such frequency combs also allow the synchronization of two independent femtosecond lasers.

The increasing research on laser cooling of atoms and molecules and many experiments with Bose–Einstein condensates have brought about some remarkable results and have considerably increased our knowledge about the interaction of light with matter on a microscopic scale and the interatomic interactions at very low temperatures. Also the realization of coherent matter waves (atom lasers) and investigations of interference effects between matter waves have proved fundamental aspects of quantum mechanics.

The largest expansion of laser spectroscopy can be seen in its possible and already realized applications to chemical and biological problems and its use in medicine as a diagnostic tool and for therapy. Also, for the solution of technical problems, such as surface inspections, purity checks of samples or the analysis of the chemical composition of samples, laser spectroscopy has offered new techniques.

In spite of these many new developments the representation of established fundamental aspects of laser spectroscopy and the explanation of the basic techniques are not changed in this new edition. The new developments mentioned above and also new references have been added. This, unfortunately, increases the number of pages. Since this textbook addresses beginners in this field as well as researchers who are familiar with special aspects of laser spectroscopy but want to have an overview on the whole field, the author did not want to change the concept of the textbook.

Many readers have contributed to the elimination of errors in the former edition or have made suggestions for improvements. I want to thank all of them. The author would be grateful if he receives such suggestions also for this new edition.

Many thanks go to all colleagues who gave their permission to use figures and results from their research. I thank Dr. H. Becker and T. Wilbourn for critical reading of the manuscript, Dr. H.J. Koelsch and C.-D. Bachem of Springer-Verlag for their valuable assistance during the editing process, and LE-TeX Jelonek, Schmidt and Vöckler for the setting and layout. I appreciate, that Dr. H. Lotsch, who has taken care for the foregoing editions, has supplied his computer files for this new edition. Last, but not least, I would like to thank my wife Harriet who made many efforts in order to give me the necessary time for writing this new edition.

Kaiserslautern, Germany Wolfgang Demtröder
April 2002

Preface to the Second Edition

During the past 14 years since the first edition of this book was published, the field of laser spectroscopy has shown a remarkable expansion. Many new spectroscopic techniques have been developed. The time resolution has reached the femtosecond scale and the frequency stability of lasers is now in the millihertz range.

In particular, the various applications of laser spectroscopy in physics, chemistry, biology, and medicine, and its contributions to the solutions of technical and environmental problems are remarkable. Therefore, a new edition of the book seemed necessary to account for at least part of these novel developments. Although it adheres to the concept of the first edition, several new spectroscopic techniques such as optothermal spectroscopy or velocity-modulation spectroscopy are added.

A whole chapter is devoted to time-resolved spectroscopy including the generation and detection of ultrashort light pulses. The principles of coherent spectroscopy, which have found widespread applications, are covered in a separate chapter. The combination of laser spectroscopy and collision physics, which has given new impetus to the study and control of chemical reactions, has deserved an extra chapter. In addition, more space has been given to optical cooling and trapping of atoms and ions.

I hope that the new edition will find a similar friendly acceptance as the first one. Of course, a texbook never is perfect but can always be improved. I, therefore, appreciate any hint to possible errors or comments concerning corrections and improvements. I will be happy if this book helps to support teaching courses on laser spectroscopy and to transfer some of the delight I have experienced during my research in this fascinating field over the last 30 years.

Many people have helped to complete this new edition. I am grateful to colleagues and friends, who have supplied figures and reprints of their work. I thank the graduate students in my group, who provided many of the examples used to illustrate the different techniques. Mrs. Wollscheid who has drawn many figures, and Mrs. Heider who typed part of the corrections. Particular thanks go to Helmut Lotsch of Springer-Verlag, who worked very hard for this book and who showed much patience with me when I often did not keep the deadlines.

Last but not least, I thank my wife Harriet who had much understanding for the many weekends lost for the family and who helped me to have sufficient time to write this extensive book.

Kaiserslautern, Germany
June 1995

Wolfgang Demtröder

Preface to the First Edition

The impact of lasers on spectroscopy can hardly be overestimated. Lasers represent intense light sources with spectral energy densities which may exceed those of incoherent sources by several orders of magnitude. Furthermore, because of their extremely small bandwidth, single-mode lasers allow a spectral resolution which far exceeds that of conventional spectrometers. Many experiments which could not be done before the application of lasers, because of lack of intensity or insufficient resolution, are readily performed with lasers.

Now several thousands of laser lines are known which span the whole spectral range from the vacuum-ultraviolet to the far-infrared region. Of particular interst are the continuously tunable lasers which may in many cases replace wavelength-selecting elements, such as spectrometers or interferometers. In combination with optical frequency-mixing techniques such continuously tunable monochromatic coherent light sources are available at nearly any desired wavelength above 100 nm.

The high intensity and spectral monochromasy of lasers have opened a new class of spectroscopic techniques which allow investigation of the structure of atoms and molecules in much more detail. Stimulated by the variety of new experimental possibilities that lasers give to spectroscopists, very lively research activities have developed in this field, as manifested by an avalanche of publications. A good survey about recent progress in laser spectroscopy is given by the proceedings of various conferences on laser spectroscopy (see "Springer Series in Optical Sciences"), on picosecond phenomena (see "Springer Series in Chemical Physics"), and by several quasi-mongraphs on laser spectroscopy published in "Topics in Applied Physics."

For nonspecialists, however, or for people who are just starting in this field, it is often difficult to find from the many articles scattered over many journals a coherent representation of the basic principles of laser spectroscopy. This textbook intends to close this gap between the advanced research papers and the representation of fundamental principles and experimental techniques. It is addressed to physicists and chemists who want to study laser spectroscopy in more detail. Students who have some knowledge of atomic and molecular physics, electrodynamics, and optics should be able to follow the presentation.

The fundamental principles of lasers are covered only very briefly because many excellent textbooks on lasers already exist.

On the other hand, those characteristics of the laser that are important for its applications in spectroscopy are treated in more detail. Examples are the frequency spectrum of different types of lasers, their linewidths, amplitude and frequency stability, tunability, and tuning ranges. The optical components such as mirrors, prisms, and gratings, and the experimental equipment of spectroscopy, for example, monochromators, interferometers, photon detectors, etc., are discussed extensively because detailed knowledge of modern spectroscopic equipment may be crucial for the successful performance of an experiment.

Each chapter gives several examples to illustrate the subject discussed. Problems at the end of each chapter may serve as a test of the reader's understanding. The literature cited for each chapter is, of course, not complete but should inspire further studies. Many subjects that could be covered only briefly in this book can be found in the references in a more detailed and often more advanced treatment. The literature selection does not represent any priority list but has didactical purposes and is intended to illustrate the subject of each chapter more thoroughly.

The spectroscopic applications of lasers covered in this book are restricted to the spectroscopy of free atoms, molecules, or ions. There exists, of course, a wide range of applications in plasma physics, solid-state physics, or fluid dynamics which are not discussed because they are beyond the scope of this book. It is hoped that this book may be of help to students and researchers. Although it is meant as an introduction to laser spectroscopy, it may also facilitate the understanding of advanced papers on special subjects in laser spectroscopy. Since laser spectroscopy is a very fascinating field of research, I would be happy if this book can transfer to the reader some of my excitement and pleasure experienced in the laboratory while looking for new lines or unexpected results.

I want to thank many people who have helped to complete this book. In particular the students in my research group who by their experimental work have contributed to many of the examples given for illustration and who have spent their time reading the galley proofs. I am grateful to colleages from many laboratories who have supplied me with figures from their publications. Special thanks go to Mrs. Keck and Mrs. Ofiiara who typed the manuscript and to Mrs. Wollscheid and Mrs. Ullmer who made the drawings. Last but not least, I would like to thank Dr. U. Hebgen, Dr. H. Lotsch, Mr. K.-H. Winter, and other coworkers of Springer-Verlag who showed much patience with a dilatory author and who tried hard to complete the book in a short time.

Kaiserslautern, Germany
March 1981

Wolfgang Demtröder

Contents

Chapter 1
Doppler-Limited Absorption and Fluorescence Spectroscopy with Lasers

In Vol. 1, Chap. 5 we presented the different realizations of tunable lasers; we now discuss their applications in absorption and fluorescence spectroscopy. First we discuss those methods where the spectral resolution is limited by the Doppler width of the molecular absorption lines. This limit can in fact be reached if the laser linewidth is small compared with the Doppler width. In several examples, such as optical pumping or laser-induced fluorescence spectroscopy, multimode lasers may be employed, although in most cases single-mode lasers are superior. In general, however, these lasers may not necessarily be frequency stabilized as long as the frequency jitter is small compared with the absorption linewidth. We compare several detection techniques of molecular absorption spectroscopy with regard to their sensitivity and their feasibility in the different spectral regions. Some examples illustrate the methods to give the reader a feeling of what has been achieved. After the discussion of Doppler-limited spectroscopy, Chaps. 2–5 give an extensive treatment of various techniques which allow sub-Doppler spectroscopy.

1.1 Advantages of Lasers in Spectroscopy

In order to illustrate the advantages of absorption spectroscopy with tunable lasers, we first compare it with conventional absorption spectroscopy, which uses incoherent radiation sources. Figure 1.1 presents schematic diagrams for both methods.

In classical absorption spectroscopy, radiation sources with a *broad emission continuum* are preferred (e.g., high-pressure Hg arcs, Xe flash lamps, etc.). The radiation is collimated by the lens L_1 and passes through the absorption cell. Behind a dispersing instrument for wavelength selection (spectrometer or interferometer), the power $P_T(\lambda)$ of the transmitted light is measured as a function of the wavelength λ (Fig. 1.1a). The power transmitted through an absorption cell with length L and absorption coefficient $\alpha(\lambda)$ is for small absorption $\alpha L \ll 1$

$$P_T = P_0 \cdot e^{-\alpha L} \approx P_0(1 - \alpha L). \tag{1.1a}$$

W. Demtröder, *Laser Spectroscopy 2*, DOI 10.1007/978-3-662-44641-6_1

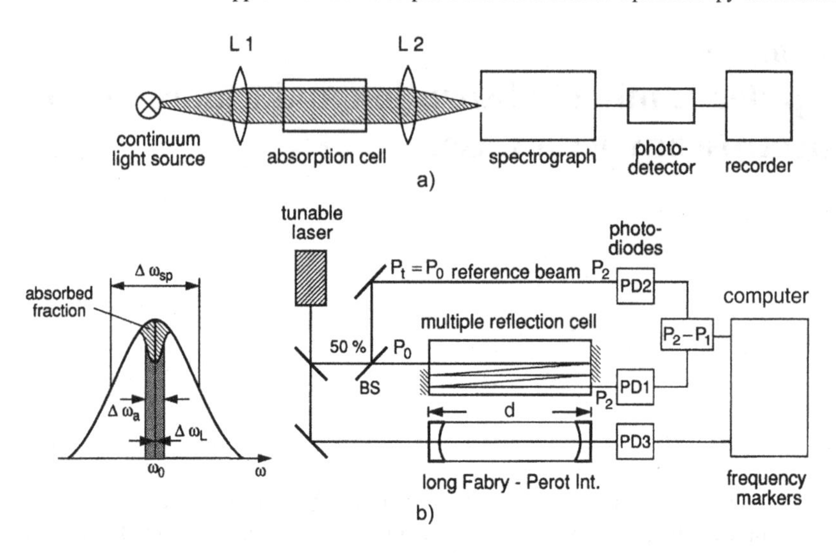

Fig. 1.1 Comparison between absorption spectroscopy with a broadband incoherent source (**a**) and with a tunable single-mode laser (**b**)

The difference

$$\Delta P = P_0 - P_T = P_0 \alpha L \tag{1.1b}$$

is proportional to the product of incident radiation power P_0 and absorption $\alpha \cdot L$.

It can be measured by comparison with a reference beam with power $P_R = P_0$, which can be realized, for instance, by shifting the absorption cell alternatively out of the light beam. This gives the absorption spectrum

$$P_A(\lambda) = a\big[P_0(\lambda) - P_T(\lambda)\big] = a\big[bP_R(\lambda) - P_T(\lambda)\big], \tag{1.2a}$$

where the constants a and b take into account wavelength-independent losses of P_R and P_T (e.g., reflections of the cell walls). Often the frequency $\omega = 2\pi\nu = 2\pi c/\lambda$ is used instead of the wavelength λ. This converts Eq. (1.2a) to

$$P_A(\omega)\,\mathrm{d}\omega = a\big[bP_R(\omega) - P_T(\omega)\big]\,\mathrm{d}\omega \tag{1.2b}$$

where $\mathrm{d}\omega$ is the spectral interval which is resolved by the spectrograph.

The *spectral resolution* is generally limited by the resolving power of the dispersing spectrometer. Only with large and expensive instruments (e.g., Fourier spectrometers) may the Doppler limit be reached [1].

The *detection sensitivity* of the experimental arrangement is defined by the minimum absorbed power that can still be detected. In most cases it is limited by the detector noise and by intensity fluctuations of the radiation source. Generally, the limit of the detectable absorption is reached at relative absorptions $\Delta P/P \geq 10^{-4}$–10^{-5}. This limit can be pushed further down only in favorable cases by using special sources and lock-in detection or signal-averaging techniques.

The absorption coefficient

$$\alpha(\lambda) = N_i \cdot \sigma_{ik}(\lambda) \tag{1.3}$$

is the product of the number density N_i [cm^{-3}] of absorbing atoms in level $|i\rangle$ and the wavelength-dependent absorption cross section $\sigma_{ik}(\lambda)$ of the absorbing transition $|i\rangle \to |k\rangle$.

The measured signal is

$$S = a\Delta P = a \cdot P_0 \cdot N_i \cdot \sigma_{ik} \cdot L. \tag{1.4a}$$

This signal should be larger than the noise equivalent power NEP (this is the input power of the detector which gives the same detector output as the noise).

This sets the sensitivity limit to $S > S_{\min} = \mathrm{NEP}$. Inserting into (1.1b) gives for the minimum measurable absorption coefficient

$$\alpha_{\min} = \mathrm{NEP}/(a P_0 L) \tag{1.4b}$$

and for the minimum detectable atomic density

$$N_i \geq \frac{\mathrm{NEP}}{a P_0 \sigma_{ik} L}. \tag{1.4c}$$

The minimum still detectable concentration N_i of absorbing molecules is determined by the noise-equivalent power NEP, the radiation input power P_0, the absorption cross section σ_{ik} of the absorbing transition and the length L of the absorption region.

In order to reach a high sensitivity one has to minimize the noise and to maximize the radiation input power P_0 and the absorption path length L. The absorption cross section σ_{ik} is a given quantity for each atomic or molecular species. If possible, one should select a transition with a large absorption cross section.

Example 1.1 For $P_0 = 10$ mW, $L = 1$ m, NEP $= 1$ μW, $a = 0.5 \to \alpha_{\min} = 2 \times 10^{-6}\ \mathrm{cm}^{-1}$ and with $\sigma_{ik} = 10^{-13}\ \mathrm{cm}^2 \to N_i \geq 10^7\ \mathrm{cm}^{-3}$. This corresponds to a gas pressure at room temperature of $p = 4 \times 10^{-8}$ Pa $= 4 \times 10^{-10}$ mb.

Contrary to radiation sources with broad emission continua used in conventional spectroscopy, tunable lasers offer radiation sources in the spectral range from the UV to the IR with extremely narrow bandwidths and with spectral power densities that may exceed those of incoherent light sources by many orders of magnitude (Vol. 1, Sects. 5.7, 5.8).

In several regards laser absorption spectroscopy corresponds to microwave spectroscopy, where klystrons or carcinotrons instead of lasers represent tunable coherent radiation sources. Laser spectroscopy transfers many of the techniques and advantages of microwave spectroscopy to the infrared, visible, and ultraviolet spectral ranges.

The advantages of absorption spectroscopy with tunable lasers may be summarized as follows:

- No monochromator is needed, since the absorption coefficient $\alpha(\omega)$ and its frequency dependence can be directly measured from the difference $\Delta P(\omega) = a[P_R(\omega) - P_T(\omega)]$ between the intensities of the reference beam with $P_R = P_2$ and transmitted beam with $P_T = P_1$ (Fig. 1.1b). The spectral resolution is higher than in conventional spectroscopy. With tunable single-mode lasers it is only limited by the linewidths of the absorbing molecular transitions. Using Doppler-free techniques (Chaps. 2–5), even sub-Doppler resolution can be achieved.
- The detection sensitivity increases with increasing spectral resolution $\omega/\Delta\omega$ as long as $\Delta\omega$ is still larger than the linewidth $\delta\omega$ of the absorption line. This can be seen as follows: The relative power attenuation per absorption path length Δx on the transition with center frequency ω_0 is, for small absorption $\alpha \cdot \Delta x \ll 1$,

$$\Delta P/P = \Delta x \cdot \int_{\omega_0 - \frac{1}{2}\Delta\omega}^{\omega_0 + \frac{1}{2}\Delta\omega} \alpha(\omega) P(\omega)\, \mathrm{d}\omega \Big/ \int_{\omega_0 - \frac{1}{2}\delta\omega}^{\omega_0 + \frac{1}{2}\delta\omega} P(\omega)\, \mathrm{d}\omega. \tag{1.5}$$

If $P(\omega)$ does not change much within the interval $\Delta\omega$, we can write

$$\int_{\omega_0 - \frac{1}{2}\Delta\omega}^{\omega_0 + \frac{1}{2}\Delta\omega} P(\omega)\, \mathrm{d}\omega = \bar{P}\Delta\omega \quad \text{and} \quad \int \alpha(\omega) P(\omega)\, \mathrm{d}\omega = \bar{P} \int \alpha(\omega)\, \mathrm{d}\omega.$$

This yields

$$\Delta P/P = \frac{\Delta x}{\Delta\omega} \int_{\omega_0 - \frac{1}{2}\delta\omega}^{\omega_0 + \frac{1}{2}\delta\omega} \alpha(\omega)\, \mathrm{d}\omega \simeq \bar{\alpha} \cdot \Delta x \frac{\delta\omega}{\Delta\omega} \quad \text{for } \Delta\omega > \delta\omega. \tag{1.6}$$

The measured relative power attenuation for an absorption pathlength Δx is therefore the product of absorption coefficient α, pathlength Δx and the ratio of absorption linewidth $\delta\omega$ and spectral resolution bandwidth $\Delta\omega$, as long as $\Delta\omega > \delta\omega$.

For $\Delta\omega \ll \delta\omega$, we obtain:

$$\frac{\Delta P(\omega)}{P(\omega)} = \Delta x \cdot \alpha(\omega). \tag{1.7}$$

This implies that the line profile $\alpha(\omega)$ of the absorption line can be measured.

Example 1.2 The spectral resolution of a 1 m spectrograph is about 0.01 nm, which corresponds at $\lambda = 500$ nm to $\Delta\omega = 2\pi \cdot 12$ GHz. The Doppler width of gas molecules with $M = 30$ at $T = 300$ K is, according to Vol. 1, (3.30a) $\delta\omega \simeq 2\pi \cdot 1$ GHz. With a single-mode laser the value of $\delta\omega$ becomes smaller than $\Delta\omega$ and the observable signal $\Delta P/P$ becomes 12 times larger than in conventional spectroscopy with the same absorption cell.

- Because of the good collimation of a laser beam, long absorption paths can be realized by multiple reflection back and forth through the multiple-path absorption cell. Disturbing reflections from cell walls or windows, which may influence the measurements, can essentially be avoided (for example, by using Brewster end windows). Such long absorption paths enable measurements of absorbing transitions even with small absorption coefficients. Furthermore, because of the higher sensitivity, lower gas pressures in the absorption cell can be used, thus avoiding pressure broadening. This is especially important in the infrared region, where the Doppler width is small and pressure broadening may become the limiting factor for the spectral resolution (Vol. 1, Sect. 3.3).

Example 1.3 With an intensity-stabilized light source and lock-in detection the minimum relative absorption that may be safely detected is about $\Delta P/P \geq 10^{-6}$, which yields the minimum measurable absorption coefficient $\alpha_{\min}$ for an absorption pathlength L,

$$\alpha_{\min} = \frac{10^{-6}}{L}\frac{\Delta\omega}{\delta\omega}\left[\mathrm{cm}^{-1}\right].$$

With conventional spectroscopy for a pathlength $L = 10$ cm and $\delta\omega = 10\Delta\omega$ we obtain $\alpha_{\min} = 10^{-6}\ \mathrm{cm}^{-1}$. With single-mode lasers one may reach $\Delta\omega < \delta\omega$ and a long absorption path with $L = 10$ m, which yields a minimum detectable absorption coefficient of $\alpha_{\min} = 10^{-9}\ \mathrm{cm}^{-1}$, i.e., an improvement of a factor of 1000!

- If a small fraction of the laser output is sent through a long Fabry–Perot interferometer with a separation d of the mirrors (Fig. 1.1b), the photodetector PD3 receives intensity peaks each time the laser frequency ν_L is tuned to a transmission maximum at $\nu = \frac{1}{2}mc/d$ (Vol. 1, Sects. 4.2–4.4). These peaks serve as accurate wavelength markers, which allow one to calibrate the separation of adjacent absorption lines. With $d = 1$ m the frequency separation $\Delta\nu_p$ between successive transmission peaks is $\Delta\nu_p = c/2d = 150$ MHz, corresponding to a wavelength separation of 10^{-4} nm at $\lambda = 550$ nm. With a semiconfocal FPI the free spectral range is $c/8d$, which gives $\Delta\nu_p = 75$ MHz for $d = 0.5$ m.
- The laser frequency may be stabilized onto the center of an absorption line. With the methods discussed in Vol. 1, Sect. 4.4 it is possible to measure the wavelength λ_L of the laser with an absolute accuracy of 10^{-8} or better. This allows determination of the molecular absorption lines with the same accuracy.
- It is possible to tune the laser wavelength very rapidly over a spectral region where molecular absorption lines have to be detected. With electro-optical components, for instance, pulsed dye lasers can be tuned over several wavenumbers within a microsecond. This opens new perspectives for spectroscopic investigations of short-lived intermediate radicals in chemical reactions. The capabilities of classical flash photolysis may be considerably extended using such rapidly tunable laser sources.

- An important advantage of absorption spectroscopy with tunable single-mode lasers stems from their capabilities to measure line profiles of absorbing molecular transitions with high accuracy. In case of pressure broadening, the determination of line profiles allows one to derive information about the interaction potential of the collision partners (Sect. 3.3 of Vol. 1 and Sect. 8.1). In plasma physics this technique is widely used to determine electron and ion densities and temperatures.
- In fluorescence spectroscopy and optical pumping experiments, the high intensity of lasers allows an appreciable population in selectively excited states to be achieved that may be comparable to that of the absorbing ground states. The small laser linewidth favors the selectivity of optical excitation and results in favorable cases in the exclusive population of single molecular levels. These advantageous conditions allow one to perform absorption and fluorescence spectroscopy of *excited* states and to transform spectroscopic methods, such as microwave or RF spectroscopy, which has until now been restricted to electronic ground states, also to excited states.
- Transient absorption and rapid relaxation processes can be studied in detail with short-pulse lasers (see Chap. 6) with time resolutions reaching down into the femtosecond range.

Since minimizing the noise is an essential task in high sensitivity spectroscopy, we will shortly discuss the different contributions to the total noise (see Vol. 1, Sect. 4.5):

(1) *Power fluctuations δP of the radiation source*

Most radiation sources show fluctuations of their output power, due to plasma instabilities in the discharge of lamps, or thermal effects which influence the density of emitting atoms, ions or molecules. Here lasers are superior to incoherent sources, because their output power can be stabilized by various techniques (see Vol. 1, Sect. 5.4.3) to a level $\delta P/P < 10^{-4}$.

(2) *Detector noise*

Because of the much higher spectral power density of most lasers compared to incoherent light sources the detector noise is generally negligible in laser spectroscopy.

(3) *Point instabilities of the laser beam*

In case of small detector areas fluctuations of the local position of the laser spot on the detector will cause fluctuations of the measured output signal. These beam instabilities can be caused by thermal effects of the laser resonator which might result in a small tilt of the resonator mirrors, or by inhomogeneous heating of the active laser medium, which leads to thermal lens effects and affects the stability of the laser beam.

(4) *Frequency fluctuations of the laser output*

In case of narrow band lasers where the laser line width is smaller than the absorption line, frequency shifts cause a change of the absorption and with it a corresponding change of the measured signal. There are several ways to stabilize the laser frequency to avoid this effect completely (see Vol. 1, Sect. 5.4.4).

(5) *Density fluctuations of the absorbing molecules*
This would affect spectroscopy with lasers in the same way as that with incoherent sources. In most cases it is negligible.

This brief overview of some of the advantages of lasers in spectroscopy will be outlined in more detail in the following chapters and several examples will illustrate their relevance.

1.2 High-Sensitivity Methods of Absorption Spectroscopy

The general method for measuring absorption spectra is based on the determination of the absorption coefficient $\alpha(\omega)$ from the spectral power density

$$P_T(\omega) = P_0 \exp[-\alpha(\omega)x], \tag{1.8}$$

which is transmitted through an absorbing path length x. In Sect. 1.1 we have seen, that the minimum detectable concentration of absorbing molecules

$$N_1 \geq \frac{\text{NEP}}{a P_0 \sigma_{ik} L} \tag{1.9}$$

is limited by the noise equivalent input power NEP, the power P_0 of the radiation source and the absorption length L. If possible one should measure the absorption on a molecular transition with a large absorption cross section σ_{ik}.

However, the spectroscopist often has only the choice to maximize L and P_0 and to minimize NEP.

The question is now, how the noise can be reduced and the absorption path length increased. For small values of αL the direct measurement of the quantity $\Delta P = P_0 - P_T$ as the small difference of two large quantities cannot be very accurate since already small fluctuations of P_0 or P_T may result in a large relative change of ΔP. Therefore, other techniques have been developed leading to an increase in sensitivity and accuracy of absorption measurements by several orders of magnitude compared to direct absorption measurements. These sensitive detection methods represent remarkable progress, since their sensitivity limits have been pushed from relative absorptions $\Delta\alpha/\alpha \simeq 10^{-5}$ down to about $\Delta\alpha/\alpha \geq 10^{-17}$. We now discuss these different methods in more detail.

1.2.1 Enhancement of Absorption in External Cavities

The first step towards higher sensitivities is the increase of the absorption path length L. This can be for instance achieved when the absorption cell is placed inside a multipass cell. One example is shown in Fig. 1.2. Two spherical mirrors with a high reflectivity R are separated by a distance d which nearly equals the mirror radius r forming a nearly spherical resonator (Vol. 1, Sect. 5.2).

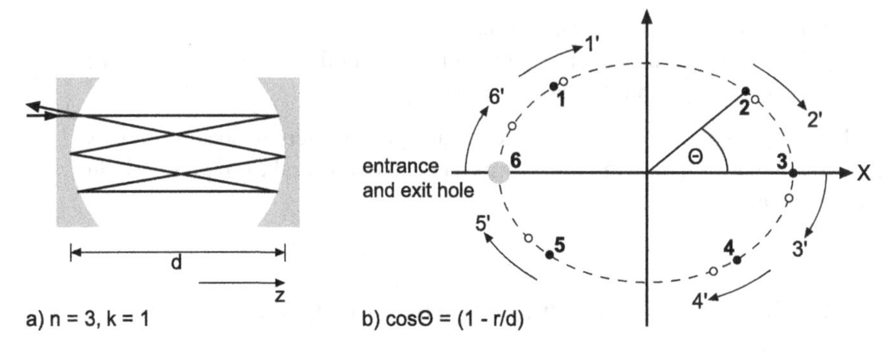

Fig. 1.2 Multipass absorption cell. (**a**) Beam geometry, (**b**) laser beam spots on a mirror surface

The laser beam enters the resonator through a small hole in one mirror and leaves the resonator after q reflections through the same hole but under a different angle. The number of transits can be chosen by adjusting the difference $(d - r) \ll d$. In Fig. 1.2b the laser spots on one mirror are shown for a given value of $d - r$. The angle θ between to adjacent spots is given by the relation

$$\cos\theta = (1 - r/d).$$

The spherical mirrors refocus the divergent laser beam at each reflection and prevent the laser spots on the mirrors to become too large. This is essential because the neighbouring spots should not overlap which would result in interference effects which give unwanted structures in the transmitted intensity when the laser wavelength is tuned.

Without absorption cell the intensity of the laser beam decreases after one roundtrip by the attenuation factor $\exp[-2(1 - R)]$. For q roundtrips with the absorption cell the transmitted intensity has decreased to $I_0 \exp[-2q(1-R) - 2q\alpha L]$.

Example 1.4 For $R = 0.99$, $\alpha = 0$ and $q = 100 \to I_T = I_0 \exp[-2] \approx 0.14 I_0$. For $R = 0.98 \to I_T = \exp[-4] \approx 0.018 I_0$. With $\alpha L = 0.01$ and $R = 0.99 \to I_T = 0.018 I_0$.

The alternative solution for a high sensitivity is the increase of the absorbed power (1.4b). This can be achieved when the absorption cell is placed inside an external resonator with a high Q-value (i.e. high reflectivity mirrors). However, now the incoming laser beam has to be mode-matched to the fundamental mode of the resonator (Fig. 1.3), which requires a proper lens system to image the incoming laser beam into the resonator. The laser power inside the resonator with mirror reflectivity R is $P_i = P_0/(1 - R)$ and the resonator eigenfrequencies of the cavity are $\nu_m = m \cdot c/d$, where d is the cavity length and m a large integer. When the laser frequency ν_L is tuned, the cavity length has to be tuned synchroneously in order to ensure that always $\nu_L = \nu_m$.

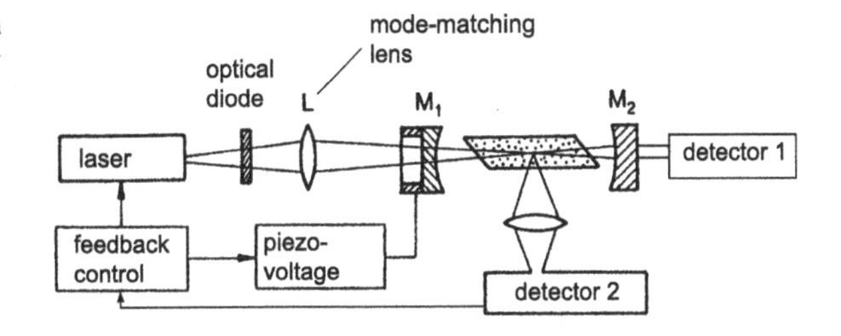

Fig. 1.3 Spectroscopy in an external enhancement cavity with mode-matching optics

Example 1.5 With $R = 0.99$ the power inside the resonator is without the absorbing sample 100 times higher than that of the incoming laser beam. For low values of αL with the absorption inside the resonator this enhancement factor is not much lower. For instance in a molecular beam with a low density of absorbing molecules typical values are $\alpha = 10^{-5}\ \text{cm}^{-1}$ and $L = 0.2\ \text{cm} \rightarrow \alpha L = 2 \times 10^{-6}$. With NEP $= 10$ nW, $P_0 = 100$ mW, $L = 100 \cdot 0.2\ \text{cm} \rightarrow \alpha_{\min} = 5 \times 10^{-9}\ \text{cm}^{-1}$.

1.2.2 Frequency Modulation

The second scheme to be treated is based on a frequency modulation of the monochromatic incident wave. It was not designed specifically for laser spectroscopy, but was taken from microwave spectroscopy where it is a standard method. The laser frequency ω_L is modulated at the modulation frequency Ω, which changes ω_L periodically from $\omega_L - \Delta\omega_L/2$ to $\omega_L + \Delta\omega_L/2$. When the laser is tuned through the absorption spectrum, the difference $\Delta P_T = P_T(\omega_L - \Delta\omega_L/2) - P_T(\omega_L + \Delta\omega_L/2)$ is detected with a lock-in amplifier (phase-sensitive detector) tuned to the modulation frequency Ω (Fig. 1.4). If the modulation sweep $\Delta\omega_L$ is sufficiently small, the first term of the Taylor expansion

$$\Delta P_T(\omega) = \frac{\mathrm{d}P_T}{\mathrm{d}\omega}\Delta\omega_L + \frac{1}{2!}\frac{\mathrm{d}^2 P_T}{\mathrm{d}\omega^2}\Delta\omega_L^2 + \dots, \tag{1.10}$$

is dominant. This term is proportional to the first derivative of the absorption spectrum $\alpha(\omega) = \frac{\Delta P(\omega)}{P_0 L}$ as can be seen from (1.1b): when $P_0 = P_R$ is independent of ω we obtain

$$\frac{\mathrm{d}\alpha(\omega)}{\mathrm{d}\omega} = -\frac{1}{P_R L}\frac{\mathrm{d}P_T}{\mathrm{d}\omega}. \tag{1.11}$$

If the laser frequency

$$\omega_L(t) = \omega_0 + a \sin\Omega t,$$

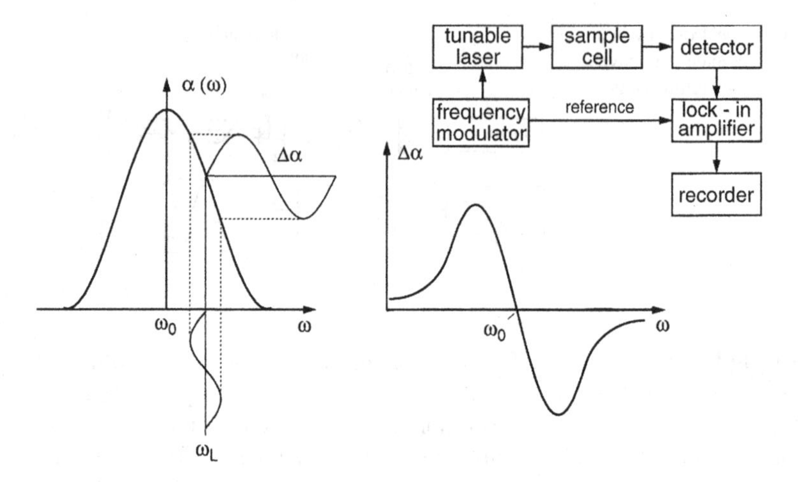

Fig. 1.4 Absorption spectroscopy with a frequency-modulated single-mode laser

is sinusoidally modulated at a modulation frequency Ω, the Taylor expansion yields

$$P_{\mathrm{T}}(\omega_{\mathrm{L}}) = P_{\mathrm{T}}(\omega_0) + \sum_n \frac{a^n}{n!} \sin^n \Omega t \left(\frac{\mathrm{d}^n P_{\mathrm{T}}}{\mathrm{d}\omega^n}\right)_{\omega_0} . \tag{1.12}$$

For $\alpha L \ll 1$, we obtain from (1.1a)

$$\left(\frac{\mathrm{d}^n P_{\mathrm{T}}}{\mathrm{d}\omega^n}\right)_{\omega_0} = -P_0 x \left(\frac{\mathrm{d}^n \alpha(\omega)}{\mathrm{d}\omega^n}\right)_{\omega_0} .$$

The terms $\sin^n \Omega t$ can be converted into linear functions of $\sin(n\Omega t)$ and $\cos(n\Omega t)$ using known trigonometric formulas.

Inserting these relations into (1.12), one finds after rearrangement of the terms the expression

$$\begin{aligned}
\frac{\Delta P_{\mathrm{T}}}{P_0} = -aL \Bigg\{ & \left[\frac{a}{4}\left(\frac{\mathrm{d}^2\alpha}{\mathrm{d}\omega^2}\right)_{\omega_0} + \frac{a^3}{64}\left(\frac{\mathrm{d}^4\alpha}{\mathrm{d}\omega^4}\right)_{\omega_0} + \dots\right] \\
& + \left[\left(\frac{\mathrm{d}\alpha}{\mathrm{d}\omega}\right)_{\omega_0} + \frac{a^2}{8}\left(\frac{\mathrm{d}^3\alpha}{\mathrm{d}\omega^3}\right)_{\omega_0} + \dots\right] \sin(\Omega t) \\
& + \left[-\frac{a}{4}\left(\frac{\mathrm{d}^2\alpha}{\mathrm{d}\omega^2}\right)_{\omega_0} + \frac{a^3}{48}\left(\frac{\mathrm{d}^4\alpha}{\mathrm{d}\omega^4}\right)_{\omega_0} + \dots\right] \cos(2\Omega t) \\
& + \left[-\frac{a^2}{24}\left(\frac{\mathrm{d}^3\alpha}{\mathrm{d}\omega^3}\right)_{\omega_0} + \frac{a^4}{384}\left(\frac{\mathrm{d}^5\alpha}{\mathrm{d}\omega^5}\right)_{\omega_0} + \dots\right] \sin(3\Omega t) \\
& + \dots \Bigg\} .
\end{aligned}$$

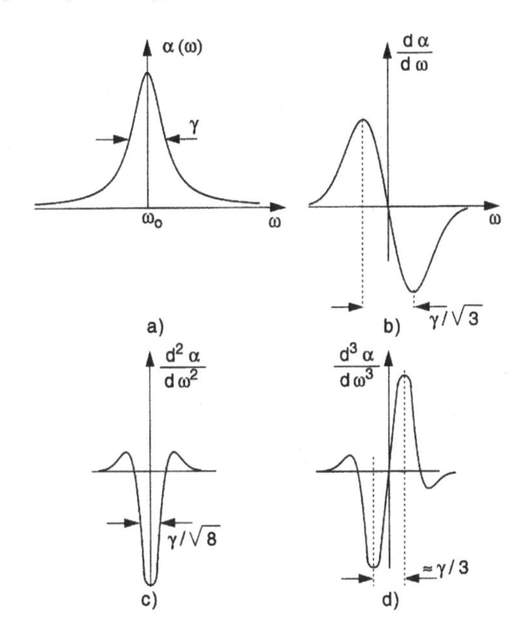

Fig. 1.5 Lorentzian line profile $\alpha(\omega)$ of halfwidth γ (FWHM) (**a**), with first (**b**), second (**c**), and third (**d**) derivatives

For a sufficiently small modulation amplitude ($a/\omega_0 \ll 1$), the first terms in each bracket are dominant. Therefore we obtain for the signal $S(n\Omega)$ behind a lock-in amplifier tuned to the frequency $n\Omega$ (Fig. 1.5):

$$S(n\Omega) = \left(\frac{\Delta P_{\text{T}}}{P_0}\right)_{n\Omega} = aL \begin{cases} b_n \sin(n\Omega t), & \text{for } n = 2m+1, \\ c_n \cos(n\Omega t), & \text{for } n = 2m. \end{cases}$$

In particular, the signals for the first three derivatives of the absorption coefficient $\alpha(\omega)$, shown in Fig. 1.5, are

$$\begin{aligned} S(\Omega) &= -aL\frac{\mathrm{d}\alpha}{\mathrm{d}\omega}\sin(\Omega t), \\ S(2\Omega) &= +\frac{a^2L}{4}\frac{\mathrm{d}^2\alpha}{\mathrm{d}\omega^2}\cos(2\Omega t), \\ S(3\Omega) &= +\frac{a^3L}{24}\frac{\mathrm{d}^3\alpha}{\mathrm{d}\omega^3}\sin(3\Omega t). \end{aligned} \tag{1.13}$$

The advantage of this "derivative spectroscopy" [2] with a frequency-modulated laser is the possibility for phase-sensitive detection, which restricts the frequency response of the detection system to a narrow frequency interval centered at the modulation frequency Ω. Frequency-independent background absorption from cell windows and background noise from fluctuations of the laser intensity or of the density of absorbing molecules are essentially reduced. Regarding the signal-to-noise ra-

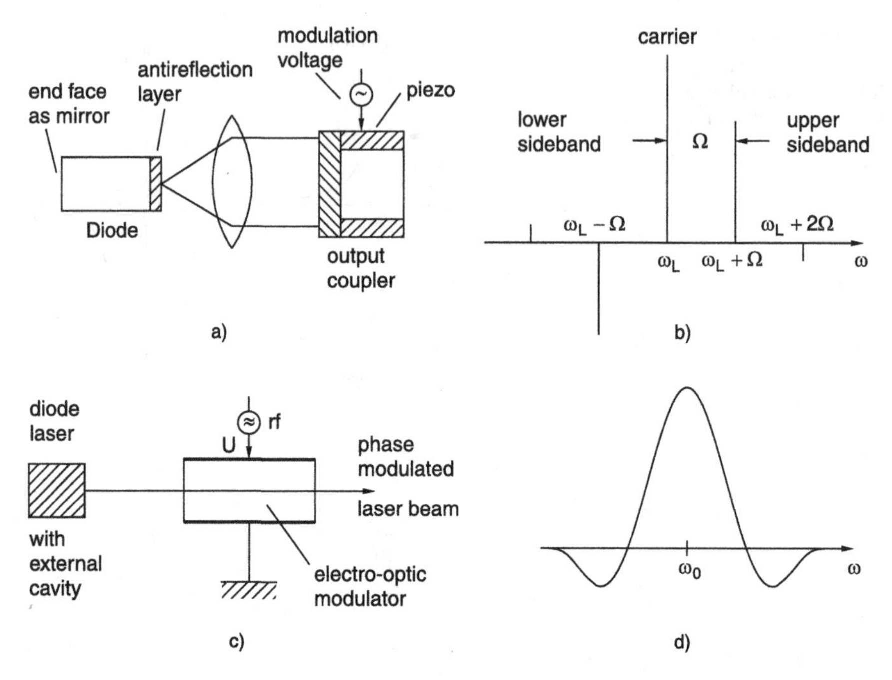

Fig. 1.6 Modulated laser sources. (**a**) Wavelength-modulation; (**b**) corresponding absorption line profiles; (**c**) phase modulation; (**d**) corresponding sideband spectrum

tio and achievable sensitivity, the *frequency* modulation technique is superior to an *intensity* modulation of the incident radiation.

There are different experimental techniques for achieving frequency modulation of single-mode lasers. One of them is based on periodically changing the length d of the laser resonator, which can be accomplished by mounting one of the resonator mirrors onto a piezo ceramic, which changes its dimensions when a voltage is applied to the piezo (Fig. 1.6a and Vol. 1, Sect. 5.4).

Because the laser frequency $\nu = q \cdot (c/2d)$, where q is an integer number, depends on the resonator length d, a change Δd causes a frequency change

$$\Delta\nu = -q\left(c/2d^2\right)\Delta d \tag{1.14a}$$

or a wavelength change (wavelength modulation)

$$\Delta\lambda = (2/q)\Delta d. \tag{1.14b}$$

The maximum signal is obtained when $\Delta\nu$ reaches the spectral width of the absorption lines which is generally the Doppler width with $\Delta\nu \approx 1$ GHz in the visible region. The length change Δd must be therefore sufficiently large to attain a modulation depth $\Delta\nu$ that reaches the Doppler width of the absorption line. The modulation frequency is limited by the mass of the mirror to a few kilohertz.

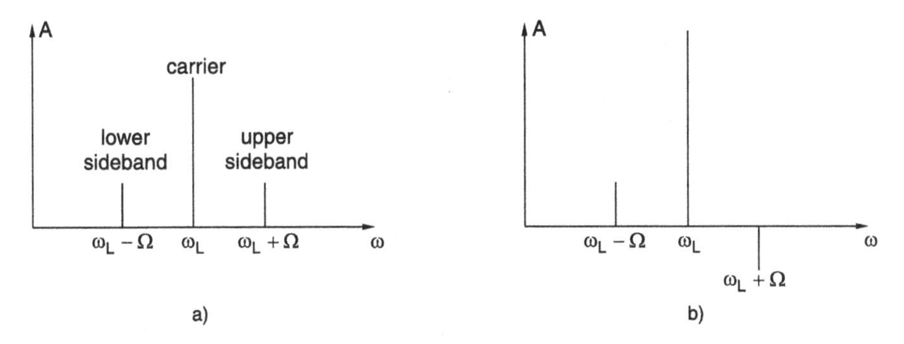

Fig. 1.7 Sideband amplitudes for: (**a**) amplitude modulation; (**b**) phase modulation

Example 1.6 For a resonator length of $d = 8$ cm, which is typical of a diode laser with an external cavity, and $\lambda = 800$ nm, the integer q becomes $q = 2 \times 10^5$. For a Doppler width of $\Delta\nu_d = 1$ GHz, a length change of $\Delta d = 2$ μm is sufficient to shift the laser frequency periodically over the absorption profile.

The technical noise, which represents the major limitation, decreases with increasing frequency. It is therefore advantageous to choose the modulation frequency as high as possible. For diode lasers this can be achieved by modulation of the diode current. For other lasers, often electro-optical modulators outside the laser resonator are used as phase modulators (Fig. 1.6), resulting in a frequency modulation of the transmitted laser beam [3].

The electro-optic crystal changes its refractive index $n(V)$ with the application of a voltage $V = V_0(1 + a \cdot \sin \Omega t)$. The amplitude of the laser wave transmitted through the phase-modulator with optical length $n \cdot L$ is

$$E = E_0 e^{i(\omega t + \phi(t))} \quad \text{with } \phi(t) = (2\pi/\lambda) n(t) \cdot L \text{ with } n = n_0(1 + b \cdot \sin \omega t). \quad (1.15)$$

While only two sidebands appear at frequencies $\omega \pm \Omega$ for amplitude modulation, for phase modulation the number of sidebands at frequencies $\omega \pm q \cdot \Omega$ depends on the modulation amplitude $\Delta\phi_m = b \cdot n_0 \cdot 2\pi/\lambda$. The amplitudes of the sidebands decrease with increasing q. For sufficiently small values of b, the higher sidebands for $q > 1$ have small intensities and can be neglected.

The phase modulation has an additional advantage: the first two sidebands at frequencies $\omega + \Omega$ and $\omega - \Omega$ have equal amplitudes but opposite phases (Fig. 1.7). A lock-in detector tuned to the modulation frequency Ω therefore receives the superposition of two beat signals between the carrier and the two sidebands, which cancel to zero if no absorption is present. Any fluctuation of the laser intensity appears equally on both signals and is therefore also cancelled. If the laser wavelength is tuned over an absorption line, one sideband is absorbed, if $\omega + \Omega$ or $\omega - \Omega$ coincides with the absorption frequency ω_0 (Fig. 1.8). This perturbs the balance and

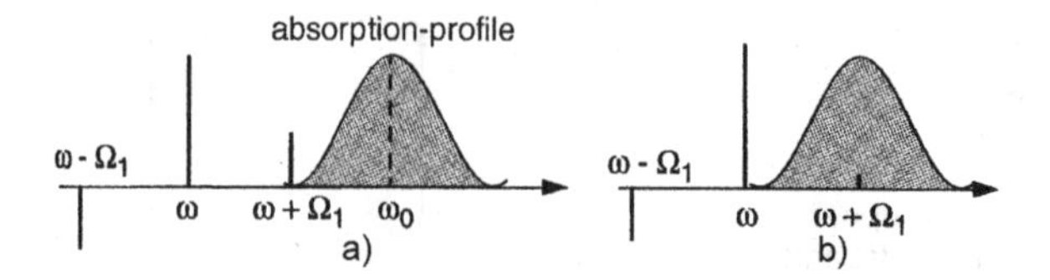

Fig. 1.8 Principle of phase-modulated absorption spectroscopy: (**a**) no absorption of carrier and sidebands, (**b**) one sideband coincides with an absorption line

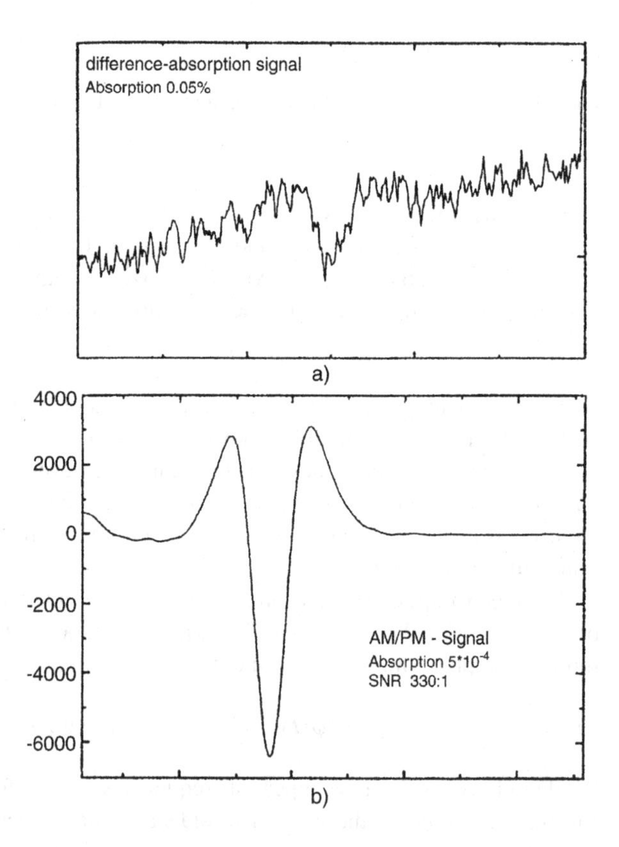

Fig. 1.9 (**a**) Water overtone absorption line measured with unmodulated (**b**) and with phase-modulated single-mode diode laser

gives raise to a signal (Fig. 1.6d) with a profile that is similar to the profile of the second derivative in Fig. 1.5c.

The sensitivity of this technique is demonstrated by Fig. 1.9, which shows an overtone absorption line of the water molecule H_2O, recorded with an unmodulated laser and with this modulation technique. The signal-to-noise ratio of the absorption measured with phase modulation is about 2 orders of magnitude higher than without modulation. The sensitivity reaches a maximum if the modulation frequency is chosen to be equal to the width of the absorption line.

If the modulation frequency Ω is chosen sufficiently high ($\Omega > 1000$ MHz), the technical noise may drop below the quantum-noise limit set by the statistical

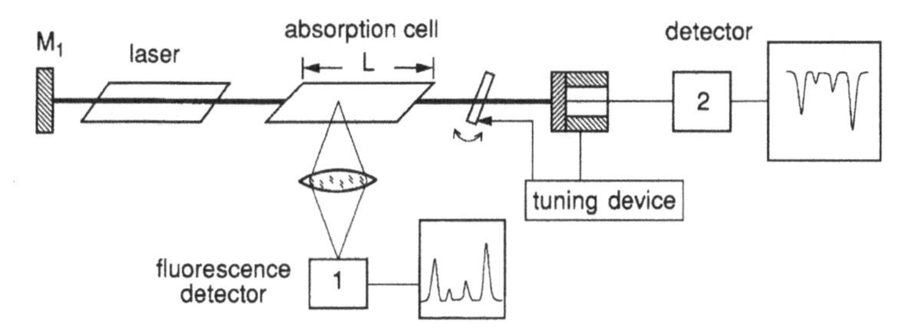

Fig. 1.10 Intracavity absorption detected either by monitoring the laser output $P(\omega_L)$ with detector 2 or the laser-induced fluorescence $I_{Fl}(\omega_L)$ with detector 1

fluctuations of detected photons. In this case, the detection limit is mainly due to the quantum limit [4]. Since lock-in detectors cannot handle such high frequencies, the signal input has to be downconverted in a mixer, where the difference frequency between a local oscillator and the signal is generated.

This new method of high-frequency modulation spectroscopy using low-frequency detection is called two-tone frequency-modulation spectroscopy [5]. Here the laser output is phase-modulated in an electro-optic $LiTaO_3$ crystal that is driven by a high-frequency voltage in the GHz range, that is amplitude-modulated at frequencies in the MHz range. The detector output is fed into a frequency mixer (downconverter) and the final signal is received with a lock-in amplifier in the kilohertz range [6, 7].

A comparison of different modulation techniques can be found in [8–10].

1.2.3 Intracavity Laser Absorption Spectroscopy ICLAS

When the absorbing sample is placed inside the laser cavity (Fig. 1.10), the detection sensitivity can be enhanced considerably, in favorable cases by several orders of magnitude. Four different effects can be utilized to achieve this "amplified" sensitivity. The first two are based on single-mode operation, the last two on multimode oscillation of the laser.

(1) Assume that the reflectivities of the two resonator mirrors are $R_1 = 1$ and $R_2 = 1 - T_2$ (mirror absorption is neglected). At the laser output power P_{out}, the power inside the cavity is $P_{int} = qP_{out}$ with $q = 1/T_2$. For $\alpha L \ll 1$, the power $\Delta P(\omega)$ absorbed at the frequency ω in the absorption cell (length L) is

$$\Delta P(\omega) = \alpha(\omega)LP_{int} = q\alpha(\omega)LP_{out} \tag{1.16}$$

and therefore q-times larger than outside the laser cavity. If the absorbed power can be measured directly, for example, through the resulting pressure increase in the absorption cell (Sect. 1.3.2) or through the laser-induced fluorescence (Sect. 1.3.1),

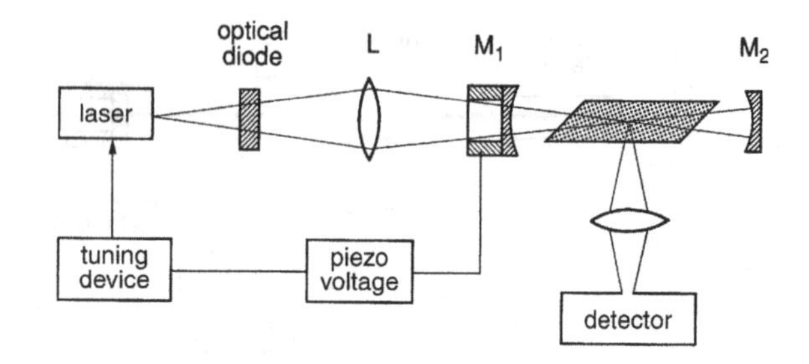

Fig. 1.11 Spectroscopy inside an external resonator, which is synchronously tuned with the laser frequency ω_L in order to be always in resonance

the signal will be q times larger than for the case of single-pass absorption outside the cavity.

Example 1.7 With a resonator transmission $T_2 = 0.02$ (a figure that can be readily realized in practice) the enhancement factor becomes $q = 50$, as long as saturation effects can be neglected and provided the absorption is sufficiently small that it does not noticeably change the laser intensity.

This q-fold amplification of the sensitivity can be also understood from the simple fact that a laser photon travels on the average q times back and forth between the resonator mirrors before it leaves the resonator. It therefore has a q-fold chance to be absorbed in the sample; in other words, the effective absorption pathlength becomes q times longer.

This sensitivity enhancement in detecting small absorptions has no direct correlation with the gain medium and can be also realized in external passive resonators. If the laser output is mode matched (Vol. 1, Sect. 5.2.3) by lenses or mirrors into the fundamental mode of the passive cavity containing the absorbing sample (Fig. 1.11), the radiation power inside this cavity is q times larger. The enhancement factor q may become larger if the internal losses of the cavity can be kept low.

Intracavity absorption cells are particularly advantageous if the absorption is monitored via the laser-induced fluorescence. Since the radiation field inside the active resonator or inside the mode-matched passive cavity is concentrated within the region of the Gaussian beam (Vol. 1, Sect. 5.9), the laser-excited fluorescence can be effectively imaged onto the entrance slit of a spectrometer with a larger efficiency than in the commonly used multipass cells. If minute concentrations of an absorbing component have to be selectively detected in the presence of other constituents with overlapping absorption lines but different fluorescence spectra, the use of a spectrometer for dispersing the fluorescence can solve the problem.

External passive resonators may become advantageous when the absorption cell cannot be placed directly inside the active laser resonator. However, there also exist some drawbacks: the cavity length has to be changed synchronously with the tunable-laser wavelength in order to keep the external cavity always in resonance. Furthermore, one has to take care to prevent optical feedback from the passive to the active cavity, which would cause a coupling of both cavities with resulting instabilities. This feedback can be avoided by an optical diode (Vol. 1, Sect. 5.2.7).

(2) Another way of detecting intracavity absorption with a very high sensitivity relies on the dependence of the single-mode laser output power on absorption losses inside the laser resonator (detector 2 in Fig. 1.10). At constant pump power, just above the threshold, minor changes of the intracavity losses may result in drastic changes of the laser output. In Vol. 1, Sect. 5.3 we saw that under steady-state conditions, the laser ouput power essentially depends on the pump power and reaches the value P_s, where the gain factor $G = \exp[-2L_1\alpha_s - \gamma]$ becomes $G = 1$. This implies that the saturated gain $g_s = 2L_1\alpha_s$ of the active medium with the length L_1 equals the total losses γ per cavity round trip (Vol. 1, Sect. 5.1).

The saturated gain $g_s = 2L_1\alpha_s$ depends on the intracavity intensity I. According to (3.61) in Vol. 1 we obtain

$$g_s = \frac{g_0}{1 + I/I_s} = \frac{g_0}{1 + P/P_s}, \tag{1.17}$$

where I_s is the saturation intensity. The gain decreases from g_0 for $P = 0$ to $g_s = g_0/2$ for $P = P_s$ (Vol. 1, Sect. 3.6). At a constant pump power, the laser power P stabilizes itself at the value where $g_s = \gamma$. With (1.17), this gives:

$$P = P_s \frac{g_0 - \gamma}{\gamma}. \tag{1.18}$$

If small additional losses $\Delta\gamma$ are introduced by the absorbing sample inside the cavity, the laser power drops to a value

$$P_\alpha = P - \Delta P = P_s \cdot \frac{g_0 - \gamma - \Delta\gamma}{\gamma + \Delta\gamma}. \tag{1.19}$$

From (1.17–1.19) we obtain for the relative change $\Delta P/P$ of the laser output power by the absorbing sample

$$\Delta P/P = \frac{g_0}{g_0 - \gamma} \frac{\Delta\gamma}{\gamma + \Delta\gamma} \simeq \frac{g_0}{\gamma} \frac{\Delta\gamma}{g_0 - \gamma}, \quad \text{for } \Delta\gamma \ll \gamma, \tag{1.20}$$

where g_0 is the unsaturated gain.

Compared with the single-pass absorption of a sample with the absorption coefficient α and the absorption pathlength L_2 outside the laser resonator where $\Delta P/P = -\alpha L_2 = -\Delta\gamma$, the intracavity absorption represents a sensitivity enhancement by the factor

$$\boxed{Q = \frac{g_0}{\gamma(g_0 - \gamma)}}. \tag{1.21}$$

At pump powers far above threshold, the unsaturated gain g_0 is large compared to the losses γ, and (1.21) reduces to

$$Q \simeq 1/\gamma, \quad \text{for } g_0 \gg \gamma.$$

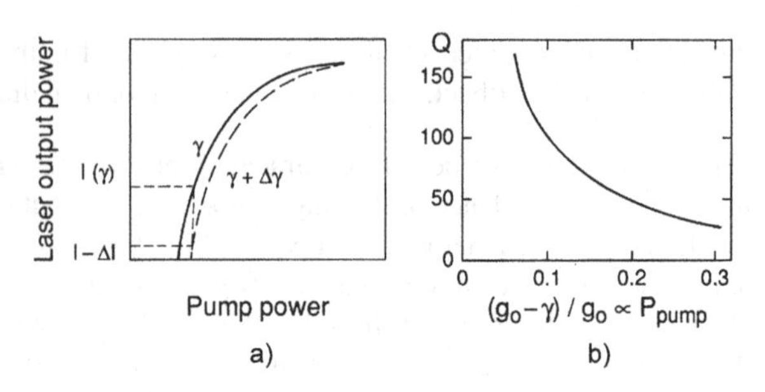

Fig. 1.12 (**a**) Laser output power $P_L(P_{pump})$ for two slightly different losses γ and $\gamma+\Delta\gamma$. (**b**) Enhancement factor Q as a function of pump power P_{pump} above threshold

If the resonator losses are mainly due to the transmission T_2 of the output mirror, the enhancement factor Q then becomes $Q = 1/\gamma = 1/T_2 = q$, which is equal to the enhancement of the previous detection method 1.

Just above threshold, however, g_0 is only slightly larger than γ and the denominator in (1.20) becomes very small, which means that the enhancement factor Q may reach very large values (Fig. 1.12). At first sight, it might appear that the sensitivity could be made arbitrarily large for $2\alpha_0 L \to \gamma$. However, there are experimental as well as fundamental limitations that restrict the maximum achievable value of Q. The increasing instability of the laser output, for instance, limits the detection sensitivity when the threshold is approached. Furthermore, just above threshold, the spontaneous radiation that is emitted into the solid angle accepted by the detector cannot be neglected. It represents a constant background power, nearly independent of γ, which puts a principal upper limit to the relative change $\Delta P/P$ and thus to the sensitivity.

Note If the gain medium is inhomogeneously broadened, the saturated gain g_s becomes

$$g_s = \frac{g_0}{\sqrt{1+I/I_s}}$$

(Vol. 1, Sect. 3.6 and Sect. 2.2). The derivation analogue to (1.17–1.20) yields

$$\frac{\Delta P}{P} = \frac{g_0^2(2\gamma\Delta\gamma + \Delta\gamma^2)}{(g_0^2-\gamma^2)(\gamma+\Delta\gamma)^2} \simeq \frac{g_0^2}{\gamma^2}\frac{\Delta\gamma}{g_0-\gamma}, \tag{1.22a}$$

instead of (1.20).

(3) In the preceding discussion of the sensitivity enhancement by intracavity absorption, we have implicitly assumed that the laser oscillates in a single mode. Even larger enhancement factors Q can be achieved, however, with lasers oscillating simultaneously in several competing modes. Pulsed or cw dye lasers without additional mode selection are examples of lasers with mode competition. As discussed in Vol. 1, Sect. 5.3), the active dye medium exhibits a broad homogeneous spectral gain profile, which allows the same dye molecules to simultaneously contribute to the gain of all modes with frequencies within the homogeneous linewidth (see the discussion in Vol. 1, Sects. 5.3, 5.8). This means that the different oscillating laser modes may share the same molecules to achieve their gains. This leads to mode competition and bring about the following mode-coupling phenomena: Assume that the laser oscillates simultaneously on N modes, which may have equal gain and equal losses, and therefore equal intensities. When the laser wavelength is tuned across the absorption spectrum of an absorbing sample inside the laser resonator, one of the oscillating modes may be tuned into resonance with an absorption line (frequency ω_k) of the sample molecules. This mode now suffers additional losses $\Delta\gamma = \alpha(\omega_k)L$, which cause the decrease ΔI of its intensity. Because of this decreased intensity the inversion of the active medium is less depleted by this mode, and the gain at ω_k *increases*. Since the other $(N-1)$ modes can participate in the gain at ω_k, their intensity will increase. This, however, again depletes the gain at ω_k and further *decreases* the intensity of the mode oscillating at ω_k. With sufficiently strong coupling between the modes, this mutual interaction may finally result in a total suppression of the absorbing mode.

Frequency fluctuations of the modes caused by external perturbations (Vol. 1, Sect. 5.4) limit the mode coupling. Furthermore, in dye lasers with standing-wave resonators, the spatial hole-burning effect (Vol. 1, Sect. 5.8) weakens the coupling between modes. Because of their slightly different wavelengths, the maxima and nodes of the field distributions in the different modes are located at different positions in the active medium. This has the effect that the volumes of the active dye from which the different modes extract their gain overlap only partly. With sufficiently high pump power, the absorbing mode has an adequate gain volume on its own and is not completely suppressed but suffers a large intensity loss.

A more detailed calculation [11, 12] yields for the relative power change of the absorbing mode

$$\frac{\Delta P}{P} = \frac{g_0 \Delta\gamma}{\gamma(g_0 - \gamma)}(1 + KN), \tag{1.22b}$$

where K $(0 \leq K \leq 1)$ is a measure of the coupling strength.

Without mode coupling ($K = 0$), (1.22b) gives the same result as (1.20) for a single-mode laser. In the case of strong coupling ($K = 1$) and a large number of modes ($N \gg 1$), the relative intensity change of the absorbing mode increases proportionally to the number of simultaneously oscillating modes.

If several modes are simultaneously absorbed, the factor N in (1.22a–1.22b) stands for the ratio of all coupled modes to those absorbed. If all modes have equal frequency spacing, their number N gives the ratio of the spectral width of the homogeneous gain profile to the width of the absorption profile.

In order to detect the intensity change of one mode in the presence of many others, the laser output has to be dispersed by a monochromator or an interferometer. The absorbing molecules may have many absorption lines within the broadband gain profile of a multimode dye laser. Those laser modes that overlap with absorption lines are attenuated or are even completely quenched. This results in "spectral holes" in the output spectrum of the laser and allows the sensitive simultaneous recording of the whole absorption spectrum within the laser bandwidth, if the laser output is photographically recorded behind a spectrograph or if an optical multichannel analyzer (Vol. 1, Sect. 4.5) is used.

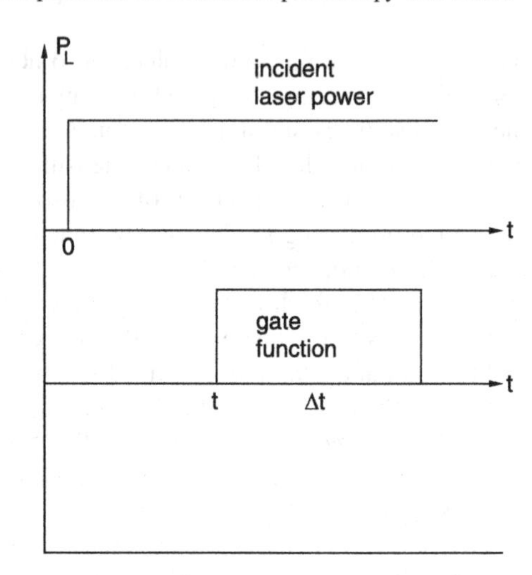

Fig. 1.13 Time profiles of incident and transmitted laser power

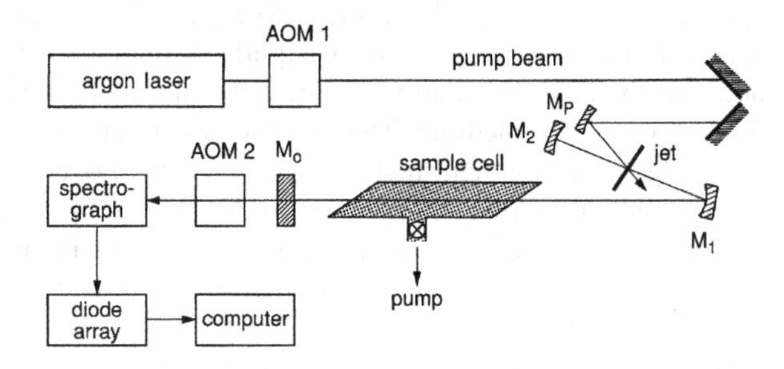

Fig. 1.14 Experimental arrangement for intracavity laser spectroscopy using a step-function pump intensity and a definite delay for the detection [13]

(4) The discussion in item 3 has assumed that the mode coupling and mode frequencies were time independent. In real laser systems this is not true. Fluctuations of mode frequencies due to density fluctuations of the dye liquid or caused by external perturbations prevent stationary conditions in a multimode laser. We can define a mean "mode lifetime" t_m, which represents the average time a specific mode exists in a multimode laser. If the measuring time for intracavity absorption exceeds the mode lifetime t_m, no quantitatively reliable information on the magnitude of the absorption coefficient $\alpha(\omega)$ of the intracavity sample can be obtained.

It is therefore better to pump the "intracavity laser" with a step-function pump laser, which starts pumping at $t = 0$ and then remains constant (Fig. 1.13). The intracavity absorption is then measured at times t with $0 < t < t_\mathrm{m}$, which are shorter than the mean mode lifetime t_m. The experimental arrangement is depicted in Fig. 1.14 [13]. The cw broadband dye laser is pumped by an argon laser. The pump beam can be switched on by the acousto-optic modulator AOM1 at $t = 0$. The dye laser output passes a second modulator AOM2, which transmits the dye laser beam for the time interval Δt at the selectable time t to the entrance slit of a high-resolution spectrograph. A photodiode array with a length D allows simultaneous recording of the spectral interval $\Delta\lambda = (\mathrm{d}x/\mathrm{d}\lambda)D$. The repetition rate f of

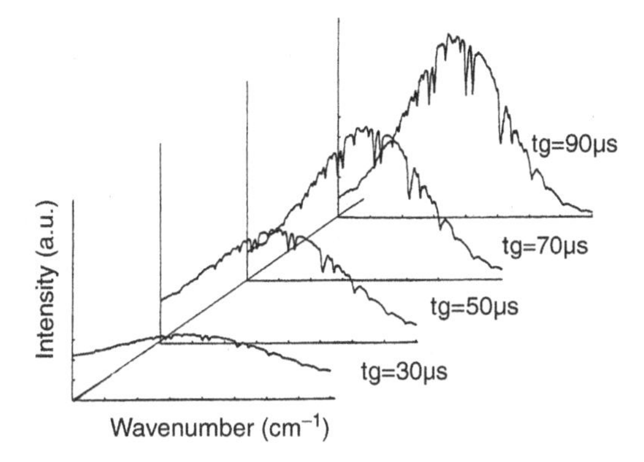

Fig. 1.15 Time evolution of the spectral profile of the laser output measured with time-resolved intracavity absorption spectroscopy [13]

the pump cycles must be smaller than the inverse delay time t between AOM1 and AOM2, i.e., $f < 1/t$.

A detailed consideration [11–17] shows that the time evolution of the laser intensity in a specific mode $q(\omega)$ with frequency ω after the start of the pump pulse depends on the gain profile of the laser medium, the absorption $\alpha(\omega)$ of the intracavity sample, and the mean mode lifetime t_m. If the broad gain profile with the spectral width $\Delta\omega_\mathrm{g}$ and the center frequency ω_0 can be approximated by the parabolic function

$$g(\omega) = g_0\left[1 - \left(\frac{\omega - \omega_0}{\Delta\omega_g}\right)^2\right],$$

the time evolution of the output power $P_\mathrm{q} = P(\omega_\mathrm{a})$ in the qth mode for a constant pump rate and times $t < t_\mathrm{m}$ is, after saturation of the gain medium

$$P_\mathrm{q}(t) = P_\mathrm{q}(0)\sqrt{\frac{t}{\pi t_\mathrm{m}}}\exp\left[-\left(\frac{\omega - \omega_0}{\Delta\omega_\mathrm{g}}\right)^2 t \Big/ t_\mathrm{m}\right]\mathrm{e}^{-\alpha(\omega_\mathrm{q})ct}. \tag{1.23}$$

The first exponential factor describes the spectral narrowing of the gain profile with increasing time t due to saturation and laser mode competition, and the second factor can be recognized as the Beer–Lambert absorption law for the transmitted laser power in the qth mode with the effective absorption length $L_\mathrm{eff} = ct$. In practice, effective absorption lengths up to 70,000 km have been realized [15]. The spectral width of the laser output becomes narrower with increasing time, but the absorption dips become more pronounced (Fig. 1.15).

Example 1.8 With typical delay times $t = 10^{-4}$ s, the effective absorption pathlength becomes $L_\mathrm{eff} = c \cdot t = 3 \times 10^8 \times 10^{-4}$ m $= 30$ km! If dips of 1 % can still be detected, this gives a sensitivity limit of $\alpha \cdot L_\mathrm{eff} = 0.01 \Rightarrow \alpha_\mathrm{min} \geq 3 \times 10^{-9}\ \mathrm{cm}^{-1}$ for the absorption coefficient $\alpha(\omega)$. For systems where $t = 10$ ms can be realized, the effective length becomes $L_\mathrm{eff} = 3 \times 10^6$ m and a limit of $\alpha_\mathrm{min} = 3 \times 10^{-11}\ \mathrm{cm}^{-1}$ is possible.

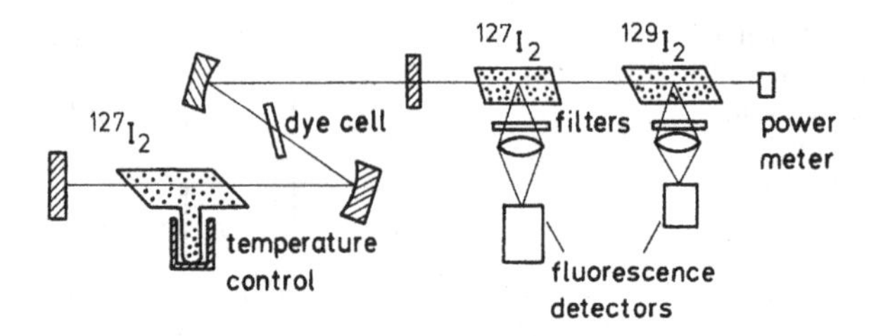

Fig. 1.16 Isotope-selective intracavity absorption spectroscopy. The frequencies ω_k absorbed by the $^{127}I_2$ isotope inside the laser cavity are missing in the laser output, which therefore does not excite any fluorescence in the same isotope outside the laser resonator [20]

In ring lasers (Vol. 1, Sect. 5.6) the spatial hole-burning effect does not occur if the laser is oscillating in unidirectional traveling-wave modes. If no optical diode is inserted into the ring resonator, the unsaturated gain is generally equal for clockwise and counterclockwise running waves. In such a bistable operational mode, slight changes of the net gain, which might be different for both waves because of their opposite Doppler shifts, may switch the laser from clockwise to counterclockwise operation, and vice versa. Such a bistable multimode ring laser with strong gain competition between the modes therefore represents an extremely sensitive detector for small absorptions inside the resonator [18].

The enhanced sensitivity of intracavity absorption may be utilized either to detect minute concentrations of absorbing components or to measure very weak forbidden transitions in atoms or molecules at sufficiently low pressures to study the unperturbed absorption line profiles. With intracavity absorption cells of less than 1 m, absorbing transitions have been measured that would demand a path length of several kilometers with conventional single-pass absorption at a comparable pressure [15, 19].

Some examples illustrate the various applications of the intracavity absorption technique.

- With an iodine cell inside the resonator of a cw multimode dye laser, an enhancement factor of $Q = 10^5$ could be achieved, allowing the detection of I_2 molecules at concentrations down to $n \leq 10^8/\mathrm{cm}^3$ [20]. This corresponds to a sensitivity limit of $\alpha L \leq 10^{-7}$. Instead of the laser output power, the laser-induced fluorescence from a second iodine cell outside the laser resonator was monitored as a function of wavelength. This experimental arrangement (Fig. 1.16) allows demonstration of the isotope-specific absorption. When the laser beam passes through two external iodine cells filled with the isotopes $^{127}I_2$ and $^{129}I_2$, tiny traces of $^{127}I_2$ inside the laser cavity are sufficient to completely quench the laser-induced fluorescence from the external $^{127}I_2$ cell, while the $^{129}I_2$ fluorescence is not affected [21]. This demonstrates that those modes of the broadband dye laser that are absorbed by the internal $^{127}I_2$ are completely suppressed.
- Detection of absorbing transitions with very small oscillator strength (Vol. 1, Sect. 2.7.2) has been demonstrated by Bray et al. [22], who measured the extremely weak $(v' = 2,\ v'' = 0)$ infrared overtone absorption band of the red-

atmospheric system of molecular oxygen and also the ($v' = 6 \leftarrow v'' = 0$) overtone band of HCl, with a cw rhodamine B dye laser (0.3 nm bandwidth) using a 97 cm-long intracavity absorption cell. Sensitivity tests indicated that even transitions with oscillator strengths down to $f \leq 10^{-12}$ could be readily detected. An example is the P(11) line in the ($2 \leftarrow 0$) band of the $b\,^1\Sigma_g^+ \leftarrow x\,^3\Sigma_g^-$ system of O_2 with the oscillator strength $f = 8.4 \times 10^{-13}$ [23]!
- The overtone spectrum $\Delta v = 6$ of SiH_4 was measured with the effective absorption pathlength $L_{eff} = 5.25$ km [13]. The spectrum shows well-resolved rotational structure, and local Fermi resonances were observed.

Although most experiments have so far been performed with dye lasers, the color-center lasers or the newly developed vibronic solid-state lasers such as the Ti:sapphire laser, with broad spectral-gain profiles (Vol. 1, Sect. 5.7.3) are equally well suited for intracavity spectroscopy in the near infrared. An example is the spectroscopy of rovibronic transitions between higher electronic states of the H_3 molecule with a color-center laser [24]. The combination of Fourier spectroscopy with ICLAS allows improved spectral resolution, while the sensitivity can also be enhanced [25, 34, 35].

Instead of absorption, weak emission lines can also be detected with the intracavity techniques [26]. If this light is injected into specific modes of the multimode laser, the intensity of these modes will increase for observation times $t < t_m$ before they can share their intensity by mode-coupling with other modes. Cavity-enhanced spectroscopy in optical fibres have been reported in [32, 33].

There are many applications of ICLAS, as for instance the in situ analysis of multi-component gases in the atmosphere [36] or the detection of gas components in flames [37].

A detailed discussion of intracavity absorption, its dynamics, and its limitations can be found in the articles of Baev et al. [15, 27], in the thesis of Atmanspacher [23], and in several review articles [13, 28–31].

1.2.4 Cavity Ring-Down Spectroscopy (CRDS)

During the last years a new, very sensitive detection technique for measuring small absorptions, cavity ring-down spectroscopy (CRDS), has been developed and gradually improved. It is based on measurements of the decay times of optical resonators filled with the absorbing species [38]. We can understand its general principle as follows: Assume a short laser pulse with input power P_0 is sent through an optical resonator with two highly reflecting mirrors (reflectivities $R_1 = R_2 = R$, and transmission $T = 1 - R - A \ll 1$, where A includes all losses of the cavity from absorption, scattering, and diffraction, except those losses introduced by the absorbing sample). The pulse will be reflected back and forth between the mirrors (Fig. 1.17), while for each round-trip a small fraction will be transmitted through the end mirror and reach the detector. The transmitted power of the first output

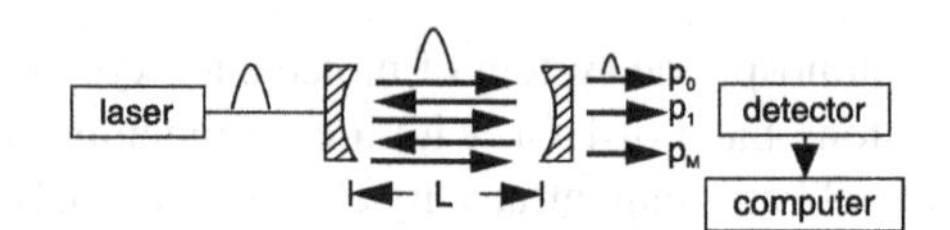

Fig. 1.17 Principle of cavity ring-down spectroscopy with pulsed lasers

pulse is

$$P_1 = T^2 e^{-\alpha L} \cdot P_0, \tag{1.24}$$

where α is the absorption coefficient of the gas sample inside the resonator with length L. For each round-trip, the power of the transmitted pulse decreases by an additional factor $R^2 \cdot \exp(-2\alpha L)$. After n round-trips, its power has decreased to

$$P_n = \left[R \cdot e^{-\alpha L}\right]^{2n} P_1 = \left[(1 - T - A)e^{-\alpha L}\right]^{2n} P_1, \tag{1.25}$$

which can be written as

$$P_n = P_1 \cdot e^{+2n(\ln R - \alpha \cdot L)}. \tag{1.26}$$

The reflectivity of the resonator mirrors is $R \geq 0.999$ and therefore $\ln R \approx R - 1 = -(T + A)$. Therefore, (1.26) can be written as

$$P_n = P_1 \cdot e^{-2n(T + A + \alpha L)}. \tag{1.27}$$

The time delay between successive transmitted pulses equals the cavity round-trip time $T_R = 2L/c$. The nth pulse therefore is detected at the time $t = 2nL/c$. If the time constant of the detector is large compared to T_R, the detector averages over subsequent pulses and the detected signals give the exponential function

$$P(t) = P_1 \cdot e^{-t/\tau_1}, \tag{1.28}$$

with the decay time

$$\tau_1 = \frac{L/c}{T + A + \alpha \cdot L}. \tag{1.29}$$

Without an absorbing sample inside the resonator ($\alpha = 0$), the decay time of the resonator will be lengthened to

$$\tau_2 = \frac{L/c}{T + A} \approx \frac{L/c}{1 - R}. \tag{1.30}$$

The absorption coefficient α can be obtained directly from the difference

$$1/\tau_1 - 1/\tau_2 = c \cdot \alpha \tag{1.31}$$

in the reciprocal decay times τ_i.

What level of accuracy can be achieved for the determination of α? Assume that the decay times τ_i can be measured within an uncertainty $\delta\tau_i$. From (1.31) we then obtain for the uncertainty $\delta\alpha$ of α

$$\frac{\partial\alpha}{\partial\tau_1} = -\frac{1}{c\tau_1^2}; \qquad \frac{\partial\alpha}{\partial\tau_2} = +\frac{1}{c\tau_2^2}.$$

With $\tau = \frac{1}{2}(\tau_1 + \tau_2)$ we obtain

$$\Rightarrow \quad \frac{\partial\alpha}{\partial\tau} = \frac{1}{c}\left(\frac{1}{\tau_2^2} - \frac{1}{\tau_1^2}\right) = \frac{\tau_1^2 - \tau_2^2}{\tau_1^2 \cdot \tau_2^2}.$$

For $\tau_1 - \tau_2 \ll \tau$ this gives

$$\delta\alpha \approx \frac{2\delta\tau}{\tau^3} \cdot (\tau_1 - \tau_2), \tag{1.32}$$

where τ is the average of τ_1 and τ_2.

In order to attain a high sensitivity for the determination of α, the decay times should be as long as possible; i.e., the reflectivity of the cavity mirrors should be as high as possible. The uncertainty $\delta\tau$ is mainly caused by the noise in the decay curves. A good signal-to-noise ratio therefore increases the accuracy of α.

Example 1.9 $R = 99.9\,\% = 0.999$; $L = d = 1$ m, $\alpha = 10^{-6}$ cm^{-1} $\Rightarrow \tau_1 = 3.03$ µs, $\tau_2 = 3.33$ µs. The difference $\Delta\tau = 0.3$ µs is rather small. If it can be measured within ± 0.03 µs then $\delta\alpha = 0.06$ µs$/(c\tau^2) = 4 \times 10^{-7}$ cm^{-1}. This means that α can only be determined with an uncertainty of 40 %.

However, for $R = 0.9999 \Rightarrow \tau_1 = 16.5$ µs and $\tau_2 = 33$ µs, $\Delta\tau = 16.5$ µs. Now $\delta\alpha = 6 \times 10^{-9}$ cm^{-1} and $\delta\alpha/\alpha = 0.6\,\%$. These figures demonstrate the importance of a high finesse cavity for the sensitivity of CRDS.

The CRDS technique uses the same principle as intra-cavity spectroscopy, namely increasing the effective absorption path length. The difference is that in CRDS the absorption coefficient is determined from a time measurement, i.e., the decay time of the "ringing cavity", while in intra-cavity spectroscopy the gain competition between different resonator modes is used as the enhancement factor.

If the reflectivity is very high, diffraction losses may become dominant, in particular for cavities with a large separation d of the mirrors. Since the TEM_{00} mode has the lowest diffraction losses, the incoming laser beam has to be mode-matched by a lens system to excite the fundamental mode of the resonator but not the higher transverse modes. Similar to intracavity absorption, this technique takes advantage of the increased effective absorption length $L_{eff} = L/(1 - R)$, because the laser pulse traverses the absorbing sample $1/(1 - R)$ times.

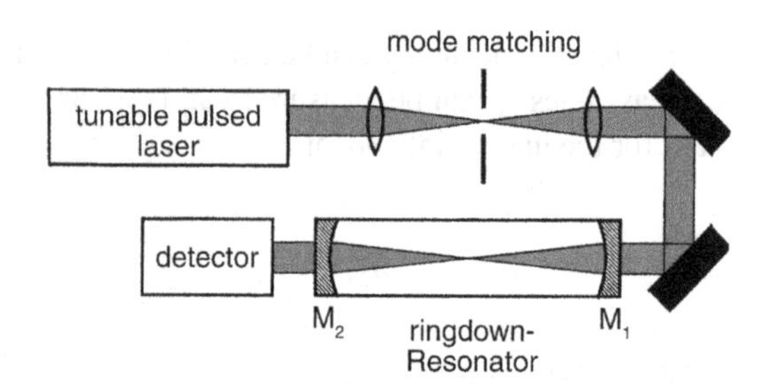

Fig. 1.18 Experimental setup with mode-matching optics

The experimental setup is shown in Fig. 1.18. The laser pulses are coupled into the resonator by carefully designed mode-matching optics, which ensure that only the TEM_{00} modes of the cavity are excited. Diffraction losses are minimized by spherical mirrors, which also form the end windows of the absorption cell. If the absorbing species are in a molecular beam inside the cavity, the mirrors form the windows of the vacuum chamber. For a sufficiently short input pulse ($T_p < T_R$), the output consists of a sequence of pulses with a time separation T_R and with exponentially decreasing intensities, which are detected with a boxcar integrator. For longer pulses ($T_p > T_R$), these pulses overlap in time and one observes a quasi-continuous exponential decay of the transmitted intensity. Instead of input pulses, the resonator can also be illuminated with cw radiation, which is suddenly switched off at $t = 0$.

If several resonator modes within the bandwidth of the laser pulse are excited, beat signals are superimposed onto the exponential decay curve. These beats are due to interference between the different modes with differing frequencies. They depend on the relative phases between the excited resonator modes. Since these phase differences vary from pulse to pulse when the cavity is excited by a train of input pulses, averaging over many excitation pulses smears out the interference pattern, and again a pure exponential decay curve is obtained.

When the laser wavelength λ is tuned over the absorption range of the molecules inside the resonator, the measured differences

$$\Delta = \frac{1}{\tau_1} - \frac{1}{\tau_2} = \frac{\tau_2 - \tau_1}{\tau_2 \cdot \tau_1} \approx \frac{\Delta\tau(\lambda)}{\tau^2} = c \cdot \alpha(\lambda)$$

yield the absorption spectrum $\alpha(\lambda)$ [39]. For illustration, the rotationally resolved vibrational overtone band $(2, 0, 5) \leftarrow (0, 0, 0)$ of the HCN molecule, obtained with cavity ring-down spectroscopy, is shown in Fig. 1.19.

The advantages of CRLS are:

(1) Intensity fluctuations of the laser play only a minor role, because the measured decay times do not depend on the laser intensity. They only increase the noise and therefore decrease the S/N-ratio.
(2) The repetition rate is several kilohertz. Therefore even within a few seconds the signal can be averaged over many excitation cycles, thus increasing the obtained S/N-ratio.
(3) With very high reflectivity mirrors long absorption paths can be realized which increases the absorption $\alpha \cdot L$ by a factor up to 10^4.

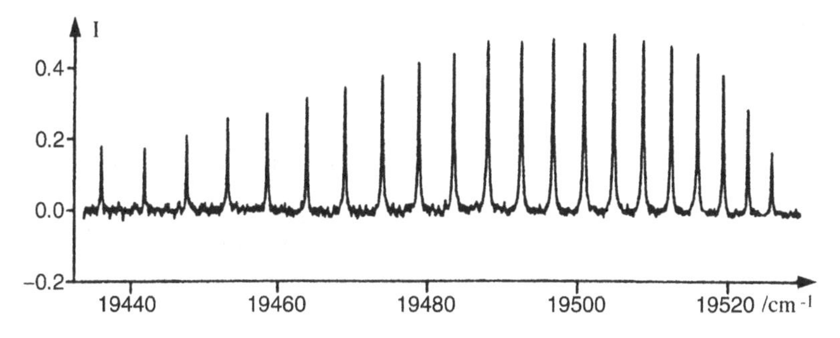

Fig. 1.19 Section of the rotational lines on the overtone band $(2, 0, 5) \leftarrow (0, 0, 0)$ of the HCN molecule, measured with CRDS [39]

The disadvantages are:

(1) Using narrow band tunable lasers the mode structure of the resonator with fixed mirror separation superimposes the absorption spectrum. When the resonator mode-frequencies are tuned synchronously with the laser frequency this problem can be avoided but the experimental expenditure becomes larger (see Vol. 1, Sect. 5.4.5).
(2) One needs really high reflectivity mirrors in order to reach a sensitivity comparable to other sensitive absorption techniques.

The possible high sensitivity of cavity ring-down spectroscopy can be achieved only if several conditions are met:

(a) The bandwidth $\delta\omega_R$ of all excited cavity modes should be smaller than the width $\delta\omega_a$ of the absorption lines. This implies that the laser pulsewidth $\delta\omega_L < \delta\omega_a$.
(b) The relaxation time T_R of the cavity must be longer than the lifetime T_{exc} of the excited molecules, which means $\delta\omega_R = 1/T_R < \delta\omega_a = 1/T_{exc}$.

Example 1.10 The resonator length is $L = 0.5$ m, the mirror reflectivity is $R = 0.995$; $\Rightarrow T_R = 3.3 \times 10^{-7}$ s, $\delta\omega_R = 3 \times 10^6$ s^{-1}. With a laser pulse duration of 10^{-8} s, the Fourier-limited laser bandwidth is $\delta\omega_L = 10^8$ s^{-1} $\rightarrow$ $\delta\nu = 1.5 \times 10^7$ s^{-1}. This is smaller than the Doppler width of absorption lines in the visible spectrum.

The frequency spacing of the longitudinal modes in a resonator with plane mirrors is 3×10^8 Hz, which shows that only one mode of the cavity is excited. In this case the resonator length L has to be synchroneously tuned when the laser wavelength is tuned (see Vol. 1, Sect. 5.5).

The major source of noise in single-shot decay is the technical noise introduced by the detection electronics and by fluctuations of the cavity length. Here an optical heterodyne detection technique can greatly enhance the signal-to-noise ratio. The experimental arrangement [40] is illustrated by Fig. 1.20. The output of a cw

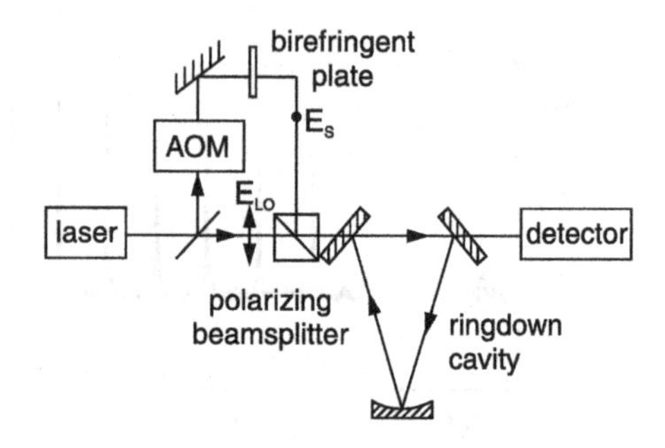

Fig. 1.20 Apparatus for heterodyne-detected cavity ring-down spectroscopy

tunable single mode diode laser is split into two parts. One part is directly sent into the cavity ring-down ring resonator and one of the TEM_{00} modes of the cavity is locked to the laser frequency. This part serves as the local oscillator. The other part is frequency shifted in an AOM (acousto-optic modulator) by one free spectral range of the cavity. It is therefore resonant with the adjacent TEM_{00} mode. It is chopped by the AOM at 40 kHz. Both parts are superimposed, pass through the ring-down cavity, and the total transmitted intensity

$$I_T \alpha \left| E_s(t) + E_{LO} \cdot e^{i(2\pi\delta\nu t+\phi)} \right|^2 = \left| E_s(t) \right|^2 + |E_{LO}|^2 + 2E_s \cdot E_{LO} \cdot \cos(2\pi\delta\nu t + \phi), \quad (1.33)$$

is measured by the detector. Since the signal enters the cavity as pulses of 12.5 μs duration at a repetition rate of 40 kHz, the transmitted signal intensity decays exponentially after the end of each pulse, while the local oscillator has constant intensity.

The interference term in (1.33) contains the product of the large amplitude E_{LO} (its transmission through the ring-down cavity is $T \approx 1$) and the small amplitude $E_s(t)$ and is, therefore, much larger than $(E_s(t))^2$. Measuring the decay time of the interference amplitude at the frequency $\delta\nu$ (= free spectral range of the cavity), therefore, gives a larger signal (which decays with 2τ) and thus a larger signal-to-noise ratio.

Because a single-mode cw laser is used, the spectral resolution is generally higher than in case of pulsed lasers. It is only limited by the linewidth of the absorption lines.

Instead of measuring the differences of decay times τ_1 and τ_2 with and without absorbing sample inside the cavity, the time-integrated transmitted intensity can be also monitored as a function of the wavelength of the incident laser pulse [44]. This method is similar to the absorption in external cavities, discussed in Sect. 1.2. An improved version of CRDS called "Cavity Leak-Out Spectroscopy (CALOS)" has been proposed and realized in several laboratories [45, 46, 48]. Here a cw laser with a frequency scan is used. Its output beam is mode-matched to the high finesse ring-down cavity (Fig. 1.21). Within each scan cycle there is a time where the laser frequency coincides with an eigen-resonance of the high finesse cavity. At this time the intra-cavity power builds up, the input beam is then shut off and the decay of the stored energy is observed for the empty cavity and the cavity filled with the absorbing gas.

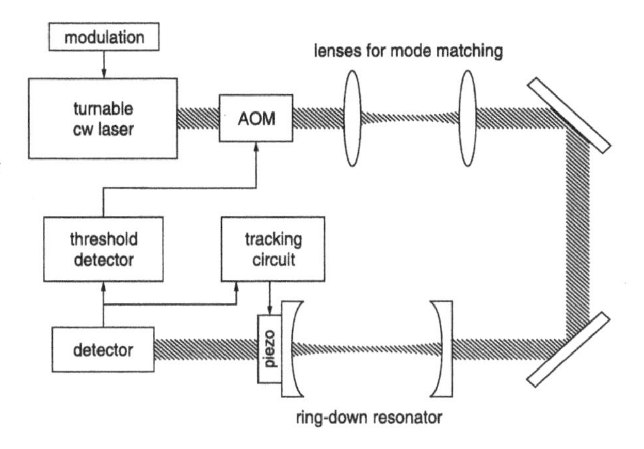

Fig. 1.21 Experimental setup of cavity leak-out spectroscopy with mode-matching optics and detector (P. Hering, Institute of Laser Medicine, University of Düsseldorf)

Either tunable cw lasers can be used or tunable sidebands of a fixed-frequency laser. Since the noise of cw-lasers is lower than for pulsed lasers, the sensitivity of CALOS is generally higher. Absorption coefficients down to $\alpha = 7 \times 10^{-11}\,\mathrm{cm}^{-1}\,\mathrm{Hz}^{-1/2}$ could be still measured and detection limits for molecular gases, relevant for medical applications, such as NO, CO, CO_2, NH_3 etc. down into the ppt-region (1 ppt means a relative concentration of 10^{-12}) have been achieved [46].

With CRDS atmospherically important radicals have been measured with high sensitivity [47].

A very sensitive modification of CRDS has been developed, where the instrumental improvements result in a new high-performance continuous-wave (cw) CRD spectrometer using a rapidly-swept cavity of simple design [49]. It employs efficient data-acquisition procedures, high-reflectivity mirrors, a low absorption flow cell, and various compact fibre-optical components in a single ended transmitter-receiver configuration suitable for remote sensing. By measuring spectroscopic features in the 1.525 μm band of C_2H_2 gas, the authors realised detection limits of 19 ntorr (2.5×10^{-11} atm) of net C_2H_2 (Doppler-limited at low pressure) and 0.37 ppbv of C_2H_2 in air (pressure-broadened at 1 atm).

Another technique uses as input an amplitude-modulated cw laser, which is split by a beam splitter into the probe beam passing the absorption cell inside the resonator and a reference beam [53]. Due to absorption and dispersion the probe beam suffers a phase shift φ of its modulation. Since the decay time τ of the resonator is related to the absorption, the phase shift depends on the decay time τ and the modulation frequency Ω by

$$\operatorname{tang} \varphi = \Omega \cdot \tau .$$

This phase-shift method is a general technique and is e.g. also applied to the measurement of lifetimes of excited atoms or molecules (see Sect. 6.3). The combination of CRDS with Fourier-spectroscopy gives for a fixed detection time a better signal-to noise ratio, because now all absorption lines within the covered spectral range are detected simultaneously. In Fig. 1.22 a possible experimental arrangement for this combined spectroscopic technique is schematically depicted. The transmitted laser

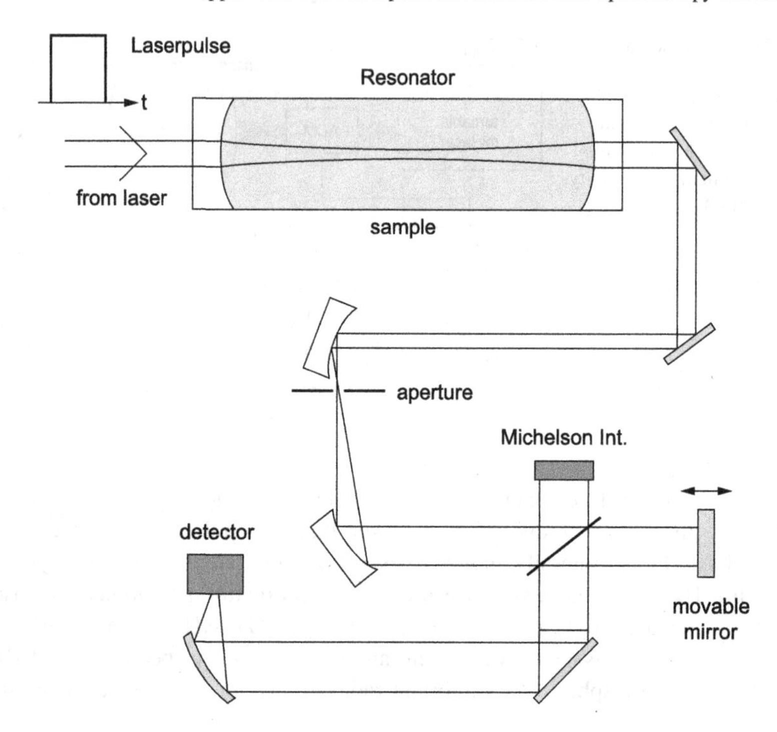

Fig. 1.22 Experimental setup for Fourier-transform spectroscopy combined with cavity ringdown spectroscopy [56]

beam is sent through a Michelson interferometer and the interference spectrum is recorded by a detector and the Fourier-transform of this spectrum is processed by a computer.

More information on CRDS and CALOS can be found in [41–43] and in several reviews [41, 45, 46, 48, 50–52, 54].

1.3 Direct Determination of Absorbed Photons

In the methods discussed in the preceding sections, the attenuation of the transmitted light (or of the laser power in intracavity spectroscopy) is monitored to determine the absorption coefficient $\alpha(\omega)$ or the number density of absorbing species. For small absorptions this means the measurement of a small difference of two large quantities, which limits, of course, the signal-to-noise ratio.

Several different techniques have been developed where the absorbed radiation power, i.e., the number of absorbed photons, can be directly monitored. These techniques belong to the most sensitive detection methods in spectroscopy, and it is worthwhile to know about them.

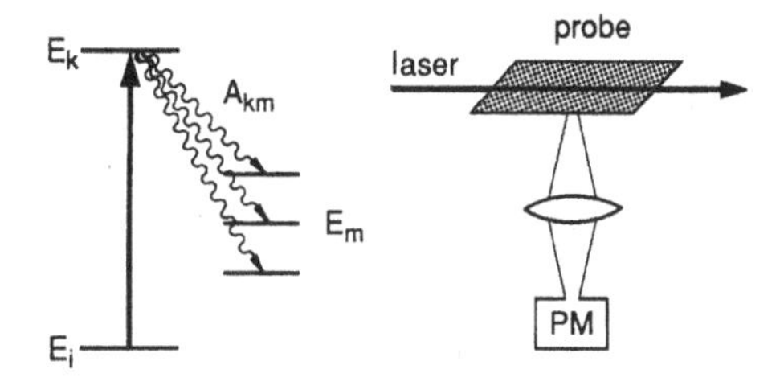

Fig. 1.23 Level scheme and experimental arrangement for fluorescence excitation spectroscopy

1.3.1 Fluorescence Excitation Spectroscopy

In the visible and ultraviolet regions a very high sensitivity can be achieved, if the absorption of laser photons is monitored through the laser-induced fluorescence (Fig. 1.23). When the laser wavelength λ_L is tuned to an absorbing molecular transition $E_i \to E_k$, the number of photons absorbed per second along the path length Δx is

$$n_a = N_i n_L \sigma_{ik} \Delta x, \tag{1.34}$$

where n_L is the number of incident laser photons per second, σ_{ik} the absorption cross section per molecule, and N_i the density of molecules in the absorbing state $|i\rangle$.

The number of fluorescence photons emitted per second from the excited level E_k is

$$n_{Fl} = N_k A_k = n_a \eta_k, \tag{1.35}$$

where $A_k = \sum_m A_{km}$ stands for the total spontaneous transition probability (Vol. 1, Sect. 2.8) to all levels with $E_m < E_k$. The quantum efficiency of the excited state $\eta_k = A_k/(A_k + R_k)$ gives the ratio of the spontaneous transition rate to the total deactivation rate, which may also include the radiationless transition rate R_k (e.g., collision-induced transitions). For $\eta_k = 1$, the number n_{Fl} of fluorescence photons emitted per second equals the number n_a of photons absorbed per second under stationary conditions.

Unfortunately, only the fraction δ of the fluorescence photons emitted into all directions can be collected, where $\delta = d\Omega/4\pi$ depends on the solid angle $d\Omega$, accepted by the fluorescence detector. Not every photon impinging onto the photomultiplier cathode releases a photoelectron; only the fraction $\eta_{ph} = n_{pe}/n_{ph}$ of these photons produces on the average n_{pe} photoelectrons. The quantity η_{ph} is called the *quantum efficiency of the photocathode* (Vol. 1, Sect. 4.5.2). The number n_{pe} of photoelectrons counted per second is then

$$n_{pe} = n_a \eta_k \eta_{ph} \delta = (N_i \sigma_{ik} n_L \Delta x) \eta_k \eta_{ph} \delta. \tag{1.36}$$

Example 1.11 Modern photomultipliers reach quantum efficiencies of $\eta_{ph} = 0.2$. With carefully designed optics it is possible to achieve a collection

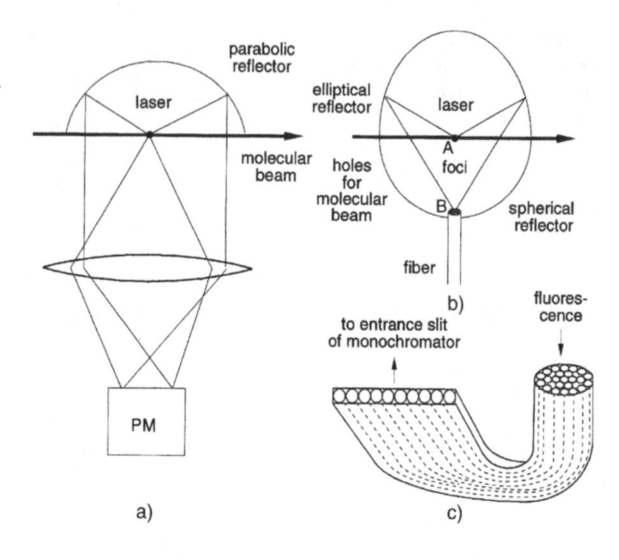

Fig. 1.24 (**a**) Parabolic optical system and (**b**) elliptical/spherical mirror system with high collection efficiency of fluorescence light. (**c**) Imaging of the fluorescence onto the entrance slit of a monochromator with a properly adjusted shape of the fiber bundle

factor $\delta = 0.1$, which implies that the collecting optics cover a solid angle $d\Omega = 0.4\pi$. Using photon-counting techniques and cooled multipliers (dark pulse rate $\leq$10 counts/s), counting rates of $n_{pe} = 100$ counts/s are sufficient to obtain a signal-to-noise ratio $S/R \sim 8$ at integration times of 1 s.

Inserting this figure for n_{PE} into (1.36) illustrates that with $\eta_k = 1$ absorption rates of $n_a = 5 \times 10^3$/s can be measured quantitatively. Assuming a laser power of 1 W at the wavelength $\lambda = 500$ nm, which corresponds to a photon flux of $n_L = 3 \times 10^{18}$/s, this implies *that it is possible to detect a relative absorption of* $\Delta P/P \leq 10^{-14}$. When placing the absorbing probe inside the cavity where the laser power is q times larger ($q \sim 10$ to 100, Sect. 1.2.3), this impressive sensitivity may be even further enhanced.

Since the attainable signal is proportional to the fluorescence collection efficiency δ, it is important to design collection optics with optimum values of δ. Two possible designs are shown in Fig. 1.24 that are particularly useful if the excitation volume is small (e.g., the crossing volume of a laser beam with a collimated molecular beam). One of these collection optics uses a parabolic mirror, which collects the light from a solid angle of nearly 2π. A lens images the light source onto the photomultiplier cathode. The design of Fig. 1.24b uses an elliptical mirror where the light source is placed in one focal point A and the polished end of a multifiber bundle into the other B. A half-sphere reflector with its center in A reflects light, which is emitted into the lower half space back into the source A, which is then further reflected by the elliptical mirror and focused into B.

The exit end of the fiber bundle can be arranged to have a rectangular shape in order to match it to the entrance slit of a spectrograph (Fig. 1.24c).

When the laser wavelength λ_L is tuned across the spectral range of the absorption lines, the total fluorescence intensity $I_{Fl}(\lambda_L) \propto n_L \sigma_{ik} N_i$ monitored as a function of laser wavelength λ_L represents an image of the absorption spectrum called the *excitation spectrum*. According to (1.36) the photoelectron rate n_{PE} is directly proportional to the absorption coefficient $N_i \sigma_{ik}$, where the proportionality factor depends on the quantum efficiency η_{ph} of the photomultiplier cathode and on the collection efficiency δ of the fluorescence photons.

Although the excitation spectrum directly reflects the absorption spectrum with respect to the line positions, the relative *intensities* of different lines $I(\lambda)$ are identical in both spectra only if the following conditions are guaranteed:

- The quantum efficiency η_k must be the same for all excited states E_k. Under collision-free conditions, i.e., at sufficiently low pressures, the excited molecules radiate before they can collide, and we obtain $\eta_k = 1$ for all levels E_k.
- The quantum efficiency η_{ph} of the detector should be constant over the whole spectral range of the emitted fluorescence. Otherwise, the spectral distribution of the fluorescence, which may be different for different excited levels E_k, will influence the signal rate. Some modern photomultipliers can meet this requirement.
- The geometrical collection efficiency δ of the detection system should be identical for the fluorescence from different excited levels. This demand excludes, for example, excited levels with very long lifetimes, where the excited molecules may diffuse out of the observation region before they emit the fluorescence photon. Furthermore, the fluorescence may not be isotropic, depending on the symmetry of the excited state. In this case δ will vary for different upper levels.

However, even if these requirements are not strictly fulfilled, excitation spectroscopy is still very useful to measure absorption lines with extremely high sensitivity, although their relative intensities may not be recorded accurately.

The technique of excitation spectroscopy has been widely used to measure very small absorptions. One example is the determination of absorption lines in molecular beams where both the pathlength Δx and the density N_i of absorbing molecules are small.

The LIF method is illustrated by Fig. 1.25, which shows a section of the excitation spectrum of the silver dimer Ag_2, taken in a collimated molecular beam under conditions comparable to that of Example 1.12.

Example 1.12 With $\Delta x = 0.1$ cm, $\delta = 0.5$, $\eta = 1$, a density $N_i = 10^7/\text{cm}^3$ of absorbing molecules yields for an absorption cross section $\sigma_{ik} = 10^{-17}$ cm^2 and an incident flux of $n_L = 10^{16}$ photons/s (=3 mW at $\lambda = 500$ nm) about 5×10^4 fluorescence photons imaged onto the photomultiplier cathode, which emits about 1×10^4 photoelectrons per second, giving rise to 10^4 counts/s at the exit of the PM.

The extremely high sensitivity of this technique has been demonstrated impressively by Fairbanks et al. [57], who performed absolute density measurements of

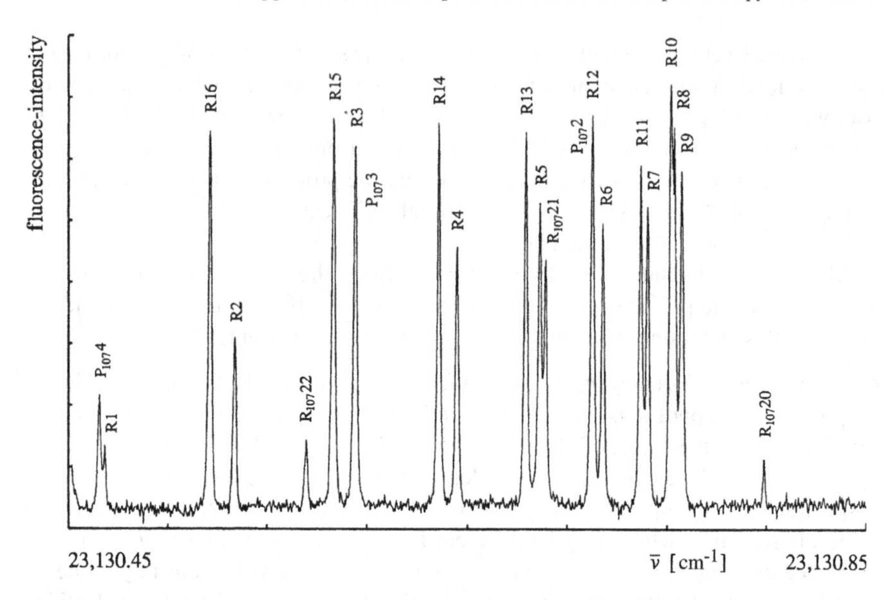

Fig. 1.25 Section of the fluorescence excitation spectrum of the $^{107}Ag^{109}Ag$ isotope, showing the band head of the $v' = 1 \leftarrow v'' = 0$ band in the $A\,^1\Sigma_u \leftarrow X\,^1\Sigma_g$ system, superimposed by some lines of the $^{107}Ag^{107}Ag$ isotope [58]

sodium vapor in the density range $N = 10^2$–10^{11} cm^{-3} using laser-excited fluorescence as a monitor. The lower detection limit of $N = 10^2$ cm^{-3} was imposed by background stray light scattered out of the incident laser beam by windows and cell walls.

Because of the high sensitivity of excitation spectroscopy, it can be successfully used to monitor minute concentrations of radicals and short-lived intermediate products in chemical reactions [59]. Besides measurements of small concentrations, detailed information on the internal state distribution $N_i(v_i'', J_i'')$ of reaction products can be extracted, since the fluorescence signal is, according to (1.36), proportional to the number N_i of absorbing molecules in the level $|i\rangle$ (Sect. 1.8.5).

Often excitation spectra of polyatomic molecules are rather complex and difficult to analyze, in particular if several absorption lines overlap with the laser line profile. In such cases "*filtered excitation spectra*" are helpful where instead of the total fluorescence from all excited levels only fluorescence lines selected through a spectrometer are detected which are emitted from a single upper level. Although this decreases the detected fluorescence intensity, it simplifies the excitation spectrum and makes the assignment much easier.

If in atoms a transition $|i\rangle \to |k\rangle$ can be selected, which represents a true two level system (i.e., the fluorescence from $|k\rangle$ terminates only in $|i\rangle$), the atom may be excited many times while it flies through the laser beam. At a spontaneous lifetime τ and a travel time T through the laser beam, a maximum of $n = T/(2\tau)$ excitation-fluorescence cycles can be achieved (photon burst). With $T = 10^{-5}$ s and $\tau = 10^{-8}$ s we can expect $n \mathrel{\hat{=}} 500$ fluorescence photons per atom! This allows single atom detection.

If molecules are diluted in solutions or in solids, single molecules can be excited if the focused laser beam diameter is smaller than the average distance between the molecules. Although the fluorescence from the upper state terminates in many rovibrational levels of the electronic ground state, these levels are rapidly quenched by collisions with the solvent molecules, which brings them back into the initial level, from where they can again be excited by the laser. This allows the detection of many photons per second from a single molecule and gives sufficient sensitivity to follow the diffusion of molecules into and out of the laser spot. This technique of "single-molecule detection" has developed into a very useful and sensitive method in chemistry and biology [60–62].

Excitation spectroscopy has its highest sensitivity in the visible, ultraviolet, and near-infrared regions. With increasing wavelength λ the sensitivity decreases for the following reasons: Eq. (1.36) shows that the detected photoelectron rate n_{PE} decreases with η_k, η_{ph}, and δ. All these numbers generally decrease with increasing wavelength. The quantum efficiency η_{ph} and the attainable signal-to-noise ratio are much lower for infrared than for visible photodetectors (Vol. 1, Sect. 4.5). By absorption of infrared photons, vibrational–rotational levels of the electronic ground state are excited with radiative lifetimes that are generally several orders of magnitude larger than those of excited *electronic* states. At sufficiently low pressures the molecules diffuse out of the observation region before they radiate. This diminishes the collection efficiency δ. At higher pressures the quantum efficiency η_k of the excited level E_k is decreased because collisional deactivation competes with radiative transitions. Under these conditions photoacoustic detection may become preferable.

1.3.2 Photoacoustic Spectroscopy

Photoacoustic spectroscopy is a sensitive technique for measuring small absorptions that is mainly applied when minute concentrations of molecular species have to be detected in the presence of other components at higher pressure. An example is the detection of spurious pollutant gases in the atmosphere. Its basic principle may be summarized as follows: The laser beam is sent through the absorber cell (Fig. 1.26). If the laser is tuned to the absorbing molecular transition $E_i \to E_k$, part of the molecules in the lower level E_i will be excited into the upper level E_k. By collisions with other atoms or molecules in the cell, these excited molecules may transfer their excitation energy $(E_k - E_i)$ completely or partly into translational, rotational, or vibrational energy of the collision partners. At thermal equilibrium, this energy is randomly distributed onto all degrees of freedom, causing an increase of thermal energy and with it a rise in temperature and pressure at a constant density in the cell.

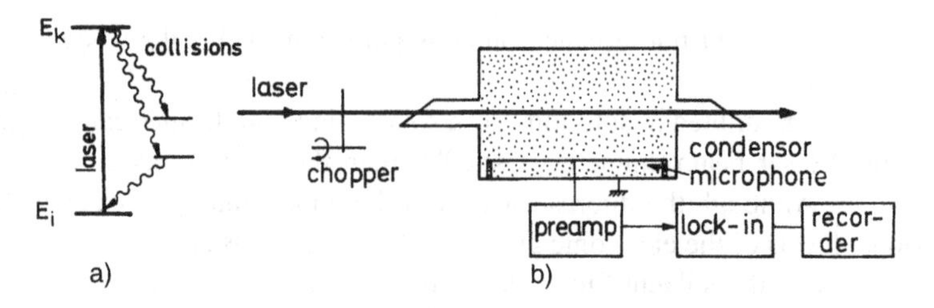

Fig. 1.26 Photoacoustic spectroscopy: (**a**) level scheme; (**b**) schematic experimental arrangement

When the laser beam is chopped at frequencies $\Omega < 1/T$, where T is the mean relaxation time of the excited molecules, periodic pressure variations appear in the absorption cell, which can be detected with a sensitive microphone placed inside the cell. The output signal S [Volt] of the microphone is proportional to the pressure change Δp induced by the absorbed radiation power $\Delta W/\Delta t$. If saturation can be neglected, the absorbed energy per cycle

$$\Delta W = N_i \sigma_{ik} \Delta x (1 - \eta_K) P_L \Delta t, \tag{1.37}$$

is proportional to the density N_i [cm^{-3}] of the absorbing molecules in level $|i\rangle$, the absorption cross section σ_{ik} [cm^2], the absorption pathlength Δx, the cycle period Δt, and the incident laser power P_L.

Contrary to LIF, the optoacoustic signal *decreases* with increasing quantum efficiency η_k (which gives the ratio of emitted fluorescence energy to absorbed laser energy) unless the fluorescence is absorbed inside the cell and contributes to the temperature rise.

Since the absorbed energy ΔW is transferred into kinetic or internal energy of all N molecules per cm^3 in the photoacoustic cell with the volume V, the temperature rise ΔT is obtained from

$$\Delta W = \frac{1}{2} f V N k \Delta T, \tag{1.38}$$

where f is the number of degrees of freedom that are accessible for each of the N molecules at the temperature T [K]. If the chopping frequency of the laser is sufficiently high, the heat transfer to the walls of the cell during the pressure rise time can be neglected. From the equation of state $pV = NVkT$, we finally obtain

$$\Delta p = N k \Delta T = \frac{2\Delta W}{fV}. \tag{1.39}$$

It is therefore advantageous to keep the volume V of the photoacoustic cell small. The output signal S from the microphone is then

$$S = \Delta p S_m = \frac{2N_i \sigma_{ik}}{fV} \Delta x (1 - \eta_k) P_L \Delta t S_m, \tag{1.40}$$

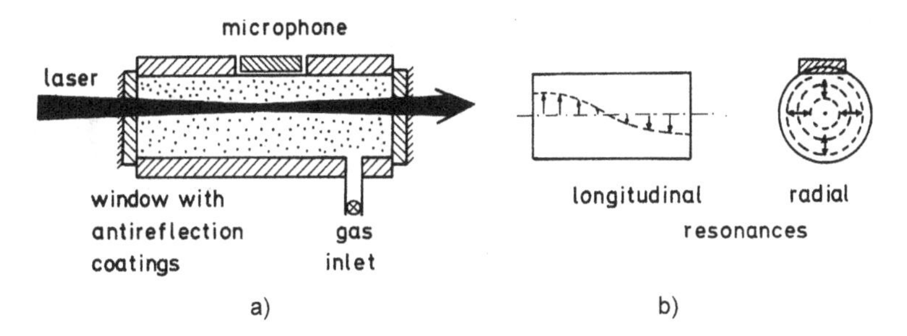

Fig. 1.27 (**a**) Spectrophone with capacitance microphone; (**b**) longitudinal and radial acoustic resonance modes

where the sensitivity S_m [Volt/Pascal] of the microphone not only depends on the characteristics of the microphone but also on the geometry of the photoacoustic cell.

With infrared lasers the molecules are generally excited into higher vibrational levels of the electronic ground state. Assuming cross sections of 10^{-18}–10^{-19} cm^2 for the collisional deactivation of the vibrationally excited molecules, the equipartition of energy takes only about 10^{-5} s at pressures around 1 mbar. Since the spontaneous lifetimes of these excited vibrational levels are typically around 10^{-2}–10^{-5} s, it follows that at pressures above 1 mbar, the excitation energy absorbed from the laser beam will be almost completely transferred into thermal energy, which implies that $\eta_k \sim 0$.

The idea of the spectraphone is very old and was demonstrated by Bell and Tyndal [63] in 1881. However, the impressive detection sensitivity obtained today could only be achieved with the development of lasers, sensitive capacitance microphones, low-noise amplifiers, and lock-in techniques. Concentrations down to the parts per billion range (ppb, or 10^{-9}) at total pressures of 1 mbar up to several atmospheres are readily detectable with a modern spectraphone (Fig. 1.27).

Modern condenser microphones with a low-noise FET preamplifier and phase-sensitive detection achieve signals of larger than 1 V/mbar ($\hat{=}10$ mV/Pa) with a background noise of 3×10^{-8} V at integration times of 1 s. This sensitivity allows detection of pressure amplitudes below 10^{-7} mbar and is, in general, not limited by the electronic noise but by another disturbing effect: laser light reflected from the cell windows or scattered by aerosols in the cell may partly be absorbed by the walls and contributes to a temperature increase. The resulting pressure rise is, of course, modulated at the chopping frequency and is therefore detected as background signal. There are several ways to reduce this phenomenon. Antireflection coatings of the cell windows or, in case of linearly polarized laser light, the use of Brewster windows minimize the reflections. An elegant solution chooses the chopping frequency to coincide with an acoustic resonance of the cell. This results in a resonant amplification of the pressure amplitude, which may be as large as 1000 fold. This experimental trick has the additional advantage that those acoustic resonances can be selected that couple most efficiently to the beam profile but are less effectively excited by heat conduction from the walls. The background

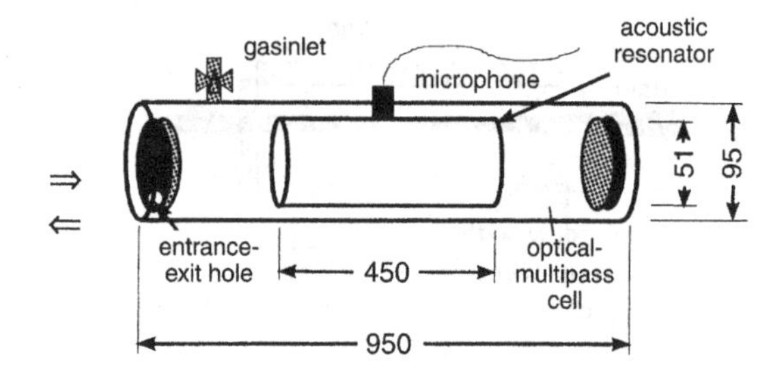

Fig. 1.28 Acoustic resonance cell inside an optical multipass cell. Dimensions are in millimeters

signal caused by wall absorption can thus be reduced and the true signal is enhanced. Figure 1.27b shows longitudinal and radial acoustic resonances in a cylindrical cell.

Example 1.13 With $N_i = 2.5 \times 10^{11}\ \text{cm}^{-3}$ ($\hat{=} 10^{-8}$ bar), $\sigma_{ik} = 10^{-16}\ \text{cm}^2$, $\Delta x = 10$ cm, $V = 50\ \text{cm}^3$, $\eta_k = 0$, $f = 6$, we obtain the pressure change $\Delta p = 1.5$ Pa ($\hat{=}0.015$ mbar) for the incident laser power $P_L = 100$ mW. With a microphone sensitivity of $S_m = 10^{-2}$ V/Pa the output signal becomes $S =$ 15 mV.

The sensitivity can be further enhanced by frequency modulation of the laser (Sect. 1.2.2) and by intracavity absorption techniques. With the spectraphone inside the laser cavity, the photoacoustic signal due to nonsaturating transitions is increased by a factor q as a result of a q-fold increase of the laser intensity inside the resonator (Sect. 1.2.3). The optoacoustic cell can be placed inside a multipath optical cell (Fig. 1.28) where an effective absorption pathlength of about 50 m can be readily realized [74].

According to (1.40) the optoacoustic signal decreases with increasing quantum efficiency because the fluorescence carries energy away without heating the gas, as long as the fluorescence light is not absorbed within the cell. Since the quantum efficiency is determined by the ratio of spontaneous to collision-induced deactivation of the excited level, it decreases with increasing spontaneous lifetime and gas pressure. Therefore, the optoacoustic method is particularly favorable to monitor vibrational spectra of molecules in the infrared region (because of the long lifetimes of excited vibrational levels) and to detect small concentrations of molecules in the presence of other gases at higher pressures (because of the large collisional deactivation rate). It is even possible to use this technique for measuring *rotational* spectra in the microwave region as well as electronic molecular spectra in the visible or ultraviolet range, where electronic states with short spontaneous lifetimes are excited. However, the sensitivity in these spectral regions is not quite as high and there are other methods that are superior.

Some examples illustrate this very useful spectroscopic technique. For a more detailed discussion of optoacoustic spectroscopy, its experimental tricks, and its

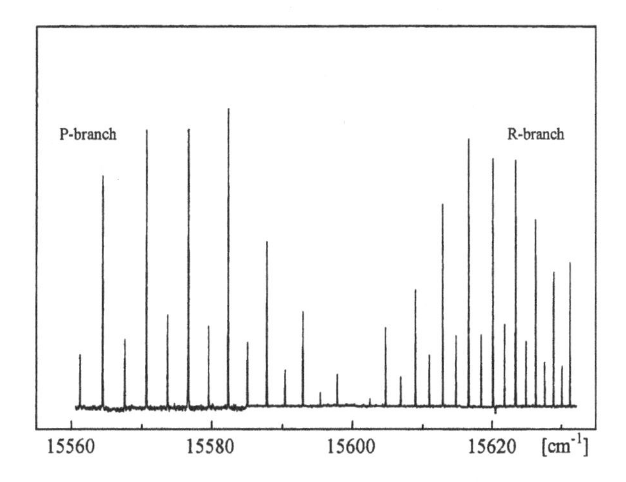

Fig. 1.29 Section of the optoacoustic overtone absorption spectrum $(5, 1, 0, 0, 0, 0) \leftarrow (0, 0, 0, 0, 0, 0)$ of acetylene around $\bar{\nu} = 15{,}600\ \text{cm}^{-1}$, corresponding to the excitation of a local mode by five vibrational quanta [74]

various applications [68], the reader is referred to recently published monographs [64, 65, 67, 80] and to conference proceedings [69–71].

Example 1.14

(a) The sensitivity of the spectraphone has been demonstrated by Kreutzer et al. [72]. At a total air pressure of 660 mbar in the absorption cell these researchers could detect concentrations of ethylene down to 0.2 ppb, of NH_3 down to 0.4 ppb, and of NO pollutants down to 10 ppb. The feasibility of determining certain important isotope abundances or ratios by simple and rapid infrared spectroscopy with the spectraphone and also the ready control of small leaks of polluting or poison gases have been demonstrated [73].

(b) The optoacoustic method has been applied with great success to high-resolution spectroscopy of rotational–vibrational bands of numerous molecules [74]. Figure 1.29 illustrates as an example a section of the visible overtone absorption spectrum of the C_2H_2 molecule where, in spite of the small absorption coefficient, a good signal-to-noise ratio could be achieved.

(c) A general technique for the optoacoustic spectroscopy of *excited* molecular vibrational states has been demonstrated by Patel [75]. This technique involves the use of vibrational energy transfer between two dissimilar molecules A and B. When A is excited to its first vibrational level by absorption of a laser photon $h\nu_1$, it can transfer its excitation energy by a near-resonant collision to molecule B. Because of the large cross section for such collisions, a high density of vibrationally excited molecules B can be achieved also for those molecules that cannot be pumped directly by existing powerful laser lines. The excited molecule B can absorb a photon $h\nu_2$ from a second, weak tunable laser, which allows spectroscopy of all accessible transitions ($v = 1 \rightarrow v = 2$). The technique has been proved for the NO molecule, where the frequency of the four transitions in the $^2\Pi_{1/2}$ and $^2\Pi_{3/2}$ subbands of ^{15}NO and the Λ doubling for the $v = 1 \rightarrow 2$ transition have been accurately measured. The following scheme illustrates the method:

$$^{14}\text{NO} + h\nu_1(\text{CO}_2\ \text{laser}) \longrightarrow {}^{14}\text{NO}^*(v = 1),$$

$$^{14}\text{NO}^*(v = 1) + {}^{15}\text{NO}(v = 0) \longrightarrow {}^{14}\text{NO}(v = 0) + {}^{15}\text{NO}^*(v = 1) + \Delta E\left(35\ \text{cm}^{-1}\right),$$

$$^{15}\text{NO}^*(v = 1) + h\nu_2(\text{spin-flip laser}) \longrightarrow {}^{15}\text{NO}^*(v = 2).$$

The last process is detected by optoacoustic spectroscopy.

(d) The application of photoacoustic detection to the visible region has been reported by Stella et al. [76]. They placed the spectraphone inside the cavity of a cw dye laser and scanned the laser across the absorption bands of the CH_4 and NH_3 molecules. The high-quality spectra with resolving power of over 2×10^5 proved to be adequate to resolve single rotational features of the very weak vibrational overtone transitions in these molecules. The experimental results are very useful for the investigation of the planetary atmospheres, where such weak overtone transitions are induced by the sun light.

(e) An interesting application of photoacoustic detection is in the measurement of dissociation energies of molecules [77]. When the laser wavelength is tuned across the dissociation limit, the photoacoustic signal drops drastically because beyond this limit the absorbed laser energy is used for dissociation. (This means that it is converted into potential energy and cannot be transferred into kinetic energy as in the case of excited-state deactivation. Only the kinetic energy causes a pressure increase.)

(f) With a special design of the spectraphone employing a coated quartz membrane for the condenser microphone, even corrosive gases can be measured [78]. This extends the applications of optoacoustic spectroscopy to aggressive pollutant gases such as NO_2 or SO_2, which are important constituents of air pollution.

Photoacoustic spectroscopy can be also applied to liquids and solids [79]. An interesting application is the determination of species adsorbed on surfaces. Optoacoustic spectroscopy allows a time-resolved analysis of adsorption and desorption processes of atoms or molecules at liquid or solid surfaces, and their dependence on the surface characteristics and on the temperature [81]. For more information see [79, 82, 83].

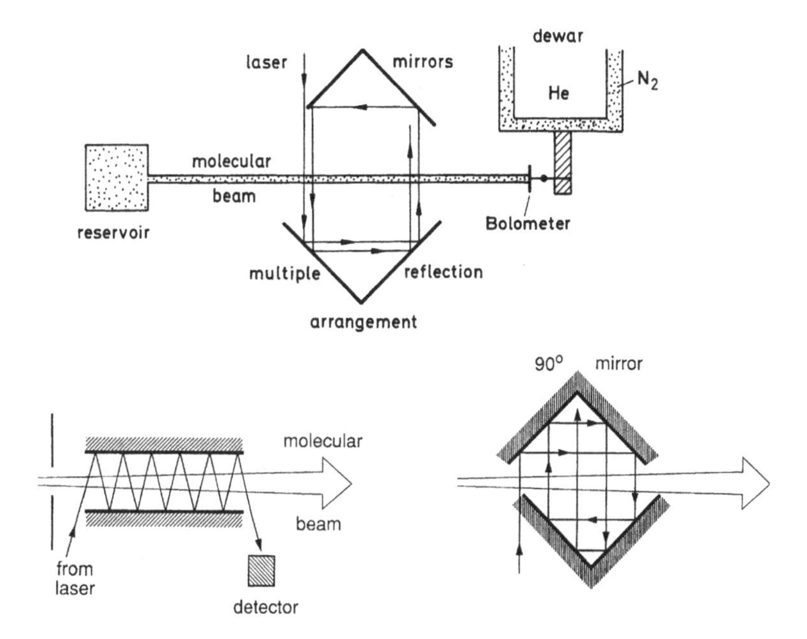

Fig. 1.30 Optothermal spectroscopy in a molecular beam with a helium-cooled bolometer as detector and two optical systems, which increase the absorption path length

1.3.3 Optothermal Spectroscopy

For the spectroscopy of vibrational–rotational transitions in molecules the laser-excited fluorescence is generally not the most sensitive tool, as was discussed at the end of Sect. 1.3.1. Optoacoustic spectroscopy, on the other hand, is based on collisional energy transfer and is therefore not applicable to molecular beams, where collisions are rare or even completely absent. For the infrared spectroscopy of molecules in a molecular beam therefore a new detection technique has been developed, which relies on the collision-free conditions in a beam and on the long radiative lifetimes of vibrational–rotational levels in the electronic ground state [84–86].

Optothermal spectroscopy uses a cooled bolometer (Vol. 1, Sect. 4.5) to detect the excitation of molecules in a beam (Fig. 1.30). When the molecules hit the bolometer they transfer their kinetic and their internal thermal energy, thereby increasing the bolometer temperature T_0 by an amount ΔT. If the molecules are excited by a tunable laser (for example, a color-center laser or a diode laser) their vibrational–rotational energy increases by $\Delta E = h\nu \gg E_{\text{kin}}$. If the lifetime τ of the excited levels is larger than the flight time $t = d/v$ from the excitation region to the bolometer, they transfer this extra energy to the bolometer. If N excited molecules hit the bolometer per second, the additional rate of heat transfer is

$$\frac{\mathrm{d}Q}{\mathrm{d}t} = N\Delta E = Nh\nu. \tag{1.41}$$

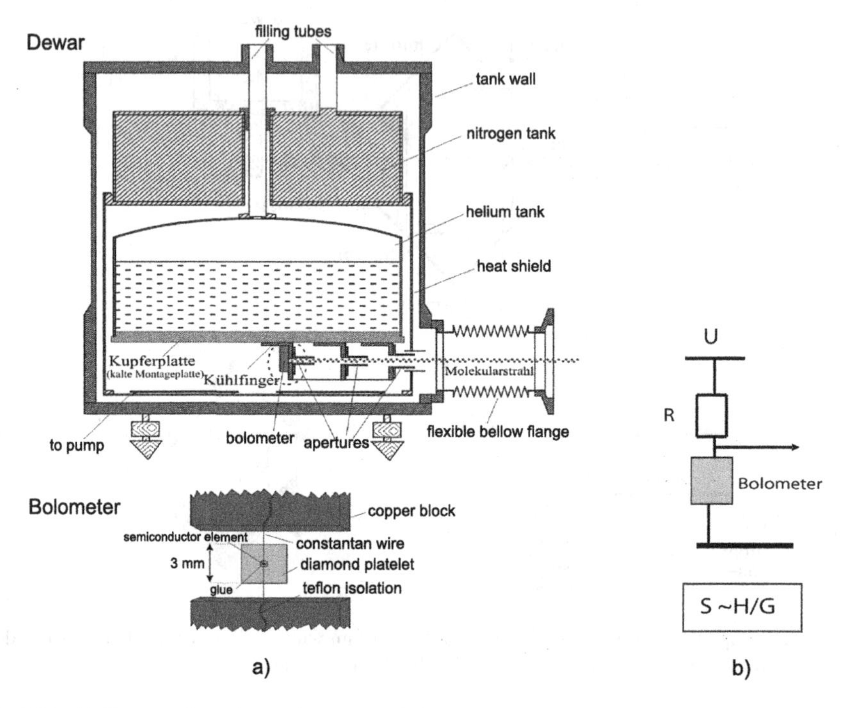

Fig. 1.31 (**a**) Construction of the bolometer cooled down to 1.5 K. (**b**) Electric circuit

With the heat capacity C of the bolometer and a heat conduction $G(T - T_0)$ the temperature T is determined by

$$Nh\nu = C\frac{\mathrm{d}T}{\mathrm{d}t} + G(T - T_0). \tag{1.42}$$

Under stationary conditions ($\mathrm{d}T/\mathrm{d}t = 0$), we obtain from (1.42) the temperature rise

$$\Delta T = T - T_0 = \frac{Nh\nu}{G}. \tag{1.43}$$

The sensitivity of the bolometer increases with decreasing heat conduction G. In general, the exciting laser beam is chopped in order to increase the signal-to-noise ratio by lock-in detection. The time constant $\tau = C/G$ of the bolometer (Vol. 1, Sect. 4.5) should be smaller than the chopping period. Therefore the bolometer must be constructed in such a way that both C and G are as small as possible.

The construction of the bolometer is shown in Fig. 1.31. The thermally isolated tank is filled with liquid nitrogen as precooler and with liquid helium as cooling medium. When the evaporated vapour above the liquid helium surface is pumped away, evaporation cooling drops the temperature down to about 1.5 K. The semiconductor detector (Fig. 1.32) with a preamplifier is mounted on the bottom of the

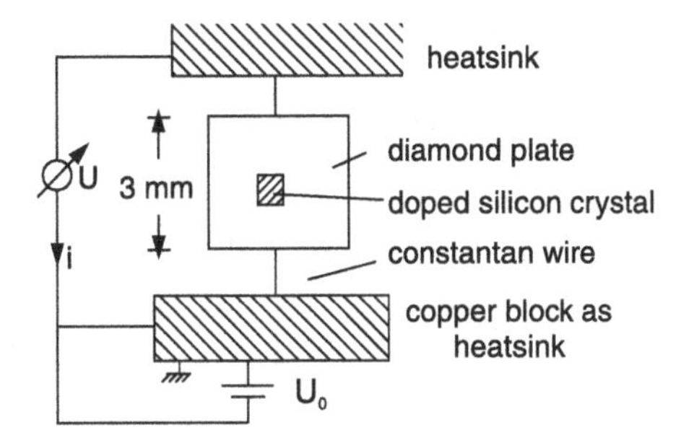

Fig. 1.32 Central part of the bolometer. The diamond plate increases the sensitivity area without contributing much to the heat capacity

liquid Helium container and has therefore the temperature of the helium tank. Apertures in front of the detector prevent heat radiation from the vacuum vessel walls, which are at room temperature, to reach the detector.

The bolometer consists of a small ($0.25 \times 0.25 \times 0.25\ \mathrm{mm}^3$) doped silicon semiconductor crystal (Fig. 1.32), which has a low heat capacity at $T = 1.5$ K.

The collimated molecular beam generally has a diameter of about 3 mm when reaching the bolometer, whereas the active area of the sensitive silicon disc in the bolometer is only $0.5 \times 0.5\ \mathrm{mm}^2$. This means that only a small fraction of the beam would hit the bolometer and contribute to the signal. In order to improve the situation the silicon disc is glued with a heat-conductive glue to a larger thin sapphire disc with an area of $3 \times 3\ \mathrm{mm}^2$. Since sapphire has a high Debye temperature, at 1.5 K only very few phonons are excited, which implies that the specific heat is very low. In spite of its larger size the sapphire disc provides only a small contribution to the total heat capacity of the bolometer, although it greatly enhances the signal (Fig. 1.32).

If a small electric current i is sent through the crystal with resistance R, a voltage $U = R \cdot i$ is generated across the bolometer. The resistance decreases with increasing temperature. The temperature change ΔT can be measured by the resulting resistance change

$$\Delta R = \frac{\mathrm{d}R}{\mathrm{d}T}\Delta T,$$

which is a function of the temperature dependence $\mathrm{d}R/\mathrm{d}T$ of the bolometer material. Large values of $\mathrm{d}R/\mathrm{d}T$ are achieved with doped semiconductor materials at low temperatures (a few Kelvin!).

Even larger values can be realized with materials around their critical temperature T_c for the transition from the superconducting to the normal conducting state. In this case, however, one always has to keep the temperature at T_c. This can be achieved by a temperature feedback control, where the feedback signal is a measure for the rate $\mathrm{d}Q/\mathrm{d}t$ of energy transfer to the bolometer by the excited molecules.

With such a detector at $T = 1.5$ K, energy transfer rates $\mathrm{d}Q/\mathrm{d}t \geq 10^{-14}$ W are still detectable [85]. This means that an absorbed laser power of $\Delta P \geq 10^{-14}$ W is measurable. In order to maximize the absorbed power, the absorption path length can be increased by an optical device consisting of two 90° reflectors, which reflect

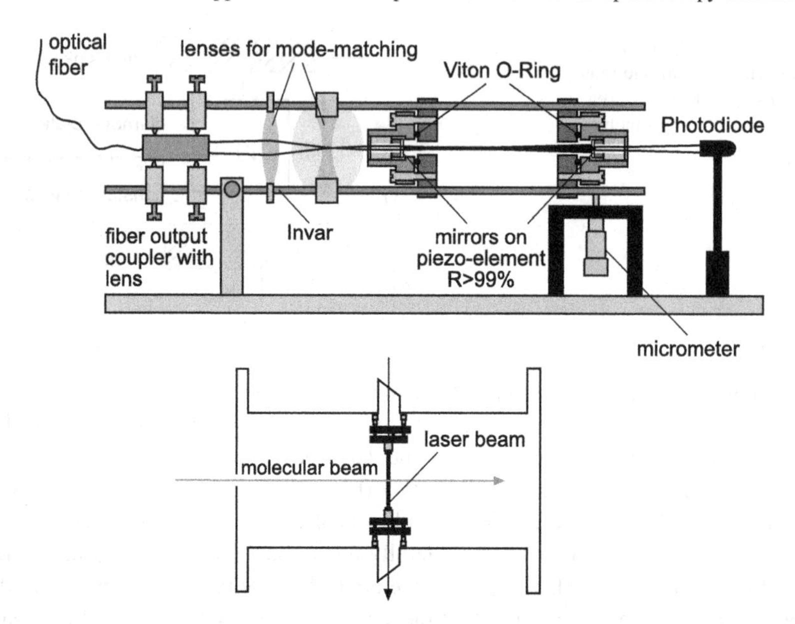

Fig. 1.33 Excitation of molecules in a collimated beam inside an optical enhancement cavity

the laser beam many times back and forth through the molecular beam (Fig. 1.30). An even higher sensitivity is achieved with an optical enhancement cavity with spherical mirrors, where the optical beam waist at the center of the cavity has to be matched to the molecular beam diameter (Fig. 1.33). Choosing the optimum values of the mirror distance d and mirror curvature radius r the beam waist area πw_0^2 can be made equal to the cross section of the molecular beam. Then all molecules pass through the laser beam. The laser beam is coupled into the resonator through a single mode optical fiber and mode-matched to the fundamental mode of the cavity by a lens system. When the laser wavelength is tuned, the enhancement cavity has to be synchronously tuned, which is accomplished by including a piezo-cylinder and an electronic feedback circuit that keeps the cavity in resonance with the laser wavelength.

The sensitivity of the optothermal technique is illustrated by the comparison of the same sections of the overtone spectrum of C_2H_4 molecules, measured with Fourier, optoacoustic, and optothermal spectroscopy, respectively (Fig. 1.34). Note the increase of spectral resolution and signal-to-noise ratio in the optothermal spectrum compared to the two other Doppler-limited techniques [87]. More examples can be found in [88].

The term photothermal spectroscopy is also used in the literature for the generation of thermal-wave phenomena in samples that are illuminated with a time-dependent light intensity [89]. In many aspects this is equivalent to photoacoustic spectroscopy. An interesting modification of this technique for the study of molecules adsorbed at surfaces is illustrated in Fig. 1.35. A small spot of a surface is irradiated by a pulsed laser beam. The absorbed power results in a tem-

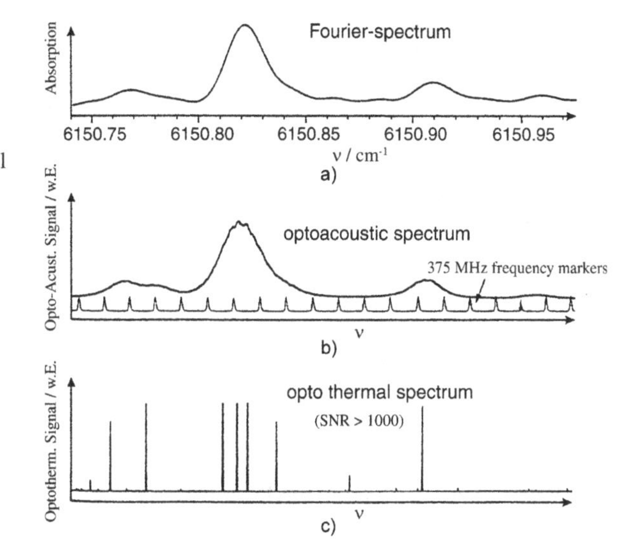

Fig. 1.34 Section of the overtone band ($\nu_5 + \nu_9$) of C_2H_4, comparison of (**a**) Fourier spectrum; (**b**) optoacoustic spectrum; (**c**) Doppler-free optothermal spectrum

perature rise of the illuminated spot. This temperature travels as a thermal shock wave through the solid. It leads to a slight time-dependent deformation of the surface from thermal expansion. This deformation is probed by the time-dependent deflection of a low-power HeNe laser beam. In the case of adsorbed molecules, the absorbed power varies when the laser wavelength is tuned over the absorption spectrum of the molecules. Time-resolved measurements of the probe-beam deflection allow one to determine the kind and amount of adsorbed molecules and to follow their desorption with time, caused by the irradiating light of the pulsed laser [90].

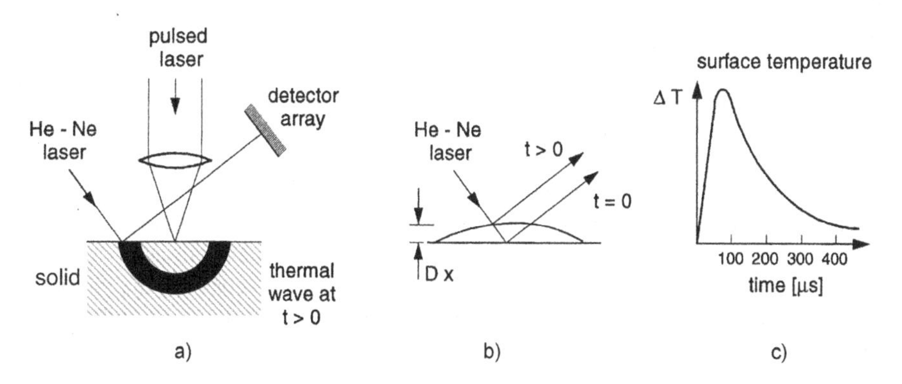

Fig. 1.35 Optothermal spectroscopy of solids and of molecules adsorbed at surfaces: (**a**) propagation of thermal wave induced by a pulsed laser; (**b**) deformation of a surface detected by the deflection of a HeNe laser beam; (**c**) time profile of surface temperature change following pulsed illumination [90]

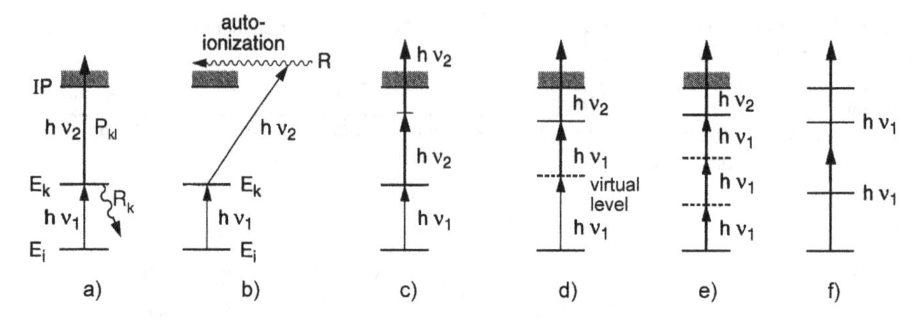

Fig. 1.36 Level schemes of ionization spectroscopy: (**a**) photoionization; (**b**) excitation of autoionizing Rydberg levels; (**c**) two-photon ionization of excited molecules; (**d**) one-photon ionization of a high lying level, excited by non-resonant two-photon process; (**e**) three-photon excitation of a level which is ionized by a fourth photon; (**f**) non-resonant two-photon ionization

1.4 Ionization Spectroscopy

1.4.1 Basic Techniques

Ionization spectroscopy monitors the absorption of photons on the molecular transition $E_i \to E_k$ by detecting the ions or electrons produced by ionization of the excited state E_k. The necessary ionization of the excited molecule may be performed by photons, by collisions, or by an external electric field.

(a) Photoionization

The excited molecules are ionized by absorption of a second photon, i.e.,

$$\mathrm{M}^*(E_k) + h\nu_2 \longrightarrow \mathrm{M}^+ + \mathrm{e}^- + E_{\mathrm{kin}}. \tag{1.44a}$$

The ionizing photon may come either from the same laser that has excited the level E_k or from a separate light source, which can be another laser or even an incoherent source (Fig. 1.36a).

The following estimation illustrates the possible sensitivity of resonant two-photon ionization spectroscopy (Fig. 1.36a). Let N_k be the density of excited molecules in level E_k, $P_{k\mathrm{I}}$ the probability per second that a molecule in level E_k is ionized by n_{L_2} photons from laser L_2 and $n_\mathrm{a} = N_i n_\mathrm{L} \sigma_{ik} \Delta x$ (1.34) the number of photons absorbed per second on the transition $E_i \to E_k$. If R_k is the total relaxation rate of level E_k, besides the ionization rate (spontaneous transitions plus collision-induced deactivation) the signal rate in counts per second for the absorption path length Δx and for n_L incident laser photons per second under steady state conditions is:

$$S_\mathrm{I} = N_k P_{k\mathrm{I}} \delta \cdot \eta = n_\mathrm{a} \frac{P_{k\mathrm{I}}}{P_{k\mathrm{I}} + R_k} \delta \cdot \eta = N_i n_\mathrm{L} \sigma_{ik} \Delta x \frac{P_{k\mathrm{I}}}{P_{k\mathrm{I}} + R_k} \delta \cdot \eta, \tag{1.44b}$$

where $P_{k\mathrm{I}} = \sigma_{k\mathrm{I}} \eta_{\mathrm{L}_2}$.

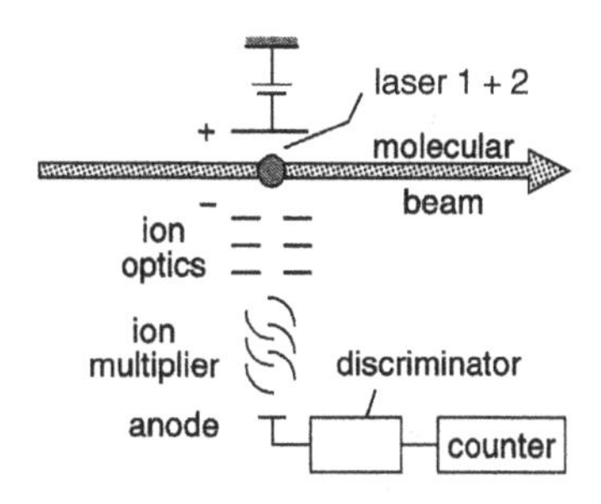

Fig. 1.37 Experimental arrangement for photoionization spectroscopy in a molecular beam

With a proper design (Fig. 1.37) the *collection* efficiency δ for the ionized electrons or ions can reach $\delta = 1$. If the electrons or ions are accelerated to several keV and detected by electron multipliers or channeltrons, a *detection* efficiency of $\eta = 1$ can also be achieved. If the ionization probability $P_{k\mathrm{I}}$ can be made large compared to the relaxation rate R_k of the level $|k\rangle$, the signal S_I then becomes with $\delta = \eta = 1$:

$$S_\mathrm{I} \sim n_\mathrm{a}. \tag{1.44c}$$

This means that every laser photon absorbed in the transition $E_i \rightarrow E_k$ *gives rise to a detected ion or electron.* It implies that *single absorbed photons* can be detected with an overall efficiency close to unity (or 100 %). In experimental practice there are, of course, additional losses and sources of noise, which limit the detection efficiency to a somewhat lower level. However, for all absorbing transitions $E_i \rightarrow E_k$ where the upper level E_k can readily be ionized, ionization spectroscopy is the most sensitive detection technique and is superior to all other methods discussed so far [93, 94].

(b) Resonant Multiphoton Ionization REMPI

In the previous section the ionization was achieved by two photons where the first photon was resonant with the excitation transition $\langle i| \rightarrow \langle k|$ as shown in Fig. 1.36a. There are several schemes where more than two photons are involved. They are used when the ionization energy is very high and a single photon from accessible lasers has not sufficient energy to ionize the excited state.

Another alternative is the resonant two-photon excitation of a high lying level which is then ionized by a third photon from the same or from another laser (Fig. 1.36d). With resonant three-photon excitation excited states with opposite parity are reached which are then ionized by a fourth photon (Fig. 1.36e). The smallest ionization probability has the non-resonant two-photon ionization (Fig. 1.36f), which can still yield enough sensitivity if pulsed lasers with sufficient output power are used.

All these multiphoton ionization processes, where at least one of the transitions is resonant with one-, two- or three-photon absorption, are subsidised under the heading *REMPI*.

A very efficient photoionization process is the excitation of high-lying Rydberg levels above the ionization limit (Fig. 1.36b), which decay by autoionization into lower levels of the ion M^+

$$M^*(E_k) + h\nu_2 \longrightarrow M^{**} \to M^+ + e^- + E_{kin}(e^-). \tag{1.44d}$$

The absorption cross section for this process is generally much larger than that of the bound–free transition described by (1.44a) (Sect. 5.4.2). Therefore lower intensities of the ionizing laser are necessary to real efficient ionization.

The excited molecule may also be ionized by a nonresonant two-photon process (Fig. 1.36c)

$$M^*(E_k) + 2h\nu_2 \longrightarrow M^+ + e^- + E_{kin}(e^-). \tag{1.44e}$$

(c) Collision-Induced Ionization

Ionizing collisions between excited atoms or molecules and electrons represent the main ionization process in gas discharges:

$$M^*(E_k) + e^- \longrightarrow M^+ + 2e^-. \tag{1.45a}$$

If the excited level E_k is not too far from the ionization limit, the molecule may also be ionized by thermal collisions with other atoms or molecules. If E_k lies above the ionization limit of the collision partners A, Penning ionization [91] becomes an efficient process and proceeds as

$$M^*(E_k) + A \longrightarrow M + A^+ + e^-. \tag{1.45b}$$

(d) Field Ionization

If the excited level E_k lies closely below the ionization limit, the molecule $M^*(E_k)$ can be ionized by an external electric dc field (Fig. 1.38a). This method is particularly efficient if the excited level is a long-lived, highly excited Rydberg state with principal quantum number n, quantum defect δ, and ionization energy Ry/n^{*2}, which can be expressed by the Rydberg constant Ry and the effective quantum number $n^* = n - \delta$. The required minimum electric field can readily be estimated from Bohr's atomic model, which gives a good approximation for atomic levels with large principal quantum number n. Without an external field the ionization energy for the outer electron at the mean radius r from the nucleus is determined by the Coulomb field of the nucleus shielded by the inner electron core.

$$IP = \int_r^\infty \frac{Z_{eff} e^2}{4\pi\epsilon_0 r^2} \mathrm{d}r = \frac{Z_{eff} e^2}{4\pi\epsilon_0 r} = \frac{\mathrm{Ry}}{(n-\delta)^2} = \frac{\mathrm{Ry}}{n^{*2}}$$

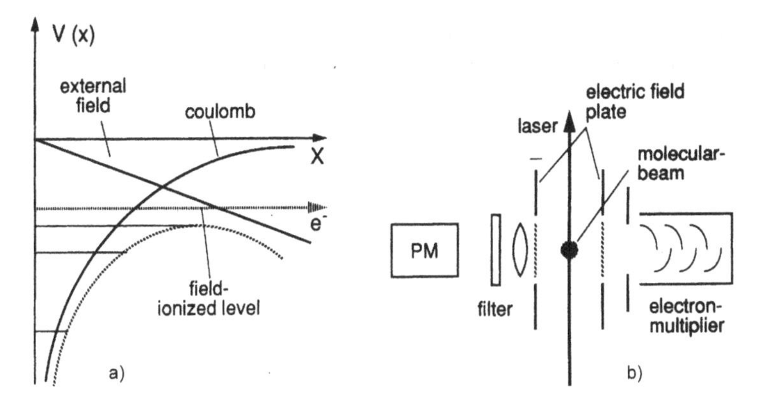

Fig. 1.38 Field ionization of highly excited levels closely below the ionization limit: (**a**) Potential diagram and (**b**) experimental arrangement for field ionization in a molecular beam. The photomultiplier monitors the LIF from the intermediate level populated in a two-step excitation process

where eZ_{eff} is the effective nuclear charge, i.e., the nuclear charge eZ partly screened by the electron cloud. If an external field $E_{\text{ext}} = -E_0 x$ is applied, the effective ionization potential is lowered to the value (see Problem 1.9)

$$IP_{\text{eff}} = IP - \sqrt{\frac{Z_{\text{eff}} e^3 E_0}{\pi \epsilon_0}}. \tag{1.46}$$

If the energy E of the excited level is above IP_{eff}, it will be field-ionized.

Techniques where the laser excites the atoms but the ionizing step is performed by field ionization have found increasing applications in the detection of Rydberg atoms in atomic beams (Sect. 9.6), in analytical chemistry for trace elements, or for small concentrations of pollutants [92].

Example 1.15 For levels 10 meV below the ionization limit, (1.46) gives $E_0 \geq 1.7 \times 10^4$ V/m for the ionizing external field. However, because of the quantum-mechanical tunnel effect the fields required for complete ionization are even lower.

1.4.2 Pulsed Versus CW Lasers for Photoionization

In case of photoionization of the excited level $|k\rangle$, the ionization probability per second

$$P_{k\text{I}} = \sigma_{k\text{I}} n_{\text{L}_2} \left[\text{s}^{-1}\right]$$

equals the product of the ionization cross section $\sigma_{k\mathrm{I}}$ [cm^2] and the photon flux density n_{L_2} [cm^{-2} s^{-1}] of the ionizing laser. We can then write (1.44b) as

$$S_\mathrm{I} = N_i \left[\frac{\sigma_{ik} n_{\mathrm{L}_1} \delta \cdot \eta}{1 + R_k/(\sigma_{k\mathrm{I}} n_{\mathrm{L}_2})} \right] \Delta x. \tag{1.47}$$

The maximum ion rate $S_\mathrm{I}^{\max}$, which is achieved if

$$\sigma_{k\mathrm{I}} n_{\mathrm{L}_2} \gg R_k, \quad \text{and} \quad \delta = \eta = 1,$$

becomes equal to the rate n_a of photons absorbed in the transition $|i\rangle \to |k\rangle$:

$$S_\mathrm{I}^{\max} = N_i \sigma_{ik} n_{\mathrm{L}_1} \Delta x = n_\mathrm{a}. \tag{1.48}$$

The following estimation illustrates under which conditions this maximum ion rate can be realized: Typical cross sections for photoionization are $\sigma_{k\mathrm{I}} \sim 10^{-17}$ cm^2. If radiative decay is the only deactivation mechanism of the excited level $|k\rangle$, we have $R_k = A_k \approx 10^8$ s^{-1}. In order to achieve $n_{\mathrm{L}_2}\sigma_{k\mathrm{I}} > A_k$, we need a photon flux $n_{\mathrm{L}_2} > 10^{25}$ cm^{-2} s^{-1} of the ionizing laser. With pulsed lasers this condition can be met readily.

Example 1.16 Excimer laser: 100 mJ/pulse, $\Delta T = 10$ ns, the cross section of the laser beam may be 1 cm^2 $\to n_{\mathrm{L}_2} = 2 \times 10^{25}$ cm^{-2}s^{-1}. With the numbers above we can reach an ionization probability of $P_{ik} = 2 \times 10^8$ s^{-1} for all molecules within the laser beam. This gives an ion rate S_I that is 2/3 of the maximum rate $S_\mathrm{I} = n_\mathrm{a}$.

The advantage of pulsed lasers is the large photon flux during the pulse time ΔT, which allows the ionization of the excited molecules *before they decay* by relaxation into lower levels where they are lost for further ionization. Their disadvantages are their large spectral bandwidth, which is generally larger than the Fourier-limited bandwidth $\Delta\nu \geq 1/\Delta T$, and their low duty cycle. At typical repetition rates of $f_\mathrm{L} = 10$ to 100 s^{-1} and a pulse duration of $\Delta T = 10^{-8}$ s, the duty cycle is only 10^{-7}–10^{-6}!

If the diffusion time t_D of molecules out of the excitation-ionization region is smaller than $1/f_\mathrm{L}$, at most the fraction $f_\mathrm{L} t_\mathrm{D}$ of molecules can be ionized, even if the ionization probability during the laser pulse ΔT approaches 100 %.

Example 1.17 Assume that the two laser beams L_1 and L_2 for excitation and ionization have the diameter $D = 1$ cm and traverse a collimated molecular beam with 1 cm^2 cross section perpendicularly. During the pulse time $\Delta T = 10^{-8}$ s the distance traveled by the molecules at the mean velocity

$\bar{v} = 500$ m/s is $d = \Delta T \bar{v} \sim 5 \times 10^{-4}$ cm. This means that all molecules in the excitation volume of 1 cm^3 can be ionized during the time ΔT. During the "dark" time $T = 1/f_L$, however, the molecules travel the distance $d = \bar{v}T \sim$ 500 cm at $f_L = 10^2$ s^{-1}. Therefore, only the fraction $1/500 = 2 \times 10^{-3}$ of all molecules in the absorbing level $|i\rangle$ are ionized in a continuous molecular beam.

There are two solutions:

(a) If pulsed molecular beams can be employed having the pulse time $\Delta T_B \leq D/\bar{v}$ and the repetition rate $f_B = f_L$, an optimum detection probability can be achieved.
(b) In continuous molecular beams the two laser beams may propagate antiparallel to the molecular beam axis. If lasers with a high-repetition frequency f_L are used (for example, copper-laser-pumped dye lasers with $f_L \leq 10^4$ s^{-1}), then the distance traveled by the molecules during the dark time is only $d = \bar{v}/f \geq 5$ cm. They can therefore still be detected by the next pulse [96, 97].

With cw lasers the duty cycle is 100 % and the spectral resolution is not limited by the laser bandwidth. However, their intensity is smaller and the laser beam must be focused in order to meet the requirement $P_{kI} > R_k$.

Example 1.18 If a cw argon laser with 10 W at $\lambda = 488$ nm ($\hat{=} 2.5 \times 10^{19}$ photon/s) is used for the ionizing step, it has to be focused down to an area of 2.5×10^{-6} cm^2, i.e., a diameter of 17 μm, in order to reach a photon flux density of $n_{L_2} = 10^{25}$ cm^{-2} s^{-1}.

Here the following problem arises: the molecules excited by laser L_1 into level $|k\rangle$ travel during their spontaneous lifetime of $\tau = 10$ ns at around thermal velocities of about $\bar{v} = 5 \times 10^4$ m/s only a distance of $d = 5$ μm before they decay into lower levels. The second laser L_2 must therefore be focused in a similar way as L_1 and its focus must overlap that of L_1 within a few microns.

A technical solution of this problem is depicted in Fig. 1.39. The dye laser L_1 is guided through a single-mode optical fiber. The divergent light emitted out of the fiber end is made parallel by a spherical lens, superimposed with the beam of the argon laser L_2 by a dichroic mirror M. Both beams are then focused by a cylindrical lens into the molecular beam, forming a rectangular "sheet of light" with a thickness of about 5–10 μm and a height of about 1 mm, adapted to the dimensions of the molecular beam [98]. All molecules in the beam traveling in the z-direction have to pass through the two laser beams. Since the transition probability for the first step $|i\rangle \rightarrow |k\rangle$ is generally larger by some orders of magnitude than that of the ionizing

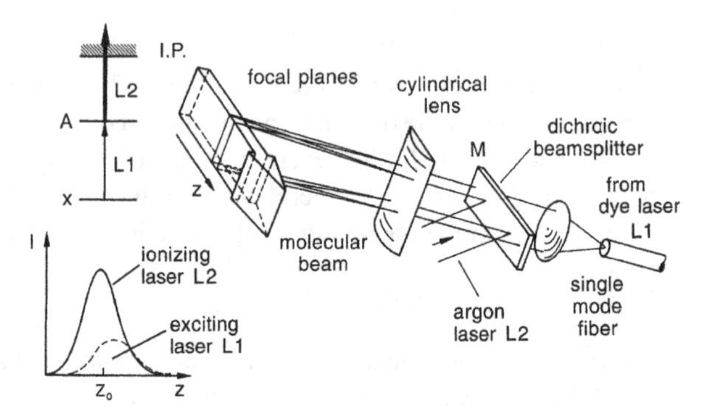

Fig. 1.39 Experimental arrangement for resonant two-photon two-color ionization with two cw lasers. The *insert* shows the optimum overlap of the Gaussian intensity profiles in the focal plane

transition, the first transition can be readily saturated (Sect. 2.1). It is therefore advantageous to adjust the relative positions of the two beams in such a way that the maximum intensity of L_2 coincides spatially with the slope of the Gaussian intensity profile of L_1 (see insert of Fig. 1.39).

Often it is possible to tune the ionizing laser L_2 to transitions from $|k\rangle$ into autoionizing Rydberg levels (Sect. 5.4). For such transitions the probability may be from two to three orders of magnitude larger than for bound–free transitions into the ionization continuum. In these cases the requirement (1.48) can be met even at much lower intensities of L_2.

1.4.3 Sensitivity of the Different Techniques

Resonant two-step ionization with two laser photons from pulsed or cw lasers represents the most versatile and sensitive detection technique. If laser L_1 excites all atoms or molecules that fly through the laser beam, single atoms or molecules can be detected [93, 99, 100] if the condition (1.48) can be fulfilled.

We have seen in Sect. 1.4.1 that resonant two-photon ionization can reach a detection efficiency of 100 %, which means that every absorbed photon can be detected. The question is now how few molecules can be still detected? According to Eq. (1.4a) the minimum number density of detectable absorbers is $N_i \geq \Delta P/(P_0 \sigma_{ik} L)$, where L is the absorption length and ΔP the absorbed power. If single absorbed photons can be detected, $\Delta P = h \cdot \nu$ and we obtain for the incident laser power $P_0 = n_{L_1} h\nu$

$$N_i \geq \frac{h \cdot \nu}{n_{L_1} h\nu \cdot \sigma_{ik} L} = \frac{1}{n_{L_1} \sigma_{ik} L}. \tag{1.49}$$

Example 1.19

(1) *RTPI in a gas cell* with an absorption length $L = 5$ cm. The incident laser has a power of $P_0 = 1$ mW at $\lambda = 500$ nm $\rightarrow n_{L_1} = P_0/h\nu =$

6.5×10^{16}/s. With an absorption cross section $\sigma_{ik} = 10^{-18}$ cm^2 we obtain: $N_i \geq 3$ cm^{-3} s^{-1}. If the cross section of the laser beam is 0.1 cm^2 the absorption volume is $V = 0.5$ cm^3 and the minimum detectable number of molecules is $N_{min} = N_i \cdot V = 1.5$ s^{-1}.

(2) *Absorption in a collimated molecular beam.* Here $L = 0.2$ cm which gives $N_1 \geq 75$ cm^{-3} s^{-1}. In order to reach single molecule detection we can focus the laser beam to a diameter of 1 mm. This gives a beam cross section of 7.7×10^{-3} cm^2 and an absorption volume of $V = 1.5 \times 10^{-3}$ cm^3 $\rightarrow N_{min} = N_i \cdot V = 0.11$ s^{-1} if every absorbed photon could be detected. Generally the collection efficiency of the ions is smaller then 100 % and the noise is larger than $N_{min} \cdot h\nu$. But still single absorbed photons per sec can be detected. This high detection efficiency has been successfully used to measure the hyperfine structure of the rare radioactive Francium isotopes and to determine their nuclear spins and magnetic moments [101].

(3) *Collisional ionization.* For collisional ionization the number of produced ions depends on the number of incident high energy particles and the collisional ionization cross section. Both quantities are generally much smaller than in case of RTPI with lasers. The scientific goal of this technique is therefore not to reach high sensitivities but to study the energy transfer in the collisions and the resultant ionic energy levels.

(4) *Field ionization* is a very efficient method to detect excited atoms or molecules. If the energy of the excited Rydberg level lies above the effective ionization potential $IP = R_y/n^{*2}$ every Rydberg level is ionized. Field ionization is therefore often used to detect the population of highly excited states.

1.4.4 Resonant Two-Photon Ionization (RTPI) Combined with Mass Spectrometry

When used in combination with mass spectrometry, RTPI allows mass- and wavelength-selective spectroscopy, even if the spectral lines of the different species overlap. This is particularly important for *molecular* isotopes with dense spectra, which overlap for the different isotopes. This is illustrated by Fig. 1.40, where the differences in the line positions in the spectra of $^{21}Li_3$ and $^{20}Li_3$ are caused partly by the different masses but mainly by the different nuclear spins. Such isotope-selective spectra give detailed information on isotope shifts of vibrational and rotational levels and facilitate the correct assignment of the spectral lines considerably. Furthermore, they yield the relative isotopic abundances.

Time-of-flight mass spectrometers with pulsed lasers are convenient since they allow the simultaneous but separate recording of the spectra of different isotopes

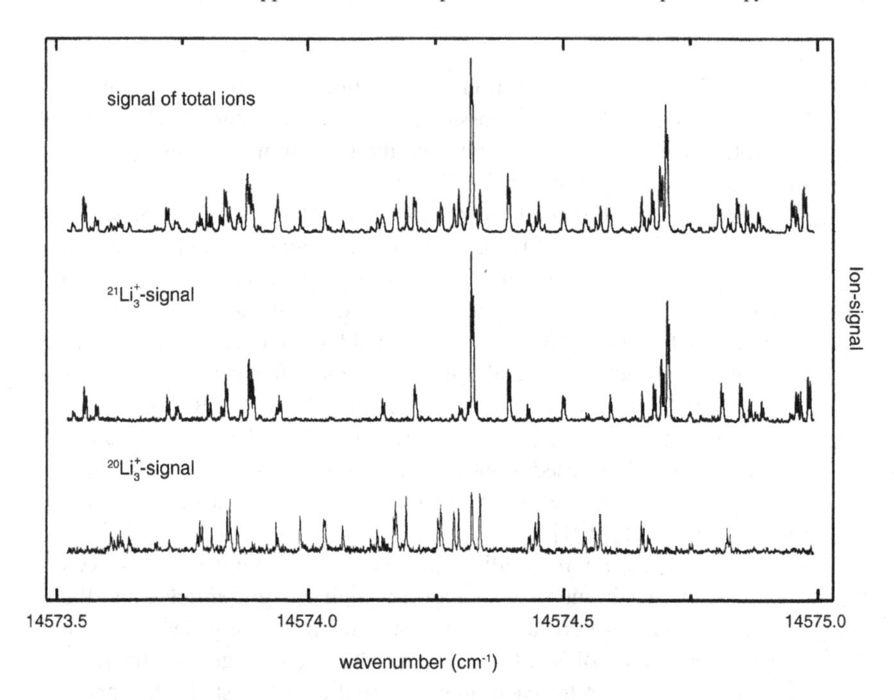

Fig. 1.40 Excitation spectra of Li_3 clusters detected by photoionization of the excited states: (**a**) no mass selection; (**b**) spectrum of the $^{21}Li_3 = {}^7Li^7Li^7Li$ isotopomer; (**c**) spectrum of $^{20}Li_3 = {}^6Li^7Li^7Li$, recorded with doubled sensitivity [95]

[103, 104]. For ionization by cw lasers, quadrupole mass spectrometers are generally used. Their disadvantage is the lower transmittance and the fact that different masses cannot be recorded simultaneously but only sequentially. At sufficiently low ion rates, delayed coincidence techniques in combination with time-of-flight spectrometers can even be utilized for cw ionization if both the photoion and the corresponding photoelectron are detected. The detected electron provides the zero point of the time scale and the ions with different masses are separated by their differences $\Delta t_a = t_{ion} - t_{el}$ in arrival times at the ion detector.

For measurements of cluster-size distributions in cold molecular beams (Sect. 4.3), or for monitoring the mass distribution of laser-desorbed molecules from surfaces, these combined techniques of laser ionization and mass spectrometry are very useful [105, 106]. For the detection of rare isotopes in the presence of other much more abundant isotopes, the double discrimination of isotope-selective excitation by the first laser L1 and the subsequent mass separation by the mass spectrometer is essential to completely separate the isotopes, even if the far wings of their absorption lines overlap [107]. The combination of resonant multiphoton ionization (REMPI) with mass spectrometry for the investigation of molecular dynamics and fragmentation is discussed in Chap. 5.

Often the laser is used to evaporate atoms or molecules from liquid or solid surfaces. These evaporated particles either leave the surface as ions which can be di-

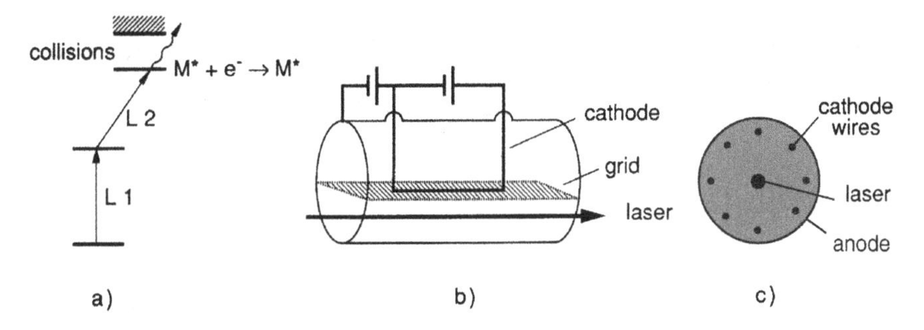

Fig. 1.41 Thermionic diode: (**a**) level scheme; (**b**) schematic arrangement; and (**c**) field-free excitation scheme, where the laser beam passes through the field-free central region in a symmetric arrangement of cathode wires

rectly detected or as neutrals which are then ionized in the ion source of the mass spectrometer by electron impact. This technique is particularly important for the mass selective detection of fragile biological molecules because the laser desorption is a "soft" evaporation technique in contrast to ion bombardment which might cause fragmentation of large molecules. Such laser mass spectrometers have been successfully used for space missions for example for the investigation of solid samples from the mars surface by the landing module *Curiosity* [108].

1.4.5 Thermionic Diode

The collisional ionization of high-lying Rydberg levels is utilized in the thermionic diode [111], which consists in its simplest form of a metallic cylindrical cell filled with a gas or vapor, a heated wire as cathode, and the walls as anode (Fig. 1.41). If a small voltage of a few volts is applied, the diode operates in the space-charge limited region and the diode current is restricted by the electron space-charge around the cathode. The laser beam passes through this space-charge region close to the cathode and excites molecules into Rydberg states, which are ionized by electron impact. Because of its much larger mass, the ion stays on the average a much longer time Δt_{ion} within the space-charge region than an electron. During this period it compensates one negative charge and therefore allows $n = \Delta t_{\text{ion}}/\Delta t_{\text{el}}$ extra electrons to leave the space-charge region. If N ions are formed per second, the diode current therefore increases by

$$\Delta i = \mathrm{e}N\Delta t_{\text{ion}}/\Delta t_{\text{el}} = \mathrm{e}MN. \tag{1.50}$$

The current magnification factor $M = \Delta t_{\text{ion}}/\Delta t_{\text{el}}$ can reach values of up to $M = 10^5$.

Sensitive and accurate measurements of atomic and molecular Rydberg levels have been performed [112–114] with thermionic diodes. With a special arrangement

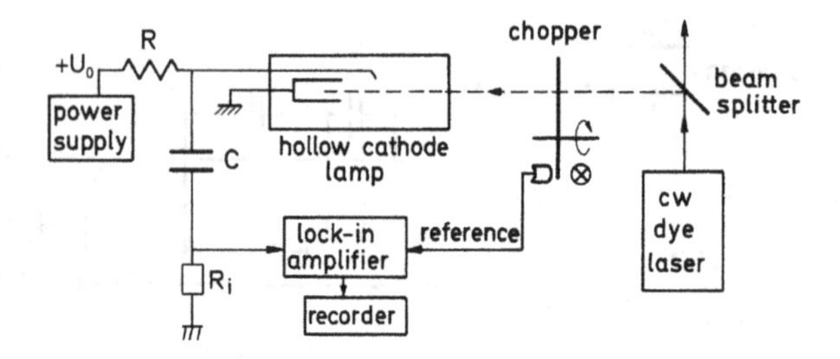

Fig. 1.42 Experimental arrangement for optogalvanic spectroscopy in a hollow cathode lamp

of the electrodes, a nearly field-free excitation zone can be realized that allows the measurement of Rydberg states up to the principal quantum numbers $n = 300$ [114] without noticeable Stark shifts.

A more detailed representation of ionization spectroscopy and its various applications to sensitive detection of atoms and molecules can be found in [92–94, 99, 100, 103].

1.5 Optogalvanic Spectroscopy

Optogalvanic spectroscopy is an excellent and simple technique to perform laser spectroscopy in gas discharges. Assume that the laser beam passes through part of the discharge volume. When the laser frequency is tuned to the transition $E_i \to E_k$ between two levels of atoms or ions in the discharge, the population densities $n_i(E_i)$ and $n_k(E_k)$ are changed by optical pumping. Because of the different ionization probabilities from the two levels, this population change will result in a change in the number of ions and free electrons and therefore a change ΔI in the discharge current that is detected as a voltage change $\Delta U = R\Delta I$ across the resistor R at a constant supply voltage U_0 (Fig. 1.42). When the laser intensity is chopped, an ac voltage is obtained, which can be directly fed into a lock-in amplifier.

Even with moderate laser powers (a few milliwatts) large signals (μV to mV) can be achieved in gas discharges of several milliamperes. Since the absorbed laser photons are detected by the optically induced current change, this very sensitive technique is called *optogalvanic spectroscopy* [115–117].

Both positive and negative signals are observed, depending on the levels E_i, E_k involved in the laser-induced transition $E_i \to E_k$. If $IP(E_i)$ is the total ionization probability of an atom in level E_i, the voltage change ΔU produced by the laser-induced change $\Delta n_i = n_{i0} - n_{i\mathrm{L}}$ is given by

$$\Delta U = R\Delta I = a\left[\Delta n_i IP(E_i) - \Delta n_k IP(E_k)\right]. \tag{1.51}$$

There are several competing processes that may contribute to the ionization of atoms in level E_i, such as direct ionization by electron impact $\mathrm{A}(E_i) + \mathrm{e}^- \to \mathrm{A}^+ + 2\mathrm{e}$, collisional ionization by metastable atoms $\mathrm{A}(E_i) + \mathrm{B}^* \to \mathrm{A}^+ + \mathrm{B} + \mathrm{e}^-$, or, in

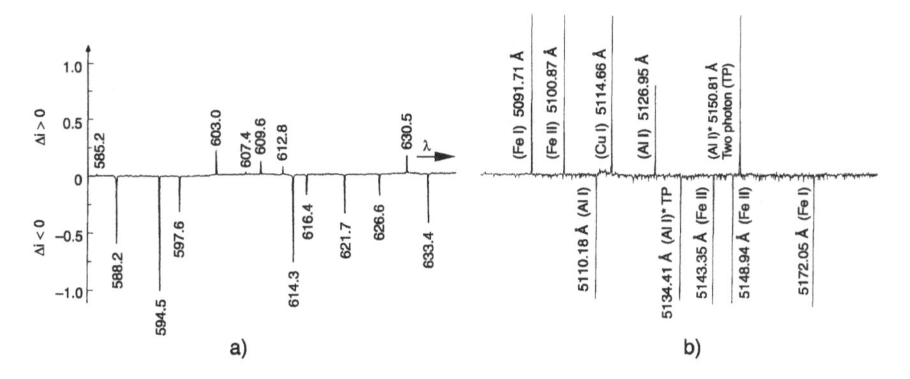

Fig. 1.43 Optogalvanic spectrum (**a**) of a neon discharge (1 mA, $p = 1$ mbar) generated with a broadband cw dye laser [117]; (**b**) of Al, Cu, and Fe vapor sputtered in a hollow cathode illuminated with a pulsed dye laser [120]

particular for highly excited levels, the direct photoionization by laser photons $A(E_i) + h\nu \rightarrow A^+ + e^-$. The change in the electron temperature caused by the atomic population change ΔN can also influence the ionization probability. The competition of these and other processes determines whether the population changes Δn_i and Δn_k cause an increase or a decrease of the discharge current. Figure 1.43a shows the optogalvanic spectrum of a Ne discharge (5 mA) recorded in a fast scan with 0.1 s time constant. The good signal-to-noise ratio demonstrates the sensitivity of the method.

In hollow cathodes the cathode material can be sputtered by ion bombardment in the discharge. The metal vapor, consisting of atoms and ions, can be investigated by optogalvanic spectroscopy. Figure 1.43b illustrates a section of the optogalvanic spectrum of aluminum, copper, and iron atoms, and ions Al^+, Fe^+, measured simultaneously in two hollow cathodes irradiated with a tunable pulsed dye laser [120].

Molecular spectra can also be measured by optogalvanic spectroscopy [121]. In particular, transitions from highly excited molecular states that are not accessible to optical excitation but are populated by electron impact in the gas discharge can be investigated. Furthermore, molecular ions and radicals can be studied. Some molecules, called *excimers* (Vol. 1, Sect. 5.7.6) are only stable in their excited states. They are therefore appropriate for this technique because they do not exist in the ground state and cannot be studied in neutral-gas cells. Examples are He_2^* or H_2^* [122, 123].

Optogalvanic spectroscopy is a suitable technique for studies of excitation and ionization processes in flames, gas discharges, and plasmas [124], which are important for the development of new energy saving light sources. Of particular interest is the investigation of radicals and unstable reaction products that are formed by electron-impact fragmentation in gas discharges. These species play an important role in the extremely rarefied plasma in molecular clouds in the interstellar medium.

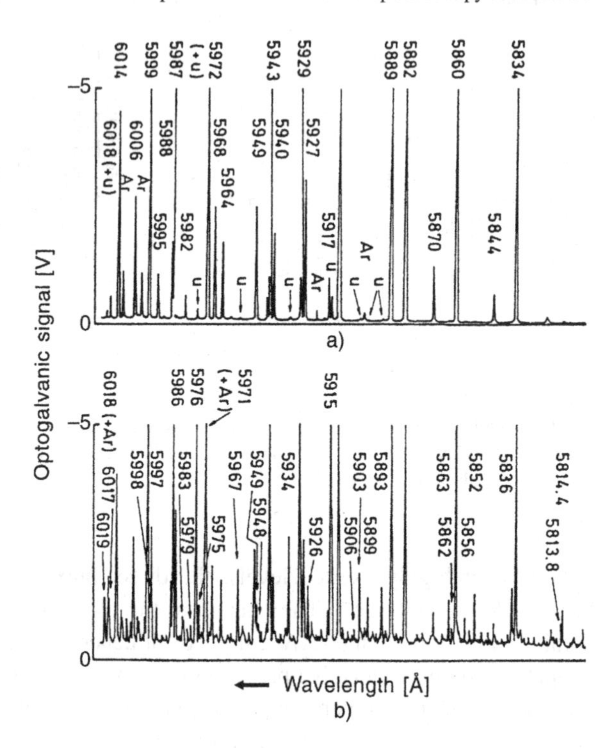

Fig. 1.44 Optogalvanic spectrum of a uranium hollow-cathode lamp filled with argon buffer gas. In the *upper spectrum* (**a**) taken at 7 mA discharge current, most of the lines are argon transitions, while in the *lower spectrum* (**b**) at 20 mA many more uranium lines appear, because of sputtering of uranium from the hollow cathode walls [125]

Besides its applications to studies of collision processes and ionization probabilities in gas discharges and combustion processes, this technique has the very useful aspect of simple wavelength calibration in laser spectroscopy [125]. A small fraction of the output from a tunable laser is split into a hollow-cathode spectral lamp and the optogalvanic spectrum of the discharge is recorded simultaneously with the unknown spectrum under investigation. The numerous lines of thorium or uranium are nearly equally distributed throughout the visible and ultraviolet spectral regions (Fig. 1.44). They are recommended as secondary wavelength standards since they have been measured interferometrically to a high precision [126, 127]. They can therefore serve as convenient absolute wavelength markers, accurate to about 0.001 cm^{-1}.

If the discharge cell has windows of optical quality, it can be placed inside the laser resonator to take advantage of the q-fold laser intensity (Sect. 1.2.3). With such an intracavity arrangement, Doppler-free saturation spectroscopy can also be performed with the optogalvanic technique (Sect. 2.2 and [128]). An increased sensitivity can be achieved by optogalvanic spectroscopy in thermionic diodes under space-charge-limited conditions (Sect. 1.4.5). Here the internal space-charge amplification is utilized to generate signals in the millivolt to volt range without further external amplification [112, 129].

For more details on optogalvanic spectroscopy see [130], the reviews [92, 115–119] and the book [131], which also give extensive reference lists.

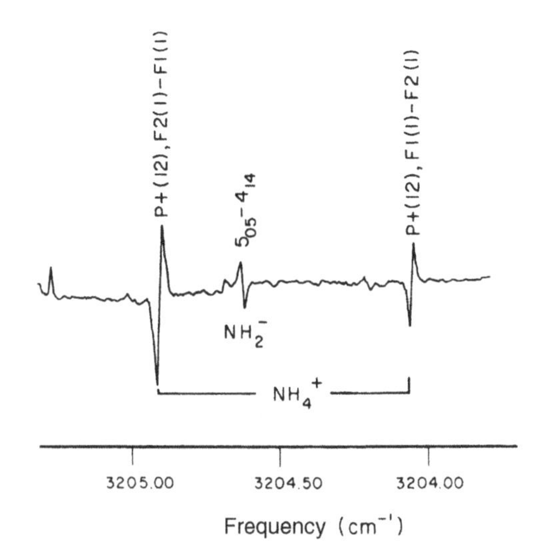

Fig. 1.45 Opposite signal phases of the derivative line profiles of negative and positive ions in velocity-modulation spectroscopy [132]

1.6 Velocity-Modulation Spectroscopy

The analysis of absorption spectra in molecular gas discharges is by no means simple, because the gas discharge produces a large variety of neutral and ionized fragments from the neutral parent molecules. The spectra of these different species may overlap, and often an unambiguous identification is not possible if the spectra are not known. An elegant technique, developed by Saykally et al. [132, 133], is very helpful in distinguishing between spectra of ionized and neutral species.

The external voltage applied to the gas discharge accelerates the positive ions toward the cathode and the negative ions toward the anode. They therefore acquire a drift velocity v_D, which causes a corresponding Doppler shift $\Delta\omega = \omega - \omega_0 = \boldsymbol{k} \cdot \boldsymbol{v}_D$ of their absorption frequency ω_0. If an ac voltage of frequency f is applied instead of a dc voltage, the drift velocity is periodically changed and the absorption frequency $\omega = \omega_0 + \boldsymbol{k} \cdot \boldsymbol{v}_D$ is modulated at the frequency f around the unshifted frequency ω_0. When the absorption spectrum is recorded with a lock-in amplifier, the spectra of the ions can be immediately distinguished from those of the neutral species. This velocity-modulation technique has the same effect as the frequency modulation discussed in Sect. 1.2.2. When the laser is scanned through the spectrum, the first derivatives of the ion lines are seen where the phase of the signals is opposite for positive and negative ions, respectively (Fig. 1.45). The two species can therefore be distinguished by the sign of the lock-in output signal.

A typical experimental arrangement [133] is depicted in Fig. 1.46. With specially designed electron-switching circuits, polarity modulation frequencies up to 50 kHz can be realized for gas discharges of 300 V and 3 A [134]. The attainable signal-to-noise ratio is illustrated by Fig. 1.47, which shows the band head of a vibrational band of the $\mathrm{A}\,^2\Pi_{1/2} \leftarrow \mathrm{X}\,^2\Sigma_g^+$ transition of the CO^+-ion [133].

This technique was first applied to the infrared region where many vibrational–rotational transitions of ions were measured with color-center lasers or diode

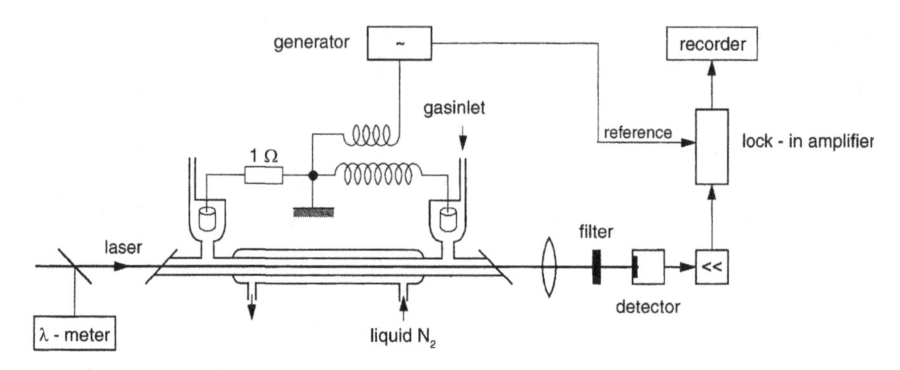

Fig. 1.46 Experimental arrangement for velocity-modulation spectroscopy [132]

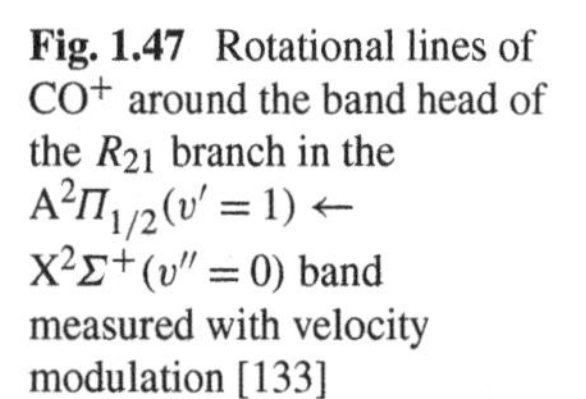

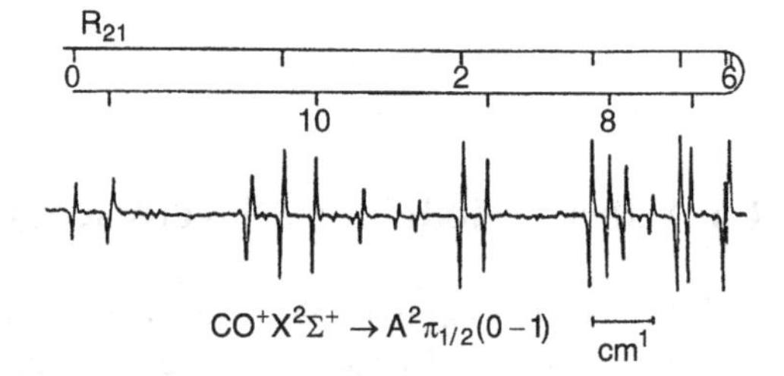

Fig. 1.47 Rotational lines of CO^+ around the band head of the R_{21} branch in the $A^2\Pi_{1/2}(v'=1) \leftarrow X^2\Sigma^+(v''=0)$ band measured with velocity modulation [133]

lasers [133, 137]. Meanwhile, electronic transitions have also been studied with dye lasers [138].

A modification of this velocity-modulation technique in fast ion beams is discussed in Sect. 4.5.

1.7 Laser Magnetic Resonance and Stark Spectroscopy

In all methods discussed in the Sects. 1.1–1.6, the laser frequency ω_L was tuned across the constant frequencies $\omega_i k$ of molecular absorption lines. For molecules with permanent magnetic or electric dipole moments, it is often preferable to tune the absorption lines by means of external magnetic or electric fields across a fixed-frequency laser line. This is particularly advantageous if intense lines of fixed-frequency lasers exist in the spectral region of interest but no tunable source with sufficient intensity is available. Such spectral regions of interest are, for example, the 3–5 μm and the 10 μm ranges, where numerous intense lines of HF, DF, CO, N_2O, and CO_2 lasers can be utilized. Since many of the vibrational bands fall into this spectral range, it is often called the *fingerprint* region of molecules.

Another spectral range of interest is the far infrared, where the rotational lines of polar molecules are found. Here a large number of lines from H_2O or D_2O lasers (125 μm) and from HCN lasers (330 μm) provide intense sources. The successful development of numerous optically pumped molecular lasers [140] has considerably increased the number of FIR lines.

1.7.1 Laser Magnetic Resonance

The molecular level E_0 with the total angular momentum J splits in an external magnetic field $\boldsymbol{B}$ into $(2J+1)$ Zeeman components. The sublevel with the magnetic quantum number M shifts from the energy E_0 at zero field to

$$E = E_0 - g\mu_0 \boldsymbol{B} M, \tag{1.52}$$

where μ_0 is the Bohr magneton, and g is the Landé factor, which depends on the coupling scheme of the different angular momenta (electronic angular momentum, electron spin, molecular rotation, and nuclear spin). The frequency ω of a transition $(v'', J'', M'') \to (v', J', M')$ is therefore tuned by the magnetic field from its unperturbed position ω_0 to

$$\omega = \omega_0 - \mu_0 \big(g' M' - g'' M''\big) \boldsymbol{B}/\hbar, \tag{1.53}$$

and we obtain on the transition $(v'', J'', M'') \to (v', J', M')$ three groups of lines with $\Delta M = M'' - M' = 0, \pm 1$, which degenerate to three single lines if $g'' = g'$ (normal Zeeman effect). The tuning range depends on the magnitude of $g'' - g'$ and is larger for molecules with a large permanent dipole moment. These are, in particular, radicals with an unpaired electron, which have a large spin moment. In favorable cases a tuning range of up to 2 cm^{-1} can be reached in magnetic fields of 2 T ($\hat{=}$20 kG).

Figure 1.48a explains schematically the appearance of resonances between the fixed frequency ω_L and the different Zeeman components when the magnetic field $\boldsymbol{B}$ is tuned. The experimental arrangement is illustrated in Fig. 1.48b. The sample is placed inside the laser cavity and the laser output is monitored as a function of the magnetic field. The cell is part of a flow system in which radicals are generated either directly in a microwave discharge or by adding reactants to the discharge close to the laser cavity. A polyethylene membrane beam splitter separates the laser medium from the sample. The beam splitter polarizes the radiation and transitions with either $\Delta M = 0$ or ± 1 can be selected by rotation of the tube about the laser axis. For illustration, Fig. 1.48c shows the laser magnetic resonance (LMR) spectrum of the CH radical with some OH lines overlapping. Concentrations of 2×10^8 molecules/cm^3 could be still detected with reasonable signal-to-noise ratio for the detector time constant of 1 s [141, 142].

The sensitivity of this intracavity technique (Sect. 1.2.3) can even be enhanced by modulating the magnetic field, which yields the first derivative of the spectrum (Sect. 1.2.2). When a tunable laser is used it can be tuned to the center ν_0 of a molecular line at zero field $\boldsymbol{B} = 0$. If the magnetic field is now modulated around zero, the phase of the zero-field LMR resonances for $\Delta M = +1$ transitions is opposite to that for $\Delta M = -1$ transitions. The advantages of this zero-field LMR spectroscopy have been proved for the NO molecule by Urban et al. [143] using a spin-flip Raman laser.

Because of its high sensitivity LMR spectroscopy is an excellent method to detect radicals in very low concentrations and to measure their spectra with high precision.

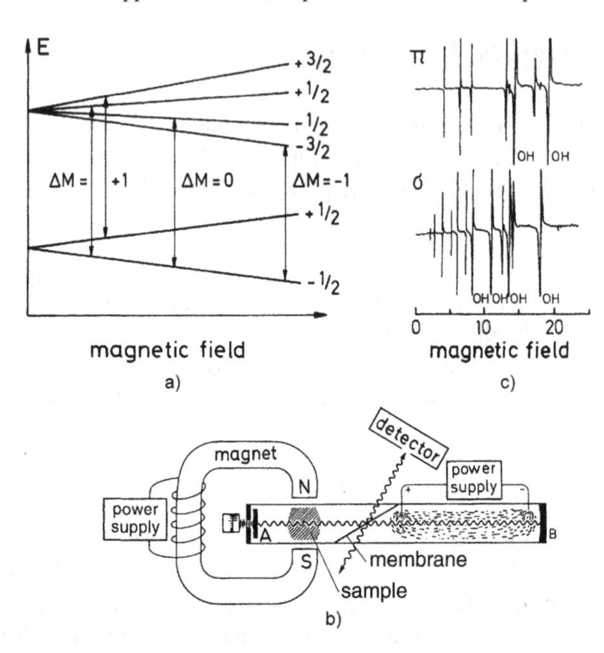

Fig. 1.48 Laser magnetic resonance spectroscopy: (**a**) level diagram; (**b**) experimental set-up with an intracavity sample; A and B are the laser resonator mirrors; (**c**) LMR spectrum of CH radicals, superimposed by some OH-lines, measured in a low-pressure oxygen–acetylene flame with an H_2O laser [142]

If a sufficient number of resonances with laser lines can be found, the rotational constant, the fine structure parameters, and the magnetic moments can be determined very accurately. The identification of the spectra and the assignment of the lines are often possible even if the molecular constants are not known beforehand [144]. Most radicals observed in interstellar space by radio astronomy [141a] have been found and measured in the laboratory with LMR spectroscopy [141b].

Often, a combination of LMR spectroscopy with a fixed-frequency laser and absorption spectroscopy at zero magnetic field with a tunable laser is helpful for the identification of spectra.

Instead of inside the laser cavity, the sample can also be placed outside between two crossed polarizers (Fig. 1.49). In a *longitudinal* magnetic field the plane of polarization of the transmitted light is turned due to the Faraday effect if its frequency ω coincides with one of the allowed Zeeman transitions. The detector receives a signal only for these resonance cases, while the nonresonant background is blocked by the crossed polarizers [145]. This technique is similar to polarization spectroscopy (Sect. 2.4). Modulation of the magnetic field and lock-in detection further enhances the sensitivity. In a transverse magnetic field, the plane of polarization of the linearly polarized incident beam is chosen 45° inclined against $\boldsymbol{B}$. Due to the Voigt effect, the plane of polarization is turned if ω_L coincides with a Zeeman transition [145].

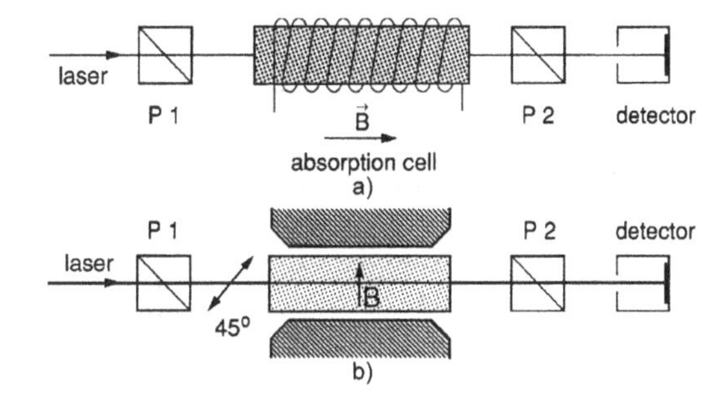

Fig. 1.49 Schematic arrangement for LMR spectroscopy using the Faraday effect in a longitudinal magnetic field (**a**), or the Voigt effect in a transverse magnetic field (**b**) [145]

If the laser beam is split into two partial beams which pass into opposite directions through the LMR cell, Doppler-free saturation spectra can be realized (see Sect. 2.2). This allows one to resolve even complex spectra of radicals or neutral molecules. The narrow spectral width of the Lamb-dips facilitates the determination of collisional broadening and the measurement of molecular transition moments.

1.7.2 Stark Spectroscopy

Analogously to the LMR technique, Stark spectroscopy utilizes the Stark shift of molecular levels in *electric* fields to tune molecular absorption lines into resonance with lines of fixed-frequency lasers. A number of small molecules with permanent electric dipole moments and sufficiently large Stark shifts have been investigated, in particular, those molecules that have rotational spectra outside spectral regions accessible to conventional microwave spectroscopy [146].

To achieve large electric fields, the separation of the Stark electrodes is made as small as possible (typically about 1 mm). This generally excludes an intracavity arrangement because the diffraction by this narrow aperture would introduce intolerably large losses. The Stark cell is therefore placed outside the resonator, and for enhanced sensitivity the electric field is modulated while the dc field is tuned. This modulation technique is also common in microwave spectroscopy. The accuracy of 10^{-4} for the Stark field measurements allows a precise determination of the absolute value for the electric dipole moment.

Figure 1.50 illustrates the obtainable sensitivity by a $\Delta M = 0$ Stark spectrum of the ammonia isotope $^{14}NH_2D$ composed of measurements with several laser lines [146]. An electric resonance signal is observed at every crossing point of the sloped energy levels with a fixed laser frequency. Since the absolute frequency of many laser lines was measured accurately within 20–40 kHz (Sect. 9.7), the absolute frequency of the Stark components at resonance with the laser line can be measured with the same accuracy. The total accuracy in the determination of the molecular parameters is therefore mainly limited by the accuracy of 10^{-4} for the electric field measurements. To date numerous molecules have been measured with laser Stark spectroscopy [146–149]. The number of molecules accessible to this technique can be vastly enlarged if tunable lasers in the relevant spectral regions

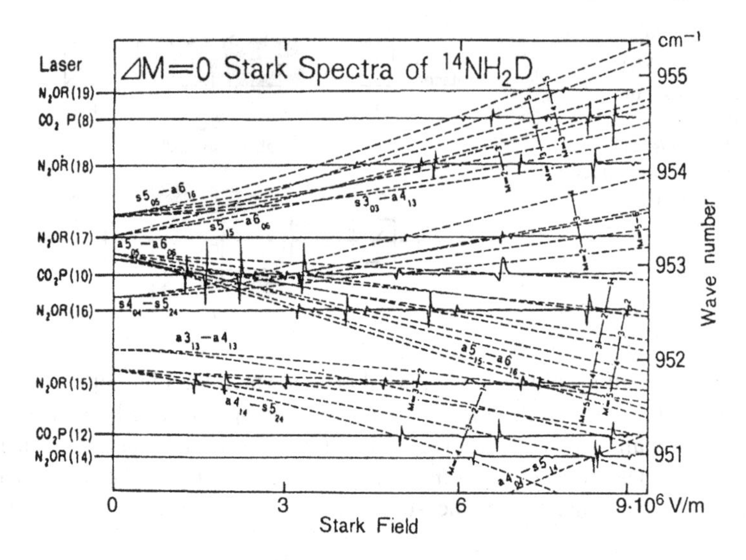

Fig. 1.50 $\Delta M = 0$ Stark spectra of $^{14}NH_2D$ in the spectral range 950–955 cm^{-1} obtained with different fixed-frequency laser lines [146]

are used, which can be stabilized close to a molecular line with sufficient accuracy and long-term stability. Stark spectroscopy with constant electric fields and tunable lasers has been performed in molecular beams at sub-Doppler resolution to measure the electric dipole moments of polar molecules in excited vibrational states [148].

An efficient way to generate coherent, tunable radiation in the far infrared is the difference frequency generation by mixing the output of an CO_2 laser kept on a selected line with the output of a tunable CO_2 waveguide laser in a MIM diode (Vol. 1, Sect. 5.8). With this technique Stark spectra of $^{13}CH_3OH$ were measured over a broad spectral range [149].

Reviews about more recent investigations in LMR and Stark-spectroscopy, including the visible and UV range, can be found in [150, 152, 153].

1.8 Laser-Induced Fluorescence

Laser-induced fluorescence (LIF) has a large range of applications in spectroscopy. First, LIF serves as a sensitive monitor for the absorption of laser photons in fluorescence excitation spectroscopy (Sect. 1.3.1). In this case, the undispersed total fluorescence from the excited level is generally monitored (see Sect. 1.3.1).

Second, it is well suited to gain information on molecular states if the fluorescence spectrum excited by a laser on a selected absorption transition is dispersed by a monochromator. The fluorescence spectrum emitted from a selectively populated rovibronic level (v_k', J_k') consists of all allowed transitions to lower levels (v_m'', J_m'') (Fig. 1.51a). The wavenumber differences of the fluorescence lines immediately yield the term differences of these terminating levels (v_m'', J_m'').

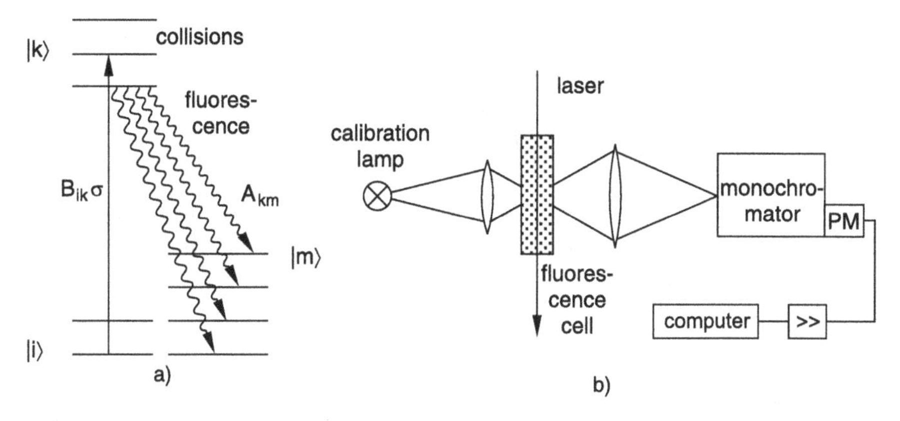

Fig. 1.51 Laser-induced fluorescence: (**a**) level scheme and (**b**) experimental arrangement for measuring LIF spectra

A third aspect of LIF is the spectroscopic study of collision processes. If the excited molecule is transferred by inelastic collisions from the primarily excited level (v_k', J_k') into other rovibronic levels, the fluorescence spectrum shows new lines emitted from these collisionally populated levels which give quantitative information on the collision cross sections (Sect. 8.4).

Another aspect of LIF concerns its application to the determination of the internal-state distribution in molecular reaction products of chemical reactions (Sects. 1.8.5 and 8.6). Under certain conditions the intensity I_{Fl} of LIF excited on the transition $|i\rangle \rightarrow |k\rangle$ is a direct measure of the population density N_i in the absorbing level $|i\rangle$.

A typical setup is shown in Fig. 1.51b. The fluorescence induced by a laser beam passing through a sample cell is imaged by a lens onto the entrance slit of a spectrometer. A lamp emitting calibration lines (for instance a thorium hollow cathode lamp) is imaged by the same optics onto the spectrometer and generates a calibration spectrum which allows the accurate wavelength-determination of the LIF-lines.

Let us now consider some basic facts of LIF in molecules. Extensive information on LIF can be found in [154–157].

1.8.1 Molecular Spectroscopy by Laser-Induced Fluorescence

Assume a rovibronic level (v_k', J_k') in an excited electronic state of a diatomic molecule has been selectively populated by optical pumping. With a mean lifetime $\tau_k = 1/\sum_m A_{km}$, the excited molecules undergo spontaneous transitions to lower levels $E_m(v_m'', J_m'')$ (Fig. 1.51). At a population density $N_k(v_k', J_k')$ the radiation power of a fluorescence line with frequency $\nu_{km} = (E_k - E_m)/h$ is given by (Vol. 1, Sect. 2.7.1)

$$P_{km} \propto N_k A_{km} \nu_{km}. \tag{1.54}$$

The spontaneous transition probability A_{km} is proportional to the square of the matrix element (Vol. 1, Sect. 2.7.2)

$$A_{km} \propto \left| \int \psi_k^* \boldsymbol{r} \psi_m \, \mathrm{d}\tau_n \, \mathrm{d}\tau_{\mathrm{el}} \right|^2 , \tag{1.55}$$

where $\boldsymbol{r}$ is the vector of the excited electron and the integration extends over all nuclear and electronic coordinates. Within the Born–Oppenheimer approximation [158, 159] the total wave function can be separated into a product

$$\psi = \psi_{\mathrm{el}} \psi_{\mathrm{vib}} \psi_{\mathrm{rot}}, \tag{1.56}$$

of electronic, vibrational, and rotational factors. If the electronic transition moment does not critically depend on the internuclear separation R, the total transition probability is then proportional to the product of three factors

$$A_{km} \propto |M_{\mathrm{el}}|^2 |M_{\mathrm{vib}}|^2 |M_{\mathrm{rot}}|^2, \tag{1.57a}$$

where the first factor

$$M_{\mathrm{el}} = \int \psi_{\mathrm{el}}^* \boldsymbol{r} \psi_{\mathrm{el}}'' \, \mathrm{d}\tau_{\mathrm{el}}, \tag{1.57b}$$

represents the electronic matrix element that depends on the coupling of the two electronic states. The second integral

$$M_{\mathrm{vib}} = \int \psi_{\mathrm{vib}}' \psi_{\mathrm{vib}}'' \, \mathrm{d}\tau_{\mathrm{vib}}, \quad \text{with } \mathrm{d}\tau_{\mathrm{vib}} = R^2 \, \mathrm{d}R, \tag{1.57c}$$

is the Franck–Condon factor, which depends on the overlap of the vibrational wave functions $\psi_{\mathrm{vib}}(R)$ in the upper and lower state. The third integral

$$M_{\mathrm{rot}} = \int \psi_{\mathrm{rot}}' \psi_{\mathrm{rot}}'' g_i \, \mathrm{d}\tau_{\mathrm{rot}}, \quad \text{with } \mathrm{d}\tau_{\mathrm{rot}} = \mathrm{d}\vartheta \, \mathrm{d}\varphi, \tag{1.57d}$$

is called the Hönl–London factor. It depends on the orientation of the molecular axis relative to the electric vector of the observed fluorescence wave. This is expressed by the factor g_i $(i = x, y, z)$, where $g_x = \sin\vartheta \cos\varphi$, $g_y = \sin\vartheta \sin\varphi$, and $g_z = \cos\vartheta$, and ϑ and φ are the polar and azimuthal angles [158].

Only those transitions for which all three factors are nonzero appear as lines in the fluorescence spectrum. The Hönl–London factor is always zero unless

$$\Delta J = J_k' - J_m'' = 0, \pm 1. \tag{1.58}$$

If a *single* upper level (v_k', J_k') has been selectively excited, each vibrational band $v_k' \to v_m''$ consists of at most *three* lines: a P *line* $(\Delta J = -1)$, a Q *line* $(\Delta J = 0)$, and an R *line* $(\Delta J = +1)$. For diatomic *homonuclear* molecules additional symmetry selection rules may further reduce the number of possible transitions. A selectively excited level (v_k', J_k') in a Π state, for example, emits on a $\Pi \to \Sigma$ transition either

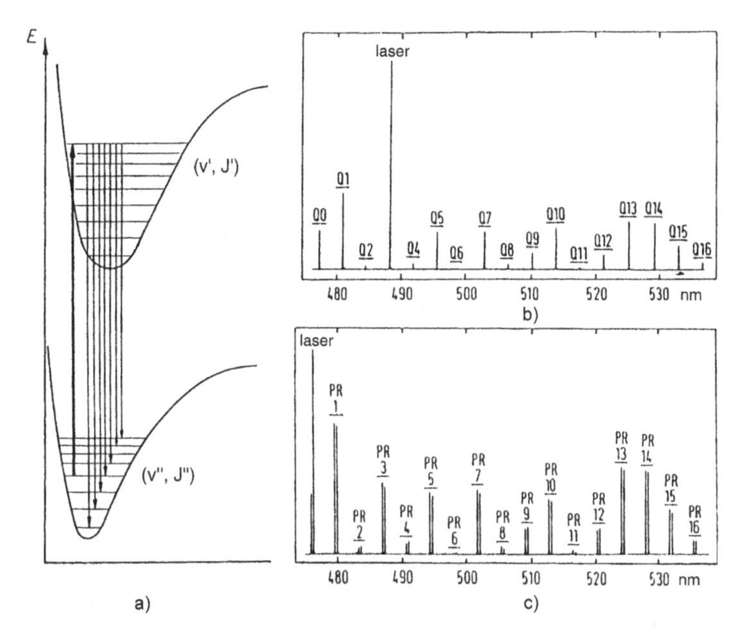

Fig. 1.52 Laser-induced fluorescence of the Na_2 molecule excited by argon laser lines: (**a**) energy level diagram; (**b**) fluorescence lines with $\Delta J = 0$ (Q-lines) emitted from the upper level ($v' = 3$, $J' = 43$) of the $B^1\Pi_u$ state, excited at $\lambda = 488$ nm; (**c**) *P* and *R* doublets, emitted from the upper level ($v' = 6$, $J' = 27$). The *numbers* indicate the vibrational quantum numbers v'' of the termination levels

only Q lines or only P and R lines, depending on the symmetry of the rotational levels, while on a $\Sigma_u \rightarrow \Sigma_g$ transition only P and R lines are allowed [159].

The fluorescence spectrum emitted from selectively excited molecular levels of a diatomic molecule is therefore very simple compared with a spectrum obtained under broadband excitation. It consists of a progression of vibrational bands where each band has at most three rotational lines. Figure 1.52 illustrates this by two fluorescence spectra of the Na_2 molecule, excited by two different argon laser lines. While the $\lambda = 488$ nm line excites a positive Λ component in the ($v' = 3$, $J' = 43$) level, which emits only Q lines, the $\lambda = 476.5$ nm line populates the negative Λ component in the ($v' = 6$, $J' = 27$) level of the $^1\Pi_u$ state, resulting in P and R lines.

1.8.2 Experimental Aspects of LIF

In atomic physics the selective excitation of single atomic levels was achieved with atomic resonance lines from hollow cathode lamps even before the invention of lasers. However, in molecular spectroscopy only fortuitous coincidences between atomic resonance lines and molecular transitions could be used in some cases.

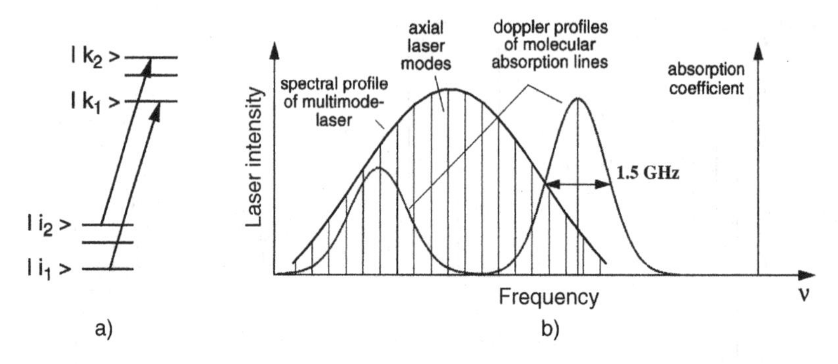

Fig. 1.53 Doppler-broadened absorption lines overlapping with the spectral line profile of a laser: (**a**) level scheme and (**b**) line profiles

Molecular lamps generally emit on many lines and are therefore not useful for the selective excitation of molecular levels.

Tunable narrow-band lasers can be tuned to every wanted molecular transition $|i\rangle \rightarrow |k\rangle$ within the tuning range of the laser. However, selective excitation of a single upper level can only be achieved if neighboring absorption lines do not overlap within their Doppler width (Fig. 1.53). In the case of atoms this can generally be achieved, whereas for molecules with complex absorption spectra, many absorption lines often overlap. In this latter case the laser simultaneously excites several upper levels, which are not necessarily energetically close to each other (Fig. 1.53a). In many cases, however, the fluorescence spectra of these levels can readily be separated by a spectrometer of medium size [160].

In order to achieve selective excitation of single levels even in complex molecular spectra, one may use collimated cold molecular beams where the Doppler width is greatly decreased and the number of absorbing levels is drastically reduced due to the low internal temperature of the molecules (Sect. 4.2).

A very elegant technique combines selective laser excitation with high-resolution Fourier transform spectroscopy of the LIF spectrum. This combination takes advantage of the simultaneous recording of all fluorescence lines and has therefore, at the same total recording time, a higher signal-to-noise ratio. It has also been applied to the spectroscopy of visible and infrared fluorescence spectra of a large variety of molecules [161–163].

When the beam of a single-mode laser passes in the z-direction through a molecular absorption cell, only molecules with velocity components $v_z = 0 \pm \gamma$ are excited if the laser is tuned to the center frequency of an absorption line with the homogeneous width γ. The fluorescence collected within a narrow cone around the z-axis then shows sub-Doppler linewidths, which may be resolved with Fourier transform spectroscopy (Fig. 1.54) [163].

The advantages of LIF spectroscopy for the determination of molecular parameters may be summarized as follows:

(a) The relatively simple structure of the fluorescence spectra allows ready assignment. The fluorescence lines can be resolved with medium-sized spectrometers.

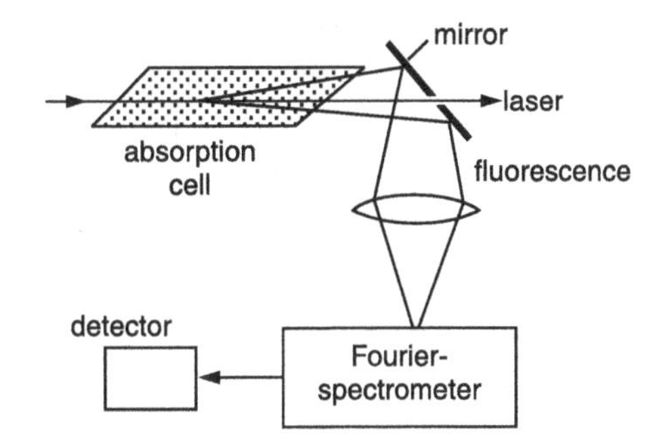

Fig. 1.54 Measurement of LIF spectrum with reduced Doppler width, excited by a single-mode laser

The demands of experimental equipment are much less stringent than those necessary for complete resolution and analysis of absorption spectra of the same molecule. This advantage still remains if a few upper levels are simultaneously populated under Doppler-limited excitation [160].

(b) The large intensities of many laser lines allow achievement of large population densities N_k in the excited level. This yields, according to (1.54), correspondingly high intensities of the fluorescence lines and enables detection even of transitions with small Franck–Condon factors (FCF). A fluorescence progression $v'_k \rightarrow v''_m$ may therefore be measured with sufficiently good signal-to-noise ratio up to very high vibrational quantum numbers v''_m. The potential curve of a diatomic molecule can be determined very accurately from the measured term energies $E(v''_m, J''_m)$ using the Rydberg–Klein–Rees (RKR) method, which is based on a modified WKB procedure [164–167]. Since the term values $E(v''_m, J''_m)$ can be immediately determined from the wave numbers of the fluorescence lines, the RKR potential can be constructed up to the highest measured level $v''_{\max}$. In some cases fluorescence progressions are found up to levels v'' just below the dissociation limit [168, 169]. This allows the spectroscopic determination of the dissociation energy by an extrapolation of the decreasing vibrational spacings $\Delta E_{\text{vib}} = E(v''_{m+1}) - E(v''_m)$ to $\Delta E_{\text{vib}} = 0$ (Birge–Sponer plot) [170–173].

(c) The relative intensities of the fluorescence lines $(v'_k, J'_k \rightarrow v''_m, J''_m)$ are proportional to the Franck–Condon factors. The comparison between calculated FCF obtained with the RKR potential from the Schrödinger equation and the measured relative intensities allows a very sensitive test for the accuracy of the potential. In combination with lifetime measurements, these intensity measurements yield absolute values of the electronic transition moment $M_{\text{el}}(R)$ and its dependence on the internuclear distance R [174].

(d) In several cases discrete molecular levels have been excited that emit continuous fluorescence spectra terminating on repulsive potentials of dissociating states [175]. The overlap integral between the vibrational eigenfunctions ψ'_{vib} of the upper discrete level and the continuous function $\psi_{\text{cont}}(R)$ of the dissociating lower state often shows an intensity modulation of the continuous fluorescence spectrum that reflects the square $|\psi'_{\text{vib}}(R)|^2$ of the upper-state wave function (Vol. 1, Fig. 2.16). If the upper potential is known, the repulsive part of

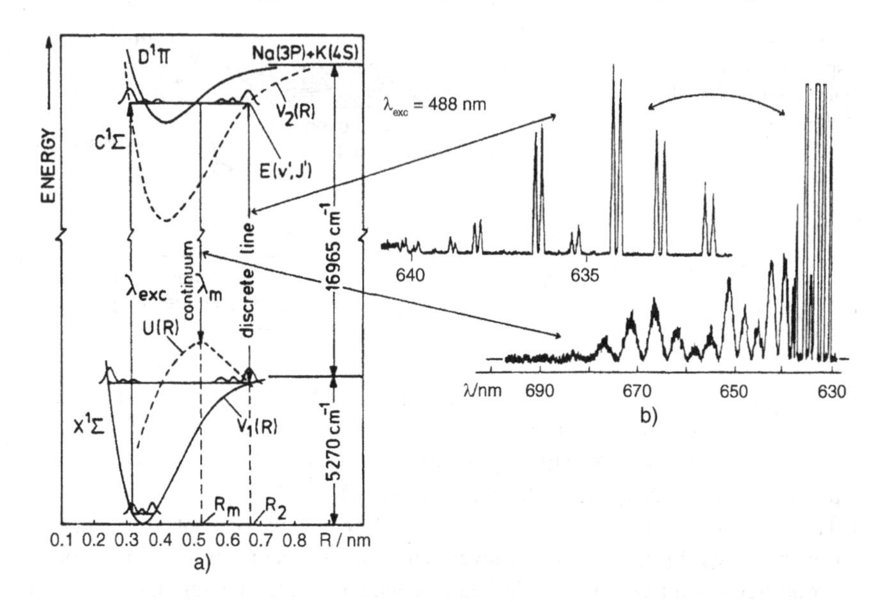

Fig. 1.55 (**a**) Schematic potential energy curve diagram illustrating bound–bound transitions between the discrete levels and bound–free transitions from an upper discrete level into the lower continuum above the dissociation energy of the electronic ground state. (**b**) Two sections of the NaK fluorescence spectrum showing both kinds of transitions [179]

the lower potential can be accurately determined [176, 177]. This is of particular relevance for excimer spectroscopy (Vol. 1, Sect. 5.7) [178].

(e) For transitions between high vibrational levels of two bound states, the main contribution to the transition probability comes from the internuclear distances R close to the classical turning points R_{min} and R_{max} of the vibrating oscillator. There is, however, a nonvanishing contribution from the internuclear distance R between R_{min} and R_{max}, where the vibrating molecule has kinetic energy $E_{kin} = E(v, J) - V(R)$. During a radiative transition this kinetic energy has to be preserved. If the total energy $E'' = E(v', J') - h\nu = V''(R) + E_{kin} = U(R)$ in the lower state is below the dissociation limit of the potential $V''(R)$, the fluorescence terminates in bound levels and the fluorescence spectrum consists of discrete lines. If it is above this limit, the fluorescence transitions end in the dissociation continuum (Fig. 1.55). The intensity distribution of these "Condon internal-diffraction bands" [179, 180] is very sensitive to the difference potential $V''(R) - V'(R)$, and therefore allows an accurate determination of one of the potential curves if the other is known [181].

1.8.3 LIF of Polyatomic Molecules

The technique of LIF is, of course, not restricted to diatomic molecules, but has also been applied to the investigation of triatomic molecules such as NO_2, SO_2, BO_2,

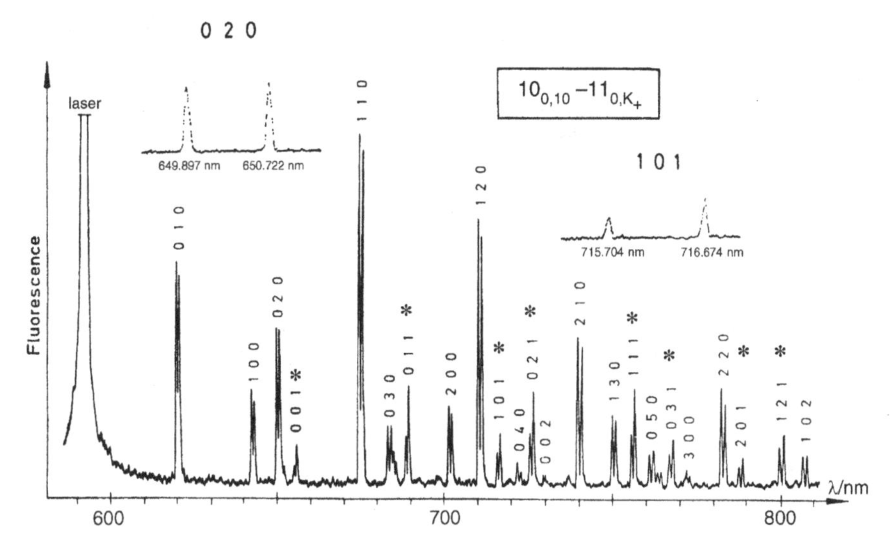

Fig. 1.56 LIF spectrum of NO_2 excited at $\lambda = 590.8$ nm. The vibrational bands terminating on ground-state vibrational levels (v_1, v_2, v_3) marked with an *asterisk* are forbidden by symmetry selection rules but are made allowable by an admixture of a perturbing level with other symmetry to the excited-state wave function [182]

NH_2, and many other polyatomic molecules. In combination with excitation spectroscopy, it allows the assignment of transitions and the identification of complex spectra. Examples of such measurements have been cited in [154–157].

The LIF of polyatomic molecules opens possibilities to recognize perturbations both in excited and in ground electronic states. If the upper state is perturbed its wave function is a linear combination of the BO wave functions of the mutually perturbing levels. The admixture of the perturber wave function opens new channels for fluorescence into lower levels with different symmetries, which are forbidden for unperturbed transitions. The LIF spectrum of NO_2 depicted in Fig. 1.56 is an example where the "forbidden" vibrational bands terminating on the vibrational levels (v_1, v_2, v_3) with an odd number v_3 of vibrational quanta in the asymmetric stretching vibrational mode in the electronic ground state are marked by an asterisk [182].

Due to nonlinear coupling between high-lying vibrational levels of polyatomic molecules, the level structure may become quite complex. The classical motion of the nuclei in such a highly vibrationally excited molecule can often no longer be described by a superposition of normal vibrations [183]. In such cases a statistical model of the vibrating molecule may be more adequate when the distribution of the energy spacings of neighboring vibrational levels is investigated. For several molecules a transition regime has been found from *classical oscillations* to *chaotic behavior* when the vibrational energy is increasing. The chaotic region corresponds to a Wigner distribution of the energy separations of neighboring vibrational levels, while the classical nonchaotic regime leads to a Poisson distribution of the vibrational energy spacings. With LIF spectroscopy such high-lying levels in the elec-

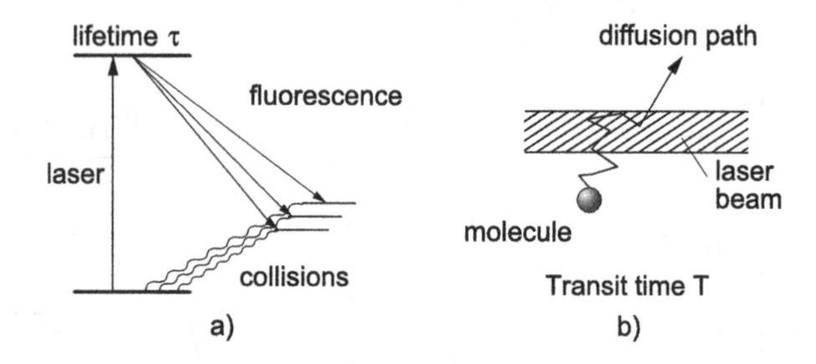

Fig. 1.57 Photon bursts. (**a**) Level scheme; (**b**) diffusion of a molecule through the laser beam

tronic ground state can be reached. The analysis of such LIF spectra therefore yields information on the dynamics of molecules in vibronically coupled levels [184–186].

1.8.4 Photon Bursts

At low gas densities each molecule can only absorb one photon, because the LIF terminates on many lower levels and only a small fraction of all excited molecules reaches the initial ground state from where the absorption started. This is different in liquids or in gases at high pressures. Here fast collisional redistribution between the different levels (v'', J'') restore thermal equilibrium and refill the level (v_i'', J_i'') depleted by the absorption of laser photons in a time T_c which is short compared to the lifetime τ of the upper laser excited state (Fig. 1.57).

If the molecule stays for a time T within the laser beam which is long compared to the lifetime τ of the upper laser-excited state $\langle k|$ the molecule can undergo $n = T/(2\tau)$ absorption-emission cycles. The laser-induced fluorescence is observed through a microscope.

Example 1.20 Typical values of lifetimes of electronically excited states are $\tau = 10^{-8}$ s. Diffusion times of a molecule through a laser beam with 0.3 cm diameter are about 1 ms. This yields $n = 5 \times 10^4$ fluorescence photons emitted by one molecule. If only 5 % of these photons (=2500 photons) can be imaged onto the detector, this is by far sufficient to be detected even by the naked eye. Therefore single molecules can be seen by this technique.

1.8.5 Determination of Population Distributions by LIF

An interesting application of LIF is the measurement of relative population densities $N(v_i'', J_i'')$ and their distribution over the different vibrational–rotational levels (v_i'', J_i'') under situations that differ from thermal equilibrium. Examples are chemical reactions of the type $AB + C \rightarrow AC^* + B$, where a reaction product AC^* with

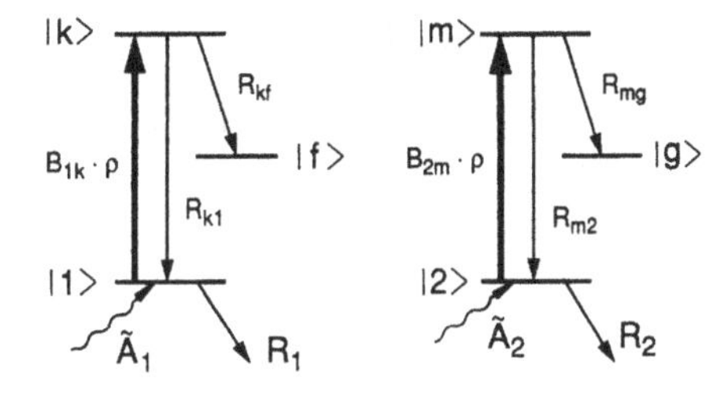

Fig. 1.58 Level scheme for measurements of population distributions in rotational–vibrational levels of molecular reaction products in their electronic ground state

internal energy is formed in a reactive collision between the partners AB and C. The measurement of the internal state distribution $N_{AC}(v'', J'')$ can often provide useful information on the reaction paths and on the potential surfaces of the collision complex (ABC)*. The initial internal state distribution $N_{AC}(v, J)$ is often far away from a Boltzmann distribution. There even exist reactions that produce population inversions allowing the realization of chemical lasers [187]. Investigations of these population distributions may finally allow the optimization and better control of these reactions.

The absorption rate of the laser tuned to the transition $\langle i| \to \langle k|$ is directly proportional to the population distribution $N(v_i'', J_i'')$. Therefore the intensity ratio of the total LIF induced by the laser tuned to the different absorption transitions yields the relative population densities of the absorbing levels. The population density $N_k(v_k, J_k)$ in the excited state $|k\rangle$ can be determined from the measurement of the fluorescence rate

$$n_{Fl} = N_k A_k V_R, \tag{1.59}$$

which represents the number of fluorescence photons n_{Fl} emitted per second from the reaction volume V_R.

In order to obtain the population densities $N_i(v_i, J_i)$ in the electronic ground state, the laser is tuned to the absorbing transitions $|i\rangle \to |k\rangle$ starting from levels (v_i'', J_i'') of the reaction products under investigation, and the total fluorescence rate (1.59) is measured for different upper levels $|k\rangle$. Under stationary conditions these rates are obtained from the rate equation

$$\frac{dN_k}{dt} = 0 = N_i B_{ik} \rho - N_k (B_{ki} \rho + A_k + R_k) \tag{1.60a}$$

which gives, according to (1.59) with $B_{ik} = B_{ki}$:

$$n_{Fl} = N_k A_k V_R = N_i A_k V_R \frac{B_{ik} \rho}{B_{ik} \rho + A_k + R_k}, \tag{1.60b}$$

where R_k is the total nonradiative deactivation rate of the level $|k\rangle$ (Fig. 1.58). If the collisional deactivation of level $|k\rangle$ is negligible compared with its radiative depopulation by fluorescence, (1.60a–1.60b) reduces to

$$n_{Fl} = N_i \cdot A_k \cdot V_R \cdot \frac{B_{ik} \rho}{A_k + B_{ik} \rho} = \frac{N_i V_R B_{ik} \rho}{1 + B_{ik} \rho / A_k}. \tag{1.61}$$

We distinguish two limiting cases:

(a) The laser intensity is sufficiently low to ensure $B_{ik}\rho \ll A_k$. Then the ratio $n_{\mathrm{Fl}}(k)/n_{\mathrm{Fl}}(m)$ of fluorescence rates observed under excitation on the two transitions $|1\rangle \rightarrow |k\rangle$ and $|2\rangle \rightarrow |m\rangle$, respectively, with the same laser intensity $I_{\mathrm{L}} = c\rho$ becomes (Fig. 1.58)

$$\frac{n_{\mathrm{Fl}}(k)}{n_{\mathrm{Fl}}(m)} = \frac{N_1}{N_2}\frac{B_{1k}}{B_{2m}} = \frac{N_1\sigma_{1k}}{N_2\sigma_{2m}} = \frac{\alpha_{1k}}{\alpha_{2m}}, \tag{1.62}$$

where σ is the optical absorption cross section and $\alpha = N\sigma$ the absorption coefficient. Therefore the ratio N_1/N_2 of the lower-level population can be obtained from the measured ratio of fluorescence rates or of absorption coefficients if the absorption cross sections are known.

(b) If the laser intensity is sufficiently high to saturate the absorbing transition we have the case $B_{ik}\rho \gg A_k$. Then (1.61) yields for $R_k < A_k$

$$\frac{n_{\mathrm{Fl}}(k)}{n_{\mathrm{Fl}}(m)} = \frac{N_1 A_k}{N_2 A_m}. \tag{1.63}$$

Under stationary conditions the population densities N_i are determined by the rate equations

$$\mathrm{d}N_i/\mathrm{d}t = 0 = \widetilde{A}_i - N_i(R_i + B_{ik}\rho) + \sum_m N_m R_{mi}, \tag{1.64}$$

where $\widetilde{A}_i$ is the feeding rate of the level $|i\rangle$ by the reaction, or by diffusion of molecules in level $|i\rangle$ into the detection volume. The rate $N_i(R_i + B_{ik}\rho)$ is the total deactivation rate of the level $|i\rangle$, whereas $\sum N_m R_{mi}$ is the sum of all transition rates from other levels $|m\rangle$ into the level $|i\rangle$.

If $\widetilde{A}_i \gg \sum N_m R_{mi}$ and the main depletion rate of N_i is due to laser absorption ($B_{ik}\rho \gg R_i$), the stationary level population becomes $N_i = \widetilde{A}_i/B_{ik}\rho$ and the ratio (1.63) of LIF rates becomes

$$\frac{n_{\mathrm{Fl}}(k)}{n_{\mathrm{Fl}}(m)} = \frac{\widetilde{A}_1 B_{2m} A_k}{\widetilde{A}_2 B_{1k} A_m}. \tag{1.65}$$

Lifetime measurements of levels $|k\rangle$ and $|m\rangle$ (see Sect. 6.3) yield the absolute values of A_k and A_m. Measuring the relative intensities of the different fluorescence transitions allows the determination of the branching ratios of these transitions from a selectively excited upper level [188].

First measurements of internal-state distributions of reaction products using LIF have been performed by Zare et al. [189–191]. One example is the formation of

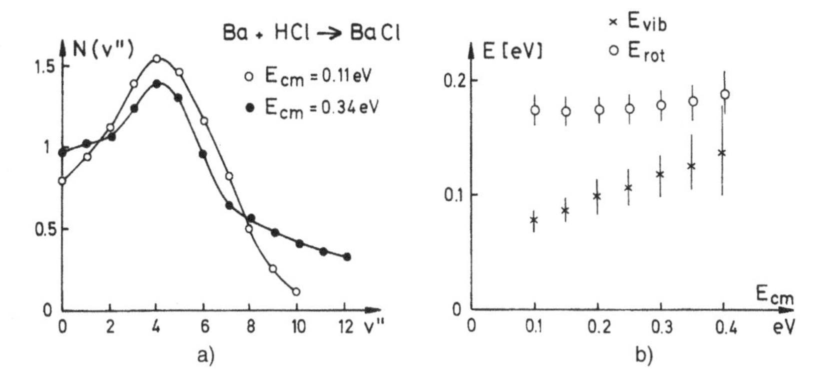

Fig. 1.59 (**a**) Vibrational level populations $N(v'')$ of BaCl for two different collision energies of the reactants Ba + HCl. (**b**) Mean vibrational and rotational energy of reactively formed BaCl as a function of the relative collision energy [191]

BaCl in the reaction of barium with halogens

$$\mathrm{Ba} + \mathrm{HCl} \rightarrow \mathrm{BaCl}^*\left(\mathrm{X}^2\Sigma^+ v'', J''\right) + \mathrm{H}. \tag{1.66}$$

Figure 1.59a exhibits the vibrational population distribution of BaCl for two different collision energies of the reactants Ba and HCl. Figure 1.59b illustrates that the total *rotational* energy of BaCl barely depends on the collision energy in the center of mass system of Ba + HCl, while the *vibrational* energy increases with increasing collision energy.

The interesting question of how the internal-state distribution of the products is determined by the internal energy of the reacting molecules can be answered experimentally with a second laser that pumps the reacting molecule into excited levels (v'', J''). The internal-state distribution of the product is measured with and without the pump laser. An example that has been studied [190] is the reaction

$$\mathrm{Ba} + \mathrm{HF}\left(v'' = 1\right) \rightarrow \mathrm{BaF}^*(v = 0\text{–}12) + \mathrm{H}, \tag{1.67}$$

where a chemical HF laser has been used to excite the first vibrational level of HF, and the internal state distribution of BaF* is measured with a tunable dye laser.

An example where LIF is used for the optimization of the production of thin amorphous layers of Si:H from the deposition of gaseous silane SiH_4 in a gas discharge is the spectroscopy of SiH radicals that are formed in the discharge [192].

Another example is the determination of vibrational–rotational population distributions in supersonic molecular beams [193], where molecules are cooled down to rotational temperatures of a few Kelvin (Sect. 4.2).

The method is not restricted to neutral molecules but can also be applied to ionic species. This is of importance when LIF is used for diagnostics of combustion processes [194, 195] or plasmas [196].

A good overview of fluorescence spectroscopy is given in [197].

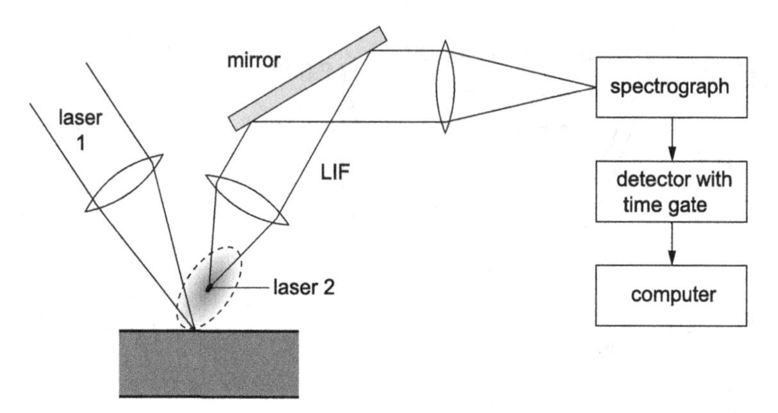

Fig. 1.60 Experimental setup for LIBS

1.8.6 Laser-Induced Breakdown Spectroscopy LIBS

This popular technique is used for the analysis of solid or liquid surfaces. Its technique can be summarized as follows (Fig. 1.60): A pulsed laser with high peak power is focussed onto the surface and evaporates electrons, neutral atoms or molecules and ions. The emitted particles form a hot plasma (up to 100,000 K), which expands and cools down, reaching soon thermal equilibrium at about 10,000 K.

With a second laser electronic transitions in the neutral atoms or the ions are excited and the laser-induced fluorescence is monitored for specific known excitation lines. With a time gate in the detector system the spectra can be taken at different times and therefore different temperatures of the expanding plasma. The intensity of the LIF gives information about the atomic composition of the evaporated material and the abundance of the excited species, if the transition probabilities are known [198, 199]. The sensitivity of the technique depends on the peak intensity and the pulse duration of the excitation laser pulses. Typical laser pulse widths range from nanoseconds to picoseconds. Recently also femtosecond lasers have been used. Since only a tiny amount of material (nanograms to picograms) is evaporated the sample is not essentially damaged by this purely optical analysis [200].

1.9 Comparison Between the Different Methods

The different sensitive techniques of Doppler-limited laser spectroscopy discussed in the previous sections supplement each other in an ideal way. In the *visible* and *ultraviolet* range, where *electronic* states of atoms or molecules are excited by absorption of laser photons, *excitation spectroscopy* is generally the most suitable technique, particularly at low molecular densities. Because of the short spontaneous lifetimes of most excited electronic states E_k, the quantum efficiency η_k reaches

100 % in many cases. For the detection of the laser-excited fluorescence, sensitive photomultipliers or intensified CCD cameras are available that allow, together with photon-counting electronics (Vol. 1, Sect. 4.5), the detection of single fluorescence photons with an overall efficiency of 10^{-3}–10^{-1} including the collection efficiency $\delta \approx 0.01$–0.3 (Sect. 1.3.1).

Excitation of very high-lying states close below the ionization limit, e.g., by ultraviolet lasers or by two-photon absorption, enables the detection of absorbed laser photons by monitoring the ions. Because of the high collection efficiency of these ions, *ionization spectroscopy* represents the most sensitive detection method, superior to all other techniques in all cases where it can be applied. Its experimental drawback is the need for two lasers, including at least one tunable laser.

In the *infrared* region, excitation spectroscopy is less sensitive because of the lower sensitivity of infrared photodetectors and because of the longer lifetimes of excited vibrational levels. These long lifetimes bring about either at low pressures a diffusion of the excited molecules out of the observation region, or at high pressures a collision-induced radiationless deactivation of the excited states. Here *photoacoustic spectroscopy* becomes superior, since this technique utilizes this collision-induced transfer of excitation energy into thermal energy. A specific application of this technique is the quantitative determination of small concentrations of molecular components in gases at higher pressures. Examples are measurements of air pollution or of poisonous constituents in auto engine exhausts, where sensitivities in the ppb range have been successfully demonstrated. For pure gases at lower pressures, where collisions are less significant, absorption spectroscopy with wavelength-modulation may be even superior to optoacoustic spectroscopy, Fig. 1.61.

For infrared spectroscopy in molecular beams optothermal spectroscopy is a very good choice (Sect. 1.3.3).

For the spectroscopy of atoms or ions in gas discharges, *optogalvanic spectroscopy* (Sect. 1.5) is a very convenient and experimentally simple alternative to fluorescence detection. In favorable cases it may even reach the sensitivity of excitation spectroscopy. For the distinction between spectra of ions and neutral species *velocity-modulation spectroscopy* (Sect. 1.6) offers an elegant solution.

Regarding detection sensitivity, LMR and Stark spectroscopy can compete with the other methods. However, their applications are restricted to molecules with sufficiently large permanent dipole moments to achieve the necessary tuning range. They are therefore mainly applied to the spectroscopy of free radicals with an unpaired electron. The magnetic moment of these radicals is predominantly determined by the electron spin and is therefore several orders of magnitude larger than that of stable molecules in $^1\Sigma$ ground states. The advantage of LMR or Stark spectroscopy is the direct determination of Zeeman or Stark splittings from which the Landé factors and therefore the coupling schemes of the different angular momenta can be deduced. A further merit is the higher accuracy of the *absolute* frequency of molecular absorption lines because the frequencies of fixed-frequency laser lines can be absolutely measured with a higher accuracy than is possible with tunable lasers.

All these methods represent modifications of *absorption* spectroscopy, whereas LIF spectroscopy is based on *emission* of fluorescence from selectively populated

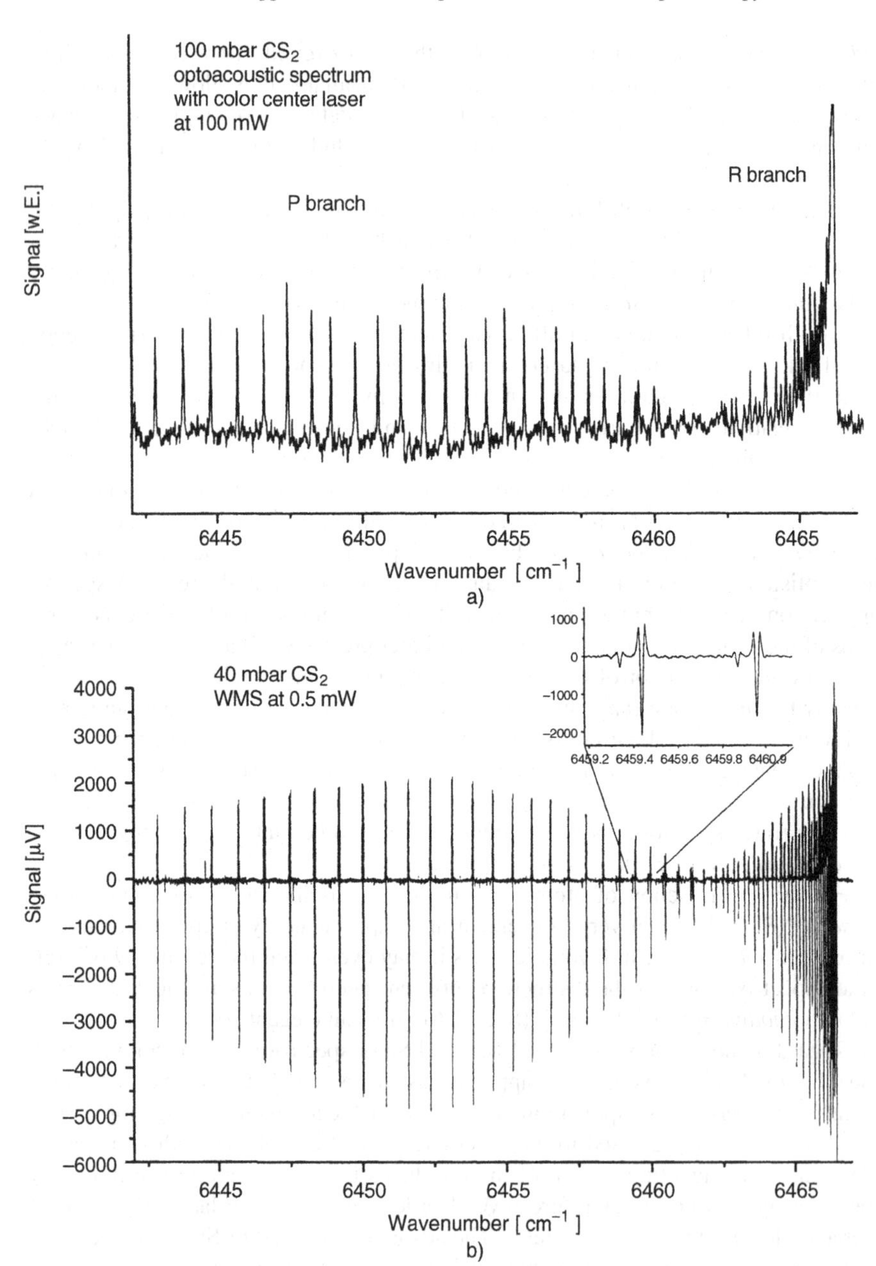

Fig. 1.61 Section of CS_2-overtone spectrum. Comparison of optoacoustic detection (**a**) and absorption spectroscopy with wavelength modulation (**b**) (G.H. Wenz, Thesis, K.L.)

upper levels. The absorption spectra depend on both the upper and lower levels. The absorption transitions start from thermally populated lower levels. If their properties are known (e.g., from microwave spectroscopy) the absorption spectra yield infor-

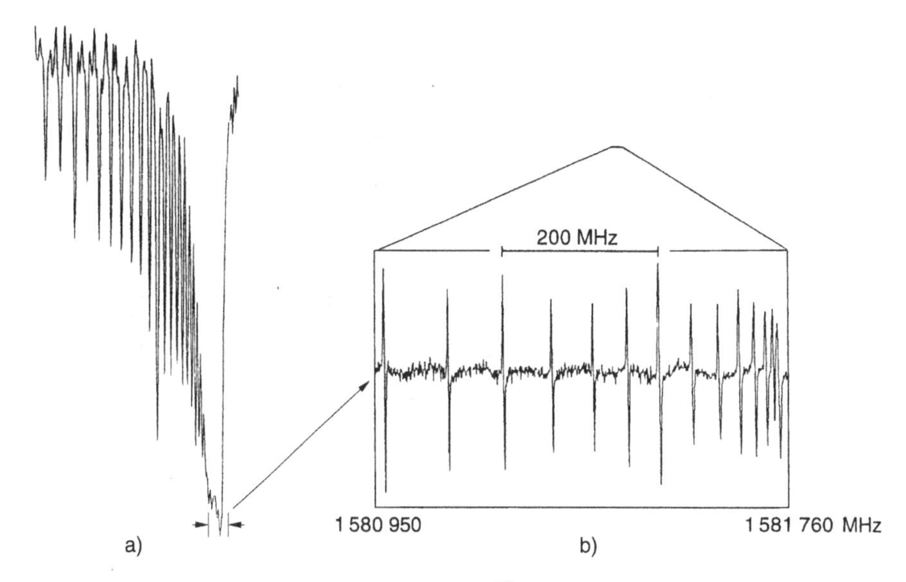

Fig. 1.62 Section of the pure rotational spectrum of $^{16}O_3$ recorded with a high-resolution Fourier spectrometer (*left spectrum*) and a small section of it, measured with a tunable far-infrared laser spectrometer (*right spectrum* with expanded frequency scale), demonstrating the superior resolution of the laser spectrometer [201]

mation on the *upper* levels. On the other hand, the LIF spectra start from one or a few upper levels and terminate on many rotational–vibrational levels of a lower electronic state. They give information about this lower state.

All these techniques may be combined with intracavity absorption when the sample molecules are placed inside the laser resonator to enhance the sensitivity. Cavity ring-down spectroscopy yields absorption spectra with a detection sensitivity that is comparable to the most advanced modulation techniques in multipass absorption spectroscopy.

A serious competitor of laser absorption spectroscopy is Fourier spectroscopy [1, 201], which offers the advantage of simultaneous recording of a large spectral interval and with it a short measuring time. In contrast, in laser absorption spectroscopy the whole spectrum must be measured sequentially, which takes much more time. However, laser techniques offer two advantages, namely, higher spectral resolution and higher sensitivity. This is demonstrated by examples in two different spectral regimes. Figure 1.62 exhibits part of the submillimeter, purely rotational spectrum of $^{16}O_3$ taken with a high-resolution ($0.003\ \mathrm{cm}^{-1}$) Fourier spectrometer (Fig. 1.62a) and a small section of Fig. 1.62a in an expanded scale, recorded with the tunable far-infrared laser spectrometer shown in Vol. 1, Fig. 5.76. Although the resolution is still limited by the Doppler width of the absorption lines, this imposes no real limitation because at $\nu = 1.5 \times 10^{12}$ Hz the Doppler width of ozone is only $\Delta\nu_D \approx 2$ MHz, compared with a resolution of 90 MHz of the Fourier spectrometer. Another example illustrates the overtone spectrum of acethylene C_2H_2 around $\lambda = 1.5\ \mu$m, recorded with a Fourier spectrometer (Fig. 1.63a) and with a color-

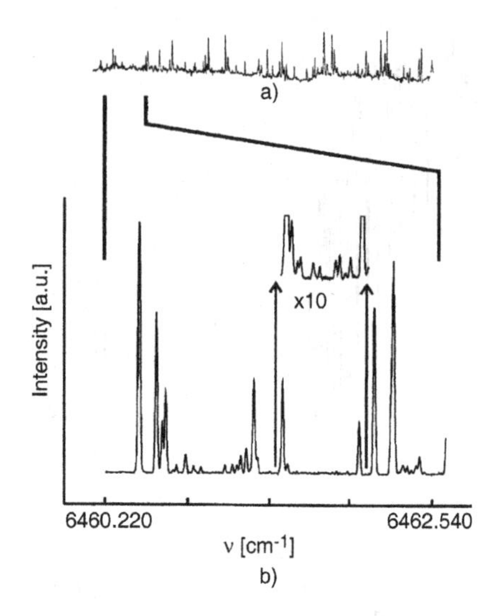

Fig. 1.63 Comparison of overtone spectra of C_2H_2 around $\lambda = 1.5\ \mu m$, measured (**a**) with a Fourier spectrometer and (**b**) with a color-center laser and intracavity optoacoustic spectroscopy [74]

center laser and photo-acoustic spectroscopy (Fig. 1.63b). Weak lines, which are barely seen in the Fourier spectrum, still have a large signal-to-noise ratio in the optoacoustic spectrum as shown by the insert in Fig. 1.63b [74].

1.10 Problems

1.1 A monochromatic laser beam is sent through a sample of diatomic molecules. The laser wavelength is tuned to a vibration–rotation transition $(v'', J'') \to (v', J')$ with an absorption cross section of $\sigma_{ik} = 10^{-18}\ \text{cm}^2$.

(a) Estimate the fraction n_i/n of molecules in the level $(v_i'' = 0, J_i'' = 20)$ for thermal equilibrium at $T = 300$ K (vibrational constant $\omega_e = 200\ \text{cm}^{-1}$, rotational constant $B_e = 1.5\ \text{cm}^{-1}$).
(b) Calculate the absorption coefficient for a total gas pressure of 10 mbar.
(c) What is the transmitted laser power P_T behind an absorption path length of 10 cm for an incident power $P_0 = 100$ mW?

1.2 A focused laser beam ($\varnothing = 0.4$ mm) with 1-mW input power at $\lambda = 623$ nm crosses a molecular beam ($\varnothing = 1$ mm) perpendicularly. The absorbing molecules with a partial flux density $N_i = n_i \cdot \bar{v} = 10^{12}/(\text{s}\,\text{cm}^2)$ and a velocity $\bar{v} = 5 \times 10^4$ cm/s have the absorption cross section $\sigma = 10^{-16}\ \text{cm}^2$.

How many photoelectrons are counted with a photomultiplier if the fluorescence emitted from the crossing volume $V_c = 10^{-4}\ \text{cm}^3$ of both the laser and molecular beams is imaged by a lens with $D = 4$ cm at a distance $L = 8$ cm from V_c onto the photocathode with quantum efficiency $\eta_{ph} = 0.2$?

1.3 A monochromatic laser beam with $P = 1$ mW is sent through a 1 m long sample cell filled with absorbing molecules. The absorbing transition has the Doppler width $\Delta\omega_D = 2\pi \times 10^9\ \mathrm{s}^{-1}$ and a peak absorption $\alpha(\omega_0) = 10^{-8}\ \mathrm{cm}^{-1}$. The laser frequency $\omega_L = \omega_0 + \Delta\omega \cdot \cos 2\pi f t$ is modulated ($\Delta\omega = 2\pi \times 10$ MHz). A detector measures the transmitted laser power P_T. Calculate the maximum ac amplitude of the detector output signal for a detector with a sensitivity of 1 V/mW. How large is the dc background signal?

1.4 How many ions per second are formed by resonant two-step photoionization if the transition $|i\rangle \to |k\rangle$ of the first step is saturated, the lifetime τ_k of the level $|k\rangle$ is $\tau_k = 10^{-8}$ s, the ionization probability by the second laser is $10^7\ \mathrm{s}^{-1}$, and the diffusion rate of molecules in the absorbing level $|i\rangle$ into the excitation volume is $\mathrm{d}N_i/\mathrm{d}t = 10^5/\mathrm{s}$?

1.5 A thermal detector (bolometer area $3 \times 3\ \mathrm{mm}^2$) is hit by a molecular beam with a density of $n = 10^8$ molecules per cm^3, having a mean velocity $\langle v\rangle = 4 \times 10^4$ cm/s and a mass $m = 28$ AMU. All impinging molecules are assumed to stick at the surface of the detector.

(a) Calculate the energy transferred per second to the bolometer.
(b) What is the temperature rise ΔT, if the heat losses of the bolometer are $G = 10^{-8}$ W/K?
(c) The molecular beam is crossed by an infrared laser ($\lambda = 1.5$ µm) with $P_0 = 10$ mW. The absorption coefficient is $\alpha = 10^{-10}\ \mathrm{cm}^{-1}$ and the absorption path length $L = 10$ cm (by the multiple-reflection techniques). Calculate the additional temperature rise of the bolometer.

1.6 The frequency ν_0 of a molecular transition to an upper level $|k\rangle$ is 10^8 Hz below the frequency ν_L of a fixed laser line. Assume that the molecules have no magnetic moment in the ground state, but $\mu = 0.5\mu_B$ ($\mu_B = 9.27 \times 10^{-24}$ J/T) in the upper state. What magnetic field $\boldsymbol{B}$ is necessary to tune the absorption line into resonance with the laser line? How many Zeeman components are observed, if the lower level has the rotational quantum number $J = 1$ and the upper $J = 2$, and the laser is (a) linearly and (b) circularly polarized?

1.7 For velocity-modulation spectroscopy an ac voltage of 2 kV peak-to-peak and $f = 1$ kHz is applied along a discharge tube with 1 m length in the z-direction.

(a) How large is the mean electric field strength?
(b) Estimate the value of Δv_t for the ion velocity $v = v_0 + \Delta v_z \sin 2\pi f t$ for ions with the mass $m = 40$ amu (1 amu $= 1.66 \times 10^{-27}$ kg) if their mean free path length is $\Lambda = 10^{-3}$ m and the density of neutral species in the discharge tube is $n = 10^{17}/\mathrm{cm}^3$.
(c) How large is the maximum modulation $\Delta\nu$ of the absorption frequency $\nu(v_t) = \nu_0 + \Delta\nu(v_t)$ for $\nu_0 = 10^{14}\ \mathrm{s}^{-1}$?

(d) How large is the ac signal for an absorbing transition with $\alpha(\nu_0) = 10^{-6}\ \mathrm{cm}^{-1}$ if the incident laser power is 10 mW and the detector sensitivity 1 V/mW?

1.8 An absorbing sample in a cell with $L = 4$ cm is placed within the laser resonator with a total resonator loss of 2 % per round trip and a transmission $T = 0.5$ % of the output coupler.

(a) Calculate the relative decrease of the laser output power of 1 mW when the laser is tuned from a frequency ν where $\alpha = 0$ to a frequency ν_0 where the sample has an absorption coefficient $\alpha = 5 \times 10^{-8}\ \mathrm{cm}^{-1}$, while the unsaturated gain $g_0 = 4 \times 10^{-2}$ of the active laser medium stays constant.
(b) How many fluorescence photons are emitted if the fluorescence yield of the sample is 0.5 and the laser wavelength is 500 nm?
(c) Could you design the optimum collection optics that image a maximum ratio of the fluorescence light onto the photomultiplier cathode of 40 mm ⌀? How many photon counts are observed if the quantum yield of the cathode is $\eta = 0.15$?
(d) Compare the detection sensitivity of (c) (dark current of the PM is 10^{-9} A) with that of method (a) if the laser output is monitored by a photodiode with a sensitivity of 10 V/W and a noise level of 10^{-9} V.

1.9 Derive Eq. (1.46).

1.10 The expanding plasma plume of evaporated material emits fluorescence when excited on an aluminium line at $\lambda = 555$ nm and on a copper line at $\lambda = 324$ nm. The intensity ratio of the total LIF excited by these lines is 1:4. The ratio of the two transition probabilities is 1:6. Calculate the relative abundance of the two atomic species.

Chapter 2
Nonlinear Spectroscopy

One of the essential advantages that single-mode lasers can offer for high-resolution spectroscopy is the possibility of overcoming the limitation set by Doppler broadening. Several techniques have been developed that are based on selective saturation of atomic or molecular transitions by sufficiently intense lasers.

The population density of molecules in the absorbing level is decreased by optical pumping. This results in a nonlinear dependence of the absorbed radiation power on the incident power. Such techniques are therefore summarized as *nonlinear spectroscopy*, which also includes methods that are based on the simultaneous absorption of two or more photons during an atomic or molecular transition. In the following sections the basic physics and the experimental realization of some important methods of nonlinear spectroscopy are discussed. At first we shall, however, treat the saturation of population densities by intense incident radiation.

2.1 Linear and Nonlinear Absorption

Assume that a monochromatic plane lightwave

$$E = E_0 \cos(\omega t - kz),$$

with the mean intensity

$$I = \frac{1}{2} c\epsilon_0 E_0^2 \; [\mathrm{W/m^2}],$$

passes through a sample of molecules, which absorb on the transition $E_i \to E_k$ $(E_k - E_i = \hbar\omega)$. The power $\mathrm{d}P$ absorbed in the volume $\mathrm{d}V = A\,\mathrm{d}z$ is then

$$\mathrm{d}P = -P_0 \alpha \,\mathrm{d}z = -A I_0 \sigma_{ik} \Delta N \,\mathrm{d}z \; [\mathrm{W}], \tag{2.1a}$$

where A is the cross-section of the illuminated area, $I_0 = P_0/A$ is the incident intensity, $\Delta N = [N_i - (g_i/g_k)N_k]$ is the difference of the population densities, and $\sigma_{ik}(\nu)$

W. Demtröder, *Laser Spectroscopy 2*, DOI 10.1007/978-3-662-44641-6_2

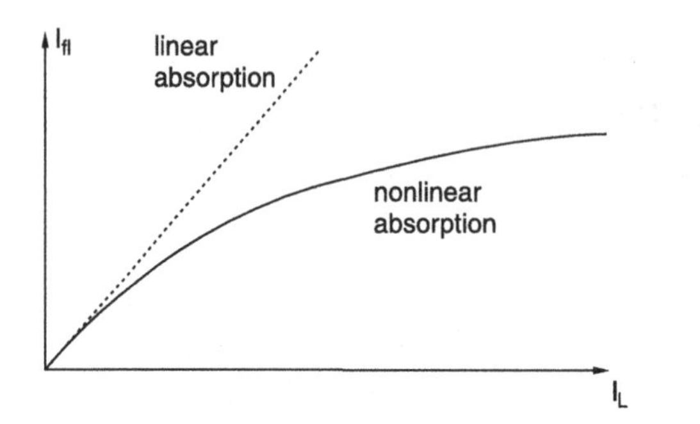

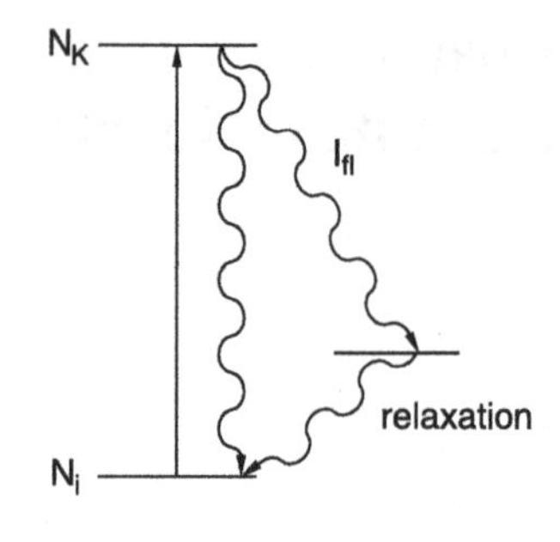

Fig. 2.1 Fluorescence intensity $I_{\mathrm{Fl}}(I_{\mathrm{L}})$ as a function of incident laser intensity for linear and nonlinear absorption

is the absorption cross section per molecule at the light frequency $\nu = \omega/2\pi$, see (5.2) in Vol. 1.

For sufficiently low incident intensities I_0, the absorption coefficient α is independent of I_0 (this implies that the population difference ΔN is not dependent on I_0!), and the absorbed power $\mathrm{d}P$ is linearly dependent on the incident power P_0. Integration of (2.1a) gives Beer's law of *linear absorption*:

$$P = P_0 \mathrm{e}^{-\alpha z} = P_0 \mathrm{e}^{-\sigma \Delta N z}. \tag{2.1b}$$

When the absorption is measured through the fluorescence intensity I_{Fl} emitted from the upper level $|k\rangle$, which is proportional to the absorbed power, we obtain the linear curve in Fig. 2.1. As the intensity of the incident radiation increases, the population N_i in the absorbing level $|i\rangle$ will be depleted when the absorption rate becomes faster than the relaxation processes that can refill it. Therefore, the absorption decreases and the curve $I_{\mathrm{Fl}}(I_0)$ in Fig. 2.1 deviates from the straight curve until it reaches a constant value (saturation). We must then generalize (2.1a) to

$$\mathrm{d}P = -P_0 \cdot \alpha(P_0)\,\mathrm{d}z = -P_0 \sigma_{ik} \Delta N(P_0)\,\mathrm{d}z \tag{2.1c}$$

where the population difference ΔN and therefore the absorption coefficient α become a function of P_0 (*nonlinear absorption*).

Example 2.1 In a first approximation we can write $\alpha = \alpha_0(1 - bI)$, which gives with a cross-section A of the incident radiation

$$\mathrm{d}P = -AI\alpha\,\mathrm{d}z = -A\big(I\alpha_0 - \alpha_0 b I^2\big)\,\mathrm{d}z. \tag{2.1d}$$

The first term describes the contribution from linear absorption, and the second from nonlinear quadratic absorption.

We will now consider the nonlinear absorption in more detail.

If the incident plane wave with the spectral energy density

$$\rho_\nu(\nu) = I_\nu(\nu)/c \; [\mathrm{W\,s^2/m^3}]$$

has the spectral width $\delta\nu_L$, the total intensity becomes

$$I = \int I_\nu(\nu)\,\mathrm{d}\nu \approx I_\nu(\nu_0)\cdot\delta\nu_L. \tag{2.2}$$

Note The difference between the spectral intensity I_ν [$\mathrm{W\,s\,m^{-2}}$] (radiation power per m^2 and frequency interval $\mathrm{d}\nu = 1\ \mathrm{s}^{-1}$) and the total intensity I [$\mathrm{W\,m^{-2}}$].

The absorbed power is then

$$\Delta P = \Delta N\,\mathrm{d}V \cdot \int I_\nu(\nu)\cdot\sigma_{ik}(\nu)\,\mathrm{d}\nu. \tag{2.3}$$

For a monochromatic laser wave tuned to the center-frequency ν_0 of an absorption line, the absorbed power is:

$$\Delta P = \Delta N\,\mathrm{d}V \cdot I(\nu_0)\cdot\sigma_{ik}(\nu_0) \tag{2.4}$$

where $\mathrm{d}V = A\,\mathrm{d}z$ is the volume of the absorbing medium traversed by the laser beam with cross-section A. If the spectral width $\delta\nu_L$ of the laser is larger than the width $\delta\nu_a$ of the absorption line, only that part of the spectral interval of the laser line inside the line width $\delta\nu_a$ of the absorption line is absorbed, and the absorbed power becomes:

$$\Delta P = \Delta N\,\mathrm{d}V \cdot I(\nu_0)\cdot\sigma(\nu_0)\cdot\delta\nu_a/\delta\nu_L. \tag{2.5}$$

This corresponds to $n_{ph} = \Delta P/h\nu$ absorbed photons. From Vol. 1, (2.15) we can deduce

$$n_{ph} = B_{ik}\cdot\rho(\nu)\Delta N\,\mathrm{d}V. \tag{2.6}$$

$B_{ik}\rho_\nu$ gives the net probability for absorbing a photon per molecule and per second within the unit volume $\mathrm{d}V = 1\ \mathrm{m}^3$, see Vol. 1, (2.15), (2.18).

The comparison of (2.3) and (2.6) yields the relation between the Einstein coefficient B_{ik} and the absorption cross-section σ_{ik}

$$\boxed{B_{ik} = \frac{c}{h\nu}\int_{\nu=0}^{\infty}\sigma_{ik}(\nu)\,\mathrm{d}\nu,} \tag{2.7}$$

where the frequencies ν outside the absorption linewidth $\delta\nu_a$ do not noticeably contribute to the integral. The absorption of the incident wave causes population changes of the levels involved in the absorbing transition. This can be described by

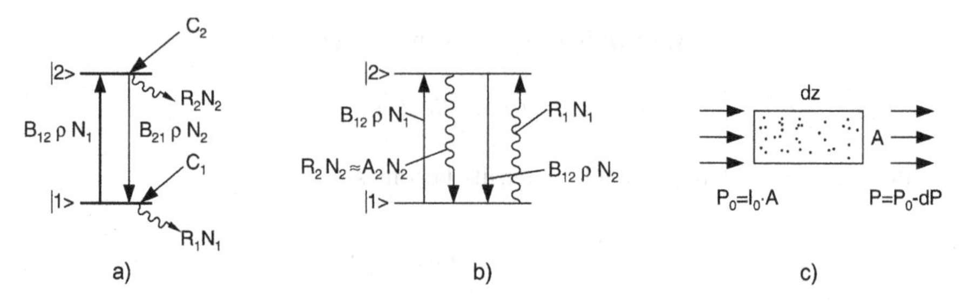

Fig. 2.2 Level diagram of an open two-level system with open relaxation channels into other levels and population paths from outside the system (**a**), a closed system (**b**), and schematic illustration of absorption (**c**)

the rate equations for the population densities N_1, N_2 of the nondegenerate levels $|1\rangle$ and $|2\rangle$ with $g_1 = g_2 = 1$ (Fig. 2.2):

$$\frac{\mathrm{d}N_1}{\mathrm{d}t} = B_{12}\rho_\nu(N_2 - N_1) - R_1 N_1 + C_1, \tag{2.8a}$$

$$\frac{\mathrm{d}N_2}{\mathrm{d}t} = B_{12}\rho_\nu(N_1 - N_2) - R_2 N_2 + C_2, \tag{2.8b}$$

where $R_i N_i$ represents the total relaxation rate (including spontaneous emission) that depopulates the level $|i\rangle$ and

$$C_i = \sum_k R_{ki} N_k + D_i, \tag{2.8c}$$

takes care of all relaxation paths from other levels $|k\rangle$ that contribute to the repopulation of level $|i\rangle$ and also of the net diffusion rate D_i of molecules in level $|i\rangle$ into the excitation volume $\mathrm{d}V$. We call the system described by (2.8a, 2.8b) an *open two-level system* because optical pumping occurs only between the two levels $|1\rangle$ and $|2\rangle$, which, however, may decay into other levels. That is, channels are open for transitions out of the system and for the population of the system from outside.

If the quantities C_i are not noticeably changed by the radiation field, we obtain from (2.8a–2.8c) under stationary conditions ($\mathrm{d}N/\mathrm{d}t = 0$) the unsaturated population difference for $\rho = 0$

$$\Delta N^0 = \Delta N(\rho = 0) = N_2^0 - N_1^0 = \frac{C_2 R_1 - C_1 R_2}{R_1 R_2}. \tag{2.9}$$

Note that $N_1^0 > N_2^0$ and therefore $\Delta N^0 < 0$.

For the saturated population difference ($\rho \neq 0$):

$$\Delta N = \frac{\Delta N^0}{1 + B_{12}\rho_\nu(1/R_1 + 1/R_2)} = \frac{\Delta N^0}{1 + S}, \tag{2.10}$$

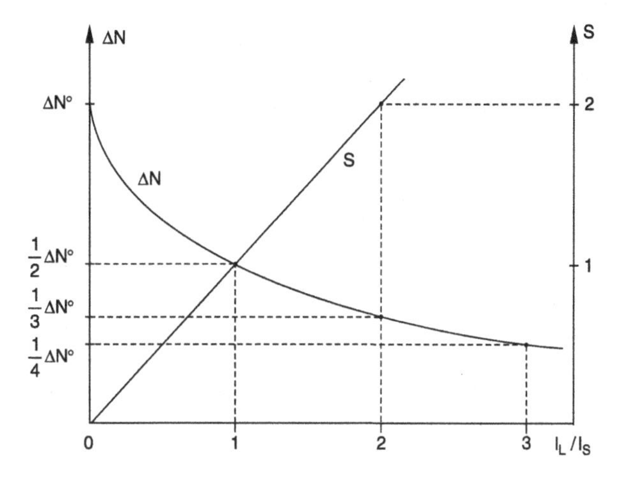

Fig. 2.3 Population difference ΔN and saturation parameter S as a function of incident laser intensity I_L

where the *saturation parameter*

$$S = \frac{B_{12}\rho_\nu}{R^*} = \frac{B_{12}I_\nu/c}{R^*} = \frac{B_{12}I}{c \cdot R_1 R_2},$$

with

$$R^* = \frac{R_1 R_2}{R_1 + R_2}, \tag{2.11}$$

gives the ratio of the induced transition probability $B_{12}\rho$ *to the "mean" relaxation probability* R^*. For $S \ll 1$ the homogeneous width of the absorption line is $\delta\nu_a = R_1 + R_2$ and therefore the absorbed intensity is $\Delta I = \Delta I_\nu(R_1 + R_2)$.

For $S \ll 1$, (2.10) can be written as

$$\Delta N \approx \Delta N^0(1 - S). \tag{2.12}$$

The spectral intensity $I_\nu = c \cdot R^*/B_{12} = I_s(\nu)$ at which the saturation parameter S becomes $S = 1$ is called the *saturation intensity*. The total saturation intensity is $I_s = \int I_s(\nu)\,d\nu \approx I_s(\nu_L)\,\delta\nu_L$. From (2.10) we derive that for $S = 1$ the population-density difference ΔN decreases to one half of its unsaturated value ΔN^0 (Fig. 2.3). The saturation power P_s is $P_s = I_s A$, where A is the cross section of the laser beam at the absorbing molecular sample.

Taking into account saturation, the power decrease of the incident light wave along the length dz of the absorption path is for $\delta\nu_L < \delta\nu_a$, according to (2.1a–2.1c and 2.10)

$$dP = -A \cdot I \cdot \sigma_{12}\frac{\Delta N^0}{1 + S}\,dz. \tag{2.13}$$

In case of incoherent light sources, such as spectral lamps, the intensity I_ν is so small that $S \ll 1$. We can then approximate (2.13) by

$$dP = -P\sigma_{12}\Delta N^0\,dz, \tag{2.14}$$

where the unsaturated population difference as given by (2.9) is independent of the intensity I, and the absorbed power is proportional to the incident power (linear absorption), i.e., the relative absorbed power $\mathrm{d}P/P$ is constant.

Higher incident intensities I_ν are achievable with lasers, where $S \ll 1$ may be no longer valid and (2.13) instead of (2.14) has to be used. Because of the decreasing population difference the absorbed power increases less than linearly with increasing incident intensity (Fig. 2.1). The decreasing relative absorption $\mathrm{d}P/P$ with increasing intensity I can readily be demonstrated when the absorbed power as a function of the incident intensity is measured via the laser-induced fluorescence. If the depopulation of the absorbing level $|1\rangle$ by absorption of laser photons becomes a noticeable fraction of the repopulation rate (Fig. 2.3), the population N_1 decreases and the laser-induced fluorescence intensity increases less than linearly with the incident laser intensity I_L.

With the Rabi flopping frequency

$$\Omega_\mathrm{R} = D_{ik} E_0/\hbar$$

which depends on the transition dipole matrix element D_{ik} and the electric field amplitude E_0 of the incident wave (see Vol. 1, (2.90)), we can relate the saturation parameter S for homogeneous line broadening (see Vol. 1, Sect. 3.1) with a line width $\gamma = \gamma_1 + \gamma_2 = R_1 + R_2$ to the Rabi flopping frequency by

$$S_{12} = \Omega_\mathrm{R}^2/\left(R^* \cdot \gamma\right) = \Omega_\mathrm{R}^2/(R_1 R_2). \tag{2.15}$$

Using the relations (2.30, 2.78 and 2.90) from Vol. 1 with the homogeneous spectral width $\delta\omega_\mathrm{a} = 2\pi\delta\nu_\mathrm{a} = \gamma = \gamma_1 + \gamma_2 = R_1 + R_2$ (see Vol. 1, Sect. 3.1.2) and the Rabi frequency Ω_{12} we obtain from (2.11)

$$S = \frac{\Omega_{12}^2}{R^*\gamma} = \frac{\Omega_{12}^2}{\gamma_1\gamma_2}, \quad \text{for } R_1 = \gamma_1,\ R_2 = \gamma_2. \tag{2.16}$$

This demonstrates that the saturation parameter can also be expressed as the square of the ratio $\Omega_{12}/\sqrt{R_1 \cdot R_2}$ of the Rabi frequency Ω_{12} at resonance ($\omega = \omega_{12}$) and the geometric mean of the relaxation rates of $|1\rangle$ and $|2\rangle$. In other words, when the atoms are exposed to light with intensity $I = I_\mathrm{s}$, their Rabi frequency is $\Omega_{12} = \sqrt{R_1 \cdot R_2}$.

The saturation parameter defined by (2.11) for the open two-level system is more general than that defined in Vol. 1, (3.67d) for a closed two-level system. The difference lies in the definition of the mean relaxation probability, which is $R = (R_1 + R_2)/2$ in the closed system but $R^* = R_1 R_2/(R_1 + R_2)$ in the open system. We can close our open system defined by the rate equations (2.8a–2.8c) by setting $C_1 = R_2 N_2$, $C_2 = R_1 N_1$, and $N_1 + N_2 = N = \text{const.}$ (see Fig. 2.2b). The rate equations then become identical to Vol. 1, (3.66) and R^* converts to R.

Regarding the saturation of the absorbing-level population N_1, there exists an important difference between a closed and an open two-level system. For the closed two-level system the stationary population density N_1 of the absorbing level $|1\rangle$ is, according to (2.8a–2.8c) and Vol. 1, (3.67) with $C_1 = R_2N_2$, $C_2 = R_1N_1$, $N_1+N_2 = N = \text{const.}$:

$$N_1 = \frac{B_{12}I/c + R_2}{2B_{12}I_\nu/c + R_1 + R_2}N, \quad \text{with } N = N_1 + N_2. \tag{2.17a}$$

N_1 can never drop below $N/2$ because

$$N_1 \geq \lim_{I\to\infty} N_1 = N/2 \to N_1 \geq N_2.$$

For our open system, however, we obtain from (2.8a–2.8c)

$$N_1 = \frac{(C_1 + C_2)B_{12}I_\nu/c + R_2C_1}{(R_1 + R_2)B_{12}I_\nu/c + R_1R_2}N. \tag{2.17b}$$

For large intensities I_ν ($S \gg 1$), the population density $N_1(I)$ approaches the limit

$$N_1(S \to \infty) = \frac{C_1 + C_2}{R_1 + R_2}N. \tag{2.17c}$$

If the repopulation rates C_1, C_2 are small compared to the depopulation rates R_1, R_2 the saturated population density N_1 may become quite small.

This case applies, for instance, to the saturation of molecular transitions in a molecular beam, where collisions are generally negligible. The excited level $|2\rangle$ decays by spontaneous emission at the rate N_2A_2 into many other rotational–vibrational levels $|m\rangle \neq |1\rangle$, and only a small fraction N_2A_{21} comes back to the level $|1\rangle$. The only repopulation mechanisms of level $|1\rangle$ are the diffusion of molecules into the excitation volume and the radiative decay rate N_2A_{21}.

For $E_2 \gg kT$, the upper level $|2\rangle$ can be populated only by optical pumping. If $|1\rangle$ is the ground state, its "lifetime" is given by the transit time $t_T = d/v$ through the excitation region of length d. We therefore have to replace in (2.8a–2.8c): $C_1 = D_1 + N_2A_{21}$, $D_2 = 0$, $R_1 = 1/t_T$, $R_2 = A_2 + 1/t_T$, and find

$$N_1 = \frac{D_1(B_{12}\rho + A_2 + 1/t_T)}{B_{12}\rho(A_2 - A_{21} + 2/t_T) + 1/t_T^2}N. \tag{2.18a}$$

Without the laser excitation ($\rho = 0$, $A_2 = A_{21} = 0$), we obtain $N_1^0 = D_1t_T$. This is the stationary population due to the diffusion of molecules into the excitation region. For large laser intensities, (2.18a) yields

$$\lim_{I\to\infty} N_1 = \frac{D_1}{A_2 - A_{21} + 2/t_T}. \tag{2.18b}$$

Example 2.2 For $d = 1$ mm, $v = 5 \times 10^4$ cm/s $\rightarrow t_T = 2 \times 10^{-6}$ s. With $D_1 = 10^{14}$ s^{-1} cm^{-3} we have the stationary population density $N_1^0 = 2 \times 10^8$ cm^{-3}. With typical figures of $A_2 = 10^8$ s^{-1}, $A_{21} = 10^7$ s^{-1} we obtain for the completely saturated population density $N_1 \approx 10^6$ cm^{-3}. The saturated population N_1 has therefore decreased to 0.5 % of its unsaturated value N_1^0.

We shall now briefly illustrate for two different situations at which intensities the saturation becomes noticeable:

(a) The bandwidth $\delta\nu_L$ of the cw laser is larger than the spectral width $\delta\nu_a$ of the absorbing transition. In this case, we get the same results both for homogeneous and inhomogeneous line broadening. The total saturation intensity $I_s = c\rho_s(\nu)\delta\nu_L$ is then according to (2.11)

$$I_s = \int I_s(\nu)\,d\nu \approx \frac{R^* c}{B_{12}} \delta\nu_L \left[\mathrm{W/m^2}\right]. \tag{2.19a}$$

Example 2.3 Saturation of a molecular transition in a molecular beam by a broadband cw laser with $\delta\nu_L = 3 \times 10^9$ s^{-1} ($\hat{=} 0.1$ cm^{-1}): With $R_1 = 1/t_T$, $R_2 = A_2 + 1/t_T$, we obtain with $B_{12} = (c^3/8\pi h\nu^3)A_{21}$ the saturation intensity

$$I_s(\delta\nu_L) = \frac{(A_2 + 1/t_T)8\pi h\nu^3 \cdot \delta\nu_L}{(t_T A_2 + 2)A_{21}c^2} \left[\mathrm{W/m^2}\right]. \tag{2.19b}$$

With the values of our previous example $A_2 = 10^8$ s^{-1}, $t_T = 2 \times 10^{-6}$ s, $\nu = 5 \times 10^{14}$ s^{-1}, $A_{21} = 10^7$ s^{-1}, and $\delta\nu_L = 3 \times 10^9$ s^{-1}, which gives

$$I_s \approx 3 \times 10^3\ \mathrm{W/m^2}.$$

If the laser beam is focused to a cross section of $A = 1$ mm^2, a laser power of 3 mW is sufficient for our example to reach the value $S = 1$ of the saturation parameter.

If the laser bandwidth $\delta\nu_L$ matches the homogeneous width $\gamma/2\pi$ of the absorption line, we find with $\gamma = A_2 + 2/t_T$ from (2.19b) the saturation intensity

$$I_s = \frac{4h\nu^3}{T A_{21} c^2}(A_2 + 1/t_T) \approx 100\ \mathrm{W/m^2} = 100\ \mathrm{\mu W/mm^2}. \tag{2.19c}$$

If the laser beam is focussed to a cross section 10×10 μm^2 the saturation power decreases to $P_s = 10$ nW!

(b) The second case deals with a cw single-mode laser with frequency $\nu = \nu_0$ tuned to the center frequency ν_0 of a homogeneously broadened atomic resonance transition. If spontaneous emission into the ground state is the only relaxation process of the upper level, the relaxation rate is $R^* = A_{21}/2$ for $S = 1$, the saturation-broadened linewidth is $\delta\nu_a = \sqrt{2}A_{21}/2\pi$, and the saturation intensity is, according to (2.11) and Vol. 1, (2.22)

$$I_s = c\rho_s\delta\nu_a = \frac{cR^* A_{21}}{\sqrt{2}\pi B_{12}} = \frac{2\sqrt{2}h\nu A_{21}}{\lambda^2}. \tag{2.20}$$

The same result could have been obtained from Vol. 1, (3.67f) $I_s = h\nu A_{21}/2\sigma_{12}$ with

$$\int \sigma_{12}\,d\nu \sim \sigma(\nu_0)\delta\nu_a = (h\nu/c)B_{12} = \left(c^2/8\pi\nu^2\right)A_{21}, \tag{2.21a}$$

which yields without saturation broadening

$$\sigma(\nu_0) \sim c^2/4\nu^2 = (\lambda/2)^2, \quad \text{and} \quad I_s = \frac{2h\nu A_{21}}{\lambda^2}. \tag{2.21b}$$

For $A_{21} = 10^8\ \text{s}^{-1}$ the relaxation due to diffusion out of the excitation volume can be neglected. At sufficiently low pressures the collision-induced transition probability is small compared to A_{21}.

Example 2.4 $\lambda = 500$ nm $\rightarrow \nu = 6 \times 10^{14}\ \text{s}^{-1}$, $A_{21} = 10^8\ \text{s}^{-1} \rightarrow I_s \approx 380\ \text{W/m}^2$. Focusing the beam to a focal area of 1 mm^2 means a saturation power of only 265 μW! With the value $A_{21} = 10^7\ \text{s}^{-1}$ of Example 2.3, the saturation intensity drops to 38 W/m^2. Focussing to $10 \times 10\ \mu\text{m}^2$ reduces the saturation power to 3.8 nW.

If collision broadening is essential, the linewidth increases and the saturation intensity increases roughly proportionally to the homogeneous linewidth. If pulsed lasers are used, the saturation peak powers are much higher because the system does generally not reach the stationary conditions. For the optical pumping time the laser pulse duration T_L often is the limiting time interval. Only for long laser pulses (e.g., copper-vapor laser-pumped dye lasers) the transit time of the molecules through the laser beam may be shorter than T_L.

2.2 Saturation of Inhomogeneous Line Profiles

In Vol. 1, Sect. 3.6 we saw that the saturation of *homogeneously broadened* transitions with Lorentzian line profiles results again in a Lorentzian profile with the halfwidth

$$\Delta\omega_s = \Delta\omega_0\sqrt{1+S_0}, \quad S_0 = S(\omega_0), \tag{2.22}$$

which is increased by the factor $(1+S_0)^{1/2}$ compared to the unsaturated halfwidth $\Delta\omega_0$. The saturation broadening is due to the fact that the absorption coefficient

$$\alpha(\omega) = \frac{\alpha_0(\omega)}{1+S(\omega)}$$

decreases by the factor $[1+S(\omega)]^{-1}$, whereas the saturation parameter $S(\omega)$ itself has a Lorentzian line profile and the saturation is stronger at the line center than in the line wings (Vol. 1, Fig. 3.24).

We will now discuss saturation of inhomogeneous line profiles. As an example we treat Doppler-broadened transitions, which represent the most important case in saturation spectroscopy.

2.2.1 Hole Burning

When a monochromatic light wave

$$E = E_0\cos(\omega t - kz), \quad \text{with } k = k_z,$$

passes through a gaseous sample of molecules with a Maxwell–Boltzmann velocity distribution, only those molecules that move with such velocities $\boldsymbol{v}$ that the Doppler-shifted laser frequency in the frame of the moving molecule $\omega' = \omega - \boldsymbol{k}\cdot\boldsymbol{v}$ with $\boldsymbol{k}\cdot\boldsymbol{v} = kv_z$ falls within the homogeneous linewidth γ around the center absorption frequency ω_0 of a molecule at rest, i.e., $\omega' = \omega_0 \pm \gamma$, can significantly contribute to the absorption. The absorption cross section for a molecule with the velocity component v_z on a transition $|1\rangle \to |2\rangle$ is without saturation ($S = 0$)

$$\sigma_{12}(\omega, v_z) = \sigma_0\frac{(\gamma/2)^2}{(\omega-\omega_0-kv_z)^2+(\gamma/2)^2}, \tag{2.23}$$

where $\sigma_0 = \sigma(\omega = \omega_0 + kv_z)$ is the maximum absorption cross section at the line center of the Doppler-shifted molecular transition.

Due to saturation, the population density $N_1(v_z)\,\mathrm{d}v_z$ decreases within the velocity interval $\mathrm{d}v_z = \gamma/k$, while the population density $N_2(v_z)\,\mathrm{d}v_z$ of the upper level $|2\rangle$ increases correspondingly (Fig. 2.4a). From (2.10) and Vol. 1, (3.72) we obtain for $S \ll 1$

$$N_i^0 - N_i = \frac{\Delta N_i^0}{1+S} \approx \Delta N_i^0(1-S).$$

The saturation parameter S is

$$S(\omega) = S_0\cdot\frac{(\gamma/2)^2}{(\omega-\omega_0)^2+(\gamma/2)^2}.$$

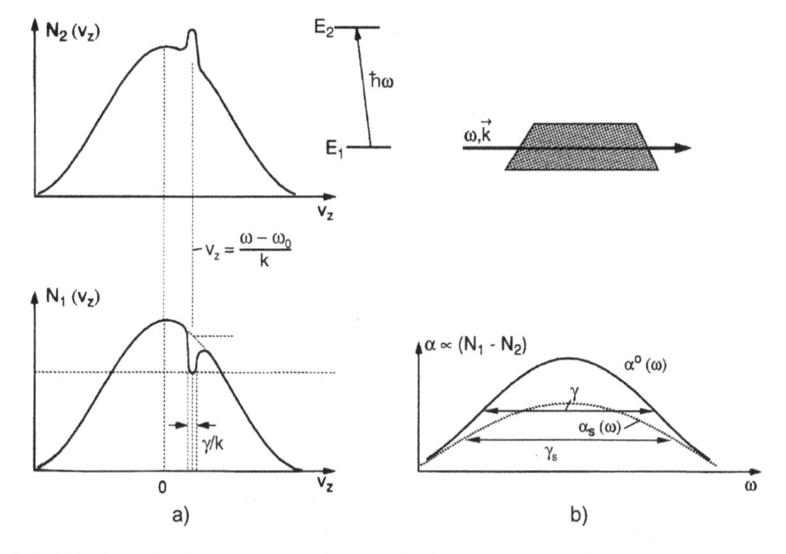

Fig. 2.4 Velocity-selective saturation of a Doppler-broadened transition: (**a**) the Bennet hole in the lower- and the Bennet peak in the upper-state population distribution $N_i(v_z)$. (**b**) Saturated absorption profile when the saturating laser is tuned across the Doppler-profile of a molecular transition (*dashed curve*)

With $\gamma_s = \gamma \cdot \sqrt{1+S}$ and $\omega_0 \to \omega_0 + kv_z$ one obtains

$$N_1(\omega, v_z) = N_1^0(v_z) - \frac{\Delta N^0}{\gamma_1 \tau}\left[\frac{S_0(\gamma/2)^2}{(\omega - \omega_0 - kv_z)^2 + (\gamma_s/2)^2}\right], \tag{2.24a}$$

$$N_2(\omega, v_z) = N_2^0(v_z) + \frac{\Delta N^0}{\gamma_2 \tau}\left[\frac{S_0(\gamma/2)^2}{(\omega - \omega_0 - kv_z)^2 + (\gamma_s/2)^2}\right], \tag{2.24b}$$

where $\gamma = \gamma_1 + \gamma_2$ denotes the homogeneous width of the transition and $\gamma_s = \gamma\sqrt{1+S_0}$. The quantity

$$\tau = \frac{1}{\gamma_1} + \frac{1}{\gamma_2} = \frac{\gamma}{\gamma_1 \cdot \gamma_2}, \tag{2.24c}$$

is called the *longitudinal relaxation time*, while

$$T = \frac{1}{\gamma_1 + \gamma_2} = \frac{1}{\gamma} \tag{2.24d}$$

is the *transverse relaxation time*.

Note that for $\gamma_1 \neq \gamma_2$ the depth of the hole in $N_1(v_z)$ and the heights of the peak in $N_2(v_z)$ are different.

Subtracting (2.24b) from (2.24a) yields for the saturated population difference

$$\Delta N(\omega_s, v_z) = \Delta N^0(v_z)\left[1 - \frac{S_0(\gamma/2)^2}{(\omega - \omega_0 - kv_z)^2 + (\gamma_s/2)^2}\right]. \tag{2.24e}$$

The velocity-selective minimum in the velocity distribution $\Delta N(v_z)$ at $v_z = (\omega - \omega_0)/k$, which is often called a *Bennet hole* [202], has the homogeneous width (according to Vol. 1, Sect. 3.6)

$$\gamma_s = \gamma\sqrt{1+S_0},$$

and a depth at the hole center at $\omega = \omega_0 + kv_z$ of

$$\Delta N^0(v_z) - \Delta N(v_z) = \Delta N^0(v_z)\frac{S_0}{1+S_0}. \tag{2.25}$$

For $S_0 = 1$ the hole depth amounts to 50 % of the unsaturated population difference. Molecules with velocity components in the interval v_z to $v_z + \mathrm{d}v_z$ give the contribution

$$\mathrm{d}\alpha(\omega, v_z)\,\mathrm{d}v_z = \Delta N(v_z)\sigma(\omega, v_z)\,\mathrm{d}v_z, \tag{2.26}$$

to the absorption coefficient $\alpha(\omega, v_z)$. The total absorption coefficient caused by all molecules in the absorbing level is then

$$\alpha(\omega) = \int \Delta N(v_z)\sigma_{12}(\omega, v_z)\,\mathrm{d}v_z. \tag{2.27}$$

Inserting $\Delta N(v_z)$ from (2.24a–2.24e), $\sigma(\omega, v_z)$ from (2.23) and $\Delta N^0(v_z)$ from Vol. 1, (3.40) yields

$$\alpha(\omega) = \frac{\Delta N\sigma_0}{v_p\sqrt{\pi}}\int\frac{\mathrm{e}^{-(v_z/v_p)^2}\,\mathrm{d}v_z}{(\omega-\omega_0-kv_z)^2+(\gamma_s/2)^2}, \tag{2.28}$$

with the most probable velocity $v_p = (2k_BT/m)^{1/2}$ and the total saturated population difference $\Delta N = \int \Delta N(v_z)\,\mathrm{d}v_z$. Despite the saturation, one obtains again a Voigt profile for $\alpha(\omega)$, similarly to Vol. 1, (3.46). The only difference is the saturation-broadened homogeneous linewidth γ_s in (2.28) instead of γ in Vol. 1, (3.33) (Fig. 2.4b).

Since for $S_0 < 1$ the Doppler width is generally large compared to the homogeneous width γ_s, the nominator in (2.28) at a given frequency ω does not vary much within the interval $\Delta v_z = \gamma_s/k$, where the integrand contributes significantly to $\alpha(\omega)$. We therefore can take the factor $\exp[-(v_z/v_p)^2]$ outside the integral. The residual integrals can be solved analytically and one obtains with $v_z = (\omega-\omega_0)/k$ and Vol. 1, (3.44) the saturated absorption coefficient

$$\alpha_s(\omega) = \frac{\alpha^0(\omega_0)}{\sqrt{1+S_0}}\exp\left\{-\left[\frac{\omega-\omega_0}{0.6\delta\omega_D}\right]^2\right\}, \tag{2.29}$$

with the unsaturated absorption coefficient

$$\alpha^0(\omega_0) = \Delta N_0\frac{\sigma_0\gamma c\sqrt{\pi}}{v_p\omega_0},$$

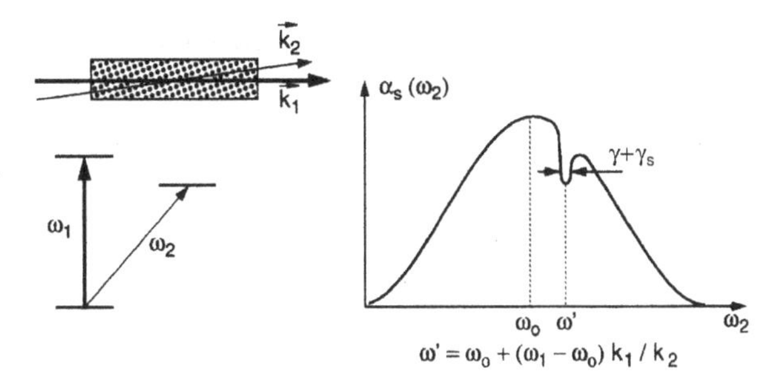

Fig. 2.5 Dip in the Doppler-broadened absorption profile $\alpha(\omega)$ burned by a strong pump laser at $\omega_p = \omega_1$ and detected by a weak probe laser at $\omega' = \omega_0 \pm (\omega_1 - \omega_0)k_1/k_2$

and the Doppler width

$$\delta\omega_D = \frac{\omega_0}{c}\sqrt{\frac{8kT\ln 2}{m}},$$

Eq. (2.29) illustrates a remarkable result: *Although at each frequency ω the monochromatic laser burns a Bennet hole into the velocity distribution $N_1(v_z)$, this hole* cannot *be detected just by tuning the laser through the absorption profile.* The absorption coefficient

$$\alpha(\omega) = \frac{\alpha^0(\omega)}{\sqrt{1+S_0}}, \tag{2.30}$$

of the inhomogeneous profile still shows a Voigt profile *without any hole* but is reduced by the constant factor $(1+S_0)^{-1/2}$, which is independent of ω (Fig. 2.4b).

Note The difference to the homogeneous absorption profile where $\alpha(\omega)$ is reduced by the *frequency-dependent* factor $(1+S(\omega))^{-1}$, see Vol. 1, (3.69) and Fig. 3.23 is the fact, that for the inhomogeneous profile the reduction factor $(1+S_0)^{-1}$ is independent of the frequency ω.

The Bennet hole can, however, be detected if *two lasers* are used:

- The saturating pump laser with the wave vector $\boldsymbol{k}_1$, which is kept at the frequency ω_1 and which burns a hole, according to (2.24a–2.24e), into the velocity class $v_z \pm \Delta v_z/2$ with $v_z = (\omega_0 - \omega_1)/k_1$ and $\Delta v_z = \gamma/k_1$ (Fig. 2.5).
- A weak probe laser with the wave vector $\boldsymbol{k}_2$ and a frequency ω tunable across the Voigt profile. This probe laser is sufficiently weak to cause no extra saturation. The absorption coefficient for the tunable probe laser is then

$$\alpha_s(\omega_1,\omega) = \frac{\sigma_0 \Delta N^0}{v_p\sqrt{\pi}} \int \frac{e^{-(v_z/v_p)^2}}{(\omega_0 - \omega - k_2 v_z)^2 + (\gamma/2)^2} \times \left[1 - \frac{S_0(\gamma/2)^2}{(\omega_0 - \omega_1 - k_1 v_z)^2 + (\gamma_s/2)^2}\right] dv_z. \tag{2.31}$$

The integration over the velocity distribution yields analogously to (2.29)

$$\alpha_s(\omega_1, \omega) = \alpha^0(\omega)\left[1 - \frac{S_0}{\sqrt{1+S_0}}\frac{(\gamma/2)^2}{(\omega-\omega')^2 + (\Gamma_s/2)^2}\right]. \tag{2.32}$$

This is an unsaturated Doppler profile $\alpha^0(\omega)$ with a saturation dip at the probe frequency

$$\omega = \omega' = \omega_0 \pm (\omega_1 - \omega_0)k_1/k_2,$$

where the + sign holds for collinear and the − sign for anticollinear propagation of the pump and probe waves.

The halfwidth $\Gamma_s = \gamma + \gamma_s = \gamma[1 + (1+S_0)^{1/2}]$ of the absorption dip at $\omega = \omega'$ equals the sum of the unsaturated homogeneous absorption width γ of the weak probe wave and that of the saturated dip (due to the strong pump). The depth of the dip at $\omega = \omega'$ is

$$\begin{aligned}\Delta\alpha(\omega') = \alpha^0(\omega') - \alpha_s(\omega') &= \alpha^0(\omega')\frac{S_0}{\sqrt{1+S_0}[(1+\sqrt{1+S_0})]^2} \\ &\approx \frac{S_0}{4}\alpha^0(\omega'), \quad \text{for } S_0 \ll 1.\end{aligned} \tag{2.32a}$$

Note In the derivation of (2.32) we have only regarded population changes due to saturation effects. We have neglected coherence phenomena that may, for instance, result from interference between the two waves. These effects, which differ for copropagating waves from those of counterpropagating waves, have been treated in detail in [203–205]. For sufficiently small laser intensities ($S \ll 1$) they do not strongly affect the results derived above, but add finer details to the spectral structures obtained.

2.2.2 *Lamb Dip*

Pump and probe waves may also be generated by a single laser when the incident beam is reflected back into the absorption cell (Fig. 2.6).

The saturated population difference in such a case of equal intensities $I_1 = I_2 = I$ of the two counterpropagating waves with wavevectors $\boldsymbol{k}_1 = -\boldsymbol{k}_2$ is then

$$\begin{aligned}\Delta N(v_z) &= \Delta N^0(v_z) \\ &\times\left[1 - \frac{S_0(\gamma/2)^2}{(\omega_0 - \omega - kv_z)^2 + (\gamma_s/2)^2} - \frac{S_0(\gamma/2)^2}{(\omega_0 - \omega + kv_z)^2 + (\gamma_s/2)^2}\right],\end{aligned} \tag{2.33}$$

where $S_0 = S_0(I)$ is the saturation parameter due to one of the running waves. Because of the opposite Doppler shifts the two waves with frequency ω burn two

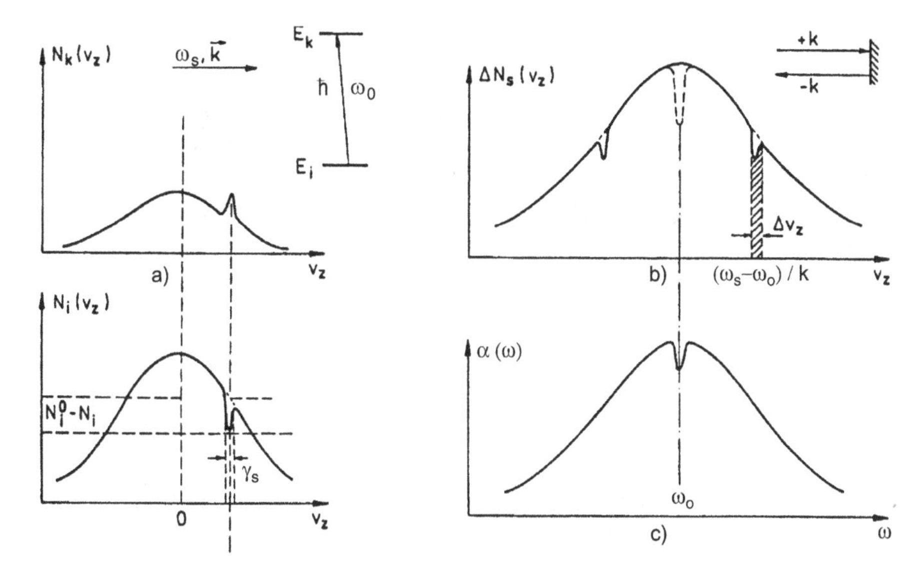

Fig. 2.6 Saturation of an inhomogeneous line profile: (**a**) Bennet hole and dip produced by a monochromatic running wave with $\omega \neq \omega_0$; (**b**) Bennet holes caused by the two counterpropagating waves for $\omega \neq \omega_0$ and for $\omega = \omega_0$ (*dashed curve*); (**c**) Lamb dip in the absorption profile $\alpha_s(\omega)$

Bennet holes at the velocity components $v_z = \pm(\omega_0 - \omega)/k$ into the population distribution $\Delta N(v_z)$ (Fig. 2.6b).

The saturated absorption coefficient then becomes

$$\alpha_s(\omega) = \int \Delta N(v_z)\big[\sigma(\omega_0 - \omega - kv_z) + \sigma(\omega_0 - \omega + kv_z)\big]\,\mathrm{d}v_z. \tag{2.34}$$

Inserting (2.33 and 2.23) into (2.34) yields in the weak-field approximation ($S_0 \ll 1$) after some elaborate calculations [203] the saturated-absorption coefficient for a sample in a standing-wave field

$$\alpha_s(\omega) = \alpha^0(\omega)\left[1 - \frac{S_0}{2}\left(1 + \frac{(\gamma_s/2)^2}{(\omega-\omega_0)^2 + (\gamma_s/2)^2}\right)\right], \tag{2.35a}$$

with

$$\gamma_s = \gamma\sqrt{1+S_0}, \quad \text{and} \quad S_0 = S_0(I, \omega_0).$$

This represents the Doppler-broadened absorption profile $\alpha^0(\omega)$ with a dip at the line center $\omega = \omega_0$ (Fig. 2.6c), which is called a *Lamb dip* after W.E. Lamb, who first explained it theoretically [206]. For $\omega = \omega_0$ the saturated absorption coefficient drops to $\alpha_s(\omega_0) = \alpha^0(\omega_0)\cdot(1 - S_0)$. The depth of the Lamb dip is $S_0 = B_{ik}I/(c\gamma_s)$, $I = I_1 = I_2$ being the intensity of one of the counterpropagating waves that form the standing wave. For $\omega_0 - \omega \gg \gamma_s$ the saturated absorption coefficient becomes $\alpha_s = \alpha_0(1 - S_0/2)$, which corresponds to the saturation by one of the two waves.

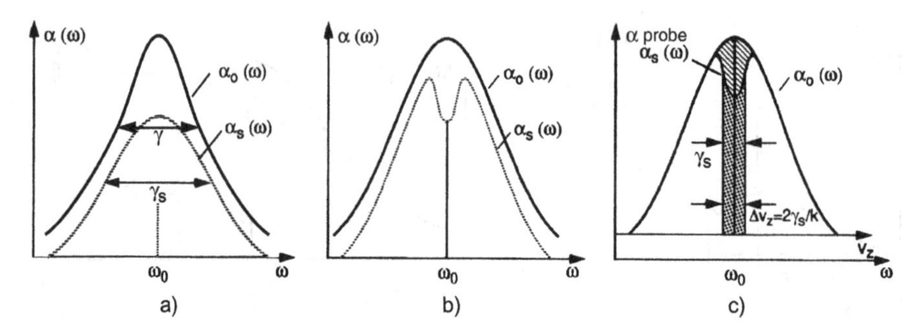

Fig. 2.7 Comparison of the saturation of a homogeneous absorption line profile (**a**) and an inhomogeneous profile (**b**) in a standing-wave field. (**c**) The traveling saturating wave is kept at $\omega = \omega_0$ and a weak probe wave is tuned across the line profile

The Lamb dip can be understood in a simple, conspicuous way: for $\omega \neq \omega_0$ the incident wave is absorbed by molecules with the velocity components $v_z = +(\omega - \omega_0 \mp \gamma_s/2)/k$, the reflected wave by other molecules with $v_z = -(\omega - \omega_0) \pm \gamma_s/2$. For $\omega = \omega_0$ both waves are absorbed by the same molecules with $v_z = (0 \pm \gamma_s/s)/k$, which essentially move perpendicularly to the laser beams. The intensity per molecule absorbed is now twice as large, and the saturation accordingly higher.

In Fig. 2.7 the differences in the saturation behavior of a homogeneous (Fig. 2.7a) and an inhomogeneous line profile are illustrated. For the inhomogeneous case two situations are illustrated:

- The absorbing sample is placed in a standing-wave field ($I_1 = I_2 = I$) and the frequency ω is tuned over the line profiles (Fig. 2.7b).
- A pump laser ($I = I_1$) is kept at the line center at ω_0 and a weak probe laser is tuned over the saturated line profiles (Fig. 2.7c).

In the first case (Fig. 2.7c), the saturation of the inhomogeneous line profile is $S_0 = B_{ik}I/(c\gamma_s)$ at the center of the Lamb dip and $S_0/2$ for $(\omega - \omega_0) \gg \gamma_s$. In the second case (Fig. 2.7c), the depth of the Bennet hole is $S_0/2 = B_{ik}I_1/(c\gamma_s)$.

For strong laser fields the approximation $S_0 \ll 1$ no longer holds. Instead of (2.35a, 2.35b) one obtains, neglecting coherent effects [203]:

$$\alpha_s(\omega) = \alpha^0(\omega)\frac{\gamma/2}{B[1-(\frac{2(\omega-\omega_0)}{A+B})^2]^{1/2}}, \tag{2.35b}$$

with

$$A = \left[(\omega-\omega_0)^2 + (\gamma/2)^2\right]^{1/2} \quad \text{and} \quad B = \left[(\omega-\omega_0)^2 + (\gamma/2)^2(1+2S)\right]^{1/2}.$$

This yields $\alpha_s(\omega_0) = \alpha^0(\omega_0)/\sqrt{1+2S}$ at the line center $\omega = \omega_0$, and $\alpha_s(\omega) = \alpha^0(\omega)/\sqrt{1+S}$ for $(\omega - \omega_0) \gg \gamma$. The maximum depth of the Lamb dip is achieved

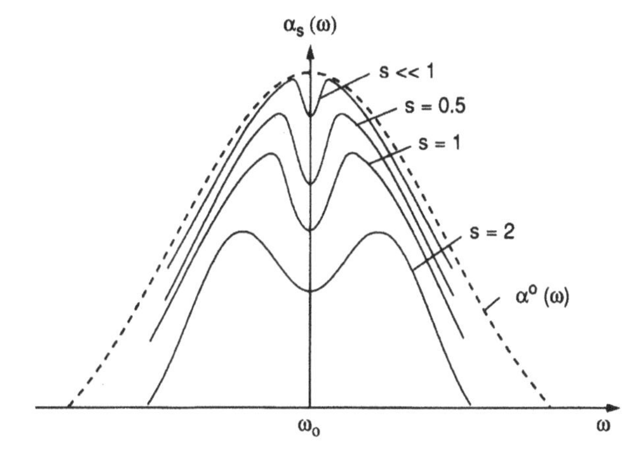

Fig. 2.8 Lamb dips for several values of the saturation parameter S_0

when

$$\frac{\alpha(\omega-\omega_0 \gg \gamma_s)-\alpha(\omega_0)}{\alpha(\omega_0)}=\frac{1}{\sqrt{1+S_0}}-\frac{1}{\sqrt{1+2S_0}},$$

becomes maximum, which occurs for $S_0 \approx 1.4$. In Fig. 2.8 the saturated absorption profile is depicted for some values of S_0.

Note The width of the Lamb dip in (2.35a) is $\delta\omega_{LD}=\gamma_s$. This corresponds, however, to the velocity interval $\Delta v_z = 2\gamma_s/k$ because the opposite Doppler shifts $\Delta\omega=(\omega_0-\omega)=\pm k v_z$ of the two Bennet holes add when the laser frequency ω is tuned.

If the intensity of the reflected wave in Fig. 2.6 is very small ($I_2 \ll I_1$), we obtain instead of (2.35a, 2.35b) a formula similar to (2.32). However, we must replace Γ_s by $\Gamma_s^* = (\gamma+\gamma_s)/2$ since the pump and probe waves are simultaneously tuned. For $S_0 \ll 1$ the result is

$$\alpha_s(\omega)=\alpha^0(\omega)\left[1-\frac{S_0}{2}\frac{(\gamma_s/2)^2}{(\omega-\omega_0)^2+(\Gamma_s^*/2)^2}\right]. \tag{2.36}$$

2.3 Saturation Spectroscopy

Saturation spectroscopy is based on the velocity-selective saturation of Doppler-broadened molecular transitions, treated in Sect. 2.2. Here the spectral resolution is no longer limited by the Doppler width but only by the much narrower width of the Lamb dip. The gain in spectral resolution is illustrated by the example of two transitions from a common lower level $|c\rangle$ to two closely spaced levels $|a\rangle$ and $|b\rangle$ (Fig. 2.9). Even when the Doppler profiles of the two transitions completely overlap, their narrow Lamb dips can clearly be separated, as long as $\Delta\omega = \omega_{ca} - \omega_{cb} > 2\gamma_s$.

Saturation spectroscopy is therefore often called *Lamb-dip spectroscopy*.

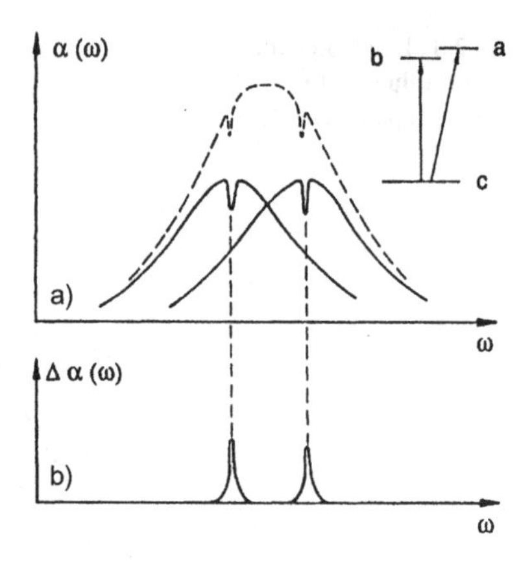

Fig. 2.9 Spectral resolution of the Lamb dips of two transitions with overlapping Doppler profiles: (**a**) direct measurement of Doppler-profiles and Lamb dips; (**b**) lock-in detection of $(\alpha^0(\omega) - \alpha_s(\omega))$

2.3.1 Experimental Schemes

A possible experimental realization of saturation spectroscopy is illustrated in Fig. 2.10. The output beam from a tunable laser is split by the beam splitter BS into a strong pump beam with the intensity I_1 and a weak probe beam with intensity $I_2 \ll I_1$, which pass through the absorbing sample in opposite directions. When the transmitted probe-beam intensity $I_{t2}(\omega)$ is measured as a function of the laser frequency ω, the detection signal $DS(\omega) \propto I_2 - I_{t2}$ shows the Doppler-broadened absorption profiles with "Lamb peaks" at their centers, because the saturated absorption shows Lamb dips at the centers of the Doppler-broadened absorption lines.

The Doppler-broadened background can be eliminated if the pump beam is chopped and the transmitted probe intensity is monitored through a lock-in amplifier that is tuned to the chopping frequency. According to (2.36), we obtain for a sufficiently weak probe intensity the Doppler-free absorption profile

$$\alpha_0 - \alpha_s = \frac{\alpha_0 S_0}{2} \frac{(\gamma_s/2)^2}{(\omega - \omega_0)^2 + (\Gamma_s^*/2)^2}. \tag{2.37}$$

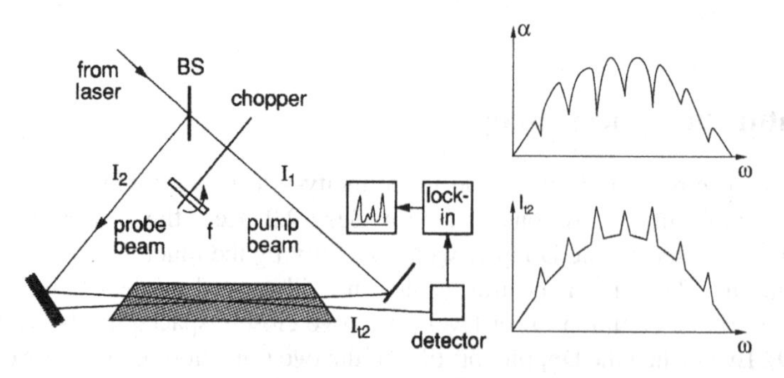

Fig. 2.10 Experimental setup for saturation spectroscopy where the transmitted probe intensity $I_2(\omega)$ is monitored

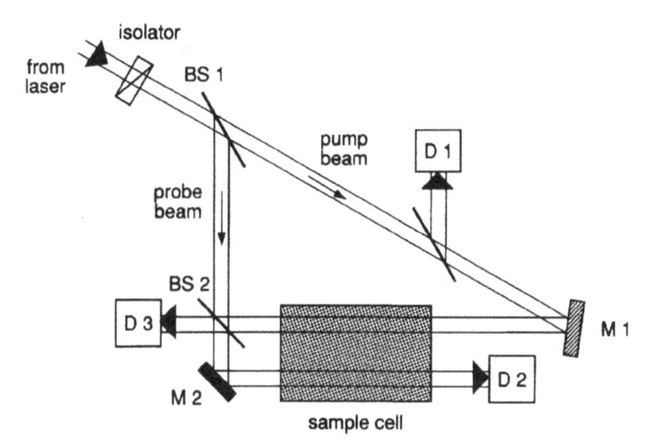

Fig. 2.11 Sensitive version of saturation spectroscopy. The probe is split by BS2 into two parallel beams, which pass through a pumped and an unpumped region, respectively. Pump and probe beams are strictly antiparallel. A Faraday isolator prevents feedback of the probe beam into the laser

To enhance the sensitivity, the probe beam can again be split into two parts. One beam passes the region of the sample that is saturated by the pump beam, and the other passes the sample cell at an unsaturated region (Fig. 2.11). The difference between the two probe-beam outputs monitored with the detectors D1 and D2 yields the saturation signal if this difference has been set to zero with the pump beam off. The pump-beam intensity measured by D3 can be used for normalization of the saturation signal. Figure 2.12 shows as an example the saturation spectrum of a mixture of different cesium isotopes contained in a glass cell heated to about 100 °C [207]. The hyperfine structure and the isotope shifts of the different isotopes can be derived from these measurements with high accuracy. The small crossing angle α [rad] of the two beams in Fig. 2.10 results in a residual Doppler width $\delta\omega_1 = \Delta\omega_D\alpha$. If the pump and probe beams are strictly anticollinear ($\alpha = 0$) the probe beam is coupled back into the laser, resulting in laser instabilities. This can be prevented by an optical isolator, such as a Faraday isolator [208], which turns the plane of polarization by 90° and suppresses the reflected beam by a polarizer (Fig. 2.11).

Instead of measuring the *attenuation* of the probe beam, the absorption can also be monitored by the laser-induced *fluorescence*, which is proportional to the ab-

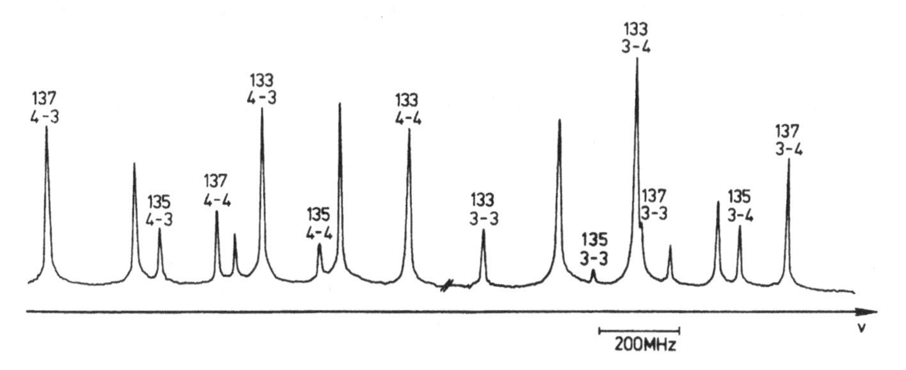

Fig. 2.12 Saturation spectrum of all hyperfine components of the $6^2S_{1/2} \rightarrow 7^2P$ transition at $\lambda = 459.3$ nm in a mixture of isotopes ^{133}Cs, ^{135}Cs, and ^{137}Cs [207]

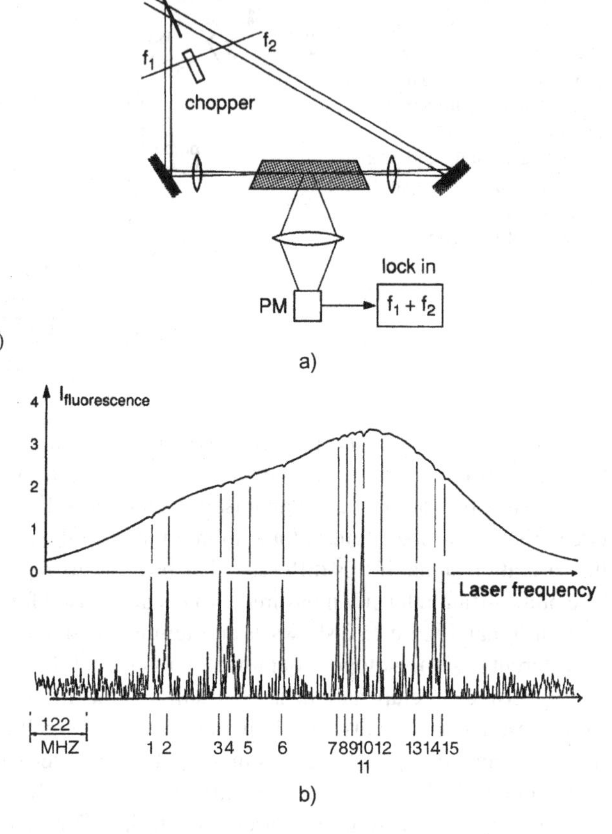

Fig. 2.13 Intermodulated fluorescence method for saturation spectroscopy at small densities of the sample molecules: (**a**) experimental arrangement; (**b**) hyperfine spectrum of the $(v''=1, J''=98) \rightarrow (v'=58, J'=99)$ line in the $X^1\Sigma_g \rightarrow {}^3\Pi_{0u}$ system of I_2 at $\lambda = 514.5$ nm, monitored at the chopping frequency f_1 of the pump beam (*upper spectrum* with the Lamb dips) and at $(f_1 + f_2)$ (*lower spectrum*) [211]

sorbed laser power. This technique is, in particular, advantageous when the density of the absorbing molecules is low and the absorption accordingly small. The change in the attenuation of the probe beam is then difficult to detect and the small Lamb dips may be nearly buried under the noise of the Doppler-broadened background. Sorem and Schawlow [209, 210] have demonstrated a very sensitive *intermodulated fluorescence technique*, where the pump beam and the probe beam are chopped at two different frequencies f_1 and f_2 (Fig. 2.13a). Assume the intensities of the two beams to be $I_1 = I_{10}(1 + \cos \Omega_1 t)$ and $I_2 = I_{20}(1 + \cos \Omega_2 t)$ with $\Omega_i = 2\pi f_i$. The intensity of the laser-induced fluorescence is then

$$I_{\text{Fl}} = C \Delta N_{\text{s}}(I_1 + I_2), \tag{2.38}$$

where ΔN_{s} is the difference between the saturated population densities of the absorbing and upper levels, and the constant C includes the transition probability and the collection efficiency of the fluorescence detector. According to (2.33), we obtain at the center of an absorption line

$$\Delta N_{\text{s}} = \Delta N^0 \big[1 - a(I_1 + I_2)\big].$$

Inserting this into (2.38) gives

$$I_{\mathrm{Fl}} = C\big[\Delta N^0(I_1 + I_2) - a\Delta N^0(I_1 + I_2)^2\big]. \tag{2.39}$$

The quadratic expression $(I_1 + I_2)^2$ contains the frequency-dependent term

$$I_{10}I_{20}\cos\Omega_1 t \cdot \cos\Omega_2 t = \frac{1}{2}I_{10}I_{20}\big[\cos(\Omega_1 + \Omega_2)t + \cos(\Omega_1 - \Omega_2)t\big],$$

which reveals that the fluorescence intensity contains linear terms, modulated at the chopping frequencies f_1 and f_2, respectively, and quadratic terms with the modulation frequencies $(f_1 + f_2)$ and $(f_1 - f_2)$, respectively. While the linear terms represent the normal laser-induced fluorescence with a Doppler-broadened excitation line profile, the quadratic ones describe the saturation effect because they depend on the decrease of the population density ΔN $(v_z = 0)$ from the simultaneous interaction of the molecules with both fields. When the fluorescence is monitored through a lock-in amplifier tuned to the sum frequency $f_1 + f_2$, the linear background is suppressed and only the saturation signals are detected. This is demonstrated by Fig. 2.13b, which shows the 15 hyperfine components of the rotational line $(v'' - 1, J'' = 98)$ $(v' = 58, J' = 99)$ in the $X^1\Sigma_g^+ \to B^3\Pi_{0u}$ transition of the iodine molecule I_2 [210, 211]. The two laser beams were chopped by a rotating disc with two rows of a different number of holes, which interrupted the beams at $f_1 = 600\ \mathrm{s}^{-1}$ and $f_2 = 900\ \mathrm{s}^{-1}$. The upper spectrum was monitored at the chopping frequency f_1 of the pump beam. The Doppler-broadened background caused by the linear terms in (2.39) and the Lamb dips both show a modulation at the frequency f_1 and are therefore recorded simultaneously. The center frequencies of the hyperfine structure (hfs) components, however, can be obtained more accurately from the intermodulated fluorescence spectrum (lower spectrum), which was monitored at the sum frequency $(f_1 + f_2)$ where the linear background is suppressed. This technique has found wide applications for sub-Doppler modulation spectroscopy of molecules and radicals at low pressures [212–214].

Example 2.5 The linewidth of the Lamb dips of the I_2 hfs components can be estimated as follows. The spontaneous lifetime of the upper level of the transition at $\lambda = 632$ nm is $\tau = 10^{-7}$ s, which gives a natural linewidth of $\delta\nu_{\mathrm{n}} = 1.5$ MHz. The vapor pressure at $T = 300$ K is $p(I) = 0.05$ mbar. The pressure broadening is then about 2 MHz. Saturation broadening at a saturation parameter $S = 3$ is $\delta\nu_{\mathrm{s}} = 2\delta\nu_{\mathrm{n}} = 3.0$ MHz. Transit time broadening for a laser beam diameter $2w = 1$ mm is $\delta\nu_{\mathrm{tr}} = 0.4v/w \approx 120$ kHz for an average velocity of $v = 300$ m/s. The total width of the Lamb dips is then $\delta\nu_{\mathrm{LD}} = \sqrt{2^2 + 3^2 + 0.1^2} \approx 3.6$ MHz.

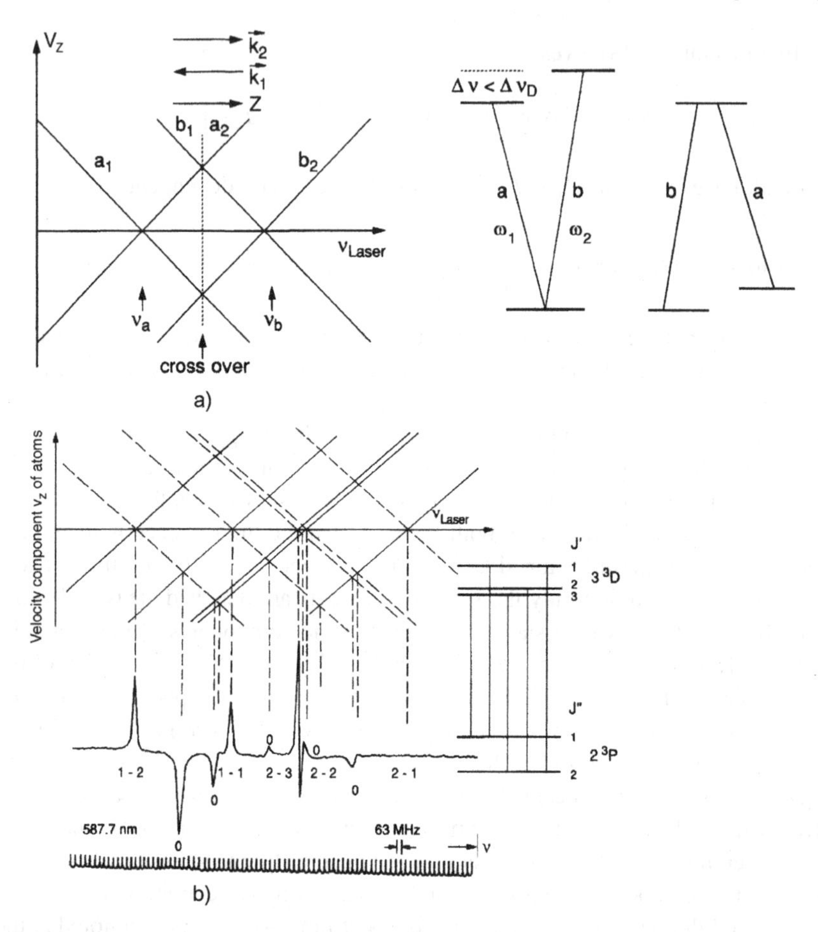

Fig. 2.14 (**a**) Generation of cross-over saturation signals; (**b**) illustration of cross-overs in the helium transition $3^3D \leftarrow 2^3P$. The cross-over signals are marked by 0 above or below the lines [215]

2.3.2 *Cross-Over Signals*

If two molecular transitions with a common lower or upper level overlap within their Doppler width, extra resonances, called *cross-over* signals, occur in the Lamb-dip spectrum. Their generation is explained in Fig. 2.14a.

Assume that for the center frequencies ω_1 and ω_2 of the two transitions $|\omega_1 - \omega_2| < \Delta\omega_D$ holds. At the laser frequency $\omega = (\omega_1 + \omega_2)/2$, the incident wave is shifted against ω_1 by $\Delta\omega = \omega - \omega_1 = (\omega_2 - \omega_1)/2$. If it saturates the velocity class $(v_z \pm \mathrm{d}v_z) = (\omega_2 - \omega_1)/2k \pm \gamma k$ on the transition 1 with center frequency ω_1, it is in resonance with molecules in this subgroup. Since the reflected wave experiences the opposite Doppler shift, when it saturates the same velocity class it is in resonance with molecules in the same subgroup in transition 2 with the center

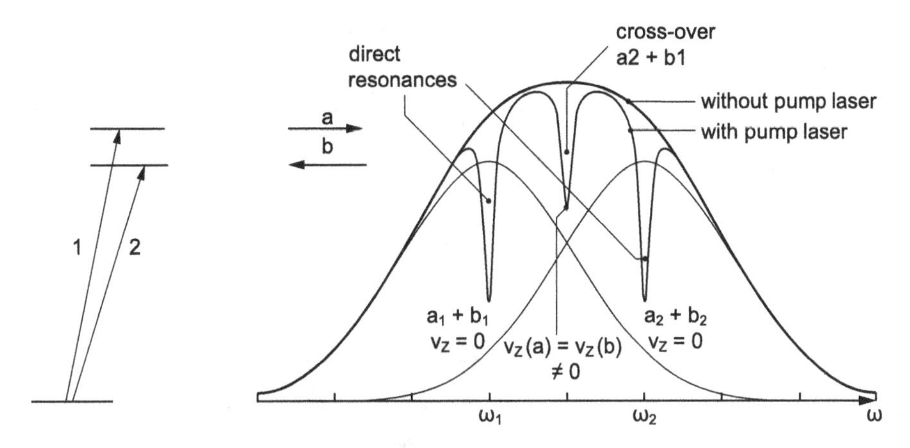

Fig. 2.15 Explanation of cross-over signals

frequency ω_2. One therefore observes, besides the saturation signals at ω_1 and ω_2 (where the velocity class $v_z = 0$ is saturated), an additional saturation signal (cross-over) at $\omega = (\omega_1 + \omega_2)/2$ because one of the two waves causes a decrease $-\Delta N_1$ of the population density N_1 in the common lower level, which is probed by the second wave on another transition. In case of a common upper level, both waves contribute at $\omega = (\omega_1 + \omega_2)/2$ to the increase ΔN_2 of the population N_2; one wave on transition a, the other on b. The sign of the cross-over signals is negative for a common lower level and positive for a common upper level. Their frequency position $\omega_c = (\omega_1 + \omega_2)/2$ is just at the center between the two transitions 1 and 2 (Fig. 2.14b).

In Fig. 2.15 the three Lamb dips (two direct resonances and 1 cross-over signal) are illustrated for the case of two transitions with a common lower level. Although these cross-over signals increase the number of observed Lamb dips and may therefore increase the complexity of the spectrum, they have the great advantage that they allow one to assign pairs of transitions with a common level. This may also facilitate the assignment of the whole spectrum; see, for example, Fig. 2.14b and [205, 214–216].

2.3.3 Intracavity Saturation Spectroscopy

When the absorbing sample is placed inside the resonator of a tunable laser, the Lamb dip in the absorption coefficient $\alpha(\omega)$ causes a corresponding peak of the laser output power $P(\omega)$ (Fig. 2.16).

The power $P(\omega)$ depends on the spectral gain profile $G(\omega)$ and on the absorption profile $\alpha(\omega)$ of the intracavity sample, which is generally Doppler broadened. The Lamb peaks therefore sit on a broad background (Fig. 2.16a). With the center frequency ω_1 of the gain profile and an absorption Lamb dip at ω_0 we obtain,

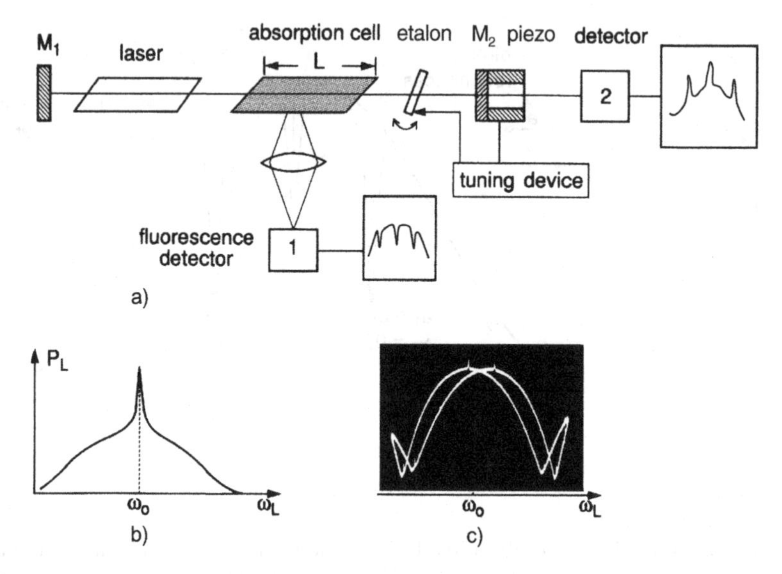

Fig. 2.16 Saturation spectroscopy inside the cavity of a laser: (**a**) experimental arrangement; (**b**) output power $P(\omega)$; (**c**) experimental detection of the Lamb peak in the output power of a HeNe laser tunable around $\lambda = 3.39\,\mu\text{m}$, caused by the Lamb dip of a CH_4 transition in a methane cell inside the laser cavity [221]

according to Vol. 1, (5.58) and (2.35a, 2.35b)

$$P_L(\omega) \propto \left\{ G(\omega - \omega_1) - \alpha^0(\omega)\left[1 - \frac{S_0}{2}\left(1 + \frac{(\gamma_s/2)^2}{(\omega-\omega_0)^2 + (\gamma_s/2)^2}\right)\right]\right\}. \tag{2.40}$$

In a small interval around ω_0 we may approximate the gain profile $G(\omega - \omega_1)$ and the unsaturated absorption profile $\alpha^0(\omega)$ by a quadratic function of ω, yielding for (2.40) the approximation

$$P_L(\omega) = A\omega^2 + B\omega + C + \frac{D}{(\omega-\omega_0)^2 + (\gamma_s/2)^2}, \tag{2.41}$$

where the constants A, B, C, and D depend on ω_0, ω_1, γ, and S_0. The derivatives

$$P_L^{(n)}(\omega) = \frac{d^n P_L(\omega)}{d\omega^n} \quad (n = 1, 2, 3, \ldots),$$

of the laser output power with respect to ω are

$$P_L^{(1)}(\omega) = 2A\omega + B - \frac{2D(\omega-\omega_0)}{[(\omega-\omega_0)^2 + (\gamma_s/2)^2]^2},$$

$$P_L^{(2)}(\omega) = 2A + \frac{6D(\omega-\omega_0)^2 - 2D(\gamma_s/2)^2}{[(\omega-\omega_0)^2 + (\gamma_s/2)^2]^3}, \tag{2.42}$$

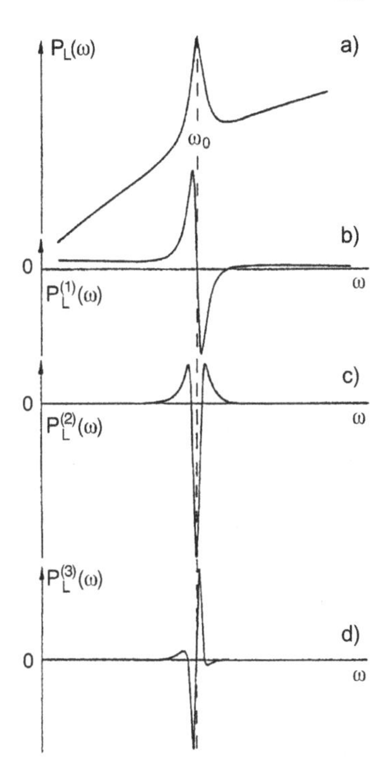

Fig. 2.17 Lamb peak at the slope of the Doppler-broadened gain profile and the first three derivatives, illustrating the suppression of the Doppler background

$$P_{\mathrm{L}}^{(3)}(\omega)=\frac{24D(\omega-\omega_0)[(\omega-\omega_0)^2-(\gamma_{\mathrm{s}}/2)^2]}{[(\omega-\omega_0)^2+(\gamma_{\mathrm{s}}/2)^2]^4}.$$

These derivatives are exhibited in Fig. 2.17, which illustrates that the broad background disappears for the higher derivatives. If the absorptive medium is the same as the gain medium, the Lamb peak appears at the center of the gain profile (Fig. 2.16b, c).

Example 2.6 The zero crossings of $P_{\mathrm{L}}^{(3)}(\omega_{\mathrm{m}})$ are at $\omega=\omega_0$ and $\omega=\omega_0\pm\gamma_{\mathrm{s}}/2$. The maxima and minima can be calculated from the condition $P_{\mathrm{L}}^{(4)}(\omega_{\mathrm{m}})=0$. The two largest central extremes are at $\omega_{\mathrm{m},1,2}=\omega_0\pm 0.16\gamma_{\mathrm{s}}$. The frequency difference between these extremes is then $\delta\omega=0.32\gamma_{\mathrm{s}}$. It is three times smaller than the width between the maximum and minimum of $P_{\mathrm{L}}^{(1)}$. The small maximum and minimum are at $\omega=\omega_0\pm 0.68\gamma_{\mathrm{s}}$.

If the laser frequency ω is modulated at the frequency Ω, the laser output

$$P_{\mathrm{L}}(\omega)=P_{\mathrm{L}}(\omega_0+a\sin\Omega t),$$

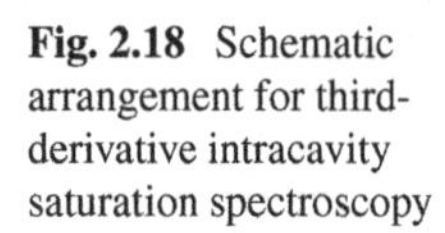

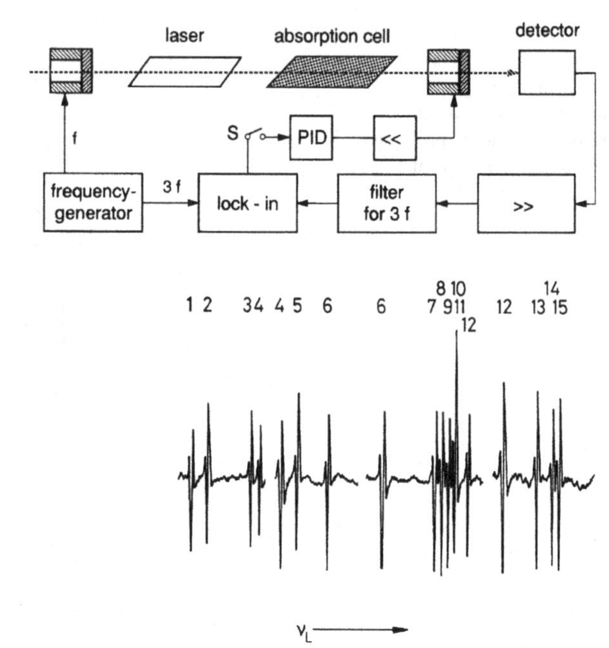

Fig. 2.18 Schematic arrangement for third-derivative intracavity saturation spectroscopy

Fig. 2.19 The third-derivative intracavity absorption spectrum of I_2 around $\lambda = 514$ nm, showing the same hfs components as Fig. 2.13 [216]

can be expanded into a Taylor series around ω_0. The derivation in Sect. 1.2.2 shows that the output $P_L(\omega, 3\Omega)$ measured with a lock-in device at the frequency 3Ω is proportional to the third derivative, see (1.13).

The experimental performance is depicted in Fig. 2.18. The modulation frequency $\Omega = 2\pi f$ is tripled by forming rectangular pulses, where the third harmonic is filtered and fed into the reference input of a lock-in amplifier that is tuned to 3Ω. Figure 2.19 illustrates this technique by the third-derivative spectrum of the same hfs components of I_2 as obtained with the intermodulated fluorescence technique in Fig. 2.13.

2.3.4 Lamb-Dip Frequency Stabilization of Lasers

The steep zero crossing of the third derivative of narrow Lamb dips gives a good reference for accurate stabilization of the laser frequency onto an atomic or molecular transition. Either the Lamb dip in the gain profile of the laser transition or Lamb dips of absorption lines of an intracavity sample can be used.

In the infrared spectral range, Lamb dips of a vibration–rotation transition of CH_4 at $\lambda = 3.39$ μm or of CO_2 around 10 μm are commonly used for frequency stabilization of the HeNe laser at 3.39 μm or the CO_2 laser. In the visible range various hyperfine components of rotational lines within the ${}^1\Sigma_g \rightarrow {}^3\Pi_{0u}$ system of the I_2 molecule are mainly chosen. The experimental setup is the same as that shown in Fig. 2.18. The laser is tuned to the wanted hfs component and then the

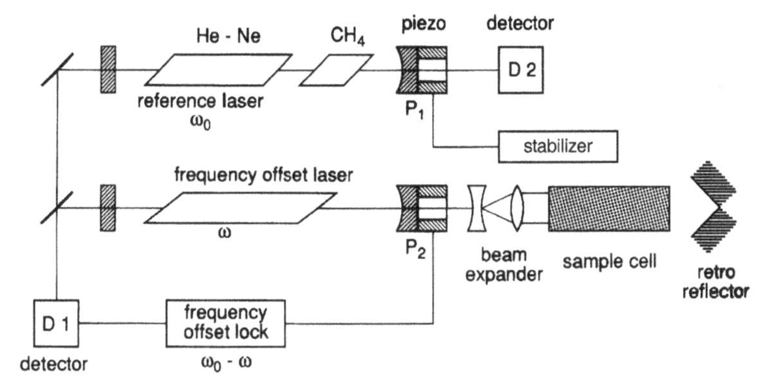

Fig. 2.20 Schematic diagram of a frequency-offset locked laser spectrometer

feedback switch S is closed in order to lock the laser onto the zero crossing of this component [217].

Using a double servo loop for fast stabilization of the laser frequency onto the transmission peak of a Fabry–Perot Interferometer (FPI) and a slow loop to stabilize the FPI onto the first derivative of a forbidden narrow calcium transition, Barger et al. constructed an ultrastable cw dye laser with a short-term linewidth of approximately 800 Hz and a long-term drift of less than 2 kHz/h [218]. Stabilities of better than 1 Hz have also been realized [219, 220].

This extremely high stability can be transferred to tunable lasers by a special *frequency-offset locking* technique [221]. Its basic principle is illustrated in Fig. 2.20. A reference laser is frequency stabilized onto the Lamb dip of a molecular transition at ω_0. The output from a second, more powerful laser at the frequency ω is mixed in detector D1 with the output from the reference laser at the frequency ω_0. An electronic device compares the difference frequency $\omega_0 - \omega$ with the frequency ω' of a stable but tunable RF oscillator, and controls the piezo P_2 such that $\omega_0 - \omega = \omega'$ at all times. The frequency ω of the powerful laser is therefore always locked to the *offset frequency* $\omega = \omega_0 - \omega'$, which can be controlled by tuning the RF frequency ω'.

For Lamb-dip spectroscopy with ultrahigh resolution, the output beam of the powerful laser is expanded before it is sent through the sample cell in order to minimize transit-time broadening (Vol. 1, Sect. 3.4). A retroreflector provides the counterpropagating probe wave for Lamb-dip spectroscopy. The real experimental setup is somewhat more complicated. A third laser is used to eliminate the troublesome region near the zero-offset frequency. Furthermore, optical decoupling elements have to be inserted to avoid optical feedback between the three lasers. A detailed description of the whole system can be found in [222].

An outstanding example of the amount of information on interactions in a large molecule that can be extracted from a high-resolution spectrum is represented by the work of Bordé et al. on saturation spectroscopy of SF_6 [223]. Many details of the various interactions, such as spin-rotation coupling, Coriolis coupling, and hyperfine structure, which are completely masked at lower resolution, can be unravelled when

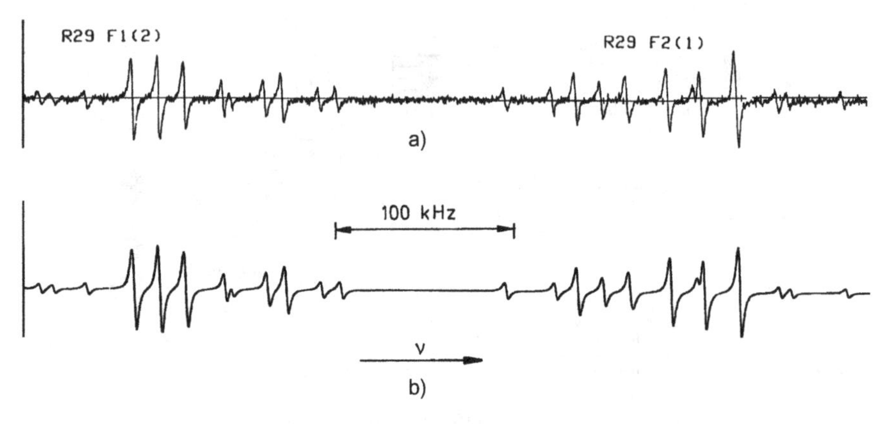

Fig. 2.21 Hyperfine and super-hyperfine structures of a rotational–vibrational transition in SF_6, showing the molecular transitions and cross-over signals: (**a**) experimental spectrum; (**b**) calculated spectrum [223]

sufficiently high resolution is achieved. For illustration Fig. 2.21 depicts a section of the saturation spectrum of SF_6 taken by this group.

Meanwhile, many complex molecular spectra have been resolved by saturation spectroscopy. One example are the ultranarrow overtone transitions of the acetylene isotopomer $^{13}C_2H_2$ [224].

2.4 Polarization Spectroscopy

While saturation spectroscopy monitors the decrease of *absorption* of a probe beam caused by a pump wave that has selectively depleted the absorbing level, the signals in polarization spectroscopy come mainly from the change of the *polarization state* of the probe wave induced by a *polarized* pump wave. Because of optical pumping, the pump wave causes not only a change in the absorption coefficient α, but also a change in the refraction index n.

This very sensitive Doppler-free spectroscopic technique has many advantages over conventional saturation spectroscopy and will certainly gain increasing attention [225, 226]. We therefore discuss the basic principle and some of its experimental modifications in more detail.

2.4.1 Basic Principle

The basic idea of polarization spectroscopy can be understood in a simple way (Fig. 2.22). The output from a monochromatic tunable laser is split into a weak probe beam with the intensity I_1 and a stronger pump beam with the intensity I_2. The probe beam passes through a linear polarizer P_1, the sample cell, and a second

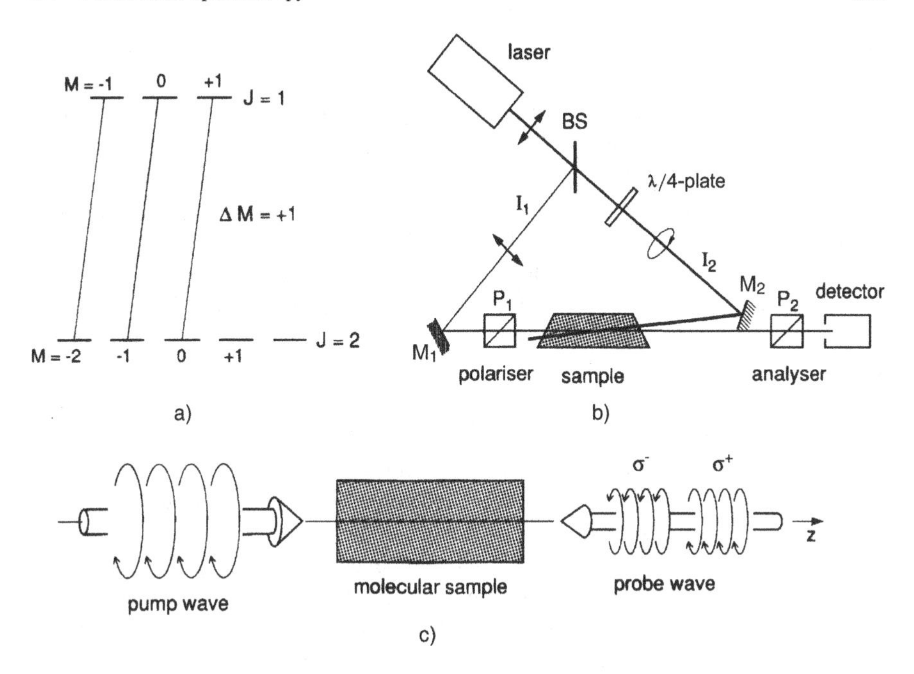

Fig. 2.22 Polarization spectroscopy: (**a**) level scheme for a P transition $J = 2 \rightarrow J = 1$; (**b**) experimental setup; (**c**) linearly polarized probe wave as superposition of σ^+ (angular momentum $+\hbar$ in z-direction) and σ^- components

linear polarizer P_2, which is crossed with P_1. Without the pump laser, the sample is isotropic and therefore does not change the polarisation of the probe beam. The detector D behind P_2 receives only a very small signal caused by the residual transmission of the crossed polarizer, which might be as small as $10^{-8} I_1$.

After having passed through a $\lambda/4$-plate, which produces a circular polarization, the pump beam travels in the opposite direction through the sample cell. When the laser frequency ω is tuned to a molecular transition $(J'', M'') \rightarrow (J', M')$, molecules in the lower level (J'', M'') can absorb the pump wave. The quantum number M that describes the projection of J onto the direction of light propagation follows the selection rule $\Delta M = +1$ for the transitions $M'' \rightarrow M'$ induced by σ^+ circularly polarized light $(M'' \rightarrow M' = M'' + 1)$. The degenerate M'' sublevels of the rotational level J'' become partially or completely depleted because of saturation. The degree of depletion depends on the pump intensity I_2, the absorption cross section $\sigma(J'', M'' \rightarrow J', M')$, and on possible relaxation processes that may repopulate the level (J'', M''). The cross section σ depends on J'', M'', J', and M'. From Fig. 2.22a it can be seen that in the case of P-transitions $(\Delta J = -1)$, not all of the M'' sublevels of the lower state are pumped. For example, from levels with $M'' = +J''$ no P-transitions with $\Delta M = +1$ are possible, while for R-transitions the levels $M' = -J'$ in the upper state are not populated. *This implies that the pumping process produces an unequal saturation and with it a nonuniform population of the M sublevels*, which is equivalent to an anisotropic distribution for the orientations of the angular momentum vector J.

Such an anisotropic sample becomes birefringent for the incident linearly polarized probe beam. Its plane of polarization is slightly rotated after having passed the anisotropic sample. This effect is quite analogous to the Faraday effect, where the nonisotropic orientation of J is caused by an external magnetic field. For polarization spectroscopy no magnetic field is needed. Contrary to the Faraday effect where all molecules are oriented, here only those molecules that interact with the monochromatic pump wave show this nonisotropic orientation. As has already been discussed in Sect. 2.2, this is the subgroup of molecules with the velocity components

$$v_z \pm \Delta v_z = (\omega_0 - \omega)/k \pm \gamma/k,$$

where Δv_z is determined by the homogeneous linewidth $\delta\omega = \gamma$.

For $\omega \neq \omega_0$ the probe wave that passes in the opposite direction through the sample interacts with a *different group* of molecules in the velocity interval $v_z \pm \Delta v_z = -(\omega_0 - \omega \pm \delta\omega)/k$, and will therefore not be influenced by the pump wave. If, however, the laser frequency ω coincides with the center frequency ω_0 of the molecular transition within its homogeneous linewidth $\delta\omega = \gamma$ (i.e., $\omega = \omega_0 \pm \delta\omega \to v_z = 0 \pm \Delta v_z$), both waves can be absorbed by the same molecules and the probe wave experiences a birefringence from the nonisotropic M distribution of the molecules in the absorbing lower rotational level J'' or in the upper level J'.

Only in this case will the plane of polarization of the probe wave be slightly rotated by $\Delta\theta$, and the detector D will receive a Doppler-free signal every time the laser frequency ω is tuned across the center of a molecular absorption line.

2.4.2 Line Profiles of Polarization Signals

Let us now discuss the generation of this signal in a more quantitative way. The linearly polarized probe wave

$$\boldsymbol{E} = \boldsymbol{E}_0 \mathrm{e}^{\mathrm{i}(\omega t - kz)}, \qquad \boldsymbol{E}_0 = \{E_0, 0, 0\},$$

can always be composed of a σ^+ and a σ^- circularly polarized component $\boldsymbol{E} = \boldsymbol{E}^+ + \boldsymbol{E}^-$ (Fig. 2.22c) where

$$E^+ = E_0^+ \mathrm{e}^{\mathrm{i}(\omega t - k^+ z)}, \qquad E_0^+ = \frac{1}{2} E_0(\hat{\boldsymbol{x}} + \mathrm{i}\hat{\boldsymbol{y}}), \tag{2.43a}$$

$$E^- = E_0^- \mathrm{e}^{\mathrm{i}(\omega t - k^- z)}, \qquad E_0^- = \frac{1}{2} E_0(\hat{\boldsymbol{x}} - \mathrm{i}\hat{\boldsymbol{y}}), \tag{2.43b}$$

where $\hat{\boldsymbol{x}}$ and $\hat{\boldsymbol{y}}$ are unit vectors in the x- and y-direction, respectively. While passing through the birefringent sample the two components experience different absorption coefficients α^+ and α^- and different refractive indices n^+ and n^- from the non-isotropic saturation caused by the σ^+-polarized pump wave. The refractive indices n^+, n^- are related to the wave numbers $k = 2\pi/\lambda$ by $n^+ = ck^+/\omega$, $n^- = ck^-/\omega$. After a path length L through the pumped region of the sample, the two components are

$$E^+ = E_0^+ \mathrm{e}^{\mathrm{i}[\omega t - k^+ L + \mathrm{i}(\alpha^+/2)L]},$$
$$E^- = E_0^- \mathrm{e}^{\mathrm{i}[\omega t - k^- L + \mathrm{i}(\alpha^-/2)L]}. \tag{2.44}$$

Because of the differences $\Delta n = n^+ - n^-$ and $\Delta\alpha = \alpha^+ - \alpha^-$ caused by the non-isotropic saturation, a phase difference

$$\Delta\phi = \left(k^+ - k^-\right)L = (\omega L/c)\Delta n,$$

has developed between the two components, and also a small amplitude difference

$$\Delta E = \frac{E_0}{2}\left[\mathrm{e}^{-(\alpha^+/2)L} - \mathrm{e}^{-(\alpha^-/2)L}\right].$$

The windows of the absorption cell with thickness d also show a small absorption and a pressure-induced birefringence from the atmospheric pressure on one side and vacuum on the other side. Their index of refraction n_w and their absorption coefficient α_w can be expressed by the complex quantity

$$n_\mathrm{w}^{*\pm} = b_\mathrm{r}^\pm + \mathrm{i}b_i^\pm, \quad \text{with } n_\mathrm{w} = b_\mathrm{r}, \quad \text{and} \quad \alpha_\mathrm{w} = 2k \cdot b_i = 2(\omega/c)b_i.$$

Behind the exit window the superposition of the σ^+ and σ^- components of the linearly polarized probe wave traveling into the z-direction yields the elliptically polarized wave

$$\begin{aligned} E(z = L) &= E^+ + E^- \\ &= \frac{1}{2}E_0 \mathrm{e}^{\mathrm{i}\omega t}\mathrm{e}^{-\mathrm{i}[\omega(nL + b_\mathrm{r})/c - \mathrm{i}\alpha L/2 - \mathrm{i}a_\mathrm{w}/2]} \\ &\quad \times \left[(\hat{x} + \mathrm{i}\hat{y})\mathrm{e}^{-\mathrm{i}\delta} + (\hat{x} - \mathrm{i}\hat{y})\mathrm{e}^{+\mathrm{i}\delta}\right], \end{aligned} \tag{2.45}$$

where $a_\mathrm{w} = 2d\alpha_\mathrm{w} = (4d \cdot \omega/c)b_i$ is the absorption in the two windows of the sample cell, and

$$n = \frac{1}{2}\left(n^+ + n^-\right); \qquad \alpha = \frac{1}{2}\left(\alpha^+ + \alpha^-\right); \qquad b = \frac{1}{2}\left(b^+ + b^-\right)$$

are the average quantities. The phase factor

$$\delta = \omega(L\Delta n + \Delta b_\mathrm{r})/2c - \mathrm{i}(L\Delta\alpha/4 + \Delta a_\mathrm{w}/2),$$

depends on the differences

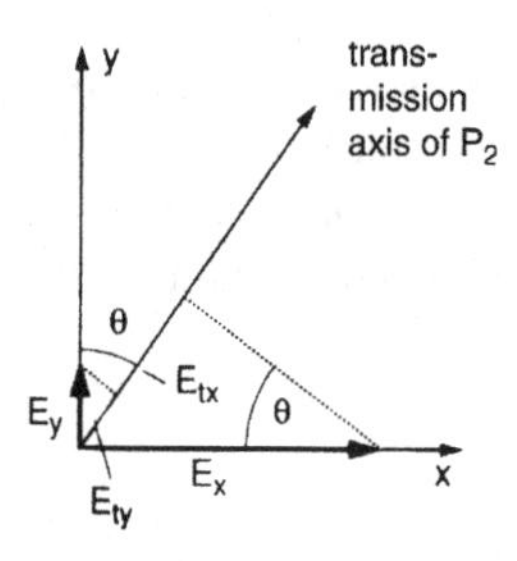

Fig. 2.23 Transmission of the elliptically polarized probe wave through the analyzer, uncrossed by the angle θ

$$\Delta n = n^+ - n^-, \qquad \Delta\alpha = \alpha^+ - \alpha^-, \qquad \Delta b_r = b_r^+ - b_r^-, \quad \text{and}$$
$$\Delta b_i = b_i^+ - b_i^- .$$

If the transmission axis of the analyzer P_2 is tilted by the small angle $\theta \ll 1$ against the y-axis (Fig. 2.23) the transmitted amplitude becomes

$$E_\mathrm{t} = E_x \sin\theta + E_y \cos\theta .$$

For most practical cases, the differences $\Delta\alpha$ and Δn caused by the pump wave are very small. Also, the birefringence of the cell windows can be minimized (for example, by compensating the air pressure by a mechanical pressure onto the edges of the windows). With

$$L\Delta\alpha \ll 1, \qquad L\Delta k \ll 1, \quad \text{and} \quad \Delta b \ll 1,$$

we can expand $\exp(\mathrm{i}\delta)$ in (2.45). We then obtain for small angles $\theta \ll 1$ ($\cos\theta \approx 1$, $\sin\theta \approx \theta$) the transmitted amplitude

$$E_\mathrm{t} = E_0 \mathrm{e}^{\mathrm{i}\omega t} \exp\left[-\mathrm{i} \left\{ \omega(nL + b_\mathrm{r})/c - \frac{\mathrm{i}}{2}(\alpha \cdot L + a_\mathrm{w}) \right\} \right] (\theta + \delta). \tag{2.46}$$

The detector signal $S(\omega)$ is proportional to the transmitted intensity

$$S(\omega) \propto I_\mathrm{T}(\omega) = c\epsilon_0 E_\mathrm{t} E_\mathrm{t}^* .$$

Even for $\theta = 0$ the crossed polarizers have a small residual transmission $I_\mathrm{t} = \xi I_0$ ($\xi \approx 10^{-6}$ to 10^{-8}). Taking this into account, we obtain with the incident probe intensity I_0 and the abbreviation $\theta' = \theta + \omega/(2c) \cdot \Delta b_\mathrm{r}$ and $a_\mathrm{w} = 2d\alpha_\mathrm{w} = 4dkb_i$, the transmitted intensity

$$\begin{aligned} I_\mathrm{t} &= I_0 \mathrm{e}^{-\alpha L - a_\mathrm{w}} \left(\xi + |\theta + \Delta|^2 \right), \\ &= I_0 \mathrm{e}^{-\alpha L - a_\mathrm{w}} \left[\xi + \theta'^2 + \left(\frac{1}{2}\Delta a_\mathrm{w} \right)^2 + \frac{1}{4}\Delta a_\mathrm{w} L \Delta\alpha + \frac{\omega}{c}\theta' L \Delta n \right. \\ &\quad \left. + \left(\frac{\omega}{2c} L \Delta n \right)^2 + \left(\frac{L\Delta\alpha}{4} \right)^2 \right]. \end{aligned} \tag{2.47}$$

The change of absorption $\Delta\alpha$ is caused by those molecules within the velocity interval $\Delta v_z = 0 \pm \gamma_s/k$ that simultaneously interact with the pump and probe waves. The line profile of $\Delta\alpha(\omega)$ is therefore, analogous to the situation in saturation spectroscopy (Sect. 2.2) a Lorentzian profile

$$\Delta\alpha(\omega) = \frac{\Delta\alpha_0}{1+x^2}, \quad \text{with } x = \frac{\omega_0 - \omega}{\gamma_s/2}, \quad \text{and} \quad \alpha_0 = \alpha(\omega_0), \tag{2.48}$$

having the halfwidth γ_s, which corresponds to the homogeneous width of the molecular transition saturated by the pump wave.

The absorption coefficient $\alpha(\omega)$ and the refractive index $n(\omega)$ are related by the Kramers–Kronig dispersion relation, see (3.36b, 3.37b) in Vol. 1, Sect. 3.1.

We therefore obtain for $\Delta n(\omega)$ the dispersion profile

$$\Delta n(\omega) = \frac{c}{\omega_0} \frac{\Delta\alpha_0 x}{1+x^2}. \tag{2.49}$$

Inserting (2.48 and 2.49) into (2.47) gives for the *circularly polarized pump beam* the line profile of the detector signal

$$S^{\mathrm{cp}} = I_t(\omega) = I_0 \mathrm{e}^{-\alpha L - a_w} \left\{ \xi + \theta'^2 + \frac{1}{4}\Delta a_w^2 + \frac{1}{2}\theta' \Delta\alpha_0 L \frac{x}{1+x^2} + \left[\frac{1}{4}\Delta\alpha_0 \Delta a_w L + \left(\frac{\Delta\alpha_0 L}{4}\right)^2\right] \frac{1}{1+x^2} + \frac{3}{4}\left(\frac{\Delta\alpha_0 x}{(1+x^2)}\right)^2 \right\}. \tag{2.50}$$

The signal contains a constant background term $\xi + \theta'^2 + \Delta a_w^2/4$ that is independent of the frequency ω. The quantity $\xi = I_T/I_0$ ($\theta = 0$, $\Delta a_w = \Delta b_r = 0$, $\Delta\alpha_0 = 0$) gives the residual transmission of the completely crossed analyzer P_2. With normal Glan–Thomson polarizers $\xi < 10^{-6}$ can be realized, with selected samples even $\xi < 10^{-8}$ can be obtained. The third term is due to the absorption part of the window birefringence. All these three terms are approximately independent of the frequency ω.

The next three terms contribute to the line profile of the polarization signal. The first frequency-dependent term in (2.50) gives the additional transmitted intensity when the angle $\theta' = \theta + (\omega/2c)\Delta b_r$ is not zero. It has a dispersion-type profile. For $\theta' = 0$ the dispersion term vanishes. The two Lorentzian terms depend on the product of Δa_w and $\Delta\alpha_0 \cdot L$ and on $(\Delta\alpha_0 L)^2$. By squeezing the windows their dichroism (that is, Δa_w) can be increased and the amplitude of the first Lorentzian term can be enlarged. Of course, the background term Δa_w^2 also increases and one has to find the optimum signal-to-noise ratio. Since in most cases $\Delta\alpha_0 L \ll 1$, the last term in (2.50), which is proportional to $\Delta\alpha_0^2 L^2$, is generally negligible. Nearly pure dispersion signals can be obtained if Δa_w is minimized and θ' is increased until the optimum signal-to-noise ratio is achieved. Therefore by controlling the birefringence of the window either dispersion-shaped or Lorentzian line profiles can be obtained for the signals.

The sensitivity of polarization spectroscopy compared with saturation spectroscopy is illustrated by Fig. 2.24a, which shows the same hfs transitions of I_2

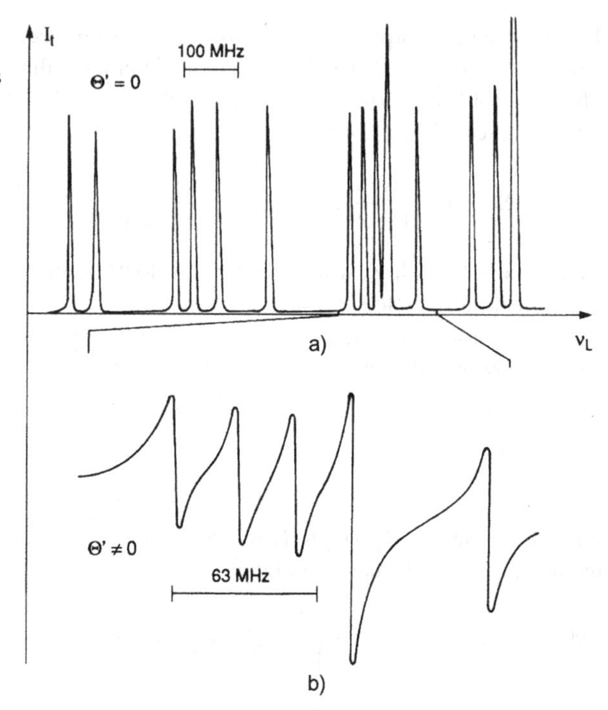

Fig. 2.24 Polarization spectrum of the same hyperfine components of I_2 as shown in Fig. 2.13 with circularly polarized pump: (**a**) with $\theta' = 0$; (**b**) enlarged section with $\theta' \neq 0$

as in Fig. 2.13 taken under comparable experimental conditions. A section of the same spectrum is depicted in Fig. 2.24b with $\theta' \neq 0$, optimized for dispersion line-profiles.

If the pump wave is *linearly polarized* with the electric field vector 45° against the x-axis, one obtains in an analogous derivation for the polarization signal instead of (2.50) the expression

$$S^{\mathrm{LP}}(\omega) = I_0 \mathrm{e}^{-\alpha L - a_\mathrm{w}} \left\{ \xi + \frac{1}{4}\theta^2 \Delta a_\mathrm{w}^2 + \left(\frac{\omega}{2c}\Delta b_\mathrm{r}\right)^2 + \frac{\Delta b_\mathrm{r}}{4}\frac{\omega}{c}\Delta\alpha_0 L \frac{x}{1+x^2} \right.$$
$$\left. + \left[-\frac{1}{4}\theta\,\Delta a_\mathrm{w} \Delta\alpha_0 L + \left(\frac{\Delta\alpha_0 L}{4}\right)^2 \right] \frac{1}{1+x^2} \right\}, \tag{2.51}$$

where $\Delta\alpha = \alpha_\parallel - \alpha_\perp$ and $\Delta b = b_\parallel - b_\perp$ are now defined by the difference of the components parallel or perpendicular to the $\boldsymbol{E}$ vector of the pump wave. Dispersion and Lorentzian terms as well as $\Delta\alpha_\mathrm{w}$ and Δb_r are interchanged compared with (2.50). If two different lasers are used as pump and probe, all probe transitions which share a common lower or upper level with the pump transition are accessible to polarisation spectroscopy. Therefore P- and R-transitions can be probed, even if the pump is stabilized on a Q-transition. In Fig. 2.25 the different possibilities are illustrated (see also Chap. 5).

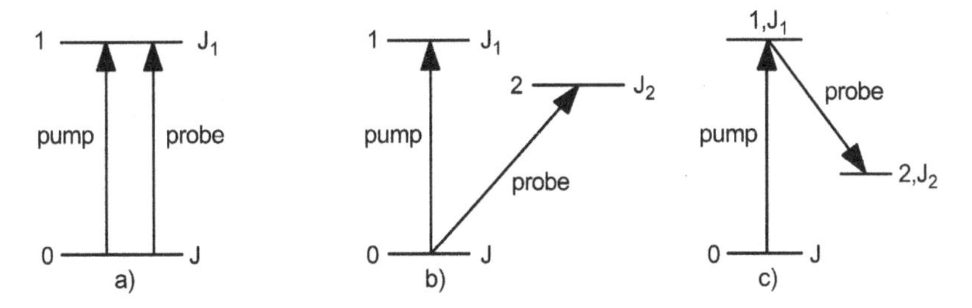

Fig. 2.25 Level scheme for possible pump-probe experiments. (**a**) Two-level scheme; (**b**) V-type scheme with three-level scheme; (**c**) Λ-type scheme

2.4.3 Magnitude of Polarization Signals

In order to understand the different magnitudes and line profiles, we have to investigate the magnitudes of the differences $\Delta\alpha_0 = \alpha^+ - \alpha^-$ for a circularly polarized pump wave and $\Delta\alpha_0 = \alpha_\parallel - \alpha_\perp$ for a linearly polarized pump wave, and their relation to the absorption cross section of the molecular transitions.

When a linearly polarized weak probe wave propagating through a ensemble of molecules is tuned to a molecular transition $|J, M\rangle \to |J_1, M \pm 1\rangle$, the difference $\Delta\alpha = \alpha^+ - \alpha^-$ of the absorption coefficients of its left- and right-hand circularly polarized components is

$$\Delta\alpha(M) = N_M\left(\sigma^+_{JJ_1M} - \sigma^-_{JJ_1M}\right), \tag{2.52}$$

where $N_M = N_J/(2J+1)$ represents the population density of one of the $(2J+1)$ degenerate sublevels $|J, M\rangle$ of a rotational level $|J\rangle$ in the lower state, and $\sigma^\pm_{JJ_1M}$ the absorption cross section for transitions starting from the level $|J, M\rangle$ and ending on $|J_1, M \pm 1\rangle$.

The absorption cross section

$$\sigma_{JJ_1M} = \sigma_{JJ_1} C(J, J_1, M, M_1) \tag{2.53}$$

can be separated into a product of two factors: the first factor is independent of the molecular orientation, but depends solely on the internal transition probability of the molecular transitions, which differs for P, Q, and R-lines [228, 229]. The second factor in (2.53) is the Clebsch–Gordan coefficient, which depends on the rotational quantum numbers J, J_1 of the lower and upper levels, and on the orientation of the molecule with respect to the quantization axis. In the case of a σ^+ pump wave, this is the propagation direction k_p; in the case of a π pump wave, it is the direction of the electric vector $\boldsymbol{E}$, see Fig. 2.26.

The total change $\Delta\alpha$ $(J \to J_1) = \alpha^+ (J \to J_1) - \alpha^- (J \to J_1)$ on a rotational transition $J \to J_1$ is due to the saturation of all allowed transitions ($M_J \to M_{J1}$ with $\Delta M = \pm 1$ for a circular pump polarization or $\Delta M = 0$ for a linearly polarized pump) between the $(2J+1)$ degenerate sublevels M in the lower level J and the

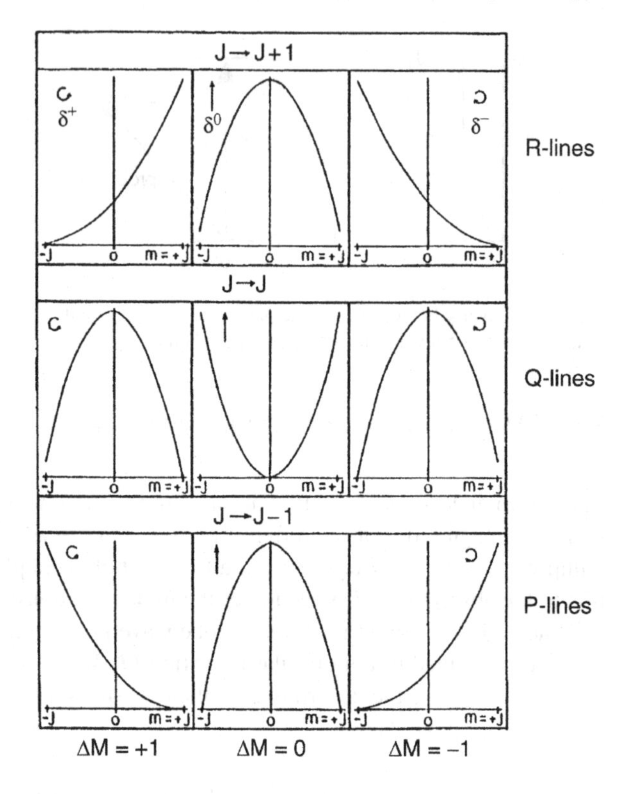

Fig. 2.26 Dependence of σ_{JJ_1M} on the orientational quantum number M for transitions with $\Delta M = 0, \pm 1$

$(2J_1 + 1)$ sublevels in the upper level J_1.

$$\Delta\alpha(J, J_1) = \sum_M N_M\left(\sigma^+_{JJ_1M} - \sigma^-_{JJ_1M}\right). \tag{2.54}$$

Because of the saturation by the pump beam, the population N_M decreases from its unsaturated value N^0_M to

$$N^S_M = \frac{N^0_M}{1+S},$$

where the saturation parameter

$$S = \frac{B_{12}\varrho_2}{R^*} = \frac{8\sigma_{JJ_1M}}{\gamma_s R^*}\frac{I_2}{\hbar\omega}, \tag{2.55}$$

depends on the absorption cross section for the pump transition, the saturated homogeneous linewidth γ_s, the mean relaxation rate R^*, which refills the saturated level population (for example, by collisions) and the number of pump photons $I_2/\hbar\omega$ per cm^2 and second [227].

From (2.52–2.54) we finally obtain for the quantity $\Delta\alpha_0$ in (2.50, 2.51)

$$\Delta\alpha_0 = \alpha_0 S_0 \Delta C^*_{JJ_1}, \tag{2.56}$$

Table 2.1 Values of $\frac{2}{3}\Delta C^*_{J_1 J_2}$ for linear pump polarization (**a**) and circular pump polarization (**b**). For the two-level system $J_1 = J_2$ and $r = (\gamma_J - \gamma_{J_2})/(\gamma_J + \gamma_{J_2})$. For the three-level system J is the rotational quantum number of the common level and $r = -1$

	$J_2 = J + 1$	$J_2 = J$	$J_2 = J - 1$
(a) Linear pump polarization			
$J_1 = J + 1$	$\frac{2J^2+J(4+5r)+5+5r}{5(J+1)(2J+3)}$	$\frac{-(2J-1)}{5(J+1)}$	$\frac{1}{5}$
$J_1 = 1$	$\frac{-(2J-1)}{5(J+1)}$	$\frac{(2J-3)(2J-1)}{5J(J+1)}$	$-\frac{2J+3}{5J}$
$J_1 = J - 1$	$\frac{1}{5}$	$-2J + \frac{3}{5J}$	$\frac{2J^2-5rJ+3}{5J(2J-1)}$
(b) Circular pump polarization			
$J_1 = J + 1$	$\frac{2J^2+J(4+r)+r+1}{(2J+3)(J+1)}$	$\frac{-1}{J+1}$	-1
$J_1 = 1$	$\frac{-1}{J+1}$	$\frac{1}{J(J+1)}$	$\frac{1}{J}$
$J_1 = J - 1$	-1	$\frac{1}{J}$	$\frac{2J^2-rJ-1}{J(2J-1)}$

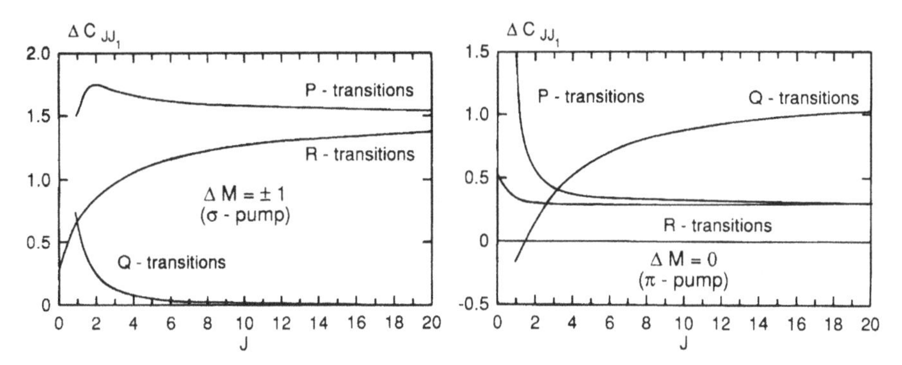

Fig. 2.27 Relative magnitude of polarization signals expressed by the factor $\Delta C^*_{JJ_1}$ as a function of rotational quantum number J for circular and linear pump polarization

where $\alpha_0 = N_J \sigma_{JJ_1}$ is the unsaturated absorption coefficient of the probe wave at the line center, and $\Delta C^*_{JJ_1}$ is a numerical factor that is proportional to the sum $\sum \Delta\sigma_{JJ_1 M} C(JJ_1 M M_1)$ with $\Delta\sigma = \sigma^+ - \sigma^-$. The numerical values of $\Delta C^*_{JJ_1}$ are compiled in Table 2.1 for P, Q, and R-transitions. Their dependence on the rotational quantum number J is illustrated by Fig. 2.27. As can be seen, for a linearly polarized pump Q-transitions have a larger cross section, while for circular pump polarization P and R-lines are favoured. This is advantageous for the assignment of molecular spectra. For illustration, in Fig. 2.28 two identical sections of the Cs_2 polarization spectrum around $\lambda = 627.8$ nm are shown, pumped with a linearly or circularly polarized pump beam, respectively. As shown below, the magnitude of the polarization signals for linearly polarized pump is large for transitions with $\Delta J = 0$, whereas for a circularly polarized pump they are maximum for $\Delta J = \pm 1$. While the Q-lines are

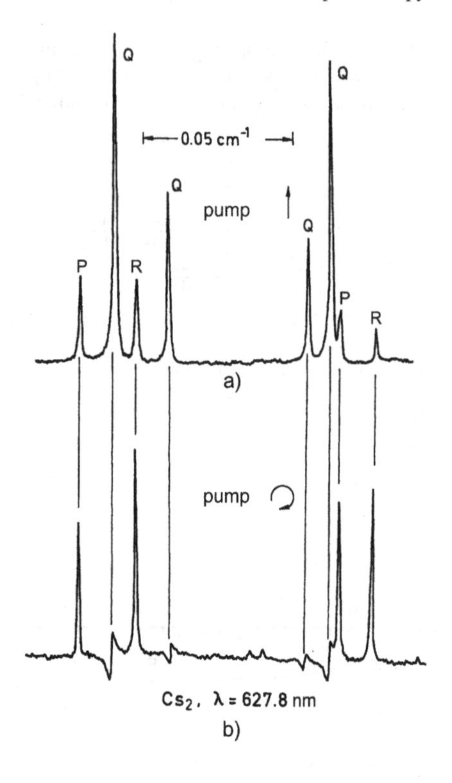

Fig. 2.28 Two identical sections of the Cs_2 polarization spectrum recorded with linear (*upper spectrum*) and circular pump polarization (*lower spectrum*)

prominent in the upper spectrum, they appear in the lower spectrum only as small dispersion-type signals.

2.4.4 Sensitivity of Polarization Spectroscopy

In the following we briefly discuss the sensitivity and the signal-to-noise ratio achievable with polarization spectroscopy. The amplitude of the dispersion signal in (2.50) for $\theta' \neq 0$ is approximately the difference $\Delta I_T = I_T(x = +1) - I_T(x = -1)$ between the maximum and the minimum of the dispersion curve. From (2.50) we obtain (Fig. 2.29)

$$\Delta S_{\max} = I_0 e^{-(\alpha L + a_w)} \cdot \theta' \Delta\alpha_0 L, \tag{2.57a}$$

while for the Lorentzian profiles $(I_T(x = 0) - I_T(x = \infty), \theta' = 0)$ we derive from (2.50)

$$\Delta S_{\max} = I_0 e^{-(\alpha L + a_w)} \cdot \frac{\Delta\alpha_0 L}{4}\left(\Delta a_w + \frac{1}{4}\Delta\alpha_0 L\right). \tag{2.57b}$$

Under general laboratory conditions the main contribution to the noise comes from fluctuations of the probe-laser intensity, while the principal limit set by shot noise

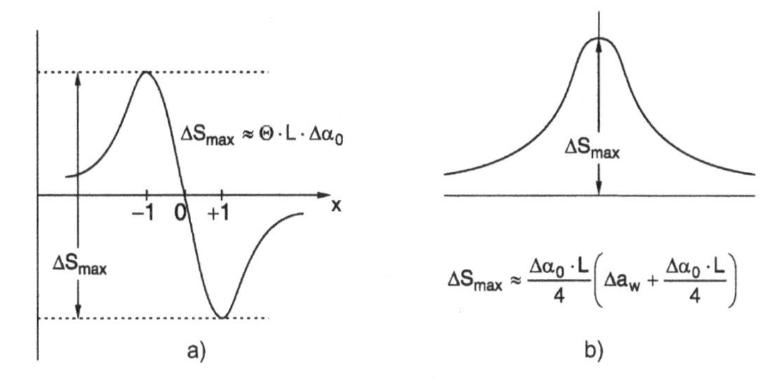

Fig. 2.29 Magnitude of signals in polarization spectroscopy: (**a**) for dispersion signals; (**b**) for Lorentzian signals

(Chap. 4) is seldom reached. The noise level is therefore essentially proportional to the transmitted intensity, which is given by the background term in (2.50).

Because the crossed polarizers greatly reduce the background level, we can expect a better signal-to-noise ratio than in saturation spectroscopy, where the full intensity of the probe beam is detected.

In the absence of window birefringence (that is, $\Delta b_r = \Delta b_i = 0$, $\theta' = \theta$) the signal-to-noise (S/N) ratio, which is, besides a constant factor a, equal to the signal-to-background ratio, becomes with (2.56) and (2.57a) for the dispersion signals

$$\frac{\mathrm{S}}{\mathrm{N}} = a\frac{\theta\alpha_0 L S_0}{\xi + \theta^2}\Delta C^*_{JJ_1}, \tag{2.58}$$

where a is a measure of the intensity stability of the probe laser, i.e., I_1/a is the mean noise of the incident probe wave with intensity I_1, and S_0 is the saturation parameter at $\omega = \omega_0$.

This ratio is a function of the uncrossing angle θ of the two polarizers (Fig. 2.23) and has a maximum for $\mathrm{d(S/N)/d}\theta = 0$, which yields $\theta^2 = \xi$ and

$$\left(\frac{\mathrm{S}}{\mathrm{N}}\right)_{\max} = a\frac{\alpha_0 L S_0 \Delta C^*_{JJ_1}}{2\sqrt{\xi}}. \tag{2.59}$$

In this case of ideal windows ($b_r = b_i = 0$) the quality of the two polarizers, given by the residual transmission ξ of the completely crossed polarizations ($\theta = 0$), limits the achievable S/N ratio.

For comparison, in saturation spectroscopy the S/N ratio is, according to (2.37), given by

$$\frac{\mathrm{S}}{\mathrm{N}} = \frac{\alpha_0 S_0 L I_0 a}{2I_0} = \frac{1}{2}a\alpha_0 L S_0. \tag{2.60}$$

Polarization spectroscopy therefore increases the S/N ratio by the factor $\Delta C^*_{JJ_1}/2\sqrt{\xi}$ for optimized dispersion-type signals.

Example 2.7 For $\Delta C^*_{JJ_1} = 0.5$, $\xi = 10^{-6}$, the S/N ratio in polarization spectroscopy with ideal cell windows becomes 500 times better than in saturation spectroscopy.

For windows with birefringence the situation is more complex: for $b_r = 0$ we obtain for the dispersion signals the optimum uncrossing angle

$$\theta' = \sqrt{\xi + \left(\frac{1}{2}\Delta a_w\right)^2}. \tag{2.61}$$

This yields the maximum obtainable signal-to-noise ratio for dispersion signals

$$\left(\frac{S}{N}\right)^{\text{disp}}_{\max} = a \cdot \alpha_0 \cdot LS_0 \cdot \Delta C^*_{JJ_1} \frac{1}{2 \cdot \sqrt{\xi + (\frac{1}{2}\Delta a_w)}}, \tag{2.62}$$

which converts to (2.59) for $\Delta a_w = 0$.

Example 2.8 For $\xi = 10^{-6}$ and $\Delta a_w = 10^{-5}$, the S/N ratio is smaller than in Example 2.7 by a factor of 0.4. For $\xi = 10^{-8}$ one reaches an improvement in the S/N compared to saturation spectroscopy by a factor of 5000 with $\Delta\alpha_w = 0$, but only 150 for $\Delta\alpha_w = 10^{-5}$.

For the Lorentzian line profiles with $\theta' = 0$ in (2.50) we obtain

$$\left(\frac{S}{N}\right)_{\max} = \frac{a}{4}\frac{\Delta\alpha_0 L(\Delta a_w + \Delta\alpha_0 L/4)}{\xi + (\Delta a_w/2)^2}. \tag{2.63}$$

Here we have to optimize Δa_w to achieve the maximum S/N ratio. The differentiation of (2.63) with respect to Δa_w gives for $\xi \ll \Delta\alpha \cdot L$ the optimum window birefringence

$$\Delta a_w \approx \frac{4\xi}{\Delta\alpha_0 \cdot L}, \tag{2.64}$$

which yields for the maximum S/N ratio for Lorentzian signals

$$\boxed{\left(\frac{S}{N}\right)_{\max} \approx a \cdot \alpha_0 L S_0 \Delta C^*_{JJ_1} \cdot \frac{\Delta\alpha_0 L}{4\xi}\left(1 + \frac{12\xi}{(\Delta\alpha_0 L)^2} - \frac{64\xi^2}{(\Delta\alpha_0 L)^3}\right).} \tag{2.65}$$

Example 2.9 With $a = 10^2$, $\xi = 10^{-6}$, $\Delta C^*_{JJ_1} = 0.5$, $\alpha_0 L = 10^{-2}$, and $S_0 = 0.1$, we obtain a signal-to-noise ratio of $5\alpha_0 \cdot L = 5 \times 10^{-2}$ for the saturation spectroscopy signal measured with the arrangement in Fig. 2.10, without lock-in detection but $S/N = 2.5 \times 10^3 \alpha_0 \cdot L = 25$ for the dispersion signal or the optimized Lorentzian profiles in polarization spectroscopy. In practice, these figures are somewhat lower because $\Delta a_w \neq 0$. One must also take into account that the length L of the interaction zone is smaller in the arrangement of Fig. 2.22 (because of the finite angle between the pump and probe beams) than for saturation spectroscopy with completely overlapping pump and probe beams (Fig. 4.12). For $\alpha_0 L = 10^{-3}$ and $\xi = 10^{-6}$, one still achieves $S/N \approx 2.5$ for dispersion and for Lorentzian signals. For $\xi = 10^{-8}$ and $\Delta a_w = 10^{-4}$, the improvement factor over saturation spectroscopy is 70 for dispersion signals. For Lorentzian signals, the optimum value of $\Delta \alpha_w$ is 4×10^{-6} for $\Delta \alpha_0 \cdot L = 10^{-2}$. With these numbers we obtain, using (2.65) an improvement factor of 1.2×10^4.

An important source of noise is the interference noise caused by the superposition of the signal wave with that part of the pump wave that is backscattered from the windows of the sample cell. Since the optical paths of the two waves through air differ considerably, every fluctuation of the air density causes a corresponding fluctuation in the phase difference between the two coherent waves, which results in the addition of random fluctuations to the signal amplitude monitored by the detector. Even without air fluctuations, periodic changes in this interference amplitude occur when the wavelength is tuned through the spectral range of interest.

This noise and the periodic false signals can be greatly reduced by implementing a fast periodic phase shift of the pump which averages over the fluctuations. To achieve this, the mirror M_2 in Fig. 2.22 reflecting thc pump beam into the sample cell is mounted on a piezo element which periodically shifts the pathlength and with it the phase of the pump wave by $\Delta\varphi > 2\pi$ when an ac voltage with a frequency f is applied to the piezo. If f is large compared to the lock-in frequency, the lock-in averages out all phase fluctuations.

2.4.5 *Advantages of Polarization Spectroscopy*

Let us briefly summarize the advantages of polarization spectroscopy, discussed in the previous sections:

- Along with the other sub-Doppler techniques, it has the advantage of high spectral resolution, which is mainly limited by the residual Doppler width due to the finite angle between the pump beam and the probe beam. This limitation corresponds to that imposed to linear spectroscopy in collimated molecular beams by the divergence angle of the molecular beam. The transit-time broadening can be reduced if the pump and probe beams are less tightly focused.

- The sensitivity is 2–3 orders of magnitude larger than that of saturation spectroscopy. It is surpassed only by that of the intermodulated fluorescence technique at very low sample pressures (Sect. 2.3.1).
- The possibility of distinguishing between P, R, and Q-lines is a particular advantage for the assignment of complex molecular spectra.
- The dispersion profile of the polarization signals allows a stabilization of the laser frequency to the line center without any frequency modulation. The large achievable signal-to-noise ratio assures an excellent frequency stability.

Meanwhile polarization spectroscopy has been applied to the measurement of many high-resolution atomic and molecular spectra. Examples can be found in [204, 230–236].

2.5 Multiphoton Spectroscopy

In this section we consider the simultaneous absorption of two or more photons by a molecule that undergoes a transition $E_i \rightarrow E_f$ with $(E_f - E_i) = \hbar \sum_i \omega_i$. The photons may either come from a single laser beam passing through the absorbing sample or they may be provided by two or more beams emitted from one or several lasers.

The first detailed theoretical treatment of two-photon processes was given in 1929 by Göppert-Mayer [237], but the experimental realization had to wait for the development of sufficiently intense light sources, now provided by lasers [238, 239].

2.5.1 Two-Photon Absorption

Two-photon absorption can be formally described by a two-step process from the initial level $|i\rangle$ via a "virtual level" $|v\rangle$ to the final level $|f\rangle$ (Fig. 2.30b). This fictitious virtual level is represented by a linear combination of the wave functions of all real molecular levels $|k_n\rangle$ that combine with $|i\rangle$ and $|f\rangle$ by allowed one-photon transitions. The excitation of $|v\rangle$ is equivalent to the sum of all off-resonance excitations of these real levels $|k_n\rangle$. The probability amplitude for a transition $|i\rangle \rightarrow |v\rangle$ is then represented by the sum of the amplitudes of all allowed transitions $|i\rangle \rightarrow |k\rangle$ with off-resonance detuning $(\omega - \omega_{ik})$. The same arguments hold for the second step $|v\rangle \rightarrow |f\rangle$.

For a molecule moving with a velocity $\boldsymbol{v}$, the probability A_{if} for a two-photon transition between the ground state E_i and an excited state E_f induced by the photons $\hbar\omega_1$ and $\hbar\omega_2$ from two light waves with the wave vectors $\boldsymbol{k}_1$, $\boldsymbol{k}_2$, the polariza-

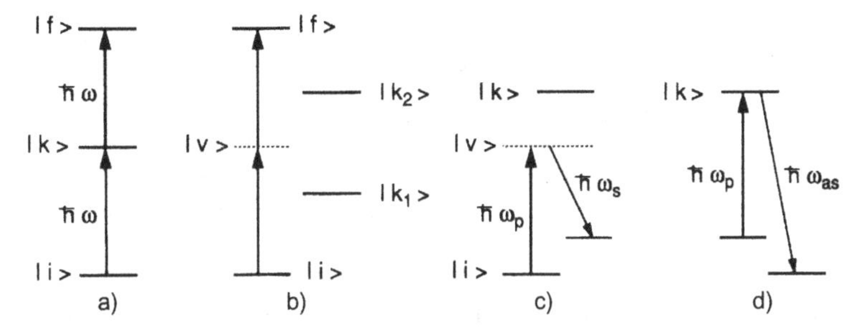

Fig. 2.30 Level schemes of different two-photon transitions: (**a**) resonant two-photon absorption with a real intermediate level $|k\rangle$; (**b**) nonresonant two-photon absorption with a virtual level $|v\rangle$; (**c**) Raman transition; (**d**) resonant anti-Stokes Raman scattering

tion unit vectors $\hat{\boldsymbol{e}}_1$, $\hat{\boldsymbol{e}}_2$, and the intensities I_1, I_2 can be written as

$$A_{if} \propto \frac{\gamma_{if} I_1 I_2}{[\omega_{if} - \omega_1 - \omega_2 - \boldsymbol{v} \cdot (\boldsymbol{k}_1 + \boldsymbol{k}_2)]^2 + (\gamma_{if}/2)^2} \times \left| \sum_k \frac{\boldsymbol{D}_{ik} \cdot \hat{\boldsymbol{e}}_1 \cdot \boldsymbol{D}_{kf} \cdot \hat{\boldsymbol{e}}_2}{\omega_{ki} - \omega_1 - \boldsymbol{v} \cdot \boldsymbol{k}_1} + \frac{\boldsymbol{D}_{ik} \cdot \hat{\boldsymbol{e}}_2 \cdot \boldsymbol{D}_{kf} \cdot \hat{\boldsymbol{e}}_1}{\omega_{ki} - \omega_2 - \boldsymbol{v} \cdot \boldsymbol{k}_2} \right|^2 . \tag{2.66}$$

The first factor gives the spectral line profile of the two-photon transition of a single molecule. It corresponds exactly to that of a single-photon transition of a moving molecule at the center frequency $\omega_{if} = \omega_1 + \omega_2 + \boldsymbol{v} \cdot (\boldsymbol{k}_1 + \boldsymbol{k}_2)$ with a homogeneous linewidth γ_{if} (Vol. 1, Sects. 3.1, 3.6). Integration over all molecular velocities v gives a *Voigt profile* with a halfwidth that depends on the relative orientations of $\boldsymbol{k}_1$ and $\boldsymbol{k}_2$. If both light waves are parallel, the Doppler width, which is proportional to $|\boldsymbol{k}_1 + \boldsymbol{k}_2|$, becomes maximum and is, in general, large compared to the homogeneous width γ_{if}. For $\boldsymbol{k}_1 = -\boldsymbol{k}_2$ the Doppler broadening vanishes and we obtain a pure Lorentzian line profile with a homogeneous linewidth γ_{if} provided that the laser linewidth is small compared to γ_{if}. This *Doppler-free two-photon spectroscopy* is discussed in Sect. 2.5.2.

Because the transition probability (2.66) is proportional to the product of the intensities $I_1 I_2$ (which has to be replaced by I^2 in the case of a single laser beam), *pulsed* lasers, which deliver sufficiently large peak powers, are generally used. The spectral linewidth of these lasers is often comparable to or even larger than the Doppler width. For nonresonant transitions $|\omega_{ki} - \omega_i| \gg \boldsymbol{v} \cdot \boldsymbol{k}_i$, and the denominators $(\omega_{ki} - \omega - \boldsymbol{k} \cdot \boldsymbol{v})$ in the sum in (2.66) can then be approximated by $(\omega_{ki} - \omega_i)$.

The second factor in (2.66) describes the transition probability for the two-photon transition. It can be derived quantum mechanically by second-order perturbation theory (see, for example, [240, 241]). This factor contains a sum of products of matrix elements $\boldsymbol{D}_{ik} \boldsymbol{D}_{kf}$ for the transitions between the initial level i and intermediate molecular levels k or between these levels k and the final state f, see Vol. 1, (2.110). The summation extends over all molecular levels k that are accessible by allowed one-photon transitions from the initial state $|i\rangle$. The denominator shows, however, that only those levels k that are not too far off resonance with one of the Doppler-shifted laser frequencies $\omega'_n = \omega_n - \boldsymbol{v} \cdot \boldsymbol{k}_n$ $(n = 1, 2)$ will mainly contribute.

Often the frequencies ω_1 and ω_2 can be selected in such a way that the virtual level is close to a real molecular eigenstate, which greatly enhances the transition probability. It is therefore generally advantageous to excite the final level E_f by two different photons with $\omega_1 + \omega_2 = (E_f - E_i)/\hbar$ rather than by two photons out of the same laser with $2\omega = (E_f - E_i)/\hbar$.

The second factor in (2.66) describes quite generally the transition probability for all possible two-photon transitions such as Raman scattering or two-photon absorption and emission. Figure 2.30 illustrates schematically three different two-photon processes. The *important point is that the same selection rules are valid for all these two-photon processes.* Equation (2.66) reveals that both matrix elements $\boldsymbol{D}_{ik}$ and $\boldsymbol{D}_{kf}$ must be nonzero to give a nonvanishing transition probability A_{if}. This means that two-photon transitions can only be observed between two states $|i\rangle$ and $|f\rangle$ that are both connected to intermediate levels $|k\rangle$ by allowed single-photon optical transitions. Because the selection rule for single-photon transitions demands that the levels $|i\rangle$ and $|k\rangle$ or $|k\rangle$ and $|f\rangle$ have opposite parity, *the two levels $|i\rangle$ and $|f\rangle$ connected by a two-photon transition* must have the same parity. In atomic two-photon spectroscopy $s \to s$ or $s \to d$ transitions are allowed, and in diatomic homonuclear molecules $\Sigma_g \to \Sigma_g$ transitions are allowed.

It is therefore possible to reach molecular states that cannot be populated by single-photon transitions from the ground state. In this regard two-photon absorption spectroscopy is complementary to one-photon absorption spectroscopy, and its results are of particular interest because they yield information about states, which often had not been found before [242]. Often excited molecular states that are accessible from the ground state by one-photon transitions are perturbed by nearby states of opposite parity, which cannot be directly observed by one-photon spectroscopy. It is generally difficult to deduce the structure of these perturbing states from the degree of perturbations, while two-photon spectroscopy allows direct access to such states.

Since the matrix elements $\boldsymbol{D}_{ik} \cdot \hat{\boldsymbol{e}}_1$ and $\boldsymbol{D}_{kf} \cdot \hat{\boldsymbol{e}}_2$ depend on the polarization characteristics of the incident radiation, it is possible to select the accessible upper states by a proper choice of the polarization. While for single-photon transitions the total transition probability (summed over all M sublevels) is *independent* of the polarization of the incident radiation, there is a distinct polarization effect in multiphoton transitions, which can be understood by applying known selection rules to the two matrix elements in (2.66). For example, two parallel laser beams, which both have right-hand circular polarization, induce two-photon transitions in atoms with $\Delta L = 2$. This allows, for instance, $s \to d$ transitions, but not $s \to s$ transitions. When a circularly polarized wave is reflected back on itself, the right-hand circular polarization changes into a left-hand one, and if a two-photon transition is induced by one photon from each wave, only $\Delta L = 0$ transitions are selected which means that now s–s transitions are allowed but not s–d transitions. Figure 2.31 illustrates the different atomic transitions that are possible by multiphoton absorption of linearly polarized light and of right or left circularly polarized light.

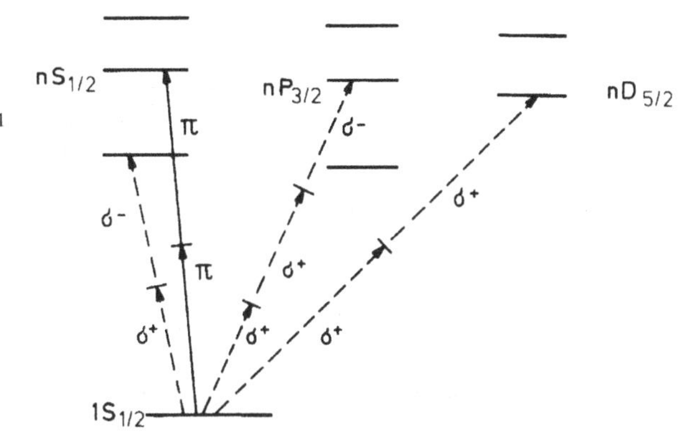

Fig. 2.31 Different two-photon transitions in an atom, depending on the polarization characteristics $\hat{e}_1$ and $\hat{e}_2$ of the two laser fields

Different upper states can therefore be selected by a proper choice of the polarization. In many cases it is possible to gain information about the symmetry properties of the upper states from the known symmetry of the ground state and the polarization of the two light waves. Since the selection rules of two-photon absorption and Raman transitions are identical, one can utilize the group-theoretical techniques originally developed for Raman scattering to analyze the symmetry properties of excited states reached by the different two-photon techniques [243, 244].

2.5.2 *Doppler-Free Multiphoton Spectroscopy*

In the methods discussed in Sects. 2.3 and 2.4, the Doppler width had been reduced or even completely eliminated by proper selection of a *velocity subgroup* of molecules with the velocity components $v_z = 0 \pm \Delta v_z$, due to selective saturation. The technique of Doppler-free multiphoton spectroscopy does not need such a velocity selection because *all* molecules in the absorbing state, regardless of their velocities, can contribute to the Doppler-free transition. Therefore the sensitivity of Doppler-free multiphoton spectroscopy is comparable to that of saturation spectroscopy in spite of the smaller transition probabilities.

While the general concepts and the transition probability of multiphoton transitions were discussed in Sect. 2.5.1, we concentrate in this subsection on *Doppler-free* multiphoton spectroscopy [245–249].

Assume a molecule moves with a velocity $\boldsymbol{v}$ in the laboratory frame. In the reference frame of the moving molecule the frequency ω of an EM wave with the wave vector $\boldsymbol{k}$ is Doppler shifted to (Vol. 1, Sect. 3.2)

$$\omega' = \omega - \boldsymbol{k} \cdot \boldsymbol{v}. \tag{2.67}$$

The resonance condition for the simultaneous absorption of two photons is

$$(E_f - E_i)/\hbar = \left(\omega'_1 + \omega'_2\right) = \omega_1 + \omega_2 - \boldsymbol{v} \cdot (\boldsymbol{k}_1 + \boldsymbol{k}_2). \tag{2.68}$$

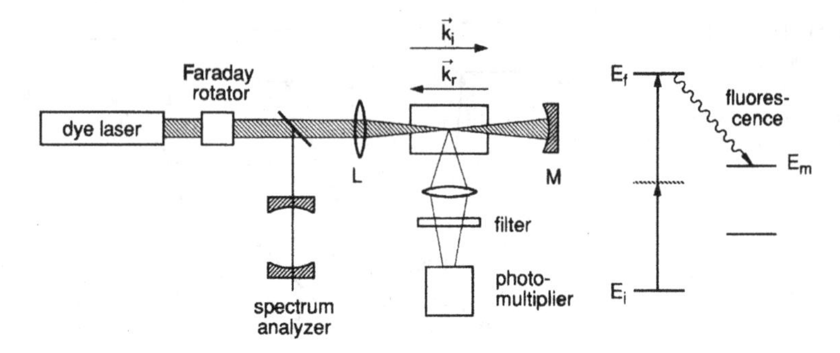

Fig. 2.32 Experimental arrangement for Doppler-free two-photon spectroscopy

If the two photons are absorbed out of two light waves with equal frequencies $\omega_1 = \omega_2 = \omega$, which travel in opposite directions, we obtain $\boldsymbol{k}_1 = -\boldsymbol{k}_2$ and (2.68) shows that the Doppler shift of the two-photon transition becomes zero. This means that *all* molecules, *independent of their velocities*, absorb at the same sum frequency $\omega_1 + \omega_2 = 2\omega$.

Although the probability of a two-photon transition is generally much lower than that of a single-photon transition, the fact that *all* molecules in the absorbing state can contribute to the signal may outweigh the lower transition probability, and the signal amplitude may even become, in favorable cases, larger than that of the saturation signals.

The considerations above can be generalized to many photons. When the moving molecule is simultaneously interacting with several plane waves with wave vectors $\boldsymbol{k}_i$ and one photon is absorbed from each wave, the total Doppler shift $\boldsymbol{v} \cdot \sum_i \boldsymbol{k}_i$ becomes zero for $\sum \boldsymbol{k}_i = 0$.

Figure 2.32 shows a possible experimental arrangement for the observation of Doppler-free two-photon absorption. The two oppositely traveling waves are formed by reflection of the output beam from a single-mode tunable dye laser. The Faraday rotator prevents feedback into the laser. The two-photon absorption is monitored by the fluorescence emitted from the final state E_f into other states E_m. From (2.66) it follows that the probability of two-photon absorption is proportional to the square I^2 of the laser intensity. Therefore, the two beams are focused into the sample cell by the lens L and the spherical mirror M.

For illustration the examples in Fig. 2.33 show the Doppler-free two-photon spectra of the $3S \to 5S$ transition in the sodium atom with resolved hyperfine structure [245].

The line profile of two-photon transitions can be deduced from the following consideration: assume the reflected beam in Fig. 2.32 to have the same intensity as the incident beam. In this case the two terms in the second factor of (2.66) become identical, while the first factor, which describes the line profile, differs for the case when both photons are absorbed out of the same beam from the case when they come from different beams. The probability for the latter case is twice as large as for the former case. This can be seen as follows: Let (a, a) be the probability for the

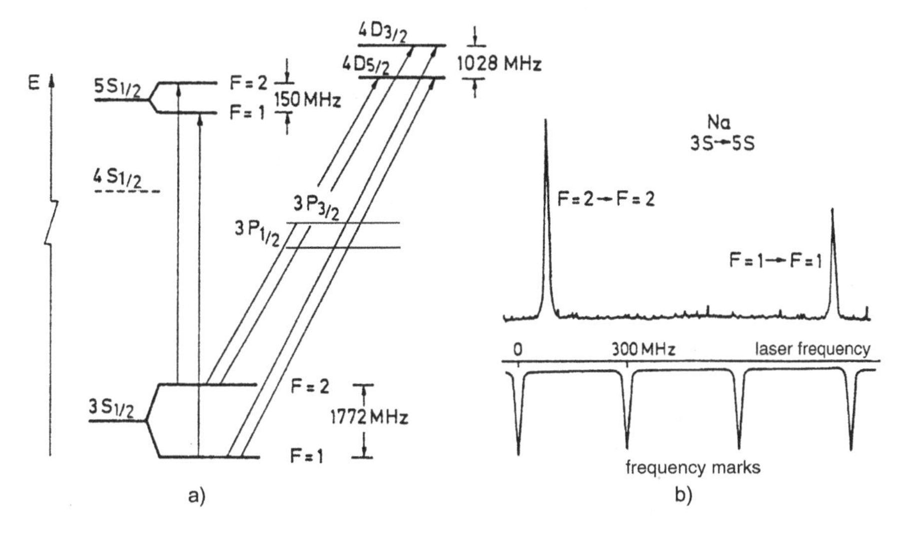

Fig. 2.33 Doppler-free two-photon spectrum of the Na atom: (**a**) level scheme of the $3S \to 5S$ and $3S \to 4D$ transitions; (**b**) $3S \to 5S$ transition for $\Delta F = 0$ with resolved hyperfine structure [245]

case that both photons are provided by the incident beam and (b, b) by the reflected beam. The total probability for a two-photon absorption with both photons of the same beam (which gives a Doppler-broadened contribution to the signal) is then $(a, a)^2 + (b, b)^2$.

The probability amplitude of the Doppler-free absorption is the sum $(a, b) + (b, a)$ of two nondistinguishable events. The total probability for this case is then $|(a, b) + (b, a)|^2$. For equal intensities of the two beams this is twice the probability $(a, a)^2 + (b, b)^2$.

We therefore obtain from (2.66) for the two-photon absorption probability

$$
\begin{aligned}
W_{if} \propto & \left| \sum_m \frac{(\boldsymbol{D}_{im}\hat{\boldsymbol{e}}_1)(\boldsymbol{D}_{mf}\hat{\boldsymbol{e}}_2)}{\omega - \omega_{im} - \boldsymbol{k}_1 \cdot \boldsymbol{v}} + \sum_m \frac{(\boldsymbol{D}_{im}\hat{\boldsymbol{e}}_2)(\boldsymbol{D}_{mf}\hat{\boldsymbol{e}}_1)}{\omega - \omega_{im} - \boldsymbol{k}_2 \cdot \boldsymbol{v}} \right|^2 I^2 \\
& \times \left[\frac{4\gamma_{if}}{(\omega_{if} - 2\omega)^2 + (\gamma_{if}/2)^2} + \frac{\gamma_{if}}{(\omega_{if} - 2\omega - 2\boldsymbol{k} \cdot \boldsymbol{v})^2 + (\gamma_{if}/2)^2} \right. \\
& \left. + \frac{\gamma_{if}}{(\omega_{if} - 2\omega + 2\boldsymbol{k} \cdot \boldsymbol{v})^2 + (\gamma_{if}/2)^2} \right].
\end{aligned}
\tag{2.69}
$$

Integration over the velocity distribution yields the absorption profile

$$
\begin{aligned}
\alpha(\omega) \propto & \, \Delta N^0 I^2 \left| \sum_m \frac{(\boldsymbol{D}_{im}\hat{\boldsymbol{e}})(\boldsymbol{D}_{mf}\hat{\boldsymbol{e}})}{\omega - \omega_{im}} \right|^2 \\
& \times \left\{ \exp\left[-\left(\frac{\omega_{if} - 2\omega}{2kv_\mathrm{p}} \right)^2 \right] + \frac{kv_\mathrm{p}}{\sqrt{\pi}} \frac{\gamma_{if}/2}{(\omega_{if} - 2\omega)^2 + (\gamma_{if}/2)^2} \right\},
\end{aligned}
\tag{2.70}
$$

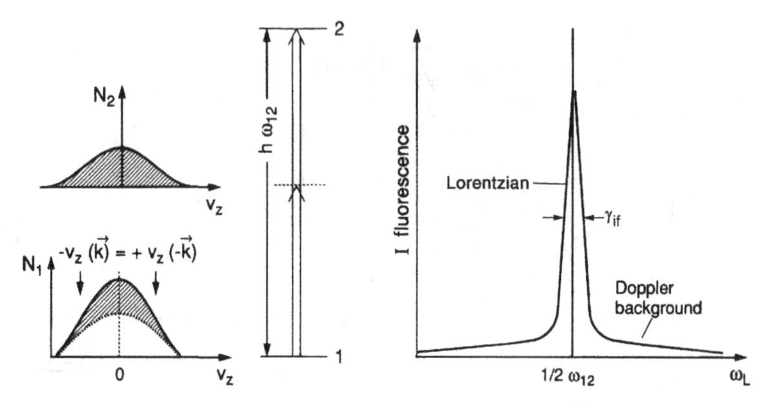

Fig. 2.34 Schematic line profile of a Doppler-free two-photon signal with (greatly exaggerated) Doppler-broadened background

where $v_\mathrm{p} = (2kT/m)^{1/2}$ is the most probable velocity, and $\Delta N_0 = N_i^0 - N_f$ the nonsaturated population difference. The absorption profile (2.70) represents a superposition of a Doppler-broadened background and a narrow Lorentzian profile with the linewidth $\gamma_{if} = \gamma_i + \gamma_f$ (Fig. 2.34).

As was shown above, the area under the Doppler profile with halfwidth $\Delta\omega_\mathrm{D}$ is half of that under the narrow Lorentzian profile with halfwidth γ_{if}. However, due to its larger linewidth its peak intensity amounts only to the fraction

$$\epsilon = \frac{\gamma_{if}\sqrt{\pi}}{2kv_\mathrm{p}} \simeq \frac{\gamma_{if}}{2\Delta\omega_\mathrm{D}},$$

of the Lorentzian peak heights.

Example 2.10 With $\gamma_{if} = 20$ MHz, $\Delta\omega_\mathrm{D} = 2$ GHz, the Doppler-free signal in Fig. 2.34 is about 200 times higher than the maximum of the Doppler-broadened background.

By choosing the proper polarization of the two laser waves, the background can often be completely suppressed. For example, if the incident laser beam has σ^+ polarization, a $\lambda/4$-plate between mirror M and the sample in Fig. 2.32 converts the reflected beam to σ^- polarization. Two-photon absorption on a $S \to S$ transition with $\Delta M = 0$ is then possible only if one photon comes from the incident wave and the other from the reflected wave because two photons out of the same beam would induce transitions with $\Delta M = \pm 2$.

For the resonance case $2\omega = \omega_{if}$, the second term in (2.69) becomes $2kv_\mathrm{p}/(\gamma_{if}\sqrt{\pi}) \gg 1$. We can neglect the contribution of the Doppler term and obtain the

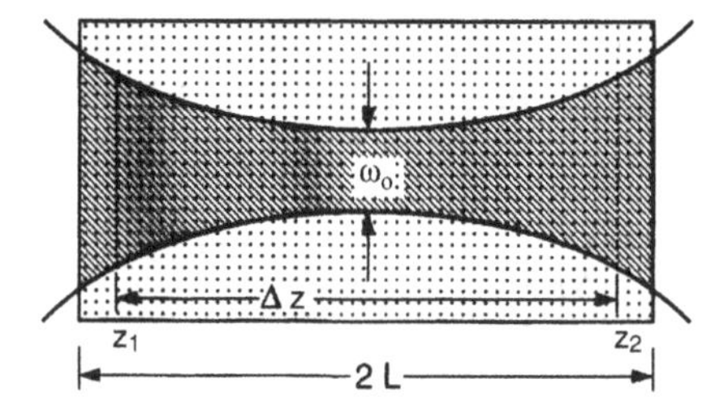

Fig. 2.35 Beam waist w_0 and Rayleigh length L for optimum focusing in two-photon spectroscopy

maximum two-photon absorption:

$$\alpha\left(\omega = \frac{1}{2}\omega_{if}\right) \propto I^2 \frac{\Delta N^0 k v_\mathrm{p}}{\sqrt{\pi}\gamma_{if}} \left| \sum_m \frac{(\boldsymbol{D}_{im} \cdot \hat{\boldsymbol{e}}) \cdot (\boldsymbol{D}_{mf} \cdot \hat{\boldsymbol{e}})}{\omega - \omega_{im}} \right|^2 . \tag{2.71}$$

Note Although the product of matrix elements in (2.71) is generally much smaller than the corresponding one-photon matrix elements, the magnitude of the Doppler-free two-photon signal (2.68) may exceed that of Doppler-free saturation signals. This is due to the fact that *all* molecules in level $|i\rangle$ contribute to two-photon absorption, whereas the signal in Doppler-free saturation spectroscopy is provided only by the subgroup of molecules out of a narrow velocity interval $\Delta v_z = 0 \pm \gamma/k$. For $\gamma = 0.01\,\Delta\omega_\mathrm{D}$, this subgroup represents only about 1 % of all molecules.

For *molecular* transitions the matrix elements D_{im} and D_{mf} are composed of three factors: the electronic transition dipole element, the Franck–Condon factor, and the Hönl–London factor (Sect. 1.7). Within the two-photon dipole approximation $\alpha(\omega)$ becomes zero if one of these factors is zero. The calculation of linestrengths for two- or three-photon transitions in diatomic molecules can be found in [243, 244].

2.5.3 Influence of Focusing on the Magnitude of Two-Photon Signals

Since the two-photon absorption probability is proportional to the square of the incident laser intensity, the signal can generally be increased by focusing the laser beam into the sample cell. However, the signal is also proportional to the number of absorbing molecules within the interaction volume, which decreases with decreasing focal volume. With pulsed lasers the two-photon transition may already be saturated at a certain incident intensity without focusing. In this case focusing will decrease the signal. The optimization of the focusing conditions can be based on the following estimation: Assume the laser beams propagate into the $\pm z$-directions. The two-photon absorption is monitored via the fluorescence from the upper level $|f\rangle$ and can be collected from a sample volume $V = \pi \cdot \int_{z_1}^{z^2} r^2(z)\,\mathrm{d}z$ (Fig. 2.35), where

$r(z)$ is the beam radius (Vol. 1, Sect. 5.3) and Δz the maximum length of the interaction volume seen by the collecting optics. The two-photon signal is then for $N_f \ll N_i$, that is, $\Delta N \approx N_i$, according to (2.71)

$$S\left(\frac{1}{2}\omega_{if}\right) \propto N_i \left| \sum_m \frac{(D_{im}\hat{\mathbf{e}}) \cdot (D_{mf}\hat{\mathbf{e}})}{\omega - \omega_{im}} \right|^2 \int_z \int_r I^2(r,z) 2\pi r \, \mathrm{d}r \, \mathrm{d}z, \tag{2.72}$$

where N_i is the density of the molecules in the absorbing level $|i\rangle$ and

$$I(r,z) = \frac{2P_0}{\pi w^2} \mathrm{e}^{-2r^2/w^2} \tag{2.73}$$

gives the radial intensity distribution of the laser beam with power P_0 in the TEM_{00} mode with a Gaussian intensity profile. The beam waist

$$w(z) = w_0 \sqrt{1 + \frac{\lambda z}{\pi w_0^2}}, \tag{2.74}$$

gives the radius of the Gaussian beam in the vicinity of the focus at $z = 0$ (Vol. 1, Sect. 5.4). The Rayleigh length $L = \pi w_0^2/\lambda$ gives that distance from the focus at $z = 0$ where the beam cross section $\pi w^2(L) = 2\pi w_0^2$ has increased by a factor of two compared to the area πw_0^2 at $z = 0$.

Inserting (2.73, 2.74) into (2.72) shows that the signal S_{if} is proportional to

$$S_{if} \propto \int_0^\infty \int_{z_1}^{z_2} \frac{1}{w^4} \mathrm{e}^{-4r^2/w^2} r \, \mathrm{d}r \, \mathrm{d}z = 4 \int_{z_1}^{z_2} \frac{\mathrm{d}z}{w^2}. \tag{2.75a}$$

S_{if} becomes maximum if the integral

$$\int_{-\Delta z/2}^{+\Delta z/2} \frac{\mathrm{d}z}{1 + (\lambda z/\pi w_0^2)^2} = 2L \arctan\left[\Delta z/(2L)\right], \tag{2.75b}$$

with $\Delta z = z_2 - z_1$ has a maximum value. Equation (2.75a–2.75b) shows that the signal cannot be increased any longer by stronger focusing when the Rayleigh length L becomes smaller than the interval Δz from which the fluorescence can be collected.

The two-photon signal can be greatly enhanced if the sample is placed either inside the laser resonator or in an external resonator, which has to be tuned synchroneously with the laser wavelength (Sect. 1.4).

2.5.4 Examples of Doppler-Free Two-Photon Spectroscopy

The first experiments on Doppler-free two-photon spectroscopy were performed on the alkali atoms [245–249] because their two-photon transitions can be induced by

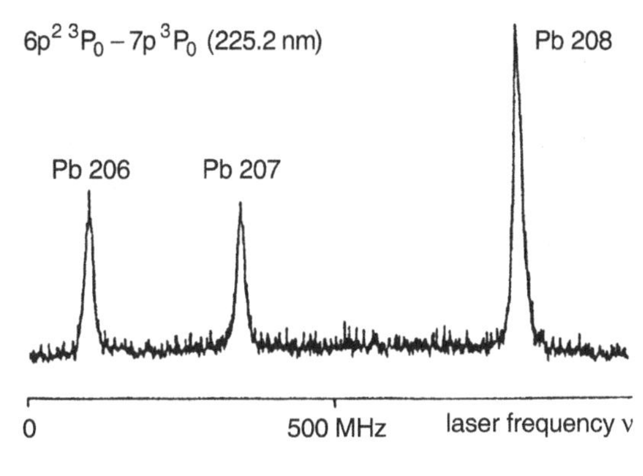

Fig. 2.36 Measurement of the isotope shift of the stable lead isotopes measured with Doppler-free two-photon spectroscopy at $\lambda_{\mathrm{exc}} = 450$ nm and monitored via fluorescence [250]

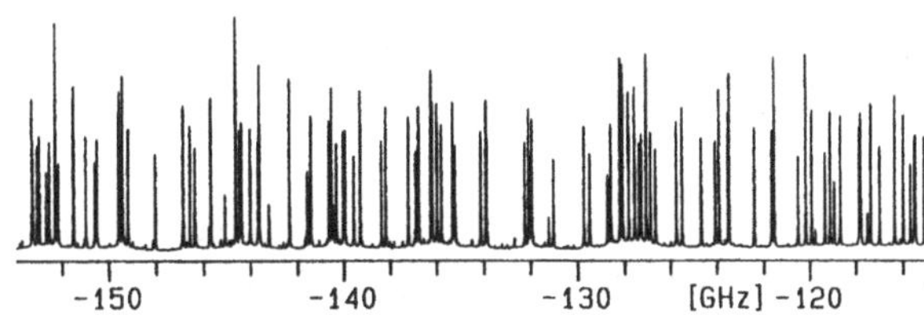

Fig. 2.37 Section of the Doppler-free two-photon excitation spectrum of the 14^1_0 Q_Q band of C_6H_6 [255]

cw dye lasers or diode lasers in convenient spectral ranges. Furthermore, the first excited P state is not too far away from the virtual level in Fig. 2.30b. This enlarges the two-photon transition probabilities for such "near-resonant" transitions. Meanwhile, there are numerous further applications of this sub-Doppler technique in atomic and molecular physics. We shall illustrate them by a few examples only.

The isotope shift between different stable lead isotopes shown in Fig. 2.36 has been determined by Doppler-free two-photon absorption of a cw dye laser beam at $\lambda = 450.4$ nm on the transition $6p^{2\,3}P_0 \rightarrow 7p\,^3P_0$ [250] using the experimental arrangement of Fig. 2.32.

Doppler-free two-photon transitions to atomic Rydberg levels [251] allow the accurate determination of quantum defects and of level shifts in external fields. Hyperfine structures in Rydberg states of two-electron atoms, such as calcium and singlet–triplet mixing of the valence state 4s and ^1D and ^3D Rydberg levels have been thoroughly studied by Doppler-free two-photon spectroscopy [252].

The application of two-photon spectroscopy to molecules has brought a wealth of new insight to excited molecular states. One example is the two-photon excitation of CO in the fourth positive system $A\,^1\Pi \leftarrow X\,^1\Sigma_g$ and of N_2 in the Lyman–Birge–Hopfield system with a narrow-band pulsed frequency-doubled dye laser. Doppler-free spectra of states with excitation energies between 8–12 eV can be measured with this technique [253].

For larger molecules, rotationally resolved absorption spectra could, for the first time, be measured, as has been demonstrated for the UV spectra of benzene C_6H_6. Spectral features, which had been regarded as true continua in former times, could now be completely resolved (Fig. 2.37) and turned out to be dense but discrete rotational-line spectra [254, 255]. The lifetimes of the upper levels could be determined from the natural linewidths of these transitions [256]. It was proven that

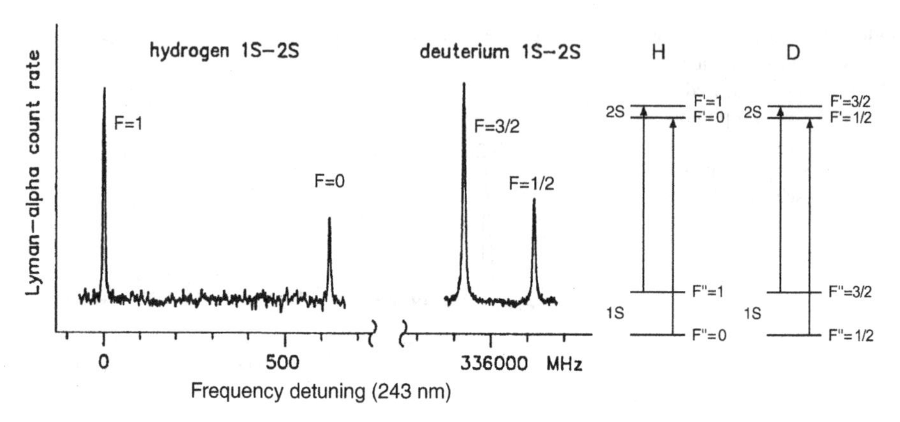

Fig. 2.38 Doppler-free two-photon transitions of hydrogen $1S$–$2S$ and deuterium $1S$–$2S$ [260]. For each of the two photons from opposite laser beams, only transitions with $\Delta F = 0$ are allowed

these lifetimes strongly decrease with increasing vibrational–rotational energy in the excited electronic state because of the increasing rate of radiationless transitions [257].

Molecular two-photon spectroscopy can also be applied in the infrared region to induce transitions between rotational–vibrational levels within the electronic ground state. One example is the Doppler-free spectroscopy of rotational lines in the ν_2 vibrational bands of NH_3 [258]. This allows the study of the collisional properties of the ν_2 vibrational manifold from pressure broadening and shifts (Vol. 1, Sect. 3.3) and Stark shifts.

Near-resonant two-photon spectroscopy of NH_3 has been reported by Winnewisser et al. [259]. Using a tunable diode laser with frequency ν_1 the two-photon excitation $\nu_1 + \nu_2$ of the $2\nu_2$ $(1, 1)$ level of NH_3 was observed with the second photon $h\nu_2$ coming from a fixed-frequency CO_2 laser.

The possibilities of Doppler-free two-photon spectroscopy for metrology and fundamental physics has been impressively demonstrated by precision measurements of the $1S$–$2S$ transition in atomic hydrogen [260–263]. Precise measurements of this one-photon forbidden transition with a very narrow natural linewidth of 1.3 Hz yield accurate values of fundamental constants and can provide stringent tests of quantum electrodynamic theory (Sect. 9.7). A comparison of the $1S$–$2S$ transition frequency with the $2S$–$3P$ frequency allows the precise determination of the Lamb shift of the $1S$ ground state [261], whereas the $2S$ Lamb shift was already measured long ago by the famous Lamb–Rutherford experiments where the RF transition between $2S_{1/2}$ and $2P_{1/2}$ were observed. Because of the isotope shift the $1S$–$2S$ transitions of ^{1}H and $^2\mathrm{H} = {}^2_1D$ differ by 2×335.49716732 GHz $= 670.99433464$ GHz (Fig. 2.38). From this measurement a value $r_D^2 - r_P^2 = 3.8212$ fm^2 for the difference of the mean square charge radii of proton and deuteron is derived. The Rydberg constant has been determined within a relative uncertainty of 10^{-10} [262–264]. The attainable precision allows a test of the limits of quantum-electrodynamics. Of particular interest is the questions

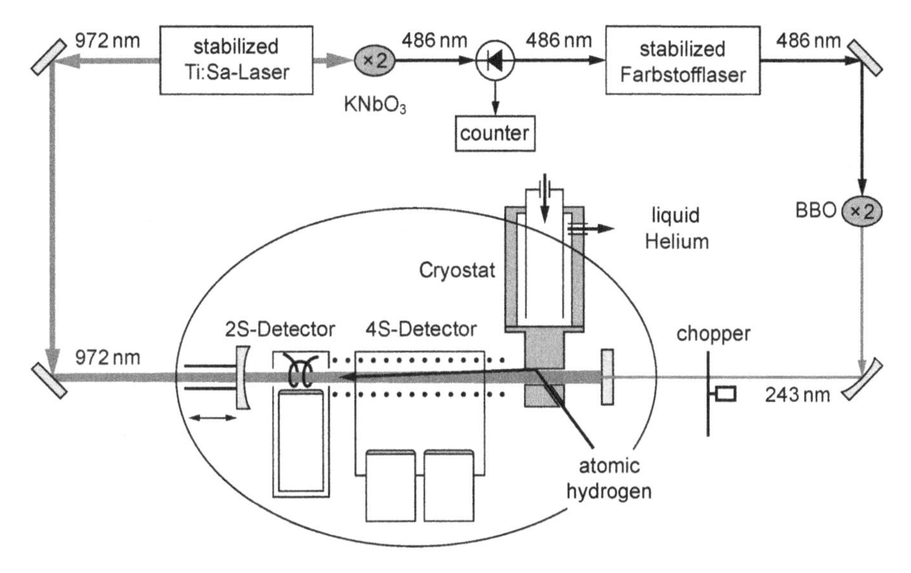

Fig. 2.39 Experimental setup for the simultaneous excitation of the two-photon transitions $1S \rightarrow 2S$ and $2S \rightarrow 4S$ in a cold atomic beam [290]

whether the fundamental physical constants such as the fine structure constant, the Rydber constant or the ratio of electron-to-proton mass are really constant or whether they might change with time. In order to test this within reasonable times the precision has to be further increased. One possible way is the increase of the interaction time of the H-atoms with the laser field. This can be achieved if the velocity of the atoms is reduced by cooling them down to very low temperatures. The experimental setup is shown in Fig. 2.39. The atoms pass through a cryostat with liquid helium and are deflected by collisions with a cold wall into the direction of the laser beam where they interact with the standing wave formed by reflection of the frequency doubled laser between the two mirrors. This results in a Doppler-free two-photon transition $1S$–$2S$ of the H-atom.

Even higher precision can be reached with the optical frequency comb (see Sect. 9.7).

2.5.5 *Multiphoton Spectroscopy*

If the incident intensity is sufficiently large, a molecule may absorb several photons simultaneously. The probability for absorption of a photon $\hbar\omega_k$ on the transition $|i\rangle \rightarrow |f\rangle$ with $E_f - E_i = \sum \hbar\omega_k$ can be obtained from a generalization of (2.66). The first factor in (2.66) now contains the product $\prod_k I_k$ of the intensities I_k of the different beams. In the case of n-photon absorption of a single laser beam this product becomes I^n. The second factor in the generalized formula (2.66) includes the sum over products of n one-photon matrix elements.

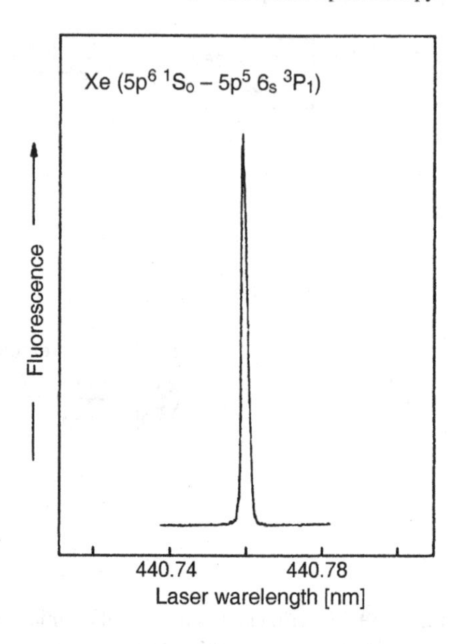

Fig. 2.40 Three-photon excited resonance fluorescence in Xe at $\lambda = 147$ nm, excited with a pulsed dye laser at $\lambda = 440.76$ nm with 80 kW peak power. The Xe pressure was 8 mtorr [265]

The three-photon absorption can be used for the excitation of high-lying molecular levels with the same parity as accessible to one-photon transitions. However, for a one-photon absorption, lasers with a wavelength $\lambda/3$ have to be available in order to reach the same excitation energy. An example of Doppler-limited collinear three-photon spectroscopy is the excitation of high-lying levels of xenon and CO with a narrow-band pulsed dye laser at $\lambda = 440$ nm (Fig. 2.40). For one-photon transitions light sources at $\lambda = 146.7$ nm in the VUV would have been necessary.

In the case of *Doppler-free* multiphoton absorption, momentum conservation

$$\sum_k \boldsymbol{p}_k = \hbar \sum_k \boldsymbol{k}_k = 0, \tag{2.76}$$

has to be fulfilled in addition to energy conservation $\sum \hbar\omega_k = E_f - E_i$. Each of the absorbed photons transfers the momentum $\hbar \boldsymbol{k}_k$ to the molecule. If (2.76) holds, the total transfer of momentum is zero, which implies that the velocity of the absorbing molecule has not changed. This means that the photon energy $\hbar \sum \omega_k$ is completely converted into excitation energy of the molecule without changing its kinetic energy. This is independent of the initial velocity of the molecule, that is, the transition is Doppler-free.

A possible experimental arrangement for Doppler-free three-photon absorption spectroscopy is depicted in Fig. 2.41. The three laser beams generated by beam splitting of a single dye laser beam cross each other under 120° in the absorbing sample.

The absorption probability increases if a two-photon resonance can be found. One example is illustrated in Fig. 2.42a, where the $4D$ level of the Na atom is excited by two-photons of a dye laser at $\lambda = 578.7$ nm and further excitation by a third photon reaching high-lying Rydberg levels nP or nF with the electronic orbital

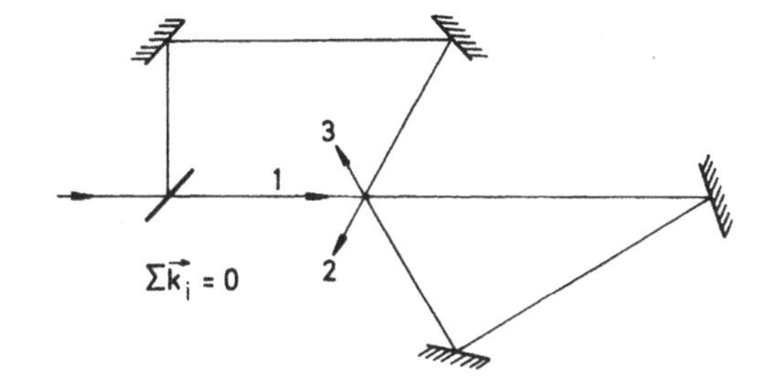

Fig. 2.41 Possible arrangements for Doppler-free three-photon spectroscopy

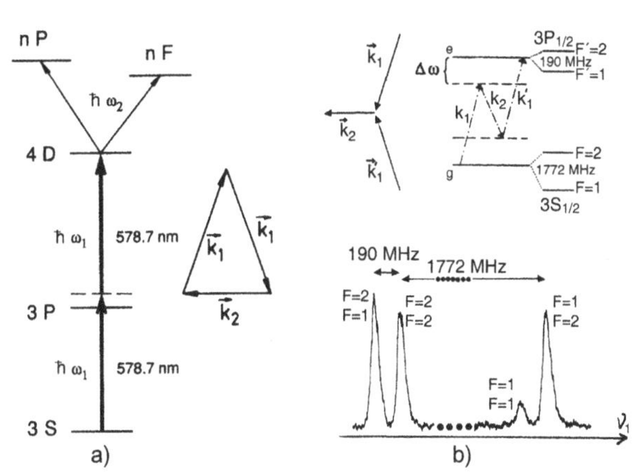

Fig. 2.42 Level schemes for Doppler-free three-photon spectroscopy of the Na atom: (**a**) stepwise excitation of Rydberg states; (**b**) Raman-type process shown for the example of the $3S$–$3P$ excitation of the Na atom. Laser 1 is tuned while L2 is kept 30 GHz below the Na D_1 line [266]

momentum $\ell = 1$ or $\ell = 3$. For Doppler-free excitation the wave-vector diagram of Fig. 2.41 has been used [266].

Three-photon excitation may also be used for a Raman-type process, depicted in Fig. 2.42b, which proceeds via two virtual levels. This Doppler-free technique was demonstrated for the $3\,^2S_{1/2} \rightarrow 3\,^2P_{1/2}$ transition of the Na atom, where the photons with the momentum $\hbar \boldsymbol{k}_1$ and $\hbar \boldsymbol{k}_1'$ are absorbed while the photon with the momentum $\hbar \boldsymbol{k}_2$ is emitted. The hyperfine structure of the upper and lower states could readily be resolved [266].

Multiphoton absorption of visible photons may result in ionization of atoms or molecules. At a given laser intensity the ion rate $N_{\text{ion}}(\lambda_L)$ recorded as a function of the laser intensity shows narrow maxima if one-, two-, or three-photon resonances occur. If, for instance, the ionization potential (IP) is smaller than $3\hbar\omega$, resonances in the ionization yield are observed, either when the laser frequency ω is in resonance with a two-photon transition between the levels $|i\rangle$ and $|f\rangle$, i.e. ($E_f - E_i = 2\hbar\omega$), or when autoionizing Rydberg states can be reached by three-photon transitions [267].

Multiphoton absorption has also been observed on transitions within the electronic ground states of molecules, induced by infrared photons of a CO_2 laser [268, 269]. At sufficient intensities this multiphoton excitation of high vibrational–rotational states may lead to the dissociation of the molecule [270].

If the first step of the multiphoton excitation can be chosen isotope selectively so that a wanted isotopomer has a larger absorption probability than the other isotopomers of the molecular species, selective dissociation may be achieved [270],

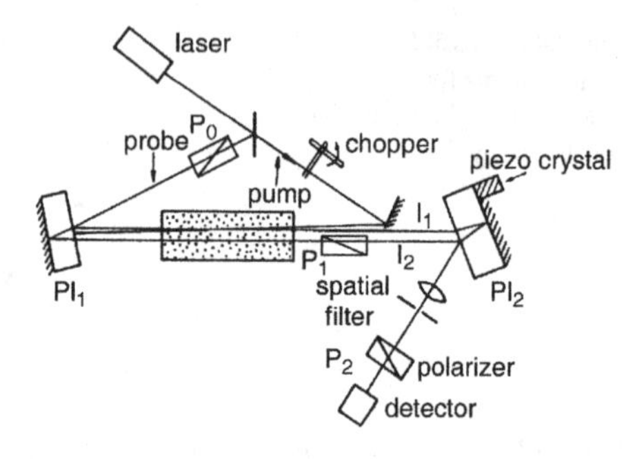

Fig. 2.43 Schematic arrangement of saturated interference spectroscopy

which can be used for isotope separation by chemical reactions with the isotope-selective dissociation products (Sect. 10.2).

2.6 Special Techniques of Nonlinear Spectroscopy

In this section we will briefly discuss some variations of saturation, polarization, or multiphoton spectroscopy that either increase the sensitivity or are adapted to the solution of special spectroscopic problems. They are often based on combinations of several nonlinear techniques.

2.6.1 Saturated Interference Spectroscopy

The higher sensitivity of polarization spectroscopy compared with conventional saturation spectroscopy results from the detection of *phase differences* rather than amplitude differences. This advantage is also used in a method that monitors the interference between two probe beams where one of the beams suffers saturation-induced phase shifts. This saturated interference spectroscopy was independently developed in different laboratories [271, 272]. The basic principle can easily be understood from Fig. 2.43. We follow here the presentation in [271].

The probe beam is split by the plane-parallel plate Pl1 into two beams. One beam passes through that region of the absorbing sample that is saturated by the pump beam; the other passes through an unsaturated region of the same sample cell. The two beams are recombined by a second plane-parallel plate Pl2. The two carefully aligned parallel plates form a Jamin interferometer [273], which can be adjusted by a piezoelement in such a way that without the saturating pump beam the two probe waves with intensities I_1 and I_2 interfere destructively.

If the saturation by the pump wave introduces a phase shift φ, the resulting intensity at the detector becomes

$$I = I_1 + I_2 - 2\sqrt{I_1 I_2}\cos\varphi. \tag{2.77}$$

The intensities I_1 and I_2 of the two interfering probe waves can be made equal by placing a polarizer P1 into one of the beams and a second polarizer P2 in front of the detector. Due to a slight difference δ in the absorptions of the two beams by the sample molecules, their intensities at the detector are related by

$$I_1 = I_2(1+\delta) \quad \text{with } \delta \ll 1 .$$

For small phase shifts φ ($\varphi \ll 1 \rightarrow \cos\varphi \approx 1 - \frac{1}{2}\varphi^2$), we can approximate (2.77) by

$$I \approx \left(\frac{1}{4}\delta^2 + \varphi^2\right) I_2 , \tag{2.78}$$

when we neglect the higher-order terms $\delta\varphi^2$ and $\delta^2\varphi^2$. The amplitude difference δ and the phase shift φ are both caused by selective saturation of the sample through the monochromatic pump wave, which travels in the opposite direction. Analogous to the situation in polarization spectroscopy, we therefore obtain Lorentzian and dispersion profiles for the frequency dependence of both quantities

$$\delta(\omega) = \frac{\delta_0}{1+x^2} , \qquad \varphi(\omega) = \frac{1}{2}\delta_0 \frac{x}{1+x^2} , \tag{2.79}$$

where $\delta_0 = \delta(\omega_0)$, $x = 2(\omega-\omega_0)/\gamma$, and γ is the homogeneous linewidth (FWHM).

Inserting (2.79) into (2.78) yields, for the total intensity I at the minimum of the interference patterns, the Lorentzian profile,

$$I = \frac{1}{4}\frac{I_2\delta_0^2}{1+x^2} . \tag{2.80}$$

According to (2.79), the phase differences $\varphi(\omega)$ depends on the laser frequency ω. However, it can always be adjusted to zero while the laser frequency is scanned. This can be accomplished by a sine-wave voltage at the piezoelement, which causes a modulation

$$\varphi(\omega) = \varphi_0(\omega) + a\sin(2\pi f_1 t) .$$

When the detector signal is fed to a lock-in amplifier that is tuned to the modulation frequency f_1, the lock-in output can drive a servo loop to bring the phase difference φ_0 back to zero. For $\varphi(\omega) \equiv 0$, we obtain from (2.78, 2.79)

$$I(\omega) = \frac{1}{4}\delta(\omega)^2 I_2 = \frac{1}{4}\frac{\delta_0^2 I_2}{(1+x^2)^2} . \tag{2.81}$$

The halfwidth of this signal is reduced from γ to $(\sqrt{2}-1)^{1/2}\gamma \approx 0.64\gamma$.

Contrary to the situation in polarization spectroscopy, where for slightly uncrossed polarizers the line shape of the polarization signal is a superposition of Lorentzian and dispersion profiles, with saturated interference spectroscopy pure Lorentzian profiles are obtained because the phase shift is compensated by the feedback control. Measuring the first derivative of the profiles, pure dispersion-type signals appear. To achieve this, the output of the lock-in amplifier that controls the

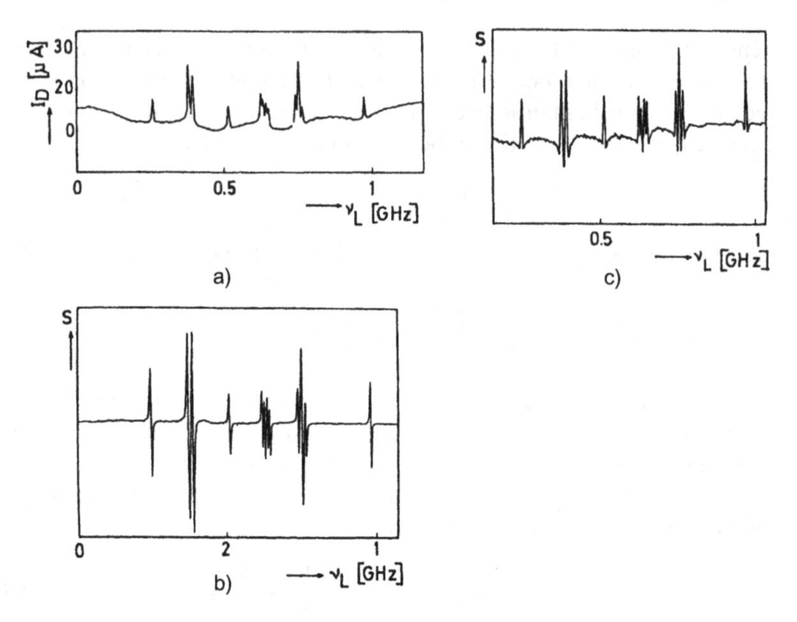

Fig. 2.44 Saturated interference spectra of I_2 at $\lambda = 600$ nm: (**a**) saturated absorption signal of the hfs components; (**b**) first-derivative spectrum of (**a**); (**c**) first derivative of saturated dispersion signal [272]

phase is fed into another lock-in that is tuned to a frequency f_2 $(f_2 \ll f_1)$ at which the saturating pump beam is chopped.

The method has been applied so far to the spectroscopy of Na_2 [271] and I_2 [272]. Figure 2.44a shows saturated absorption signals in I_2 obtained with conventional saturation spectroscopy using a dye laser at $\lambda = 600$ nm with 10 mW pump power and 1 mW probe power. Figure 2.44b displays the first derivative of the spectrum in Fig. 2.44a and Fig. 2.44c the first derivative of the saturated interference signal.

The sensitivity of the saturated interference technique is comparable to that of polarization spectroscopy. While the latter can be applied only to transitions from levels with a rotational quantum number $J \geq 1$, the former works also for $J = 0$. An experimental drawback may be the critical alignment of the Jamin interferometer and its stability during the measurements.

2.6.2 Doppler-Free Laser-Induced Dichroism and Birefringence

A slight modification of the experimental arrangement used for polarization spectroscopy allows the simultaneous observation of saturated absorption *and* dispersion [274]. While in the setup of Fig. 2.22 the probe beam had linear polarization, here a circularly polarized probe and a linearly polarized pump beam are used (Fig. 2.45). The probe beam can be composed of two components with linear polarization parallel and perpendicular to the pump beam polarization. Due to anisotropic saturation

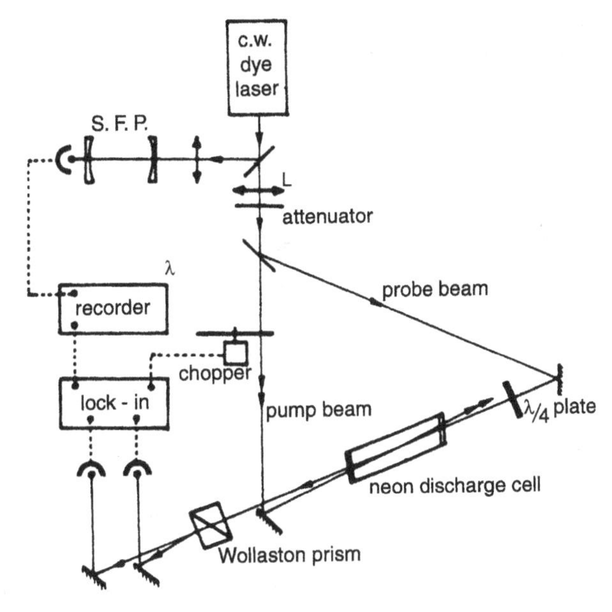

Fig. 2.45 Experimental arrangement for observation of Doppler-free laser-induced dichroism and birefringence [274]

by the pump, the absorption coefficients $\alpha_{\|}$ and $\alpha_{\perp}$ and the refractive indices $n_{\|}$ and $n_{\perp}$ experienced by the probe beam are different for the parallel and the perpendicular polarizations. This causes a change of the probe beam polarization, which is monitored behind a linear analyzer rotated through the angle β from the reference direction π. Analogous to the derivation in Sect. 2.4.2, one can show that the transmitted intensity of a circularly polarized probe wave with incident intensity I is for $\alpha L \ll 1$ and $\Delta n(L/\lambda) \ll 1$

$$I_{\mathrm{t}}(\beta) = \frac{I}{2}\left(1 - \frac{\alpha_{\|} + \alpha_{\perp}}{2} L - \frac{L}{2}\Delta\alpha\cos 2\beta - \frac{\omega L}{c}\Delta n \sin 2\beta\right), \tag{2.82}$$

with $\Delta\alpha = \alpha_{\|} - \alpha_{\perp}$ and $\Delta n = n_{\|} - n_{\perp}$.

The difference of the two transmitted intensities

$$\Delta_1 = I_{\mathrm{t}}(\beta = 0^\circ) - I_{\mathrm{t}}(\beta = 90^\circ) = I L \Delta\alpha/2, \tag{2.83}$$

gives the pure dichroism signal (anisotropic saturated absorption), while the difference

$$\Delta_2 = I_{\mathrm{t}}(45^\circ) - I_{\mathrm{t}}(-45^\circ) = I(\omega L/c)\Delta n, \tag{2.84}$$

yields the pure birefringence signal (saturated dispersion). A birefringent Wollaston prism behind the interaction region allows the spatial separation of the two probe beam components with mutually orthogonal polarizations. The two beams are monitored by two identical photodiodes. After a correct balance of the output signals, a differential amplifier records directly the desired differences Δ_1 and Δ_2 if the axes of the birefringent prism have suitable orientations.

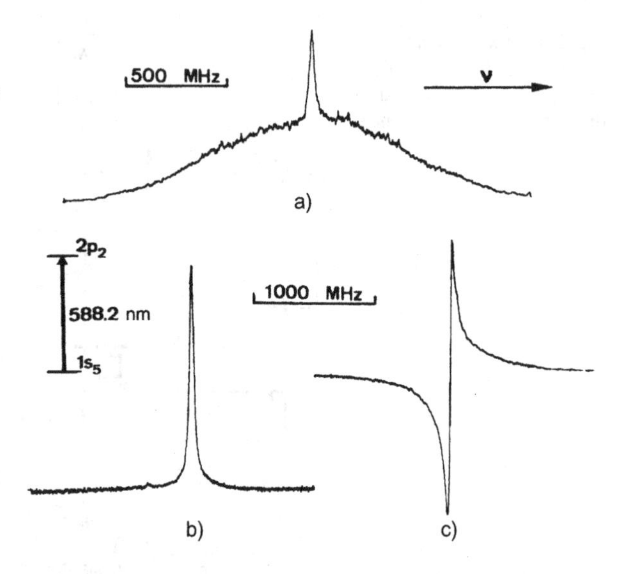

Fig. 2.46 Comparison of different techniques for measuring the neon transition $1s_2$–$2p_2$ at $\lambda = 588.2$ nm: (**a**) intracavity saturation spectroscopy (Lamb peak of the laser output $I_L(\omega)$ with Doppler-broadened background); (**b**) laser-induced dichroism; and (**c**) laser-induced birefringence [274]

Figure 2.46 illustrates the advantages of this technique. The upper spectrum represents a Lamb peak in the intracavity saturation spectrum of the neon line ($1s \to 2p$) at $\lambda = 588.2$ nm (Sect. 2.3.3). Due to the collisional redistribution of the atomic velocities, a broad and rather intense background appears in addition to the narrow peak. This broad structure is not present in the dichroism and birefringent curves (Fig. 2.46b, c). This improves the signal-to-noise ratio and the spectral resolution.

2.6.3 *Heterodyne Polarization Spectroscopy*

In most detection schemes of saturation or polarization spectroscopy the intensity fluctuations of the probe laser represent the major contribution to the noise. Generally, the noise power spectrum $P_{\text{Noise}}(f)$ shows a frequency-dependence, where the spectral power density decreases with increasing frequency (e.g., $1/f$-noise). It is therefore advantageous for a high S/N ratio to detect the signal S behind a lock-in amplifier at high frequencies f.

This is the basic idea of heterodyne polarization spectroscopy [275, 276], where the pump wave at the frequency ω_p passes through an acousto-optical modulator, driven at the modulation frequency f, which generates sidebands at $\omega = \omega_p \pm 2\pi f$ (Fig. 2.47). The sideband at $\omega_t = \omega_p + 2\pi f$ is sent as a pump beam through the sample cell, while the probe beam at frequency ω_p is split off the laser beam in front of the modulator. Otherwise the setup is similar to that of Fig. 2.22.

The signal intensity transmitted by the analyzer P_3 is described by (2.50). However, the quantity $x = 2(\omega - \omega_0 - 2\pi f)/\gamma$ now differs from (2.48) by the frequency shift f, and the amplitude of the polarization signal is modulated at the difference frequency f between pump and probe beam. The signal can therefore be detected

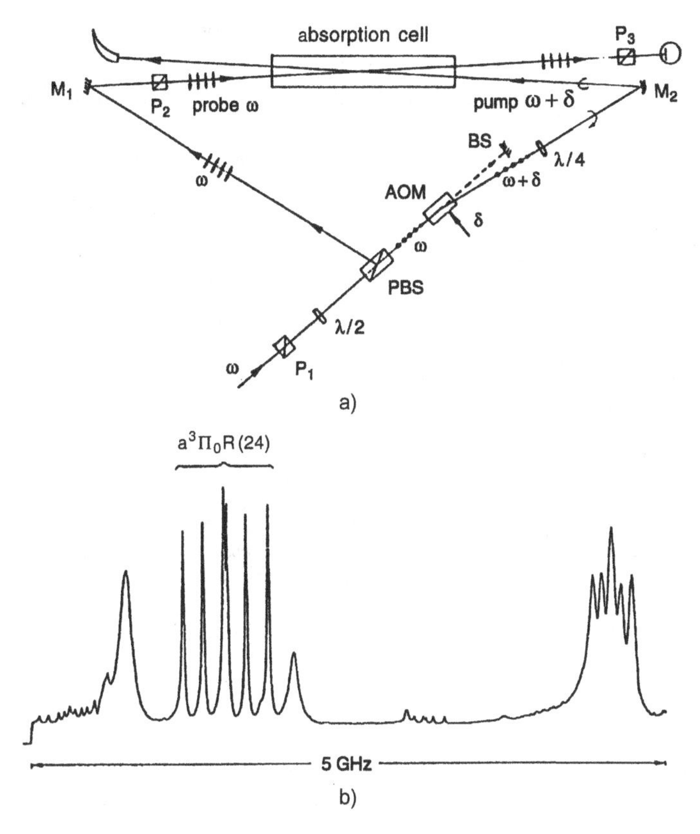

Fig. 2.47 (**a**) Experimental arrangement of heterodyne polarization spectroscopy; (**b**) section of the Na_2 polarization spectrum showing the hyperfine splitting of the R(24) rotational line of the spin-forbidden transition $X^1\Sigma_g \rightarrow {}^3\Pi_u$ [276]

through a lock-in device tuned to this frequency, which is chosen in the MHz range. No further chopping of the pump beam is necessary.

2.6.4 Combination of Different Nonlinear Techniques

A combination of Doppler-free two-photon spectroscopy and polarization spectroscopy was utilized by Grützmacher et al. [277] for the measurement of the Lyman-λ line profile in a hydrogen plasma at low pressure.

A famous example of such a combination is the first precise measurement of the Lamb shift in the $1S$ ground state of the hydrogen atom by Hänsch et al. [261]. Although new and more accurate techniques have been developed by Hänsch and coworkers (Sect. 9.7), the "old" technique is quite instructive and shall therefore be briefly discussed here.

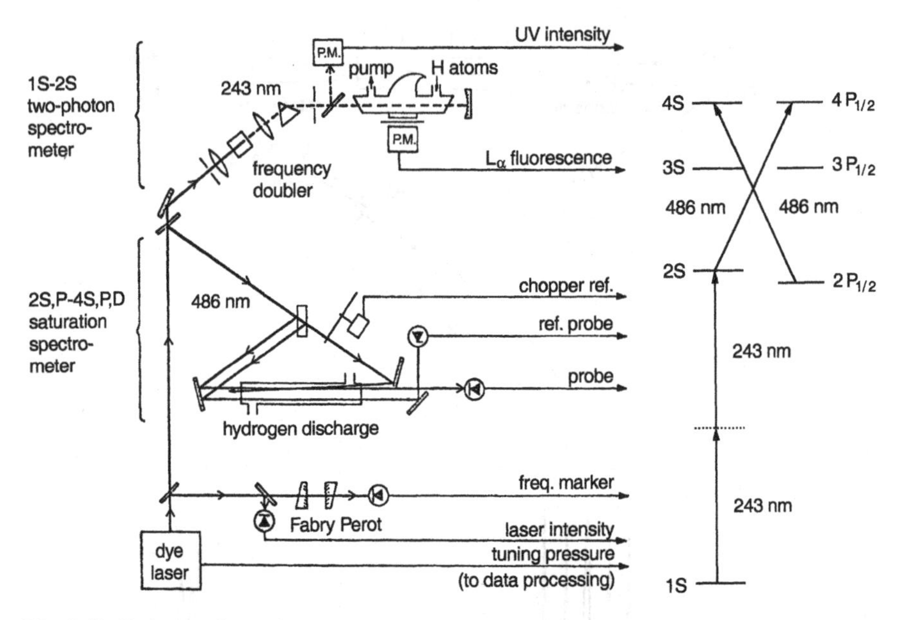

Fig. 2.48 Early experimental setup for measurements of the Lamb shift in the $1S$ state and of the fine structure in the $^2P_{1/2}$ state of the H atom by combination of Doppler-free two-photon and saturation spectroscopy [261]

The experimental arrangement is shown in Fig. 2.48. The output of a tunable dye laser at $\lambda = 486$ nm is frequency-doubled in a nonlinear crystal. While the fundamental wave at 486 nm is used for Doppler-free saturation spectroscopy [261] or polarization spectroscopy [278] of the Balmer transition $2S_{1/2}$–$4P_{1/2}$, the second harmonics of the laser at $\lambda = 243$ nm induce the Doppler-free two-photon transition $1S_{1/2} \rightarrow 2S_{1/2}$. In the simple Bohr model [279], both transitions should be induced at the same frequency since in this model $\nu(1S–2S) = 4\nu(2S–4P)$. The measured frequency difference $\Delta\nu = \nu(1S–2S) - 4\nu(2S–4P)$ yields the Lamb shift $\delta\nu_\mathrm{L}(1S) = \Delta\nu - \delta\nu_\mathrm{L}(2S) - \Delta\nu_\mathrm{fs}(4S_{1/2}–4P_{1/2}) - \delta\nu_\mathrm{L}(4S)$. The Lamb shift $\delta\nu_\mathrm{L}(2S)$ is known and $\Delta\nu_\mathrm{fs}(4S_{1/2}–4P_{1/2})$ can be calculated within the Dirac theory. The frequency markers of the FPI allow the accurate determination of the hfs splitting of the $1S$ state and the isotope shift $\Delta\nu_\mathrm{Is}(^1\mathrm{H}–^2\mathrm{H})$ between the $1S$–$2S$ transitions of hydrogen ^{1}H and deuterium ^{2}H (Fig. 2.38).

The more recent version of this precision measurement of the $1S$–$2S$ transition is shown in Fig. 2.49. The hydrogen atoms are formed in a microwave discharge and effuse through a cold nozzle into the vacuum, forming a collimated beam of color H-atoms. The laser beam is sent through the nozzle and is reflected back anticollinear to the atomic beam axis. The metastable $2S$ atoms fly through an electric field where the $2S$ state is mixed with the $2P$ state. The $2P$ atoms emit Lyman-α fluorescence, which is detected by a solar blind photomultiplier.

If the microwave discharge is pulsed the Lyman α-radiation emitted from the excited H-atoms can be detected behind a time gate as a function of the time de-

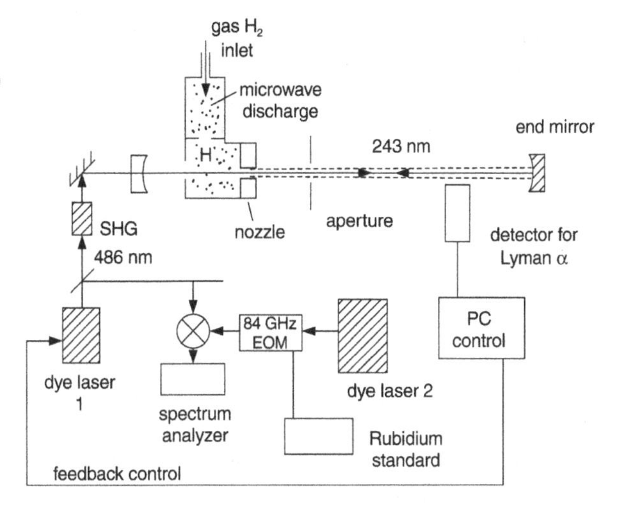

Fig. 2.49 Experimental setup for the precise determination of H($1S \to 2S$) frequency [280]

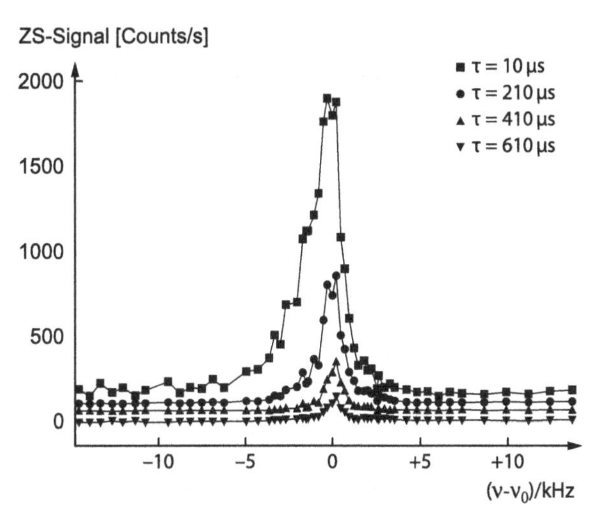

Fig. 2.50 Spectral line profiles of the $1S$–$2S$ transition measured with Doppler-free two-photon excitation in an atomic hydrogen beam for different atomic velocities

lay between excitation and detection. This allows the selection of different velocity groups of H-atoms. When the laser frequency is tuned over the $1S \to 2S$ transition the spectral profile of the two-photon transition depends on the second order Doppler-effect (because the first order effect is eliminated by the Doppler-free excitation). This results in a slight shift and asymmetry of the line profiles which become smaller and more symmetric with decreasing atomic velocity (Fig. 2.50). The reason for the asymmetry is the fact, that slow atoms have a longer interaction time with the laser wave and therefore give a larger contribution to the signal.

The achieved relative accuracy of determining the line centre is about 1×10^{-14}. This allows a very accurate determination of the $1S \to 2S$ transition from which the mean charge radius of the proton can be deduced. Of fundamental importance is the check, whether the fundamental constants change with time. These precision experiments allow a very small upper limit for a possible variation with time for the values of the fine-structure constant or the Rydberg constant. Tuning the laser frequency over both transitions of hydrogen and deuterium atoms gives an accurate value of the isotope shift of the deuterium atom. The achievable $1S$–$2S$ linewidth is about 1 kHz, limited by transit-time broadening. The uncertainty of determining the line center is below 30 Hz!

2.7 Conclusion

The few examples shown above illustrate that nonlinear spectroscopy represents an important branch of laser spectroscopy of atoms and molecules. Its advantages are the Doppler-free spectral resolution if narrow-band lasers are used and the possibilities to reach high-lying states by multiphoton absorption with pulsed or cw lasers. Because of its relevance for molecular physics, numerous books and reviews cover this field. The references [203, 205, 281–289] represent only a small selection.

In combination with double-resonance techniques, nonlinear spectroscopy has contributed particularly to the assignment of complex spectra and has therefore considerably increased our knowledge about molecular structure and dynamics. This subject is covered in Chap. 5.

2.8 Problems

2.1

(a) A collimated sodium beam is crossed by the focussed beam (focal area $A = 0.2 \times 0.01\ \text{cm}^2$) of a single-mode cw dye laser, tuned to the hyperfine component $(F' = 2 \leftarrow F'' = 1)$ of the D_2 transition $3\,^2P_{3/2} \leftarrow 3\,^2S_{1/2}$ of Na. Calculate the saturation intensity I_s if the mean velocity of sodium atoms is $\bar{v} = 5 \times 10^4\ \text{cm/s}$. The lifetime τ_K of the upper level is $\tau_K = 16$ ns and the residual Doppler width can be neglected.

(b) How large is I_s in a sodium cell at $P_{\text{Na}} = 10^{-6}$ mbar with $P_{\text{Ar}} = 10$ mbar additional argon pressure? The pressure broadening is 25 MHz/mbar for Na–Ar collisions.

2.2 A pulsed dye laser with the pulse length $\Delta T = 10^{-8}$ s and with a peak power of $P = 1$ kW at $\lambda = 600$ nm illuminates sample in a cell at $p = 1$ mbar and $T = 300$ K. A rectangular intensity profile is assumed with a laser-beam cross section of $1\ \text{cm}^2$. Which fraction of all N_i in the absorbing lower level $|i\rangle$ is excited when the laser is tuned to a weak absorbing transition $|i\rangle \rightarrow |k\rangle$ with the absorption cross section $\sigma_{ik} = 10^{-18}\ \text{cm}^2$? The laser bandwidth is assumed to be 3 times the Doppler width.

2.3 In an experiment on polarization spectroscopy the circularly polarized pump laser causes a change $\Delta\alpha = \alpha^+ - \alpha^- = 10^{-2}\alpha_0$ of the absorption coefficient. By which angle is the plane of polarization of the linearly polarized probe laser beam at $\lambda = 600\ \mu\text{m}$ tuned after passing through the pumped region with length L, if the absorption without pump laser $\alpha_0 L = 5 \times 10^{-2}$?

2.4 Estimate the fluorescence detection rate (number of detected fluorescence photons/s) on the Na transition $5s \to 3p$, obtained in the Doppler-free free-photon experiment of Fig. 2.32, when a single-mode dye laser is tuned to $\nu/2$ of the transition $3s \to 5s$ ($\nu = 1 \times 10^{15}\ \text{s}^{-1}$) in a cell with a Na density of $n = 10^{12}\ \text{cm}^{-3}$. The laser power is $P = 100$ mW, the beam is focused to the beam waist $w_0 = 10^{-2}$ cm and a length $L = 1$ cm around the focus is imaged with a collection efficiency of 5 % onto the fluorescence detector. The absorption cross section is $\sigma = a \cdot I$ with $a = 10^{-10}\ \text{W}^{-1}$ and the transition probability of the $5s \to 3p$ transition is $A_{ki} = 0.2(A_k + R_{\text{coll}})$.

2.5 The saturation spectrum of the Na D_1 transition $3\,^2S_{1/2} \to 3\,^2P_{1/2}$ shows the resolved hyperfine components. Estimate the relative magnitude of the cross-over signal between the two transitions $3\,^2S_{1/2}(F'' = 1) \to 3\,P_{1/2}$ ($F' = 1$ and $F' = 2$) sharing the same lower level, if the laser intensity is 2 times the saturation intensity I_s for the transition to $F' = 1$.

2.6 Which fraction of H atoms in the $1\,^2S_{1/2}$ ground state that can be excited by a Doppler-free two-photon transition into the $2\,^2S_{1/2}$ state in a collimated H atomic beam with $\overline{v} = 10^3$ m/s, when a laser with $I = 10^3\ \text{W/cm}^2$ and a rectangular beam cross section of $1 \times 1\ \text{mm}^2$ crosses the atomic beam perpendicularly and the absorption probability is $P_{if} = (\sigma_0 \cdot I)^2/(\gamma \cdot h\nu)^2$ where $\sigma_0 = 10^{-18}\ \text{cm}^2$ and γ is the linewidth.

Chapter 3
Laser Raman Spectroscopy

For many years Raman spectroscopy has been a powerful tool for the investigation of molecular vibrations and rotations. In the pre-laser era, however, its main drawback was a lack of sufficiently intense radiation sources. The introduction of lasers, therefore, has indeed revolutionized this classical field of spectroscopy. Lasers have not only greatly enhanced the sensitivity of spontaneous Raman spectroscopy but they have furthermore initiated new spectroscopic techniques, based on the stimulated Raman effect, such as coherent anti-Stokes Raman scattering (CARS) or hyper-Raman spectroscopy. The research activities in laser Raman spectroscopy have recently shown an impressive expansion and a vast literature on this field is available. In this chapter we summarize only briefly the basic background of the Raman effect and present some experimental techniques that have been developed for Raman spectroscopy of gaseous media. For more thorough studies of this interesting field the textbooks and reviews given in [291–302] and the conference proceedings [303, 304] are recommended. More information on Raman spectroscopy of liquids and solids can be found in [301, 305–308].

3.1 Basic Considerations

Raman scattering may be regarded as an inelastic collision of an incident photon $\hbar\omega_i$ with a molecule in the initial energy level E_i (Fig. 3.1a). Following the collision, a photon $\hbar\omega_s$ with lower energy is detected and the molecule is found in a higher-energy level E_f

$$\hbar\omega_i + M(E_i) \to M^*(E_f) + \hbar\omega_s, \quad \text{with } \hbar(\omega_i - \omega_s) = E_f - E_i > 0. \tag{3.1a}$$

The energy difference $\Delta E = E_f - E_i$ may appear as vibrational, rotational, or electronic energy of the molecule.

If the photon $\hbar\omega_i$ is scattered by a vibrationally excited molecule, it may gain energy and the scattered photon has a higher frequency ω_{as} (Fig. 3.1c), where

W. Demtröder, *Laser Spectroscopy 2*, DOI 10.1007/978-3-662-44641-6_3

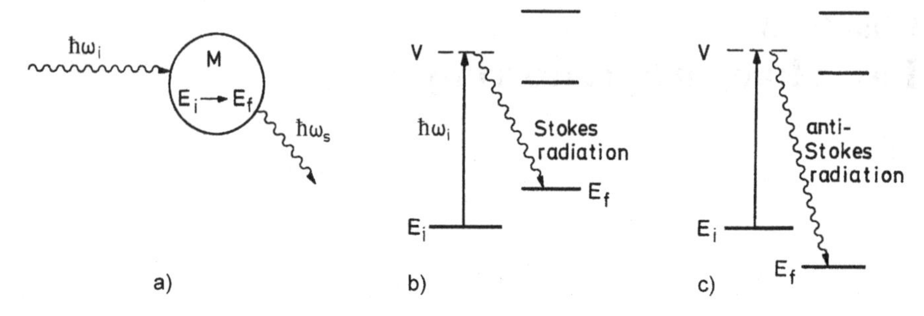

Fig. 3.1 Schematic level diagram of Raman scattering

$$\hbar\omega_{\mathrm{as}} = \hbar\omega_{\mathrm{i}} + E_{\mathrm{i}} - E_{\mathrm{f}}, \quad \text{with } E_{\mathrm{i}} > E_{\mathrm{f}}. \tag{3.1b}$$

This "superelastic" photon scattering is called *anti-Stokes radiation.*

In the energy level scheme (Fig. 3.1b), the intermediate state $E_{\mathrm{v}} = E_{\mathrm{i}} + \hbar\omega_{\mathrm{i}}$ of the system "during" the scattering process is often formally described as a *virtual* level, which, however, is not necessarily a "real" stationary eigenstate of the molecule. If the virtual level coincides with one of the molecular eigenstates, one speaks of the *resonance Raman effect.*

A classical description of the *vibrational Raman effect* (which was the main process studied before the introduction of lasers) has been developed by Placek [298]. It starts from the relation

$$\boldsymbol{p} = \boldsymbol{\mu}_0 + \tilde{\alpha}\boldsymbol{E}, \tag{3.2}$$

between the electric field amplitude $\boldsymbol{E} = \boldsymbol{E}_0 \cos\omega t$ of the incident wave and the dipole moment $\boldsymbol{p}$ of a molecule. The first term $\boldsymbol{\mu}_0$ represents a possible *permanent* dipole moment while $\tilde{\alpha}\boldsymbol{E}$ is the *induced* dipole moment. The polarizability is generally expressed by the tensor (α_{ij}) of rank two, which depends on the molecular symmetry. Dipole moment and polarizability are functions of the coordinates of the nuclei and electrons. However, as long as the frequency of the incident radiation is far off resonance with electronic or vibrational transitions, the nuclear displacements induced by the polarization of the electron cloud are sufficiently small. Since the electronic charge distribution is determined by the nuclear positions and adjusts "instantaneously" to changes in these positions, we can expand the dipole moment and polarizability into Taylor series in the normal coordinates q_n of the nuclear displacements

$$\begin{aligned}\boldsymbol{\mu} &= \boldsymbol{\mu}(0) + \sum_{n=1}^{Q}\left(\frac{\partial\boldsymbol{\mu}}{\partial q_n}\right)_0 q_n + \cdots,\\ \alpha_{ij}(q) &= \alpha_{ij}(0) + \sum_{n=1}^{Q}\left(\frac{\partial\alpha_{ij}}{\partial q_n}\right)_0 q_n + \cdots,\end{aligned} \tag{3.3}$$

where $Q = 3N - 6$ (or $3N - 5$ for linear molecules) gives the number of normal vibrational modes for a molecule with N nuclei, and $\mu(0) = \mu_0$ and $\alpha_{ij}(0)$ are the

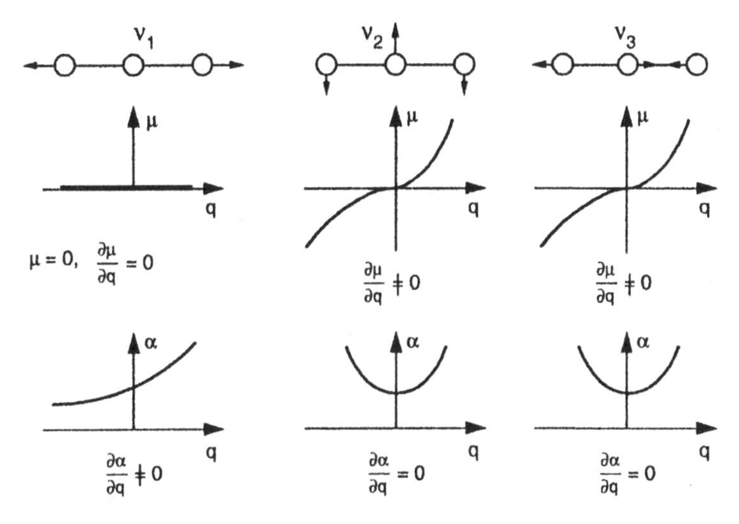

Fig. 3.2 Dependence $\partial\mu/\partial q$ of dipole moment and $\partial\alpha/\partial q$ of polarizability on the normal vibrations of the CO_2 molecule

dipole moment and the polarizability at the equilibrium configuration $q_n = 0$. For small vibrational amplitudes the normal coordinates $q_n(t)$ of the vibrating molecule can be approximated by

$$q_n(t) = q_{n0}\cos(\omega_n t), \tag{3.4}$$

where q_{n0} gives the amplitude, and ω_n the vibrational frequency of the nth normal vibration. Inserting (3.4 and 3.3) into (3.2) yields the total dipole moment

$$\begin{aligned} \boldsymbol{p} = \boldsymbol{\mu}_0 + \sum_{n=1}^{Q}\left(\frac{\partial\boldsymbol{\mu}}{\partial q_n}\right)_0 q_{n0}\cos(\omega_n t) + \alpha_{ij}(0)\boldsymbol{E}_0\cos(\omega t) \\ + \frac{1}{2}\boldsymbol{E}_0\sum_{n=1}^{Q}\left(\frac{\partial\alpha_{ij}}{\partial q_n}\right)_0 q_{n0}\big[\cos(\omega+\omega_n)t + \cos(\omega-\omega_n)t\big]. \end{aligned} \tag{3.5}$$

The second term describes the infrared spectrum, the third term the Rayleigh scattering, and the last term represents the Raman scattering. In Fig. 3.2 the dependence of $\partial\boldsymbol{\mu}/\partial q$ and $\partial\alpha/\partial q$ is shown for the three normal vibrations of the CO_2 molecule. This illustrates that $\partial\boldsymbol{\mu}/\partial q \neq 0$ for the bending vibration ν_2 and for the asymmetric stretch ν_3. These two vibrational modes are called "infrared active." The polarizability change is $\partial\alpha/\partial q \neq 0$ only for the symmetric stretch ν_1, which is therefore called "Raman active."

Since an oscillating dipole moment is a source of new waves generated at each molecule, (3.5) shows that an elastically scattered wave at the frequency ω of the incident wave is produced (Rayleigh scattering) as are inelastically scattered components with the frequencies $\omega - \omega_n$ (*Stokes waves*) and superelastically scattered waves with the frequencies $\omega + \omega_n$ (*anti-Stokes components*). The microscopic contributions from each molecule add up to macroscopic waves with intensities that depend on the population $N(E_i)$ of the molecules in the initial level E_i, on the intensity of the incident radiation, and on the expression $(\partial\alpha_{ij}/\partial q_n)q_n$, which describes the dependence of the polarizability components on the nuclear displacements.

Although the classical theory correctly describes the frequencies $\omega \pm \omega_n$ of the Raman lines, it fails to give the correct intensities and a quantum mechanical treatment is demanded. The expectation value of the component α_{ij} of the polarizability tensor is given by

$$\langle\alpha_{ij}\rangle_{ab} = \int u_b^*(q)\alpha_{ij}u_a(q)\,\mathrm{d}q, \tag{3.6}$$

where the functions $u(q)$ represent the molecular eigenfunctions in the initial level a and the final level b. The integration extends over all nuclear coordinates. This shows that a computation of the intensities of Raman lines is based on the knowledge of the molecular wave functions of the initial and final states. In the case of vibrational–rotational Raman scattering these are the rotational–vibrational eigenfunctions of the electronic ground state.

For small displacements q_n, the molecular potential can be approximated by a harmonic potential, where the coupling between the different normal vibrational modes can be neglected. The functions $u(q)$ are then separable into a product

$$u(q) = \prod_{n=1}^{Q} w_n(q_n, v_n), \tag{3.7}$$

of vibrational eigenfunction of the nth normal mode with v_n vibrational quanta. Using the orthogonality relation

$$\int w_n w_m\,\mathrm{d}q = \delta_{nm}, \tag{3.8}$$

of the functions $w_n(q_n)$, one obtains from (3.6 and 3.3)

$$\langle\alpha_{ij}\rangle_{ab} = (\alpha_{ij})_0 + \sum_{n=1}^{Q}\left(\frac{\partial\alpha_{ij}}{\partial q_n}\right)_0 \int w_n(q_n, v_a)q_n w_n(q_n, v_b)\,\mathrm{d}q_n. \tag{3.9}$$

The first term is a constant and is responsible for the Rayleigh scattering. For nondegenerate vibrations the integrals in the second term vanish unless $v_a = v_b \pm 1$. In these cases it has the value $[\frac{1}{2}(v_a + 1)]^{1/2}$ [308]. The basic intensity parameter of vibrational Raman spectroscopy is the derivative $(\partial\alpha_{ij}/\partial q)$, which can be determined from the Raman spectra.

The intensity of a Raman line at the Stokes or anti-Stokes frequency $\omega_s = \omega \pm \omega_n$ is determined by the population density $N_i(E_i)$ in the initial level $E_i(v, J)$, by the intensity I_L of the incident pump laser, and by the Raman scattering cross section $\sigma_R(i \to f)$ for the Raman transition $E_i \to E_f$:

$$I_s = N_i(E_i)\sigma_R(i \to f)I_L. \tag{3.10}$$

At thermal equilibrium the population density $N_i(E_\mathrm{i})$ follows the Boltzmann distribution

$$N_i(E_\mathrm{i}, v, J) = \frac{N}{Z} g_i \mathrm{e}^{-E_\mathrm{i}/kT}, \quad \text{with } N = \sum N_i. \tag{3.11a}$$

The statistical weight factors g_i depend on the vibrational state $v = (n_1 v_1, n_2 v_2, \ldots)$, the rotational state with the rotational quantum number J, the projection K onto the symmetry axis in the case of a symmetric top, and furthermore on the nuclear spins I of the N nuclei. The partition function

$$Z = \sum_i g_i \mathrm{e}^{-E_\mathrm{i}/kT}, \tag{3.11b}$$

is a normalization factor, which makes $\sum N_i(v, J) = N$, as can be verified by inserting (3.11b) into (3.11a).

In case of Stokes radiation the initial state of the molecules may be the vibrational ground state, while for the emission of anti-Stokes lines the molecules must have initial excitation energy. Because of the lower population density in these excited levels, the intensity of the anti-Stokes lines is lower by a factor $\exp(-\hbar\omega_v/kT)$.

Example 3.1 $\hbar\omega_v = 1000\ \mathrm{cm}^{-1}$, $T = 300\ \mathrm{K} \to kT \sim 250\ \mathrm{cm}^{-1} \to \exp(-E_\mathrm{i}/kT) \approx \mathrm{e}^{-4} \approx 0.018$. With comparable cross sections σ_R the intensity of anti-Stokes lines is therefore lower by two orders of magnitude compared with those of Stokes lines.

The scattering cross section depends on the matrix element (3.9) of the polarizability tensor and furthermore contains the ω^4 frequency dependence derived from the classical theory of light scattering. One obtains [309] analogously to the two-photon cross section (Sect. 2.5)

$$\sigma_\mathrm{R}(i \to f) = \frac{8\pi\omega_\mathrm{s}^4}{9\hbar c^4} \left| \sum_j \frac{\langle\alpha_{ij}\rangle \hat{\boldsymbol{e}}_\mathrm{L} \langle\alpha_{jf}\rangle \hat{\boldsymbol{e}}_\mathrm{s}}{\omega_{ij} - \omega_\mathrm{L} - \mathrm{i}\gamma_j} + \frac{\langle\alpha_{ji}\rangle \hat{\boldsymbol{e}}_\mathrm{L} \langle\alpha_{jf}\rangle \hat{\boldsymbol{e}}_\mathrm{s}}{\omega_{jf} - \omega_\mathrm{L} - \mathrm{i}\gamma_j} \right|^2, \tag{3.12}$$

where $\hat{\boldsymbol{e}}_\mathrm{L}$ and $\hat{\boldsymbol{e}}_\mathrm{s}$ are unit vectors representing the polarization of the incident laser beam and the scattered light. The sum extends over all molecular levels j with homogeneous width γ_j accessible by single-photon transitions from the initial state i. We see from (3.12) that the initial and final states are connected by *two-photon* transitions, which implies that both states have the same parity. For example, the vibrational transitions in homonuclear diatomic molecules, which are forbidden for single-photon infrared transitions, are accessible to Raman transitions.

The matrix elements $\langle\alpha_{ij}\rangle$ depend on the symmetry characteristics of the molecular states. While the theoretical evaluation of the magnitude of $\langle\alpha_{ij}\rangle$ demands a knowledge of the corresponding wave functions, the question whether $\langle\alpha_{ij}\rangle$ is zero or not depends on the symmetry properties of the molecular wave functions

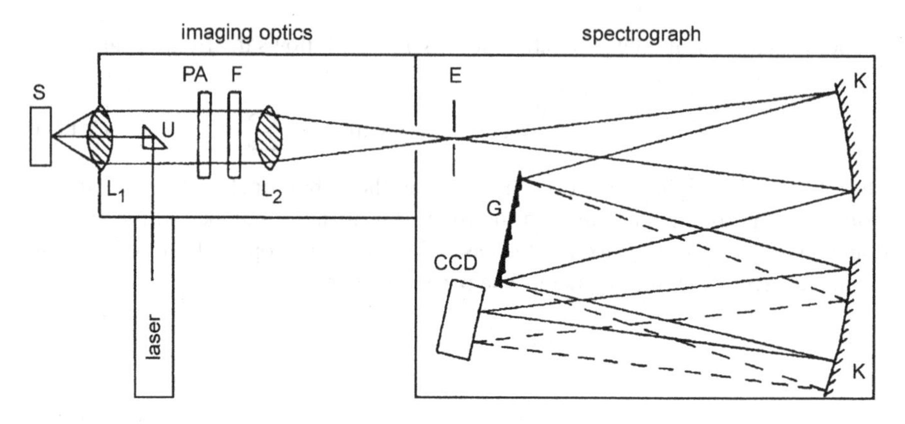

Fig. 3.3 Schematic arrangement for Raman spectroscopy

for the states $|i\rangle$ and $|f\rangle$ and can therefore be answered by group theory without explicitly calculating the matrix elements (3.9).

According to (3.12), the Raman scattering cross section increases considerably if the laser frequency ω_L matches a transition frequency ω_{ij} of the molecule (resonance Raman effect) [310, 311]. With tunable dye lasers and optical frequency doubling this resonance condition can often be realized. The enhanced sensitivity of resonant Raman scattering can be utilized for measurements of micro-samples or of very small concentrations of molecules in solutions, where the absorption of the pump wave is small in spite of resonance with a molecular transition.

If the frequency difference $\omega_L - \omega_s$ corresponds to an electronic transition of the molecule, we speak of electronic Raman scattering [312, 313], which gives complementary information to electronic-absorption spectroscopy. This is because the initial and final states must have the same parity, and therefore a direct dipole-allowed electronic transition $|i\rangle \rightarrow |f\rangle$ is not possible.

In paramagnetic molecules Raman transitions between different fine-structure components (spin-flip Raman transitions) can occur [299]. If the molecules are placed in a longitudinal magnetic field parallel to the laser beam, the Raman light is circularly polarized and is measured as σ^+ light for $\Delta M = +1$ and as σ^- for $\Delta M = -1$ transitions.

The spontaneous Raman spectrum can be measured with a grating spectrograph (Fig. 3.3). The laser is reflected by a prism and focussed onto the sample S. The scattered Raman light is collected by the lens L_1 and focussed by L_2 onto the entrance slit of the spectrograph. Since the intensity of the Raman lines is generally very small, precautions have to be taken to supress the scattered laser light, which might be stronger than the Raman light by several orders of magnitude. Often a double spectrograph is used or a combination of spectral filters and spectrograph or interferometers and spectrograph.

The spatial resolution can be greatly enhanced by confocal microscopy. Here the exciting laser beam is focused by a short focal lens onto the sample and the Raman light is collected by the same lens and imaged onto a small aperture, for instance an

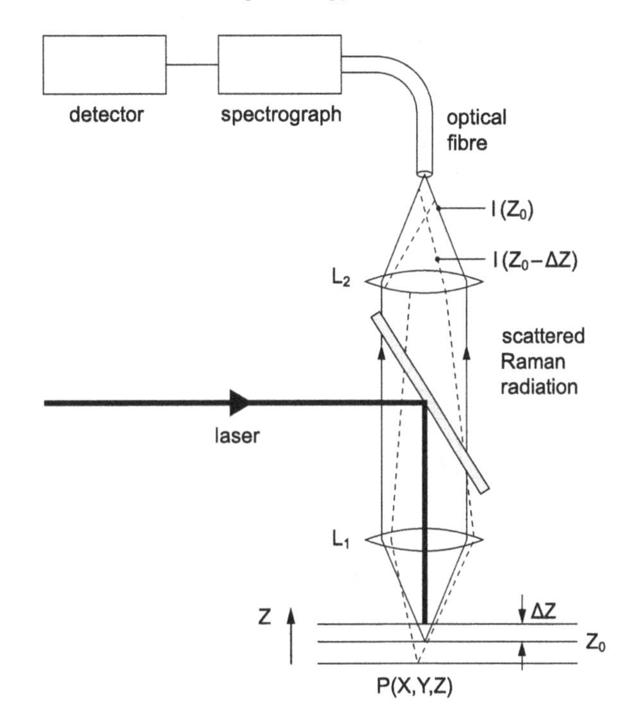

Fig. 3.4 Spatially resolved Raman-spectroscopy with confocal microscopy

optical fibre with small diameter (Fig. 3.4). Light from other points than the focal point are not transmitted by the aperture. A lateral resolution of a few microns can be obtained. The axial resolution depends on the focal length of lens L_1. The light from layers above or below the focal plane has a larger diameter and therefore only a small fraction passes the aperture.

3.2 Experimental Techniques of Linear Laser Raman Spectroscopy

The scattering cross sections in spontaneous Raman spectroscopy are very small, typically on the order of 10^{-30} cm^2. The experimental problems of detecting weak signals in the presence of intense background radiation are by no means trivial. The achievable signal-to-noise ratio depends both on the pump intensity and on the sensitivity of the detector. Recent years have brought remarkable progress on the source as well as on the detector side [314]. The incident light intensity can be greatly enhanced by using multiple reflection cells, intracavity techniques (Sect. 1.2.3), or a combination of both. Figure 3.5 depicts as an example of such advanced equipment a Raman spectrometer with a multiple-reflection Raman cell inside the resonator of an argon laser. The laser can be tuned by the Brewster prism with reflecting backside (LP + M) to the different laser lines [315]. A sophisticated system of mirrors CM collects the scattered light, which is further imaged by the lens L_1 onto the entrance

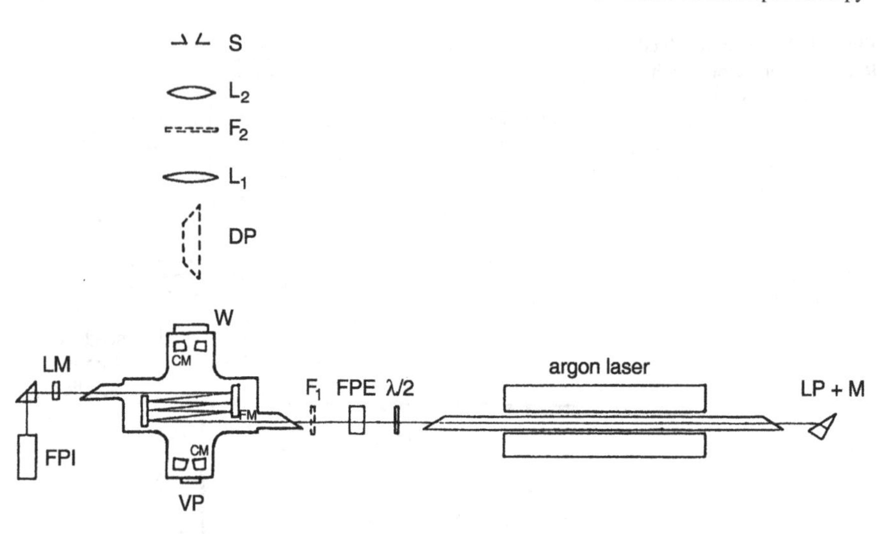

Fig. 3.5 Experimental arrangement for intracavity Raman spectroscopy with an argon laser: CM, multiple reflection four-mirror system for efficient collection of scattered light; LM, laser-resonator mirror; DP, Dove prism, which turns the image of the horizontal interaction plane by 90° in order to match it to the vertical entrance slit S of the spectrograph; FPE, Fabry–Perot etalon to enforce single-mode operation of the argon laser; LP, Littrow prism for line selection [315]

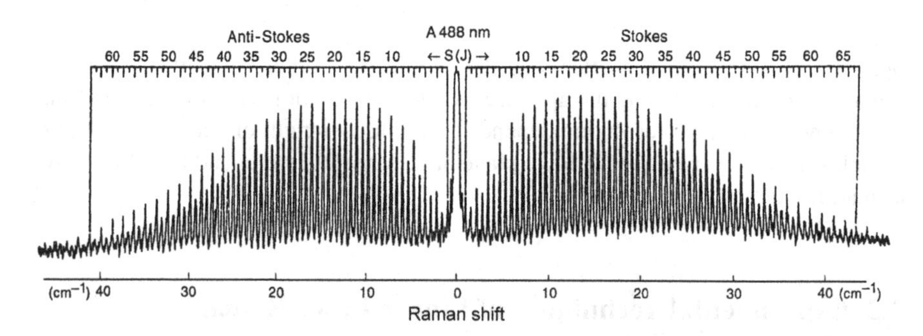

Fig. 3.6 Rotational Raman spectrum of C_2N_2 excited with the 488 nm line of the argon laser in the experimental setup of Fig. 3.5 and recorded on a photographic plate with 10 min exposure time [315]

slit S of the spectrometer. A Dove prism DP [316] turns the image of the line source by 90° to make it parallel to the entrance slit. Figure 3.6, which shows the pure rotational Raman spectrum of C_2N_2, illustrates the sensitivity that can be obtained with this setup [315].

In earlier days of Raman spectroscopy the photographic plate was the only detector used to record the Raman spectra. The introduction of sensitive photomultipliers and, in particular, the development of image intensifiers and optical multichannel analyzers with cooled photocathodes (Vol. 1, Sect. 4.5) have greatly enhanced the detection sensitivity. Image intensifiers and instrumentation such as optical multi-

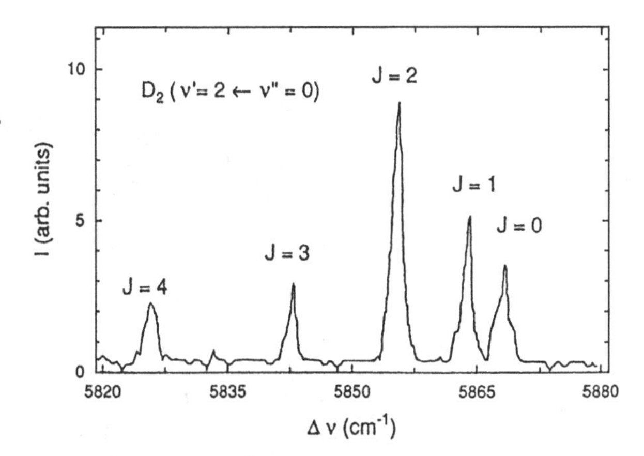

Fig. 3.7 Rotationally resolved Q-branch in the ($v' = 2 \leftarrow v'' = 0$) overtone spectrum of the D_2 molecule, measured with the sample inside the resonator of a 250 W argon laser at $\lambda = 488$ nm [318]

channel analyzers (OMAs) or CCD arrays (Vol. 1, Sect. 4.5.3) allow simultaneous recording of extended spectral ranges with sensitivities comparable to those of photomultipliers [316].

The third experimental component that has contributed to the further improvement of the quality of Raman spectra is the introduction of digital computers to control the experimental procedure, to calibrate the Raman spectra, and to analyze the data. This has greatly reduced the time spent for preparing the data for the interpretation of the results [317].

Because of the increased sensitivity of an intracavity arrangement, even weak vibrational overtone bands with $\Delta v > 1$ can be recorded with rotational resolution. For illustration, Fig. 3.7 shows the rotationally resolved Q-branch of the D_2 molecule for the transitions ($v' = 2 \leftarrow v'' = 0$) [318]. The photon counting rate for the overtone transitions was about 5000 times smaller than those for the fundamental ($v' = 1 \leftarrow v'' = 0$) band. This overtone Raman spectroscopy can also be applied to large molecules, as has been demonstrated for the overtone spectrum of the torsional vibration of CH_3CD_3 and C_2H_6, where the torsional splittings could be measured up to the 5th torsional level [319].

Just as in absorption spectroscopy, the sensitivity may be enhanced by difference laser Raman spectroscopy, where the pump laser passes alternately through a cell containing the sample molecules dissolved in some liquid and through a cell containing only the liquid. The basic advantages of this difference technique are the cancellation of unwanted Raman bands of the solvent in the spectrum of the solution and the accurate determination of small frequency shifts due to interactions with the solvent molecules.

In the case of strongly absorbing Raman samples, the heat production at the focus of the incident laser may become so large that the molecules under investigation may thermally decompose. A solution to this problem is the rotating sample technique [320], where the sample is rotated with an angular velocity Ω. If the interaction point with the laser beam is R centimeters away from the axis, the time T spent by the molecules within the focal region with diameter d [cm] is $T = d/(R\Omega)$. This technique which allows much higher input powers and therefore better signal-to-

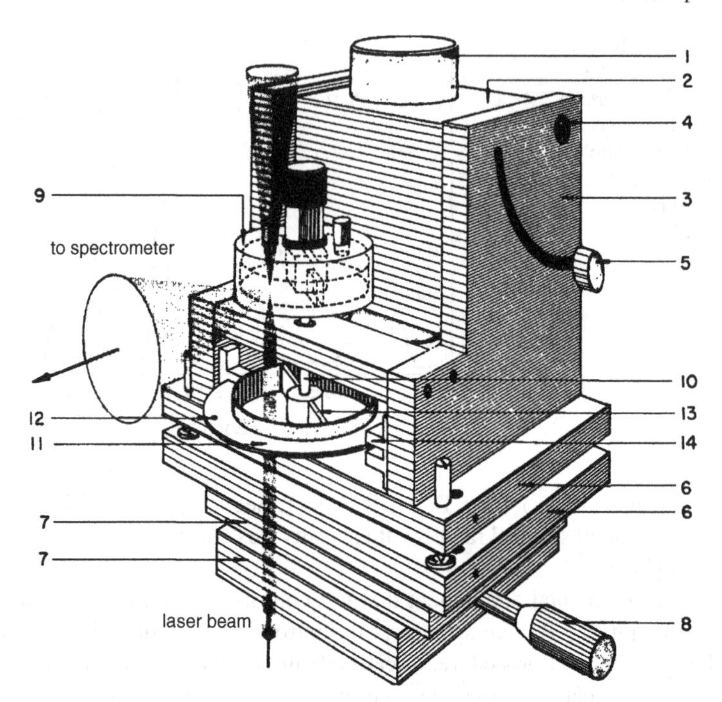

Fig. 3.8 Rotating sample cell used for difference Raman spectroscopy: *1*, motor; *2*, motor block; *3*, side parts; *4*, motor axis; *5*, set screw; *6*, kinematic mount; *7*, x–y precision ball glider; *8*, adjustment screw; *9*, divided liquid cell for difference Raman spectroscopy; *10*, axis for trigger wheel; *11*, trigger wheel; *12*, trigger hole; *13*, bar; *14*, optoelectronic array consisting of a photodiode and transistor [320]

noise ratios, can be combined with the difference technique by mounting a cylindrical cell onto the rotation axis. One half of the cell is filled with the liquid Raman sample in solution, the other half is only filled with the solvent (Fig. 3.8).

A larger increase of sensitivity in linear Raman spectroscopy of liquids has been achieved with optical-fiber Raman spectroscopy. This technique uses a capillary optical fiber with the refractive index n_f, filled with a liquid with refractive index $n_e > n_f$. If the incident laser beam is focused into the fiber, the laser light as well as the Raman light is trapped in the core due to internal reflection and therefore travels inside the capillary. With sufficiently long capillaries (1–30 m) and low losses, very high spontaneous Raman intensities can be achieved, which may exceed those of conventional techniques by factors of 10^3 [321]. Figure 3.9 shows schematically the experimental arrangement where the fiber is wound on a drum. Because of the increased sensitivity, this fiber technique also allows one to record second- and third-order Raman bands, which facilitates complete assignments of vibrational spectra [322].

The sensitivity of Raman spectroscopy in the gas phase can be greatly enhanced by combination with one of the detection techniques discussed in Chap. 1. For example, the vibrationally excited molecules produced by Raman–Stokes scattering can

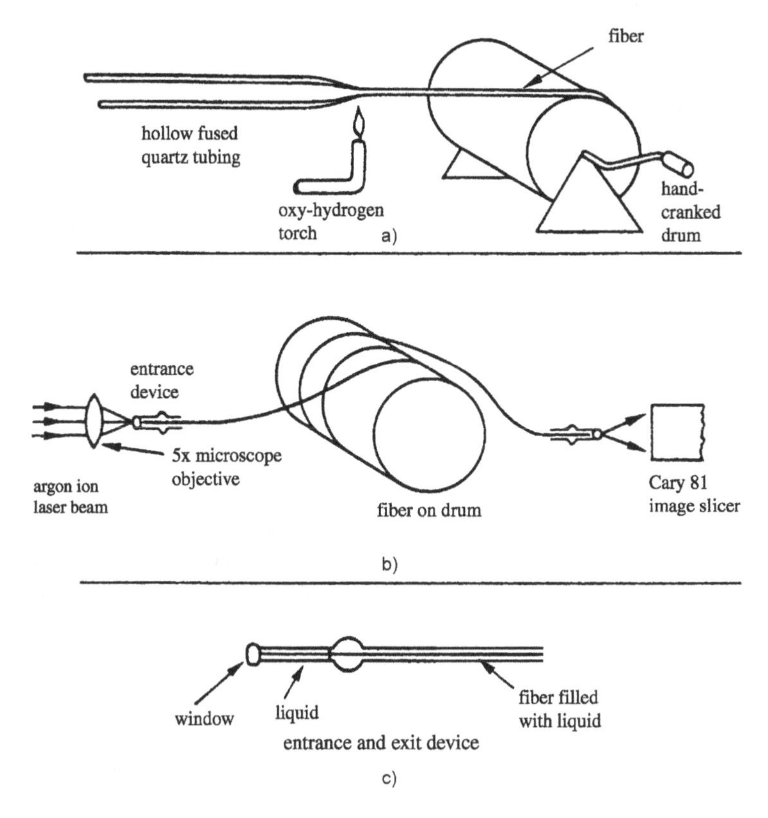

Fig. 3.9 Raman spectroscopy of liquid samples in a thin capillary fiber: (**a**) production of the fiber; (**b**) incoupling of an argon laser beam with a microscope objective into the fiber and imaging of the outcoupled radiation into a spectrometer; (**c**) fiber with liquid [320]

be selectively detected by resonant two-photon ionization with two visible lasers or by UV ionization with a laser frequency ω_{UV}, which can ionize molecules in level E_{f} but not in E_{i} (Fig. 3.10).

The combination of Raman spectroscopy with Fourier-transform spectroscopy [323] allows the simultaneous detection of larger spectral ranges in the Raman spectra.

The information obtained from linear Raman spectroscopy is derived from the following experimental data:

- The linewidth of the scattered radiation, which represents for gaseous samples a convolution of the Doppler width, collisional broadening, spectral profile of the exciting laser, and natural linewidth, depending on the lifetimes of molecular levels involved in the Raman transition.
- The degree of polarization ρ of the scattered light, defined as

$$\rho = \frac{I_{\parallel} - I_{\perp}}{I_{\parallel} + I_{\perp}}, \tag{3.13}$$

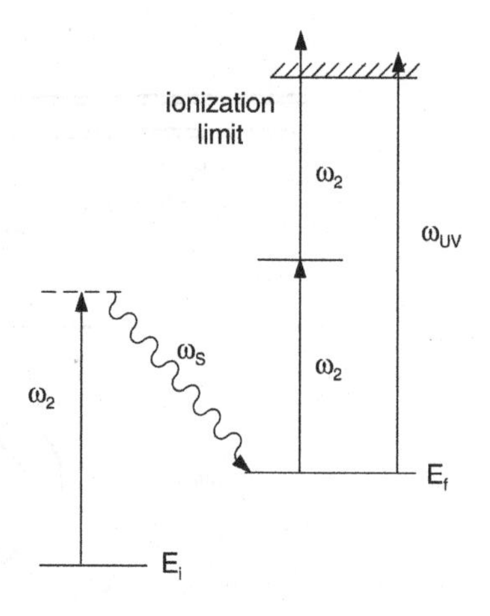

Fig. 3.10 Detection of Raman–Stokes scattering by photoionization of the excited level E_f either by one UV photon ($IP(E_\text{f}) < \hbar\omega_\text{UV} < IP(E_\text{i})$) or by resonant two-photon ionization

where $I_\parallel$ and $I_\perp$ are the intensities of the scattered light with a polarization parallel and perpendicular, respectively, to that of a linearly polarized excitation laser. A more detailed calculation shows that for statistically oriented molecules the degree of polarization

$$\rho = \frac{3\beta^2}{45\overline{\alpha}^2 + 4\beta^2}, \tag{3.14}$$

depends on the mean value $\overline{\alpha} = (\alpha_{xx} + \alpha_{yy} + \alpha_{zz})/3$ of the diagonal components of the polarizability tensor $\tilde{\alpha}$ and on the anisotropy

$$\begin{aligned}\beta^2 = \frac{1}{2}\big[&(\alpha_{xx} - \alpha_{yy})^2 + (\alpha_{yy} - \alpha_{zz})^2 + (\alpha_{zz} - \alpha_{xx})^2 \\ &+ 6\big(\alpha_{xy}^2 + \alpha_{xz}^2 + \alpha_{zx}^2\big)\big].\end{aligned} \tag{3.15}$$

Measurements of ρ and β therefore allow the determination of the polarizability tensor [324].

It turns out that

$$\begin{aligned}\overline{\alpha_{xx}^2} = \overline{\alpha_{yy}^2} = \overline{\alpha_{zz}^2} &= \frac{1}{45}\big(45\overline{\alpha^2} + 4\beta^2\big), \\ \overline{\alpha_{xy}^2} = \overline{\alpha_{xz}^2} = \overline{\alpha_{yz}^2} &= \frac{1}{15}\beta^2.\end{aligned} \tag{3.16}$$

With the experimental arrangement of Fig. 3.11, where the exciting laser is polarized in the x-direction and the Raman light is observed in the y-direction without polarizer ($\mu_x + \mu_z$), the measured intensity becomes

$$I_{x,(x+z)} = \frac{\omega^4 \cdot I_0}{16\pi^2\varepsilon_0^2 c^4}\big(\alpha_{xx}^2 + \alpha_{zx}^2\big). \tag{3.17}$$

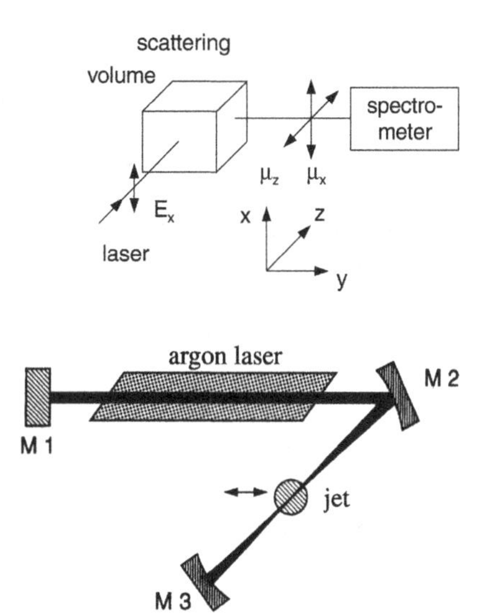

Fig. 3.11 Possible scattering geometry for measurements of the components α_{xx} and α_{zx} of the polarizability tensor

Fig. 3.12 Intracavity Raman spectroscopy of molecules in a cold jet with spatial resolution

- The intensity of the Raman lines is proportional to the product of the Raman scattering cross section σ_R, which depends according to (3.12) on the matrix elements $\langle\alpha_{ij}\rangle$ of the polarizability tensor and the density N_i of molecules in the initial state. If the cross sections σ_R have been determined elsewhere, the intensity of the Raman lines can be used for measurements of the population densities $N(v, J)$. Assuming a Boltzmann distribution (3.11a), the temperature T of the sample can be derived from measured values of $N(v, J)$. This is frequently used for the determination of unknown temperature profiles in flames [325] or of unknown density profiles in liquid or gaseous flows [326] at a known temperature (Sect. 3.5).

  One example is intracavity Raman spectroscopy of molecules in a supersonic jet, demonstrated by van Helvoort et al. [327]. If the intracavity beam waist of an argon-ion laser is shifted to different locations of the molecular jet (Fig. 3.12), the vibrational and rotational temperatures of the molecules (Sect. 4.2) and their local variations can be derived from the Raman spectra.

More details of recent techniques in linear laser Raman spectroscopy can be found in [301, 328].

3.3 Nonlinear Raman Spectroscopy

When the intensity of the incident light wave becomes sufficiently large, the induced oscillation of the electron cloud surpasses the linear range assumed in Sect. 3.1. This implies that the induced dipole moments p of the molecules are no longer proportional to the electric field $\boldsymbol{E}$ and we have to generalize (3.2). The function $\boldsymbol{p}(\boldsymbol{E})$ can be expanded into a power series of E^n $(n = 0, 1, 2, \dots)$, which is generally written as

$$p(E) = \mu + \tilde{\alpha} E + \tilde{\beta} E \cdot E + \tilde{\gamma} E \cdot E \cdot E, \tag{3.18a}$$

where $\tilde{\alpha}$ is the polarizability tensor, $\tilde{\beta}$ is named *hyper-polarizability*, and $\tilde{\gamma}$ is called the *second hyper-polarizability*. The quantities α, β, and γ are tensors of rank two, three, and four, respectively.

In component notation ($i = x, y, z$) (3.18a) can be written as

$$p_i(E) = \mu_i + \sum_k \alpha_{ik} E_k + \sum_k \sum_j \beta_{ikj} E_k E_j + \sum_k \sum_j \sum_l \gamma_{ikjl} E_k E_j E_l. \tag{3.18b}$$

This gives for the polarization $P = Np$ of a medium with N oriented dipoles

$$P_i(E) = \epsilon_0 \left(\chi_i + \sum_k \chi_{ik} E_k + \sum_{k,j} \chi_{ikj} E_k E_j + \cdots \right), \tag{3.18c}$$

which corresponds to Vol. 1, (6.3) discussed in the section on nonlinear optics and frequency conversion, if we define the susceptibilities $\chi_i = N\mu_i/\epsilon_0$, $\chi_{ik} = N\alpha_{ik}/\epsilon_0$, and so on.

For sufficiently small electric field amplitudes E the nonlinear terms in (3.18a) can be neglected, and we then obtain (3.2) for the linear Raman spectroscopy.

3.3.1 Stimulated Raman Scattering

If the incident laser intensity I_L becomes very large, an appreciable fraction of the molecules in the initial state E_i is excited into the final state E_f and the intensity of the Raman-scattered light is correspondingly large. Under these conditions we have to consider the simultaneous interaction of the molecules with *two* EM waves: the laser wave at the frequency ω_L and the Stokes wave at the frequency $\omega_S = \omega_L - \omega_V$ or the anti-Stokes wave at $\omega_a = \omega_L + \omega_V$. Both waves are coupled by the molecules vibrating with the frequencies ω_V. This parametric interaction leads to an energy exchange between the pump wave and the Stokes or anti-Stokes waves. This phenomenon of *stimulated* Raman scattering, which was first observed by Woodbury et al. [329] and then explained by Woodbury and Eckhardt [330], can be described in classical terms [331, 332].

The Raman medium is taken as consisting of N harmonic oscillators per unit volume, which are independent of each other. Because of the combined action of the incident laser wave and the Stokes wave, the oscillators experience a driving force F that depends on the total field amplitude E

$$E(z,t) = E_L e^{i(\omega_L t - k_L z)} + E_S e^{i(\omega_S t - k_S z)}, \tag{3.19}$$

where we have assumed plane waves traveling in the z-direction. The potential energy W_{pot} of a molecule with induced dipole moment $\boldsymbol{p} = \alpha \boldsymbol{E}$ in an EM field with amplitude $\boldsymbol{E}$ is, according to (3.2, 3.3) with $\mu = 0$

$$W_{\text{pot}} = -\boldsymbol{p} \cdot \boldsymbol{E} = -\alpha(q)E^2. \tag{3.20}$$

The force $\boldsymbol{F} = -\operatorname{grad} W_{\text{pot}}$ acting on the molecule gives

$$\boldsymbol{F}(z,t) = +\frac{\partial}{\partial q}\{[\alpha(q)]E^2\} = \left(\frac{\partial \alpha}{\partial q}\right)_0 E^2(z,t). \tag{3.21}$$

The equation of motion for the molecular oscillator with oscillation amplitude q, mass m, and vibrational eigenfrequency ω_{v} is then

$$\frac{\partial^2 q}{\partial t^2} - \gamma \frac{\partial q}{\partial t} + \omega_{\text{v}}^2 q = \left(\frac{\partial \alpha}{\partial q}\right)_0 E^2/m, \tag{3.22}$$

where γ is the damping constant that is responsible for the linewidth $\Delta\omega = \gamma$ of spontaneous Raman scattering. Inserting the complex ansatz

$$q = \frac{1}{2}\left(q_{\text{v}} \mathrm{e}^{\mathrm{i}\omega t} + q_{\text{v}}^* \mathrm{e}^{-\mathrm{i}\omega t}\right), \tag{3.23}$$

into (3.22) we get with the field amplitude (3.19)

$$\left(\omega_{\text{v}}^2 - \omega^2 + \mathrm{i}\gamma\omega\right) q_{\text{v}} \mathrm{e}^{\mathrm{i}\omega t} = \frac{1}{2m}\left(\frac{\partial \alpha}{\partial q}\right)_0 E_{\text{L}} E_{\text{S}} \mathrm{e}^{\mathrm{i}[(\omega_{\text{L}} - \omega_{\text{S}})t - (k_{\text{L}} - k_{\text{S}})z]}. \tag{3.24}$$

Comparison of the time-dependent terms on both sides of (3.24) shows that $\omega = \omega_{\text{L}} - \omega_{\text{S}}$. The molecular vibrations are therefore driven at the difference frequency $\omega_{\text{v}} = \omega_{\text{L}} - \omega_{\text{S}}$. Solving (3.24) for q_{v} yields

$$q_{\text{v}} = \frac{(\partial\alpha/\partial q)_0 E_{\text{L}} E_{\text{S}}}{2m\left[\omega_{\text{v}}^2 - (\omega_{\text{L}} - \omega_{\text{S}})^2 + \mathrm{i}(\omega_{\text{L}} - \omega_{\text{S}})\gamma\right]} \mathrm{e}^{-\mathrm{i}(k_{\text{L}} - k_{\text{S}})z}. \tag{3.25}$$

The induced oscillating molecular dipoles $\boldsymbol{p}(\omega, z, t)$ result in a macroscopic polarization $P = Np$. According to (3.5), the polarization $P_{\text{S}} = P(\omega_{\text{S}})$ at the Stokes frequency ω_{S}, which is responsible for Raman scattering, is given by

$$P_{\text{S}} = \frac{1}{2} N \left(\frac{\partial \alpha}{\partial q}\right)_0 qE. \tag{3.26}$$

Inserting q from (3.23, 3.25) and E from (3.19) yields the *nonlinear* polarization

$$P_{\text{S}}^{\text{NL}}(\omega_{\text{S}}) = N \frac{(\partial\alpha/\partial q)_0^2 E_{\text{L}}^2 E_{\text{S}}}{4m\left[\omega_{\text{v}}^2 - (\omega_{\text{L}} - \omega_{\text{S}})^2 + \mathrm{i}\gamma(\omega_{\text{L}} - \omega_{\text{S}})\right]} \mathrm{e}^{-\mathrm{i}(\omega_{\text{s}} t - k_{\text{s}} z)}. \tag{3.27}$$

This shows that a *polarization wave* travels through the medium with an amplitude proportional to the product $E_{\text{L}}^2 \cdot E_{\text{S}}$. It has the same wave vector k_{S} as the Stokes wave and can therefore amplify this wave. The amplification can be derived from the wave equation

$$\Delta E = \mu_0 \sigma \frac{\partial}{\partial t} E + \mu_0 \epsilon \frac{\partial^2}{\partial t^2} E + \mu_0 \frac{\partial^2}{\partial t^2} \left(P_S^{NL} \right), \tag{3.28}$$

for waves in a medium with the conductivity σ, where P_S^{NL} acts as the driving term.

For the one-dimensional problem ($\partial/\partial y = \partial/\partial x = 0$) with the approximation $d^2E/dz^2 \ll k\,dE/dz$ and with (3.26), the equation for the Stokes wave becomes

$$\frac{dE_S}{dz} = -\frac{\sigma}{2}\sqrt{\mu_0/\epsilon}\,E_S + N\frac{k_S}{2\epsilon}\left(\frac{\partial \alpha}{\partial q}\right)_0 q_v E_L. \tag{3.29}$$

Substituting q_v from (3.25) gives the final result for the case $\omega_v = \omega_L - \omega_S$

$$\frac{dE_S}{dz} = \left[-\frac{\sigma}{2}\sqrt{\mu_0/\epsilon} + N\frac{(\partial\alpha/\partial q)_0^2 E_L^2}{4m\epsilon i\gamma(\omega_L - \omega_S)}\right] E_S = (-f + g)E_S. \tag{3.30}$$

Integration of (3.30) yields

$$E_S = E_S(0)e^{(g-f)z}. \tag{3.31a}$$

The Stokes wave is amplified if the gain g exceeds the losses f. The amplification factor g depends on the square of the laser amplitude E_L and on the term $(\partial\alpha/\partial q)_0^2$. Stimulated Raman scattering is therefore observed only if the incident laser intensity exceeds a threshold value that is determined by the nonlinear term $(\partial\alpha_{ij}/\partial q)_0$ in the polarization tensor of the Raman-active normal vibration and by the loss factor $f = \frac{1}{2}\sigma(\mu_0/\epsilon)^{1/2}$.

According to (3.31a–3.31b), the Stokes intensity increases exponentially with the length z of the interaction zone. If, however, the pump wave is absorbed by the medium, the pump intensity at the position z decreases to

$$I_L(z) = I_L(0)\cdot e^{-\alpha\cdot z}$$

and therefore the gain factor g in (3.30) is smaller. There is an effective length $z = L_{eff}$ for the optimum intensity I_s of the Stokes wave, which is given by

$$L_{eff} = (1/\alpha_L)\left[1 - e^{-\alpha\cdot L}\right]$$

and the amplitude of the Stokes wave becomes for a medium with length L

$$E_s = E_s(0)\cdot e^{(g\cdot L_{eff} - f\cdot L)}. \tag{3.31b}$$

While the intensity of anti-Stokes radiation is very small in spontaneous Raman scattering due to the low thermal population density in excited molecular levels (Sect. 3.1), this is not necessarily true in stimulated Raman scattering. Because of the strong incident pump wave, a large fraction of all interacting molecules is excited

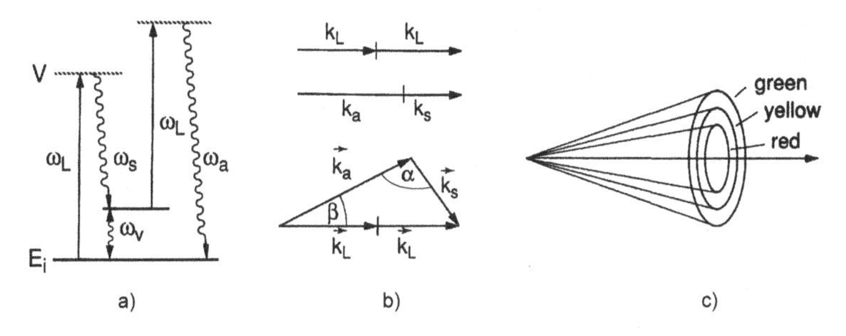

Fig. 3.13 Generation of stimulated anti-Stokes radiation: (**a**) term diagram illustrating energy conservation; (**b**) vector diagram of momentum conservation for the collinear and noncollinear case; (**c**) radiation cone for different values of $\boldsymbol{k}_S$ showing red, yellow, and green rings of anti-Stokes radiation excited by a ruby laser at 694 nm

into higher vibrational levels, and strong anti-Stokes radiation at frequencies $\omega_L + \omega_v$ has been found.

According to (3.26), the driving term in the wave equation (3.28) for an anti-Stokes wave at $\omega_a = \omega_L + \omega_V$ is given by

$$P_{\omega_a}^{NL} = \frac{1}{2} N \left(\frac{\partial \alpha}{\partial q} \right)_0 q_v E_L e^{i[(\omega_L + \omega_v)t - k_L z]} . \tag{3.32}$$

For small amplitudes $E_a \ll E_L$ of the anti-Stokes waves, we can assume that the molecular vibrations are independent of E_a and can replace q_v by its solution (3.25). This yields an equation for the amplification of E_a that is analogous to (3.29) for E_S

$$\begin{aligned} \frac{dE_a}{dz} &= -\frac{f}{2} E_a e^{i(\omega_a t - k_a z)} \\ &\quad + N_v \left[\frac{\omega_a \sqrt{\mu_0/\epsilon}}{8 m_v} \left(\frac{\partial \alpha}{\partial q} \right)_0^2 \right] E_L^2 E_S^* e^{i(2k_L - k_S - k_a)z} , \end{aligned} \tag{3.33}$$

where N_v is the density of vibrationally excited molecules. This shows that, analogously to sum- or difference-frequency generation (Vol. 1, Sect. 5.8), a macroscopic wave can build up only if the phase-matching condition

$$k_a = 2k_L - k_S , \tag{3.34}$$

can be satisfied. In a medium with normal dispersion this condition cannot be met for collinear waves. From a three-dimensional analysis, however, one obtains the vector equation

$$2\boldsymbol{k}_L = \boldsymbol{k}_S + \boldsymbol{k}_a , \tag{3.35}$$

which reveals that the anti-Stokes radiation is emitted into a cone whose axis is parallel to the beam-propagation direction (Fig. 3.13). The apex angle β of this cone is obtained by multiplying (3.35) with $\boldsymbol{k}_a$

$$2\boldsymbol{k}_L \cdot \boldsymbol{k}_a = 2k_L k_a \cos\beta = \boldsymbol{k}_S \cdot k_a \cos\alpha + k_a^2 . \tag{3.36a}$$

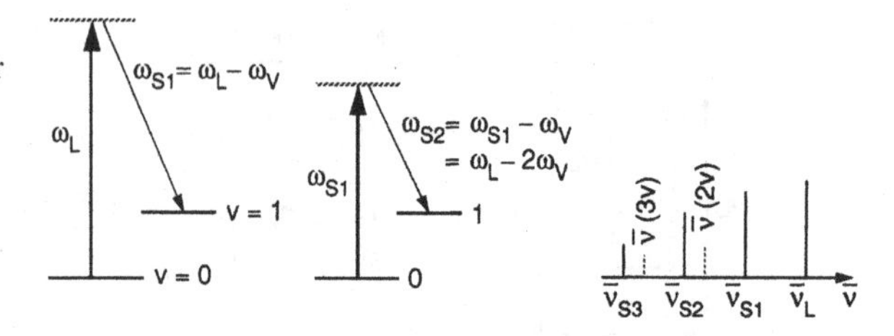

Fig. 3.14 Level diagram for the generation of higher-order Stokes sidebands, which differ from the vibrational overtone frequencies

With $k = n \cdot \omega/c$ this can be written as

$$\cos\beta = \frac{n(\omega_S)\omega_S\cos\alpha + n(\omega_a)\omega_a}{2n(\omega_L)\omega_L}. \tag{3.36b}$$

For $n(\omega_S) = n(\omega_a) = n(\omega_L)$ the collinear case can be realized ($\alpha = \beta = 0$). This is exactly what has been observed [329, 332].

Let us briefly summarized the differences between the linear (spontaneous) and the nonlinear (induced) Raman effect:

- While the intensity of spontaneous Raman lines is proportional to the incident pump intensity, but lower by several orders of magnitude compared with the pump intensity, the stimulated Stokes or anti-Stokes radiation depend in a nonlinear way on I_p but have intensities comparable to that of the pump wave.
- The stimulated Raman effect is observed only above a threshold pump intensity, which depends on the gain of the Raman medium and the length of the pump region.
- Most Raman-active substances show only one or two Stokes lines at the frequencies $\omega_S = \omega_L - \omega_V$ in stimulated emission. At higher pump intensities, however, lines at the frequencies $\omega = \omega_L - n\omega_V$ $(n = 1, 2, 3)$, which do not correspond to overtones of vibrational frequencies, have been observed besides these Stokes lines. Because of the anharmonicity of molecular vibrations, the vibrational levels in an unharmonic potential have energies $E_v = \hbar\omega_v(n + \frac{1}{2}) - x_k\hbar(n + \frac{1}{2})^2$ and therefore the spontaneous Raman lines from vibrational overtones are shifted against ω_L by $\Delta\omega = n\omega_v - (n^2 + n)x_k$, where x_k represent the anharmonicity constants. In Fig. 3.14 is illustrated that these higher-order Stokes lines are generated by consecutive Raman processes, induced by the pump wave, the Stokes wave, etc.
- The linewidths of spontaneous and stimulated Raman lines depend on the linewidth of the pump laser. For narrow linewidths, however, the width of the stimulated Raman lines becomes smaller than that of the spontaneous lines, which are Doppler-broadened by the thermal motion of the scattering molecules. A Stokes photon $\hbar\omega_s$, which is scattered into an angle ϕ against the incident laser beam by a molecule moving with the velocity $\boldsymbol{v}$, has a Doppler-shifted frequency

$$\begin{aligned}\omega_s &= \omega_L - \omega_V - (\boldsymbol{k}_L - \boldsymbol{k}_S)\cdot\boldsymbol{v}\\ &= \omega_L - \omega_v - \big[1 - (k_S/k_L)\cos\phi\big]\boldsymbol{k}_L\cdot\boldsymbol{v}.\end{aligned} \tag{3.37}$$

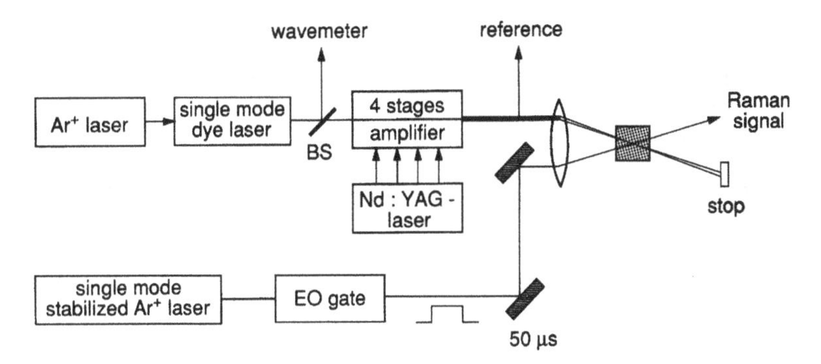

Fig. 3.15 Schematic diagram of a stimulated Raman spectrometer with pulsed, amplified cw pump laser and a single-mode cw probe laser [337]

In the case of spontaneous Raman scattering we have $0 \leq \phi \leq 2\pi$, and the spontaneous Raman lines show a Doppler width that is $(k_S/k_L) = (\omega_S/\omega_L)$ times that of fluorescence lines at ω_L. For induced Raman scattering $\boldsymbol{k}_S \parallel \boldsymbol{k}_L \to \cos\phi = 1$, and the bracket in (3.37) has the value $(1 - k_S/k_L) \ll 1$, if $\omega_V \ll \omega_L$.

- The main merit of the stimulated Raman effect for molecular spectroscopy may be seen in the much higher intensities of stimulated Raman lines. During the same measuring time one therefore achieves a much better signal-to-noise ratio than in linear Raman spectroscopy. The experimental realization of stimulated Raman spectroscopy is based on two different techniques:

  (a) Stimulated Raman gain spectroscopy (SRGS), where a strong pump laser at ω_L is used to produce sufficient gain for the Stokes radiation at ω_S according to (3.31a–3.31b). This gain can be measured with a weak tunable probe laser tuned to the Stokes wavelengths [333] (Fig. 3.15).
  (b) Inverse Raman spectroscopy (IRS), where the attenuation of the weak probe laser at ω_1 is measured when the strong pump laser at ω_2 is tuned through Stokes or anti-Stokes transitions [334].

Several high-resolution stimulated Raman spectrometers have been built [333–335] that are used for measurements of linewidths and Raman line positions in order to gain information on molecular structure and dynamics. A good compromise for obtaining a high resolution and a large signal-to-noise ratio is to use a pulsed pump laser and a single-mode cw probe laser (quasi-cw spectrometer [335]). The narrow-band pulsed laser can be realized by pulsed amplification of a single-mode cw laser (Vol. 1, Sect. 5.7). For illustration Fig. 3.15 shows a typical quasi-cw stimulated Raman spectrometer, where the wavelengths of the tunable dye laser are measured with a traveling Michelson wavemeter (Vol. 1, Sect. 4.4)

There is another important application of stimulated Raman scattering in the field of *Raman lasers*. With a tunable pump laser at the frequency ω_L, intense coherent radiation sources at frequencies $\omega_L \pm n\omega_V$ $(n = 1, 2, 3, \dots)$ can be realized that cover the UV and infrared spectral range if visible pump lasers are used (Vol. 1, Sect. 5.8).

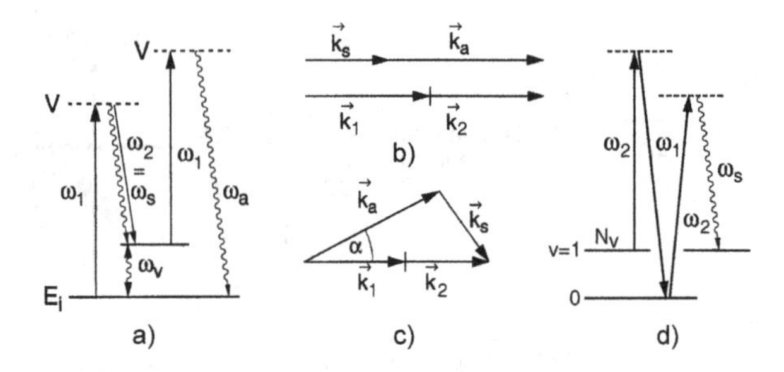

Fig. 3.16 (**a**) Level diagram of CARS; (**b**) vector diagrams for phase matching in gases with negligible dispersion and (**c**) in liquids or solids with noticeable dispersion; (**d**) generation of coherent Stokes radiation at $\omega_S = 2\omega_2 - \omega_1$

More details on the stimulated Raman effect and reference to experiments in this field can be found in [301, 336–342].

3.3.2 Coherent Anti-Stokes Raman Spectroscopy

In Sect. 3.3.1 we discussed the observation that a sufficiently strong incident pump wave at the frequency ω_L can generate an intense Stokes wave at $\omega_S = \omega_L - \omega_V$. Under the combined action of both waves, the nonlinear polarization P_{NL} of the medium is generated that contains contributions at the frequencies $\omega_V = \omega_L - \omega_S$, $\omega_S = \omega_L - \omega_V$, and $\omega_a = \omega_L + \omega_V$. These contributions act as driving forces for the generation of new waves. Provided that the phase-matching conditions $2\boldsymbol{k}_L = \boldsymbol{k}_S + \boldsymbol{k}_a$ can be satisfied, a strong anti-Stokes wave $E_a \cos(\omega_a t - \boldsymbol{k}_a \cdot \boldsymbol{r})$ is observed in the direction of $\boldsymbol{k}_a$.

Despite the enormous intensities of stimulated Stokes and anti-Stokes waves, stimulated Raman spectroscopy has been of little use in molecular spectroscopy. The high threshold, which, according to (3.30), depends on the molecular density N, the incident intensity $I \propto E_L^2$, and the square of the small polarizability term $(\partial\alpha_{ij}/\partial q)$ in (3.27), limits stimulated emission to only the strongest Raman lines in materials of high densities N.

The recently developed technique of coherent anti-Stokes Raman spectroscopy (CARS), however, combines the advantages of signal strength obtained in stimulated Raman spectroscopy with the general applicability of spontaneous Raman spectroscopy [336–347]. In this technique *two* lasers are needed. The frequencies ω_1 and ω_2 of the two incident laser waves are chosen such that their difference $\omega_1 - \omega_2 = \omega_V$ coincides with a Raman-active vibration of the molecules under investigation. These two incident waves correspond to the pump wave $(\omega_1 = \omega_L)$ and the Stokes wave $(\omega_2 = \omega_S)$ in stimulated Raman scattering. The advantage is that the Stokes wave at ω_2 is already present and does not need to be generated in the medium. These two waves are considered in (3.19).

Because of the nonlinear interaction discussed in Sect. 3.3.1, new Stokes and anti-Stokes waves are generated (Fig. 3.16d). The waves ω_1 and ω_2 produce a large population density of vibrationally excited molecules by stimulated Raman scattering. These excited molecules act as the nonlinear medium for the generation of

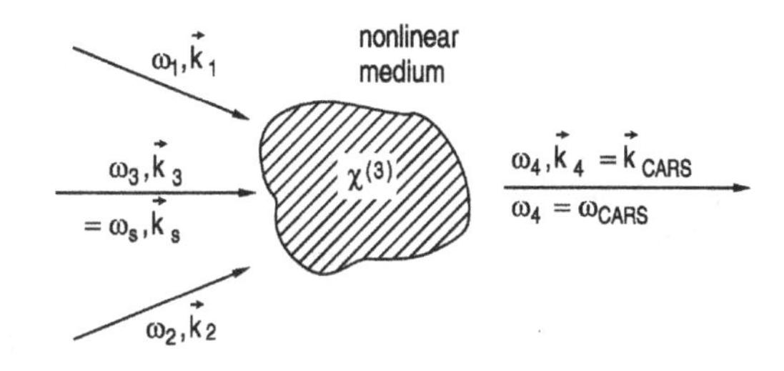

Fig. 3.17 CARS as a four-wave mixing process

anti-Stokes radiation at $\omega_a = 2\omega_1 - \omega_2$ by the incident wave with frequency ω_1. In a similar way, a Stokes wave with frequency $\omega_s = 2\omega_2 - \omega_1$ is generated by the incident waves at ω_1 and ω_2 (Fig. 3.16d). Since four waves are involved in the generation of the anti-Stokes wave, CARS is called a *four-wave parametric mixing process* (Fig. 3.17).

It can be derived from (3.33) that the power S of the CARS signal (which is proportional to the square of the amplitude E_a)

$$S \propto N^2 I_1^2 I_2, \tag{3.38}$$

increases with the square N^2 of the molecular density N and is proportional to the product $I_1(\omega_1)^2 I_2(\omega_2)$ of the pump laser intensities. It is therefore necessary to use either high densities N or large intensities I. If the two pump beams are focused into the sample, most of the CARS signal is generated within the local volume where the intensities are maximum. CARS spectroscopy therefore offers high spatial resolution.

If the incident waves are at *optical* frequencies, the difference frequency $\omega_R = \omega_1 - \omega_2$ is small compared with ω_1 in the case of rotational–vibrational frequencies ω_R. In gaseous Raman samples the dispersion is generally negligible over small ranges $\Delta\omega = \omega_1 - \omega_2$ and satisfactory phase matching is obtained for *collinear* beams. The Stokes wave at $\omega_S = 2\omega_2 - \omega_1$ and the anti-Stokes wave at $\omega_a = 2\omega_1 - \omega_2$ are then generated in the same direction as the incoming beams (Fig. 3.16b). In liquids, dispersion effects are more severe and the phase-matching condition can be satisfied over a sufficiently long coherence length only, if the two incoming beams are crossed at the phase-matching angle (Fig. 3.16c).

In the collinear arrangement the anti-Stokes wave at $\omega_a = 2\omega_1 - \omega_2$ ($\omega_a > \omega_1$!) is detected through filters that reject both incident laser beams as well as the fluorescence that may be generated in the sample. Figure 3.18 illustrates an early experimental setup used for rotational–vibrational spectroscopy of gases by CARS [348]. The two incident laser beams are provided by a Q-switched ruby laser and a tunable dye laser that was pumped by this ruby laser. Because the gain of the anti-Stokes wave depends quadratically on the molecular density N (see (3.38)), megawatt-range power levels of the incident beams are required for gaseous samples, while kilowatt powers are sufficient for liquid samples [349].

The most common pump system for pulsed CARS experiments are two dye lasers pumped by the same pump laser (N_2 laser, excimer laser, or frequency-doubled Nd:YAG laser). This system is very flexible because both frequencies ω_1 and ω_2

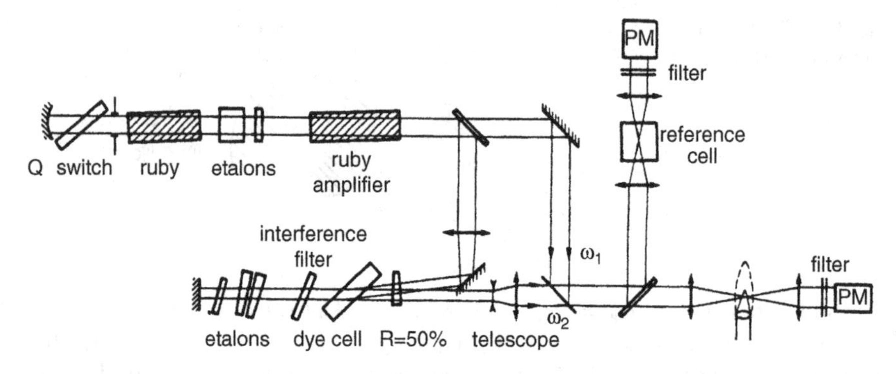

Fig. 3.18 Experimental setup of an early CARS experiment in gases using a ruby laser and a dye laser pumped by the ruby laser [348]

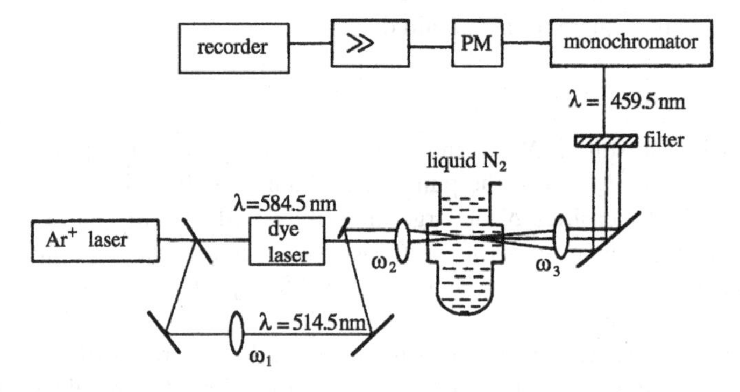

Fig. 3.19 Experimental setup for cw CARS in liquids [349]

can be varied over large spectral ranges. Since both the frequency and intensity fluctuations of the dye lasers result in strong intensity fluctuations of the CARS signal, the stability of the dye lasers needs particular attention. With compact and stable systems the signal fluctuations can be reduced below 10 % [350].

In addition, many CARS experiments have been performed with cw dye lasers with liquid samples as well as with gaseous ones. An experimental setup for cw CARS of liquid nitrogen is shown in Fig. 3.19, where the two incident collinear pump waves are provided by the 514.5 nm argon laser line (ω_1) and a cw dye laser (ω_2) pumped by the same argon laser [349].

The advantage of cw CARS is its higher spectral resolution because the bandwidth $\Delta\nu$ of single-mode cw lasers is several orders of magnitude below that of pulsed lasers. In order to obtain sufficiently high intensities, intracavity excitation has been used. A possible experimental realization (Fig. 3.20) places the sample cell inside the ring resonator of an argon ion laser, where the cw dye laser is coupled into the resonator by means of a prism [334]. The CARS signal generated in the sample cell at the beam waist of the resonator is transmitted through the dichroic mirror

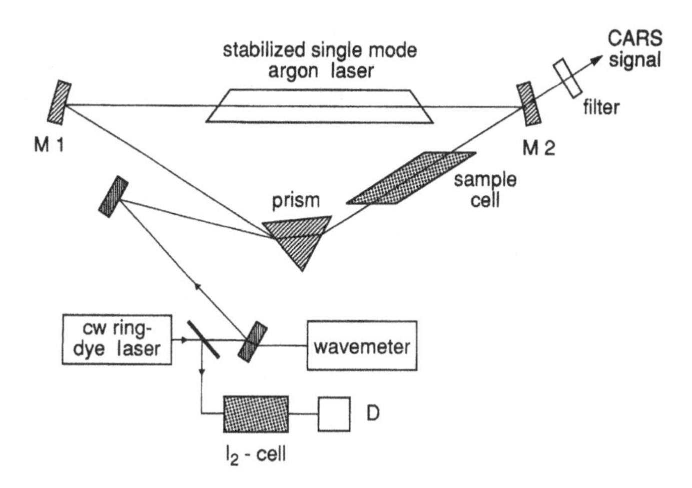

Fig. 3.20 Schematic arrangement of a cw CARS spectrometer with intracavity excitation of the sample [334]

M2 and is spectrally purified by a filter or a prism and a monochromator before it reaches a photomultiplier.

High-resolution CARS can be also performed with injection-seeded pulsed dye lasers [334, 351]. If the radiation of a single-mode cw dye laser with frequency ω is injected into the cavity of a pulsed dye laser that has been mode matched to the Gaussian beam of the cw laser (Vol. 1, Sect. 5.8), the amplification of the gain medium is enhanced considerably at the frequency ω and the pulsed laser oscillates on a single cavity mode at the frequency ω. Only some milliwatts of the cw laser are needed for injection, while the output of the single-mode pulsed laser reaches several kilowatts, which may be further amplified (Vol. 1, Sect. 5.5). Its bandwidth $\Delta\nu$ for pulses of duration Δt is only limited by the Fourier limit $\Delta\nu = 1/(2\pi\,\Delta t)$.

3.3.3 Resonant CARS and BOX CARS

If the frequencies ω_1 and ω_2 of the two incident laser waves are chosen to match a molecular transition, one or even two of the virtual levels in Fig. 3.16 coincide with a real molecular level. In this case of *resonant CARS*, the sensitivity may be increased by several orders of magnitude. Because of the larger absorption of the incident waves, the absorption path length must be sufficiently short or the density of absorbing molecules must be correspondingly small [352, 353].

A certain disadvantage of collinear CARS in gases is the spatial overlap of the two parallel incident beams with the signal beam. This overlap must be separated by spectral filters. This disadvantage can be overcome by the BOX CARS technique [354], where the pump beam of laser L_1 ($\boldsymbol{k}_1, \omega_1$) is split into two parallel beams that are focused by a lens into the sample (Fig. 3.21b), where the directions of the

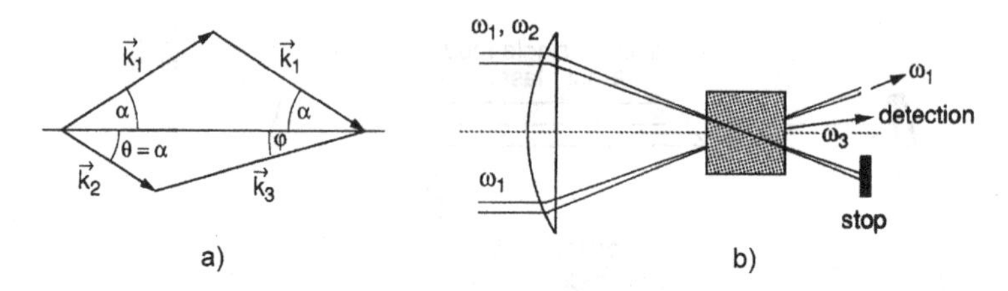

Fig. 3.21 Wave vector diagram for BOX CARS (**a**) and experimental realization (**b**)

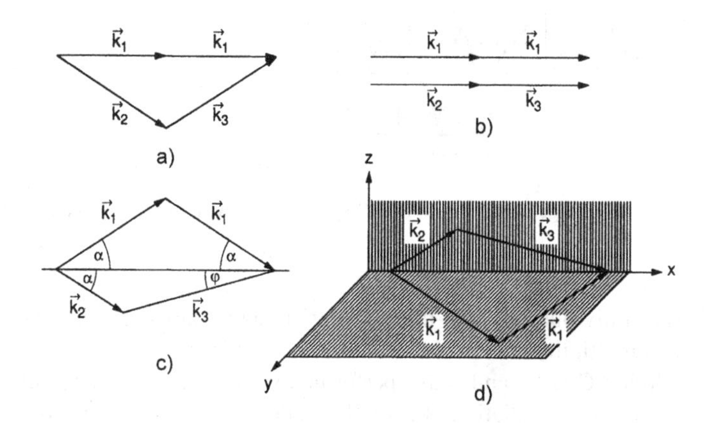

Fig. 3.22 Comparison of phase-matching diagrams for CARS: (**a**) for the general case $\boldsymbol{k}_1 \nparallel \boldsymbol{k}_2$; (**b**) the collinear case $\boldsymbol{k}_1 \parallel \boldsymbol{k}_2$; (**c**) for BOX CARS; and (**d**) for folded BOX CARS

three incoming beams match the vector diagram of Fig. 3.21a. The CARS signal beam can now be separated by geometrical means (beam stop and apertures). For comparison the vector diagrams of the phase-matching condition (3.35) are shown in Fig. 3.22 for the general case ($\boldsymbol{k}_1 \neq \boldsymbol{k}_2$), the collinear CARS arrangement, and for BOX CARS, where the vector diagram has the form of a box. From Fig. 3.21a with the relation $|\boldsymbol{k}| = n \cdot \omega / c$, the phase-matching conditions

$$\begin{aligned} n_2\omega_2 \sin\theta &= n_3\omega_3 \sin\varphi, \\ n_2\omega_2 \cos\theta + n_3\omega_3 \cos\varphi &= 2n_1\omega_1 \cos\alpha, \end{aligned} \tag{3.39}$$

can be derived, which yield for $\theta = \alpha$ the relation

$$\sin\varphi = \frac{n_2\omega_2}{n_3\omega_3} \sin\alpha, \tag{3.40}$$

between the angle φ of the CARS signal beam $\boldsymbol{k}_\mathrm{a} = \boldsymbol{k}_3$ against the z-direction and the angle α of the incident beams.

The CARS signal is generated within the common overlap volume of the three incident beams. This BOX-CAR technique considerably increases the spatial resolution, which may reach values below 1 mm.

In another beam configuration (folded BOX CARS), the pump beam with ω_1 is split into two parallel beams, which are directed by the focusing lens in such a way that the wave vectors $\boldsymbol{k}_2$ and $\boldsymbol{k}_3 = \boldsymbol{k}_{as}$ are contained in a plane orthogonal to that of the two $\boldsymbol{k}_1$ vectors (Fig. 3.22d). This has the advantage that neither of the two pump beams overlaps the signal beam at the detector [355]. If the Raman shifts are small, spectral filtering becomes difficult and the advantage of this folded BOX CAR technique is obvious.

The advantages of CARS may be summarized as follows:

- The signal levels in CARS may exceed those obtained in spontaneous Raman spectroscopy by a factor of 10^4–10^5.
- The higher frequency $\omega_3 > \omega_1, \omega_2$ of the anti-Stokes waves allows one to use filters that reject the incident light as well as fluorescence light.
- The small beam divergence allows one to place the detector far away from the sample, which yields excellent spatial discrimination against fluorescence or thermal luminous background such as occurs in flames, discharges, or chemiluminescent samples.
- The main contribution to the anti-Stokes generation comes from a small volume around the focus of the two incident beams. Therefore very small sample quantities (microliters for liquid samples or millibar pressures for gaseous samples) are required. Furthermore, a high spatial resolution is possible, which allows one to probe the spatial distribution of molecules in definite rotational–vibrational levels. The measurements of local temperature variations in flames from the intensity of anti-Stokes lines in CARS is an example where this advantage is utilized.
- The high spatial resolution is utilized in CARS microscopy, which is mainly applied to the investigation of biological cells and their compositions (see Sect. 3.4.3).
- A high spectral resolution can be achieved without using a monochromator. The Doppler width $\Delta\omega_D$, which represents a principal limitation in 90° spontaneous Raman scattering, is reduced to $[(\omega_2 - \omega_1)/\omega_1]\Delta\omega_D$ for the collinear arrangement used in CARS. While a resolution of 0.3 to 0.03 cm^{-1} is readily obtained in CARS with pulsed lasers, even linewidths down to 0.001 cm^{-1} have been achieved with single-mode lasers.

The main disadvantages of CARS are the expensive equipment and strong fluctuations of the signals caused by instabilities of intensities and alignments of the incident laser beams. The sensitivity of detecting low relative concentrations of specified sample molecules is mainly limited by interference with the nonresonant background from the other molecules in the sample. This limitation can be overcome, however, by resonant CARS.

3.3.4 Hyper-Raman Effect

The higher-order terms $\beta\boldsymbol{E}\boldsymbol{E}$, $\gamma\boldsymbol{E}\boldsymbol{E}\boldsymbol{E}$ in the expansion (3.18a) of $\boldsymbol{p}(\boldsymbol{E})$ represent the hyper-Raman effect. Analogous to (3.3), we can expand β in a Taylor series in

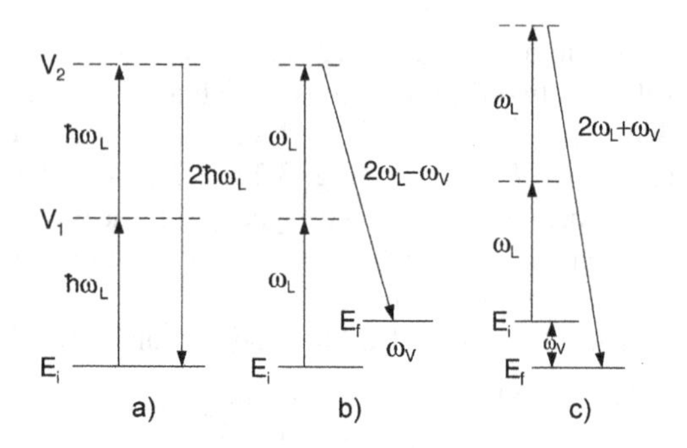

Fig. 3.23 Hyper-Rayleigh scattering (**a**), Stokes hyper-Raman (**b**), and anti-Stokes hyper-Raman scattering (**c**)

the normal coordinates $q_n = q_{n0}\cos(\omega_n t)$

$$\beta = \beta_0 + \sum_{n=1}^{2Q}\left(\frac{\partial\beta}{\partial q_n}\right)_0 q_n + \cdots . \tag{3.41}$$

Assume that two laser waves $E_1 = E_{01}\cos(\omega_1 t - k_1 z)$ and $E_2 = E_{02}\cos(\omega_2 t - k_2 t)$ are incident on the Raman sample. From the third term in (3.18a) we then obtain with (3.41) contributions to $p(E)$ due to β_0

$$\beta_0 E_{01}^2\cos(2\omega_1 t), \quad \text{and} \quad \beta_0 E_{02}^2\cos(2\omega_2 t), \tag{3.42}$$

which give rise to *hyper-Rayleigh scattering* at frequencies $2\omega_1$, $2\omega_2$ and $\omega_1 + \omega_2$ (Fig. 3.23a). The term $(\partial\beta/\partial q_n)q_{n0}\cos(\omega_n t)$ of (3.41) inserted into (3.18a) yields contributions

$$p^{\mathrm{HR}} \propto \left(\frac{\partial\beta}{\partial q}\right)_0 q_{n0}\big[\cos(2\omega_1 \pm \omega_n)t + \cos(2\omega_2 \pm \omega_n)t\big], \tag{3.43}$$

which are responsible for *hyper-Raman scattering* (Fig. 3.23b, c) [356].

Since the coefficients $(\partial\beta/\partial q)_0$ are very small, one needs large incident intensities to observe hyper-Raman scattering. Similar to second-harmonic generation (Vol. 1, Sect. 5.8), hyper-Rayleigh scattering is forbidden for molecules with a center of inversion. The hyper-Raman effect obeys selection rules that differ from those of the linear Raman effect. It is therefore very attractive to molecular spectroscopists since molecular vibrations can be observed in the hyper-Raman spectrum that are forbidden for infrared as well as for linear Raman transitions. For example, spherical molecules such as CH_4 have no pure rotational Raman spectrum but a hyper-Raman spectrum, which was found by Maker [357]. A general theory for rotational and rotational–vibrational hyper-Raman scattering has been worked out by Altmann and Strey [358].

The intensity of hyper-Raman lines can be considerably enhanced when the molecules under investigation are adsorbed at a surface [359], because the surface lowers the symmetry and increases the induced dipole moments.

Similar to the induced Raman effect, the hyper-Raman effect can also be used to generate coherent radiation in spectral ranges where no intense lasers exist. One example is the generation of tunable radiation around 16 μm by the stimulated hyper-Raman effect in strontium vapor [360].

3.3.5 Summary of Nonlinear Raman Spectroscopy

In the previous subsections we briefly introduced some nonlinear techniques of Raman spectroscopy. Besides stimulated Raman spectroscopy, Raman gain spectroscopy, inverse Raman spectroscopy, and CARS, several other special techniques such as the Raman-induced Kerr effect [361] or coherent Raman ellipsometry [362] also offer attractive alternatives to conventional Raman spectroscopy.

All these nonlinear techniques represent coherent third-order processes analogous to saturation spectroscopy, polarization spectroscopy, or two-photon absorption (Chap. 2), because the magnitude of the nonlinear signal is proportional to the third power of the involved field amplitudes (3.18a–3.18c).

The advantages of these nonlinear Raman techniques are the greatly increased signal-to-noise ratio and thus the enhanced sensitivity, the higher spectral and spatial resolution, and in the case of the hyper-Raman spectroscopy, the possibility of measuring higher-order contributions of molecules in the gaseous, liquid, or solid state to the susceptibility.

Additionally, there are several good books and reviews on nonlinear Raman spectroscopy. For more thorough information the reader is therefore referred to [301, 328, 334, 338, 339, 345, 347, 363].

3.4 Special Techniques

In this section we briefly discuss some special techniques of linear and nonlinear Raman spectroscopy that have particular advantages for different applications. These are the resonance Raman effect, surface-enhanced Raman signals, Raman microscopy, and time-resolved Raman spectroscopy.

3.4.1 Resonance Raman Effect

The Raman scattering cross section can be increased by several orders of magnitude if the excitation wavelength matches an electronic transition of the molecule, that is, when it coincides with or comes close to a line in an electronic absorption band. In this case, the denominator in (3.12) becomes very small for the neighboring lines in this band, and several terms in the sum (3.12) give large contributions to the signal.

The Raman lines appear predominantly at those downward transitions that have the largest Franck–Condon factors, that is, the largest overlap of the vibrational wavefunctions in the upper and lower state (Fig. 3.24). Unlike the nonresonant case, in spontaneous resonance Raman scattering a larger number of Raman lines may

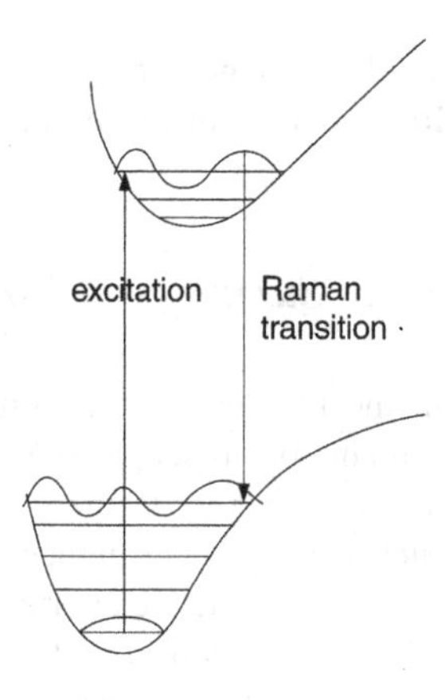

Fig. 3.24 Resonance Raman effect

appear that are shifted against the excitation line by several vibrational quanta. They correspond to Raman transitions terminating on higher vibrational levels of the ground electronic state. This is quite similar to laser-excited fluorescence and opens the possibility to determine the anharmonicity constants of the molecular potential curve or potential surface of the lower electronic state.

In stimulated Raman scattering the strongest Raman transition will have the largest gain and reaches threshold before the other transitions can develop. Just above threshold we therefore expect only a single Raman line in the stimulated Raman spectrum, while at higher pump powers more lines will appear.

Resonance Raman scattering is particularly advantageous for samples with small densities, for example, for gases at low pressures, where the absorption of the incident radiation is not severe and where nonresonant Raman spectroscopy might not be sufficiently sensitive.

If the excited state lies above the dissociation limit of the upper electronic state, the scattered Raman light shows a continuous spectrum. The intensity profile of this spectrum yields information on the repulsive part of the potential in the upper state.

3.4.2 Surface-Enhanced Raman Scattering

The intensity of Raman scattered light may be enhanced by several orders of magnitude if the molecules are adsorbed on a surface [364]. There are several mechanisms that contribute to this enhancement. Since the amplitude of the scattered radiation is proportional to the induced dipole moment

$$\boldsymbol{p}_{\mathrm{ind}} = \alpha \cdot \boldsymbol{E},$$

the increase of the polarizability α by the interaction of the molecule with the surface is one of the causes for the enhancement. In the case of metal surfaces, the electric field $\boldsymbol{E}$ at the surface may also be much larger than that of the incident radiation, which also leads to an increase of the induced dipole moment. Both effects depend on the orientation of the molecule relative to the surface normal, on its distance from the surface, and on the morphology, in particular the roughness of the surface. Small metal clusters on the surface increase the intensity of the molecular Raman lines. The frequency of the exciting light also has a large influence on the enhancement factor. In the case of metal surfaces it becomes maximum if it is close to the plasma frequency of the metal.

Because of these dependencies, surface-enhanced Raman spectroscopy has been successfully applied for surface analysis and also for tracing small concentrations of adsorbed molecules [364].

One example is the detection of explosive molecules such as TNT on porous gold surfaces where a detection sensitivity down to 0.1 ppm could be achieved [365].

3.4.3 Raman Microscopy

For the nondestructive investigation of very small samples, for example, parts of living cells or inclusions in crystals, the combination of microscopy and Raman spectroscopy turns out to be a very useful technique. The laser beam is focused into the sample and the Raman spectrum that is emitted from the small focal spot is monitored through the microscope with subsequent spectrometer. This also applies to measurements of phase transitions in molecular crystals under high pressures. These pressures, which reach up to several gigapascals, can be realized with moderate efforts in a small volume between two diamonds with area A that are pressed together by a force F producing a pressure $p = F/A$. For example, with $F = 10^3$ Pa and $A = 10^{-6}$ cm^2, one obtains a pressure of 10^9 Pa. The phase transitions lead to a frequency shift of molecular vibrations, which are detected by the corresponding shift of the Raman lines [366].

One example of the application of this technique is the investigation of different phases of "fluid inclusions" in a quartz crystal found in the Swiss Alps. Their Raman spectra permitted determination of CO_2 as a gaseous inclusion and water as a liquid inclusion, and they showed that the mineral thought to be $CaSO_4$ was in fact $CaCO_3$ [367].

A typical experimental arrangement for Raman microscopy is shown in Fig. 3.25. The output beam of an argon laser or a dye laser is focused by a microscope objective into the microsample. The backscattered Raman light is imaged onto the entrance slit of a double or triple monochromator, which effectively supresses scattered laser light. A CCD camera at the exit of the monochromator records the wanted spectral range of the Raman radiation [364, 368, 369].

3.4.4 Time-Resolved Raman Spectroscopy

Both linear and nonlinear Raman spectroscopy can be combined with time-resolved detection techniques when pumping with short laser pulses [370]. Since Raman spectroscopy allows the determination of molecular parameters from measurements

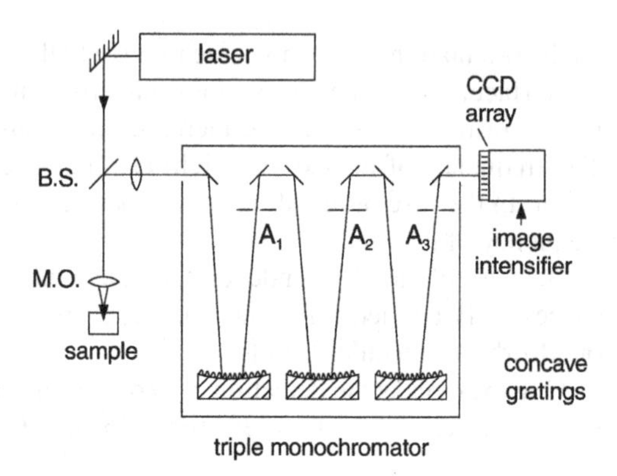

Fig. 3.25 Raman microscopy with suppression of scattered laser light by a triple monochromator with three apertures A_i

of frequencies and populations of vibrational and rotational energy levels, time-resolved techniques give information on energy transfer between vibrational levels or on structural changes of short-lived intermediate species in chemical reactions. One example is the vibrational excitation of molecules in liquids and the collisional energy transfer from the excited vibrational modes into other levels or into translational energy of the collision partners. These processes proceed on picosecond to femtosecond time scales [369, 371].

Time-resolved Raman spectroscopy has proved to be a very useful tool to elucidate fast processes in biological molecules, for instance, to follow the fast structural changes during the visual process where, after photoexcitation of rhodopsin molecules, a sequence of energy transfer processes involving isomerization and proton transfer takes place. This subject is treated in more detail in Chap. 6 in comparison with other time-resolved techniques.

3.5 Applications of Laser Raman Spectroscopy

The primary object of Raman spectroscopy is the determination of molecular energy levels and transition probabilities connected with molecular transitions that are not accessible to infrared spectroscopy. Linear laser Raman spectroscopy, CARS, and hyper-Raman scattering have very successfully collected many spectroscopic data that could not have been obtained with other techniques. Besides these basic applications to molecular spectroscopy there are, however, a number of scientific and technical applications of Raman spectroscopy to other fields, which have become feasible with the new methods discussed in the previous sections. We can give only a few examples.

Since the intensity of spontaneous Raman lines is proportional to the density $N(v_i, J_i)$ of molecules in the initial state (v_i, J_i), Raman spectroscopy can provide information on the population distribution $N(v_i, J_i)$, its local variation, and on concentrations of molecular constituents in samples. This allows one, for instance, to probe the temperature in flames or hot gases from the rotational Raman spectra [372, 373, 377, 378] and to detect deviations from thermal equilibrium.

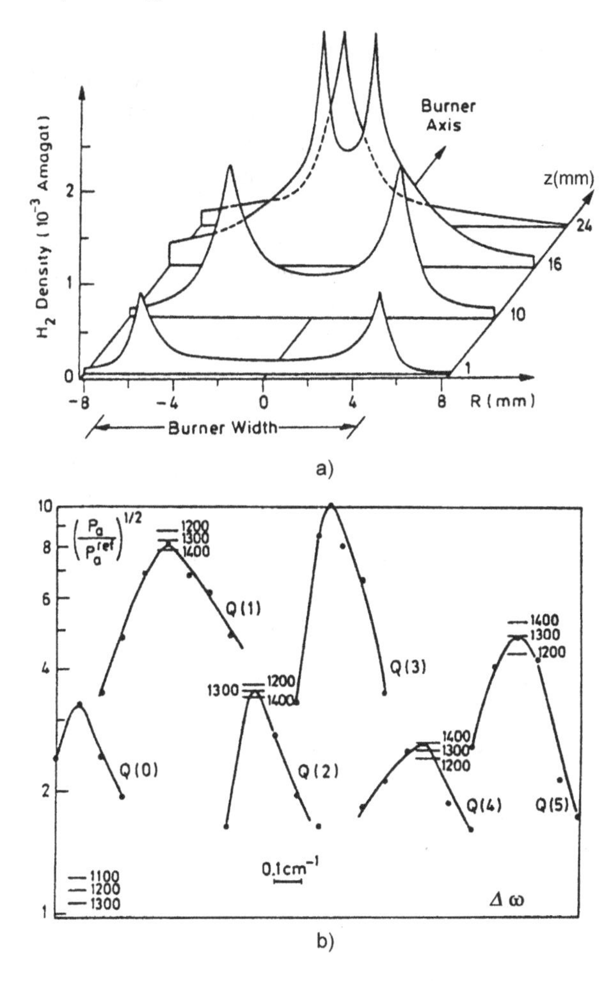

Fig. 3.26 Determination of density profiles of H_2 molecules in a flame. R is the distance from the burner axis, z the distance along this axis. The profiles in (**a**) have been obtained from the spatial variations of the Q line intensities and spectral profiles shown in (**b**). The relative intensities of $Q(J)$ furthermore allow the determinations of the temperature profiles. The *numbers with the horizontal lines* give the expected signal heights fot the labeled temperatures T in [K] [348]

With CARS the spatial resolution is greatly increased, in particular if BOX CARS is used. The focal volume from which the signal radiation is generated can be made smaller than 0.1 mm^3 [344]. The local density profiles of reaction products formed in flames or discharges can therefore be accurately probed without disturbing the sample conditions. The intensity of the stimulated anti-Stokes radiation is proportional to N^2 (3.31a–3.31b). Figure 3.26 shows for illustration the H_2 distribution in a horizontal Bunsen flame, measured from the CARS spectrum of the Q branch in H_2. The H_2 molecules are formed by the pyrolysis of hydrocarbon molecules [348]. Another example is the measurement of CARS spectra of water vapor in flames, which allowed one to probe the temperature in the postflame region of a premixed CH_4 air flame [373].

One definite advantage of CARS is the fact that the detector can be far away from the sample (remote Raman-spectroscopy) because the intensity of the signal does not decrease as $1/r^2$ as for spontaneous Raman spectroscopy or fluorescence detection, due to the good collimation of the anti-Stokes beam. This remote sensing can be used for instance for safety inspection where explosives should be detected at large distances up to 50 m [374]. For safety inspections at airports a small portable Raman spectrometer is useful [375]. One important field of applications is forensic science where the sensitivity and the simultaneous spectral information can be used to identify hidden drugs, explosives or other dangerous material [376].

With a detection sensitivity of 10 to 100 ppm, CARS is not as good as some other techniques in monitoring pollutant gases at low concentrations (Sect. 1.2), but its advantage is the capability to examine a large number of species quickly by tuning the dye lasers. The good background rejection allows the use of this technique under conditions of bright background radiation where other methods may fail [377]. Examples are temperature and concentration measurements of molecular nitrogen, oxygen, and methane in a high-temperature furnace up to 2000 K [378], where the thermal radiation is much stronger than laser-induced fluorescence. Therefore CARS is the best choice because the detector can be placed far from the furnace.

A further example of the scientific application of CARS is the investigation of cluster formation in supersonic beams (Sect. 4.3), where the decrease in the rotational and vibrational temperatures during the adiabatic expansion (Sect. 4.2) and the degree of cluster formation in dependence on the distance from the nozzle can be determined [379].

CARS has been successfully used for the spectroscopy of chemical reactions (Sect. 8.4). The BOX CARS technique with pulsed lasers offers spectral, spatial, and time-resolved investigations of collision processes and reactions, not only in laboratory experiments but also in the tougher surroundings of factories, in the reaction zone of car engines, and in atmospheric research (Sect. 10.2 and [380, 381]).

The detection sensitivity of CARS ranges from 0.1–100 ppm ($\hat{=}10^{-7} - 10^{-4}$ relative concentrations) depending on the Raman cross sections. Although other spectroscopic techniques such as laser-induced fluorescence or resonant two-photon ionization (Sect. 1.2) may reach higher sensitivities, there are enough examples where CARS is the best or even the only choice, for instance, when the molecules under investigation are not infrared active or have no electronic transitions within the spectral range of available lasers.

An interesting application of Raman spectroscopy in medicine is the in vivo detection of breast cancer [382]. This technique is very promising because the spatial and spectral resolution allows the localization of spots with a different molecular composition and is able to distinguish between benign and malign cancer cells.

Further information on laser Raman spectroscopy can be found in [383–386].

3.6 Problems

3.1 What is the minimum detectable concentration N_i of molecules with a Raman scattering cross section $\sigma = 10^{-30}$ cm^2, if the incident cw laser has 10 W output power at $\lambda = 500$ nm that is focused into a scattering volume of 5 mm $\times$ 1 mm^2, which can be imaged with 10 % collection efficiency onto a photomultiplier with the quantum efficiency $\eta = 25$ %? The multiplier dark current is 10 photoelectrons per second and a signal-to-noise ratio of 3:1 should be achieved.

3.2 The bent water molecule H_2O has three normal vibrations. Which of these are Raman active and which are infrared active? Are there also vibrations that are both infrared as well as Raman active?

3.3 A small molecular sample of 10^{21} molecules in a volume of 5 mm × 1 mm^2 is illuminated by 10 W of argon laser radiation at $\lambda = 488$ nm. The Raman cross section is $\sigma = 10^{-29}$ cm^2 and the Stokes radiation is shifted by 1000 cm^{-1}. Calculate the heat energy $\mathrm{d}W_{\mathrm{H}}/\mathrm{d}t$ generated per second in the sample, if the molecules do not absorb the laser radiation or the Stokes radiation. How much is $\mathrm{d}W_{\mathrm{H}}/\mathrm{d}t$ increased if the laser wavelength is close to resonance of an absorbing transition, causing an absorption coefficient $\alpha = 10^{-1}$ cm^{-1}?

3.4 Estimate the intensity of Raman radiation emerging out of the endface of an optical fiber of 100 m length and 0.1 mm ∅, filled with a Raman-active medium with a molecular density $N_i = 10^{21}$ cm^{-3} and a Raman scattering cross section of $\phi_{\mathrm{R}} = 10^{-30}$ cm^2, if the laser radiation (1 W) and the Raman light are both kept inside the fiber by total internal reflection.

3.5 The two parallel incident laser beams with a Gaussian intensity profile are focused by a lens with $f = 5$ cm in a BOX CARS arrangement into the sample. Estimate the spatial resolution, defined by the halfwidth $S_{\mathrm{A}}(z)$ of the CARS signal, when the beam diameter of each beam at the lens is 3 mm and their separation is 20 mm.

Chapter 4
Laser Spectroscopy in Molecular Beams

For many years molecular beams were mainly employed for scattering experiments. The combination of new spectroscopic methods with molecular beam techniques has brought about a wealth of new information on the structure of atoms and molecules, on details of collision processes, and on fundamentals of quantum optics and the interaction of light with matter.

There are several aspects of laser spectroscopy performed with molecular beams that have contributed to the success of these combined techniques. First, the spectral resolution of absorption and fluorescence spectra can be increased by using collimated molecular beams with reduced transverse velocity components (Sect. 4.1). Second, the internal cooling of molecules during the adiabatic expansion of supersonic beams compresses their population distribution into the lowest vibrational–rotational levels. This greatly reduces the number of absorbing levels and results in a drastic simplification of the absorption spectrum (Sect. 4.2).

The low translational temperature achieved in supersonic beams allows the generation and observation of loosely bound van der Waals complexes and clusters (Sect. 4.3). The collision-free conditions in molecular beams after their expansion into a vacuum chamber facilitates saturation of absorbing levels, since no collisions refill a level depleted by optical pumping. This makes Doppler-free saturation spectroscopy feasible even at low cw laser intensities (Sect. 4.4).

New techniques of high-resolution laser spectroscopy in beams of positive or negative ions have been developed. These techniques are discussed in Sects. 4.5 and 4.6.

Several examples illustrate the advantages of molecular beams for spectroscopic investigation. The wide, new field of laser spectroscopy of collision processes in crossed molecular beams is discussed in Chap. 8.

A detailed discussion on the formation and characteristic features of molecular beams can be found in the extensive literature [387].

W. Demtröder, *Laser Spectroscopy 2*, DOI 10.1007/978-3-662-44641-6_4

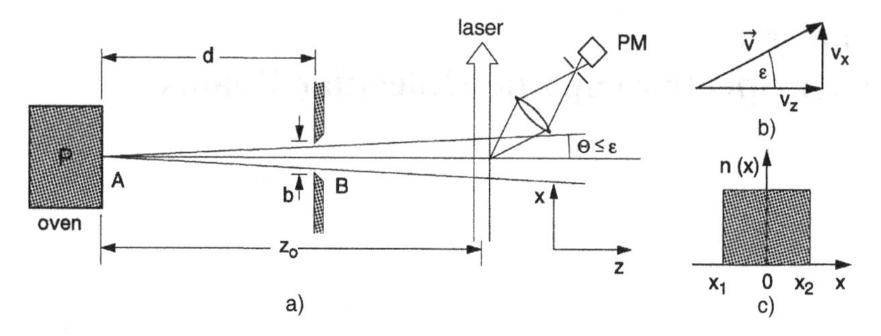

Fig. 4.1 Laser excitation spectroscopy with reduced Doppler width in a collimated molecular beam: (**a**) schematic experimental arrangement; (**b**) collimation ratio; (**c**) density profile $n(x)$ in a collimated beam effusing from a point source A

4.1 Reduction of Doppler Width

Let us assume molecules effusing into a vacuum tank from a small hole A in an oven that is filled with a gas or vapor at pressure p (Fig. 4.1). The molecular density behind A and the background pressure in the vacuum tank are sufficiently low to assure a large mean free path of the effusing molecules, such that collisions can be neglected. The number $N(\theta)$ of molecules that travel into the cone $\theta \pm \mathrm{d}\theta$ around the direction θ against the symmetry axis (which we choose to be the z-axis) is proportional to $\cos\theta$. A slit B with width b, at a distance d from the point source A, selects a small angular interval $-\epsilon \le \theta \le +\epsilon$ around $\theta = 0$ (Fig. 4.1). The molecules passing through the slit B, which is parallel to the y-axis, form a molecular beam in the z-direction, collimated with respect to the x-direction. The collimation ratio is defined by (Fig. 4.1b)

$$\frac{v_x}{v_z} = \tan\epsilon = \frac{b}{2d}. \tag{4.1}$$

If the source diameter is small compared with the slit width b and if $b \ll d$ (which means $\epsilon \ll 1$), the flux density behind the slit B is approximately constant across the beam diameter, since $\cos\theta \simeq 1$ for $\theta \ll 1$. For this case, the density profile of the molecular beam is illustrated in Fig. 4.1c.

The density $n(v)\,\mathrm{d}v$ of molecules with velocities $v = |\boldsymbol{v}|$ inside the interval v to $v + \mathrm{d}v$ in a molecular beam at thermal equilibrium, which effuses with the most probable velocity $v_\mathrm{p} = (2kT/m)^{1/2}$ into the z-direction, can be described at the distance $r = (z^2 + x^2)^{1/2}$ from the source A as

$$n(v, r, \theta)\,\mathrm{d}v = C\frac{\cos\theta}{r^2} n v^2 \mathrm{e}^{-(v/v_\mathrm{p})^2}\,\mathrm{d}v, \tag{4.2}$$

where the normalization factor $C = (4/\sqrt{\pi})v_\mathrm{p}^{-3}$ assures that the total density n of the molecules is $n = \int n(v)\,\mathrm{d}v$.

Note The mean flux density is $N = \overline{nv} = \int v n(v)\,\mathrm{d}v$.

If the collimated molecular beam is crossed perpendicularly with a monochromatic laser beam with frequency ω propagating into the x-direction, the absorption probability for each molecule depends on its velocity component v_x. In Vol. 1, Sect. 3.2 it was shown that the center frequency of a molecular transition, which is ω_0 in the rest frame of the moving molecule, is Doppler shifted to a frequency ω_0' according to

$$\omega_0' = \omega_0 - \boldsymbol{k} \cdot \boldsymbol{v} = \omega_0 - k v_x, \quad k = |\boldsymbol{k}|. \tag{4.3}$$

Only those molecules with velocity components v_x in the interval $\mathrm{d}v_x = \delta\omega_\mathrm{n}/k$ around $v_x = (\omega - \omega_0)/k$ essentially contribute to the absorption of the monochromatic laser wave, because these molecules are shifted into resonance with the laser frequency ω within the natural linewidth $\delta\omega_\mathrm{n}$ of the absorbing transition.

When the laser beam in the x–z-plane ($y = 0$) travels along the x-direction through the molecular beam, its power decreases as

$$P(\omega) = P_0 \exp\left[-\int_{x_1}^{x_2} \alpha(\omega, x)\,\mathrm{d}x\right]. \tag{4.4}$$

The absorption within the distance $\Delta x = x_2 - x_1$ inside the molecular beam is generally extremely small. Typical figures for $\Delta P(\omega) = P_0 - P(x_2, \omega)$ range from 10^{-4} to 10^{-15} of the incident power. We can therefore use the approximation $\mathrm{e}^{-x} \simeq 1 - x$ and obtain with the absorption coefficient

$$\alpha(\omega, x) = \int n(v_x, x)\sigma(\omega, v_x)\,\mathrm{d}v_x, \tag{4.5}$$

the spectral profile of the absorbed power

$$\Delta P(\omega) = P_0 \int_{-\infty}^{+\infty} \left[\int_{x_1}^{x_2} n(v_x, x)\sigma(\omega, v_x)\,\mathrm{d}x\right] \mathrm{d}v_x. \tag{4.6}$$

With $v_x = (x/r)v \to \mathrm{d}v_x = (x/r)\,\mathrm{d}v$ and $\cos\theta = z/r$, we derive from (4.2) for the molecular density

$$n(v_x, x)\,\mathrm{d}v_x = Cn \frac{z}{x^3} v_x^2 \exp\left[-(r v_x / x v_\mathrm{p})^2\right] \mathrm{d}v_x. \tag{4.7}$$

The absorption cross section $\sigma(\omega, v_x)$ describes the absorption of a monochromatic wave of frequency ω by a molecule with a velocity component v_x. Its spectral profile is represented by a Lorentzian (Vol. 1, Sect. 3.6), namely

$$\sigma(\omega, v_x) = \sigma_0 \frac{(\gamma/2)^2}{(\omega - \omega_0 - k v_x)^2 + (\gamma/2)^2} = \sigma_0 L(\omega - \omega_0, \gamma). \tag{4.8}$$

Inserting (4.7) and (4.8) into (4.6) yields the absorption profile

$$\Delta P(\omega) = a_1 \int_{-\infty}^{+\infty} \left[\int_{x_1}^{x_2} \Delta\omega_0^2 \frac{\exp[-c^2 \Delta\omega_0^2 (1 + z^2/x^2)/\omega_0^2 v_\mathrm{p}^2]}{(\omega - \omega_0 - k v_x)^2 + (\gamma/2)^2}\,\mathrm{d}x\right] \mathrm{d}\Delta\omega_0,$$

with $a_1 = P_0 n\sigma_0 \gamma c^3 z/(\sqrt{\pi} v_p^3 \omega_0^3)$, and $\Delta\omega_0 = \omega_0' - \omega_0 = v_x\omega_0/c$, where $\omega_0' = \omega_0 + kv_x$ is the Doppler-shifted eigenfrequency ω_0. The integration over $\Delta\omega_0$ extends from $-\infty$ to $+\infty$ since the velocities v are spread from 0 to ∞.

The integration over x is analytically possible and yields with $x_1 = -r\sin\epsilon$, $x_2 = +r\sin\epsilon$

$$\Delta P(\omega) = a_2 \int_{-\infty}^{+\infty} \frac{\exp[-(\frac{c(\omega-\omega_0')}{\omega_0' v_p \sin\epsilon})^2]}{(\omega-\omega_0')^2 + (\gamma/2)^2} \mathrm{d}\omega_0', \quad \text{with } a_2 = a_1\left(\frac{c\gamma}{2z\omega_0}\right)^2. \tag{4.9}$$

This represents a *Voigt profile*, that is, a convolution product of a Lorentzian function with halfwidth γ and a Doppler function. A comparison with Vol. 1, (3.33) shows, however, that the Doppler width is reduced by the factor $\sin\epsilon = v_x/v = b/2d$, which equals the collimation ratio of the beam. The collimation of the molecular beam therefore reduces the Doppler width $\Delta\omega_0$ of the absorption lines to the width

$$\boxed{\Delta\omega_D^* = \Delta\omega_D \sin\epsilon,} \quad \text{with } \Delta\omega_D = 2\omega_0(v_p/c)\sqrt{\ln 2}, \tag{4.10}$$

where $\Delta\omega_D$ is the corresponding Doppler width in a gas at thermal equilibrium.

Example 4.1 Typical figures of $b = 1$ mm and $d = 5$ cm yield a collimation ratio $b/(2d) = 1/100$. This brings the Doppler width $\Delta\nu_0 = \Delta\omega_0/2\pi \approx$ 1500 MHz down to $\Delta\nu_D^* = \Delta\omega_D^*/2\pi \approx 15$ MHz, which is of the same order of magnitude as the natural linewidth γ of many molecular transitions.

Note For larger diameters of the oven hole A, the density profile $n(x)$ of the molecular beam is no longer rectangular but decreases gradually beyond the limiting angles $\theta = \pm\epsilon$. For $\Delta\omega_D^* > \gamma$, the absorption profile is then altered compared to that in (4.9), while for $\Delta\omega_D^* \ll \gamma$ the difference is negligible because the Lorentzian profile is dominant in the latter case [388].

The technique of reducing the Doppler width by the collimation of mo lecular beams was employed before the invention of lasers to produce light sources with narrow emission lines [389]. Atoms in a collimated beam were excited by electron impact. The fluorescence lines emitted by the excited atoms showed a reduced Doppler width if observed in a direction perpendicular to the atomic beam. However, the intensity of these atomic beam light sources was very weak and only the application of intense monochromatic, tunable lasers has allowed one to take full advantage of this method of *Doppler-free spectroscopy*.

A typical laser spectrometer for sub-Doppler excitation spectroscopy in a collimated molecular beam is shown in Fig. 4.2. The laser wavelength λ_L is controlled by a computer, which also records the laser-induced fluorescence $I_{Fl}(\lambda_L)$. Spectral regions in the UV can be covered by frequency-doubling the visible laser frequency

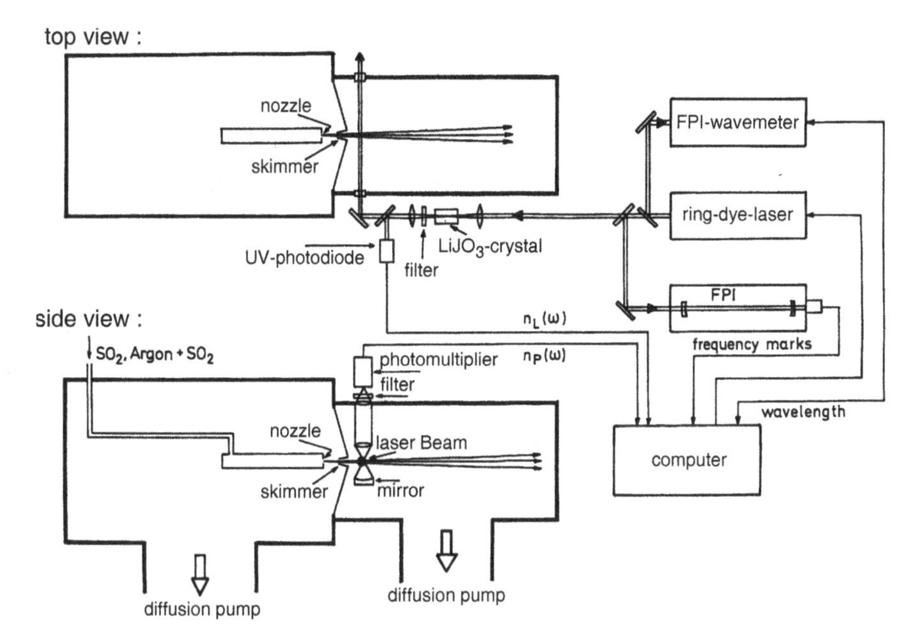

Fig. 4.2 Laser spectrometer for sub-Doppler excitation spectroscopy in a collimated molecular beam

in a nonlinear optical crystal, such as $LiIO_3$. For effective collection of the fluorescence, the optical system of Fig. 1.24 can be utilized. The transmission peaks of a long FPI give frequency marks separated by the free spectral range $\delta\nu = c/2d$ of the FPI (Vol. 1, Sect. 4.2.7). The absolute laser wavelength is measured by a wavemeter described in Vol. 1, Sect. 4.4.

The achievable spectral resolution is demonstrated by Fig. 4.3, which shows a small section of the Na_2 spectrum of the $A\,^1\Sigma_u \leftarrow X\,^1\Sigma_g$ system. Because of spin–orbit coupling with a $^3\Pi_u$ state, some rotational levels of the $A\,^1\Sigma_u$ state are mixed with levels of the $^3\Pi_u$ state and therefore show hyperfine splittings [390].

Particularly for polyatomic molecules with their complex visible absorption spectra, the reduction of the Doppler width is essential for the resolution of single lines [392]. This is illustrated by a section from the excitation spectrum of the SO_2 molecule, excited with a single-mode frequency-doubled dye laser tunable around $\lambda = 304$ nm (Fig. 4.4b). For comparison the same section of the spectrum as obtained with Doppler-limited laser spectroscopy in an SO_2 cell is shown in Fig. 4.4a [391].

The possibilities of molecular beam spectroscopy can be enhanced by allowing for spectrally resolved fluorescence detection or for resonant two-photon ionization in combination with a mass spectrometer. Such a molecular beam apparatus is shown in Fig. 4.5. The photomultiplier PM1 monitors the total fluorescence $I_{Fl}(\lambda_L)$ as a function of the laser wavelength λ_L (excitation spectrum, Sect. 1.3). Photomultiplier PM2 records the dispersed fluorescence spectrum excited at a fixed laser

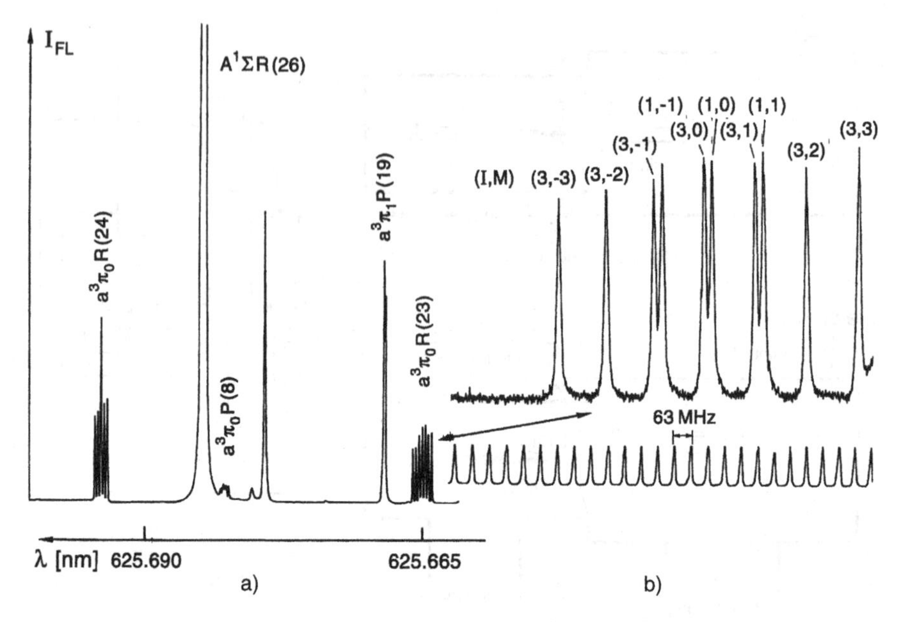

Fig. 4.3 (**a**) Hyperfine structure of rotational lines in the $A\,^1\Sigma_u \leftarrow X\,^1\Sigma_g^+$ system of Na_2, caused by spin–orbit coupling between the $A\,^1\Sigma_u$ and $a\,^3\Pi_u$ state [390]; and (**b**) enlarged scan of the HF multiplet of the R(23) line

wavelength, where the laser is stabilized onto a selected molecular absorption line. In a second crossing point of the molecular beam with two laser beams within the ion chamber of a quadrupole mass spectrometer, the molecules are selectively excited by laser L_1 and the excited molecules are ionized by L_2. The ions are extracted into the mass spectrometer and detected by an ion detector, such as a channeltron or an ion multiplier. This allows the selection of spectra of specific molecules (for example, isotopomers) when several species are present in the molecular beam.

The excitation spectrum can be further simplified and its analysis facilitated by recording a *filtered excitation spectrum* (Fig. 4.6b). The monochromator is set to a selected vibrational band of the fluorescence spectrum while the laser is tuned through the absorption spectrum. Then only transitions to those upper levels appear in the excitation spectrum that emit fluorescence into the selected band. These are levels with a certain symmetry, determined by the selected fluorescence band.

The selective excitation of single upper levels, which is possible in molecular beams with sufficiently good collimation, results even in polyatomic molecules in astonishingly simple fluorescence spectra. This is, for example, demonstrated in Fig. 1.56 for the NO_2 molecule, which is excited at a fixed wavelength $\lambda_L = 592$ nm. The fluorescence spectrum consists of readily assigned vibrational bands that are composed of three rotational lines (strong P and R lines, and weak Q lines).

Instead of measuring the fluorescence intensity $I_{Fl}(\lambda_L)$, the excitation spectrum can also be monitored via resonant two-photon ionization (RTPI). This is illustrated by Fig. 4.7, which shows a RTPI spectrum of a band head of the Cs_2 molecule,

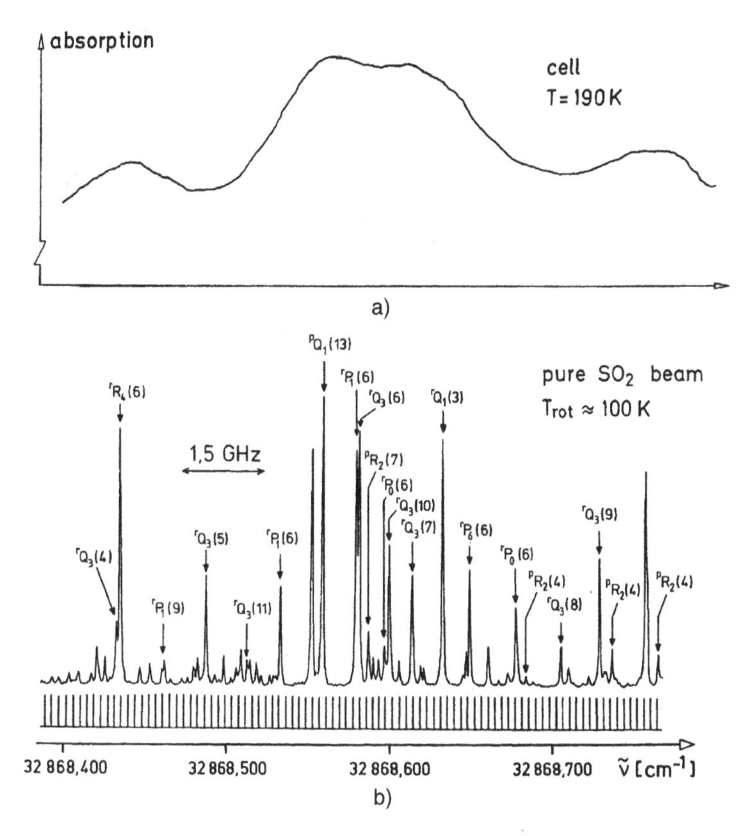

Fig. 4.4 Section of the excitation spectrum of SO_2: (**a**) taken in a SO_2 cell at 0.1 mbar with Doppler-limited resolution; (**b**) taken in a collimated SO_2 beam [391]

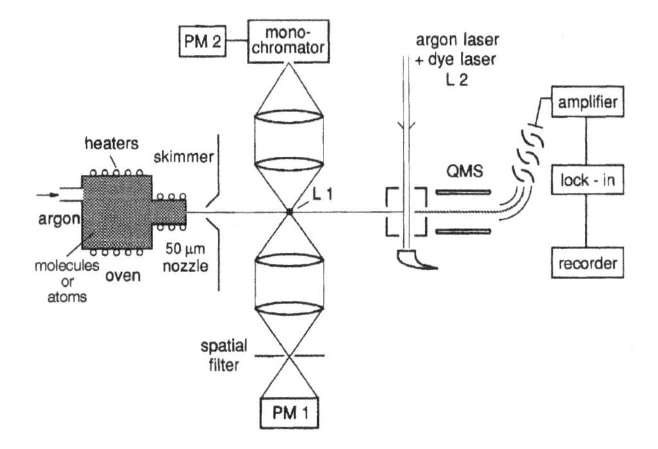

Fig. 4.5 Experimental setup for sub-Doppler spectroscopy in a collimated molecular beam. Photomultiplier PM1 monitors the total undispersed fluorescence, while PM2 behind a monochromator measures the dispersed fluorescence spectrum. The mass-specific absorption can be monitored by resonant two-color two-photon ionization in the ion source of a mass spectrometer

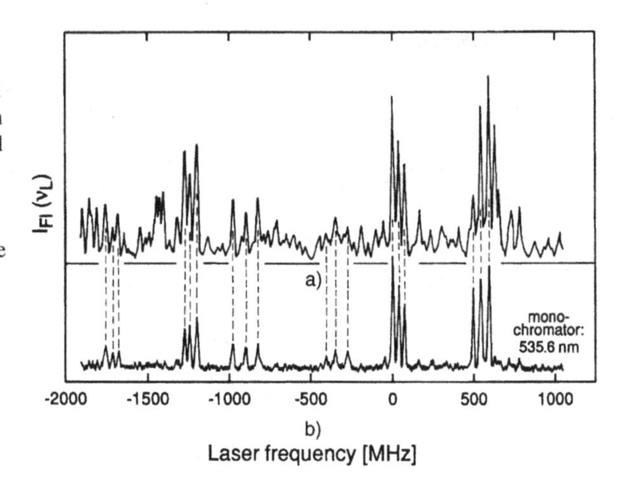

Fig. 4.6 Section of the spectrum of NO_2 excited at $\lambda_{ex} = 488$ nm in a collimated NO_2 beam with a collimation ratio of $\sin\epsilon = 1/80$: (**a**) total fluorescence monitored; and (**b**) filtered excitation spectrum, where instead of the total fluorescence only the fluorescence band at $\lambda = 535.6$ nm for the lower vibrational level (0, 10) was monitored by PM2 behind a monochromator [393]

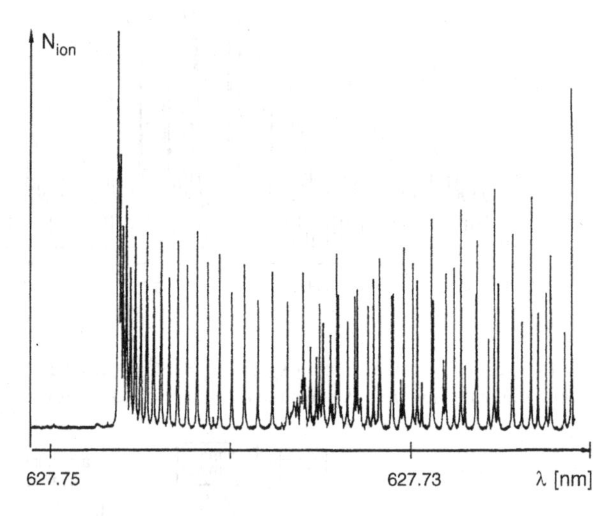

Fig. 4.7 Absorption spectrum of the (0–0) band head of the $C\,^1\Pi_u \leftarrow X\,^1\Sigma_g^+$ system of Cs_2, monitored by resonant two-photon ionization spectroscopy in a cold collimated Ar beam seeded with cesium

excited by a tunable cw dye laser and ionized by a cw argon laser [394] using the arrangement of Fig. 4.5.

Besides these three examples, a large number of atoms and molecules have been studied in molecular beams with high spectral resolution. For atoms mainly hyperfine structure splittings, isotope shifts, and Zeeman splittings have been investigated by this technique, because these splittings are generally so small that they may be completely masked in Doppler-limited spectroscopy [393, 394]. An impressive illustration of the sensitivity of this technique is the measurement of nuclear-charge radii and nuclear moments of stable and radioactive unstable isotopes through the resolution of optical hfs splittings and isotope shifts (Fig. 4.8) performed by several groups [395, 397]. Even spurious concentrations of short-lived radioactive isotopes could be measured in combination with an on-line mass separator (Fig. 4.8). From the hfs-splittings the properties of the radioactive nuclei, such as the charge radius, the nuclear spin and the spatial neutron distribution can be deduced. E. Otten,

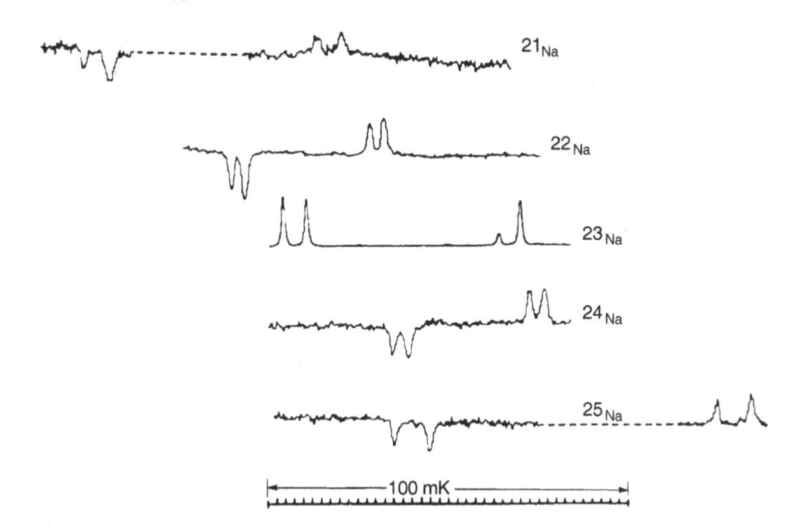

Fig. 4.8 Hyperfine structure and isotope shifts of radioactive Na-isotope [396]

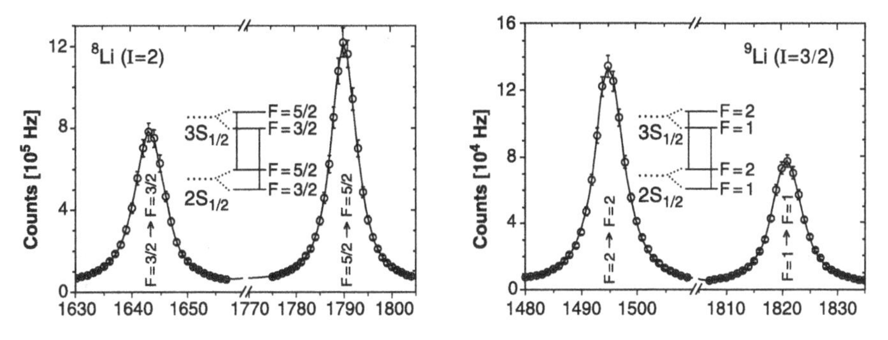

Fig. 4.9 Hyperfine structure of the radioactive lithium isotopes ^{8}Li and ^{9}Li [399]

H.J. Kluge and their groups have performed precision measurements of several short lived isotopes. One example, shown in Fig. 4.9 is the hfs of the radioactive isotopes ^{8}Li and ^{9}Li from which the spatial distribution of the additional neutrons can be inferred [398].

For molecules the line densities are much higher, and often the rotational structure can only be resolved by sub-Doppler spectroscopy. Limiting the collimation angle of the molecular beam below 2×10^{-3} rad, the residual Doppler width can be reduced to values below 500 kHz. Such high-resolution spectra with linewidths of less than 150 kHz could be, for instance, achieved in a molecular iodine beam since the residual Doppler width of the heavy I_2 molecules, which is proportional to $m^{-1/2}$, is already below this value for a collimation ratio $\epsilon \leq 4 \times 10^{-4}$ [400]. At such small linewidths the transit-time broadening from the finite interaction time of the molecules with a focused laser beam is no longer negligible, since the spontaneous lifetime already exceeds the transit time.

More examples of sub-Doppler spectroscopy in atomic or molecular beams can be found in reviews on this field by Jacquinot [396], and Lange et al. [401], in the two-volume edition on molecular beams by Scoles [402], as well as in [403–406].

4.2 Adiabatic Cooling in Supersonic Beams

For effusive beams discussed in the previous section, the pressure in the reservoir is so low that the mean free path Λ of the molecules is large compared with the diameter a of the hole A. This implies that collisions during the expansion can be neglected. Now we will treat the case where $\Lambda \ll a$. This means that the molecules suffer many collisions during their passage through A and in the spatial region behind A. In this case the expanding gas may be described by a hydrodynamic-flow model [407]: the expansion occurs so rapidly that essentially no heat exchange occurs between the gas and the walls, the expansion is adiabatic, and the enthalpy per mole of the expanding gas is conserved.

The total energy E of a mole with mass M is the sum of the internal energy $U = U_{\text{trans}} + U_{\text{rot}} + U_{\text{vib}}$ of a gas volume at rest in the reservoir, its potential energy pV, and the kinetic-flow energy $\frac{1}{2}Mu^2$ of the gas expanding with the mean flow velocity $u(z)$ in the z-direction into the vacuum. Energy conservation demands for the total energy before and after the expansion

$$U_0 + p_0 V_0 + \frac{1}{2}Mu_0^2 = U + pV + \frac{1}{2}Mu^2. \tag{4.11}$$

If the mass flow $\mathrm{d}M/\mathrm{d}t$ through A is small compared to the total mass of the gas in the reservoir, we can assume thermal equilibrium inside the reservoir, which implies $u_0 = 0$. Since the gas expands into the vacuum chamber, the pressure after the expansion is small ($p \ll p_0$). Therefore we may approximate (4.11) by setting $p = 0$, which yields

$$U_0 + p_0 V_0 = U + \frac{1}{2}Mu^2. \tag{4.12}$$

This equation illustrates that a "cold beam" with small internal energy U is obtained, if most of the initial energy $U_0 + p_0 V_0$ is converted into kinetic-flow energy $\frac{1}{2}Mu^2$. The flow velocity u may exceed the local velocity of sound $c(p, T)$. In this case a supersonic beam is produced. In the limiting case of total conversion we would expect $U = 0$, which means $T = 0$. We will later discuss several reasons why this ideal case cannot be reached in reality.

The decrease of the internal energy means also a decrease of the relative velocities of the molecules. In a microscopic model this may be understood as follows (Fig. 4.10): during the expansion the faster molecules collide with slower ones flying ahead and transfer kinetic energy.

The energy transfer rate decreases with decreasing relative velocity and decreasing density and is therefore important only during the first stage of the expan-

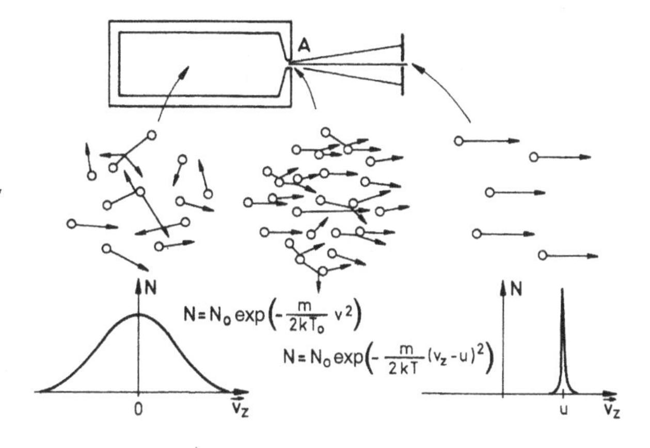

Fig. 4.10 Molecular model of adiabatic cooling by collisions during the expansion from a reservoir with a Maxwellian velocity distribution into the directed molecular flow with a narrow distribution around the flow velocity u [409]

sion. Head-on collisions with impact parameter zero narrow the velocity distribution $n(v_\parallel)$ of velocity components $v_\parallel = v_z$ parallel to the flow velocity $\boldsymbol{u}$ in the z-direction. This results in a modified Maxwellian distribution

$$n(v_z) = C_1 \exp\left(-\frac{m(v_z - u)^2}{2kT_\parallel}\right), \tag{4.13}$$

around the flow velocity u. This distribution may be characterized by the *translational temperature* $T_\parallel$, which is a measure of the width of the distribution (4.13).

For collisions with nonzero impact parameter both collision partners are deflected. If the deflection angle is larger than the collimation angle ϵ, these molecules can no longer pass the collimating aperture B in Fig. 4.1. The aperture causes for effusive as well as for supersonic beams a reduction of the transverse velocity components. Along the beam axis z the width of the distribution $n(v_x)$ measured within a fixed spatial interval Δx, which can be set by the optics for LIF detection, decreases proportionally to $\Delta x/z$. This is often named *geometrical cooling* because the reduction of the width of $n(v_x)$ is not caused by collision but by a pure geometrical effect. The transverse velocity distribution

$$n(v_x) = C_2 \exp\left(-\frac{mv_x^2}{2kT_\perp}\right) = C_2 \exp\left(-\frac{mv^2 \sin^2\epsilon}{2kT_\perp}\right), \tag{4.14}$$

is often characterized by a *transverse temperature* $T_\perp$, which is determined by the velocity distribution $n(v)$, the collimation ratio $\tan\epsilon = v_x/v_z = b/2d$, or $\sin\epsilon = v_x/v$ and the distance z from the nozzle.

The reduction of the velocity distributions $n(v_x)$ and $n(v_z)$ can be measured with different spectroscopic techniques. The first method is based on measurements of the Doppler profiles of absorption lines (Fig. 4.11). The beam of a single-mode dye laser is split into one beam that crosses the molecular beam perpendicularly, and another that is directed anticollinearly to the molecular beam. The maximum ω_m of the Doppler-shifted absorption profile yields the most probable velocity $v_p = (\omega_0 - \omega_m)/k$, while the absorption profiles of the two arrangements give the distribution $n(v_\parallel)$ and $n(v_\perp)$ [408, 409].

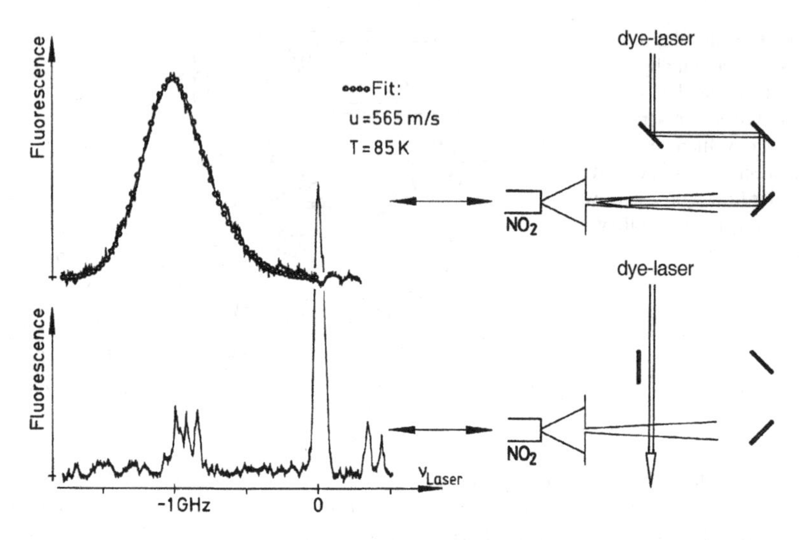

Fig. 4.11 Determination of the velocity distributions $n(v_\parallel)$ and $n(v_\perp)$ by measuring the Doppler profile of absorption lines in a thermal, effusive NO_2 beam [409]

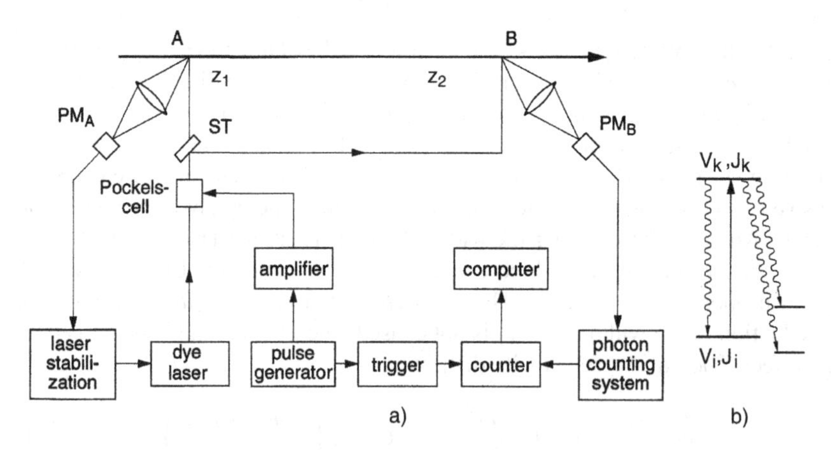

Fig. 4.12 Quantum-state-specific determination of the velocity distribution $n(i, v_\parallel)$ of molecules in an absorbing level $|i\rangle = (v_i, J_i)$ by time-of-flight measurements

The second method is based on a time-of-flight measurement. The laser beam is again split, but now both partial beams cross the molecular beam perpendicularly at different positions z_1 and z_2 (Fig. 4.12). When the laser is tuned to a molecular transition $|i\rangle \to |k\rangle$ the lower level $|i\rangle = (v_i, J_i)$ is partly depleted due to optical pumping. The second laser beam therefore experiences a smaller absorption and produces a smaller fluorescence signal. In the case of molecules even small intensities are sufficient to saturate a transition and to completely deplete the lower level (Sect. 2.1).

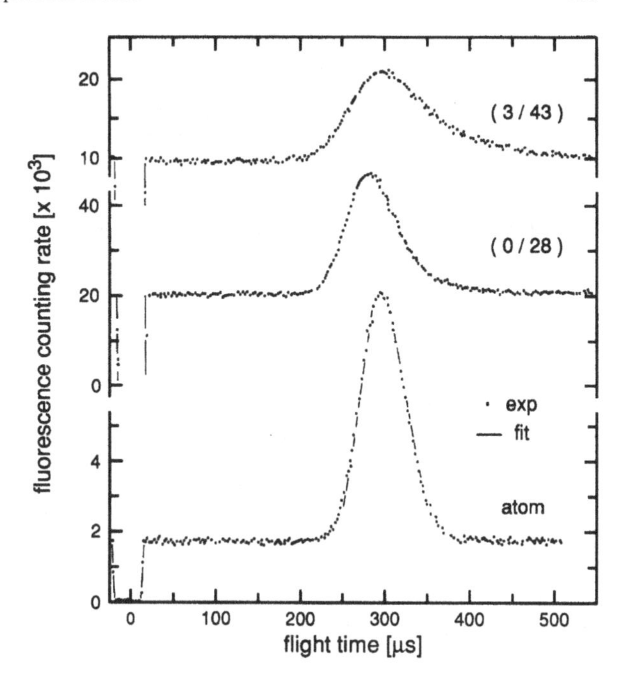

Fig. 4.13 Time-of-flight distribution of Na atoms and Na_2 molecules in two different vibrational–rotational levels ($v'' = 3$, $J'' = 43$) and ($v'' = 0$, $J'' = 28$) [410]

If the first laser beam is interrupted at the time t_0 for a time interval Δt that is short compared to the transit time $T = (z_2 - z_1)/v$ (this can be realized by a Pockels cell or a fast mechanical chopper), a pulse of molecules in level $|i\rangle$ can pass the pump region without being depleted. Because of their velocity distribution, the different molecules reach z_2 at different times $t = t_0 + T$. The time-resolved detection of the fluorescence intensity $I_{\mathrm{Fl}}(t)$ induced by the second, noninterrupted laser beam yields the distribution $n(T) = n(\Delta z/v)$, which can be converted by a Fourier transformation into the velocity distribution $n(v)$. Figure 4.13 shows as an example the velocity distribution of Na atoms and Na_2 molecules in a sodium beam in the intermediate range between effusive and supersonic conditions. If the molecules Na_2 had been formed in the reservoir before the expansion, one would expect the relation $v_p(\mathrm{Na}) = \sqrt{2} v_p(\mathrm{Na}_2)$ because the mass $m(\mathrm{Na}_2) = 2m(\mathrm{Na})$. The result of Fig. 4.13 proves that the Na_2 molecules have a larger most probable velocity v_p. This implies that most of the dimers are formed during the adiabatic expansion [410].

The laser spectroscopic techniques provide much more detailed information about the state-dependent velocity distribution than measurements with mechanical velocity selectors. Note that in Fig. 4.13 not only $v_p(\mathrm{Na}_2) > v_p(\mathrm{Na})$ but the velocity distribution of the Na_2 molecules differs for different vibration–rotation levels (v, J). This is due to the fact that molecules are being formed by stabilizing collisions during the adiabatic expansion. Molecules in lower states have suffered more collisions with atoms of the "cold bath." Their distribution $n(v)$ becomes narrower and their most probable velocity v_p more closely approaches the flow velocity u.

The main advantage of cold molecular beams for molecular spectroscopy is the decrease of rotational energy U_{rot} and vibrational energy U_{vib}, which results in

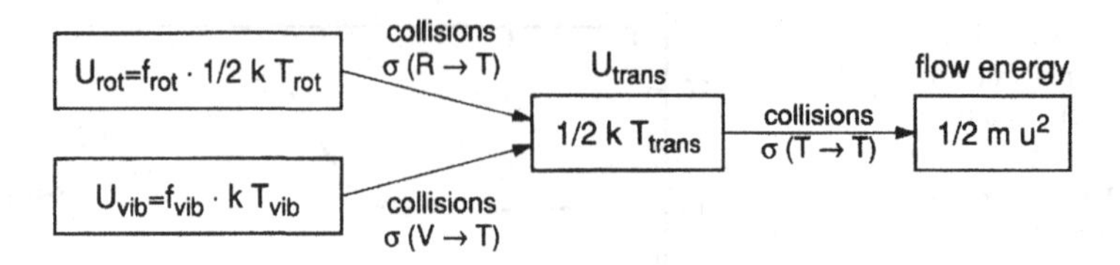

Fig. 4.14 Energy transfer diagram for adiabatic cooling in supersonic molecular beams

a compression of the population distribution $n(v, J)$ into the lowest vibrational and rotational levels. This energy transfer proceeds via collisions during the adiabatic expansion (Fig. 4.14). Since the cross sections for collisional energy transfer $U_{\text{rot}} \to U_{\text{trans}}$ are generally smaller than for elastic collisions ($U_{\text{trans}} \to U_{\text{trans}}$), the rotational energy of the molecules before the expansion cannot be completely transferred into flow energy during the short time interval of the expansion, where collisions are important. This implies that after the expansion $U_{\text{rot}} > U_{\text{trans}}$, which means that the translational energy of the relative velocities (internal kinetic energy) has cooled faster than the rotational energy. The rotational degrees of freedom are not completely in thermal equilibrium with the translation. However, for molecules with sufficiently small rotational constants, the cross sections $\sigma_{\text{rot–rot}}$ for collisional redistribution within the manifold of rotational levels may be larger than $\sigma_{\text{rot–trans}}$. In these cases it is often possible to describe the rotational population $n(J)$ approximately by a Boltzmann distribution

$$n(J) = C_2(2J+1)\exp\left(-\frac{E_{\text{rot}}}{kT_{\text{rot}}}\right), \tag{4.15}$$

where $C_2 = n_{\text{v}}/Z$ is a constant, depending on the partition function Z and the total population density n_{v} in a vibrational level. This defines a *rotational temperature* T_{rot}, which is higher than the translational temperature $T_{\|}$ defined by (4.13).

The rotational temperature can be determined experimentally by measuring relative intensities of the absorption lines

$$I_{\text{abs}} = C_1 n_i(v_i, J_i) B_{\text{ik}} \rho_L, \tag{4.16a}$$

for different rotational lines in the same vibrational band, starting from different rotational levels $|J_i\rangle$ in the lower state. If the laser intensity $I_{\text{L}} = \rho_{\text{L}} c$ is sufficiently low, saturation can be neglected and the line intensities (monitored, for instance, by laser-induced fluorescence, LIF) are proportional to the unsaturated population densities $n_1(v_i J_i)$.

For unperturbed transitions the relative transition propabilities of different rotational lines in the same vibrational bands are given by the corresponding Hönl–London factors [411] and can thus be readily calculated. In case of perturbed spectra this may be no longer true. Here the following alternatives can be used:

- If the laser intensity is sufficiently large the molecular transitions are saturated. In the case of complete saturation ($S \gg 1$, Sect. 2.1) every molecule in the absorbing level $|i\rangle$ passing through the laser beam will absorb a photon. The intensity of the absorption line is then *independent* of the transition probability B_{ik} but only depends on the molecular flux $N(v_i, J_i) = u(v_i, J_i) \times n(v_i, J_i)$. We then obtain instead of (4.16a)

$$I_{abs} = C_2 u(v_i, J_i) n(v_i, J_i). \tag{4.16b}$$

- Another possibility for the determination of T_{rot} in the case of perturbed spectra is based on the following procedure: at first the relative intensities $I_{th}(v, J)$ are measured in an effusive thermal beam with a sufficiently low pressure p_0 in the reservoir. Here the population distribution $n(v, J)$ of the total density n can still be described by a Boltzmann distribution

$$n_{th}(v_i, J_i) = (g_i/Z) \cdot n \exp(-E_i/kT_0),$$

with the partition function $Z = \sum_i g_i e^{-E_i/kT_0}$ and the reservoir temperature T_0 since cooling is negligible. Then the changes in the relative intensities are observed while the pressure p_0 is increased and adiabatic cooling starts. This yields the dependence $T_{rot}(p_0, T_0)$ of rotational temperatures in the supersonic beam on the reservoir parameters p_0 and T_0.

If the intensities in the supersonic beam are named $I_s(v, J)$, we obtain

$$\frac{I_s(v_i, J_i)/I_s(v_k, J_k)}{I_{th}(v_i, J_i)/I_{th}(v_k, J_k)} = \frac{n_s(v_i, J_i)/n_s(v_k, J_k)}{n_{th}(v_i, J_i)/n_{th}(v_k, J_k)}. \tag{4.17}$$

Since the relative Boltzmann distribution in the thermal beam is

$$\frac{n_{th}(v_i, J_i)}{n_{th}(v_k, J_k)} = \frac{g_i}{g_k} e^{-(E_i - E_k)/kT_0}, \tag{4.18}$$

we can determine the rotational and vibrational temperatures T_{rot} and T_{vib} with (4.15)–(4.18) from the equation

$$n(v_i J_i) = (g_i/Z) \cdot n\, e^{-E_{vib}/kT_{vib}} e^{-E_{rot}/kT_{rot}}, \tag{4.19}$$

where $g_i = g_{vib} g_{rot}$ is the statistical weight factor.

The cross sections $\sigma_{vib-trans}$ or $\sigma_{vib-rot}$ are generally much smaller than $\sigma_{rot-trans}$. This implies that the cooling of vibrational energy is less effective than that of E_{rot}. Although the population distribution $n(v)$ deviates more or less from a Boltzmann distribution, it is often described by the *vibrational temperature* T_{vib}. From the discussion above we can deduce the relation

$$T_{trans} < T_{rot} < T_{vib}. \tag{4.20}$$

The lowest translational temperatures $T_\parallel < 1$ K can be reached with supersonic beams of noble gas atoms. The reason for this fact is the following: If two atoms A recombine during the expansion to form a dimer A_2, the binding energy is transferred to a third collision partner. This results in heating of the cold beam and prevents the translational temperature from reaching its lowest possible value. Since

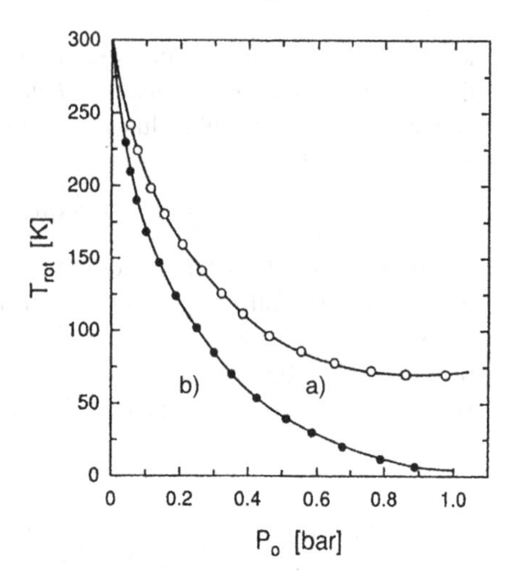

Fig. 4.15 Rotational temperature T_{rot} of NO_2 molecules in a pure NO_2 beam (*a*) and in an argon beam seeded with 5 % NO_2 (*b*) as a function of the pressure p_0 in the reservoir

the binding energy of noble gas atoms is very small, this heating effect is generally negligible in beams of noble gas atoms.

In order to reach low values of T_{rot} for molecules, it is advantageous to use noble gas atomic beams that are "seeded" with a few percent of the wanted molecules. The cold bath of the atoms acts as a heat sink for the transfer of rotational energy of the molecules to translational energy of the atoms. This effect is demonstrated by Fig. 4.15, which shows the rotational temperature T_{rot} as a function of the pressure p_0 in the reservoir for a pure NO_2 beam and an argon beam seeded with 5 % NO_2 molecules.

Example 4.2 In a molecular beam of 3 % NO_2 diluted in argon one measures at a total pressure of $p_0 = 1$ bar in the reservoir and a nozzle diameter $a = 100\ \mu m$

$$T_{trans} \approx 1\ K, \qquad T_{rot} \approx 5\text{–}10\ K, \qquad T_{vib} \approx 50\text{–}100\ K.$$

Because of the small cross sections $\sigma_{vib\text{–}trans}$, vibrational cooling in seeded noble gas beams is not as effective. Here a beam of cold inert molecules such as nitrogen N_2 or SF_6 seeded with the molecules M under investigation may be better for cooling T_{vib} by vibrational–vibrational energy transfer [412].

The reduction of T_{rot} and T_{vib} results in a drastic simplification of the molecular absorption spectrum because only the lowest, still populated levels contribute to the absorption. Transitions from low rotational levels become stronger, those from higher rotational levels are nearly completely eliminated. Even complex spectra, where several bands may overlap at room temperature, reduce at sufficiently low rotational temperatures in cold beams to a few rotational lines for each band, which are grouped around the band head. This greatly facilitates their assignments and allows

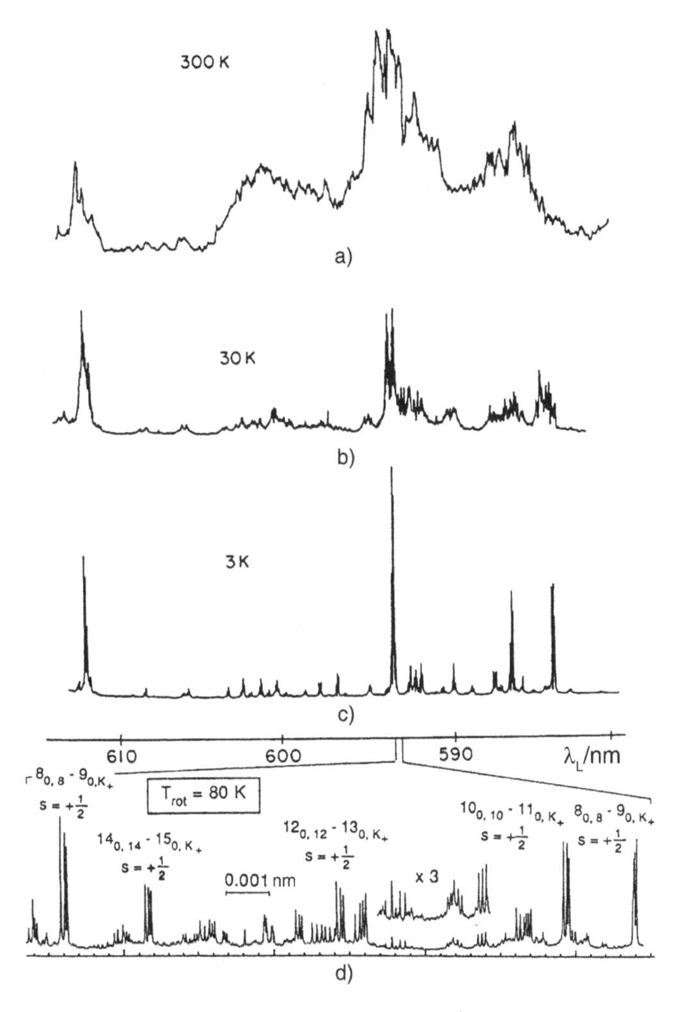

Fig. 4.16 Section of the excitation spectrum of NO_2 obtained under different experimental conditions: (**a**) in a vapor cell at $T = 300$ K, $p(NO_2) = 0.05$ mbar; (**b**) in a pure NO_2 beam at $T_{rot} = 30$ K; (**c**) in a supersonic argon beam seeded with 5 % NO_2 at $T_{rot} = 3$ K, where (**a–c**) were excited with a dye laser with 0.05 nm bandwidth [413]; (**d**) 0.01 nm section of (**b**) recorded with a single-mode dye laser (1 MHz bandwidth) [414]

one to determine the band origins more reliably. For illustration, Fig. 4.16 depicts the same section of the visible NO_2 spectrum recorded under different experimental conditions. While in the congested spectrum at room temperature no spectral lines can be recognized, the spectrum at $T_{rot} = 3$ K in a cold He jet seeded with NO_2 [413] clearly demonstrates the cooling effect and shows the well-separated manifold of vibronic bands. The spectrum in Fig. 4.16d shows one band recorded with sub-Doppler resolution at $T_{rot} = 80$ K, where rotational levels up to $J = 12$ are populated.

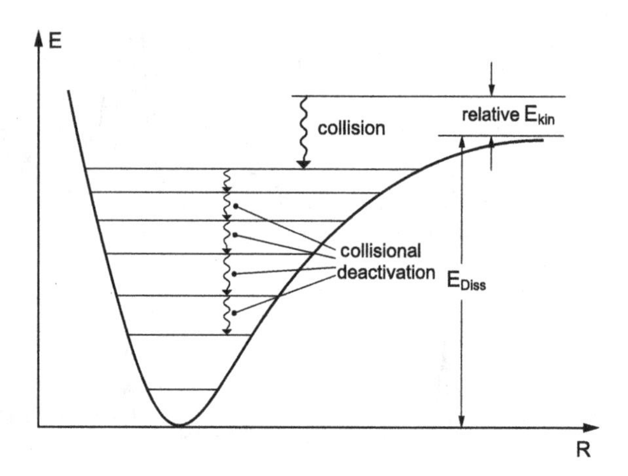

Fig. 4.17 Potential diagram for the explanation of the formation of molecular complexes in cold molecular beams by collisional deactivation

In addition, a large number of molecules has been investigated in cold molecular beams. Even large biomolecules become accessible to laser spectroscopic techniques [413–418].

Rotational temperatures $T_{\mathrm{rot}} < 1$ K have been achieved with pulsed supersonic beams. A valve between the reservoir and nozzle opens for times $\Delta t \approx 0.1$–1 ms with repetition frequencies f adapted to that of the pulsed lasers. Pressures p_0 up to 100 bar are used, which demand only modest pumping speeds because of the small duty cycle $\Delta t \cdot f \ll 1$.

4.3 Formation and Spectroscopy of Clusters and Van der Waals Molecules in Cold Molecular Beams

Because of their small relative velocities Δv (Fig. 4.10), atoms A or molecules M with mass m may recombine to bound systems A_n or M_n ($n = 2, 3, 4, \dots$) if the small translational energy $\frac{1}{2}m\overline{\Delta v^2}$ of their relative motion can be transferred to a third collision partner (which may be another atom or molecule or the wall of the nozzle). This results in the formation of loosely bound atomic or molecular complexes (e.g., NaHe, I_2He_4) or clusters that are bound systems of n equal atoms or molecules (e.g., Na_n, Ar_n, $(H_2O)_n$, where $n = 2, 3, \dots$) (Fig. 4.17).

In a thermodynamic model condensation takes place if the vapor pressure of the condensating substance falls below the total local pressure. The vapor pressure

$$p_s = A\mathrm{e}^{-B/T},$$

in the expanding beam decreases exponentially with decreasing temperature T, while the total pressure p_t decreases because of the decreasing density in the expanding beam and the falling temperature (Fig. 4.18). If sufficient three-body collisions occur in regions where $p_s \le p_t$ recombination can take place.

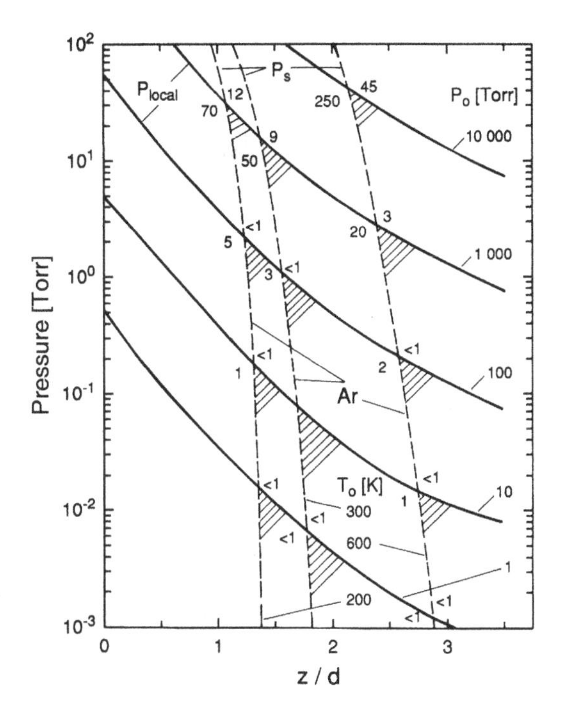

Fig. 4.18 Vapor pressure p_s of argon and local total pressure p_{loc} as a function of normalized distance $z^* = z/d$ from the nozzle in units of the nozzle diameter d for different stagnation pressures p_0 in the reservoir. Condensation can take place in the hatched areas. The numbers of three-body collisions at the points where $p_s = p_{loc}$ are also given [419]

Clusters represent a transition regime between molecules and small liquid drops or small solid particles, and they have therefore found increasing interest [420–423]. Laser spectroscopy has contributed in an outstanding way to the investigation of cluster structures and dynamics. A typical experimental arrangement for the study of small metal clusters is shown in Fig. 4.5. The metal vapor is produced in an oven and mixed with argon. The mixture expands through a small nozzle ($\sim$50 μm diameter) and cools to rotational temperatures of a few Kelvin. Resonant two-photon ionization with a tunable dye laser and an argon-ion laser converts the clusters $A_n(i)$ in a specified level $|i\rangle$ into ions, which are monitored behind a quadrupole mass spectrometer in order to select the wanted cluster species A_n [424]. If the wavelength of the ionizing laser is chosen properly, fragmentation of the ionized clusters A_n^+ can be avoided or at least minimized, and the measured mass distribution $N(A_n^+)$ represents the distribution of the neutral clusters $N(A_n)$, which can be studied as a function of the oven parameters (p_0, T_0), seed-gas concentration, and the nozzle diameter [425]. This gives information about the nucleation process of clusters during the expansion of a supersonic beam.

The unimolecular dissociation of cluster ions after photoexcitation can be used to determine their binding energy as a function of cluster size [426] with the apparatus shown schematically in Fig. 4.19. The clusters, formed during the adiabatic expansion through the nozzle, are photoionized by a pulsed UV laser L_1 in region 1. After acceleration the ions fly through a field-free region 2 and pass deflection plates at a time t that depends on their mass. When the wanted cluster mass arrives at the deflection region the deflecting voltage is switched off, allowing this species

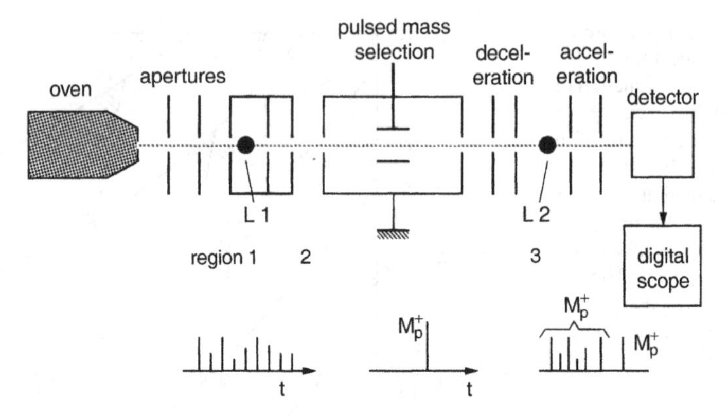

Fig. 4.19 Time-of-flight molecular beam apparatus for the measurement of cluster fragmentation

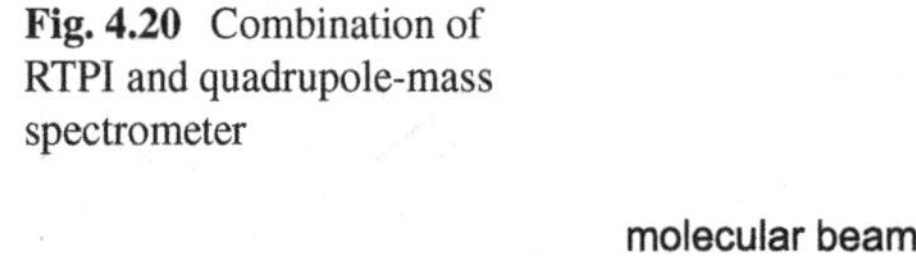

Fig. 4.20 Combination of RTPI and quadrupole-mass spectrometer

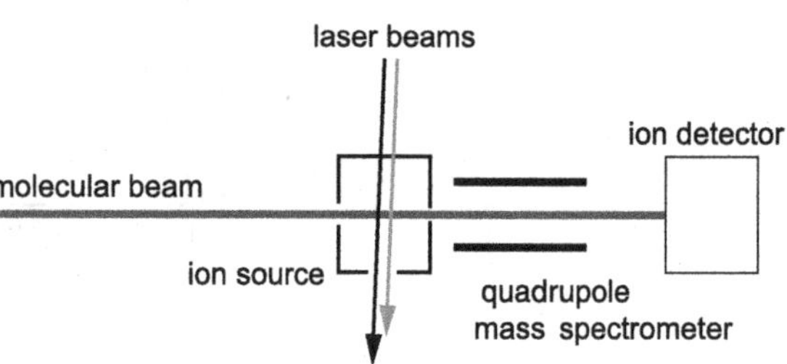

to fly straight on. In region 3 the related cluster ions are excited by a second tunable laser L_2, which results in their fragmentation. The fragments are detected by a time-of-flight mass spectrometer.

Clusters are often *floppy systems*, which do not have a rigid molecular geometry. Their spectroscopy therefore offers the possibility to study new behavior of quantum systems between regular, "well-behaved" molecules with normal vibrations and classically chaotic systems with irregular movements of the nuclei [427].

The most intensively studied clusters are alkali-metal clusters [428–431], where the stability and the ionization energies have been measured as have the electronic spectra and their transition from localized molecular orbitals to delocalized band structure of solids [432, 433].

Molecular clusters are also formed during the adiabatic expansion of free jets. Examples are the production of benzene clusters $(C_6H_6)_n$ and their analysis by two-photon ionization in a reflectron [434], or the determination of the structure and ionization potential of benzene–argon complexes by two-color resonance-enhanced two-photon ionization techniques [435]. Since clusters are formed with a broad mass distribution, it is advantageous to select individual masses by a mass spectrometer. For cw-spectroscopy a quadrupole mass filter is the best choice (Fig. 4.20) while for pulsed excitation a time of flight mass spectrometer is preferable (Fig. 4.19).

The structure of molecular complexes in their electronic ground state can be obtained from direct IR laser absorption spectroscopy in pulsed supersonic-slit jet ex-

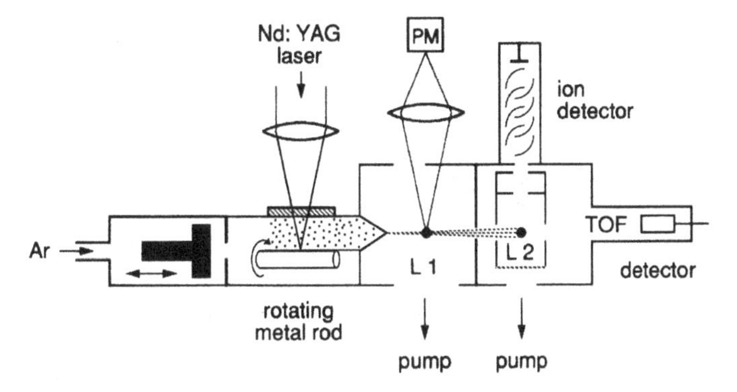

Fig. 4.21 Production of cold metal clusters by laser evaporation of metal vapors

pansions [436]. This allows one to follow the formation rate of clusters and complexes during the adiabatic expansion [438]. Selective photodissociation of van der Waals clusters by infrared lasers may be used for isotope separation [439].

An interesting technique for the production of metal clusters has been developed by Smalley et al. [440]. A slowly rotating metal rod is irradiated by pulses from a Nd:YAG laser (Fig. 4.21). The metal vapor produced by evaporation of material in the focal spot on the rod surface is mixed with a noble gas, which is let into the evaporation chamber through a pulsed nozzle synchronized with the laser pulses. The resulting mixture of noble gas and metal vapor expands through a narrow nozzle. In the resulting supersonic pulsed jet, metal clusters are formed, which can be analyzed by their fluorescence induced by laser L_1 or by two-photon ionization with L_2. The mass distribution can be measured with a time-of-flight spectrometer.

With this technique metal clusters of materials with high melting points could be produced, which are more difficult to realize by evaporation in hot furnaces. A famous example of carbon clusters formed by this technique for the first time were the fullerenes C_{60}, C_{70}, etc. [441].

An elegant technique for studying van der Waals complexes at low temperatures was developed by Toennies and coworkers [442]. A beam of large He clusters (10^4–10^5 He atoms) passes through a region with a sufficient vapor pressure of atoms or molecules. The He droplets pick up a molecule which either sticks to the surface or diffuses into the central part of the droplet, where it is cooled down to a low temperature of 100 mK up to a few Kelvin (see Fig. 4.22). Since the interaction with the He atoms is very small, the spectrum of this trapped molecule does not differ much from that of a free cold molecule. However, unlike cooling during the adiabatic expansion of a supersonic jet, where $T_{vib} > T_{rot} > T_{trans}$, in this case $T_{rot} = T_{vib} = T_{He}$ [443–445]. This implies that all molecules are at their lowest vibration-rotational levels and the absorption spectrum becomes considerably simplified.

At temperatures below 2 K, the inner part of the He droplet becomes superfluid. In this case the molecules imbedded in the droplet can freely rotate, which shows up in their rotational absorption spectrum. For illustration, Fig. 4.23 shows a section

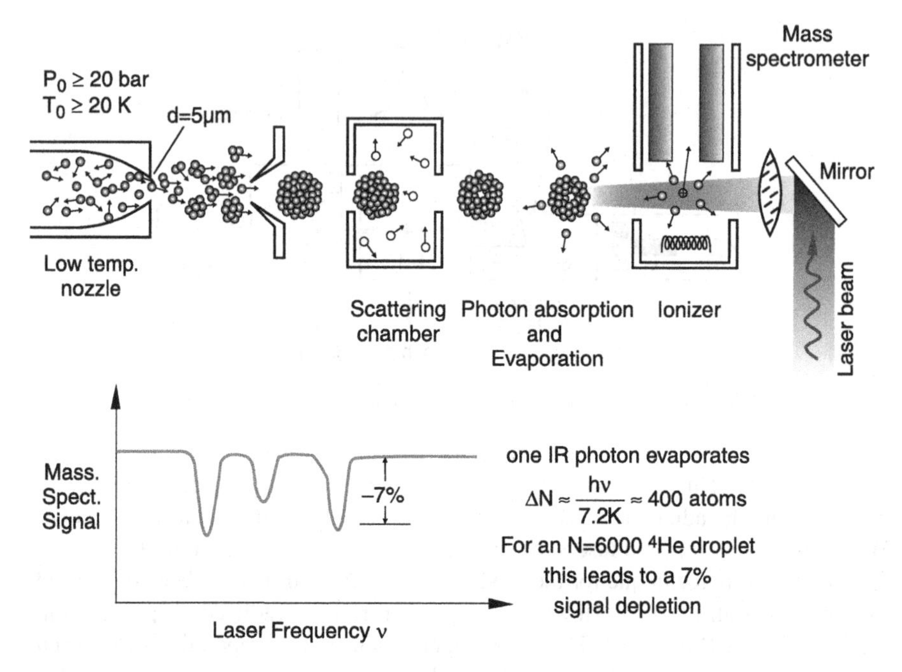

Fig. 4.22 Experimental setup for the production of He droplets, pick-up of molecules and IR-spectroscopy of molecules in He droplets, using depletion spectroscopy (P. Toennies, http://wwwuser.gwdg.de/mpisfto/)

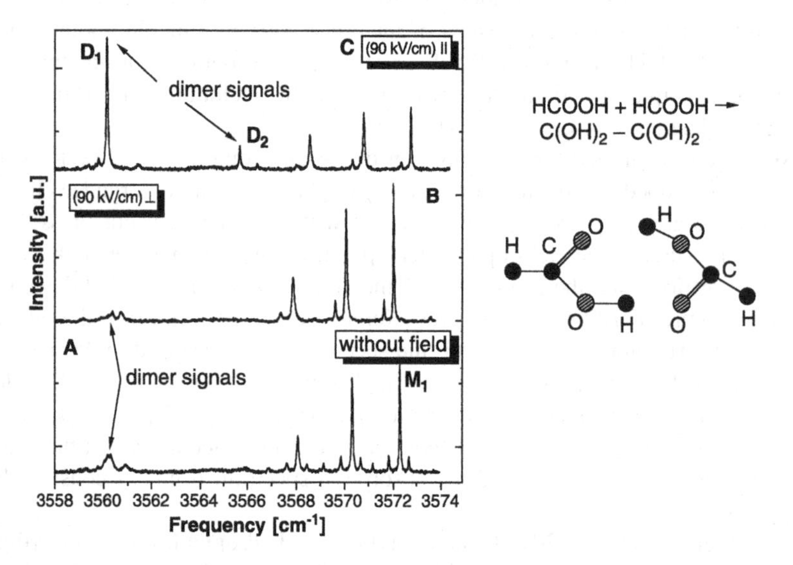

Fig. 4.23 Section of the formic acid HCOOH and its dimer in cold He droplets with and without an external electric field [446]

of the infrared spectrum of formic acid molecules and their dimers [446]. High-resolution spectroscopy of small molecules such as NO has been performed for He droplets, where the hyperfine structure could be resolved [447].

Interesting studies of high-spin alkali clusters attached to He droplets have been performed by several groups [448]. The cold He droplets pass through an alkali vapor where they pick up alkali atoms. On the surfaces of the droplets the atoms can diffuse and recombine into dimers, trimers and larger clusters. If they recombine into their electronic ground state, the large binding energy releases a correspondingly large recombination energy, which results in the heating of the droplet and the evaporation of the alkali atoms. Therefore they can be attached to the surface only if they recombine into high-spin states which have a much smaller binding energy.

This technique provides access to the spectroscopy of high-spin states which are not readily formed in the gas phase, and furthermore it allows these clusters with the helium droplets to be investigated by observing their level shifts.

A similar technique, where alkali molecules and clusters in high spin states are formed, was invented by Scoles and coworkers [450]. Here the He-clusters pass through a region with high alkali vapor pressure. The alkali atoms condense at the surface of the He clusters. Alkali atoms meet and recombine by migration along the surface. Since the singlet states of alkali dimers have a large binding energy, which is transferred to the He cluster, this leads to the evaporation of many He atoms and may completely destroy the cluster. However, the formation of triplet states with a much smaller binding energy will evaporate only a few He atoms. The alkali dimers and multimers in high spin states rapidly adjust their temperature to that of the He cluster. They can be studied by laser spectroscopy, thus giving access to states that are difficult to produce in normal gas phase spectroscopy [451].

Van der Waals diatomics formed through the weak interaction between a noble gas atom and an alkali atom have also been studied in noble gas beams seeded with alkali vapor [452].

4.4 Nonlinear Spectroscopy in Molecular Beams

The residual Doppler width from the finite collimation ratio ϵ of the molecular beam can be completely eliminated when nonlinear Doppler-free techniques are applied. Since collisions can generally be neglected at the crossing point of the molecular and laser beam, the lower molecular level $|i\rangle$ depleted by absorption of laser photons can be only refilled by diffusion of new, unpumped molecules into the interaction zone and by the small fraction of the fluorescence terminating on the initial level $|i\rangle$. The saturation intensity I_s is therefore lower in molecular beams than in gas cells (Example 2.3).

A possible arrangement for saturation spectroscopy in a molecular beam is depicted in Fig. 4.24. The laser beam crosses the molecular beam perpendicularly and is reflected by the mirror M1. The incident and the reflected beam can only be ab-

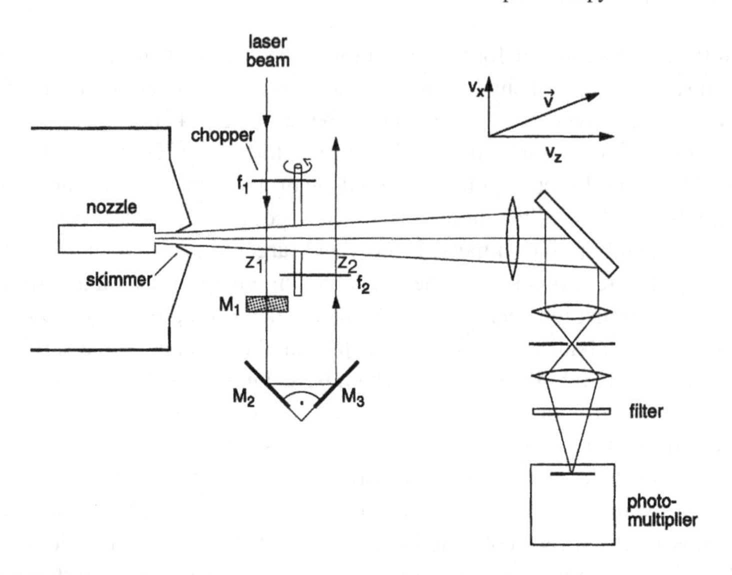

Fig. 4.24 Experimental setup for saturation spectroscopy in a collimated molecular beam

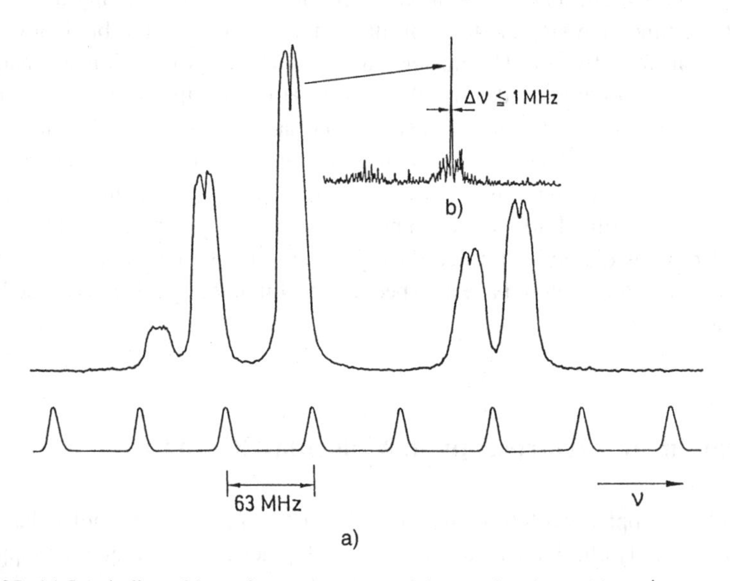

Fig. 4.25 (**a**) Lamb dips of hyperfine components of the rotational transition $J' = 1 \leftarrow J'' = 0$ in a collimated NO_2 beam. The residual Doppler width is 15 MHz. (**b**) The *insert* shows one component with suppression of the Doppler background by chopping the laser beams and using lock-in techniques [453]

sorbed by the same molecules within the transverse velocity group $v_x = 0 \pm \gamma k$ if the laser frequency $\omega_L = \omega_0 \pm \gamma$ matches the molecular absorption frequency ω_0 within the homogeneous linewidth γ. When tuning the laser frequency ω_L one observes narrow Lamb dips (Fig. 4.25) with a saturation-broadened width γ_s at the center of broader profiles with a reduced Doppler-width $\epsilon \Delta\omega_D$, from the collimation ratio $\epsilon \ll 1$ of the molecular beam (Sect. 4.1).

It is essential that the two laser beams cross the molecular beam exactly perpendicularly; otherwise, the opposite Doppler shifts $\pm\delta\omega_D$ observed for the two beams result in a broadening of the Lamb dips for $2\delta\omega_D \leq \gamma$, while no Lamb dips can be observed for $2\delta\omega_D \gg \gamma$!

Example 4.3 In a supersonic beam with $u = 10^3$ m/s, $\epsilon = 10^{-2}$, the residual Doppler width for a visible transition with $\nu = 6 \times 10^{14}$ is $\epsilon \Delta\omega_D \sim 2\pi \cdot 20$ MHz. For a crossing angle of 89° the Doppler shift between the two beams is $2\delta\omega_D = 2\pi \cdot 60$ MHz. In this case the Doppler shift $2\delta\omega_D$ is larger than the residual Doppler width, and even in the linear spectrum with reduced Doppler width the lines would be doubled. No Lamb dips can be observed with opposite laser beams.

If the width γ_s of the Lamb dips is very narrow, the demand for exactly perpendicular crossing becomes very stringent. In this case, an arrangement where the mirror M1 is removed and replaced by the retroreflectors M2, M3 in Fig. 4.24 is experimentally more convenient. The two laser beams intersect the molecular beam at two closely spaced locations z_1 and z_2. This arrangement eliminates the reflection of the laser beam back into the laser.

The Doppler-broadened background with the residual Doppler width from the divergence of the molecular beam can completely be eliminated by chopping the two laser beams at two different frequencies f_1, f_2, and monitoring the signal at the sum frequency $f_1 + f_2$ (intermodulated fluorescence, Sect. 2.3.1). This is demonstrated by the insert in Fig. 4.25. The linewidth of the Lamb dips in Fig. 4.25 is below 1 MHz and is mainly limited by frequency fluctuations of the cw single-mode dye laser [453].

Additionally, several experiments on saturation spectroscopy of molecules and radicals in molecular beams have been reported [454, 455] where finer details of congested molecular spectra, such as hyperfine structure or Λ-doubling can be resolved. Another alternative is Doppler-free two-photon spectroscopy in molecular beams, where high-lying molecular levels with the same parity as the absorbing ground state levels are accessible [456].

The improvements in the sensitivity of CARS (Sect. 3.3) have made this nonlinear technique an attractive method for the investigation of molecular beams. Its spectral and spatial resolution allow the determination of the internal-state distributions of molecules in effusive or in supersonic beams, and their dependence on the location with respect to the nozzle (Sect. 3.5). An analysis of rotationally-resolved CARS spectra and their variation with increasing distance z from the nozzle allows the determination of rotational and vibrational temperatures $T_{rot}(z)$, $T_{vib}(z)$, from which the cooling rates can be obtained [457]. With cw CARS realized with focused cw laser beams the main contribution to the signal comes from the small focal volume, and a spatial resolution below 1 mm^3 can be achieved [458].

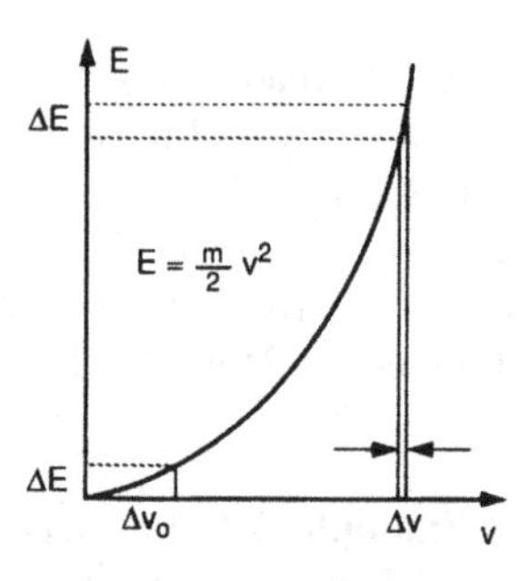

Fig. 4.26 Acceleration cooling

Another example of the application of CARS is the investigation of cluster formation in a supersonic beam. The formation rate of clusters can be inferred from the increasing intensity $I(z)$ of characteristic cluster bands in the CARS spectra.

The advantage of CARS compared to infrared absorption ion spectroscopy is the higher sensitivity and the fact that nonpolar molecules, such as N_2, can also be studied [459]. With pulsed CARS short-lived transient species produced by photo dissociation in molecular beams can also be investigated [460, 461].

4.5 Laser Spectroscopy in Fast Ion Beams

In the examples considered so far, the laser beam was crossed *perpendicularly* with the molecular beam, and the reduction of the Doppler width was achieved through the limitation of the maximum transverse velocity components v_x by geometrical apertures. One therefore often calls this reduction of the transverse velocity components geometrical cooling. Kaufmann [462] and Wing et al. [463] have independently proposed another arrangement, where the laser beam travels *collinearly* with a fast ion or atom beam and the narrowing of the *longitudinal* velocity distribution is achieved by an acceleration voltage (*acceleration cooling*). This fast-ion-beam laser spectroscopy (FIBLAS) can be understood as follows: Assume that two ions start from the ion source (Fig. 4.26) with different thermal velocities $v_1(0)$ and $v_2(0)$. After being accelerated by the voltage U their kinetic energies are

$$E_1 = \frac{m}{2}v_1^2 = \frac{1}{2}mv_1^2(0) + eU,$$

$$E_2 = \frac{m}{2}v_2^2 = \frac{1}{2}mv_2^2(0) + eU.$$

Subtracting the first equation from the second yields

$$v_2^2 - v_1^2 = v_2^2(0) - v_1^2(0) \quad \Rightarrow \quad \Delta v = v_1 - v_2 = \frac{v_0}{v}\Delta v_0,$$

with

$$v = \frac{1}{2}(v_1 + v_2), \quad \text{and} \quad v_0 = \frac{1}{2}\big[v_1(0) + v_2(0)\big].$$

Since $E_{\text{th}} = (m/2)v_0^2$ and $v = (2eU/m)^{1/2}$, we obtain for the final velocity spread

$$\Delta v = \Delta v_0 \sqrt{E_{\text{th}}/eU}. \tag{4.21}$$

For $E_{\text{th}} \ll eU \Rightarrow \Delta v \ll \Delta v_0$.

Example 4.4 $\Delta E_{\text{th}} = 0.1$ eV, $eU = 10$ keV $\rightarrow \Delta v = 3 \times 10^{-3} \Delta v_0$. This means that the Doppler width of the ions in the ion source has been decreased by acceleration cooling by a factor of 300! If the laser crosses the ion beam perpendicularly, the transverse velocity components for ions with $v = 3 \times 10^5$ m/s at a collimation ratio $\epsilon = 10^{-2}$ are $v_x = v_y \leq 3 \times 10^3$ m/s. This would result in a residual Doppler width of $\Delta \nu \sim 3$ GHz, which illustrates that for fast beams the longitudinal arrangement is superior to the transverse one.

This reduction of the velocity spread results from the fact that *energies* rather than velocities are added (Fig. 4.26). If the energy $eU \gg E_{\text{th}}$, the velocity change is mainly determined by U, but is hardly affected by the fluctuations of the initial thermal velocity. This implies, however, that the acceleration voltage has to be extremely well stabilized to take advantage of this acceleration cooling.

A voltage change ΔU results in a change Δv in the velocity v. From $(m/2)v^2 = e \cdot U$ we obtain

$$\Delta v = (e/m \cdot v)\Delta U.$$

Since $\nu = \nu_0(1 + v/c)$, the frequency change becomes

$$\Delta \nu = \frac{\nu_0}{c} \Delta v = \frac{\nu_0}{c} \frac{e}{mv} \Delta U = \nu_0 \sqrt{\frac{eU}{2mc^2} \frac{\Delta U}{U}}. \tag{4.22}$$

Example 4.5 At the acceleration voltage $U = 10$ kV, which is stable within ± 1 V, an absorption line of neon ions ($m = 21$ AMU) at $\nu_0 = 5 \times 10^{14}$ suffers Doppler broadening from the voltage instability according to (4.22) of $\Delta \nu \approx 25$ MHz.

A definite advantage of this coaxial arrangement of laser beam and ion beam in Fig. 4.27 is the longer interaction zone between the two beams because the laser-induced fluorescence can be collected by a lens from the path length Δz of several centimeters, compared to a few millimeters in the perpendicular arrangement. This increases the sensitivity because of the longer absorption path. Furthermore, transit-time broadening for an interaction length L of 10 cm becomes

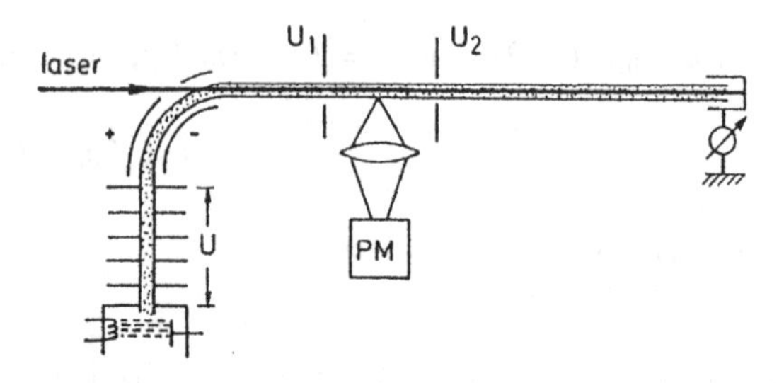

Fig. 4.27 Collinear laser spectroscopy in ion beams

$\delta\nu_{tr} \approx 0.4v/L \approx 2$ MHz, while for the perpendicular intersection of a laser beam with diameter $2w = 1$ mm, the transit time broadening $\delta\nu_{tr} = 400$ MHz gives a non-negligible contribution.

A further advantage of collinear laser spectroscopy is the possibility of "electric Doppler tuning". The absorption spectrum of the ions can be scanned across a *fixed* laser frequency ν_0 simply by tuning the acceleration voltage U. This allows one to use high-intensity, fixed-frequency lasers, such as the argon-ion laser. Because of their high gain the interaction zone may even be placed inside the laser cavity. Instead of tuning the acceleration voltage U (which influences the beam collimation), the velocity of the ions in the interaction zone with the laser beam is tuned by retarding or accelerating potentials U_1 and U_2 (Fig. 4.28).

Example 4.6 A voltage shift of $\Delta U = 100$ V at $U = 10$ kV causes for H_2^+ ions a relative frequency shift of $\Delta\nu/\nu \approx 1.5 \times 10^{-5}$. At an absorption frequency of $\nu = 6 \times 10^{14}$ s^{-1} this gives an absolute shift of $\Delta\nu \approx 10$ GHz.

If the ion beam passes through a differentially pumped alkali-vapor cell the ions can suffer charge-exchange collisions, where an electron is transferred from an alkali atom to the ion. Because such charge-exchange collisions show very large cross sections they occur mainly at large impact parameters, and the transfer of energy and

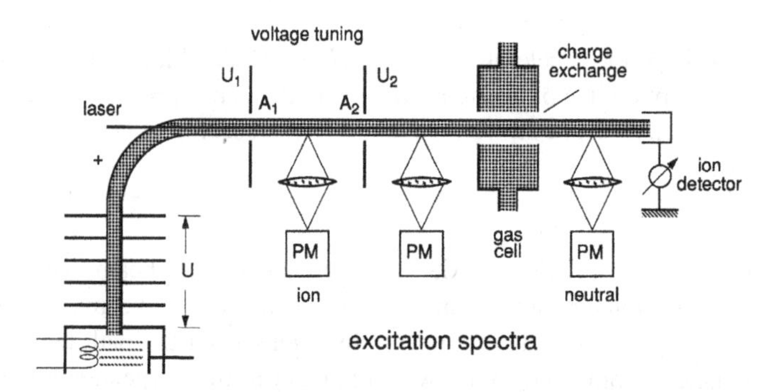

Fig. 4.28 Arrangement for "electric Doppler tuning" applied to the spectroscopy of fast ionized or neutral species

momentum is small. This means that the fast beam of neutralized atoms has nearly the same narrow velocity distribution as the ions before the collisions. The charge exchange produces neutral atoms or molecules in highly excited states. This offers the possibility to investigate electronically excited atoms or neutral molecules and to study their structure and dynamics.

With this technique it is, for instance, possible to investigate atomic or molecular Rydberg states and excimers (molecular dimers that are stable in excited states but are unstable in their electronic ground state, see Vol. 1, Sect. 5.7) in more detail. Examples are high-resolution studies of fine structure and barrier tunneling in excited triplet states of He_2 [464, 465].

4.6 Applications of FIBLAS

Some special techniques and possible applications of fast-ion-beam laser spectroscopy (FIBLAS) are illustrated by four different groups of experiments.

4.6.1 Spectroscopy of Radioactive Elements

The first group comprises high-resolution laser spectroscopy of short-lived radioactive isotopes with lifetimes in the millisecond range. The ions are produced by nuclear reactions induced by bombardment of a thin foil with neutrons, protons, γ-quanta, or other particles inside the ion source of a mass spectrometer. They are evaporated and enter after mass selection the interaction zone of the collinear laser [466].

Precision measurements of hyperfine structure and isotope shifts yield information on nuclear spins, quadrupole moments, and nuclear deformations. The results of these experiments allow tests of nuclear models of the spatial distribution of protons and neutrons in highly deformed nuclei [467]. In Fig. 4.8 the hyperfine spectra of different Na isotopes are depicted, which had been produced by spallation of aluminum nuclei by proton bombardment according to the reaction $^{27}\mathrm{Al}(p, 3p, xn)$ $^{25-x}\mathrm{Na}$ [468] and in Fig. 4.29 the hfs of 6 isotopes of the titanium ion Ti^+ illustrates the good signal-to-noise ratio. Such precision measurements have been performed in several laboratories for different families of isotopes [466–469].

4.6.2 Photofragmentation Spectroscopy of Molecular Ions

Besides excitation spectroscopy of bound–bound transitions, photofragmentation spectroscopy has gained increasing interest. Here predissociating upper levels of parent molecular ions M^+, which decay into neutral and ionized fragments, are

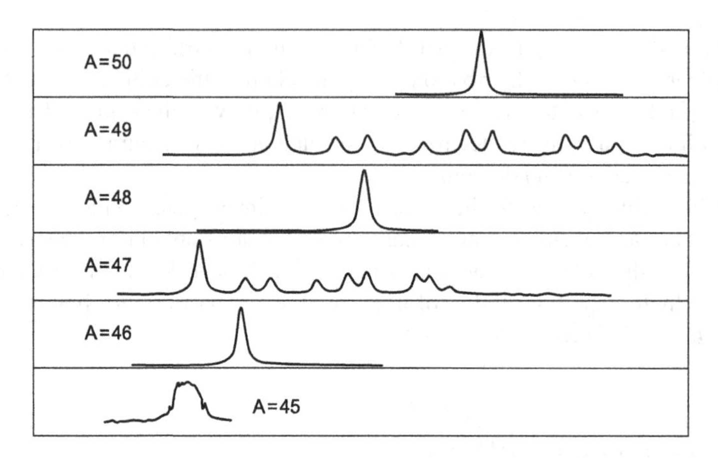

Fig. 4.29 Hyperfine spectra of 6 isotopes of the Titan ions for the determination of nuclear spins, magnetic moments and isotope shifts [468]

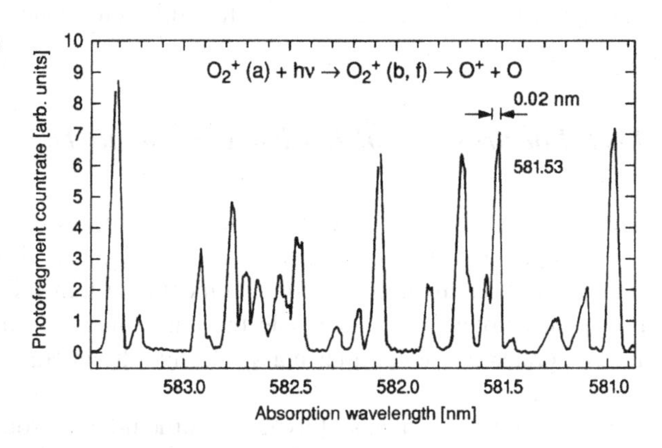

Fig. 4.30 Dependence of the O^+ photofragment signal on the absorption wavelength of O_2^+, obtained by Doppler tuning at a fixed laser wavelength [470]

excited. The ionized fragments can be detected with a mass spectrometer, while the neutral fragments need to be ionized by laser photons or by electron impact.

For illustration, Fig. 4.30 shows the number of O^+ ions formed in the photodissociation reaction

$$O_2^+ + h\nu \rightarrow O_2^{+*} \rightarrow O^+ + O,$$

as a function of the absorption wavelength $\lambda = c/\nu$ [470].

With a properly selected polarization of the laser, the photofragments are ejected into a direction perpendicular to the ion beam direction. Their transverse energy distribution can be measured with a position-sensitive detector, because their impact position x, y at the ion detector centered around the position $x = y = 0$ of the parent ion beam is given by $x = (v_x/v_z)z$, where z is the distance between the excitation zone and the detector [471, 472].

A special version of ion spectroscopy is the Coulomb-explosion technique (Fig. 4.31). A collimated beam of molecular ions with several MeV kinetic energy

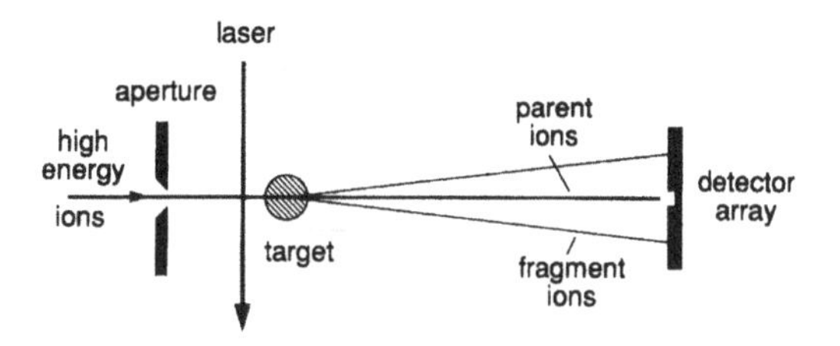

Fig. 4.31 Schematic illustration of Coulomb-explosion technique [473]

pass through a thin foil where all valence electrons are stripped. By Coulomb explosion the fragments are ejected and are detected by a position-sensitive detector. The geometry and structure of the original parent ion M^+ in its electronic ground state can be inferred from the measured pattern. Excitation of the ions M^+ by a laser just before they enter the foil allows the determination of the molecular structure in excited states [473].

If the kinetic energy distribution of fragments produced by direct photodissociation of molecular ions is measured, the form of the repulsive potential may be deduced. This can be realized with the apparatus shown in Fig. 4.32. In the long interaction zone with the collinear laser beam, part of the parent ions are photodissociated. Both the parent ions and the fragment ions are deflected by a quadrupole field. The fragments are separated and their energies are measured by two 180° energy analyzers. The parent ions M^+ are mass and energy selected before they enter the interaction zone [474, 475].

Of particular interest is the multiphoton dissociation by infrared lasers, which can be studied in more detail by such an arrangement. One example is the dissociation of SO_2^+ ions induced by multiphoton absorption of CO_2 laser photons. The relative probability for the two channels

$$SO_2^+ \begin{cases} \to SO^+ + O, \\ \to S^+ + O_2, \end{cases}$$

depends on the wavelength and on the intensity of the CO_2 laser [476].

The combination of fast ion beam photofragmentation with field-dissociation spectroscopy opens interesting new possibilities for studying long-range ion–atom interactions. This has been demonstrated by Bjerre and Keiding [477], who measured the O^+–O potential in the internuclear distance range 1–2 nm from electric field-induced dissociation of selectively laser-excited O_2^+ ions in a fast beam.

4.6.3 Saturation Spectroscopy in Fast Beams

For the elimination of the residual Doppler width, the FIBLAS technique allows an elegant realization of saturation spectroscopy with a single fixed-frequency laser (Fig. 4.33). The ions are accelerated by the voltage U, which is tuned to a value

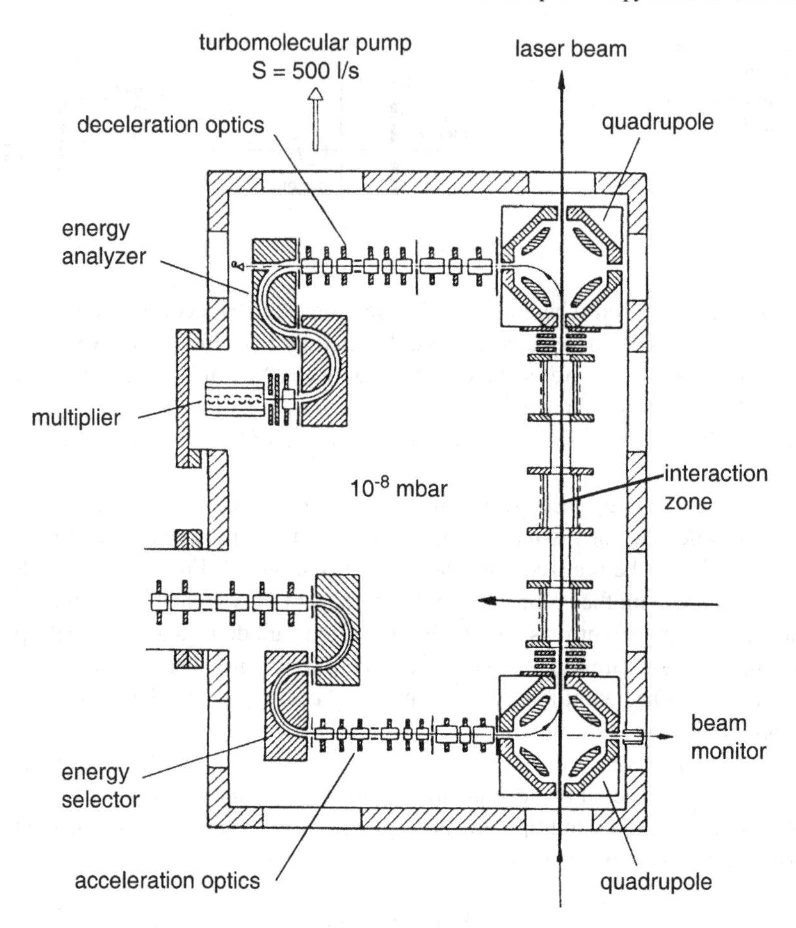

Fig. 4.32 Experimental setup for photofragmentation spectroscopy with energy and mass selection of the parent ions and the fragment ions [474]

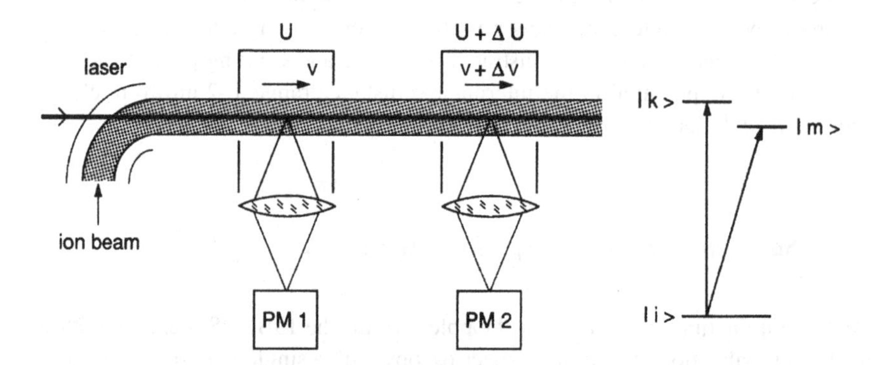

Fig. 4.33 Saturation spectroscopy in fast ion beams

where the laser radiation is absorbed in the first part of the interaction zone on a transition $|i\rangle \leftarrow |k\rangle$ at the fixed laser frequency

$$\nu_L = \nu_0\sqrt{1 + (2eU)/(mc^2)}. \tag{4.23}$$

In the second part of the interaction zone an additional voltage ΔU is applied, which changes the velocity of the ions. If the laser-induced fluorescence is monitored by PM2 as a function of ΔU, a Lamb dip will be observed at $\Delta U = 0$ because the absorbing level $|i\rangle$ has already been partly depleted in the first zone.

If several transitions are possible from level $|i\rangle$ with frequencies ω within the Doppler tuning range

$$\Delta\nu(\Delta U) = \nu_0(1 \pm \sqrt{1 + \Delta U/U}), \tag{4.24}$$

a Lamb-dip spectrum of all transitions starting from the level $|i\rangle$ is obtained when U is kept constant and ΔU is tuned [480].

Fast ion and neutral beams are particularly useful for very accurate measurements of lifetimes of highly excited ionic and neutral molecular levels (Sect. 6.3).

4.7 Spectroscopy in Cold Ion Beams

Although the velocity spread of ions in fast beams is reduced by acceleration cooling, their internal energy (E_{vib}, E_{rot}, E_e), which they acquired in the ion source, is generally not decreased unless the ions can undergo radiative transitions to lower levels on their way from the ion source to the laser interaction zone. Therefore, other techniques have been developed to produce "cold ions" with low internal energies. Three of them are shown in Fig. 4.34.

In the first method (Fig. 4.34a), a low current discharge is maintained through a glass nozzle between a thin tungsten wire acting as a cathode and an anode ring on the vacuum side. The molecules are partly ionized or dissociate in the discharge during their adiabatic expansion into the vacuum [481]. If the expanding beam is crossed by a laser beam just behind the nozzle, excitation spectroscopy of cold molecular ions or short-lived radicals can be performed [482]. Erman et al. [483] developed a simple hollow-cathode supersonic beam arrangement, which allows sub-Doppler spectroscopy of cold ions (Fig. 4.34b).

Instead of gas discharges, electrons emitted from hot cathodes can be used for ionization (Fig. 4.34c). With several cathodes arranged cylindrically around a cylindrical grid acting as an anode, a large electron current can be focused into the cold molecular beam. Because of the low electron mass, electron impact ionization at electron energies closely above ionization threshold does not much increase the rotational energy of the ionized molecules, and rotationally cold molecular ions can be formed from cold neutral molecules. Rotational temperatures of about 20 K have been reached, for instance, when supersonically cooled neutral triacetylene molecules were ionized by 200 eV electrons in a seeded free jet of helium [484].

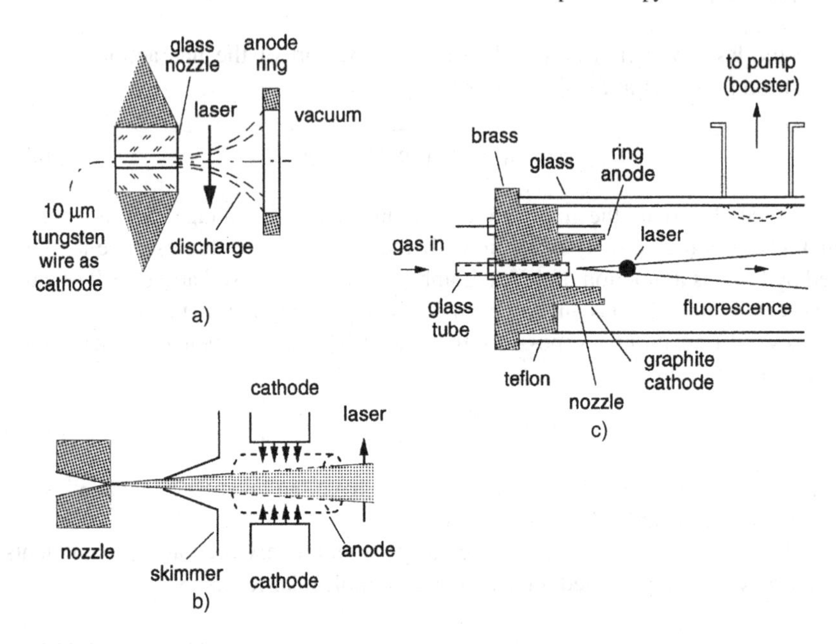

Fig. 4.34 Three possible arrangements for producing cold molecular ions

The vibrational energy depends on the Franck–Condon factors of the ionizing transitions. If the electron beam is modulated, lock-in detection allows separation of the spectra of neutral and ionized species [485]. When pulsed lasers are used, the electron gun can also be pulsed in order to reach high peak currents.

Cold ions can also be formed by two-photon ionization directly behind the nozzle of a supersonic neutral molecular beam [486]. These cold ions can then be further investigated by one of the laser spectroscopic techniques discussed above. The combination of pulsed lasers and pulsed nozzles with time-of-flight spectrometers gives sufficiently large signals to study not only molecular excitation but also the different fragmentation processes [487–489].

4.8 Laser Photo-Detachment in Molecular Beams

The collinear arrangement of laser- and ion-beam can be also used for the spectroscopy of negative ions [490], which play an important role in the higher atmosphere and also for many chemical reactions. Although meanwhile hundreds of different stable negative ions are known, spectra with rotational resolution have been only measured for a few of them. The collinear arrangement allows, due to the spectral narrowing (see Sect. 4.5) a high spectral resolution. In photo-detachment spectroscopy the extra electron is removed by absorption of a photon, i.e. the negative ion is photo-ionized, leaving a neutral species according to the scheme:

$$\mathrm{M}^- + h\nu \rightarrow \mathrm{M} + \mathrm{e}^- + E_{\mathrm{kin}}$$

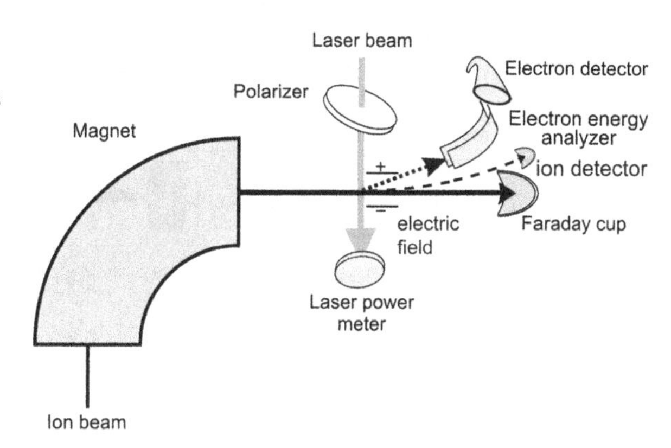

Fig. 4.35 A typical laser photodetachment electron spectroscopy setup. Electrons are energy analyzed and guided into a detector. The laser beam can be polarized to examine the angular distribution [492]

The kinetic energy of the electrons is

$$E_{\text{kin}} = h\nu - (E_{\text{ion}} - E_k),$$

where E_{ion} is the ionization energy of the negative ion in its ground state and E_k the energy of level $|k\rangle$ in which the neutral molecule ends. The correlation between the electrons influences strongly the binding energy. Since the binding energy of the electron in negative ions is generally small, infrared or visible lasers can be used for photo-detachment spectroscopy. The experimental arrangement is shown in Fig. 4.35. The ions are mass-selected by a magnet. The laser beam crosses the ion beam perpendicular. The energy spectrum of the photo-ionized electrons and the rate of neutral molecules is measured as a function of the laser wavelength. The remaining negative ions which are not photo-ionized, can be separated from the neutrals by an electric field. The neutrals and the remaining negative ions can be detected by Faraday cups.

If the binding energy differs for different isobars of a species, photo-detachment can be used for efficient selection of the wanted isobar.

The photo-detachment probability might depend on the polarization state of the laser. Detachment with linear polarized lasers can lead to interference effects between electrons ejected into different directions but collected again by the electric extraction field. This effect differs for circular polarization.

A collinear scheme for ion- and laser beams enlarges the interaction time between the laser and the negative ions. The synchronization of the laser pulses with the pulsed ion source ensures an optimum time overlap between the ion pulses and the laser pulses. The electrons are focussed by a weak magnetic field onto the electron detector. The remaining negative ions are separated from the neutrals by a perpendicular electric field.

With this technique the photo-detachment spectrum of NO^- ions has been measured [493]. Here the laser beam is reflected by a gold mirror into the anti-collinear direction of the ion beam (Fig. 4.36). The assembly is located in an electrode arrangement of three electrodes creating an electric field that bumps the ion packet over the mirror and returns it to the incident beam axis.

More information on this field can be found in [490–495].

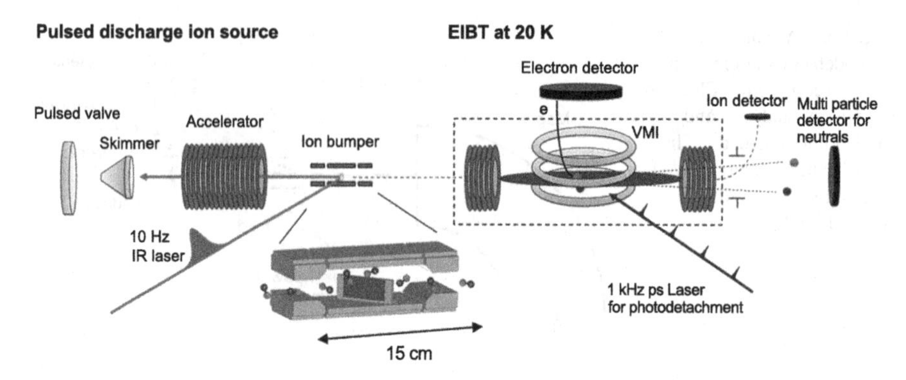

Fig. 4.36 Schematic of the reflective beam blocker setup. Ions are produced in a pulsed discharge and accelerated to 7 keV. A gold mirror is placed in the beam path to illuminate the fast moving ion packet with a 10 Hz IR laser pulse. A set of electrostatic electrodes is used to bump the ion beam over the reflective beam blocker. On the way to the ion detector the ion beam enters an electrostatic ion beam trap (EIBT) dedicated to photoelectron-photofragment coincidence (PPC) experiments [493]

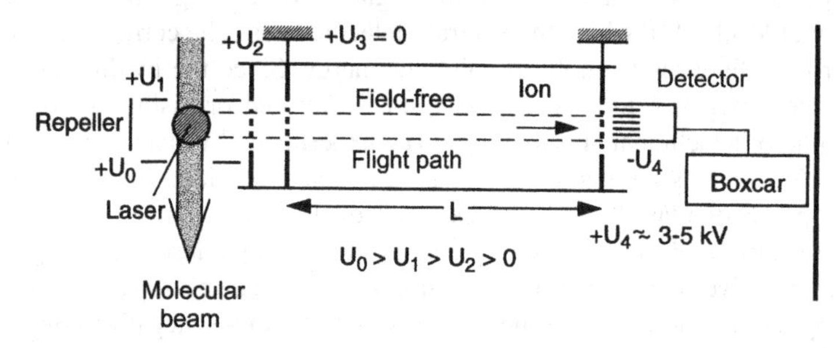

Fig. 4.37 Mass-selective laser spectroscopy in a molecular beam using time-of-flight mass spectrometry

4.9 Combination of Molecular Beam Laser Spectroscopy and Mass Spectrometry

The combination of laser- and mass-spectrometry has brought a wealth of new information on the structure and dynamics of molecules [494]. If, for instance a mixture of different isotopomers of a molecular species (these are molecules with different atomic isotopes) is present, the absorption spectra of the different isotopomers might overlap which impedes the analysis of the spectrum. Therefore the separation of these isotopomers by a mass spectrometer will facilitate the unambiguous analysis. The isotope shift of spectral lines gives additional information on the molecular structure.

For pulsed laser excitation time of flight mass spectrometers are the optimum choice [495]. A schematic sketch of their design is shown in Fig. 4.37. By resonant two-step excitation and ionization in a molecular beam molecular ions are generated which are extracted by electric fields, pass a field-free region and are then acceler-

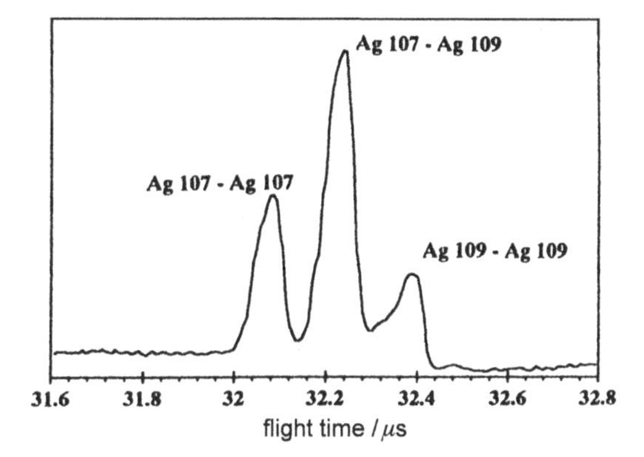

Fig. 4.38 Flight times of the different isotopomers of the silber-dimer Ag_2 in a time-of-flight mass spectrometer

ated onto the detector. Their kinetic energy $E_{\text{kin}} = (m/2)v^2 = q \cdot U$ is proportional to the total acceleration voltage U and their charge q. Measuring the time of flight T through the field-free region with length L allows the determination of their mass

$$m = 2qU/v^2 = 2qUT^2/L^2.$$

The time resolved measurement of the different arrival times of the ions yields the masses m.

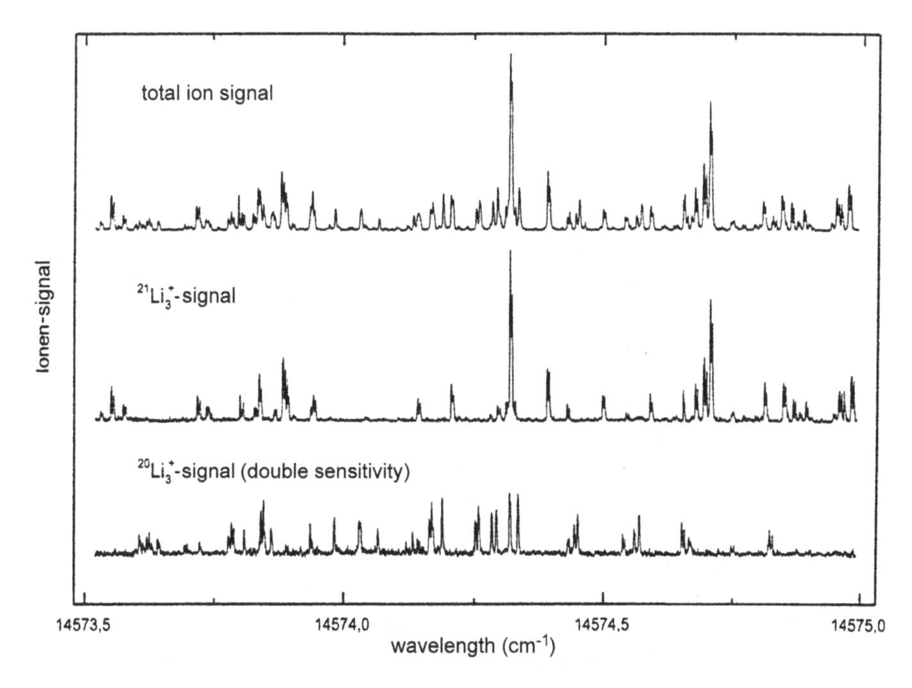

Fig. 4.39 Section of the excitation spectrum of the Li_3-trimer. *Upper spectrum*: without mass selection; *below*: isotope-selective spectra of $^{21}Li_3$ and $^{20}Li_3$ [497]

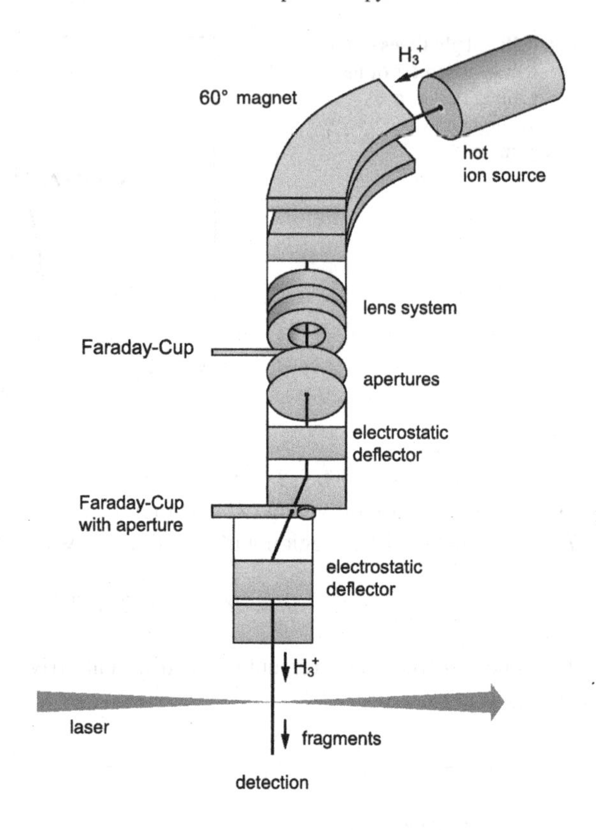

Fig. 4.40 Experimental arrangement for the study of photo-dissociation of H_3^+-ions [498]

In Fig. 4.38 the time of flight spectrum of the silver Ag_2 dimers $^{107}Ag^{107}Ag$, $^{107}Ag^{109}Ag$ and $^{109}Ag^{109}Ag$ is depicted [496]. When the time window of the detector is kept fixed on the arrival time of a specific isotopomer and the laser wavelength is tuned over the spectral range of an electronic transition, the spectra of the different isotopomers can be selectively recorded. In Fig. 4.39 such excitation spectra of different Li_2 isotopomers are recorded [497] which illustrate the simplification of the spectrum compared with the spectrum without mass selection which represents a superposition of the spectra of all isotopomers.

The formation and photo-dissociation of H_3^+-ions plays an important role in astrophysics to explain the development and the collapse of molecular clouds. Therefore laboratory experiments are demanded to measure absolute cross sections for these processes. In Fig. 4.40 a possible experimental arrangement for such measurements is shown [498]. The ions are created in a special ion source, are accelerated and mass selected. The transmitted H_3^+-ions are excited by one- or two-laser beams and the fragments are detected.

These few examples will have demonstrated that the combination of pulsed lasers, pulsed molecular beams, and time-of-flight mass spectrometry represents a powerful technique for studying the selective excitation, ionization, and fragmentation of wanted molecules out of a large variety of different molecules or species in a molecular beam [487–489, 500–504]. The technique, refined by Boesl

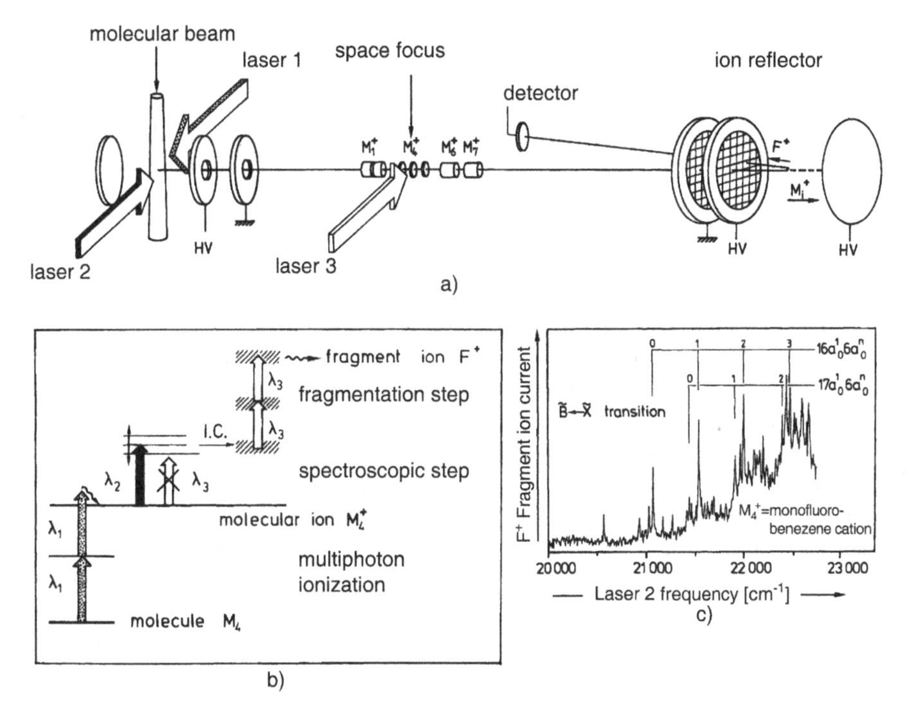

Fig. 4.41 Spectroscopy of fragmentation products produced by photoionization and excitation of large molecules: (**a**) experimental setup; (**b**) level scheme; and (**c**) spectrum of the fragment ion [487]

et al. [487] is illustrated by Fig. 4.41: rotationally and vibrationally cold neutral parent molecules M in a supersonic molecular beam pass through the ion source of a time-of-flight mass spectrometer. A pulsed laser L_1 forms molecular ions M^+ by resonant enhanced multiphoton ionization. By selecting special intermediate states of M, the molecular ion M^+ can often be preferentially prepared in a selected vibrational level.

In the second step, after a time delay Δt that is long compared with typical lifetimes of excited states of M^+ but shorter than the time of flight out of the excitation region, a pulsed tunable dye laser L_2 excites the molecular ions M^+ from their electronic ground state into selected electronic states of interest.

The detection of spectroscopic excitation is performed by photofragmentation of $(M^+)^*$ with a pulsed laser L_3 according to the dissociation process

$$\left(M^+\right)^* + h\nu_3 \rightarrow \sum_i M_i^+ + \sum_k M_k$$

where ionized as well as neutral fragments can be formed. For selective excitation of the target species and discrimination against secondary fragment ions F^+ from unwanted molecular ions and other fragment ions produced by laser L_1, the laser L_3

crosses the time-of-flight spectrometer (reflectron [501]) at the space focus in the field-free drift region, where ions of the same mass (in the example of Fig. 4.41b this is M_4^+) are compressed to ion bunches, while ion clouds of different masses are already separated by several microseconds due to their different flight times. Choosing the correct time delay of L_3 allows the mass-selective excitation of M_4^+ at the space focus.

If the wavelength λ_3 of laser L_3 is chosen properly, the ions M_4^+ cannot be fragmented by L_3 if they are in their electronic ground state. Therefore, the excitation by L_2 can be monitored by L_3.

Behind the space focus, the whole assembly of molecular primary and secondary ions passes a field-free drift region and enters the ion mirror of the reflectron. This mirror has two functions: by setting the potential of the mirror end plate lower than the potential in the ion source where all primary ions are formed, these primary ions hit the end plate and are eliminated. All secondary fragment ions F_i^+ formed at the space focus have considerably less kinetic energy (with the neutral fragments carrying away the residual kinetic energy). They are reflected in the ion mirror and reach the ion detector. By choosing the right reflecting field strength, secondary fragment ions F_i^+ (within the mass range of interest) are focused in time and appear in a narrow time window (for example, $\Delta t = 10$ ns) at the detector. This allows an excellent discrimination against most sources of noise, which may produce signals at different times or in very wide time ranges.

The spectroscopic technique described above is applicable to most ionic states, but is particularly useful for nonfluorescing or nonpredissociating molecular ion states, such as those of many radical cations. Because of the considerably lower energies of their first excited electronic states in comparison to their neutral parent molecules, internal conversion is enhanced, thus suppressing fluorescence. Typical examples are the cations of all mono- and of many di- and trihalogenated benzenes as well as of the benzene cation itself. For illustration, the UV/VIS spectrum of the monofluorobenzene cation shown in Fig. 4.41c was measured by the method described above and revealed for the first time vibrational resolution for this molecular cation.

The whole experimental arrangement is very flexible due to an uncomplicated mechanical setup; it allows some more ion optical variations as well as laser excitation schemes as presented above and thus several additional possibilities to perform spectroscopy and analysis of ionized and of neutral molecules [487].

The detailed process of the decomposition of photoionized mass-selected metastable alkali clusters has been investigated in a tandem time-of-flight mass spectrometer [503]. It was found that the ion clusters dissociate through the evaporation of either a single neutral atoms or a dimer. The neutral clusters formed during the adiabatic expansion from a nozzle into vacuum are collimated by a skimmer, photoionized by laser 1 and pass with a selectable delay through the beam of the fragmentation laser 2. A pulsed mass selector transmits only ions with the target mass (Fig. 4.42).

Further information on molecular multiphoton ionization and ion fragmentation spectroscopy can be found in [504].

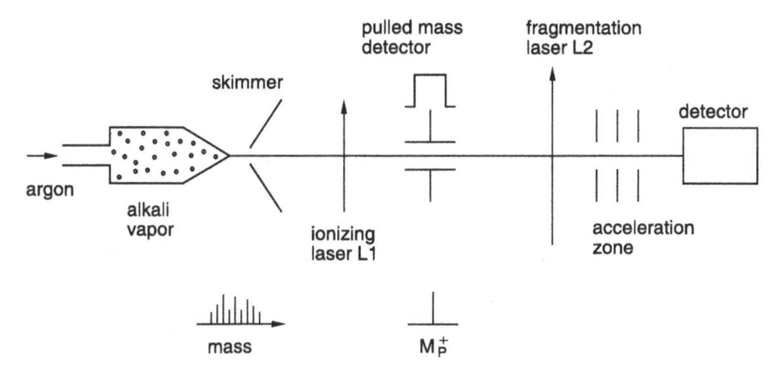

Fig. 4.42 Setup for studying the decomposition of mass-selected cluster ions

4.10 Problems

4.1 A collimated effusive molecular beam with a rectangular density profile behind the collimating aperture has a thermal velocity distribution at $T = 500$ K. Calculate the intensity profile $\alpha(\omega)$ of an absorption line, centered at ω_0 for molecules at rest, if the beam of a weak tunable monochromatic laser crosses the molecular beam under 45° against the molecular beam axis

(a) for negligible divergence of the molecular beam, and
(b) for a collimation angle of $\epsilon = 5°$.

4.2 A monochromatic laser beam ($\lambda = 500$ nm) in the x-direction crosses a supersonic divergent atomic beam perpendicularly to the beam axis at a distance $d = 10$ cm away from the nozzle. Calculate the halfwidth Δx of the spatial fluorescence distribution $I(x)$, if the laser frequency is tuned to the center of an atomic transition with a homogeneous linewidth of $\Delta\nu_h = 10$ MHz (saturation effects shall be negligible).

4.3 A laser beam from a tunable monochromatic laser is directed along the z-axis against a collimated thermal molecular beam with a Maxwell–Boltzmann velocity distribution. Calculate the spectral profiles of the absorption $\alpha(\omega)$

(a) for a weak laser (no saturation);
(b) for a strong laser (complete saturation $s \gg 1$), where the saturated homogeneous linewidth is still small compared to the Doppler width.

4.4 The beam of a monochromatic cw laser is split into two beams that intersect an atomic beam perpendicularly at the positions z_1 and $z_2 = z_1 + d$. The laser frequency is tuned to the center of the absorption transition $|k\rangle \leftarrow |i\rangle$ where the laser depletes the level $|i\rangle$. Calculate the time profile $I(z_2, t)$ of the LIF signal measured at z_2, if the first beam at z_1 is interrupted for a time interval $\Delta t = 10^{-7}$ s, which is short compared to the mean transit $\bar{t} = d/\bar{v}$ with $d = 0.4$ m for a supersonic beam

(a) with the velocity distribution $N(v) = C \cdot e^{-\frac{m}{2kT}(v-\bar{u})^2}$;
(b) with the approximate velocity distribution $N(v) = a(u - 10|u - v|)$, $u = \bar{v} = 10^3$ m/s, $0.9u \le v \le 1.1u$.

4.5 Calculate the population distribution $N(v'')$ and $N(J'')$ of vibrational and rotational levels of Na_2 molecules in a supersonic beam with $T_{vib} = 100$ K and $T_{rot} = 10$ K. Which fraction of all molecules is in the levels ($v'' = 0$, $J'' = 20$) and ($v'' = 1$, $J'' = 20$)? At which value of J'' is the maximum of $N(J'')$? The rotational constant is $B_e = 0.15\ \mathrm{cm}^{-1}$, and the vibrational constant is $\omega_e = 150\ \mathrm{cm}^{-1}$.

Chapter 5
Optical Pumping and Double-Resonance Techniques

Optical pumping means selective population or depletion of atomic or molecular levels by absorption of radiation, resulting in a population change ΔN in these levels, which causes a noticeable deviation from the thermal equilibrium population. With intense atomic resonance lines emitted from hollow-cathode lamps or from microwave discharge lamps, optical pumping had successfully been used for a long time in *atomic* spectroscopy, even before the invention of the laser [505, 506]. However, the introduction of lasers as very powerful pumping sources with narrow linewidths has substantially increased the application range of optical pumping. In particular, lasers have facilitated the transfer of this well-developed technique to *molecular* spectroscopy. While early experiments on optical pumping of molecules [507, 508] were restricted to accidental coincidences between molecular absorption lines and atomic resonance lines from incoherent sources, the possibility of tuning a laser to the desired molecular transition provides a much more selective and effective pumping process. It allows, because of the larger intensity, a much larger change $\Delta N_i = N_{i0} - N_i$ of the population density in the selected level $|i\rangle$ from its unsaturated value N_{i0} at thermal equilibrium to a nonequilibrium value N_i.

This change ΔN_i of the population density can be probed by a second EM wave, which may be a radio frequency (RF) field, a microwave, or another laser beam. If this "probe wave" is tuned into resonance with a molecular transition sharing one of the two levels $|i\rangle$ or $|k\rangle$ with the pump transition, the pump laser and the probe wave are simultaneously in resonance with the coupled atomic or molecular transitions (Fig. 5.1). This situation is therefore called *optical–RF, optical–microwave or optical–optical double resonance.*

Even this double-resonance spectroscopy has already been applied to the study of atomic transitions before lasers were available. In these pre-laser experiments incoherent atomic resonance lamps served as pump sources and a radio frequency field provided probe transitions between Zeeman levels of optically excited atomic states [509]. However, with tunable lasers as pump sources, these techniques are no longer restricted to some special favorable cases, and the achievable signal-to-noise ratio of the double-resonance signals may be increased by several orders of magnitude [510].

W. Demtröder, *Laser Spectroscopy 2*, DOI 10.1007/978-3-662-44641-6_5

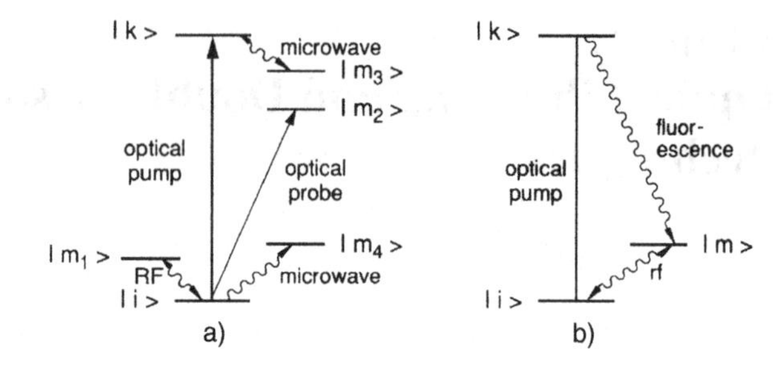

Fig. 5.1 Schematic level schemes of optical pumping and double-resonance transitions

In this chapter we shall treat the most important laser double-resonance techniques by illustration with several examples. While the pump transition is always induced by a pulsed or cw laser, the probe field may be provided by any coherent source in the spectral range between the RF region and the ultraviolet.

5.1 Optical Pumping

The effect of optical pumping on a molecular system depends on the characteristics of the pump laser, such as intensity, spectral bandwidth, and polarization, and on the linewidth and the transition probability of the absorbing transition. If the bandwidth $\Delta\omega_{\mathrm{L}}$ of the pump laser is larger than the linewidth $\Delta\omega$ of the molecular transition, all molecules in the absorbing level $|i\rangle$ can be pumped. In case of dominant Doppler broadening this means that molecules within the total velocity range can simultaneously be pumped into a higher level $|k\rangle$. If the laser bandwidth $\Delta\omega_{\mathrm{L}}$ is small compared to the inhomogeneous width $\Delta\omega$ of a molecular transition, only a subgroup of molecules with a matching absorption frequency $\omega = \omega_0 - \boldsymbol{k}\cdot\boldsymbol{v} = \omega_{\mathrm{L}} \pm \Delta\omega_{\mathrm{L}}/2$ is pumped (Sect. 2.2).

There are several different aspects of optical pumping that are related to a number of spectroscopic techniques based on optical pumping. The *first aspect* concerns the *increase or decrease of the population* in selected levels. At sufficiently high laser intensities the molecular transition can be saturated. This means that a maximum change $\Delta N = N_{is} - N_{i0}$ of the population densities can be achieved, where ΔN is negative for the lower level and positive for the upper level of the transition (Sect. 2.1). In case of *molecular* transitions, where only a small fraction of all excited molecules returns back into the initial level $|i\rangle$ by fluorescence, this level may be depleted rather completely.

Since the fluorescent transitions must obey certain selection rules, it is often possible to populate a selected level $|m\rangle$ by fluorescence from the laser-pumped upper level (Fig. 5.1b). Even with a weak pumping intensity large population densities in the level $|m\rangle$ may be achieved. In the pre-laser era, the term "optical pumping" was used for this special case because this scheme was the only way to achieve an appreciable population change with incoherent pumping sources.

Excited molecular levels $|k\rangle$ with $E_k \gg kT$ are barely populated at thermal equilibrium. With lasers as pumping sources large population densities N_k can be

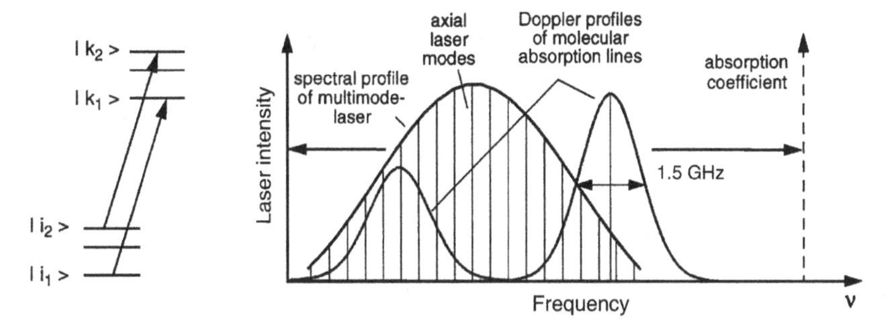

Fig. 5.2 Overlap of several Doppler-broadened absorption lines with the laser line profile leads to simultaneous optical pumping of several levels

achieved, which may become comparable to those of the absorbing ground states. This opens several possibilities for new experimental techniques:

- The selectively excited molecular levels emit fluorescence spectra that are much simpler than those emitted in gas discharges, where many upper levels are populated. These laser-induced fluorescence spectra are therefore more readily assignable and allow the determination of molecular constants for all levels in lower states into which fluorescence transitions terminate (Sect. 1.8).
- A sufficiently large population of the upper state furthermore allows the measurement of absorption spectra for transitions from this state to still higher-lying levels (excited-state spectroscopy, stepwise excitation) (Sect. 5.4). Since all absorbing transitions start from this selectively populated level, the absorption spectrum is again much simpler than in gas discharges.

The selectivity of optical pumping depends on the laser bandwidth and on the line density of the absorption spectrum. If several absorption lines overlap within their Doppler width with the spectral profile of the laser, more than one transition is simultaneously pumped, which means that more than one upper level is populated (Fig. 5.2). In such cases of dense absorption spectra, optical pumping with narrow-band lasers in collimated cold molecular beams can be utilized to achieve the wanted selectivity for populating a single upper level (Sects. 4.3, 5.5).

The situation is different when narrow-band lasers are used as pumping sources: If the beam of a *single-mode* laser with the frequency ω is sent in the z-direction through an absorption cell, only molecules within the velocity group $v_z = (\omega - \omega_0 \pm \gamma)/k$ can absorb the laser photons $\hbar\omega$ (Sect. 2.2) on the transition $|i\rangle \to |k\rangle$ with $E_k - E_i = \hbar\omega_0$ and homogeneous linewidth γ. Therefore only molecules within this velocity group are excited. This implies that the absorption of a tunable narrow-band probe laser by these excited molecules yields a Doppler-free double-resonance signal.

A further important aspect of optical pumping with a *polarized* laser is the selective population or depletion of degenerate M sublevels $|J, M\rangle$ of a level with angular momentum $\boldsymbol{J}$. These sublevels differ by the projection $M\hbar$ of $\boldsymbol{J}$ onto the quantization axis. Atoms or molecules with a nonuniform population density $N(J, M)$

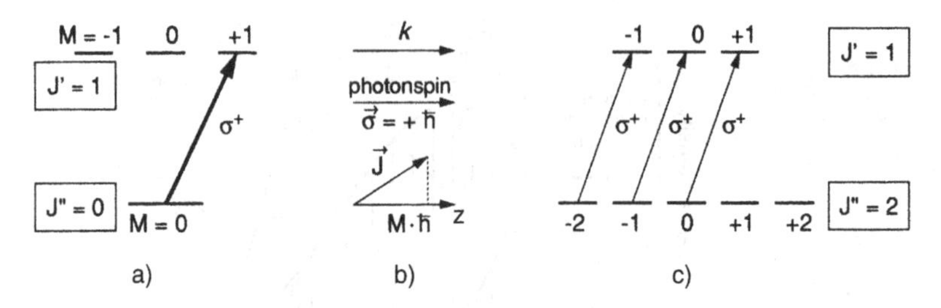

Fig. 5.3 (**a**) Orientation in the upper state produced by optical pumping with σ^+ light on an R transition $J'' = 0 \to J' = 1$. (**b**) Classical model of orientation where J precesses around the z-direction with a projection $M\hbar$. (**c**) Orientation in the lower state caused by partial depletion for the example of a P pump transition $J'' = 2 \to J' = 1$

of these sublevels are oriented because their angular momentum $\boldsymbol{J}$ has a preferential spatial distribution while under thermal equilibrium conditions $\boldsymbol{J}$ points into all directions with equal probability, that is, the orientational distribution is uniform. The highest degree of orientation is reached if only one of the $(2J + 1)$ possible M sublevels is selectively populated.

By choosing the appropriate polarization of the pump laser, it is possible to achieve equal population densities within the pairs of sublevels $|\pm M\rangle$ with the same value of $|M|$ while levels with different $|M|$ may have different populations. This situation is called *alignment*.

Note that orientation or alignment can be produced in both the upper state of a pump transition due to a M-selective population as well as in the lower state because of the corresponding M-selective depletion (Fig. 5.3).

Example 5.1 Let us illustrate the above consideration by some specific examples: if the pump beam propagating into the z-direction has σ^+ polarization (left-circularly polarized light inducing transitions $\Delta M = +1$), we choose the direction of the $\boldsymbol{k}$ vector (i.e., the z-axis) as the quantization axis. The photon spin $\boldsymbol{\sigma} = +\hbar\boldsymbol{k}/k$ points into the propagation direction and absorption of these photons induces transitions with $\Delta M = +1$. For a transition $J'' = 0 \to J' = 1$ optical pumping with σ^+ light can only populate the upper sublevel with $M = +1$. This generates orientation of the atoms in the upper state since their angular momentum precesses around the $+z$-direction with the projection $+\hbar$ (Fig. 5.3a). Optical pumping with σ^+ light on a P-transition $J'' = 2 \to J' = 1$ causes M-selective depletion of the lower state and therefore orientation of molecules in the lower state (Fig. 5.3c).

Example 5.2 Optical pumping with linearly polarized light (π-polarization) propagating into the z-direction can be regarded as a superposition of σ^+ and σ^- light (Sect. 2.4). This means that $\Delta M = \pm 1$ transitions can simultane-

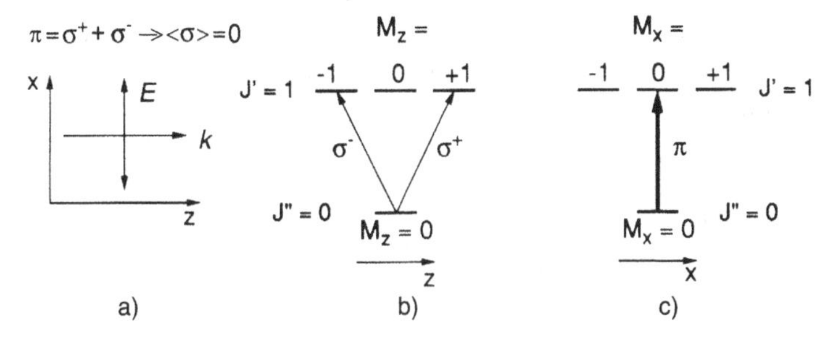

Fig. 5.4 Alignment in the upper level of a R-transition $J'' = 0 \to J' = 1$ by a linearly polarized pump wave: (**a**) direction of $\boldsymbol{E}$ and $\boldsymbol{k}$; (**b**) level scheme with z-quantization axis; and (**c**) with x-quantization axis

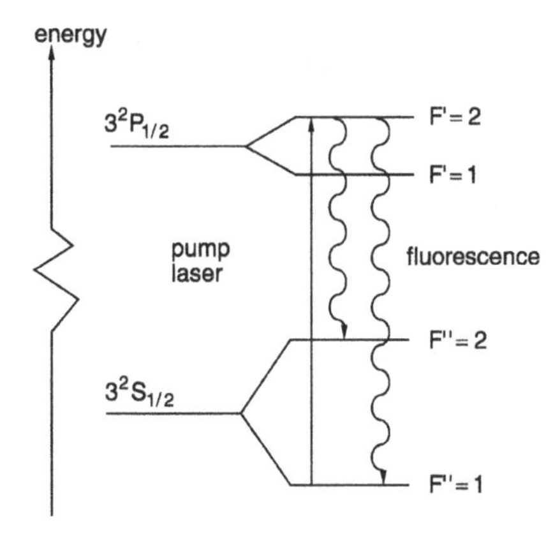

Fig. 5.5 Level scheme for nuclear spin polarization by optical pumping of the Na atom with nuclear spin $I = \frac{3}{2}\hbar$

ously be pumped with equal probability. The upper state becomes aligned because both sublevels $M = \pm 1$ are equally populated (Fig. 5.4).

Note If the direction of the $\boldsymbol{E}$ vector of the linearly polarized light is selected as the quantization axis (which we chose as the x-axis), the two sublevels with $M_z = \pm 1$ now transform into the sublevel $M_x = 0$ (Fig. 5.4). The linearly polarized pump induces transitions with $\Delta M_x = 0$ and again produces alignment since only the component $M_x = 0$ is populated. Of course, the selection of the quantization axis cannot change the physical situation but only its description.

If the laser bandwidth is small enough to resolve the hyperfine structure, a specific hfs component can be selectively populated. This can result in a nuclear spin orientation as illustrated in Fig. 5.5 for the Na atom with a nuclear spin $I = 3/2$. The laser selectively populates the $F' = 2$ component in the upper $3^2P_{1/2}$ state. The fluorescence from this state terminates at both the $F'' = 1$ and $F'' = 2$ components in the $3^2S_{1/2}$ state. The $F = 1$ component is again excited into the $3^2P_{1/2}$ state.

After a few absorption–emission cycles, the $F'' = 1$ component will be completely depleted and the $F'' = 2$ component is selectively populated.

For a quantitative treatment of optical pumping we consider a pump transition between the levels $|J_1 M_1\rangle \rightarrow |J_2 M_2\rangle$. Without an external magnetic field all $(2J + 1)$ sublevels $|M\rangle$ are degenerate and their population densities at thermal equilibrium are, without the pump laser

$$N^0(J, M) = \frac{N^0(J)}{2J + 1}, \quad \text{with } N^0(J) = \sum_{M=-J}^{+J} N^0(M, J). \tag{5.1}$$

The decrease of $N_1^0(J, M)$ by optical pumping $P_{12} = P(|J_1 M_1\rangle \rightarrow |J_2 M_2\rangle)$ can be described by the rate equation

$$\frac{\mathrm{d}}{\mathrm{d}t} N_1(J_1, M_1) = \sum_{M_2} P_{12}(N_2 - N_1) + \sum_k (R_{k1} N_k - R_{1k} N_1). \tag{5.2}$$

Included are the optical pumping (induced absorption and emission) and all relaxation processes that refill level $|1\rangle$ from other levels $|k\rangle$ or that deplete $|1\rangle$. The optical pumping probability

$$P_{12} \propto \left| \langle J_1 M_1 | \boldsymbol{D} \cdot \boldsymbol{E} | J_2 M_2 \rangle \right|^2, \tag{5.3}$$

is proportional to the square of the transition matrix element (Vol. 1, Sect. 2.8.2) and depends on the scalar product $\boldsymbol{D} \cdot \boldsymbol{E}$, of transition dipole moment $\boldsymbol{D}$, and electric field vector $\boldsymbol{E}$, that is, on the polarization of the pump laser. Molecules that are oriented with their transition dipole moment parallel to the electric field vector have the highest optical pumping probability.

When the laser intensity is sufficiently high, the population difference ΔN^0 decreases according to (2.10) to its saturated value

$$\Delta N^{\mathrm{s}} = \frac{\Delta N^0}{1 + S}.$$

Since the transition probability and therefore also the saturation parameter S have a maximum value for molecules with $\boldsymbol{D} \parallel \boldsymbol{E}$, the population density N will decrease with increasing laser intensity for molecules with $\boldsymbol{D} \parallel \boldsymbol{E}$ more than for those with $\boldsymbol{D} \perp \boldsymbol{E}$. This means that the degree of orientation decreases with increasing saturation (Fig. 5.6).

The pump rate

$$P_{12}(N_2 - N_1) = \sigma_{12} N_{\mathrm{ph}}(N_2 - N_1), \tag{5.4}$$

is proportional to the optical absorption cross section σ_{12} and the photon flux rate N_{ph} [number of photons/cm^2 s]. The absorption cross section can be written as a product

$$\sigma(J_1 M_1, J_2 M_2) = \sigma_{J_1 J_2} \cdot C(J_1 M_1, J_2 M_2), \tag{5.5}$$

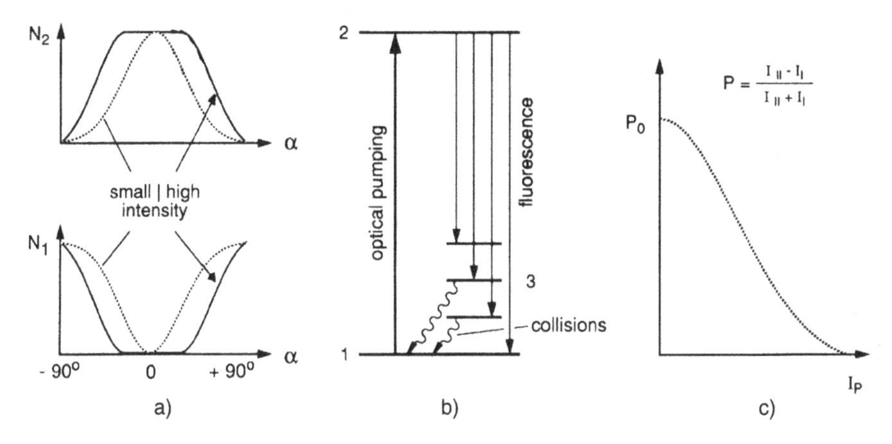

Fig. 5.6 (**a**) Decrease of molecular orientation with increasing pump intensity from saturation in the lower and upper state; (**b**) schematic level scheme monitored through the polarization P_0 of laser-induced fluorescence (**c**)

of two factors. The cross section $\sigma_{J_1J_2}$ is independent of the molecular orientation and is essentially equal to the product of the electronic transition probability times the Franck–Condon factor times the Hönl–London factor. The second factor is the Clebsch–Gordan coefficient $C(J_1M_1J_2M_2)$, which depends on the rotational quantum numbers *and on the molecular orientation* in both levels of the pump transition [511]. In the case of molecules the orientation is partly canceled by rotation. The maximum achievable degree of orientation obtained with optical pumping depends on the orientation of the transition-dipole vector with respect to the molecular rotation axis and is different for P, Q, or R-transitions in diatomic molecules [512]. For molecules the maximum orientation is therefore generally smaller than in case of atoms.

The coupling of the molecular angular momentum $\boldsymbol{J}$ with a possible nuclear spin $\boldsymbol{I}$ leads to a precession of $\boldsymbol{J}$ around the total angular momentum $\boldsymbol{F} = \boldsymbol{J} + \boldsymbol{I}$, which further reduces the molecular orientation [513]. A careful analysis of experiments on optical pumping of molecules gives detailed information on the various coupling mechanisms between the different angular momenta in selected molecular levels [514].

Another aspect of optical pumping is related to the coherent excitation of two or more molecular levels. This means that the optical excitation produces definite phase relations between the wave functions of these levels. This leads to interference effects, which influence the spatial distribution and the time dependence of the laser-induced fluorescence. This subject of coherent spectroscopy is covered in Chap. 7.

A thorough theoretical treatment of optical pumping can be found in the review of Happer [515] and [516]. Specific aspects of optical pumping by lasers with particular attention to problems arising from the spectral intensity distribution of the pump laser and from saturation effects were treated in [517, 518]. Applications of optical pumping methods to the investigation of small molecules were discussed in [508].

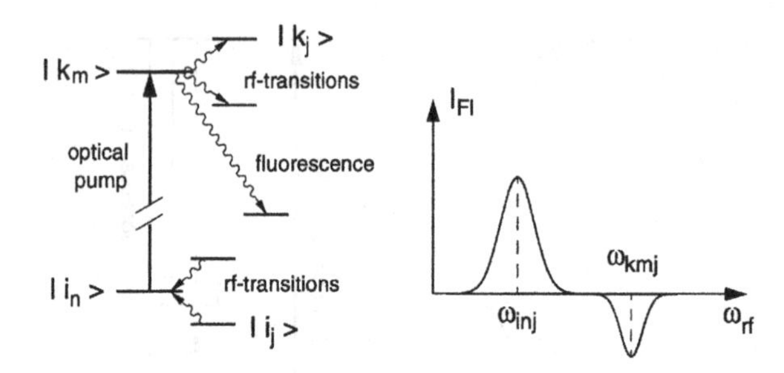

Fig. 5.7 Optical–radio frequency double resonance for lower and upper level of the optical transition

5.2 Optical–RF Double-Resonance Technique

The combination of laser-spectroscopic techniques with molecular beams and RF spectroscopy has considerably enlarged the application range of optical–RF double-resonance schemes. This optical–RF double-resonance method has now become a very powerful technique for high-precision measurements of electric or magnetic dipole moments, of Landé factors, and of fine or hyperfine splitting in atoms and molecules. It is therefore used in many laboratories.

5.2.1 Basic Considerations

Two different levels $|i\rangle$ and $|k\rangle$ that are connected by an optical transition may be split into closely spaced sublevels $|i_n\rangle$ and $|k_m\rangle$. A narrow-band laser that is tuned to the transition $|i_n\rangle \rightarrow |k_m\rangle$ between specific sublevels selectively depletes the level $|i_n\rangle$ and increases the population of the level $|k_m\rangle$ (Fig. 5.7). Examples of this situation are hyperfine components of two rotational–vibrational levels in two different electronic states of a molecule or Zeeman sublevels of atomic electronic states.

If the optically pumped sample is placed inside an RF field with the frequency ω_{rf} tuned into resonance with the transition $|i_j\rangle \rightarrow |i_n\rangle$ between two sublevels of the lower state, the level population $N(i_n)$ that was depleted by optical pumping will increase again. This leads to an increased absorption of the optical pump beam, which may be monitored by the corresponding increase of the laser-induced fluorescence intensity. Measuring $I_{\mathrm{Fl}}(\omega_{\mathrm{rf}})$ while ω_{rf} is tuned yields a double-resonance signal at $\omega_{\mathrm{rf}} = \omega_{inj} = [E(i_n) - E(i_j)]/\hbar$ (Fig. 5.7).

Each absorbed RF photon leads to an extra absorbed optical photon of the pump beam. This optical–RF double resonance therefore yields an internal energy amplification factor $V = \omega_{\mathrm{opt}}/\omega_{\mathrm{rf}}$ for the detection of an RF transition. With $\omega_{\mathrm{opt}} = 3 \times 10^{15}$ Hz and $\omega_{\mathrm{rf}} = 10^7$ Hz, we obtain $V = 3 \times 10^8$! Since optical photons can be detected with a much higher efficiency than RF quanta, this inherent energy amplification results in a corresponding increase in the detection sensitivity.

RF transitions between sublevels of the upper state result in a change of the polarization and the spatial distribution of the laser-induced fluorescence. They can

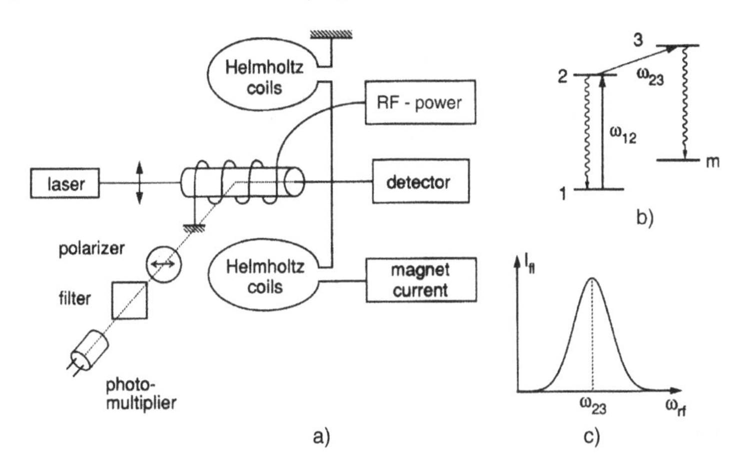

Fig. 5.8 Experimental arrangement for optical–RF double-resonance spectroscopy (**a**), level scheme (**b**) and double-resonance signal monitored through the LIF (**c**)

therefore be monitored through polarizers in front of the photomultiplier. Since the RF transitions deplete the optically pumped upper level, the RF double-resonance signal at $\omega_2 = \omega_{kmj} = [E(k_j) - E(k_m)]/\hbar$ has an opposite sign to that of the lower state at $\omega_1 = \omega_{inj}$ (Fig. 5.7).

In case of Zeeman sublevels or hfs levels, the allowed RF transitions are *magnetic* dipole transitions. Optimum conditions are then achieved if the sample is placed at the maximum of the magnetic field amplitude of the RF field. This can be realized, for instance, inside a coil that is fed by an RF current. For *electric* dipole transitions (for example, between Stark components in an external dc electric field) the electric amplitude of the RF field should be maximum in the optical pumping region.

A typical experimental arrangement for measuring RF transitions between Zeeman levels in the upper state of the optical transition is shown in Fig. 5.8. A coil around the sample cell provides the RF field, while the dc magnetic field is produced by a pair of Helmholtz coils. The fluorescence induced by a polarized dye laser beam is monitored by a photomultiplier through a polarizer as a function of the radio frequency $\omega_{\rm rf}$ [514].

Instead of tuning $\omega_{\rm rf}$ one may also vary the dc magnetic or electric field at a fixed value of $\omega_{\rm rf}$ thus tuning the Zeeman or Stark splittings into resonance with the fixed radio frequency (Sect. 1.7). This has the experimental advantage that the RF coil can be better impedance-matched to the RF generator.

The fundamental advantage of the optical–RF double-resonance technique is its high spectral resolution, which is not limited by the optical Doppler width. Although the optical excitation may occur on a Doppler-broadened transition, the optical–RF double-resonance signal $I_{\rm Fl}(\omega_{\rm rf})$ is measured at a low radio frequency $\omega_{\rm rf}$, and the Doppler width, which is according to Vol. 1, (3.43) porportional to the frequency, is reduced by the factor $(\omega_{\rm rf}/\omega_{\rm opt})$. This makes the residual Doppler width of the double-resonance signal completely negligible compared with other broadening effects such as collisional or saturation broadening. In the absence of these additional

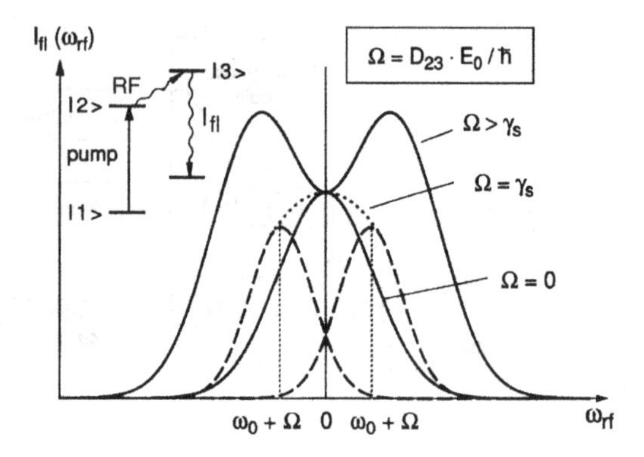

Fig. 5.9 Saturation broadening and Rabi splitting of double-resonance signals with increasing RF power P

line-broadening effects, the halfwidth of the double-resonance signal for the RF transition $|2\rangle \rightarrow |3\rangle$

$$\Delta\omega_{23} = (\Delta E_2 + \Delta E_3)/\hbar, \tag{5.6}$$

is essentially determined by the energy level widths ΔE_i of the corresponding levels $|2\rangle$ and $|3\rangle$, which are related to their spontaneous lifetime τ_i by $\Delta E_i = \hbar/\tau_i$. For transitions between sublevels in the ground state, the radiative lifetimes may be extremely long and the linewidth is only limited by the transit time of the molecules through the RF field. Sub-kilohertz resonances have been observed in the RF-optical double resonance spectroscopy of rare-earth ions [519].

With increasing RF intensity, however, saturation broadening is observed (Vol. 1, Sect. 3.6) and the double-resonance signal may even exhibit a minimum at the center frequency ω_{23} (Fig. 5.9). This can readily be understood from the semiclassical model of Vol. 1, Sect. 2.7: for large RF field amplitudes E_{rf} the Rabi flopping frequency Vol. 1, (2.90)

$$\Omega = \sqrt{(\omega_{23} - \omega_{rf})^2 + D_{23}^2 E_{rf}^2/\hbar},$$

becomes comparable to the natural linewidth $\delta\omega_n$ of the RF transition. The resulting modulation of the time-dependent population densities $N_2(t)$ and $N_3(t)$ causes a splitting of the line profile into two components with $\omega = \omega_{23} \pm \Omega$, which can be observed for $\Omega > \delta\omega_n$ (Fig. 5.9). If the halfwidth Δ_{23} of the double-resonance signal is plotted against the RF power P_{rf}, the extrapolation toward $P_{rf} = 0$ yields the linewidth $\gamma = 1/\tau$ of those levels of which the natural linewidth represents the dominant contribution to broadening; it allows one to determine the natural lifetime τ.

The achievable accuracy is mainly determined by the signal-to-noise ratio of the double-resonance signal, which limits the accuracy of the exact determination of the center frequency ω_{23}. However, the *absolute* accuracy in the determination of the RF frequency ω_{23} is generally several orders of magnitude higher than in conventional optical spectroscopy, where the level splitting $\Delta E = \hbar\omega_{23} = h(\nu_{31} - \nu_{21})$ is indirectly deduced from a small difference $\Delta\lambda$ between two directly measured wavelengths $\lambda_{31} = c/\nu_{31}$ and $\lambda_{21} = c/\nu_{21}$.

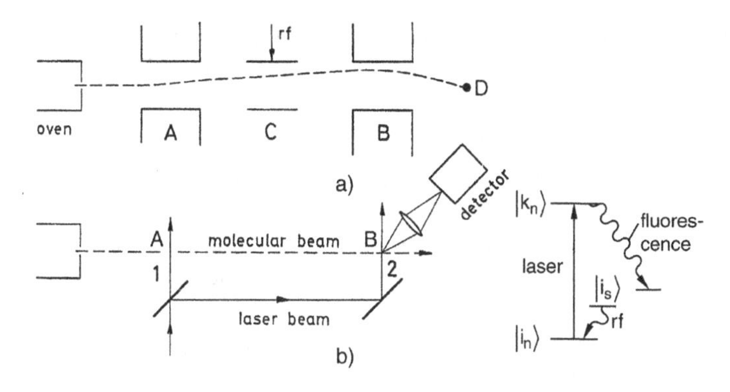

Fig. 5.10 Comparison between the conventional Rabi method (**a**) and its laser version (**b**)

5.2.2 Laser–RF Double-Resonance Spectroscopy in Molecular Beams

The *Rabi technique* of radio frequency or microwave spectroscopy in atomic or molecular beams [520–523] has made outstanding contributions to the accurate determination of ground state parameters, such as the hfs splittings in atoms and molecules, small Coriolis splitting in rotating and vibrating molecules, or the narrow rotational structures of weakly bound van der Waals complexes [524]. Its basic principle is illustrated in Fig. 5.10. A collimated beam of molecules with a permanent dipole moment is deflected in a static inhomogeneous magnetic field A but is deflected back onto the detector D in a second static field B with opposite field gradient. An RF field C is applied between A and B; which induces transitions between the molecular levels $|i_n\rangle$ and $|i_j\rangle$. Since the magnetic moment generally differs for the two levels, the deflection in B will change after such an RF transition and the detector output $S(\omega_{\mathrm{rf}})$ will show a resonance signal for the resonance radio frequency $\omega_{\mathrm{rf}} = [E(i_n) - E(i_j)]/\hbar$. For the measurement of Zeeman or Stark components an additional dc magnetic or electric field is applied in the RF region C. A famous example of such a device is the cesium clock [525].

The technique is restricted to atoms or molecules with a sufficiently large difference of the dipole moments in the two levels since the change of deflection in the inhomogeneous fields must be detectable. Furthermore, the detection of neutral particles with a universal ionization detector is generally not very sensitive except for those molecules (for example, alkali atoms or dimers) that can be monitored with a Langmuir–Taylor detector.

The laser version of the Rabi method (Fig. 5.10b) overcomes both limitations. The two magnets A and B are replaced by the two partial beams 1 and 2 of a laser that cross the molecular beam perpendicularly at the positions A and B. If the laser frequency ω_{L} is tuned to the molecular transition $|i_n\rangle \rightarrow |k_n\rangle$, the lower level $|i_n\rangle$ is partly depleted due to optical pumping in the first crossing point A. Therefore the absorption of the second beam in the crossing point B is decreased, which can be monitored through the laser-induced fluorescence. The RF transition $|i_j\rangle \to |i_n\rangle$ induced in C increases the population density $N(i_n)$ again and with it the fluorescence signal in B.

This laser version has the following advantages:

- It is not restricted to molecules with a permanent dipole moment but may be applied to all molecules that can be excited by existing lasers.
- Even at moderate laser powers the optical transition may become saturated (Sect. 2.1) and the lower level $|i_n\rangle$ can be appreciably depleted. This considerably increases the population difference $\Delta N = N(i_j) - N(i_n)$ and thus the absorption of the RF field on the transition $|i_j\rangle \to |i_n\rangle$, which is proportional to ΔN. In the conventional Rabi technique the population $N(E)$ follows a Boltzmann distribution, and for $\Delta E \ll kT$ the difference ΔN becomes very small.
- The detection of the RF transitions through the change in the LIF intensity is much more sensitive than the universal ionization detector. Using resonant two-photon ionization (Sect. 1.3), the sensitivity may further be enhanced.
- In addition, the signal-to-noise ratio is higher because only molecules in the level $|i_n\rangle$ contribute to the RF–double-resonance signal, while for the conventional Rabi technique the difference in deflection is generally so small that molecules in other levels also reach the detector and the signal represents a small difference of two large background currents.

These advantages allow the extension of the laser Rabi technique to a large variety of different problems [526], which shall be illustrated by three examples:

Example 5.3 (Measurements of the hfs in the $^1\Sigma_g^+$ state of Na_2) The small magnetic hfs in the $^1\Sigma$ state of a homonuclear diatomic molecule is caused by the interaction of the nuclear spins I with the weak magnetic field produced by the rotation of the molecule. For the Na_2 molecular ground state the hyperfine splittings are smaller than the natural linewidth of the optical transitions. They could nevertheless be measured by the laser version of the Rabi technique [527]. A polarized argon laser beam at $\lambda = 476.5$ nm crosses the sodium beam and excites the Na_2 molecules on the transition $X\,^1\Sigma_g^+(v'' = 0, J'' = 28) \to B\,^1\Pi_u(v' = 3, J' = 27)$. The hfs splittings are much smaller than the laser linewidth. Therefore all hfs components are pumped by the laser. However, the optical transition probability depends on the hfs components of the lower and upper states. Therefore the depletion of the lower levels $|i_n\rangle$ differs for the different hfs components. The RF transitions change the population distribution and affect the monitored LIF signal. The hfs constant for the level $(v'' = 0, J'' = 28)$ was determined to be 0.17 ± 0.03 kHz, while the quadrupole coupling constant was found to be $\text{eQq} = 463.7 \pm 0.9$ kHz.

With a tunable single-mode cw dye laser any wanted transition can be selected and the hfs constants can be obtained for arbitrary rotational lev-

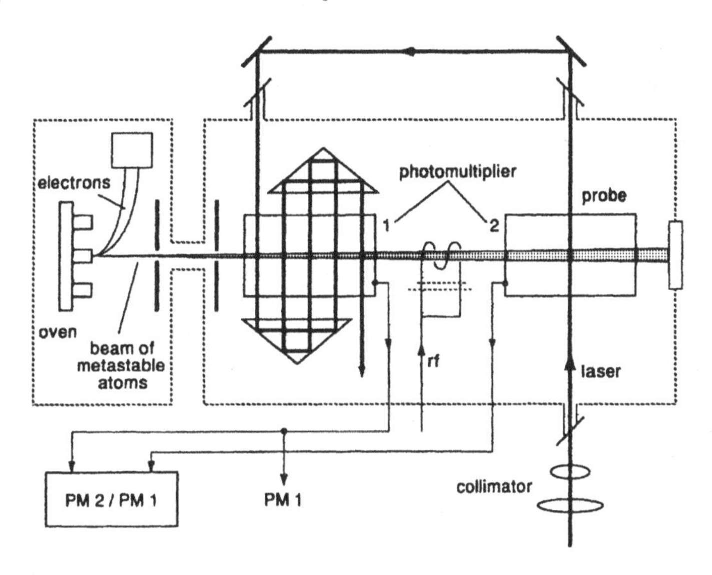

Fig. 5.11 Atomic beam resonance apparatus with combined electron-impact and laser pumping for the sensitive detection of optical–RF double resonance in highly excited states [529]

els [528]. The hyperfine splittings are very small compared to the natural linewidth of the optical transition $X\,^1\Sigma_g \rightarrow B\,^1\Pi_u$, which is about 20 MHz because of the short lifetime $\tau = 7$ ns of the upper levels. At larger laser intensities optical pumping of the overlapping hfs components generates a coherent superposition of the lower-state hfs components. This results in a drastic change of line profiles and center frequencies of the RF double-resonance signals with increasing laser intensity. One therefore has to measure at different laser intensities I_L and to extrapolate to $I_L \rightarrow 0$.

Example 5.4 (Hyperfine structure of highly excited atomic levels) The combination of electron-impact excitation with the laser version of the Rabi technique allows measurements of hfs splittings in higher excited states. Penselin and coworkers [529, 531] have developed the atomic beam RF resonance apparatus shown in Fig. 5.11. Metals with a high condensation temperature are evaporated from a source heated by an electron gun and are further excited by electron impact into metastable states. Behind a collimating aperture optical pumping with a single-mode dye laser in a multiple-path arrangement results in a selective depletion of single hfs sublevels of the metastable states, which are refilled by RF transitions. In order to enhance the sensitivity, the laser beam is reflected by a prism arrangement back and forth and crosses the atomic beam several times. With difference and ratio recording the influence of fluctuations of the laser intensity or of the atomic-beam intensity could be minimized. The high sensitivity allowed a good signal-to-noise ratio even in cases where the population density of the metastable levels was less than 1 % of the ground-state population.

An interesting application of optical-RF double resonance is the realization of sensitive magnetometers. Here a cell with rubidium vapor at room temperature is placed in a magnetic field and the RF is tuned to transitions between Zeeman components in the $^2S_{1/2}$ state. Since the Landé factors are known, the magnetic field strength can be obtained by measuring the radiofrequency [530].

5.3 Optical–Microwave Double Resonance

Microwave spectroscopy has contributed in an outstanding way to the precise determination of molecular parameters, such as bond lengths and bond angles, nuclear equilibrium configurations of polyatomic molecules, fine and hyperfine splittings, or Coriolis interactions in rotating molecules. Its application was, however, restricted to transitions between thermally populated levels, which generally means transitions within electronic ground states [533].

At room temperatures ($T = 300$ K $\rightarrow kT \sim 250$ cm^{-1}) the ratio $h\nu/kT$ is very small for typical microwave frequencies $\nu \sim 10^{10}$ Hz ($\hat{=}0.3$ cm^{-1}). For the thermal population distribution

$$N_k/N_i = (g_k/g_i)\mathrm{e}^{-h\nu_{ik}/kT},$$

the power of a microwave absorbed along the path length Δx through the sample is

$$\Delta P = -P_0\sigma_{ik}\big[N_i - (g_i/g_k)N_k\big]\Delta x \approx -P_0 N_i \sigma_{ik}\Delta x h\nu_{ik}/kT. \tag{5.7}$$

Since $h\nu/kT \ll 1$, the absorbed power is small, because induced absorption and emission nearly balance one another. Furthermore, the absorption coefficient σ_{ik}, which scales with ν^3, is many orders of magnitude smaller than in the optical region.

Optical–microwave double resonance (OMDR) can considerably improve the situation and extends the advantages of microwave spectroscopy to excited vibrational or electronic states, because selected levels in these states can be populated by optical pumping. Generally dye lasers or tunable diode lasers are used for optical pumping. However, even fixed frequency lasers can often be used. Many lines of intense infrared lasers (for example, CO_2, N_2O, CO, HF, and DF lasers) coincide with rotational–vibrational transitions of polyatomic molecules. Even for lines that are only close to molecular transitions the molecular lines may be tuned into resonance by external magnetic or electric fields (Sect. 1.6). The advantages of this OMDR may be summarized as follows:

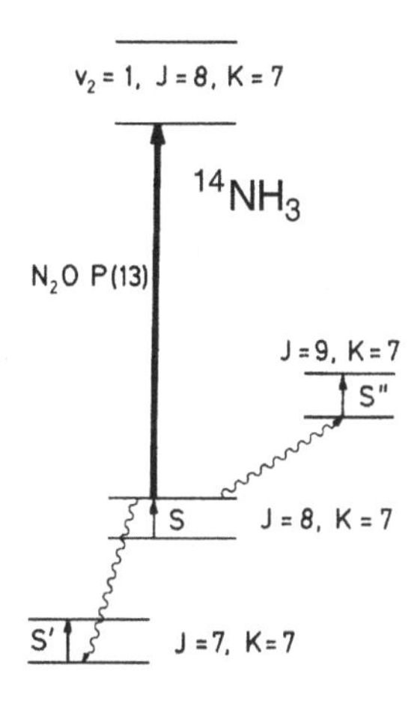

Fig. 5.12 Infrared–microwave double resonance in NH_3. The microwave transitions between the inversion doublets can start from the laser-pumped level (signal S) or from levels where the depletion has been transferred by collisions (secondary DR signals S′, S″) [534]

- Optical pumping of a level $|i\rangle$ increases the population difference $\Delta N = N_i - N_k$ for a microwave transition by several orders of magnitude compared to the value $N_i h\nu/kT$ of (5.7).
- The selective population or depletion of a single level simplifies the microwave spectrum considerably. The difference of the microwave spectra with and without optical pumping yields directly those microwave transitions that start from one of the two levels connected by the pump transition.
- The detection of the microwave transition does not rely on the small absorption of the microwave but can use the higher sensitivity of optical or infrared detectors.

Some examples illustrate these advantages.

Example 5.5 (Laser version of the cesium clock) The cesium clock provides our present time or frequency standard that is based on the hfs transition $7^2S_{1/2}$ $(F = 3 \rightarrow F = 4)$ in the electronic ground state of Cs. The accuracy of the frequency standard depends on the achievable signal-to-noise ratio and on the symmetry of the line profile of the microwave transition. With optical pumping and probing (see Fig. 5.10b), the signal-to-noise ratio can be increased by more than one order of magnitude. A further advantage is the absence of the two inhomogeneous magnetic fields, which may cause stray fields in the RF zone [532].

Example 5.6 Figure 5.12 illustrates the level scheme for the enhancement of microwave signals on transitions between the inversion doublets of NH_3 in rotational levels of the vibrational ground state and of the excited state $v_2' = 1$ by infrared pumping with an N_2O-laser line. This pumping causes selective depletion of the upper inversion component of the $(J'' = 8,\ K'' = 7)$ level and an increased population of the lower inversion component in the upper

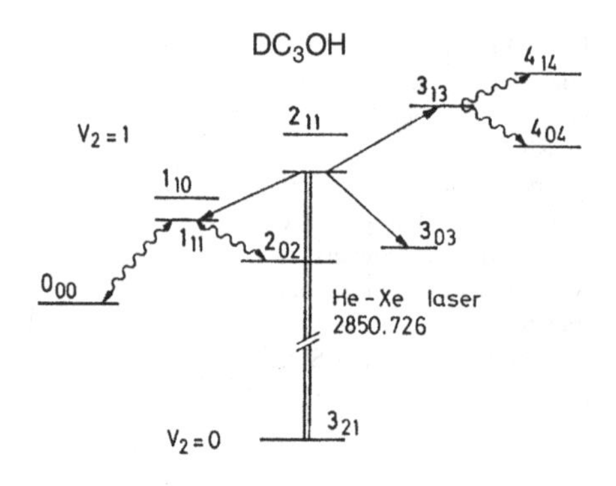

Fig. 5.13 Infrared–microwave double resonance in the vibrationally excited $v_2 = 1$ state of DCCCOH. The *solid arrows* indicate the microwave transitions, the *wavy arrows* secondary MW transitions [535]

vibrational level. The selective change ΔN of the populations is partly transferred by collisions to neighboring levels (Sect. 8.2), resulting in secondary collision-induced double-resonance signals [534].

Example 5.7 Figure 5.13 gives an example of microwave spectroscopy in an excited vibrational state, where the ($N_{\mathrm{Ka,Kc}} = 2_{1,2}$) rotational level in the excited vibrational state $v_2 = 1$ of DCCCHO has been selectively populated by infrared pumping with a HeXe laser [535]. The solid arrows represent the direct microwave transitions from the pumped level, while the wavy arrows correspond to "triple-resonance" transitions starting from levels that have been populated either by the first microwave quantum or by collision-induced transitions from the laser-pumped level.

Further examples and experimental details on infrared–microwave double-resonance spectroscopy can be found in [536–539].

The electronically excited states of most molecules are far less thoroughly investigated than their ground states. On the other hand, their level structure is generally more complex because of interactions between electron and nuclear motions, which are more pronounced in excited states (breakdown of the Born–Oppenheimer approximation, perturbations). It is therefore most desirable to apply spectroscopic techniques that are sensitive and selective and that facilitate assignment. This is just what the optical–microwave double-resonance technique can provide.

Example 5.8 One example of this technique is the excited-state spectroscopy of the molecule BaO [540]. In the volume that is crossed by an atomic barium beam and a molecular O_2 beam, BaO molecules are formed in various rovibronic levels (v'', J'') of their electronic $X\,^1\Sigma_g$ ground state by the reaction $Ba + O_2 \rightarrow BaO + O$ (Fig. 5.14). With a dye laser tuned to

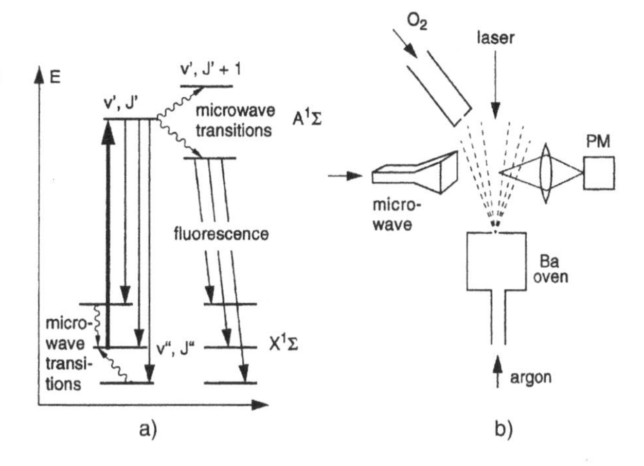

Fig. 5.14 Level scheme (**a**) and experimental arrangement (**b**) for detection of optical–microwave DR in BaO molecules [540]

the transitions $A\,^1\Sigma(v', J') \leftarrow X\,^1\Sigma(v'', J'')$ different levels (v', J') in the excited A state are selectively populated. Their quantum numbers (v', J') can be determined from the LIF spectrum (Sect. 1.7).

When the crossing volume is irradiated with a microwave produced by a tunable clystron, transitions $J'' \to J'' \pm 1$ or $J' \to J' \pm 1$ between adjacent rotational levels in the X or the A state can be induced by tuning the microwave to the corresponding frequencies.

Since optical pumping by the laser has *decreased* the population $N(v'', J'')$ below its thermal equilibrium value, microwave transitions $J'' \to J'' \pm 1$ *increase* $N(v'', J'')$, which result in a corresponding increase of the total fluorescence intensity. The transitions $J' \to J' \pm 1$, on the other hand, *decrease* the population $N(v', J')$ in the optically excited levels. Therefore they decrease the fluorescence intensity emitted from (v', J'), but generate new lines in the fluorescence spectrum originating from the levels $(J' \pm 1)$. They can be separated by a monochromator or by interference filters.

The use of OMDR spectroscopy allows very accurate measurements of rotational spacings in both the ground state and in electronically exited states [541]. The sensitivity is sufficiently high to measure even very small concentrations of molecules or radicals that are formed as intermediate products in chemical reactions. This has been proven, for example, for the case of NH_2 radicals in a discharge flow system [542].

Example 5.9 An impressive example of the OMDR version of the Rabi technique is the measurement of the hyperfine splittings in the electronic ground state of CaCl [543], where linewidths as small as 15 kHz could be achieved for the OMDR signals. They were only limited by the transit time of

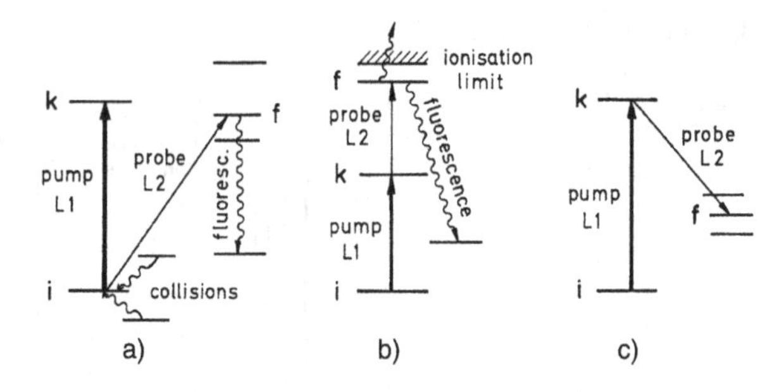

Fig. 5.15 Different schemes for OODR: (**a**) V-type OODR; (**b**) stepwise excitation; (**c**) Λ-type OODR

the molecules flying through the microwave zone. With an additional dc electric field in the microwave region, the resulting Stark splittings of the OMDR signals were observed and the electric dipole moments could be accurately determined [544, 545].

Besides the precise determination of rotational level spacings in excited states, another application of OMDR is the investigation of atomic or molecular Rydberg states. If a state with principal quantum number n is excited by one UV photon or two visible photons, microwave transitions to neighboring Rydberg states are induced. Energy shifts of these levels by external fields or by perturbations due to interactions between different Rydberg levels or autoionization processes can be measured with high accuracy [546].

5.4 Optical–Optical Double Resonance

The optical–optical double-resonance (OODR) technique is based on the simultaneous interaction of a molecule with two optical waves that are tuned to two molecular transitions sharing a common level. This may be the lower or the upper level of the pump transition. The three possible level schemes of OODR are depicted in Fig. 5.15.

The "V-type" double-resonance method of Fig. 5.15a (named after its V-shaped appearance in the level diagram) may be regarded as the inversion of laser-induced fluorescence (Fig. 5.16). With the LIF technique one *excited* level $|2\rangle$ is selectively populated. The resulting fluorescence spectrum corresponds to all allowed optical transitions from this level into the lower levels $|m\rangle$. The analysis of the LIF spectrum mainly yields information on the *lower* state. With V-type OODR, on the other hand, a single *lower* level $|i\rangle$ is selectively *depleted*. The OODR transitions, monitored through the difference in the absorption of the probe wave with and without the pump, start from the lower level $|1\rangle$ and end in all upper levels $|m\rangle$, which can be reached by probe transitions from $|1\rangle$. They give mainly information about these *upper* levels.

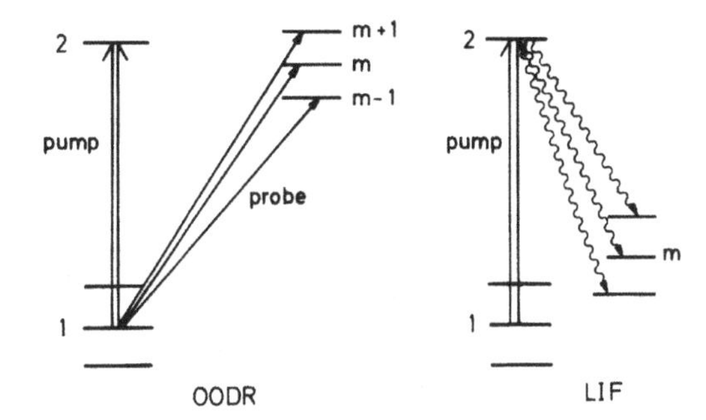

Fig. 5.16 V-type OODR compared with LIF

The second OODR scheme (Fig. 5.15b) represents stepwise excitation of high-lying levels via a common intermediate level $|2\rangle$, which is the upper level of the pump transition but the lower level of the probe transition. This scheme allows the investigation of higher levels (e.g., Rydberg states), where the absorption of the probe can be monitored by either LIF from these levels, or by the ions produced by the absorption of a third photon.

The last scheme in Fig. 5.15c, named Λ-type OODR, represents a stimulated resonant Raman process where the molecules are coherently transferred from level $|1\rangle$ to levels $|m\rangle$ by absorption of the pump laser and stimulated emission by the probe laser.

We will now discuss these three schemes in more detail.

5.4.1 Simplification of Complex Absorption Spectra

The V-type OODR scheme can be utilized for the simplification and analysis of infrared, visible, or UV molecular spectra. This is particularly helpful if the spectra are perturbed and might not show any regular pattern. Assume the pump laser L_1, with intensity I_1, which has been tuned to the transition $|1\rangle \rightarrow |2\rangle$ is chopped at the frequency f_1. The population densities N_1, N_2 then show a corresponding modulation

$$\begin{aligned} N_1(t) &= N_1^0\{1 - aI_1[1 + \cos(2\pi f_1 t)]\}, \\ N_2(t) &= N_2^0\{1 + bI_1[1 + \cos(2\pi f_1 t)]\}, \end{aligned} \tag{5.8}$$

which has an opposite phase for the two levels. The modulation amplitudes a and b depend on the transition probability of the pump transition and on possible relaxation processes such as spontaneous emission, collisional relaxation, or diffusion of molecules into or out of the pump region (Sect. 2.1).

The LIF intensity $I_{\text{Fl}}(\lambda_2)$ induced by the tunable probe laser L_2 will be modulated at the frequency f_1 if the wavelength λ_2 coincides with an absorbing transition $|1\rangle \rightarrow |m\rangle$ from the optically pumped lower level $|1\rangle$ to an upper level $|m\rangle$ or with a downward transition $|2\rangle \rightarrow |m\rangle$ from the upper pump level $|2\rangle$ to a lower level $|m\rangle$.

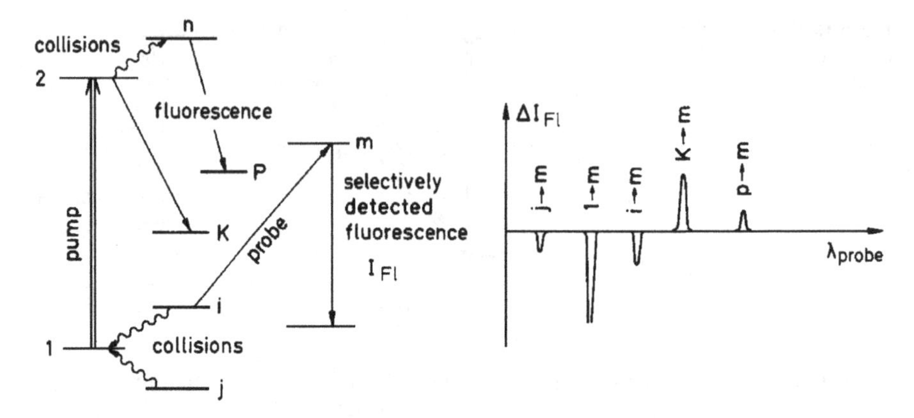

Fig. 5.17 OODR signals with opposite phase for probe transitions starting from the lower or the upper pump level, respectively

If the probe laser-induced fluorescence $I_{\text{Fl}}(\lambda_2)$ is monitored through a lock-in amplifier at the frequency f_1, one obtains negative OODR signals for all transitions $|1\rangle \rightarrow |m\rangle$ and positive signals for the transitions $|2\rangle \rightarrow |m\rangle$ (Fig. 5.17). From the phase of the lock-in signal it is therefore, in principle, possible to decide which of the two possible types of probe transitions is detected. Since this double-resonance technique selectively detects transitions that start from or terminate at levels labeled by the pump lever, it is often called *labeling spectroscopy*.

In reality, the situation is generally more complex. If the OODR experiments are performed on molecules in a cell that have a thermal velocity distribution and that may suffer collisions, double-resonance signals are also observed for probe transitions starting from other levels than $|1\rangle$ or $|2\rangle$. This is due to the following facts:

(a) Even with a narrow-band pump laser several absorbing transitions may simultaneously be pumped if the absorption profiles overlap within their Doppler width (Fig. 5.2).
(b) Collisions may transfer the modulation of the population density N_1 to neighboring levels if the collisional time is shorter than the chopping period $1/f_1$ of the pump laser (Sect. 8.3). This causes additional double-resonance signals, which may impede the analysis (Fig. 5.18). On the other hand, they also can give much information on cross sections for collisional population and energy transfer (Chap. 8).
(c) The fluorescence from the level $|2\rangle$ into all accessible lower levels $|m\rangle$ is also modulated at f_1 and leads to a corresponding modulation of the population densities N_m.

In order to avoid such secondary OODR signals, sub-Doppler excitation under collision-free conditions has to be realized. This can be achieved in a collimated molecular beam that is intersected by the two lasers L_1 and L_2 either at two differ-

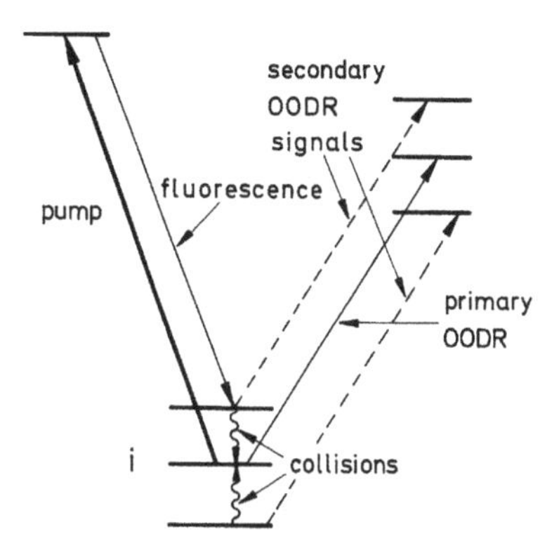

Fig. 5.18 Secondary OODR signals induced by collisions or by fluorescence from the optically pumped levels

ent positions z_1 and z_2 (Fig. 5.19) or with two overlapping laser beams (Fig. 5.20). In the first arrangement the probe laser-induced fluorescence $I_{\text{Fl}}(\lambda_2)$ can be imaged separately onto the detector, and chopping of the pump laser L_1 with phase-sensitive detection of $I_{\text{Fl}}(\lambda_2)$ yields the wanted OODR spectrum. In the second arrangement with superimposed pump and probe beams, the detected fluorescence $I_{\text{Fl}} = I_{\text{Fl}}(L_1) + I_{\text{Fl}}(L_2)$ is the sum of pump and probe laser-induced fluorescence intensities, which are both modulated at the chopping frequency f_1. In order to eliminate the large background of $I_{\text{Fl}}(L_1)$ one can use a double-modulation technique, where L_1 is modulated at f_1, L_2 at f_2, and the fluorescence is monitored at the sum frequency $f_1 + f_2$. This selects the OODR signals from the background, as can be seen from the following consideration.

The fluorescence intensity $I_{\text{Fl}}(\lambda_2)$ induced by the probe laser with intensity $I_2 = I_{20}(1 + \cos 2\pi f_2 t)$ on a transition starting from the lower level $|i\rangle$ of the pump transition is

$$I_{\text{Fl}}(\lambda_2) \propto N_i I_2 = N_{i0}\big\{1 - a I_{10}\big[1 + \cos(2\pi f_1 t)\big]\big\} I_{20}\big[1 + \cos(2\pi f_2 t)\big]. \quad (5.9)$$

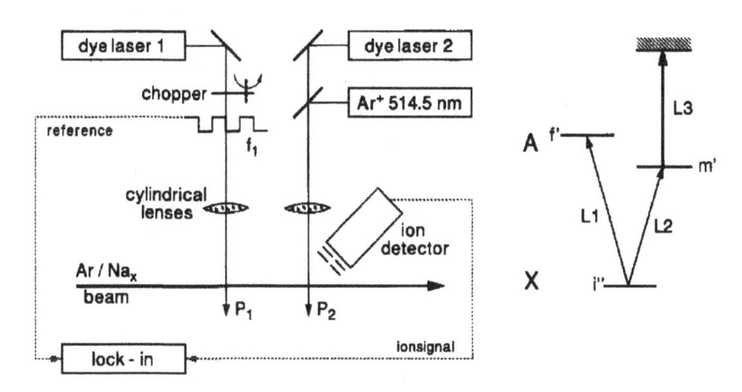

Fig. 5.19 OODR spectroscopy in a molecular beam with spatially separated pump and probe beams. The probe transition L_2 can be monitored by ionizing the excited levels m' with L_3

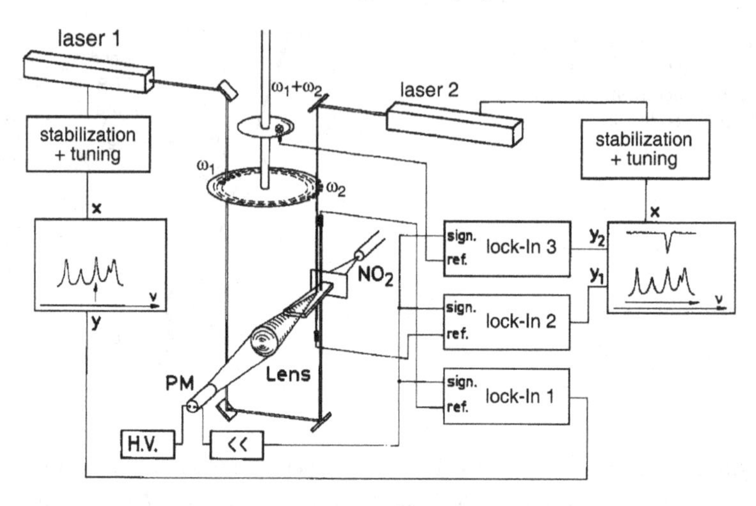

Fig. 5.20 OODR spectroscopy in a collimated molecular beam with overlapping pump and probe beams and detection at the sum frequency

The nonlinear term

$$
\begin{aligned}
&I_{10}I_{20}\cos(2\pi f_1 t)\cos(2\pi f_2 t) \\
&\quad = \frac{1}{2} I_{10}I_{20}\big[\cos 2\pi(f_1 + f_2)t + \cos 2\pi(f_1 - f_2)t\big],
\end{aligned}
\tag{5.10}
$$

which represents the OODR signal, is modulated at the frequencies $(f_1 + f_2)$ and $(f_1 - f_2)$ whereas the linear terms, representing the pump laser-induced fluorescence or the fluorescence induced by the probe laser on transitions from nonmodulated levels only contain the frequencies f_1 or f_2.

The application of OODR in molecular beams to the analysis of complex perturbed spectra is illustrated by Fig. 5.21, which shows a section of the visible NO_2 spectrum around $\lambda = 488$ nm. In spite of the small residual Doppler width of 15 MHz, not all lines are fully resolved and the analysis of the spectrum turns out to be very difficult, because the upper state is heavily perturbed. If in the OODR experiment the pump laser L_1 is kept on line 1, one obtains the two OODR signals as shown in the upper left part of Fig. 5.21, which proves that lines 1 and 4 in the lower spectrum share a common lower level. The right part of Fig. 5.21 also shows that the lines 2 and 5 start from a common lower level as well as the lines 3 and 6. The whole lower spectrum consists of two rovibronic transitions $J'', K'' = (10, 5) \to (J', K') = (11, 5)$ (each with three hfs components), which end in two closely spaced rotational levels with equal quantum numbers (J', K') that belong to two different vibronic states coupled by a mutual interaction [547].

This V-type OODR has been used in many laboratories. Further examples can be found in [548–552].

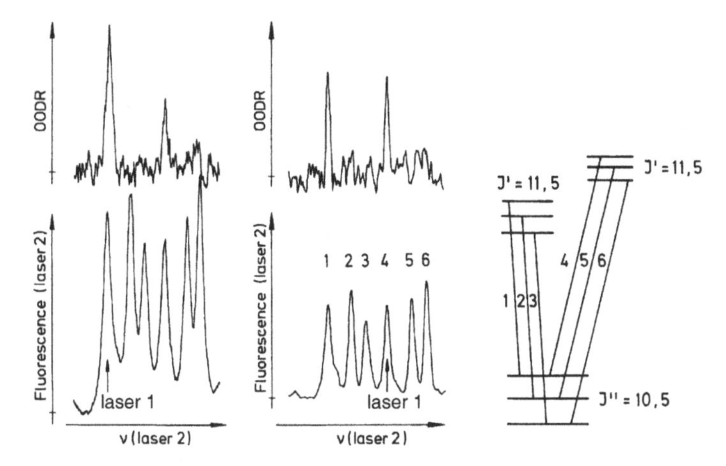

Fig. 5.21 Section of the linear excitation spectrum of NO_2 around $\lambda = 488$ nm (*lower spectrum*) and OODR signals obtained with the pump laser kept on transition 1 or 4, respectively [547]

5.4.2 Stepwise Excitation and Spectroscopy of Rydberg States

When the excited state $|2\rangle$ has been selectively populated by optical pumping with laser L_1, transitions to still higher levels $|m\rangle$ can be induced by a second tunable laser L_2 (Fig. 5.15b). This *two-step excitation* may be regarded as the special resonance case of the more general two-photon excitation with two different photons $\hbar\omega_2$ (Sect. 2.5). Because the upper levels $|m\rangle$ must have the same parity as the initial level $|1\rangle$, they cannot be reached by an allowed one-photon transition. The one-step excitation by a frequency-doubled laser with the photon energy $2\hbar\omega = \hbar(\omega_1 + \omega_2)$ therefore excites, in the same energy range, levels of opposite symmetry compared to those levels reached by the two-step excitation.

With two visible lasers, levels $|m\rangle$ with excitation energies up to 6 eV can be reached. Optical frequency doubling of both lasers allows even the population of levels up to 12 eV. This makes the Rydberg levels of most atoms and molecules accessible to detailed investigations. The population of Rydberg levels of species M can be monitored either by their fluorescence or by detecting the ions M^+ or the electrons e^- that are produced by photoionization, field ionization, collisional ionization, or autoionization of the Rydberg levels.

Rydberg states have some remarkable characteristics (Table 5.1). Their spectroscopic investigation allows one to study fundamental problems of quantum optics (Sect. 9.5), nonlinear dynamics, and chaotic behavior of quantum systems (see below). Therefore detailed studies of atomic and molecular Rydberg states have found increasing interest [553–566].

The term value T_n of a Rydberg level with principal quantum number n is given by the Rydberg formula

$$T_n = IP - \frac{R}{(n-\delta)^2} = IP - \frac{R}{n^{*2}}, \tag{5.11}$$

Table 5.1 Characteristic properties of Rydberg atoms

Property	n-dependence	Numerical values for		
		H ($n=2$)	H ($n=50$)	Na (10d)
Binding energy	$-R\cdot n^{-2}$	4 eV	5.4 meV	0.14 eV
$E(n+1)-E_n$	$\frac{R}{n^2}-\frac{R}{(n+1)^2}\propto\frac{1}{n^3}$	$\frac{5}{36}R\approx 2$ eV	0.2 meV $\hat{=}$ 2 cm^{-1}	$\approx$1.5 meV
Mean radius	$a_0 n^2$	$4a_0\approx 0.2$ nm	132 nm	7 nm
Geometrical cross section	$\pi a_0^2 n^4$	$\approx 1.2\times 10^{-16}$	5×10^{-10}	$\approx 10^{-12}$ [cm^2]
Spontaneous lifetime	$\propto n^3$	5×10^{-9}	1.5×10^{-4}	$\approx 10^{-6}$ [s]
Critical field	$E_c=\pi\epsilon_0 R^2 e^{-3} n^{-4}$	5×10^9	5×10^3	3×10^6 [V/m]
Orbital period	$T_n\propto n^3$	10^{-15}	2×10^{-11}	2×10^{-13} [s]
Polarizability	$\alpha\propto n^7$	10^{-6}	10^4	20 [s^{-1} V^{-2} m^2]

where IP is the ionization potential, R the Rydberg constant, and $\delta(n,\ell)$ the quantum defect that depends on n and on the angular momentum $\ell\hbar$ of the Rydberg electron. The quantum defect describes the deviation of the real potential (from the nucleus, including its shielding by the core electrons) from a pure Coulomb potential $V=Z_{\mathrm{eff}}e^2/(4\pi\epsilon_0 r)$. With the effective quantum number $n^*=n-\delta$, (5.11) can be formally written as Rydberg levels in a Coulomb potential.

The investigation of atomic Rydberg states started with alkali atoms since they can be readily prepared in cells or atomic beams. Because of their relatively low ionization energy, their Rydberg states can be reached by stepwise excitation with two dye lasers in the visible spectral region. The first step generally uses the resonance transitions $nS\rightarrow nP$ to the first excited state, which has a large transition probability and therefore can be already saturated at laser intensities of $I(L_1)\leq 0.1$ W/cm^2! The second step requires higher intensities $I(L_2)\approx 1$–100 W/cm^2 (because the transition probability decreases with n^{-3} [559]), which, however, can be readily realized with cw dye lasers. For illustration, Fig. 5.22 exhibits the term diagram and measured Rydberg series $3p^2P\rightarrow nS$, and $3p^2P\rightarrow nD$ of Na atoms excited via the $3p^2P_{3/2}$ state.

Example 5.10 From (5.11) one can calculate that the term energy of Na$^*(n=40)$ is only about 0.005 eV below the ionization limit, and Fig. 5.23b shows that an external field with $E=100$ V/cm is sufficient to ionize this state.

The external field E [V/m] decreases the ionization threshold of a Rydberg state down to the "appearance potential" AP of the ions (Fig. 5.24):

$$AP=IP-\sqrt{e^3E/\pi\epsilon_0}, \tag{5.12}$$

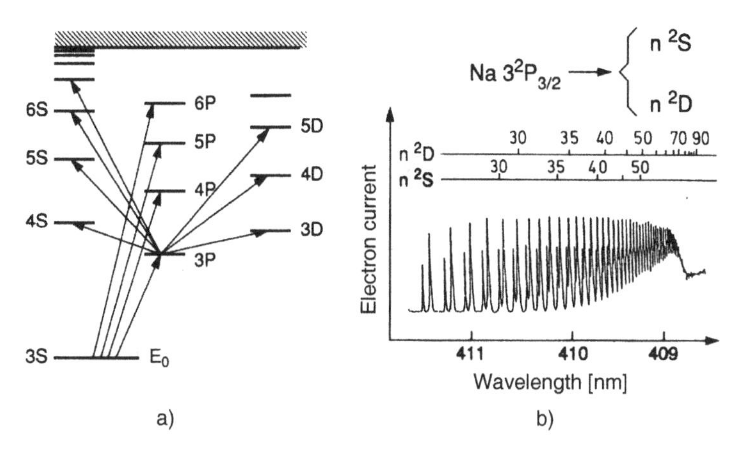

Fig. 5.22 Level scheme for two-step excitation of Rydberg levels of alkali atoms (**a**) and Rydberg series Na $3\,^2P_{3/2} \rightarrow n\,^2S, n\,^2D$ measured by field ionization of the Rydberg states (**b**) [557]

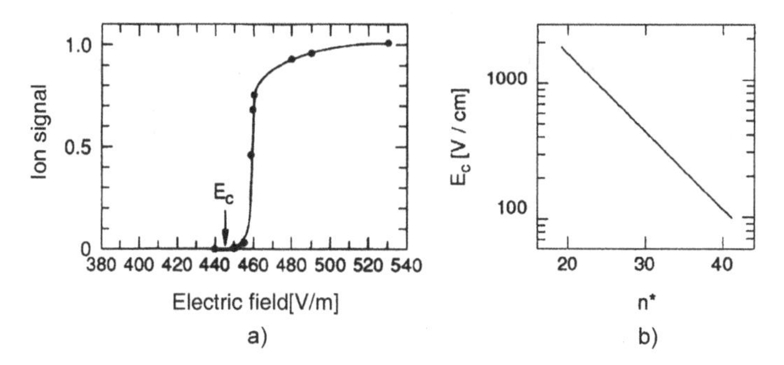

Fig. 5.23 Field ionization of atomic Rydberg levels: (**a**) ionization rate of the Na $(31\,^2S)$ level; (**b**) dependence of the threshold field E_c on the effective principal quantum number n^* for Na$(n^* S)$ Rydberg levels [557]

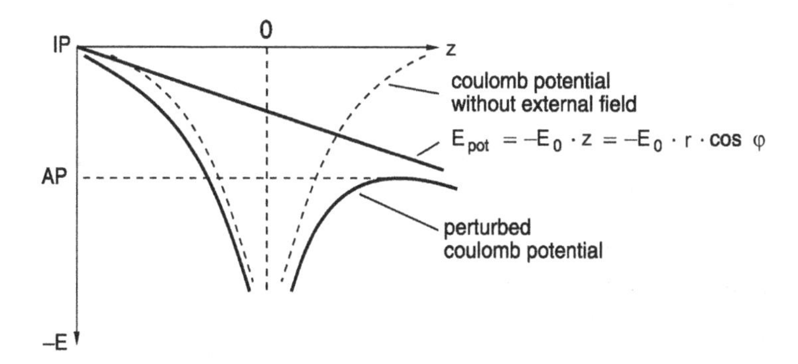

Fig. 5.24 Coulomb potential with external field. AP = appearance potential, r = distance of Rydberg electron from the atomic nucleus

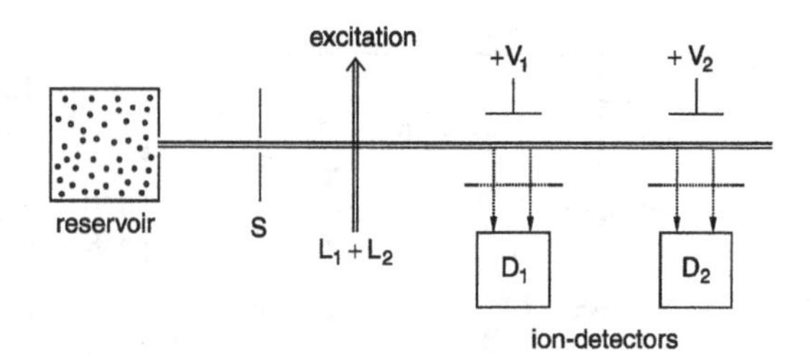

Fig. 5.25 Selective detection of Rydberg atoms in specific Rydberg states

which can be readily derived from the maximum energy of the superposition of Coulomb potential and external potential $V_{\text{ext}} = -Ez$ of the electric field in the z-direction.

Example 5.11 For $E = 1000$ V/cm the decrease $\Delta IP = IP - AP$ of the effective ionization potential is $\Delta IP = 20$ meV $\hat{=}$ 194 cm^{-1}. Rydberg levels with $n^* \geq 24$ are already field ionized.

The strong dependence of the field-ionization probability on the principal quantum number n can be utilized to detect atoms in specific Rydberg states n. When the Rydberg atoms in an atomic beam initially fly through a region with an electric field E_1 and then through a field with $E_2 > E_1$, the Rydberg atoms at all levels $n > n_1$ are field-ionized in the first field, while the atoms in level $n_1 - 1$, which pass through the first field unaffected, are then ionized in the second field if E_2 is properly chosen (Fig. 5.25).

Collisional deactivation of Rydberg levels under "field-free" conditions can be studied in the thermionic heat pipe (Sect. 1.4.5) with shielding grids or wires [561]. Since the excitation occurs in a field-free region, the effects of collisional broadening and shifts (Vol. 1, Sect. 3.3) can be separately investigated up to very high quantum numbers [560–562]. In addition, the influence of the hyperfine energy of core electrons on the term values of Rydberg levels of the Sr atom could be studied in a heat pipe up to $n = 300$ [563].

When Rydberg levels are selectively excited under collision-free conditions in an atomic beam, it is found that after a time, which is short compared to the spontaneous lifetime, the population of neighboring Rydberg levels increases. The reason for this surprising result is the following: because of the large dipole moment of Rydberg atoms, the weak thermal radiation emitted by the walls of the apparatus is sufficient to induce transitions between the optically pumped level $|n, \ell\rangle$ and neighboring Rydberg levels $|n + \Delta n, \ell \pm 1\rangle$ [564].

The interaction of the Rydberg atoms with the thermal radiation field results in a small shift of the Rydberg levels (AC-Stark shift), which only amounts to $\Delta\nu/\nu \approx 2 \times 10^{-12}$ for rubidium. It has recently been measured with extremely well stabilized lasers [565]. In order to eliminate the influence of the thermal radiation field one has to enclose the interaction zone of the laser and atomic beam by walls cooled to a few degrees Kelvin.

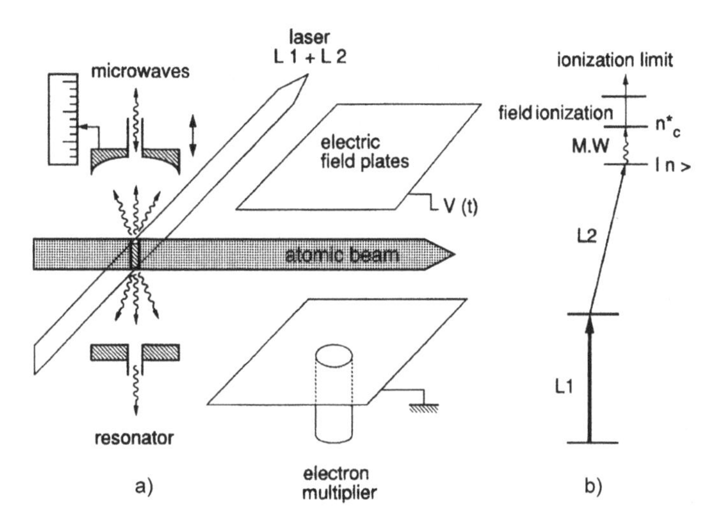

Fig. 5.26 Sensitive detection of microwave radiation by microwave-photon ionization of Rydberg levels

On the other hand, the large dipole moment of Rydberg atoms offers the possibility to use them as sensitive detectors for microwave and submillimeter-wave radiation [566]. For the detection of radiation with frequency ω, a Rydberg level $|n\rangle$ is selectively excited by stepwise excitation with lasers in an external electric dc field. The field strength is adjusted in such a way that the energy E_n of the Rydberg level $|n\rangle$ is just below the critical value E_c for field ionization, but $E_n + \hbar\omega$ is just above. Every absorbed microwave photon $\hbar\omega$ then produces an ion that can be detected with 100 % efficiency (Fig. 5.26).

There is another interesting aspect of Rydberg levels: the Coulomb energy of a Rydberg electron in level $|n\rangle$ from the electric field of the core decreases with $1/n^{*2}$. For sufficiently large principle quantum numbers n^* the Zeeman energy in an external magnetic field $\boldsymbol{B}$ may become larger than the Coulomb energy.

The Lorentz force $\boldsymbol{F} = q(\boldsymbol{v} \times \boldsymbol{B})$ causes the electron to precess around the magnetic field direction $\boldsymbol{B}$, and even if the total energy of the electron lies above the field-free ionization limit, the electron cannot escape except into the direction of $\boldsymbol{B}$. This leads to relatively long lifetimes of such autoionizing states. The corresponding classical trajectories of the electron in such states are complicated and may even be chaotic. At present, investigations in several laboratories are attempting to determine how the chaotic behavior of the classical model is related to the term structure of the quantum states [567]. The question whether "quantum chaos" really exists is still matter of controversy [569–572].

Example 5.12 For $n = 100$ the Coulomb binding energy of the electron to the proton in a hydrogen Rydberg atom is $E_{\text{coul}} = \text{Ry}/n^2 = 1.3 \times 10^{-3}$ eV.

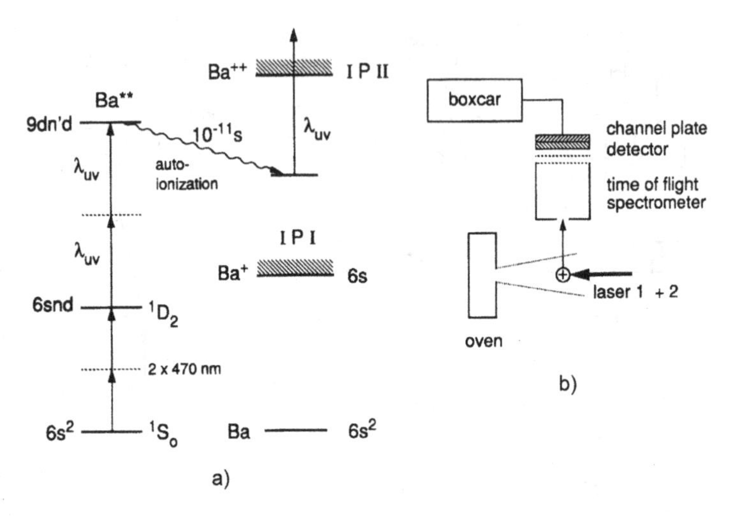

Fig. 5.27 Double excited planetary atoms: (**a**) level diagram for two-step excitation of two electrons in the Ba atom with subsequent autoionization to Ba^{++}; (**b**) experimental arrangement [574]

The magnetic energy of the electron with an average orbital angular momentum of $50\hbar$ in a field of $B = 1$ T is $E_{\text{mag}} = \ell\mu_B = 2.8 \times 10^{-3}$ eV, where $\mu_B = 5.6 \times 10^{-5}$ eV/T is the Bohr magneton.

Until now detailed experiments on Rydberg atoms in crossed electric and magnetic fields have been performed on alkali atoms [568] and on Rydberg states of the H atom [573], which are excited either by direct two-photon transitions or by stepwise excitation via the 2^2p state. Since the ionization energy of H is 13.6 eV, one needs photon energies above 6.7 eV ($\lambda \leq 190$ nm), which can be produced by frequency doubling of UV lasers in gases or metal vapors [574, 575].

In the examples given so far only one single Rydberg electron was excited. Special laser excitation techniques allow the simultaneous excitation of two electrons into high-lying Rydberg states [576]. The total energy of this doubly excited system is far above the ionization energy. The correlation between the two electrons causes an exchange of energy and results in auto-ionization (Fig. 5.27). The population of the doubly-excited Rydberg state is achieved by two two-photon transitions: at first a one-electron Rydberg state is excited by a two-photon transition induced by a visible laser. A successive two-photon transition with UV photons excites a second electron from the core into a Rydberg state [577, 578]. Such doubly excited Rydberg atoms, which are named *planetary atoms* [579], offer the unique possibility to study the correlations between two electrons that are both in defined Rydberg orbits (n_1, ℓ_1, S_1) and (n_2, ℓ_2, S_2) by measuring the autoionization time for different sets of quantum numbers.

Molecular Rydberg series are by far more complex than those of atoms. The reason is that many more electronic states exist in molecules, where furthermore

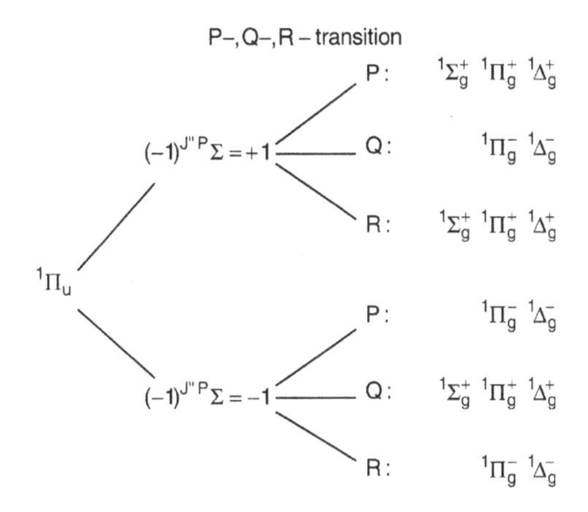

Fig. 5.28 Molecular Rydberg states of homonuclear alkali dimers, accessibly by one-photon transitions from the intermediate $B\,^1\Pi_u$ state

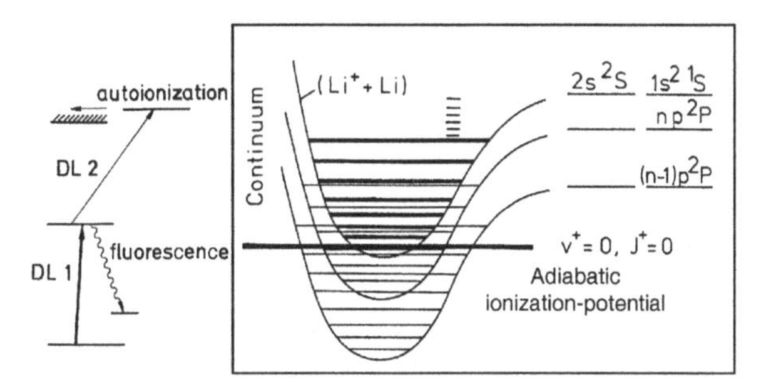

Fig. 5.29 Autoionization of molecular Rydberg levels (n^*, v^*, J^*) above the energy of the lowest level of the molecular ion M^+

each of these states has a manifold of vibrational–rotational levels. Often vibronic levels of different electronic Rydberg states are coupled by several interactions (vibronic, spin-orbit, Coriolis interaction, etc.). This interaction shifts their energy, and such perturbed Rydberg series often show an irregular structure that is not easy to analyze [580].

Here, the stepwise excitation is, in particular, very helpful since only those Rydberg levels are excited that are connected by allowed electric dipole transitions to the known intermediate level, populated by the pump. This shall be illustrated by the example of a relatively simple molecule, the lithium dimer Li_2 [581, 582]. When the pump laser selectively populates a level (v'_k, J'_k) in the $B\,^1\Pi_u$ state, all levels $(v^*, J^* = J'_k \pm 1$ or $J^* = J'_k)$ in the Rydberg states $ns\,^1\Sigma$ with $\ell = 0$ and $nd\,^1\Delta$, $nd\,^1\Pi$, or $nd\,^1\Sigma$ with $\ell = 2$ and $\lambda = 2, 1, 0$ are accessible by probe laser transitions with $\Delta J = +1$ (R-lines), $\Delta J = 0$ (Q-lines), or $\Delta J = -1$ (P-lines). This is shown in (Fig. 5.28), where all possible transitions from the two Λ components of the $B\,^1\Pi_u$ state with parity -1 and $+1$ into the different Rydberg states are compiled.

The excitation of Rydberg levels below the ionization potential *IP* can be monitored via the probe laser-induced fluorescence. There are, however, also many Rydberg levels above *IP* where higher vibrational or rotational levels are excited (Fig. 5.29). If the Rydberg electron can gain enough energy from the vibration or

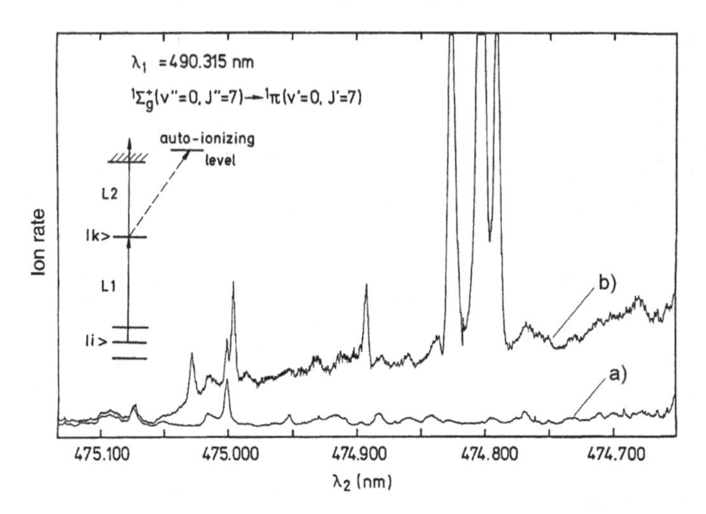

Fig. 5.30 Direct ionization and autoionization resonances just above the ionization threshold observed through the ion yield $N_{\text{ion}}(\lambda_2)$ in Li_2 when the first laser excites the level $|k\rangle \hat{=} (v' = 2, J' = 7)$ in the $B\,^1\Pi_u$ state: (**a**) with L_1 off; (**b**) with L_1 on

rotation of the molecular core it can leave the core. This *autoionization* demands a coupling between the kinetic energy of the nuclei and the energy of the Rydberg electron, which implies a breakdown of the adiabatic Born–Oppenheimer approximation [580]. This coupling is generally weak and the autoionization is therefore much slower than that in doubly excited atoms caused by electron–electron correlation. The autoionizing states are therefore "long-living" states and the corresponding transitions remarkably narrow. Since the spontaneous lifetimes increase proportionally to n^3, the autoionization may still be faster than the radiative decay of the levels. The ions produced by autoionization can therefore be used for sensitive detection of transitions to Rydberg levels above the ionization limit of molecules. When the wavelength λ (L_2) of the second laser is tuned while L_1 is kept at a fixed wavelength λ_1, one observes sharp intense resonances in the ion yield $N_{\text{ion}}(\lambda_2)$ superimposed on a continuous background of direct photoionization (Fig. 5.30). The analysis of such spectra demands the application of quantum defect theory [583–585].

A common experimental arrangement for the study of molecular Rydberg states is depicted in Fig. 5.31. The output beams of two pulsed narrow-band dye lasers, pumped by the same excimer laser, are superimposed and cross the molecular beam perpendicularly. The fluorescence emitted from the intermediate level (v', J') or from the Rydberg levels (v^*, J^*) can be monitored by a photomultiplier. The ions produced by autoionization (or for levels slightly below *IP* by field ionization) are extracted by an electric field and are accelerated onto an ion multiplier or channel plate. This allows the detection of single ions. In order to avoid electric Stark shifts of the Rydberg levels during their excitation, the extraction field is switched on only after the end of the laser pulse. Experimental details and more infor-

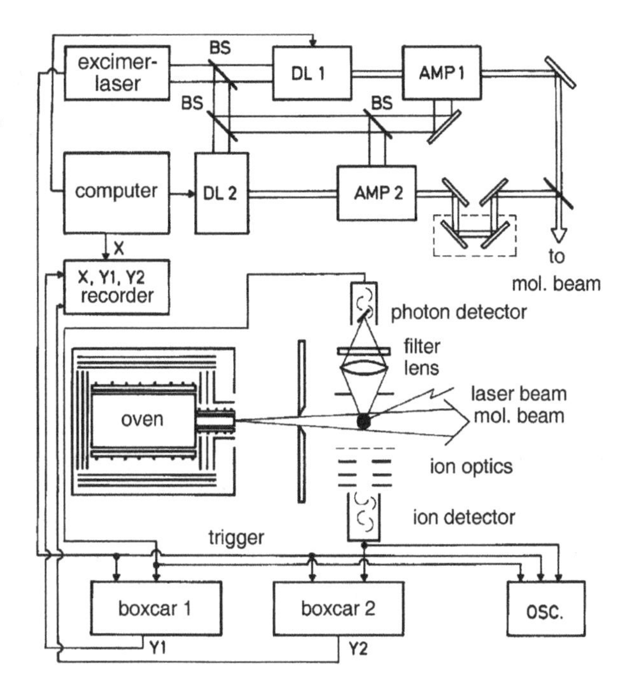

Fig. 5.31 Experimental arrangement for the measurement of molecular Rydberg levels with two-step excitation in a molecular beam [582]

mation on the structure and analysis of molecular Rydberg spectra can be found in [582–586].

A powerful technique for studying molecular ion states and ionization potentials of molecules is the zero-kinetic energy (ZEKE) electron spectroscopy where the photoelectrons with a very small kinetic energy, produced by laser excitation of ionic states just at the ionization limit, are selectively detected [588]. When the laser is tuned through the spectral range of interest, the excitation threshold for a new state is marked by the appearance of ZEKE electrons. Because of their very small kinetic energy, the electrons can be extracted by a small electric field and collected onto a detector with high efficiency. This ZEKE electron spectroscopy in combination with laser excitation and molecular beam techniques is a rapidly developing field of molecular Rydberg state spectroscopy [587–591].

From the optically pumped atomic or molecular Rydberg levels neighboring levels can be reached by microwave transitions, as was mentioned above. This "triple resonance" (two-step laser excitation plus microwave) is a very accurate method to measure quantum defects, fine-structure splitting, and Zeeman and Stark splitting in Rydberg states [592].

The Stark splitting and the field ionization of very high Rydberg levels provide sensitive indicators for measuring weak electric fields. These few examples demonstrate the variety of information obtained from Rydberg state spectroscopy. More examples can be found in Sect. 9.5 and in the literature on laser spectroscopy of Rydberg states [593–597].

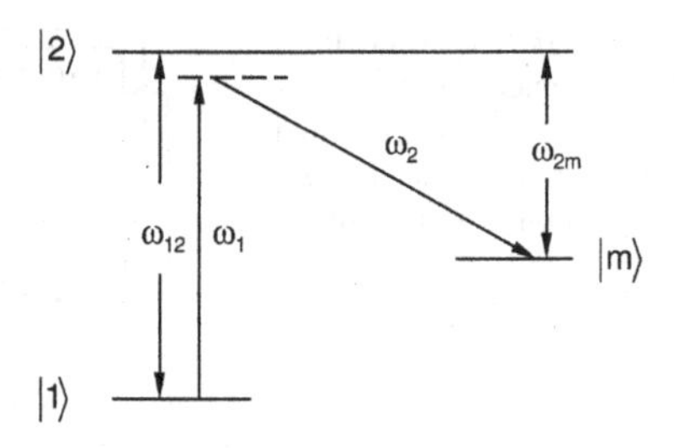

Fig. 5.32 Stimulated emission spectroscopy

5.4.3 Stimulated Emission Pumping

In the Λ-type OODR scheme (Fig. 5.32) the probe laser induces downward transitions from the upper level $k = |2\rangle$ of the pump transition to lower levels $f = |m\rangle$. This process, which is called *stimulated emission pumping* (SEP), may be regarded as a resonantly induced Raman-type transition. In case of monochromatic pump and probe lasers tuned to the frequencies ω_1 and ω_2, respectively, the resonance condition for a molecule moving with velocity $\boldsymbol{v}$ is

$$\omega_1 - \boldsymbol{k}_1 \cdot \boldsymbol{v} - (\omega_2 - \boldsymbol{k}_2 \cdot \boldsymbol{v}) = (E_m - E_1)/\hbar \pm \Gamma_{m1}, \tag{5.13}$$

where ω_i, $\boldsymbol{k}_i$ are frequency and wave vector of laser i, and $\Gamma_{m1} = \gamma_1 + \gamma_m$ is the sum of the homogeneous widths of initial and final levels of the Raman transition.

Note that the laser frequency ω_1 does not necessarily coincide with $\omega_{12} = (E_2 - E_1)/\hbar$ when a virtual level is reached by L_1. In the case where $\omega_1 = \omega_{12}$ we have resonance-enhanced Raman transitions.

In any case, the probe laser is in resonance if its frequency is

$$\omega_2 = \omega_1 + (\boldsymbol{k}_2 - \boldsymbol{k}_1) \cdot \boldsymbol{v} - (E_m - E_1)/\hbar \pm \Gamma_{m1}. \tag{5.14}$$

For a collinear arrangement of the two laser beams $\boldsymbol{k}_1 \parallel \boldsymbol{k}_2$ the Doppler shift becomes very small for $|k_1| - |k_2| \ll |k_1|$, and the width of the OODR signal $S(\omega_2)$ is only determined by the sum of the level widths of the initial and final levels, while the width of the upper level $|2\rangle$ does not enter into the signal width. If the two levels $|1\rangle$ and $|m\rangle$ are, for example, vibrational–rotational levels in the electronic ground state of a molecule, their spontaneous lifetimes are very long (even infinite for homonuclear diatomics!) and the level widths are only determined by the transit time of the molecule through the laser beams. In such cases, the linewidth of the OODR signal may become smaller than the natural linewidth of the optical transitions $|1\rangle \to |2\rangle$ (Sect. 9.4).

A more thorough consideration starts similar to the discussion of saturation spectroscopy in Sect. 2.2 from the absorption coefficient of the probe wave

$$\alpha(\omega) = \int_{-\infty}^{+\infty} \sigma(v_z, \omega) \Delta N(v_z) \, \mathrm{d}v_z, \tag{5.15}$$

with $\Delta N = (N_m - N_2)$. The population density $N_2(v_z)$ in the common upper level $|2\rangle$ is altered through optical pumping with the laser L_1 from its thermal equilibrium value $N_2^0(v_z)$ to

$$N_2^s(v_z) = N_2^0(v_z) + \frac{1}{2}\frac{[N_1(v_z) - N_2(v_z)]\gamma_2 S_0}{(\omega - \omega_{12} - k_1 v_z)^2 + (\Gamma_{12}/2)^2(1+S)}, \tag{5.16}$$

where S is the saturation parameter of the pump transition, see (2.7). Inserting (5.16) into (5.15) and integrating over all velocity components v_z yields [598]

$$\alpha(\omega) = \alpha_0(\omega)\left(1 - \frac{1}{2}\frac{N_2^0 - N_1^0}{N_2^0 - N_m^0}\frac{k_2 S}{k_1\sqrt{1+S}}\frac{\gamma_1 \Delta\Gamma}{[\Omega_2 \pm (k_2/k_1)\Omega_1]^2 + (\Delta\Gamma)^2}\right), \tag{5.17}$$

with $\Omega_1 = \omega_1 - \omega_2$, $\Omega_2 = \omega_{2\mathrm{m}}$, and $\Delta\Gamma = \Gamma_{2\mathrm{m}} + (k_2/k_1)\Gamma_{12}\sqrt{1+S}$.

This represents a Doppler profile $\alpha_0(\omega)$ with a narrow dip at the frequency

$$\omega_2 = (k_2/k_1)(\omega_1 - \omega_{12}) + \omega_{2\mathrm{m}}.$$

If the second term in the bracket of (5.17) becomes larger than 1, $\alpha(\omega)$ becomes negative, that is, the probe wave is amplified instead of attenuated (Raman gain, see Vol. 1, Sect. 5.7).

Chopping of the pump laser and detection of the probe laser Raman signal behind a lock-in amplifier yields a narrow OODR signal with a linewidth

$$\Delta\Gamma = \gamma_m + \left[(k_2/k_1)\gamma_1 + (1 \mp k_2/k_1)\gamma_2\right]\sqrt{1+S}, \tag{5.18}$$

where the $-$ sign in the bracket of (5.18) stands for collinear, and the $+$ sign for anticollinear propagation of the two laser beams. This shows that for the collinear arrangement the level width γ_2 of the common upper level only enters with the fraction $(1 - k_2/k_1)$, which becomes very small for $k_2/k_1 \sim 1$ [599].

This Λ-type OODR can be used to probe high-lying vibrational–rotational levels close to the dissociation limit of the electronic ground state. As an example, Fig. 5.33 shows the term diagram of the Cs_2 molecule where a high-lying vibrational level ($v' = 50$, $J' = 48$) in the $D\,^1\Sigma_u$ state is optically pumped at the inner turning point of the vibrating molecule and the probe laser induces downward transition with large Franck–Condon factors from the outer turning point into levels (v_3, J_3) closely below the dissociation energy of the $X\,^1\Sigma_g^+$ ground state. These levels show perturbations by hyperfine mixing with triplet levels induced by indirect nuclear spin–nuclear spin interactions, which are mediated by the electron spin. This OODR spectroscopic technique therefore can give very detailed information about the interaction of atoms at large internuclear distances, which are gaining increasing interest after the realization of Bose–Einstein condensation of alkali atoms, where such interactions cause spin-flips resulting in a loss of atoms from the condensate (see Sect. 9.1.10).

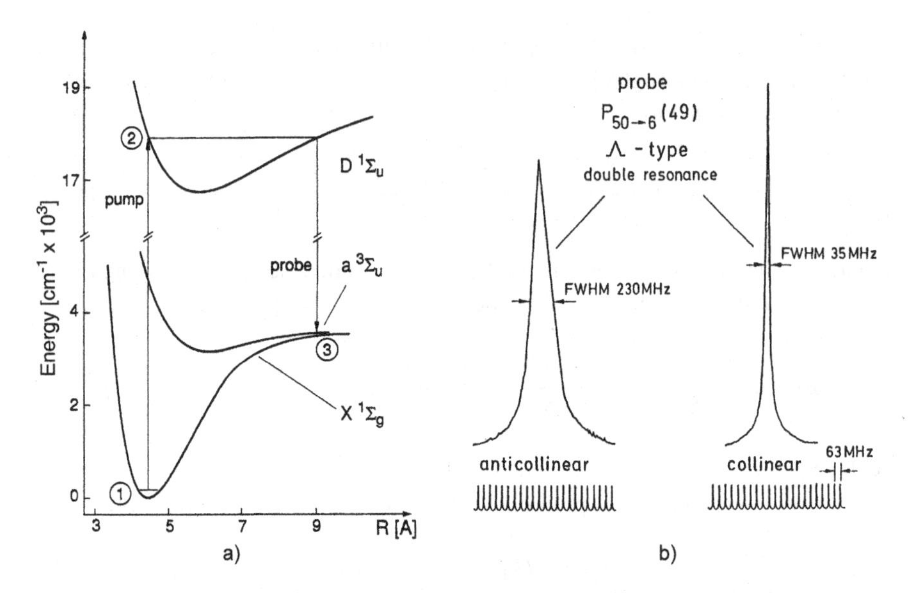

Fig. 5.33 Λ-type OODR spectroscopy illustrated for the example of the Cs_2 molecule: (**a**) term diagram; (**b**) two OODR signals of the same transition measured with collinear and anticollinear propagation of pump and probe beams. For this example the homogeneous width of the upper (predissociating) level is $\gamma_2 \sim 100$ MHz [605]

The linewidth of the OODR signal for collinear propagation is much smaller than the level width of the predissociating upper level.

The downward transitions can be probed by the change in transmission of the probe wave by OODR-polarization spectroscopy or by ion-dip spectroscopy (Sect. 5.5.1).

Often the super-high spectral resolution of the induced resonant Raman transitions obtained with single-mode lasers is not necessary if the levels $|m\rangle$ are separated by more than one Doppler width. Then pulsed lasers can be used for stimulated emission pumping [600]. Many experiments on high vibrational levels in the electronic ground state of polyatomic molecules have been performed so far by SEP with pulsed lasers. Compilations may be found in [601–604].

The stimulated emission pumping technique allows one to selectively populate high vibrational levels of molecules that are going to collide with other atoms or molecules. The investigation of the cross section for reactive collisions as a function of the excitation energy gives very valuable information on the reaction mechanism and the intermolecular potential (Chap. 8). Therefore, the SEP technique is a useful method for the selective population transfer into high-lying levels that are not accessible by direct absorption from the ground state.

Up to now we have only considered the population changes ΔN_i under SEP. There are, however, also coherent effects and an even more efficient scheme for population transfer is provided by the coherent stimulated rapid adiabatic passage (STIRAP) method, explained in Chap. 7.

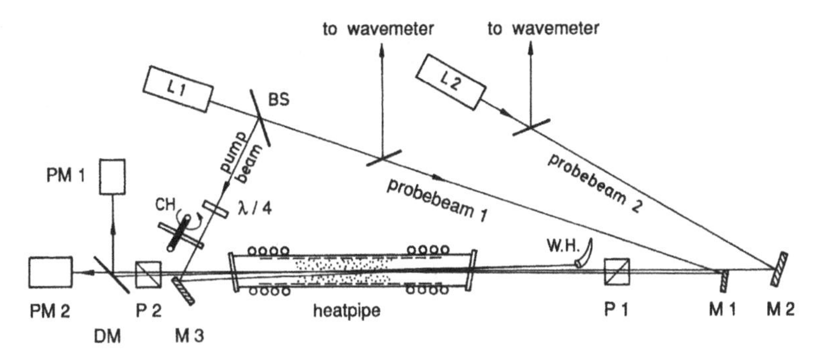

Fig. 5.34 Experimental arrangement for OODR Λ-type double resonance

5.5 Special Detection Schemes of Double-Resonance Spectroscopy

Because of their great advantages for the analysis of molecular spectra, a large variety of different detection schemes for double-resonance spectroscopy have been developed. In this section we present four examples.

5.5.1 OODR-Polarization Spectroscopy

Similar to polarization spectroscopy with a single tunable laser (Sect. 2.4), in OODR-polarization spectroscopy the sample cell is placed between two crossed polarizers and the transmitted probe laser intensity $I_T(\lambda_2)$ is measured as a function of the probe laser wavelength λ_2.

The optical pumping of the molecules is now performed with a separate laser beam, which is sent through the sample cell anticollinearly to the probe beam for V-type OODR, but collinearly for Λ-type OODR (Fig. 5.34). In order to keep the pump transition on the wanted selected transition, at first an ordinary Doppler-free polarization spectrum of the pump laser must be recorded. Therefore, the pump laser beam is split into a pump and a probe beam, and the spectrum is recorded while laser L_2 is switched off. Now the pump laser is stabilized onto the wanted transition and the second (weak) probe laser L_2 is simultaneously sent through the cell.

There are several experimental tricks to separate the two probe beams of L_1 and L_2: for Λ-type OODR these two probe beams travel into opposite directions and can therefore be detected by two different detectors. In case of V-type OODR the pump beam L_1 and the probe beam L_2 are anticollinear, which means that the two probe beams are collinear. If their wavelengths differ sufficiently they can be separated by a prism behind the analyzer. Although this is efficient for separating both probe beams, it has the disadvantage that the refraction angle of the prism changes with wavelength and the detector must be realigned when the probe wavelength is scanned over a larger spectral range. If the wavelengths λ_1, λ_2 of the two probe

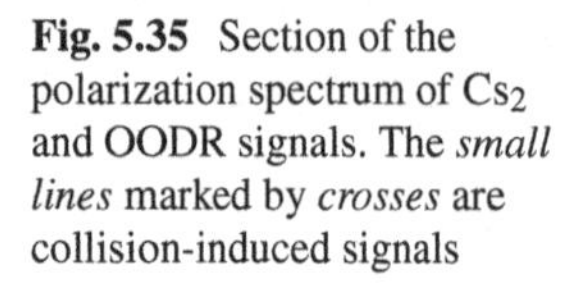

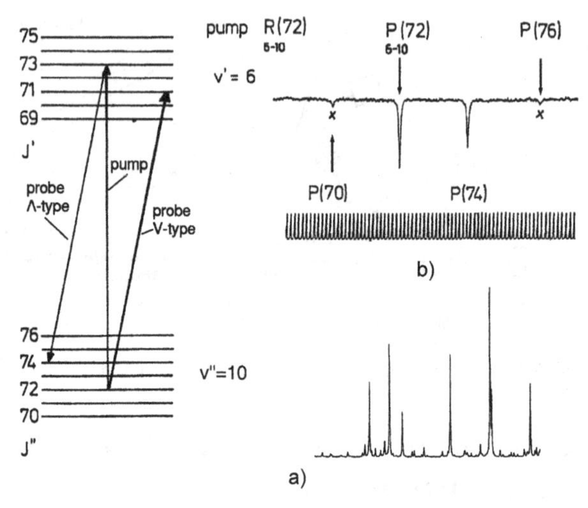

Fig. 5.35 Section of the polarization spectrum of Cs_2 and OODR signals. The *small lines* marked by *crosses* are collision-induced signals

beams differ sufficiently, the two probe beams can be separated by a dichroic mirror (Fig. 5.34).

The sum-frequency method is easier with respect to the optical arrangement but with more electronics involved. The pump beam L_1 and the probe beam L_2 are chopped at two different frequencies with the same chopper (Fig. 5.20). Both probe beams are detected by the same detector. The output signals are fed parallel into two lock-in detectors tuned to f_1 and $f_1 + f_2$, respectively. While the lock-in at f_1 records the polarization spectrum of L_1 if λ_1 is tuned, the lock-in at $(f_1 + f_2)$ records the OODR spectrum induced by both lasers L_1 and L_2 at a fixed wavelength λ_1 while λ_2 is tuned.

For illustration, Fig. 5.35 displays a section of the polarization spectrum of Cs_2 with V-type and Λ-type OODR signals of Cs_2 transitions with a common upper or lower level, respectively [605, 606].

OODR polarization spectroscopy is particularly useful for assigning complex spectra of larger molecules. Since many organic molecules such as benzene or naphthalene have no absorption bands in the visible and only start to absorb in the UV, the output frequencies of the two visible lasers used for OODR have to be doubled. In order to achieve sufficiently high intensities for the nonlinear spectroscopy, frequency doubling is required in external cavities. In Fig. 5.36 the complete experimental setup for OODR polarization spectroscopy in the UV is shown [607], while Fig. 5.37 illustrates the advantage of OODR. In the lower trace a small section of the dense Doppler-free polarization spectrum of the naphthalene molecule around 310 nm is shown. The upper spectrum shows some OODR lines together with their level schemes and assignments. The pump is stabilized onto the $^rP_0(58)$ transition while the probe laser is tuned. Both V-type and Λ-type double resonance signals can be identified. In principle, the signs of the V-type and Λ-type signals should be opposite, but in this case the lock-in detector was set to quadrature and the magnitude but not the sign of the signals was detected, because this gave a better signal-to-noise ratio.

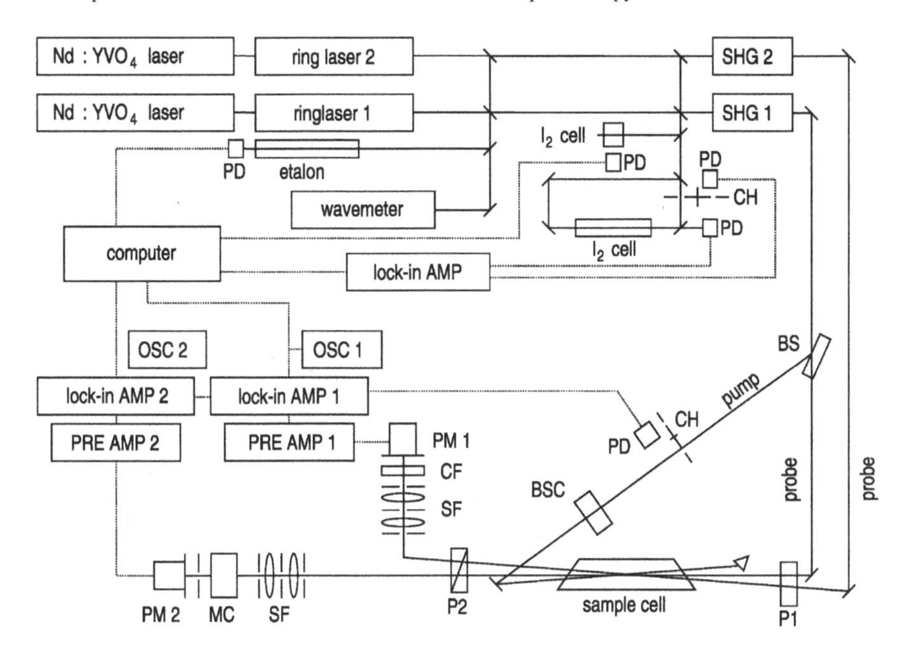

Fig. 5.36 Experimental setup for optical–optical double resonance polarization spectroscopy in the UV [607]

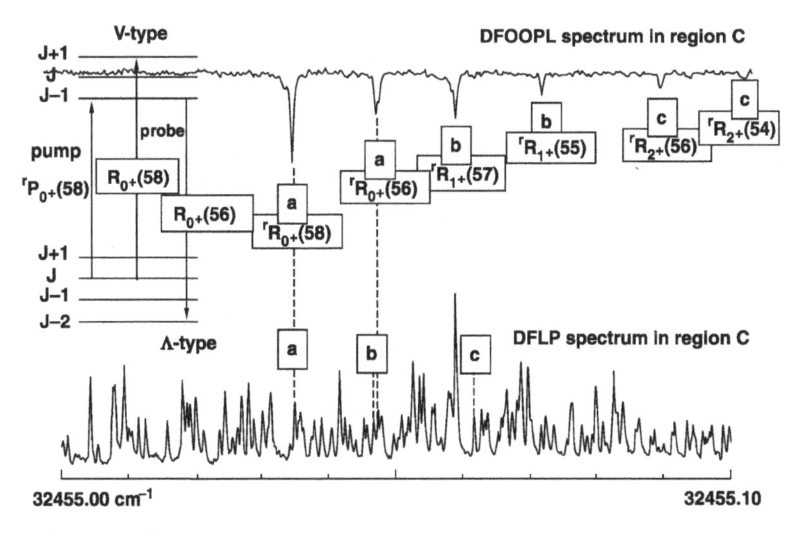

Fig. 5.37 Section of the OODR spectrum (*upper trace*) and the polarization spectrum (*lower trace*) of naphthalene [607]

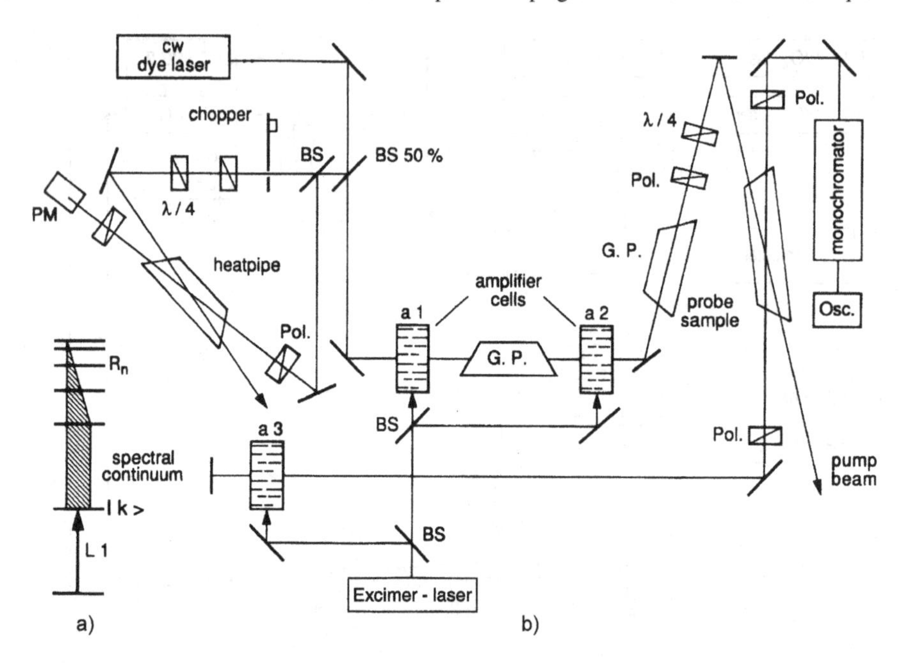

Fig. 5.38 Polarization labeling method: (**a**) term diagram; and (**b**) experimental arrangement

5.5.2 Polarization Labeling

Often it is advantageous to gain an overlook over a wider range of the OODR spectra with lower resolution before Doppler-free scans over selected sections of this range are performed. Here, the polarization labeling technique, first introduced by Schawlow and his group [608, 609], turns out to be very useful. A polarized pump laser L_1 orients molecules in a selected lower level $|i\rangle$ or an upper level $|k\rangle$ and therefore *labels* those levels. Instead of a single-mode probe laser, a linearly polarized spectral continuum is now sent through the sample between two crossed polarizers. Only those wavelengths λ_{im} or λ_{km} are affected in their polarization characteristics that correspond to molecular transitions starting from the labeled levels $|i\rangle$ or $|k\rangle$. These wavelengths are transmitted through the crossed analyzer, are separated by a spectrograph, and are simultaneously recorded by an optical multichannel analyzer or a CCD camera (Vol. 1, Sect. 4.5). The spectral resolution is limited by the spectrograph and the spectral range by its dispersion and by the length of the diode array of the OMA.

A typical experimental arrangement used in our lab for studying molecular Rydberg states is exhibited in Fig. 5.38. The pump laser is a single-mode cw dye laser, which is amplified in two pulsed amplifier cells pumped by a nitrogen or excimer laser. The spectral continuum is provided by the broad fluorescence excited in a dye cell by part of the excimer laser beam. Since this spectral continuum is only formed within the small pump region in the dye cell, it can be transformed

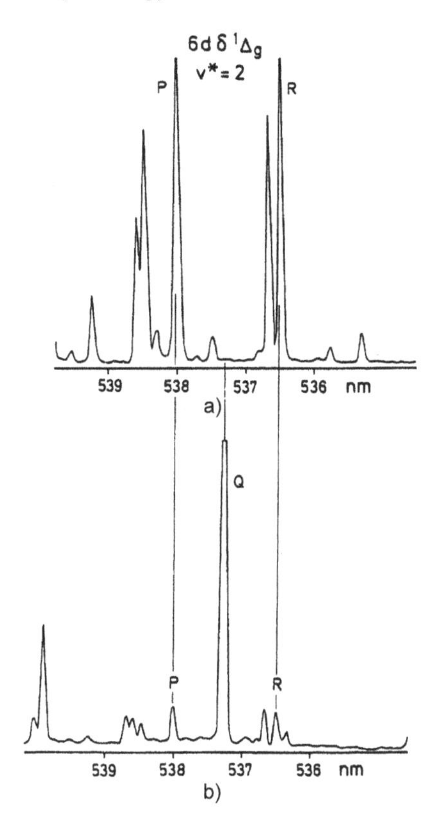

Fig. 5.39 Section of *polarization-labeling* spectrum of Li_2 Rydberg transitions: (**a**) pump laser being circularly polarized; and (**b**) linear polarization. Note the difference in intensities for Q and P, R transitions in both spectra

into a parallel beam by a lens and is sent antiparallel to the pump beam through a heat pipe containing lithium vapor. The OMA system behind the spectrograph is gated in order to make the detector sensitive only for the short time of the probe pulse.

Figure 5.39 illustrates a section of the polarization labeling spectrum of Rydberg transitions in Li_2 starting from the intermediate level $B\,^1\Pi_u$ ($v' = 1$, $J' = 27$), which is labeled by the pump laser [610]. In the upper spectrum the pump laser was circularly polarized, enhancing P and R transitions, while in the lower spectrum the pump laser was linearly polarized with an angle of 45° between the electric vectors of pump and probe. The Q-lines are now pronounced, while the P and R lines are weakened (Sect. 2.4). This demonstrates the advantage of polarization labelling for the analysis of the spectra.

5.5.3 *Microwave–Optical Double-Resonance Polarization Spectroscopy*

A very sensitive and accurate double-resonance technique is microwave–optical double-resonance polarization spectroscopy (MOPS), developed by Ernst et al.

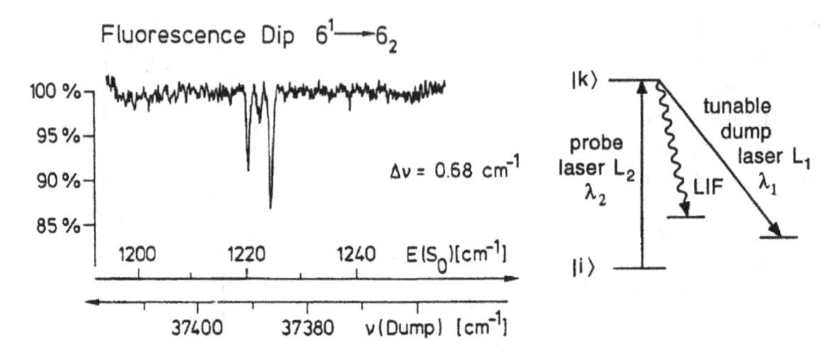

Fig. 5.40 Fluorescence-dip spectroscopy of C_6H_6 molecules. The upper rotational level $J' = 6$, $K' = 6$, $\ell'_6 = -1$ of the $6^1 v$ vibrational state has been selectively excited by L_2 and the fluorescence is observed while the wavelength λ_1 of the dump laser L_1 is tuned [614]

[611]. This technique detects microwave transitions in a sample between crossed polarizers through the change in transmission of a polarized optical wave. The sensitivity of the method has been demonstrated by measurements of the hfs of rotational transitions in the electronic ground state of CaCl molecules that were produced by the reaction $2Ca + Cl_2 \rightarrow 2CaCl$ in an argon flow. In spite of the small concentrations of CaCl reaction products and the short absorption pathlength in the reaction zone, a good signal-to-noise ratio could be achieved at linewidths of 1–2 MHz [612]. A more recent example is the application of this technique to Na_3 clusters in a cold molecular beam, where the hfs in the electronic groundstate of Na_3 has been measured [613].

5.5.4 Hole-Burning and Ion-Dip Double-Resonance Spectroscopy

Instead of keeping the pump laser at a fixed wavelength λ_1 and tuning the probe laser wavelength λ_2, one may also fix the probe transition $|i\rangle \rightarrow |k\rangle$ while the pump laser is scanned. The probe laser-induced fluorescence intensity I_{Fl} excited at λ_2 then exhibits a dip when the tuned pump radiation coincides with a transition that shares the common lower level $|i\rangle$ or upper level $|k\rangle$ with the probe laser (V-type or Λ-type OODR) (Fig. 5.40). With pulsed lasers, the time delay Δt between pump and probe laser pulses can be varied in order to study the relaxation of the hole that was burned into the population N_i by the pump or N_k by the dump laser [614].

If the absorption of the probe laser is monitored via one- or two-photon ionization of the upper probe transition (Fig. 5.41b), the ion signal shows a dip when the tunable pump beam burns a hole into the common lower level population N_i (*ion-dip spectroscopy*). This opens another way to detect stimulated emission pumping. Now three lasers are employed (Fig. 5.41b). The first laser L_1 is kept on a transition $|i\rangle \rightarrow |k\rangle$. The molecules in the excited level $|k\rangle$ are ionized by a second laser L_2. In favorable cases photons from L_1 may also be used for ionization, thus making

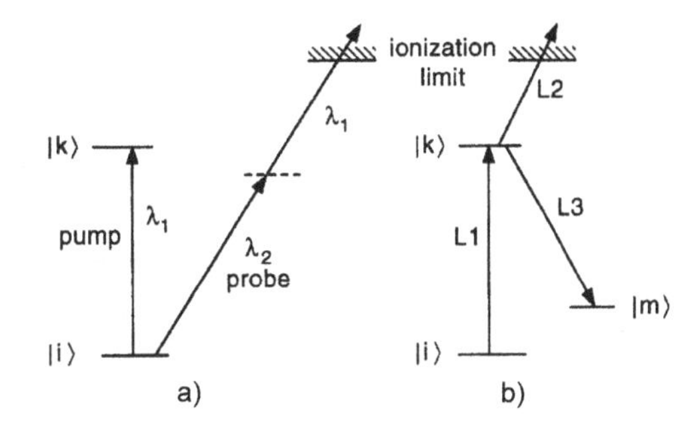

Fig. 5.41 Ion-dip spectroscopy. The ion dips are monitored by the depletion of the lower level $|i\rangle$ (**a**) on the upper level $|k\rangle$ by stimulated emission pumping (**b**)

L_2 unnecessary. A third tunable laser L_3 induces downward transitions $|k\rangle \rightarrow |m\rangle$, which cause a depletion of level $|k\rangle$. A dip in the ion signal $S(\lambda_3)$ is observed any time the wavelength λ_3 of the stimulated emission laser coincides with a transition $|k\rangle \rightarrow |m\rangle$ [614, 615].

5.5.5 Triple-Resonance Spectroscopy

For some spectroscopic problems it is necessary to use three lasers in order to populate molecular or atomic states that cannot be reached by two-step excitation. One example is the investigation of high-lying vibrational levels in excited electronic states, which give information about the interaction potential between excited atoms at large internuclear separations. This potential $V(R)$ may exhibit a barrier or hump, and the molecules in levels above the true dissociation energy $V(R = \infty)$ may tunnel through the potential barrier. Such a triple resonance scheme is illustrated in Fig. 5.42a for the Na_2 molecule. A dye laser L_1 excites the selected level (v', J')

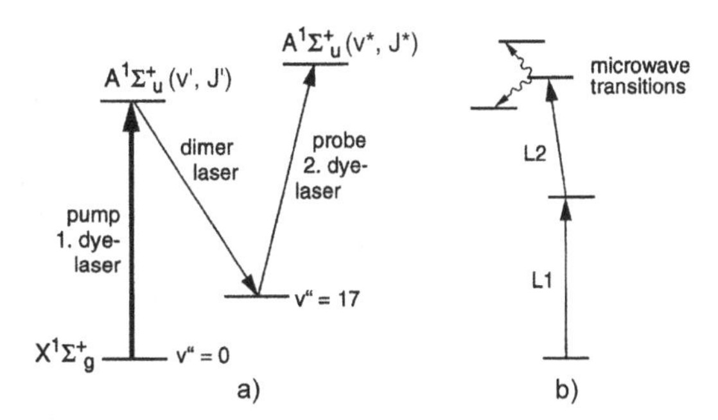

Fig. 5.42 Triple-resonance spectroscopy: (**a**) population of high vibrational levels in the ground state by stimulated emission pumping and absorption of a third laser radiation, resulting in the population of levels just below and above the dissociation energy of the excited $A\,^1\Sigma_u$ state of Na_2 [617]; (**b**) stepwise excitation of Rydberg levels by OODR and microwave-induced transitions to neighboring levels

in the $A\,^1\Sigma_u$ state. If a sufficiently large population $N(v', J')$ can be achieved, population inversion with respect to high vibrational levels (v'', J'') of the electronic ground state is reached and laser oscillation starts (Na_2 dimer laser), which populates the lower level (v'', J''). From this level a third laser induces transitions into levels (v^*, J^*) close to the dissociation limit of the $A\,^1\Sigma_u$ state. These levels could not have been populated from low vibrational levels (v'', J'') of the special state because the corresponding Franck–Condon factors are too small [616]. With this technique the very last bound level in the $X\,^1\Sigma_g^+$ ground state of Na_2 molecules could be measured [617]. These measurements yield the scattering length [618] for Na–Na collisions, an important quantity for the achievement of Bose–Einstein condensates (Sect. 9.1.10).

A second example, mentioned already, is microwave spectroscopy of Rydberg levels that have been excited by resonant two-step absorption of two dye lasers (Fig. 5.41b). The high accuracy of microwave spectroscopy allows the precise determination of finer details, such as field-induced energy shifts of Rydberg levels, broadening of Rydberg transitions by blackbody radiation, or other effects that might not be resolvable with optical spectroscopy.

5.5.6 Photoassociation Spectroscopy

The recently developed technique of photoassociation spectroscopy is a modification of stimulated emission spectroscopy (see Sect. 5.4.3) that is applied to systems of cold colliding atoms. Its principle is illustrated in Fig. 5.43. A pair of colliding cold atoms A + A or A + B approaching each other in a cold gas (e.g., a Bose–Einstein condensate, see Sect. 9.1.10) are excited by a narrow-band laser L_1 into a stable electronic state of the dimer A_2 or AB at a distance R that corresponds to the outer turning point of a high vibration level $|v'\rangle$ in the excited state. A second laser L_2 induces transitions from the inner turning point down into bound vibrational levels $|v'', J''\rangle$ of the electronic ground state. By tuning the wavelength λ_1 of L_1, the vibrational levels in the upper electronic state can be selected within a wide range, thus optimizing the Franck–Condon factors for downward transitions into wanted vibrational levels $|v''\rangle$ of the ground state. At very low temperatures, even the highest vibration level $|v''_{max}\rangle$ of the ground state will be stable. Precision spectroscopy of the rotational and hyperfine structure of such levels closely below the dissociation energy gives more insight into the complicated interactions between electrons and nuclei at large internuclear distances R, where even retardation effects have to be taken into account and where rotational spacings, hfs and fine structure are of the same order of magnitude. Taking into account all of these effects enables the very accurate construction of the potential curve $E_{pot}(R)$ at large values of R.

From measurements of the absorption probability of transitions from very high levels of the ground state the upper state potential and the dependence of the transition moment on the inter-nuclear distance R can be obtained with high precision.

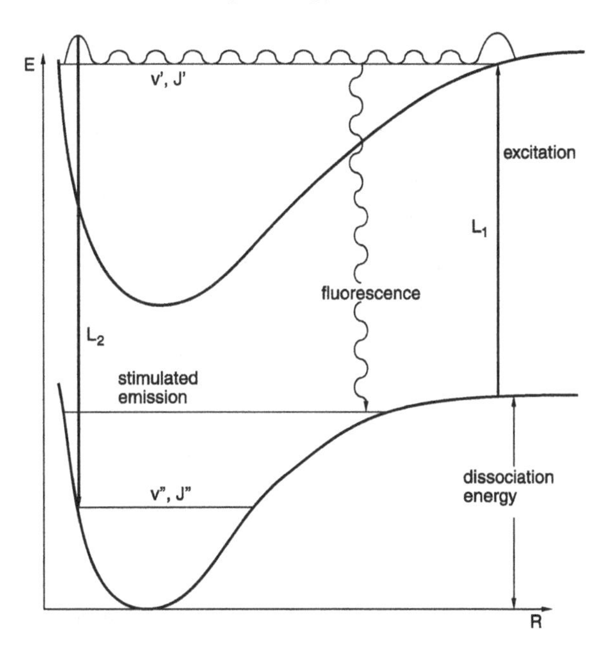

Fig. 5.43 Level scheme for photoassociation spectroscopy

Generally the triplet ground state of alkali dimers is not directly accessible by absorption spectroscopy. Photo-association spectroscopy of cold colliding Li atoms wit a relative angular momentum $l = 1$ (p-scattering) allows transitions from the $^3\Sigma_g$ state into an excited triplet state. Their measurements give precise values of high vibrational levels and allow the determination of the triplet potential. From measurements of rotational levels also the spin–spin and spin–rotation interaction can be deduced [620].

By exciting a pair of metastable He* atoms in the metastable state $2\,^3S_1$ with a narrow-band laser at $\lambda = 1083$ nm, bound vibrational levels of the helium dimer He_2^* in the electronic $\mathrm{O^+}_u$ state $(2\,^3S_1 + 2\,^3P_0)$ could be populated in a magnetic trap [619]. The inner turning point at $R_1 = 150a_0$ and the outer turning point at $1150a_0 = 57.5$ nm show that these excited dimers are indeed giant molecules in terms of their size.

Infrared-UV double resonance has proved to be very useful for the solution of the conformational structure of large bio-molecules [621]. Here the absorption of an infrared photon on a selected vibrational transition is detected by two-photon ionization of the excited vibrational level. In Fig. 5.44 the infrared UV-double resonance spectra for the two lowest energy conformations of the MAPE molecule (2-methylamino 1-phenyl ethanol) [621]. The lower level of the 4 IR-transitions A; B; C and D can be labelled by tuning the IR-laser on the corresponding transitions which depletes the lower levels and weakens the selected line in the absorption spectrum.

More information can be found in [622–624].

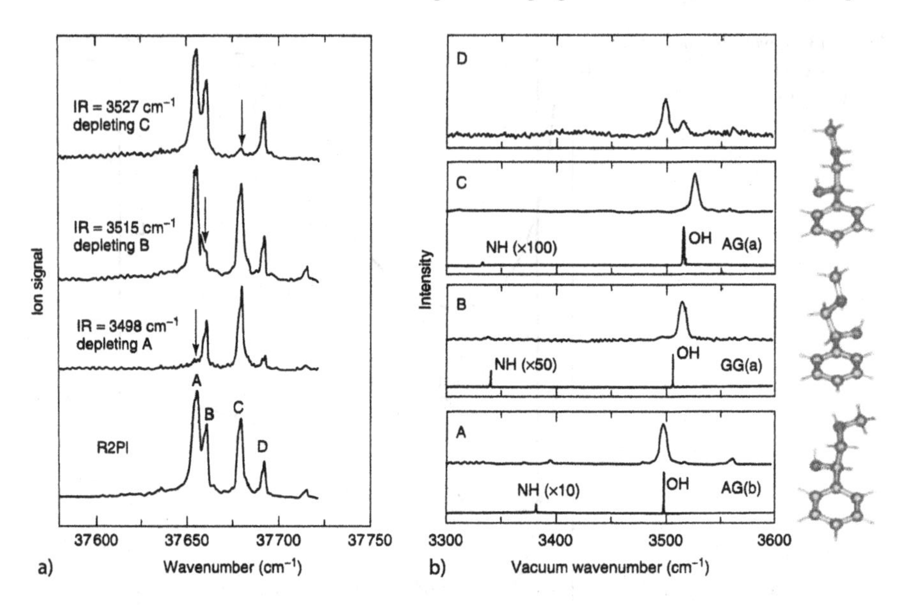

Fig. 5.44 IR-UV spectra of the lowest energy conformers of 2-methylamino 1-phenyl ethanol, MAPE (**b**), and the UV-R2PI and IR-UV hole-burn spectra of the origin bands associated with the three lowest energy conformers of MAPE (**a**). The AG and GG nomenclature specifies anti or gauche conformations around the CCCN and OCCN dihedral angles of the ethanolamine side chain. Note the slight shifts in the displaced OH bands, reflecting the conformational changes in the OH $\rightarrow$ NH_2 hydrogen-bond distances and angles [621]

5.6 Problems

5.1 A linearly polarized laser beam passes in z-direction through a sample cell. Its wavelength is tuned to a transition $|1\rangle \rightarrow |2\rangle$ with $J_1 = 0$, $J_2 = 1$. With the laser off, the upper level $|2\rangle$ is assumed to be empty. What is the relative population $N_{21}(M_2 = \pm 1)/N_{20}(M_2 = 0)$ of the M_2 levels and the corresponding degree of alignment, if the excitation rate is $10^7\ \mathrm{s}^{-1}$, the effective lifetime is $\tau_2 = 10^{-8}$ s, and the collision-induced mixing rate $(M_2 = \pm 1) \leftarrow M_2 = 0$ is $5 \times 10^6\ \mathrm{s}^{-1}$?

5.2 Assume the absorption cross section $\sigma = 10^{-13}\ \mathrm{cm}^2$ for a linearly polarized laser with $\boldsymbol{E} \parallel \boldsymbol{D}$ and $\lambda = 500$ nm. What is the angular distribution $N_2(\alpha)$ of excited atoms with the dipole transition moment $\boldsymbol{D}$ forming the angle α against $\boldsymbol{E}$ on the transition $J_1 = 0 \rightarrow J_2 = 1$ for the laser intensity I?

Calculate the saturation parameter $S(\alpha = 0)$ for $I = 10^4\ \mathrm{W/m}^2$ and a spontaneous transition probability $A_{21} = 5 \times 10^6\ \mathrm{s}^{-1}$.

5.3

(a) Molecules with a magnetic moment $\mu = 0.8\mu_B$ ($\mu_B = 9.27 \times 10^{-24}$ J/T) pass with $\boldsymbol{v} = 500$ m/s in the z-direction through an inhomogeneous magnetic field ($\mathrm{d}B/\mathrm{d}x = 1$ T/m, $L_B = 0.1$ m) of a Rabi molecular beam apparatus. Calculate the deflection angle of the molecules with mass $m = 50$ AMU (1 AMU $= 1.66 \times 10^{-27}$ kg).

(b) If the molecules enter the magnetic field in a nearly parallel beam with a divergence angle $\epsilon = 1°$ and a uniform density $N_0(x)$, they arrive at the detector 50 cm downstream of the end of the magnetic field with a transverse density distribution $N(x)$. What is the approximate profile of $N(x)$?

(c) Assume 60 % of all molecules to be in level $|1\rangle$ with $\mu_1 = 0.8\mu_B$ and 40 % in level $|2\rangle$ with $\mu_2 = 0.9\mu_B$. What is the profile $N(x)$ with $N = N_1 + N_2$? How does $N(x)$ change, if an RF transition before the inhomogeneous magnet equalizes the population N_1 and N_2?

5.4 In an optical–microwave double-resonance experiment the decrease of the population N_i caused by excitation from level $|i\rangle$ into level $|k\rangle$ by a cw laser is 20 %. The lifetime of the upper level is $\tau_k = 10^{-8}$ s and its collisional deactivation rate is 5×10^7 s^{-1}. The collisional refilling rate of level $|i\rangle$ is 5×10^7 s^{-1}. Compare the magnitude of the microwave signals $|k\rangle \rightarrow |m\rangle$ between rotational levels in the upper electronic state with that on the transitions $|i\rangle \rightarrow |n\rangle$ in the ground state, if without optical pumping a thermal ground-state population distribution is assumed and the cross section σ_{km} and σ_{in} for the microwave transitions are equal.

5.5 In a V-type OODR scheme molecules in a sample cell are excited by a chopped pump laser at $\lambda = 500$ nm on the transition $|i\rangle \rightarrow |k\rangle$ with the absorption cross section $\sigma_{ik} = 10^{-14}$ cm^2. What pump intensity is required in order to achieve a modulation of 50 % of the lower-level population N_i, if the relaxation rate into the level $|i\rangle$ is $10^7 \cdot (N_{10} - N_1)$ s^{-1}?

5.6 A potassium atom is excited into a Rydberg level ($n = 50$, $\ell = 0$) with the quantum defect $\delta = 2.18$.

(a) Calculate the ionization energy of this level.

(b) What is the minimum electric field required for field ionization (the tunnel effect is neglected)?

(c) What frequency ω_{rf} is required for resonant RF transitions between levels ($n = 50, \ell = 0$) $\rightarrow$ ($n = 50, \ell = 1, \delta = 1.71$)?

5.7 In a Λ-type OODR scheme a transition is induced between the ground-state levels $|1\rangle$ and $|3\rangle$ of the Cs_2 molecule. The radiative lifetime of these levels is $\tau_1 = \tau_3 = \infty$. The upper level $|2\rangle$ has a lifetime of $\tau_2 = 10$ ns. Calculate the linewidth of the OODR signal

(a) for collinear, and

(b) for anticollinear propagation of pump and probe beams,

if the laser linewidth is $\delta\nu_L \leq 1$ MHz, the transit time of the molecules through the focused laser beam is 30 ns, and the wavelengths are $\lambda_{pump} = 580$ nm, $\lambda_{probe} = 680$ nm.

Chapter 6
Time-Resolved Laser Spectroscopy

The investigation of fast processes, such as electron motions in atoms or molecules, radiative or collision-induced decays of excited levels, isomerization of excited molecules, or the relaxation of an optically pumped system toward thermal equilibrium, opens the way to study in detail the dynamic properties of excited atoms and molecules. A thorough knowledge of dynamical processes is of fundamental importance for many branches of physics, chemistry, or biology. Examples are predissociation rates of excited molecules, femtosecond chemistry, or the understanding of the visual process and its different steps from the photoexcitation of rhodopsin molecules in the retina cells to the arrival of electrical nerve pulses in the brain.

In order to study these processes experimentally, one needs a sufficiently good time resolution, which means that the resolvable minimum time interval Δt must still be shorter than the time scale T of the process under investigation. While the previous chapters emphasized the high *spectral* resolution, this chapter concentrates on experimental techniques that allow high *time* resolution.

The development of ultrashort laser pulses and of new detection techniques that allow a very high time resolution has brought about impressive progress in the study of fast processes. The achievable time resolution has been pushed recently into the attosecond range (1 as $= 10^{-18}$ s). Spectroscopists can now quantitatively follow up ultrafast processes, which could not be resolved ten years ago.

The *spectral* resolution $\Delta\nu$ of most time-resolved techniques is, in principle, confined by the Fourier limit $\Delta\nu = a/\Delta T$, where ΔT is the duration of the short light pulse and the factor $a \simeq 1$ depends on the profile $I(t)$ of the pulse. The spectral bandwidth $\Delta\nu$ of such *Fourier-limited pulses* is still much narrower than that of light pulses from incoherent light sources, such as flashlamps or sparks. Some time-resolved coherent methods based on regular trains of short pulses even circumvent the Fourier limit $\Delta\nu$ of a single pulse and simultaneously reach extremely high spectral and time resolutions (Sect. 7.4).

We will at first discuss techniques for the generation and detection of short laser pulses before their importance for different applications is demonstrated by some examples. Methods for measuring lifetimes of excited atoms or molecules and of fast relaxation phenomena are presented. These applications illustrate the relevance

W. Demtröder, *Laser Spectroscopy 2*, DOI 10.1007/978-3-662-44641-6_6

of pico- and femtosecond molecular physics and chemistry for our understanding of fundamental dynamical processes in molecules.

The special aspects of time-resolved coherent spectroscopy are covered in Sects. 7.2, 7.4. For a more extensive representation of the fascinating field of time-resolved spectroscopy some monographs [625–628], reviews [629–632], and conference proceedings [633–635] should be mentioned.

6.1 Generation of Short Laser Pulses

For incoherent, pulsed light sources (for example, flashlamps or spark discharges) the duration of the light pulse is essentially determined by that of the electric discharge. For a long time, microsecond pulses represented the shortest available pulses. Only recently could the nanosecond range be reached by using special discharge circuits with low inductance and with pulse-forming networks [636, 637].

For laser pulses, on the other hand, the time duration of the pulse is not necessarily limited by the duration of the pump pulse, but may be much shorter. Before we present different techniques to achieve ultrashort laser pulses, we will discuss the relations between the relevant parameters of a laser that determine the time profile of a laser pulse.

6.1.1 Time Profiles of Pulsed Lasers

In active laser media pumped by pulsed sources (for example, flashlamps, electron pulses, or pulsed lasers), the population inversion necessary for oscillation threshold can be maintained only over a time interval ΔT that depends on the duration and power of the pump pulse. A schematic time diagram of a pump pulse, the population inversion, and the laser output is shown in Fig. 6.1. As soon as threshold is reached, laser emission starts. If the pump power is still increasing, the gain becomes high and the laser power rises faster than the inversion, until the increasingly induced emission reduces the inversion to the threshold value.

The time profile of the laser pulse is not only determined by the amplification per round trip $G(t)$ (Vol. 1, Sect. 5.2) but also by the relaxation times τ_i, τ_k of the upper and lower laser levels. If these times are short compared to the rise time of the pump pulse, quasi-stationary laser emission is reached, where the inversion $\Delta N(t)$ and the output power $P_L(t)$ have a smooth time profile, determined by the balance between pump power $P_p(t)$, which creates the inversion, and laser output power $P_L(t)$, which decreases it. Such a time behavior, which is depicted in Fig. 6.1a, can be found, for instance, in many pulsed gas lasers such as the excimer lasers (Vol. 1, Sect. 5.7).

In some pulsed lasers (for example, the N_2 laser) the lower laser level has a longer effective lifetime than the upper level [638]. The increasing laser power $P_L(t)$ decreases the population inversion by stimulated emission. Since the lower level is not

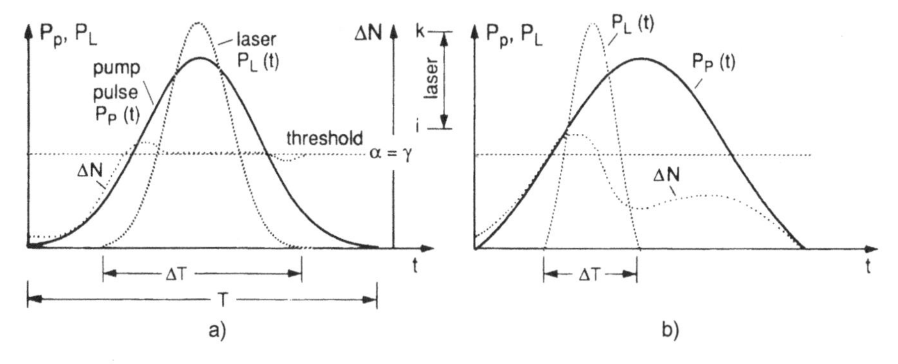

Fig. 6.1 Time profiles of the pump power $P_p(t)$, the inversion density $\Delta N(t)$, and laser output power $P_L(t)$: (**a**) for sufficiently short lifetime τ_i of the lower laser level; and (**b**) for a self-terminating laser with $\tau_i^{eff} > \tau_k^{eff}$

Fig. 6.2 Schematic representation of spikes in the emission of a flashlamp-pumped solid-state laser with long relaxation times τ_i, τ_k

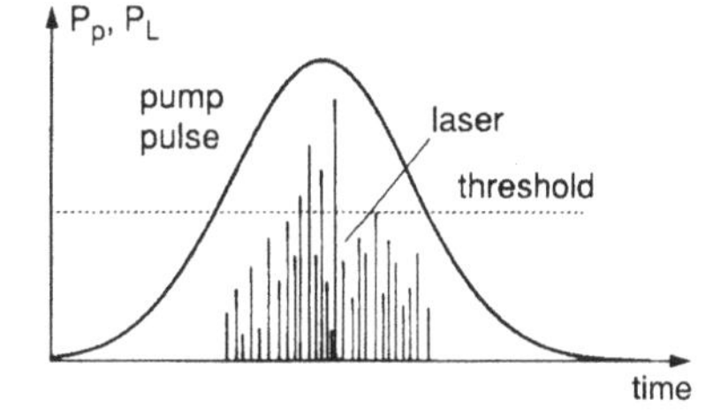

sufficiently quickly depopulated, it forms a bottleneck for maintaining threshold inversion. The laser pulse itself limits its duration and it ends before the pump pulse ceases (self-terminating laser, Fig. 6.1b).

If the relaxation times τ_i, τ_k are long compared to the rise time of the pump pulse, a large inversion ΔN may build up before the induced emission is strong enough to deplete the upper level. The corresponding high gain leads to a large amplification of induced emission and the laser power P_L may become so high that it depletes the upper laser level faster than the pump can refill it. The inversion ΔN drops below threshold and the laser oscillation stops long before the pump pulse ends. When the pump has again built up a sufficiently large inversion, laser oscillations starts again. In this case, which is, for instance, realized in the flashlamp-pumped ruby laser, the laser output consists of a more or less irregular sequence of "spikes" (Fig. 6.2) with a duration of $\Delta T \simeq 1$ μs for each spike, which is much shorter than the pump–pulse duration $T \simeq 100$ μs to 1 ms [639, 640].

For time-resolved laser spectroscopy, pulsed dye lasers are of particular relevance due to their continuously tunable wavelength. They can be pumped by flashlamps ($T \simeq 1$ μs to 1 ms), by other pulsed lasers, for example, by copper-vapor lasers ($T \simeq 50$ ns), excimer lasers ($T \simeq 15$ ns), nitrogen lasers ($T = 2$–10 ns), or frequency-doubled Nd:YAG lasers ($T = 5$–15 ns). Because of the short relaxation times τ_i, $\tau_k (\simeq 10^{-11}$ s), no spiking occurs and the situation of Fig. 6.1a is realized (Vol. 1, Sect. 5.7). The dye laser pulses have durations between 1 ns to 500 μs, depending

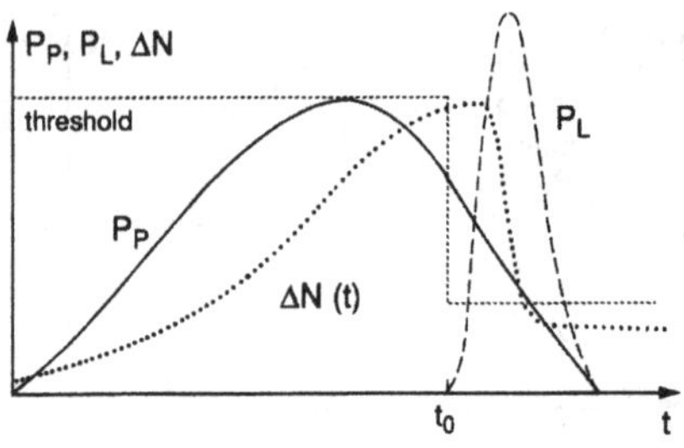

Fig. 6.3 Pump power $P_P(t)$, resonator losses $\gamma(t)$, inversion density $\Delta N(t)$, and laser output power $P_L(t)$ for a Q-switched laser

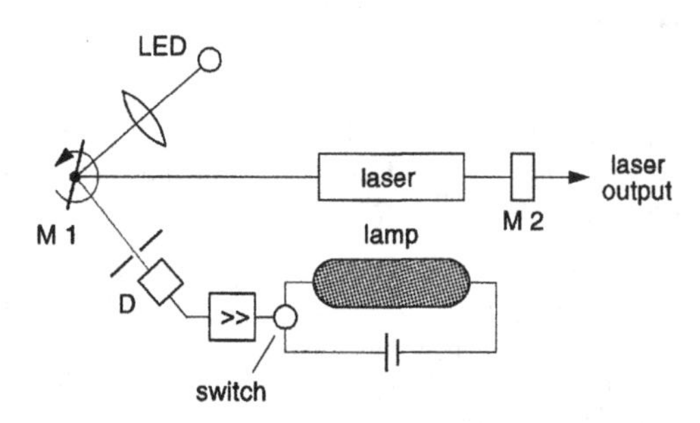

Fig. 6.4 Experimental realization of Q-switching by a rapidly spinning cavity mirror M1

on the pump pulses; typical peak powers range from 1 kW to 10 MW and pulse repetition rates from 1 Hz to 15 kHz [641].

6.1.2 Q-Switched Lasers

In order to obtain a single, powerful pulse of a flashlamp-pumped laser instead of the irregular sequence of many spikes, the technique of Q-switching was developed. Q-switching is based on the following principle: Until a selected time t_0 after the start of the pump pulse at $t = 0$, the cavity losses of a laser are kept so high by a closed "optical switch" inside the laser resonator that the oscillation threshold cannot be reached. Therefore a large inversion ΔN is built up by the pump (Fig. 6.3). If the switch is opened at $t = t_0$ the losses are suddenly lowered (that is, the quality factor or Q-value of the cavity (Vol. 1, Sect. 5.1) jumps from a low to a high value). Because of the large amplification $G \propto B_{ik}\rho\Delta N$ for induced emission, a quickly rising intense laser pulse develops, which depletes in a very short time the whole inversion that had been built up during the time interval t_0. This converts the energy stored in the active medium into a giant light pulse [640, 642, 643]. The time profile of the pulse depends on the rise time of Q-switching. Typical durations of these giant pulses are 1–20 ns and peak powers up to 10^9 W are reached, which can be further increased by subsequent amplification stages.

Such an optical switch can be realized, for instance, if one of the resonator mirrors is mounted on a rapidly spinning motor shaft (Fig. 6.4). Only at that time t_0 where the surface normal of the mirror coincides with the resonator axis is the incident light reflected back into the resonator, giving a high Q-value of the laser

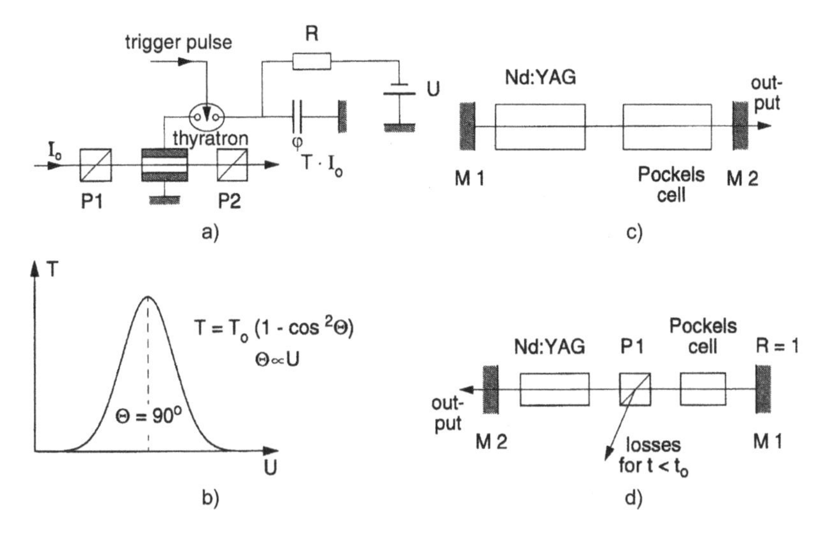

Fig. 6.5 Q-switching with a Pockels cell inside the laser resonator: (**a**) Pockel's cell between two crossed polarizers, (**b**) transmission $T(\theta) \propto U$; (**c**), (**d**) possible experimental arrangements

cavity [644]. The optimum time t_0 can be selected by imaging the beam of a light-emitting diode (LED) after reflection at the spinning mirror onto the detector D, which provides the trigger signal for the flashlamp of the Q-switched laser. This technique, however, has some disadvantages: the spinning mirror is not very stable and the switching time is not sufficiently short. Therefore other Q-switching methods have been developed that are based on electro-optical or acousto-optical modulators [645].

One example is given in Fig. 6.5, where a Pockels cell between two crossed polarizers acts as a Q-switch [646]. The Pockels cell consists of an optically anisotropic crystal that turns the plane of polarization of a transmitted, linearly polarized wave by an angle $\theta \propto |E|$ when an external electric field $\boldsymbol{E}$ is applied. Since the transmittance of the system is $T = T_0(1 - \cos^2\theta)$, it can be changed by the voltage U applied to the electrodes of the Pockels cell. If for times $t < t_0$ the voltage is $U = 0$, the crossed polarizers have the transmittance $T = 0$ for $\theta = 0$. When at $t = t_0$ a fast voltage pulse $U(t)$ is applied to the Pockels cell, which causes a rotation of the polarization plane by $\theta = 90°$, the transmittance rises to its maximum value T_0. The linearly polarized lightwave can pass the switch, is reflected back and forth between the resonator mirrors M1 and M2, and is amplified until the inversion ΔN is depleted below threshold. The experimental arrangement shown in Fig. 6.5d works differently and needs only one polarizer P1, which is used as polarizing beam splitter. For $t < t_0$ a voltage U is kept at the Pockels cell, which causes circular polarization of the transmitted light. After reflection at M1 the light again passes the Pockels cell, reaches P1 as linearly polarized light with the plane of polarization turned by $\theta = 90°$, and is totally reflected by P1 and therefore lost for further amplification. For $t > t_0$ U is switched off, P1 now transmits, and the laser output is coupled out through M2.

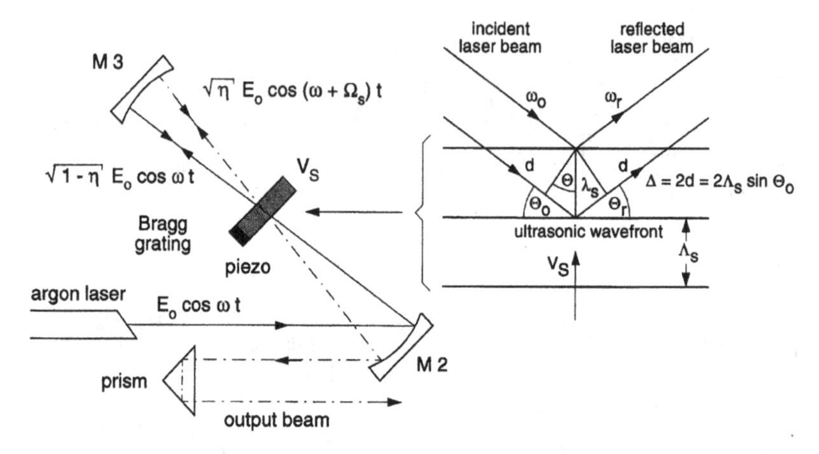

Fig. 6.6 Cavity dumping of a cw argon laser by a pulsed acoustic wave. The *insert* shows the Bragg reflection of an optical wave at a running ultrasonic wave with wavelength Λ_s

The optimum choice of t_0 depends on the duration T of the pump pulse $P_p(t)$ and on the effective lifetime τ_k of the upper laser level. If $\tau_k \gg T \approx t_0$, only a small fraction of the energy stored in the upper laser level is lost by relaxation and the giant pulse can extract nearly the whole energy. For example, with the ruby laser ($\tau_k \approx 3$ ms) the switching time t_0 can be chosen close to the end of the pump pulse ($t_0 = 0.1$–1 ms), while for Nd:YAG-lasers ($\tau_k \approx 0.2$ ms) the optimum switching time t_0 lies before the end of the pump pulse. Therefore only part of the pump energy can be converted into the giant pulse [640, 643, 647].

6.1.3 Cavity Dumping

The principle of Q-switching can also be applied to cw lasers. Here, however, an inverse technique, called *cavity dumping*, is used. The laser cavity consists of highly reflecting mirrors in order to keep the losses low and the Q-value high. The intracavity cw power is therefore high because nearly no power leaks out of the resonator. At $t = t_0$ an optical switch is activated that couples a large fraction of this stored power out of the resonator. This may again be performed with a Pockels cell (Fig. 6.5d), where now M1 and M2 are highly reflecting, and P1 has a large transmittance for $t < t_0$, but reflects the light out of the cavity for a short time Δt at $t = t_0$.

Often an acousto-optic switch is used, for example, for argon lasers and cw dye lasers [648]. Its basic principle is explained in Fig. 6.6. A short ultrasonic pulse with acoustic frequency f_s and pulse duration $T \gg 1/f_s$ is sent at $t = t_0$ through a fused quartz plate inside the laser resonator. The acoustic wave produces a time-dependent spatially periodic modulation of the refractive index $n(t, z)$, which acts as a Bragg grating with the grating constant $\Lambda = c_s/f_s$, equal to the acoustic wavelength Λ where c_s is the sound velocity. When an optical wave $E_0 \cos(\omega t - \boldsymbol{k} \cdot \boldsymbol{r})$ with the

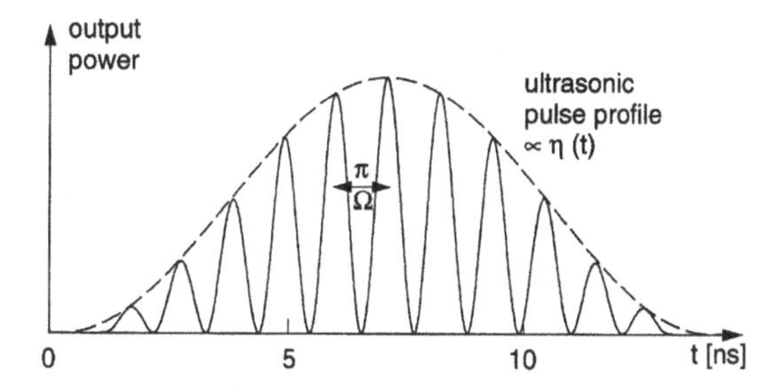

Fig. 6.7 Intensity profile of a cavity-dumped laser pulse, showing the intensity modulation at twice the ultrasonic frequency Ω

wavelength $\lambda = 2\pi/k$ passes through the Bragg plate, the fraction η of the incident intensity I_0 is diffracted by an angle θ determined by the Bragg relation:

$$2\Lambda \sin\theta = \lambda/n. \tag{6.1}$$

The fraction η depends on the modulation amplitude of the refractive index and thus on the power of the ultrasonic wave.

When the optical wave is reflected at an acoustic wave front moving with the velocity v_s, its frequency ω suffers a Doppler shift, which is, according to (6.1) with $c = \lambda \cdot \omega/2\pi$ and $v_s = \Lambda\Omega/2\pi$

$$\Delta\omega = 2\frac{nv_s}{c}\omega\sin\theta = 2n\frac{\Lambda\Omega}{\lambda\omega}\omega\sin\theta = \Omega, \tag{6.2}$$

and turns out to be equal to the acoustic frequency $\Omega = 2\pi f_s$. The amplitude of the deflected fraction is $E_1 = \sqrt{\eta}E_0\cos(\omega+\Omega)t$, that of the unaffected transmitted wave $E_2 = \sqrt{1-\eta}E_0\cos\omega t$. After reflection at the mirror M3 the fraction $\sqrt{1-\eta}E_1$ is transmitted and the fraction $\sqrt{\eta}E_2$ is deflected by the Bragg plate into the direction of the outcoupled beam. This time, however, the reflection occurs at receding acoustic wavefronts and the Doppler shift is $-\Omega$ instead of $+\Omega$. The total amplitude of the extracted wave is therefore

$$\begin{aligned} E_c &= \sqrt{\eta}\sqrt{1-\eta}E_0\left[\cos(\omega+\Omega)t + \cos(\omega-\Omega)t\right] \\ &= 2\sqrt{\eta(1-\eta)}E_0\cos\Omega t\cos\omega t. \end{aligned} \tag{6.3}$$

The average output power $P_c \propto E_c^2$ of the light pulse is then with $\omega \gg \Omega$ and $\langle\cos^2\omega t\rangle = 0.5$:

$$P_c(t) = \eta(t)\left[1-\eta(t)\right]P_0\cos^2\Omega t, \tag{6.4}$$

where the time-dependent efficiency $\eta(t)$ is determined by the time profile of the ultrasonic pulse (Fig. 6.7) and P_0 is the intracavity power. During the ultrasonic pulse the fraction $2\eta(1-\eta)$ of the optical power $\frac{1}{2}\epsilon_0 E_0^2$, stored within the laser resonator, can be extracted in a short light pulse, which is still modulated at twice the acoustic frequency Ω. With $\eta = 0.3$ one obtains an extraction efficiency of $2\eta(1-\eta) = 0.42$.

The repetition rate f of the extracted pulses can be varied within wide limits by choosing the appropriate repetition rate of the ultrasonic pulses. Above a critical frequency f_c, which varies for the different laser types, the peak power of the extracted pulses decreases if the time between two successive pulses is not sufficiently long to recover the inversion and to reach sufficient intracavity power.

The technique of cavity dumping is mainly applied to gas lasers and cw dye lasers. One achieves pulse durations $\Delta T = 10$–100 ns, pulse-repetition rates of 0–4 MHz, and peak powers that may be 10–100 times higher than for normal cw operation with optimized transmission of the output coupler. The average power depends on the repetition rate f. Typical values for $f = 10^4$–4×10^6 Hz are 0.1–40 % of the cw output power. The disadvantage of the acoustic cavity dumper compared to the Pockels cell of Fig. 6.5 is the intensity modulation of the pulse at the frequency 2Ω.

Example 6.1 With an argon laser, which delivers 3 W cw power at $\lambda = 514.5$ nm, the cavity dumping yields a pulse duration $\Delta T = 10$ ns. At a repetition rate $f = 1$ MHz peak powers of 60 W are possible. The on/off ratio is then $f \Delta T = 10^{-2}$ and the average power $P \approx 0.6$ W is 20 % of the cw power.

6.1.4 Mode Locking of Lasers

Without frequency-selective elements inside the laser resonator, the laser generally oscillates simultaneously on many resonator modes within the spectral gain profile of the active medium (Vol. 1, Sect. 5.3). In this "multimode operation" no definite phase relations exist between the different oscillating modes, and the laser output equals the sum $\sum_k I_k$ of the intensities I_k of all oscillating modes, which are more or less randomly fluctuating in time (Vol. 1, Sect. 5.3.4).

If coupling between the phases of these simultaneously oscillating modes can be established, a coherent superposition of the mode *amplitudes* may be reached, which leads to the generation of short output pulses in the picosecond range. This mode coupling or *mode locking* has been realized by optical modulators inside the laser resonator (*active* mode locking) or by saturable absorbers (*passive* mode locking) or by a combined action of both locking techniques [640, 649–653].

(a) Active Mode Locking

When the intensity of a monochromatic lightwave

$$E = A_0 \cos(\omega_0 t - kx),$$

is modulated at the frequency $f = \Omega/2\pi$ (for example, by a Pockels cell or an acousto-optic modulator), the frequency spectrum of the optical wave contains, besides the carrier at $\omega = \omega_0$, sidebands at $\omega = \omega_0 \pm \Omega$ (that is, $\nu = \nu_0 \pm f$) (Fig. 6.8).

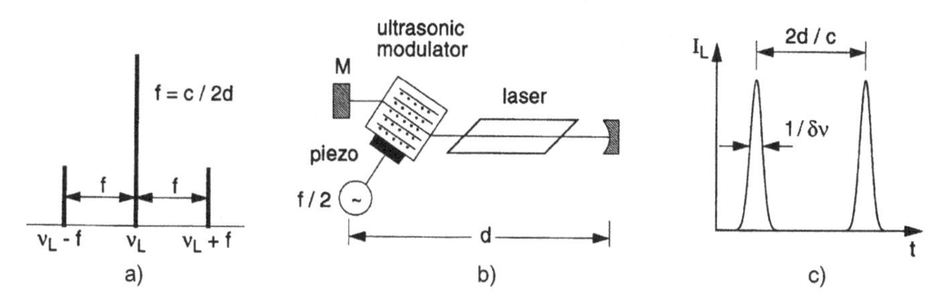

Fig. 6.8 Active mode locking: (**a**) sideband generation; (**b**) experimental arrangement with a standing ultrasonic wave inside the laser resonator; (**c**) idealized output pulses

If the modulator is placed inside the laser resonator with the mirror separation d and the mode frequencies $\nu_m = \nu_0 \pm m \cdot c/2d$ $(m = 0, 1, 2, \ldots)$, the sidebands coincide with resonator mode frequencies if the modulation frequency f equals the mode separation $\Delta\nu = c/2d$. The sidebands can then reach the oscillation threshold and participate in the laser oscillation. Since they pass the intracavity modulator they are also modulated and new sidebands $\nu = \nu_0 \pm 2f$ are generated. This continues until all modes inside the gain profile participate in the laser oscillation. There is, however, an important difference from normal multimode operation: the modes do not oscillate independently, but are phase-coupled by the modulator. At a certain time t_0, the amplitudes of all modes have their maximum at the location of the modulator and this situation is repeated after each cavity round-trip time $T = 2d/c$ (Fig. 6.8c). We will discuss this in more detail: The modulator has the time-dependent transmission

$$T = T_0\big[1 - \delta(1 - \cos\Omega t)\big] = T_0\big[1 - 2\delta\sin^2(\Omega/2)t\big], \tag{6.5}$$

with the modulation frequency $f = \Omega/2\pi$ and the modulation amplitude $2\delta \le 1$. Behind the modulator the field amplitude A_k of the kth mode becomes

$$A_k(t) = T A_{k0}\cos\omega_k t = T_0 A_{k0}\big[1 - 2\delta\sin^2(\Omega/2)t\big]\cos\omega_k t. \tag{6.6}$$

This can be written with $\sin^2 x/2 = \frac{1}{2}(1 - \cos x)$ as

$$A_k(t) = T_0 A_{k0}\left[(1-\delta)\cos\omega_k t + \frac{1}{2}\delta\big[\cos(\omega_k + \Omega)t + \cos(\omega_k - \Omega)t\big]\right]. \tag{6.7}$$

For $\Omega = \pi c/d$, the sideband $\omega_k + \Omega$ corresponds to the next resonator mode and generates the amplitude

$$A_{k+1} = \frac{1}{2}A_0 T_0\delta\cos\omega_{k+1},$$

which is further amplified by stimulated emission, as long as ω_{k+1} lies within the gain profile above threshold. Since the amplitudes of all three modes in (6.7) achieve their maxima at times $t = q2d/c$ $(q = 0, 1, 2, \ldots)$, their phases are coupled by the modulation. A corresponding consideration applies to all other sidebands generated by modulation of the sidebands in (6.7).

Within the spectral width $\delta\nu$ of the gain profile

$$N = \frac{\delta\nu}{\Delta\nu} = 2\delta\nu\frac{d}{c},$$

oscillating resonator modes with the mode separation $\Delta\nu = c/2d$ can be locked together. The superposition of these N phase-locked modes results in the total amplitude

$$A(t) = \sum_{k=-m}^{+m} A_k \cos(\omega_0 + k\Omega)t, \quad \text{with } N = 2m+1. \tag{6.8}$$

For equal mode amplitudes $A_k = A_0$, (6.8) gives the total time-dependent intensity

$$I(t) \propto A_0^2 \frac{\sin^2(\frac{1}{2}N\Omega t)}{\sin^2(\frac{1}{2}\Omega t)} \cos^2 \omega_0 t. \tag{6.9}$$

If the amplitude A_0 is time independent (cw laser), this represents a sequence of equidistant pulses with the separation

$$T = \frac{2d}{c} = \frac{1}{\Delta\nu}, \tag{6.10}$$

which equals the round-trip time through the laser resonator. The pulse width

$$\Delta T = \frac{2\pi}{(2m+1)\Omega} = \frac{2\pi}{N\Omega} = \frac{1}{\delta\nu}, \tag{6.11}$$

is determined by the number N of phase-locked modes and is inversely proportional to the spectral bandwidth $\delta\nu$ of the gain profile above threshold (Fig. 6.9).

The peak power of the pulses, which can be derived from the intensity maxima in (6.9) at times $t = 2\pi q/\Omega = q(2d/c)$ $(q = 0, 1, 2, \ldots)$, is proportional to N^2. The pulse energy is therefore proportional to $N^2\Delta T \propto N$. In between the main pulses $(N-2)$ small maxima appear, which decrease in intensity as N increases.

Note For equal amplitudes $A_k = A_0$ the time-dependent intensity $I(t)$ in (6.9) corresponds exactly to the spatial intensity distribution $I(x)$ of light diffracted by a grating with N grooves that are illuminated by a plane wave. One has to replace Ωt by the phase difference ϕ between neighboring interfering partial waves, see (4.28) in Vol. 1, Sect. 4.1.3 and compare Fig. 4.21 in Vol. 1 and Fig. 6.9.

In real mode-locked lasers the amplitudes A_k are generally not equal. Their amplitude distribution A_k depends on the form of the spectral gain profile. This modifies (6.9) and gives slightly different time profiles of the mode-locked pulses, but does not change the principle considerations.

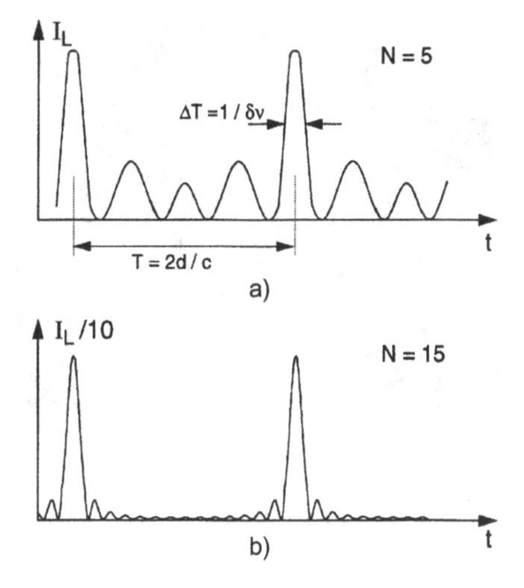

Fig. 6.9 Schematic profile of the output of a mode-locked laser: (**a**) with 5 modes locked; (**b**) with 15 modes locked

For *pulsed* mode-locked lasers the envelope of the pulse heights follows the time profile $\Delta N(t)$ of the inversion, which is determined by the pump power $P_{\text{p}}(t)$. Instead of a continuous sequence of equal pulses one obtains a finite train of pulses (Fig. 6.10).

For many applications a single laser pulse instead of a train of pulses is required. This can be realized with a synchronously triggered Pockels cell outside the laser resonator, which transmits only one selected pulse out of the pulse train. It is triggered by a mode-locked pulse just before the maximum of the train envelope, then it opens for a time $\Delta t < 2d/c$ and transmits only the next pulse following the trigger pulse [654]. Another method of single-pulse selection is cavity dumping of a mode-locked laser [655]. Here the trigger signal for the intracavity Pockels cell (Fig. 6.5) is synchronized by the mode-locked pulses and couples just one mode-locked pulse out of the cavity (Fig. 6.10b).

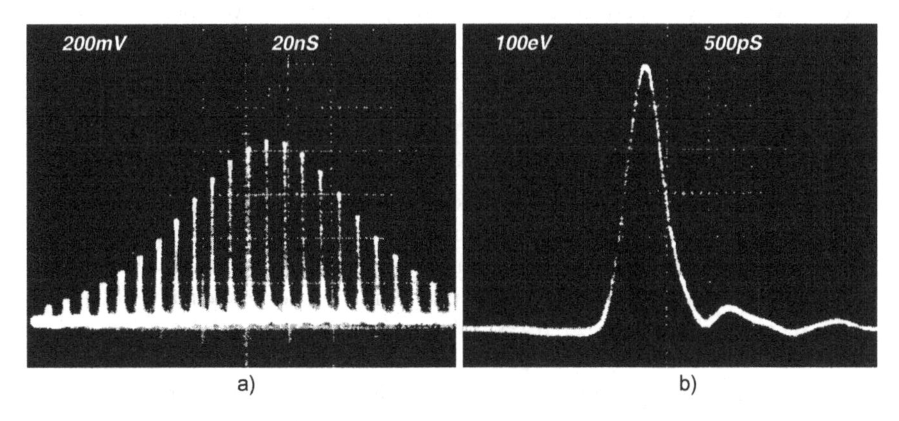

Fig. 6.10 (**a**) Pulse train of a mode-locked pulsed Nd:YAG laser; and (**b**) single pulse selected by a Pockel's cell. Note the different time scales in (**a**) (2 ns/div) and (**b**) (500 ps/div) [654]

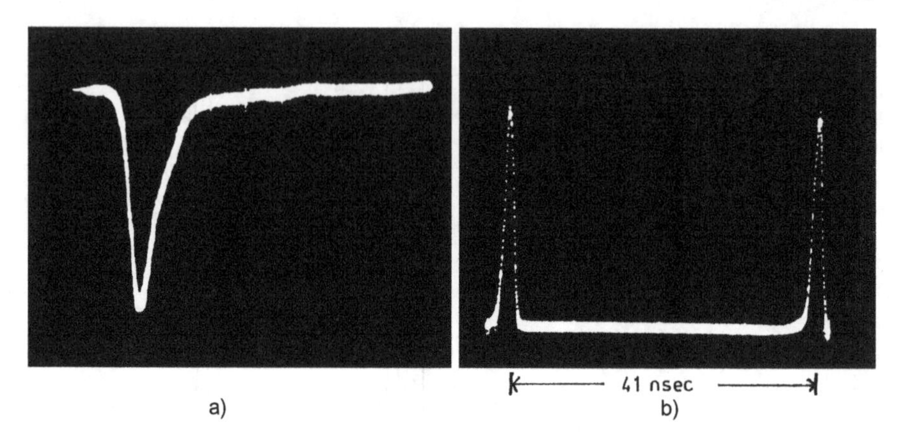

Fig. 6.11 Measured pulses of a mode-locked argon laser at $\lambda = 488$ nm: (**a**) monitored with a fast photodiode and a sampling oscilloscope (500 ps/div). The small oscillations after the pulse are cable reflections; (**b**) the attenuated scattered laser light was detected by a photomultiplier (single-photon counting) and stored in a multichannel analyzer. The time resolution is limited by the pulse rise times of the photodiode and photomultiplier, respectively [656]

Example 6.2

(a) The Doppler-broadened gain profile of the HeNe laser at $\lambda = 633$ nm has the spectral bandwidth $\delta\nu \approx 1.5$ GHz. Therefore, mode-locked pulses with durations down to $\Delta T \approx 500$ ps can be generated.

(b) Because of the higher temperature in the discharge of an argon-ion laser the bandwidth at $\lambda = 514.5$ nm is about $\delta\nu = 5$–7 GHz and one would expect pulses down to 150 ps. Experimentally, 200 ps pulses have been achieved. The apparently longer pulse width in Fig. 6.11 is limited by the time resolution of the detectors.

(c) The actively mode-locked Nd:glass laser [651, 652] delivers pulses at $\lambda = 1.06$ μm with durations down to 5 ps with high peak power ($\geq 10^{10}$ W), which can be frequency doubled or tripled in nonlinear optical crystals with high conversion efficiency. This yields powerful short light pulses in the green or ultraviolet region.

(d) Because of the large bandwidth $\delta\nu$ of their spectral gain profile, dye lasers, Ti:sapphire, and color-center lasers (Vol. 1, Sect. 5.7) are the best candidates for generating ultrashort light pulses. With $\delta\nu = 3 \times 10^{13}$ Hz (this corresponds to $\delta\lambda \sim 30$ nm at $\lambda = 600$ nm), pulse widths down to $\Delta T = 3 \times 10^{-14}$ s should be possible. This can, indeed, be realized with special techniques (Sect. 6.1.5). With active mode locking, however, one only reaches $\Delta T \geq 10$–50 ps [653]. This corresponds to the transit time of a light pulse through the modulator, which imposes a lower limit, unless new techniques are used.

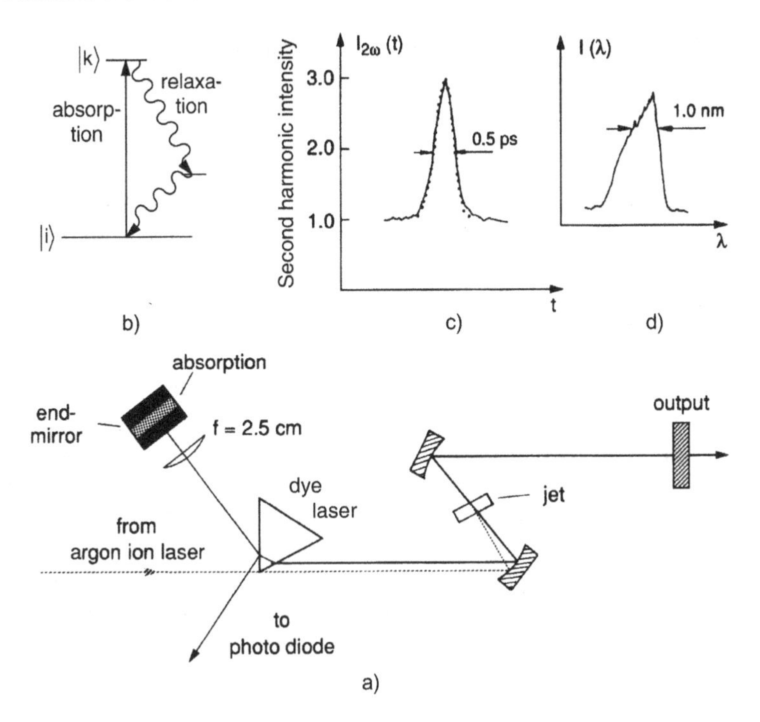

Fig. 6.12 Passive mode locking of a cw dye laser: (**a**) experimental arrangement; (**b**) level scheme of absorber; (**c**) time profile; and (**d**) spectral profile of a mode-locked pulse [652]

(b) Passive Mode Locking

Passive mode locking is a technique that demands less experimental effort than active mode-locking; it can be applied to pulsed as well as to cw lasers. Pulse widths below 1 ps have been realized. Its basic principles can be understood as follows: Instead of the active modulator, a saturable absorber is put inside the laser resonator, close to one of the end mirrors (Fig. 6.12). The absorbing transition $|k\rangle \leftarrow |i\rangle$ takes place between the levels $|i\rangle$ and $|k\rangle$ with short relaxation times τ_i, τ_k. In order to reach oscillation threshold in spite of the absorption losses the gain of the active medium must be correspondingly high. In the case of a pulsed pump source, the emission of the active laser medium at a time shortly before threshold is reached consists of fluorescence photons, which are amplified by induced emission. The peak power of the resulting photon avalanches (Vol. 1, Sect. 5.2) fluctuates more or less randomly. Because of nonlinear saturation in the absorber (Sect. 2.1), the most intense photon avalanche suffers the lowest absorption losses and thus experiences the largest net gain. It therefore grows faster than other competing weaker avalanches, saturates the absorber more, and increases its net gain even more. After a few resonator round trips this photon pulse has become so powerful that it depletes the inversion of the active laser medium nearly completely and therefore suppresses all other avalanches.

This nonlinear interaction of photons with the absorbing and the amplifying media leads under favorable conditions to mode-locked laser operation starting from a statistically fluctuating, unstable threshold situation. After this short unstable transient state the laser emission consists of a stable, regular train of short pulses with the time separation $T = 2d/c$, as long as the pump power remains above threshold (which is now lower than at the beginning because the absorption is saturated).

This more qualitative representation illustrates that the time profile and the width ΔT of the pulses is determined by the relaxation times of absorber and amplifier. In order to suppress the weaker photon avalanches reliably, the relaxation times of the absorber must be short compared with the resonator round-trip time. Otherwise, weak pulses, which pass the absorber shortly after the stronger saturating pulse, would take advantage of the saturation and would experience smaller losses. The recovery time of the amplifying transition in the active laser medium, on the other hand, should be comparable to the round-trip time in order to give maximum amplification of the strongest pulse, but minimum gain for pulses in between. A more thorough analysis of the conditions for stable passive mode locking can be found in [640, 657, 658].

The Fourier analysis of the regular pulse train yields again the mode spectrum of all resonator modes participating in the laser oscillation. The coupling of the modes is achieved at the fixed times $t = t_0 + q2d/c$ when the saturating pulse passes the absorber. This explains the term *passive mode locking*. Different dyes can be employed as saturable absorbers. The optimum choice depends on the wavelength. Examples are methylene blue, diethyloxadicarbocyanine iodide DODCI, or polymethinpyrylin [659], which have relaxation times of 10^{-9}–10^{-11} s.

Also semiconductors can be used as saturable absorbers. In Fig. 6.13 the principle operation is illustrated. The absorbed photon leads to the excitation of an electron from the valence band into the conduction band and the creation of a hole in the valence band. The thermalization of electrons and holes (Fig. 6.13b and c) and their recombination limits the saturation time and restores the absorption before the next laser pulse again saturates the transition.

Passive mode locking can also be realized in cw lasers. However, the smaller amplification restricts stable operation to a smaller range of values for the ratio of absorption to amplification, and the optimum conditions are more critical than in pulsed operation [660, 661]. With passively mode-locked cw dye lasers pulse widths down to 0.5 ps have been achieved [662].

More detailed representations of active and passive mode locking can be found in [640, 663, 664].

(c) Synchronous Pumping with Mode-Locked Lasers

For synchronous pumping the mode-locked pump laser L_1, which delivers short pulses with the time separation $T = 2d_1/c$, is employed to pump another laser L_2 (for example, a cw dye laser or a color-center laser). This laser L_2 then operates in a pulsed mode with the repetition frequency $f = 1/T$. An example, illustrated by Fig. 6.14, is a cw dye laser pumped by an acousto-optically mode-locked argon laser.

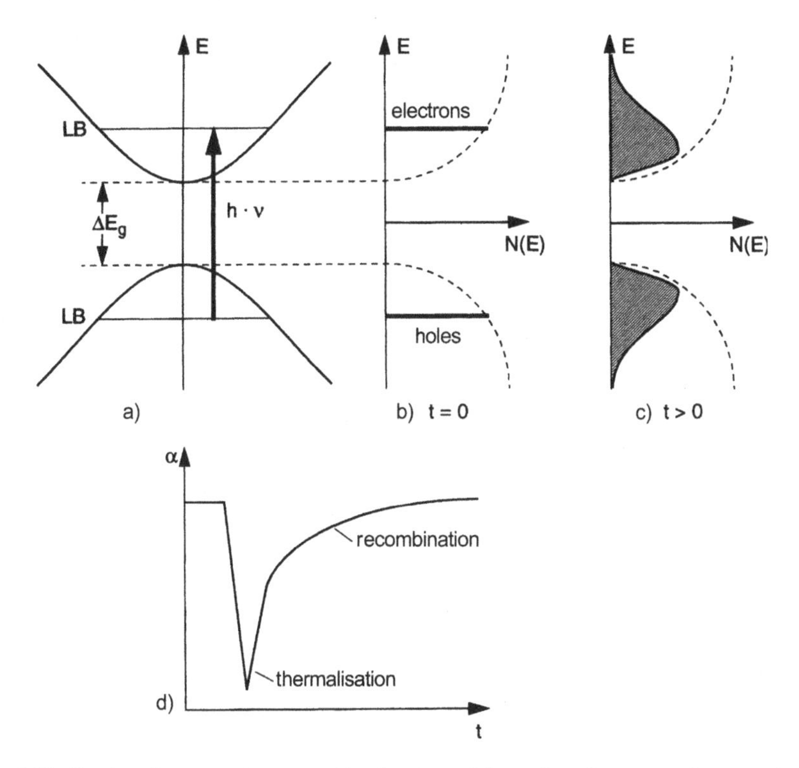

Fig. 6.13 Semiconductors as fast saturable absorbers: (**a**) transitors between valence- and conduction-band; (**b**) energy distribution of electrons and holes before laser excitation; (**c**) time dependence of the absorption coefficient

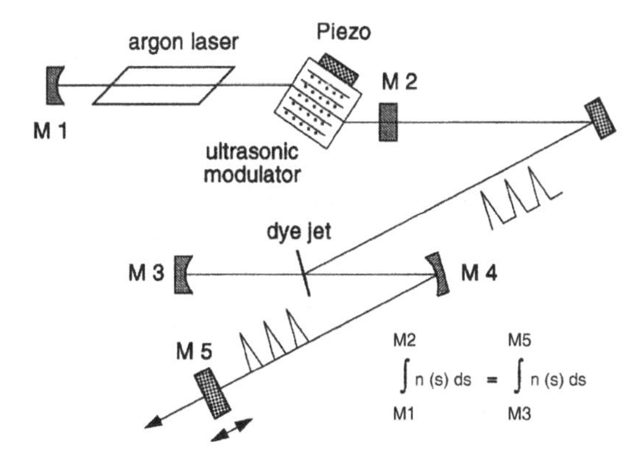

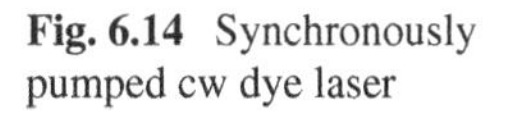
Fig. 6.14 Synchronously pumped cw dye laser

The optimum gain for the dye-laser pulses is achieved if they arrive in the active medium (dye jet) at the time of maximum inversion $\Delta N(t)$ (Fig. 6.15). If the optical cavity length d_2 of the dye laser is properly matched to the length d_1 of the pump laser resonator, the round-trip times of the pulses in both lasers become equal and the arrival times of the two pulses in the amplifying dye jet are synchro-

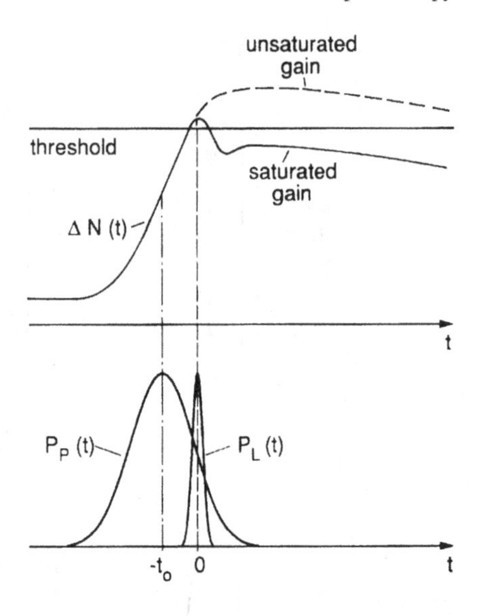

Fig. 6.15 Schematic time profiles of argon laser pump pulse $P_p(t)$, inversion $\Delta N(t)$ in the dye jet, and dye laser pulse $P_L(t)$ in a synchronously pumped cw dye laser

nized. Because of saturation effects, the dye-laser pulses become much shorter than the pump pulses, and pulse widths below 1 ps have been achieved [665–667]. For the experimental realization of accurate synchronization one end mirror of the dye-laser cavity is placed on a micrometer carriage in order to adjust the length d_2. The achievable pulse width ΔT depends on the accuracy $\Delta d = d_1 - d_2$ of the optical cavity length matching. A mismatch of $\Delta d = 1$ μm increases the pulsewidth from 0.5 to 1 ps [668].

For many applications the pulse repetition rate $f = c/2d$ (which is $f = 150$ MHz for $d = 1$ m) is too high. In such cases the combination of synchronous pumping and cavity dumping (Sect. 6.1.2) is helpful, where only every kth pulse ($k \geq 10$) is extracted due to Bragg reflection by an ultrasonic pulsed wave in the cavity dumper. The ultrasonic pulse now has to be synchronized with the mode-locked optical pulses in order to assure that the ultrasonic pulse is applied just at the time when the mode-locked pulse passes the cavity dumper (Fig. 6.16b).

The technical realization of this synchronized system is shown in Fig. 6.16a. The frequency $\nu_s = \Omega/2\pi$ of the ultrasonic wave is chosen as an integer multiple $\nu_s = q \cdot c/2d$ of the mode-locking frequency. A fast photodiode, which detects the mode-locked optical pulses, provides the trigger signal for the RF generator for the ultrasonic wave. This allows the adjustment of the phase of the ultrasonic wave in such a way that the arrival time of the mode-locked pulse in the cavity dumper coincides with its maximum extraction efficiency. During the ultrasonic pulse only one mode-locked pulse is extracted. The extraction repetition frequency $\nu_e = (c/2d)/k$ can be chosen between 1 Hz to 4 MHz by selecting the repetition rate of the ultrasonic pulses [669].

There are several versions of the experimental realizations of mode-locked or synchronously pumped lasers. Table 6.1 gives a short summary of typical operation parameters of the different techniques. More detailed representations of this subject can be found in [663–670].

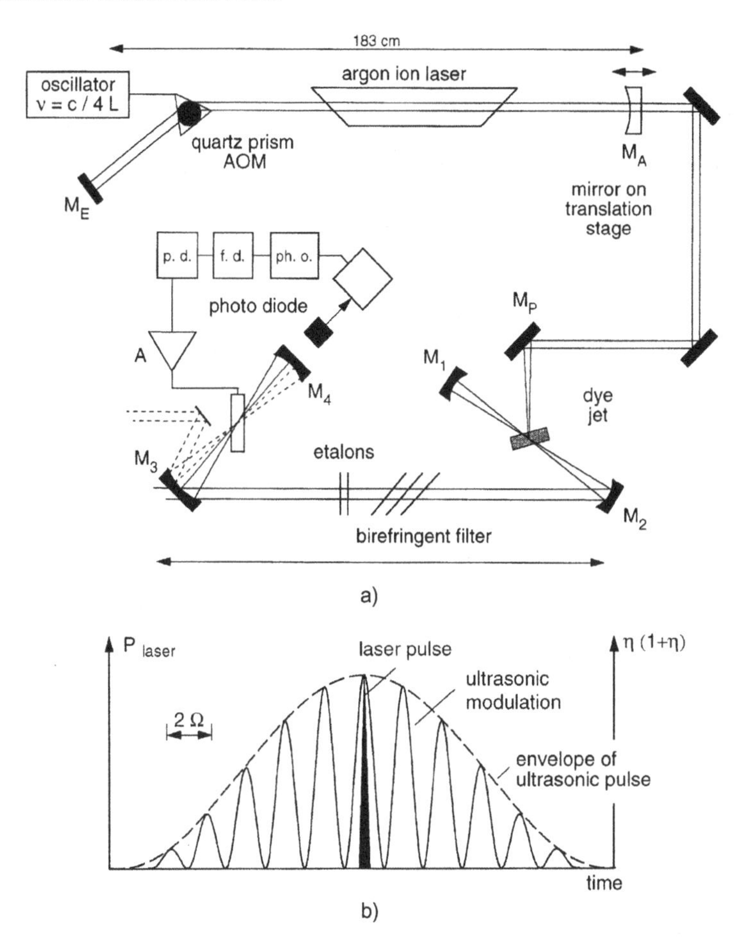

Fig. 6.16 Cavity-dumped synchronously pumped dye laser system with synchronization electronics [Courtesy Spectra-Physics]: (**a**) experimental setup; (**b**) correct synchronization time of a mode-locked pulse with the time of maximum output coupling of the cavity dumping pulse

6.1.5 Generation of Femtosecond Pulses

In the last sections it was shown that passive mode locking or synchronous pumping allows the generation of light pulses with a pulse width below 1 ps. Recently some new techniques have been developed that generate still shorter pulses. The shortest light pulses reported up to now are only 5 fs long [690]. At $\lambda = 600$ nm this corresponds to less than 3 oscillation periods of visible light! For illustration Fig. 6.17a shows such femtosecond pulses where the electric field amplitude $E(t)$ has its maximum at the peak of the pulse envelope $A(t)$ and the optical frequency $\nu(t)$ is constant during the pulse. In Fig. 6.17b a chirped pulse is shown where the optical frequency increases during the pulse (up-chirp). The Fourier-transform of $E(t)$ gives the frequency spectrum $E(\omega)$ of the pulses. Generation of higher harmonics of such

Table 6.1 Summary of different mode-locking techniques. With active mode locking of cw lasers an average power of 1 W can be achieved

Technique	Mode locker	Laser	Typical pulse duration	Typical pulse energy
Active mode locking	Acousto-optic modulator Pockels cell	Argon, cw HeNe, cw Nd:YAG, pulsed	300 ps 500 ps 100 ps	10 nJ 0.1 nJ 10 μJ
Passive mode locking	Saturable absorber	Dye, cw Nd:YAG	1 ps 1–10 ps	1 nJ 1 nJ
Synchronous pumping	Mode-locked pump laser and matching of resonator length	Dye, cw Color center	1 ps 1 ps	10 nJ 10 nJ
Colliding pulse mode locking CPM	Passive mode locking and eventual synchronous pumping	Ring dye laser	<100 fs	≈ 1 nJ
Kerr lens mode locking	Optical Kerr effect	Ti:sapphire	<10 fs	≈ 10–100 nJ

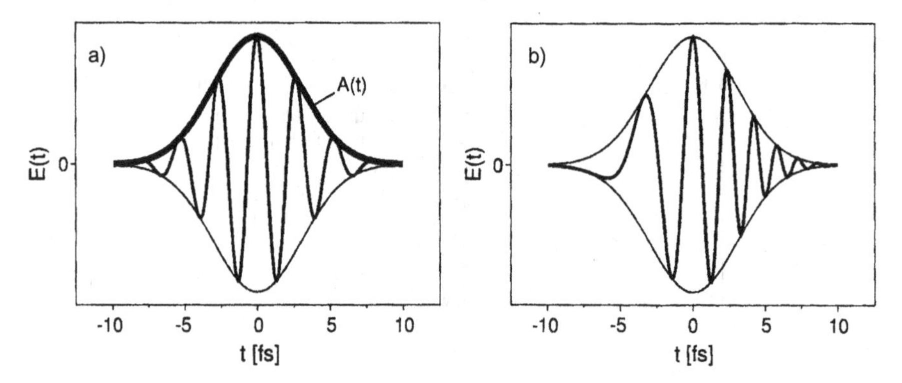

Fig. 6.17 Envelope $A(t)$ and electric field amplitude $E(t)$ of a femtosecond pulse. (**a**) Electric field maximum at pulse maximum. (**b**) Up chirped pulse where the optical frequency $\nu(t)$ increases with time

short pulses produces even shorter pulses in the VUV with pulsewidths of down to 60 attoseconds (0.06 fs). We will now discuss some of these new techniques.

(a) The Colliding Pulse Mode-Locked Laser

A cw ring dye laser pumped by an argon laser can be passively mode locked by an absorber inside the ring resonator. The mode-locked dye laser pulses travel into both directions in the ring, clockwise and counterclockwise (Fig. 6.18). If the absorber, realized by a thin dye jet, is placed at a location where the path length A1–A2 between the amplifying jet and the absorbing jet is just one-quarter of the total ring length L, the net gain per round-trip is maximum when the counterpropagating

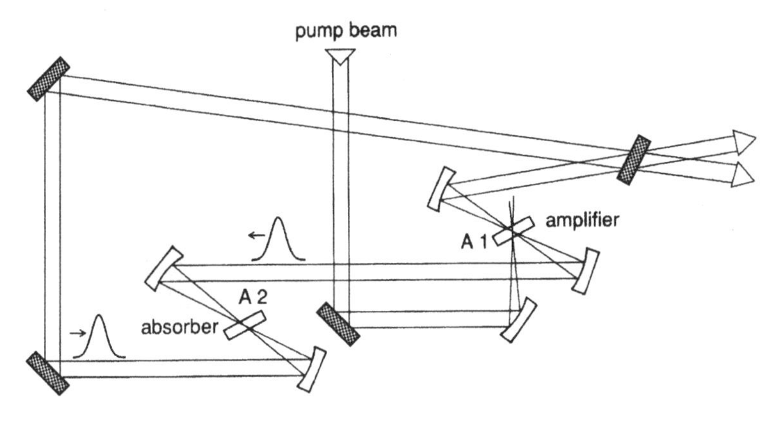

Fig. 6.18 CPM ring dye laser. The distance A1–A2 is one-quarter the total round-trip length L

pulses collide within the absorber. This can be seen as follows: For this situation the time separation $\Delta t = T/2$ between the passage of successive pulses (one clockwise and the next counter-clockwise) through the amplifier achieves the maximum value of one-half of the round-trip time T. This means that the amplifying medium has a maximum time to recover its inversion after it was depleted by the previous pulse.

The total pulse intensity in the absorber where the two pulses collide is twice that of a single pulse. This means larger saturation and less absorption. Both effects lead to a maximum net gain if the two pulses collide in the absorber jet.

At a proper choice of the amplifying gain and the absorption losses, this situation will automatically be realized in the passively mode-locked ring dye laser. It leads to an energetically favorable stable operation, which is called *colliding-pulse mode* (CPM) *locking*, and the whole system is termed a *CPM laser*. This mode of operation results in particularly short pulses down to 50 fs. There are several reasons for this pulse shortening:

(i) The transit time of the light pulses through the thin absorber jet ($d < 100\,\mu$m) is only about 400 fs. During their superposition in the absorber the two colliding light pulses form, for a short time, a standing wave, which generates, because of saturation effects, a spatial modulation $N_i(z)$ of the absorber density N_i and a corresponding refractive index grating with a period of $\lambda/2$ (Fig. 6.19). This grating causes a partial reflection of the two incident light pulses. The reflected part of one pulse superimposes and interferes with the oppositely traveling pulse, resulting in a coupling of the two pulses. At $t = t_0$ the constructive interference is maximized (additive-pulse mode locking).

(ii) The absorption of both pulses has a minimum when the two pulse maxima just overlap. At this time the grating is most pronounced and the coupling is maximized. The pulses are therefore shortened for each successive round-trip until the shortening is compensated by other phenomena, which cause a broadening of the pulses. One of these effects is the dispersion due to the dielectric layers on the resonator mirrors, which causes a different round-trip time for the different wave-

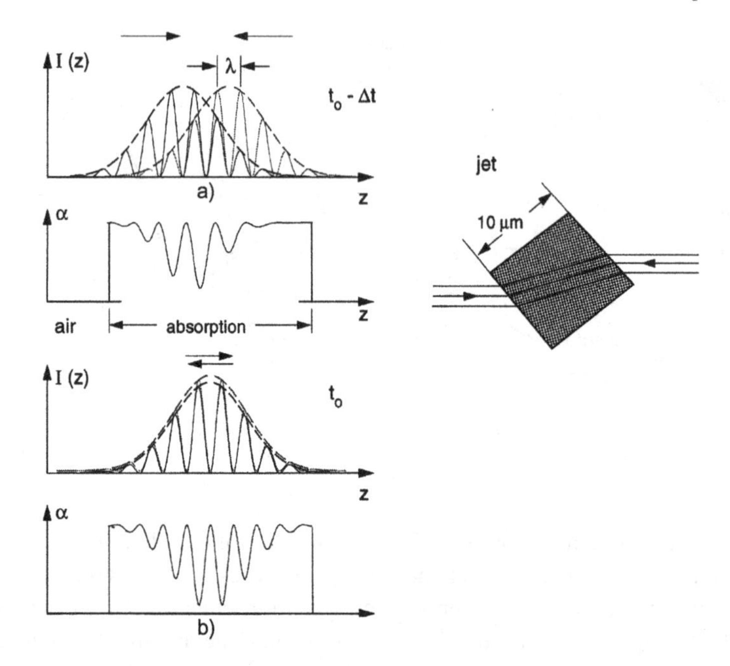

Fig. 6.19 Density grating generated in the absorber jet by the colliding optical pulses at two different times: (**a**) partial overlap at time $t_0 - \Delta t$; (**b**) complete overlap at $t = t_0$

Fig. 6.20 Compensation of mirror dispersion by prisms within the cavity

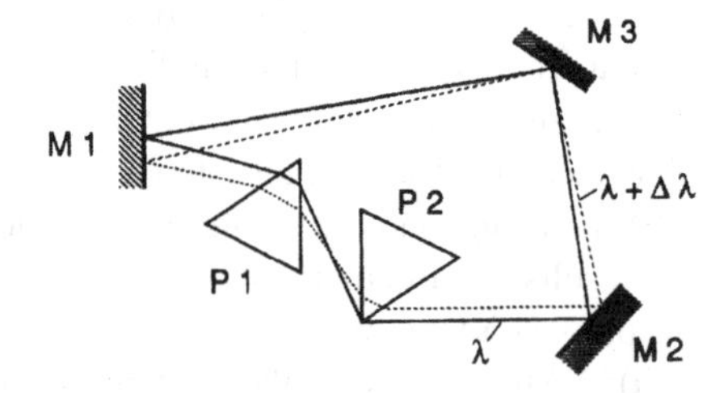

lengths contained in the short pulse. The shorter the pulses the broader their spectral profile $I(\lambda)$ becomes and the more serious are dispersion effects.

The mirror dispersion can be compensated up to the first order by inserting into the ring resonator (Fig. 6.20) dispersive prisms, which introduce different optical path lengths $d_p n(\lambda)$ [673]. This dispersion compensation can be optimized by shifting the prisms perpendicularly to the pulse propagation, thus adjusting the optical path length $d_p n(\lambda)$ of the pulses through the prisms.

In principle, the lower limit ΔT_{min} of the pulse width is given by the Fourier limit $\Delta T_{min} = a/\delta\nu$, where $a \sim 1$ is a constant that depends on the time profile of the pulse (Sect. 6.2.2). The larger the spectral width $\delta\nu$ of the gain profile is, the smaller ΔT_{min} becomes. In reality, however, the dispersion effects, which increase with $\delta\nu$, become more and more important and prevent reaching the principal lower limit of ΔT_{min}. In Fig. 6.21 the achievable limit ΔT_{min} is plotted against the spec-

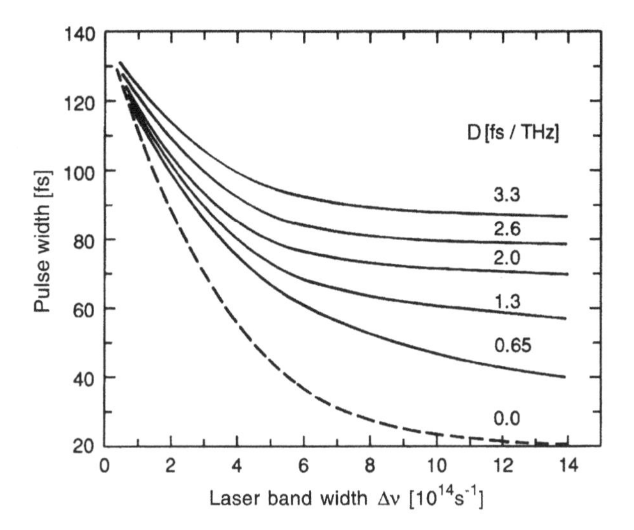

Fig. 6.21 Theoretical lower limit of pulse width ΔT as a function of the spectral bandwidth $\Delta\nu$ of a mode-locked laser for different values of the dispersion parameter D [fs/THz] [751]

tral bandwidth for different dispersions [fs/cm], where the dashed curve gives the dispersion-free Fourier limit of the pulse width ΔT [673, 674].

Pulse widths below 100 fs can be reached with this CPM technique [675, 676]. If the CPM ring dye laser is synchronously pumped by a mode-locked argon laser stable operation over many hours can be realized [677]. Using a novel combination of saturable absorber dyes and a frequency-doubled mode-locked Nd:YAG laser as a pump, pulse widths down to 39 ps at $\lambda = 815$ nm have been reported [678].

(b) Kerr Lens Mode Locking

For a long time dye solutions were the favorite gain medium for the generation of femtosecond pulses because of their broad spectral gain region. Meanwhile, different solid state gain materials have been found with very broad fluorescence bandwidths, which allow, in combination with new nonlinear phenomena, the realization of light pulses down to 5 fs.

For solid-state lasers typical lifetimes of the upper laser level range from 10^{-6} s to 10^{-3} s. This is much longer than the time between successive pulses in a mode-locked pulse train, which is about 10–20 ns. Therefore the saturation of the amplifying medium cannot recover within the time between two pulses and the amplifying medium therefore cannot contribute to the mode locking by dynamic saturation as in the case of CPM mode locking discussed before. One needs a fast saturable absorber, where the saturation can follow the short pulse profile of the mode-locked pulses. Such passively mode-locked solid-state lasers might not be completely stable with regard to pulse stability and pulse intensities and they generally do not deliver pulses below 1 ps.

The crucial breakthrough for the realization of ultrafast pulses below 100 fs was the discovery of a fast pulse-forming mechanism in 1991, called *Kerr lens mode*

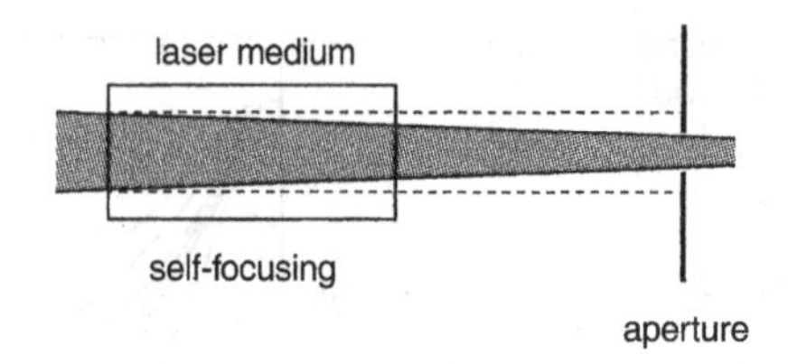

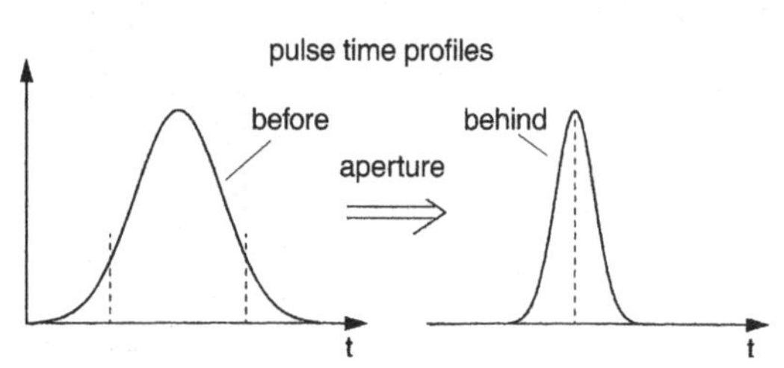

Fig. 6.22 Kerr lens mode locking

locking (KLM), which can be understood as follows: For large incident intensities I the refractive index n of a medium depends on the intensity. One can write

$$n(\omega, I) = n_0(\omega) + n_2(\omega)I.$$

This intensity-dependent change of the refractive index is caused by the nonlinear polarization of the electron shell induced by the electric field of the optical wave and is therefore called the optical Kerr effect.

Because of the radial intensity variation of a Gaussian laser beam, the refractive index of the medium under the influence of a laser beam shows a radial gradient with the maximum value of n at the central axis. This acts as a focusing lens and leads to focusing of the incident laser beam, where the focal length depends on the intensity. Since the central part of the pulse time-profile has the largest intensity, it is also focused more strongly than the outer parts, where the intensity is lower. A circular aperture at the right place inside the laser resonator transmits only this central part, that is, it cuts away the leading and trailing edges and the transmitted pulse is therefore shorter than the incident pulse (Fig. 6.22).

Example 6.3 For sapphire Al_2O_3 $n_2 = 3 \times 10^{-16}$ cm^2/W. At the intensity of 10^{14} W/cm^2 the refractive index changes by $\Delta n = 3 \times 10^{-2} n_0$. For a wave with $\lambda = 1000$ nm this leads to a phase shift of $\Delta\Phi = (2\pi/\lambda)\Delta n = 300 \cdot 2\pi$ after a pathlength of 1 cm, which results in a radius of curvature of the phase-front $R = 4$ cm with the corresponding focal length of the Kerr lens.

Often the laser medium itself acts as Kerr medium and forms an additional lens inside the laser resonator. This is shown schematically in Fig. 6.23, where the lenses with focal lengths f_1 and f_2 are in practice curved mirrors [679]. Without the Kerr lens the resonator is stable if the distance between the two lenses is $f_1 + f_2$. With the Kerr lens this distance has to be modified to $f_1 + f_2 + \delta$, where the quantity δ

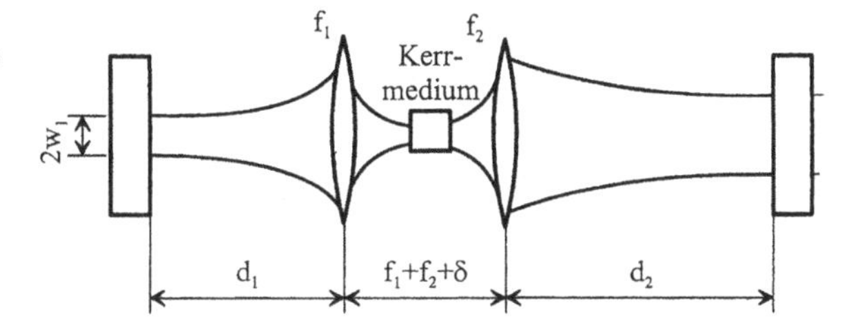

Fig. 6.23 Schematic illustration of Kerr lens mode locking inside the laser resonator [679]

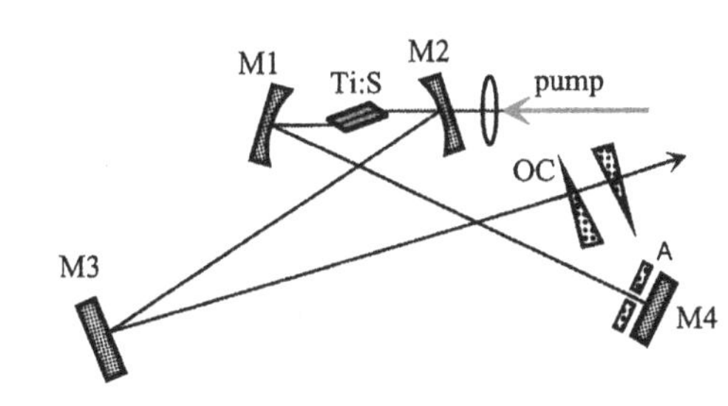

Fig. 6.24 Schematic diagram of the mirror dispersion controlled (MDC) Ti:sapphire oscillator used for soft-aperture and hard-aperture mode locking [679]

depends on the focal length of the Kerr lens, and therefore on the pulse intensity. If the distance between the two lenses is $f_1 + f_2 + \delta$ the resonator is only stable for values of δ within the limits

$$0 < \delta < \delta_1, \quad \text{or} \quad \delta_2 < \delta < \delta_1 + \delta_2,$$

where

$$\delta_1 = \frac{f_2^2}{d_2 - f_2}, \qquad \delta_2 = \frac{f_1^2}{d_1 - f_1}. \tag{6.12}$$

Choosing the right value of δ makes the resonator stable only for the time interval around the pulse maximum.

In Fig. 6.24 the experimental setup for a femtosecond Ti:sapphire laser with Kerr lens mode locking is shown, where the amplifying laser medium acts simultaneously as a Kerr lens. The folded resonator is designed in such a way that only the most intense part of the pulse is sufficiently focused by the Kerr lens to always pass for every round-trip through the spatially confined active region pumped by the argon-ion laser. Here the active gain medium defined by the pump focus acts as spatial "soft aperture." A mechanical aperture A in front of mirror M4 in Fig. 6.24 realizes "hard aperture" Kerr lens mode locking. Output pulses with pulsewidths below 10 fs have been demonstrated with this design.

A Kerr-lens mode-locked Ti:Sapphire laser with a threshold that is ten times less than in conventional Kerr-lens lasers has been reported by Fujimoto and his group [679]. The schematic diagram of this design with an astigmatically compensated folded cavity is shown in Fig. 6.25. The dispersion of the cavity is compensated by a prism pair and the output coupler has a transmission of 1 %. The low threshold permits the use of low-power inexpensive pump lasers.

An alternative method for the realization of KLM uses the birefringent properties of the Kerr medium, which turns the plane of polarization of the light wave

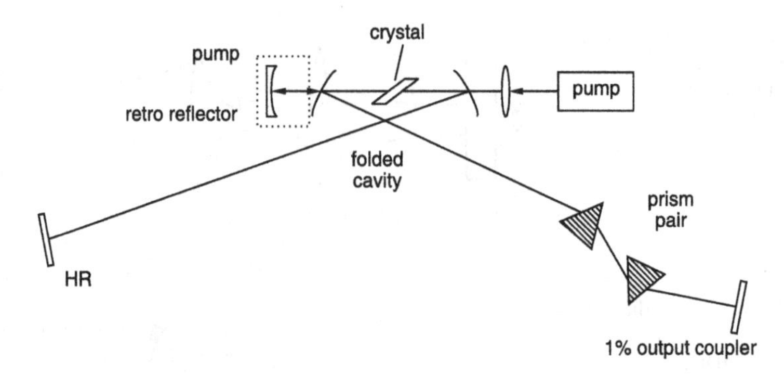

Fig. 6.25 Schematic diagram of the ultralow-threshold Ti:Al_2O_3 laser with Kerr-lens mode-locking [680]

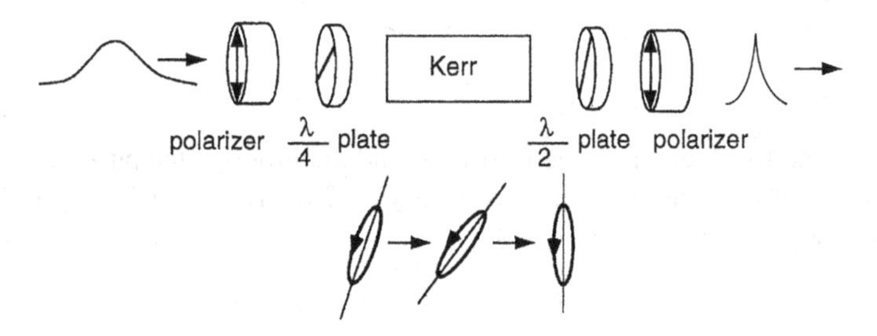

Fig. 6.26 Polarization-additive pulse mode-locking [681]

passing through the Kerr medium. This is illustrated in Fig. 6.26. The incident wave passes through a linear polarizer and is then elliptically polarized by a $\lambda/4$-plate. The Kerr medium causes a time and intensity-dependent nonlinear polarization rotation. A $\lambda/2$-plate and a linear polarizer behind the Kerr medium can be arranged in such a way that the transmission reaches its maximum at the peak of the incident pulse, thus shortening the pulse width [681]. This device acts similarly to a passive saturable absorber and is particularly useful for fiber lasers with ultrashort pulses.

In Table 6.1 the different techniques for the generation of short laser pulses and their typical parameters are compared. It shows that pulse widths below 1 ps can be achieved with the CPM technique and with Kerr lens mode-locking. In the next section we will discuss how short laser pulses can be further compressed by nonlinear effects in optical fibres.

6.1.6 Optical Pulse Compression

Since the principle lower limit $\Delta T_{\text{min}} = 1/\delta\nu$ of the optical pulse is given by the spectral bandwidth $\delta\nu$ of the gain medium, it is desirable to make $\delta\nu$ as large as possible. The idea of spectral broadening of optical pulses by *self-phase modulation* in optical fibers with subsequent pulse compression represented a breakthrough

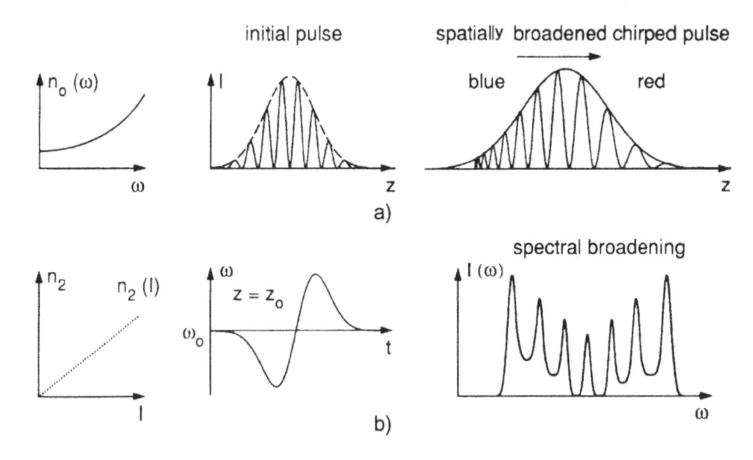

Fig. 6.27 Spatial and spectral broadening of a pulse in a medium with normal linear (**a**) and nonlinear (**b**) refractive index

for achieving pulsewidths of only a few femtoseconds. The method is based on the following principle: When an optical pulse with the spectral amplitude distribution $E(\omega)$ propagates through a medium with refractive index $n(\omega)$, its time profile will change because the group velocity

$$v_g = \frac{d\omega}{dk} = \frac{d}{dk}(v_{ph}k) = v_{ph} + k\frac{dv_{ph}}{dk}, \tag{6.13}$$

which gives the velocity of the pulse maximum, shows a dispersion

$$\frac{dv_g}{d\omega} = \frac{dv_g}{dk}\Big/\frac{d\omega}{dk} = \frac{1}{v_g}\frac{d^2\omega}{dk^2}. \tag{6.14}$$

For $d^2\omega/dk^2 \neq 0$, the velocity differs for the different frequency components of the pulse, which means that the shape of the pulse will change during its propagation through the medium (GVD = group velocity dispersion) (Fig. 6.27a). For negative dispersion ($dn/d\lambda < 0$), for example, the red wavelengths have a larger velocity than the blue ones, that is, the pulse becomes spatially broader. A quantitative description starts from the wave equation for the pulse envelope [682, 683]

$$\frac{\partial E}{\partial z} + \frac{1}{v_g}\frac{\partial E}{\partial t} + \frac{i}{2v_g^2}\frac{\partial v_g}{\partial \omega}\frac{\partial^2 E}{\partial t^2} = 0, \tag{6.15}$$

which can be derived from the general wave equation with the slowly varying envelope approximation ($\lambda\partial^2 E/\partial z^2 \ll \partial E/\partial z$) [640].

For the time profile of a pulse with half-width $\tau(L)$ after travelling over a pathlength L through the dispersive medium one obtains from (6.15)

$$\tau(L) = \tau_0\sqrt{1 + \left(L^2/L_D^2\right)}$$

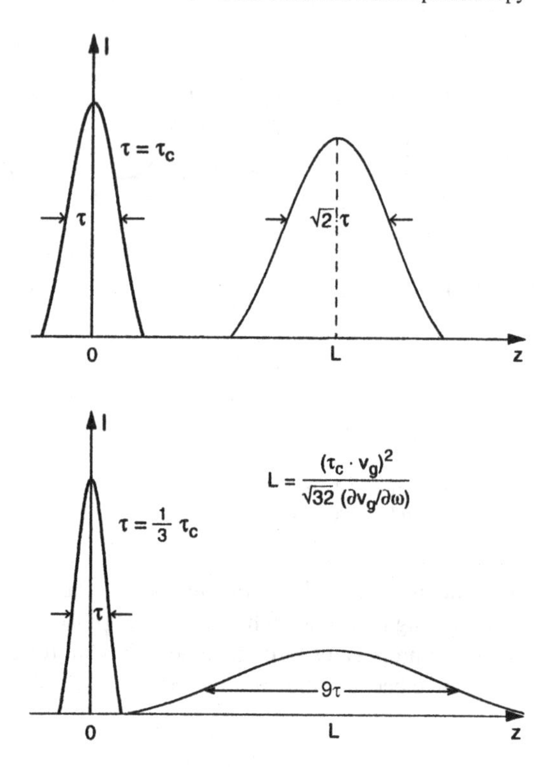

Fig. 6.28 Schematic illustration of the broadening of short light pulses due to group velocity dispersion GVD

where

$$\tau_0 = \tau(L = 0) \quad \text{and} \quad L_\text{D} = \tau_0^2/\beta_2 \quad \text{with } \beta_2 = (\partial v_\text{g}/\partial \omega)/v_\text{g}^2.$$

Is the dispersion length. After a pathlength $L = L_\text{D}$ the pulse-width has broadened by a factor $\sqrt{2}$. This shows that the broadening becomes the larger the smaller the pulse-width τ_0 of the initial pulse is. This is illustrated by Fig. 6.28, which shows schematically the broadening of two pulses with different initial pulse widths. For

$$L = (3)^{1/2} L_\text{D} = 3^{1/2} (\tau_0 \cdot v_\text{g})^2/(\partial v_\text{g}/\partial \omega)$$

the pulse width has doubled.

Example 6.4 $L = 0.2$ m, $v_\text{g} = 10^8$ m/s, $\partial v_\text{g}/\partial \omega = 10^{-8}$ m $\rightarrow \tau_c \approx 1$ ps. For $\tau_0 = \tau_c$ the pulse broadens by a factor $2^{1/2}$ after a path length $L = 20$ cm. For $\tau_0 = 0.3\tau_c$ it broadens after 20 cm already by a factor 9.

If the optical pulse of a mode-locked laser is focused into an optical fiber, the intensity I becomes very high. The amplitude of the forced oscillations of the electrons in the fiber material under the influence of the optical field increases with the

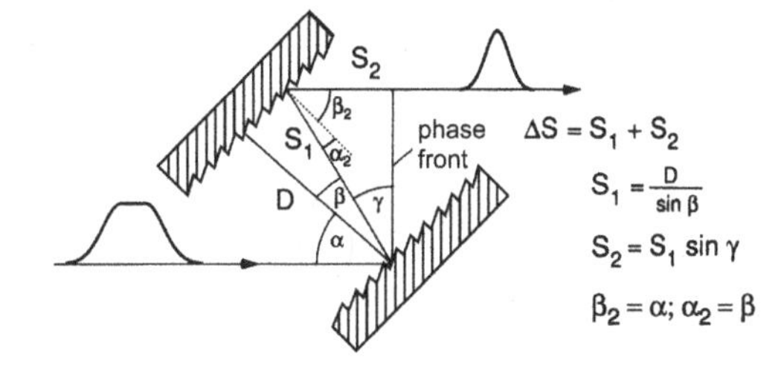

Fig. 6.29 Pulse compression by a grating pair

field amplitude and the refractive index becomes intensity dependent:

$$n(\omega, I) = n_0(\omega) + n_2 I(t), \tag{6.16}$$

where $n_0(\omega)$ describes the linear dispersion (see Sect. 6.1.5). The phase $\phi = \omega t - kz$ of the optical wave $E = E_0 \cos(\omega t - kz)$ with $k = n\omega/c$

$$\phi = \omega t - \omega n \frac{z}{c} = \omega\left(t - n_0 \frac{z}{c}\right) - AI(t), \quad \text{with } A = n_2 \omega \frac{z}{c}, \tag{6.17}$$

now depends on the intensity

$$I(t) = c\epsilon_0 \int \left|E_0(\omega, t)\right|^2 \cos^2(\omega t - kz)\, \mathrm{d}\omega. \tag{6.18}$$

Since the momentary optical frequency

$$\omega = \frac{\mathrm{d}\phi}{\mathrm{d}t} = \omega_0 - A\frac{\mathrm{d}I}{\mathrm{d}t}, \tag{6.19}$$

is a function of the time derivative $\mathrm{d}I/\mathrm{d}t$, (6.19) illustrates that the frequency decreases at the leading edge of the pulse ($\mathrm{d}I/\mathrm{d}t > 0$), while at the trailing edge ($\mathrm{d}I/\mathrm{d}t < 0$) ω increases (*self-phase modulation*). The leading edge is red shifted, the trailing edge blue shifted. This frequency shift during the pulse time τ is called "chirp". The *spectral profile* of the pulse becomes broader (Fig. 6.27b).

The linear dispersion $n_0(\lambda)$ causes a spatial *broadening, the intensity-dependent refractive index $n_2 I(t)$ a* spectral *broadening*. The spatial broadening of the pulse (which corresponds to a broadening of its time profile) is proportional to the length of the fiber and depends on the spectral width $\Delta\omega$ of the pulse and on its intensity.

The compression of this spectrally and spatially broadened pulse can now be achieved by a grating pair with the separation D that has a larger pathlength for red wavelengths than for blue ones and therefore delays the leading red edge of the pulse compared to its blue trailing edge. This can be seen as follows [685]: the optical pathlength $S(\lambda)$ between two phase-fronts of a plane wave before and after the grating is, according to Fig. 6.29

$$S(\lambda) = S_1 + S_2 = \frac{D}{\cos\beta}(1 + \sin\gamma), \quad \text{with } \gamma = 90° - (\alpha + \beta). \tag{6.20}$$

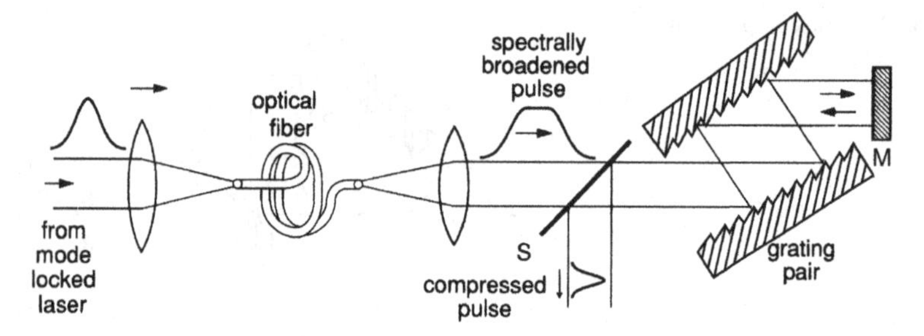

Fig. 6.30 Experimental arrangement for the generation of femtosecond pulses by self-phase-modulation with subsequent pulse compression by a grating pair [686]

This transforms with $\cos(\alpha+\beta)=\cos\alpha\cos\beta-\sin\alpha\sin\beta$ into

$$S(\lambda)=D\left[\frac{1}{\cos\beta}+\cos\alpha-\sin\alpha\tan\beta\right].$$

With the grating equation $d(\sin\alpha-\sin\beta)=\lambda$ of a grating with groove separation d and its dispersion $\mathrm{d}\beta/\mathrm{d}\lambda=1/(d\cos\beta)$ for a given angle of incidence α (Vol. 1, Sect. 4.1.3), we obtain the spatial dispersion

$$\frac{\mathrm{d}S}{\mathrm{d}\lambda}=\frac{\mathrm{d}S}{\mathrm{d}\beta}\frac{\mathrm{d}\beta}{\mathrm{d}\lambda}=\frac{-D\lambda}{d^2\cos^3\beta}=\frac{-D\lambda}{d^2[1-(\lambda/d-\sin\alpha)^2]^{3/2}}. \tag{6.21}$$

This shows that the dispersion is proportional to the grating separation D and increases with λ. By choosing the correct value of D one can just compensate the chirp of the pulse generated in the optical fiber and obtain a compressed pulse.

Note The diffraction by the grating causes an additional phase shift which amounts in the first diffraction order to 2π per grating groove. In reference [687] it is shown, that in spite of this phase shift the delay time of the pulse by the diffraction grating pair is $\Delta\tau=S(\omega)/c$.

A typical experimental arrangement is depicted in Fig. 6.30 [686]. The optical pulse from the mode-locked laser is spatially and spectrally broadened in the optical fiber and then compressed by the grating pair. The dispersion of the grating pair can be doubled if the pulse is reflected by the mirror M and passes the grating pair again. Pulse widths of 16 fs have been obtained with such a system [688].

With a combination of prisms and gratings (Fig. 6.31) not only the quadratic but also the cubic term in the phase dispersion

$$\begin{aligned}\phi(\omega)=\phi(\omega_0)&+\left(\frac{\partial\phi}{\partial\omega}\right)_{\omega_0}(\omega-\omega_0)+\frac{1}{2}\left(\frac{\partial^2\phi}{\partial\omega^2}\right)_{\omega_0}(\omega-\omega_0)^2\\&+\frac{1}{6}\left(\frac{\partial^3\phi}{\partial\omega^3}\right)_{\omega_0}(\omega-\omega_0)^3,\end{aligned} \tag{6.22}$$

can be compensated [689]. This allows one to reach pulsewidths of 6 fs.

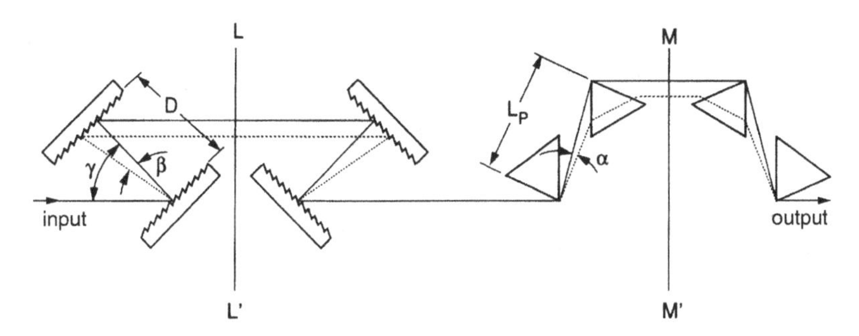

Fig. 6.31 Sequence of grating pairs and prism pairs for the compensation of quadratic and cubic phase dispersion. LL′ and MM′ are two phase-fronts. The *solid line* represents a reference path and the *dashed line* illustrates the paths for the wave of wavelength λ, which is diffracted by an angle β at the first grating and refracted by an angle α against the reference path in a prism [689]

More information and a detailed discussion of the different compression techniques can be found in [691].

6.1.7 Sub 10 fs Pulses with Chirped Laser Mirrors

The development of broadband saturable semiconductor absorber mirrors and of dispersion-engineered chirped multilayer dielectric mirrors has allowed the realization of self-starting ultrashort laser pulses, which routinely reach sub-10 fs pulsewidths and peak powers above the megawatt level.

Femtosecond pulse generation relies on a net negative intracavity group-delay dispersion (GDD). Since solid-state gain media introduce a frequency-dependent positive dispersion, this must be overcompensated by media inside the laser cavity that have a correspondingly large negative dispersion. We saw in Sect. 6.1.5 that intracavity prisms can be used as such compensators. However, here the GDD shows a large wavelength dependence and for very short pulses (with a corresponding broad spectral range) this results in asymmetric pulse shapes with broad pedestals in the time domain. The invention of chirped dielectric low-loss laser mirrors has brought a substantial improvement [692].

A chirped mirror is a dielectric mirror with many alternate layers with low and high indices of refraction. The reflectivity r at each boundary is $r = (n_\mathrm{r} - n_\mathrm{l})/(n_\mathrm{r} + n_\mathrm{l})$. If the optical thicknesses of these layers are always the same, we have the Bragg reflector shown in Fig. 6.32a. Such a mirror will, however, not generate a chirp in the reflected pulse. To achieve such a chirp the thicknesses of the layers must slowly vary from the first to the last layer, as shown in Fig. 6.32b. Two different wavelengths within the spectral bandwidth of the pulse now penetrate different path lengths into the stack and therefore suffer a different delay (positive group delay dispersion). When they are superimposed again after reflection, the reflected pulse has become broader because of the chirp. Proper engineering of the chirped mirror

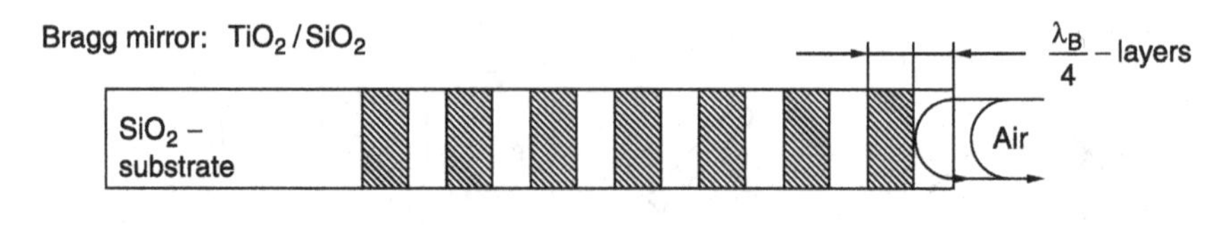

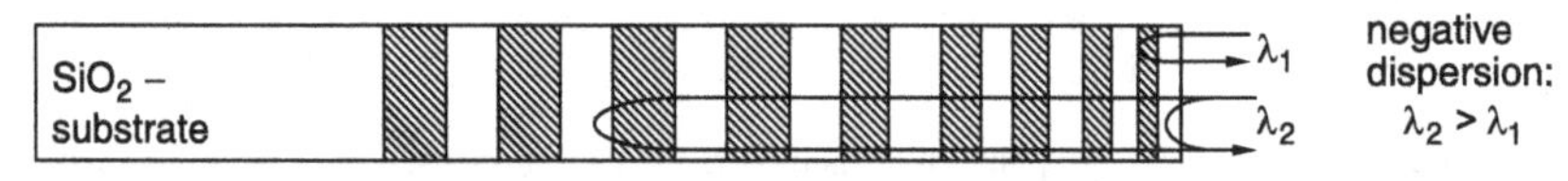

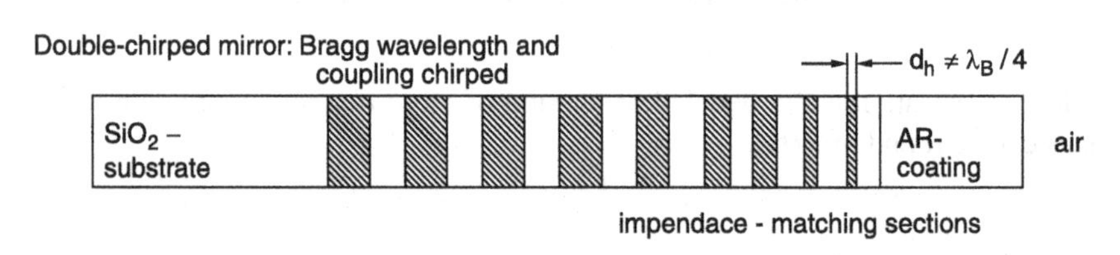

Fig. 6.32 Chirped mirrors: (**a**) Bragg mirror with no chirp; (**b**) simple chirped mirror for one wavelength; (**c**) double-chirped mirror with matching sections to avoid residual reflections [695]

allows good compensation of the negative group dispersion of the laser cavity. In the mirror in Fig. 6.32a only the Bragg wavelength is linearly chirped. For the double-chirped mirror in Fig. 6.32b, the spectral width for which dispersion compensation can be reached is greatly increased. This also means that very short pulses can be properly chirped. In order to avoid undesired reflections from the front surface or from sections inside the mirror that cause oscillations in the group dispersion, an antireflection coating and matching sections between the Bragg layers are necessary [695].

Instead of varying the thicknesses of the layers, one can also produce chirped mirrors by smoothly altering the indices of refraction and the difference between them $(n_r - n_l)$ (Fig. 6.33).

These mirrors may be regarded as one-dimensional holograms that are generated when a chirped and an unchirped laser pulse from opposite directions are superimposed in a medium where they generate a refractive index pattern proportional to their total intensity [693]. When a chirped pulse is reflected by such a hologram, it becomes compressed, similar to the situation with phase-conjugated mirrors.

In practice, such mirrors are produced by evaporation techniques controlled by a corresponding computer program. In Fig. 6.33 the variation of the refractive index for the different dielectric layers is shown for a mirror with negative GDD, and in Fig. 6.34 the reflectivity and the group delay is plotted as a function of wavelength for mirrors with negatively and positively chirped graded-index profiles. In combi-

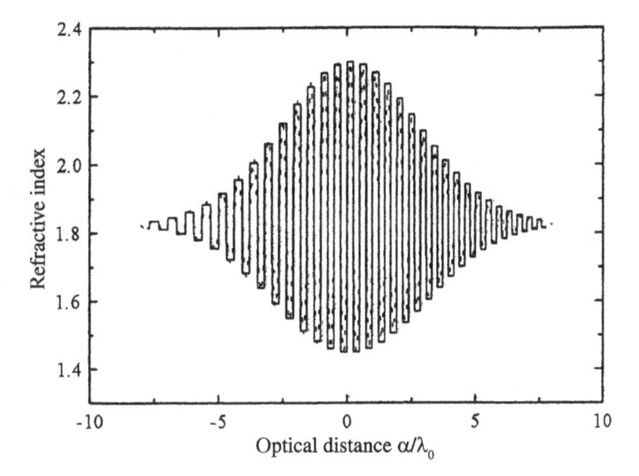

Fig. 6.33 Refractive index profile of a discrete-valued chirped dielectric mirror

nation with Kerr lens mode locking such chirped mirrors allow the generation of femtosecond pulses down to 4 fs [679].

In Fig. 6.34 the measured group delay dispersion (GDD) is shown as realized with a double-chirped mirror, compared with the wanted one. This illustrates that for the spectral range 1000–1200 nm the match is quite good while for $\lambda < 1000$ nm and $\lambda > 1200$ nm still large deviations appear [693b, 693c].

Another alternative for the generation of ultrafast pulses is the passive mode locking by fast semiconductor saturable absorbers in front of chirped mirrors (Fig. 6.35) in combination with Kerr lens mode locking [694]. The recovery time of the saturable absorber must be generally faster then the laser pulse width. This is provided by KLM, which may be regarded as artificial saturable absorber that is as fast as the Kerr nonlinearity following the laser intensity. Since the recovery time in a semi-

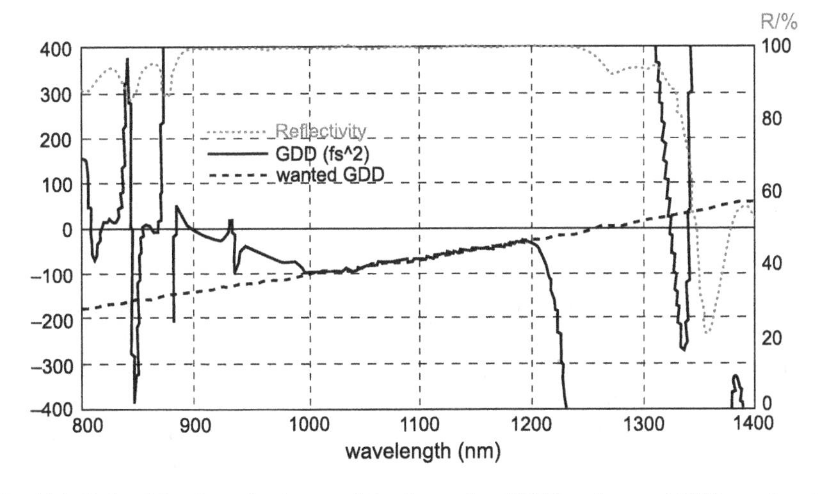

Fig. 6.34 Reflectivity R, realized group delay despersion (GDD) and wanted GDD as a function of λ for a chirped mirror [693b]

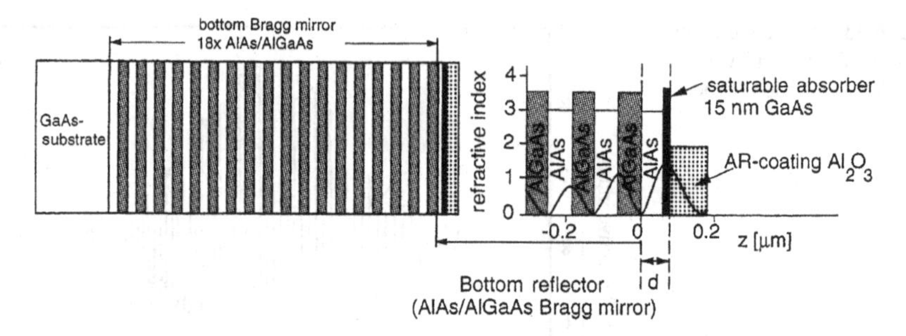

Fig. 6.35 Semiconductor Bragg mirror with a 15 nm saturable absorber layer of GaAs [694]

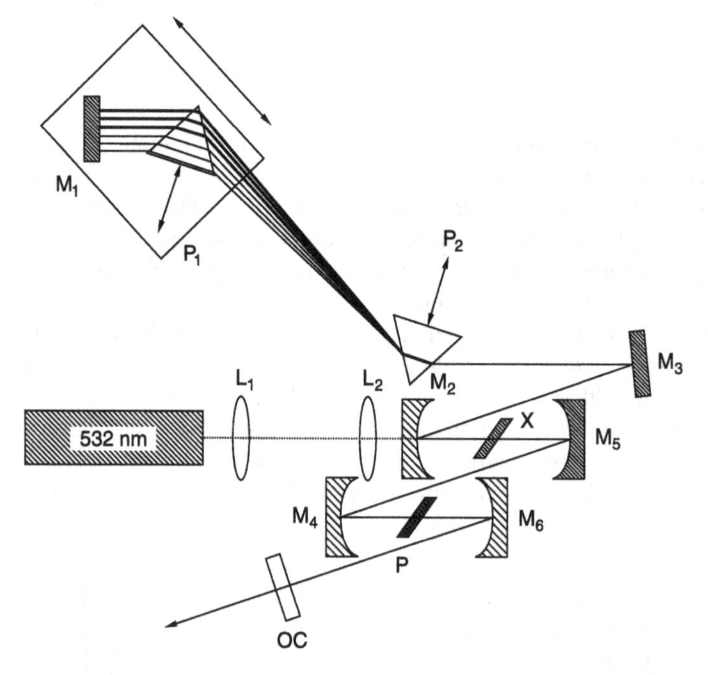

Fig. 6.36 Setup of the double-Z cavity for an ultrashort pulse Ti:sapphire laser. The two prisms P_1 and P_2 and eight bounces on double-chirped mirrors M_2–M_6 provide flat dispersion. A second focus in a BK7-plate (P) leads to enhanced SPM, and the laser generates significantly wider spectra [695]

conductor material is given by the relaxation of the excited electrons into the initial state in the valence band, these materials represent fast saturable absorbers in the sub-picosecond range, but do not reach the 10 fs limit. Nevertheless, soliton-like pulses down to 13 fs have been achieved.

A combination of saturable semiconductor media in front of a chirped mirror and KLM techniques can be used for reliable operation of sub-10 fs pulses.

A possible setup for generating the shortest pulses without pulse compression is shown in Fig. 6.36. It consists of five chirped mirrors M_2–M_6 and two prisms P_1, P_2.

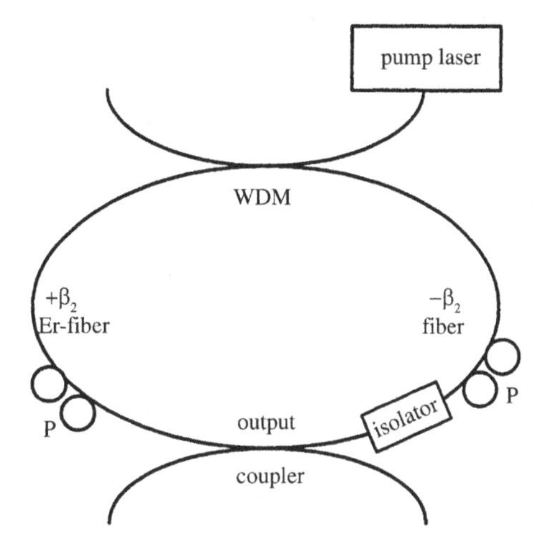

Fig. 6.37 Basic principle of fiber ring laser with positive and negative dispersion parts of the fiber. The pump laser is coupled into the fiber ring by a wavelength-division multiplexed coupler (WDM)

This setup provides a flat dispersion by the mirrors and dispersion compensation by the prisms. The curved mirrors focus the pump beam into the laser crystal X. The glass plate P in the second focus causes self-phase modulation which results in a significantly wider spectrum of the laser emission and thus a shorter pulse [695].

6.1.8 Fiber Lasers

The production of optical fibres which are doped with rear earth atoms has considerably pushed the development of fibre lasers. The large length of the amplifying medium allows lasing at low pump energies and the broad spectral gain profile enables the realization of widely tunable lasers and the generation of very short pulses. The advantages of such fibre lasers are the compact design with integrated optical components, their reliability and the inherent alignment which makes their daily use easy.

The principle design of fibre lasers is shown in Fig. 6.37. The pump laser is coupled into the fibre ring through an optical fibre. The wavelength-dependent coupling is realized by contact between the kernels of the two fibres where most of the cladding has been removed at the contact point. The fibre ring consists alternately of parts with negative and positive dispersion, which can be realized by a corresponding concentration of impurity atoms, such as erbium ions. The unidirectional operation of the ring laser is enforced by an optical diode which has high losses for the unwanted direction. Fibre loops P within the ring ensure a selected polarization of the laser light. Turning of these loops turns the plane of polarization into any wanted direction.

Instead of the ring fibre laser also a linear fibre laser such as the design shown in Fig. 6.38, can be realized. The light amplification takes place in the erbium doped fibre ring. The saturable absorber in front of the end mirror causes passive mode

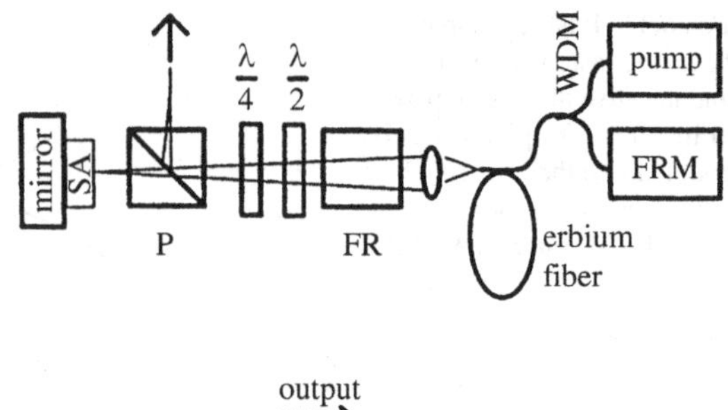

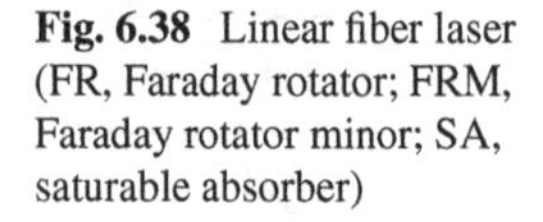
Fig. 6.38 Linear fiber laser (FR, Faraday rotator; FRM, Faraday rotator minor; SA, saturable absorber)

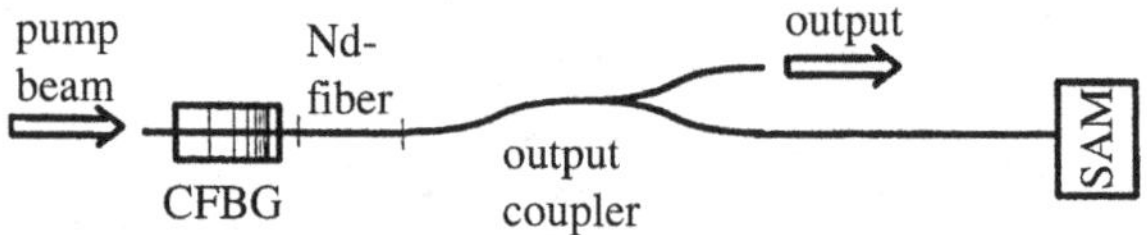

Fig. 6.39 Fully integrated passively mode-locked Nd-fiber laser (CFBG, chirped fiber Bragg grating for dispersion compensation; SAM, saturable absorber minor) [709]

locking and produces short pulses with a width of a few femtoseconds. The outcoupling of these pulses be either realized through the polarizing beam splitter P or by a fibre switch where the two fibres have close contact.

While in Fig. 6.38 the optical components are partly outside the fibre. A completely integrated design is shown in Fig. 6.39 where the saturable absorber in integrated into the fibre end. The resonator end mirror is also an integrated Bragg reflector consisting of alternating layers with high and low refractive index.

With such fibre lasers [698] pulses below 100 fs can be realized as a matter of routine, and with special designs also below 10 fs [699]. With cw fibre lasers pumped by diode lasers output powers in the range of several kilowatts are achieved [700, 701].

6.1.9 Soliton Lasers

In the Sect. 6.1.7 we discussed the self-phase modulation of an optical pulse in a fiber because of the intensity-dependent refractive index $n = n_0 + n_2 I(t)$. While the resultant spectral broadening of the pulse leads in a medium with normal negative dispersion ($\mathrm{d}n_0/\mathrm{d}\lambda < 0$) to a spatial broadening of the pulse, an anomalous linear dispersion ($\mathrm{d}n_0/\mathrm{d}\lambda > 0$) would result in a pulse compression. Such anomalous dispersion can be found in fused quartz fibers for $\lambda > 1.3\,\mu\mathrm{m}$ [696, 697]. For a suitable choice of the pulse intensity the dispersion effects caused by $n_0(\lambda)$ and by $n_2 I(t)$ may cancel, which means that the pulse propagates through the medium without changing its time profile. Such a pulse is named a fundamental soliton [702, 703].

Such a pulse propagation is illustrated in Fig. 6.40, which shows the spectral and time profiles of a pulse propagating through a medium in a linear dispersive medium without self-phase modulation (Fig. 6.40a) and in a nonlinear medium (Fig. 6.40b) with negative dispersion and positive SPM.

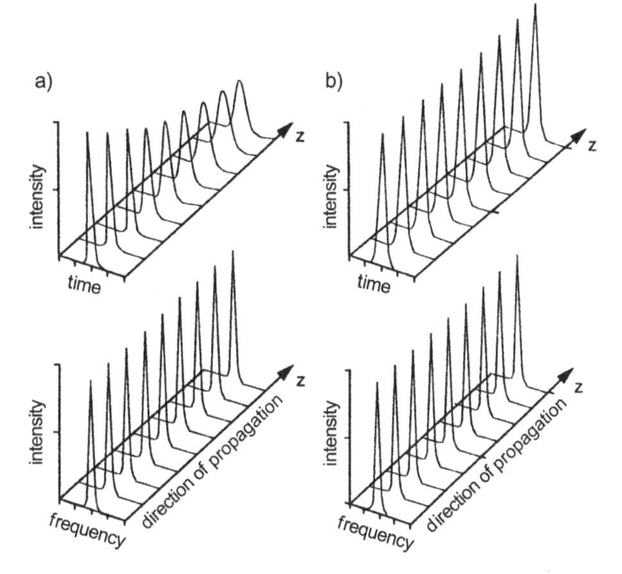

Fig. 6.40 Comparison of spectral and time profile for the propagation of short pulses: (**a**) linear propagation in a medium without self-phase modulation, (**b**) fundamental soliton for negative dispersion and positive SPM

Introducing the refractive index $n = n_0 + n_2 I$ into the wave equation (6.15) yields stable solutions that are called *solitons* of order N. While the fundamental soliton ($N = 1$) has a constant time profile $I(t)$, the higher-order solitons show an oscillatory change of their time profile $I(t)$: the pulsewidth decreases at first and then increases again. After a path length z_0, which depends on the refractive index of the fiber and on the pulse intensity, the soliton recovers its initial form $I(t)$ [704, 705].

Optical solitons in fused quartz fibers can be utilized to achieve stable femtosecond pulses in broadband infrared lasers, such as the color-center laser or the Ti:sapphire laser. Such a system is called a *soliton laser* [706–713]. Its experimental realization is shown in Fig. 6.41.

A KCl:T1° color-center laser with end mirrors M0 and M1 is synchronously pumped by a mode-locked Nd:YAG laser. The output pulses of the color-center laser at $\lambda = 1.5$ nm pass the beam splitter S. A fraction of the intensity is reflected by S and is focused into an optical fiber where the pulses propagate as solitons, because the dispersion of the fiber at 1.5 μm is $\mathrm{d}n/\mathrm{d}\lambda > 0$. The pulses are compressed, are reflected by M5, pass the fiber again, and are coupled back into the laser resonator. If the length of the fiber is adjusted properly, the transit time along the path M0–S–M5–S–M0 just equals the round-trip time $T = 2d/c$ through the laser resonator M0–M1–M0. In this case compressed pulses are always injected into the laser resonator at the proper times $t = t_0 + q2d/c$ ($q = 1, 2, \ldots$) to superimpose the pulses circulating inside the laser cavity. This injection of shortened pulses leads to a decrease of the laser pulse width until other broadening mechanisms, which increase with decreasing pulse width, compensate the pulse shortening.

In order to match the phase of the reflected fiber light pulse to that of the cavity internal pulse, the two resonator path lengths have to be equal within a small

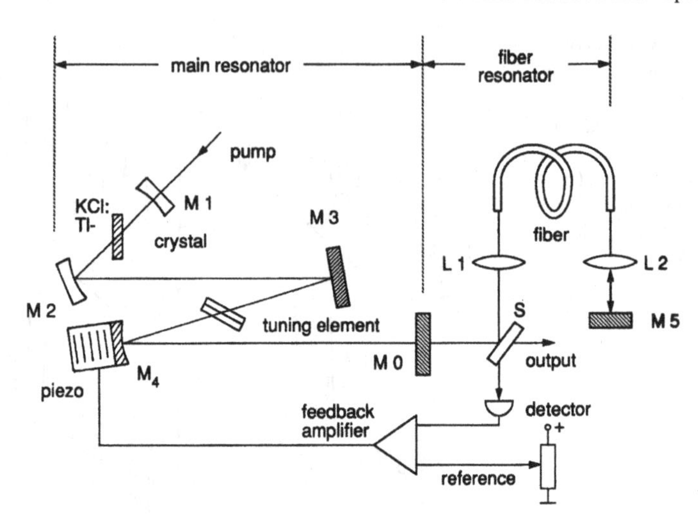

Fig. 6.41 Soliton laser [713]

fraction of a wavelength. The output power, transmitted through S to the detector, critically depends on the proper matching of both cavity lengths and can therefore be used as feedback control for stabilizing the length of the laser resonator, which is controlled by the position of M4 mounted on a piezo cylinder. It turns out that the best stabilization can be achieved with solitons of order $N \geq 2$ [707].

With such a KCl:T1° color-center soliton laser, stable operation with pulse widths of 19 fs was demonstrated [712]. This corresponds at $\lambda = 1.5$ μm to only four oscillation periods of the infrared wave. More about soliton lasers can be found in [706–714].

The fabrication of rare-earth-doped optical fibers with a wide bandwidth gain have pushed the development of optical fiber amplifiers. This large bandwidth together with the low pump power requirements facilitated the realization of passively mode-locked femtosecond fiber lasers. The advantages of such fiber lasers are their compact setup with highly integrated optical components, their reliability, and their pre-alignment, which makes their daily operation more convenient [709].

The basic principle of a fiber ring laser is schematically shown in Figs. 6.37–6.39. The pump laser is coupled into the fiber ring laser through a fiber splice and the output power is extracted through a second splice. The fiber ring consists of negative $(-\beta_2)$ and positive $(+\beta_2)$ dispersion parts with an erbium amplifying medium where the dispersion can be controlled by small concentrations of dopants. The isolator enforces unidirectional operation and fiber loops preserve the polarization state. Instead of ring fiber lasers, linear fiber lasers have also been realized, as shown in Fig. 6.38. The saturable absorber in front of the end mirror allows passive mode locking and results in femtosecond operation. An example of a fully integrated fiber femtosecond laser is given in Fig. 6.39, where the saturable absorber is butted directly to a fiber end, and a chirped fiber Bragg grating (CFBG) is used for dispersion compensation.

Output pulses with about 70 μJ pulse energy and pulse widths below 100 fs have been generated with such fiber lasers [710].

Soliton ring fiber lasers can be also realized with active mode locking by polarization modulation [711], or by additive-pulse mode locking (APM). In the latter technique the pulse is split into the two arms of an interferometer and the coherent superposition of the self-phase modulated pulses results in pulse shortening [681].

The tuning range of an erbium soliton laser could be greatly enlarged by Raman shifting [715]. Such a system, which delivers 170 fs bandwidth-limited pulses with pulse energies of 24 pJ, could be tuned from 1000 nm to 1070 nm. After amplification in an erbium fiber amplifier, output pulses with 74 fs width and 8 nJ could be obtained. Frequency-doubling in a periodically poled lithium-niobate crystal shifts the infrared soliton laser wavelength into the visible range [716].

6.1.10 Wavelength-Tunable Ultrashort Pulses

Up to now we have only discussed femtosecond lasers that can deliver short pulses with a fixed wavelength. For time-resolved spectroscopy the tunability of the wavelength provides many advantages, and so much effort has been expended in the development of sources of ultrashort pulses with a wide tuning range. There are several ways to realize them.

Similar to cw lasers, active media with a broad spectral gain profile can be used for pulsed lasers. With wavelength-selective optical elements inside the laser resonator, the laser wavelength can be tuned across the whole gain profile. However, the drawback is the widening of the pulse length ΔT with decreasing spectral width $\Delta \nu$ due to the principal Fourier limit $\Delta T > 2\pi/\Delta\nu$.

Therefore, a new principle was developed for the generation of ultrashort tunable pulses which is based on parametric oscillators and amplifiers (see Vol. 1, Sect. 5.8.8).

In the parametric process (Fig. 6.42a), a pump photon $\hbar\omega_p$ is split in a nonlinear optical crystal into a signal photon $\hbar\omega_s$ and an idler photon $\hbar\omega_i$, such that energy and momentum are conserved:

$$\omega_p = \omega_s + \omega_i, \tag{6.23a}$$

$$\boldsymbol{k}_p = \boldsymbol{k}_s + \boldsymbol{k}_i. \tag{6.23b}$$

In the parametric amplifier two laser beams are focused into the nonlinear optical crystal: the seed laser beam with frequency ω_s and a strong pump beam with ω_p. Due to the parametric process in the nonlinear crystal, a new wave is generated with a difference frequency $\omega_i = \omega_p - \omega_s$ (Fig. 6.42b). Most of the pump photons split up into a signal and an idler photon where the signal photons superimpose the seed photons and amplify the weak seed beam.

There are important differences to lasers, where the amplification occurs in media with inverted populations, and the gain depends on the time interval in which the

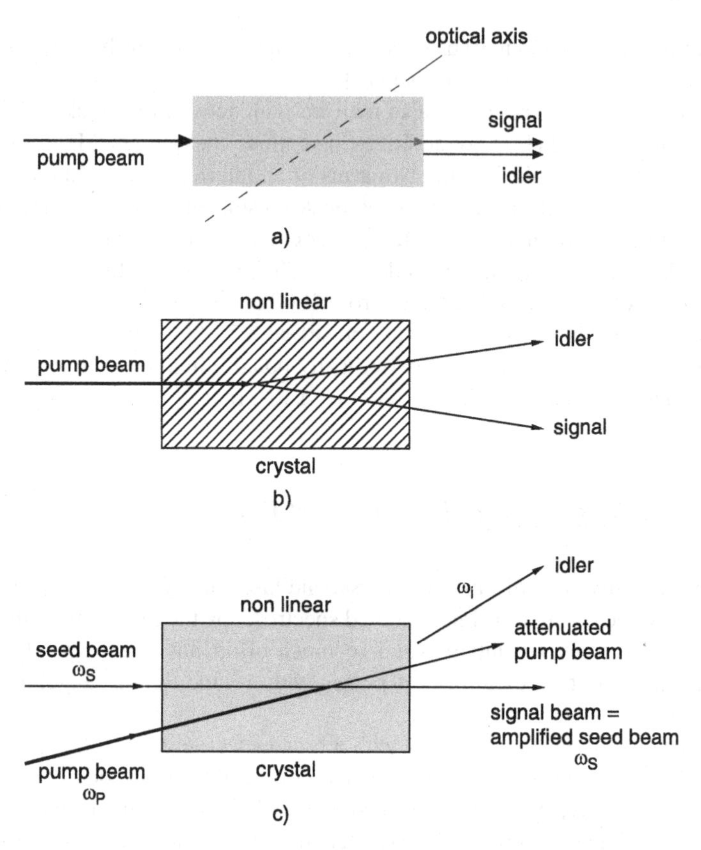

Fig. 6.42 Schematic diagram of a parametric process (**a**) collinear optical parametric amplifier, (**b**) parametric process in a nonlinear optical crystal, (**c**) noncollinear optical parametric amplifier NOPA

inversion can be maintained. For conventional laser pulse amplification, the width of the amplified pulse therefore depends on the width of the pump pulse (see Fig. 6.1), although the time profile of the amplified pulse does not necessarily follow that of the pump pulse. For parametric amplification, on the other hand, there is no population inversion in the gain medium. The amplification is due to the nonlinear interaction between pump and seed pulses, and the time profile of the amplified seed pulse is the convolution of the seed pulse and pump pulse profiles.

In order to ensure that this superposition is always in phase, the *phase velocities* of the interacting waves have to be matched. This corresponds exactly to the phase-matching condition for the frequency doubling in nonlinear crystals and is equivalent to momentum conservation (6.23b) for the interacting photons. This condition can be realized in birefringent crystals. However, there is an additional condition: if the three short pulses are to have maximum overlap during their passage through the crystal and therefore an optimum interaction, their *group velocities* must also be matched. This cannot generally be achieved for the collinear propagation of seed

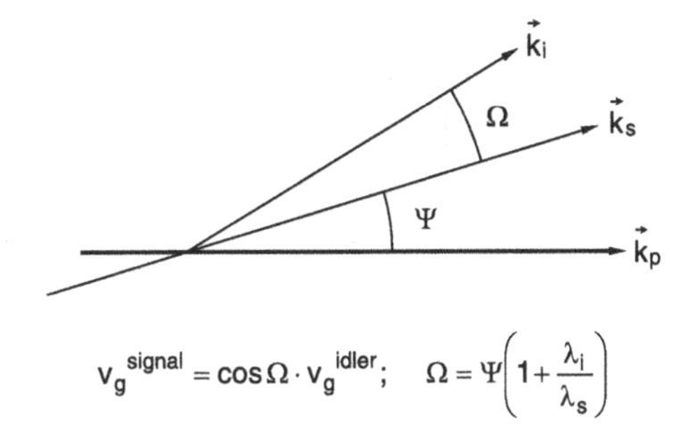

Fig. 6.43 Group-velocity matching in a noncollinear optical parametric amplifier (NOPA)

and pump beams in birefringent crystals. With a group velocity mismatch, the amplified signal pulse becomes broader and the amplification factor decreases from its optimum value. The noncollinear arrangement shown in Fig. 6.43, called NOPA (noncollinear optical parametric amplifier), solves this problem. With the angle ψ between the pump and seed beams, momentum conservation (6.23b) gives for the angle Ω between signal and idler beams the relation

$$\Omega = \psi(1 + \lambda_i/\lambda_s). \tag{6.24}$$

For a properly chosen value of ψ, the group velocities

$$v_g^{idler} = \cos\Omega \cdot v_g^{signal} = \cos(\psi + \psi\lambda_i/\lambda_s)v_g^{signal} \tag{6.25}$$

of idler and signal waves in the selected direction become equal. The phase-matching condition for the phase velocities can be fulfilled by choosing the proper angle with respect to the optical axis of the nonlinear crystal.

Wavelength tuning over a larger spectral range can be achieved in the following way (Fig. 6.44).

A small part of the pump beam at λ_p is focused into a CaF_2 plate, where it generates a spectral continuum in the focus. Most of the pump beam is frequency doubled in a nonlinear crystal. The continuum emission from this small bright spot in the CaF_2 plate is collected by a parabolic mirror and imaged into a nonlinear BBO crystal, where it represents the seed beam which overlaps with the frequency-doubled beam of the pump laser at $\lambda_p/2$. The angle between the optical axis and the seed beam allows phase-matching for sum or difference frequency generation only for a specific wavelength of the seed beam. For this orientation of the crystal, the phase and group velocities of the interacting waves must be equal. Changing this angle allows continuous tuning of the output wavelength [716]. For a pump wavelength $\lambda_p/2 = 387$ nm and a seed continuum between 500 nm and 800 nm, the output wavelength can be tuned across the whole visible range [718].

Such a widely tunable NOPA consists essentially of three parts (Fig. 6.45):

(a) The generation of a spectral continuum for the seed beam.
(b) The parametric amplifier, which is the nonlinear optical crystal where pump beam and seed beam overlap and the amplified output beam is generated at the sum or difference frequency due to parametric interaction.

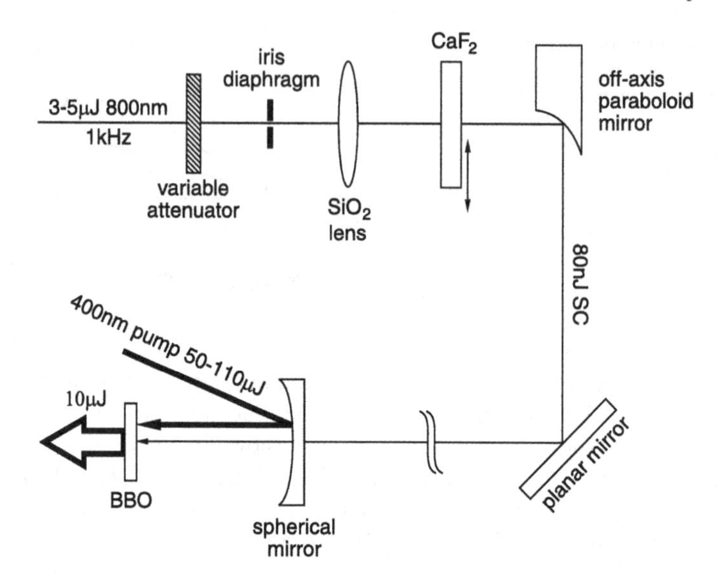

Fig. 6.44 Tunable optical parametric amplifier with a white light source and a blue pump laser [716]

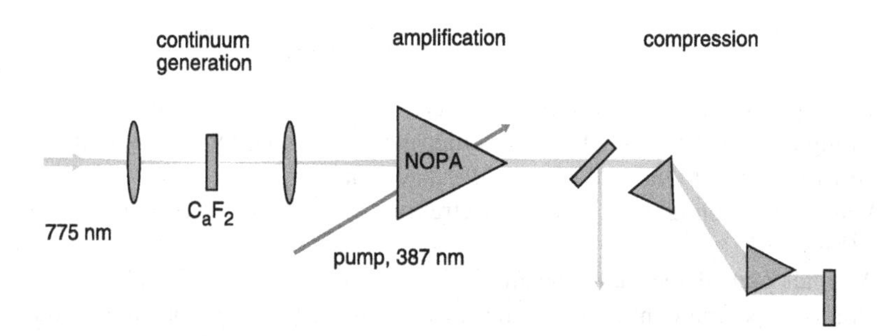

Fig. 6.45 Principle design of a NOPA [724]

(c) The pulse compressor, which may consist of a prism or grating arrangement (see the next section).

With such a system, pulses with widths of less than 20 fs could be generated with wavelengths that are continuously tunable in the visible between 470 nm and 750 nm and in the near infrared between 865 nm and 1600 nm [719].

The real experimental setup is somewhat more complicated, because of the many lenses, mirrors, beam-splitters and apertures that improve the quality of the beam. This is illustrated by Fig. 6.46, which shows a NOPA built in Freiburg [720].

For many spectroscopic experiments, more than one laser is necessary [721, 722]. Here a design which offers three independently tunable phase-coherent sources is very helpful [723]. Its principle is illustrated in Fig. 6.47.

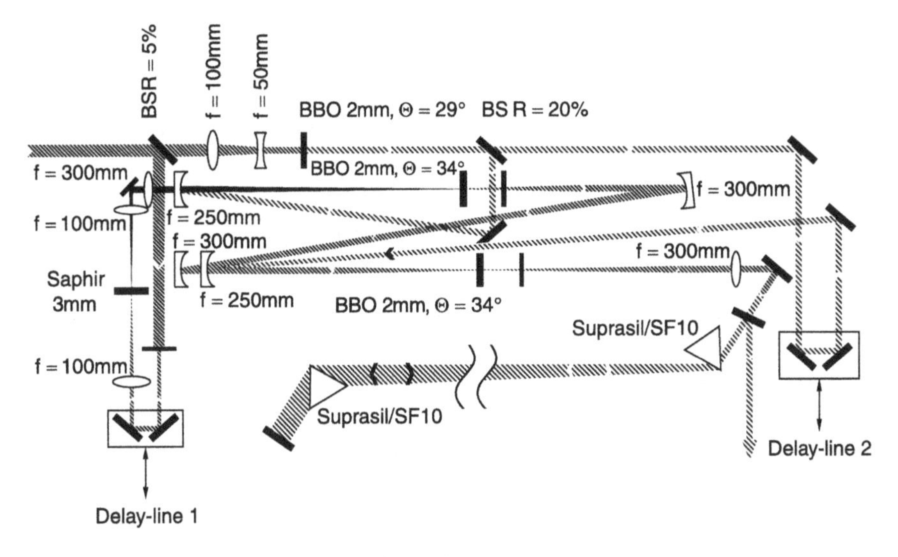

Fig. 6.46 More detailed drawing of a NOPA [720]

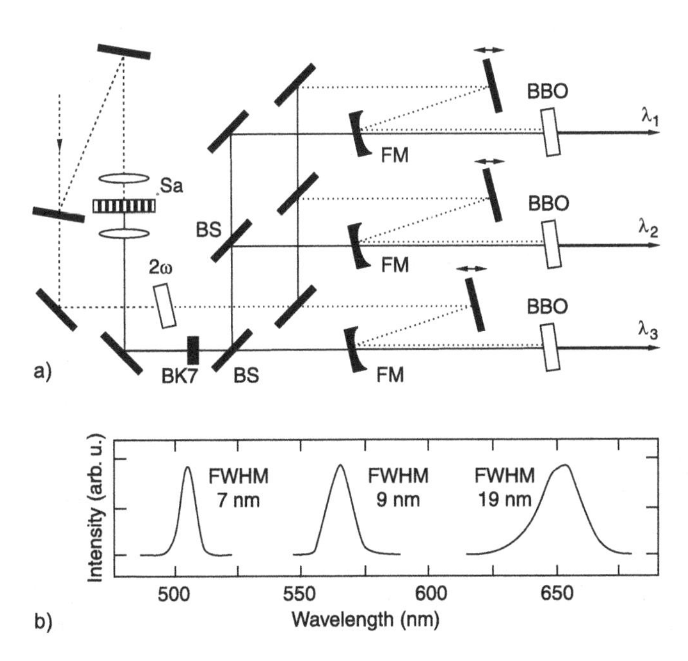

Fig. 6.47 (**a**) Three phase-locked NOPAs: *dashed*, 775 nm pump beam; *dotted*, 387 nm pump beam; *solid*, continuum; Sa: sapphire; BK7: glass substrate; BS: broadband beam. (**b**) The output pulses of the three NOPAs [723]

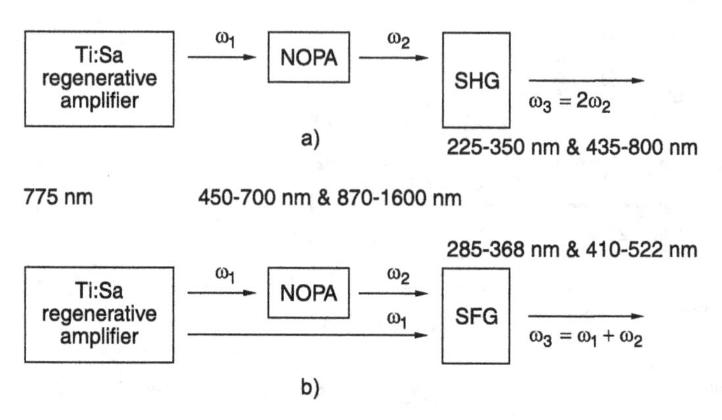

Fig. 6.48 Wavelength-ranges available with NOPAs pumped by a Ti:sapphire regenerative amplifier based on second harmonic frequency generation SHG of the NOPA output (**a**) and sum-frequency generation SFG (**b**) [716]

A small fraction of the pump beam generates a broad spectral continuum in a sapphire plate. The larger fraction is frequency doubled and sent by beam splitters into three BBO crystals for parametric amplification. The seed beam collimated from the focal spot in the sapphire plate is also imaged into the three BBO crystals, where they are superimposed on the frequency-doubled pump beams. The three parametric amplifiers can be independently tuned by changing the orientations of the crystals. The total tuning range is only limited by the spectral bandwidth of the seed beam continuum.

The wavelength ranges available with the second harmonic and the sum-frequency generation based on parametric device schemes are illustrated by the schematic design in Fig. 6.48 [724].

6.1.11 Shaping of Ultrashort Light Pulses

For many applications a specific time profile of short laser pulses is desired. One example is the coherent control of chemical reactions (see Sect. 10.2). Recently, some techniques have been developed that allow such pulse shaping and that work as follows: The output pulses of a femtosecond laser are reflected by an optical diffraction grating. Because of the large spectral bandwidth of the pulse, the different wavelengths are diffracted into different directions (Fig. 6.49). The divergent beam is made parallel by a lens and is sent through a liquid crystal display (LCD), which is formed by a two-dimensional thin array of liquid crystal pixels with transparent electrodes. The different wavelengths are now spatially separated. If a voltage is applied to a pixel, the liquid crystal changes its refractive index and with it the phase of the transmitted partial wave. Therefore the phase front of the transmitted pulse differs from that of the incident pulse, depending on the individual voltages applied to the different pixels. A second lens recombines the dispersed partial waves, and

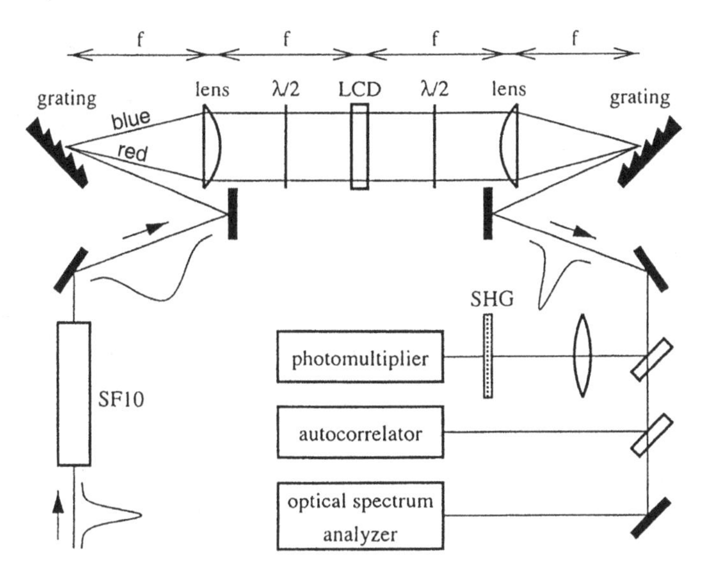

Fig. 6.49 Schematic experimental setup for pulse shaping of femtosecond pulses [727]

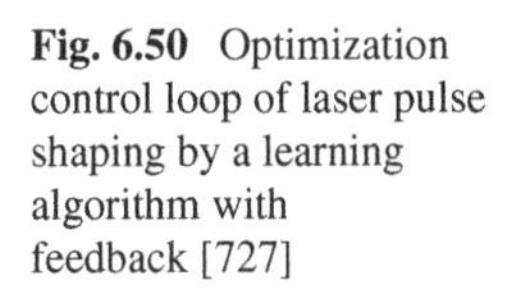

Fig. 6.50 Optimization control loop of laser pulse shaping by a learning algorithm with feedback [727]

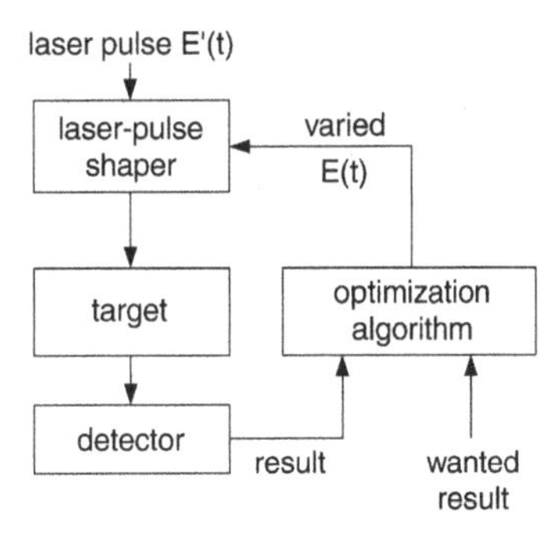

reflection by a second grating again overlaps the different wavelengths. This results in a light pulse with a time profile that depends on the phase differences between its spectral components, which in turn can be controlled by the LCD, driven by a special computer program (Fig. 6.50) [725, 726]. A self-learning algorithm can be incorporated into the closed loop, which compares the output pulse form with the wanted one and tries to vary the voltage at the different pixels in such a way that the wanted pulse form is approximated [727]. More details can be found in [728].

6.1.12 Generation of High-Power Ultrashort Pulses

The peak powers of ultrashort light pulses, which are generated by the techniques discussed in the previous section, are for many applications not high enough. Examples where higher powers are required are nonlinear optics and the generation of

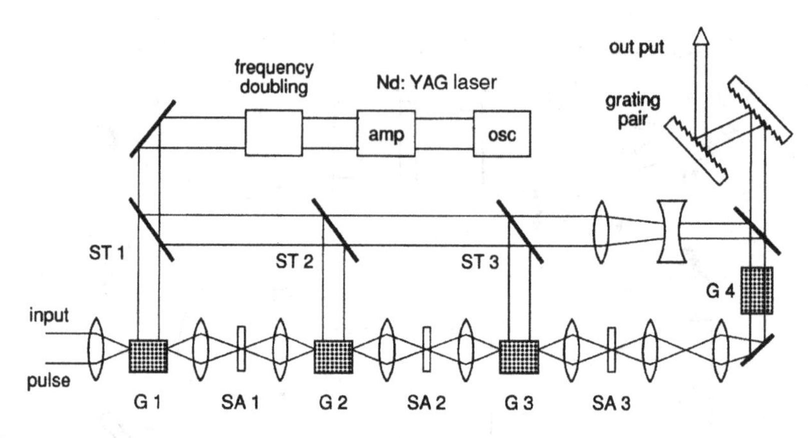

Fig. 6.51 Amplification of ultrashort light pulses through a chain of amplifier dye cells G1–G4, pumped by a frequency-doubled pulsed Nd:YAG laser. Saturable absorbers SA1–SA3 are placed between the amplifier cells in order to prevent feedback by reflection and to suppress amplified spontaneous emission

VUV ultrashort pulses, multiphoton ionization and excitation of multiply charged ions, the generation of high-temperature plasmas for optical pumping of X-ray lasers, or industrial applications in short-time material processing. Therefore, methods must be developed that increase the energy and peak power of ultrashort pulses. One solution of this problem is the amplification of the pulses in dye cells that are pumped by pulsed, powerful lasers, such as excimer, Nd:YAG, or Nd:glass lasers (Fig. 6.51). The dispersion of the amplifying cells leads to a broadening of the pulse, which can, however, be compensated by pulse compression with a grating pair. In order to suppress optical feedback by reflection from cell windows and amplification of spontaneous emission, saturable absorber cells are placed between the amplifying states, which are saturated by the wanted high-power pulses but suppress the weaker fluorescence [729–733].

If a laser beam with intensity I_0 passes through an amplifier cell of length L and with a gain coefficient $-\alpha$ ($\alpha < 0$), the output intensity becomes

$$I_{\text{out}} = I_0 \mathrm{e}^{-\alpha L}. \tag{6.26}$$

With increasing intensity, saturation starts and the gain coefficient decreases to

$$\alpha(I) = \frac{\alpha_0}{1+S} = \frac{\alpha_0}{1+I/I_\text{s}} \tag{6.27}$$

where $\alpha_0 = \alpha(I = 0)$ is the small signal gain for $I \to 0$ and I_s the saturation intensity for which $S = 1$ (see Sect. 2.1). Equations (6.26) and (6.27) yield with $I_{\text{in}} = I(z = 0)$ and $I_{\text{out}} = I(z = L)$ the equation

$$\frac{\mathrm{d}I}{\mathrm{d}z} = -\alpha \cdot I = -I \cdot \frac{\alpha_0}{1+I/I_\text{s}}, \tag{6.28}$$

$$\Rightarrow \quad \frac{\mathrm{d}I}{I_s} + \frac{\mathrm{d}I}{I} = -\alpha_0 \,\mathrm{d}z. \tag{6.29}$$

Integration from $z = 0$ to $z = L$ gives

$$I_{\text{out}} = I_{\text{in}} - I_s \big(\ln(I_{\text{out}}/I_{\text{in}}) + \alpha_0 \cdot L\big). \tag{6.30}$$

The unsaturated gain ($I \ll I_s$) would be

$$G_0 = \frac{I_{\text{out}}}{I_{\text{in}}} \approx \mathrm{e}^{-\alpha_0 L}.$$

With small saturation we obtain

$$I_{\text{out}} = I_{\text{in}} + I_s(-\alpha_0 + \alpha) \cdot L \tag{6.31}$$

and the output intensity becomes:

$$I_{\text{out}} = I_{\text{in}} + I_s \cdot \ln(G_0/G). \tag{6.32}$$

The higher saturation intensity I_s is, the larger becomes the maximum output intensity. The amplified intensity therefore depends on the incident intensity I_{in}, the small signal gain G_0, the saturated gain G and the saturation intensity I_s. If the amplifying medium is completely saturated, the gain drops to $G = 1$ and I_{out} is limited to the maximum value

$$I_{\text{out}}^{\max} = I_{\text{in}}. \tag{6.33}$$

In order to achieve a larger amplification, several amplifier stages are necessary.

A serious limitation is the low repetition rate of most pump lasers used for the amplifier chain. Although the input pulse rate of the pico- or femtosecond pulses from mode-locked lasers may be many megahertz, most solid-state lasers used for pumping only allow repetition rates below 1 kHz. Copper-vapor lasers can be operated up to 20 kHz. Recently, a multi-kilohertz Ti:Al_2O_3 amplifier for high-power femtosecond pulses at $\lambda = 764$ nm has been reported [734].

Over the past ten years new concepts have been developed that have increased the peak power of short pulses by more than four orders of magnitude, reaching the terawatt (10^{12} W) or even the petawatt (10^{15} W) regime [733, 734–745]. One of these methods is based on chirped pulse amplification, which works as follows (Fig. 6.52): The femtosecond output pulse from the laser oscillator is stretched in time by a large factor, for example, 10^4. This means that a 100 fs pulse becomes 1 ns long and its peak power decreases by the same factor. This long pulse is now amplified by a factor up to 10^{10}, which increases its energy, but, because of the pulse stretching, the peak power is far smaller than for the case where the initial short pulse had been amplified accordingly. This prevents destruction of the optics by peak powers that exceed the damage threshold. Finally, the amplified pulse is again compressed before it is sent to the target.

We will now discuss the different components of this process in more detail. The oscillator consists of one of the femtosecond devices discussed previously. They

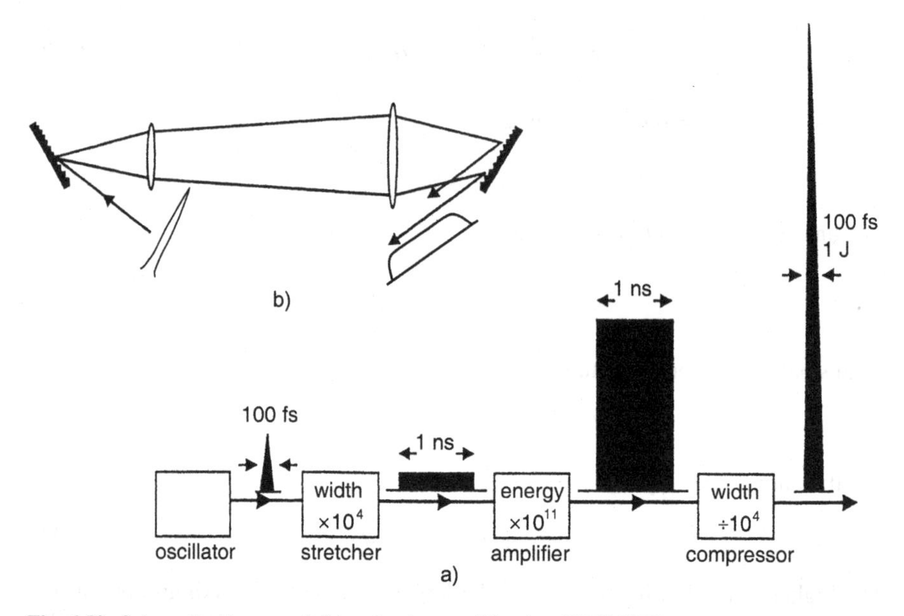

Fig. 6.52 Schematic diagram of chirped pulse amplification (CPA) [745]

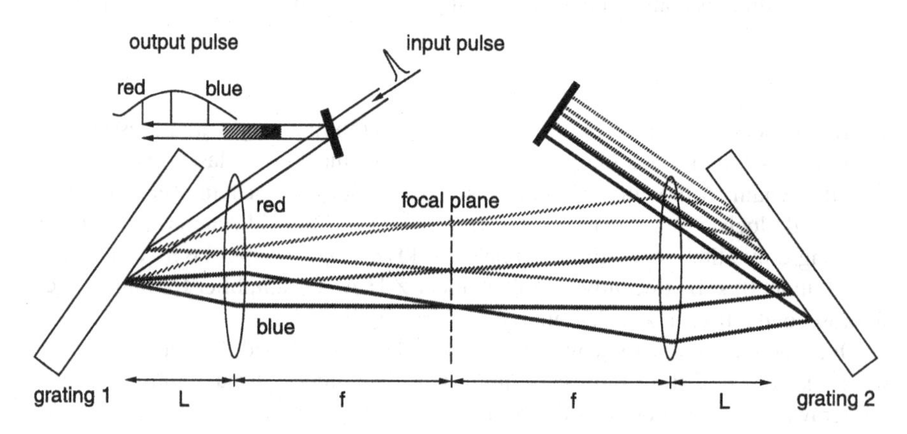

Fig. 6.53 Design for stretching femtosecond pulses and generation of a frequency chirp

deliver chirped pulses where the frequency of the light wave varies during the pulse width Δt. Such a chirped pulse can be stretched by a grating pair, where the two gratings, however, are not parallel as for pulse compression, but are tilted against each other (Fig. 6.53). This increases the path difference between the blue and the red components in the pulse and stretches the pulse length. Another design of an aberration-free pulse stretcher with two curved mirrors and a grating is described in [733] and is depicted in Fig. 6.54.

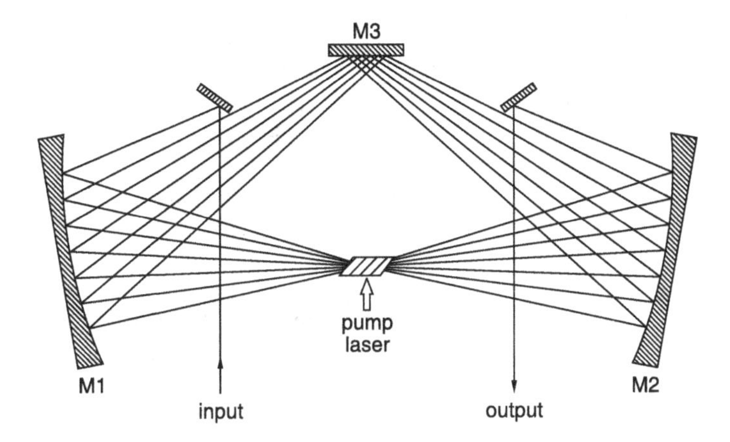

Fig. 6.54 Regenerative amplifier for short light pulses

The amplification is performed in a multipass amplifier system. Here the stretched laser pulse is sent many times through a gain medium, which is pumped by a nanosecond pump pulse from a Nd:YAG laser (regenerative amplifier, Fig. 6.54). For amplification of Ti:sapphire pulses a highly doped Ti:sapphire crystal serves as the gain medium. After each transit of the stretched pulse through the gain medium, which depletes the inversion, the pump pulse regenerates it again. Often the system is designed in such a way that the different transits pass at slightly different locations through the amplifying medium. The number of transits depends on the geometrical mirror arrangement and is limited by the duration of the pump pulse.

A second amplifier stage can be used for higher amplification. The different stages are separated by an optical diode, which prevents back reflections into the previous stage. A Pockels cell (PC) selects the amplified pulses with a repetition rate of a few kilohertz (limited by the pump laser) from the many more nonamplified pulses of the oscillator (at a repetition rate of about 80 MHz). The upper pulse energy that can be extracted from the amplifying medium is given by the saturation fluence, which depends on the emission cross section of the medium. For Ti:sapphire, for example, the highest achievable intensity is about 100 TW/cm^2. Figure 6.55 shows the total setup for pulse stretching, amplification and compression.

During the multipass transits the spatial mode quality of the pulse might be deteriorated, which means that the pulse cannot be tightly focused into the target. If the multipass design forms a true resonant cavity and the incoming pulses of the pump laser and the oscillator are both carefully mode-matched to the fundamental Gaussian mode of this cavity, the resonator will only support the TEM_{00} modes and the system acts as a spatial filter, because all other transverse modes will not be amplified. Such a regenerative amplifier preserves the spatial Gaussian pulse profile, which allows one to achieve diffraction-limited tight focusing and correspondingly high intensities in the focal plane. After amplification the pulse with energy W is

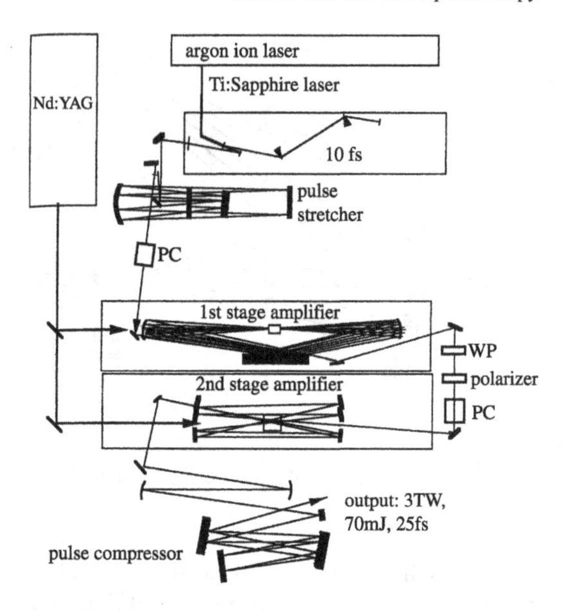

Fig. 6.55 Schematic diagram of the oscillator–amplifier system of a 3-TW 10 Hz Ti:sapphire CPA laser [733]

compressed again, thus producing pulses with a duration $\tau = 20$–100 fs and very high peak powers $P = W/\tau$ up to several terawatts.

Example 6.5 Assume the laser oscillator emits pulses with $\Delta t = 20$ fs and a pulse energy of 1 nJ. The peak power of these pulses is then 50 kW. At a repetition rate of 100 MHz the average power is 0.1 W. Stretching the pulse width by a factor of 10^4 increases the pulse width to 200 ps and reduces the peak power to 5 W. For a mirror separation of 10 cm in the regenerative amplifier, the round trip time for the pulses is $T = 0.7$ ns. If the width of the pump pulse is 7 ns, 10 roundtrips can be amplified; i.e., the amplified pulse makes 20 transits through the amplifying medium during the pump pulse. With an amplification factor of 10^4, the peak power of the amplified pulses becomes 50 kW. In a second amplification stage with $G = 10^3$, the peak power increases to 50 MW and the pulse energy to 10 mJ. Compressing the pulse again by a factor of 10^4 gives a pulse peak power of 5×10^{11} W = 0.5 terawatts.

Most of the experiments on femtosecond pulses performed up to now have used dye lasers, Ti:sapphire lasers, or color-center lasers. The spectral ranges were restricted to the regions of optimum gain of the active medium. New spectral ranges can be covered by optical mixing techniques (Vol. 1, Sect. 5.8). One example is the generation of 400 fs pulses in the mid-infrared around $\lambda = 5$ μm by mixing the output pulses from a CPM dye laser at 620 nm with pulses of 700 nm in a $LiIO_3$ crystal [737]. The pulses from the CPM laser are amplified in a six-pass dye amplifier pumped at a repetition rate of 8 kHz by a copper-vapor laser. Part of the amplified

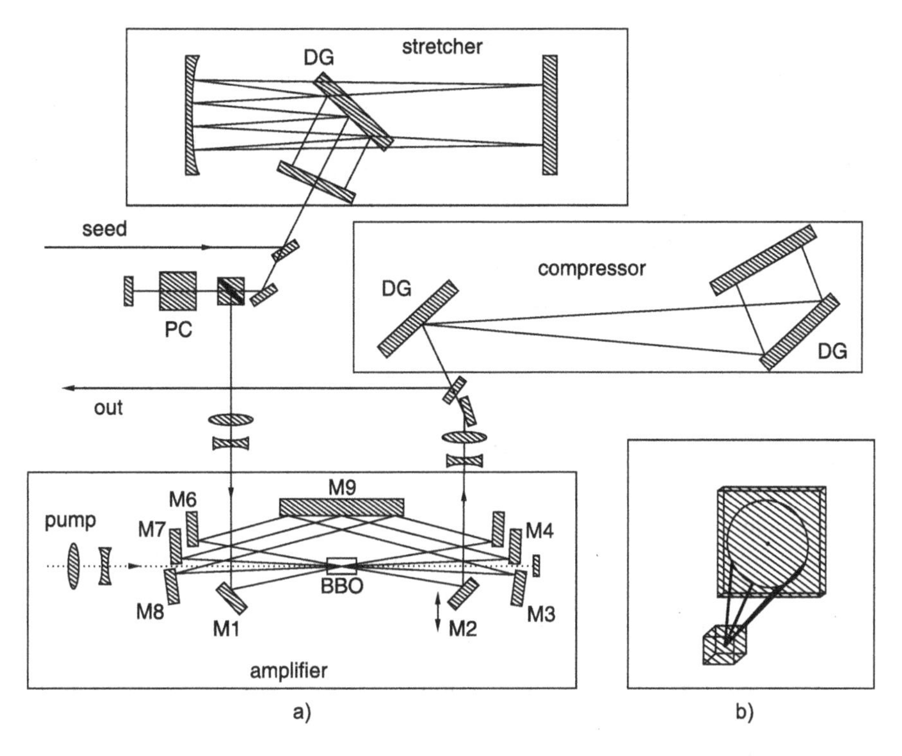

Fig. 6.56 Experimental setup for parametric amplification of ultrashort pulses, including a pulse stretcher before and a pulse compressor after amplification [739]

beam is focused into a traveling-wave dye cell [738], where an intense femtosecond pulse at $\lambda = 700$ nm is generated. Both laser beams are then focused into the nonlinear $LiIO_3$ mixing crystal, which delivers output pulses with 10 mJ energy and 400 fs pulsewidth.

A second method is based on parametric amplification. Instead of a gain medium with population inversion, here the parametric interaction between a pump wave and a signal wave in a nonlinear optical crystal is used to amplify optical pulses (see Sect. 6.1.10). The advantage of using this technique is the higher gain that results. It is possible to reach a higher amplification after just a single amplifier stage than achieved in a regenerative amplifier, where the gain is limited by saturation effects.

The nonlinear medium, however, cannot store energy as is done in the usual gain media with an inverted population. The pump, signal and idler waves must obey the phase-matching condition, but they also should overlap in time during their passage through the interaction zone in the nonlinear medium. Therefore they should have the same pulse width. A picosecond pump laser can be used for the amplification of femtosecond pulses. The femtosecond signal pulse is stretched before the parametric amplification in order to match the width of the pump pulse (Fig. 6.56), and it is then compressed after the amplification stage [739].

Another method uses nanosecond pump pulses. The femtosecond signal pulse is again stretched into the subnanosecond range and sent through several amplification stages. The pump pulses for these stages are delayed for each stage by a specific time interval to achieve optimum overlap with the signal pulse [740]. Instead of several nonlinear crystals, the signal pulses can be sent through the same crystal several times in an arrangement similar to that of a regenerative amplifier. However, the difference here is that parametric interaction is used instead of the gain medium with population inversion. Since the amplification is higher, fewer amplification stages are required, and it was shown that just a single stage of such a "regenerative parametric stage" can yield peak powers in the terawatt range [741].

6.1.13 Reaching the Attosecond Range

There are a lot of processes in nature that occur at timescales below 1 fs. Examples include the movement of electrons in atoms or molecules, the excitation of electrons from inner shells of atoms, and the subsequent decay of electrons from higher energy states into this hole in the core which results in X-ray emission or Auger processes. This decay may proceed within 10^{-16} s. Many more electronic processes proceed on an attosecond timescale, such as the rearrangement of the electron shell after optical excitation. Even shorter decay times can be found for excited atomic nuclei. If such processes need to be investigated with high time resolution, the probe must be faster than the investigated process [742]. Therefore, new techniques have been developed over the past few years that allow time resolutions in the attosecond range (1 attosecond $= 10^{-18}$ s) [743, 744].

One of these techniques is based on the generation of high harmonics of a powerful femtosecond laser pulse [746].

The minimum time duration of a pulse with the carrier frequency $\nu = c/\lambda$ is

$$\Delta T = \lambda/c = 1/\nu.$$

For $\lambda = 700$ nm this gives $\Delta T = 2.3 \times 10^{-15}$ s $= 2.3$ fs. In order to realize pulses in the attosecond range (1 as $= 10^{-18}$ s) one has to choose shorter wavelengths. This can be achieved by generation of high harmonics. For the 50th harmonics at $\lambda = 14$ nm one obtains already a lower limit for the pulse width of $\Delta T = 46$ as.

If the output pulse of a powerful femtosecond laser is focused into a gas jet of noble gas atoms, high harmonics of the fundamental laser wavelength are generated due to the nonlinear interaction of the laser field with the electrons of the atoms. The strong laser field, which can surpass the Coulomb field within the atom by several orders of magnitude, leads to extreme enharmonic movement of the electrons, which are accelerated back and forth with the optical frequency of the laser field. Since accelerated charges radiate, the periodically changing acceleration of the electron results in emission at frequencies $\omega_n = n\omega$ where the integer n can have values of up to 350! For a fundamental wavelength $\lambda = 700$ nm, the har-

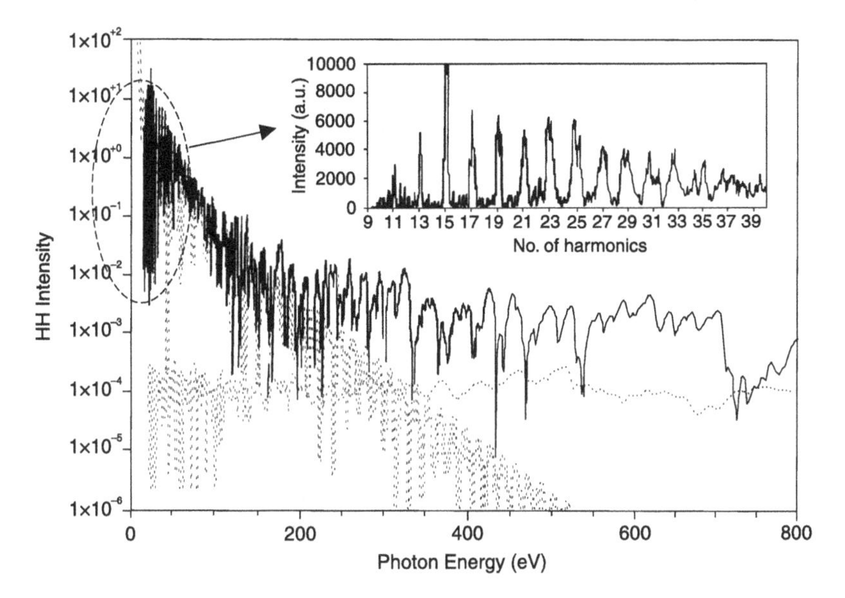

Fig. 6.57 Spectral intensity distribution of the higher harmonics produced by focussing a femtosecond terawatt laser at $\lambda = 720$ nm into a neon jet [749]

monic $n\omega$ with $n = 350$ has a wavelength $\lambda = 2$ nm in the X-ray region. This corresponds to a photon energy of 500 eV. The intensity distribution of the higher harmonics, as generated by a 5 fs pulse at $\lambda = 720$ nm with an output peak power of 0.2 terawatt, is shown in Fig. 6.57 on a logarithmic scale. One can see that the harmonic intensity at $n\omega$ decreases with increasing n up to $n = 80$ (225 eV), while for higher harmonics it stays approximately constant. Because the power of the nth harmonic at $\omega_n = n\omega$ is proportional to $I(\omega)^n$, the pulse width of the harmonic is much narrower than that of the pulse at ω. Only the central part of the pulse $I(\omega)$ around the peak intensity essentially contributes to the higher harmonics generation.

Another way of producing attosecond pulses with high-energy photons is based on plasma generation when a high-intensity laser beam is focused onto the surface of a solid material. If a 5 fs pulse at $\lambda = 780$ nm with an intensity $I = 10^{21}$ W/cm^2 is focused onto a solid target, the free electrons in the resulting plasma emit attosecond pulses with photon energies of 20–70 eV [747].

To illustrate atomic dynamics, which require subfemtosecond resolution, Fig. 6.58 shows the time-resolved field ionization of neon atoms in the optical field of a 5 fs laser pulse which consists of only three optical cycles within the pulse half width [750]. The whole process proceeds within about 6 fs, but one can clearly see peaks in the ionization probability at times of maxima in the optical field, which means that the time resolution is below 1 fs.

Such femto- to attosecond X-ray pulses can be used for the generation of Laue diagrams of vibrational or electronical excited molecules. The exposure time must

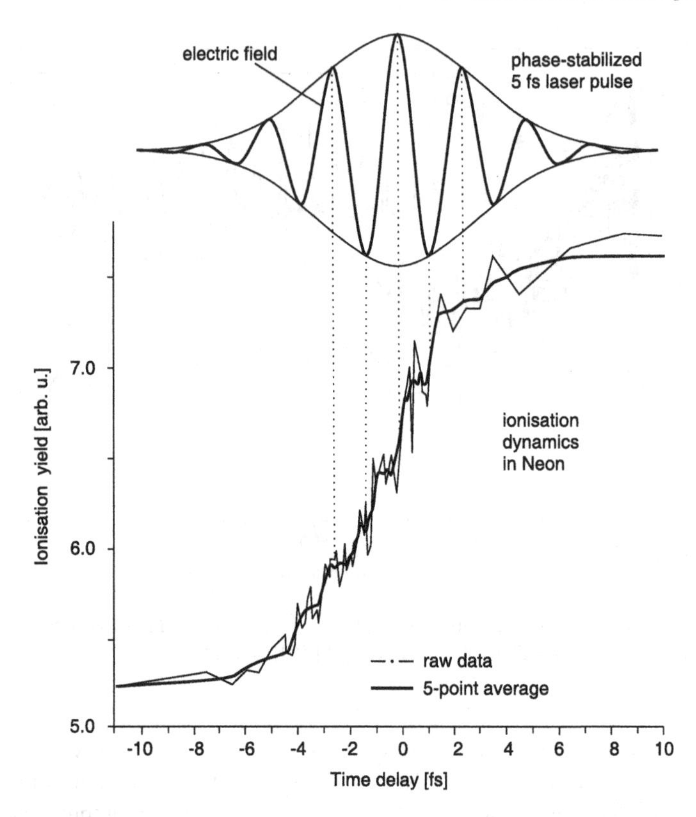

Fig. 6.58 Real-time observation of the optical field ionization of neon atoms with attosecond time resolution [748]

be short compared to the vibration period in order to elucidate the structures of molecules in specific excited states excited by optical or UV femtosecond pulses. If the time delay between the two pulses is varied, structural changes in the excited molecule during its vibrations or after the rearrangement of its electron shell following excitation can be tracked.

The optical phase of the carrier wave in a linearly polarized femtosecond pulse can be measured by the photoelectron rate (Fig. 6.59). If the electrons are produced by the nth harmonic of the visible femtosecond pulse, the rate is proportional to the $2n$th power of the visible field amplitude. The amplitude depends strongly on the phase of the optical wave relative to the envelope maximum of the pulse. Measurements of this photo-electron rate as a function of the phase shift of the field amplitude against the pulse maximum allows the determination of the phase and the pulse width of the high-harmonic attosecond pulse [754]. There are many more applications of attosecond pulses; these can be found, for example, in the publications of the groups of P. Corkum at the NRC in Ottawa [753] and F. Krausz at the MPI for Quantum Optics in Garching, who have pioneered this field [754].

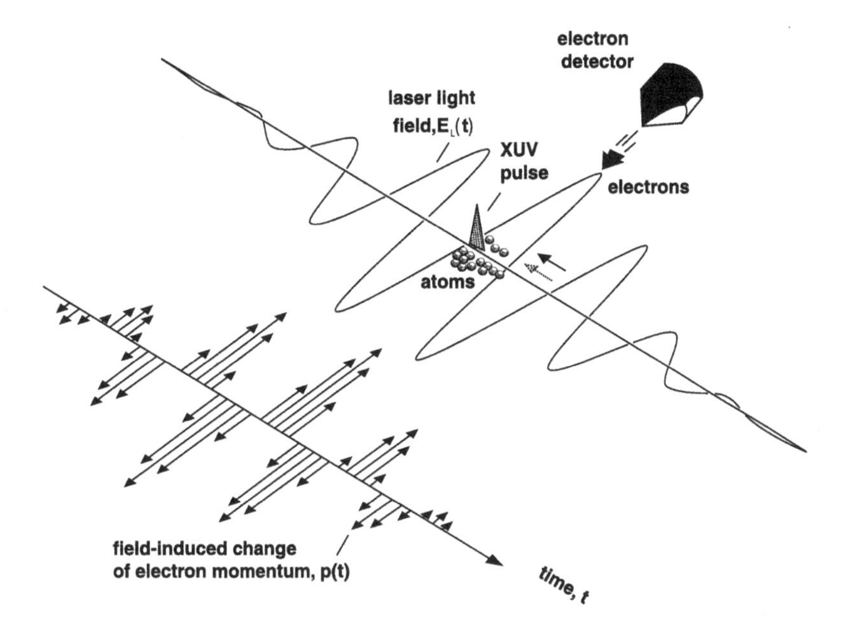

Fig. 6.59 Measuring the absolute phase of an optical field by photoelectron spectroscopy [749]

For more experimental details and special experimental setups for the generation of femtosecond and attosecond pulses, the reader is referred to the literature [749–751, 753–755].

6.1.14 Summary of Short Pulse Generation

There are several ways of generating short pulses. One is based on mode-locking lasers which have gain media with a broad spectral range. While the first experiments relied on dye lasers or Nd:YAG-lasers, the Ti:Sapphire laser has now become the most attractive choice. Some of the most commonly used materials are listed in Table 6.1 and [756]. Laser pulses down to 4 fs have been demonstrated using Kerr lens mode-locking and chirped mirrors.

Another method uses optical pulse compression in optical fibers, where the intensity-dependent part of the refractive index causes a frequency chirp and increases the spectral profile and the time duration of the pulse. Subsequent pulse compression, achieved with a pair of optical gratings or by using prisms, leads to a drastic shortening of the pulse.

A third way of generating short pulses is based on the parametric interaction of a signal pulse with a short pump pulse in a nonlinear crystal. This method allows the generation of short pulses with a wide tuning range of their center wavelength if the signal pulse comes from a point-like white light source which is generated by focusing a high-power laser pulse into a glass or CaF_2^- disk.

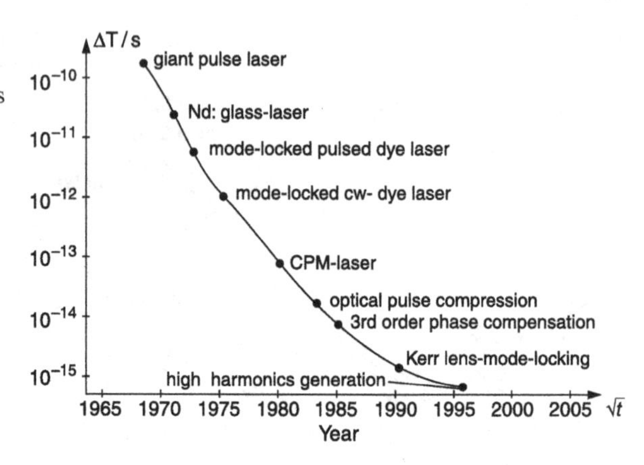

Fig. 6.60 Historical development of the progress in generating ultrashort pulses

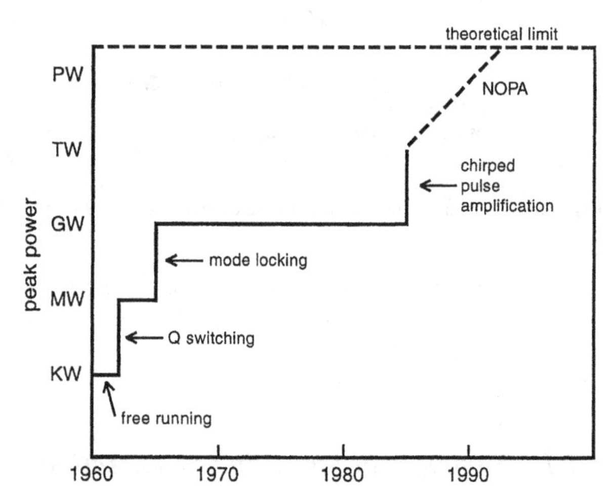

Fig. 6.61 The progress in achievable peak powers with the invention of different techniques for laser pulse generation [745]

The path to the shorter and shorter pulses and higher peak powers achieved over the last few decades is illustrated on a logarithmic scale in Figs. 6.60 and 6.61.

Note The extrapolation into the future may be erroneous [751].

A good review on femtosecond pulses can be found in [752].

6.2 Measurement of Ultrashort Pulses

During recent years the development of fast photodetectors has made impressive progress. For example, PIN photodiodes (Vol. 1, Sect. 4.5) are available with a rise time of 20 ps [760]. The easiest and cheapest way to achieve measurements of pulses with $\Delta T > 10^{-10}$ s is through detection by photodiodes, CCD detectors or photomultipliers, as discussed in Vol. 1, Sect. 4.5. However, until now the only detector

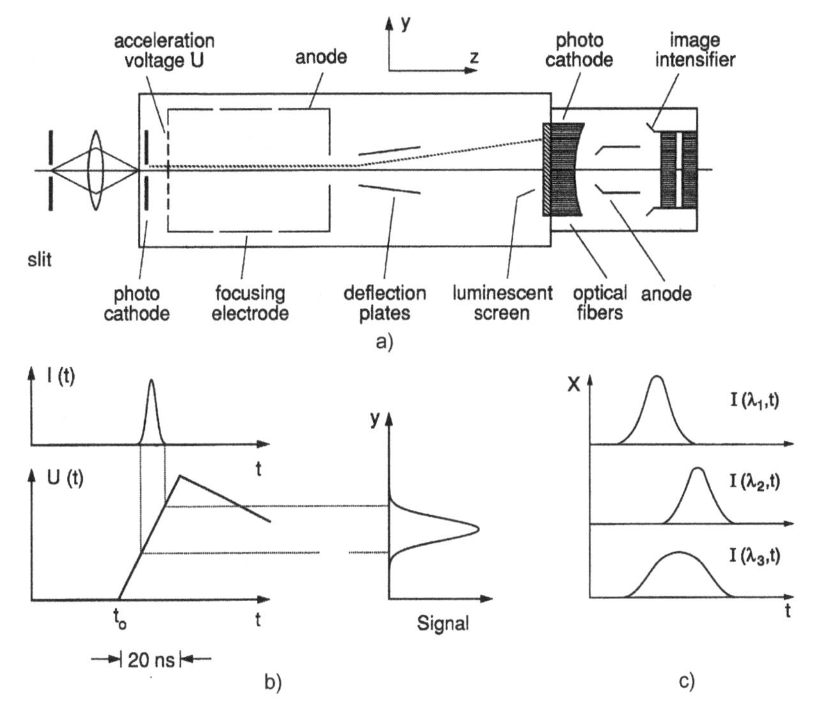

Fig. 6.62 Streak camera. (**a**) Design and (**b**) schematic diagram of the relation between the time profile $I(t)$ and the spatial distribution $S(y)$ at the output plane; (**c**) spectrally resolved time profiles $I(\lambda, t)$

that reaches a time resolution slightly below 1 ps is the streak camera [761]. Femtosecond pulses can be measured with optical correlation techniques, even if the detector itself is much slower. Since such correlation methods represent the standard technique for measuring of ultrashort pulses, we will discuss them in more detail. A short laser pulse is completely characterized by the amplitude, frequency, polarization and phase of the electric field. The more of these quantities can be measured, the better can the pulse be characterized. In this section we will introduce different techniques which are capable to measure some or even all of these characteristic features.

6.2.1 Streak Camera

The basic principle of a streak camera is schematically depicted in Fig. 6.62. The optical pulse with the time profile $I(t)$ is focused onto a photocathode, where it produces a pulse of photoelectrons $N_{\mathrm{PE}}(t) \propto I(t)$. The photoelectrons are extracted into the z-direction by a plane grid at the high voltage U. They are further accelerated and imaged onto a luminescent screen at $z = z_{\mathrm{S}}$. A pair of deflection plates can de-

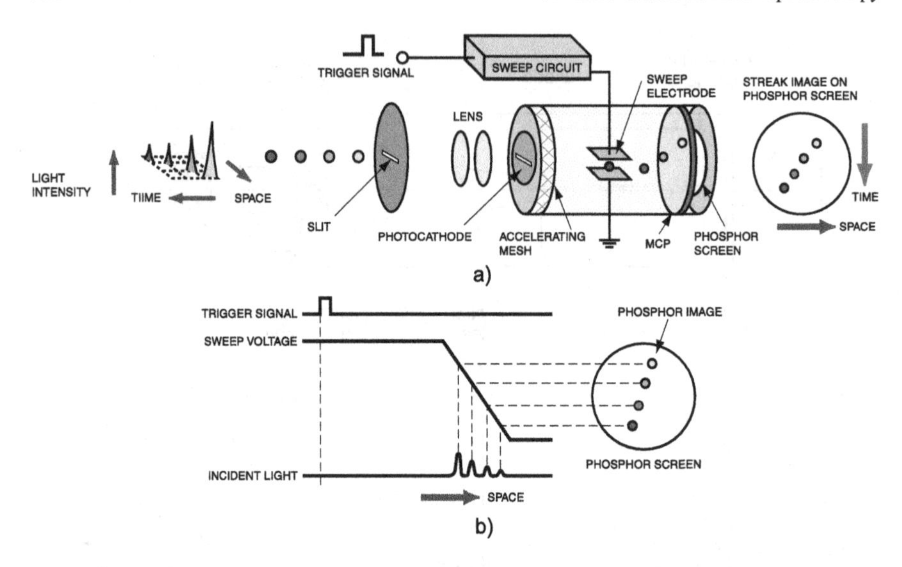

Fig. 6.63 (**a**) Operating principle of the streak tube [762]; (**b**) operation timing (at time of sweep)

flect the electrons into the y-direction. If a linear voltage ramp $U_y(t) = U_0 \cdot (t - t_0)$ is applied to the deflection plates, the focal point of the electron pulse $(y_s(t), z_s)$ at the screen in the plane $z = z_s$ depends on the time t when the electrons enter the deflecting electric field. The spatial distribution $N_{PE}(y_s)$ therefore reflects the time profile $I(t)$ of the incident light pulse (Fig. 6.62b).

When the incident light is imaged onto a slit parallel to the x-direction the electron optics transfer the slit image to the luminescent screen S. This allows the visualization of an intensity–time profile $I(x,t)$, which might depend on the x-direction. For example, if the optical pulse is at first sent through a spectrograph with a dispersion $\mathrm{d}\lambda/\mathrm{d}x$, the intensity profile $I(x,t)$ reflects for different values of x the time profiles $I(\lambda,t)$ of the different spectral components of the pulse. The distribution $N_{PE}(x_s, y_s)$ on the screen S then yields the time profiles $I(\lambda,t)$ of the spectral components (Fig. 6.62c). The screen S is generally a luminescent screen (like the oscilloscope screen), which can be viewed by a video camera either directly or after amplification through an image intensifier. Often microchannel plates are used instead of the screen.

The start time t_0 for the voltage ramp $U_y = (t - t_0)U_0$ is triggered by the optical pulse. Since the electronic device that generates the ramp has a finite start and rise time, the optical pulse must be delayed before it reaches the cathode of the streak camera. This assures that the photoelectrons pass the deflecting field during the linear part of the voltage ramp. The optical delay can be realized by an extra optical path length such as a spectrograph (Sect. 6.4.1).

In Fig. 6.63 the design and operation principle of a modern streak camera is illustrated for the example of 4 short light pulses which are monitored on the phosphor screen. The trigger signal for the spatial deflection which starts the sweep voltage is

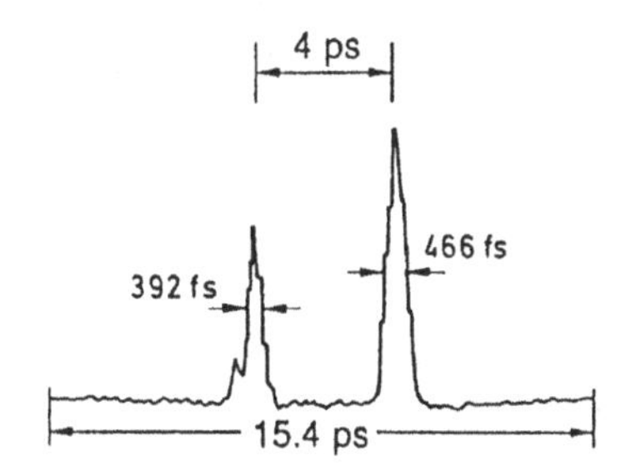

Fig. 6.64 Streak image of two subpicosecond pulses separately by 4 ps, measured with the femtosecond streak camera [761]

provided by the first optical pulse. This fast spatial deflection of the streak camera converts the time resolution into a space resolution.

In commercial streak cameras the deflection speed can be selected between 1 cm/50 ps to 1 cm/10 ns. With a spatial resolution of 0.1 nm, a time resolution of 1 ps is achieved. A femtosecond streak camera has been developed [761] that has a time resolution selectable between 200 fs to 8 ps over a spectral range 200–850 nm. Figure 6.64 illustrates this impressive resolution by showing the streak camera screen picture of two femtosecond pulses which are separated by 4 ps. Recent designs even reach a resolution of 200 fs. The spatial separation depends on the sweep voltage ramp. More details can be found in [760, 761, 763, 764].

6.2.2 Optical Correlator for Measuring Ultrashort Pulses

For measurements of optical pulse widths below 1 ps the best choice is a correlation technique that is based on the following principle: the optical pulse with the intensity profile $I(t) = c\epsilon_0|E(t)|^2$ and the halfwidth ΔT is split into two pulses $I_1(t)$ and $I_2(t)$, which travel different path lengths s_1 and s_2 before they are again superimposed (Fig. 6.65). For a path difference $\Delta s = s_1 - s_2$ the pulses are separated by the time interval $\tau = \Delta s/c$ and the coherent superposition of their ampli-

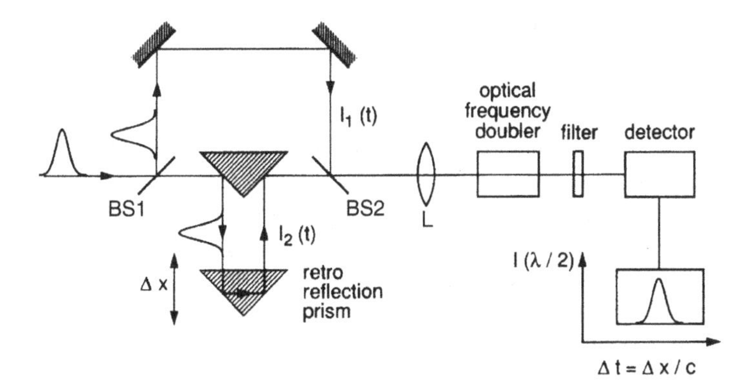

Fig. 6.65 Optical correlator with translation-retroreflecting prism and second-harmonic generation

tudes $E_i(t)$ yields the total intensity at the measuring time t

$$\begin{aligned} I(t,\tau) &= c\epsilon_0\big[E_1(t)+E_2(t-\tau)\big]^2 \\ &= I_1(t)+I_2(t-\tau)+2c\epsilon_0 E_1(t)\cdot E_2(t-\tau). \end{aligned} \tag{6.34}$$

A *linear* detector has the output signal $S_L(t)=aI(t)$. If the time constant T of the detector is large compared to the pulse length ΔT, the output signal is

$$S_L(\tau)=a\big\langle I(t,\tau)\big\rangle=\frac{a}{T}\int_{-T/2}^{+T/2} I(t,\tau)\,\mathrm{d}t. \tag{6.35}$$

For strictly monochromatic cw light ($E_0(t)=\text{const}$) and $E_1=E_2=E_0\cos\omega\omega t$, the integration yields

$$S_L(\tau)=2a\left\{\langle I_0\rangle+\frac{c\epsilon_0}{T}\int_{-T/2}^{+T/2} E_0^2\cos\omega t\cos\omega(t-\tau)\,\mathrm{d}t\right\} \tag{6.36}$$

and the signal becomes for $T\to\infty$

$$S_L(\tau)=2aI_0(1+\cos\omega\tau) \tag{6.37}$$

which represents on oscillatory function of the delay time τ with the oscillation period $\Delta\tau=\pi/\omega=\lambda/2c$ (two-beam interference, see Vol. 1, Sect. 4.2).

Mode-locked pulses of duration ΔT and spectral width $\Delta\omega\approx 1/\Delta T$ are composed of many modes with different frequencies ω. The oscillations of these modes have different periods $\Delta\tau(\omega)=\pi/\omega\ll\Delta T$, and after a time $t\geq\tau_c$ their phases differ by more than π and their amplitudes cancel. In other words, the coherence time is $\tau_c\leq\Delta T$ and interference structure can only be observed for delay times $\tau\leq\Delta T$.

The signal output of the linear detector is now

$$S_L=2a\left\{\langle I_0\rangle+\frac{c\epsilon_0}{\Delta\omega T}\int_{\omega_0-\Delta\omega/2}^{\omega_0+\Delta\omega/2}\int_{-T/2}^{+T/2} E^2(t)\cos\omega t\cos\omega(t-\tau)\,\mathrm{d}t\,\mathrm{d}\omega\right\}. \tag{6.38}$$

For $T>\Delta T=1/\Delta\omega$ the integral vanishes and the signal does not depend on τ.

A linear detector with a time constant $T\gg\tau$ would therefore give an output signal that is independent of τ and that yields no information on the time profile $I(t)$! This is obvious, because the detector measures only the integral over $I_1(t)+I_2(t+\tau)$; that is, the sum of the energies of the two pulses, which is independent of the delay time τ as long as $T>\tau$. Therefore *linear detectors* with time resolution T cannot be used for the measurement of time profiles of ultrashort pulses with $\Delta T<T$.

If, however, the two noncollinear pulses are focused into a nonlinear optical crystal which doubles the optical frequency, the intensity of the second harmonic

$I(2\omega) \propto (I_1 + I_2)^2$ is proportional to the *square* of the incident intensity (Vol. 1, Sect. 5.7), and the measured averaged signal $S(2\omega, \tau) \propto I(2\omega, \tau)$ becomes

$$\langle S_{\mathrm{NL}}(2\omega, \tau)\rangle \propto \frac{1}{T}\int_{-T/2}^{+T/2} \left|\left[E_1(t) + E_2(t-\tau)\right]^2\right|^2 \mathrm{d}t. \tag{6.39}$$

With

$$E_1 = E(t)\mathrm{e}^{\mathrm{i}[\omega t + \phi(t)]}; \qquad E_2 = E(t-\tau)\mathrm{e}^{\mathrm{i}[\omega(t-\tau)+\phi(t-\tau)]}$$

where $E(t)$ is the slowly varying (compared to the inverse optical frequency) envelope of the pulse, ω is its center frequency and $\phi(t)$ is the slowly varying phase (e.g., for a chirped pulse), we obtain for the detector signal after the frequency doubling crystal

$$\begin{aligned} &S_{\mathrm{NL}}(2\omega, \tau) \\ &\quad = C \cdot \int \left|\left\{E(t)\mathrm{e}^{\mathrm{i}[\omega t + \phi(t)]} + E(t-\tau)\mathrm{e}^{\mathrm{i}[\omega(t-\tau)+\phi(t-\tau)]}\right\}^2\right|^2 \mathrm{d}t \\ &\quad = C \cdot \left[A_1 + 4A_2(\tau) + 4A_3(\tau)\,\mathrm{Re}\left\{\mathrm{e}^{\mathrm{i}(\omega\tau + \Delta\phi)}\right\} + 2A_4(\tau)\,\mathrm{Re}\left\{\mathrm{e}^{2\mathrm{i}(\omega\tau+\Delta\phi)}\right\}\right] \\ &\quad \cong A_1 + 4A_2 + 4A_3 \cos\Delta\phi \cos\omega\tau + 2A_4 \cos(2\Delta\phi)\cos(2\omega\tau). \end{aligned} \tag{6.40}$$

With $\Delta\phi = \phi(t+\tau) - \phi(t)$ and

$$A_1 = \int_{-\infty}^{+\infty} \left[E^4(t) + E^4(t-\tau)\right] \mathrm{d}t \quad \text{(constant background)}$$

$$A_2(\tau) = \int_{-\infty}^{+\infty} E^2(t) \cdot E^2(t-\tau)\,\mathrm{d}t \quad \text{(pulse envelope)}$$

$$A_3(\tau) = \int_{-\infty}^{+\infty} E(t) \cdot E(t-\tau) \cdot \left[E^2(t) + E^2(t-\tau)\right] \mathrm{d}t \quad \text{(interference term at } \omega)$$

$$A_4(\tau) = \int_{-\infty}^{+\infty} E^2(t) \cdot E^2(t-\tau)\,\mathrm{d}t \quad \text{(interference term at } 2\omega).$$

If the detector only monitors the intensity, not the phases, and if its time constant T is large compared to the pulse width ΔT ($T \gg \Delta T$), the terms $4A_3 \cos\Delta\phi \cdot \cos\omega\tau$ and $2A_4 \cos(2\Delta\phi) \cdot \cos(2\omega\tau)$ in (6.40) average to zero and (6.40) reduces to

$$S_{\mathrm{NL}}(2\omega, \tau) \approx 2\int I^2(t)\,\mathrm{d}t + 4\int I(t) \cdot I(t-\tau)\,\mathrm{d}t. \tag{6.41}$$

The first integral is independent of τ and gives a constant background when the delay time τ is varied. The second integral, however, does depend on τ. It gives information on the intensity profile $I(t)$ of the pulse because it represents the convolution of the intensity profile $I(t)$ with the time-delayed profile $I(t+\tau)$ of the same pulse (intensity autocorrelation).

Note The difference between linear detection (6.35) and nonlinear detection (6.41). With linear detection the *sum* $I_1(t) + I_2(t - \tau)$ is measured, which is independent of τ as long as $\tau < T$. The nonlinear detector measures the signal $S(2\omega, \tau)$ that contains the *product* $I_1(t)I_2(t - \tau)$, which does depend on τ as long as τ is smaller than the maximum width of the pulses.

Note Shifting the time from t to $t + \tau$ changes the product $I(t) \cdot I(t - \tau)$ into $I(t + \tau) \cdot I(t) = I(t) \cdot I(t + \tau)$.

All these devices, which are called *optical correlators*, measure the correlation between the field amplitude $E(t)$ or the intensity $I(t)$ at the time t and its values $E(t+\tau)$ or $I(t+\tau)$ at a later time. These correlations are mathematically expressed by normalized correlation functions of order k. The normalized first order correlation function

$$G^{(1)}(\tau) = \frac{\int_{-\infty}^{+\infty} E(t) \cdot E(t+\tau)\,\mathrm{d}t}{\int_{-\infty}^{+\infty} E^2(t)\,\mathrm{d}t} = \frac{\langle E(t) \cdot E(t+\tau)\rangle}{\langle E^2(t)\rangle}, \tag{6.42}$$

describes the correlation between the field amplitudes at times t and $t + \tau$. From (6.42) we obtain $G^{(1)}(0) = 1$, and for pulses with a finite pulse duration ΔT (6.42) yields $G^{(1)}(\infty) = 0$.

The normalized second-order correlation function

$$G^{(2)}(\tau) = \frac{\int I(t) \cdot I(t+\tau)\,\mathrm{d}t}{\int I^2(t)\,\mathrm{d}t} = \frac{\langle I(t) \cdot I(t+\tau)\rangle}{\langle I^2(t)\rangle}, \tag{6.43}$$

describes the *intensity* correlation, where again $G^{(2)}(0) = 1$. The correlation signal (6.39) after the optical frequency doubler can be written in normalized form in terms of $G^{(2)}(\tau)$ for $I_1 = I_2 = I/2$ as

$$S_{\mathrm{NL}}(2\omega, \tau) \propto \left[G^{(2)}(0) + 2G^{(2)}(\tau)\right] = \left[1 + 2G^{(2)}(\tau)\right]. \tag{6.44}$$

Note $S_{\mathrm{NL}}(2\omega, \tau)$ is symmetric, i.e., $S(\tau) = S(-\tau)$. This implies that a possible asymmetry of the pulse time profile does not show up in the signal S_{NL}.

If I_1 and I_2 come from the same pulse, the correlation is also called "*autocorrelation*" in contrast to the situation where I_1 and I_2 come from two different pulses.

There are two different techniques that are used to measure the time profiles and optical oscillations of ultrashort pulses: noncollinear intensity correlation and interferometric autocorrelation. While the former measures the envelope of the pulse, the latter can even measure the optical oscillations within the pulse envelope. Combined with the spectral resolution, the time profiles of the different spectral components within the optical pulse spectrum can be simultaneously measured by the *FROG* technique. The relative phases of these spectral components are observable using the *SPIDER* technique (see Sect. 6.2.4).

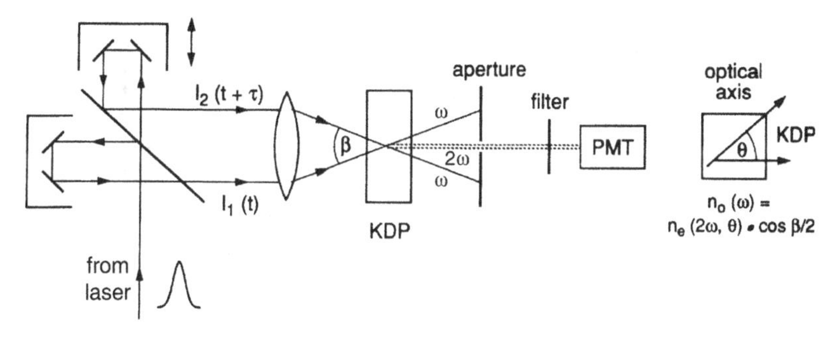

Fig. 6.66 Background-free measurement of the second-order correlation function $G^{(2)}(\tau)$ by choosing the phase-matching condition properly (SF, spectral filter neglecting scattered light of the fundamental wave at ω; KDP, potassium-dihydrogen-phosphate crystal for frequency doubling)

For sub-femtosecond pulses a new techniques has been developed which is called *CRAB* (*complete reconstruction of attosecond bursts*). Instead of using a nonlinear interaction for generating the signal, a weak femtosecond pulse is used to probe the attosecond pulse by measuring the energy spectrum of photo-electrons produced by photo-ionization of atoms by the attosecond pulse (see Fig. 6.59 and Sect. 6.2.5).

(a) Noncollinear Intensity Correlation

The intensity profile $I(t)$ of the pulses is measured using the intensity correlation method. For a collinear geometry, the intensity correlation (6.44) yields the normalized signal $S(2\omega, \tau = 0) = 3$ for completely overlapping pulses ($\tau = 0$) and the background signal $S(2\omega, \tau = \infty) = 1$ for completely separated pulses ($\tau \gg \Delta T$). The signal-to-background ratio is therefore 3:1.

The τ-independent background in (6.39) can be suppressed when the two beams are focused into the doubling crystal under different angles $\pm\beta/2$ against the z-direction (Fig. 6.66) where the signal $S(2\omega)$ is detected. If the phase-matching conditions for the doubling crystal (Vol. 1, Sect. 5.7) are chosen in such a way that for two photons out of the same beam no phase matching occurs, but only for one photon out of each beam, such that the term A_1 in (6.40) does not contribute to the signal (background-free detection) [765, 766]. In another method of background-free pulse measurement the polarization plane of one of the two beams in Fig. 6.66 is turned in such a way that a properly oriented doubling crystal (generally, a KDP crystal) fulfills the phase-matching condition only if the two photons each come from a different beam [767]. In this noncollinear scheme no interference occurs (A_3 and $A_4 = 0$) and the measured signal equals the envelope of the pulse profile in Fig. 6.71. For methods with background suppression no signal is obtained for $\tau \gg \Delta T$.

In the methods discussed above, one of the retroreflectors is mounted on a translational stage moved via micrometer screws by a step motor while the signal $S(2\omega, \tau)$ is recorded. Since τ must be larger than ΔT, the translational move should be at

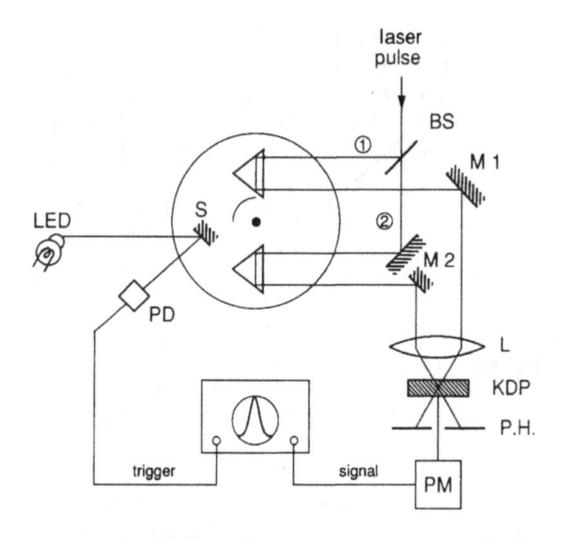

Fig. 6.67 Rotating autocorrelator allowing direct observations of the correlation signal $S(2\omega, \tau)$ on a scope that is triggered by the output signal of the photodiode PD

least $\Delta S = \frac{1}{2}c\tau \geq \frac{1}{2}c\Delta T$. For pulses of 10 ps this means $\Delta S \geq 1.5$ mm. With a rotating correlator (Fig. 6.67) the signal $S(2\omega, \tau)$ can be directly viewed on a scope, which is very useful when optimizing the pulse width. Two retroreflecting prisms are mounted on a rotating disc. During a certain fraction ΔT_{rot} of the rotation period T_{rot} the reflected beams reach the mirrors M1 and M2 and are focused into the KDP crystal. The viewing oscilloscope is triggered by a pulse obtained by reflecting the light of a LED onto the photodetector PD. A compact autocorrelator for measuring the intensity profile of femtosecond pulses from tunable sources is shown in Fig. 6.68. This device allows online measurements of pulses with 100 Hz to 10 kHz repetition rates in the wavelength range 420–1460 nm. The time delay is controlled by a piezo translator [769].

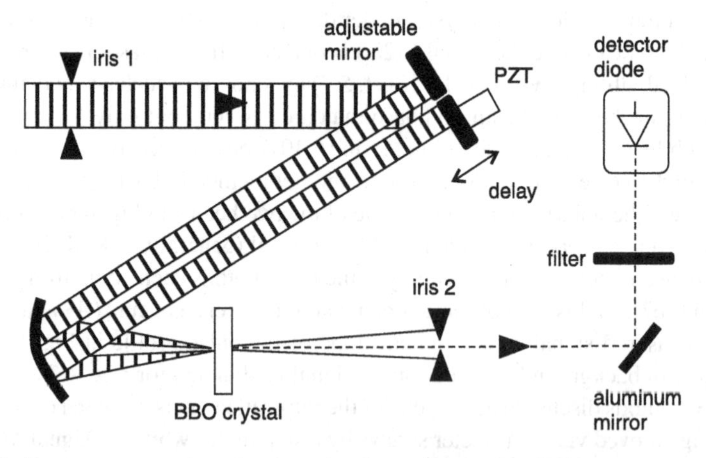

Fig. 6.68 Compact autocorrelator for measuring femtosecond pulses [769]

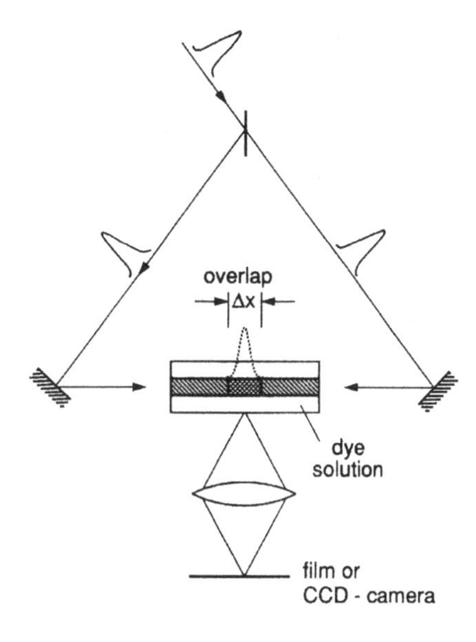

Fig. 6.69 Measurement of short pulses via two-photon induced fluorescence

Instead of using optical frequency doubling other nonlinear effects can also be used, such as two-photon absorption in liquids or solids, which can be monitored by the emitted fluorescence. If the optical pulse is again split into two pulses traveling in the opposite $\pm z$-directions through the sample cell (Fig. 6.69), the spatial intensity profile $I_{\mathrm{Fl}}(z) \propto I^2(\omega, \tau)$ can be imaged by magnifying optics onto a vidicon or an image intensifier. Since a pulse width $\Delta T = 1$ ps corresponds to a path length of 0.3 mm, this technique, which is based on the spatial resolution of the fluorescence intensity, is limited to pulse widths $\Delta T \geq 0.3$ ps. For shorter pulses the delay time τ between the pulses has to be varied and the total fluorescence

$$I_{\mathrm{Fl}}(\tau) = \int I(z, \tau)\,\mathrm{d}z, \tag{6.45}$$

has to be measured as a function of τ [768, 770].

(b) Interferometric Autocorrelation

In interferometric autocorrelation the coherent superposition of the two collinear partial beams is realized. The basic principle is shown in Fig. 6.65. The incoming laser pulse is split by the beamsplitter BS1 into two parts, which travel through two different pathlengths and are then collinearly superimposed at BS2. When they are focused by the lens L into a nonlinear optical crystal, the output signal (6.39) is generated at 2ω. Instead of the delay line arrangement in Fig. 6.65 a Michelson interferometer in Fig. 6.70 can also be used. The second harmonics are detected by a photomultiplier, while the fundamental wavelength is rejected by a filter.

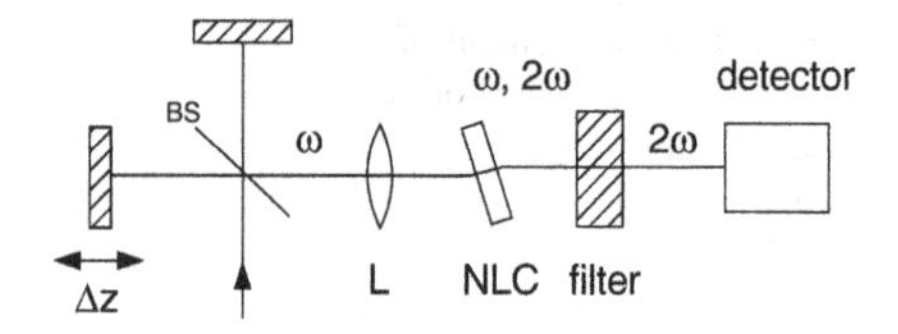

Fig. 6.70 Michelson interferometer for interferometric autocorrelation

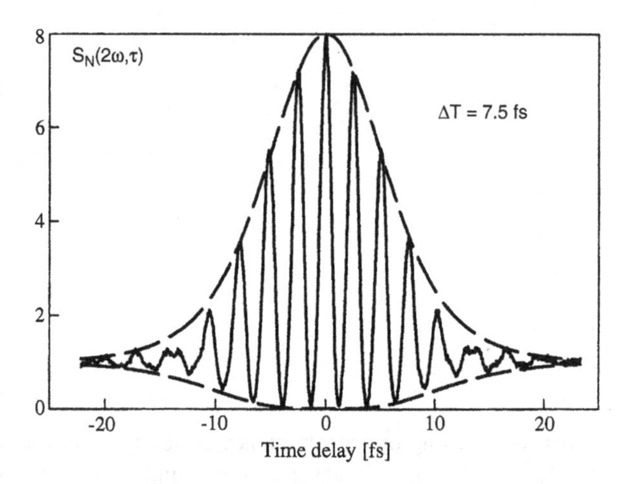

Fig. 6.71 Interferometric autocorrelation trace of a 7.5 fs pulse with upper and lower envelopes [759]

The nonlinear crystal can be omitted, if the detector itself has a nonlinear response. This is, for instance, the case for a semiconductor detector with a band gap $\Delta E > h\nu$, where only two-photon absorption contributes to the signal.

In interferometric autocorrelation, averaging is not complete (unlike in intensity correlation), and the phases of the electric fields have to be taken into account. Here all of the terms A_1–A_4 in (6.40) can contribute to the signal. If a filter which rejects the fundamental frequency ω and transmits only the doubled frequency 2ω is inserted behind the frequency-doubling crystal, the third term in (6.40) with A_3 is suppressed.

A typical signal as a function of the delay time τ is shown in Fig. 6.71.

The upper envelope of this interference pattern is obtained if the phase $\omega_0\tau$ is replaced by the constant phase 2π, the lower envelope for $\omega_0\tau = \pi$. The maximum signal is, according to (6.40), $S_{\mathrm{N}}^{\max}(2\omega, \tau) = 8$, while the background $S(2\omega, \infty) = 1$. The signal-to-background ratio of 8:1 is therefore larger than that for intensity correlation. For $\omega_0\tau = \pi$, the minimum value is $S_{\mathrm{N}}^{\min}(2\omega, \tau) = 0$ (Fig. 6.71).

For illustration, Fig. 6.72 shows a 12 fs pulse measured with intensity correlation (a) and interferometric autocorrelation (b).

It is important to note that the profile $S(\tau)$ of the correlation signal depends on the time profile $I(t)$ of the light pulse. It gives the correct pulse width ΔT only if an assumption is made about the pulse profile. For illustration, Fig. 6.73a depicts the signal $S(2\omega, \tau)$ of Fourier-limited pulses with the Gaussian profile $I(t) = I_0 \exp(-t^2/0.36\Delta T^2)$ with and without background suppression. From the halfwidth $\Delta\tau$ of the signal the halfwidth ΔT of the pulses can only be derived if

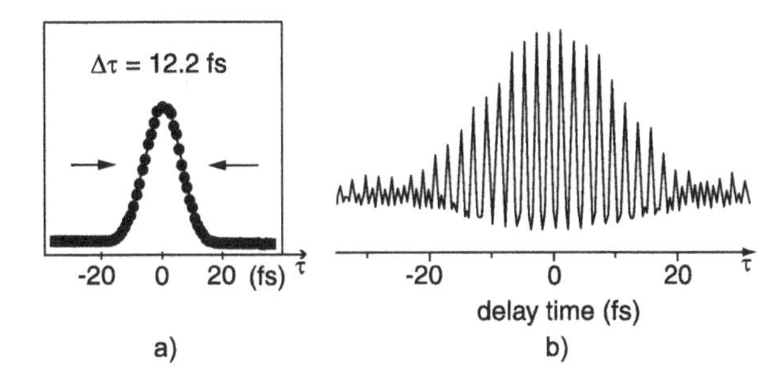

Fig. 6.72 A femtosecond pulse with time duration $\Delta T = 12.2$ fs: (**a**) measured with intensity correlation; (**b**) measured with interferometric autocorrelation [771]

Table 6.2 Ratios $\Delta\tau/\Delta T$ of the width $\Delta\tau$ of the autocorrelation profile and ΔT of the pulse $I(t)$, and products $\Delta\nu \cdot \Delta T$ of spectral width $\Delta\nu$ and duration ΔT of pulses with different profiles $I(t)$

Pulse profile	Mathematical expression for $I(t)$	$\Delta\tau/\Delta T$	$\Delta\nu \cdot \Delta T$
Rectangular	$\begin{cases} I_0 & \text{for } 0 \le t \le \Delta T, \\ 0 & \text{elsewhere} \end{cases}$	1	0.886
Gaussian	$I_0 \exp[-t^2/(0.36\Delta T^2)]$	$\sqrt{2}$	0.441
Sech^2	$\mathrm{sech}^2(t/0.57\Delta T)$	1.55	0.315
Lorentzian	$[1+(2t/\Delta T)^2]^{-1}$	2	0.221

the pulse profile $I(t)$ is known. In Table 6.2 the ratio $\Delta\tau/\Delta T$ and the products $\Delta T \cdot \Delta\nu$ are compiled for different pulse profiles $I(t)$, while Fig. 6.73a–c illustrates the corresponding profiles and contrasts of $G^{(2)}(\tau)$. Even noise pulses and continuous random noise result in a maximum of the correlation function $G^{(2)}(\tau)$ at $\tau = 0$ (Fig. 6.73d), and the contrast becomes $G^{(2)}(0)/G^{(1)}(\infty) = 2$ [765, 773]. For the determination of the real pulse profile one has to measure the function $G^{(2)}(\tau)$ over a wider range of delay times τ. Generally, a model profile is assumed and the calculated functions $G^{(2)}(\tau)$ and even $G^{(3)}(\tau)$ are compared with the measured ones [774].

In Fig. 6.74 the power density spectrum and the interferometric autocorrelation signal of a femtosecond laser pulse is compared.

In Fig. 6.75 a chirped hyperbolic secant pulse

$$E(t) = \left[\mathrm{sech}(t/\Delta T)\right]^{(1-\mathrm{i}a)} = \left(\frac{2}{\mathrm{e}^{t/\Delta T} + \mathrm{e}^{-t/\Delta T}}\right)^{(1-\mathrm{i}a)}, \tag{6.46}$$

is shown for $a = 2$ and $\Delta T = 10$ fs. Chirped pulses result in a more complex autocorrelation signal.

With interferometric autocorrelation the chirp of a pulse and the resulting change in its time profile can be determined. This is illustrated by the example of a chirped pulse with a Gaussian profile

$$E(t) = E_0 \exp\left[-(1+\mathrm{i}a)(t/\Delta T)^2\right] \tag{6.47}$$

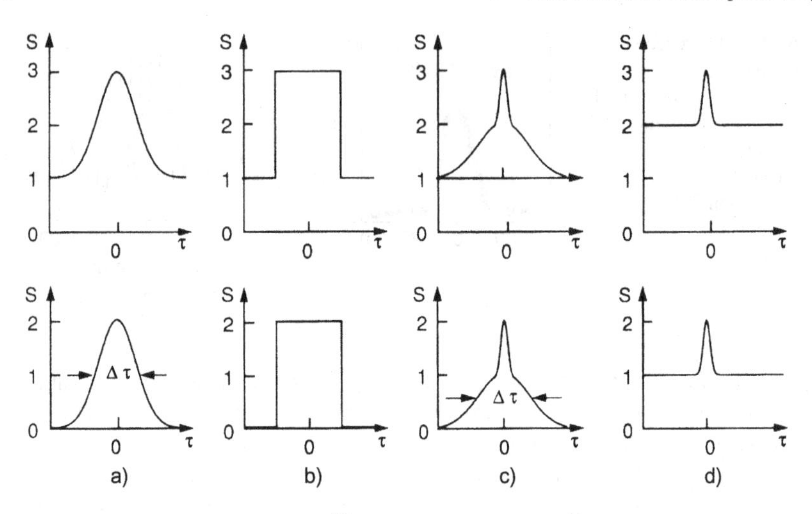

Fig. 6.73 Autocorrelation signal $S \propto G^{(2)}(\tau)$ for different pulse profiles without background suppression (*upper part*) and with background suppression (*lower part*): (**a**) Fourier-limited Gaussian pulse; (**b**) rectangular pulse; (**c**) single noise pulse; and (**d**) continuous noise

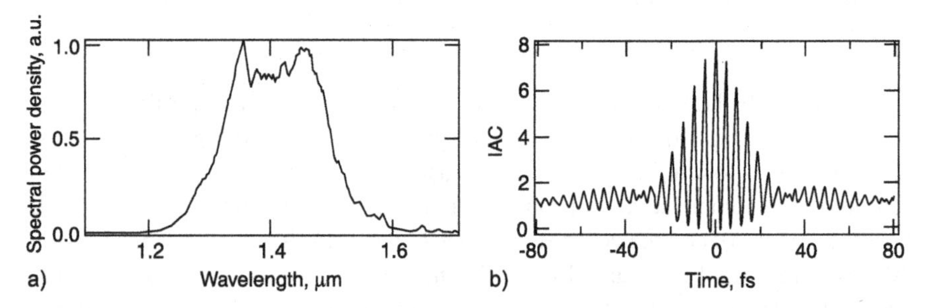

Fig. 6.74 Femtosecond laser pulse. (**a**) Optical power spectrum; (**b**) interferometric autocorrelation of the same pulse [828]

where ΔT is the width of the unchirped pulse ($a = 0$). The interferometric autocorrelation gives for such a pulse:

$$G_2(\tau) = 1 + 2\mathrm{e}^{-(\tau/\Delta T)^2} + 4\mathrm{e}^{-[\frac{3+a^2}{4}(\tau/\Delta T)^2]} \cdot \cos\frac{a}{2}\left(\frac{\tau}{\Delta T}\right)^2 \cdot \cos\omega\tau$$
$$+ 2\mathrm{e}^{-(1+a^2)(\tau/\Delta T)^2} \cdot \cos 2\omega\tau. \tag{6.48}$$

In Fig. 6.75b, the upper and lower envelopes of the autocorrelation signal are plotted as a function of the normalized delay time $\tau/\Delta T$ for different values of the chirp parameter a in (6.47) [775].

The drawback of the techniques discussed so far is their lack of phase measurements for the different spectral components within the pulse spectral profile. This can be overcome by the *frequency-resolved optical gating* technique (FROG) [776].

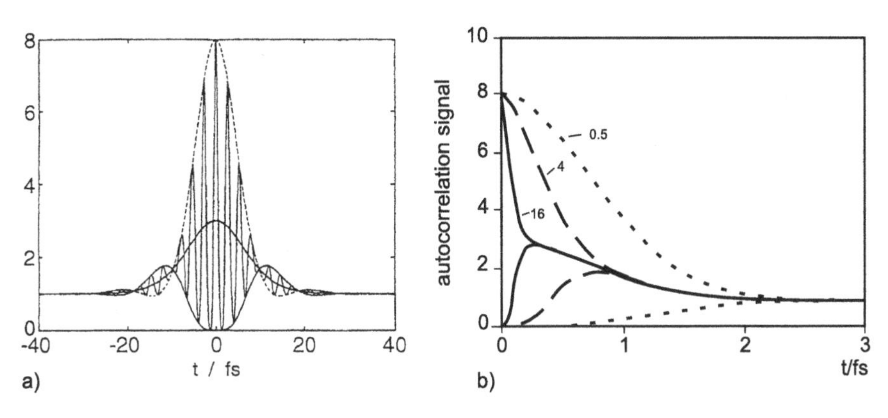

Fig. 6.75 (**a**) Interferometric autocorrelation of a pulse with $\Delta T = 10$ fs and a chirp of $a = 2$ (*dashed line*). The *solid line* shows the pulse profile obtained by intensity correlation. (**b**) Upper and lower envelopes of a chirped Gaussian pulse [675] for various chirp parameters a [775]

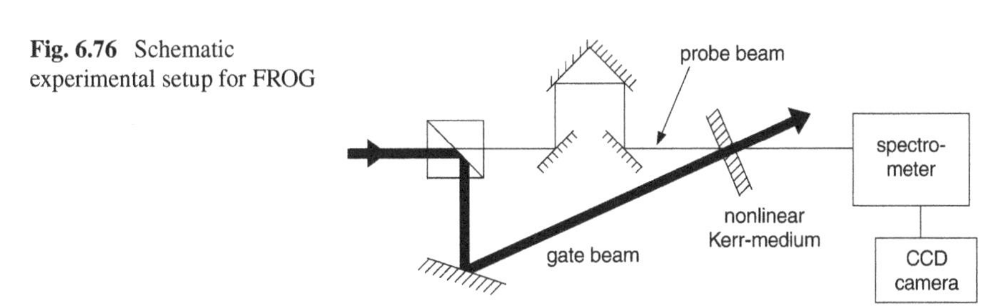

Fig. 6.76 Schematic experimental setup for FROG

6.2.3 FROG Technique

We have seen that the second-order autocorrelations are symmetric and therefore do not provide any information on possible pulse asymmetries. Here the FROG technique is useful, since it allows measurements of the third-order autocorrelation. Its basic features are depicted in Fig. 6.76. As in the other autocorrelation techniques the incoming pulse is split by a polarizing beamsplitter into two partial beams with amplitudes E_1 and E_2. The probe pulse with amplitude $E_1(t)$ passes through a shutter (Kerr cell), that is opened by the delayed gate pulse, which is a replica $E_2(t-\tau)$ of the signal pulse. The signal transmitted by the Kerr gate is then

$$\boldsymbol{E}_s(t,\tau) \propto \boldsymbol{E}(t) \cdot g(t-\tau), \tag{6.49}$$

where g is the gate function $g(t-\tau) \propto I_2(t-\tau) \propto E_2^2(t-\tau)$.

If the transmitted pulse is sent through a spectrometer, where it is spectrally dispersed, a CCD camera will record the time dependence of the spectral components, which gives the two-dimensional function

$$I_t(\omega,\tau) = \left| \int_{-\infty}^{+\infty} \boldsymbol{E}(t) \cdot g(t-\tau) \mathrm{e}^{\mathrm{i}\omega t} \right|. \tag{6.50}$$

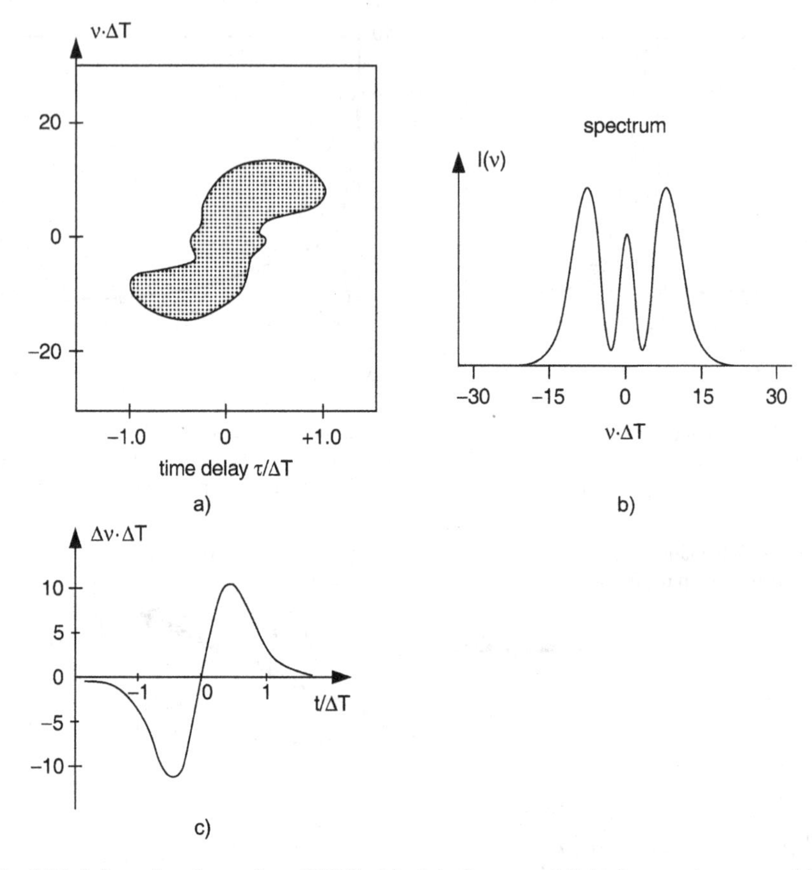

Fig. 6.77 Information drawn from FROG: (**a**) plot of measured light frequencies versus delay time τ in units of pulse length ΔT; (**b**) frequency spectrum of (**a**); (**c**) frequency chirp [777]

In Fig. 6.77 this two-dimensional function is illustrated for a Gaussian pulse profile without frequency chirp and for a chirped pulse.

The integration of (6.49) over the delay time, which can be experimentally achieved by opening the gate for a time larger than all relevant delay times, yields the time profile of the pulse

$$\boldsymbol{E}(t) = \int_{-\infty}^{+\infty} \boldsymbol{E}_s(t, \tau)\,\mathrm{d}\tau. \tag{6.51}$$

The two functions $E_s(t, \omega)$ and $E_s(t, \tau)$ form a Fourier pair, related to each other by

$$\boldsymbol{E}_s(t, \omega_\tau) = \frac{1}{2\pi} \int_{-\infty}^{+\infty} \boldsymbol{E}_s(t, \tau)\mathrm{e}^{-\mathrm{i}\omega_\tau \cdot \tau}\,\mathrm{d}\tau. \tag{6.52}$$

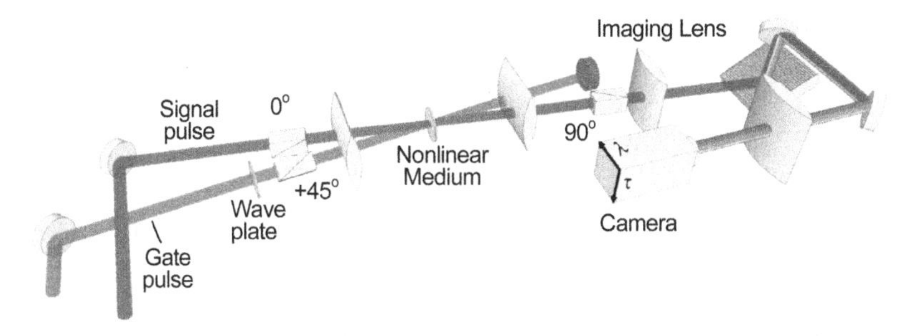

Fig. 6.78 The schematic of single-shot polarization-gate (PG) cross-correlation frequency-resolved optical gating (XFROG) setup. The *orange beam* and *red beam* represent the known reference pulse and unknown pulse, respectively. The experimental apparatus is essentially a PG FROG setup with the known reference pulse replacing one of the replicas. The apparatus becomes a Blind PG FROG setup when the known reference is replaced by an unknown one [779]

The measured spectrogram $S_E(\omega, \tau)$ can be expressed by

$$I_E(\omega, \tau) = \left| \int_{-\infty}^{+\infty} \int_{-\infty}^{+\infty} \boldsymbol{E}_s(t, \omega_\tau) \mathrm{e}^{-\mathrm{i}\omega t + \mathrm{i}\omega_\tau \cdot \tau} \, \mathrm{d}\omega_\tau \, \mathrm{d}t \right|^2 . \tag{6.53}$$

The unknown signal $E(t, \omega)$ can be extracted by a two-dimensional (the two dimensions are t and τ) phase retrieval. The reconstruction of the pulse $E(t, \omega)$ yields the instantaneous frequency as a function of time and the pulse spectrum, shown in Fig. 6.77b [778]. In the FROG-technique gate pulse and signal pulse come from the same source and therefore have the same but unknown pulse profile. This autocorrelation technique is therefore a self-referencing method. Instead of splitting the incoming pulse into two parts one can use as gate pulse a pulse from a second source with known pulse profile (cross correlation FROG = XFROG). The deconvolution of the measured signal with the known gate pulse profile therefore gives the wanted profile of the signal pulse. If the pulse profile of the reference pulse is unknown the technique is called "*Blind FROG*." In this case the correct deconvolution of the two pulses is obtained in an iterative process until the convolution agrees with the measured signal.

In Fig. 6.78 a schematic experimental setup is shown for XFROG with a polarization gate, where the reference pulse changes the polarization dependent transmission of a nonlinear medium for the signal pulse [779]. The gate pulse comes from another source than the signal pulse.

A simplified version of FROG, called *GRENOULLE* (*grating-eliminated no-nonsense observation of ultra-fast laser-light E-fields*), is an optical-gating (FROG) technique. It [784] is a second-harmonic-generation (SHG) frequency-resolved optical gating technique, and is able to measure the intensity, the spatial chirp, the phase and the tilt of the phasefront. The beam spitter, the translational stage and their combining optics used in FROG is replaced by a Fresnel biprism, which simplifies the optical settings and realignments considerably. The setup and its comparison, with the conventional FROG arrangement are shown schematically in Fig. 6.79 [784, 787].

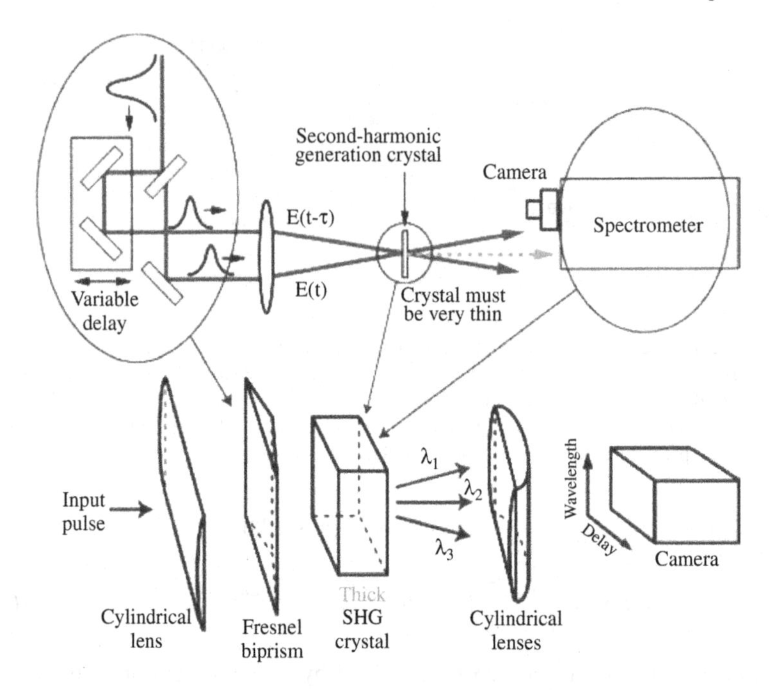

Fig. 6.79 An SHG FROG device (*above*) and the simpler version, GRENOUILLE (*below*) [787]

For more information on the FROG technique and the recent literature on this subject, see [778].

6.2.4 SPIDER Technique

The FROG method provides information on the time-dependent frequency spectrum of a short pulse but cannot measure the phases of these spectral components. A newly developed technique is helpful in this case; this method is called SPIDER (*S*pectral *P*hase *I*nterferometry for *D*irect *F*ield *R*econstruction). It uses the interference structure generated when two spatially separated pulses are superimposed [788]. Similar to the autocorrelation method, the two pulses are generated from the input pulse that is to be measured, using a beam splitter and a delay line which changes the time delay between the two pulses. The second pulse is therefore a copy of the first pulse with a time delay τ. The electric field amplitude

$$E(x) = \sqrt{I(x)}\mathrm{e}^{\mathrm{i}\phi(x)}$$

of the first pulse interferes with the field

$$E(x+\Delta x) = \sqrt{I(x+\Delta x)}\mathrm{e}^{\mathrm{i}\phi(x+\Delta x)}$$

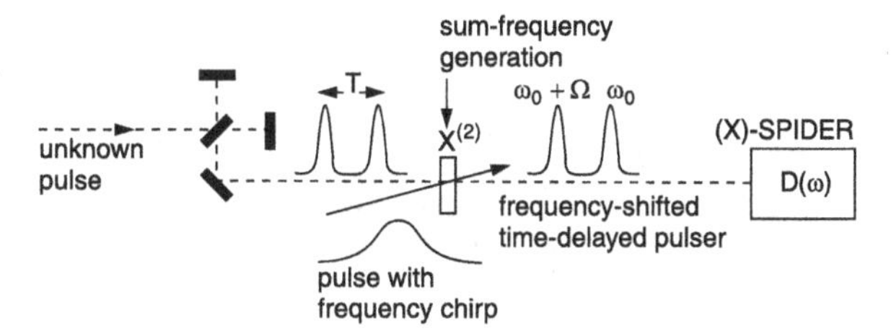

Fig. 6.80 Schematic representation of the SPIDER technique [789]

of the second pulse. The detector measures the square of the total amplitude and delivers the signal

$$S(x,\Delta x)=I(x)+I(x+\Delta x)+2\sqrt{I(x)}\sqrt{I(x+\Delta x)}\cos\{\phi(x)-\phi(x+\Delta x)\}.$$

The intensity measurement at the location x is therefore related to the phase difference $\Delta\phi=\phi(x)-\phi(x+\Delta x)$ between the phases of the wavefront at the locations x and $x+\Delta x$.

These two pulses are superimposed in a nonlinear crystal with a third pulse with a large frequency chirp where the sum frequencies $\omega_1+\omega_3$ and $\omega_2+\omega_3$ are generated (Fig. 6.80). This third pulse, which is generated from the input pulse by beam-splitting, is sent through a dispersive medium where a frequency chirp is produced which makes the pulse much broader than the two other pulses. Because of the chirp, the superposition of the third pulse with the first pulse gives another sum frequency $\omega_1+\omega_3$, and than $\omega_2+\omega_3$ with the second delayed pulse. When the frequency ω_3 of the chirped pulse changes during the delay time Δt by $\Delta\omega=\Omega$, the detector receives the signal

$$\begin{aligned}S(\omega)&=I(\omega+\omega_3)+I(\omega+\omega_3+\Omega)\\&\quad+2\sqrt{I(\omega+\omega_3)}\sqrt{I(\omega+\omega_3+\Omega)}\cos\{\phi(\omega+\omega_3)-\phi(\omega+\omega_3+\Omega)\}.\end{aligned}$$

This signal is measured behind a spectrograph as a function of the delay Δt between pulses 1 and 2. The measured frequency shift $\Omega=\phi\Delta t$ of the sum frequency gives the phase $\phi(t)$ of the unknown input pulse and its time profile $I(t)$. Figure 6.81 illustrates the principle of the SPIDER technique using a schematic diagram.

One drawback of the FROG and SPIDER techniques is the fact that the pulse measurement does not take place at the location of the sample where the short pulse is used to investigate time-dependent processes in atoms or molecules. This is where the pulse profile and its phases actually need to be determined. In order to remove this drawback, E. Riedle and his group [789] extended the SPIDER technique into the ZAP-SPIDER method (Zero *A*dditional *P*hase *SPIDER*), which is illustrated in Fig. 6.82. Here the unknown pulse is sent directly into the nonlinear crystal, where it is superimposed by two chirped pulses from slightly different directions and which have a delay Δt. The pulses at the sum frequency generated in the nonlinear crystal have different frequencies ω_s and $\omega_s+\Omega$, and are emitted in different directions because of the phase-matching condition for sum frequency generation. One of these sum frequency pulses is sent through a variable delay line

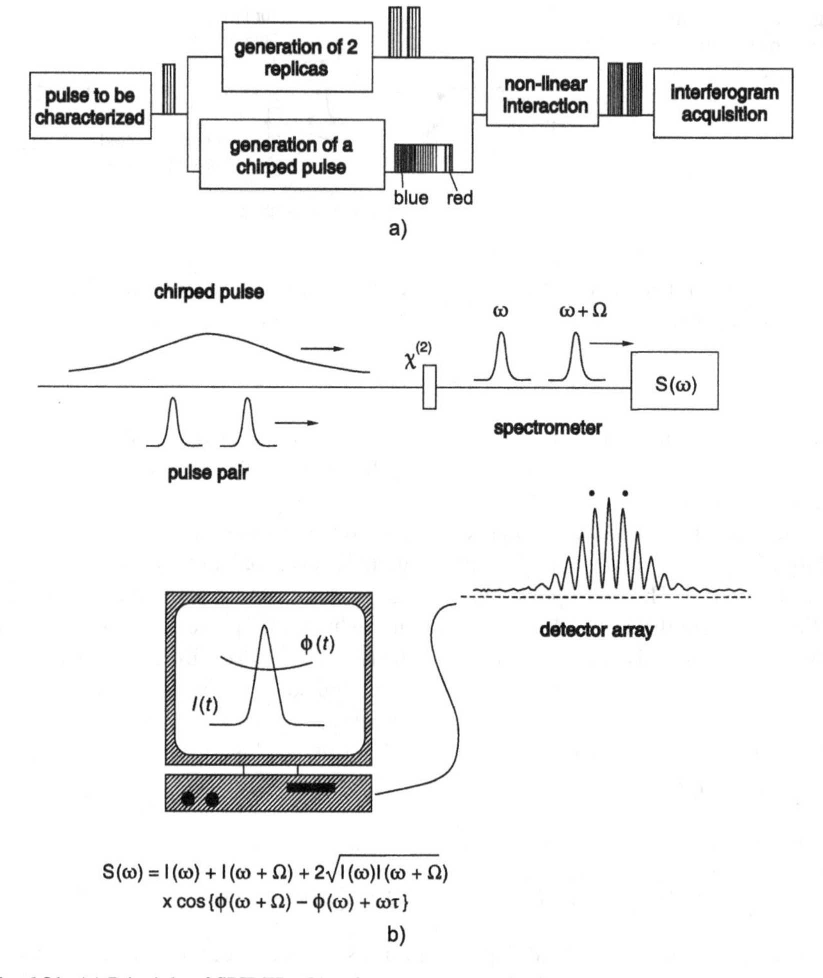

Fig. 6.81 (**a**) Principle of SPIDER; (**b**) pulse sequence, profiles and measured signal [772]

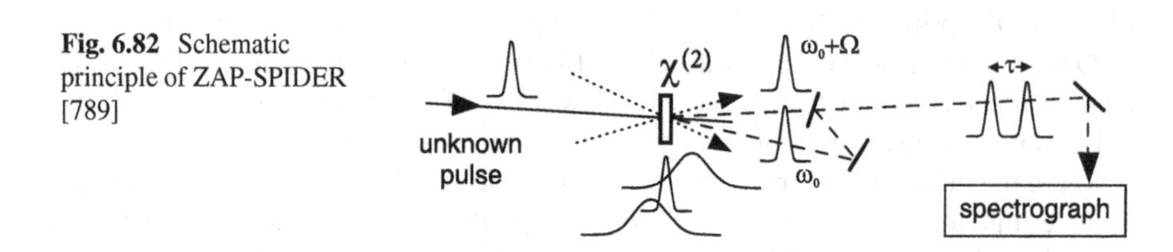

Fig. 6.82 Schematic principle of ZAP-SPIDER [789]

and is then superimposed onto the other pulse. This gives rise to an interference pattern, which depends on the relative phase between the two pulses and contains all of the information on the time profiles and phases of the unknown pulse. The spectrum of the two pulses can be measured using a spectrograph. The experimental setup is shown in Fig. 6.83: the chirp is produced by sending part of the pulse from

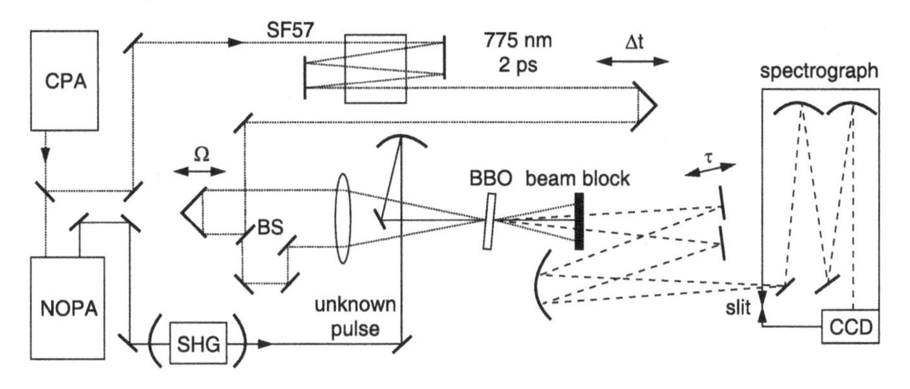

Fig. 6.83 Experimental setup for the ZAP-SPIDER technique [789]

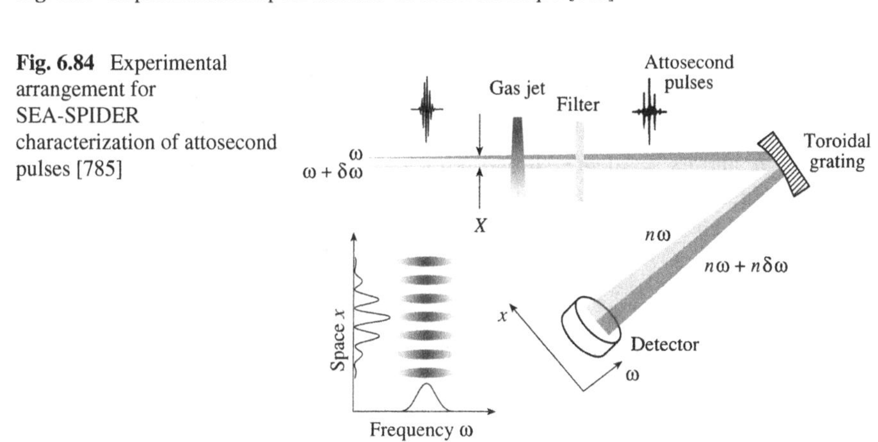

Fig. 6.84 Experimental arrangement for SEA-SPIDER characterization of attosecond pulses [785]

a femtosecond laser through a dispersive SF57 glass block which also stretches the pulsewidth to 2 ps. The pulse delay can be controlled with a retroreflector on a translational stage. This pulse is again split by the beam splitter BS into two pulses which can be separately delayed and are then sent into the BBO nonlinear crystal, where they are superimposed onto the unknown pulse and the sum frequency is generated.

For the measurements of pulse profiles and phases of high harmonics a XUV version of SPIDER has been developed [785, 790]. This involves the mixing of two pulses which are replica of each other, where one of the pulses is frequency-shifted by a small amount $\delta\omega$. This is achieved by mixing the two pulses with a third frequency chirped pulse. The frequency shift $\delta\omega$ is then dependent on the time delay τ between the two pulses [785]. The pulses pass through an atomic beam of noble gas atoms where high harmonics at $n\omega$ and $n(\omega + \delta\omega)$ are generated (Fig. 6.84). They are filtered against the fundamental pulses and are diffracted by a torodial grating which separates the different frequencies spatially before they fall onto the detector. The frequency dependent pattern of the spatially separated harmonics is measured as a function of the delay time τ. The Fourier-transform of the measured

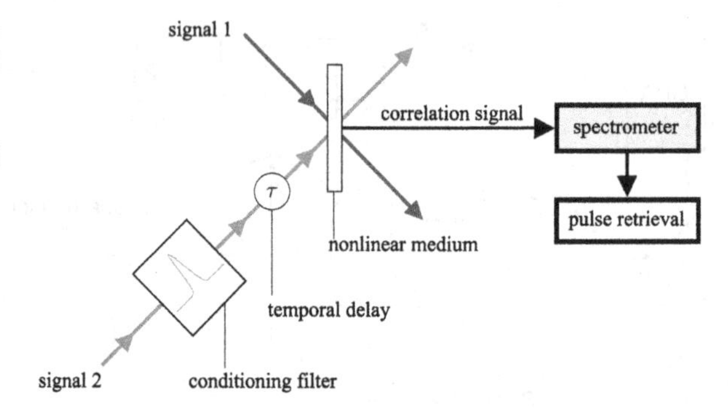

Fig. 6.85 Optical schematic of the very advanced method for phase and intensity retrieval of E-fields (VAMPIRE) technique [791]

signal $S(\omega)$ gives the time profiles of the pulses. The signal of the two interferring pulses is then

$$S(\omega, \tau) = \left|E(\omega) + E(\omega + \delta\omega)\right| e^{i\omega\tau}.$$

The principle of this method is depected in Fig. 6.85. The signal pulse at the frequency ω and its frequency-shifted replica at $\omega + \delta\omega$ pass with a time delay τ through a beam of noble gas atoms. The high harmonics at $n\omega$ and $n(\omega + \delta\omega)$ are diffracted by a concave grating and reach the detector.

6.2.5 CRAB- and VAMPIRE-Techniques

The basic principle of CRAB (*complete reconstruction of attosecond bursts*) is shown in Fig. 6.59.

The few-cycle optical pulses produce photoelectrons by ionizing atoms in a collimated atomic beam (see [744] and [749]). The photoelectrons are detected on both sides of the atomic beam perpendicular to the laser and atomic beams in the direction of the alternating electric field vector of the optical wave. Since the photoelectrons are generated by multiphoton absorption (for $h\nu = 1.8$ eV, ten photons must be absorbed to ionize the neon atoms) the number of photoelectrons depends strongly on the electric field strength. Therefore, nearly all photoelectrons are generated close to the maximum of the periodically changing field $\boldsymbol{E}(t)$. In a few-cycle pulse, $E(t)$ depends on the phase of the optical field relative to the envelope of the pulse. The largest value of $E(t)$ is achieved if the maximum of E coincides with the maximum of the pulse envelope. The number of photo-electrons are therefore different for the two sides, because the electric field amplitudes E and $-E$ have different time shifts against the pulse maximum and therefore different amplitudes. Measurements of the ratio N_1/N_2 of the numbers of photoelectrons emitted to both sides can determine the phase of the optical field.

Further information can be obtained from measurements of the energy spectrum of the photo-electrons. A weak low frequency laser field is applied to the interaction region with a delay τ. The electrons gain momentum in this field and their final energy spectrum depends on the delay time τ of the probe field against the attosecond high intensity ionizing field. The weak probe field causes a phase modulation of the electron wave packet and influences their trajectory.

In order to remove the ambiguities of pulse analysis for the FROG technique, a self referenced two-pulse measurement technique was invented [791], which is called VAMPIRE (*very advanced method for phase and intensity retrieval*).

Its basic principle [779] is illustrated in Fig. 6.85. The signal beam 1 interferes in a nonlinear medium with the signal beam 2, specially prepared by a conditional filter which consists of a spectral and a temporal phase modulator. The spectral phase modulator can be realized by an unbalanced Mach–Zehnder interferometer (see Vol. 1, Sect. 4.2.4) with a dispersive element placed in one of the arms. The spectral phase modulator breaks the symmetry of the measured signal and therefore avoids ambiguities caused by symmetric VAMPIRE spectrograms.

6.2.6 Comparison of the Different Techniques

All techniques discussed so far have their advantages and limitations. Table 6.2 compiles which combinations of the electric field contribute to the measured signal for the main different variants of FROG. In the normal FROG-technique the pulse interferes with the time-shifted version of itself. It therefore represents an autocorrelation method, which gives no information on the phase of the optical wave. The two interfering pulses can generate in a nonlinear crystal second harmonics or they can produce an optical grating where diffraction occurs or third harmonic generation is used.

Blind FROG recovers the signal from mixing two pulses with unknown pulse profiles in a nonlinear medium, while XFROG is a cross-correlation technique which uses a known gate profile and therefore simplifies the analysis of the signal. If the gate pulse is polarized, a polarization dependent gate gives additional information on the characteristics of the signal pulse.

The main advantage of SPIDER is its capability to measure the phase A drawback is the fact that properly chirped pulses are required.

Note If the signal consists of a strictly equidistant pulse train, the combination of FROG and SPIDER yields correct information about the pulse length, the frequency chirp and CRAB allows to measure the phases of the optical field. In reality, however, there are random fluctuations of the pulse width and the time separation between successive pulses. In such cases the analysis of the measured signal does not necessarily give the correct information about the pulse duration or of its phases. As has been discussed in detail in [782] it turns out that SPIDER underestimates the average pulse duration but gives a good value of the average phase, while FROG retrieves the correct average pulse duration but not the real pulse profile.

6.3 Lifetime Measurement with Lasers

Measurements of lifetimes of excited atomic or molecular levels are of great interest for many problems in atomic, molecular, or astrophysics, as can be seen from the following three examples:

(i) From the measured lifetimes $\tau_k = 1/A_k$ of levels $|k\rangle$, which may decay by fluorescence into the lower levels $|m\rangle$, the absolute transition probability $A_k = \sum_m A_{km}$ can be determined (Vol. 1, Sect. 2.8.1). Knowing the lifetime of level $|k\rangle$ the relative intensities I_{km} of transitions $|k\rangle \to |m\rangle$ allow the determination of the absolute transition probabilities A_{km}. This yields the transition dipole matrix elements $\langle k|\boldsymbol{r}|m\rangle$ (Vol. 1, Sect. 2.8.2). The values of these matrix elements are sensitively dependent on the wave functions of the upper and lower states. Lifetime measurements therefore represent crucial tests for the quality of computed wave functions and can be used to optimize models of the electron distribution in complex atoms or molecules.

(ii) The intensity decrease $I(\omega, z) = I_0 \mathrm{e}^{-\alpha(\omega) z}$ of light passing through absorbing samples depends on the product $\alpha(\omega) z = N_i \sigma_{ik}(\omega) z$ of the absorber density N_i and the absorption cross section σ_{ik}. Since σ_{ik} is proportional to the transition probability A_{ik} (Vol. 1, Table 2.2), it can be determined from lifetime measurements (see item (i)). Together with measurements of the absorption coefficient $\alpha(\omega)$ the density N_i of the absorbers can be determined. This problem is very important for testing of models of stellar atmospheres [793]. A well-known example is the measurement of absorption profiles of Fraunhofer lines in the solar spectrum. They yield density and temperature profiles and the abundance of the elements in the sun's atmosphere (photosphere and chromosphere). The knowledge of transition probabilities allows absolute values of these quantities to be determined.

(iii) Lifetime measurements are not only important to gain information on the dynamics of excited states but also for the determination of absolute cross sections for quenching collisions. The probability R_{kn} per second for the collision-induced transition $|k\rangle \to |n\rangle$ in an excited atom or molecule A

$$R_{kn} = \frac{1}{\bar{v}} \int_0^\infty N_\mathrm{B}(v) \sigma_{kn}(v) v \, \mathrm{d}v = N_\mathrm{B} \langle \sigma_{kn}^\mathrm{coll} \cdot v \rangle \approx N_\mathrm{B} \langle \sigma_{kn}^\mathrm{coll} \rangle \cdot \bar{v}, \tag{6.54}$$

depends on the density N_B of the collision partners B, the collision cross section $\sigma_{kn}^\mathrm{coll}$, and the mean relative velocity $\bar{v}$. The total deactivation probability P_k of an excited level $|k\rangle$ is the sum of radiative probability $A_k = \sum_m A_{km} = 1/\tau^\mathrm{rad}$ and the collisional deactivation probability R_k. Since the measured effective lifetime is $\tau_k^\mathrm{eff} = 1/P_k$, we obtain the equation

$$\frac{1}{\tau_k^\mathrm{eff}} = \frac{1}{\tau_k^\mathrm{rad}} + R_k, \quad \text{with } R_k = \sum_n R_{kn}, \tag{6.55}$$

where the summation extends over all levels $|n\rangle$ that can be populated by collisional transitions $|k\rangle \to |n\rangle$.

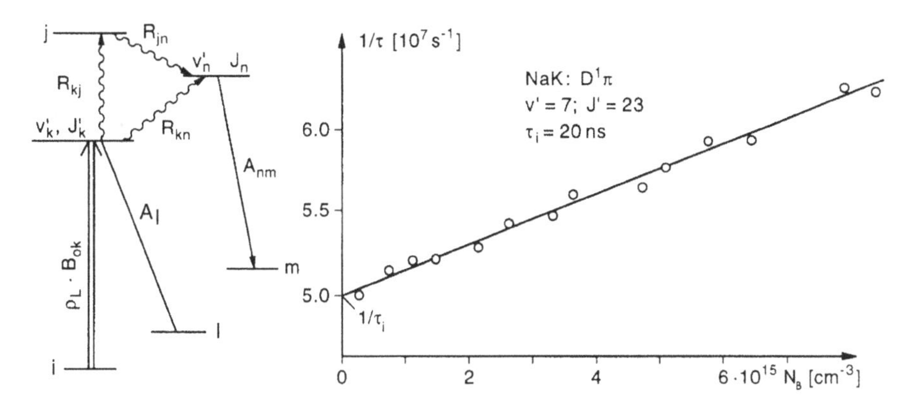

Fig. 6.86 Collisional depopulation of the excited level (v_k', J_k') of a molecule and example of a Stern–Vollmer plot for the NaK level $D^1\Pi_u$ ($v' = 7$, $J' = 13$) depopulated by collisions with argon atoms at the density N_B

In a gas cell at the temperature T the mean relative velocity between collision partners A and B with masses M_A, M_B is

$$\bar{v} = \sqrt{8kT/\pi\mu}, \quad \text{with } \mu = \frac{M_\mathrm{A} M_\mathrm{B}}{M_\mathrm{A} + M_\mathrm{B}}. \tag{6.56}$$

Using the thermodynamic equation of state $p = N \cdot kT$, we can replace the density N_B in (6.54) by the pressure p and obtain the *Stern–Vollmer equation*:

$$\frac{1}{\tau_k^\mathrm{eff}} = \frac{1}{\tau_k^\mathrm{rad}} + b\sigma_k p, \quad \text{with } b = (8/\pi\mu kT)^{1/2}. \tag{6.57}$$

It represents a straight line when $1/\tau^\mathrm{eff}$ is plotted versus p (Fig. 6.86). The slope $\tan\alpha = b\sigma_k$ yields the total quenching cross section σ_k and the intersect with the axis $p = 0$ gives the radiative lifetime $\tau_k^\mathrm{rad} = \tau_k^\mathrm{eff}\,(p = 0)$.

In the following subsections we will discuss some experimental methods of lifetime measurements [794, 795]. Nowadays lasers are generally used for the selective population of excited levels. In this case, the induced emission, which contributes to the depletion of the excited level, has to be taken into account if the exciting laser is not switched off during the fluorescence detection. The rate equation for the time-dependent population density of the level $|k\rangle$, which gives the effective lifetime τ_k^eff, is then

$$\frac{\mathrm{d}N_k}{\mathrm{d}t} = +N_i B_{ik}\rho_\mathrm{L} - N_k(A_k + R_k + B_{ki}\rho_\mathrm{L}), \tag{6.58}$$

where ρ_L is the spectral energy density of the exciting laser, which is tuned to the transition $|i\rangle \to |k\rangle$. The solution $N_k(t) \propto I_\mathrm{Fl}(t)$ of (6.58) depends on the time profile $I_\mathrm{L}(t) = c\rho_\mathrm{L}(t)$ of the excitation laser.

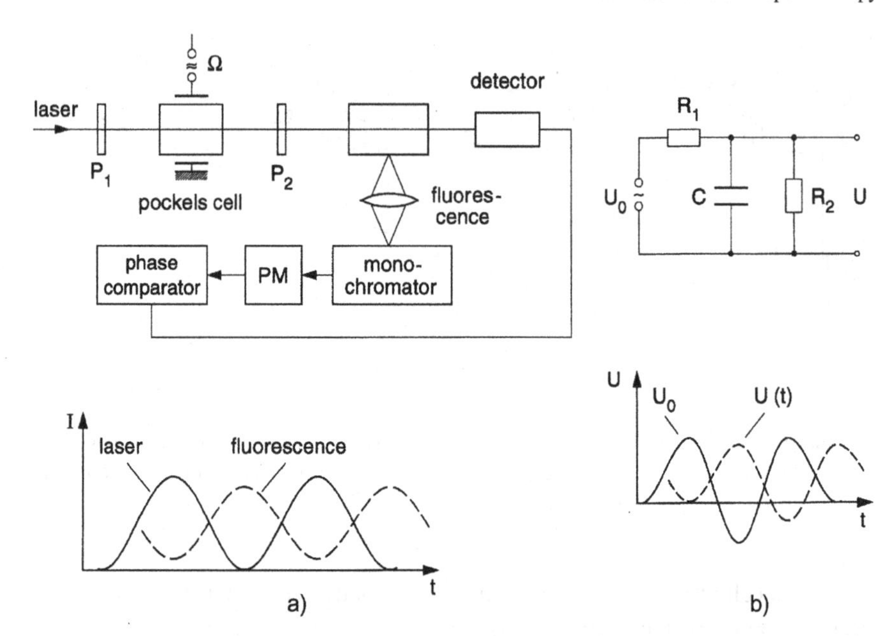

Fig. 6.87 Phase-shift method for the measurement of excited-state lifetimes: (**a**) experimental arrangement; and (**b**) equivalent electric network

6.3.1 Phase-Shift Method

If the laser is tuned to the center frequency ω_{ik} of an absorbing transition $|i\rangle \rightarrow |k\rangle$, the detected fluorescence intensity I_{Fl} monitored on the transition $|k\rangle \rightarrow |m\rangle$ is proportional to the laser intensity I_{L} as long as saturation can be neglected. In the *phase-shift method* the laser intensity is modulated at the frequency $f = \Omega/2\pi$ (Fig. 6.87a) according to

$$I_{\mathrm{L}}(t) = \frac{1}{2} I_0 (1 + a \sin \Omega t) \cos^2 \omega_{ik} t \quad \text{with } |a| \leq 1. \tag{6.59}$$

Inserting (6.59) with $I_{\mathrm{L}}(t) = c\rho_{\mathrm{L}}(t)$ into (6.58) yields the time-dependent population density $N_k(t)$ of the upper level, and therefore also the fluorescence power $P_{\mathrm{Fl}}(t) = N_k(t) A_{km}$ emitted on the transition $|k\rangle \rightarrow |m\rangle$. The result is

$$P_{\mathrm{Fl}}(t) = b \left[1 + \frac{a \sin(\Omega t + \phi)}{[1 + (\Omega \tau_{\mathrm{eff}})^2]^{1/2}} \right] \cos^2 \omega_{km} t, \tag{6.60}$$

where the constant $b \propto N_i \sigma_{ik} I_{\mathrm{L}} V$ depends on the density N_i of the absorbing molecules, the absorption cross section σ_{ik}, the laser intensity I_{L}, and the excitation volume V seen by the fluorescence detector. Since the detector averages over the optical oscillations ω_{km} we obtain $\langle \cos^2 \omega_{km} \rangle = 1/2$, and (6.60) gives similarly to (6.59) a sinewave-modulated function with a reduced amplitude and the phase shift ϕ against the exciting intensity $I_L(t)$. This phase shift depends on the modulation frequency Ω and the effective lifetime τ_{eff}. The evaluation yields

$$\tan\phi = \Omega\tau_{eff}. \tag{6.61}$$

According to (6.55–6.58) the effective lifetime is determined by the inverse sum of all deactivation processes of the excited level $|k\rangle$. In order to obtain the spontaneous lifetime $\tau_{spont} = 1/A_k$ one has to measure $\tau_{eff}(p, I_L)$ at different pressures p and different laser intensities I_L, and extrapolate the results towards $p \to 0$ and $I_L \to 0$. The influence of induced emission is a definite drawback of the phase-shift method. It can be eliminated if ϕ is measured at different intensities I_L with the extrapolation $\phi(I_L \to 0)$.

Note This problem of exciting atoms with sine wave-modulated light and determining the mean lifetime of their exponential decay from measurements of the phase shift ϕ is mathematically completely equivalent to the well-known problem of charging a capacitor C from an ac source with the voltage $U_0(t) = U_1 \sin\Omega t$ through the resistor R_1 with simultaneous discharging through a resistor R_2 (Fig. 6.87b). The equation corresponding to (6.58) is here

$$C\frac{\mathrm{d}U}{\mathrm{d}t} = \frac{U_0 - U}{R_1} - \frac{U}{R_2}, \tag{6.62}$$

which has the solution

$$U = U_2 \sin(\Omega t - \phi), \quad \text{with } \tan\phi = \Omega\frac{R_1R_2C}{R_1 + R_2}, \tag{6.63}$$

where

$$U_2 = U_0\frac{R_2}{[(R_1 + R_2)^2 + (\Omega CR_1R_2)^2]^{1/2}}.$$

A comparison with (6.60) shows that the mean lifetime τ corresponds to the time constant $\tau = RC$ with $R = R_1R_2/(R_1 + R_2)$ and the laser intensity to the charging current $I(t) = (U_0 - U)/R_1$.

Equation (6.61) anticipates a pure exponential decay. This is justified if a single upper level $|k\rangle$ is selectively populated. If several levels are simultaneously excited the fluorescence power $P_{Fl}(t)$ represents a superposition of decay functions with different decay times τ_k. In such cases the phase shifts $\phi(\Omega)$ and the amplitudes $a/(1 + \Omega^2\tau^2)^{1/2}$ have to be measured for different modulation frequencies Ω. The mathematical analysis of the results allows one to separate the contributions of the simultaneously excited levels to the decay curve and to determine the different lifetimes of these levels [796]. A better solution is, however, if the fluorescence is dis-

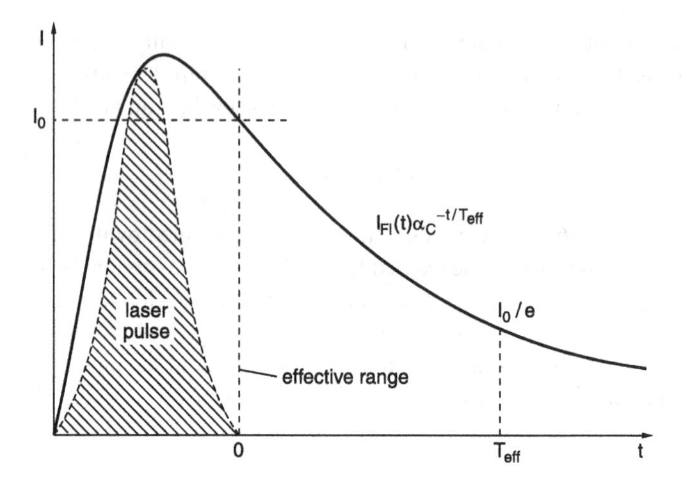

Fig. 6.88 Lifetime measurement after pulse excitation

persed by a monochromator and the detector selectively monitors a single transition from each of the different excited levels $|k_n\rangle$ separately.

6.3.2 Single-Pulse Excitation

The molecules are excited by a short laser pulse. The trailing edge of this pulse should be short compared with the decay time of the excited level, which is directly monitored after the end of the excitation pulse (Fig. 6.88). Either the time-resolved LIF on transitions $|k\rangle \rightarrow |m\rangle$ to lower levels $|m\rangle$ is detected or the time-dependent absorption of a second laser, which is tuned to the transition $|k\rangle \rightarrow |j\rangle$, to higher levels $|j\rangle$.

The time-dependent fluorescence can be viewed either with an oscilloscope or may be recorded by a transient recorder. Another method is based on a boxcar integrator, which transmits the signal through a gate that opens only during a selected time interval Δt (Vol. 1, Sect. 4.5.6). After each successive excitation pulse the delay ΔT of the gate is increased by T/m. After m excitation cycles the whole time window T has been covered (Fig. 6.89). The direct observation of the decay curve on an oscilloscope has the advantage that nonexponential decays can be recognized immediately. For sufficiently intense fluorescence one needs only a single excitation pulse, although generally averaging over many excitation cycles will improve the signal-to noise ratio.

This technique of single-pulse excitation is useful for low repetition rates. Examples are the excitation with pulsed dye lasers pumped by Nd:YAG or excimer lasers [794, 797].

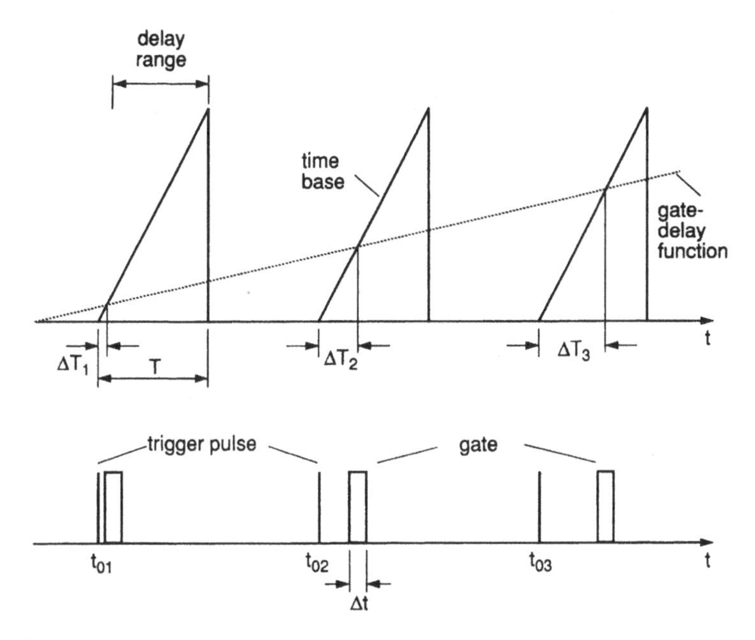

Fig. 6.89 Lifetime measurements with a gated boxcar system with successively increasing gate delay time

6.3.3 Delayed-Coincidence Technique

As in the previous method the delayed-coincidence technique also uses short laser pulses for the excitation of selected levels. However, here the pulse energy is kept so low that the detection probability P_D of a fluorescence photon per laser excitation pulse remains small ($P_\mathrm{D} \leq 0.1$). If $P_\mathrm{D}(t)\,\mathrm{d}t$ is the probability of detecting a fluorescence photon in the time interval t to $t + \mathrm{d}t$ after the excitation then the mean number $n_\mathrm{Fl}(t)$ of fluorescence photons detected within this time interval for N excitation cycles ($N \gg 1$) is

$$n_\mathrm{Fl}(t)\,\mathrm{d}t = N P_\mathrm{D}(t)\,\mathrm{d}t. \tag{6.64}$$

The experimental realization is shown schematically in Fig. 6.90. Part of the laser pulse is send to a fast photodiode. The output pulse of this diode at $t = t_0$ starts a time-amplitude converter (TAC), which generates a fast-rising voltage ramp $U(t) = (t - t_0)U_0$. A photomultiplier with a large amplification factor generates for each detected fluorescence photon an output pulse that triggers a fast discriminator. The normalized output pulse of the discriminator stops the TAC at time t. The amplitude $U(t)$ of the TAC output pulse is proportional to the delay time $t - t_0$ between the excitation pulse and the fluorescence photon emission. These pulses are stored in a multichannel analyzer. The number of events per channel gives the number of fluorescence photons emitted at the corresponding delay time.

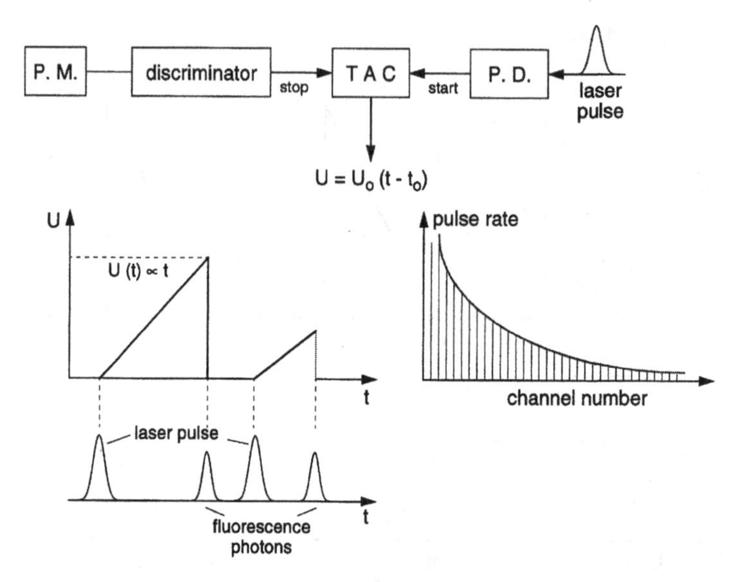

Fig. 6.90 Basic principle of lifetime measurements with the delayed-coincidence single-photon counting technique

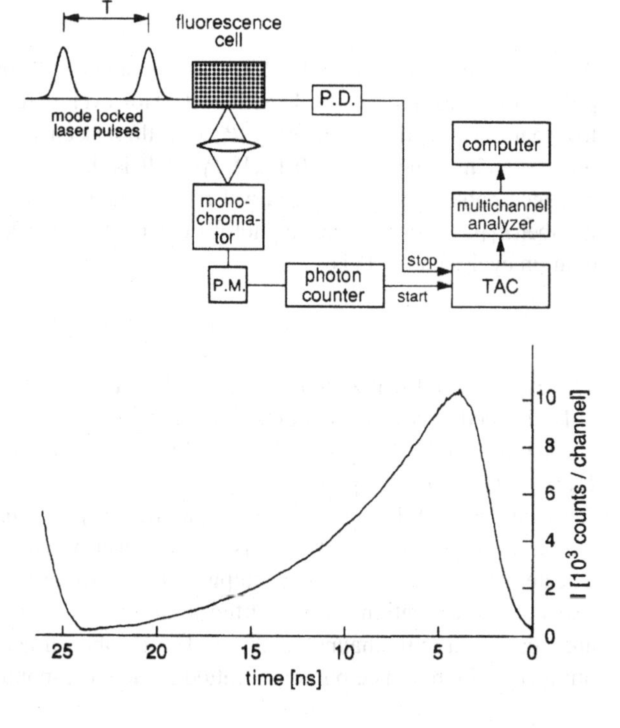

Fig. 6.91 Experimental arrangement for lifetime measurements with the delayed-coincidence single-photon counting technique and decay curve of the Na_2 $(B^1\Pi_u\ v'=6, J'=27)$ level [656]

The repetition rate f of the excitation pulses is chosen as high as possible since the measuring time for a given signal-to-noise-ratio is proportional to $1/f$. An upper limit for f is determined by the fact that the time $T = 1/f$ between two successive laser pulses should be at least three times the lifetime τ_k of the measured level $|k\rangle$. This technique is therefore ideally suited for excitation with mode-locked or cavity-dumped lasers. There is, however, an electronic bottleneck: the input pulse rate of a TAC is limited by its dead time τ_D and should be smaller than $1/\tau_D$. It is therefore advantageous to invert the functions of the start and stop pulses. The fluorescence pulses (which have a much smaller rate than the excitation pulses) now act as start pulses and the next laser pulse stops the TAC. This means that the time $(T - t)$ is measured instead of t. Since the time T between successive pulses of a mode-locked laser is very stable and can be accurately determined from the mode-locking frequency $f = 1/T$, the time interval between successive pulses can be used for time calibration of the detection system [656]. In Fig. 6.91 the whole detection system is shown together with a decay curve of an excited level of the Na_2 molecule, measured over a period of 10 min. More information about the delayed-coincidence method can be found in [798].

6.3.4 Lifetime Measurements in Fast Beams

The most accurate method for lifetime measurements in the range of 10^{-7}–10^{-9} s is based on a modern version of an old technique that was used by W. Wien 70 years ago [799]. Here a time measurement is reduced to measurements of a pathlength and a velocity.

The atomic or molecular ions produced in an ion source are accelerated by the voltage U and focused to form an ion beam. The different masses are separated by a magnet (Fig. 6.92) and the wanted ions are excited at the position $x = 0$ by a cw laser beam. The LIF is monitored as a function of the variable distance x between the excitation region and the position of a special photon detector mounted on a precision translational drive. Since the velocity $v = (2eU/m)^{1/2}$ is known from the measured acceleration voltage U, the time $t = x/v$ is determined from the measured positions x.

The excitation intensity can be increased if the excitation region is placed inside the resonator of a cw dye laser that is tuned to the selected transition. Before they reach the laser beam, the ions can be preexcited into highly excited long-living levels by gas collisions in a differentially pumped gas cell (Fig. 6.93a). This opens new transitions for the laser excitation and allows lifetime measurements of high-lying ionic states even with visible lasers [800].

The ions can be neutralized by charge-exchange collisions in differentially pumped alkali-vapor cells. Since charge exchange occurs with large collision cross sections at large impact parameters (grazing collisions), the momentum transfer is very small and the velocity of the neutrals is nearly the same as that of the ions. With this technique lifetimes of highly excited neutral atoms or molecules can be measured with high accuracy.

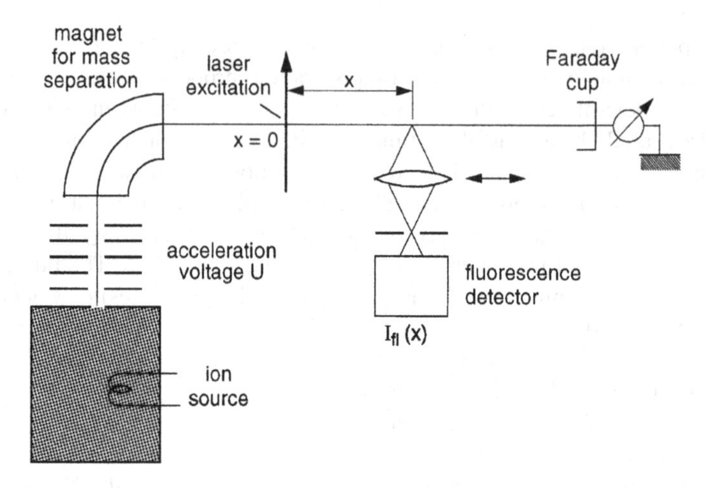

Fig. 6.92 Lifetime measurements of highly excited levels of ions or neutral atoms and molecules in a fast beam

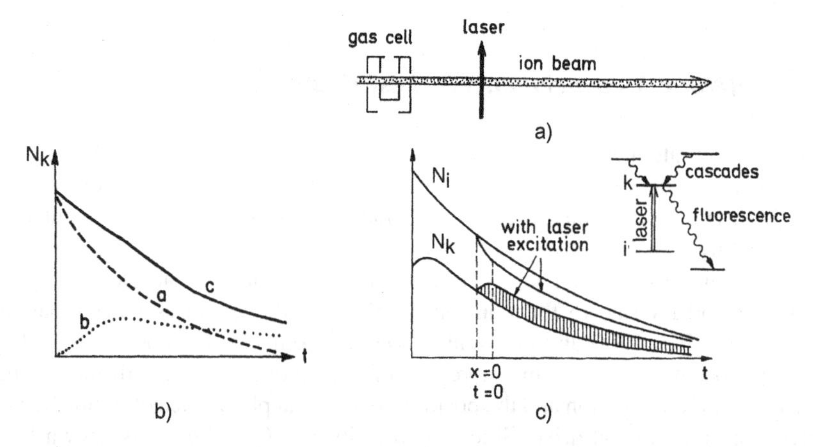

Fig. 6.93 Cascade-free lifetime measurements despite the simultaneous excitation of many levels: (**a**) preexcitation by collisions in a gas cell with subsequent laser excitation; (**b**) decay of level $|k\rangle$ without cascading (*curve a*), its feeding by cascades (*curve b*), and resulting population $N_k(t)$ with cascading and decaying (*curve c*). (**c**) The fluorescence $I(x,\lambda)$ is measured alternately with and without selective laser excitation

Collisional preexcitation has the drawback that several levels are simultaneously excited, which may feed by cascading fluorescence transitions the level $|k\rangle$ whose lifetime is to be measured. These cascades alter the time profile $I_{\mathrm{Fl}}(t)$ of the level $|k\rangle$ and falsify the real lifetime τ_k (Fig. 6.93b). This problem can be solved by a special measurement cycle: for each position x the fluorescence is measured alternately with and without laser excitation (Fig. 6.93c). The difference of both counting rates yields the LIF without cascade contributions. In order to eliminate fluctuations of the laser intensity or the ion beam intensity a second detector is installed at the fixed

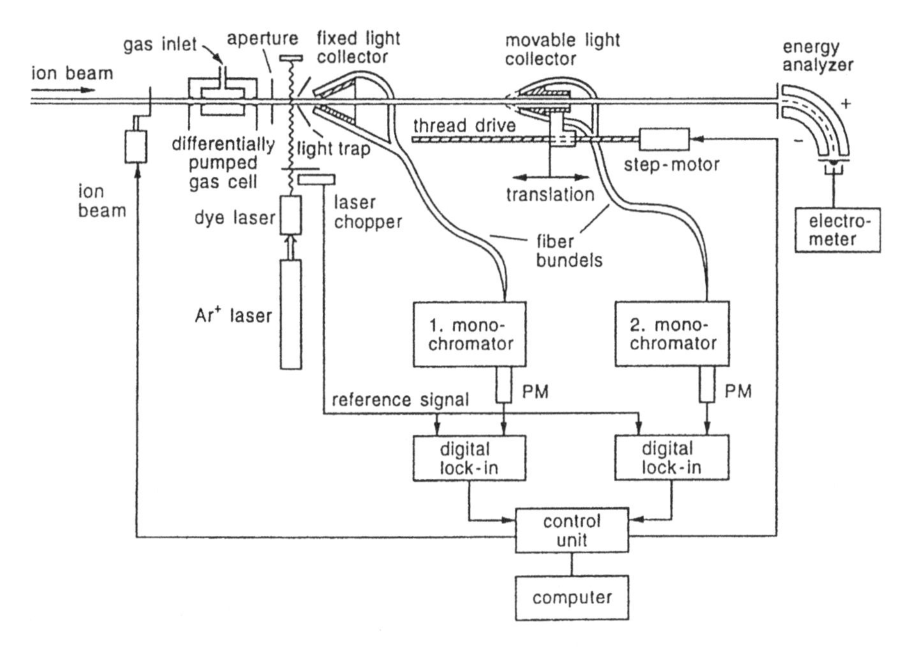

Fig. 6.94 Experimental arrangement for cascade-free lifetime measurements in fast beams of ions or neutrals with fluorescence collection by conically shaped optical-fiber bundles

position x_0 (Fig. 6.94). The normalized ratios $S(x)/S(x_0)$, which are independent of these fluctuations, are then fed into a computer that fits them to a theoretical decay curve [801].

The time resolution Δt of the detectors is determined by their spatial resolution Δx and the velocity v of the ions or neutrals. In order to reach a good time resolution, which is *independent* of the position x of the detector, one has to take care that the detector collects the fluorescence only from a small path interval Δx, but still sees the whole cross section of the slightly divergent ion beam. This can be realized by specially designed bundles of optical fibers, which are arranged in a conical circle around the beam axis (Fig. 6.94), while the outcoupling end of the fiber bundle has a rectangular form which is matched to the entrance slit of a spectrograph.

Lifetimes of atoms and ions have been measured very accurately with this technique. More experimental details and different versions of this laser beam method can be found in the extensive literature [800–804].

Example 6.6 Ne ions (23 atomic mass units, AMU) accelerated by $U = 150$ kV have a velocity $v = 10^6$ m/s. In order to reach a time resolution of 1 ns the spatial resolution of the detector must be $\Delta x = 1$ mm.

High-lying excited states of atoms and ions, which are of interest in astrophysics, can be produced by laser ablation from surfaces. The expanding plasma cloud,

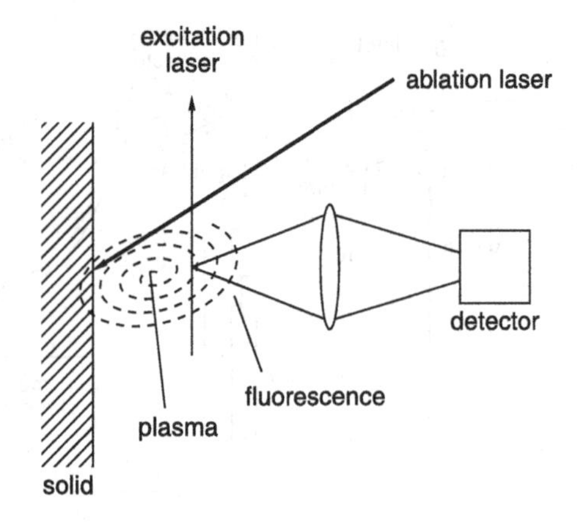

Fig. 6.95 Laser ablation from surfaces and laser excitation of the plasma plume with fluorescence detection

which consists of hot atoms and ions, is irradiated by a second laser, and the time-resolved fluorescence of the excited states is monitored. Since many of these states can only be reached by VUV lasers, the excitation must be performed in a vacuum chamber and VUV-sensitive detectors (for instance solar blind multipliers) must be used [805]. A typical experimental setup is shown in Fig. 6.95. The excitation laser can be time-delayed with respect to the ablation laser by a electronic delay unit which allows delay times of $\Delta\tau > 1$ ns.

6.4 Spectroscopy in the Pico-to-Attosecond Range

For measurements of very fast relaxation processes that demand a time resolution below 10^{-10} s most detectors (except the streak camera) are not fast enough. Here the pump-and-probe technique is the best choice. It is based on the following principle shown in Fig. 6.96.

The molecules under investigation are excited by a fast laser pulse on the transition $|0\rangle \to |1\rangle$. A probe pulse with a variable time delay τ against the pump pulse probes the time evolution of the population density $N_1(t)$. The time resolution is only limited by the pulse width ΔT of the two pulses, but not by the time constants of the detectors!

In early experiments of this kind a fixed-frequency mode-locked Nd:glass or Nd:YAG laser was used. Both pulses came from the same laser and fortuitious coincidences of molecular transitions with the laser wavelength were utilized [806]. The time delay of the probe pulse is realized, as shown in Fig. 6.97, by beam splitting and a variable path-length difference. Since the pump and probe pulses coincide with the same transition $|i\rangle \to |k\rangle$, the absorption of the probe pulse measured as a function of the delay time τ, in fact, monitors the time evolution of the population difference $[N_k(t) - N_i(t)]$. A larger variety of molecular transitions becomes accessible if the Nd:YAG laser wavelength is Raman shifted (Vol. 1, Sect. 6.8) into spectral regions of interest [807].

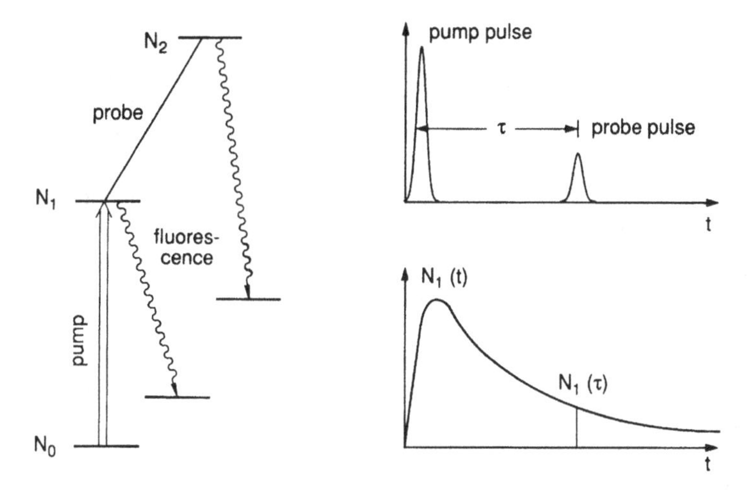

Fig. 6.96 Pump-and-probe technique

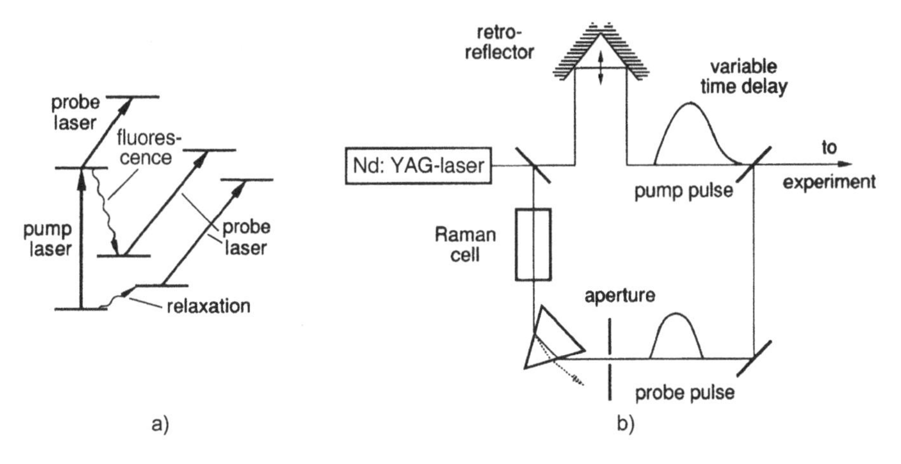

Fig. 6.97 Pump-and-probe technique for the measurements of ultrafast relaxation processes

A broader application range is opened by a system of two independently tunable mode-locked dye lasers, which have to be pumped by the same pump laser in order to synchronize the pump and probe pulses [808]. For studies of vibrational levels in the electronic ground states of molecules the difference frequency generation of these two dye lasers can be used as a tunable infrared source for direct excitation of selected levels on infrared-active transitions. Raman-active vibrations can be excited by spontaneous or stimulated Raman transitions (Chap. 3). Another useful short-pulse source for these experiments is a three-wavelength Ti:sapphire laser, where two of the wavelengths can be indepently tuned [811].

In addition, short-pulse tunable optical parametric oscillators have been realized, where the pump wavelength and the signal or idler waves can be used for pump-and-probe experiments [809]. The wide tuning range allows more detailled investigations compared to the restricted use of fixed frequency lasers [810]. Here the different femtosecond NOPAs (see Sect. 6.1.10) in combination with frequency doubling or sum-frequency generation provide the spectroscopist with widely tunable ultrafast intense radiation sources which can be tuned across the infrared and visible to the UV region. These open the door to studies of fast dynamical processes in chemistry and biology (see Sect. 10.3).

Recently it has become possible to phase-lock two different femtosecond lasers. This opens many possibilities for spectroscopic applications. One example is the use of an infrared femtosecond pulse to excite nuclear vibrations of molecules and a second UV femtosecond pulse for electronic excitation. This allows one to study the influence of fast changes of the electron cloud on the nuclear oscillation period [812].

These developments widen the range of applications considerably. We will now give some examples of applications of the pump-and-probe technique.

6.4.1 *Pump-and-Probe Spectroscopy of Collisional Relaxation in Liquids*

Because of the high molecular density in liquids, the average time τ_c between two successive collisions of a selectively excited molecule A with other molecules A of the same kind or with different molecules B is very short (10^{-12}–10^{-11} s). If A has been excited by absorption of a laser photon its excitation energy may be rapidly redistributed among other levels of A by collisions, or it may be transferred into internal energy of B or into translational energy of A and B (temperature rise of the sample). With the pump-and-probe technique this energy transfer can be studied by measurements of the time-dependent population densities $N_m(t)$ of the relevant levels of A or B. The collisions not only change the population densities but also the phases of the wave functions of the coherently excited levels (Vol. 1, Sect. 2.10). These phase relaxation times are generally shorter than the population relaxation times.

Besides excitation and probing with infrared laser pulses, the CARS technique (Sect. 3.5) is a promising technique to study these relaxation processes. An example is the measurement of the dephasing process of the OD stretching vibration in heavy water D_2O by CARS [813]. The pump at $\omega = \omega_L$ is provided by an amplified 80 fs dye laser pulse from a CPM ring dye laser. The Stokes pulse at ω_s is generated by a synchronized tunable picosecond dye laser. The CARS signal at $\omega_{as} = 2\omega_L - \omega_s$ is detected as a function of the time delay between the pump and probe pulses.

Another example is the deactivation of high vibrational levels in the S_0 and S_1 singlet states of dye molecules in organic liquids pumped by a pulsed laser (Fig. 6.98). The laser populates many vibrational levels in the excited S_1 singlet state, which are accessible by optical pumping on transitions starting from thermally populated levels in the electronic ground state S_0. These excited levels $|v'\rangle$

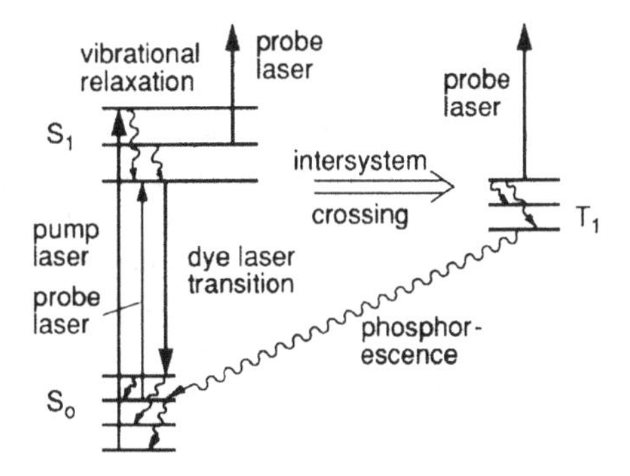

Fig. 6.98 Measurements of fast relaxation processes in excited and ground states

rapidly relax by inelastic collisions into the lowest vibrational level $|v' = 0\rangle$ of S_1, which represents the upper level of the dye laser transition. This relaxation process can be followed up by measuring the time-dependent absorption of a weak probe laser pulse on transitions from these levels into higher excited singlet states.

Fluorescence and stimulated emission on transitions ($v' = 0 \rightarrow v'' > 0$) lead to a fast rise of the population densities $N(v'')$ of high vibrational levels in the S_0 state. This would result in a self-termination of the laser oscillation if these levels were not depopulated quickly enough by collisions. The relaxation of $N(v'')$ toward the thermal equilibrium population $N_0(v'')$ can again be probed by a weak visible probe laser. Polarization spectroscopy (Sect. 2.4) with femtosecond pulses allows one to separately determine the decay times τ_{vib} of population redistribution and the dephasing times [814].

Of particular importance for dye laser physics is the intersystem crossing of dye molecules from the excited S_1 state into levels of the triplet state T_1. Because the population of these long-living triplet levels represents a severe loss for the dye laser radiation because of absorption on electronic transitions to higher triplet states, the time-dependent triplet concentration and possible quenching processes by triplet-quenching additives have been investigated in detail [815]. Furthermore, spin-exchange and transitions by collisions between excited S_1 dye molecules and triplet O_2 molecules or between T_1 dye molecules and excited O_2 ($^1\Delta$) molecules play a crucial role in photodynamical processes in cancer cells (Sect. 10.5).

6.4.2 Electronic Relaxation in Semiconductors

A very interesting problem is concerned with the physical limitations of the ultimate speed of electronic computers. Since any *bit* corresponds to a transition from a nonconducting to a conducting state of a semiconductor, or vice versa, the relaxation time of electrons in the conduction band and the recombination time certainly impose a lower limit for the minimum switching time. This electronic relaxation can be measured with the pump-and-probe technique. The electrons are excited by a femtosecond laser pulse from the upper edge of the valence band into levels with energies $E = \hbar\omega - \Delta E$ (ΔE: band gap) in the conduction band, from where they relax into the lower edge of the conduction band before they recombine with holes in the valence band. Since the optical reflectivity of the semiconductor sample depends on the energy distribution $N(E)$ of the free conduction electrons, the reflection of a weak probe laser pulse can be used to monitor the distribution $N(E)$ [816]. Because of their fast relaxation semiconductors can be used as saturable absorbers for passive mode locking in femtosecond lasers [694]. In this case, a thin semiconductor sheet is placed in front of a resonator mirror (see Sect. 6.1.12). The characteristic time scales for interband and intraband electron relaxation are again measured with the pump-and probe technique [817]. Of particular interest for the magnetic storage of information is the timescale of magnetization or demagnetization of thin magnetic films. It turns out that femtosecond laser pulses focused onto the film can demagnetize a small local area within about 100 fs [818].

6.4.3 Femtosecond Transition State Dynamics

The pump-and-probe technique has proved to be very well suited for studying short-lived transient states of molecular systems that had been excited by a short laser pulse before they dissociate:

$$\mathrm{AB} + h\nu \longrightarrow [\mathrm{AB}]^* \longrightarrow \mathrm{A}^* + \mathrm{B}.$$

An illustrative example is the photodissociation of excited NaI molecules, which has been studied in detail by Zewail et al. [819].

The adiabatic potential diagram of NaI (Fig. 6.99) is characterized by an avoided crossing between the repulsive potential of the two interacting neutral atoms Na + I and the Coulomb potential of the ions $\mathrm{Na}^+ + \mathrm{I}^-$, which is mainly responsible for the strong binding of NaI at small internuclear distances R. If NaI is excited into the repulsive state by a short laser pulse at the wavelength λ_1, the excited molecules start to move toward larger values of R with a velocity $v(R) = [(2/\mu)(E - E_{\mathrm{pot}}(R)]^{1/2}$.

Example 6.7 For $E - E_{\mathrm{pot}} = 1000\ \mathrm{cm}^{-1}$ and $\mu = m_1 m_2/(m_1 + m_2) =$ 19.5 AMU $\rightarrow v \approx 10^3$ m/s. The time $\Delta T = \Delta R/v$ of passing through an interval of $\Delta R = 0.1$ nm is then $\Delta T = 10^{-13}\ \mathrm{s} = 100$ fs.

When the excited system [NaI]* reaches the avoided crossing at $R = R_\mathrm{c}$ it may either stay on the potential $V_1(R)$ and oscillate back and forth between R_1 and R_2, or it may tunnel to the potential curve $V_0(R)$, where it separates into Na + I.

The time behavior of the system can be probed by a probe pulse with the wavelength λ_2 tuned to the transition from $V_1(R)$ into the excited state $V_2(R)$ that dissociates into $\mathrm{Na}^* + \mathrm{I}$. At the fixed wavelength $\lambda_2 = 2\pi c/\omega_2$ the dissociating system

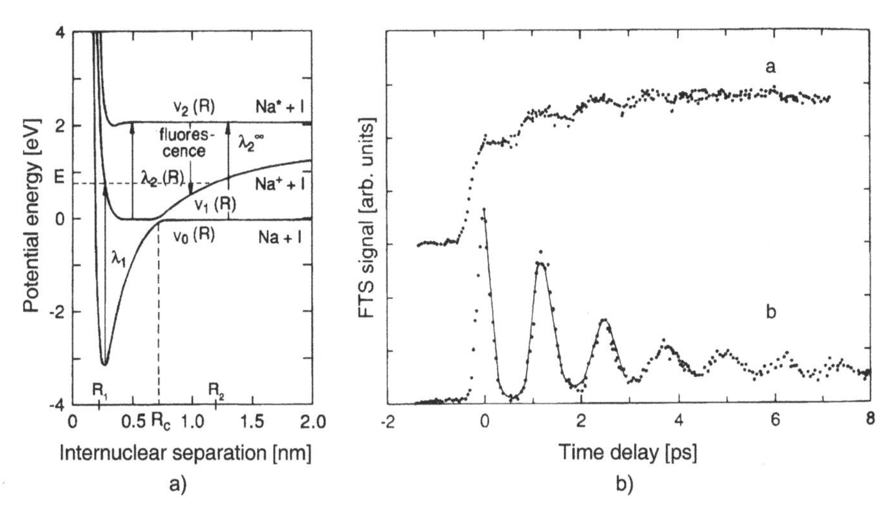

Fig. 6.99 (**a**) Potential diagram of NaI with the pump transition at λ_1 and the tunable probe pulse at $\lambda_2(R)$. (**b**) Fluorescence intensity $I_{\mathrm{Fl}}(\Delta t)$ as a function of the delay time Δt between pump and probe pulses: (*curve a*) with λ_2 tuned to the atomic Na* transition and (*curve b*) λ_2 tuned to $\lambda_2(R)$ with $R < R_c$ [819]

absorbs the probe pulse only at that distance R where $V_1(R) - V_2(R) = \hbar\omega_2$. If λ_2 is tuned to the sodium resonance line $3s \rightarrow 3p$, absorption occurs for $R = \infty$.

Since the dissociation time is very short compared to the lifetime of the excited sodium atom Na*(3p), the dissociating (NaI)* emits nearly exclusively at the atomic resonance fluorescence. The atomic fluorescence intensity $I_{\mathrm{Fl}}(\mathrm{Na}^*, \Delta t)$, monitored in dependence on the delay time Δt between the pump-and-probe pulse, gives the probability for finding the excited system [NaI]* at a certain internuclear separation R, where $V_1(R) - V_2(R) = \hbar\omega_2$ (Fig. 6.99b).

The experimental results [820] shown in Fig. 6.99b reflect the oscillatory movement of Na*I(R) on the potential $V_1(R)$ between R_1 and R_2, which had been excited at the inner turning point R_1 by the pump pulse at $t = 0$. This corresponds to a periodic change between the covalent and ionic potential. The damping is due to the leakage into the lower-state potential around the avoided crossing at $R = R_c$. If λ_2 is tuned to the atomic resonance line, the accumulation of Na*(3p) atoms can be measured when the delay between the pump-and-probe pulse is increased.

6.4.4 Real-Time Observations of Molecular Vibrations

The time scale of molecular vibrations is on the order of 10^{-13}–10^{-15} s. The vibrational frequency of the H_2 molecule, for example, is $\nu_{\mathrm{vib}} = 1.3 \times 10^{14}\ \mathrm{s}^{-1} \rightarrow T_{\mathrm{vib}} = 7.6 \times 10^{-15}$ s, that of the Na_2 molecule is $\nu_{\mathrm{vib}} = 4.5 \times 10^{12}\ \mathrm{s}^{-1} \rightarrow T_{\mathrm{vib}} = 2 \times 10^{-13}$ s, and even the heavy I_2 molecule still has $T_{\mathrm{vib}} = 5 \times 10^{-13}$ s. With

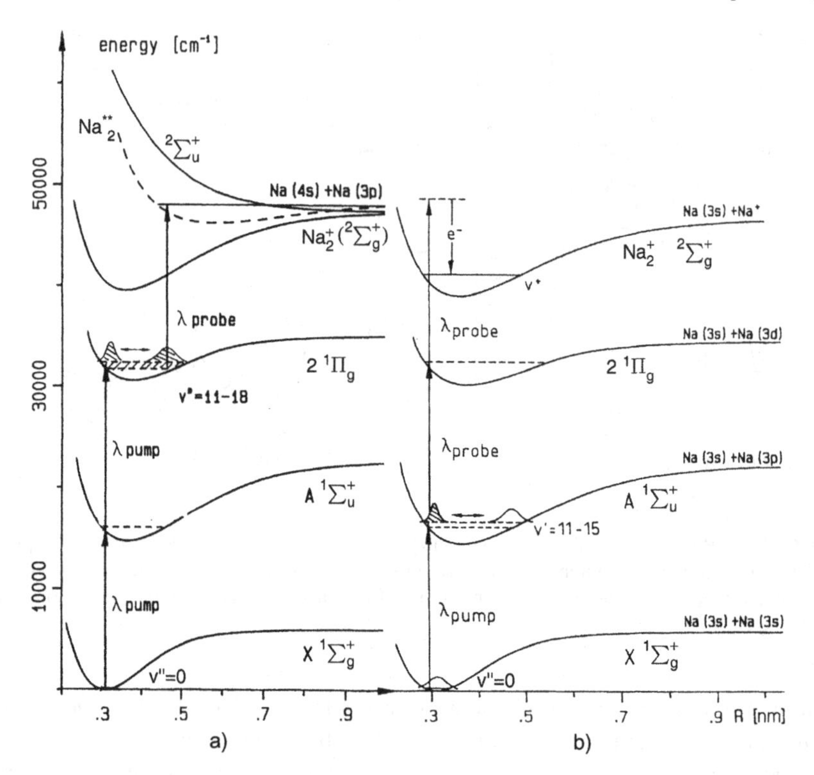

Fig. 6.100 Femtosecond spectroscopy of Na_2: (**a**) potential curve diagram illustrating the preparation of a vibrational wave packet in the $2^1\Pi_g$ state of Na_2 due to coherent simultaneous two-photon excitation of vibrational levels ($v' = 11$–18). Further excitation by a third photon results in production of $Na_2^{**} \to Na^+ + Na^*$ from the outer turning point of the wave packet. (**b**) One-photon excitation of a vibrational wave packet in the $A^1\Sigma_u$ state with subsequent two-photon ionization from the *inner* turning point [822]

conventional techniques one always measures a time average over many vibrational periods.

With femtosecond pump-and-probe experiments "fast motion pictures" of a vibrating molecule may be obtained, and the time behavior of the wave packets of coherently excited and superimposed molecular vibrations can be mapped. This is illustrated by the following examples dealing with the dynamics of molecular multiphoton ionization and fragmentation of Na_2, and its dependence on the phase of the vibrational wave packet in the intermediate state [821]. There are two pathways for photoionization of cold Na_2 molecules in a supersonic beam (Fig. 6.100):

(i) One-photon absorption of a femtosecond pulse ($\lambda = 672$ nm, $\Delta T = 50$ fs, $I = 50\ \mathrm{GW/cm^2}$) leads to simultaneous coherent excitation of vibrational levels $v' = 11$–15 in the $A^1\Sigma_u$ state of Na_2 at the inner part of the potential $V_1(R)$. This generates a vibrational wave packet, which oscillates at a frequency of $3 \times 10^{12}\ \mathrm{s^{-1}}$ back and forth between the inner and outer turning point. Resonant

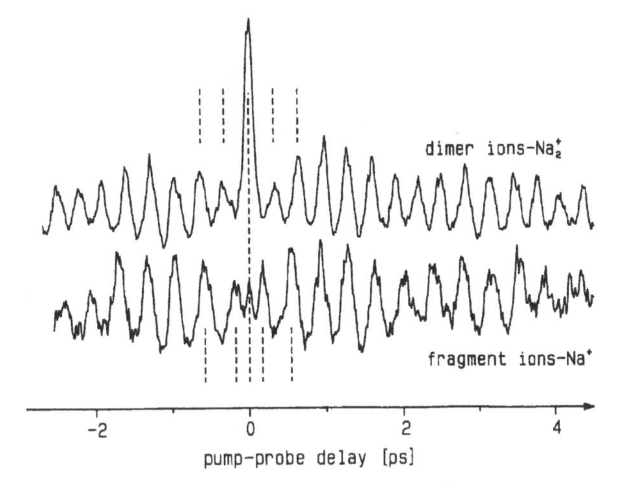

Fig. 6.101 Observed ion rates $N(Na_2^+)$ (*upper trace*) and $N(Na^+)$ (*lower trace*) as a function of the delay time Δt between the pump and probe pulses [822]

enhanced two-photon ionization of the excited molecules by the probe pulse has a larger probability at the inner turning point than at the outer turning point, because of favorable Franck–Condon factors for transitions from the $A^1\Sigma_u$ state to the near-resonant intermediate state $2^1\Pi_g$, which enhances ionizing two-photon transitions at small values of the internuclear distance R (Fig. 6.100b). The ionization rate $N(Na_2^+, \Delta t)$ monitored as a function of the delay time Δt between the weak pump pulse and the stronger probe pulse, yields the upper oscillatory function of Fig. 6.101 with a period that matches the vibrational wave-packet period in the $A^1\Sigma_u$ state.

(ii) The second possible competing process is the two-photon excitation of wavepackets of the $v' = 11$–18 vibrational levels in the $2^1\Pi_g$ state of Na_2 by the pump pulse, with subsequent one-photon excitation into a doubly excited state of Na_2^{**}, which autoionizes according to

$$Na_2^{**} \longrightarrow Na_2^+ + e^- \longrightarrow Na^+ + Na^* + e^-,$$

and results in the generation of Na^+ ions. The number $N(Na^+, \Delta t)$ of atomic ions Na^+, measured as a function of the delay time Δt between pump and probe pulses, shows again an oscillatory structure (Fig. 6.101, lower trace), but with a time shift of half a vibrational period against the upper trace. In this case, the ionization starts from the *outer* turning point of the $2^1\Pi_g$ levels and the oscillatory structure shows a 180° shift and slightly different oscillation period, which corresponds to the vibrational period in the $2^1\Pi_g$ state.

The photoelectrons and ions and their kinetic energies can be measured with two time-of-flight mass spectrometers arranged into opposite directions perpendicular to the molecular and the laser beams [823, 824].

An interesting application is the laser induced isomerization of molecules which is shown schematically for the example of the stilben molecule in Fig. 6.102.

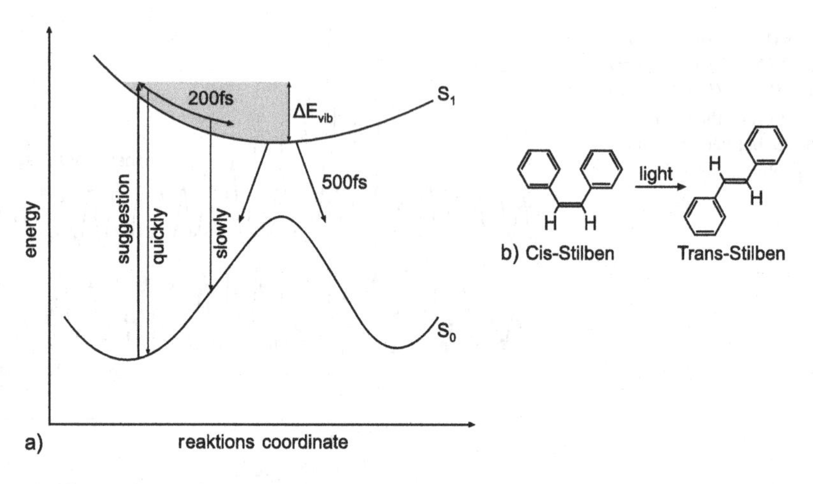

Fig. 6.102 (**a**) Schematic potential diagram for illustration of the isomerization of Stilben; (**b**) transition from Cis- to trans-Stilben

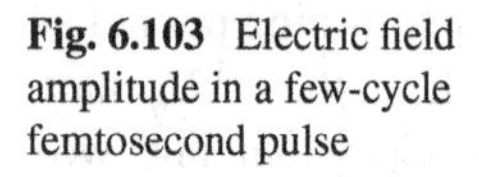

Fig. 6.103 Electric field amplitude in a few-cycle femtosecond pulse

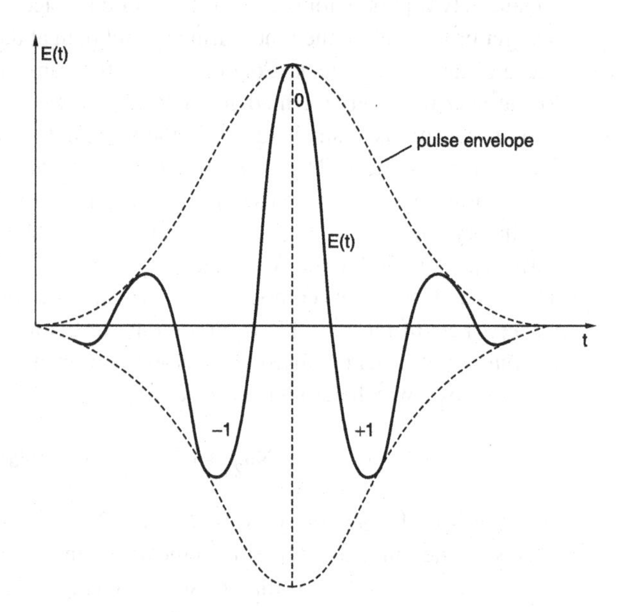

6.4.5 Attosecond Spectroscopy of Atomic Inner Shell Processes

For inner-shell spectroscopy, where the energy separation of atomic states ranges from several hundred eV up to keV, the higher harmonics generated by high-intensity femtosecond pulses can be used [825]. Unfortunately these high harmonic pulses with pulse widths in the attosecond range appear as pulse trains with a repetition frequency twice that of the optical frequency of the generating pulse, i.e., a time separation of about 2.3 fs because they are generated always at the maxima

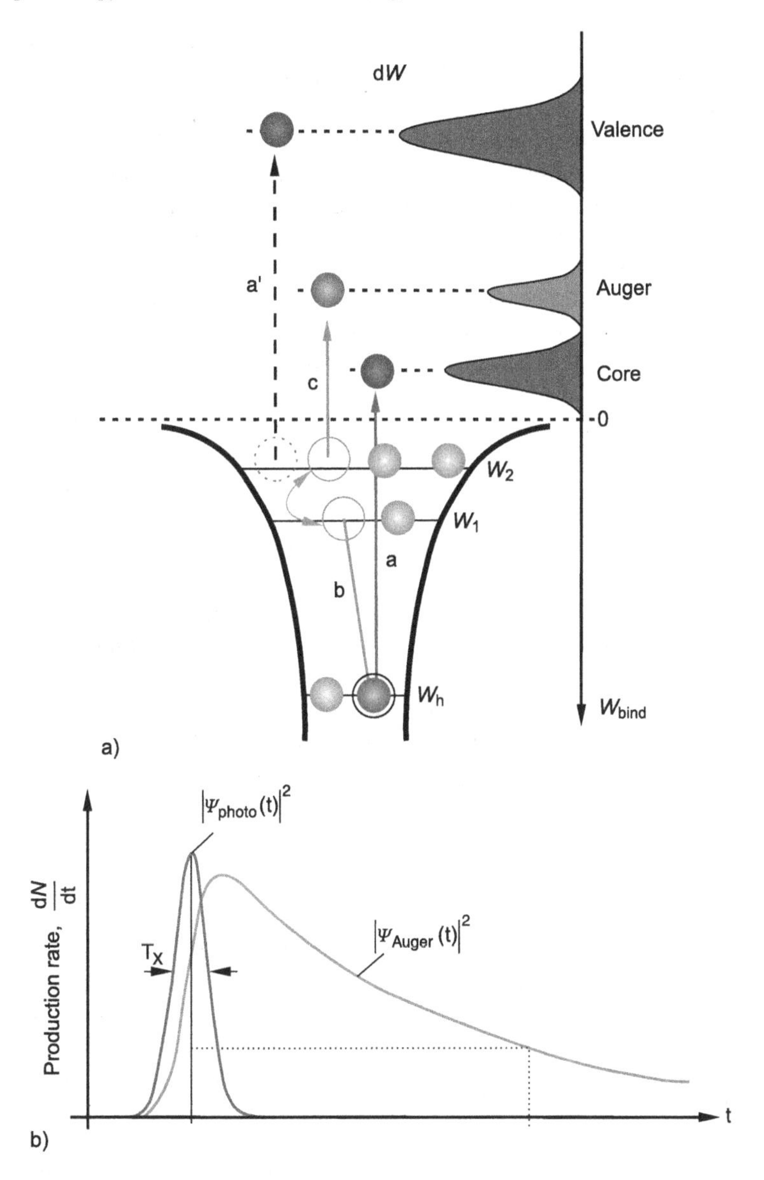

Fig. 6.104 (**a**) Schematic illustration of inner shell excitation by a VUV pulse (a) and relaxation (b) into the vacancy, resulting in the emission of an Auger electron (c) or the excitation into a higher state (a') [826]. (**b**) Exitation probability $|\Psi_{\text{photo}}(t)|^2$ and Auger process as a function of time

of the electric field amplitude of the fundamental optical wave. The interpretation of results from time-resolved spectroscopy obtained through atomic excitation by such pulse trains is not unambiguous. Therefore, it is desirable to generate single pulses instead of pulse trains. This is possible if a pulse of less than 5 fs is used for the generation of higher harmonics. Using phase control of such a pulse, one maximum of the field amplitude $E(t)$ can be shifted into the maximum of the pulse envelope. The initial and the following maximum then have a lower amplitude (Fig. 6.103). Since the intensity of the nth harmonic scales with the $2n$th power $E^{2n}(t)$ of the optical electric field in the femtosecond pulse, only this largest field maximum E_0 essentially contributes to the generation of higher harmonics. With $n = 15$, for example, the generation of XUV by the adjacent field extrema E_{-1} and E_{+1} in Fig. 6.103, which reach only 60 % of the central maximum, has a probability of 10^{-7} compared to that for the central maximum.

Using such short attosecond XUV pulses, the temporal evolution of the Auger process after inner shell excitation can be followed with the pump-and-probe technique.

An XUV pulse excites an electron from the inner shell E_i of an atom (Fig. 6.104), producing an inner shell vacancy which is rapidly filled by an electron from a higher shell with energy E_k. The energy difference $\Delta E = E_k - E_i$ is either carried away by an XUV photon or it is transferred to another electron in an outer shell E_m with $IP - E_m < \Delta E$ (Auger electron). The delay time of this Auger electron with respect to the excitation pulse corresponds exactly to the lifetime of the inner shell vacancy. The emitted Auger electron leaves a vacancy in the outer shell which can be detected by the corresponding decrease in the photoionization with visible light. The delay time can be measured with attosecond time resolution by using a fraction of the visible femtosecond pulse with variable delay with respect to the excitation pulse, in the same approach as used for photoionization with femtosecond pulses, as discussed in Sect. 6.1.13 [826].

6.4.6 Transient Grating Techniques

If two light pulses of different propagation directions overlap in an absorbing sample, they produce an interference pattern because of the intensity-dependent saturation of the population density (Fig. 6.105). When a probe pulse is sent through the overlap region in the sample, this interference pattern shows up as periodic change of the sample transmission and therefore acts as a grating that produces diffraction orders of the probe beam. The grating vector is $k_G = k_2 - k_1$, and the grating period depends on the angle Θ between the two pump beams. The grating amplitude can be inferred from the relative intensity of the different diffraction orders. This gives information on the saturation intensities. The grating will fade away if the delay times τ_1 and τ_2 are larger than the relaxation time of the sample molecules. Therefore this technique of transient gratings gives information on the dynamics of the sample [827].

Most experiments have been performed in solid or liquid samples where the relaxation times are in the range of femto- to picoseconds [824].

There are numerous other examples where pico- and femtosecond spectroscopy have been applied to problems in atomic and molecular physics. Some of them are discussed in Sects. 7.7 and 10.2.

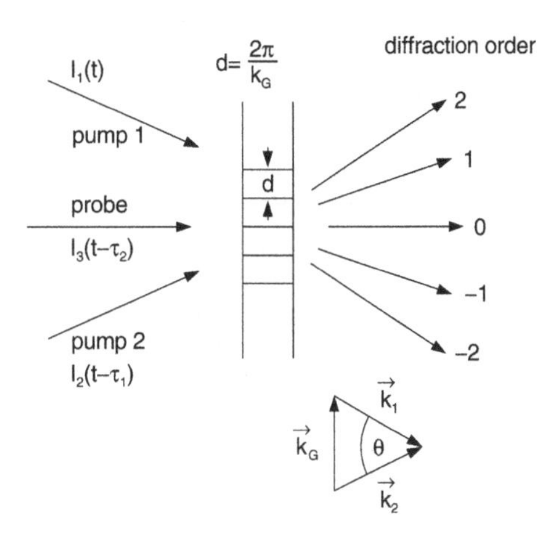

Fig. 6.105 Schematic diagram of a transient grating experiment

In particular the high peak powers now available by the invention of new techniques (Fig. 6.61) allow a new class of experiments in nonlinear physics. Examples are the generation of high harmonics up to the 60th overtone, which generates XUV frequencies with $\lambda = 13$ nm from the fundamental wave with $\lambda = 800$ nm, or multiphoton ionization to produce highly charged ions from neutral atoms. The electric fields of such high power laser pulses exceed the inner-atomic field strength resulting in field ionization. The behavior of atoms and molecules in such strong electric ac fields have brought some surprises and many theoreticians are working on adequate models to describe such extreme situations.

For more information on femtosecond lasers and spectroscopy see [626, 828].

6.5 Problems

6.1 A Pockels cell inside a laser resonator is used as a Q-switch. It has a maximum transmission of 95 % for the applied voltage $U = 0$. What voltage U is required to prevent lasing before the gain $G_\alpha = \exp(\alpha L)$ of the active medium exceeds the value $G_\alpha = 10$, when the "half-wave voltage" of the Pockels cell is 2 kV? What is the effective amplification factor G_{eff} immediately after the opening of the Pockels cell if the total cavity losses are 30° per round trip?

6.2 What is the actual time profile of mode-locked pulses from a cw argon laser if the gain profile is Gaussian with a halfwidth of 8 GHz (FWHM)?

6.3 An optical pulse with the Gaussian intensity profile $I(t)$, center wavelength $\lambda_0 = 600$ nm, and initial halfwidth $\tau = 500$ fs propagates through an optical fiber with refractive index $n = 1.5$.

(a) How large is its initial spatial extension?
(b) How long is the propagation length z_1 after which the spatial width of the pulse has increased by a factor of 2 from linear dispersion with $\mathrm{d}n/\mathrm{d}\lambda = 10^3$ per cm?

(c) How large is its spectral broadening at z_1 if its peak intensity is $I_\mathrm{p} = 10^{13}\ \mathrm{W/m^2}$ and the nonlinear part of the refractive index is $n_2 = 10^{-20}\ \mathrm{m^2/W}$?

6.4 Calculate the separation D of a grating pair that just compensates a spatial dispersion $\mathrm{d}S/\mathrm{d}\lambda = 10^5$ for a center wavelength of 600 nm, a groove spacing of $d = 1\ \mu\mathrm{m}$ and angle of incidence $\alpha = 30°$.

6.5 Calculate the pressure p of the argon buffer gas at $T = 500$ K that decreases the radiative lifetime $\tau = 16$ ns of an excited Na_2 level to 8 ns, if the total quenching cross section is $\sigma = 10^{-14}\ \mathrm{cm^2}$.

Chapter 7
Coherent Spectroscopy

This chapter provides an introduction to different spectroscopic techniques that are based either on the coherent excitation of atoms and molecules or on the coherent superposition of light scattered by molecules and small particles. The coherent excitation establishes definite phase relations between the amplitudes of the atomic or molecular wave functions; this, in turn, determines the total amplitudes of the emitted, scattered, or absorbed radiation.

Either two or more molecular levels of a molecule are excited coherently by a spectrally broad, short laser pulse (level-crossing and quantum-beat spectroscopy) or a whole ensemble of many atoms or molecules is coherently excited simultaneously into identical levels (photon-echo spectroscopy). This coherent excitation alters the spatial distribution or the time dependence of the total, emitted, or absorbed radiation amplitude, when compared with incoherent excitation. Whereas methods of incoherent spectroscopy measure only the total intensity, which is proportional to the population density and therefore to the square $|\psi|^2$ of the wave function ψ, the coherent techniques, on the other hand, yield additional information on the amplitudes and phases of ψ.

Within the density-matrix formalism (Vol. 1, Sect. 2.9) the coherent techniques measure the off-diagonal elements ρ_{ab} of the density matrix, called the *coherences*, while incoherent spectroscopy only yields information about the diagonal elements, representing the time-dependent population densities. The off-diagonal elements describe the atomic dipoles induced by the radiation field, which oscillate at the field frequency ω and which represent radiation sources with the field amplitude $A_k(\boldsymbol{r}, t)$. Under coherent excitation the dipoles oscillate with definite phase relations, and the phase-sensitive superposition of the radiation amplitudes A_k results in measurable interference phenomena (quantum beats, photon echoes, free induction decay, etc.).

After switching off the excitation sources, the phase relations between the different oscillating atomic dipoles are altered by different relaxation processes, which *perturb* the atomic dipoles. We may classify these processes into two categories:

W. Demtröder, *Laser Spectroscopy 2*, DOI 10.1007/978-3-662-44641-6_7

- Population-changing processes: the decay of the population density in the excited level $|2\rangle$

$$N_2(t) = N_2(0)\mathrm{e}^{-t/\tau_{\mathrm{eff}}},$$

by spontaneous emission or by inelastic collisions decreases the intensity of the radiation with the time constant $T_1 = \tau_{\mathrm{eff}}$, often called the *longitudinal relaxation time*.

- Phase-perturbing collisions (Vol. 1, Sect. 3.3), or the different Doppler shifts of the frequencies of emitting atoms with their individual velocity $\boldsymbol{v}_k$ result in a change of the relative phases of the atomic dipoles, which also affects the total amplitudes of the superimposed, emitted partial waves. The time constant T_2 of this *phase decay* is called the *transverse relaxation time*. Generally, the phase decay is faster than the population decay and therefore $T_2 < T_1$. While phase-perturbing collisions result in a homogeneous line broadening $\gamma_2^{\mathrm{hom}} = 1/T_2^{\mathrm{hom}}$, the velocity distribution, which gives rise to Doppler broadening, adds the inhomogeneous contribution $\gamma_2^{\mathrm{inhom}} = 1/T_2^{\mathrm{inhom}}$ to the line profile, which has the total spectral width

$$\Delta\omega = 1/T_1 + 1/T_2^{\mathrm{hom}} + 1/T_2^{\mathrm{inhom}}. \tag{7.1}$$

The techniques of coherent spectroscopy that are discussed below allow the elimination of the inhomogeneous contribution and therefore represent methods of "Doppler-free" spectroscopy, *although the coherent excitation may use spectrally broad radiation*. This is an advantage compared with the nonlinear Doppler-free techniques discussed in Chap. 2, where narrow-band single-mode lasers are required.

The combination of ultrafast light pulses with coherent spectroscopy allows, for the first time, the direct measurements of wave packets of coherently excited molecular vibrations and their decay (Sects. 7.4–7.8).

A rapidly expanding area of coherent spectroscopy is heterodyne and correlation spectroscopy, based on the interference between two coherent light waves. These two light waves may be generated by two stabilized lasers, or by one laser and the Doppler-shifted laser light scattered by a moving particle (atom, molecule, dust particle, microbes, living cells, etc.). The observed frequency distribution of the *beat spectrum* allows a spectral resolution down into the millihertz range (Sect. 7.8).

Correlation fluorescence spectroscopy in combination with confocal microscopy has proved to be a very versatile tool for detection and temporal investigations of biomolecules in living cells and of the interaction of different molecules.

In the following sections the most important techniques of coherent spectroscopy are discussed in order to illustrate more quantitatively the foregoing statements.

7.1 Level-Crossing Spectroscopy

Level-crossing spectroscopy is based on the measured change in the spatial intensity distribution or the polarization characteristics of fluorescence that is emitted from

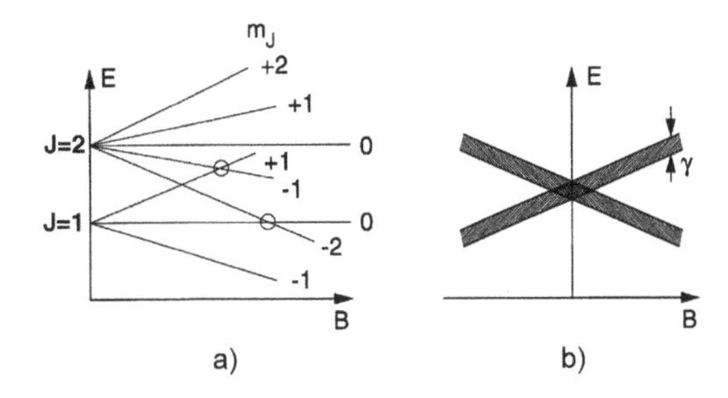

Fig. 7.1 Schematic diagram of level crossing: (**a**) general case and (**b**) Hanle effect

coherently excited levels when these levels cross under the influence of external magnetic or electric fields. Examples are fine or hyperfine levels with different Zeeman shifts, which may cross at a certain value B_c of the magnetic field B (Fig. 7.1). A special case of level crossing is the *zero-field level crossing* (Fig. 7.1b) that occurs for a degenerate level with total angular momentum $J > 0$. For $B \neq 0$, the $(2J + 1)$ Zeeman components split, which changes the polarization characteristics of the emitted fluorescence. This phenomenon was observed as early as 1923 by *W. Hanle* [831] and is therefore called the Hanle effect.

7.1.1 Classical Model of the Hanle Effect

A typical experimental arrangement for level-crossing spectroscopy is depicted in Fig. 7.2. Atoms or molecules in a homogeneous magnetic field $\boldsymbol{B} = \{0, 0, B_z\}$ are excited by the polarized optical wave $\boldsymbol{E} = E_y \cos(\omega t - kx)$ and tuned to the transi-

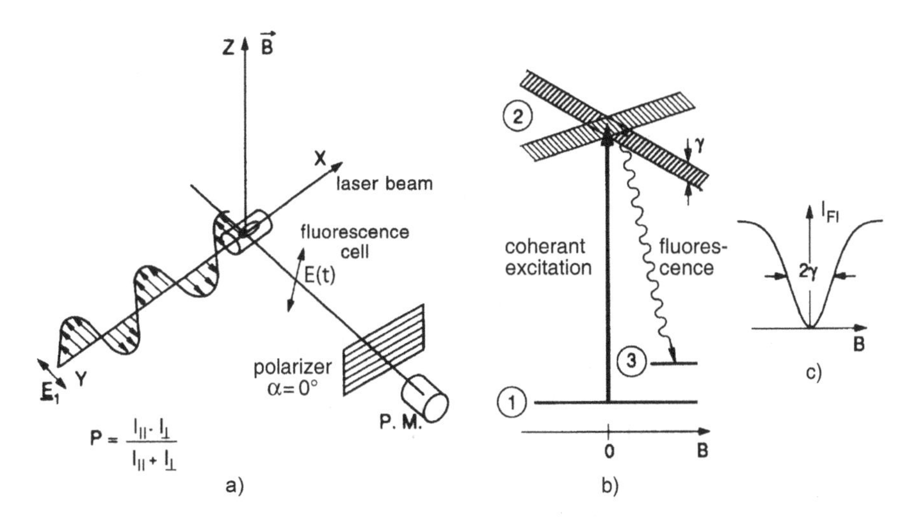

Fig. 7.2 Level-crossing spectroscopy: (**a**) experimental arrangement; (**b**) level scheme; and (**c**) Hanle signal

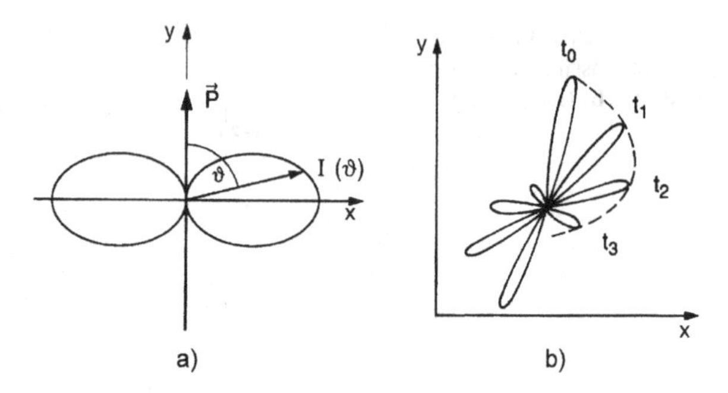

Fig. 7.3 Spatial intensity distribution of the radiation emitted by a classical oscillating dipole: (**a**) stationary case for $B = 0$; and (**b**) time sequence of intensity distributions of a dipole with decay constant γ, precessing in a magnetic field $\boldsymbol{B} = \{0, 0, B_z\}$

tion $|1\rangle \rightarrow |2\rangle$. The fluorescence $I_{\mathrm{Fl}}(B)$ emitted from the excited level $|2\rangle$ into the y-direction is observed behind a polarizer as a function of the magnetic field B.

In a classical, vivid description, the excited atom is represented by a damped oscillator that oscillates in the y-direction with a spatial emission characteristics $I(\vartheta) \propto \sin^2 \vartheta$, depending on the angle ϑ between observation direction and dipole axis (Fig. 7.3a). If the atom with an excited-state lifetime $\tau_2 = 1/\gamma_2$ is excited at $t = t_0$ by a short laser pulse, the amplitude of the emitted radiation is for $B = 0$:

$$\boldsymbol{E}(t) = \boldsymbol{E}_0 \mathrm{e}^{-(\mathrm{i}\omega + \gamma/2)(t - t_0)}. \tag{7.2}$$

With the angular momentum $\boldsymbol{J}$ in level $|2\rangle$, the atomic dipole with the magnetic dipole moment $\mu = g_J \mu_0$ (g_J: Landé factor; μ_0: Bohr's magneton) will precess for $B \neq 0$ around the z-axis with the precession frequency

$$\Omega_{\mathrm{p}} = g_J \mu_0 B/\hbar. \tag{7.3}$$

Together with the dipole axis the direction of maximum emission, which is perpendicular to the dipole axis, also precesses with Ω_{p} around the z-axis while the amplitude of the dipole oscillation decreases as $\exp[-(\gamma/2)t]$ (Fig. 7.3b). If the fluorescence is observed in the y-direction (Fig. 7.2a) behind a polarizer with the transmission axis tilted by the angle α against the x-axis, the measured intensity $I_{\mathrm{Fl}}(t)$ becomes

$$I_{\mathrm{Fl}}(B, \alpha, t) = I_0 \mathrm{e}^{-\gamma(t - t_0)} \sin^2\bigl[\Omega_{\mathrm{p}}(t - t_0)\bigr] \cos^2 \alpha. \tag{7.4}$$

This intensity can either be monitored in a time-resolved fashion after pulsed excitation in the constant magnetic field B (*quantum beats*, Sect. 7.2), or the time-integrated fluorescence intensity is measured as a function of B (level crossing), where the excitation may be pulsed or cw.

In the case of cw excitation the fluorescence intensity $I(t)$ at any time t is the result of all atoms excited within the time interval between $t_0 = -\infty$ and $t_0 = t$. It can therefore be written as

$$I(t, B, \alpha) = C I_0 \cos^2 \alpha \int_{t_0 = -\infty}^{t} \mathrm{e}^{-\gamma(t - t_0)} \sin^2\bigl[\Omega_{\mathrm{p}}(t - t_0)\bigr] \mathrm{d}t_0. \tag{7.5}$$

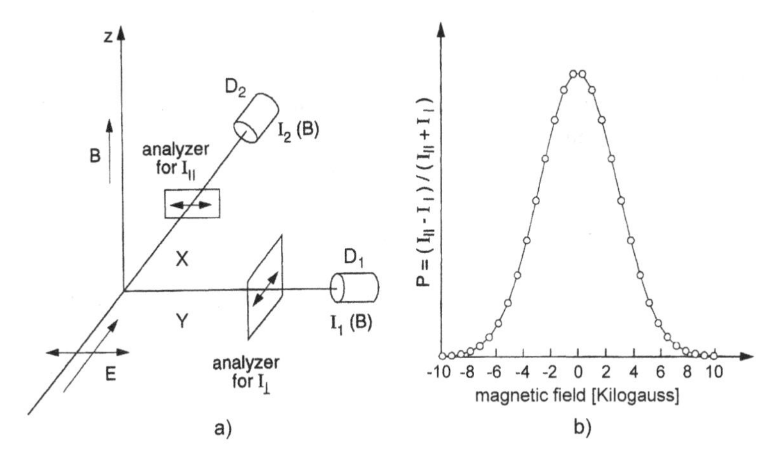

Fig. 7.4 Hanle signal of the fluorescence $I_{\mathrm{Fl}}(B)$ from laser-excited $\mathrm{Na}_2^*(B^1\Pi_u, v' = 10, J' = 12)$ molecules: (**a**) schematic experimental setup for difference and ratio recording; and (**b**) measured polarization $P(B)$ [835]

The transformation $(t - t_0) \to t'$ of the variable t_0 shifts the integration limits to the interval from $t' = \infty$ to $t' = 0$, which shows that the integral is independent of t. It can be solved with the substitution $2\sin^2 x = 1 - \cos 2x$, which yields the Hanle signal, observed by the detector D_1 in Fig. 7.4a:

$$I(B, \alpha) = C\frac{I_0 \cos^2\alpha}{2\gamma}\left(1 - \frac{\gamma^2}{\gamma^2 + 4\Omega_\mathrm{p}^2}\right). \tag{7.6a}$$

Inserting (7.3) we obtain for $\alpha = 0$

$$I(B, \alpha = 0) = C\frac{I_0}{2\gamma}\left(1 - \frac{1}{1 + (2g\mu_0 B/\hbar\gamma)^2}\right). \tag{7.7}$$

This intensity gives, as a function of the applied magnetic field, a Lorentzian-type signal (Fig. 7.2c) with the halfwidth (FWHM)

$$\Delta B_{1/2} = \frac{\hbar\gamma}{g\mu_0} = \frac{\hbar}{g\mu_0\tau_\mathrm{eff}}, \quad \text{with } \tau_\mathrm{eff} = 1/\gamma. \tag{7.8}$$

From the measured halfwidth $\Delta B_{1/2}$ the product $g_J \tau_\mathrm{eff}$ of Landé factor g_J of the excited level $|2\rangle$ times its effective lifetime τ_eff can be derived. For *atomic* states the Landé factor g_J is generally known, and the measured value of $\Delta B_{1/2}$ determines the lifetime τ_eff. Measurements of $\tau_\mathrm{eff}(p)$ as a function of pressure in the sample cell then yield by extrapolation $p \to 0$ the radiative lifetime τ_n (Sect. 6.3). The Hanle effect therefore offers, like other Doppler-free techniques, an alternative method for the measurement of atomic lifetimes from the width $\Delta B_{1/2}$ of the signal [832].

In case of excited *molecules* the angular momentum coupling scheme is often not known, in particular if hyperfine structure or perturbations between different electronic states affect the coupling of the various angular momenta. The total angular momentum $\boldsymbol{F} = \boldsymbol{N} + \boldsymbol{S} + \boldsymbol{L} + \boldsymbol{I}$ is composed of the molecular-rotation angular momentum $\boldsymbol{N}$, the total electron spin $\boldsymbol{S}$, the electronic orbital momentum $\boldsymbol{L}$, and the nuclear spin $\boldsymbol{I}$. The measured width $\Delta B_{1/2}$ can then be used to determine the Landé factor g and thus the coupling scheme, if the lifetime τ is already known from other independent measurements, for example, those discussed in Sect. 6.3 [833].

Note That the total intensity of the fluorescence is independent of the magnetic field, although this field does change the spatial distribution and the polarization characteristics of the fluorescence. It is therefore advantageous, to use the experimental setup in Fig. 7.4a, where two detectors monitor the fluorescence emitted into the x- and y-direction respectively. The fluorescence intensity detected in the x-direction through a polarizer with an angle β against the y-direction is

$$\begin{aligned} I_2(B, \beta) &= C \cdot I_0 \cos^2 \beta \int_{t_0=-\infty}^{+t} \mathrm{e}^{-\gamma(t-t_0)} \cos^2\left[\Omega_\mathrm{p}(t - t_0)\right] \mathrm{d}t_0 \\ &= C \frac{I_0 \cos^2 \beta}{2\gamma} \left(1 + \frac{\gamma^2}{\gamma^2 + 4\Omega_\mathrm{p}^2}\right). \end{aligned} \tag{7.6b}$$

With $\alpha = \beta = 0$ the difference signal becomes

$$I_2 - I_1 = \frac{C I_0 \gamma}{\gamma^2 + 4\Omega_\mathrm{p}^2}$$

and gives twice the signal magnitude without the constant background. The sum of the two measured signals

$$I_1 + I_2 = C \frac{I_0}{\gamma}$$

is independent of B and gives just twice the background term. The ratio

$$R = (I_1 - I_2)/(I_1 + I_2) = \frac{1}{1 + (2\Omega_\mathrm{p}/\gamma)^2}$$

is furthermore independent of the incident intensity and therefore does not suffer from intensity fluctuations. It has a better signal-to noise ratio compared to the measurement in only one direction.

For illustration, a molecular Hanle signal is depicted in Fig. 7.4 that was obtained when Na_2 molecules were excited in a magnetic field B into the upper level $|2\rangle = B^1\Pi_u$ ($v' = 10, J' = 12$) by an argon laser at $\lambda = 475$ nm [835]. For this Na_2 level with $S = 0$ Hund's coupling case a_α applies, and the Landé factor

$$g_F = \frac{F(F+1) + J(J+1) - I(I+1)}{2J(J+1)F(F+1)} \tag{7.9}$$

decreases strongly with increasing rotational quantum number J [836]. For larger values of J one therefore needs much larger magnetic fields than for atomic-level crossings.

7.1.2 Quantum-Mechanical Models

The quantum-mechanical treatment of level-crossing spectroscopy [834, 837] starts with the Breit formula

$$I_{\mathrm{Fl}}(2 \to 3) = C\left|\langle 1|\boldsymbol{\mu}_{12} \cdot \boldsymbol{E}_1|2\rangle\right|^2\left|\langle 2|\boldsymbol{\mu}_{23} \cdot \boldsymbol{E}_2|3\rangle\right|^2, \tag{7.10}$$

for the fluorescence intensity excited on the transition $|1\rangle \to |2\rangle$ by a polarized light wave with the electric field vector $\boldsymbol{E}_1$ and observed on the transition $|2\rangle \to |3\rangle$ behind a polarizer with its axis parallel to $\boldsymbol{E}_2$ [838]. The spatial intensity distribution and the polarization characteristics of the fluorescence depend on the orientation of the molecular transition dipoles $\boldsymbol{\mu}_{12}$ and $\boldsymbol{\mu}_{23}$ relative to the electric vectors $\boldsymbol{E}_1$ and $\boldsymbol{E}_2$ of absorbed and emitted radiation.

The level $|2\rangle$ with the total angular momentum $\boldsymbol{J}$ has $(2J+1)$ Zeeman components with the magnetic quantum number M, which are degenerate at $B=0$. The wave function of $|2\rangle$

$$\psi_2 = \sum_k c_k \psi_k \mathrm{e}^{-\mathrm{i}\omega_k t}, \tag{7.11}$$

is a linear combination of the wave functions $\psi_k \exp(-\mathrm{i}\omega_k t)$ of all coherently excited Zeeman components. The products of matrix elements in (7.10) contain the interference terms $c_{M_i} c_{M_k} \psi_i \psi_k \exp[\mathrm{i}(\omega_k - \omega_i)t]$.

For $B=0$ all frequencies ω_k are equal. The interference terms become time independent and describe together with the other constant terms in (7.10) the spatial distribution of I_{Fl}. However, for $B \neq 0$ the phase factors $\exp[\mathrm{i}(\omega_k - \omega_i)t]$ are time dependent. Even when all levels (J, M) for a given J have been excited coherently at time $t=0$ and their wave functions $\psi_k(t=0)$ all have the same phase $\psi(0)=0$, the phases of ψ_k develop differently in time because of the different frequencies ω_k. The magnitudes and the signs of the interference terms change in time, and for $(\omega_i - \omega_k) \gg 1/\tau$ (which is equivalent to $g_J \mu_0 B/\hbar \gg \gamma$) the interference terms average to zero and the observed time-averaged intensity distribution becomes isotropic.

Note Although the magnetic field alters the spatial distribution and the polarization characteristics of the fluorescence, it does not change the total fluorescence intensity!

Example 7.1 The quantum-mechanical description may be illustrated by a specific example, taken from [839], where optical pumping on the transition

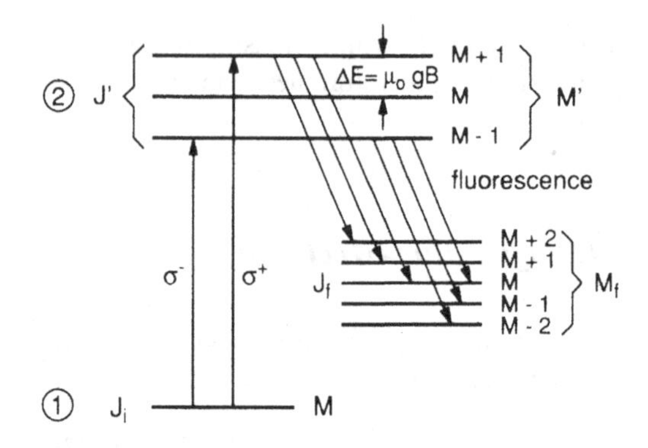

Fig. 7.5 Level scheme for optical pumping of Zeeman levels by light linearly polarized in the y-direction. Each of the two excited Zeeman levels can decay in three final-state sublevels. Superposition of the two different routes $(J'', M'') \rightarrow (J_f, M_f)$ generates interference effects [839]

$(J, M) \rightarrow (J', M')$ is achieved by linearly polarized light with a polarization vector in the x–y-plane, perpendicular to the field direction along the z-axis (Fig. 7.5). The linearly polarized light may be regarded as a superposition of left- and right-hand circularly polarized components σ^+ and σ^-. The excited-state wave function for the excitation by light polarized in the x-direction is

$$|2\rangle_x = (-1/\sqrt{2})\big(a_{M+1}|M+1\rangle + a_{M-1}|M-1\rangle\big), \tag{7.12}$$

and for excitation with y-polarization

$$|2\rangle_y = (-\mathrm{i}\sqrt{2})\big(a_{M+1}|M+1\rangle - a_{M-1}|M-1\rangle\big). \tag{7.13}$$

The coefficients a_M are proportional to the matrix element of the transition $(J, M) \rightarrow (J', M')$. The optical pumping process generates a coherent superposition of the eigenstates ($M' = M \pm 1$) as long as the spectral width of the pump radiation is broader than the level splitting.

The time development of the excited-state wave function is described by the time-dependent Schrödinger equation

$$-\frac{\hbar}{\mathrm{i}}\frac{\partial \psi_2}{\partial t} = \mathcal{H}\psi_2, \tag{7.14}$$

where the operator $\mathcal{H}$ has the eigenvalues $E_M = E_0 + \mu_0 g M B$. Including the spontaneous emission in a semiclassical way by the decay constant γ (Vol. 1, Sect. 2.7.5), we may write the solution of (7.14)

$$\psi_2(t) = \mathrm{e}^{-(\gamma/2)t}\exp(-\mathrm{i}\mathcal{H}t/\hbar)\psi_2(0), \tag{7.15}$$

where the operator $\exp(-\mathrm{i}\mathcal{H}t)$ is defined by its power-series expansion. From (7.12)–(7.15) we find for excitation with y-polarization

$$\psi_2(t) = \frac{-\mathrm{i}}{\sqrt{2}}\mathrm{e}^{-(\gamma/2)t}\exp(-\mathrm{i}\mathcal{H}t/\hbar)\big[a_{M+1}|M+1\rangle - a_{M-1}|M-1\rangle\big]$$

$$
\begin{aligned}
&= \frac{-\mathrm{i}}{\sqrt{2}} \mathrm{e}^{-(\gamma/2)t} \mathrm{e}^{-\mathrm{i}(E_0+\mu_0 g M B)t/\hbar} \\
&\quad \times \left[a_{M+1} \mathrm{e}^{-\mathrm{i}\mu_0 g B t/\hbar} |M+1\rangle - a_{M-1} \mathrm{e}^{(\mathrm{i}\mu_0 g B)t/\hbar} |M-1\rangle \right] \\
&= \mathrm{e}^{-(\gamma/2)t} \mathrm{e}^{-\mathrm{i}(E_0+\mu_0 g M B)t/\hbar} \\
&\quad \times \left[|\psi_2\rangle_x \sin(\mu_0 g B t/\hbar) + |\psi_2\rangle_y \cos(\mu_0 g B t/\hbar) \right].
\end{aligned} \tag{7.16}
$$

This shows that the excited wave function under the influence of the magnetic field B changes continuously from $|\psi_2\rangle_x$ to $|\psi_2\rangle_y$, and back. If the fluorescence is detected through a polarizer with the transmission axis parallel to the x axis, only light from the $|\psi_2\rangle_x$ component is detected. The intensity of this light is

$$
I(E_x, t) = C \left| \psi_{2x}(t) \right|^2 = C \mathrm{e}^{-\gamma t} \sin^2(\mu_0 g B t/\hbar), \tag{7.17}
$$

which gives, after integration over all time intervals from $t = -\infty$ to the time t_0 of observation, the Lorentzian intensity profile (7.7) with $\gamma = 1/\tau$ observed in the y-direction behind a polarizer in the x-direction (Fig. 7.2)

$$
I_y(E_x) = \frac{C\tau I_0}{2} \frac{(2\mu_0 g \tau B/\hbar)^2}{1 + (2\mu_0 g \tau B/\hbar)^2}, \tag{7.18}
$$

which turns out to be identical with the classical result (7.7).

In Fig. 7.6 the level scheme of the hyperfine levels in the $5P_{3/2}$ level of the Rubidium atom is shown with level crossings at $B = 0$ and $B \neq 0$. Therefore level crossing spectroscopy of the Rb-atom will give several signals at different magnetic fields B.

Measurements of the Hanle effect and level crossings at magnetic fields $B \neq 0$ showed for higher intensities of the excitation laser nonlinear Zeeman shifts [840].

7.1.3 Experimental Arrangements

Level-crossing spectroscopy was used in atomic physics even before the invention of lasers [831, 842–844]. These investigations were, however, restricted to atomic resonance transitions that could be excited with intense hollow-cathode or microwave atomic-resonance lamps. Only a very few molecules have been studied, where accidental coincidences between atomic resonance lines and molecular transitions were utilized [836].

Optical pumping with tunable lasers or even with one of the various lines of fixed-frequency lasers has largely increased the application possibilities of level-crossing spectroscopy to the investigation of molecules and complex atoms. Because of the

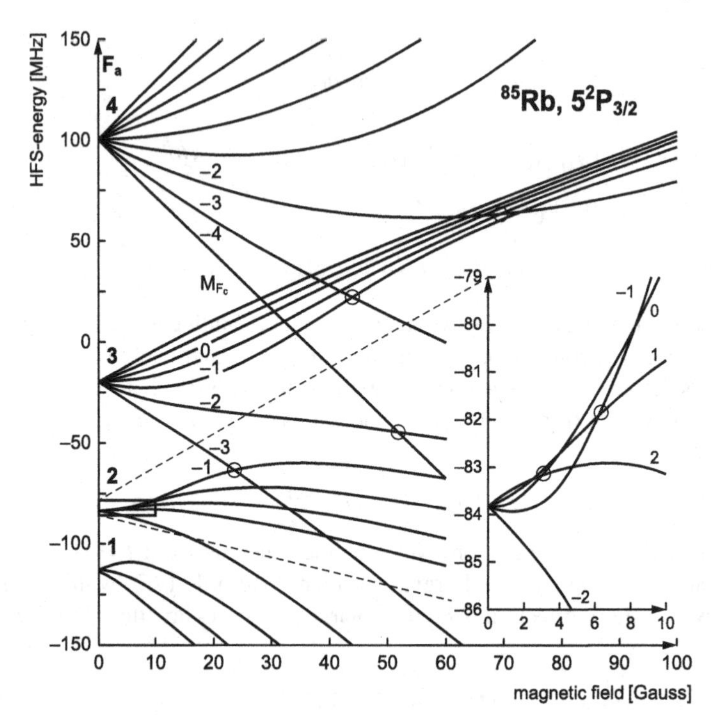

Fig. 7.6 Hanle-effect and level crossings at $B \neq 0$ in the $5^2P_{3/2}$-state of the ^{85}Rb-isotope. The *insert* shows the nonlinear Zeeman-shifts

higher laser intensity the population density in the excited state is much larger, and therefore the signal-to-noise ratio is higher. In combination with two-step excitation (Sect. 5.4), Landé factors and lifetimes of high-lying Rydberg levels can be studied with this technique. Furthermore, lasers have introduced new versions of this technique such as stimulated level-crossing spectroscopy [845].

Level-crossing spectroscopy with lasers has some definite experimental advantages. Compared with other Doppler-free techniques it demands a relatively simple experimental arrangement. Neither single-mode lasers and frequency-stabilization techniques nor collimated molecular beams are required. The experiments can be performed in simple vapor cells, and the experimental expenditure is modest. In many cases no monochromator is needed since sufficient selectivity in the excitation process can be achieved to avoid simultaneous excitation of different molecular levels with a resulting overlap of several level-crossing signals.

There are, of course, also some disadvantages. One major problem is the change of the absorption profile with the magnetic field. The laser bandwidth must be sufficiently large in order to assure that all Zeeman components can absorb the radiation independent of the field strength B. On the other hand, the laser bandwidth should not be too large, to avoid simultaneous excitation of different, closely-spaced transitions. This problem arises particularly in *molecular* level-crossing spectroscopy, where several molecular lines often overlap within their Doppler widths. In such cases a compromise has to be found for an intermediate laser bandwidth, and the fluorescence may have to be monitored through a monochromator in order to discriminate against other transitions. Because of the high magnetic fields required for Hanle signals from short-lived molecular levels with high rotational quantum numbers and therefore with small Landé factors, careful magnetic shielding of the photomultiplier is essential to avoid a variation of the multiplier gain factor with the magnetic field strength.

The level-crossing signal may only be a few percent of the field-independent background intensity. In order to improve the signal-to-noise ratio, either the field is modulated or the polarizer in front of the detector is made to rotate and the signal is recovered by lock-in detection.

Since the total fluorescence intensity is independent of the magnetic field, ratio recording can be used to eliminate possible field-dependent absorption effects. If the signals of the two detectors D_1 and D_2 in Fig. 7.4 are $S_1 = I_{\|}(B) \cdot f(B)$; $S_2 = I_{\perp}(B) \cdot f(B)$, the ratio

$$R = \frac{S_1 - S_2}{S_1 + S_2} = \frac{I_{\|} - I_{\perp}}{I_{\|} + I_{\perp}},$$

no longer depends on the field-dependent absorption $f(B)$ [835].

7.1.4 Examples

A large number of atoms and molecules have been investigated by level-crossing spectroscopy using laser excitation. A compilation of the measurements up to 1975 can be found in the review of Walther [846], up to 1990 in [847], and up to 1997 in [848].

The iodine molecule has been very thoroughly studied with electric and magnetic level-crossing spectroscopy. The hyperfine structure of the rotational levels affects the profile of the level-crossing curves [849]. A computer fit to the non-Lorentzian superposition of all Hanle curves from the different hfs levels allows simultaneous determination of the Landé factor g and the lifetime τ [850]. Because of different predissociation rates the effective lifetimes of different hfs levels differ considerably.

In larger molecules the phase-coherence time of excited levels may be shorter than the population lifetime because of perturbations between closely spaced levels of different electronic states, which cause a dephasing of the excited-level wave functions. One example is the NO_2 molecule, where the width of the Hanle signal turns out to be more than one order of magnitude larger than expected from independent measurements of population lifetime and Landé factors [851, 852]. This discrepancy is explained by a short intramolecular decay time (dephasing time), but a much larger radiative lifetime [853].

The energy levels $E_k(B)$, split by a magnetic field, can be made to recross again at $B \neq 0$ by an additional electric field. Such Zeeman–Stark recrossings allow the determination of magnetic and electric moments of selectively excited rotational levels. In the case of perturbed levels these moments are altered, and the

g-values may vary strongly from level to level even if the level-energy separation is small [854].

An external electric field may also prevent level crossings caused by the magnetic field. Such anti-crossing effects in Rydberg states of Li atoms in the presence of parallel magnetic and electric fields have been studied in the region where the Stark and Zeeman components of adjacent principal quantum numbers n overlap. The experimental results give information on core effects and interelectronic coupling [855, 856].

By stepwise excitation with two or three lasers, highly excited states and Rydberg levels of atoms and molecules can be investigated. These techniques allow measurement of the natural linewidth, the fine structure, and hfs parameters of high-lying Rydberg states. Most experiments have been performed on alkali atoms. For example, Li atoms in an atomic beam were excited by a frequency-doubled pulsed dye laser and level-crossing signals of the excited Rydberg atoms were detected by field ionization [857].

An example of a more recent experiment is the study of hyperfine structure of highly excited levels of neutral copper atoms, which yielded the magnetic-dipole and electric-quadrupole interaction constants. The experimental results allowed a comparison with theoretical calculations based on multiconfiguration Hartree–Fock methods [858]. Measurements of lifetimes [859] and of state multipoles using level-crossing techniques can be found in [860].

Since molecular radicals have an unpaired valence electron they possess a large magnetic and in case of heteropolar molecules also an electric dipole moment. They therefore allow magnetic or electric level crossing spectroscopy with high sensitivity. This was proved by Cahn et al. [841] who measured the crossings of rotational levels of the BaF radical in an external magnetic field. Because of the long lifetime of rotational levels in the electronic ground state, the observed level crossing signals had an extremely narrow linewidth and can be used for high precision experiments, such as the search for parity violation.

7.1.5 Stimulated Level-Crossing Spectroscopy

So far we have considered level crossing monitored through spontaneous emission. A level-crossing resonance can also manifest itself as a change in absorption of an intense monochromatic wave tuned to the molecular transition when the absorbing levels cross under the influence of external fields. The physical origin of this stimulated level-crossing spectroscopy is based on saturation effects and may be illustrated by a simple example [845].

Consider a molecular transition between the two levels $|a\rangle$ and $|b\rangle$ with the angular momenta $J = 0$ and $J' = 1$ (Fig. 7.7). We denote the center frequencies of the $\Delta M = +1, 0, -1$ transitions by ω_+, ω_0, and ω_- and the corresponding matrix elements by μ_+, μ_0, and μ_-, respectively. Without an external field the M sublevels are degenerate and $\omega_+ = \omega_- = \omega_0$. The monochromatic wave

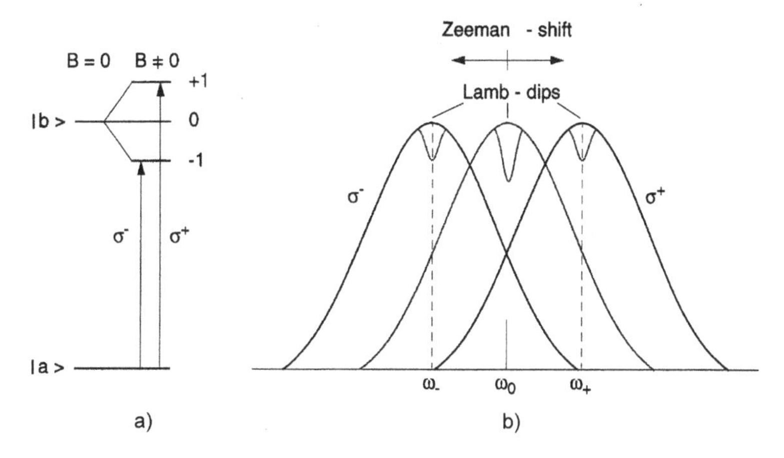

Fig. 7.7 Stimulated level-crossing spectroscopy with a common lower level: (**a**) level scheme; and (**b**) saturation holes in the Doppler-broadened population distribution with and without magnetic field

$\boldsymbol{E} = \boldsymbol{E}_0 \cos(\omega t - kx)$, linearly polarized in the y-direction, induces transitions with $\Delta M = 0$ without an external field. The saturated absorption of the laser beam is then, according to (2.30),

$$\alpha_s(\omega) = \frac{\alpha_0(\omega_0)}{\sqrt{1+S_0}} \mathrm{e}^{-[(\omega-\omega_0)/\Delta\omega_D]^2}, \tag{7.19}$$

where $\alpha_0 = (N_a - N_b)|S\mu|^2\omega/(\hbar\gamma^2)$ is the unsaturated absorption coefficient, and $S_0 = E_0^2|\mu|^2/(\hbar^2\gamma_s)$ is the saturation parameter at the line center (Vol. 1, Sect. 3.6). If an external electric or magnetic field is applied in the z-direction, the degenerate levels split and the laser beam, polarized in the y-direction, induces transitions $\Delta M = \pm 1$ because it can be composed of $\sigma^+ + \sigma^-$ contributions (see previous section). If the level splitting $(\omega_+ - \omega_-) \gg \gamma$, the absorption coefficient is now the sum of two contributions:

$$\alpha_s(\omega) = \frac{\alpha_0^+}{(1+S_0^+)^{1/2}} \mathrm{e}^{-[(\omega-\omega_+)/\Delta\omega_D]^2} + \frac{\alpha_0^-}{(1+S_0^-)^{1/2}} \mathrm{e}^{-[(\omega-\omega_-)/\Delta\omega_D]^2}. \tag{7.20}$$

For a $J = 1 \to 0$ transition $|\mu_+|^2 = |\mu_-|^2 = \frac{1}{2}|\mu_0|^2$. Neglecting the difference in the absorption coefficients $\alpha(\omega_+)$ and $\alpha(\omega_-)$ for $\omega_+ - \omega_- \ll \Delta\omega_D$, we may approximate (7.20) for $\hbar(\omega_+ - \omega_-) \gg \gamma$ by

$$\alpha_s(\omega) = \frac{\alpha_0^0}{(1+\frac{1}{2}S_0)^{1/2}} \mathrm{e}^{-[(\omega-\omega_0)/\Delta\omega_D]^2},$$

$$\text{with } S_0^+ \approx S_0^- = \frac{1}{2}S_0, \text{ and } \alpha_0^0 = \alpha_0^+ + \alpha_0^-, \tag{7.21}$$

which differs from (7.19) by the factor $\frac{1}{2}$ in the denominator. The difference of the absorption coefficient with and without a field (that is, for $\omega^+ - \omega^- > \gamma$) is

$$\Delta\alpha(\omega) = \alpha_0^0 \mathrm{e}^{-[(\omega-\omega_0)/\Delta\omega_\mathrm{D}]^2} \left(\frac{1}{\sqrt{1+\frac{1}{2}S^2}} - \frac{1}{\sqrt{1+S^2}} \right), \tag{7.22}$$

where the saturation parameter

$$S = S_0 \frac{1}{(w - w_0)^2 + (\gamma_\mathrm{s})^2}$$

has a Lorentzian line profile (see Vol. 1, Sect. 3.6). For $S \ll 1$, this becomes

$$\Delta\alpha(\omega) \approx \frac{1}{4} S^2(\omega) \alpha_0^0 \mathrm{e}^{-[(\omega-\omega_0)/\Delta\omega_\mathrm{D}]^2} \tag{7.23}$$

(see Vol. 1, Sect. 3.6).

This demonstrates that the effect of level splitting on the absorption appears only in the saturated absorption and disappears for $S \to 0$. The absorption frequency is changed by altering the magnetic field. The absorption coefficient $\alpha_\mathrm{s}(\omega_0, B)$, measured as a function of the magnetic field B while the laser is kept at ω_0, has a maximum at $B = 0$, and the transmitted laser intensity $I_t(B)$ shows a corresponding "Lamb dip." Although saturation effects may influence the line shape of the level-crossing signal, for small saturation it may still be essentially Lorentzian. The advantage of stimulated versus spontaneous level crossing is the larger signal-to-noise ratio and the fact that level crossings in the ground state can also be detected.

Most experiments on stimulated level crossing are based on intracavity absorption techniques because of their increased sensitivity (Sect. 1.2.3). Luntz and Brewer [861] demonstrated that even such small Zeeman splittings as in the molecular $^1\Sigma$ ground state can be precisely measured. They used a single-mode HeNe laser oscillating on the 3.39 μm line, which coincides with a vibration–rotation transition in the $^1\Sigma$ ground state of CH_4. Level crossings were detected as resonances in the laser output when the CH_4 transition was tuned by an external magnetic field. The rotational magnetic moment of the $^1\Sigma$ state of CH_4 was measured as $0.36 \pm 0.07\mu_\mathrm{N}$. In addition, Stark-tuned level-crossing resonances in the excited vibrational level of CH_4 have been detected with this method [862].

A number of stimulated level-crossing experiments have been performed on the active medium of gas lasers, where the gain of the laser transition is changed when sublevels of the upper or lower laser level cross each other. The whole gain tube is, for instance, placed in a longitudinal magnetic field, and the laser output is observed as a function of the magnetic field. Examples are the observation of stimulated hyperfine level crossings in a Xe laser [863], where accurate hyperfine splittings could be determined, or the measurement of Landé factors of atomic laser levels with high precision, as the determination of $g(^2P_4) = 1.3005 \pm 0.1$ % in neon by Hermann et al. [864].

Two-photon induced level crossing [865], which relies on the OODR scheme of Raman-type transitions (Fig. 7.8), has been performed with the two neon transitions

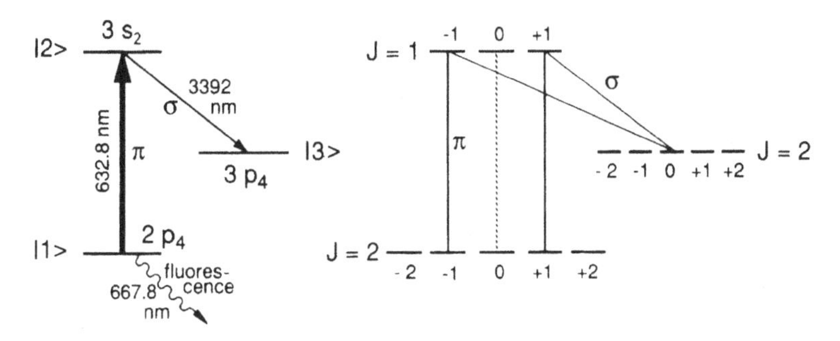

Fig. 7.8 Level scheme for two-photon stimulated Hanle effect. Only the sublevels $M = \pm 1$ of the lower state $|1\rangle$ are pumped by Raman-type transitions

at $\lambda_1 = 632.8$ nm and $\lambda_2 = 3.39$ μm, which have the common upper $3s_2$ level. Here the Paschen notation of neon levels is used [866]. A HeNe laser in an external magnetic field B is simultaneously oscillating on both transitions. Each level splits into $(2J + 1)$ Zeeman components, where J is the quantum number of the total angular momentum, which is the sum of the angular momenta of the core and the excited electron. The Hanle signal $S(B)$ is monitored via the fluorescence from the $3p_4$ level at $\lambda = 667.8$ nm.

There is one important point to note. The width $\Delta B_{1/2}$ of the level-crossing signal reflects the average width $\gamma = \frac{1}{2}(\gamma_1 + \gamma_2)$ of the two crossing levels. If these levels have a smaller width than the other level of the optical transition, level-crossing spectroscopy allows a higher spectral resolution than, for example, saturation spectroscopy, where the limiting linewidth $\gamma = \gamma_a + \gamma_b$ is given by the sum of upper and lower level widths. Examples are all cw laser transitions, where the upper level always has a longer spontaneous lifetime than the lower level (otherwise inversion could not be maintained). Level-crossing spectroscopy of the upper level then yields a higher spectral resolution than the natural linewidth of the fluorescence between both levels. This is, in particular, true for level-crossing spectroscopy in electronic ground states, where the spontaneous lifetimes are infinite and other broadening effects, such as transit-time broadening or the finite linewidth of the laser, limit the resolution [867].

7.2 Quantum-Beat Spectroscopy

Quantum-beat spectroscopy represents not only a beautiful demonstration of the fundamental principles of quantum mechanics, but this Doppler-free technique has also gained increasing importance in atomic and molecular spectroscopy. Whereas commonly used spectroscopy in the frequency domain yields information on the stationary states $|k\rangle$ of atoms and molecules, which are eigenstates of the total Hamiltonian

$$\mathcal{H}\psi_k = E_k\psi_k, \tag{7.24a}$$

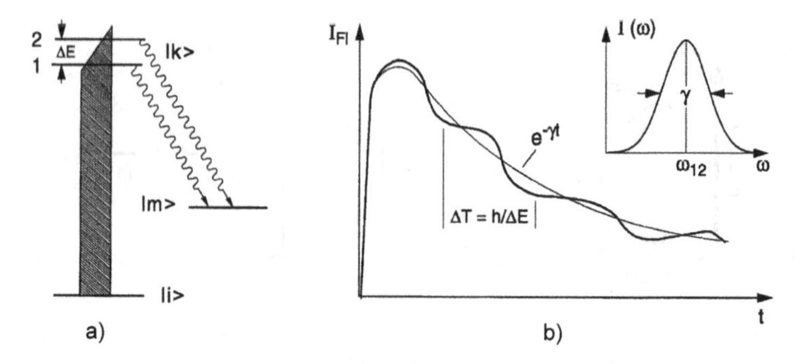

Fig. 7.9 (**a**) Level scheme illustrating the coherent excitation of levels $|1\rangle$ and $|2\rangle$ by a short pulse. (**b**) Quantum beats observed in the fluorescence decay of two coherently excited levels. *Insert:* Fourier spectrum $I(\omega)$ of (**b**) with $\omega_{12} = \Delta E/\hbar$

time-resolved spectroscopy with sufficiently short laser pulses characterizes nonstationary states

$$\big|\psi(t)\big\rangle = \sum c_k|\psi_k\rangle \mathrm{e}^{-\mathrm{i}E_k t/\hbar}, \tag{7.24b}$$

which can be described as a coherent time-dependent superposition of stationary eigenstates. Because of the different energies E_k this superposition is no longer independent of time. With time-resolved spectroscopy the time-dependence of $|\psi(t)\rangle$ can be measured in the form of a signal $S(t)$. The Fourier transformation of this time-dependent signal $S(t)$ yields the spectral information on the spectral components $c_k|\psi_k\rangle$ and their energies E_k. This is illustrated by the following subsections.

7.2.1 Basic Principles

If two closely spaced levels $|1\rangle$ and $|2\rangle$ are simultaneously excited from a common lower level $|i\rangle$ at time $t = 0$ by a short laser pulse with the pulse width $\Delta t < \hbar/(E_2 - E_1)$ (Fig. 7.9a), the wave function of the "coherent superposition state $|1\rangle + |2\rangle$" at $t = 0$

$$\psi(t=0) = \sum_k c_k\psi_k(0) = c_1\psi_1(0) + c_2\psi_2(0), \tag{7.25a}$$

is represented by a linear combination of the wave functions ψ_k $(k = 1, 2)$ of the "unperturbed" levels $|k\rangle$. The probability for the population of the level $|k\rangle$ at $t = 0$ is given by $|c_k|^2$. If the population N_k decays with the decay constant $\gamma_k = 1/\tau_k$ into a lower level $|m\rangle$, the time-dependent wave function of the coherent superposition state becomes

$$\psi(t) = \sum_k c_k\psi_k(0)\mathrm{e}^{-(\mathrm{i}\omega_{km}+\gamma_k/2)t}, \quad \text{with } \omega_{km} = (E_k - E_m)/\hbar. \tag{7.25b}$$

If the detector measures the total fluorescence emitted from both levels $|k\rangle$, the time-dependent signal $S(t)$ is

$$S(t) \propto I(t) = C\left|\langle\psi_m|\boldsymbol{\epsilon}\cdot\boldsymbol{\mu}|\psi(t)\rangle\right|^2. \tag{7.26}$$

Here C is a constant factor depending on the experimental arrangement, $\boldsymbol{\mu} = e\cdot\boldsymbol{r}$ is the dipole operator, and $\boldsymbol{\epsilon}$ gives the polarization direction of the emitted light. Inserting (7.25a) into (7.26) yields for equal decay constants $\gamma_1 = \gamma_2 = \gamma$ of both levels

$$I(t) = C\mathrm{e}^{-\gamma t}(A + B\cos\omega_{21}t), \tag{7.27a}$$

with

$$\begin{aligned} A &= c_1^2\left|\langle\psi_m|\boldsymbol{\epsilon}\cdot\boldsymbol{\mu}|\psi_1\rangle\right|^2 + c_2^2\left|\langle\psi_m|\boldsymbol{\epsilon}\cdot\boldsymbol{\mu}|\psi_2\rangle\right|^2, \\ B &= 2c_1c_2\left|\langle\psi_m|\boldsymbol{\epsilon}\cdot\boldsymbol{\mu}|\psi_1\rangle\right|\cdot\left|\langle\psi_m|\boldsymbol{\epsilon}\cdot\boldsymbol{\mu}|\psi_2\rangle\right|. \end{aligned} \tag{7.27b}$$

This represents an exponential decay $\exp(-\gamma t)$ superimposed by a modulation with the frequency $\omega_{21} = (E_2 - E_1)/\hbar$, which depends on the energy separation ΔE_{21} of the two coherently excited levels (Fig. 7.9b). This modulation is called *quantum beats*, because it is caused by the interference of the time-dependent wave functions of the two coherently excited levels.

The physical interpretation of the quantum beats is based on the following fact. When the molecule has reemitted a photon, there is no way to distinguish between the transitions $1 \to m$ or $2 \to m$ if the total fluorescence is monitored. As a general rule, in quantum mechanics the total probability amplitude of two indistinguishable processes is the sum of the two corresponding amplitudes and the observed signal is the square of this sum. This quantum-beat interference effect is analogous to Young's double-slit interference experiment. The quantum beats disappear if the fluorescence from only one of the upper levels is selectively detected.

The Fourier analysis of the time-dependent signal (7.27a–7.27b) yields a Doppler-free spectrum $I(\omega)$, from which the energy spacing ΔE as well as the width γ of the two levels $|k\rangle$ can be determined, even if ΔE is smaller than the Doppler width of the detected fluorescence (Fig. 7.9c). Quantum-beat spectroscopy therefore allows Doppler-free resolution [868].

7.2.2 Experimental Techniques

The experimental realization generally uses short-pulse lasers, such as pulsed dye lasers (Vol. 1, Sect. 5.7), or mode-locked lasers (Sect. 6.1). The time response of the detection system has to be fast enough to resolve the time intervals $\Delta t < \hbar/(E_2 - E_1)$. Fast transient digitizers or boxcar detection systems (Vol. 1, Sect. 4.5) meet this requirement.

When atoms, ions, or molecules in a fast beam are excited and the fluorescence intensity is monitored as a function of the distance z downstream of the excitation point, the time resolution $\Delta t = \Delta z/v$ is determined by the particle velocity v and the resolvable spatial interval Δz from which the fluorescence is collected [869]. In this case, detection systems can be used that integrate over the intensity and measure the quantity

$$I(z)\Delta z = \left[\int_{t=0}^{\infty} I(t,z)\,\mathrm{d}t\right]\Delta z.$$

The excitation can even be performed with cw lasers since the bandwidth necessary for coherent excitation of the two levels is assured by the short interaction time $\Delta t = d/v$ of a molecule with velocity v passing through a laser beam with the diameter d.

Example 7.2 With $d = 0.1$ cm and $v = 10^8$ cm/s $\to t \to 10^{-9}$ s. This allows coherent excitation of two levels with a separation up to 1000 MHz.

Figure 7.10 illustrates as an example quantum beats measured by Andrä et al. [869] in the fluorescence of $^{137}Ba^+$ ions following the excitation of three hfs levels in the $6p^2P_{3/2}$ level. Either a tunable dye laser beam crossed perpendicularly with the ion beam or the beam of a fixed-frequency laser crossed under the tilting angle θ with the ion beam can be used for excitation. In the latter case, Doppler tuning of the ion transitions is achieved by tuning the ion velocity (Sect. 4.5) or the tilting angle θ, since the absorption frequency is $\omega = \omega_L - |\boldsymbol{k} \parallel \boldsymbol{v}|\cos\theta$. The lower spectra in Fig. 7.10 are the Fourier transforms of the quantum beats, which yield the hfs transitions depicted in the energy level diagrams.

By choosing the correct angle θ_i either the excitation $6s^2S_{1/2}(F'' = 1) \to 6p^2P_{3/2}(F' = 0, 1, 2)$ can be selected (upper part in Fig. 7.10) or the excitation $F'' = 2 \to F' = 1, 2, 3$ (lower part).

Because of sub-Doppler resolution, quantum-beat spectroscopy has been used to measure fine or hyperfine structure and Lamb shifts of excited states of neutral atoms and ions [870].

If the Zeeman sublevels $|J, M = \pm 1\rangle$ of an atomic level with $J = 1$ are coherently excited by a pulsed laser, the fluorescence amplitudes are equal for the transitions $(J = 1, M = +1 \to J = 0)$ and $(J = 1, M = -1 \to J = 0)$. One therefore observes Zeeman quantum beats with a 100 % modulation of the fluorescence decay (Fig. 7.11).

Quantum beats can be observed not only in emission but also in the transmitted intensity of a laser beam passing through a coherently prepared absorbing sample. This has first been demonstrated by Lange et al. [872, 873]. The method is based on time-resolved polarization spectroscopy (Sect. 2.4) and uses the pump-and-probe technique discussed in Sect. 6.4. A polarized pump pulse orientates atoms in a cell placed between two crossed polarizers (Fig. 7.12) and generates a coherent superposition of levels involved in the pump transition. This results in an oscillatory time dependence of the transition dipole moment with an oscillation period $\Delta T = 1/\Delta\nu$

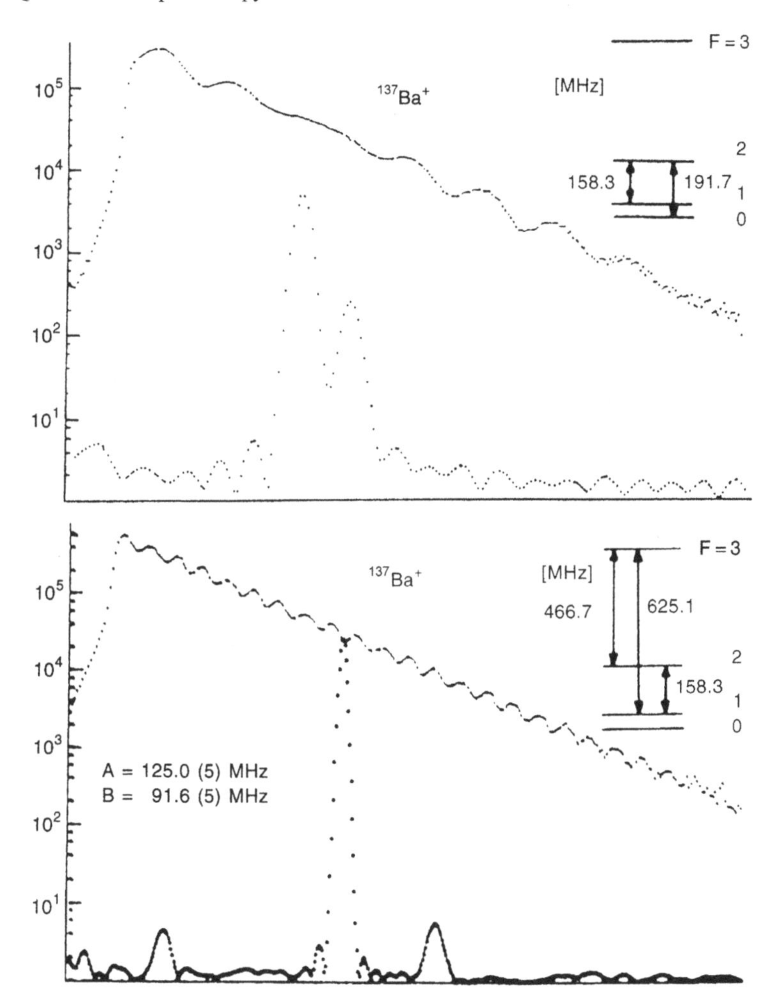

Fig. 7.10 Observed quantum beats in the fluorescence of ^{137}Ba$^+$ ions following an excitation of different sublevel groups at $\lambda = 455.4$ nm in a fast ion beam, and the corresponding Fourier transform spectra. The level schemes represent the hfs of the emitting upper level $6p^2P_{3/2}$ with the measured beat frequencies [869]

that is determined by the splitting $\Delta\nu$ of the sublevels. When a probe pulse with variable delay Δt propagates through the sample, its transmittance $I_T(\Delta t)$ through the crossed analyzer shows this oscillation.

This time-resolved polarization spectroscopy has, quite similar to its cw counterpart, the advantage of a zero-background method, avoiding the problem of finding a small signal against a large background. In contrast to cw polarization spectroscopy, no narrow-band single-frequency lasers are required and a broadband laser source can be utilized, which facilitates the experimental setup considerably. The *time* resolution of the pump-and-probe technique is *not* limited by that of the detec-

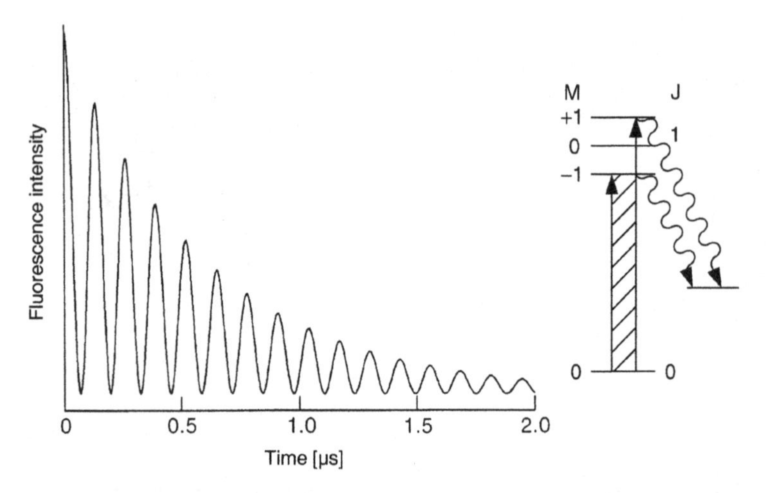

Fig. 7.11 Zeeman quantum beats observed in the fluorescence of Yb atoms in a magnetic field after pulsed excitation at $\lambda = 555.6$ nm [871]

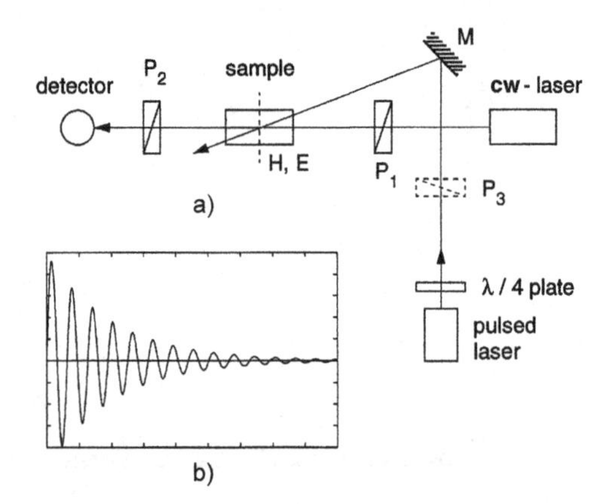

Fig. 7.12 Quantum-beat spectroscopy of atomic or molecular ground states measured by time-resolved polarization spectroscopy: (**a**) experimental arrangement; and (**b**) Zeeman quantum beat signal of the Na $3^2S_{1/2}$ ground state recorded by a transient digitizer with a time resolution of 100 ns. (Single pump pulse, time scale 1 μs/div, magnetic field $B = 1.63 \times 10^{-4}$ T) [872]

tor (Sect. 6.4). With a mode-locked cw dye laser a time resolution in the picosecond range can be achieved [874].

The *spectral* resolution of the Fourier-transformed spectrum is not limited by the bandwidth of the lasers or the Doppler width of the absorbing transition, but only by the homogeneous width of the levels involved. The amplitude of the quantum-beat signal is damped by collisions and by diffusion of the oriented atoms out of the interaction zone. If the diffusion time represents the limiting factor, the damping may be decreased by adding a noble gas, which slows down the diffusion. This decreases the spectral linewidth in the Fourier spectrum until collisional broadening becomes dominant. The decay time of the quantum-beat signal yields the phase relaxation time T_2 [875].

The basic difference of stimulated quantum beats in emission or absorption is illustrated by Fig. 7.13. The V-type scheme of Fig. 7.13a creates coherences in the

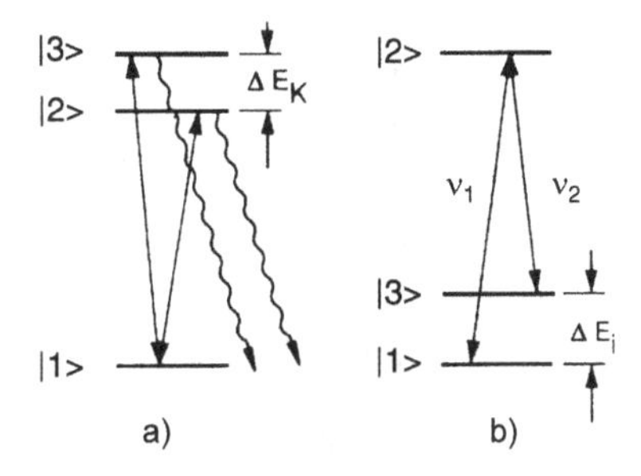

Fig. 7.13 Preparation of coherences: (**a**) in the excited state; and (**b**) in the ground state of an atomic system

excited state that can be observed by stimulated emission. The Λ-type scheme of Fig. 7.13b, on the other hand, describes coherence in the ground state, which can be monitored by absorption [873].

The time development of these coherences corresponds to a time-dependent susceptibility $\chi(t)$ of the sample, which affects the polarization characteristics of the probe pulse and appears as quantum beats of the transmitted probe pulse intensity.

An interesting technique for measuring hyperfine splittings of excited atomic levels by quantum-beat spectroscopy has been reported by Leuchs et al. [876]. The pump laser creates a coherent superposition of HFS sublevels in the excited state that are photoionized by a second laser pulse with variable delay. The angular distribution of photoelectrons, measured as a function of the delay time, exhibits a periodic variation because of quantum beats, reflecting the hfs splitting in the intermediate state.

7.2.3 Molecular Quantum-Beat Spectroscopy

Because quantum-beat spectroscopy offers Doppler-free spectral resolution, it has gained increasing importance in molecular physics for measurements of Zeeman and Stark splittings or of hyperfine structures and perturbations in excited molecules. The time-resolved measured signals yield not only information on the dynamics and the phase development in excited states but allow the determination of magnetic and electric dipole moments and of Landé g-factors.

One example is the measurement of hyperfine quantum beats in the polyatomic molecule propynal HC≡CCHO by Huber and coworkers [877]. In order to simplify the absorption spectrum and to reduce the overlap of absorbing transitions from different lower levels, the molecules were cooled by a supersonic expansion (Sect. 4.2). The Fourier analysis of the complex beat pattern (Fig. 7.14) showed that several upper levels had been excited coherently. Excitation with linear and circular polarization with and without an external magnetic field, allowed the analysis of this complex pattern, which is due to singlet–triplet mixing of the excited levels [877, 878].

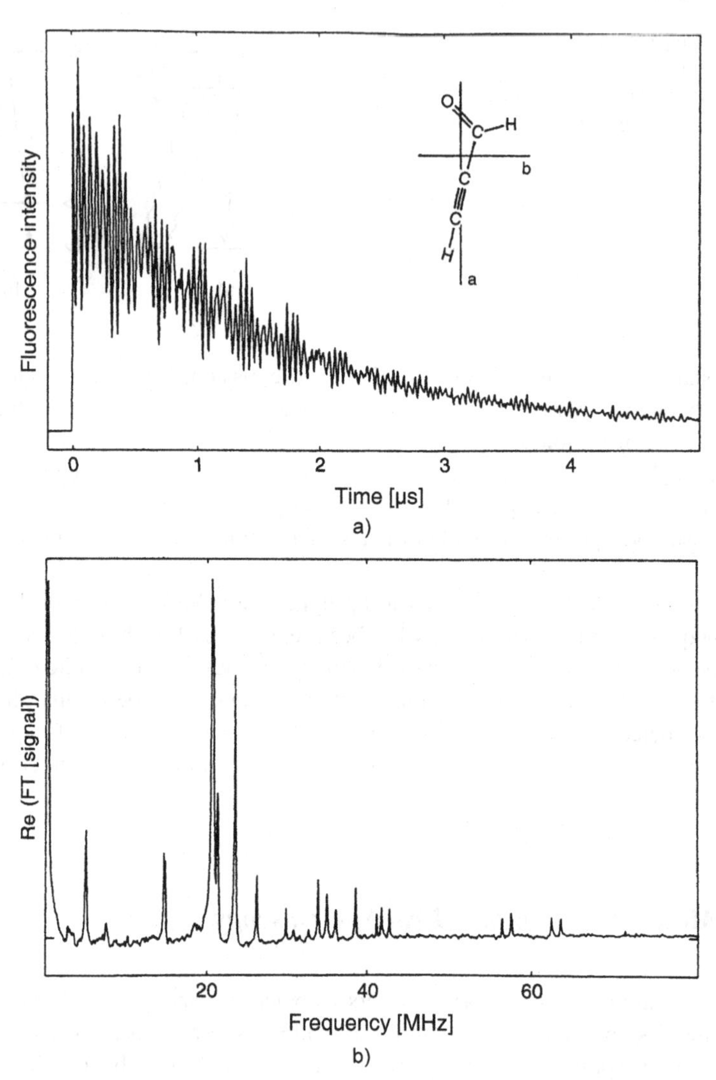

Fig. 7.14 Complex quantum-beat decay of at least seven coherently excited levels of propynal (**a**) with the corresponding Fourier transform spectrum (**b**) [877]

Many other molecules, such as SO_2 [880], NO_2 [881], or CS_2 [882], have been investigated. A fine example of the capabilities of molecular quantum-beat spectroscopy is the determination of the magnitude and orientation of excited-state electric dipole moments in the vibrationless S_1 state of planar propynal [883].

More examples and experimental details as well as theoretical aspects of quantum-beat spectroscopy can be found in several reviews [868, 878, 884], papers [871–885, 887], and a book [886].

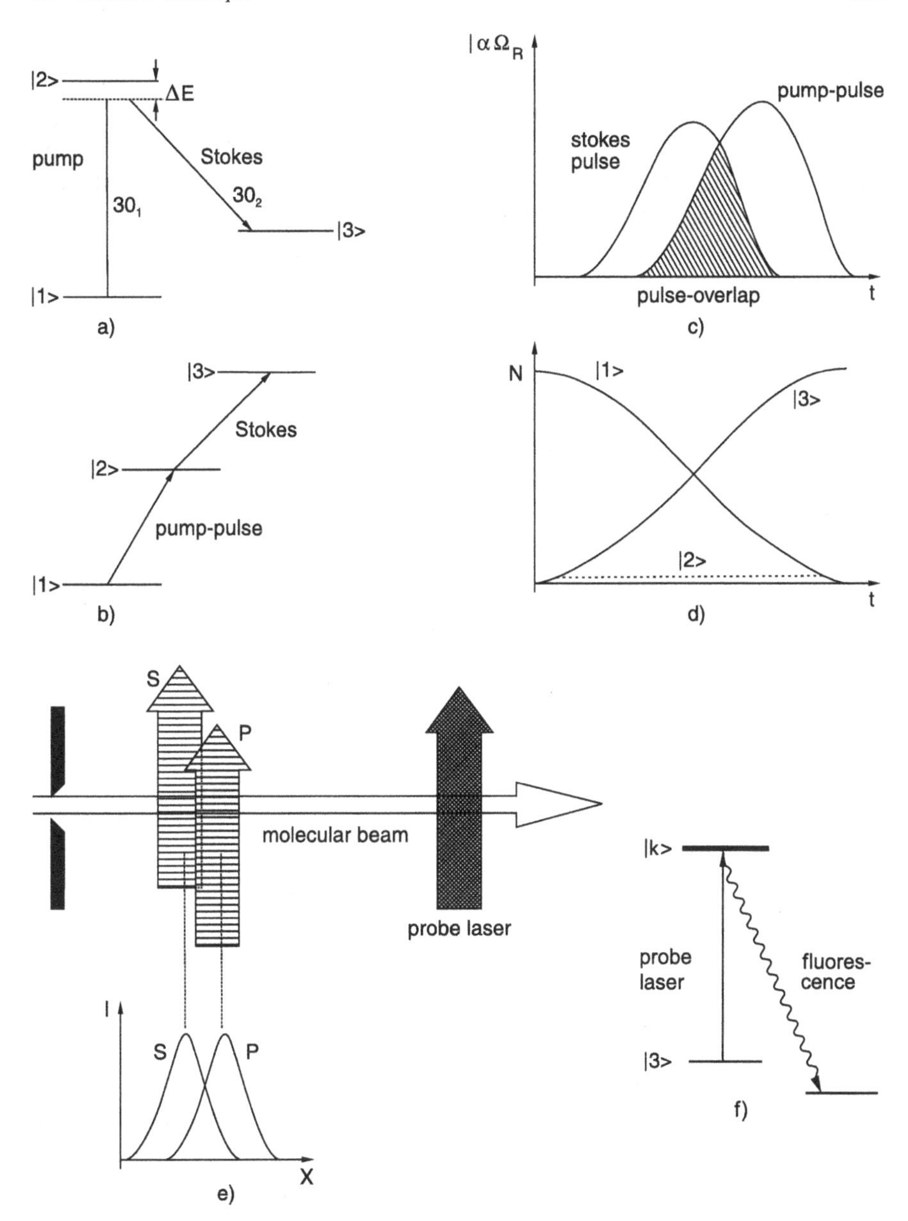

Fig. 7.15 STIRAP: (**a**) Λ-level scheme; (**b**) coherent excitation of high-lying levels; (**c**) spatial overlay of pump and Stokes pulses; (**d**) populations ($N(t)$) of the three levels; (**e**) experimental arrangement; (**f**) measurement of the population N_3 by LIF [888]

7.3 STIRAP Technique

The acronym STIRAP stands for *STI*mulated *R*aman *A*diabatic *P*assage [888]. Its principle is illustrated in Fig. 7.15. The molecules under investigation are irradiated by two laser pulses: a pump laser L_1 and a Stokes laser L_2, which generate a stimulated Raman process transfering the molecule from level $|1\rangle$ into level $|3\rangle$ (Fig. 7.15a). The frequency ν_1 of the pump laser is slightly detuned from resonance with the transition $|1\rangle \rightarrow |2\rangle$; the difference frequency $\nu_1 - \nu_2$ is, however, exactly matched to the energy difference $E_3 - E_1$. The time sequence of the two pulses is counterintuitive and differs from normal double resonance: the Stokes laser pulse is applied first, and then the pump laser. What is the advantage of this arrangement?

The Stokes laser generates a coherent superposition of the wavefunctions of levels $|2\rangle$ and $|3\rangle$. The states $|2\rangle$ and $|3\rangle$ are, however, not occupied before the pump pulse arrives. The wavefunction oscillates between levels $|2\rangle$ and $|3\rangle$ with the Rabi frequency Ω which depends on the intensity of the Stokes pulse and its detuning from resonance. Now the pump pulse comes with a time delay Δt with respect to the Stokes pulse, where Δt is smaller than the width of the Stokes pulse, which means that the two pulses still overlap (Fig. 7.15b). This places the molecule at a coherent superposition of levels $|1\rangle$ and $|2\rangle$ and $|2\rangle$ and $|3\rangle$. If the delay Δt, the detuning $\Delta\nu$ and the intensities of the two lasers are correctly chosen, the population in level $|1\rangle$ can be completely transferred into level $|3\rangle$ without creating a population in level $|2\rangle$ (Fig. 7.15c). The coherently excited levels $|1\rangle$ and $|2\rangle$ are described by the wavefunction

$$\Phi = \sin\theta|1\rangle + \cos\theta|2\rangle .$$

The transfer can be regarded as adiabatic if the adiabasy-condition

$$\hbar\,\mathrm{d}\theta/\mathrm{d}t \ll |\varepsilon_+ - \varepsilon_-|$$

is fullfilled, where

$$\varepsilon_\pm = \pm\frac{1}{2}\hbar\left(\Delta^2 + \Omega^2\right)^{1/2}$$

is the energy detuning of the pump laser from resonance.

Contrary to stimulated emission pumping with the time sequence pump pulse–Stokes pulse, where a maximum of 50 % of the population N_1 can be transferred to $|3\rangle$ (because a maximum population difference $N_2 - N_1 = 0 \rightarrow N_2 = N_1(0)/2$ and $N_3 - N_2 = 0 \rightarrow N_3 = N_2 = N_1(0)/2$ can be reached), a transfer efficiency of 100 % can be achieved with the SIRAP technique [889]. The population N_3 can be monitored through the laser-induced fluorescence induced by a weak third laser (probe laser in Fig. 7.15e).

The transfer efficiency was checked for several molecules [890]. The technique is very useful for generating molecules in definite quantum states. If these molecules are reactants for reactive collisions, the initial conditions for the reaction are known. Changing the populated state then gives information on the dependence of the reaction probability on the initial states of the reactants (see Sect. 8.4).

7.4 Excitation and Detection of Wave Packets in Atoms and Molecules

In the previous section we saw that the coherent excitation of several eigenstates by a short laser pulse leads to an excited nonstationary state

$$|\psi(t)\rangle = \sum c_k|\psi_k\rangle \exp(-\mathrm{i}E_k t/\hbar),$$

which is described by a linear combination of stationary wave functions $|\psi_k\rangle$. Such a superposition is called a *wave packet*. Whereas quantum-beat spectroscopy gives information on the time development of this wave packet, it does not tell about the spatial localization of the system characterized by the wave packet $|\psi(x,t)\rangle$. This localization aspect is discussed in the present section.

We shall start with wave packets in atomic Rydberg states [891]: if a short laser pulse of duration τ brings free atoms from a common lower state into high Rydberg states (Sect. 5.4), all accessible levels within the energy interval $\Delta E = \hbar/\tau$ can be excited simultaneously. The total wave function describing the coherent superposition of excited Rydberg levels is the linear combination

$$\psi(x,t) = \sum_{n,\ell,m} a_{n,\ell,m} R_{n,\ell}(r) Y_{\ell,m}(\theta) \exp(-\mathrm{i}E_n t/\hbar), \tag{7.28}$$

of stationary Rydberg state hydrogen-like functions with the principal quantum number n, angular momentum quantum number ℓ, and magnetic quantum number m. This superposition (7.28) represents a localized, nonstationary wave packet.

Depending on the kind of preparation, different types of Rydberg wave packets can be formed:

Radial wave packets are localized only with respect to the radial electronic coordinate r. They consist of a superposition, (7.28), with several different values of n but only a few values of ℓ or m.

Angular Rydberg wave packets, on the other hand, are formed by a superposition, (7.28), with many different values of ℓ and m but only a single fixed value of n.

When the Rydberg levels $|n\ell m\rangle$ are excited by a short laser pulse with the duration τ from the ground state $|i\rangle$, where the electron is localized within a few Bohr radii around the nucleus, the excitation process is fast compared to the oscillation period of a radial Rydberg wave packet, provided that $\tau \ll \hbar/(E_n - E_{n-1})$. This fast excitation corresponds to a vertical transition in the potential diagram of Fig. 7.16a. A delayed probe pulse transfers the Rydberg electron into the ionization continuum (Fig. 7.16b). The probability of photoionization depends strongly on the radial coordinate r. Only in the near-core region (small r) can the Rydberg electron absorb a photon because the coupling to the nucleus can take the recoil and helps to satisfy energy and angular momentum conservation. For large values of r the electron moves like a free particle, which has a very small probability of absorbing a visible photon. Therefore the number of photoelectrons detected as a function of the

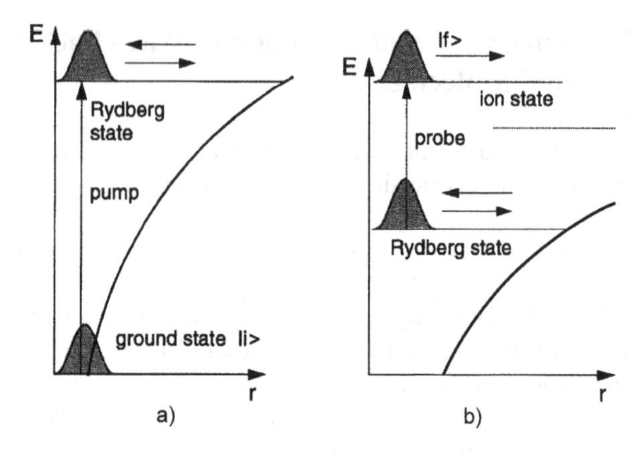

Fig. 7.16 (**a**) Excitation of a radial Rydberg wave packet by a short pump pulse from the ground state $|i\rangle$ into a Rydberg state at its inner turning point; and (**b**) its detection by a delayed photoionizing probe pulse [891]

delay time Δt between the pump and probe pulses will exhibit maxima each time the Rydberg electron is close to the nucleus, that is, if Δt is a multiple of the mean classical orbit time. Experiments performed on sodium Rydberg atoms confirm this oscillatory photoionization yield $N_{\mathrm{PE}}(\Delta t)$ [892].

A second example, which was discussed in Sect. 6.4.4, concerns wave packets of molecular vibrations studied with femtosecond time resolution [822]. While stationary spectroscopy with narrow-band lasers excites the molecule into vibrational eigenstates corresponding to a time average over many vibrational periods, the excitation by short pulses produces nonstationary wave packets composed of all vibrational eigenfunctions within the energy range $\Delta E = h/\tau$. For high vibrational levels these wave packets represent the classical motion of the vibrating nuclei. As outlined in Sect. 6.4.4, the pump-and-probe technique with femtosecond resolution allows the real-time observation of the motion of vibrational wave packets. This is illustrated by Fig. 7.17, which shows schematically the level diagram of the I_2 molecule. A short pump pulse ($\Delta T \approx 70$ fs) at $\lambda_1 = 620$ nm excites at least two vibrational levels v' in the $B^3\Pi_{ou}$ state simultaneously (coherently) from the $v'' = 0$ vibrational level of the X-ground state. The probe pulse at $\lambda_2 = 310$ nm excites the molecules further into a higher Rydberg state, where the excitation probability has a maximum at the inner turning point, since the Franck–Condon factor has here its maximum. The fluorescence from this state is monitored as a function of the delay time Δt between pump and probe pulse. The measured signal $I_{\mathrm{Fl}}(\Delta t)$ shows fast oscillations with a frequency $\nu = (\nu_1 + \nu_2)/2$, which equals the mean vibrational frequency of the coherently excited vibrational levels in the $B^3\Pi_{ou}$ state. Due to anharmonicity of the ${}^3\Pi_{ou}$ potential, the two vibrational frequencies ν_1 and ν_2 differ. The slowly varying envelope of the signal in Fig. 7.17 reflects the difference $\nu_1 - \nu_2$. After the recurrence time $\Delta T = 1/(\nu_1 - \nu_2)$, the two coherently excited vibrations are again "in phase." The Fourier transform of this beat signal gives the absolute vibrational energy E_{vib} and the separation ΔE_{vib} of the two coherently excited levels (Fig. 7.17c). This demonstrates that the "real time" vibrational motion of the molecule can be viewed with such a "stroboscopic" method. More examples can be found in [893–897].

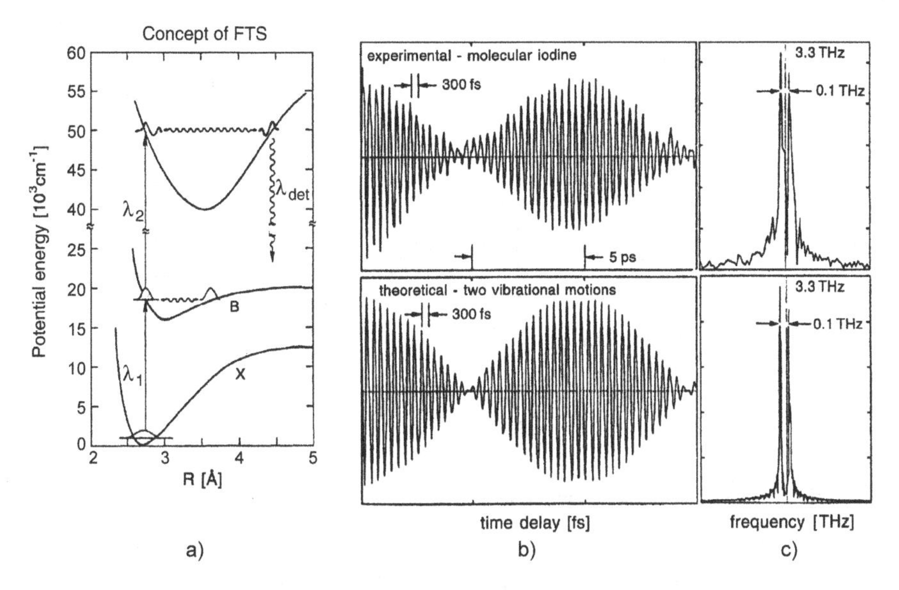

Fig. 7.17 Vibrational motion of the I_2 molecule: (**a**) potential curve of the ground state, the excited $^3\Pi_{ou}$ state reached by the pump laser, and a higher excited state of I_2 populated by the probe laser. The probe laser-induced fluorescence is used to monitor the vibrational motion of the I_2 molecule in the $B^3\Pi_{ou}$ state depicted by the probability $\mathcal{P}(R)$ in a superpositional vibrational level v'. (**b**) Probe-laser induced fluorescence intensity as a function of the delay time between probe pulse and pump pulse, showing the oscillation of the wave packet. The short period gives the mean vibrational period in the two states, the long one represents the recurrence time. (**c**) Fourier spectrum of (**b**) [895]

7.5 Coherent Control

Coherent control of a quantum system tries to transfer a selected quantum state into another wanted quantum state by optimizing the temporal and spectral profile of the exciting laser pulse. One example is the coherent control of a molecule where the wavefunctions of short lived excited molecular states are controlled which affects the different decay channels into which the excited molecule is transferred. The phase of this wavefunction depends on the characteristic features of the exciting laser pulse. Its spectral and time profile can be formed within wide limits by the technique of coherent control. The different possible decay channels interfere with each other and the goal of coherent control is constructive interference for the wanted channel and destructive interference for the unwanted channels, which can be briefly explained as follows (Fig. 7.18): A femtosecond-laser pulse is sent onto an optical grating. Here the different frequency components of the pulse are dispersed and are diffracted by the grating into different directions. The total diffracted light is collimated by a lens into a parallel beam which passes through a phase plate. This plate consists of many liquid crystals which can be individually controlled by an external voltage which orientates the molecules in the liquid crystal. This changes the refractive index and with it the phase of the transmitted light. While the phase-front

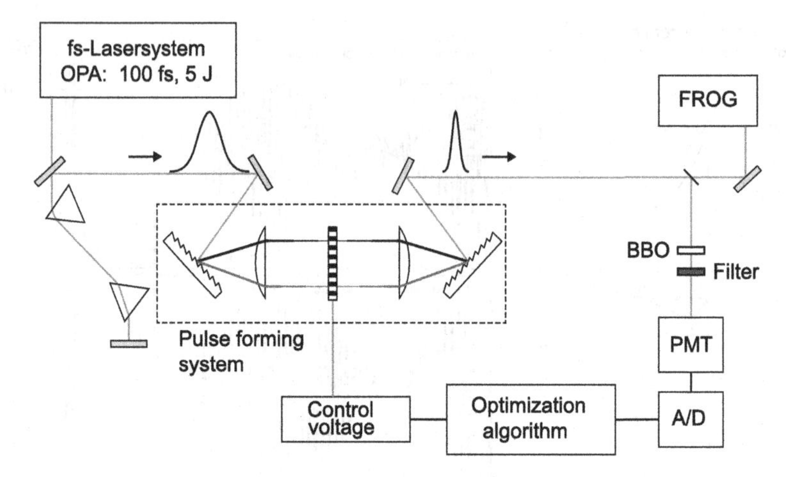

Fig. 7.18 Experimental setup for pulse profile optimization

of the incoming light is a plane, that of the transmitted light is a complex surface with a form that depends on the voltages applied to the individual liquid crystals. The transmitted light is focused by a lens onto a second grating which recombines the different optical frequencies. This superposition results in a pulse with a time profile, which depends on the individual phases of the frequency components. The pulses are sent to the sample molecules and the different decay channels of the excited molecules are monitored. By an optimization feedback algorithm the voltages to the individual liquid crystals are now changed in such a way, that the wanted decay channel is maximized and the other unwanted channels suppressed.

This technique is called *coherent control*, because the time-dependent phase of the wavefunction of the coherently excited state is controlled in order to optimize the result of the molecular decay [898, 899].

7.6 Optical Pulse-Train Interference Spectroscopy

Let us consider atoms with optical transitions between a single level $|i\rangle$ and a state $|k\rangle$ that is split into two sublevels $|k_1\rangle$ and $|k_2\rangle$ (Fig. 7.9). If the atoms are irradiated by a short laser pulse of duration $\tau < \hbar/\Delta E = \hbar/(E_{k1} - E_{k2})$ and mean optical frequency $\omega = (E_i - E_k)/\hbar$, an induced dipole moment is produced, which oscillates at the frequency ω. The envelope of the damped oscillation shows a modulation with the beat frequency $\Delta\omega = \Delta E/\hbar$ (quantum beats, Sect. 7.2).

If the sample is now exposed not to a single pulse but to a regular train of pulses with the repetition frequency f such that $\Delta\omega = q \cdot 2\pi f$ $(q \in N)$, the laser pulses always arrive "in phase" with the oscillating dipole moments. With this synchronization the contributions of subsequent pulses in the regular pulse train add up coherently in phase and a macroscopic oscillating dipole moment is produced in the sample, where the damping between successive pulses is always compensated by the next pulse [901].

The regular pulse train of a mode-locked laser with the pulse-repetition frequency f corresponds in the frequency domain to a spectrum consisting of the carrier frequency $\nu_0 = \omega/2\pi$ and sidebands at $\nu_0 \pm q \cdot f$ ($q \in N$) (Sect. 6.1.4). If a molecular sample with a level scheme depicted in Fig. 7.13b and a sublevel splitting $\Delta\nu = 2qf$ in the lower state is irradiated by such a pulse train, where the frequency ν_0 is chosen as $\nu_0 = (\nu_1 + \nu_2)/2$, the two frequencies $\nu_{1,2} = \nu_0 \pm qf$ are absorbed on the two molecular transitions ν_1, ν_2. This may be regarded as the superposition of two Raman processes ($|1\rangle \rightarrow |2\rangle \rightarrow |3\rangle$: Stokes process) and ($|3\rangle \rightarrow |2\rangle \rightarrow |1\rangle$: anti-Stokes process), where the population of the two sublevels $|1\rangle$ and $|3\rangle$ oscillates periodically with the frequency $\Delta\nu = \Delta E/h$. The level splitting ΔE can be obtained much more accurately from this oscillation frequency than from the difference of the two optical frequencies ν_1, ν_2.

The experimental arrangement, which is similar to that of time-resolved polarization spectroscopy (Fig. 7.12), is depicted in Fig. 7.19. The pulse train is provided by a synchronously pumped, mode-locked cw dye laser. A fraction of each pulse is split by the beam splitter BS and passes through an optical delay line. Pump and probe pulses propagate either in the same direction or in opposite ones through the sample cell, which is placed between two crossed polarizers. Either the transmitted probe-pulse intensity or the laser-induced fluorescence is monitored as a function of pulse-repetition frequency f and delay time Δt. For the resonance case $f = \Delta\nu/q$, the signal $S(f)$ becomes maximum (Fig. 7.19b). The halfwidth of $S(f)$ is determined by the homogeneous level widths of the levels $|1\rangle$ and $|3\rangle$ in Fig. 7.13b. If these are, for example, hfs levels in the electronic ground state, their spontaneous lifetime is very long (Vol. 1, Problem 3.2b) and the halfwidth of $S(f)$ is mainly limited by collisional broadening and by the interaction time of the atoms with the laser. The latter can be increased by adding noble gases as buffer gas, which increases the time for diffusion of the pumped atoms out of the laser beam. As can be seen from Fig. 7.19d, linewidths down to a few hertz can be achieved. The center frequency of these signals $S(f)$ can be measured within uncertainties below 1 Hz, because the repetition frequency can be determined very precisely by digital counters [902]. The accuracy is mainly limited by the achievable signal-to-noise ratio and by asymmetries of the line profile $S(f)$.

This method allows the measurement of atomic level splittings by electronic counting of the pulse-repetition frequency! If the delay time Δt of the probe pulse is continuously varied at a fixed repetition frequency f, the oscillating atomic dipole moment becomes apparent through the time-dependent probe-pulse transmission $I_T(\Delta t)$, as shown in Fig. 7.19c for the Cs atom. The fast oscillation corresponds to the hfs splitting of the $7^2S_{1/2}$ ground state, the slowly damped oscillation to that of the excited state (quantum beats in transmission, Sect. 7.2.2). The Fourier transform of the quantum-beat signal yields the Doppler-free absorption spectrum of the D_2 line with the hfs splittings in the upper and lower states, and the decay time of the $^2P_{3/2}$ state.

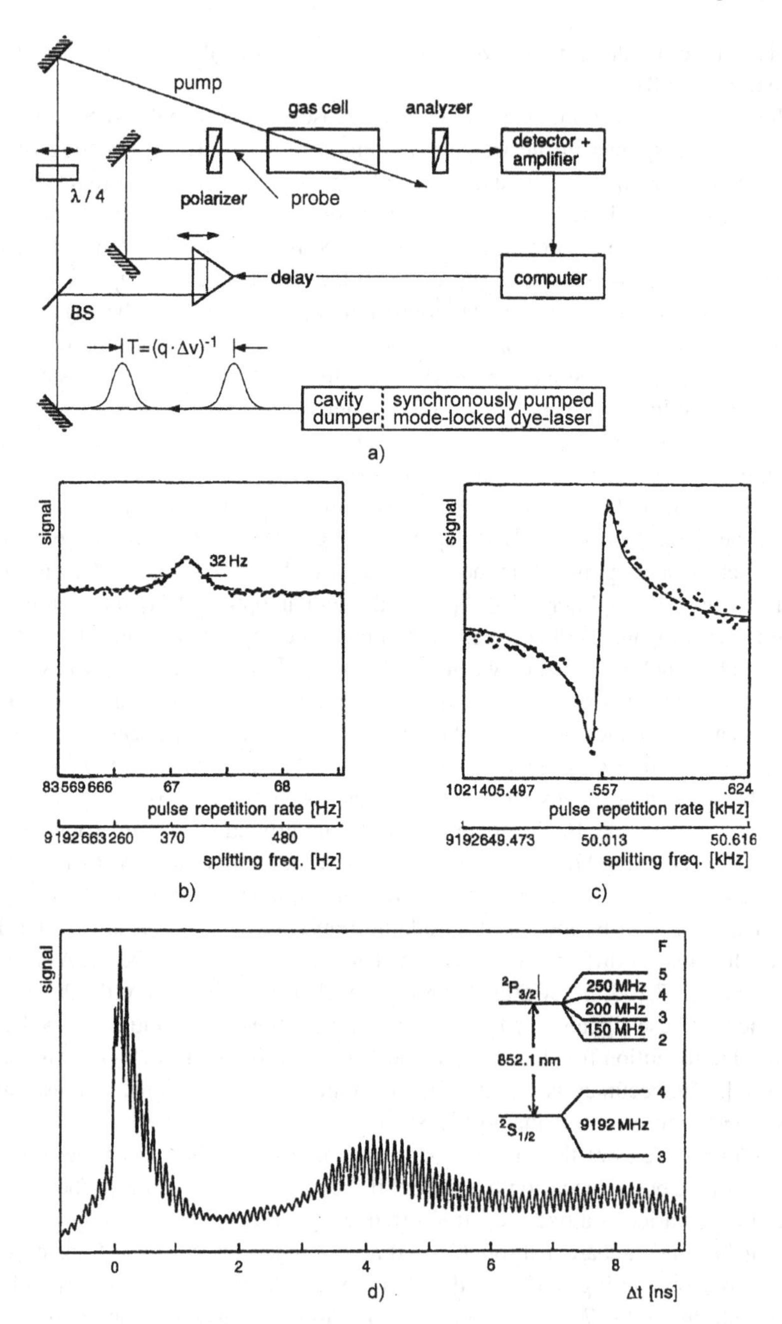

Fig. 7.19 Measurements of the hfs splittings in the $7^2S_{1/2}$ ground state of Cs atoms with the pulse-train interference method. Excitation occurs on the D_2 line at $\lambda = 852.1$ nm with a repetition rate $f = \Delta\nu/q$ with $q = 110$. (**a**) Experimental arrangement; (**b**) transmission of the probe pulse as a function of f; (**c**) fluorescence intensity $I_{\mathrm{Fl}}(f)$ as a function of f with a modulated small external magnetic field; and (**d**) as a function of the delay time Δt at a fixed repetition frequency f [902]

7.7 Photon Echoes

Assume that N atoms have simultaneously been excited by a short laser pulse from a lower level $|1\rangle$ into an upper level $|2\rangle$. The total fluorescence intensity emitted on the transition $|2\rangle \to |1\rangle$ is given by (Vol. 1, Sect. 2.7.4)

$$I_{\mathrm{Fl}} = \sum_N \hbar\omega A_{21} = \frac{\omega^4}{3\pi_0 c^3}\frac{g_1}{g_2}\left|\sum_N \langle D_{21}\rangle\right|^2, \tag{7.29}$$

where D_{21} is the dipole matrix element of the transition $|2\rangle \to |1\rangle$, and g_1, g_2 are the weight factors of levels $|1\rangle$, $|2\rangle$, respectively, see Vol. 1, (2.50) and Table 2.2. The sum extends over all N atoms.

If the atoms are excited *incoherently*, no definite phase relations exist between the wave functions of the N excited atoms. The cross terms in the square of the sum (7.29) average to zero, and we obtain in the case of identical atoms

$$\left|\sum_N \langle D_{12}\rangle\right|^2 = \sum_N \left|\langle D_{12}\rangle\right|^2 = N\left|\langle D_{12}\rangle\right|^2 \to I_{\mathrm{Fl}}^{\mathrm{incoh}} = N\hbar\omega A_{21}. \tag{7.30}$$

The situation is drastically changed, however, under *coherent* excitation, where definite phase relations are established between the N excited atoms at the time $t = 0$ of excitation. If all N excited atomic states are *in phase* we obtain

$$\left|\sum_N \langle D_{12}\rangle\right|^2 = \left|N\langle D_{12}\rangle\right|^2 \to I_{\mathrm{Fl}}^{\mathrm{coh}} = N^2\left|\langle D_{12}\rangle\right|^2 = N \cdot I_{\mathrm{Fl}}^{\mathrm{incoh}}. \tag{7.31}$$

This implies that the fluorescence intensity $I_{\mathrm{Fl}}(t)$ at times $t \le T_{\mathrm{c}}$, where all excited atoms oscillate still *in phase*, is N times larger than in the incoherent case (Dicke super-radiance) [903].

This phenomenon of super-radiance is used in the photon-echo technique for high-resolution spectroscopy to measure population and phase decay times, expressed by the *longitudinal* and *transverse* relaxation times T_1 and T_2, see (7.1). This technique is analogous to the spin-echo method in nuclear magnetic resonance (NMR) [904]. Its basic principle may be understood in a simple model, transferred from NMR to the optical region [905].

Corresponding to the magnetization vector $\boldsymbol{M} = \{M_x, M_y, M_z\}$ in NMR spectroscopy, we introduce for our optical two-level system the *pseudopolarization vector*

$$\boldsymbol{P} = \{P_x, P_y, P_3\}, \quad \text{with } P_3 = D_{12}\Delta N, \tag{7.32}$$

with the two components P_x, P_y of the atomic polarization, but a third component P_3 representing the product of the transition dipole moment D_{12} and the population density difference $\Delta N = N_1 - N_2$. Instead of the Bloch equation in NMR

$$\frac{\mathrm{d}\boldsymbol{M}}{\mathrm{d}t} = \boldsymbol{M} \times \boldsymbol{\Omega}, \tag{7.33a}$$

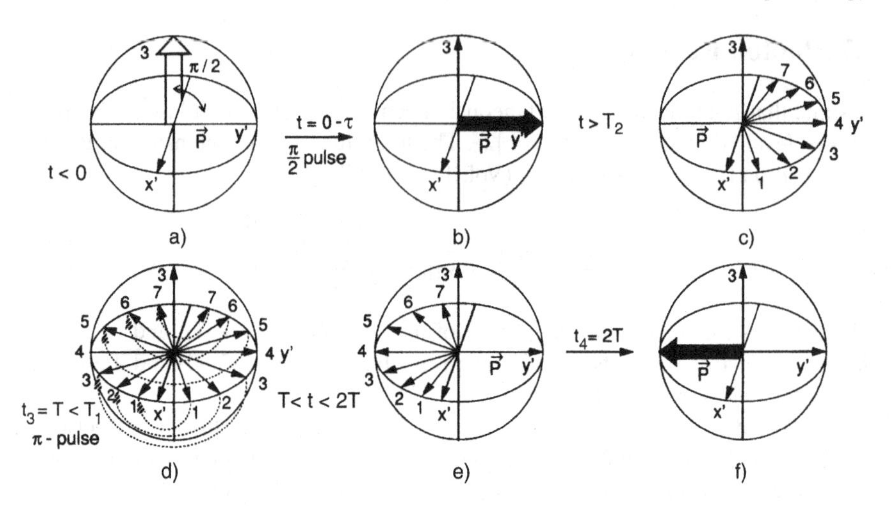

Fig. 7.20 Time development of the pseudopolarization vector and generation of a photon echo observed at $t = 2\tau$ after applying a $\pi/2$-pulse at $t = 0$ and a π-pulse at $t = \tau$ (see text)

for the time variation of the magnetization $\boldsymbol{M}$ under the influence of a magnetic RF field with frequency Ω, one obtains from Vol. 1, (2.80–2.85) the *optical Bloch equation*

$$\frac{\mathrm{d}\boldsymbol{P}}{\mathrm{d}t} = \boldsymbol{P} \times \boldsymbol{\Omega} - \{P_x/T_2;\, P_y/T_2;\, P_3/T_1\}, \tag{7.33b}$$

for the pseudopolarization vector under the influence of the optical field. The bracket contains the damping terms from phase relaxation and population relaxation that were neglected in (7.33a). The vector

$$\boldsymbol{\Omega} = \big\{(D_{12}/2\hbar),\, A_0,\, \Delta\omega\big\}, \tag{7.34}$$

is named the *optical nutation*. Its components represent the transition dipole D_{12}, the amplitude A_0 of the optical wave, and the frequency difference $\Delta\omega = \omega_{12} - \omega$ between the atomic resonance frequency $\omega_{12} = (E_2 - E_1/\hbar)$ and the optical field frequency ω. The time development of $\boldsymbol{P}$ describes the time-dependent polarization of the atomic system. This is illustrated by Fig. 7.20, which shows $\boldsymbol{P}$ in a $\{x, y, 3\}$ coordinate system, where the axis 3 represents the population difference $\Delta N = N_1 - N_2$. For times $t \le 0$ all atoms are in their ground state where they are randomly oriented. This implies $P_x = P_y = 0$ and $\Delta N = N_1$. At $t = 0$ an optical $\pi/2$ pulse is applied that excites the atoms into level $|2\rangle$. For a proper choice of pulse intensity and pulse duration τ it is possible to achieve, at the end of the pump pulse ($t = \tau$), equal populations $N_1 = N_2 \to \Delta N(t = \tau) = 0$. This means the probabilities $|a_i|^2$ to find the system in level $|i\rangle$ change from $|a_1|^2 = 1$, $|a_2|^2 = 0$ before the pulse to $|a_1|^2 = |a_2|^2 = 1/2$ after the pulse. This implies $P_3(t = 0) = 0$, which means that the pseudopolarization vector now lies in the x–y-plane. Since such a pulse changes the phase of the probability amplitudes $a_i(t)$ by $\pi/2$, it is

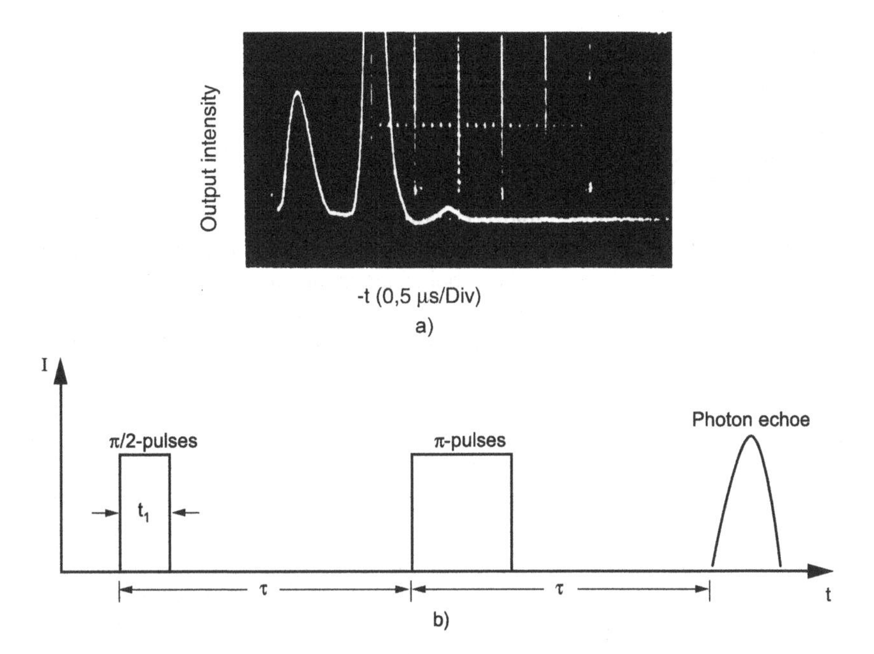

Fig. 7.21 (**a**) Pulse sequence in photon echoe experiments. (**b**) Time sequence of pulses for photon echoes. Oscilloscope trace of $\pi/2$- and π-pulses and a photon echo observed from SF_6 molecules that were excited by two CO_2 laser pulses [907]

called a $\pi/2$-*pulse* (Vol. 1, Sect. 2.7.6). At $t = \tau$ all induced atomic dipoles oscillate in phase, resulting in the macroscopic polarization $\boldsymbol{P}$, which we have assumed to point into the y-direction (Fig. 7.20b).

Because of the finite linewidth $\Delta\omega$ of the transition $|1\rangle \rightarrow |2\rangle$ (for example, the Doppler width in a gaseous sample), the frequencies $\omega_{12} = (E_1 - E_2)/\hbar$ of the atomic transitions of our N dipoles are distributed within the interval $\Delta\omega$. This causes the phases of the N oscillating dipoles to develop in time at different rates after the end of the $\pi/2$-pulse at $t > \tau$. After a time $t > T_2$, which is large compared to the phase relaxation time T_2, the phases are again randomly distributed (Fig. 7.20c,d).

If a second laser pulse that has the proper intensity and duration to invert the phase of the induced polarization (π-pulse) is applied to the sample at a time $t_3 = T < T_1$, it causes a reversal of the phase development for each dipole (Fig. 7.20d–f). This means that after a time $t_4 = 2T$ all dipoles are again in phase (Fig. 7.20f). As discussed above, while these excited atoms are in phase they emit a superradiant signal at the time $t = 2T$ that is called *photon* echo (Fig. 7.21).

In the ideal case the magnitude of the photon echo is $N_2(2T)$ times larger than the incoherent fluorescence (which is emitted at all times $t > 0$), where $N_2(2T)$ is the number density of excited atoms at $t = 2T$.

There are, however, two relaxation processes that prevent the original state, as prepared just after the first $\pi/2$-pulse at $t = 0$, from being completely reestablished at the echo time $t = 2T$. Because of spontaneous or collision-induced decay, the population of the upper state decreases to

$$N_2(2T) = N_2(0)e^{-2T/T_1}. \tag{7.35}$$

This means that the echo amplitude decreases because of population decay with the longitudinal relaxation constant T_1.

A second, generally more rapid relaxation is caused by phase-perturbing collisions (Vol. 1, Sect. 3.3), which change the phase development of the atoms and therefore prevent all atoms from being again in phase at $t = 2T$. Because such phase-perturbing collisions give rise to homogeneous line broadening (Vol. 1, Sect. 3.5), the phase relaxation time due to these collisions is called T_2^{hom} in contrast to the inhomogeneous phase relaxation caused, for instance, by the different Doppler shifts of moving atoms in a gas (Doppler broadening).

The important point is that the *inhomogeneous* phase relaxation, which occurs between $t = 0$ and $t = 2T$ because of the random Doppler-shifted frequencies of atoms with different velocities, does *not* prevent a complete restoration of the initial phases by the π-pulse. If the velocity of a single atom does not change within the time $2T$, the different phase development of each atom between $t = 0$ and $t = T$ is exactly reversed by the π-pulse. This means that even in the presence of inhomogeneous line broadening, the homogeneous relaxation processes (that is, the homogeneous part of the broadening) can be measured with the photon-echo method. *This technique therefore allows Doppler-free spectroscopy.*

The production of the coherent state by the first pulse must, of course, be faster than these homogeneous relaxation processes. This implies that the laser pulse has to be sufficiently intense. From Vol. 1, (2.93) we obtain the condition

$$D_{12}E_0 > \pi\hbar\left(1/T_1 + 1/T_2^{\text{hom}}\right), \tag{7.36}$$

for the product of laser field amplitude E_0 and transition matrix element D_{12}. With typical relaxation times of 10^{-6} to 10^{-9} s the condition (7.36) requires power densities in the range kW/cm^2 to MW/cm^2, which can be readily achieved with pulsed or mode-locked lasers. With increasing delay time T between the first $\pi/2$-pulse and the second π-pulse the echo intensity I_e decreases exponentially as

$$I_e(2T) = I_e(0)\exp\left(-2T^{\text{hom}}\right), \quad \text{with } 1/T^{\text{hom}} = \left(1/T_1 + 1/T_2^{\text{hom}}\right). \tag{7.37}$$

From the slope of a logarithmic plot of $I_e(2T)$ versus the delay time T the homogeneous relaxation times can be obtained.

The qualitative presentation of photon echoes, discussed above, may be put on a more quantitative base by using time-dependent perturbation theory. We outline briefly the basic considerations, which can be understood from the treatment in Vol. 1, Sect. 2.9. For a more detailed discussion see [905].

A two-level system can be represented by the time-dependent wave function (Vol. 1, (2.58))

$$\psi(t) = \sum_{n=1}^{2} a_n(t)u_n e^{-E_n t/\hbar}. \tag{7.38}$$

Before the first light pulse is applied, the system is in the lower level $|1\rangle$, which means $|a_1| = 1$ and $|a_2| = 0$. The harmonic perturbation (Vol. 1, (2.46))

$$V = -D_{12}E_0 \cos\omega t, \quad \text{with } \hbar\omega, = E_2 - E_1 \tag{7.39}$$

produces a linear superposition

$$\psi(t) = \cos\left(\frac{D_{12}E_0}{2\hbar}t\right)u_i e^{-iE_1t/\hbar} + \sin\left(\frac{D_{12}E_0}{2\hbar}t\right)u_2 e^{-iE_2t/\hbar}. \tag{7.40}$$

If this perturbation consists of a short intense light pulse of duration τ such that

$$(D_{12}E_0/\hbar)\tau = \pi/2, \tag{7.41}$$

the total wave function becomes, with $\cos(\pi/4) = \sin(\pi/4) = 1/\sqrt{2}$,

$$\psi(t) = \frac{1}{\sqrt{2}}\left(u_1 e^{-iE_1t/\hbar} + u_2 e^{-iE_2t/\hbar}\right). \tag{7.42}$$

After the time T, the phases have developed to $E_nT/\hbar$. If now a second π-pulse with $|(D_{12}E_0/\hbar)|\tau = \pi$ is applied, the wave functions of the upper and lower states are just interchanged, so that for time $t = T$

$$\begin{aligned} e^{-iE_1T/\hbar}u_1 &\to e^{-iE_1T/\hbar}u_2 e^{-iE_2(t-T)/\hbar}, \\ e^{-iE_2T/\hbar}u_2 &\to e^{-iE_2T/\hbar}u_1 e^{-iE_1(t-T)/\hbar}. \end{aligned} \tag{7.43}$$

The total wave function therefore becomes

$$\psi(t,T) = \frac{1}{\sqrt{2}}\left(u_2 e^{-i\omega_k(t-2T)/2} - u_1 e^{+i\omega_k(t-2T)/2}\right), \tag{7.44}$$

and the dipole moment for each atom is

$$D_{12} = \langle\psi^*|e\boldsymbol{r}|\psi\rangle = -\langle u_2^*|e\boldsymbol{r}|u_1\rangle e^{-i\omega_k(t-2T)}. \tag{7.45}$$

If the different atoms have slightly different absorption frequencies $\omega = (E_2 - E_1)/\hbar$ (for example, because of the different velocities in a gas), the phase factors for different atoms are different for $t \neq 2T$, and the macroscopic fluorescence intensity is the incoherent superposition (7.30) from all atomic contributions. However, for $t = 2T$ the phase factor is zero for all atoms, which implies that all atomic dipole moments are in phase, and super-radiance is observed.

Photon echoes were first observed in ruby crystals using two ruby laser pulses with variable delay [906]. The application of this technique to gases started with CO_2 laser pulses incident on a SF_6 sample. The collision-induced homogeneous relaxation time T_2^{hom} has been measured from the time decay of the echo amplitude. Figure 7.21 is an oscilloscope trace of the $\pi/2$- and π-pulses, which shows the echo obtained from a SF_6 cell at the pressure of 0.01 mb [907] as the small third pulse.

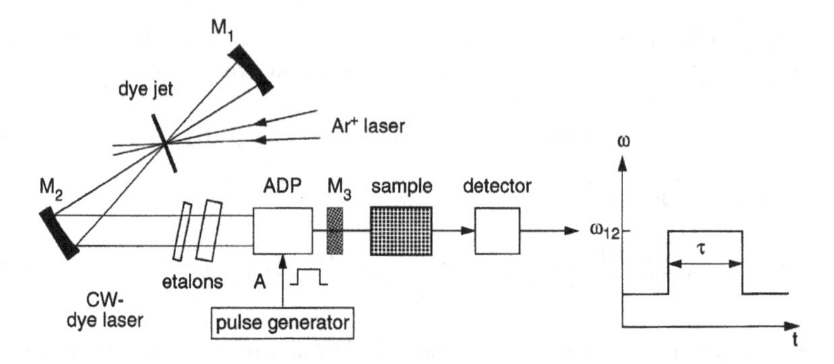

Fig. 7.22 Schematic of the laser-frequency switching apparatus for observing photon echoes and coherent optical transients. The intracavity ADP crystal is oriented in such a way that it changes the refractive index n and therefore the laser frequency without altering the polarization characteristics of the laser beam when a voltage is applied

Fig. 7.23 (**a**) Stark switching technique for the case of a Doppler-broadened molecular transition. (**b**) Infrared photon echo for a $^{13}CH_3F$ vibration–rotation transition. The molecules are switched twice into resonance with a cw CO_2 laser by the two Stark pulses shown in the *lower trace*. The third pulse is the photon echo [908]

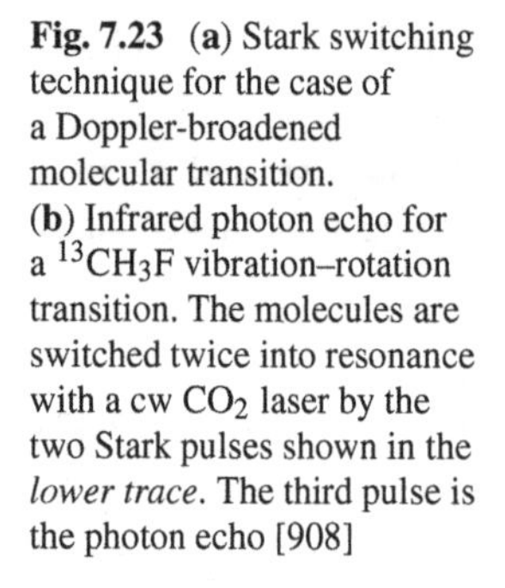

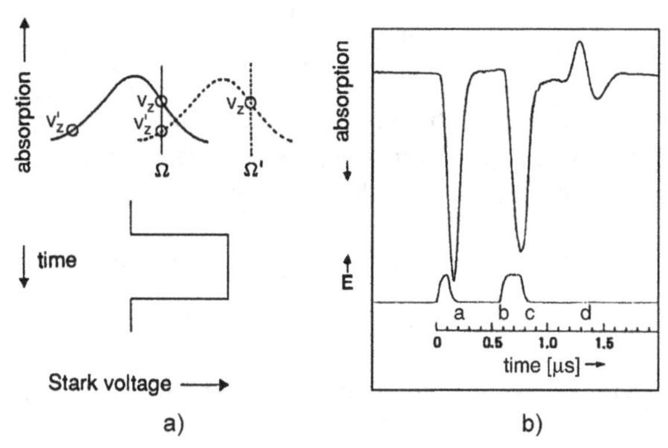

For sufficiently large transition-dipole matrix elements D_{12}, photon echoes can be also observed with cw lasers if the molecules are tuned for a short time interval into resonance with the laser frequency. There are two possible experimental realizations: the first uses an electro-optical pulsed modulator inside the laser cavity, which shifts the laser frequency $\omega \neq \omega_{12}$ for short time intervals τ into resonance ($\omega = \omega_{12}$) with the molecules. The second method shifts the absorption frequency ω_{12} of the molecules by a pulsed electric field (Stark shifting) into resonance with the fixed laser frequency ω_L. Figure 7.22 shows an example of an experimental arrangement [908] where a stable, tunable cw dye laser is used, and frequency switching is achieved with an ammonium dihydrogen phosphate (ADP) crystal driven by a sequence of low-voltage pulses. It produces a variation of the refractive index n and therefore a shift of the laser wavelength $\lambda = c/(2nd)$ in a resonator with mirror separation d.

Figure 7.23 illustrates the Stark switching technique, which can be applied to all those molecules that show a sufficiently large Stark shift [908]. In the case of Doppler-broadened absorption lines the laser of fixed frequency initially excited molecules of velocity v_z. A Stark pulse, which abruptly shifts the molecular absorption profile from the solid to the dashed curve, causes the velocity group v'_z to come into resonance with the laser frequencies Ω. We have assumed that the Stark shift of the molecular eigenfrequencies is larger than the homogeneous linewidth, but smaller than the Doppler width. With two Stark pulses the group v'_z emits an echo. This is shown in Fig. 7.23b, where the CH_3F molecules are switched twice into resonance with a cw CO_2 laser by 60 V/cm Stark pulses [909]. A more detailed discussion of photon echoes can be found in [905, 910–912].

7.8 Optical Nutation and Free-Induction Decay

If the laser pulse applied to the sample molecules is sufficiently long and intense, a molecule (represented by a two-level system) will be driven back and forth between the two levels at the Rabi flopping frequency (Vol. 1, (2.96)). The time-dependent probability amplitudes $a_1(t)$ and $a_2(t)$ are now periodic functions of time and we have the situation depicted in Vol. 1, Fig. 2.23. Since the laser beam is alternately absorbed (induced absorption $E_1 \to E_2$) and amplified (induced emission $E_2 \to E_1$), the intensity of the transmitted beam displays an oscillation. Because of relaxation effects this oscillation is damped and the transmitted intensity reaches a steady state determined by the ratio of induced to relaxation transitions. According to Vol. 1, (2.96) the flopping frequency depends on the laser intensity and on the detuning $(\omega_{12} - \omega)$ of the molecular eigenfrequency ω_{12} from the laser frequency ω. This detuning can be performed either by tuning of the laser frequency ω (Fig. 7.22) or by Stark tuning of the molecular eigenfrequencies ω_{12} (Fig. 7.23).

We consider the case of a gaseous sample with Doppler-broadened transitions [912], where the molecular levels are shifted by a pulsed electric field. The Stark shift of the molecular eigenfrequencies is larger than the homogeneous linewidth but smaller than the Doppler width (Fig. 7.23a). During the steady-state absorption preceding the Stark pulse, the single-mode cw laser excites only a narrow velocity group of molecules around the velocity component v_{z1} within the Doppler line shape. Sudden application of a Stark field shifts the transition frequency ω_{12} of this subgroup out of resonance with the laser field and another velocity subgroup around v_{z2} is shifted into resonance. At the end of the Stark pulse the first group of molecules is again switched into resonance. We can now realize two different situations:

(a) The two Stark pulses act for the optical excitation as $\pi/2$ and π-pulses. This generates a photon echo (Fig. 7.23b).
(b) The pulses are longer than the Rabi oscillation period. Then the molecules of each subgroup start their damped oscillation with the Rabi frequency $\Omega = D_{12}E_0/\hbar$ at the beginning of the excitation, giving the optical nutation patterns in Fig. 7.24. When the Stark pulse terminates at $t = \tau$, the initial velocity group is switched back into resonance and generates the second nutation pattern in Fig. 7.24a.

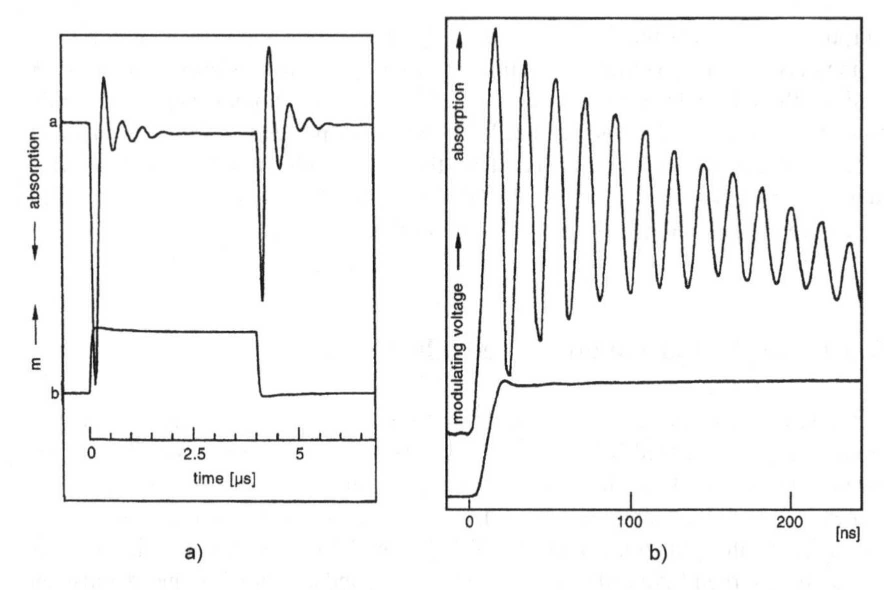

Fig. 7.24 (**a**) Optical nutation in $^{13}CH_3F$ observed with CO_2 laser excitation at $\lambda = 9.7$ μm. The Rabi oscillations appear because the Stark pulse (*lower trace*) is longer than in Fig. 7.23. (**b**) Optical free-induction decay in I_2 vapor following resonant excitation with a cw dye laser at $\lambda = 589.6$ nm. At the time $t = 0$ the laser is frequency-shifted with the arrangement depicted in Fig. 7.22 by $\Delta\omega = 54$ MHz out of resonance with the I_2 transition. The slowly varying envelope is caused by a superposition with the optical nutation of molecules in the velocity group $v_z = (\omega - \omega_0)/k$, which are now in resonance with the laser frequency ω. Note the difference in time scales of (**a**) and (**b**) [705]

The amplitude $A(t)$ of this delayed nutation depends on the population of the first subgroup v_{z1} at the time τ where the Stark pulse ends. This population had partially been saturated before $t = 0$, but collisions during the time τ try to refill it. Berman et al. [913] have shown that

$$A(\tau) \propto N_1(\tau) - N_2(\tau) = \Delta N_0 + \left[\Delta N_0 - \Delta N(0)\right] e^{-\tau/T_1}, \qquad (7.46)$$

where ΔN_0 is the unsaturated population difference in the absence of radiation, and $\Delta N(0)$, $\Delta N(\tau)$ are the saturated population difference at $t = 0$ or the partially refilled difference at $t = \tau$. The dependence of $A(\tau)$ on the length τ of the Stark pulse therefore allows measurement of the relaxation time T_1 for refilling the lower level and depopulating the upper one.

After the end of the laser pulse the induced dipole moment of the coherently prepared molecular system performs a damped oscillation at the frequency ω_{12}, where the damping is determined by the sum of all relaxation processes (spontaneous emission, collisions, etc.) that affect the phase of the oscillating dipole.

This *optical induction free decay* can be measured with a beat technique: at time $t = 0$ the frequency ω of a cw laser is switched from $\omega = \omega_{12}$ to $\omega' \neq \omega_{12}$ out of resonance with the molecules. The superposition of the damped wave at ω_{12} emitted by the coherently prepared molecules with the wave at ω' gives a beat signal at the difference frequency $\Delta\omega = \omega_{12} - \omega'$, which is detected [909]. If $\Delta\omega$ is smaller than the Doppler width, the laser at ω' interacts with another velocity subgroup of molecules and produces optical nutation, which superimposes the free-induction decay and which is responsible for the slowly varying envelope in Fig. 7.24b.

These experiments give information on the polarizability of excited molecules (from the amplitude of the oscillation), on the transverse phase relaxation times T_2 (which depend on the cross sections of phase-perturbing collisions, Vol.1 , Sect. 3.3), and on the population decay time T_1. For more details see [911, 913].

7.9 Self-Induced Transparency

One characteristic feature of optical pulses is the pulse area

$$A = \kappa \iint \left[\boldsymbol{D}_{ik} \cdot \boldsymbol{E}\left(z, t'\right)/\hbar\right]^2 \mathrm{d}t' = \Omega_0 \cdot \tau .$$

Here $\boldsymbol{D}_{ik}$ is the matrix transition moment, $\boldsymbol{E}$ is the electric field vector, Ω_0 the Rabi-frequency in case of resonance $\omega = \omega_0$ and τ is the pulse length. For $A = \pi$ the pulse is called a π-pulse. When optical π-pulses with a wavelength λ matching the energy difference ΔE_{ik} between two atomic levels $|i\rangle$ and $|k\rangle$ pass through a system of two-level atoms with the absorption coefficient α, they produce a population inversion. This causes for a following weak optical pulse amplification instead of absorption without the π-pulse.

An incident pulse with the pulse area $A = 2\pi$ propagates over considerable distances without noticeable attenuation. The time profile of a pulse with length τ approaches after a pathlength $\Delta z = m/\alpha$ ($m = 1; 2; 3; \ldots$) a hyperbolic secant profile with the electric field amplitude

$$E(z, t) = (2E_0/\alpha \cdot z)\, \mathrm{sech}\left[(t - z/v_g)/\tau_0\right] \cdot \mathrm{e}^{\mathrm{i}z/4z_0)} .$$

Here v_g is the group velocity, $z_0 = \tau_0^2/D_v$ and D_v is the group-velocity dispersion. For $z_0 \to \infty$ i.e. $D_v \to 0$ the pulse form remains unchanged. Such pulses are called *solitons* (Fig. 7.25).

The reason for this absorption-free propagation can be understood as follows: The first half of the 2π-pulse is absorbed and creates a population inversion. Therefore the second half of the pulse causes stimulated emission which just compensates the absorption of the first half (self-phase-modulation). The absorption of the first half pulse area and the amplification of the second half restores the 2π-pulse. But this restored pulse is delayed for each cycle by $\tau/2$ against a weak pulse starting at the same time. The group velocity of solitons is therefore smaller than for weak pulses.

Since short pulses are spectrally broad, one might assume, that the pulse width increases due to linear dispersion. However, for solitons this linear dispersion is exactly compensated by the nonlinear self-phase-modulation.

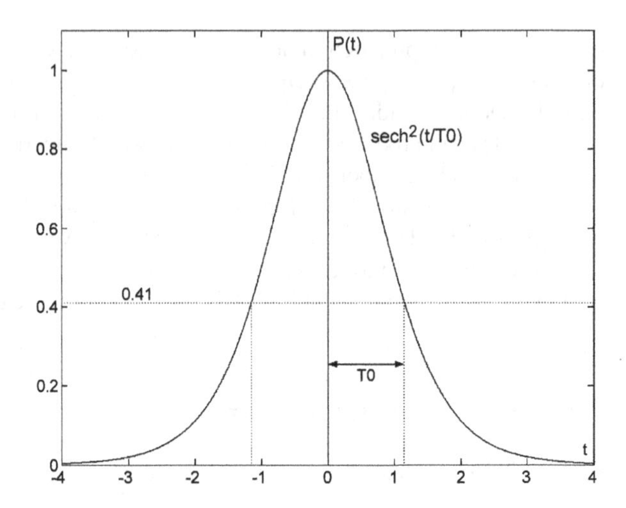

Fig. 7.25 Time profile $I(t/\tau_0)$ of a soliton

The following equation for the change of the pulse area A during the pulse propagation through the sample with absorption coefficient α can be derived [914]:

$$\mathrm{d}A(z)/\mathrm{d}z = -(\alpha/2)\sin\bigl(A(z)\bigr).$$

For $A = m \cdot 2\pi$ the pulse area A does not change because $\mathrm{d}A/\mathrm{d}z = 0$. It turns out, however, that for $m > 1$ the pulse splits into m 2π-pulses, which propagate as hyperbolic secant-pulses through the medium. The total pulse area A is still preserved.

For $A < \pi \sin A > 0$ which means that $\mathrm{d}A/\mathrm{d}z < 0$. The pulse is absorbed and after a pathlength $z = 2/\alpha$ the pulse amplitude decreases to $1/2$ of its initial value.

Summary Self-induced transparency has 3 main features:

1. *Loss-less propagation through media.* This means the energy of the pulse is preserved. There is no spontaneous emission, because the Rabi frequency Ω_R is large compared to the inverse spontaneous lifetime of the atoms: $\Omega_\mathrm{R} \gg 1/\tau_\mathrm{spontan}$.
2. *Pulse reshaping.* For $A = m \cdot 2\pi$ the shape of the pulse is preserved. For $\pi < A < 2\pi$ is $\sin A < 0$ and $\mathrm{d}A/\mathrm{d}z > 0$. The pulse area increases until $A = 2\pi$.
3. *Decreased pulse velocity* v.

After a pathlength L through a medium with refractive index n the pulse travelling with the velocity $v < c$ is delayed by

$$\Delta t = L/v - L/(c/n).$$

In Fig. 7.26 a comparison between absorption, stimulated emission and self-induced transparency is illustrated. More information about self-induced transparency can be found in [914–916].

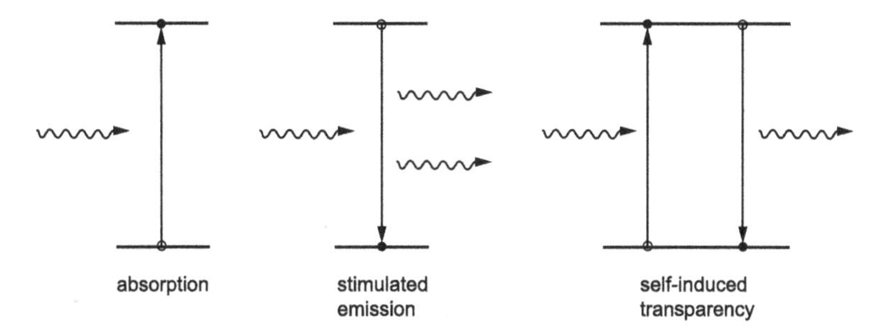

Fig. 7.26 Comparison of absorption, stimulated emission and self-induced transparency

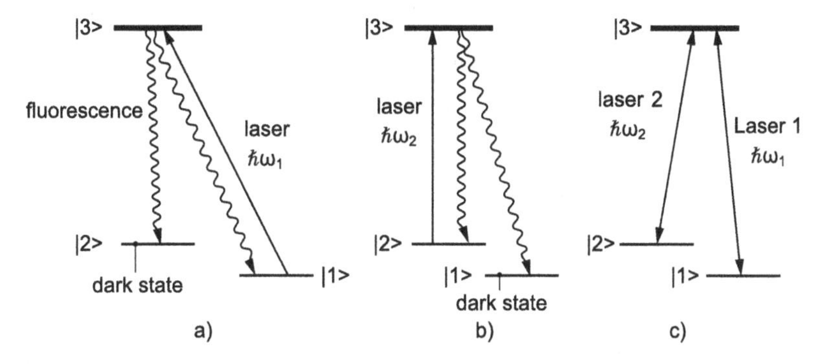

Fig. 7.27 (**a**) and (**b**) Normal incoherent dark states; (**c**) coherent dark states

7.10 Coherent Dark States and Dark Resonances

Let us consider atoms (for instance Na-atoms) with two sublevels $|1\rangle$ and $|2\rangle$ in the electronic ground state and a level $|3\rangle$ in the excited state (Fig. 7.27a). If a narrow band laser is tuned tot he transition $|1\rangle \rightarrow |3\rangle$ the atoms are excited into the level $|3\rangle$ which can decay by fluorescence into the levels $|1\rangle$ and $|2\rangle$. All atoms in level $|1\rangle$ can again be excited into level $|3\rangle$. This leads after a few excitation cycles to a complete depopulation of $|1\rangle$ because all atoms are now in level $|2\rangle$ (optical pumping). The frequency ω_1 can no longer be absorbed. For this frequency the atoms appear as non existing. The populated level $|2\rangle$ is called a normal dark state, because it does not absorbed the laser on he frequency ω_1 and therefore does not contribute to the fluorescence in spite of its population. The absorption cell appears dark. If the laser is tuned to the frequency ω_2 of the transition $|2\rangle \rightarrow |3\rangle$ (Fig. 7.27b) the level $1 >$ becomes a dark state.

Different from these normal dark states are the coherent dark states. Here two lasers are used (Fig. 7.27c). If laser 1 excites the atoms from $|1\rangle$ into $|3\rangle$ they can be transferred by laser 2 into level $|2\rangle$. From here they can be pumped again by laser 2 into level $|3\rangle$ and further by laser 1 into $|1\rangle$. For sufficiently large laser intensities the stimulated processes are fast compared with the spontaneous emission and the Raby oscillations dominate which periodically alter the populations of $|1\rangle$ and $|2\rangle$.

With the Rabi-oscillation frequencies $\Omega_{13} = D_{13}E_{01}/\hbar$ and $\Omega_{23} = D_{23}E_{02}/\hbar$ in the electromagnetic fields $E_i = E_{0i}\cos(\omega_i t)$ an antisymmetric coherent superposition of the two coherently coupled levels $|1\rangle$ and $|2\rangle$

$$|as\rangle_{\text{coherent}} = \big(\Omega_2|1\rangle - \Omega_1|2\rangle\big)/\Omega \quad \text{with} = \big(\Omega_1^2 + \Omega_2^2\big)^{1/2}$$

is created which is called a coherent dark state. This state is *not* an eigenstate of the atom because it is not constant in time. The phase of its wavefunction is periodically changing due to the different energies of the two levels

$$|as\rangle_{\text{coherent}} = \Omega_2|1\rangle \exp\big[-\mathrm{i}(E_1/\hbar)t\big] - \Omega_2|2\rangle \exp\big[-(E_2/\hbar)t\big].$$

Why is this a dark state? This can be seen as follows [917].

The transition amplitude for the dipole transition from the dark state $|as\rangle_{\text{coherent}}$ to the upper state $|3\rangle$ is

$$\langle 3|D_{as,3} \cdot E|as\rangle = (\Omega_1 \cdot \Omega_2/\Omega)\big\{\exp\big[\mathrm{i}\omega_1 t - \mathrm{i}(E_1/\hbar)t\big] - \exp[\mathrm{i}\omega_2 t - \mathrm{i}(E_2/\hbar)t\big\}$$

where

$$D_{as,3} = \mathrm{e} \cdot \int \psi_3 r \psi_{as}\, \mathrm{d}\tau$$

is the transition dipole moment for the transition $|as\rangle \to |3\rangle$. If the frequency difference $\omega_1 - \omega_2$ equals the difference $(E_1 - E_2)/\hbar$, this expression becomes identical zero. The physical reason is the destructive interference of the two transition amplitudes of the transitions $|1\rangle \to |3\rangle$ and $|2\rangle \to |3\rangle$.

Such a dark state is created by optical pumping. After a few excitation cycles the action of stimulated and spontaneous processes reaches the "correct" mixture of the levels $|1\rangle$ and $|2\rangle$. Then a steady state condition is reached where the level populations do not change any longer, because now no more transitions to level $|3\rangle$ occur. The population ratio N_1/N_2 depends on the ratio of the transition probabilities $(D_{1,3}/D_{2,3})^2$. This coherent dark state $|as\rangle$ is called a *trapping state*.

When the frequency of one laser is kept constant and the frequency of the other laser is tuned over the resonance, a very narrow dip in the fluorescence intensity is observed for $\omega_1 - \omega_2 = (E_1 - E_2)/\hbar$. This dark resonance is extremely narrow, because the lifetimes of the ground state levels $|1\rangle$ and $|2\rangle$ are nearly infinitely long. The resonance width is therefore only determined by collision processes and by the diffusion of the atoms out of the laser beams.

Instead of using two lasers one laser would be sufficient if ist output is modulated at a frequency $f = (E_1 - E_2)/h$. This creates sidebands in the frequency spectrum of the laser and one sideband replaces the second laser.

There are a couple of interesting applications where these narrow dark resonances are used for precision spectroscopy. One example is the realization of a very sensitive magnetometer [918]. When cesium atoms are brought into an external magnetic field the detection of dark resonances on transitions between selected Zeeman components can be measured with very high precision. This allows for example

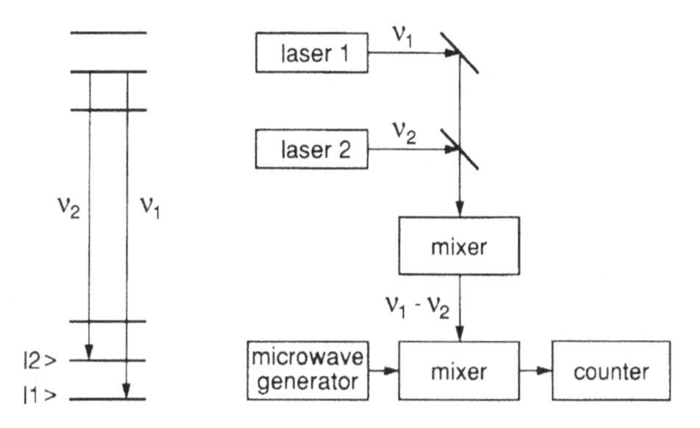

Fig. 7.28 Heterodyne spectroscopy with two lasers, stabilized onto two molecular transitions. The difference frequency, generated in a nonlinear crystal, is either measured or a second downconversion is used by mixing with a microwave

the determination of weak magnetic fields within an uncertainty of 10 pT(10^{-11} T). A possible medical application is the possibility of measuring the periodically modulated magnetic field of the heart beats in the human body.

Up to now the hyperfine transition in the ground state of the Cs-atom at 9.192 GHz represents the accepted frequency standard. An alternative to the cesium atomic fountain is the dark resonance of Cs atoms in a cell when a coherent dark state of the hyperfine levels is realized where the optical transition is excited by a frequency modulated laser with a modulation frequency which matches the hyperfine splitting in the Cs ground state. This modulation frequency can be used for the stabilization of the microwave which modulates the laser output. Since the dark resonance is very narrow, the uncertainty of the stable frequency is small.

7.11 Heterodyne Spectroscopy

Heterodyne spectroscopy uses two cw lasers with the frequencies ω_1 and ω_2 ($\Delta\omega = \omega_1 - \omega_2 \ll \omega_1, \omega_2$), which are stabilized onto two molecular transitions sharing a common level (Fig. 7.28). Measurements of the difference frequency of the two lasers then immediately yields the level splitting $\Delta E = E_1 - E_2 = \hbar\Delta\omega$ of the molecular levels $|1\rangle$ and $|2\rangle$.

For sufficiently low difference frequencies ($\Delta\nu = \Delta\omega/2\pi \leq 10^9$ Hz) fast photodiodes or photomultipliers can be used for detecting $\Delta\omega$. The two laser beams are superimposed onto the active area of the detector. The output signal S of the photodetector is proportional to the incident intensity, averaged over the time constant τ of the detector. For $\Delta\omega \ll 2\pi/\tau \ll \omega = (\omega_1 + \omega_2)/2$ we obtain the time-averaged output signal

$$\langle S\rangle \propto \left\langle (E_1 \cos\omega_1 t + E_2 \cos\omega_2 t)^2 \right\rangle = \frac{1}{2}\left(E_1^2 + E_2^2\right) + E_1 E_2 \cos(\omega_1 - \omega_2)t, \quad (7.47)$$

which contains, besides the constant term $(E_1^2 + E_2^2)/2$, an ac term with the difference frequency $\Delta\omega$. Fast electronic counters can directly count frequencies up to $\Delta\nu \simeq 10^9$ Hz. For higher frequencies a mixing technique can be used. The superimposed two laser beams are focused onto a nonlinear crystal, which generates the difference frequency $\Delta\omega = \omega_1 - \omega_2$ (Vol. 1, Sect. 5.8). The output of the crystal is then mixed with a microwave ω_{MW} in a Schottky diode or a point-contact MIM diode (Vol. 1, Sects. 4.5.2c and 5.8.5). For a proper choice of ω_{MW} the difference frequency $\Delta\omega - \omega_{MW}$ falls into a frequency range $< 10^9$ Hz and thus can be counted directly. Often the output of the two lasers and the microwave can be mixed in the same device [920].

The accuracy of stabilizing the two lasers onto molecular transitions increases with decreasing linewidth. Therefore, the narrow Lamb dips of Doppler-broadened molecular transitions measured with saturation spectroscopy (Sect. 2.2) are well suited [921]. This was proved by Bridges and Chang [922] who stabilized two CO_2 lasers onto the Lamb dips of different rotational lines within the vibrational transitions $(00^\circ1) \rightarrow (10^\circ0)$ at 10.4 μm and $(00^\circ1) \rightarrow (02^\circ0)$ at 9.4 μm. The superimposed beams of the two lasers were focused into a GaAs crystal, where the difference frequency was generated.

Ezekiel and coworkers [923] used the reduction of the Doppler width in a collimated molecular beam (Sect. 4.1) for accurate heterodyne spectroscopy. The beams of two argon lasers intersect the collimated beam of I_2 molecules perpendicularly. The laser-induced fluorescence is utilized to stabilize the laser onto the centers of two hfs components of a visible rotational transition. The difference frequency of the two lasers then yields the hfs splittings.

Instead of two different lasers, a single laser can be employed when its output is amplitude-modulated at the variable frequency f. Besides the carrier frequency ν_0 two tunable sidebands $\nu_0 \pm f$ appear. If the carrier frequency ν_0 is stabilized onto a selected molecular line, the sidebands can be tuned across other molecular transitions (*sideband spectroscopy*) [924, 925]. The experimental expenditure is smaller since only one laser has to be stabilized. The sideband tuning can be achieved with acousto-optical modulators at relatively small RF powers. The difference frequency $\Delta\omega$ is, however, restricted to achievable modulation frequencies $(\Delta\omega \leq 2\pi f_{max})$.

Using this technique, molecular ions and short-lived radicals have been successfully investigated [926, 927].

Since the radiation of even a stabilized single-mode laser is not strictly represented by a monochromatic wave but actually has a finite line width due to frequency and phase fluctuations (see Vol. 1, Sect. 5.6), the different frequency components within the laser bandwidth can interfere and produce beat signals $S(\nu - \nu_0)$, which can be monitored with a spectrum analyzer. This allows the determination of the laser line profile with high accuracy [928].

7.12 Correlation Spectroscopy

Correlation spectroscopy is based on the correlation between the measured frequency spectrum $S(\omega)$ of the photodetector output and the frequency spectrum $I(\omega)$ of the incident light intensity. This light may be the direct radiation of a laser or the light scattered by moving particles, such as molecules, dust particles, or microbes (homodyne spectroscopy). In many cases the direct laser light and the scattered light are superimposed on the photodetector, and the beat spectrum of the coherent superposition is detected (heterodyne spectroscopy) [929, 930].

Correlations in the fluorescence of molecules excited by pulsed or cw lasers are often measured, since these provide information on the dynamics of molecular processes. This branch of fluorescence correlation spectroscopy has undergone rapid development [932] because of its many advantages compared to conventional spectroscopy; for instance its wide dynamic range, covering rates from μs^{-1} to s^{-1}. This will be discussed in Sect. 7.12.4.

7.12.1 Basic Considerations

We assume that the incident light wave has the amplitude $E(\boldsymbol{r}, t)$, which does not vary over the area of the photocathode. The probability for the emission of one photoelectron per unit time

$$\mathcal{P}^{(1)}(t) = \frac{c\epsilon_0 \eta A}{h\nu} E^*(t)E(t) = \frac{\eta A}{h\nu} I(t), \tag{7.48}$$

where $\eta(\lambda)$ is the quantum efficiency of the photocathode, A its surface area and $I(t) = c\epsilon_0 E^*(t)E(t)$ is the incident intensity. The measured photocurrent $i(t)$ at the detector output generated by $n(t)$ photoelectrons per second is then

$$i(t) = ean(t) = ea\mathcal{P}^{(1)}(t) = \frac{ea\eta A}{h\nu} I(t), \tag{7.49}$$

where a is the amplification factor of the detector, which is $a = 1$ for a vacuum photocell, but $a \simeq 10^6$–10^8 for a photomultiplier.

The joint probability per unit time that one photoelectron is emitted at time t *and also* another electron at time $t + \tau$ is described by the product

$$\mathcal{P}^{(2)}(t, t+\tau) = \mathcal{P}^{(1)}(t) \cdot \mathcal{P}^{(1)}(t+\tau) = \frac{\eta^2 A^2}{(h\nu)^2} I(t) \cdot I(t+\tau). \tag{7.50}$$

Most experimental arrangements generally detect the time-averaged photocurrent. For the description of the measured signals, therefore, the normalized correlation functions

$$G^{(1)}(\tau) = \frac{\langle E^*(t) \cdot E(t+\tau)\rangle}{\langle E^*(t) \cdot E(t)\rangle}, \tag{7.51}$$

$$G^{(2)}(\tau) = \frac{\langle E^*(t) \cdot E(t) \cdot E^*(t+\tau) \cdot E(t+\tau)\rangle}{[\langle E^*(t) \cdot E(t)\rangle]^2} = \frac{\langle I(t) \cdot I(t+\tau)\rangle}{\langle I\rangle^2}, \tag{7.52}$$

of first and second order are introduced to describe the correlation between the field amplitudes and the intensities, respectively, at the times t and $t + \tau$. The first-order correlation function $G^{(1)}(t)$ is identical to the normalized mutual coherence function, defined by Vol. 1, (2.119).

The Fourier transform of $G^{(1)}(\tau)$

$$F(\omega)=\frac{1}{2\pi}\int_{-\infty}^{\infty} G^{(1)}(\tau)\mathrm{e}^{\mathrm{i}\omega\tau}\,\mathrm{d}\tau, \tag{7.53}$$

can be calculated by inserting (7.51) into (7.53). This yields, as was first shown by Wiener [935]

$$F(\omega)=\frac{E^*(\omega)\cdot E(\omega)}{\langle E^*\cdot E\rangle}=\frac{I(\omega)}{\langle I\rangle}. \tag{7.54}$$

The Fourier transform of the first-order correlation function $G^{(1)}(\tau)$ represents the normalized frequency spectrum of the incident light-wave intensity $I(\omega)$ (Wiener–Khintchine theorem) [930, 935].

Example 7.3 For completely uncorrelated light we have $G^{(1)}(\tau)=\delta(0-\tau)$, where δ is the Kronnecker symbol, that is, $\delta(x)=1$ for $x=0$, and 0 elsewhere. For a strictly monochromatic light wave $E=E_0\cos\omega t$, the first-order correlation function $G^{(1)}(\tau)=\cos(\omega t)$ becomes a periodic function of time τ oscillating with the period $\Delta\tau=2\pi/\omega$ between the values $+1$ and -1 (Vol. 1, Sect. 2.8.4).

The second-order correlation function is a constant $G^{(2)}(\tau)=1$ for cases of both completely uncorrelated light and strictly monochromatic light.

Inserting $G^{(1)}(\tau)=\delta(0-\tau)$ into (7.53) yields $F(\omega)=1$ and therefore a constant intensity $I(\omega)=\langle I\rangle$ (white noise) for completely uncorrelated light, while $G^{(1)}(\tau)=\cos(\omega\tau)$ gives $F(\omega)=\cos^2\omega t+\frac{1}{2}$, and therefore $I(\omega)=\langle I\rangle(\cos^2\omega t+\frac{1}{2})$.

We must now discuss how the measured frequency spectrum $i(\omega)$ of the photocurrent i is related to the wanted frequency spectrum $I(\omega)$ of the incident light wave.

Note That the photocurrent $i=ne=e\eta I$ is proportional to the incident intensity I, where n is the rate of photoelectrons and η the quantum efficiency of the photocathode.

Similar to (7.51) we define the correlation function

$$C(\tau)=\frac{\langle i(t)\cdot i(t+\tau)\rangle}{\langle i^2\rangle}, \tag{7.55}$$

of the time-resolved photocurrent $i(t)$. The Fourier transform of (7.55) yields, analogously to (7.53), the power spectrum $P(\omega)$ of the measured photocurrent

$$P(\omega) = \frac{1}{2\pi} \int_{-\infty}^{\infty} C(\tau) e^{i\omega\tau} \, d\tau. \tag{7.56}$$

The correlation function (7.55) of the photocurrent is determined by the two contributions

(a) The statistical process of emission of photoelectrons from the photocathode, resulting in a statistical fluctuation of the photocurrent, even when the incident light wave would consist of a completely regular and noise-free photon flux.
(b) The amplitude fluctuations of the incident light wave that are caused by the characteristics of the photon source or by the objects that scatter light onto the detector.

Let us consider these contributions for two different cases:

(1) for a constant light intensity I described by the probability function

$$\mathcal{P}(I) = \delta\big(I - \langle I \rangle\big), \tag{7.57}$$

where the mean photoelectron number per unit time is $\langle n \rangle$. Then the probability $\mathcal{P}(n, dt)$ of detecting n photoelectrons per second is given by the Poisson distribution

$$\mathcal{P}(n) = \frac{1}{n!} \langle n \rangle^n e^{-\langle n \rangle}, \tag{7.58}$$

with the mean square fluctuation

$$\big\langle (\Delta n)^2 \big\rangle = \langle n \rangle, \tag{7.59}$$

of the rate $n = i_{ph}/e$ of the photoelectron emission [936, 937]. Here, the fluctuations arise solely from the statistical nature of photoelectron emission, not from fluctuations of the incident light.

(2) The light intensity scattered by statistically fluctuating particles can generally be described by a Gaussian intensity distribution [938]

$$\mathcal{P}(I)\, dI = \frac{1}{\langle I \rangle} e^{-I/\langle I \rangle} \, dI. \tag{7.60}$$

If a quasi-monochromatic light wave $I(\omega)$ with a statistically fluctuating intensity distribution (7.60) falls onto the photocathode, the probability of detecting n photoelectrons within the time interval dt is not described by (7.58) but by the Bose–Einstein distribution

$$\mathcal{P}(n) = \frac{1}{1 + \langle n \rangle (1 + 1/\langle n \rangle)^n}, \tag{7.61}$$

which leads to a mean-square deviation of the photoelectron rate of $\langle (\Delta n)^2 \rangle = \langle n \rangle^2$ instead of (7.59).

Including both contributions (a) and (b), the mean-square deviation of the photoelectron rate becomes [929]

$$\left\langle (\Delta n)^2 \right\rangle = \langle n \rangle + \langle n \rangle^2, \tag{7.62}$$

and the correlation function n for the photoelectron emission

$$\begin{aligned} C(\tau) &= e\langle n \rangle \delta(0-\tau) + \langle e \cdot n \rangle^2 = \langle i \rangle \delta(0-\tau) + i(t) \cdot i(t+\tau) \\ &= \langle i \rangle \delta(0-\tau) + \langle i \rangle^2 G^{(2)}(\tau) \\ &= e\eta I \delta(0-\tau) + e^2\eta^2 \langle I \rangle^2 G^{(2)}(\tau). \end{aligned} \tag{7.63}$$

This shows that the autocorrelation function $C(\tau)$ of the photoelectron current is directly related to the second-order correlation function $G^{(2)}(\tau)$ of the light field.

As proved by Siegert [939] for optical fields with a Gaussian intensity distribution (7.60), the second-order correlation function $G^{(2)}(\tau)$ is related to $G^{(1)}(\tau)$ by the Siegert relation

$$G^{(2)}(\tau) = \left[G^{(2)}(0)\right]^2 + \left|G^{(1)}(\tau)\right|^2 = 1 + \left|G^{(1)}(\tau)\right|^2. \tag{7.64}$$

The procedure to obtain information on the spectral distribution of the light incident onto the detector is then as follows: from the time-resolved measured photocurrent $i(t)$ the correlation function $C(\tau)$ can be derived (7.63), which yields $G^{(2)}(\tau)$ and with (7.64) $G^{(1)}(\tau)$. The Fourier transform of $G^{(1)}(\tau)$ then gives the wanted intensity spectrum $I(\omega)$ according to (7.53) and (7.54).

In Fig. 7.29 an electronic device (digital clipping correlation) is shown, which measures the photoelectron statistics.

Example 7.4 A time-dependent optical field with the amplitude

$$E(t) = E_0 \mathrm{e}^{-\mathrm{i}\omega_0 t - (\gamma/2)t}, \tag{7.65}$$

has, according to (7.51), the first-order autocorrelation function

$$G^{(1)}(\tau) = \mathrm{e}^{-\mathrm{i}\omega_0 \tau} \mathrm{e}^{-(\gamma/2)\tau}. \tag{7.66}$$

The Fourier transformation (7.53), (7.54) yields the spectral distribution

$$I(\omega) = \frac{\langle I \rangle}{2\pi} \int_{-\infty}^{+\infty} \mathrm{e}^{\mathrm{i}(\omega-\omega_0)\tau - (\gamma/2)\tau} \, \mathrm{d}\tau = \frac{\langle I \rangle \gamma / 2\pi}{(\omega-\omega_0)^2 + (\gamma/2)^2}, \tag{7.67}$$

of the Lorentzian line profile (Vol. 1, Sect. 3.1). Inserting (7.66) into (7.63) yields the correlation function $C(\tau)$ of the photoelectron current which is,

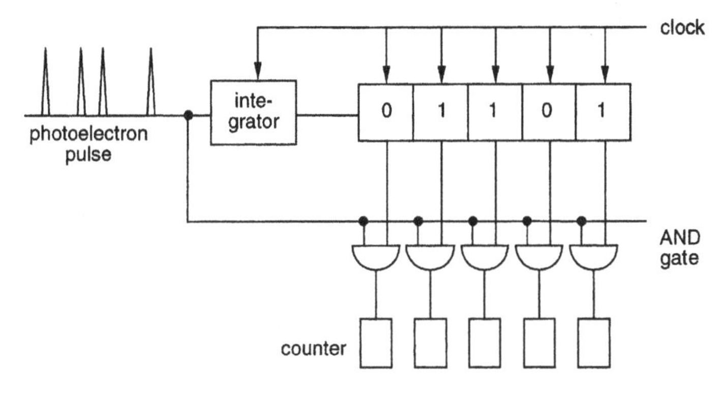

Fig. 7.29 Measurement of photoelectron statistics with a digital clipping correlator [940]

according to (7.56), related to the power spectrum $P_i(\omega)$ of the photocurrent

$$P_i(\omega \geq 0) = \frac{e}{\pi}\langle i\rangle + 2\langle i\rangle^2\delta(\omega) + 2\langle i\rangle^2 \frac{\gamma/2\pi}{\omega^2 + (\gamma/2)^2}. \qquad (7.68)$$

Equation (7.68) reveals that the power spectrum has a peak at $\omega = 0$ $[\delta(\omega = 0) = 1]$, giving the dc part $2\langle i\rangle^2$. The first term $(e/\pi)\langle i\rangle$ represents the shot-noise term, and the third term describes a Lorentzian frequency distribution peaked at $\omega = 0$ with the total power $(2/\pi\gamma)\langle i\rangle^2$. This represents the light-beating spectrum, which gives information on the intensity profile $I(\omega)$ of the incident light wave.

For more examples see [929, 930].

The correlation function $C(\tau) = \langle i(t)\rangle\langle(t + \tau)\rangle/\langle i\rangle^2$ can be directly measured with a digital correlator, which measures the photoelectron statistics. A simple version of several possible realizations is depicted in Fig. 7.29. The time is divided into equal sections Δt by an internal clock. If the number $N \cdot \Delta t_i$ of photoelectrons measured within the ith time interval Δt_i exceeds a given number N_m, the correlator gives a normalized output pulse, counted as "one." For $N\Delta t_i < N_m$, the output gives "zero." The output pulses are transferred to a "shift register" and to "AND gates," which open for "one" and close for "zero", and are finally stored in counters (Malvern correlator [941, 942]).

7.12.2 Homodyne Spectroscopy

We will illustrate homodyne spectroscopy (self-beating spectroscopy) with some examples.

(1) The output of a laser does not represent a strictly monochromatic wave (even if the laser frequency is stabilized) because of frequency and phase fluctuations

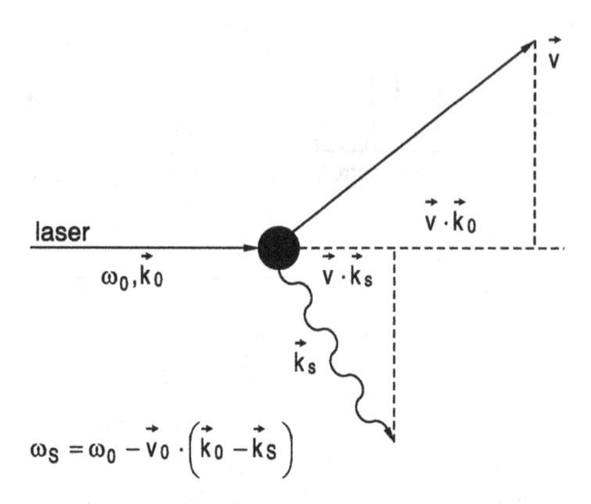

Fig. 7.30 Doppler shift of light scattered by a moving particle

(Vol. 1, Sect. 5.6). Its intensity profile $I(\omega)$ with the linewidth $\Delta\omega$ can be detected by homodyne spectroscopy. The different frequency contributions inside the line profile $I(\omega)$ interfere, giving rise to beat signals at many different frequencies $\omega_i - \omega_k < \Delta\omega$ [929]. If a photodetector is irradiated by the attenuated laser beam, the frequency distribution of the photocurrent (7.68) can be measured with an electronic spectrum analyzer. This yields, according to the discussion above, the spectral profile of the incident light. In the case of narrow spectral linewidths this correlation technique represents the most accurate measurement for line profiles [940].

(2) The light scattered elastically by small particles contains information on the size, structure, and movement of the particles. The most simple case that can be treated theoretically is the model of homogeneous spherical particles.

We will distinguish three different situations:

(a) Randomly distributed *static* scatterers. In this case the correlation function is time independent, and one obtains for the normalized correlation function

$$G^{(1)} = \frac{\langle E(t)E(t+\tau)\rangle}{|E(t)|^2} = 1.$$

The intensity spectrum of the elastically scattered radiation from N scatterers is

$$I(\omega) = N|E|^2\delta(\omega - \omega_0)$$

where ω_0 is the frequency of the laser radiation.

(b) Spherical scatterers with constant velocity $\boldsymbol{v}$

$$G^{(1)}(\tau) = \mathrm{e}^{\mathrm{i}\boldsymbol{q}\cdot\boldsymbol{v}\tau},$$

where $\boldsymbol{q} = \boldsymbol{k}_0 - \boldsymbol{k}_s$ is the wave vector difference between incident and scattered light. The intensity of the scattered light is

$$I(\omega) = N|E|^2\delta(\omega - \omega_0 + \boldsymbol{q}\cdot\boldsymbol{v})$$

where $\boldsymbol{q}\cdot\boldsymbol{v}$ represents the Doppler shift of the light scattered by moving particles (Fig. 7.30).

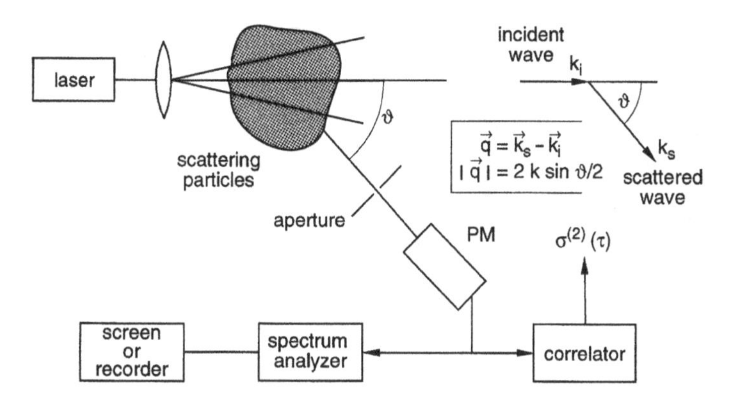

Fig. 7.31 Schematic experimental setup for measuring the autocorrelation function of scattered light (homodyne spectroscopy), with a correlator as an alternative to an electronic spectrum analyzer

(c) Spherical scatterers with translational diffusion

$$G^{(1)}(\tau) = e^{iD_T q^2 |\tau|},$$

$$I(\omega) = N|E|^2 \frac{D_\tau q^2/\pi}{(\omega - \omega_0)^2 + (D_T q^2)^2},$$

where D_T is the diffusion coefficient (see also Vol. 1, Sect. 5.6).

When monochromatic light is scattered by moving particles that show thermal motion, the field amplitudes $E(\omega)$ show a Gaussian distribution. The experimental arrangement for measuring the homodyne spectrum is shown in Fig. 7.31. The power spectrum $P(\omega)$ of the photocurrent (7.68), which is related to the spectral distribution $I(\omega)$, is measured either directly by an electronic spectrum analyzer, or with a correlator, which determines the Fourier transform of the autocorrelation function $C(\tau) \propto \langle i(t)\rangle\langle i(t+\tau)\rangle$. According to (7.63), $C(\tau)$ is related to the intensity correlation function $G^{(2)}(\tau)$, which yields $G^{(1)}(\tau)$ (7.64), and $I(\omega)$.

(3) Another example of an application of homodyne spectroscopy is the measurement of the size distribution of small particles in the nanometer range that are dispersed in liquids or gases and fly through a laser beam. The intensity I_s of the scattered light depends in a nonlinear way on the size and the refractive index of the particles. For small homogeneous spheres with diameters d, which are small compared to the wavelength ($d \ll \lambda$), the relation $I_s \propto d^6$ holds because the amplitude is a coherent superposition of the contributions from all atoms and is therefore proportional to d^3. In Fig. 7.32 the measured intensity distribution of laser light scattered by a mixture of latex spheres with $d = 22.8$ nm and $d = 5.7$ nm (small squares) is compared with the size distribution obtained from electron microscopy, which can be used for calibration [943].

(4) A further example is the light scattering by a liquid sample when the temperature is changed around the critical temperature and the sample undergoes a phase

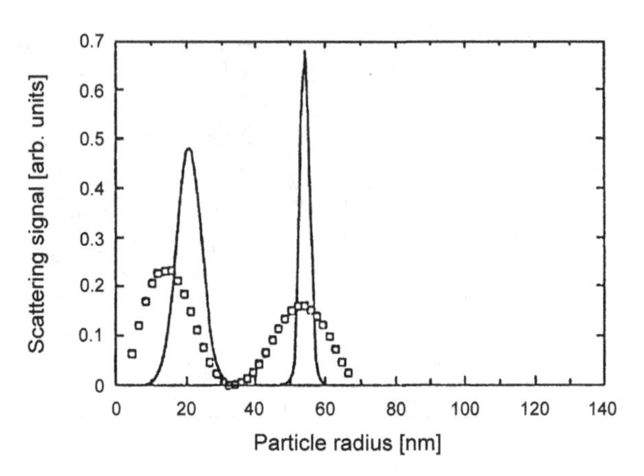

Fig. 7.32 Intensity distribution of the light intensity scattered by a mixture of two kinds of homogeneous small latex balls with diameters $d = 22.8$ nm and $d = 5.7$ nm (*squares*), compared with the size distribution of the balls measured by electron microscopy (*solid line*) [943]

transition [944]. At such a phase transition long-range and short-range order generally change. This affects the correlation between the molecules.

A famous example for the application of intensity-correlation interferometry in astronomy is the Hanbury Brown–Twiss interferometer sketched in Fig. 7.33. In its original form it was intended to measure the degree of spatial coherence of starlight (Vol. 1, Sect. 2.8) [945] from which diameters of stars could be determined. In its modern version it measures the degree of coherence and the photon statistics of laser radiation in the vicinity of the laser threshold [946].

7.12.3 Heterodyne Correlation Spectroscopy

For heterodyne correlation spectroscopy the scattered light that is to be analyzed is superimposed on the photocathode by part of the direct laser beam (Fig. 7.34). Assume the scattered light has the amplitude E_s at the detector and a frequency distribution around ω_s, whereas the direct laser radiation $E_L = E_0 \exp(-i\omega_0 t)$ acts as monochromatic local oscillator with constant amplitude E_0. The total amplitude

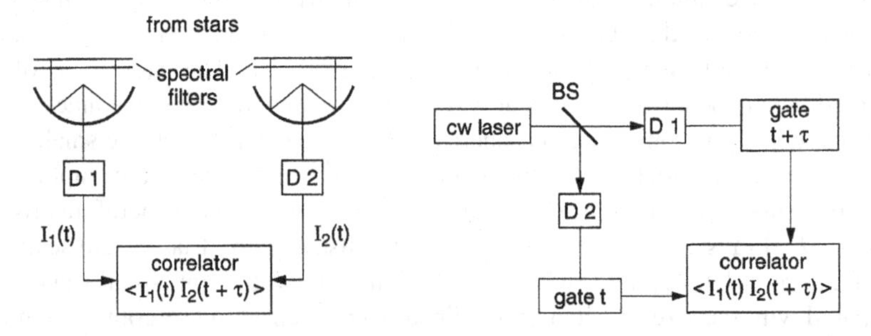

Fig. 7.33 (**a**) Basic principle of the Hanbury Brown–Twiss intensity-correlation interferometer in astronomy; (**b**) its application to measurements of spectral characteristics and statistics of laser radiation

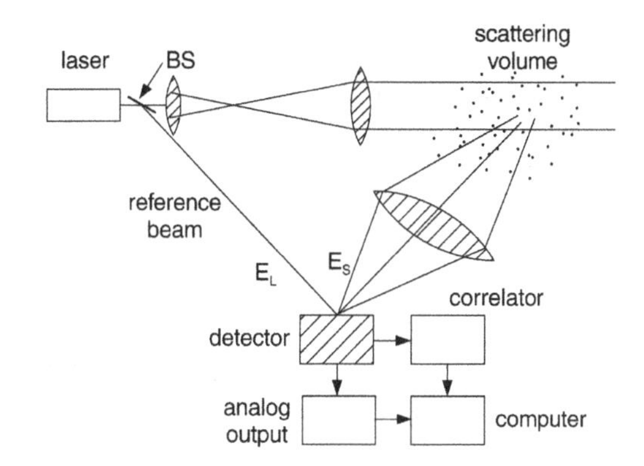

Fig. 7.34 Schematic setup for heterodyne correlation spectroscopy

is then

$$E(t) = E_0 \exp(-\mathrm{i}\omega_0 t) + E_\mathrm{s}(t). \tag{7.69}$$

The autocorrelation function of the photocurrent is, according to (7.55), (7.63)

$$C(\tau) = e \cdot \eta\delta(0-\tau)\langle E^*(t) \cdot E(t)\rangle + e^2\eta^2 E^*(t)E(t)E^*(t+\tau)E(t+\tau). \tag{7.70a}$$

Inserting (7.69) into (7.70a), one obtains 16 terms, where three are time-independent. We note, however, that for $E_\mathrm{s} \ll E_0$ the terms with E_s^2 can be neglected. Furthermore, the time average $\langle E_\mathrm{L} E_\mathrm{s}\rangle$ is zero. Therefore (7.70a) reduces to

$$C(\tau) = e\eta I_\mathrm{L}\delta(0-\tau) + e^2\eta^2 I_\mathrm{L}^2 + e^2\eta^2 I_\mathrm{L}\langle I_\mathrm{S}\rangle\left(\mathrm{e}^{\mathrm{i}\omega_0\tau}G^{(1)} + c.c.\right). \tag{7.70b}$$

The power spectrum of the photocurrent i is then obtained from (7.56) as

$$P_i(\omega) = \frac{e}{2\pi} i_\mathrm{L} + i_\mathrm{L}^2\delta(0-\omega) + \frac{i_\mathrm{L}}{2\pi}\langle i_\mathrm{s}\rangle \int_{+\infty}^{-\infty} \mathrm{e}^{\mathrm{i}\omega\tau}\left[\mathrm{e}^{\mathrm{i}\omega_0\tau}G_\mathrm{s}^{(1)}(\tau) + \mathrm{c.c.}\right]\mathrm{d}\tau. \tag{7.71}$$

The first term represents the shot noise, the second is the dc term, and the third term gives the heterodyne beat spectrum with the difference $(\omega - \omega_0)$ and sum $(\omega + \omega_0)$ frequencies. The detector is not fast enough to detect the sum frequency. The output signal of the third term therefore contains only the difference frequency $(\omega - \omega_0)$. If this difference frequency lies in an inconvenient frequency range, the local oscillator can be shifted with an acousto-optic modulator to the frequency $\omega_\mathrm{L} \pm \Delta\omega$ in order to bring the difference frequency $\omega_\mathrm{L} \pm \Delta\omega - \omega_\mathrm{s}$ into an easily accessible range [947].

Example 7.5 When we assume the correlation function

$$G_\mathrm{s}^{(1)}(\tau) = \mathrm{e}^{-(\mathrm{i}\omega_\mathrm{s}+\gamma)\tau},$$

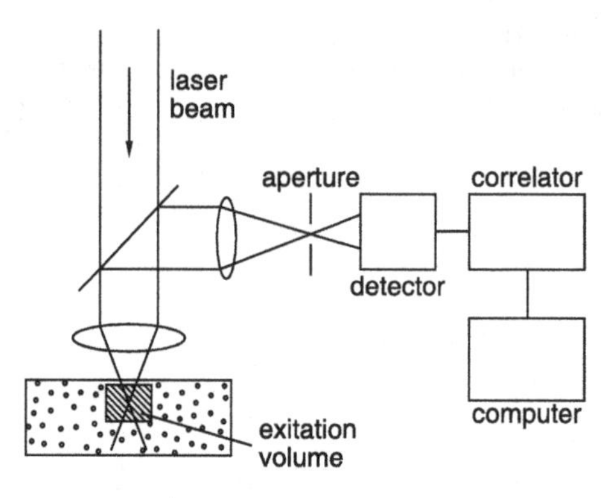

Fig. 7.35 Correlation spectroscopy of single-particle detection with a confocal arrangement

for the scattered light, the power spectrum of (7.71) becomes

$$P_i(\omega) = \frac{e i_L}{\pi} + i_L^2 \delta(0-\omega) + \frac{(\gamma/\pi)\langle i_s \rangle}{(\omega_s - \omega_L)^2 + \gamma^2}. \tag{7.72}$$

The heterodyne spectrum is a Lorentzian like the homodyne spectrum (7.67), but its maximum is shifted from ω_L to $\omega = (\omega_s - \omega_L)$.

7.12.4 Fluorescence Correlation Spectroscopy and Single Molecule Detection

The combination of confocal microscopy, laser excitation, and time-resolved fluorescence measurements allows the detection of single molecules or microparticles and the investigation of their movements in gases or liquids [948] (see Sect. 10.1.2). The exciting laser is focused by a microscope into the sample. If the focal diameter is smaller than the average distance between the molecules diluted in a liquid or a gas, a single molecule is excited (see Figs. 10.43 and 10.44). The laser-induced fluorescence or the light scattered by the microparticle is collected by the same microscope and imaged onto a small aperture in front of the detector. This aperture defines the volume from which light can be detected (Fig. 7.35). The lifetime of the excited state is generally much smaller than the diffusion time of the molecule through the laser focus. In the case of microparticles the scattered light has no time delay against the excitation light. In this case the number of scattered photons depends on the transit time of the particle through the laser focus. Each particle moving through the excitation region gives a small burst pulse of scattered light. The random time sequence of these pulses yields information on the diffusion time, the transport coefficients, and the correlation between the movement of different particles. Measuring the fluorescence intensity signal $F(t)$ and $F(t+\tau)$ allows the determination of the first-order correlation function.

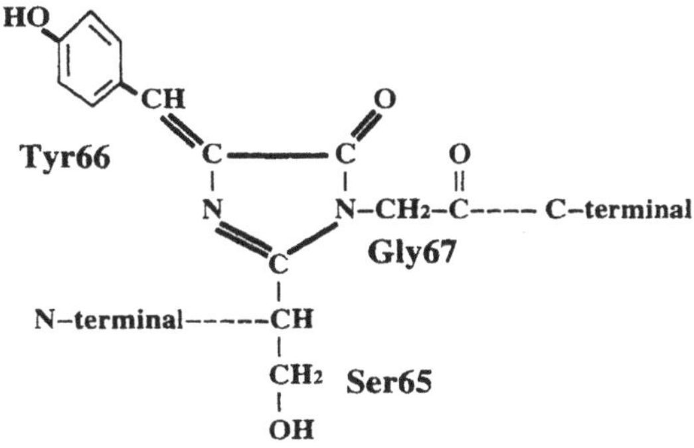

Fig. 7.36 The chemical structure of the fluorophore in the green fluorescent protein [931]

A particular useful molecule for single molecule detection is the green fluorescent protein GFP (Figs. 7.36 and 10.6), which absorbs in the blue spectral region, has a high quantum efficiency and emits fluorescence in the green spectral range. It readily attaches to other biological molecules such as bacteria or viruses and can be therefore used to follow up the path of such particles in liquids and their entrance into biological cells [949, 950].

Assume the laser beam in the z-direction is focused into a sample of N diffusing particles in the observation volume $V_o = w^2 \Delta z$, where w is the laser beam waist and Δz the Rayleigh length of the focal region. In this case, the intensity correlation function

$$G(\tau) = \langle I(t) \cdot I(t+\tau) \rangle / I^2$$

becomes

$$G(\tau) = \frac{\tau_D}{N(\tau + \tau_D)} \sqrt{\frac{1}{1 + \frac{\tau w^2}{\tau_D \Delta z^2}}}$$

with a diffusion time

$$\tau_D = w^2/4D,$$

where D is the diffusion constant. The experimental conditions are chosen such that $N \ll 1$, which implies that the probability P of finding more than one particle during the observation time within the observation volume is $P \ll 1$. For these small numbers, a Poisson distribution $P_p(N)$ is valid.

Example 7.6 The probability of finding one particle is $P(1) = (P/1)\mathrm{e}^{-1}$ and two particles is $(P^2/2)\mathrm{e}^{-2}$; with $P = 10^{-2}$ we obtain $P(1) = 3.7 \times 10^{-3}$ and $P(2) = 6.8 \times 10^{-6}$.

For every 550 measurements where only one particle is present, there is one with two particles.

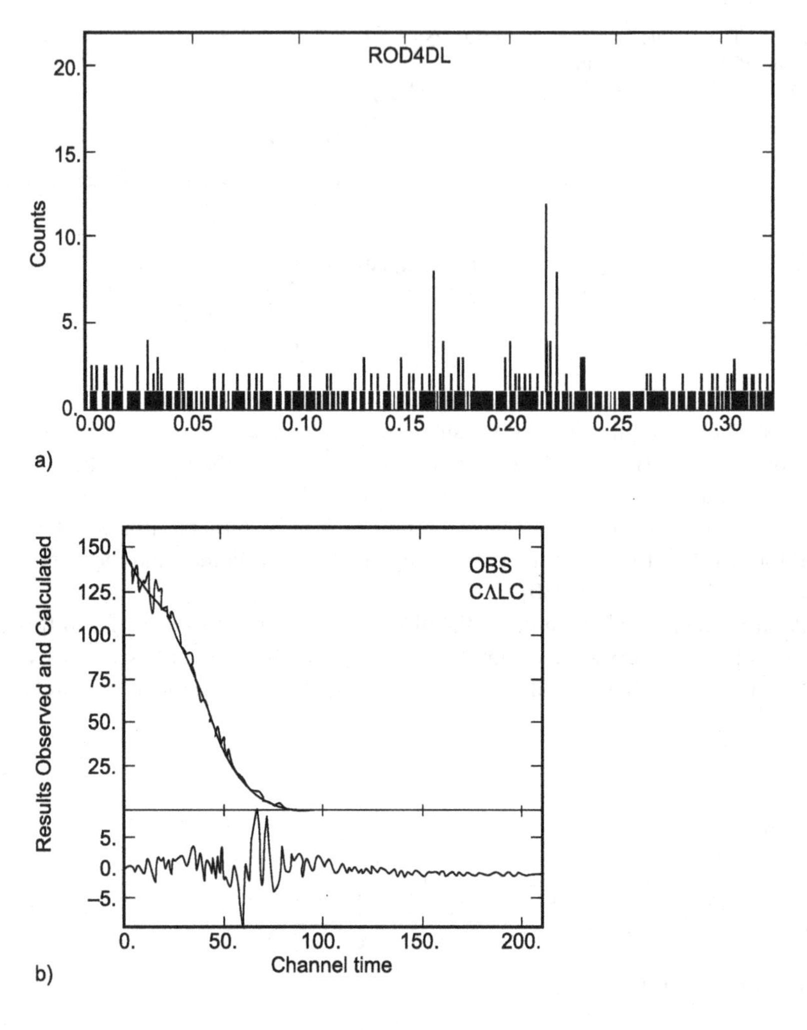

Fig. 7.37 (**a**) Fluorescence signals of single Rhodamin 6G molecules diffusing through the laser beam in the observation volume; (**b**) autocorrelation function of (**a**) [952]

Figure 7.37a shows a typical record of the detected photons scattered from single rhodamine 6G molecules diffusing through the observation volume [952]. The small signals are dark current pulses. The different magnitudes of the real signals are due to the fact that the molecules diffuse through different spots of the radial Gaussian laser beam profile. The corresponding autocorrelation function, depicted in Fig. 7.37b, gives a value of $N = 5 \times 10^{-3}$ and a diffusion time of $\tau_D = 40$ µs.

In Fig. 7.38 the auto-correlation function $G(\tau)$ of dye molecules in a liquid sample excited by polarized light is shown [934]. The step increase for short delay times τ is due to the "anti-bunching effect". When a molecule is excited at time t it can be only again excited after a is has returned into the initial ground state by spontaneous

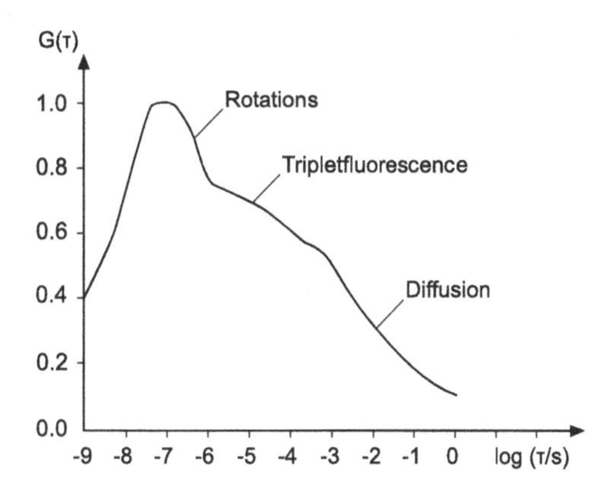

Fig. 7.38 Autocorrelation curve of dye molecules in a liquid

emission or by collision induced transitions. If the delay time T becomes smaller than the effective lifetime of the excited level the probability for fluorescence decreases and with it the correlation function. The rotation of the molecules decreases the initial orientation of the molecules generated by the excitation with linearly polarized light. This causes a decrease of the correlation function with a decay time of about 100 ns. Many organic molecules undergo radiationless transitions from the initially excited singlet state (lifetimes several ns) into the triplet state which has much larger lifetimes (several μs) and delays the fluorescence.

The diffusion of the excited molecules out of the observation region follows for a cylindrical volume with radius w the equation

$$G^{(2)}(\tau) = \frac{1}{\langle N\rangle(1+\tau/\tau_0)}\left[\frac{1}{1+w^2\tau/(\Delta z^2\tau_D)}\right]^{1/2}$$

where N is the number of molecules in the observation volume V and τ_D is the diffusion time (average time the molecules spend inside the observation volume), which is related to the diffusion coefficient D by $\tau_D = w^2/4D$. The fluctuation due to diffusion of molecules into or out of the observation volume are much slower and cause a decay of $G(\tau)$ with a time constant of about 1 ms.

Example 7.7 For rhodamine 6G molecules in water the diffusion coefficient is $D = 2.8 \times 10^{-10}$ m^2/s. In a cylindrical observation volume $V = 3 \times 10^{-13}$ m^3 with $w = 10$ μm and a length $z = 1$ mm the radial diffusion time becomes $\tau_D = w^2/4D = 89$ ms. Since the axial diffusion time is much longer, it does not play a significant role fort he fluctuations.

Since the last factor in () is nearly 1 the correlation function becomes at a concentration of molecules $n = N/V = 10^{13}$ m^{-3}

$$G^{(2)}(\tau) = 1+0.33/(1+11\tau); \qquad G(0) = 1.33; \qquad G(\tau = 0.09\text{ s}) = 1.16.$$

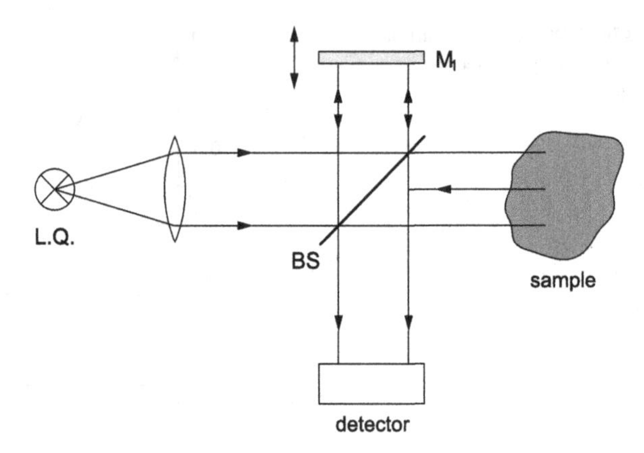

Fig. 7.39 Principle of optical coherence tomography

7.13 Optical Coherence Tomography

Optical coherence tomography is a technique which allows the investigation of the structure of materials such as biological tissue. Its basic principle is shown in Fig. 7.39. The output beam of a light source (light emitting diode or a broadband diode laser) is imaged by a lens into a wide parallel beam, which is split by the beam splitter BS into one beam which is sent to the sample, from where it is backscattered, and a second beam which is reflected by the mirror M1. The two beams are superimposed at the beamsplitter and the detector measures the interference pattern. The mirror M1 can be moved and the interferogram is measured as a function of the path-difference between the two beams. The systems acts as a Michelson interferometer. The interference pattern reflects the intensity distribution of the back-scattered light which depends on the optical properties of the sample. The coherence length L_c of the light source should be larger than the maximum path difference Δs. It is, however, preferable to choose a light source with a broad spectral profile $\Delta \nu$ which implies a small coherence length $L_c = c/\Delta \nu$, because then only back-scattered light from a thin layer $\Delta d = Lc/2$ of the sample is monitored. The distance d of this layer from the surface of the sample can be chosen by the path difference Δs selected by the position of the movable mirror M1. It is therefore possible to inspect the different layers of the sample. The axial spatial resolution $\Delta z = L_c/2 = c/2\Delta \nu$ is determined by the spectral width of the incident light, while the lateral resolution is given by the cross section of the light beam incident on the sample. One can achieve an axial resolution of a few μm.

The backscattered light depends on the structure of the sample material. For instance have cancer cells in a biological tissue a different backscattering coefficient than normal cells. They cause a different amplitude and phase of the backscattered light.

The penetration depth depends on the absorption and scattering coefficients of the sample. It can be maximized by choosing the optimum wavelength. For a wavelength-specific measurement of the spectral characteristic of the sample the broad spectral range of the light source can be dispersed by a prism (Fig. 7.40) and the interference can be measured by a CCD-detector as a function of the wavelength.

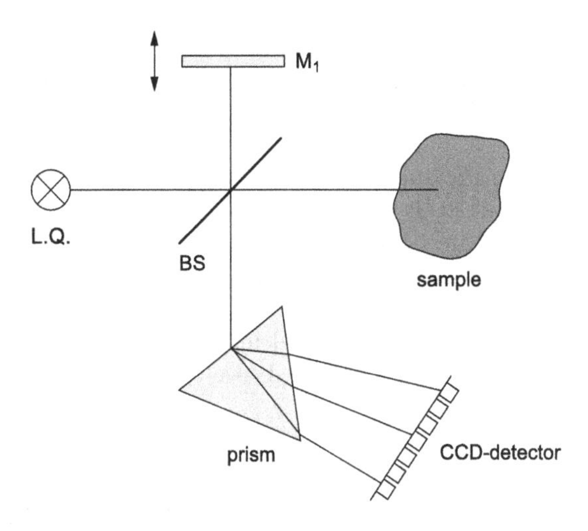

Fig. 7.40 Optimal coherence tomography with spectral resolution

Summary The main advantages of OCT can be summarized as follows:

1. The possibility to investigate the morphology of living tissue with microscopic resolution while optical microscopy needs thin cuts of tissue which can be only prepared as frozen material.
2. There is no preparation of the sample necessary
3. Instant direct infomation about the morphology of living tissue or other materials
4. No ionizing radiation is used in contrast to X-ray inspection.

Applications to medical problems and material sciences are discussed in Chap. 10.

This chapter can only give a brief survey of some aspects of coherent spectroscopy. For more details, the reader is referred to the literature. Good representations of the fundamentals of coherent spectroscopy and its various applications can be found in several textbooks [930, 953–955] and conference proceedings [957].

7.14 Problems

7.1 The Zeeman components of an excited atomic level with the quantum numbers ($J=1, S=0, L=1, I=0$) and a radiative lifetime $\tau = 15$ ns cross at zero magnetic field. Calculate the Landé g-factor and the halfwidth $\Delta B_{1/2}$(HWHM) of the Hanle signal. Compare this with the halfwidth measured for a molecular level with ($J=20, \Lambda=1, F=21, S=0, I=0, \tau=15$ ns).

7.2 What is the width $\Delta B_{1/2}$ of the Hanle signal from the crossing of the two Zeeman components of one of the hyperfine levels $F=3$ or $F=2$ in the ground state $5^2S_{1/2}$ of the rubidium atom $^{85}_{37}$Rb with a nuclear spin $I=5/2$, if the mean transit time of the atoms in a buffer gas through the excitation region is $T=0.1$ s? Explain why this method allows the realization of a sensitive magnetometer (see also [867]).

7.3 The two Zeeman components of an atomic level ($L=1, S=1/2, J=1/2$) are coherently excited by a short laser pulse. Calculate the quantum-beat period of the emitted fluorescence in a magnetic field of 10^{-2} Tesla.

7.4 In a femtosecond pump-and-probe experiment the pump pulse excites the vibrational levels $v'=10$–12 in the $A^1\Sigma_u$ state of Na_2 coherently. The vibrational spacings are $109\ cm^{-1}$ and $108\ cm^{-1}$. The probe laser pulse excites the molecules only from the inner turning point into a higher Rydberg state. If the fluorescence $I_{Fl}(\Delta t)$ from this state is observed as function of the delay time Δt between the pump and probe pulses, calculate the period ΔT_1 of the oscillating signal and the period ΔT_2 of its modulated envelope.

7.5 Two lasers are stabilized onto the Lamb dips of two molecular transitions. The width $\Delta\nu$ of the Lamb dip is 10 MHz and the rms fluctuations of the two laser frequencies is $\delta\nu = 0.5$ MHz. How accurately measured is the separation $\nu_1 - \nu_2$ of the two transitions, if the signal-to-noise ratio of the heterodyne signal of the two superimposed laser beams is 50?

7.6 Particles in a liquid show a random velocity distribution with $\sqrt{\langle v^2 \rangle} = 1\ mm/s$. What is the spectral line profile of the beat signal, if the direct beam of a HeNe laser at $\lambda = 630$ nm and the laser light scattered by the moving particles is superimposed on the photocathode of the detector?

Chapter 8
Laser Spectroscopy of Collision Processes

The two main sources of information about atomic and molecular structure and interatomic interactions are provided by spectroscopic measurements and by the investigation of elastic, inelastic, or reactive collision processes. For a long time these two branches of experimental research developed along separate lines without a strong mutual interaction. The main contributions of classical spectroscopy to the study of collision processes have been the investigations of collision-induced spectral line broadening and line shifts (Vol. 1, Sect. 3.3).

The situation has changed considerably since lasers were introduced to this field. In fact, laser spectroscopy has already become a powerful tool for studying various kinds of collision processes in more detail. The different spectroscopic techniques presented in this chapter illustrate the wide range of laser applications in collision physics. They provide a better knowledge of the interaction potentials and of the different channels for energy transfer in atomic and molecular collisions, and they give information that often cannot be adequately obtained from classical scattering experiments without lasers.

The high spectral resolution of various Doppler-free techniques discussed in Chaps. 2–4 has opened a new dimension in the measurement of collisional line broadening. While in Doppler-limited spectroscopy small line-broadening effects at low pressures are completely masked by the much larger Doppler width, Doppler-free spectroscopy is well suited to measure line-broadening effects and line shifts in the kilohertz range. This allows detection of *soft collisions* at large impact parameters of the collision partners that probe the interaction potential at large internuclear separations and that contribute only a small line broadening.

Some techniques of laser spectroscopy, such as the method of separated fields (*optical Ramsey fringes*, Sect. 9.4), coherent transient spectroscopy (Sect. 7.6), or polarization spectroscopy (Sect. 2.4) allow one to distinguish between phase-changing, velocity-changing, or orientation-changing collisions.

The high *time resolution* that is achievable with pulsed or mode-locked lasers (Chap. 6) opens the possibility for studying the dynamics of collision processes and relaxation phenomena. The interesting questions of how and how fast the excitation energy that is selectively pumped into a polyatomic molecule by absorption of laser

W. Demtröder, *Laser Spectroscopy 2*, DOI 10.1007/978-3-662-44641-6_8

photons, is redistributed among the various degrees of freedom by intermolecular or intramolecular energy transfer can be addressed by femtosecond laser spectroscopy.

One of the attractive goals of laser spectroscopy of reactive collision processes is the basic understanding of chemical reactions. The fundamental question in laser chemistry of how the excitation energy of the reactants influences the reaction probability and the internal state distribution of the reaction products can, at least partly, be answered by detailed laser-spectroscopic investigations. Section 8.4 treats some experimental techniques in this field.

The most detailed information on the collision process can be obtained from laser spectroscopy of *crossed-beam* experiments, where the initial quantum states of the collision partners before the collision are marked, and the scattering angle as well as the internal energy of the reactants is measured. In such an "ideal scattering experiment" all relevant parameters are known (Sect. 8.5).

The new and interesting field of *light-assisted collisions* (often called optical collisions), where absorption of laser photons by a collision pair results in an effective excitation of one of the collision partners, is briefly treated in the last section of this chapter. For further studies of the subject covered in this chapter, the reader is referred to books [958–960], reviews [961–967], and conference proceedings [968–971].

8.1 High-Resolution Laser Spectroscopy of Collisional Line Broadening and Line Shifts

In Vol. 1, Sect. 3.3. we discussed how elastic and inelastic collisions contribute to the broadening and shifts of spectral lines. In a semiclassical model of a collision between partners A and B, the particle B travels along a definite path $\boldsymbol{r}(t)$ in a coordinate system with its origin at the location of A. The path $\boldsymbol{r}(t)$ is completely determined by the initial conditions $\boldsymbol{r}(0)$ and $(\mathrm{d}\boldsymbol{r}/\mathrm{d}t)_0$ and by the interaction potential $V(\boldsymbol{r}, E_\mathrm{A}, E_\mathrm{B})$, which may depend on the internal energies E_A and E_B of the collision partners. In most models a spherically symmetric potential $V(r)$ is assumed, which may have a minimum at $r = r_0$ (Fig. 8.1). If the impact parameter b is large compared to r_0 the collision is classified as a *soft* collision, while for $b \leq r_0$ *hard* collisions occur.

For soft collisions B passes only through the long-range part of the potential and the scattering angle θ is small. The shift ΔE of the energy levels of A or B during the collision is accordingly small. If one of the collision partners absorbs or emits radiation during a soft collision, its frequency distribution will be only slightly changed by the interaction between A and B. Soft collisions therefore contribute to the kernel of a collision-broadened line, that is, the spectral range around the line center.

For *hard* collisions, on the other hand, the collision partners pass through the short-range part of their interaction potential, and the level shift ΔE during the collision is correspondingly larger. Hard collisions therefore contribute to the line wings (Vol. 1, Fig. 3.1) [972].

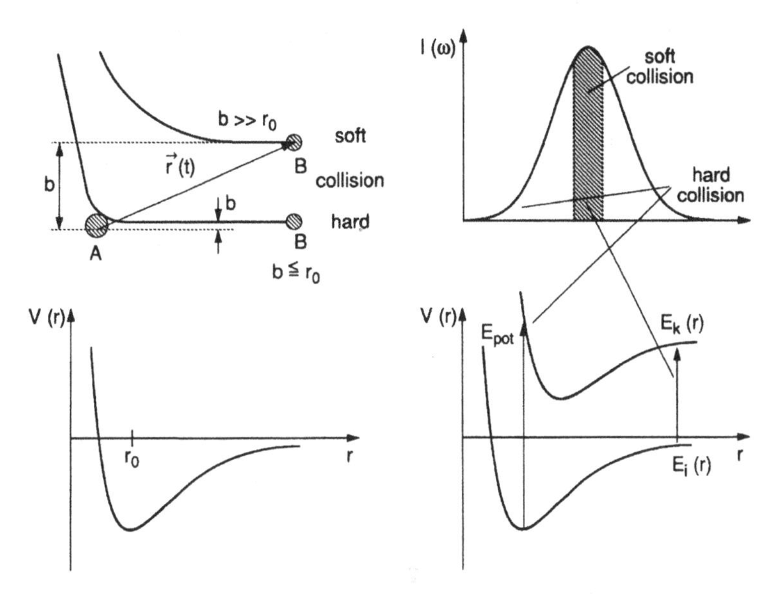

Fig. 8.1 Interaction potential $V(r)$ and semiclassical model for soft collisions with impact parameter $b \gg r_0$ and hard collisions ($b < r_0$)

8.1.1 Sub-Doppler Spectroscopy of Collision Processes

In Doppler-limited spectroscopy the effect of collisions on the line kernel is generally completely masked by the much larger Doppler width. Any information on the collision can therefore be extracted only from the line wings of the Voigt profile (Vol. 1, Sect. 3.2), by a deconvolution of the Doppler-broadened Gaussian profile and the Lorentzian profile of collisional broadening [973]. Since the collisional linewidth increases proportionally to the pressure, reliable measurements are only possible at higher pressures where collisional broadening becomes comparable to the Doppler width. At such high pressures, however, many-body collisions may no longer be negligible, since the probability that N atoms are found simultaneously within a volume $V \approx r^3$ increases with the Nth power of the density. This implies that not only two-body collisions between A and B but also many-body collisions $\mathrm{A} + N \cdot \mathrm{B}$ may contribute to line profile and line shift of transitions in A. In such cases the line profile no longer yields unambiguous information on the interaction potential $V(\mathrm{A}, \mathrm{B})$ [974].

With techniques of sub-Doppler spectroscopy, even small collisional broadening effects can be investigated with high accuracy. One example is the measurement of pressure broadening and shifts of narrow Lamb dips (Sect. 2.2) of atomic and molecular transitions, which is possible with an accuracy of a few kilohertz if stable lasers are used. The most accurate measurements have been performed with stabilized HeNe lasers on the transitions at 633 nm [975] and 3.39 μm [976]. When the laser frequency ω is tuned across the absorption profiles of the absorbing sample inside the laser resonator, the output power of the laser $P_\mathrm{L}(\omega)$ exhibits sharp

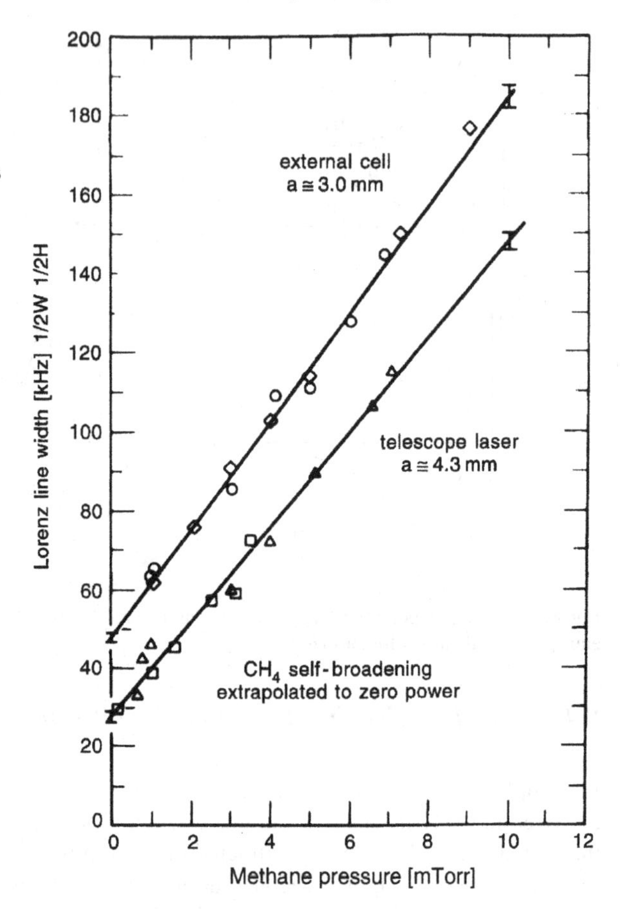

Fig. 8.2 Linewidth of the Lamb peak in the output power of a HeNe laser at $\lambda = 3.39\ \mu m$ with an intracavity CH_4 absorption cell and different beam waists of the expanded laser beam, causing a different transit-time broadening [976]

Lamb peaks (inverse Lamb dips) at the line centers of the absorbing transitions (Sect. 2.3). The line profiles of these peaks are determined by the pressure in the absorption cell, by saturation broadening, and by transit-time broadening (Vol. 1, Sect. 3.4). Center frequency ω_0, linewidth $\Delta\omega$, and line profile $P_L(\omega)$ are measured as a function of the pressure p (Fig. 8.2). The slope of the straight line $\Delta\omega(p)$ yields the line-broadening coefficient [977], while the measurement of $\omega_0(p)$ gives the collision-induced line shift.

A more detailed consideration of collisional broadening of Lamb dips or peaks must also take into account velocity-changing collisions. In Sect. 2.2 it was pointed out that only molecules with the velocity components $v_z = 0 \pm \gamma/k$ can contribute to the simultaneous absorption of the two counterpropagating waves. The velocity vectors $\boldsymbol{v}$ of these molecules are confined to a small cone within the angles $\beta \leq \pm\epsilon$ around the plane $v_z = 0$ (Fig. 8.3a), where

$$\sin\beta = v_z/|\boldsymbol{v}| \quad \Rightarrow \quad \sin\varepsilon \leq \gamma/(k \cdot v), \quad \text{with } v = |\boldsymbol{v}|. \tag{8.1}$$

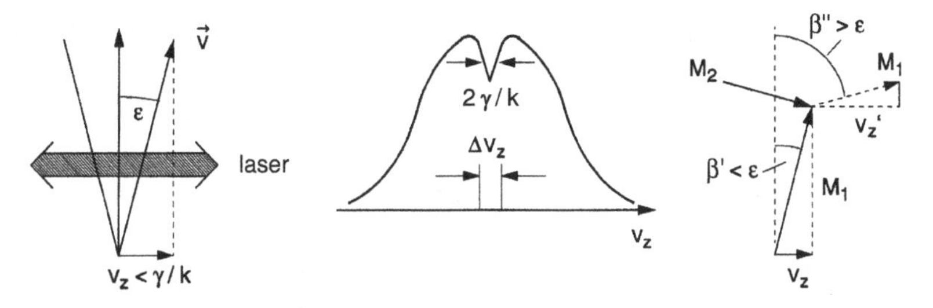

Fig. 8.3 Only molecules with velocity vectors within the angular range $\beta \leq \epsilon$ around the plane $z = 0$ contribute to the line profile of the Lamb dip. Velocity-changing collisions that increase β to values $\beta > \epsilon$ push the molecules out of resonance with the laser field

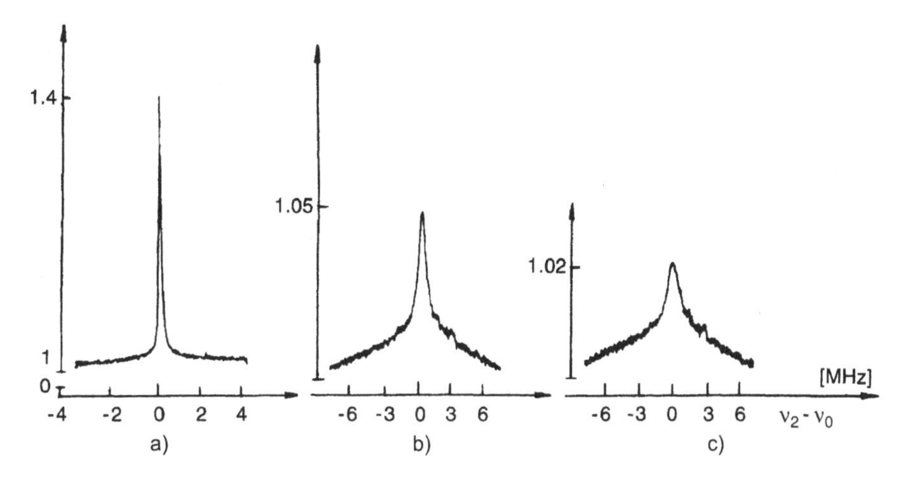

Fig. 8.4 Line profiles of Lamb peaks of a HeNe laser at $\lambda = 3.39$ μm with intracavity CH_4 absorption cell: (**a**) pure CH_4 at 1.4 mbar; (**b**) addition of 30 mbar He; and (**c**) 79 mbar He [978]

During a collision a molecule is deflected by the angle $\theta = \beta'' - \beta$ (Fig. 8.3c). If $v \cdot \sin\theta < \gamma/k$, the molecule after the collision is still in resonance with the standing light wave inside the laser resonator. Such soft collisions with deflection angles $\theta < \epsilon$ therefore do not appreciably change the absorption probability of a molecule. Because of their statistical phase jumps (Vol. 1, Sect. 3.3) they do, however, contribute to the linewidth. The line profile of the Lamb dip broadened by soft collisions remains Lorentzian.

Collisions with $\theta > \epsilon$ may shift the absorption frequency of the molecule out of resonance with the laser field. After a hard collision the molecule, therefore, can only contribute to the absorption in the line wings.

The combined effect of both kinds of collisions gives a line profile with a kernel that can be described by a Lorentzian profile slightly broadened by soft collisions. The wings, however, form a broad background caused by velocity-changing collisions. The whole profile cannot be described by a single Lorentzian function. In Fig. 8.4 such a line profile is shown for the Lamb peak in the laser output $P_L(\omega)$ at

$\lambda = 3.39$ µm with a methane cell inside the laser resonator for different pressures of CH_4 and He [978].

8.1.2 Combination of Different Techniques

Often the collisional broadening of the Lamb dip and of the Doppler profile can be measured simultaneously. A comparison of both broadenings allows the separate determination of the different contributions to line broadening. For phase-changing collisions there is no difference between the broadenings of the two different line profiles. However, velocity-changing collisions do affect the Lamb-dip profile (see above), but barely affect the Doppler profile because they mainly cause a redistribution of the velocities but do not change the temperature.

Since the homogeneous width γ of the Lamb-dip profile increases with pressure p, the maximum allowed deflection angle ϵ in (8.1) also increases with p. A comparison of pressure-induced effects on the kernel and on the background profile of the Lamb dips and on the Doppler profile therefore yields more detailed information on the collision processes. Velocity-selective optical pumping allows the measurement of the shape of velocity-changing collisional line kernels over the full thermal range of velocity changes [979].

Collisions may also change the orientation of atoms and molecules (Sect. 5.1), which means that the orientational quantum number M of an optically pumped molecule is altered. This can be monitored by polarization spectroscopy (Sect. 2.4). The orientation of molecules within the velocity interval $\Delta v_z = (\omega_\text{p} \pm \gamma)/k$ induced by the polarized pump radiation with frequency ω_p determines the polarization characteristics of the transmitted probe laser wave at frequency ω and therefore the detected signal $S(\omega)$. Any collision that alters the orientation, the population density N_i of molecules in the absorbing level $|i\rangle$, or the velocity v of absorbing molecules affects the line profile $S(\omega)$ of the polarization signal. The velocity-changing collisions have the same effect on $S(\omega)$ as on the Lamb-dip profiles in saturation spectroscopy. The orientation-changing collisions decrease the magnitude of the signal $S(\omega)$, while inelastic collisions result in the appearance of new *satellite polarization signals* in neighboring molecular transitions (Sect. 8.2). Orientation-changing collisions may also be detected with OODR saturation spectroscopy (Sect. 5.4) if the Lamb-dip profile is measured in dependence on the pressure for different angles α between the polarization planes of the linearly polarized pump and probe waves.

A commonly used technique for the investigation of depolarizing collisions in *excited* states is based on the orientation of atoms or molecules by optical pumping with a polarized laser and the measurement of the degree of polarization $P = (I_\parallel - I_\perp)/(I_\parallel + I_\perp)$ of the fluorescence emitted from the optically pumped or the collisionally populated levels (Sect. 8.2).

Because of the large mean radius $\langle r_n \rangle \propto n^2$ of the Rydberg electron, Rydberg atoms or molecules have very large collision cross sections. Therefore optical transitions to Rydberg states show large collisional broadening, which can be studied with

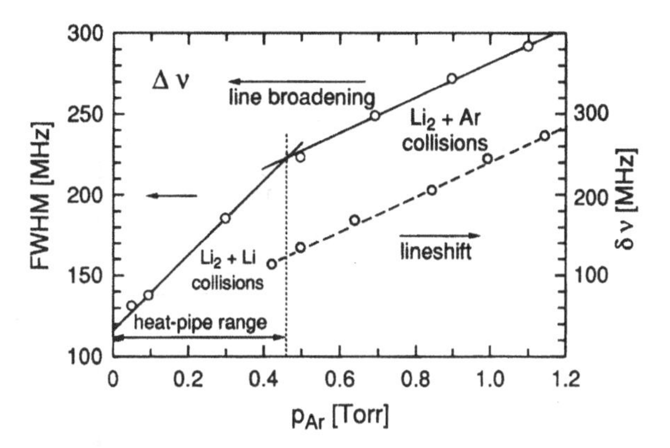

Fig. 8.5 Pressure broadening (*left scale*) and shift (*right scale*) of a Doppler-free rotational line in the Rydberg system $B^1\Sigma_u \to 6d\delta^1\Delta_g$ of the Li_2 molecule in a heat pipe for $Li_2^* + Li$ collisions at $p < 0.4$ mbar and for $Li_2^* + Ar$ collisions for $p > 0.4$ mbar [981]

Doppler-free two-photon spectroscopy or with two-step excitation (Sect. 5.4). For illustration, Fig. 8.5 illustrates pressure broadening and shifts of a rotational transition to a Rydberg level of the Li_2 molecule measured with Doppler-free OODR polarization spectroscopy (Sect. 5.5) in a lithium/argon heat pipe [980], where the intermediate level $B(v', J')$ was pumped optically by a circularly polarized pump laser. For the chosen temperature and pressure conditions the argon is confined to the cooled outer parts of the heat pipe, and the center of the heat pipe contains pure lithium vapor (98 % Li atoms and 2 % Li_2 molecules) with a total vapor pressure $p(\mathrm{Li}) = p(\mathrm{Ar})$ up to argon pressures of 0.7 mbar. The observed pressure broadening and shift in this range $p < 0.7$ mbar are therefore caused by $Li_2^* + Li$ collisions.

For $p(\mathrm{Ar}) > 0.7$ mbar (0.5 torr) the argon begins to diffuse into the central part, if the temperature and thus the lithium vapor pressure remains constant while $p(\mathrm{Ar})$ increases. The slope of the curve $\Delta\omega(p)$ yields for $p > 0.7$ mbar the cross section for $Li_2^* + Ar$ collisions. For the example depicted in Fig. 8.5 the cross sections for line broadening are $\sigma(\mathrm{Li}_2^* + \mathrm{Li}) = 60\ \mathrm{nm}^2$ and $\sigma(\mathrm{Li}_2^* + \mathrm{Ar}) = 41\ \mathrm{nm}^2$, whereas the line shifts are $\partial\nu/\partial p = -26$ MHz/mbar for $Li_2^* + Ar$ collisions [981]. Similar measurements have been performed on Sr Rydberg atoms [982], where the pressure shift and broadenings of Rydberg levels $R(n)$ for the principle quantum numbers n in the range $8 \le n \le 35$ were observed.

8.2 Measurements of Inelastic Collision Cross Sections of Excited Atoms and Molecules

Inelastic collisions transfer the internal energy of an atom or molecule A either into internal energy of the collision partner B or into relative kinetic energy of both partners. In the case of atoms, either *electronic* internal energy can be transferred to the collision partner or the magnetic energy of spin-orbit interaction. In the first case this leads to collision-induced transitions between different electronic states, and in the second case to transitions between atomic fine-structure components [982]. For *molecules* many more possibilities for energy transfer by inelastic collisions exist,

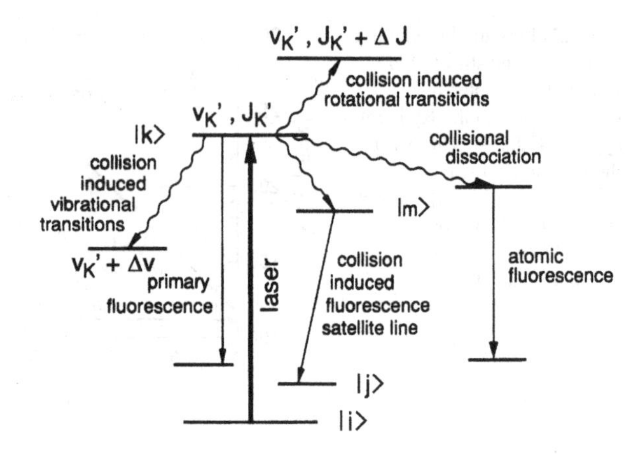

Fig. 8.6 Schematic term diagram for illustration of different possible inelastic collisional transitions of an optically pumped level $|k\rangle = (v'_k, J'_k)$ of a molecule M*

such as rotational–vibrational or electronic energy transfer, and collision-induced dissociation.

A large variety of different spectroscopic techniques has been developed for the detailed investigation of these various inelastic collisions. They are illustrated in the following sections.

8.2.1 Measurements of Absolute Quenching Cross Sections

In Sect. 6.3 we saw that the effective lifetime $\tau_k^{\text{eff}}(N_\text{B})$ of an excited level $|k\rangle$ of an atom or molecule A depends on the density N_B of the collision partners B. From the slope of the Stern–Volmer plot

$$1/\tau_k^{\text{eff}} = 1/\tau_k^{\text{rad}} + \sigma_k^{\text{total}} \overline{v} N_\text{B}, \tag{8.2}$$

the total deactivation cross section σ_k^{total} can be obtained, see (6.57). Since the collisional deactivation diminishes the fluorescence intensity emitted by $|k\rangle$, the inelastic collisions are called *quenching collisions* and σ_k^{total} is named the *quenching cross section.*

Several possible decay channels contribute to the depopulation of level $|k\rangle$, and the quenching cross section can be written as the sum

$$\sigma_k^{\text{total}} = \sum_m \sigma_{km} = \sigma_k^{\text{rot}} + \sigma_k^{\text{vib}} + \sigma_k^{\text{el}}, \tag{8.3}$$

over all collision-induced transitions $|k\rangle \rightarrow |m\rangle$ into other levels $|m\rangle$, which might be rotational, vibrational, or electronic transitions (Fig. 8.6).

While measurements of the effective lifetime $\tau_k^{\text{eff}}(N_\text{B})$ yield *absolute* values of the *total quenching cross section* σ_k^{total}, the different contributions in (8.3) have to be determined by other techniques, such as LIF spectroscopy. Even if only their *relative* magnitudes can be measured, this is sufficient to obtain, together with the absolute value of σ^{total}, their absolute magnitudes.

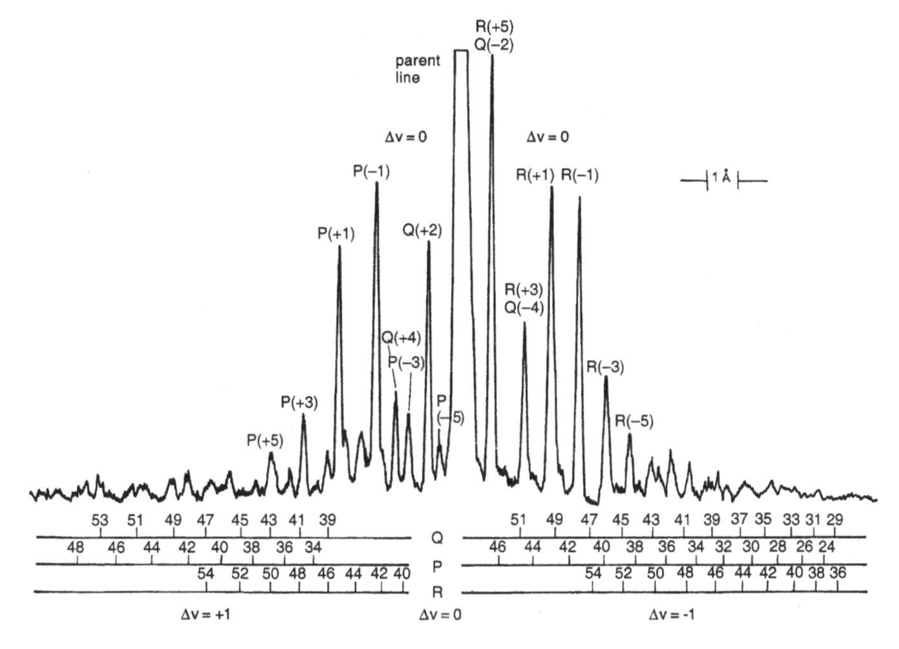

Fig. 8.7 Collision-induced satellite lines $Q(\Delta J')$, $R(\Delta J')$, and $P(\Delta J')$ in the LIF spectrum of Na_2 under excitation of the rotational level $B^1\Pi_u(v' = 6, J' = 43)$. The parent line is recorded with one-twentieth the sensitivity. The high satellite lines are related to $\Delta v = 0$ transitions and partly superimpose the weaker satellites from $\Delta v = \pm 1$ collision-induced transitions [987]

8.2.2 Collision-Induced Rovibronic Transitions in Excited States

When the level $|k\rangle = |v', J_k'\rangle$ of an excited molecule M^* has been selectively populated by optical pumping, inelastic collisions $M^* + B$, which occur during the lifetime τ_k, will transfer $M^*(k)$ into other levels $|m\rangle = |v_k' + \Delta v, J_k' + \Delta J\rangle$ of the same or of another electronic state:

$$M^*\left(v_k', J_k'\right) + B \rightarrow M^*\left(v_k' + \Delta v, J_k' + \Delta J\right) + B^* + \Delta E_{\text{kin}}. \tag{8.4}$$

The difference $\Delta E = E_k - E_m$ of the internal energies before and after the inelastic collision is transferred either into internal energy of B or into translation energy of the collisions partners.

Molecules in the collisionally populated levels $|m\rangle$ can decay by the emission of fluorescence or by further collisions. In the LIF spectrum new lines then appear besides the *parent lines*, which are emitted from the optically pumped level (Fig. 8.7). These new lines, called *collision-induced satellites* contain *the complete information on the collision process that has generated them.* Their wavelength λ allows the assignment of the upper level $|m\rangle = (v_k' + \Delta v, J_k' + \Delta J)$, their intensities yield the collisional cross sections σ_{km}, and their degree of polarization compared with that of the parent line gives the cross section for depolarizing, that is, orientation-changing collisions. This can be seen as follows: Assume the upper level $|k\rangle$ is optically pumped on the transition $|i\rangle \rightarrow |k\rangle$ (Fig. 8.6) with a pump rate $N_i P_{ik}$ and collisions induce transitions $|k\rangle \rightarrow |m\rangle$. The rate equation for the population densities N_k, N_m can be written as

$$\frac{\mathrm{d}N_k}{\mathrm{d}t} = N_i P_{ik} - N_k\left(A_k + \sum_m R_{km}\right) + \sum_n N_n R_{nk}, \tag{8.5a}$$

$$\frac{\mathrm{d}N_m}{\mathrm{d}t} = N_k R_{km} - N_m\left(A_m + \sum_m R_{mn}\right) + \sum_n N_n R_{nm}, \tag{8.5b}$$

where the last two terms describe the collisional depopulation $|m\rangle \rightarrow |n\rangle$ and the repopulation $|n\rangle \rightarrow |k\rangle$ or $|n\rangle \rightarrow |m\rangle$ from other levels $|n\rangle$.

Under stationary conditions (optical pumping with a cw laser) $(\mathrm{d}N_k/\mathrm{d}t) = (\mathrm{d}N_m/\mathrm{d}t) = 0$. If the thermal population N_n is small (that is, $E_n \gg kT$) the last terms in (8.5a–8.5b) demand at least two successive collisional transitions $|k\rangle \rightarrow |n\rangle \rightarrow |k\rangle$ or $|k\rangle \rightarrow |n\rangle \rightarrow |m\rangle$, respectively, during the lifetime τ, which means a high collision rate. At sufficiently low pressures they can therefore be neglected.

From (8.5a–8.5b) the stationary population densities N_k, N_m are then obtained as

$$\begin{aligned} N_k &= \frac{N_i P_{ik}}{A_k + \sum R_{km}}, \\ N_m &= \frac{N_k R_{km} + \sum N_n R_{nm}}{A_m + \sum R_{mn}} \simeq N_k \frac{R_{km}}{A_m}. \end{aligned} \tag{8.6}$$

The ratio of the fluorescence intensities of satellite line $|m\rangle \rightarrow |j\rangle$ to parent line $|k\rangle \rightarrow |i\rangle$

$$\frac{I_{mj}}{I_{ki}} = \frac{N_m A_{mj} h\nu_{mj}}{N_k A_{ki} h\nu_{ki}} = R_{km} \frac{A_{mj}\nu_{mj}}{A_m A_{ki}\nu_{ki}}, \tag{8.7}$$

directly yields the probability R_{km} for collision-induced transitions $|k\rangle \rightarrow |m\rangle$, if the relative radiative transition probabilities A_{mj}/A_m and A_{ki}/A_k are known. Measurements of the spontaneous lifetime τ_k and τ_m (Sect. 6.3) allow the determination of $A_k = 1/\tau_k$ and $A_m = 1/\tau_m$. The absolute values of A_{mj} and A_{ki} can then be obtained by measuring the *relative* intensities of all fluorescence lines emitted by $|m\rangle$ and $|k\rangle$ under collision-free conditions.

The probability R_{km} is related to the collision cross section σ_{km} by

$$R_{km} = (N_\mathrm{B}/\overline{v}) \int \sigma_{km}(v_\mathrm{rel}) v_\mathrm{rel}\,\mathrm{d}v, \tag{8.8}$$

where N_B is the density of collision partners B, and v_rel the relative velocity between M and B.

When the experiments are performed in a cell at temperature T, the velocities follow a Maxwellian distribution, and (8.8) becomes

$$R_{km} = N_\mathrm{B}\left(\frac{8kT}{\pi\mu}\right)^{-1/2} \langle\sigma_{km}\rangle, \tag{8.9}$$

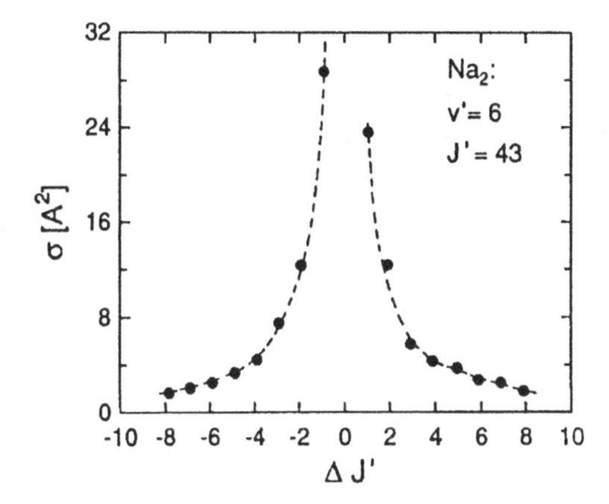

Fig. 8.8 Absolute integral cross sections $\sigma(\Delta J')$ for collision-induced rotational transitions when $Na_2^*(B^1\Pi_u, v'=6, J'=43)$ collides with He-atoms at $T=500$ K [987]

where $\mu = m_M m_B/(m_M + m_B)$ is the reduced mass, and $\langle\sigma_{km}\rangle$ means the average of $\sigma_{km}(v)$ over the velocity distribution. The cross sections σ_{km} obtained in this way represent integral cross sections, integrated over all scattering angles θ.

Such determinations of rotationally inelastic integral cross sections σ_{km} for collision-induced transitions in excited molecules obtained from measurements of satellite lines in the fluorescence spectrum have been reported for a large variety of different molecules, such as I_2 [983, 984], Li_2 [985, 986], Na_2 [987], or NaK [988]. For illustration, the cross section $\sigma(\Delta J)$ for the transition $J \to J + \Delta J$ in excited Na_2^* molecules induced by collisions $Na_2^* + He$ are plotted in Fig. 8.8. They rapidly decrease from a value $\sigma(\Delta J = \pm 1) \approx 0.3$ nm^2 to $\sigma(\Delta J = \pm 8) \approx 0.02$ nm^2. This decrease is essentially due to energy and momentum conservation, since the energy difference $\Delta E = E(J \pm \Delta J) - E(J)$ has to be transferred into the kinetic energy of the collision partners. The probability for this energy transfer is proportional to the Boltzmann factor $\exp[-\Delta E/(kT)]$ [989].

For interaction potentials $V(R)$ of the collision pair M + B with spherical symmetry, which only depend on the internuclear distance R, no internal angular momentum of M can be transferred. The absolute values of the cross sections $\sigma(\Delta J)$ are therefore a measure for the *nonspherical* part of the potential $V(\mathrm{M}, \mathrm{B})$. This potential can be represented by the expansion

$$V(R, \theta, \phi) = V_0(R) + \sum_{\ell,m} a_\ell m Y_\ell^m(\phi), \tag{8.10}$$

where the Y_ℓ^m denote the spherical surface functions. A homonuclear diatomic molecule M in a Σ state has a cylindrically symmetric electron cloud. Its interaction potential must be independent of ϕ and furthermore has a symmetry plane perpendicular to the internuclear axis. For such symmetric states (8.10) reduces to

$$V(R, \theta) = a_0 V_0(R) + a_2 P_2(\cos\theta) + \cdots, \tag{8.11}$$

where the coefficient a_2 of the Legendre polynomical $P_2(\cos\theta)$ can be determined from $\sigma(\Delta J)$, while a_0 is related to the value of the elastic cross section [990, 991].

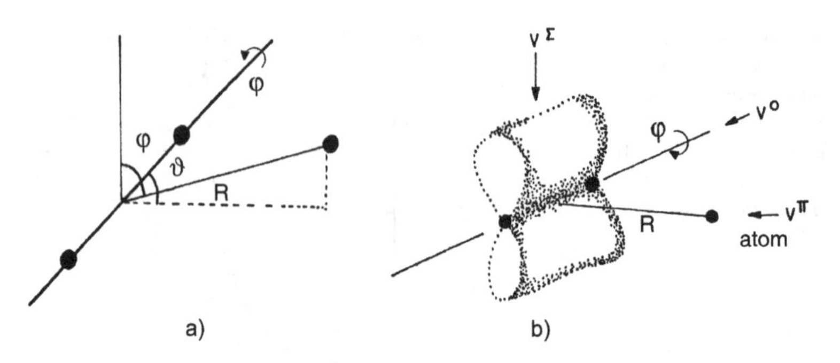

Fig. 8.9 (**a**) Illustration of the variables R, θ, and φ in the interaction potential $V(R,\theta,\varphi)$. (**b**) Schematic electron cloud distribution for excited homonuclear diatomic molecules in a Π state. Also shown are the two directions of φ for a Σ-type and a Π-type interaction potential

In a Π state, however, the electron cloud possesses an electronic angular momentum and the charge distribution no longer has cylindrical symmetry (Fig. 8.9b). The interaction potential $V(R,\theta,\phi)$ depends on all three variables. Since the two Λ components of a rotational level in a Π state ($\Lambda = 1$) correspond to different charge distributions, different magnitudes of the cross sections $\sigma(J, \pm\Delta J)$ for collision-induced rotational transitions are expected for the two Λ components of the same rotational level $|J\rangle$ [992, 993]. Such an asymmetry, which can be recognized in Fig. 8.7 (transitions with $\Delta J =$ even reach other Λ components than those with $\Delta J =$ odd), has indeed been observed [986, 992, 994].

In addition to LIF resonant two-photon ionization (Sect. 1.4) can also be used for the sensitive detection of collision-induced rotational transitions. This method represents an efficient alternative to LIF for those electronic states that do not emit detectable fluorescence because there are no allowed optical transitions into lower states. An illustrative example is the detailed investigation of inelastic collisions between excited N_2 molecules and different collision partners [995]. A vibration–rotation level (v', J') in the a $^1\Pi_g$ state of N_2 is selectively populated by two-photon absorption (Fig. 8.10). The collision-induced transitions to other levels $(v' + \Delta v, j' + \Delta J)$ are monitored by resonant two-photon ionization (REMPI, Sect. 1.2) with a pulsed dye laser. The achievable good signal-to-noise ratio is demonstrated by the collisional satellite spectrum in Fig. 8.10b, where the optically pumped level was $(v' = 2, J' = 7)$. This level is ionized by the $P(7)$ *parent line* in the spectrum, which has the signal height 7.25 on the scale of Fig. 8.10b.

If the parent level $|k\rangle = (v_k, J_k)$ of the optically pumped molecule has been oriented by polarized light, this orientation is partly transferred by collisions to the levels $(v_k + \Delta v, J_k + \Delta J)$. This can be monitored by measuring the polarization ratio $R_p = (I_\parallel - I_\perp)/(I_\parallel + I_\perp)$ of the collisional satellite lines in the fluorescence [996].

The cross sections for collision-induced vibrational transitions are generally much smaller than those for rotational transitions. This is due to energy conservation (if $\Delta E_{vib} \gg kT$) and also to dynamical reasons (the collision has to be sufficiently nonadiabatic, that is, the collision time should be comparable or shorter than the vibrational period) [991]. The spectroscopic detection is completely anal-

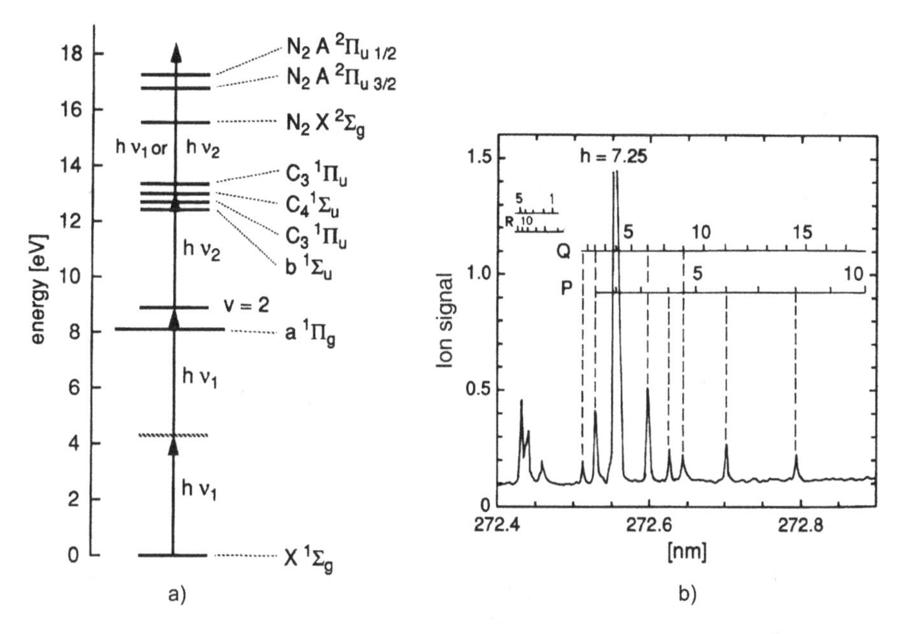

Fig. 8.10 Term diagram for selective excitation of the level ($v' = 2$, $J' = 7$) in the $a^1\Pi_g$ state of N_2 with the detection of collisional transitions by resonant two-photon ionization [995]

ogous to that for rotational transitions. In the LIF spectrum new collision-induced bands $(v_k + \Delta_k) \rightarrow (v_m)$ appear (Fig. 8.7), with a rotational distribution that reflects the probabilities of collision-induced transitions $(v_k, J_k) \rightarrow (v_k + \Delta_\mathrm{v}, J_k + \Delta J)$ [985, 987, 997, 998].

Vibrational transitions within the electronic ground state can be studied by time-resolved infrared spectroscopy or by pump–probe measurements (Sect. 8.3.1).

8.2.3 Collisional Transfer of Electronic Energy

Collisions may also transfer electronic energy. For instance, the electronic energy of an excited atom A^* or molecule M^* can be converted in a collision with a partner B into translational energy E_kin or, with higher probability, into internal energy of B [965].

For collisions at thermal energies the collision time $T_\mathrm{coll} = d/\overline{v}$ is long compared to the time for an electronic transition. The interaction $V(\mathrm{A}^*, \mathrm{B})$ or $V(\mathrm{M}^*, \mathrm{B})$ can then be described by a potential. Assume that the two potential curves $V(\mathrm{A}_i, \mathrm{B})$ and $V(\mathrm{A}_k, \mathrm{B})$ cross at the energy $E(R_\mathrm{c})$ (Fig. 8.11). If the relative kinetic energy of the collision partners is sufficiently high to reach the crossing point, the collision pair may jump over to the other potential curve [999]. In Fig. 8.11, for instance, a collision $\mathrm{A}_i + \mathrm{B}$ can lead to electronic excitation $|i\rangle \rightarrow |k\rangle$ if $E_\mathrm{kin} > E_2$, while for a collisional deexcitation $|k\rangle \rightarrow |i\rangle$ only the kinetic energy $E_\mathrm{kin} > E_1$ is required.

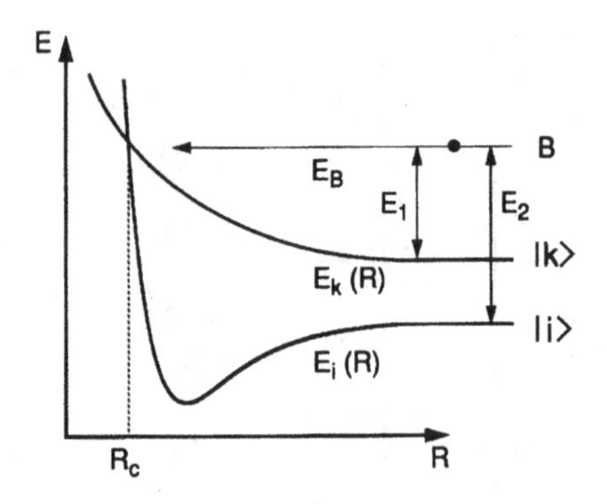

Fig. 8.11 Potential diagram for collision-induced transitions between the two electronic states $|i\rangle$ and $|k\rangle$ that can occur at the crossing point at $R = R_c$ of the two potential curves $E_i R = M_i + B$ and $E_k R = M_k + B$

The cross sections of electronic energy transfer $A^* + B \to A + B^* + \Delta E_{kin}$ are particularly large in cases of energy resonance, which means $\Delta E(A^* - A) \simeq \Delta E(B^* - B) \to \Delta E_{kin} \le kT$. A well-known example is the collisional excitation of Ne atoms by metastable He* atoms, which represents the main excitation mechanism in the HeNe laser.

The experimental proof for such electronic energy transfer ($E \to E$ transfer) is based on the selective excitation of A by a laser and the spectrally resolved detection of the fluorescence from B* [1000, 1001].

In collisions between excited atoms and molecules either the atom A* or the molecule M* may be electronically excited. Although the two cases

$$M(v_i'', J_i'') + A^* \to M^*(v', J') + A, \tag{8.12a}$$

and

$$M^*(v_k', J_k') + A \to M(v'', J'') + A^*, \tag{8.12b}$$

represent inverse processes, their collision cross sections may differ considerably. This was demonstrated by the example $M = Na_2$, $A = Na$. Either the Na atoms were selectively excited into the $3P$ state and the fluorescence of Na_2^* was measured with spectral and time resolutions [1002], or Na_2 was excited into a definite level (v_k', J_k') in the $^1A\Sigma_u$ state and the energy transfer was monitored by time-resolved measurements of the shortened lifetime $\tau_k(v_k', J_k')$ and the population $N_{3p}(t)$ of the Na* $(3P)$ level detected through the atomic fluorescence [1003]. There are two distinct processes that lead to the formation of Na*:

(a) direct energy transfer according to (8.12b), or
(b) collision-induced dissociation $Na_2^* + Na \to Na^* + Na + Na$.

This collision-induced dissociation of electronically excited molecules plays an important role in chemical reactions and has therefore been studied for many molecules [1004–1006]. The collision cross sections for this process may be sufficiently large to generate inversion between atomic levels. Laser action based on *dissociation pumping* has been demonstrated. Examples are the powerful iodine laser [1007] and the Cs laser [1008].

The transfer of electronic energy into rovibronic energy has much larger cross sections than the electronic to translational energy ($E \to T$) transfer [1009]. These processes are of crucial importance in photochemical reactions of larger molecules and detailed studies with time-resolved laser spectroscopy have brought much more insight into many biochemical processes [1010–1012].

One example of a state-selective experimental technique well suited for studies of electronic to vibrational ($E \to V$) transfer processes is based on CARS (Sect. 4.4). This has been demonstrated by Hering et al. [1013], who studied the reactions

$$\mathrm{Na}^*(3P) + \mathrm{H}_2(v'' = 0) \to \mathrm{Na}(3S) + \mathrm{H}_2(v'' = 1, 2, 3), \tag{8.13}$$

where Na* was excited by a dye laser and the internal state distribution of $\mathrm{H}_2(v'', J'')$ was measured with CARS.

An interesting phenomenon based on collisions between excited atoms and ground-state atoms is the macroscopic diffusion of optically pumped atoms realized in the *optical piston* [1014]. It is caused by the difference in cross sections for velocity-changing collisions involving excited atoms A* or ground-state atoms A, respectively, and results in a spatial separation between optically pumped and unpumped atoms, which can be therefore used for isotope separation [1015].

8.2.4 Energy Pooling in Collisions Between Excited Atoms

Optical pumping with lasers may bring an appreciable fraction of all atoms within the volume of a laser beam passing through a vapor cell into an excited electronic state. This allows the observation of collisions between two *excited* atoms, which lead to many possible excitation channels where the sum of the excitation energies is accumulated in one of the collision partners. Such *energy-pooling* processes have been demonstrated for Na* + Na*, where reactions

$$\mathrm{Na}^*(3P) + \mathrm{Na}^*(3P) \to \mathrm{Na}^{**}(nL) + \mathrm{Na}(3S), \tag{8.14}$$

have been observed [1016], leading to the excitation of high-lying levels $|n, L\rangle$ (Fig. 8.12).

Measurements of the fluorescence intensity $I_{\mathrm{Fl}}(n, L)$ emitted by the levels Na** ($n, L = 4D$ or $5S$) yields the collision rate

$$k_{n,L} = N^2(\mathrm{Na}^* 3p) \cdot \sigma_{n,L} \overline{v} \left[1/\mathrm{s\,cm}^3\right],$$

of optically pumped Na atoms, which increases with the square N^2 of the density of excited Na* atoms. This density cannot be determined directly from the measured Na*(3P) fluorescence because radiation trapping falsifies the results [1017]. The attenuation of the transmitted laser beam is a better measure because every absorbed photon produces one excited Na*(3P) atom.

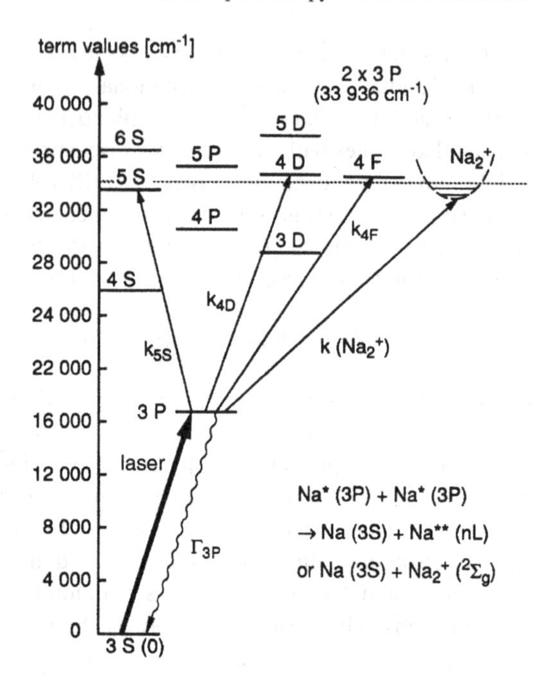

Fig. 8.12 Level diagram of the Na atom with schematic illustration of the different excitation channels in collisions $Na^*(3P) + Na^*(3P)$ [1018]

Since the sum of the excitation energies of the two colliding $Na^*(3P)$ atoms is higher than the ionization limit of the Na_2 molecule, associative ionization

$$Na^*(3P) + Na^*(3P) \underset{k_{Na_2^+}}{\rightarrow} Na_2^+ + e^-, \tag{8.15}$$

can be observed. The measurement of Na_2^+ ions gives the reaction rate $k(Na_2^+)$ for this process [1018].

A further experimental step that widens the application range is the excitation of two different species A_1 and A_2 with two dye lasers. An example is the simultaneous excitation of $Na^*(3P)$ and $K^*(4P)$ in a cell containing a mixture of sodium and potassium vapors. Energy pooling can lead to the excitation of high-lying Na^{**} or K^{**} states, which can be monitored by their fluorescence [1019]. When both lasers are chopped at two different frequencies f_1 and f_2 (where $1/f$ is long compared to the collisional transfer time), lock-in detection of the fluorescence emitted from the highly excited levels at f_1, f_2 or $f_1 + f_2$ allows the distinction between the energy pooling processes $Na^* + Na^*$, $K^* + K^*$, and $Na^* + K^*$ leading to the excitation of the levels $|n, L\rangle$ of Na^{**} or K^{**}. For further examples see [1020, 1021].

8.2.5 Spectroscopy of Spin-Flip Transitions

Collision-induced transitions between fine-structure components where the relative orientation of the electron spin with respect to the orbital angular momentum is

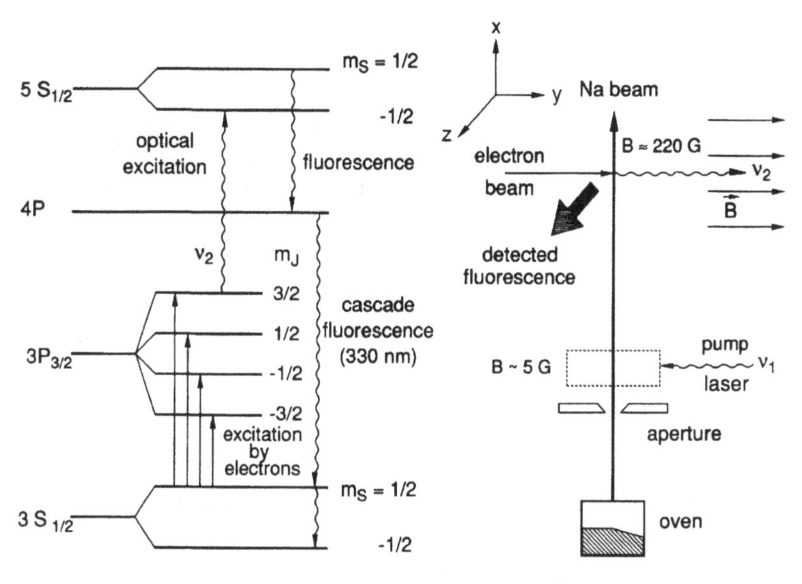

Fig. 8.13 The relative populations of Zeeman levels $|M_j\rangle$ in the $3^2P_{3/2}$ state of Na are probed by laser excitation of the $5^2S_{1/2}$ state, which is monitored via the cascade fluorescence into the $3S_{1/2}$ state [1030]

changed have been studied in detail by laser-spectroscopic techniques [1022]. One of the methods often used is *sensitized fluorescence*, where one of the fine-structure components is selectively excited and the fluorescence of the other component is observed as a function of pressure [1023]. Either pulsed excitation and time-resolved detection is used [1024] or the intensity ratio of the two fine-structure components is measured under cw excitation [1025].

Of particular interest is the question into which of the fine-structure components of the atoms a diatomic molecule dissociates after excitation into predissociating states. The investigation of this question gives information on the recoupling of angular momenta in molecules at large intermolecular distances and about crossings of potential curves [1026]. Early measurements of these processes in cells often gave wrong answers because collision-induced fine-structure transitions and also radiation trapping had changed the original fine-structure population generated by the dissociation process. Therefore, laser excitation in molecular beams has to be used to measure the initial population of a certain fine-structure component [1027].

Collision-induced spin-flip transitions in molecules transfer the optically excited molecule from a singlet into a triplet state, which may drastically change its chemical reactivity [1028]. Such processes are important in dye lasers and have therefore been intensively studied [1029]. Because of the long lifetime of the lowest triplet state T_0, its population $N(t)$ can be determined by time-resolved measurements of the absorption of a dye laser tuned to the transition $T_0 \to T_1$.

A laser-spectroscopic technique for investigations of spin-flip transitions under electron impact excitation of Na atoms [1030] is shown in Fig. 8.13. Opti-

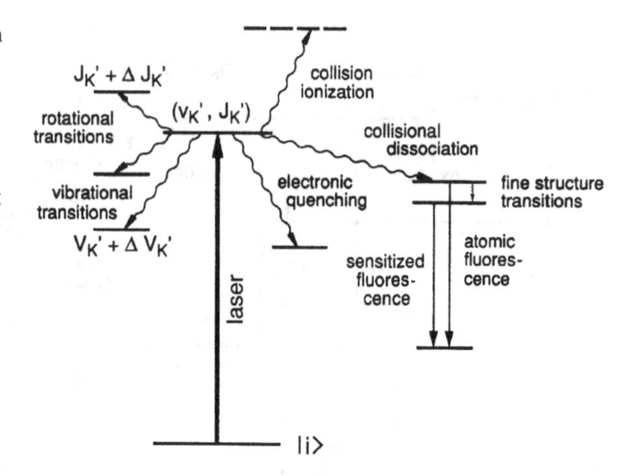

Fig. 8.14 Schematic diagram of all possible collision-induced transitions from a molecular level (v_k', J_k') in an excited electronic state selectively populated by optical pumping

cal pumping with σ^+ light on the $3S_{1/2} \rightarrow 3P_{1/2}$ transition in an external weak magnetic field orients the Na atoms into the $3S_{1/2}(m_s = +1/2)$ level. Starting from this level the Zeeman components $3P_{3/2}(M_J)$ are populated by electron impact. The population N_M is probed by a cw dye laser tuned to the transitions $3^2P_{3/2}(M_J) \rightarrow 5^2S_{1/2}(M_s = \pm 1/2)$, which can be monitored by the cascade fluorescence $5S_{1/2} \rightarrow 4P \rightarrow 3S_{1/2}$.

Figure 8.14 summarizes all possible collision-induced transitions from a selectively excited molecular level $|v_k', J_k'\rangle$ to other molecular or atomic levels.

8.3 Spectroscopic Techniques for Measuring Collision-Induced Transitions in the Electronic Ground State of Molecules

In the electronic ground states of molecules collision-induced transitions represent, for most experimental situations, the dominant mechanism for the redistribution of energy, because the radiative processes are generally too slow. In cases where a nonequilibrium distribution has been produced (for example, by chemical reactions or by optical pumping), these collisions try to restore thermal equilibrium. The relaxation time of the system is determined by the absolute values of collision cross sections.

For most infrared molecular lasers, such as the CO_2 and the CO laser, or for *chemical lasers*, such as the HF or HCl lasers, collisional energy transfer between vibrational–rotational levels of the lasing molecules plays a crucial role for the generation and maintenance of inversion and gain. These lasers are therefore called *energy-transfer lasers* [1031]. Also in many visible molecular lasers, which oscillate on transitions between optically pumped levels (v', J') of the excited electronic state and high vibrational levels of the electronic ground state, collisional deactivation of the lower laser level is essential because the radiative decay is generally too slow. Examples are dye lasers [1032] or dimer lasers, such as the Na_2 laser or the I_2 laser [1033].

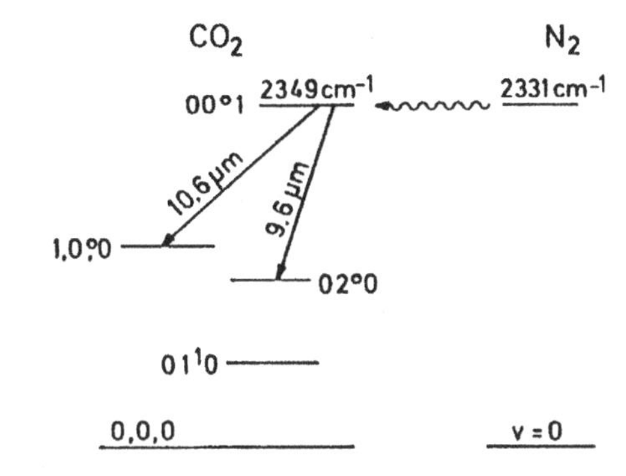

Fig. 8.15 Vibrational energy transfer from N_2 ($v = 1$) to the upper vibrational level ($v_1 = 0$, $v_2 = 0$, $l_{vib} = 0$, $v_3 = 1$) of the CO_2 laser transitions

The internal energy $E_i = E_{vib} + E_{rot}$ of a molecule $M(v_i, J_i)$ in its electronic ground state may be transferred during a collision with another molecule AB

$$M(E_i) + AB(E_m) \rightarrow M(E_i - \Delta E_1) + AB^*(E_m + \Delta E_2) + \Delta E_{kin}, \qquad (8.16)$$

into vibrational energy of AB* ($V \rightarrow V$ transfer), rotational energy ($V \rightarrow R$ transfer), electronic energy ($V \rightarrow E$ transfer), or translational energy ($V \rightarrow T$ transfer). In collisions of M with an atom A only the last two processes are possible.

The experiments show that the cross sections are much larger for $V \rightarrow V$ or $V \rightarrow R$ transfer than for $V \rightarrow T$ transfer. This is particularly true when the vibrational energies of the two collision partners are near resonant.

A well-known example of such a near-resonant $V \rightarrow V$ transfer is the collisional excitation of CO_2 molecules by N_2 molecules

$$CO_2(0,0,0) + N_2(v = 1) \rightarrow CO_2(0,0,1) + N_2(v = 0), \qquad (8.17)$$

which represents the main excitation mechanism of the upper laser level in the CO_2 laser (Fig. 8.15) [1034].

The experimental techniques for the investigation of inelastic collisions involving molecules in their electronic ground state generally differ from those discussed in Sect. 8.2. The reasons are the long spontaneous lifetimes of ground-state levels and the lower detection sensitivity for infrared radiation compared to those for the visible or UV spectrum. Although infrared fluorescence detection has been used, most of the methods are based on absorption measurements and double-resonance techniques.

8.3.1 Time-Resolved Infrared Fluorescence Detection

The energy transfer described by (8.16) can be monitored if M* is excited by a short infrared laser pulse and the fluorescence of AB* is detected by a fast cooled infrared detector (Vol. 1, Sect. 4.5) with sufficient time resolution. Such measurements have been performed in many laboratories [963]. For illustration, an experiment carried

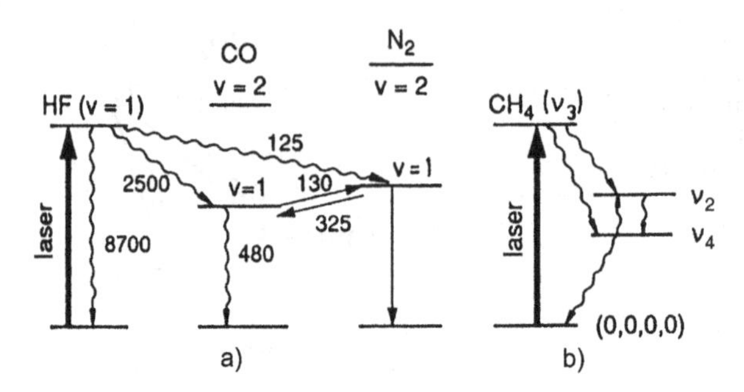

Fig. 8.16 Fluorescence detection of vibrational energy transfer from an optically pumped molecular level: (**a**) intermolecular transfer HF* → CO, N_2. The mean numbers of collisions are given for the different collision-induced transitions. (**b**) *Intra*molecular transfer $CH_4^*(\nu_3) \rightarrow CH_4^*(\nu_2) + \Delta E_{kin}$ [1035]

out by Green and Hancock [1035] is explained by Fig. 8.16a: a pulsed HF laser excites hydrogen fluoride molecules into the vibrational level $v = 1$. Collisions with other molecules AB (AB = CO, N_2) transfer the energy to excited vibrational levels of AB*. The infrared fluorescence emitted by AB* and HF* has to be separated by spectral filters. If two detectors are used, the decrease of the density $N(HF^*)$ of vibrationally excited HF molecules and the build-up and decay of $N(AB^*)$ can be monitored simultaneously.

For larger molecules M two different collisional relaxation processes have to be distinguished: collisions M* + AB may transfer the internal energy of M* to AB* (*inter*molecular energy transfer), or may redistribute the energy among the different vibrational modes of M* (*intra*molecular transfer) (Fig. 8.16b). One example is the molecule M = C_2H_4O, which can be excited on the C–H stretching vibration by a pulsed parametric oscillator at 3000 cm^{-1} [1036]. The fluorescence emitted from other vibrational levels which are populated by collision-induced transitions is detected through spectral filters. Further examples can be found in several reviews [1037–1039].

8.3.2 Time-Resolved Absorption and Double-Resonance Methods

While collision-induced transitions in excited electronic states can be monitored through the satellite lines in the *fluorescence* spectrum (Sect. 8.2.2), inelastic collisional transfer in electronic ground states of molecules can be studied by changes in the *absorption* spectrum. This technique is particularly advantageous if the radiative lifetimes of the investigated rotational–vibrational levels are so long that fluorescence detection fails because of intensity problems.

A successful technique for studying collision-induced transitions in electronic ground states is based on time-resolved double resonance [1040]. The method is explained by Fig. 8.17. A pulsed laser L_1 tuned to an infrared or optical transition

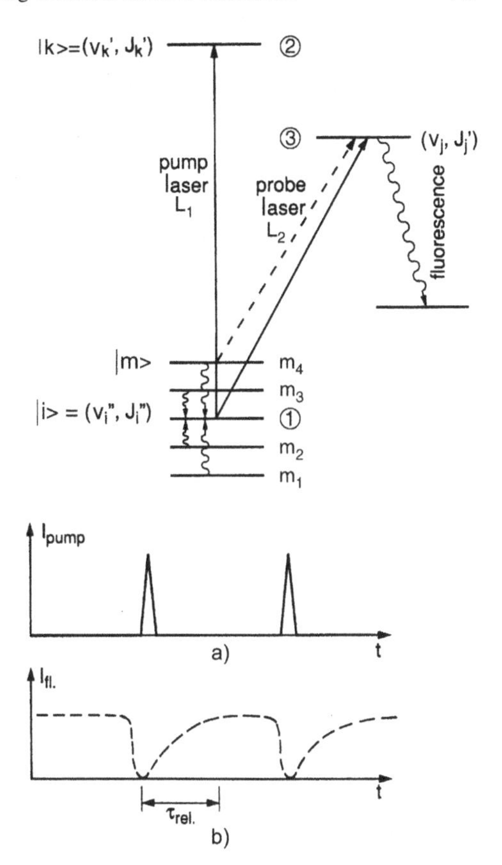

Fig. 8.17 (**a**) Level diagram for the determination of the refilling rate of a lower level (v_i'', J_i'') depleted by a pump laser pulse via measurements of the time-dependent probe laser absorption; (**b**) time dependence of pump pulses and level population $N_i(t)$ monitored via the probe laser-induced fluorescence

$|i\rangle \to |k\rangle$ depletes the lower level $|1\rangle = (v_i'', J_i'')$. The depleted level is refilled by collisional transitions from other levels. A second weak probe laser L_2 is tuned to another transition $|1\rangle \to |3\rangle = (v_j', J_j')$ starting from the depleted lower level $|i\rangle$. If the pump and probe laser beams overlap within a path length Δz in the absorbing sample, the absorption of the probe laser radiation can be measured by monitoring the transmitted probe laser intensity

$$I(\Delta z, t) = I_0 \mathrm{e}^{-\alpha \Delta z} \approx I_0\left[1 - N_i(t)\sigma_{ij}^{\mathrm{abs}} \Delta z\right]. \tag{8.18}$$

The time-resolved measurement of $I(\Delta z, t)$ yields the time dependence of the population density $N_i(t)$. Using a cw probe laser the absorption can be measured by the transmitted intensity $I(\Delta z, t)$, or by the time-dependent fluorescence induced by the probe laser, which is proportional to $N_j(t) \propto N_i(t) \cdot P(L_2)$. If a pulsed probe laser is used, the time delay Δt between pump and probe is varied and the measured values of $\alpha(\Delta t)$ yield the refilling time of level $|1\rangle$.

Without optical pumping the population densities are time independent at thermal equilibrium. From the rate equation

$$\frac{\mathrm{d}N_i^0}{\mathrm{d}t} = 0 = -N_i^0 \sum_m R_{im} + \sum_m N_m^0 R_{mi}, \tag{8.19}$$

we obtain the condition of *detailed balance*

$$N_i^0 \sum_m R_{im} = \sum_m N_m^0 R_{mi}, \tag{8.20}$$

where R_{im} are the relaxation probabilities for transitions $|i\rangle \to |m\rangle$. If the population N_i^0 is depleted to the saturated value $N_i^s < N_i^0$ by optical pumping at $t = 0$, the right side of (8.20) becomes larger than the left side and collision-induced transitions $|m\rangle \to |i\rangle$ from neighboring levels $|m\rangle$ refill $N_i(t)$. The time dependence $N_i(t)$ after the end of the pump pulse is obtained from

$$\frac{\mathrm{d}N_i}{\mathrm{d}t} = \sum_m N_m R_{mi} - N_i(t) \sum_m R_{im}. \tag{8.21a}$$

If we assume that optical pumping of level $|i\rangle$ does not essentially affect the population densities N_m $(m \neq i)$, we obtain from (8.20), (8.21a)

$$\frac{\mathrm{d}N_i}{\mathrm{d}t} = \left[N_i^0 - N_i(t)\right] \sum_m R_{im} = \left[N_i^0 - N_i(t)\right] K_i, \tag{8.21b}$$

with the relaxation rate

$$K_i = \sum_m R_{im} = N_\mathrm{B} \bar{v} \sigma_i^\mathrm{total} = N_\mathrm{B} \sqrt{\frac{8kT}{\pi\mu}} \langle \sigma_i^\mathrm{total} \rangle, \tag{8.22}$$

where N_B is the density of collision partners, $\mu = M_\mathrm{A} \cdot M_\mathrm{B}/(M_\mathrm{A} + M_\mathrm{B})$ the reduced mass and $\sigma_i^\mathrm{total} = \sum_m \sigma_{im}$ the total collision cross section.

The relaxation constant K_i depends on the total cross section $\langle \sigma_i^\mathrm{total} \rangle = \sum_m \langle \sigma_{im} \rangle$ averaged over the thermal velocity distribution and on the temperature T. Integration of (8.21a–8.21b) gives

$$N_i(t) = N_i^0 + \left[N_i^s(0) - N_i^0\right] \mathrm{e}^{-K_i t}. \tag{8.23}$$

This shows that after the end of the pump pulse at $t = 0$ the population density $N_i(t)$ returns exponentially from its saturated valued $N_i^s(0)$ back to its equilibrium value N_i^0 with the time constant

$$\tau = (K_i)^{-1} = \left(\overline{v}_\mathrm{rel} \langle \sigma_i^\mathrm{total} \rangle N_\mathrm{B}\right)^{-1}, \tag{8.24}$$

which depends on the averaged total refilling cross section $\langle \sigma_i^\mathrm{total} \rangle$ and on the number density N_B of collision partners.

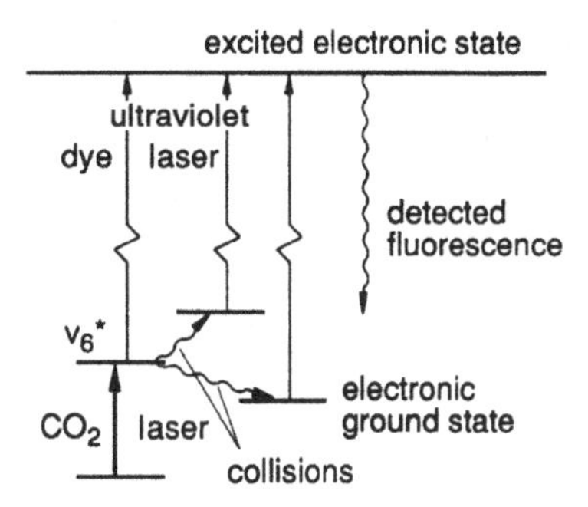

Fig. 8.18 Term diagram of IR–UV double resonance for measurements of collision-induced intramolecular vibrational transitions in the D_2CO molecule [1041]

One example of this pump-and-probe technique is the investigation of collision-induced vibrational–rotational transitions in the different isotopes HDCO and D_2CO of formaldehyde by an infrared–UV double resonance [1041]. A CO_2 laser pumpes the ν_6 vibration of the molecule (Fig. 8.18). The collisional transfer into other vibrational modes is monitored by the fluorescence intensity induced by a tunable UV dye laser with variable time delay.

The time resolution of the pump-and-probe technique is not limited by the rise time of the detectors. It can therefore be used in the pico- and femto-second range (Sect. 6.4) and is particularly advantageous for the investigation of ultrafast relaxation phenomena, such as collisional relaxation in liquids and solids [1042, 1043]. It is also useful for the detailed *real-time* study of the formation and dissociation of molecules where the collision partners are observed during the short time interval when forming or breaking a chemical bond [1044].

In Fig. 8.19 the experimental setup for such time resolved double resonance experiments is schematically shown. The laser beam is split by the beam splitter BS into a pump pulse and a probe pulse which travel along different pathlengths before they enter the sample cell. The probe pulse is sent through a Raman shifter where its wavelength can be tuned to different transitions of the sample molecule. The time delay between pump- and probe pulse can be tuned by a movable retro-reflector.

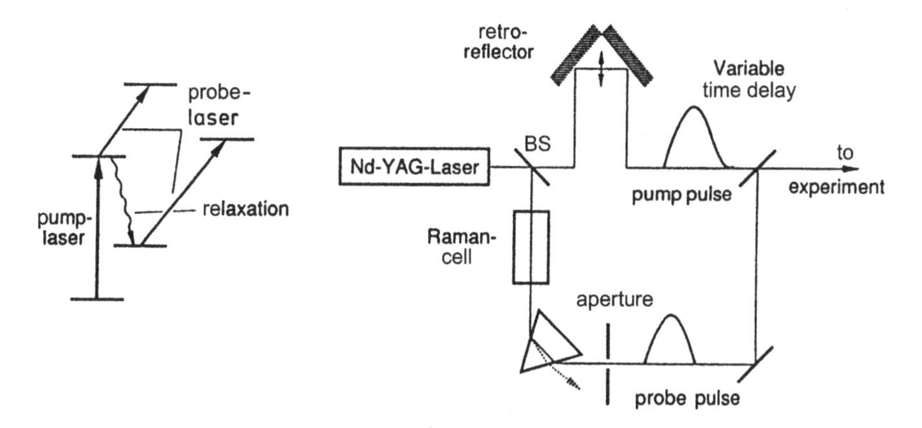

Fig. 8.19 Pump- and probe-technique for the measurement of ultrafast processes

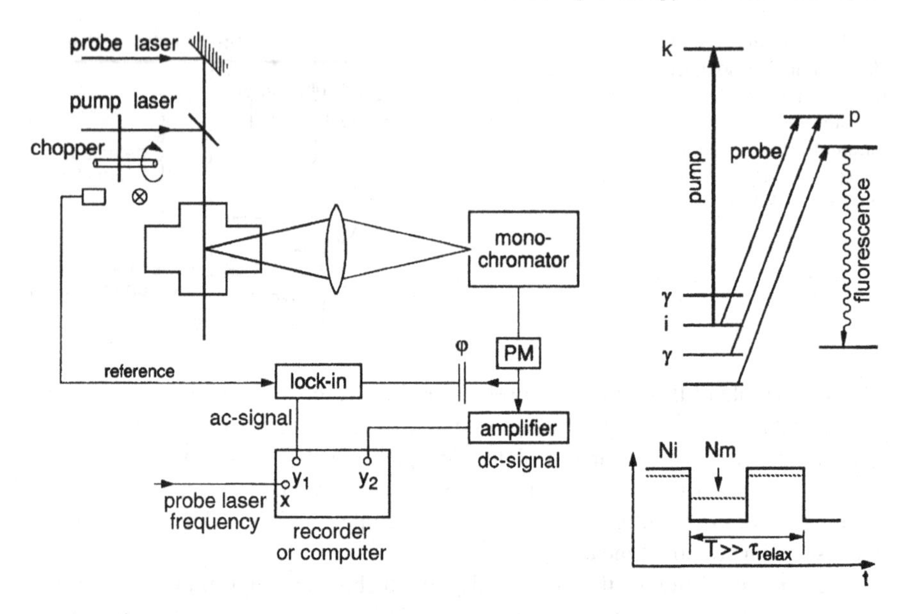

Fig. 8.20 Experimental arrangement and term diagram for measurements of individual collision-induced transitions in molecular electronic ground states with cw lasers

8.3.3 Collision Spectroscopy with Continuous-Wave Lasers

State-selective measurements of the different contributions R_{im} in (8.3, 8.22) can be performed without time resolution if the absolute value of $\sigma_i^{\text{total}} = \sum \sigma_{im}$ is known from time resolved investigations, discussed in the previous section. The pump laser may be a cw dye laser that is tuned to the wanted transition $|i\rangle \rightarrow |k\rangle$ (Fig. 8.20) and is chopped at a frequency $f \ll 1/\tau_i = K_i$. This always guarantees quasi-stationary population densities N_i^s during the on phase and N_i^0 during the off phase of the pump laser.

In this case the deviation $N_i^s - N_i^0$ from thermal equilibrium during the on phase does affect via collisional transfer the population densities N_m of neighboring levels $|m\rangle$ (8.21a). Under stationary conditions we obtain in analogy to (8.19, 8.21a–8.21b) for the on-phase of the pump laser

$$\frac{\mathrm{d}N_m}{\mathrm{d}t} = 0 = \sum_j (N_j R_{jm} - N_m R_{mj}) - \left(N_m - N_m^0\right) K_m, \tag{8.25a}$$

and for the off-phase

$$\frac{\mathrm{d}N_m^0}{\mathrm{d}t} = \sum_j N_j^0 R_{jm} - N_m^0 R_{mj} = 0, \tag{8.25b}$$

where the right side of (8.25a–8.25b) represents the relaxation of N_m toward the equilibrium population N_m^0. Inserting the deviations $\Delta N = N - N^0$ into

(8.25a–8.25b) yields with (8.19)

$$0=\sum_j(\Delta N_j R_{jm}-\Delta N_m R_{mj})-\Delta N_m K_m \quad \Rightarrow \quad \Delta N_m=\frac{\sum \Delta N_j R_{jm}}{\sum R_{mj}+K_m}. \tag{8.26}$$

This equation relates the population change ΔN_m with that of all other levels $|j\rangle$, where the sum over j also includes the optically pumped level $|i\rangle$. For $j \neq i$ the quantities ΔN_j and R_{jm} are proportional to the density N_B of the collision partners. At sufficiently low pressures we can neglect all terms $\Delta N_j \cdot R_{jm}$ for $j \neq i$ in (8.26) and obtain for the limiting case $p \to 0$ the relation

$$\frac{\Delta N_m}{\Delta N_i}=\frac{R_{im}}{\sum_j R_{mj}+K_m}. \tag{8.27}$$

The relaxation constant K_m can be obtained from a time-resolved measurement (Sect. 8.3.2). The signals from the chopped cw laser experiments give the following information: the ratio $\Delta N_m/\Delta N_i$ is directly related to the ratio of the ac absorption signals of the probe laser tuned to the transition $|i\rangle \to |p\rangle$ and $|m\rangle \to |p\rangle$, respectively, when the lock-in detector is tuned to the chopping frequency f of the pump laser. This means that the ratio

$$\frac{S_m^{\mathrm{ac}}}{S_i^{\mathrm{ac}}}=\frac{B_{mp}\Delta N_m}{B_{ip}\Delta N_i}, \tag{8.28}$$

of the ac signals is proportional to the ratio of the population changes and depends furthermore on the Einstein coefficients B_{mp} and B_{ip}.

If the time-averaged dc part of the probe laser signal

$$S_m^{\mathrm{dc}}=C B_{mp}\frac{N_m^s+N_m^0}{2}, \tag{8.29}$$

for a square-wave modulation of the pump laser is also monitored, the relative changes $\Delta N_m/N_m$ of the population densities N_m can be obtained from the relations

$$\frac{\Delta N_m/N_m^0}{\Delta N_i/N_i^0}=\frac{S_m^{\mathrm{ac}}}{S_m^{\mathrm{dc}}}\frac{S_i^{\mathrm{dc}}}{S_i^{\mathrm{ac}}}. \tag{8.30}$$

With $N_i^0/N_m^0=(g_i/g_m)\exp(-\Delta E/kT)$, the absolute values of ΔN_m can be determined from (8.30).

8.3.4 Collisions Involving Molecules in High Vibrational States

Molecules in high vibrational levels $|v\rangle$ of the electronic ground state generally show much larger collision cross sections. If the kinetic energy E_kin of the collision partners exceeds the energy difference $(E_\mathrm{D}-E_\mathrm{v})$ between the dissociation

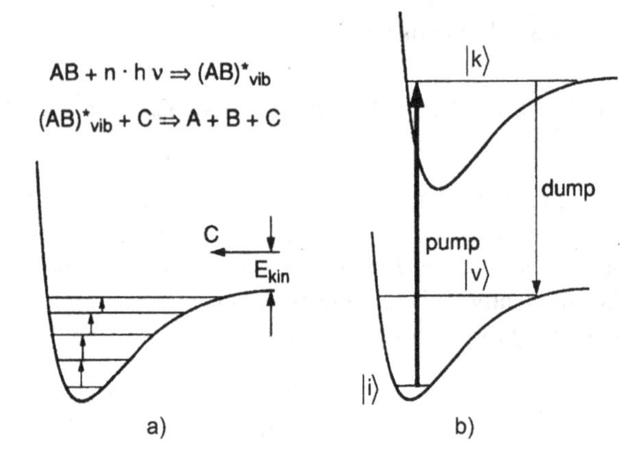

Fig. 8.21 Collision-induced dissociation of high vibrational levels populated either by multiple IR-photon absorption (**a**) or by stimulated emission pumping (**b**)

energy E_D and the vibrational energy E_v, dissociation can take place (Fig. 8.21a). Since the resulting fragments often have a larger reactivity than the molecules, the probability for the initiation of chemical reactions generally increases with increasing vibrational energy. The investigation of collisions involving vibrationally excited molecules is therefore of fundamental interest. Several spectroscopic methods have been developed for achieving a sufficient population of high vibrational levels: optical pumping with infrared lasers with one- or multiple-photon absorption is a possibility for infrared-active vibrational transitions [1045]. For homonuclear diatomics or for infrared-inactive modes, this method is not applicable.

Here, stimulated emission pumping is a powerful technique to achieve a large population in selected vibrational levels [1046, 1047]. While the pump laser is kept on the transition $|i\rangle \to |k\rangle$, the probe laser is tuned to the downward transition $|k\rangle \to |v\rangle$ (Fig. 8.21b). Proper selection of the upper level $|k\rangle$ allows one to reach sufficiently large Franck–Condon factors for the transition $|k\rangle \to |v\rangle$. With pulsed lasers a considerable fraction of the initial population N_i in the level $|i\rangle$ can be transferred into the final level $|v\rangle$.

With a coherent stimulated Raman process (STIRAP) (see Sect. 7.3), nearly the whole initial population N_i may be transferred into the final level $|v\rangle$ [1048]. Here no exact resonance with the intermediate excited level $|k\rangle$ is wanted in order to avoid transfer losses by spontaneous emission from level $|k\rangle$. The population transfer can be explained by an adiabatic passage between *dressed states* (that is, states of the molecule plus the radiation field) [1049, 1050].

Investigations of collision processes with molecules in these selectively populated levels $|v\rangle$ yields the dependence of collision cross sections on the vibrational energy for collision-induced dissociation as well as for energy transfer into other bound levels of the molecule. Since the knowledge of this dependence is essential for a detailed understanding of the collision dynamics, a large number of theoretical and experimental papers on this subject have been published. For more information the reader is referred to some review articles and the literature given therein [1039, 1045, 1051–1053].

8.4 Spectroscopy of Reactive Collisions

A detailed understanding of reactive collisions is the basis for optimization of chemical reactions that is not dependent on "trial-and-error methods." Laser-spectroscopic techniques have opened a large variety of possible strategies to reach this goal [1054]. Two aspects of spectroscopic investigations for reactive collisions shall be emphasized:

(a) By selective excitation of a reactant the dependence of the reaction probability on the internal energy of the reactants can be determined much more accurately than by measuring the temperature dependence of the reaction. This is because, in the latter case, internal energy and translational energy both change with temperature.
(b) The spectroscopic assignment of the reaction products and measurements of their internal energy distributions allow the identification of the different reaction pathways and their relative probabilities under definite initial conditions for the reactants.

The experimental conditions for the spectroscopy of reactive collisions are quite similar to those for the study of inelastic collisions. They range from a determination of the velocity-averaged reaction rates under selective excitation of reactants in cell experiments to a detailed state-to-state spectroscopy of reactive collisions in crossed molecular beams (Sect. 8.5). Some examples shall illustrate the state of the art:

The first experiments on state-selective reactive collisions were performed for the experimentally accessible reactions

$$\begin{aligned} \mathrm{Ba}+\mathrm{HF}(v=0,1) &\rightarrow \mathrm{BaF}(v=0\text{–}12)+\mathrm{H}, \\ \mathrm{Ba}+\mathrm{CO_2} &\rightarrow \mathrm{BaO}+\mathrm{CO}, \\ \mathrm{Ba}+\mathrm{O_2} &\rightarrow \mathrm{BaO}+\mathrm{O}, \end{aligned} \tag{8.31}$$

where the internal state distribution $N(v'', J'')$ of the reaction products was measured by LIF in dependence on the vibrational energy of the halogen molecule [1055, 1056].

The development of new infrared lasers and of sensitive infrared detectors has allowed investigations of reactions where the reaction-product molecules have known infrared spectra but do not absorb in the visible. An example is the endothermic reaction

$$\mathrm{Br}+\mathrm{CH_3F}(\nu_3) \rightarrow \mathrm{HBr}+\mathrm{CH_2F}, \tag{8.32}$$

where the CF stretching vibration ν_3 of CH_3F is excited by a CO_2 laser [1057]. The reaction probability increases strongly with increasing internal energy of CH_3F. If a sufficiently large concentration of excited $CH_3F(\nu_3)$ molecules is reached, energy-pooling collisions

$$\mathrm{CH_3F}^*(\nu_3)+\mathrm{CH_3F}^*(\nu_3) \rightarrow \mathrm{CH_3F}^{**}(2\nu_3)+\mathrm{CH_3F}, \tag{8.33}$$

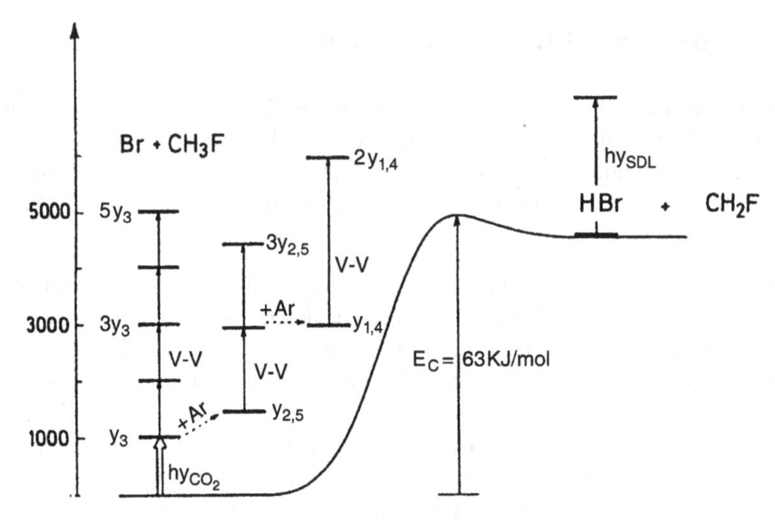

Fig. 8.22 Term diagram of the endothermic reaction $Br + CH_3F^*(\nu^3) \rightarrow HBr + CH_2F$ [1057]

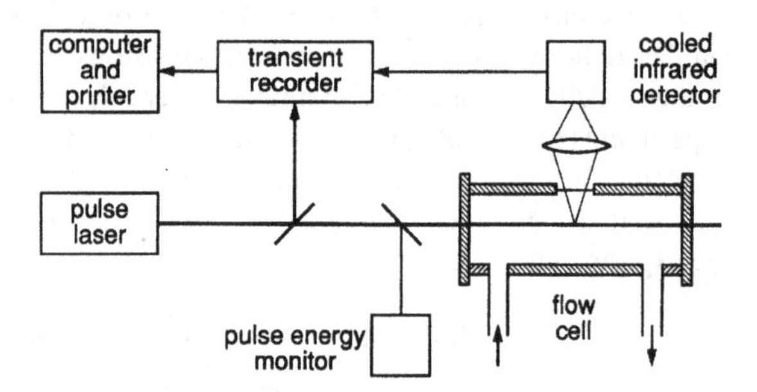

Fig. 8.23 Experimental arrangement for studies of infrared LIF in chemical reactions with spectral and time resolution

result in an increase of E_{vib} in one of the collision partners. The term diagram relevant for (8.32) is shown in Fig. 8.22. If the sum of internal plus kinetic energy is larger than the reaction barrier, excited product molecules may appear that can be monitored by their infrared fluorescence.

A typical experimental arrangement for such investigations is depicted in Fig. 8.23. In a flow system, where the reactive collisions take place, the levels (v, J) of the reactant molecules are selectively excited by a pulsed infrared laser. The time-dependent population of excited levels in the reactants or the product molecules are monitored through their fluorescence, detected by fast, cooled infrared detectors (Vol. 1, Sect. 4.5).

A longstanding controversy existed on the detailed reaction mechanism of the elementary exchange reaction of hydrogen

$$H_a + H_bH_c \rightarrow H_aH_b + H_c, \tag{8.34a}$$

where accurate ab initio calculations of the H_3 potential surface have been performed. The experimental test of theoretical predictions is, however, very demanding. First, isotope substitution has to be performed in order to distinguish between elastic and reactive scattering. Instead of (8.34a), the reaction

$$H + D_2 \rightarrow HD + D, \tag{8.34b}$$

is investigated. Second, all electronic transitions fall into the VUV range. The excitation of H or D_2 and the detection of HD or D require VUV lasers. Third, hydrogen atoms have to be produced since they are not available in bottles.

The first experiments were carried out in 1983 [1058, 1059]. The H atoms were produced by photodissociation of HJ molecules in an effusive beam using the fourth harmonics of Nd:YAG lasers. Since the dissociated iodine atom is found in the two fine-structure levels $I(P_{1/2})$ and $I(P_{3/2})$, two groups of H atoms with translational energies $E_{\text{kin}} = 0.55$ eV or 1.3 eV in the center-of-mass system $H + D_2$ are produced. If the slower H atoms collide with D_2 they can reach vibrational–rotational excitation energies in the product molecule up to $(v = 1,\ J = 3)$, while the faster group of H atoms can populate levels of HD up to $(v = 3,\ J = 8)$. The internal-state distribution of the HD molecules can be monitored either by CARS (Sect. 3.3) or by resonant multiphoton ionization [1058]. Because of their fundamental importance, these measurements have been repeated by several groups with other spectroscopic techniques that have improved signal-to-noise ratios [1060].

An infrared laser-based method for investigating the nascent state distribution of the reaction products HF from the reaction $F + H_2 \rightarrow HF + H$ in crossed molecular beams under single collision conditions has been presented by the group of D.J. Nesbit [1061]. The experimental setup (Fig. 8.24) consisted of a pulsed supersonic discharge source of F atoms, which collide with H_2 molecules in a second pulsed jet source. The product $HF(v, J)$ is probed in the intersection region by the absorption of a single-mode tunable IR laser, where full vibration–rotation resolution can be achieved (Fig. 8.25).

Ion–molecule reactions, which play an important role in interstellar clouds and in many chemical and biological processes, have increasingly caught the attention of spectroscopists. As a recently investigated example, the charge-exchange reaction

$$N^+ + CO \rightarrow CO^+ + N,$$

is mentioned, where the rotational distribution $N(v, J)$ of CO^+ for different vibrational levels v has been measured by infrared fluorescence (Fig. 8.26) in dependence on the kinetic energy of the N^+ ions [1062].

Photochemical reactions are often initiated by direct photodissociation or by collision-induced dissociation of a laser-excited molecule, where radicals are formed as intermediate products, which further react by collisions. The dynamics of photodissociation after excitation of the parent molecule by a UV laser has therefore been studied thoroughly [1063]. While the first experiments were restricted to measurements of the internal-state distribution of the dissociation products, later more refined arrangements also allowed the determination of the angular distribution and of the orientation of the products for different polarizations of the photodissociating

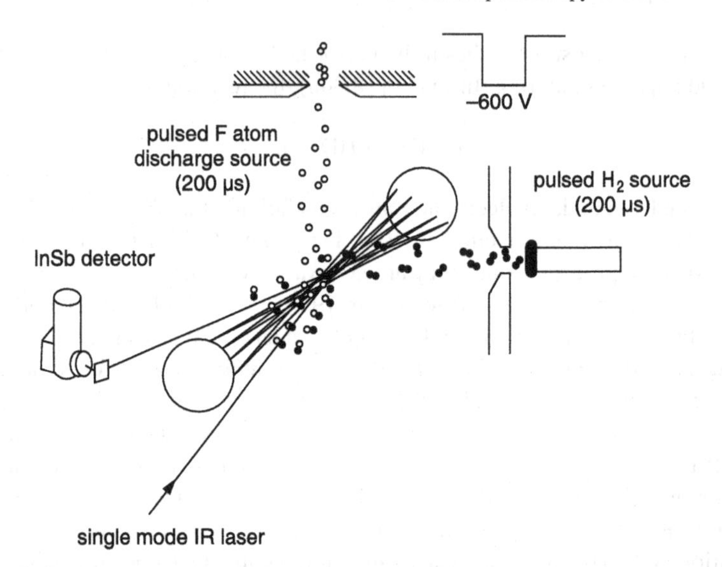

Fig. 8.24 Schematic diagram of the crossed-jet direct-absorption reactive scattering experiment. Fluorine atoms produced in a discharge pulsed jet expansion are intersected at 90° 4.5 cm downstream with a pulse of supersonically cooled H_2. Tunable single-mode IR laser light is multipassed perpendicular to the collision plane and probes HF(v, J) products by direct absorption [1061]

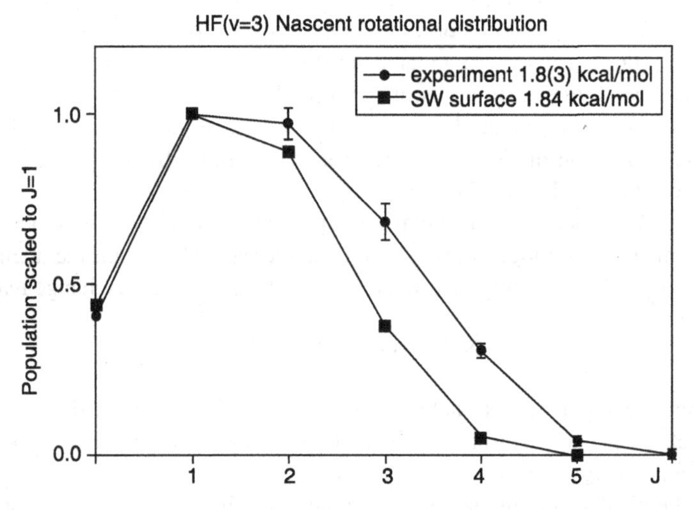

Fig. 8.25 Rotational distribution of HF(v = 3) reaction products from the reaction $F + H_2 \rightarrow HF + H$ [1061]

laser [1064–1066]. The technique is illustrated by the example

$$\mathrm{ICN} \underset{248\ \mathrm{nm}}{\overset{h\nu}{\rightarrow}} \mathrm{CN} + \mathrm{I}$$

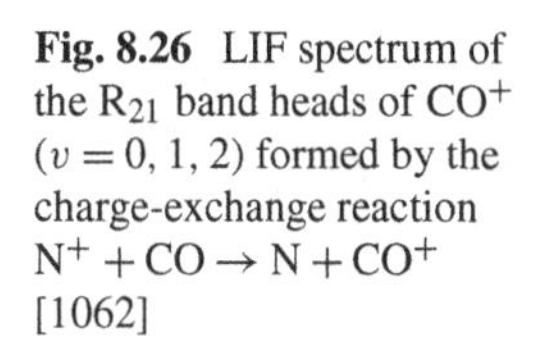

Fig. 8.26 LIF spectrum of the R_{21} band heads of CO^+ ($v = 0, 1, 2$) formed by the charge-exchange reaction $N^+ + CO \rightarrow N + CO^+$ [1062]

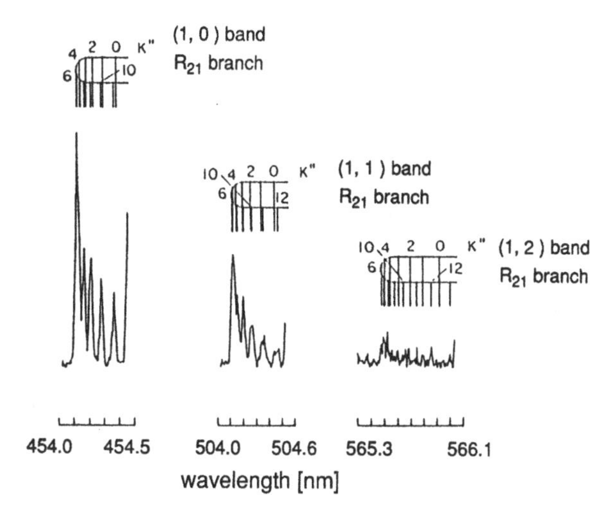

which was investigated with the apparatus depicted in Fig. 8.27. The ICN molecules are dissociated by the circularly polarized radiation of a KrF laser at $\lambda = 248$ nm. The orientation of the CN fragments is monitored through the fluorescence induced by a polarized dye laser. The dye laser is tuned through the $B^2\Sigma \leftarrow X^2\Sigma^+$ systen of CN. This circular polarization is periodically switched between σ^+ and σ^- by a photoelastic polarization modulator (PEM). The corresponding change in the fluorescence intensity is a measure of the orientation of the CN fragments [1067].

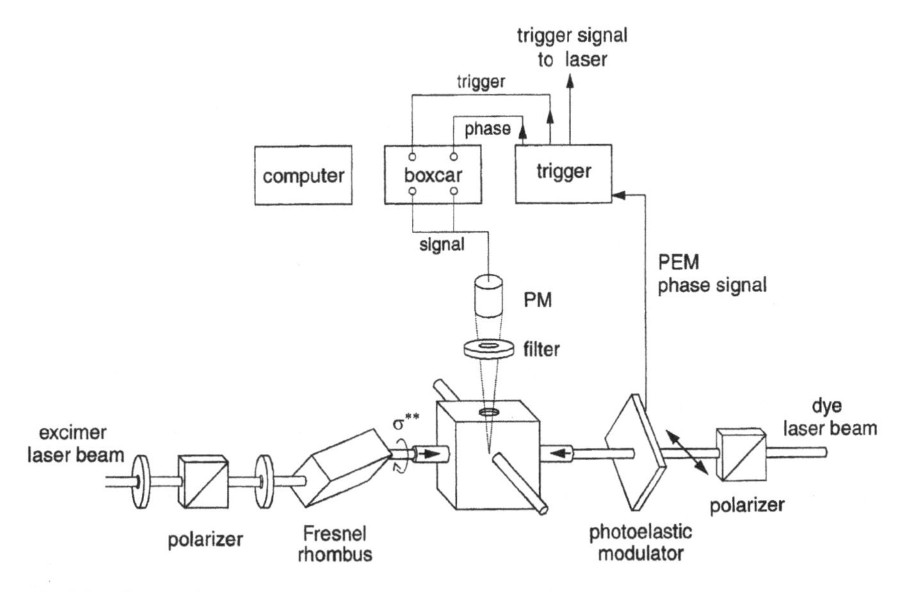

Fig. 8.27 Schematic experimental arrangement for measurements of the fragment orientation after photodissociation of the parent molecule [1067]

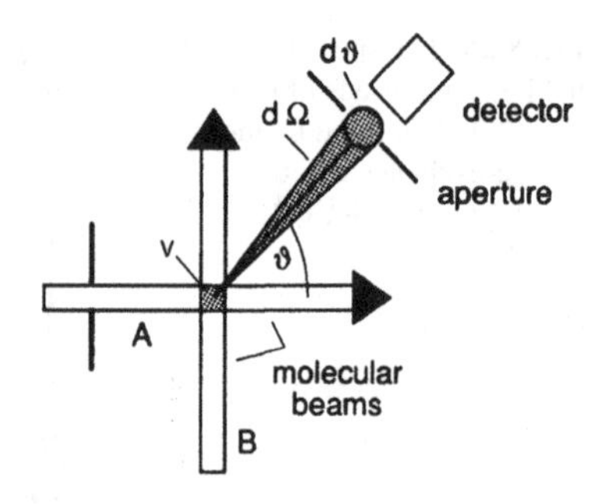

Fig. 8.28 Schematic diagram for the measurement of differential cross sections in crossed molecular beams

A famous example is the photodissociation of H_2O leading to a preferentially populated Λ component of the OH fragment, which forms the basis of the astronomical OH maser observed in interstellar clouds [1068]. There are many more examples of laser spectroscopy of chemical reactions, some of which can be found in [1069–1073].

8.5 Spectroscopic Determination of Differential Collision Cross Sections in Crossed Molecular Beams

The techniques discussed in Sects. 8.2–8.4 allowed the measurement of absolute rate constants of selected collision-induced transitions, which represent *integral* inelastic cross sections integrated over the angular distribution and averaged over the thermal velocity distribution of the collision partners. Much more detailed information on the interaction potential can be extracted from measured *differential* cross sections, obtained in crossed molecular beam experiments [958, 1074, 1075].

Assume that atoms or molecules A in a collimated beam collide with atoms or molecules B within the interaction volume V formed by the overlap of the two beams (Fig. 8.28). The number of particles A scattered per second by the angle θ into the solid angle $\mathrm{d}\Omega$ covered by the detector D is given by

$$\frac{\mathrm{d}N_\mathrm{A}(\theta)}{\mathrm{d}t}\,\mathrm{d}\Omega = n_\mathrm{A} v_\mathrm{r} n_\mathrm{B} V \frac{\mathrm{d}\sigma(\theta)}{\mathrm{d}\Omega}\,\mathrm{d}\Omega, \tag{8.35}$$

where n_A is the density of the incident particles A, n_B the density of particles B, v_r is the relative velocity, and $(\mathrm{d}\sigma(\theta)/\mathrm{d}\Omega)$ the differential scattering cross section. The scattering angle ϑ is related to the impact parameter b and the interaction potential. It can be calculated for a given potential $V(r)$ [1076]. Differential cross sections probe definite local parts of the potential, while integral cross sections only reflect the global effect of the potential on the deflected particles, averaged over all impact parameters. We will discuss which techniques can be used to measure differential cross sections for elastic, inelastic, and reactive collisions.

With classical techniques, the energy loss of the particle A during an inelastic collision is determined by measuring its velocity before and after the collision with

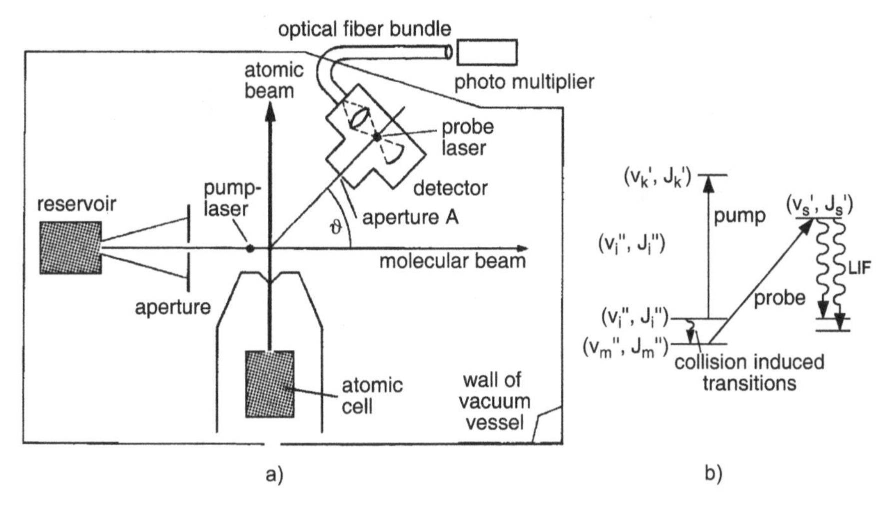

Fig. 8.29 (**a**) Experimental arrangement for the spectroscopic determination of state-to-state differential cross sections [1078]. (**b**) Level scheme for pump-probe experiment

velocity selectors or by time-of-flight measurements [991]. Since the velocity resolution $\Delta v/v$ and therefore the minimum detectable energy loss ΔE_{kin} is restricted by technical limitations, this method can be applied to only a limited number of real problems.

For polar molecules with an electric dipole moment, Rabi spectrometers with electrostatic quadrupole deflection fields offer the possibility to detect rotationally inelastic collisions since the focusing and deflecting properties of the quadrupole field depend on the quantum state $|J, M\rangle$ of a polar molecule. This technique is, however, restricted to molecules with a large electric dipole moment and to small values of the rotational quantum number J [1076].

Applications of laser-spectroscopic techniques can overcome many of the above-mentioned limitations. The energy resolution is higher by several orders of magnitude than that for time-of-flight measurements. In principle, molecules in arbitrary levels $|v, J\rangle$ can be investigated, as long as they have absorption transitions within the spectral range of available lasers. The essential progress of laser spectroscopy is, however, due to the fact that besides the scattering angle θ, the initial and final quantum states of the scattered particle A can also be determined. In this respect, laser spectroscopy of collisions in crossed beams represents the *ideal complete scattering experiment* in which all the relevant parameters are measured.

The technique is illustrated by Fig. 8.29. A collimated supersonic beam of argon, which contains Na atoms and Na_2 molecules, is crossed perpendicularly with a noble gas beam [1077]. Molecules of Na_2 in the level $|v''_m J''_m\rangle$ that have been scattered by the angle ϑ are monitored by a *quantum-state-specific detector*. It consists of a cw dye laser focused into a spot behind the aperture A and tuned to the transition $|v''_m, J''_m\rangle \rightarrow |v'_j, J'_j\rangle$, and an optical system of a mirror and lens, which collects the LIF through an optical fiber bundle onto a photomultiplier. The entire detector can be turned around the scattering center.

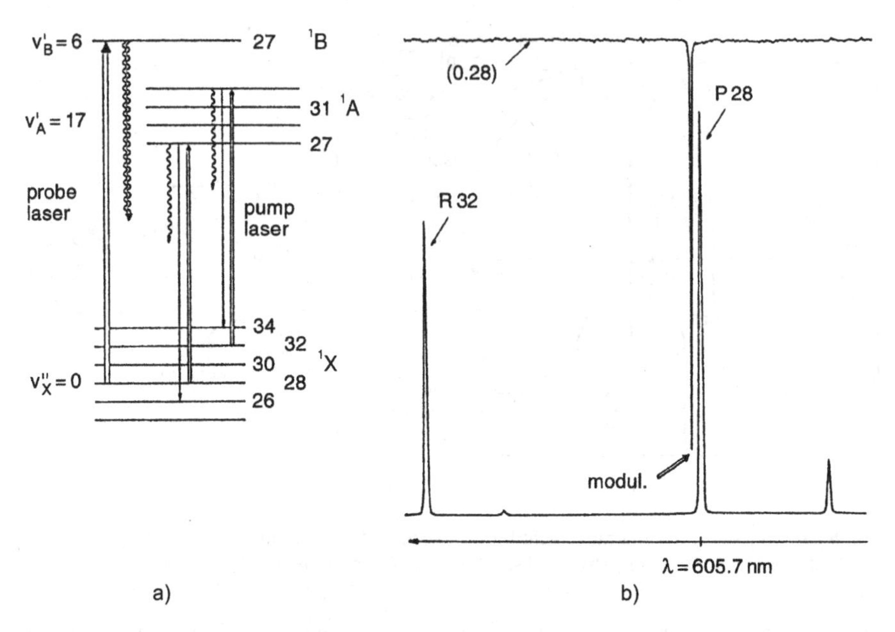

Fig. 8.30 (**a**) Level diagram of Na_2 for optical depletion of the levels $(v'', J'') = (0, 28)$ and optical probing of collisionally populated levels $(v'', J'' + \Delta J)$. (**b**) The measured spectrum in the *right part* gives the experimental proof that a level (v'', J'') can be completely depleted by optical pumping. The *lower spectrum* shows the pump laser-induced fluorescence I_{Fl} (L_1) when the laser L_1 is tuned, the *upper trace* the probe laser-induced fluorescence, where the probe laser was stabilized onto the transition $X(v'' = 0, J'' = 28) \rightarrow B(v' = 6, J' = 27)$. When the pump laser is tuned over the transition starting from $X(v'' = 0, J'' = 28)$ the probe laser-induced fluorescence drops nearly to zero. The *lower spectrum* has been shifted by 2 mm to the right [1079]

We define θ as the scattering angle in the center-of-mass system (while ϑ is that in the laboratory system). The detected intensity of the LIF is a measure of the scattering rate $[\mathrm{d}N(v''_m, J''_m, \theta)/\mathrm{d}t]\,\mathrm{d}\Omega$ which has two contributions:

(a) the elastically scattered molecules $(v''_m, J''_m) \rightarrow (v''_m, J''_m, \theta)$, and
(b) the sum of all inelastically scattered molecules which have suffered collision-induced transitions $\sum_n [(v''_n, J''_n) \rightarrow (v''_m, J''_m)]$.

In order to select molecules with definite initial and final states, an OODR method is used (Sect. 5.4). Shortly before they reach the scattering volume, the molecules pass through the beam of a pump laser that induces transitions $(v''_i, J''_i) \rightarrow (v'_k, J'_k)$ and depletes the lower level (v''_i, J''_i) by optical pumping (Fig. 8.30). If the scattering rate is measured by the detector alternately with the pump laser on and off, the difference of the two signals just gives the contribution of those molecules with initial state (v''_i, J''_i), final state (v''_m, J''_m), and scattering angle θ [1078].

Since the scattering angle θ is related to the impact parameter b, the experiment indicates which impact parameters contribute preferentially to collision-induced vibrational or rotational transitions, and how the transferred angular momentum ΔJ

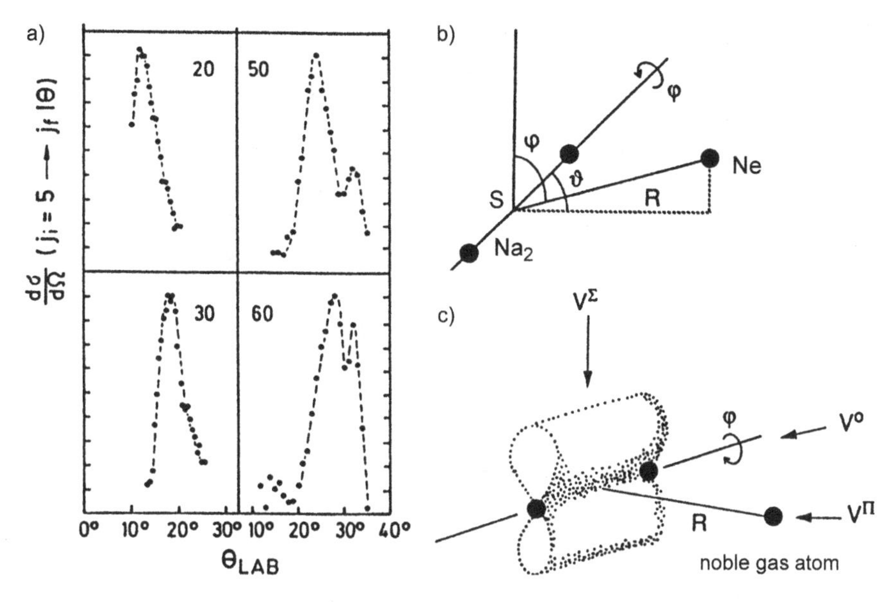

Fig. 8.31 (**a**) Differential cross sections for collision-induced notational transitions $J'' \to J'' + \Delta J$ for different values of ΔJ, as a function of the scattering angle in the center of mass system; (**b**, **c**) dependence of interaction potential in the orientation of the molecule against the line between atom and molecule

depends on impact parameters, initial state (v_i'', J_i''), and collision partner B [1079]. In Fig. 8.31 some differential cross sections $\sigma(\theta)$ for inelastic Na_2 + Ne collisions are plotted versus the scattering angle θ. Note that angular momentum transfer of up to $\Delta J = 55\hbar$ has been observed. The analysis of the measured data yields very accurate interaction potentials.

Of particular interest is the dependence of collision cross sections on the vibrational state, since the probability of inelastic and reactive collisions increases with increasing vibrational excitation [1080]. Stimulated emission pumping or STIRAP (see Sect. 7.3) before the scattering center allows a large population transfer into selected, high vibrational levels of the electronic ground state. For some molecules, such as Na_2 or I_2, an elegant realization of this technique is based on a molecular beam dimer laser [1081], which uses the molecules in the primary beam as an active medium pumped by an argon laser or a dye laser (Fig. 8.32). Threshold pump powers below 1 mW (!) could indeed be achieved.

Optical pumping also allows measurements of differential cross sections for collisions where one of the partners is electronically excited. The interaction potential $V(A^*, B)$ could be determined in this way for alkali–noble gas partners, which represent exciplexes since they are bound in an excited state but dissociate in the unstable ground state. One example is the investigation of K + Ar collision partners [1082]. In the crossing volume of a potassium beam and an argon beam, the potassium atoms are excited by a cw dye laser into the $4P_{3/2}$ state. The scattering rate for K atoms is measured versus the scattering angle θ with the pump laser on and

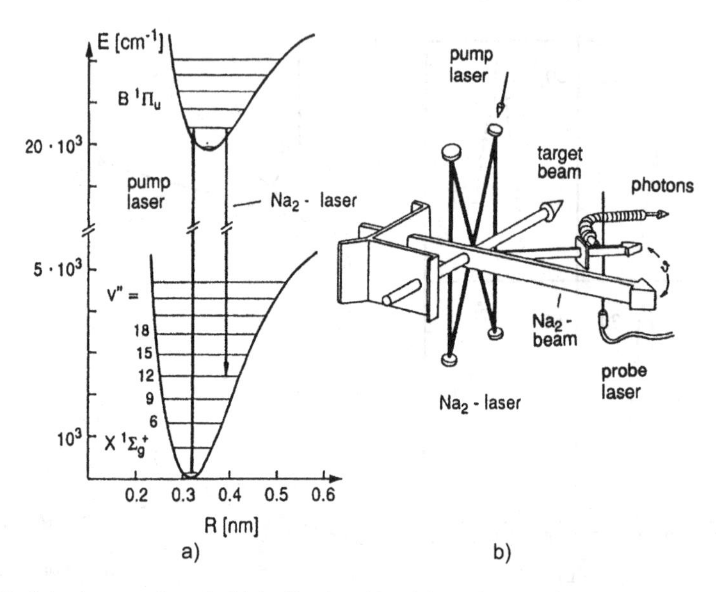

Fig. 8.32 Selective pumping of a high vibrational level (v, J) by stimulated emission pumping in Na_2 with a Na_2 dimer laser: (**a**) level scheme; (**b**) experimental realization [1081]

off. The signal difference yields the contribution of the excited atoms to the elastic scattering if the fraction of excited atoms can be determined, which demands a consideration of the hyperfine structure. In order to reach the maximum concentration of excited K atoms, either a single-mode circularly polarized dye laser must be used, which is tuned to the transition $F'' = 2 \rightarrow F' = 3$ between selected hfs components, or a broad-band laser, which pumps all allowed hfs transitions simultaneously.

An example of measurements of *inelastic* collisions between excited atoms A^* and molecules M in crossed beams is the experiment by Hertel et al. [1083], where laser-excited $Na^*(3P)$ atoms collide with molecules such as N_2 or CO and the energy transfer

$$A^*M(v'' = 0) \rightarrow A + M(v'' > 0) + \Delta E_{kin}, \tag{8.36}$$

is studied. If the scattering angle θ and the velocity of the scattered Na atoms are measured, kinematic considerations (energy and momentum conservation) allow the determination of the fraction of the excitation energy that is converted into internal energy of M and that is transferred into kinetic energy ΔE_{kin}. The experimental results represent a crucial test of ab initio potential surfaces $V(r, \theta, \phi)$ for the interaction potential $Na^* - N_2$ $(v = 1, 2, 3, \ldots)$ [1084].

Scattering of electrons, fast atoms, or ions with laser-excited atoms A^* can result in elastic, inelastic, or superelastic collisions. In the latter case, the excitation energy of A^* is partly converted into kinetic energy of the scattered particles. Orientation of the excited atoms by optical pumping with polarized lasers allows investigations of the influence of the atomic orientation on the differential cross sections for $A^* + B$ collisions, which differs for collisions with electrons or ions from the case of neutral atoms [1085]. An example of the measurement of differential cross sections of the reactive collisions $Na^* + HF$ is given in [1086].

Measurements of differential cross-sections for reactive collisions allow the determination of reaction rates and the internal state distributions of reaction products as well as their dependence on scattering angle and internuclear distances during the reaction. This gives a much more detailed insight into the time sequence of the reaction during the approach of the reactants. Therefore, many experiments have been performed where the quantum states of the reactants have been selected by laser excitation, the product state distribution by LIF and the scattering angle by using crossed beam arrangements and angular resolved detection. One example is the observation of resonances in the cross-section for the reaction $F + HD \rightarrow FH + D$ or $FD + H$, where a virtual level of the transition state opens up the specific reaction channel for the formation of $HF(v = 3)$ in a specific quantum state [1087].

From the measured angular distribution of the reaction products IF formed in the reaction $F + C_2H_5I \rightarrow IF + C_2H_5$, it can be concluded that the transition complex during this reaction has a lifetime of at least 2–3 rotational periods [1088].

8.6 Photon-Assisted Collisional Energy Transfer

If two collision partners A and M absorb a photon $h\nu$ at a relative distance R_c, one of the partners may remain in an excited state after the collision:

$$A + M + h\nu \rightarrow A^* + M. \tag{8.37}$$

This reaction can be regarded as the three step process:

$$A + M \rightarrow (AM), \tag{8.38}$$

$$(AM) \rightarrow h\nu \rightarrow (AM)^*, \tag{8.39}$$

$$(AM)^* \rightarrow A^* + M. \tag{8.40}$$

The two collision partners form a collision complex that is excited by absorption of the photon $h\nu$ at a relative distance R_c, where the energy difference $\Delta E = E(AM)^* - E(AM)$ between the two potential curves just equals the photon energy $h \cdot \nu$ (Fig. 8.33a) and the Franck–Condon factor has a maximum value [1089]. This complex decays after a short time into $A^* + M$.

If the potential is not monotonic, there are generally two trajectories with different impact parameters that lead to the same deflection angle θ in the center of mass system (Fig. 8.33b).

In a similar way collisions with excited atoms can be studied. The energy transfer by inelastic collisions between excited atoms A^* and ground-state atoms B

$$A^* + B \rightarrow B^* + A + \Delta E_{kin}, \tag{8.41}$$

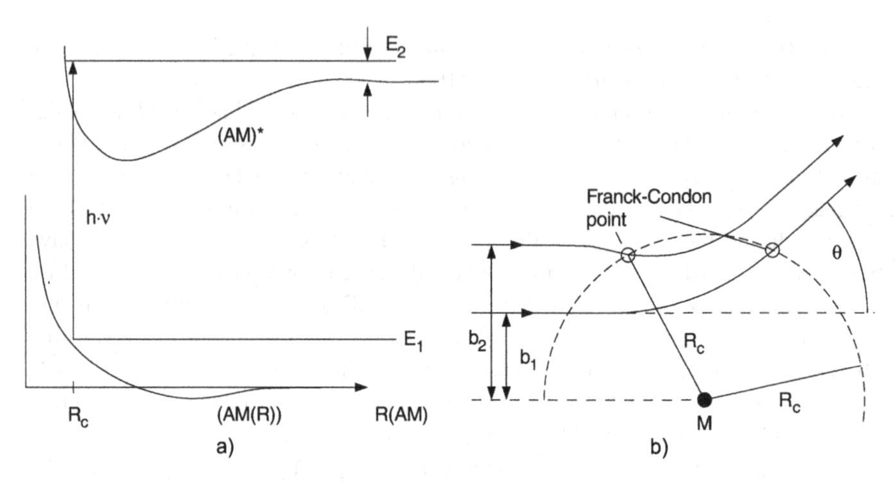

Fig. 8.33 Schematic diagram of "optical collisions": (**a**) potential curves of ground and excited state of the collision complex; (**b**) classical trajectories in the center-of-mass system for two different impact parameters b_1, b_2

is governed by the conservation of energy and momentum. The energy difference $\Delta E_{\text{int}} = E(\text{A}^*) - E(\text{B}^*)$ is transferred into kinetic energy ΔE_{kin} of the collision partners. For $\Delta E_{\text{kin}} \gg KT$, the cross section σ for the reaction (8.41) becomes very small, while for near-resonant collisions ($\Delta E_{\text{kin}} \ll KT$), σ may exceed the gas kinetic cross section by several orders of magnitude.

If such a reaction (8.41) proceeds inside the intense radiation field of a laser, a photon may be absorbed or emitted during the collision, which may help to satisfy energy conservation for small ΔE_{kin} even if ΔE_{int} is large. Instead of (8.41), the reaction is now

$$\text{A}^* + \text{B} \pm \hbar\omega \rightarrow \text{B}^* + \text{A} + \Delta E_{\text{kin}}. \tag{8.42}$$

For a suitable choice of the photon energy $\hbar\omega$, the cross section for a nonresonant reaction (8.41) can be increased by many orders of magnitude through the help of the photon, which makes the process near resonant. Such *photon-assisted* collisions will be discussed in this section.

In the *molecular* model of Fig. 8.34a the potential curves $V(R)$ are considered for the collision pairs $\text{A}^* + \text{B}$ and $\text{A} + \text{B}^*$. At the critical distance R_c the energy difference $\Delta E = V(\text{AB}^*) - V(\text{A}^*\text{B})$ may be equal to $\hbar\omega$. Resonant absorption of a photon by the collision pair $\text{A}^* + \text{B}$ at the distance R_c results in a transition into the upper potential curve $V(\text{AB}^*)$, which separates into $\text{A} + \text{B}^*$. The whole process (8.42) then leads to an energy transfer from A^* to B^*, where the initial and final kinetic energy of the collision partners depends on the slope $\text{d}V/\text{d}R$ of the potential curves and the internuclear distance R_c where photon absorption takes place.

If we start with $\text{A} + \text{B}^*$ the inverse process of stimulating photon emission will result in a transition from the upper into the lower potential, which means an energy transfer from B^* to A^* for the separated atoms.

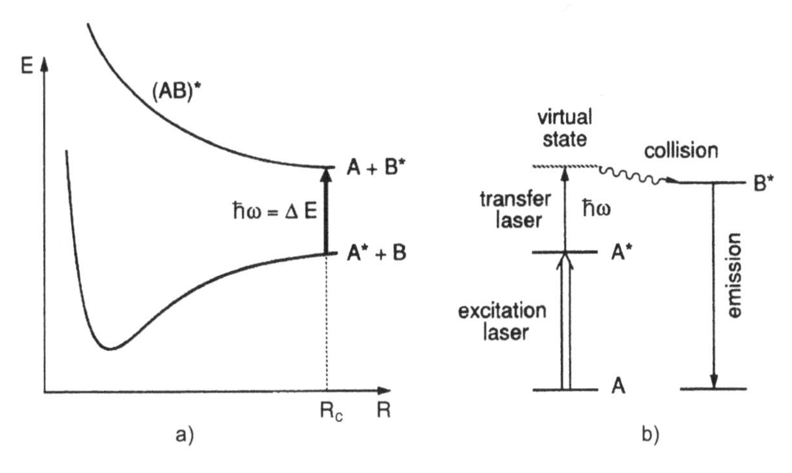

Fig. 8.34 Photon-assisted collisional energy transfer: (**a**) molecular model; (**b**) dressed-atom model

For the experimental realization of such photon-assisted collisional energy transfer one needs two lasers: the pump laser L_1 excites the atoms A into the excited state A^* and the transfer laser L_2 induces the transition between the two potential curves $V(A^*B) \rightarrow V(AB^*)$.

In the *atomic model* (Fig. 8.34b, often called the *dressed-atom model*) [1090] the excited atom A^* absorbes a photon $\hbar\omega$ of the transfer laser, which excites A^* further into a *virtual state* $A^* + \hbar\omega$ that is near resonant with the excited state of B^*. The collisional energy transfer $A^* + \hbar\omega \rightarrow B^* + \Delta E_{\text{kin}}$ with $\Delta E_{\text{kin}} \ll \hbar\omega$ then proceeds with a much higher probability than the off-resonance process without photon absorption [1091].

The first experimental demonstration of photon-assisted collisional energy transfer was reported by Harris and coworkers [1092], who studied the process

$$\begin{aligned}\text{Sr}^*\left(5^1P\right) + \text{Ca}\left(4^1S\right) + \hbar\omega &\rightarrow \text{Sr}\left(5^1S\right) + \text{Ca}\left(5^1D\right),\\ &\rightarrow \text{Sr}\left(5^1S\right) + \text{Ca}\left(4p^{21}S\right).\end{aligned}$$

The corresponding term diagram is depicted in Fig. 8.35. The strontium atoms are excited by a pulsed pump laser at $\lambda = 460.7$ nm into the level $5s5p\,^1P_1^0$. During the collision with a Ca atom in its ground state the collision pair $\text{Sr}(5^1P_1^0)$–$\text{Ca}(4^1S)$ absorbs a photon $\hbar\omega$ of the transfer laser at $\lambda = 497.9$ nm. After the collision the excited $\text{Ca}^*(4p^{21}S)$ atoms are monitored through their fluorescence at $\lambda = 551.3$ nm.

It is remarkable that the transitions $4s^{21}S_1 \rightarrow 4p^{21}S$ and $4s^{21}S \rightarrow 4p^{21}D$ of the Ca atom do not represent allowed dipole transitions and are therefore forbidden for the isolated atom. Regarding the photon-absorption probability, the absorption by a collision pair with subsequent energy transfer to the atom is therefore called *collision-induced absorption* or collision-aided radiative excitation [1093], where a *dipole-allowed* transition of the molecular collision pair A^*B or AB^* takes place.

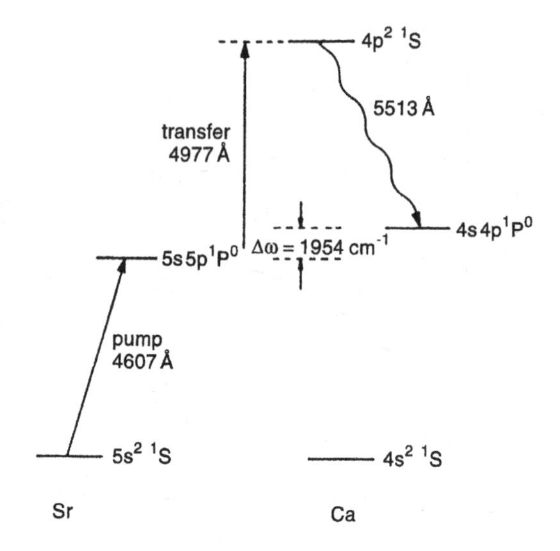

Fig. 8.35 Term diagram for photon-assisted collisional energy transfer from $Sr^*(^1P^0)$ to $Ca(4p^{2\,1}S)$ [1092]

The molecular transition has a dipole transition moment $\mu(R)$ that depends on the internuclear separation $R(A^*B)$ and that approaches zero for $R \to \infty$ [1094]. It is caused by the induced dipole–dipole interaction from the polarizability of the collision partners.

Such collision-induced absorption of radiation is important not only for the initiation of chemical reactions (Sect. 10.2), but also plays an important role in the absorption of infrared radiation in planetary atmospheres and interstellar molecular clouds. By the formation of H_2–H_2 collision pairs, for instance, vibration–rotational transitions within the electronic ground state of H_2 become allowed although they are forbidden in the isolated homonuclear H_2 molecule [1095, 1096].

An interesting aspect of collision-aided radiative excitation is its potential for optical cooling of vapors. Since the change in kinetic energy of the collision partners per absorbed photon can be much larger than that transferred by photon recoil (Sect. 9.1), only a few collisions are necessary for cooling to low temperatures compared with a few thousand for recoil cooling [1093].

Detailed investigations of the transfer cross section and its dependence on the wavelength λ_2 of the transfer laser L_2 have been performed by Toschek and coworkers [1097], who studied the reaction $Sr(5s5p) + Li(2s) + \hbar\omega \to Sr(5s^2) + Li(4d)$. The experimental setup is shown in Fig. 8.36. A mixture of strontium and lithium vapors is produced in a heat pipe [980]. The beams of two dye lasers pumped by the same N_2 laser are superimposed and focused into the center of the heat pipe. The first laser L_1 at $\lambda_1 = 407$ nm excites the Sr atoms. The pulse of the second laser L_2 is delayed by a variable time interval Δt and its wavelength λ_2 is tuned around $\lambda_1 = 700$ nm. The intensity $I_{Fl}(\lambda_2, \Delta t)$ of the fluorescence emitted by the excited Li^* atoms is monitored in the transition $Li(d^2D \to 2p^2P)$ at $\lambda = 610$ nm as a function of the wavelength λ_2 and the delay time Δt. For the resonance case λ_2^{res} (that is, $\Delta E_{kin} = 0$ in (8.42)) energy transfer cross sections up to $\sigma_T > 2 \times 10^{-13}\, I \cdot cm^2$ (where I is measured in MW/cm^2) have been determined. For sufficiently high

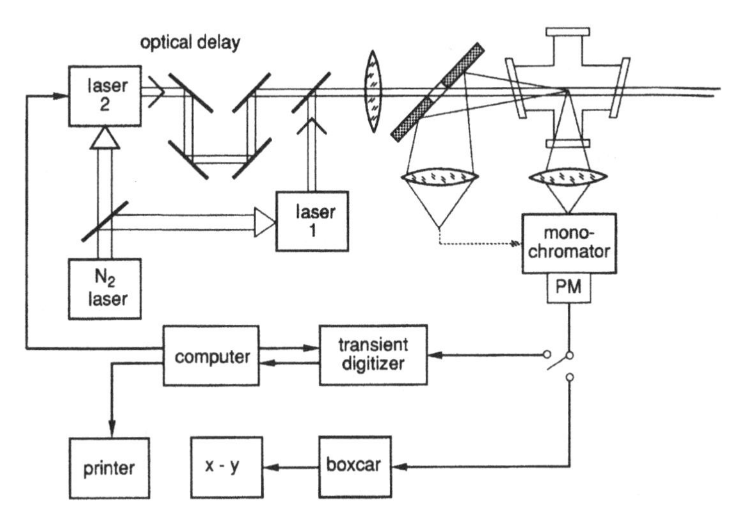

Fig. 8.36 Experimental setup for studies of photon-assisted collisional energy transfer [1097]

intensities I of the transfer laser, the photon-assisted collisional transfer cross sections σ_T exceed the gas-kinetic cross sections $\sigma \approx 10^{-15}$ cm^2 by 2–3 orders of magnitude.

From the angular distribution of K(^2P) atoms formed in the optical collision

$$\mathrm{K}\big((4s)^{22}P\big) + \mathrm{Ar} + h\nu \rightarrow \mathrm{K}\big((4p)^{22}P\big) + \mathrm{Ar},$$

the repulsive potential curves of the unstable molecule KAr($X^2\Sigma$) and KAr($B^2\Sigma$) could be obtained with an accuracy of about 1 % [1098].

Example 8.1 For saturation of the pump transition even moderate intensities ($I_p < 10^3$ kW/cm^2) are sufficient (Sect. 2.1). Assume an output power of 50 kW of the transfer laser L_2 and a beam diameter of 1 mm in the focal plane, which yields $I_{\mathrm{transf}} = 6.7$ MW/cm^2, energy transfer cross sections of $\sigma_T \approx 1.3 \times 10^{-12}$ cm^2, which is about 1000 times larger than the gas-kinetic cross section. The energy transfer $R_T = \sigma n \overline{v} h\nu_T$, where n is the density of collision partners, colliding with the excited atoms. With $n = 10^{16}$ cm^3, $\overline{v} = 10^5$ cm/s and $h\nu_T = 2$ eV, we obtain $R_T = 10^9$ eV/s.

For the photoassociation of the collision pair Na + Kr a crossed beam experiment was performed (Fig. 8.37). The Na-atoms were excited by a laser into the $3p$ state. These excited atoms collide with Kr-atoms. The collision pair was further excited by a second laser anticollinear with the first laser into a bound Rydberg state of the NaKr dimer which could decay by fluorescence into a bound $A^2\Pi$ state of the NaKr molecule [1099].

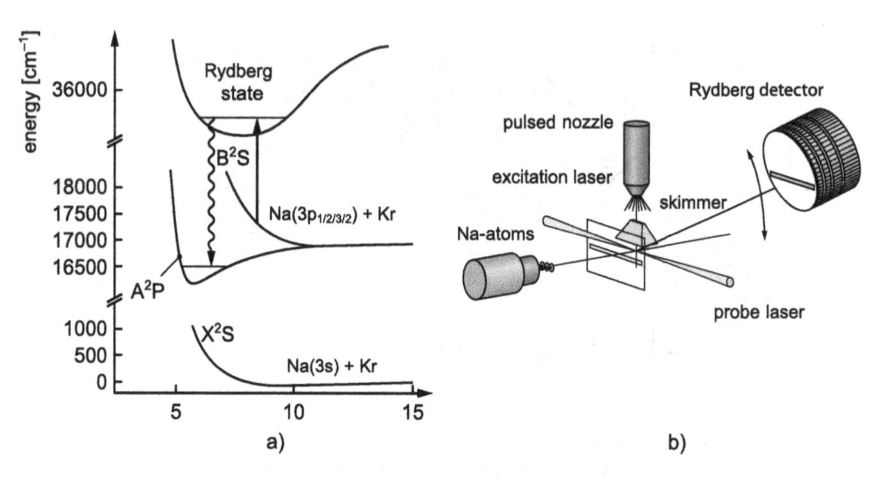

Fig. 8.37 Photoassociation of collision pairs. (**a**) Energy level scheme for the excitation of a collision pair into bound Rydberg states, (**b**) experimental arrangement with crossed atomic beams [1099]

Further information on this interesting field and its various potential application can be found in [1100, 1101]. A compendium of data on Atomic Collisions up to 1994 is compiled in [1103]. A chronicle of fifty years on the dynamics of molecular processes is given in [1104].

8.7 Problems

8.1 Assume an inelastic cross section of $\sigma^{\text{inel}} = 10^{-15}$ cm^2 for collisions between excited atoms A* and partners B, and a velocity-changing cross section $\sigma_v(\Delta v) = \sigma_0 \exp(-\Delta v^2/\langle v^2\rangle)$ for velocity changes Δv of atom A. Discuss for $\sigma_0 = 10^{-14}$ cm^2 the line profile of the Lamb dip in the transition A $\rightarrow$ A* if the radiative lifetime is $\tau_{\text{rad}} = 10^{-8}$ s, the pressure of partners B is 1 mbar, the temperature $T = 300$ K, the saturation parameter $S = 1$, and the reduced mass $\mu(\text{AB}) = 40$ AMU.

8.2 The effective lifetime of an excited molecular level is $\tau_{\text{eff}}(p = 5\ \text{mbar}) = 8 \times 10^{-9}$ s and $\tau_{\text{eff}}(p = 1\ \text{mbar}) = 12 \times 10^{-9}$ s for molecules with the mass $M = 43$ AMU in a gas cell with argon buffer gas at $T = 500$ K. Calculate the radiative lifetime, the collision-quenching cross section, and the homogeneous linewidth $\Delta\nu(p)$.

8.3 A single-mode laser beam with a Gaussian profile ($w = 1$ mm) and a power of 10 mW, tuned to the center ω_0 of the sodium line ($3^2S_{1/2} \rightarrow 3^2P_{1/2}$) passes through a cell containing sodium vapor at $p = 10^{-3}$ mbar at a temperature of 450 K. The absorption cross section is $\sigma_{\text{abs}} = 5 \times 10^{-11}$ cm^2, the natural linewidth is $\delta\nu_n = 10$ MHz, the collision cross section $\sigma_{\text{col}} = 10^{-12}$ cm^2 and the Doppler width $\delta\nu_{\text{D}} = 1$ GHz. Calculate the saturation parameter S, the absorption coefficient $\alpha_s(\omega_0)$, and the mean density of ground-state and excited atoms that collide in an energy-pooling collision with $\sigma_{\text{ep}} = 3 \times 10^{-14}$ cm^2. Which Rydberg levels are accessible by energy-pooling collisions Na$^*(3P)$ + Na$^*(3P)$ if the relative kinetic energy is $E_{\text{kin}} \leq 0.1$ eV?

8.4 A square-wave chopped cw laser depopulates the level (v_i'', J_i'') in the electronic ground state of the molecule M with $m = 40$ AMU. What is the minimum chopping period T to guarantee quasi-stationary conditions for level $|i\rangle$ and its neighboring levels $|J_i \pm \Delta J\rangle$, if the total refilling cross section of level $|i\rangle$ is $\sigma = 10^{-14}$ cm^2 and the individual cross sections $\sigma_{ik} = (3 \times 10^{-16}\ \Delta\text{J})$ cm^2 for transitions between neighboring levels $|J\rangle$ and $|J \pm \Delta J\rangle$ of the same vibrational level at a pressure of 1 mbar and a temperature of 300 K?

Chapter 9
New Developments in Laser Spectroscopy

During the last few years several new ideas have been born and new spectroscopic techniques have been developed that not only improve the spectral resolution and increase the sensitivity for investigating single atoms but that also allow several interesting experiments for testing fundamental concepts of physics. In the historical development of science, experimental progress in the accuracy of measurements has often brought about a refinement of theoretical models or even the introduction of new concepts [1105]. Examples include A. Einstein's theory of special relativity based on the interferometric experiments of Michelson and Morley [1106], M. Planck's introduction of quantum physics for the correct explanation of the measured spectral distribution of blackbody radiation, the introduction of the concept of electron spin after the spectroscopic discovery of the fine structure in atomic spectra [1107], the test of quantum electrodynamics by precision measurements of the Lamb shift [1108] or the still unresolved problem of a possible time dependence of physical fundamental constants, which might be solved by performing extremely accurate optical frequency measurements. In this chapter some of these new and exciting developments are presented.

9.1 Optical Cooling and Trapping of Atoms

In order to improve the accuracy of spectroscopic measurements of atomic energy levels, all perturbing effects leading to broadening or shifts of these levels must be either eliminated or should be sufficiently well understood to introduce appropriate corrections. One of the largest perturbing effects is the thermal motion of atoms. In Chap. 4 we saw that in collimated atomic beams the velocity components v_x and v_y, which are perpendicular to the beam axes, can be drastically reduced by collimating apertures (geometrical cooling). The component v_z can be compressed by adiabatic cooling into a small interval $v_z = u \pm \Delta v_z$ around the flow velocity u. If this reduction of the velocity distribution is described by a *translational temperature* T_{trans}, values of $T_{\mathrm{trans}} < 1$ K can be reached. However, the molecules still have a nearly

W. Demtröder, *Laser Spectroscopy 2*, DOI 10.1007/978-3-662-44641-6_9

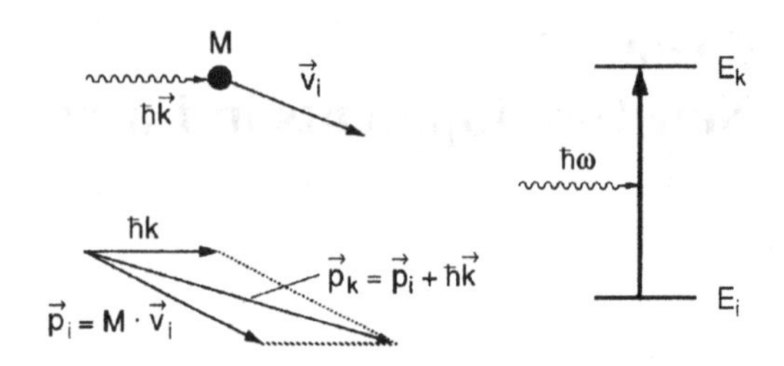

Fig. 9.1 Atomic recoil for absorption and emission of a photon

uniform but large velocity u, and broadening effects such as transit-time broadening or shifts caused by the second-order Doppler effect (see below) are not eliminated.

In this section we discuss the new technique of *optical cooling*, which decreases the velocity of atoms to a small interval around $v = 0$. Optical cooling down to "temperatures" of a few microKelvin has been achieved; by combining optical and evaporative cooling even the nanoKelvin range was reached. This brought the discovery of quite new phenomena, such as Bose–Einstein condensation or atom–lasers, and atomic fountains [1109–1111].

9.1.1 Photon Recoil

Let us consider an atom with rest mass M in the energy level E_i that moves with the velocity $\boldsymbol{v}$. If this atom absorbs a photon of energy $\hbar\omega_{ik} \simeq E_k - E_i$ and momentum $\hbar\boldsymbol{k}$, it is excited into the level E_k. Its momentum changes from $\boldsymbol{p}_i = M\boldsymbol{v}_i$ before the absorption to

$$\boldsymbol{p}_k = \boldsymbol{p}_i + \hbar\boldsymbol{k}, \tag{9.1}$$

after the absorption (the *recoil effect*, Fig. 9.1).

The relativistic energy conservation demands

$$\hbar\omega_{ik} = \sqrt{p_k^2c^2 + \left(M_0c^2 + E_k\right)^2} - \sqrt{p_i^2c^2 + \left(M_0c^2 + E_i\right)^2}. \tag{9.2}$$

When we extract $(M_0c^2 + E_k)$ from the first root and $(M_0c^2 + E_i)$ from the second, we obtain by a Taylor expansion the power series for the resonant absorption frequency

$$\omega_{ik} = \omega_0 + \boldsymbol{k} \cdot \boldsymbol{v}_i - \omega_0\frac{v_i^2}{2c^2} + \frac{\hbar\omega_0^2}{2Mc^2} + \cdots. \tag{9.3}$$

The first term represents the absorption frequency $\omega_0 = (E_k - E_i)$ of an atom at rest if the recoil of the absorbing atom is neglected. The second term describes the linear Doppler shift (first-order Doppler effect) caused by the motion of the atom at the time of absorption. The third term expresses the quadratic Doppler effect (second-order Doppler effect). Note that this term is independent of the direction of the velocity $\boldsymbol{v}$. It is therefore *not* eliminated by the "Doppler-free" techniques described in Chaps. 2–5, which only overcome the *linear* Doppler effect.

Example 9.1 A parallel beam of Ne ions accelerated by 10 keV moves with the velocity $v_z = 3 \times 10^5$ m/s. When the beam is crossed perpendicularly by a single-mode laser beam tuned to a transition with $\lambda = 500$ nm, even ions with $v_x = v_y = 0$ show a quadratic relativistic Doppler shift of $\Delta\nu/\nu = 5 \times 10^{-7}$, which yields an absolute shift of $\Delta\nu = 250$ MHz. This should be compared with the linear Doppler shift of 600 GHz, which appears when the laser beam is parallel to the ion beam (Example 4.6).

The last term in (9.3) represents the recoil energy of the atom from momentum conservation. The energy $\hbar\omega$ of the absorbed photon has to be larger than that for recoil-free absorption by the amount

$$\Delta E = \frac{\hbar^2\omega_0^2}{2Mc^2} \quad \Rightarrow \quad \frac{\Delta E}{\hbar\omega} = \frac{1}{2}\frac{\hbar\omega}{Mc^2}. \tag{9.4}$$

When the excited atom in the state E_k with the momentum $\boldsymbol{p}_k = M\boldsymbol{v}_k$ emits a photon, its momentum changes to

$$\boldsymbol{p}_i = \boldsymbol{p}_k - \hbar\boldsymbol{k}$$

after the emission. The emission frequency becomes analogous to (9.3)

$$\omega_{ik} = \omega_0 + \boldsymbol{k}\boldsymbol{v}_k - \frac{\omega_0 v_k^2}{2c^2} - \frac{\hbar\omega_0^2}{2Mc^2}. \tag{9.5}$$

The difference between the resonant absorption and emission frequencies is

$$\Delta\omega = \omega_{ik}^{\text{abs}} - \omega_{ki}^{\text{em}} = \frac{\hbar\omega_0^2}{Mc^2} + \frac{\omega_0}{2c^2}\left(v_k^2 - v_i^2\right) \approx \frac{\hbar\omega_0^2}{Mc^2}, \tag{9.6}$$

since the second term can be written as $(\omega_0/Mc^2)\cdot(E_k^{\text{kin}} - E_i^{\text{kin}})$ and $\Delta E^{\text{kin}} \ll \hbar\omega$. It can be therefore neglected for atoms with thermal velocities.

The relative frequency shift between absorbed and emitted photons because of recoil

$$\frac{\Delta\omega}{\omega} = \frac{\hbar\omega_0}{Mc^2}, \tag{9.7}$$

equals the ratio of photon energy to rest-mass energy of the atom.

For γ-quanta in the MeV range this ratio may be sufficiently large that $\Delta\omega$ becomes larger than the linewidth of the absorbing transition. This means that a γ-photon emitted from a free nucleus cannot be absorbed by another identical nucleus at rest. The recoil can be greatly reduced if the atoms are embedded in a rigid crystal structure below the Debye temperature. This recoil-free absorption and emission of γ-quanta is called the *Mössbauer effect* [1112].

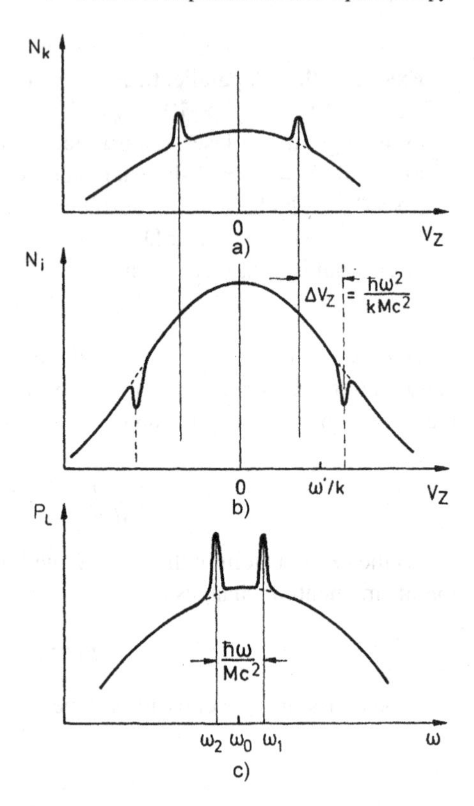

Fig. 9.2 Generation of the recoil doublet of Lamb dips: (**a**) population peaks in the upper-state population for $\omega \neq \omega_0$; (**b**) Bennet holes in the lower-state population; (**c**) recoil doublet in the output power $P_L(\omega)$ of the laser

In the *optical region* the recoil shift $\Delta\omega$ is extremely small and well below the natural linewidth of most optical transitions. Nevertheless, it has been measured for selected narrow transitions [1113, 1114].

9.1.2 Measurement of Recoil Shift

When the absorbing molecules with the resonant absorption frequency ω_0 are placed inside the laser resonator, the standing laser wave of frequency $\omega \neq \omega_0$ burns two Bennet holes into the population distribution $N_i(v_z)$ (Fig. 9.2b and Sect. 2.2), which, according to (9.3), appear at the velocity components

$$v_{zi} = \pm\big[\omega' - \hbar\omega_0^2/\big(2Mc^2\big)\big]/k, \quad \text{with } \omega' = \omega - \omega_0\big(1 - v^2/2c^2\big). \tag{9.8}$$

The corresponding peaks in the population distribution $N_k(v_z)$ of molecules in the upper level $|k\rangle$ are shifted due to photon recoil (Fig. 9.2a). They show up, according to (9.5), at the velocity components

$$v_{zk} = \pm\big[\omega' + \hbar\omega_0^2/\big(2Mc^2\big)\big]/k. \tag{9.9}$$

For the example, illustrated by Fig. 9.2, we have chosen

$$\omega < \omega_0 \quad \Rightarrow \quad \omega' < 0.$$

The two holes in the ground-state population coincide for $v_{zi} = 0$, which gives $\omega' = \hbar\omega_0^2/2Mc^2$. This happens, according to (9.8), for the laser frequency

$$\omega = \omega_1 = \omega_0\left[\left(1 - v^2/c^2\right) + \hbar\omega_0/\left(2Mc^2\right)\right], \qquad (9.10a)$$

while the two peaks in the upper level population coincide for another laser frequency

$$\omega = \omega_2 = \omega_0\left[\left(1 - v^2/c^2\right) - \hbar\omega_0/\left(2Mc^2\right)\right]. \qquad (9.10b)$$

The absorption of the laser is proportional to the population difference $\Delta N = N_i - N_k$. This difference has two maxima at ω_1 and ω_2. The laser output therefore exhibits two Lamb peaks (*inverse Lamb dips*) (Fig. 9.2c) at the laser frequencies ω_1 and ω_2, which are separated by twice the recoil energy

$$\Delta\omega = \omega_1 - \omega_2 = \hbar\omega^2/Mc^2. \qquad (9.11)$$

Example 9.2

(a) For the transition at $\lambda = 3.39$ μm in the CH_4 molecule ($M = 16$ AMU) the recoil splitting amounts to $\Delta\omega = 2\pi \cdot 2.16$ kHz, which is still larger than the natural linewidth of this transition [1114, 1115].

(b) For the calcium intercombination line ${}^1S_0 \rightarrow {}^3P_1$ at $\lambda = 657$ nm we obtain with $M = 40$ AMU the splitting $\Delta\omega = 2\pi \cdot 23.1$ kHz [1116].

(c) For a rotation–vibration transition in SF_6 at $\lambda = 10$ μm the frequency ω is one-third that of CH_4 but the mass 10 times larger than for CH_4. The recoil splitting therefore amounts only to about 0.02 kHz and is not measurable [1118].

Since such small splittings can only be observed if the width of the Lamb peaks is smaller than the recoil shift, all possible broadening effects, such as pressure broadening and transit-time broadening, must be carefully minimized. This can be achieved in experiments at low pressures and with expanded laser beam diameters [1117]. An experimental example is displayed in Fig. 9.3.

The velocities of very cold atoms are very small, i.e., the linear and the quadratic Doppler effects both become small and the recoil term becomes significant. It turns out that for cold Ca atoms at $T = 10$ μK, the recoil effect leads to large asymmetries in the Lamb dips of absorption spectra taken with short pulses [1119]. These are not found in experiments performed at room temperature, where the broad Doppler background masks these asymmetries, and they are based on the fundamental asymmetry between absorption and stimulated emission with short pulses.

The transit-time broadening can greatly be reduced by the optical Ramsey method of separated fields. The best resolution of the recoil splittings has indeed

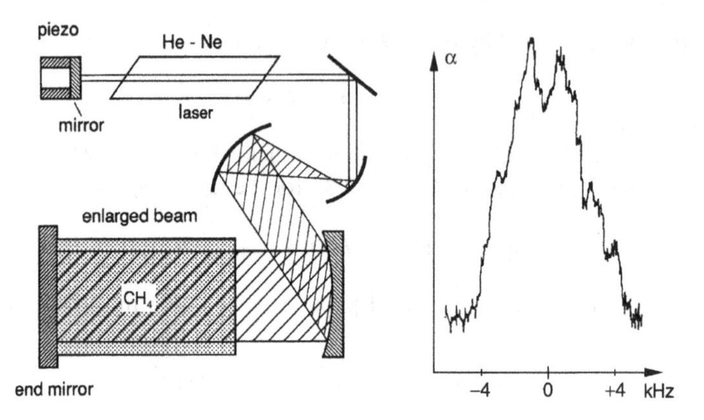

Fig. 9.3 Schematic experimental arrangement for measuring the recoil splitting and the measured signal of the recoil doublet of the hyperfine component $8 \rightarrow 7$ of the $(P(7), \nu_3)$ transition at $\lambda = 3.39\ \mu\mathrm{m}$ in methane [1115]

been achieved with this technique (Sect. 9.4). The transit time can also be increased if only molecules with small transverse velocity components contribute to the Lamb dip. If the laser intensity is kept so small that saturation of the molecular transition is reached only for molecules that stay within the laser beam for a sufficiently long time, that is, molecules with small components v_x, v_y, the transit-time broadening is greatly reduced [1115].

9.1.3 Optical Cooling by Photon Recoil

Although the recoil effect is very small when a single photon is absorbed, it can be used effectively for optical cooling of atoms by the cumulative effect of many absorbed photons. This can be seen as follows: When atom A stays for the time T in a laser field that is in resonance with the transition $|i\rangle \rightarrow |k\rangle$, the atom may absorb and emit a photon $\hbar\omega$ many times, provided the optical pumping cycle is short compared to T and the atom behaves like a true two-level system. This means that the emission of fluorescence photons $\hbar\omega$ by the excited atom in level $|k\rangle$ brings the atom back only to the initial state $|i\rangle$, but never to other levels. With the saturation parameter $S = B_{ik}\rho(\omega_{ik})/A_{ik}$, the fraction of excited atoms becomes

$$\frac{N_k}{N} = \frac{S}{1+2S}.$$

The fluorescence rate is $N_k A_k = N_k/\tau_k$. Since N_k can never exceed the saturated value $N_k = (N_i + N_k)/2 = N/2$, the minimum recycling time for the saturation parameter $S \rightarrow \infty$ (Vol. 1, Sect. 3.6) is $\Delta T = 2\tau_k$.

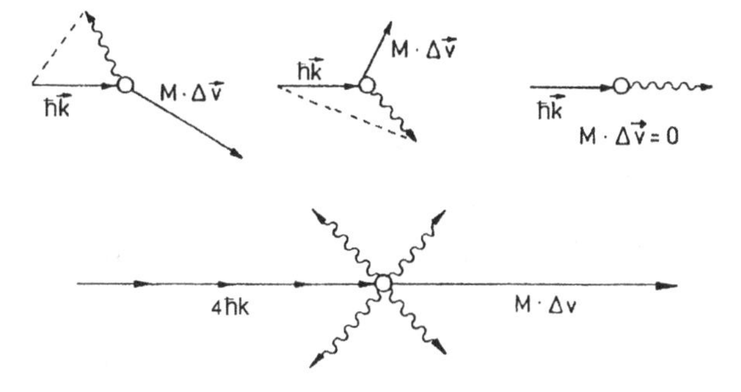

Fig. 9.4 Recoil momentum of an atom for a fixed direction of the absorbed photons but different directions of the emitted fluorescence photons

Example 9.3 When an atom passes with a thermal velocity $v = 500$ m/s through a laser beam with 2 mm diameter, the transit time is $T = 4$ µs. For a spontaneous lifetime $\tau_k = 10^{-8}$ s the atom can undergo $q \le (T/2)/\tau_k = 400$ absorption–emission cycles during its transit time.

When a laser beam is sent through a sample of absorbing atoms, the LIF is generally isotropic, that is, the spontaneously emitted photons are randomly distributed over all spatial directions. Although each emitted photon transfers the momentum $\hbar\boldsymbol{k}$ to the emitting atoms, the time-averaged momentum transfer tends to zero for sufficiently large values of $q = T/2\tau_k$.

The *absorbed* photons, however, all come from the same direction. Therefore, the momentum transfer for q absorptions adds up to a total recoil momentum $\boldsymbol{p} = q\hbar\boldsymbol{k}$ (Fig. 9.4). This changes the velocity $\boldsymbol{v}$ of an atom which flies against the beam propagation by the amount $\Delta v = \hbar k/M$ per absorption. For q absorption–emission cycles we get

$$\Delta v = q\frac{\hbar k}{M} = q\frac{\hbar\omega}{Mc}. \tag{9.12a}$$

Atoms in a collimated atomic beam can therefore be slowed down by a laser beam propagating anticollinearly to the atomic beam [1120]. This can be expressed by the "cooling force"

$$\boldsymbol{F} = M\frac{\Delta\boldsymbol{v}}{\Delta T} = \frac{\hbar\boldsymbol{k}}{\tau_k}\frac{S}{1+2S}. \tag{9.12b}$$

Note That the *recoil energy* transferred to the atom

$$\Delta E_{\text{recoil}} = q\frac{\hbar^2\omega^2}{2Mc^2}, \tag{9.13}$$

is still very small.

Example 9.4

(a) For Na atoms with $M = 23$ AMU, which absorb photons $\hbar\omega \simeq 2$ eV on the transition $3^2S_{1/2} \rightarrow 3^2P_{3/2}$, (9.12a, 9.12b) gives $\Delta v = 3$ cm/s per photon absorption. In order to reduce the initial thermal velocity of $v =$ 600 m/s at $T = 500$ K to 20 m/s (this would correspond to a temperature $T = 0.6$ K for a thermal distribution with $\overline{v} = 20$ m/s), this demands $q = 2 \times 10^4$ absorption–emission cycles. At the spontaneous lifetime $\tau_k =$ 16 ns a minimum cooling time of $T = 2 \times 10^4 \cdot 2 \cdot 16 \times 10^{-9}$ s $\simeq 600$ µs is required. This gives a negative acceleration of $a = -10^6$ m/s^2, which is 10^5 times the gravity acceleration on earth $g = 9.81$ m/s^2! During the time T the atom has traveled the distance $\Delta_z = v_0T - \frac{1}{2}aT^2 \simeq 18$ cm. During this deceleration path length it always has to remain within the laser beam. The total energy transferred by recoil to the atom is only -2×10^{-2} eV. It corresponds to the kinetic energy $\frac{1}{2}Mv^2$ of the atom and is very small compared to $\hbar\omega = 2$ eV.

(b) For Mg atoms with $M = 24$ AMU, which absorb on the singlet resonance transition at $\lambda = 285.2$ nm, the situation is more favorable because of the higher photon energy $\hbar\omega \simeq 3.7$ eV and the shorter lifetime $\tau_k = 2$ ns of the upper level. One obtains: $\Delta v = -6$ cm/s per absorbed photon; $q = 1.3 \times 10^4$. The minimum cooling time becomes $T = 3 \times 10^{-5}$ s and the deceleration path length $\Delta z \simeq 1$ cm.

(c) In [1121] a list of other atoms can be found that were regarded as possible candidates for optical cooling. Some of them have since been successfully tried.

The following remarks may be useful:

(a) Without additional tricks, this optical-cooling method is restricted to true two-level systems. Therefore, *molecules cannot be cooled with this technique*, because after their excitation into level $|k\rangle$ they return by emission of fluorescence into many lower vibrational–rotational levels and only a small fraction goes back into the initial level $|i\rangle$. Thus, only one optical pumping cycle is possible. There have been, however, other cooling mechanisms for molecules proposed and partly realized (see Sect. 9.1.5).

(b) The sodium transition $3S$–$3P$, which represents the standard example for optical cooling, is in fact a multi level system because of its hyperfine structure (Fig. 9.5). However, after optical pumping with circularly polarized light on the hfs transition $3^2S_{1/2}(F'' = 2) \rightarrow 3^2P_{3/2}(F' = 3)$, the fluorescence can only reach the initial lower level $F'' = 2$. A true two-level system would be realized if any overlap of the pump transition with other hfs components could be avoided (see below).

(c) Increasing the intensity of the pump laser decreases the time ΔT for an absorption–emission cycle. However, for saturation parameters $S > 1$ this decrease is small and ΔT soon reaches its limit $2\tau_k$. On the other hand, the in-

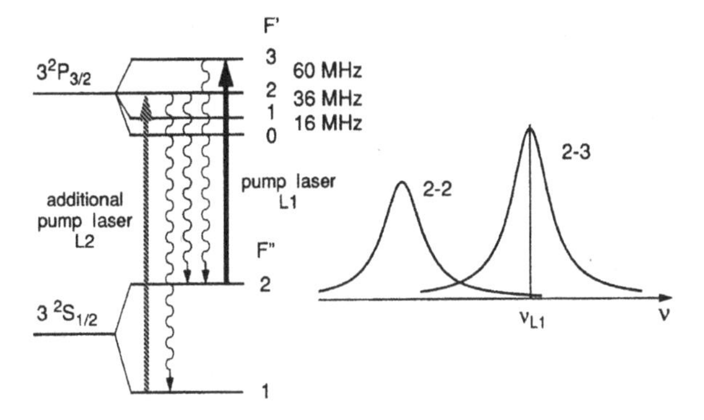

Fig. 9.5 Level diagram of the Na $3^2S_{1/2} \rightarrow 3^2P_{3/2}$ transition with hyperfine splittings. Optical pumping on the hfs component $F'' = 2 \rightarrow F' = 3$ represents a true two-level system, provided any overlap between the hfs components can be avoided. The additional pump laser L_2 is necessary in order to pump atoms transferred into level $F'' = 1$ by spectral overlap between the components $2 \rightarrow 3$ and $2 \rightarrow 2$ back into level $F'' = 2$

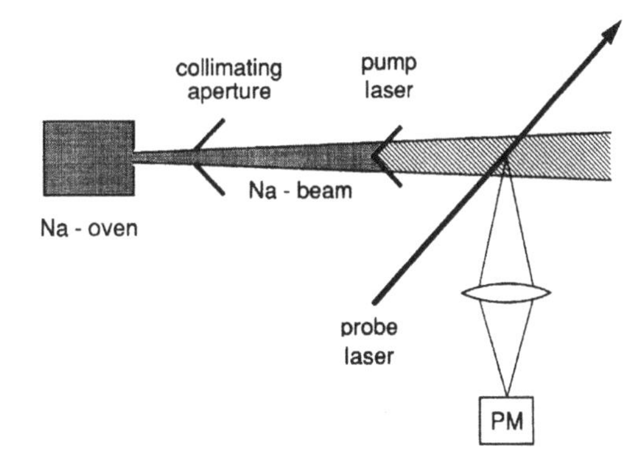

Fig. 9.6 Simplified experimental realization for the deceleration of atoms in a collimated beam by photon recoil

duced emission increases at the expense of spontaneous emission. Since the emitted induced photon has the same $\boldsymbol{k}$-vector as the absorbed induced photon, the net momentum transfer is zero. The total deceleration rate therefore has a maximum at the optimum saturation parameter $S \simeq 1$.

9.1.4 Experimental Arrangements

For the experimental realization of optical cooling, which uses a collimated beam of atoms and a counterpropagating cw laser (dye laser or diode laser, Fig. 9.6) the following difficulties have to be overcome: during the deceleration time the Doppler-shifted absorption frequency $\omega(t) = \omega_0 + \boldsymbol{k} \cdot \boldsymbol{v}(t)$ changes with the decreasing velocity v, and the atoms would come out of resonance with the monochromatic laser.

Three solutions have been successfully tried: either the laser frequency

$$\omega(t) = \omega_0 + k \cdot v(t) \pm \gamma, \tag{9.14}$$

is synchronously tuned with the changing velocity $v(t)$ in order to stay within the linewidth γ of the atomic transition [1122, 1123], or the absorption frequency of the atom is appropriately altered along the deceleration path [1124, 1125]. A third alternative uses a broadband laser for cooling. We will discuss these methods briefly: When the laser frequency ω for a counterpropagating beam $\boldsymbol{k}$ antiparallel to $\boldsymbol{v}$ is kept in resonance with the constant atomic frequency ω_0, it must have the time dependence

$$\omega(t) = \omega_0 \left(1 - \frac{v(t)}{c}\right) \quad \Rightarrow \quad \frac{\mathrm{d}\omega}{\mathrm{d}t} = -\frac{\omega_0}{c}\frac{\mathrm{d}v}{\mathrm{d}t}. \tag{9.15}$$

The velocity change per second for the optimum deceleration with a pump cycle period $T = 2\tau$ is, according to (9.12a, 9.12b)

$$\frac{\mathrm{d}v}{\mathrm{d}t} = \frac{\hbar\omega_0}{2Mc\tau}. \tag{9.16}$$

Inserting this into (9.15) yields for the necessary time dependence of the laser frequency

$$\omega_\mathrm{L}(t) = \omega(0)(1 + \alpha t), \quad \text{with } \alpha = \frac{\hbar\omega_0}{2Mc^2\tau} \ll 1. \tag{9.17}$$

This means that the pump laser should have a linear frequency chirp.

Example 9.5 For Na atoms with $v(0) = 1000\ \mathrm{m/s}$, insertion of the relevant numbers into (9.17) yields $\mathrm{d}\omega/\mathrm{d}t = 2\pi \cdot 1.7\ \mathrm{GHz/ms}$, which means that fast frequency tuning in a controlled way is required in order to fulfill (9.17).

An experimental realization of the controlled frequency chirp uses amplitude modulation of the laser with the modulation frequency Ω_1. The sideband at $\omega_\mathrm{L} - \Omega_1$ is tuned to the atomic transition and can be matched to the time-dependent Doppler shift by changing $\Omega_1(t)$ in time. In order to compensate for optical pumping into other levels than $F'' = 2$ by overlap of the laser with other hfs components, the transition $F'' = 1 \rightarrow F' = 2$ is simultaneously pumped (Fig. 9.5). This can be achieved if the pump laser is additionally modulated at the second frequency Ω_2, where the sideband $\omega_\mathrm{L} + \Omega_2$ matches the transition $F'' = 1 \rightarrow F' = 2$ [1123, 1126].

For the second method, where the laser frequency ω_L is kept constant, the atomic absorption frequency must be altered during the deceleration of the atoms. This can be realized by Zeeman tuning (Fig. 9.7). In order to match the Zeeman shift to the changing Doppler shift $\Delta\omega(z)$, the longitudinal magnetic field must have the z-dependence

$$B = B_0\sqrt{1 - 2az/v_0^2}, \tag{9.18}$$

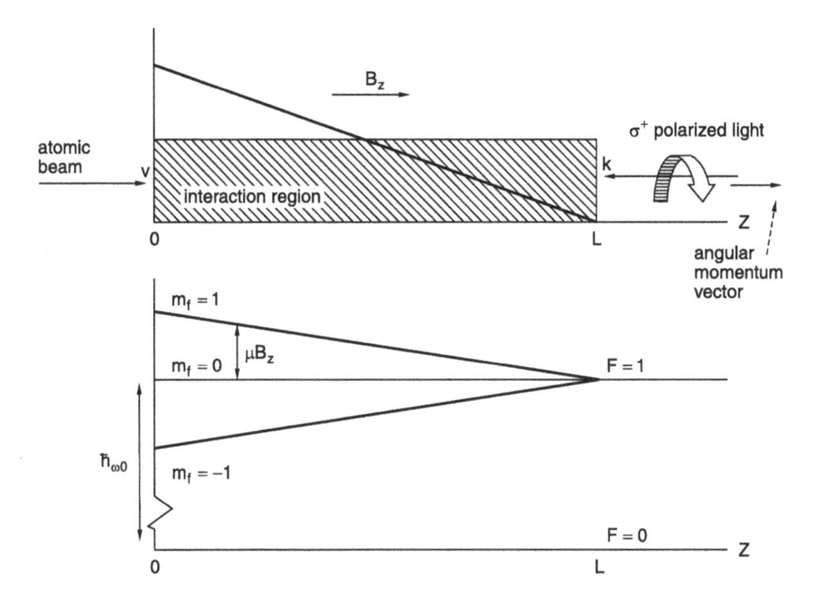

Fig. 9.7 Level diagram for laser cooling by Zeeman tuning

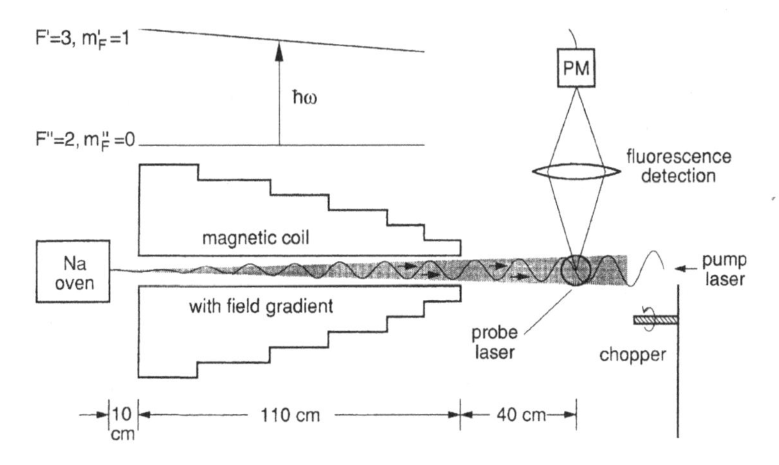

Fig. 9.8 Laser cooling of atoms in a collimated beam with a fixed laser frequency and Zeeman tuning of the atomic absorption frequency [1125]

for atoms that enter the field at $z = 0$ with the velocity v_0 and experience the negative acceleration a [m/s^2] by photon recoil [1125]. This field dependence $B(z)$ can be realized by a proper choice of the windings $N_W(z)$ per centimeters of the magnetic field coil (Fig. 9.8).

Most optical cooling experiments have been performed up to now on alkali atoms, such as Na or Rb, using a single-mode cw dye laser. The velocity decrease of the atoms is monitored with the tunable probe laser L_2, which is sufficiently weak

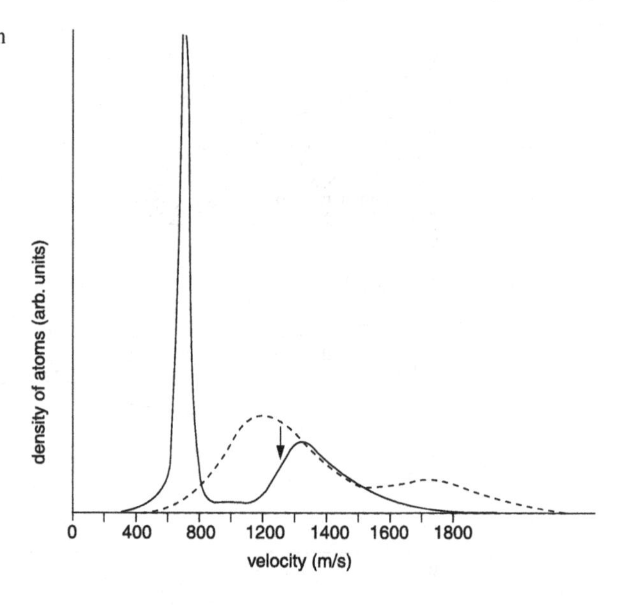

Fig. 9.9 Velocity distribution before (*dashed*) and after (*solid*) Zeeman cooling. The *arrow* indicates the highest velocity resonant with the slowing laser. (The extra bump at 1700 m/s is from $F=1$ atoms, which are optically pumped into $F=2$ during the cooling process.) [W. Phillips, Nobel Lecture 1995]

that it does not affect the velocity distribution. The probe-laser-induced fluorescence $I_{\mathrm{Fl}}(\omega_2)$ is measured as a function of the Doppler shift. Experiments have shown that the atoms could be completely stopped and their velocity could even be reversed [1122]. An example of the compression of the thermal velocity distribution into a narrow range around $v=200$ m/s is illustrated in Fig. 9.9 for Na atoms.

A favorable alternative to dye lasers is a GaAs diode laser, which can cool rubidium or cesium atoms [1127–1129] and also metastable noble gas atoms, such as He* or Ar* [1130]. The experimental expenditures are greatly reduced since the GaAs laser is much less expensive than an argon laser plus dye-laser combination. Furthermore, the frequency modulation is more readily realized with a diode laser than with a dye laser.

A very interesting alternative laser for optical cooling of atoms in a collimated beam is the *modeless* laser [1131], which has a broad spectral emission (without mode structure, when averaged over a time of $T>10$ ns, with an adjustable bandwidth and a tunable center frequency). Such a laser can cool all atoms regardless of their velocity if its spectral width $\Delta\omega_{\mathrm{L}}$ is larger than the Doppler shift $\Delta\omega_{\mathrm{D}}=v_0k$ [1132].

With the following experimental trick it is possible to compress the velocity distribution $N(v_z)$ of atoms in a beam into a small interval Δv_z around a wanted final velocity v_{f}. The beam from the modeless laser propagates anticollinearly to the atomic beam and cools the atoms (Fig. 9.10). A second single-mode laser intersects the atomic beam under a small angle α against the beam axis. If it is tuned to the frequency

$$\omega_2=\omega_0+kv_{\mathrm{f}}\cos\alpha,$$

it accelerates the atoms as soon as they have reached the velocity v_{f}. This second laser therefore acts as a barrier for the lower velocity limit of cooled atoms [1133].

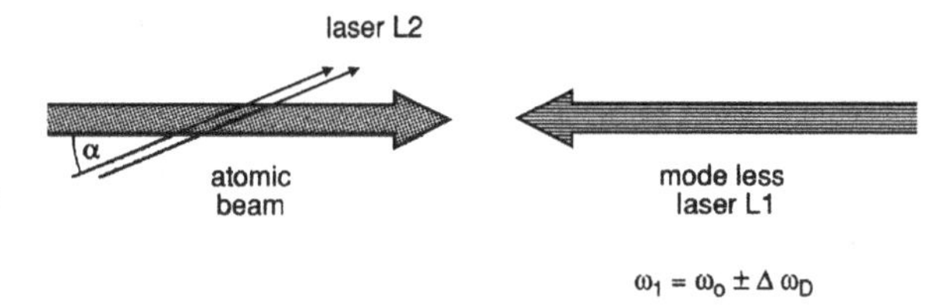

Fig. 9.10 Cooling of all atoms with a counterpropagating modeless laser. A cooling stop at a selectable velocity v_f can be realized with a second copropagating single-mode laser [1133]

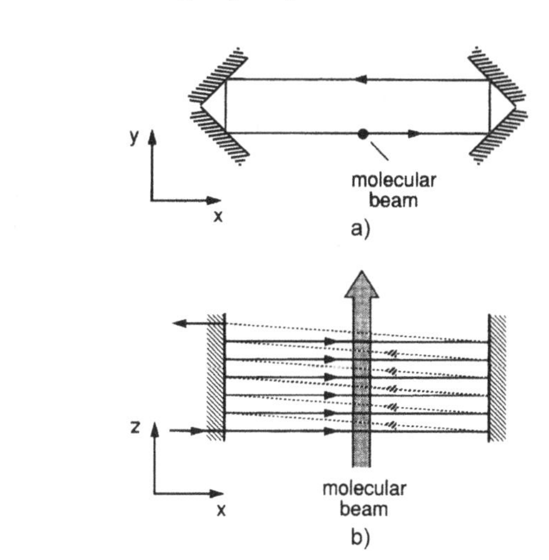

Fig. 9.11 Deflection of atoms in a collimated atomic beam using a multiple-path geometry. The molecular beam travels into the z-direction: (**a**) view into the z-direction; (**b**) view into the y-direction. On the *dashed* return paths the laser beam does not intersect the atomic beam

The photon recoil can be used not only for the deceleration of collimated atomic beams but also for the deflection of the atoms, if the laser beam crosses the atomic beam perpendicularly [1134–1136]. In order to increase the transferred photon momentum, and with it the deflection angle, an experimental arrangement is chosen where the laser beam crosses the atomic beam many times in the same direction (Fig. 9.11). The deflection angle δ per absorbed photon, which is given by $\tan\delta = \hbar k/(mv_z)$, increases with decreasing atomic velocity v_z. Optically cooled atoms can therefore be deflected by large angles. Since the atomic absorption frequency differs for different isotopes, this deflection can be used for spatial isotope separation [1137] if other methods cannot be applied (Sect. 10.1.6).

An interesting application of atomic deflection by photon recoil is the collimation and focusing of atomic beams with lasers [1138]. Assume atoms with the velocity $\boldsymbol{v} = \{v_x \ll v_z, 0, v_z\}$ pass through a laser resonator, where an intense standing op-

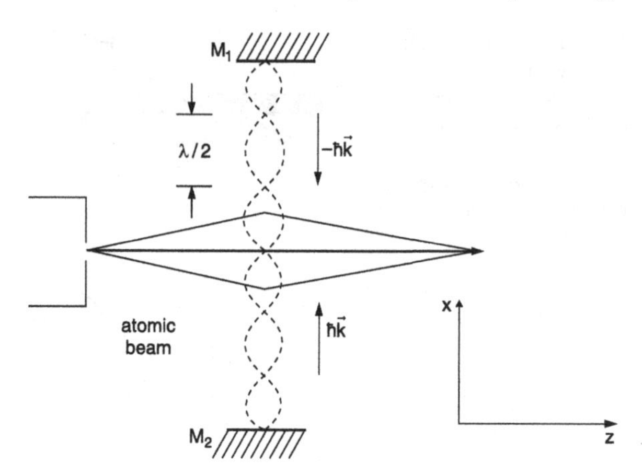

Fig. 9.12 Standing optical wave acting as a lens for a divergent atomic beam

tical wave in the $\pm x$-direction is present. If the laser frequency ω_L is kept slightly below the atomic resonance frequency ω_0 ($\gamma > \omega_0 - \omega_L > 0$), atoms with transverse velocity components v_x are always pushed back to the z-axis by photon recoil because that part of the laser wave with the $\boldsymbol{k}$-vector antiparallel to v_x always has a larger absorption probability than the component with $\boldsymbol{k}$ parallel to v_x. The velocity component v_x is therefore reduced and the atomic beam is collimated.

If the atoms have been optically cooled before they pass the standing laser wave, they experience a large collimation in the maximum of the standing wave, but are not affected in the nodes. The laser wave acts like a transmission grating that "channels" the transmitted atoms and focuses the different channels (Fig. 9.12) [1139].

A schematic diagram of an apparatus for optical cooling of atoms, deflection of the slowed-down atoms, and focusing is depicted in Fig. 9.13.

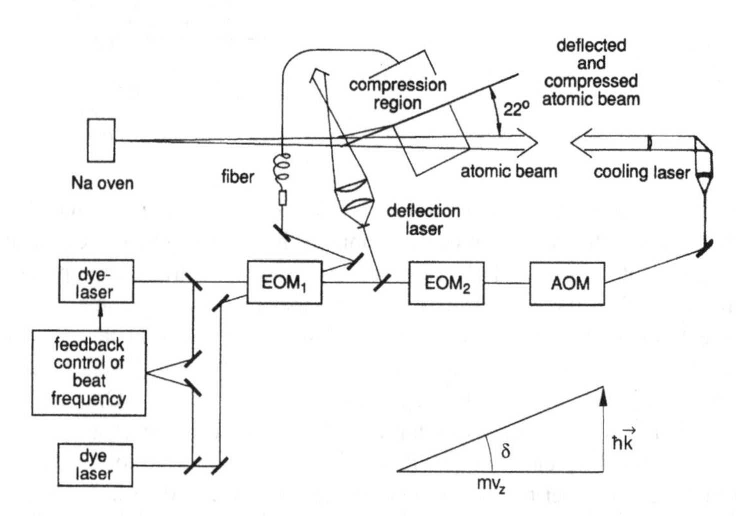

Fig. 9.13 Cooling, deflection and compression of atoms by photon recoil. The electro-optic modulators (EOM) and the acousto-optic modulator (AOM) serve for sideband generation and frequency tuning of the cooling laser sideband [1136]

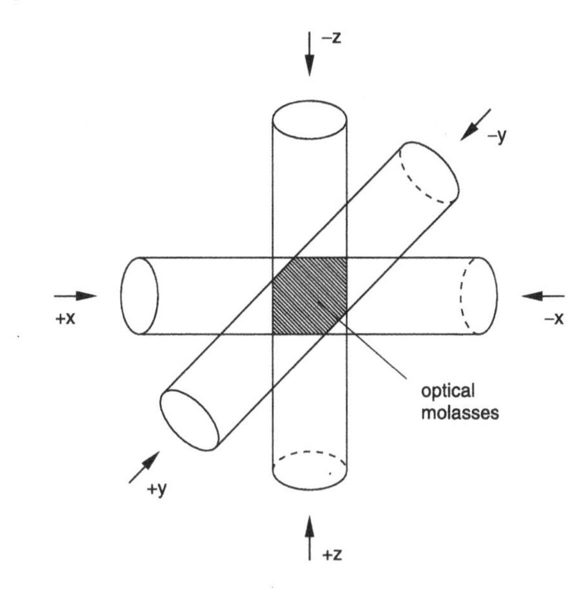

Fig. 9.14 Optical molasses with six pairwise counterpropagating laser beams

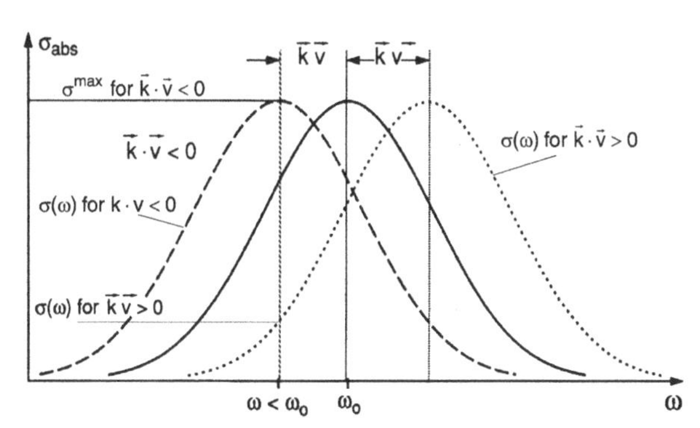

Fig. 9.15 For $\omega < \omega_0$, the absorption probability is larger for $\boldsymbol{k} \cdot \boldsymbol{v} < 0$ than for $\boldsymbol{k} \cdot \boldsymbol{v} > 0$

9.1.5 Three-dimensional Cooling of Atoms; Optical Molasses

Up to now we have only considered the cooling of atoms that all move into one direction. Therefore only one velocity component has been reduced by photon recoil. For cooling of atoms in a thermal gas where all three velocity components $\pm v_x$, $\pm v_y$, $\pm v_z$ have to be reduced, six laser beams propagating into the $\pm x$-, $\pm y$-, $\pm z$-directions are required [1140]. All six beams are generated by splitting a single laser beam (Fig. 9.14). If the laser frequency is tuned to the red side of the atomic resonance ($\Delta\omega = \omega - \omega_0 < 0$), a repulsive force is always acting on the atoms, because for atoms moving toward the laser wave ($\boldsymbol{k} \cdot \boldsymbol{v} < 0$) the Doppler-shifted absorption frequency is shifted toward the resonance frequency ω_0, whereas for the counterpropagating wave ($\boldsymbol{k} \cdot \boldsymbol{v} > 0$) it is shifted away from resonance (Fig. 9.15). For a quantitative description we denote the absorption rate $R \propto \sigma(\omega)$ of an atom with $\boldsymbol{k} \cdot \boldsymbol{v} > 0$ by $R^+(v)$, and for $\boldsymbol{k} \cdot \boldsymbol{v} < 0$ by $R^-(v)$. The net recoil force component

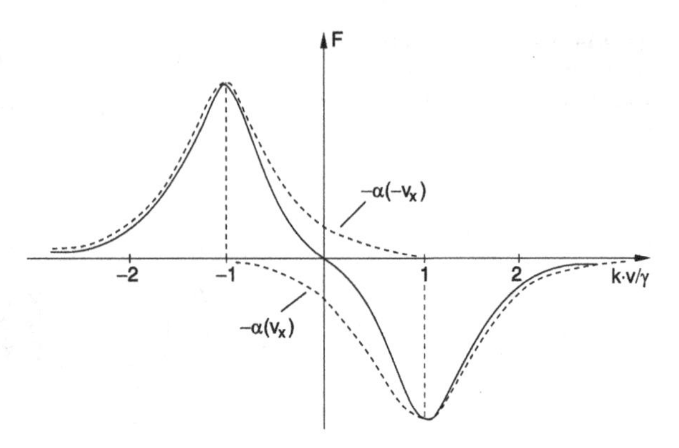

Fig. 9.16 Frictional force in an optical molasses (*solid curve*) for a red detuning $\delta = -\gamma$. The *dotted curve* shows the absorption profiles by a single atom moving with $v_x = \pm\gamma/k$ for a single laser beam propagating into the x-direction

F_i $(i = x, y, z)$ is then

$$F_i = \left[R^+(v_i) - R^-(v_i)\right] \cdot \hbar k. \tag{9.19}$$

For a Lorentzian absorption profile with FHWM γ, the frequency dependence of the absorption rate is (Fig. 9.15)

$$R^\pm(v) = \frac{R_0}{1 + (\frac{\omega_L - \omega_0 \mp kv}{\gamma/2})^2}. \tag{9.20}$$

Inserting (9.20) into (9.19) yields for $\boldsymbol{k} \cdot \boldsymbol{v} \ll \omega_L - \omega_0 = \delta$ the net force (Fig. 9.16)

$$F_i = -a \cdot v_i, \quad \text{with } a = R_0 \frac{16\delta\hbar k^2}{\gamma^2[1 + (2\delta/\gamma)^2]^2}. \tag{9.21}$$

An atom moving within the overlap region of the six running laser waves therefore experiences a force $F_i(v_i) = -av_i$ $(i = x, y, z)$ that damps its velocity. From the relation $\mathrm{d}v/\mathrm{d}t = F/m \Rightarrow \mathrm{d}v/v = -a/m\,\mathrm{d}t$ we obtain the time-dependent velocity

$$v = v_0 \mathrm{e}^{-(a/m)t}. \tag{9.22}$$

The velocity decreases exponentially with a damping time $t_D = m/a$.

Example 9.6

(a) For rubidium atoms ($m = 85$ AMU) the wavenumber is $k = 8 \times 10^6\ \mathrm{m}^{-1}$. At a detuning $\delta = \gamma$ and an absorption rate $R_0 = \gamma/2$ we obtain: $a = 4 \times 10^{-21}$ Ns/m. This gives a damping time $t = 35$ μs.

(b) For Na atoms with $\delta = 2\gamma$ one obtains $a = 1 \times 10^{-20}$ Ns/m and $t_{\mathrm{D}} = 2.3$ µs. The atoms move in this light trap like particles in viscous molasses and therefore this atomic trapping arrangement is often called *optical molasses*.

Note These optical trapping methods reduce the velocity components (v_x, v_y, v_z) to a small interval around $v = 0$. However, they do *not* compress the atoms into a spatial volume, except if the dispersion force for the field gradients $\nabla I \neq 0$ is sufficiently strong. This can be achieved by electric or magnetic field gradients, as discussed in Sect. 9.1.7.

9.1.6 Cooling of Molecules

The optical cooling techniques discussed so far are restricted to true two-level systems because the cooling cycle of induced absorption and spontaneous emission has to be performed many times before the atoms come to rest. In molecules the fluorescence from the upper excited level generally ends in many rotational–vibrational levels in the electronic ground state that differ from the initial level. Therefore most of the molecules cannot be excited again with the same laser. They are lost for further cooling cycles.

Cold molecules are very interesting for several scientific and technical applications. One example is chemical reactions initiated by collisions between cold molecules where the collision time is very long and the reaction probability might become larger by several orders of magnitude. In addition, interactions of cold molecules with surfaces where the sticking coefficient will be 100 % open up new insights into molecule–surface interactions and reactions between cold adsorbed molecules. Finally, the possibility to reach Bose–Einstein condensation of molecular gases opens new fascinating aspects of collective molecular quantum phenomena.

There have been several proposals how molecules might be cooled in spite of the above-mentioned difficulties [1141–1143]. An optical version of these proposals is based on a frequency comb laser, which oscillates on many frequencies, matching the relevant frequencies of the transitions from the upper to the lower levels with the highest transition probabilities [1143]. In this case, the molecules can be repumped into the upper level from many lower levels, thus allowing at least several pumping cycles.

A very interesting optical cooling technique starts with the selective excitation of a collision pair of cold atoms into a bound level in an upper electronic state (Fig. 9.17). While this excitation occurs at the outer turning point of the upper-state potential, a second laser dumps the excited molecule down into a low vibrational level of the electronic ground state by stimulated emission pumping (photo-induced association). In favorable cases the level $v = 0$ can be reached. If the colliding atoms

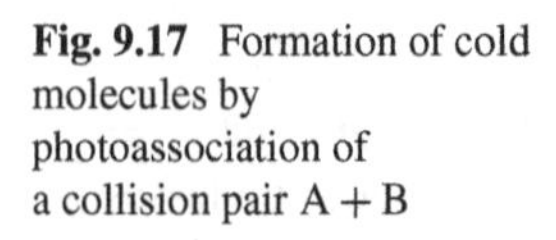

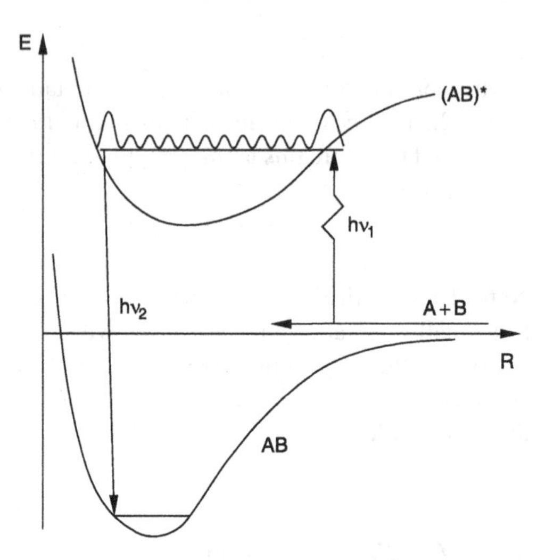

Fig. 9.17 Formation of cold molecules by photoassociation of a collision pair A + B

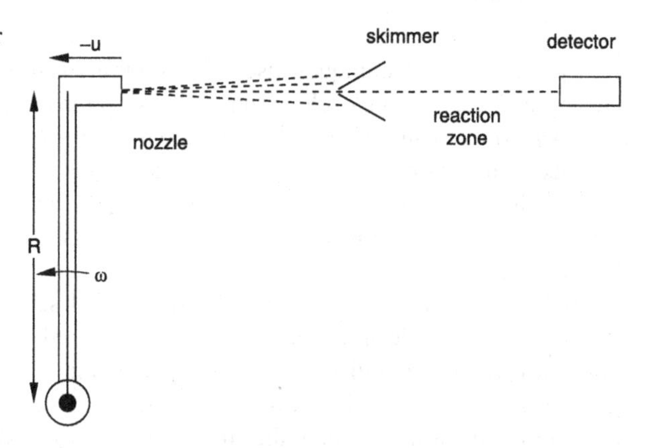

Fig. 9.18 Rotating nozzle for producing a beam of slow molecules

are sufficiently cold, the angular momentum of their relative movement is zero (S-wave scattering). Therefore the final ground state has the rotational quantum number $J = 0$ if the two lasers transfer no angular momentum to the molecule (either two transitions with $\Delta J = 0$ or the absorbing and emission transition as R-transitions with $\Delta J = +1$ in absorption and -1 in emission) [1144]. The kinetic energy of the molecule formed by photoassociation is always smaller than that of the colliding atoms because of momentum conservation [622].

A promising nonoptical technique relies on cooling of molecules by collisions with cold atoms. If the gas mixture of atoms and molecules can be trapped in a sufficiently small volume long enough to achieve thermal equilibrium between atoms and molecules, the optically cooled atoms act as a heat sink for the molecules, which will approach the same temperature as the atoms (sympathetic cooling) [1145].

An interesting proposal that could be realized uses a cold supersonic molecular beam with flow velocity u, which expands through a rotating nozzle (Fig. 9.18). We saw in Chap. 4 that in supersonic beams the velocity spread around the flow velocity u may become very small. The translational temperature in the frame moving with the velocity u can be as small as 0.1 K. If the nozzle moves with the speed $-v$, the molecules have the velocity $u - v$ in the laboratory frame. Tuning the angular velocity ω of the nozzle rotating on a circle with radius R makes it possible to reach any molecular velocity $v_m = u - \omega R$ between u and 0 in the laboratory frame. Since the beam must be collimated in order to reduce the other velocity components, there is only a small time interval per rotation period where the nozzle is in line with the collimating apertures. The cold molecules therefore appear as pulses behind the apertures.

An elegant technique has been developed in several laboratories, where cold helium clusters moving through a gas cell of atoms or molecules pick up these molecules, which then can diffuse into the interior of the helium cluster. The molecules then acquire the low temperature of the cluster. The binding energy is taken away by evaporation of He atoms from the cluster surface [1146, 1147].

At sufficiently low temperatures, the He droplet becomes superfluid [1146]. The molecules inside the droplet can then freely rotate. Because the superfluid He droplet represents a quantum fluid with discrete excitation energies, the energy transfer between the molecule and its superfluid surroundings is limited and the molecular spectra show sharp lines. At the critical temperature between normal fluidity and superfluidity, the lines start to become broader. Laser spectroscopy of these molecules therefore gives direct information on the interaction of molecules and surroundings under different conditions from measured differences between the rotational spectrum of a free molecule and that for a molecule inside a cold He droplet. This has been verified for instance by M. Havenith and her group [1146, 1147].

Injecting different molecules into the He droplet gives the possibility to study chemical reactions between very cold molecules.

9.1.7 Optical Trapping of Atoms

The effectiveness of the optical molasses for cooling atoms anticipates that the atoms are trapped within the overlap region of the six laser beams for a sufficiently long time. This demands that the potential energy of the atoms shows a sufficiently deep minimum at the center of the trapping volume, that is, restoring forces must be present that will bring escaping atoms back to the center of the trapping volume.

We will briefly discuss the two most commonly used trapping arrangements. The first is based on induced electric dipole forces in inhomogeneous electric fields and the second on magnetic dipole forces in magnetic quadrupole fields. Letokhov proposed [1150, 1151] to use the potential minima of a three-dimensional standing optical field composed by the superposition of three perpendicular standing waves for spatial trapping of cooled atoms. It turns out that atoms can be indeed trapped this way, but only after they have been cooled down to very low temperatures. However, Ashkin and Gordon calculated [1152] that the dispersion forces in focused Gaussian beams could be employed for trapping atoms.

(a) Induced Dipole Forces in a Radiation Field

When an atom with the polarizability α is brought into an inhomogeneous electric field $\boldsymbol{E}$, a dipole moment $\boldsymbol{p} = \alpha \boldsymbol{E}$ is induced and the force

$$\boldsymbol{F} = -(\boldsymbol{p} \cdot \text{grad})\boldsymbol{E} = -\alpha(\boldsymbol{E} \cdot \nabla)\boldsymbol{E} = -\alpha\left[\left(\nabla\frac{1}{2}E^2\right) - \boldsymbol{E} \times (\nabla \times \boldsymbol{E})\right], \quad (9.23)$$

acts onto the induced dipole. The same relation holds for an atom in an optical field. However, when averaging over a cycle of the optical oscillation the last term in (9.23) vanishes, and we obtain for the mean dipole force [1153]

$$\langle \boldsymbol{F}_\mathrm{D} \rangle = -\frac{1}{2}\alpha\nabla\left(E^2\right). \quad (9.24)$$

The polarizability $\alpha(\omega)$ depends on the frequency ω of the optical field. It is related to the refractive index $n(\omega)$ of a gas with the atomic density N (Vol. 1, Sect. 2.6.3) by $\alpha \approx \epsilon_0(\epsilon - 1)/N$. With $(\epsilon - 1) = n^2 - 1 \approx 2(n - 1)$ we obtain

$$\alpha(\omega) = \frac{2\epsilon_0[n(\omega) - 1]}{N}. \quad (9.25)$$

Inserting (2.50) of Vol. 1 for $n(\omega)$, the polarizability $\alpha(\omega)$ becomes

$$\alpha(\omega) = \frac{e^2}{2m_\mathrm{e}\omega_0}\frac{\Delta\omega}{\Delta\omega^2 + (\gamma_\mathrm{s}/2)^2}, \quad (9.26)$$

where m_e is the electron mass, $\Delta\omega = \omega - (\omega_0 + \boldsymbol{k} \cdot \boldsymbol{v})$ is the detuning of the field frequency ω against the Doppler-shifted eigenfrequency $\omega_0 + \boldsymbol{kv}$ of the atom, and $\gamma_\mathrm{s} = \delta\omega_\mathrm{n}\sqrt{1 + S}$ is the saturation-broadened linewidth characterized by the saturation parameter S (Vol. 1, Sect. 3.6).

For $\Delta\omega \ll \gamma_\mathrm{s}$ the polarizability $\alpha(\omega)$ increases nearly linearly with the detuning $\Delta\omega$. From (9.24) and (9.26) it follows that in an intense laser beam ($S \gg 1$) with the intensity $I = \epsilon_0 c E^2$ the force $\boldsymbol{F}_\mathrm{D}$ on an induced atomic dipole is

$$\boldsymbol{F}_\mathrm{D} = -a\Delta\omega\nabla I, \quad \text{with } a = \frac{e^2}{m\epsilon_0 c\gamma^2\omega_0 S}. \quad (9.27)$$

This reveals that in a homogeneous field (for example, an extended plane wave) $\nabla I = 0$ and the dipole force becomes zero. For a Gaussian beam with the beam waist w propagating in the z-direction, the intensity $I(r)$ in the x–y-plane is, according to Vol. 1, (5.32)

$$I(r) = I_0\mathrm{e}^{-2r^2/w^2}, \quad \text{with } r^2 = x^2 + y^2.$$

The intensity gradient $\nabla I = -(4r/w^2)I(r)\hat{\boldsymbol{r}}$ points into the radial direction and the dipole force $\boldsymbol{F}_\mathrm{D}$ is then directed toward the axis $r = 0$ for $\Delta\omega < 0$ and radially outwards for $\Delta\omega > 0$.

Fig. 9.19 Longitudinal and transverse forces on an atom in the focus of a Gaussian beam

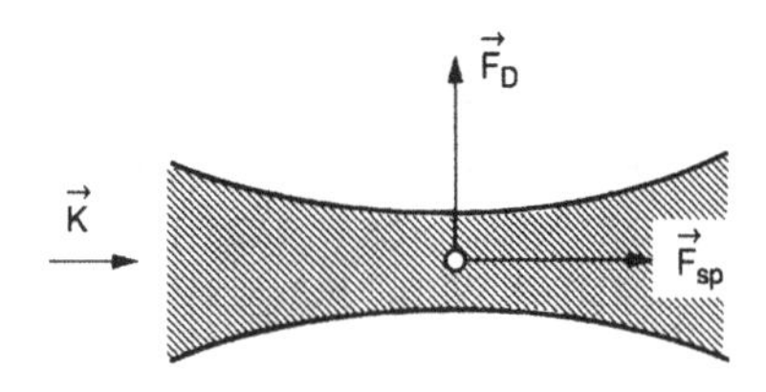

For $\Delta\omega < 0$ the z-axis of an intense Gaussian laser beam with $I(r = 0) = T_0$ represents a minimum of the potential energy

$$E_{\text{pot}} = \int_0^{\infty} F_{\text{D}}\,\text{d}r = +a\Delta\omega I_0, \tag{9.28}$$

where atoms with sufficiently low radial kinetic energy may be trapped. In the focus of a Gaussian beam we have an intensity gradient in the r-direction as well as in the z-direction. If the two forces are sufficiently strong, atoms can be trapped in the focal region.

Besides this dipole force in the r- and z-directions the recoil force in the $+z$-direction acts onto the atom (Fig. 9.19). In a standing wave in $\pm z$-direction the radial force and the recoil force may be sufficiently strong to trap atoms in all directions. For more details see [1154–1156].

Example 9.7 In the focal plane of a Gaussian laser beam with $P_{\text{L}} = 200$ mW focused down to a beam waist of $w = 10$ μm ($I_0 = 1.2 \times 10^9$ W/m^2), the radial intensity gradient is $\partial I/\partial r = 2r/w^2 I_0 \text{e}^{-2r^2/w^2}$, which gives for $r = w : (\partial I/\partial r)_{r=w} = 2I_0/w \cdot \text{e}^2$. With the number above one obtains: $(\partial I/\partial r)_{r=w} = 2.4 \times 10^{13}$ W/m^3 and the radial dipole force acting on a Na atom is for $\Delta\omega = -|\gamma| = -2\pi \times 10^7$ s^{-1}, $S = 0$, and $r = w$

$$\boldsymbol{F}_{\text{D}} = +a\Delta\omega \frac{4r}{w^2} I_0 \hat{\boldsymbol{r}}_0 = 1.5 \times 10^{-16}\ \text{N}.$$

The axial intensity gradient is for a focusing lens with $f = 5$ cm and a beam diameter $\partial I/\partial z = 4.5 \times 10^5$ W/m^3. This gives an axial dipole force of $F_{\text{D}}(z) = 3 \times 10^{-24}$ N, while the recoil force is about

$$\boldsymbol{F}_{\text{recoil}} = 3.4 \times 10^{-20}\ \text{N}.$$

In the axial direction the recoil force is many orders of magnitude larger than the axial dipole force. The potential minimum with respect to the radial dipole force is $E_{\text{pot}} \simeq -5 \times 10^{-7}$ eV. In order to trap atoms in this minimum their radial kinetic energy must be smaller than 5×10^{-7} eV, which corresponds to the "temperature" $T \simeq 5 \times 10^{-3}$ K.

Example 9.8 Assume a standing laser wave with $\lambda = 600$ nm, an average intensity $I = 10\ \mathrm{W/cm^2}$, and a detuning of $\Delta\omega = \gamma = 60$ MHz. The maximum force acting on an atom because of the intensity gradient $\nabla I = 6 \times 10^{11}\ \mathrm{W/m^3}$ between maxima and nodes of the field becomes, according to (9.27), with a saturation parameter $S = 10$: $F_\mathrm{D} = 10^{-17}$ N. The trapping energy is then 1.5×10^{-5} eV$\hat{\approx} T = 0.15$ K.

Example 9.8 demonstrates that the negative potential energy in the potential minima (the nodes of the standing wave for $\Delta\omega > 0$) is very small. The atoms must be cooled to temperatures below 1 K before they can be trapped.

Another method of trapping cooled atoms is based on the net recoil force in a three-dimensional light trap, which can be realized in the overlap region of six laser beams propagating into the six directions $\pm x$, $\pm y$, $\pm z$.

(b) Magneto-Optical Trap

A very elegant experimental realization for cooling and trapping of atoms is the magneto-optical trap (MOT), which is based on a combination of optical molasses and an inhomogeneous magnetic quadrupole field (Fig. 9.21). Its principle can be understood as follows: In a magnetic field the atomic energy levels E_i experience Zeeman shifts

$$\Delta E_i = -\boldsymbol{\mu}_i \cdot \boldsymbol{B} = -\mu_\mathrm{B} \cdot g_F \cdot m_F \cdot B, \tag{9.29}$$

which depend on the Lande g-factor g_F, Bohr's magneton μ_B, the quantum number m_F of the projection of the total angular momentum F onto the field direction, and on the magnetic field B.

In the MOT the inhomogeneous field is produced by two equal electric currents flowing into opposite directions through two coils with radius R and distance $D = R$ (anti-Helmholtz arrangement) (Fig. 9.20). If we choose the z-direction as the symmetry axis through the center of the coils, the magnetic field around $z = 0$ in the middle of the arrangement can be described by the linear dependence

$$B = bz, \tag{9.30}$$

where the constant b depends on the current through the coil and the size of the anti-Helmholtz coils. The Zeeman splittings of the transition from $F = 0$ to $F = 1$ are shown in Fig. 9.21b. Atoms in the center of this MOT are exposed to the six red-tuned laserbeams of the optical molasses (Fig. 9.21e). Let us at first only consider the two beams in the $\pm z$-direction, where the laserbeam in $+z$-direction is σ^+ polarized. Then the reflected beam in the $-z$ direction is σ^- polarized. For an atom at $z = 0$ where the magnetic field is zero, the absorption rates are equal for both laser beams, which means that the average momentum transferred to the atom is zero. For an atom at $z > 0$, however, the σ^--beam is preferentially absorbed because here the

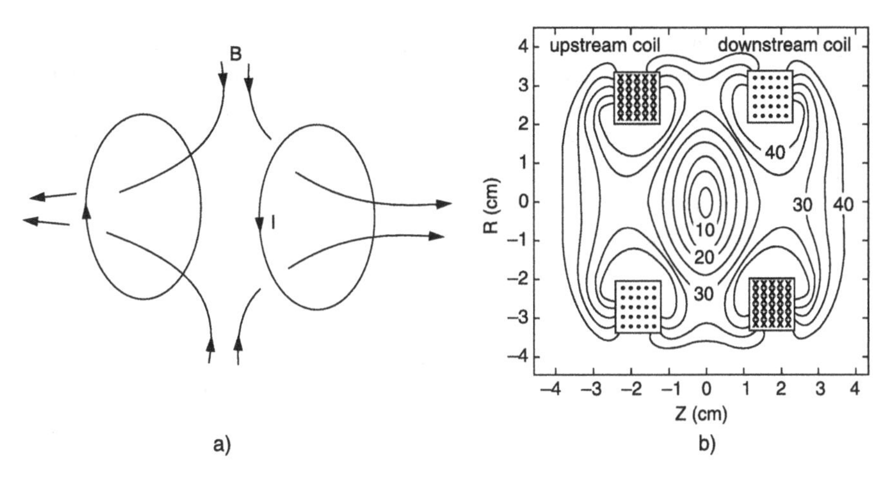

Fig. 9.20 Magnetic field of the MOT: (**a**) magnetic field lines; (**b**) equipotential lines

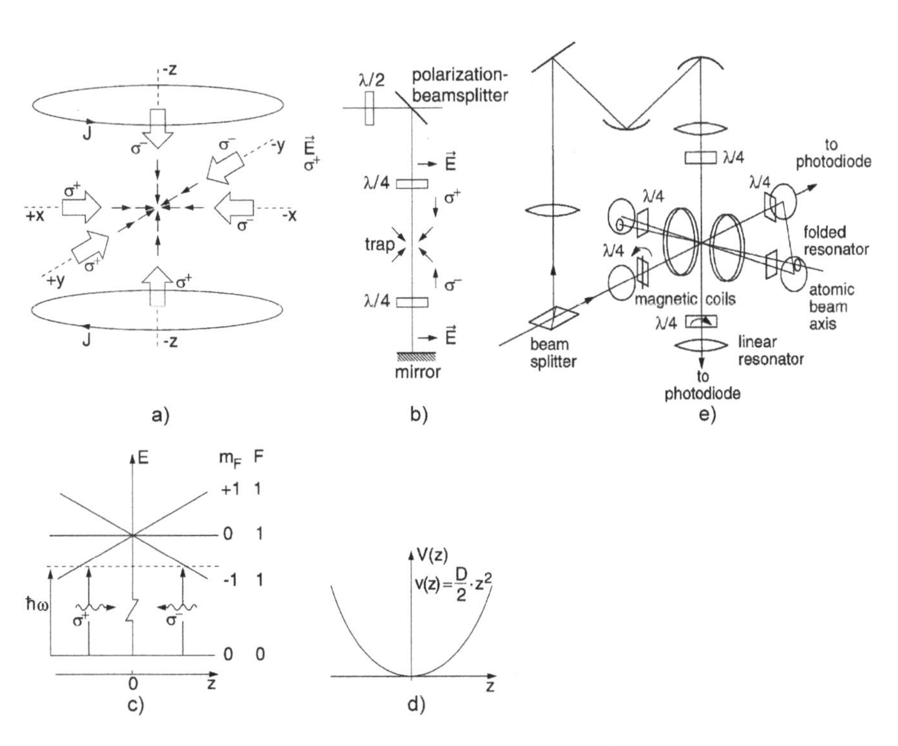

Fig. 9.21 Magneto-optical trap. (**a**) Optical molasses inside the center of an inhomogeneous magnetic field; (**b**) preparation of one of the three laser beams; (**c**) level scheme illustrating the principle of the MOT; (**d**) potential of the MOT; (**e**) schematic experimental setup

frequency difference $\omega_L - \omega_0$ is smaller than for the σ^+-beam. This means that the atom experiences a net momentum transfer into the $-z$-direction, back to the center.

In a similar way an atom at $z < 0$ shows a preferential absorption of σ^+-light and gets a net momentum in the $+z$-direction. This shows that the atoms in the MOT are compressed toward the trap center.

We will now discuss this spatially dependent restoring force more quantitatively. From the above discussion the net force

$$\boldsymbol{F}(z) = R_{\sigma^+}(z)\hbar\boldsymbol{k}_{\sigma^+} + R_{\sigma^-}(z)\hbar\boldsymbol{k}_{\sigma^-}, \tag{9.31}$$

is determined by the difference of the absorption rates R_{σ^+}, R_{σ^-} (note that the wave vectors are antiparallel). For a Lorentzian absorption profile with halfwidth γ, the absorption rates become

$$R_{\sigma^\pm} = \frac{R_0}{1 + [\frac{\omega_L - \omega_0 \pm \mu b z/\hbar}{\gamma/2}]}. \tag{9.32}$$

Around $z = 0$ ($\mu b z \ll \hbar\gamma$) this expression can be expanded as a power series of $\mu b z/\hbar\gamma$.

Taking only the linear term into account yields with $\delta = \omega_L - \omega_0$:

$$F_z = -D \cdot z, \quad \text{with } D = R_0 \mu \cdot b \frac{16k \cdot \delta}{\gamma^2(1 + (2\delta/\gamma)^2)^2}. \tag{9.33}$$

We therefore obtain a restoring force that increases linearly increasing with z. The potential around the center of the MOT can be then described (because of $F_z = -\partial V/\partial z$) as the harmonic potential

$$V(z) = \frac{1}{2} D z^2. \tag{9.34}$$

The atoms oscillate like harmonic oscillators around $z = 0$ and are spatially stabilized.

Note There is a second force

$$F_\mu = -\boldsymbol{\mu} \cdot \text{grad}\, \boldsymbol{B},$$

which acts on atoms with a magnetic moment in an inhomogeneous magnetic field.

Inserting the numbers for sodium atoms, it turns out that this force is negligibly small compared to the recoil force at laser powers in the milliwatt range. At very low temperatures, however, this force is essential to trap the atoms after the laser beams have been shut off (see Sect. 9.1.10).

In the discussion above we have neglected the velocity-dependent force in the optical molasses (Sect. 9.1.5).

The total force acting on an atom in the MOT

$$F_z = -Dz - av,$$

results in a damped oscillation around the center with an oscillation frequency

$$\Omega_0 = \sqrt{D/m}, \tag{9.35}$$

and a damping constant

$$\beta = a/m.$$

So far we have only considered the movement in the z-direction. The anti-Helmholtz coils produce a magnetic quadrupole field with three components. From Maxwell's equation div $B = 0$ and the condition $\partial B_x/\partial x = \partial B_y/\partial y$, which follows from the rotational symmetry of the arrangement, we obtain the relations

$$\frac{\partial B_x}{\partial x} = \frac{\partial B_y}{\partial y} = -\frac{1}{2}\frac{\partial B_z}{\partial z}. \tag{9.36}$$

The restoring forces in the x- and y-directions are therefore half of the forces in z-directions. The trapped thermal cloud of atoms fills an ellipsoidal volume.

Instead of using counter-propagating laser beams with σ^+ and σ^- polarization, one can also cool the atoms if the two beams have the same polarization but slightly different frequencies $\omega^+ = \omega_0 + \Delta\omega$ and $\omega^- = \omega_0 - \Delta\omega$, which guarantees that the atoms moving out of the trap are pushed back. These frequencies can be produced as the two sidebands generated by acousto-optic modulation of the incident laser beam tuned to the center frequency ω_0 of the atomic transition.

Example 9.9 With a magnetic field gradient of 0.04 T/m a sodium atom at $z = 0$ in a light trap with two counterpropagating σ^+-polarized laser beams L^+, L^- in the $\pm z$-direction with $I_+ = 0.8 I_{\text{sat}}$, $\omega_+ = \omega_0 - \gamma/2$ and $I_- = 0.15 I_{\text{sat}}$, $\omega_- = \omega_0 + \gamma/10$, the negative acceleration of a Na atom moving away from $z = 0$ reaches a value of $a = -3 \times 10^4\ \text{m/s}^2$ ($\hat{=} 3 \times 10^3$ g!), driving the atom back to $z = 0$.

Generally, the MOT is filled by slowing down atoms in an atomic beam (see Sect. 9.1.3). Spin-polarized cold atoms can also be produced by optical pumping in a normal vapor cell and trapped in a magneto-optic trap. This was demonstrated by Wieman and coworkers [1158b], who captured and cooled 10^7 Cs atoms in a low-pressure vapor cell by six orthogonal intersecting laser beams. A weak magnetic field gradient regulates the light pressure in conjunction with the detuned laser frequency to produce a damped harmonic motion of the atoms around the potential minimum. This arrangement is far simpler than an atomic beam. Effective kinetic temperatures of 1 μK have been achieved for Cs atoms. For more details on MOT and their experimental realizations see [1110, 1111] and [1160].

9.1.8 Optical Micro-traps

Besides the possibility to store atoms in a MOT there are several proposals for different realizations of very small traps for cold atoms which have been meanwhile experimentally verified. They use either electric or magnetic fields or the electromagnetic fields in laser beams for storing atoms. The goal of these micro-traps with dimensions of a few micrometres is the realization of small trapping volumes which yields for a given number of trapped atoms a high atom density and facilitates the achievement of Bose–Einstein condensation (see Sect. 9.1.10). The principle of all traps is based on the forces acting on atoms with a dipole moment in inhomogeneous fields. Choosing an appropriate geometry of these fields it is possible to create a potential minimum within a small volume where the atoms are trapped. These potential minima are generally too shallow for trapping atoms at room temperature and the atoms therefore have to be cooled before they can be trapped.

We will now present some examples for often used micro-traps.

(a) Atom Trap in the Evanescent Field of a Laser Wave

The first example is a micro-trap within the evanescent field of a laser wave closely above the horizontal surface of a transparent solid at $z = 0$, illustrated in Fig. 9.22 [1161].

A laser beam is sent through a prism and is totally reflected at the prism basis if the inclination angle $\theta > \theta_t$ exceeds the limiting angel θ_t for total reflection. The electric field of the totally reflected wave does not suddenly drops to zero at the reflecting boundary between the medium with refractive index n_1 and air ($n_2 < n_1$) but penetrates into the air with an exponentially decreasing amplitude $E(z) = E_0 \cdot \exp(-z/z_e)$ (*evanescent wave*). After a pathlength

$$z_e = \lambda/\left[2\pi n_1\sqrt{(\sin^2\theta - n_2^2)}\right]. \tag{9.37}$$

The amplitude has decreased to $1/e$ of its value at the boundary surface. Just above the surface therefore a large gradient of the electric field strength

$$E = E_0 e^{-z/z}\cos(\omega t - kx) \tag{9.38}$$

is realized. If the laser frequency ω_L is shifted against the eigenfrequency ω_0 of the atoms by the amount $\Delta\omega = (\omega_L - \omega_0)$ the atoms experience a potential

$$V(z) = \frac{1}{2}\hbar \cdot \Delta\omega \ln\left[\left(1 + 2\Omega^2(z)\right)/\left(4\Delta\omega^2 + \gamma^2\right)\right] \tag{9.39}$$

which can be positive or negative depending on the sign of $\Delta\omega$. The Rabi-oscillation frequency $\Omega = M_{ik} \cdot E/\hbar$ depends on the intensity of the laser wave and on the homogeneous linewidth γ of the atomic transition $E_i \to E_k$. The potential therefore

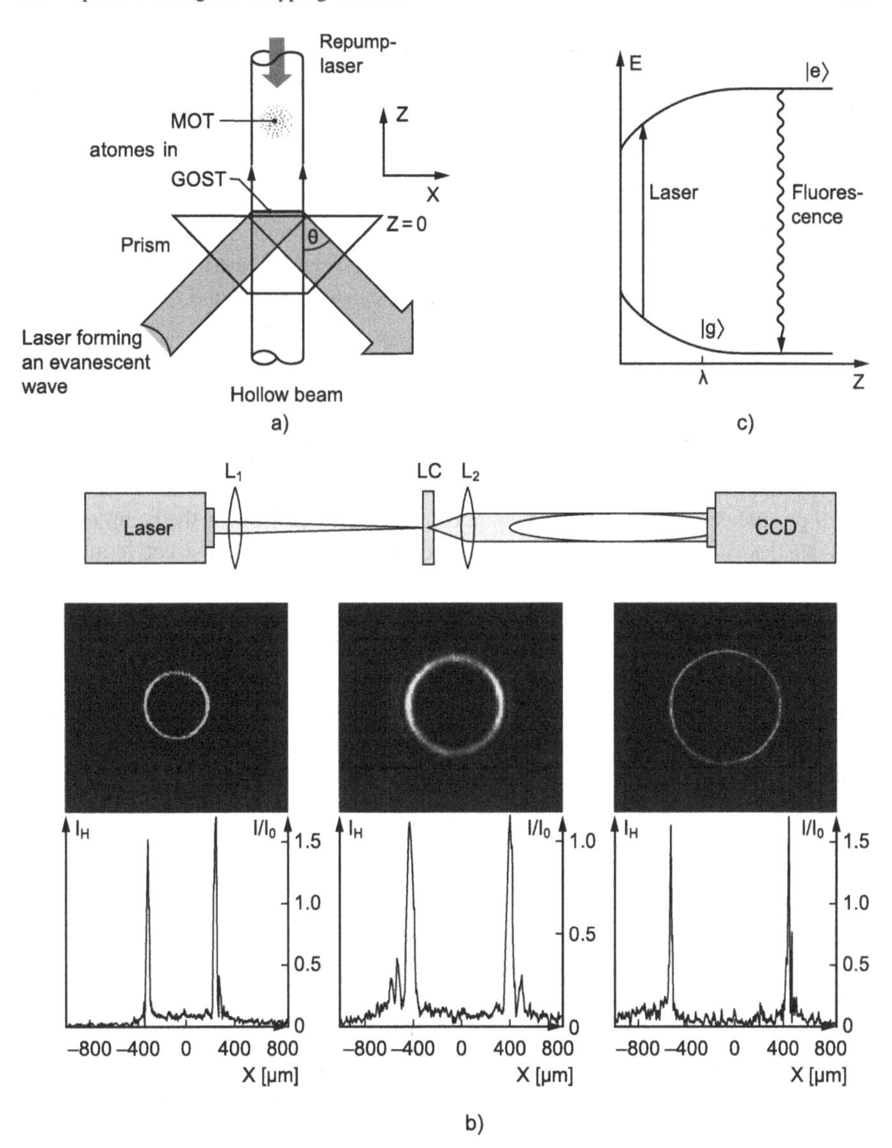

Fig. 9.22 (**a**) Scheme of a gravito-optical surface trap (GOST) with an evanescent wave and a hollow laser beam, (**b**) production of a hollow beam, (**c**) level scheme for optical cooling in the GOST

only depends on the laser detuning $\Delta\omega/\gamma$ and on the square $E^2(z)$ of the laser wave amplitude.

Because of the field gradient $\partial V/\partial z$ in the z-direction the atoms with the electric dipole moment μ experience a force in the $\pm z$-direction

$$F_z = -\mu \cdot \partial V/\partial z$$

which depends on the sign of the detuning $\Delta\omega$. For a repulsive force, acting into the $+z$-direction there is an equilibrium with the gravitational force $-m \cdot g$ for the distance z_e from the surface where

$$m \cdot g = -\mu \cdot \partial V / \partial z.$$

Here the potential has a minimum and the atoms can be trapped. For atoms falling from above onto the surface the evanescent wave acts like a trampoline. If their velocity is not too high, they are reflected and they oscillate around the equilibrium distance z_e.

In order to reach a stabilization in the x- and y-directions an additional laser with a donut intensity profile (x, y) travelling into the z-direction is superimposed (Fig. 9.22b). Its radial intensity gradient generates a stabilizing force in x- and y-directions [1162, 1163]. It is produced by sending a laser beam with a Gaussian beam profile through the focusing lens L_1 and an array of liquid crystals (Fig. 9.22c), which produces a phase shift, that changes across the diameter of the beam and results in a destructive interference at the center of the beam and constructive interference at the edge.

This works very well as can be seen from the radial intensity profiles in Fig. 9.22d. Because the trap is based on a combination of gravitations force and optical force field, it is often named GOST (gravito-optical surface trap).

In order to trap atoms at a temperature T by an inhomogeneous magnetic field the amount of negative potential energy $\mu \cdot B_{max}$ at the minimum of the trap should be larger than ηkT

$$\mu \cdot B_{max} > \eta kT.$$

This makes the total energy negative, i.e. $E_{total} = \eta kT - \mu B_{max} < 0$. The factor $\eta \approx 3$–5 ensures that also atoms with a kinetic energy well above the mean thermal energy (maximum of the Boltzmann distribution) are still trapped and the percentage of atoms leaving the trap is sufficiently small. At thermal energies the potential depth of the trap is not sufficient to keep atoms in the trap. Therefore the atoms have to be cooled down in a MOT to temperatures of a few μK. They can be then transported into the trap by photon recoil of a blue detuned repump-laser (Fig. 9.22a).

For further cooling of the atoms in the trap the different dependence of the atomic energy levels in the lower and upper state from the distance to the surface is used. For the lower state the force on between atom and surface is repulsive, in the upper state it is attractive. If the laser is red shifted against the atomic resonance it excites atoms close to the surface, which emit fluorescence on the average farther away from the surface. There the atom loses energy for each absorption-emission cycle.

(b) Joffe–Pritchard Traps

While the atomic trap based on the evanescent light wave uses a combination of laser field, static magnetic field and gravitational force the Joffe–Pritchard trap only

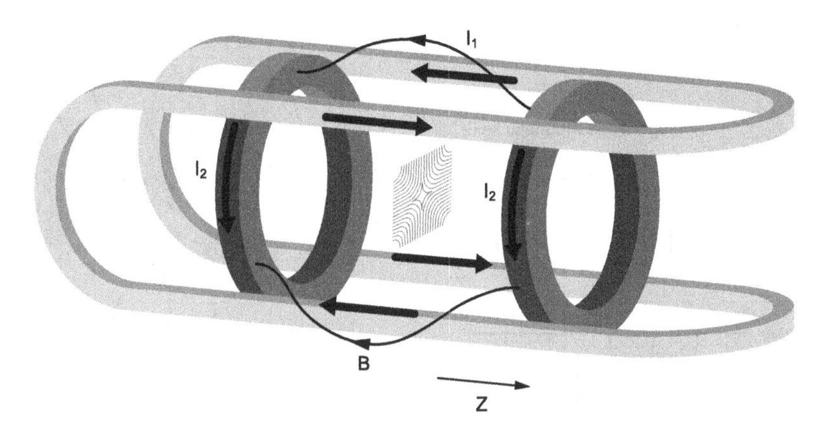

Fig. 9.23 Joffé–Pritchard trap

needs static magnetic fields: The field of a pair of Helmholtz coils superimposed by the magnetic field of 4 linear wires through which an electric current is sent (Fig. 9.23).The Helmholtz coils produce a static homogeneous field in z-direction which ensures that the magnetic field in the center of the trap where the atoms are stored is never zero. The 4 wires (Joffe-bars) generate a two-dimensional quadrupole field. The curved end pieces cause an axial stabilization of the atoms [1164].

Since the conducting wires path can be produced as conductor paths on a microchip a micro-trap can be realized. In order to understand the trapping characteristics of such traps we will consider the form of magnetic fields for various wire configurations and start with the magnetic field

$$B_{\mathrm{w}} = (\mu_0 I)/(2\pi r) \quad \text{with } r = \left(x^2 + y^2\right)^{1/2}$$

at a distance r from a straight wire in z-direction carrying the electric current I.

When the homogeneous magnetic field of a Helmholtz coil pair in y-direction is superimposed (Fig. 9.24a) the total field is

$$B_{\mathrm{total}} = \{0, B_0, 0\} + (\mu_0 I)/\left(x^2 + y^2\right)\{x, -y, 0\}.$$

The magnetic field lines are shown in Fig. 9.24b. On the line $x_{\mathrm{c}} = 0$, $y_{\mathrm{c}} = \mu_0 I/(2\pi B_0 I)$ at a distance

$$r_{\mathrm{c}} = (\mu_0 I)/(2\pi B_0)$$

from the wire the magnetic field of the wire becomes $B_{\mathrm{w}} = -B_0$, i.e. the two field just cancel. In the vicinity of this line $r = r_{\mathrm{c}}$ the total magnetic field can be described approximately as a linear quadrupole field, which means that $B_{\mathrm{total}} \sim (y - y_{\mathrm{c}})$. The potential energy of this field has a minimum at $y = y_{\mathrm{c}}$ and stabilizes the atoms in the radial direction but not in the axial direction. With increasing current I through the wire the line $r = r_{\mathrm{c}}$ i.e. $B = 0$ shifts towards larger distances r. It is therefore

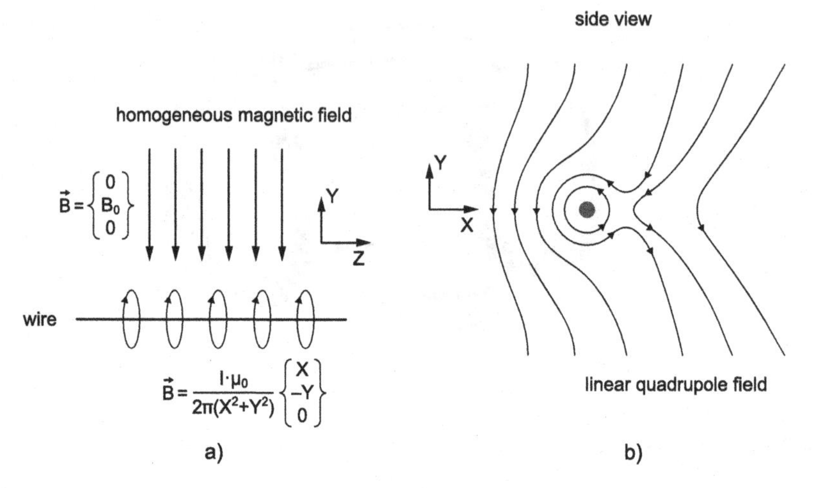

Fig. 9.24 (**a**) The two magnetic fields of a Joffé–Pritchard trap, (**b**) field lines of the total field

possible to decrease the volume of the trap to realize a micro-trap by decreasing the current I.

The radial force on an atom is

$$F_r = \partial B/\partial r = -(\mu_0 I)/(2\pi r^2).$$

Since r is very small, large forces can be realized even with moderate fields B, i.e. small currents I through the wires.

The direction of the magnetic moment against the quantization axis is characterized by the magnetic quantum number m_F where F is the quantum number of the total angular momentum (including nuclear spin). Atoms can be trapped around the potential minimum if their magnetic moment $\boldsymbol{\mu}$ is parallel to $\boldsymbol{B}$, because the potential energy is

$$E_{pot} = -\boldsymbol{\mu} \cdot \boldsymbol{B} = g_F m_F \mu_B |\boldsymbol{B}|$$

where g_F is the Lande-factor and μ_B is the Bohr magneton. Atoms with $g_F m_F > 0$ ($\boldsymbol{\mu}$ parallel to $\boldsymbol{B}$ = *low field seekers*) are pulled towards regions with low magnetic field and can be therefore trapped in a volume around the potential minimum (Fig. 9.25). Stabilization in the radial direction is, however, only guaranteed, if the magnetic moment μ always points during the motion of the Atom in the direction of the field gradient. This implies that the Lamor frequency

$$\omega_L = g_F \mu |\boldsymbol{B}|$$

must be large compared to the relative change $(\mathrm{d}\omega_L/\mathrm{d}t)/\omega_L$ of the Lamor frequency ω_L. This is no longer the case for the line $B = 0$ because there is $\omega_L = 0$: This could cause spin flips of the atoms which would then be pushed out of the trapping volume. This drawback can be overcome by a superposition of another axial magnetic field B_a which avoids that $B = 0$ within the trapping volume.

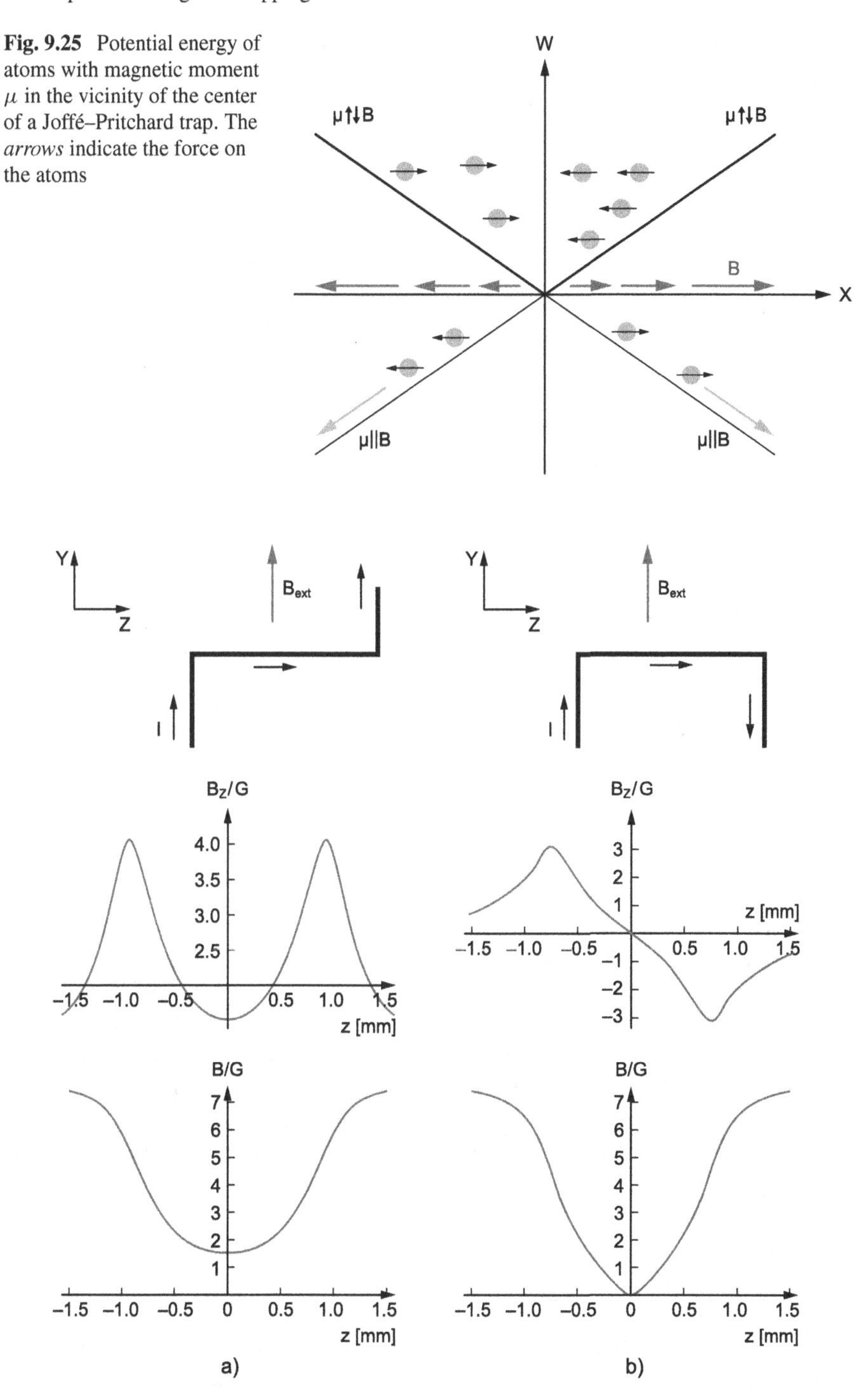

Fig. 9.25 Potential energy of atoms with magnetic moment μ in the vicinity of the center of a Joffé–Pritchard trap. The *arrows* indicate the force on the atoms

Fig. 9.26 Joffé trap (**a**) with Z-shaped electric wire, (**b**) with U-shaped wire with the corresponding magnetic field

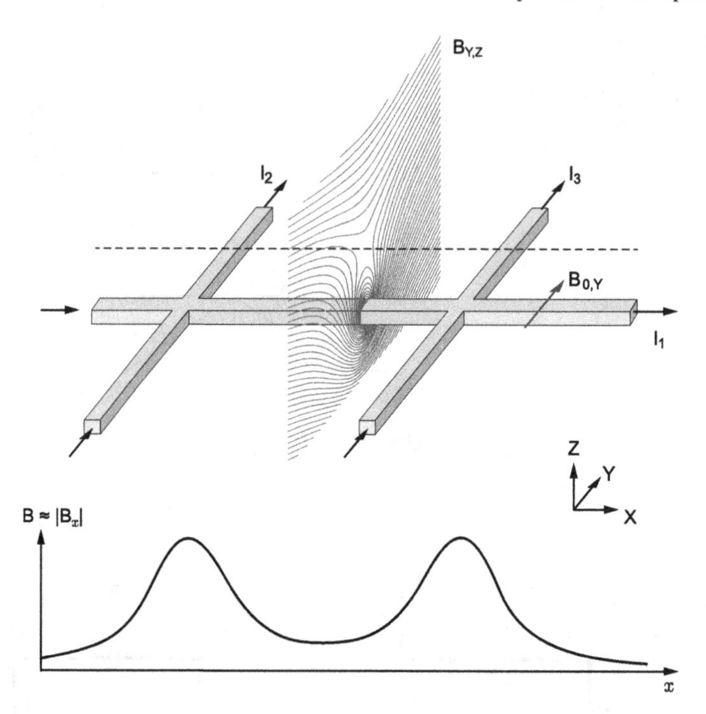

Fig. 9.27 Magnetic trap with crossed wires and a superimposed external magnetic field

The resulting force on the atoms is in the vicinity of y_c given by

$$F = \mu(\partial B/\partial y)_{y=y_c} = \mu_0 I/(2\pi r^3)y.$$

The potential around its minimum is approximately a harmonic potential with a linear restoring force. The atoms perform a radial harmonic oscillation.

The stabilization in the axial direction can be obtained by vertically bending the two end parts of the wire (Fig. 9.26) giving it a Z-form (Fig. 9.26a) or a U-form (Fig. 9.26b). The calculation gives the result that in both cases a minimum of the total magnetic field strength occurs. For atoms with $\boldsymbol{\mu}$ parallel to $\boldsymbol{B}$ (low field seekers) the field minimum is also a minimum of the potential energy, which means that the atoms can be trapped, where the trapping volume is smaller for the U-configuration.

(c) Other Micro-traps

Another realization of a micro trap which consists of crossed linear conducting wires and as superimposed homogeneous magnetic field B_{0y} in y-direction is shown in Fig. 9.27. The total magnetic field is similar to that shown in Fig. 9.26. The trap can be also realized as a micro trap with conducting wires deposited on a chip. The trap has a potential minimum in the mind between the two wires in y-direction slightly above the wire in x-direction (see also Fig. 9.26b). The trap can be further simplified by reducing the arrangement to two crossed straight wires in x- and y-directions with a superimposed homogeneous magnetic field in y-direction. In Fig. 9.27b the two field components B_x and B_z are plotted as a function of x [1164].

A further realization of a micro-trap uses a specially formed magnetic field of a permanent magnet which has a minimum in the mid of the arrangement.

9.1.9 Optical Cooling Limits

The lowest achievable temperatures of the trapped atoms can be estimated as follows: because of the recoil effect during the absorption and emission of photons, each atom performs a statistical movement comparable to the Brownian motion. If the laser frequency ω_L is tuned to the resonance frequency ω_0 of the atomic transition, the net damping force becomes zero. Although the time average $\langle v \rangle$ of the atomic velocity approaches zero, the mean value of $\langle v^2 \rangle$ increases, analogous to the random-walk problem [1166, 1167]. The optical cooling for $\omega - \omega_0 < 0$ must compensate this "statistical heating" caused by the statistical photon scattering. If the velocity of the atoms has decreased to $v < \gamma/k$, the detuning $\omega - \omega_0$ of the laser frequency must be smaller than the homogeneous linewidth of the atomic transition γ in order to stay in resonance. This yields a lower limit of $\hbar\Delta\omega < k_B T_{min}$, or with $\Delta\omega = \gamma/2$

$$T_{min} = T_D = \hbar\gamma/2k_B = \hbar/(2\tau \cdot k_B) \quad \text{(Doppler limit)}, \tag{9.40}$$

if the recoil energy $E_r = \hbar\omega^2(2Mc^2)^{-1}$ is smaller than the uncertainty $\hbar\gamma$ of the homogeneous linewidth.

Example 9.10

(a) For Mg^+ ions with $\tau = 2$ ns $\to \gamma/2\pi = 80$ MHz (9.40) yields $T_D = 2$ mK.
(b) For Na atoms with $\gamma/2\pi = 10$ MHz $\to T_D = 240$ μK.
(c) For Rb atoms with $\gamma/2\pi = 5.6$ MHz $\Rightarrow T_D = 140$ μK.
(d) For Cs atoms with $\gamma/2\pi = 5.0$ MHz $\Rightarrow T_D = 125$ μK.
(e) For Ca atoms cooled on the narrow intercombination line at $\lambda = 657$ nm with $\gamma/2\pi = 20$ kHz the Doppler limit $T_D = 480$ nK is calculated from (9.40).

Meanwhile, experiments have shown that, in fact, temperatures lower than this calculated Doppler limit can be reached [1168–1170]. How is this possible?

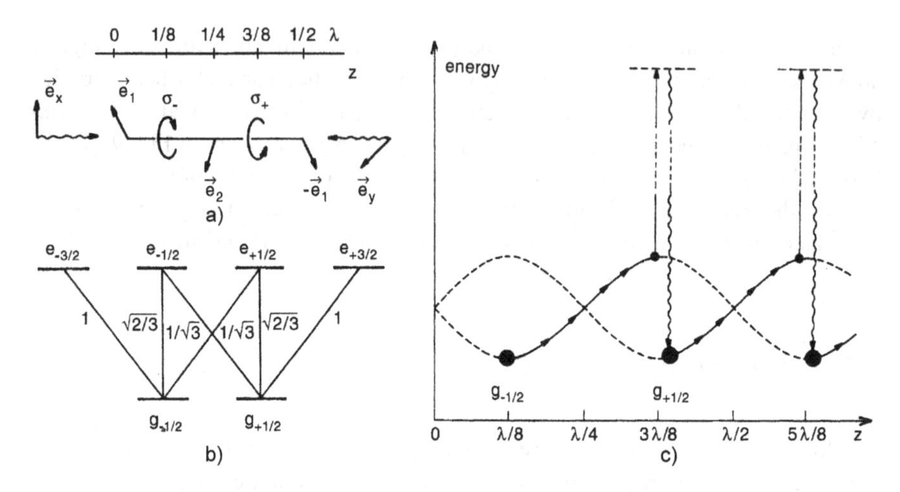

Fig. 9.28 Schematic diagram of polarization gradient (Sisyphus) cooling: (**a**) two counter-propagating linearly polarized waves with orthogonal polarization create a standing wave with z-dependent polarization. (**b**) Atomic level scheme and Clebsch–Gordan coefficients for a $J_g = 1/2 \leftrightarrow J_e = 3/2$ transition. (**c**) Atomic Sisyphus effect in the lin ⊥ lin configuration [1169]

The experimental results can be explained by the following model of *polarization gradient cooling* [1169–1171]. If two orthogonally polarized light waves travel anti-collinearly in $\pm z$-directions through the atoms of the optical molasses in a magnetic field, the total field amplitude acting on the atoms is

$$\boldsymbol{E}(z,t) = \boldsymbol{E}_1 \hat{\boldsymbol{e}}_x \cos(\omega t - kz) + E_2 \hat{\boldsymbol{e}}_y \cos(\omega t + kz). \tag{9.41}$$

This field shows a z-dependent elliptical polarization: for $z = 0$ it has linear polarization along the direction $\hat{\boldsymbol{e}}_1 = (\hat{\boldsymbol{e}}_x + \hat{\boldsymbol{e}}_y)$ (assuming $E_1 = E_2$), for $z = \lambda/8$ it has elliptical σ^- polarization, for $z = \lambda/4$ again linear polarization along $\hat{\boldsymbol{e}}_2 = (\hat{e_x} - \hat{\boldsymbol{e}}_y)$, for $z = 3\lambda/8$ elliptical σ^+ polarization, etc. (Fig. 9.28a).

For an atom at rest with the level scheme of Fig. 9.28b, the energies and the populations of the two ground-state sublevels $g_{-1/2}$, $g_{+1/2}$ depend on the location z because the energy depends on the amplitude of the standing laser field (dynamical Stark-shift). For example, for $z = \lambda/8$ the atom rests in a σ^- light field and is therefore optically pumped from both ground-state level vía the $e_{-3/2}$ and $e_{-1/2}$ excited levels into the $g_{-1/2}$ level, giving the stationary population probabilities $|(g_{-1/2})|^2 = 1$ and $|(g_{+1/2})|^2 = 0$, while for $z = (3/8)\lambda$ the atom is pumped by σ^+ light into the $g_{+1/2}$-level.

The electric field $\boldsymbol{E}(z,t)$ of the standing light wave causes a shift and broadening of the atomic Zeeman levels (*ac Stark shift*) that depends on the saturation parameter, which in turn depends on the transition probability, the polarization of $\boldsymbol{E}$, and the frequency detuning $\omega_L - \omega_0$. It differs for the different Zeeman transitions. Since the σ^- transition starting from $g_{-1/2}$ is three times as intense as that from $g_{+1/2}$ (Fig. 9.28b), the light shift Δ_- of $g_{-1/2}$ is three times the shift Δ_+ of $g_{+1/2}$. If the atom is moved to $z = (3/8)\lambda$, the situation is reversed, since now pumping occurs on a σ^+ transition.

The z-dependent energy shift of the ground-state sublevel therefore follows the curve of Fig. 9.28c. For those z values where a linearly polarized light field is present, the transition probabilities and light shifts are equal for the two sublevels.

The cooling now proceeds as follows: the important point is that the optical pumping between the sublevels takes a certain time τ_p, depending on the absorption probability and the spontaneous lifetime of the upper level. Suppose the atom starts at $z = \lambda/8$ and moves to the right with such a velocity that it travels a distance of $\lambda/4$ during the time τ_p. Then it always climbs up the potential $E^-_{pot}(g_{-1/2})$. When the optical pumping takes place at $t = 3\lambda/8$ (which is the maximum transition probability for σ^+-light!), it is transferred to the minimum $E^+_{pot}(g_{+1/2})$ of the $g_{+1/2}$ potential and can again climb up the potential. During the optical pumping cycle it absorbs less photon energy than it emits, and the energy difference must be supplied by its kinetic energy. This means that its velocity decreases. The process is reminiscent of the Greek myth of Sisyphus, the king of Corinth, who was punished by the gods to roll a heavy rock uphill. Just before he reached the top, the rock slipped from his grasp and he had to begin again. Sisyphus was condemned to repeat this exhausting task for eternity. Therefore polarization gradient cooling is also called *Sisyphus cooling* [1172].

Because the population density (indicated by the magnitude of the dots in Fig. 9.28c) is larger in the minimum than in the maxima of the potentials $E_{pot}(z)$, on the average the atom climbs uphill more than downhill. It transfers part of its kinetic energy to photon energy and is therefore cooled.

Depending on the polarization of the two counterpropagating laser beams the lin ⊥ lin configuration can be used with two orthogonal linear polarizations or the σ^+–σ^- configuration with a circularly polarized σ^+ wave, superimposed by the reflected σ^- wave. With Sisyphus cooling temperatures as low as 5–10 μK can be achieved.

The lower limit for the achieved temperature is given by the recoil energy that each atom acquires when it emits or absorbs a photon. This recoil limit is reached when the thermal energy kT equals the recoil energy $p^2/2M = \hbar^2 k^2/2M$

$$k_B T_{recoil} = \hbar^2 k^2/2M = h^2/\left(\lambda^2 \cdot 2M\right) \quad \text{(recoil limit)}, \tag{9.42}$$

where M is the mass of the atom and $\hbar \boldsymbol{k}$ the momentum of the photon.

Example 9.11 For Na atoms $m = 23$ AMU, $\lambda = 589$ nm $\to T_{recoil} = 1.1$ μK. This is 220 times lower than the Doppler limit.

The recoil limit can be overcome by a recently discovered cooling scheme called Raman cooling (Fig. 9.29). Here a stimulated Raman scattering process is used, which traverses from level 1 via a virtual level down to level 3. If the levels 1 and 3

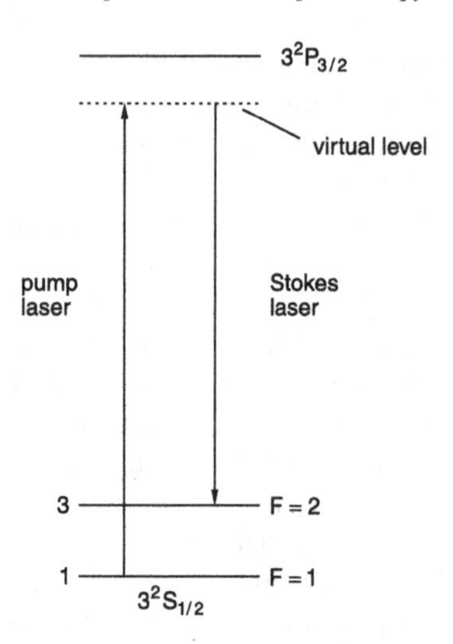

Fig. 9.29 Level scheme for Raman cooling

are hyperfine components of the electronic ground-state, the energy difference between 1 and 3 is very small compared to $h\nu$. Since the energies of the exciting photon and Raman photon are nearly equal, the absorption and emission recoils cancel each other out because both photons travel in the same direction. Since the Stokes photon has a slightly smaller momentum than the incident photon, this small difference results in a cooling of the translational energy of the atom. Because no fluorescence photon is emitted, the statistically varying recoil present for Sysiphos cooling is missing here, so the cooling limit can be pushed further down.

Review articles about laser cooling can be found in [1173–1175].

9.1.10 Bose–Einstein Condensation

At sufficiently low temperatures where the de Broglie wavelength

$$\lambda_{\mathrm{DB}} = \frac{h}{m \cdot v}, \tag{9.43}$$

becomes larger than the mean distance $d = n^{-1/3}$ between the atoms in the cold gas with density n, a phase transition takes place for bosonic particles with integer total spin. More and more particles occupy the lowest possible energy state in the trap potential and are then indistinguishable, which means that all these atoms in the same energy state are described by the same wave function (note that for bosons the Pauli exclusion principle does not apply). Such a situation of a macroscopic state occupied by many indistinguishable particles is called a *Bose–Einstein condensate* (BEC, Nobel prize in physics 2001 for E. Cornell, W. Ketterle, and C. Wiemann).

More detailed calculations show that BEC is reached if

$$n \cdot \lambda_{\mathrm{DB}}^3 > 2.612. \tag{9.44}$$

With $v^2 = 3k_B T/m$ we obtain the de Broglie wavelength

$$\lambda_{DB} = \frac{h}{\sqrt{3mK_B T}}, \tag{9.45}$$

and the condition (9.45) for the critical density becomes

$$n > 13.57(m \cdot k_B T)^{3/2}/h^3.$$

This gives the critical temperature T_c, where Bose–Einstein-condensation starts

$$T_c = (0.08h^2 n^{2/3})/(m \cdot k_B). \tag{9.46}$$

The critical temperature increases with the atomic density as $n^{2/3}$. The minimum density for BEC depends on the atomic mass and on the temperature and decreases with $T^{3/2}$ [1110, 1176].

Example 9.12 For Na atoms at a temperature of 10 μK the critical density would be $n = 6 \times 10^{14}/\text{cm}^3$, which is at present not achievable. For experimentally realized densities of 10^{12} cm^{-3} the atoms have to be cooled below 100 nK in order to reach BEC.

For rubidium atoms BEC was observed at $T = 170$ nK and a density of 3×10^{12} cm^{-3}.

9.1.11 Evaporative Cooling

The numerical examples illustrates that the usual techniques of optical cooling are not sufficient to reach BEC at realistic atomic densities unless one uses the experimentally difficult Raman-cooling or the coherent generation of dark states (see Sect. 7.10).

Fortunately there exists a very old but efficient method for cooling liquids, namely evaporation cooling [1177] which is used in daily life when cooling hot coffee in a cup by blowing over its surface. For achieving BEC the technique is applied as follows: The atoms are optically precooled in a MOT.

These optically cooled atoms are then transferred from the MOT to a purely magnetic trap (Fig. 9.30), where the restoring force

$$\boldsymbol{F} = \mu \,\text{grad}\, \boldsymbol{B},$$

keeps the cold atoms in a cloud around the minimum of the potential energy $W = -\mu \cdot B$ at the center of the magnetic trap. The total energy of the atoms is $W = W_{pot} + W_{kin}$. The atoms with the highest kinetic energy occupy the highest energy levels of the trap. Now the particles with the highest energy are removed

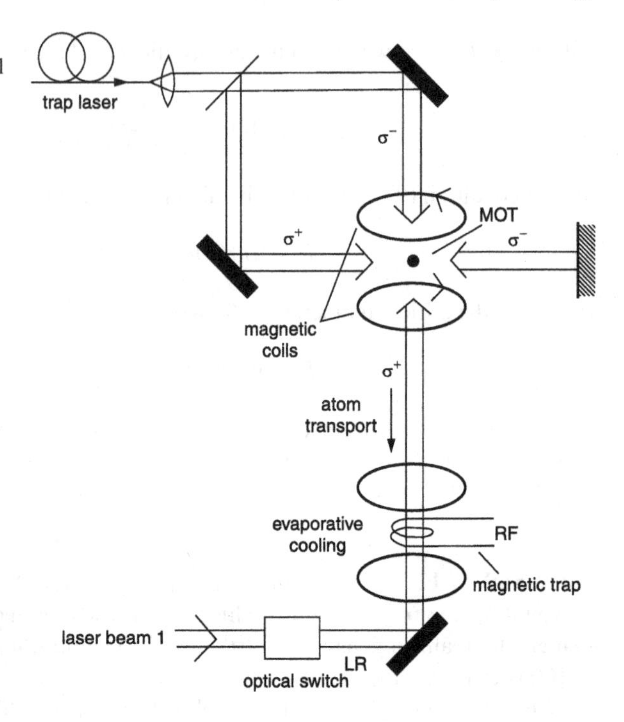

Fig. 9.30 Schematic experimental setup for optical cooling in the MOT; atom transport is achieved in the magnetic trap by switching off laser beam 1

from the trap, thus perturbing the equilibrium velocity distribution. This is achieved by a radio-frequency field that induces spin flips and thus reverses the sign of the force, changing it from a restoring to a repellent force. This is shown in Fig. 9.31b in a potential diagram. If the remaining atoms suffer sufficient mutual collisions, a new equilibrium state is reached with a lower temperature. Again, the upper part of the Maxwellian velocity distribution is thrown out of the trap. Lowering the radio frequency continuously results always in a loss of the atoms with the highest kinetic energy, which reach the largest distance from the trap center and therefore have the largest Zeeman splittings.

The cooling process must be slow enough to maintain thermal equilibrium, but should be sufficiently fast in order not to lose particles from the trap by collision-induced spin flips [1178].

The phase transition to BEC manifests itself by the sudden increase of the atomic density (Fig. 9.32), which can be monitored by measuring the absorption of an expanded probe laser beam (Fig. 9.33). As example of BEC in a magnetic trap Fig. 9.34 shows the spatial confinement of the atomic cloud in a Z-shaped Joffe–Pritchard micro-trap (see Sect. 9.1.8b), where a homogeneous magnetic field in y-direction is superimposed to the field produced by the current through the wire. The volume of the atomic cloud is an ellipsoid with rotational symmetry around the z-axis.

Meanwhile BEC in various trap configurations has been obtained in many labs worldwide. After the first realization at JILA in Boulder and at MIT several labs had reported BEC only a few months later, e.g. by G. Rempe University of Konstanz,

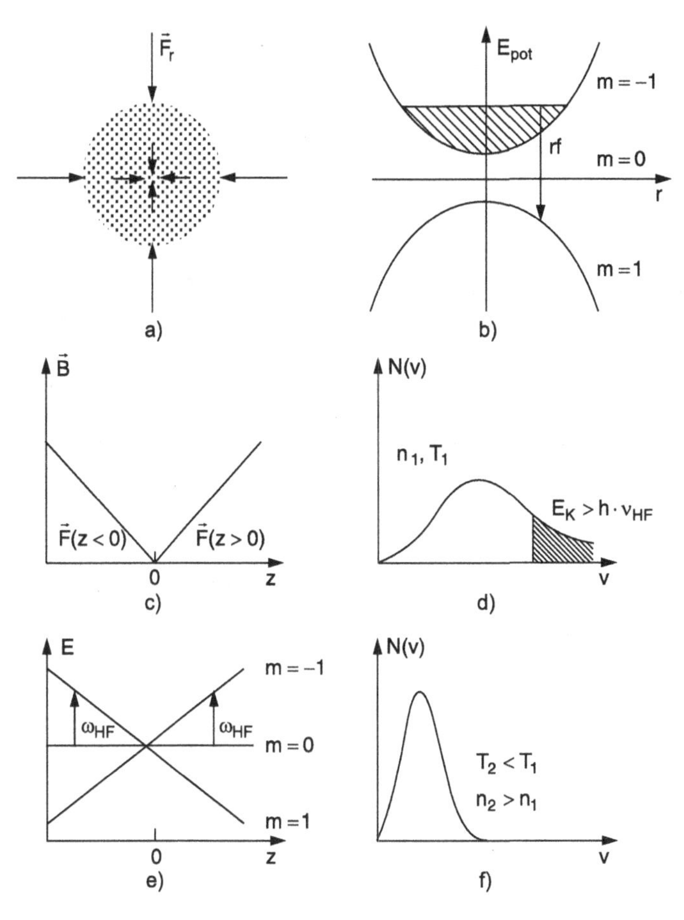

Fig. 9.31 Evaporative cooling of a cold atom cloud in a magnetic field. (**a**) Optical molasses; (**b**) trap potential; (**c**) magnetic field $B(z)$; (**d**) Boltzmann distribution before evaporative cooling; (**e**) HF transitions; (**f**) $N(v)$ after evaporation cooling

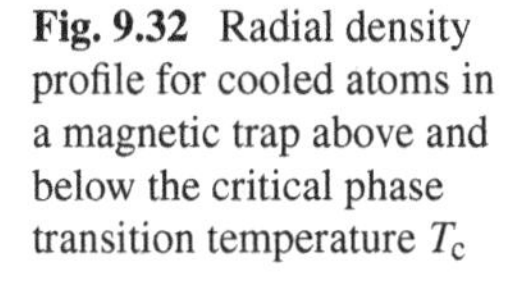

Fig. 9.32 Radial density profile for cooled atoms in a magnetic trap above and below the critical phase transition temperature T_c

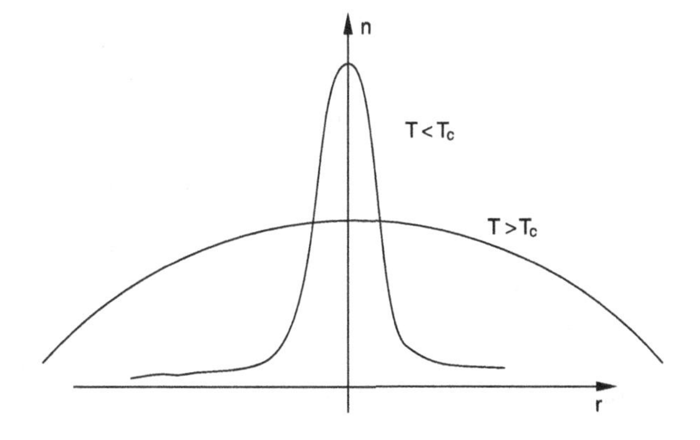

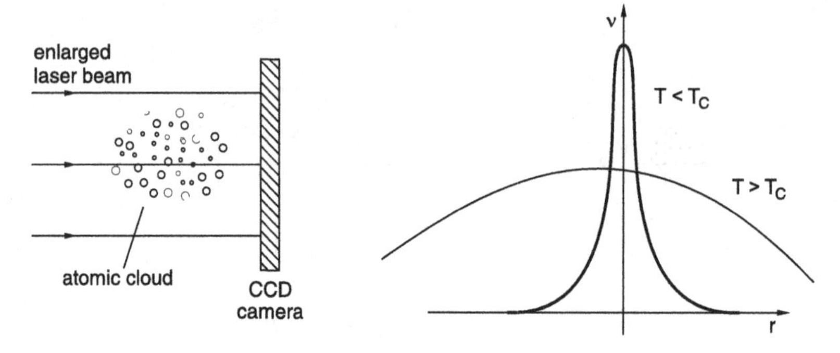

Fig. 9.33 (**a**) Measurement of spatial extension and radial density profile of the cold atomic cloud through the absorption of an enlarged weak laser beam. (**b**) Radial density profiles above and below the critical phase transition temperature T_c

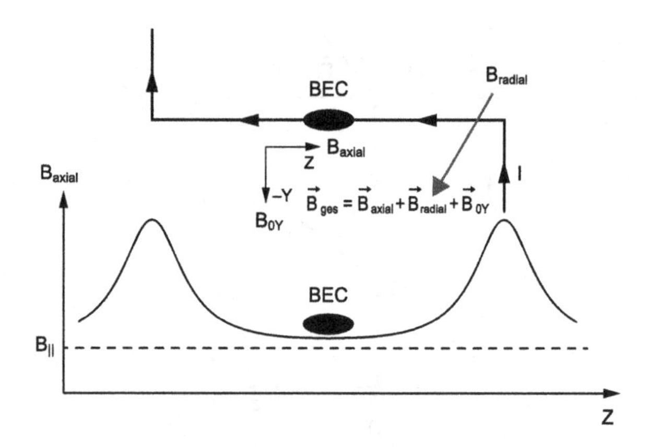

Fig. 9.34 BEC in a Joffé–Pritchard trap with Z-shaped wires

T.W. Hänsch LMU München, W. Ertmer University Hannover, D. Kleppner MIT, R.G. Hulet at Rice University and many others.

If the atoms are sufficiently cold, they can be trapped in the potential minima of a three-dimensional standing laser wave field (optical lattice). The idea has been already proposed by Lethokov [1184], but could not be realized at that time because the lowest obtainable temperatures were still too high. Nowadays many experiments have been performed with cold atoms or molecules in optical lattices (see Sect. 9.1.16).

The BEC can be kept only for a limited time. There are several loss mechanisms that result in a decay of the trapped particle density. These are spin-flip collisions, three-body recombination where molecules are formed, collisions with excited atoms where the excitation energy may be converted into kinetic energy, and collisions with rest gas atoms or molecules. The background pressure therefore has to be as low as possible (typical pressures are 10^{-10} to 10^{-11} mbar).

The scattering length a of the atoms in the BEC determines the elastic cross section $\sigma_{el} = 8\pi a^2$. The value of a is positive for a repulsive mean potential of the condensate. For negative scattering lengths the condensate will finally collapse.

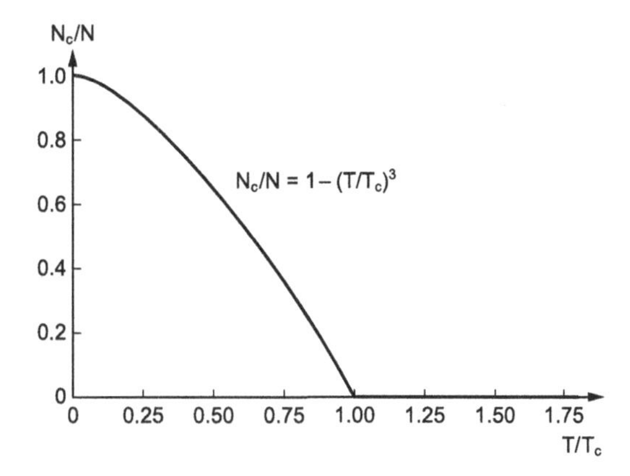

Fig. 9.35 Fraction N_c/N of condensed atoms as a function of the normalized temperature T/T_c

It is interesting to note that the value of the scattering length a for atoms A can be obtained from laser spectroscopy by measuring the energy and the vibrational wave function of the last bound vibrational level in the ground-state potential of the molecule A_2 [1194].

9.1.12 Properties of the Bose–Einstein Condensate

When the temperature T decreases below the critical temperature T_c BEC starts at first only for a small fraction of all atoms in the magnetic trap, i.e. not all atoms are immediately transferred into the coherent state of BEC. With decreasing temperature the fraction of the condensed atoms increases (Fig. 9.35). The BEC-state is separated from the normal state with $T > T_c$ by an energy gap similar to the energy gap between the Cooper pair state in supra-conductivity and the normal state. One therefore has to pump energy into the condensate in order to convert the assemble into the normal state. Also the energy distribution changes from a Maxwell–Boltzmann distribution above T_c to a distribution

$$N_c = N_t\left(1 - (T/T_c)^3\right)$$

of the condensed atoms below T_c, where N_t is the total number of atoms in the trap. Only at $T = 0$ all atoms are in the coherent BEC state. In Fig. 9.36 the density distribution $N(E)$ is shown for three different temperatures.

Since the atomic density in a BEC is very high, collisions between atoms take place. Because of the low temperature (a few nano-Kelvin) the relative velocity is extremely small. Since the angular momentum $L = q\hbar$ $(q = 1; 2; 3\ldots)$ is quantized, the relative velocity is not high enough to transfer even the minimum amount $\Delta L = md \cdot \Delta v$ of the angular momentum $|\boldsymbol{L}| = L = m \cdot v \cdot d$ (d = mean distance between the collision partners) during a collision. Therefore only S-scattering with $\Delta L = 0$ can occur which means that only elastic collisions can happen, which only change the velocity of the collision partners but not their internal energy. These elas-

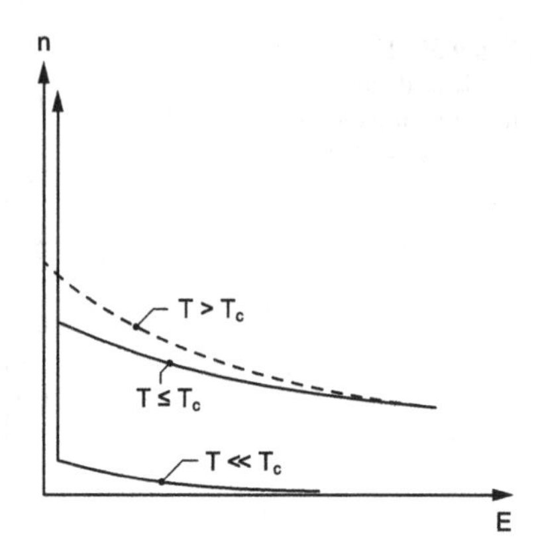

Fig. 9.36 Density distribution $n(E)$ above and below the critical temperature T_c

tic collision are responsible for the evaporative cooling which would not be possible without any collisions because they restore thermal equilibrium.

The question is, which role collision play in a BEC. The atoms are exposed to a laser field which excites them into higher electronic states. In collisions between excited atoms and ground state atoms the excitation energy can be transferred either into excitation energy of the other atom or into translational energy. The latter case will increase the velocity of atoms which then cannot be kept by the trapping potential and leave the trap. Another possibility is the associative recombination of a collision pair (see Fig. 9.17 and Sect. 9.1.6), where a stable dimer molecule is formed which, however, cannot be kept in the magnetic trap because its spin is different from that of the atoms.

All atoms for which BEC has achieved, have a magnetic moment which is due to the vector sum of electron spin and nuclear spin. Collisions between such atoms can flip the spin and therefore the direction of the magnetic moment. This will lead to an escape of the atoms from the trap and will decrease the density of the trapped atoms. Spin flips during a collision are due either to the direct dipole–dipole interaction between the collision partners or to an indirect interaction between the nuclear spins which is mediated by the electron spins, since they are coupled through the hyperfine-interaction with the nuclear spins.

The potential energy of these interactions can be positive of negative which is determined by the sign of the scattering length a [1180]. For $a < 0$ the interaction potential is attractive, for $a > 0$ repulsive. Any BEC can only be stable for $a > 0$ because for an attractive potential the atoms would condense into a solid [1181, 1182]. However, it could be shown that for a Li BEC the condensate was stable below a critical density even for $a < 0$ [1185].

Strontium atoms have a large scattering length ($a = +800a_0$) and suffer therefore large collision losses. Nevertheless BEC of ^{86}Sr atoms has been achieved at low atomic densities of $3 \times 10^{12}\ \text{cm}^{-3}$ where the condensate was stable for several seconds [1186].

A detailed discussion of the role of collisions in a BEC can be found in [1180, 1185].

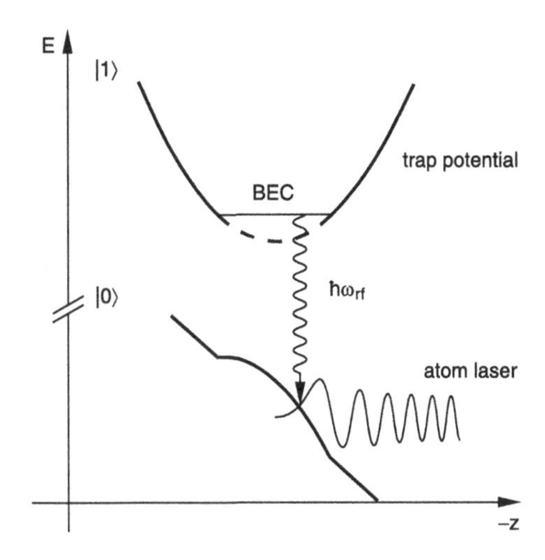

Fig. 9.37 Generation of an atom laser by extracting atoms out of a BEC

Many features of a BEC are similar to those of supra fluids. For example it is possible to create vortices by stirring the BEC with a laser beam for inducing the transfer of angular momentum $L = q \cdot \hbar$. The rotational energy of such a vortex is

$$E_{\text{rot}} = \pi N L^2 \hbar^2 / m \left(\ln(1.46b/\xi)\right),$$

where N is the atomic density, b the distance from the vortex center and ξ the extension of the vortex.

9.1.13 Atom Lasers

BEC represents a macroscopic quantum phenomenon similar to supra-conductivity or supra-liquidity. It opens the way for many new investigations which give insight into the collective behaviour of atoms or molecules. All atoms in a BEC are coherent, i.e. they are described by the same wave function. If atoms of a BEC are released from the trap they form a coherent flux of atoms. This can be realized either by switching off the trap field which lets the atoms freely fall in the gravitational field, or by inducing spin flips by rf-radiation (Fig. 9.37). Such a coherent atomic beam called *atom laser* or *BOSER* (in analogy to the LASER which represents a coherent beam of photons) reaches atomic densities which are larger by several orders of magnitude compared with an incoherent atomic beam [1179] similar to a laser where the photon density can be higher by many orders of magnitude than in an incoherent light beam. The divergence of such a coherent beam of atoms is very small. The brightness of the atom laser defined as the integrated flux of atoms per source area, divided by the velocity spread $\Delta v_x \cdot \Delta v_y \cdot \Delta v_z$ is very high and can reach values of 2×10^{24} atoms $(\text{s}^2\,\text{m}^{-5})$ and is therefore 4 orders of magnitude higher than in supersonic beams [1176].

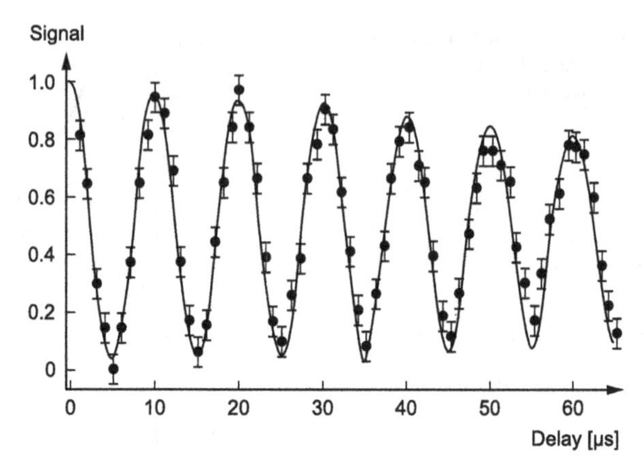

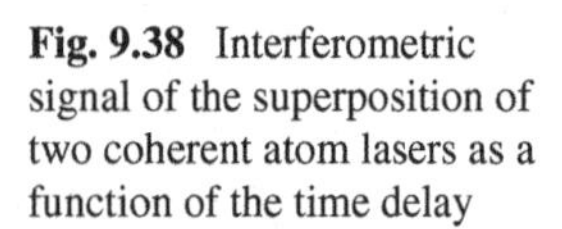
Fig. 9.38 Interferometric signal of the superposition of two coherent atom lasers as a function of the time delay

A quasi-continuous atom laser has been realized in the group of Th.W. Hänsch by using a weak radio-frequency field with a small coupling strength as output coupler for the BEC. Atoms could be continuously extracted from the BEC over a time interval up to 100 ms [1187].

The duration was limited by the finite number of atoms in the BEC and could be extended with improved techniques for increasing the number of trapped atoms.

Such atom beams with high brightness and small velocity spread are very useful for investigations of collision processes and chemical reactions at very low temperatures.

It could be proved that the superposition of two coherent atom laser beams which were released from two separate atomic traps results in interference structures similar to the superposition of coherent light waves [1179]. In Fig. 9.38 the total intensity of the superimposed two beams from two different BEC traps is shown as a function of the delay time between the two beams. The active medium of the laser which acts as amplifier corresponds to the BEC of the BOSER which is the reservoir for the atom laser and acts as amplifier because nearly all atoms in the BEC are in their ground state and feed the atom laser as long as the trapping potential is switched off.

There are, however, also principal differences between photon laser and atom laser:

- *Photons* can be generated and destroyed through emission and absorption which is not possible for *atoms*.
- Atoms in a collimated beam are deflected because of the larger mass in the gravitational field of the earth, while the deflection of photons is negligibly small.
- Atoms can collide with each other which diminishes the density and limits the extension length of an atom laser, while the collision cross section for photon–photon collisions is exceedingly small. One might compare the loss through collision in an atom laser with the spontaneous emission which also represents a loss mechanism.
- The BE-condensate is in thermal equilibrium at very low temperatures while the active medium of a laser extremely deviates from thermal equilibrium at high

temperatures because the inverted population is opposite to the Boltzmann distribution.

A more detailed discussion of the comparison between photon laser and atom laser can be found in [1183].

Since photons are also Bosons with spin $1\hbar$ the question arises whether a BEC can be realized with photons. Since their mass $m = h\nu/c^2$ is much smaller than that of atoms, BEC should start already at much higher temperatures according to Eq. (9.46). However, because there exist no photons at rest a photon condensate has to be defined in another way, namely as an ensemble of non-distinguishable photons, which are all in the same quantum state, i.e. in the same mode of the radiation field inside a resonator with a high Q-value. A first demonstration of such a photon condensate inside an optical micro-resonator was published by Weitz and coworkers in Bonn [1188].

9.1.14 Production and Trapping of Cold Fermi-Gases

Fermions (particles with a spin of $n \cdot \hbar/2$ and $n =$ odd integer) play an important role in many fields of physics, because all stable elementary particles, such as electron, proton and neutron are Fermi particles with a spin of $\hbar/2$. Also atoms with a total angular momentum of $n \cdot \hbar/2$ ($n =$ odd) are Fermions, such as the lithium isotope ^{6}Li with a nuclear spin $I = 1 \cdot \hbar$ and an electron spin $s = \hbar/2$, resulting in a total angular momentum of $F = 3/2 \cdot \hbar$ or $F = 1/2 \cdot \hbar$.

For Fermions the Pauli principle holds that any quantum state can be only occupied by at most one Fermion. Therefore a direct Bose–Einstein condensation of Fermions is not possible. However, two Fermions may couple to a Fermi-pair which is a Boson such as in supra-conductivity where two electrons couple to a Cooper pair and the condensation is realized by the Bosonic Fermi-pairs. Similar processes occur in supra-fluidity where two ^{3}He atoms (Fermions) couple to a Boson.

The question is now, whether such pairing can also happen for atoms in the gas phase, if the gas is cooled down to sufficiently low temperatures. The experimental possibilities of spectroscopic investigations of cold Fermi-pairs might elucidate many unclear aspects of supra-conductivity in solids or of supra-fluidity [1189].

The difference between Bose- and Fermi-gases becomes significant only at low temperatures when Bosons form a BEC, where all particles are in the lowest energy state while Fermions occupy all states up the Fermi energy.

Most experiments for achieving cooling of Fermi gases have been performed with alkali atoms. There are only two stable Fermionic alkali isotopes namely ^{6}Li and ^{40}K.

The techniques for cooling and trapping are the same for Fermions and Bosons: Optical cooling by laser beams and trapping in magnetic traps. In order to reach sufficiently low temperatures where BEC starts, evaporation cooling is used for Bosons (see Sect. 9.1.11). This technique cannot be used for Fermions because of the following reason: In order to maintain thermal equilibrium during the cooling process, elastic collisions are necessary. Such collisions are not possible with Fermions, because the Pauli principle does not allow Fermions in the same quantum state to come closely together, which is necessary for a collision. This can be illustrated by observing a cold gas cloud of Bosons and compare its size with a cloud of Fermions under identical conditions. While at higher temperatures no difference is observed, for temperatures below the Fermi temperature the size of the Fermi gas cloud is much larger than that of the Bose gas. The reason for this difference is the "*Fermi-pressure*" due to the Pauli-principle, which prevents a collapse of the Fermi-gas cloud and is, for example, responsible for the stability of white dwarfs stars.

In order to reach very low temperatures also for Fermi gases one has to look for other cooling mechanisms. One possibility is the *sympathetic cooling* where a mixture of Bose- and Fermi-gases (e.g. ^{6}Li and ^{7}Li) is used. The Bosons are cooled by one of the above discussed methods (e.g. by evaporation cooling) and elastic collisions lead to thermal equilibrium between Bose particles and Fermi particles and establish equal temperatures for the two species. However, I order to realize the formation of Fermion-pairs. Still lower temperatures are necessary. A promising technique uses a mixture of Fermions in different spin states which are then not identical and thus the Pauli principle does not exclude that pairs of Fermions can be formed. Such Fermions must furthermore show an attractive potential to allow the formation of stable pairs. With a tunable external magnetic field the potential energies of the atoms can be shifted in such a way, that they coincide with the energy levels of the stable pair. This leads to the formation of stable diatomic molecules (see next section).

The first condensation of a Fermi gas was realized by the research group of Deborah Jin [1191] using ^{40}K isotopes in a double MOT, where temperatures below 400 nK were reached. Here the atoms are precooled in a first MOT down to 150 μK by laser cooling. Then they are pushed by a recoil laser beam into a second MOT, where an ultrahigh vacuum was maintained in order to avoid collision with the residual gas. The atoms are then trapped in a Joffe–Pritchard trap and are pumped by optical pumping into two different hyperfine levels. Atoms in different levels are no longer identical and can collide with each other. Therefore evaporation cooling can be applied which finally brings the gas to temperatures below the critical temperature for BEC.

Further details about cooling of Fermions can be found in [1192].

9.1.15 BEC of Molecules

Recently the Bose–Einstein condensation of diatomic molecules has been reported by several groups [1195–1197, 1199–1201]. At first sight this appears astonishing, because molecules do not represent two-level systems and therefore cannot be optically cooled like atoms. One must therefore look for other cooling schemes. We will discuss some of them here.

A promising way to produce cold molecules is the photoassociation of cold atoms in a Bose–Einstein condensate, which was discussed in Sect. 5.5.6. Here cold

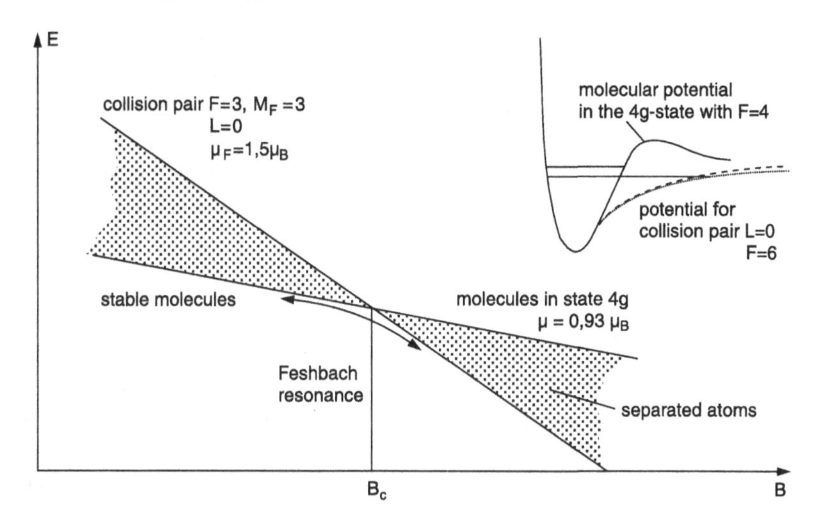

Fig. 9.39 Potential energy of the collision pair $\mathrm{Cs}(F=3, M_\mathrm{F}=3)+\mathrm{Cs}(F=3, M_\mathrm{F}=3)$ and the molecule Cs_2 as a function of an external magnetic field [1201]

molecules in stable higher vibrational levels of the electronic ground state can be produced with very low translational energy.

Another method is based on the sympathetic cooling of molecules in a cold atomic gas which does not react with the molecules but cools their translational energy down to the atomic temperature.

A very elegant method that was finally successfully used to generate molecular BECs uses the difference in the potential energy of atoms with electron spin and molecules in a magnetic field [1199]. In Fig. 9.39, the energy of a collision pair Cs + Cs with parallel electron spins and that of the molecule Cs_2 is plotted as a function of the magnetic field strength B. The field dependence $E(B)$ is stronger for the collision pair with the atoms in hyperfine levels with the quantum numbers $F=3$; $M_F=3$; $L=0$ and the magnetic moment $\mu=1.5\mu_\mathrm{B}$ than for the molecule in the $4g$ hyperfine level with $F=4$ and $\mu=0.93\mu_\mathrm{B}$. At a critical field strength B_c the two lines cross each other. This means that a molecular level exists which has the same energy as the collision pair (Feshbach resonance; see the inset in Fig. 9.39). On the left side of the Feshbach resonance, the molecular state has a lower energy and thus stable molecules can be formed. If the magnetic field is slowly (adiabatically) scanned from higher field strengths across the Feshbach resonance to lower values, stable molecules can be produced. This is monitored by the corresponding decrease in the atomic density in the BEC. When the field strength is again increased to higher values above B_c, the atomic density increases again, which means that the molecules dissociate.

Since the dimer Cs_2 has an integer total spin, it is a boson and can form a Bose–Einstein condensate [1200]. Spectroscopy of these dimers, which represent huge molecules with an internuclear separation of about 100 nm (!) gives information on the energy of the last bound molecular levels directly below the dissociation energy. Often such levels are only stabilized by the centrifugal barrier for states with rotational quantum numbers $J > 0$. The binding energies of these states allow more detailed analysis of the different contributions to the long-range forces of weakly bound systems [1201].

Feshbach-like resonances in a molecular cesium BEC indicate that even molecular dimers $(Cs_2)_2$ can be formed from the atomic dimers Cs_2 when the magnetic field is tuned over molecular resonances [1202a].

9.1.16 Atoms and Molecules in Optical Lattices

Sufficiently cold atoms can be trapped in optical lattices, formed by a three-dimensional overlap of standing laser waves in x-, y-, and z-directions. If the kinetic energy of the atoms is lower than the potential energy minima of such lattices they are trapped in these minima and cannot leave their fixed positions. Such a state is called a *Mott insulator* (in analogy to the situation in solid state physics explained by Sir Nevil Mott where a Mott insulator describes a situation where electrical conductivity should be possible according to the band structure but the repulsive interaction between the electrons prevents electron transport). The potential well depth can be tuned by changing the intensity of the three laser beams.

The research group of Th.W. Hänsch succeeded in 2002 to trap for the first time cold atoms in such an optical lattice. They cooled at first the atoms below the critical temperature fort BE condensation. Then the well depth of the optical lattice was increased. This transferred the coherent state of the atoms in the free BEC, where all atoms are in the same state i.e. described by the same wavefunction and are therefore not distinguishable, into the incoherent Mott state, where each atom sits on its separate location and can be therefore distinguished from the other atoms. Decreasing the well depth brings the atoms again back into the coherent BEC state. Just by changing the well depth of the optical lattice switches the atomic ensemble from a coherent into an incoherent state and back [1204].

Example 9.13 In a standing light wave with $\lambda = 600$ nm and a mean intensity of 10 W cm^2 the intensity gradient between maxima and nodes is grad $I = 6 \times 10^{11}$ W/cm^3. For a detuning $\Delta\omega = \gamma = 6 \times 10^7$ Hz and a saturation parameter $S = 10$ the dipole force acting on an atom is $F_D = 10^{-17}$ N. The well depth is about 10^{-5} eV corresponding to a temperature of 100 mK.

By choosing a suitable atom density it is possible to place one atom on each potential minimum of the lattice. Such atoms isolated from their surrounding are excellent candidates for precise atomic clocks if one chooses a narrowband forbidden electronic transition to a metastable atomic state as clock transition. Since the atoms are (besides their zero-point motion) at rest no Doppler-effect contributes to line broadening or shift. There are intense investigations to realize an atomic clock on narrow transitions of the strontium atom which should be more stable by at least one order of magnitude than other competing atomic clocks. Single atoms on selected sites of the optical lattice can be excited by absorption of optical photons and their fluorescence can be observed [1204].

If more than one atom is trapped in a potential minimum, molecule formation can be studied and its dependence on the barrier height. This gives a handle to control chemical reactions by optical means.

It is also possible to trap Fermi-particles in an optical lattice. Such a system has many similarities with a correlated electron gas in a solid.

Recently a mixture of Fermions and Bosons could be trapped in an optical lattice and the different behaviour of these particles could be explored when the barrier height was altered.

Mott insulators give valuable information for many problems in solid state physics. The minima of the optical lattice are separated by $\lambda/2$. This means that the lattice constant is about 500 times larger than in a solid crystal. While for solid crystals X-rays are necessary for structure characterization, with optical lattices visible lasers can be used to obtain diffraction patterns due to multi-beam interference of light, scattered by the different atoms on well-defined places.

Since the barrier heights between the potential minima can be readily changed, diffusion process can be studied in detail by optical spectroscopy. Non-occupied minima correspond to vacancies in solid crystals and their influence on the interaction between neighbouring atoms gives information on vacancy effects in solids.

Also the influence of magnetic fields on the behaviour of trapped atoms has been studied. One example is the *Meissner effect*, which was first discovered for supraconducting solids and has now been investigated for atoms in optical lattices in external magnetic fields [1205].

9.1.17 Applications of Cooled Atoms and Molecules

Optical cooling and deflection of atoms open new areas of atomic and molecular physics. Collisions at very small relative velocities, where the deBroglie wavelength $\lambda_{DB} = h/(mv)$ is large, can now be studied. They give information about the long-range part of the interaction potential, where new phenomena arise, such as retardation effects and magnetic interactions from electron or nuclear spins [1202b,c]. One example is the study of collisions between Na atoms in their $3\,^2S_{1/2}$ ground state. The interaction energy depends on the relative orientation of the two electron spins $S = \frac{1}{2}$. The atoms with parallel spins form a Na_2 molecule in a $^3\Sigma_u$ state, while atoms with antiparallel spins form a $Na_2(^1\Sigma_g^+)$ molecule. At large internuclear distances ($r > 1.5$ nm) the energy differences between the $^3\Sigma_u$ and $^1\Sigma_g$ potentials become comparable to the hyperfine splitting of the $Na(3\,^2S_{1/2})$ atoms. This has been experimentally demonstrated for Cs_2-molecular [1206]. The interaction between the nuclear spins and the electron spins leads to a mixing of the $^3\Sigma_u$ and $^1\Sigma_u$ states, which corresponds in the atomic model of colliding atoms to *spin-flip* collisions (Fig. 9.40).

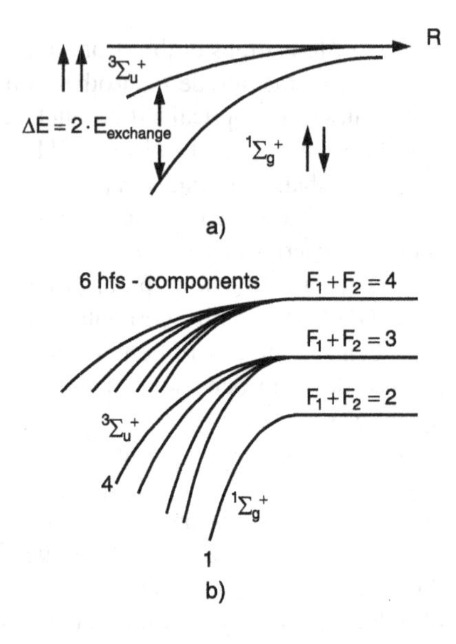

Fig. 9.40 Interaction between two Na atoms at large internuclear distances R for different spin orientations: (**a**) without hyperfine structure and (**b**) including the nuclear spins $I = (3/2)\hbar$, which gives three dissociation limits

Collisions between cold atoms in a trap can be studied experimentally by measuring the loss rate of trapped atoms under various trap conditions (temperature, magnetic-field gradients, light intensity, etc.). It turns out that the density of excited atoms cannot be neglected compared with the density of ground-state atoms, and the interaction between excited- and ground-state atoms plays an essential role. For collisions at very low temperatures the absorption and emission of photons during the collisions is important, because the collision time $\tau_c = R_c/v$ becomes very long at low relative atomic velocities v. The two dominant energy-transfer processes are collision-induced fine-structure transitions in the excited state and radiative redistribution, where a photon is absorbed by an atom at the position $\boldsymbol{r}_1$ in the trap potential $V(\boldsymbol{r})$ and another photon with a slightly different energy is reemitted after the atom has moved to another position $\boldsymbol{r}_2$.

Another application is the deflection of atoms by photon recoil. For sufficiently good beam collimation, the deflection from single photons can be detected. The distribution of the transverse-velocity components contains information about the statistics of photon absorption [1207]. Such experiments have successfully demonstrated the antibunching characteristics of photon absorption [1208]. The photon statistic is directly manifest in the momentum distribution of the deflected atoms [1209]. Optical collimation by radial recoil can considerably decrease the divergence of atomic beams and thus the beam intensity. This allows experiments in crossed beams that could not be performed before because of a lack of intensity.

A very interesting application of cold trapped atoms is their use for an optical frequency standard [1210]. They offer two major advantages: reduction of the Doppler effect and prolonged interaction times on the order of 1 s or more. Optical frequency standards may be realized either by atoms in optical traps or by atomic fountains [1211].

For the realization of an atomic fountain, cold atoms are released in the vertical direction out of an atomic trap. They are decelerated by the gravitational field and return back after having passed the culmination point with $v_z = 0$.

Example 9.14 Assume the atoms start with $v_{0z} = 5$ m/s. Their upward flight time is then $t = v_{0z}/g = 0.5$ s, their path length is $z = v_0 t - gt^2/2 = 1.25$ m, and their total flight time is 1 s. Their transit time through a laser beam with the diameter $d = 1$ cm close to the culmination point is $T_{tr} = 90$ ms, and the maximum transverse velocity is $v \leq 0.45$ m/s. The transit-time broadening is then less than 10 Hz.

There are many possible applications of cold trapped molecules.

One example is the spectroscopy of highly forbidden transitions, which becomes possible because of the long interaction time. Another aspect is a closer look at the chemistry of cold trapped molecules, where the reaction rates and the molecular dynamics are dominated by tunneling and a manipulation of molecular trajectories seems possible. Experiments on testing time-reversal symmetry via a search for a possible electric dipole moment of the proton or the electron [1212] are more sensitive when cold molecules are used [1213, 1214].

9.2 Spectroscopy of Single Ions

During recent years it has become possible to perform detailed spectroscopic investigations of single ions that are stored in electromagnetic (EM) traps and cooled by special laser arrangements. This allows tests of fundamental problems of quantum mechanics and electrodynamics and, furthermore, opens possibilities for precise frequency standards.

9.2.1 Trapping of Ions

Since ions show stronger interactions with EM fields than neutral atoms, which experience only a weak force because of their polarizability, they can be stored more effectively in EM traps. Therefore trapping of ions was achieved long before neutral particles were trapped [1215, 1216]. Two different techniques have been developed to store ions within a small volume: in the *radio frequency (RF) quadrupole trap* [1216, 1217, 1242] the ions are confined within a hyperbolic electric dc field superimposed by a RF field, while in the *Penning trap* [1220] a dc magnetic field with a superimposed electric field of hyperbolic geometry is used to trap the ions.

The EM quadrupole trap (the Paul trap, which won the Nobel prize for Wolfgang Paul in 1989) is formed by a ring electrode with a hyperbolic surface and a ring

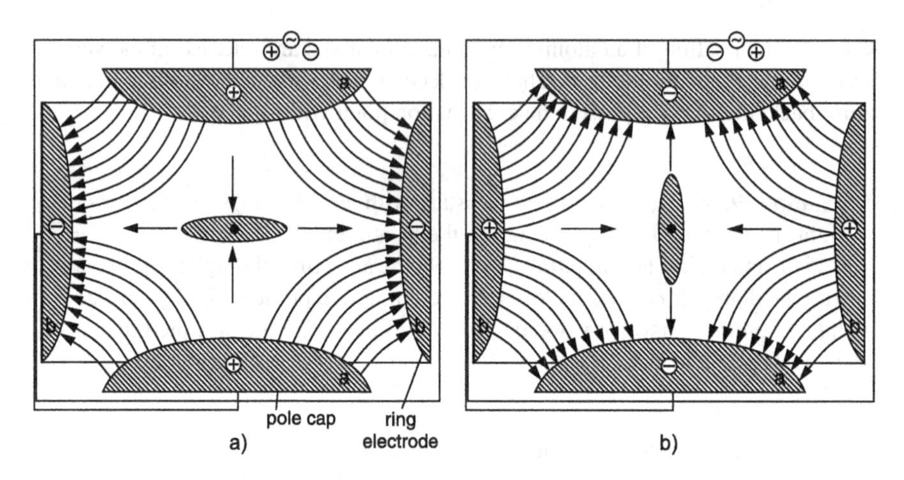

Fig. 9.41 Paul trap: field lines and extension of the ion cloud for two opposite phases of the electric RF field

radius of r_0 as one pole, and two hyperbolic caps as the second pole (Fig. 9.41). The whole system has cylindrical symmetry around the z-axis. The distance $2z_0$ between the two pole caps (which are at equal potential) is adjusted to be $2z_0 = r_0\sqrt{2}$.

With the voltage U between the caps and the ring electrode, the electric potential $\phi(r, z)$ is for points within the trap [1217],

$$\phi = \frac{U}{2r_0^2}(r^2 - 2z^2), \quad \text{with } r^2 = x^2 + y^2. \tag{9.47}$$

The applied voltage $U = U_0 + V_0 \cos(\omega_{\mathrm{RF}} t)$ is a superposition of the dc voltage U_0 and the RF voltage $V_0 \cos(\omega_{\mathrm{RF}} t)$. In Fig. 9.41 the electric field lines and the extension of the trapped ion cloud are shown for two opposite phases of the voltage U. The equation of motion

$$m\ddot{\boldsymbol{r}} = -q \cdot \operatorname{grad}\phi, \tag{9.48}$$

for an ion with charge q inside the trap can be derived from (9.47). One obtains

$$\frac{\mathrm{d}^2}{\mathrm{d}t^2}\begin{pmatrix} x \\ y \\ z \end{pmatrix} + \frac{\omega_{\mathrm{RF}}^2}{4}(a + b\cos\omega_{\mathrm{RF}} t)\begin{pmatrix} x \\ y \\ -2z \end{pmatrix} = 0. \tag{9.49}$$

The two parameters

$$a = \frac{4qU_0}{mr_0^2\omega_{\mathrm{RF}}^2}, \qquad b = \frac{4qV_0}{mr_0^2\omega_{\mathrm{RF}}^2}, \tag{9.50}$$

are determined by the dc voltage U_0, the RF amplitude V_0, and the angular frequency ω_{RF} of the RF voltage, respectively. Because of the cylindrical symmetry, the same equation holds for the x- and y-components, while for the movement in z-direction $-2z$ appears in (9.49) because $r_0^2 = 2z_0^2$.

The equation of motion (9.49) is known as *Mathieu's differential equation*. It has stable solutions only for certain values of the parameters a and b [1221]. Charged particles that enter the trap from outside cannot be trapped. Therefore, the ions have to be produced inside the trap. This is generally achieved by electron-impact ionization of neutral atoms.

The stable solutions of (9.49) can be described as a superposition of two components: A periodic "micromovement" of the ions with the RF ω_{RF} around a "guiding center" that itself performs slower harmonic oscillations with the frequency Ω in the x–y-plane and with the frequency $\omega_z = 2\Omega$ in the z-direction (secular motion) [1217]. The x- and z-components of the ion motion are

$$x(t) = x_0\left[1 + (b/4)\cos(\omega_{\mathrm{RF}} t)\right]\cos\Omega t, \tag{9.51a}$$

$$z(t) = z_0\left[1 + \sqrt{2(\omega_z/\omega_{\mathrm{RF}})}\cos(\omega_{\mathrm{RF}} t)\right]\cos(2\Omega t). \tag{9.51b}$$

The Fourier analysis of $x(t)$ and $z(t)$ yields the frequency spectrum of the ion movement, which contains the fundamental frequency ω_{RF} and its harmonics $n\omega_{\mathrm{RF}}$, as well as the sidebands at $n\omega_{\mathrm{RF}} \pm m\Omega$.

Unlike the rf fields of the Paul trap, the Penning trap uses only dc fields: a dc electric field between a ring electrode and the pole caps (Fig. 9.42a) and a dc magnetic field in the z-direction. Two different designs are shown in Fig. 9.42a and b. The ionic motion is complex. It consists of three components. There is cyclotron resonance motion in circles around a point that moves with the motion of the magnetron in a circle around the magnetic field lines (Fig. 9.42c). Superimposed on this is an axial oscillation in the z-direction. The superposition of all three components leads to the ion path depicted in Fig. 9.42d.

The RF quadrupole trap and the Penning trap can be used not only to trap ions but also to perform very precise measurements of their masses [1219]. Unlike other mass spectrometers, the signal is not obtained from mass-selected particles impinging on the detector, but from an induction voltage picked up by external electrodes from the ion motion. Fourier analysis of this signal gives the frequencies of the three components. Since the cyclotron resonance frequency

$$\omega_{\mathrm{c}} = q \cdot B/m$$

depends on the mass of the ion, the measured frequency ω_{c} yields the ion mass m directly if the magnetic field B is known.

The trapped ions can be monitored either by laser-induced fluorescence [1222, 1223] or by the RF voltage that is induced in an outer RF circuit by the motion of the ions [1217]. The LIF detection is very sensitive for a true two-level system, where the fluorescence photon rate R of a single ion with a spontaneous upper-level lifetime τ_k may reach $R = (2\tau_k)^{-1}$ [s^{-1}] (Sect. 9.1.4). Even for a three-level system, this can be achieved if a second laser is used that refills the ground-state level, depleted by optical pumping (Fig. 9.5). For $\tau_k = 10^{-8}$ s, this implies that for sufficiently large laser intensities a single ion emits up to 5×10^7 fluorescence photons per second, which allows the detection of a single stored ion [1225, 1226].

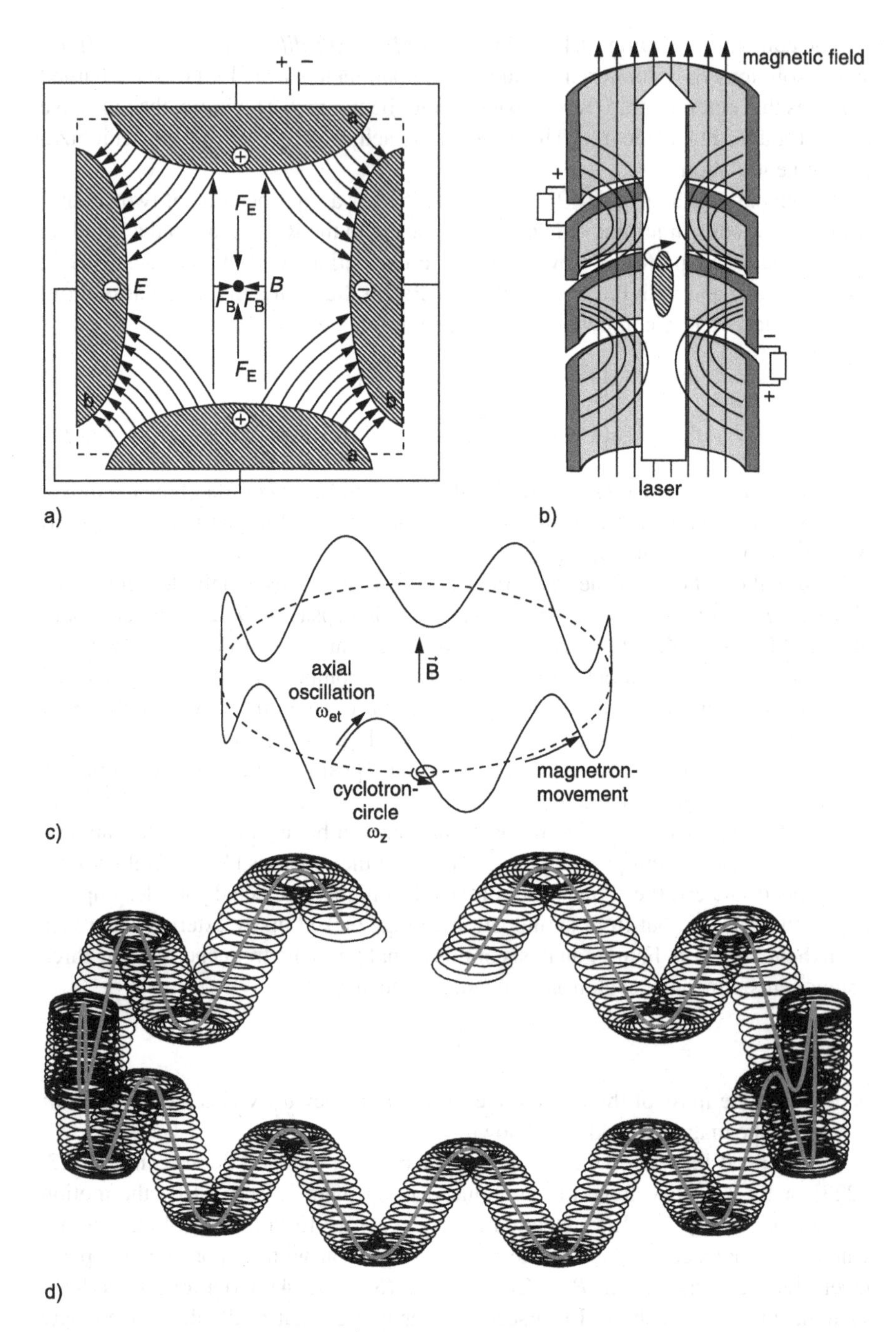

Fig. 9.42 Penning trap. (**a**) Electric field lines, magnetic field and forces on an ion; (**b**) experimental realization; (**c**) simplified path of an ion in the Penning trap; (**d**) real path of an ion [G. Werth, Mainz]

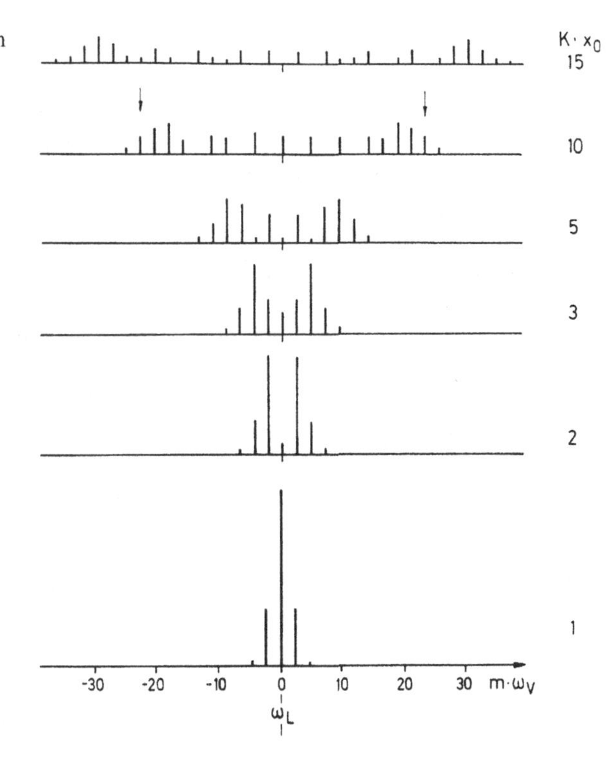

Fig. 9.43 Sideband spectrum of an oscillating ion as a function of the oscillating amplitude

9.2.2 *Optical Sideband Cooling*

Assume that an ion in the Paul trap (absorption frequency ω_0 for $v = 0$) performing a harmonic motion in the x-direction with the velocity $v_x = v_0 \cos \omega_\mathrm{v} t$, is irradiated by a monochromatic wave propagating in the x-direction. In the frame of the oscillating ion, the laser frequency is modulated at the oscillation frequency due to the oscillating Doppler shift. If the linewidth γ of the absorbing transition is smaller than ω_v, the absorption spectrum of the oscillating ion consists of discrete lines at the frequencies $\omega_\mathrm{m} = \omega_0 \pm m\omega_\mathrm{v}$. The relative line intensities are given by the mth-order Bessel function $J_\mathrm{m}(v_0\omega_0/c\omega_\mathrm{m})$ [1227, 1228], which depend on the velocity amplitude v_0 of the ion (Fig. 9.43).

If the laser is tuned to the frequency $\omega_\mathrm{L} = \omega_0 - m\omega_\mathrm{v}$ of a lower sideband, the atom can only absorb during that phase of the oscillation when the atom moves *toward* the laser wave ($\boldsymbol{k} \cdot \boldsymbol{v} < 0$). If its spontaneous lifetime is large compared to the oscillation period $T = 2\pi/\omega_\mathrm{v}$, the fluorescence is uniformly averaged over all phases of the oscillation and its frequency distribution is symmetric to ω_0. On the average, the atom therefore loses more energy by emission than it gains by absorption. This energy difference is taken from its kinetic energy, resulting in an average decrease of its oscillation energy by $m\hbar\omega_\mathrm{v}$ per absorption–emission cycle. As depicted in Fig. 9.43, the absorption spectrum narrows down to an interval around ω_0. This has the unwanted effect that the efficiency of sideband cooling at $\omega_L = \omega_0 - m\omega_v$ decreases with decreasing oscillation energy.

In a quantum-mechanical model, the ion confined to the trap can be described by a nearly harmonic oscillator with vibrational energy levels determined by the trap potential. Optical cooling corresponds to a compression of the population probability into the lowest levels.

Optical sideband cooling is quite analogous to the Doppler cooling by photon recoil discussed in Sect. 9.1. The only difference is that the confinement of the ion within the trap leads to discrete energy levels of the oscillating ion, whereas the translational energy of a free atom corresponds to a continuous absorption spectrum within the Doppler width.

Optical sideband cooling was demonstrated for Mg^+ ions in a Penning trap [1227, 1228] and for Ba^+ ions in an RF quadrupole trap [1229]. The Mg^+ ions were cooled below 0.5 K on the $3s^2S_{1/2}$–$3p^2P_{3/2}$ transition by a frequency-doubled dye laser at $\lambda = 560.2$ nm. The vibrational amplitude of the oscillating ions decreased to a few microns. With decreasing temperature the ions are therefore confined to a smaller and smaller volume around the trap center.

The cooling can be monitored with a weak probe laser that is tuned over the absorption profile. Its intensity must be sufficiently small to avoid heating of the ions for frequencies $\omega_{\text{probe}} > \omega_0$.

Recently, single Ba^+ ions could be trapped in a Paul trap, cooled down to the milliKelvin range, and observed with a microscope by their LIF. Since Ba^+ represents a three-level system (Fig. 9.44), two lasers have to be used that are tuned to the transitions $6p\,^2S_{1/2} \to 6p\,^2P_{1/2}$ (pump transition for cooling) and $5d\,^2D_{3/2} \to 6p\,^2P_{1/2}$ (repumping to avoid optical pumping into the $5d\,^2D_{3/2}$ level).

Any change in the number N of trapped ions (from a change in electron impact ionization, or from collisions with residual gas molecules, which may throw an ion out of the trap) appears as a step in the LIF signal (Fig. 9.45).

Imaging of the spatial distribution of the LIF source by a microscope in combination with an image intensifier (Vol. 1, Sect. 4.5) allows the measurement of the average spatial probability of a single ion as a function of its "temperature" [1223]. The spectrum of single ions in a trap has recently been observed [1224].

9.2.3 Direct Observations of Quantum Jumps

In quantum mechanics the probability $\mathcal{P}_1(t)$ of finding an atomic system at the time t in the quantum state $|1\rangle$ is described by time-dependent wave functions. If one wants to find out whether a system is with certainty $[\mathcal{P}_1(t) = 1]$ in a well-defined quantum state $|1\rangle$ one has to perform a measurement that, however, changes this state of the system. Many controversial opinions have been published on whether it is possible to perform experiments with a single atom in such a way that its initial state and a possible transition to a well-defined final state can be unambiguously determined.

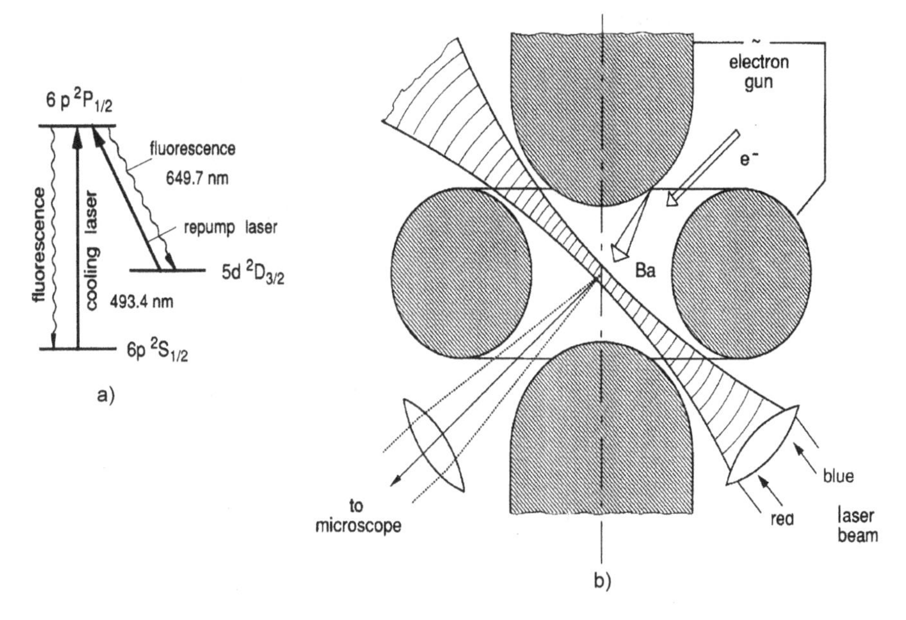

Fig. 9.44 (**a**) The Ba^+ ion as a three-level system. The second laser at $\lambda = 649.7$ nm pumps the $5d\,^2D_{3/2}$ atoms back into the $6p\,^2P_{1/2}$ level. (**b**) Experimental arrangement for trapping and cooling of Ba^+ ions in a Paul trap [1223]

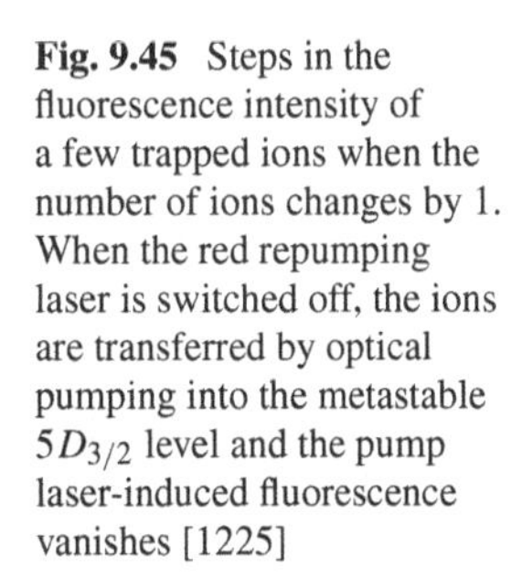

Fig. 9.45 Steps in the fluorescence intensity of a few trapped ions when the number of ions changes by 1. When the red repumping laser is switched off, the ions are transferred by optical pumping into the metastable $5D_{3/2}$ level and the pump laser-induced fluorescence vanishes [1225]

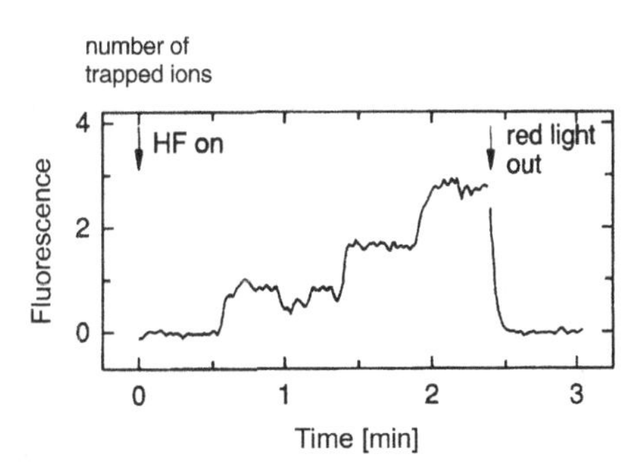

Laser-spectroscopic experiments with single ions confined in a trap have proved that such information can be obtained. The original idea was proposed by Dehmelt [1230] and has since been realized by several groups [1231–1233]. It is based on the coupling of an intense allowed transition with a weak dipole-forbidden transition via a common level. For the example of the Ba^+ ion (Fig. 9.46) the metastable $5\,^2D_{5/2}$ level with a spontaneous lifetime of $\tau = (32 \pm 5)$ s can serve as a "shelf state." Assume that the Ba^+ ion is cooled by the pump laser at $\lambda = 493.4$ nm and the population leaking by fluorescence into the $5\,^2D_{3/2}$ level is pumped back into the $6\,^2P_{1/2}$ level by the second laser at $\lambda = 649.7$ nm. If the pump transition is saturated, the fluorescence rate is about 10^8 photons per second with the lifetime $\tau(6\,^2P_{1/2}) = 8$ ns. If the metastable $5\,^2D_{5/2}$ level is populated (this can be reached,

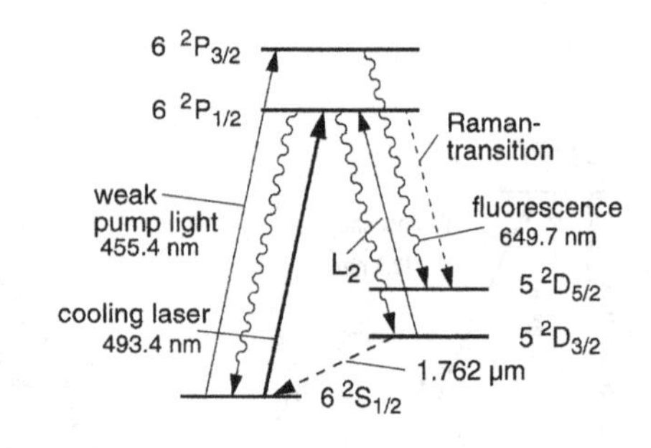

Fig. 9.46 More detailed level scheme of Ba^+ including the fine-structure splitting. The "dark level" $5\,^2D_{5/2}$ can be populated by off-resonance Raman transitions $6\,^2S_{1/2} \to 6\,^2P_{3/2} \to 5\,^2P_{5/2}$ induced by the cooling laser

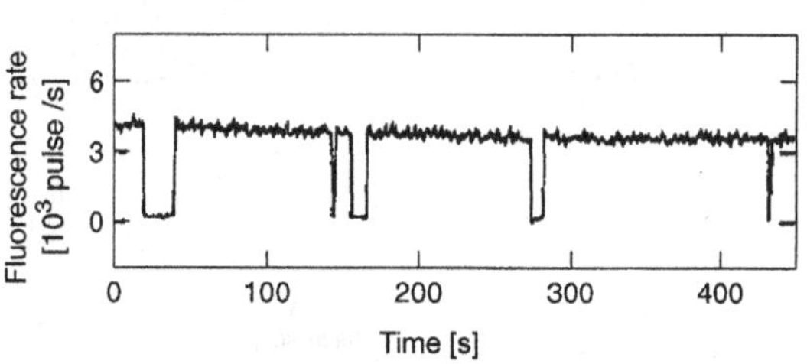

Fig. 9.47 Experimental demonstration of quantum jumps of a single ion [1232]

for example, by exciting the $6\,^2P_{3/2}$ level with a weak laser at $\lambda = 455$ nm, which decays by fluorescence into the $5\,^2D_{5/2}$ level, or even without any further laser by a nonresonant Raman transition induced by the cooling laser), the ion is, on the average, for $\tau(5\,^2D_{5/2}) = 32$ s not in its ground state $6\,^2S_{1/2}$ and therefore cannot absorb the pump radiation at $\lambda = 493$ nm. The fluorescence rate becomes zero but jumps to its value of 10^8 photons/s as soon as the $5\,^2D_{5/2}$ level returns by emission of a photon at $\lambda = 1.762$ μm back into the $6\,^2S_{1/2}$ level.

The allowed transition $6\,^2S_{1/2}$–$6\,^2P_{1/2}$ serves as an amplifying detector for a single quantum jump on the dipole-forbidden transition $5\,^2D_{5/2} \to 6\,^2S_{1/2}$ [1232]. In Fig. 9.47 the statistically occurring quantum jumps can be seen by the in-and-out phases of the fluorescence. The effective lifetime of the $5\,^2D_{5/2}$ level can be shortened by irradiating the Ba^+ ion with a third laser at $\lambda = 614.2$ nm, which induces the transitions $5\,^2D_{5/2} \to 6\,^2P_{3/2}$ where the upper level decays into the $6\,^2S_{1/2}$ ground state.

Similar observations of quantum jumps have been made for Hg^+ ions confined in a Penning trap [1233].

Of fundamental interest are measurements of the photon statistics in a three-level system, which can be performed by observing the statistics of quantum jumps. While the durations Δt_i of the "on-phases" or the "off-phases" show an exponential distribution, the probability $P(m)$ of m quantum jumps per second exhibits a Poisson distribution (Fig. 9.48). In a two-level system the situation is different. Here, a second fluorescence photon can be emitted after a first emission only, when the upper state has been reexcited by absorption of a photon. The distribution $P(\Delta T)$ of the time intervals ΔT between successive emission of fluorescence photons shows a *sub-Poisson distribution* that tends to zero for $\Delta T \to 0$ (*photon antibunching*), because at least half of a Rabi period has to pass after the emission of a photon before a second photon can be emitted [1234].

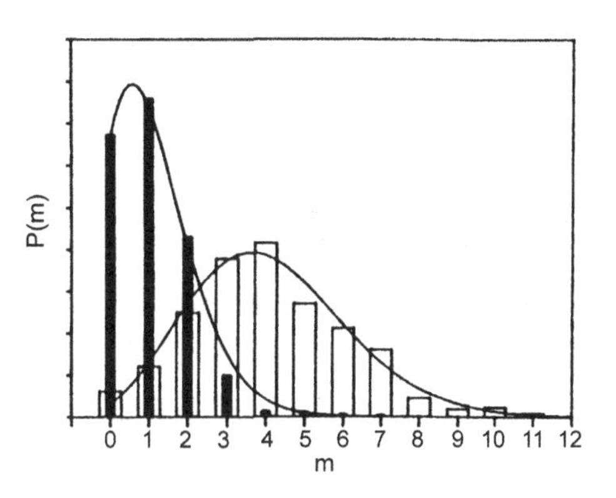

Fig. 9.48 Distribution $P(m)$ of m quantum jumps per second measured over a period of 150 s (*black bars*) and 600 s (*open bars*). The *curves* are fits of Poisson distributions with two adapted parameters [1232]

9.2.4 Formation of Wigner Crystals in Ion Traps

If several ions are trapped in an ion trap and are cooled by optical sideband cooling, a "phase transition" may occur at the temperature T_c where the ions arrange into a stable, spatially symmetric configuration like in a crystal [1235–1238]. The distances between the ions in this *Wigner crystal* are about 10^3–10^4 times larger than those in an ordinary ion crystal such as NaCl. Wigner crystals of electrons, where the electrons are located at certain regular positions in an external field, were first proposed by E. Wigner in 1934.

This phase transition from the statistically distributed ions in a "gaseous" cloud to an ordered Wigner crystal can be monitored by the change in the fluorescence intensity. This is observed as a function of the detuning $\Delta\omega$ of the cooling laser. With decreasing red detuning the temperature of the ions is lowered until the phase transition occurs at the critical temperature T_c. The distribution $I_{Fl}(\Delta\omega)$ has a completely different shape for the ordered state of the Wigner crystal than for the disordered ion cloud (Fig. 9.49a). For very small detunings the cooling rate becomes smaller then the heating rate and the crystal "melts" again.

When the detuning $\Delta\omega$ of the cooling laser or its intensity is changed, one observes a typical hysteresis (Fig. 9.49b). At about 160 μW laser power and a fixed detuning of -120 MHz the fluorescence intensity increases by a factor of 4, indicating a phase transition to the ordered Wigner crystal. With further increases of the laser power, the system suddenly jumps back at $P_L \simeq 400$ μW into the disordered state. Similar hysteresis curves can be found when the amplitude b of the trap's RF voltage is changed [1238]. Using a microscope in combination with a sensitive image intensifier the location of the ions can be made visible on a screen (Fig. 9.50) and the transition from the disordered state of the ion cloud to the ordered Wigner crystal can be directly observed [1235, 1237].

Similar to the situation for a coupled pendulum, normal vibrations can be excited in a Wigner crystal. For example, the two-ion crystal has two normal vibrations where the two ions oscillate in the ion trap potential either in phase or with opposite phases. For the in-phase oscillations, the Coulomb repulsion between the ions does not influence the oscillation frequency because the distance between the

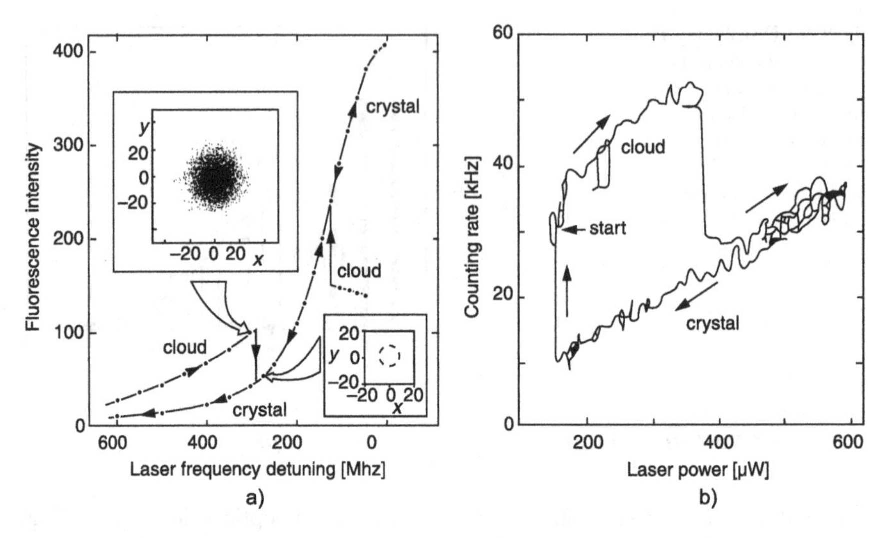

Fig. 9.49 (**a**) Fluorescence intensity as a function of laser frequency detuning. (**b**) Hysteresis curve around the phase transition from the ion cloud to the ordered Wigner crystal [1235]

Fig. 9.50 Photograph of a Wigner crystal of seven trapped ions, taken with a microscope and an image intensifier. The distance between the ions is about 20 μm [1236]

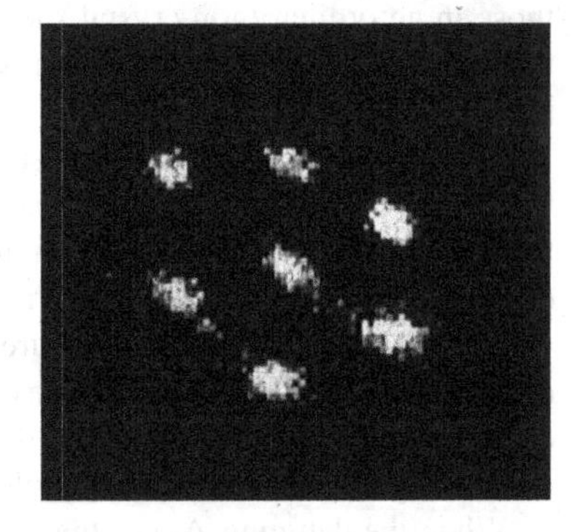

ions does not change. The restoring force is solely due to the trap potential. One obtains two degenerate oscillations in the x- and y-directions with the frequencies $\Omega_x = \Omega_y$ (9.51a), and an oscillation in the z-direction with $\Omega_z \neq \Omega_x$. For the oscillation with opposite phases the vibrational frequency is additionally determined by the Coulomb repulsion. One obtains for the three components $\Omega_{2x} = \sqrt{3}\Omega_{1x} = \Omega_{2y}$ and $\Omega_{2z} = \sqrt{3}\Omega_{1z}$ [1239].

These vibrational modes can be excited if an additional ac voltage with the proper frequency is applied to the trap electrodes. The excitation leads to heating of the Wigner crystal, which causes a decrease in the laser-induced fluorescence. By choosing the proper intensity and the detuning $\Delta\omega$ of the cooling laser, this heating can be kept stable. Such measurements allow the study of many-particle effects with samples of a selected small number of ions. They may give very useful information on solid-state problems [1240].

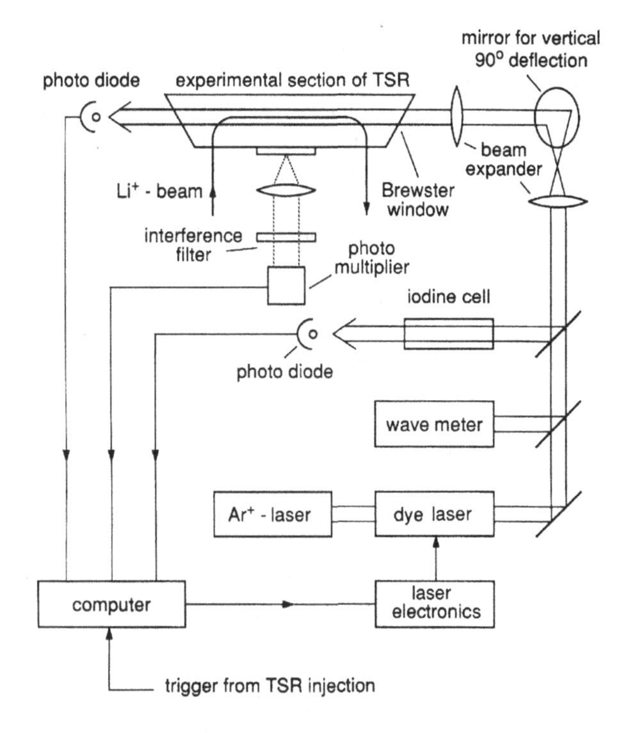

Fig. 9.51 Experimental setup for laser spectroscopy in the ion test storage ring TSR in Heidelberg [1246]

Since the ions in a Wigner crystal form a regular lattice, Bragg reflection of incident radiation can be observed. In contrast to solid crystals where the distance between neighboring atoms is a few Angstroms and Bragg reflection is only possible with X-rays, here the distances are of the order of the wavelength of visible light, and Bragg diffraction and reflection with visible laser light can even be observed for Wigner crystals that only consist of a small number of ions. Long-range order has been observed in spherical ion clouds consisting of up to 5×10^4 ions, while bulk behavior was found for a larger number 2.7×10^5 ions [1241].

9.2.5 *Laser Spectroscopy of Ions in Storage Rings*

In recent years, increasing efforts have been made to apply laser-spectroscopic techniques to high-energy ions in storage rings (Fig. 9.51) [1242]. Laser cooling of these ions has become an effective alternative to electron cooling or stochastic cooling [1243]. Unlike optical cooling of neutral atom beams, where the atoms can be stopped completely, in this case the velocity distribution of the ions around the mean velocity v_m is compressed. The "cooled" ions then all move with the same velocity v_m (see Sect. 4.2). Laser cooling can reduce the transitional temperature of the relative velocities from 300 K down to below 5 K. Beam cooling in storage rings not only results in a much better beam quality, but may also lead to a condensation of ion beams resulting in *one-dimensional Wigner crystals* [1244a]. Possible candidates for such cooling experiments are metastable Li^+ ions, where the cooling transition $1s2s\,^3S_1 \rightarrow 1s2p\,^3P_1$ with $\lambda = 548.5$ nm lies conveniently within the tuning range of cw dye lasers. At 100 keV kinetic energy excited Li^+ ions in the $2p\,^3P_1$ level with a spontaneous lifetime of $\tau = 43$ ns travel about 8 cm. This allows 100 absorption–emission cycles per ion round-trip over the interaction length of 8 m between laser beam and ion beam.

In spite of large efforts, one-dimensional Wigner crystals in high-energy storage rings have not yet been realized. However, phase transitions and ordered structures of laser-cooled Mg^+ ions stored in a small RF quadrupole storage ring with a diameter of 115 mm have been observed recently [1244b]. Cooling of the ions results in a linear or a helical structure of ions aligned along the center line of the quadrupole field.

There are, however, other interesting experiments that were successful. For instance, in the electron cooler ions and atoms move with nearly equal velocities and form an unusual kind of a plasma. Here laser spectroscopy may help to better understand the physics of highly ionized plasmas [1245].

An interesting aspect is the spontaneous radiative recombination of electrons and ions, which can be enhanced by irradiation with resonant laser light. This has been demonstrated in the test storage ring in Heidelberg [1246]. A proton beam of $E_k =$ 21 MeV is superimposed by a cold electron beam. If a pulsed, tunable dye laser beam propagates antiparallel to the ion beam through a linear section of the storage ring, the laser wavelength of 450.5 nm is Doppler shifted into resonance with the downward transition $H^+ + e^- \rightarrow H(2s)$ from the ionization limit to the $2s$ state. The enhancement factor G, which represents the ratio of induced-to-spontaneous recombinations, depends on the laser wavelength and reaches values of $G \simeq 50$ for the resonance wavelength $\lambda = 450.5$ nm.

Further experiments of fundamental interest are precision measurements of transition frequencies of very fast moving ions, allowing tests of special relativity.

9.2.6 *Quantum Computer with Stored Ions*

Ions can be stored in linear Paul traps in a regular row with constant distances depending on the well depth of the trap. They form a one-dimensional Wigner crystal (Fig. 9.52). In order to trap them the ions must be sufficiently cold. If single ions should be excited by a laser, the distance must be larger than the wavelength of the laser and the amplitude of their thermal motion must be small compared to the distance to the neighbouring ion. Since the ions in the harmonic potential of the trap can only occupy the energy levels of the harmonic oscillator a low temperature means that they occupy only the lowest energy level in the trap.

Such a one-dimensional chain of ions where each ion can be selectively excited, can be used as the basis of a quantum computer [1247, 1248]. This can be seen as follows: Every computer is based on the two bit states "0" and "1". In the atomic quantum computer these bit states correspond to an atom in the ground state and in a long living excited state, e.g. a metastable electronic state or an excited hyperfine

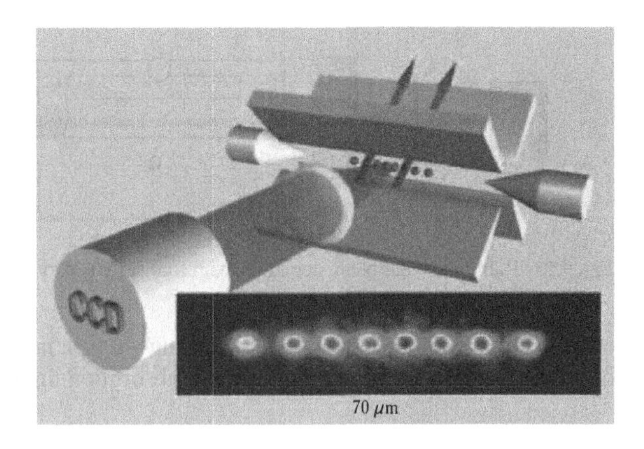

Fig. 9.52 Linear Ion-trap with 8 stored ions

level of the electronic ground state with lifetimes longer than 1 s. Such long-living states can be populated by optical excitation via short living states (see Fig. 9.46).

The basic idea of a quantum computer needs the realization of *entangled states*. Such states are the coherent superposition of real atomic states. An entangled state of an atom with two hyperfine levels $|0\rangle$ and $|1\rangle$ is for instance the state

$$|e\rangle = c_0|0\rangle + c_1|1\rangle.$$

This entangled state is named "*quantum bit or q-bit*." It replaces the bits "0" and "1" of the classical computer.

There are several ways to realize such entangled states with techniques of laser spectroscopy (see Sect. 7.9). A promising method is the linear ion chain: The ions haver in addition to their internal energy structure a further degree of freedom due to the oscillatory motion in the Paul trap, which causes the possible population of different trap states and the Coulomb interaction between the ions. If one ion is electronically excited by absorption of a laser photon, the Coulomb interaction with the neighbouring ions changes slightly. This changes the position of these ions and therefore their potential energy in the trap. If the laser beam propagates in the direction of the ion chain, the recoil of an ion, which has absorbed a photon shifts its position and its kinetic energy. Due to the Coulomb interaction this energy change is partly transferred to the neighbouring ions. Therefore the excitation of a single ions results in a coherent change of the states of the neighbouring ions and creates an entangled state of the ion ensemble.

For more details see some recent publications [1249–1252].

9.3 Optical Ramsey Fringes

In the previous sections we discussed how the interaction time of atoms or ions with laser fields can be greatly increased by cooling and trapping. Optical cooling cannot be applied to molecules, which do not represent two-level systems. Here another technique has been developed that increases the interaction time of atoms

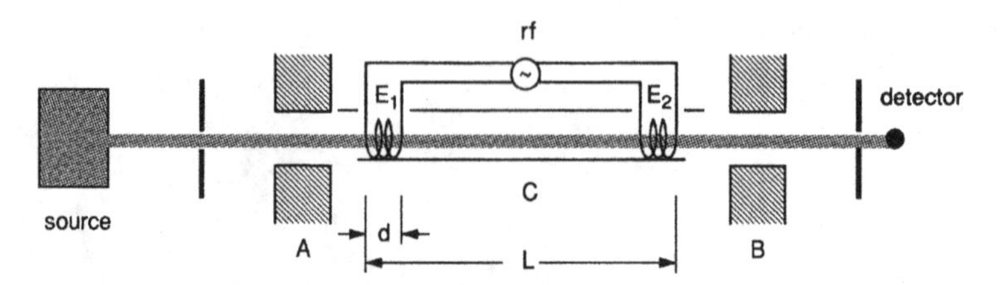

Fig. 9.53 Rabi molecular beam apparatus with Ramsey's separated fields

or molecules with EM fields by increasing the spatial interaction zone and thus decreasing the transit-time broadening of absorption lines (Vol. 1, Sect. 3.4).

9.3.1 Basic Considerations

The problem of transit-time broadening was recognized many years ago in electric or magnetic resonance spectroscopy in molecular beams [1253]. In these Rabi experiments [1254], the natural linewidth of the radio frequency or microwave transitions is extremely small because the spontaneous transition probability is, according to Vol. 1, (2.22), proportional to ω^3. The spectral widths of the microwave or RF lines are therefore determined mainly by the transit time $\Delta T = d/\overline{v}$ of molecules with the mean velocity $\overline{v}$ through the interaction zone in the C field (Fig. 5.10a) with length d.

A considerable reduction of the time-of-flight broadening could be achieved by the realization of Ramsey's ingenious idea of separated fields [1255]. The molecules in the beam pass two phase-coherent fields that are spatially separated by the distance $L \gg d$, which is large compared with the extension d of each field (Fig. 9.53). The interaction of the molecules with the first field creates a dipole moment of each molecular oscillator with a phase that is dependent on the interaction time $\tau = d/\overline{v}$ and the detuning $\Omega = \omega_0 - \omega$ of the radiofrequency ω from the center frequency ω_0 of the molecular transition (Vol. 1, Sect. 2.8). After passing the first interaction zone, the molecular dipole precesses in the field-free region at its eigenfrequency ω_0. When it enters the second field, it therefore has accumulated the phase angle $\Delta\varphi = \omega_0 T = \omega_0 L/\overline{v}$. During the same time T the field phase has changed by ωT. The relative phase between the dipole and the field has therefore changed by $(\omega_0 - \omega)T$ during the flight time T through the field-free region.

The interaction between the dipole and the second field with the amplitude $E_2 = E_0 \cos\omega t$ depends on their relative phases. The observed signal is related to the power absorbed by the molecular dipoles in the second field and is therefore proportional to $E_2^2 \cos[(\omega - \omega_0)T]$. When we assume that all N molecules passing the field per second have the same velocity v, we obtain the signal

$$S(\omega) = aNE_2^2 \cos\big[(\omega_0 - \omega)L/v\big], \tag{9.52}$$

where the constant a depends on the geometry of the beam and the field.

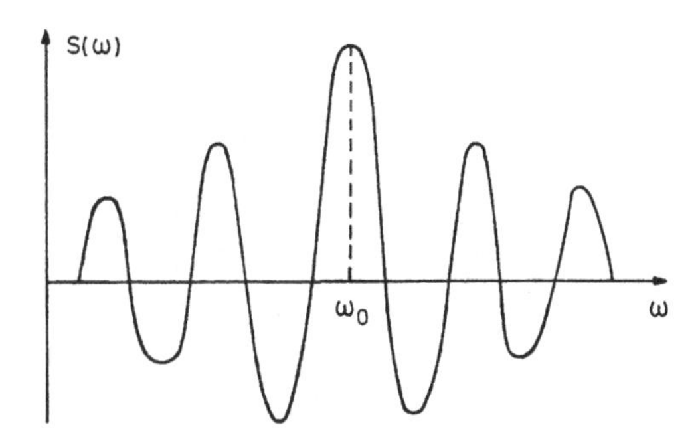

Fig. 9.54 Signal power absorbed by the molecules in the second field as a function of detuning $\Omega = \omega - \omega_0$ (Ramsey fringes) for a narrow velocity distribution $N(v)$

Measured as a function of the field frequency ω, this signal exhibits an oscillatory pattern called *Ramsey fringes* (Fig. 9.54). The full halfwidth of the central fringe, which is $\delta\omega = \pi(v/L)$, decreases with increasing separation L between the fields.

This interference phenomenon is quite similar to the well-known Young's interference experiment (Vol. 1, Sect. 2.9.2), where two slits are illuminated by coherent light and the superposition of light from both slits is observed as a function of the optical path difference Δs. The number of maxima observed in the two-slit interference pattern depends on the coherence length ℓ_c of the incident light and on the slit separation. The fringes can be seen if $\Delta s \le \ell_c$.

A similar situation is observed for the Ramsey fringes. Since the velocities of the molecules in the molecular beam are not equal but follow a Maxwellian distribution, the phase differences $(\omega_0 - \omega)L/v$ show a corresponding distribution. The interference pattern is obtained by integrating the contributions to the signal from all molecules $N(v)$ with the velocity v

$$S = C \int N(v) E^2 \cos\left[(\omega_0 - \omega)L/v\right] \mathrm{d}v. \tag{9.53}$$

Similar to Young's interference with partially coherent light, the velocity distribution will smear out the interference pattern for the higher-order fringes, that is, large $(\omega_0 - \omega)$, but will essentially leave the central fringe narrow for small $(\omega_0 - \omega)$. With a halfwidth Δv of the velocity distribution $N(v)$, this restricts the maximum field separation to about $L \le v^2/(\omega_0 \Delta v)$, since for larger L the higher interference orders for fast molecules overlap with the first order of slow molecules. Using supersonic beams with a narrow velocity distribution (Sect. 4.1), larger separations L are allowed. In general, however, only the zeroth order of the Ramsey interference is utilized in high-resolution spectroscopy and the "velocity averaging" of the higher orders has the advantage that it avoids the overlap of different orders for two closely spaced molecular lines.

The extension of Ramsey's idea to the optical region seems quite obvious if the RF fields in Fig. 9.53 are replaced by two phase-coherent laser fields. However, the transfer from the RF region, where the wavelength λ is larger than the field extension d, to the optical range where $\lambda \ll d$, causes some difficulties [1256]. Molecules with slightly inclined path directions traverse the standing optical fields at different phases (Fig. 9.55). Consider molecules starting from a point $x = 0$, $z = 0$ at

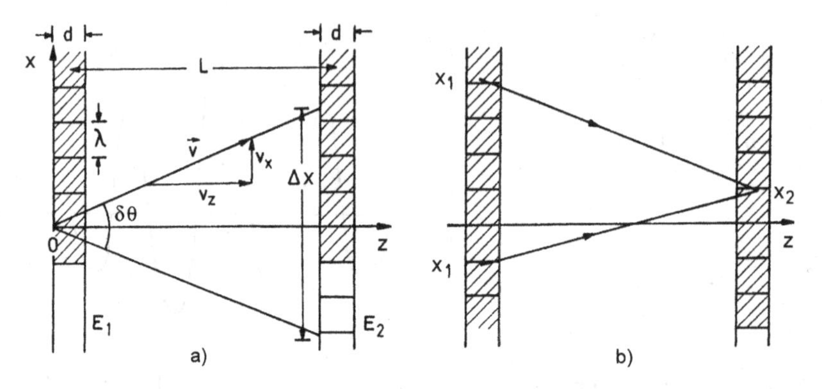

Fig. 9.55 Molecules starting from the same point with different values of v_x (**a**) or from different points x_1, x_2 in the first zone (**b**) experience different phase differences in the second field

the beginning of the first field. Only those molecules with flight directions within a narrow angular cone $\delta\theta \le \lambda/2d$ around the z-axis experience phases at the end of the first field differing by less than π. These molecules, however, traverse the *second* field at a distance L downstream within the extension $\Delta x = L\delta\theta \le L\lambda/2d$, where the phase φ of the optical field has a spatial variation up to $\Delta\varphi \le L\pi/d$. If the method of separated fields is to increase the spectral resolution, L has to be large compared to d, which implies that $\Delta\varphi \gg \pi$. Although these molecules have experienced nearly the same phase in the first field, they *do not* generate observable Ramsey fringes when interacting with the second field because their interaction phases are all different, and the total signal is obtained by summing over all molecules present at the time t in the second field. This implies averaging over their different phases, which means that the Ramsey fringes are washed out. The same is true for molecules starting from different points $(x_1, z = 0)$ in the first interaction zone that arrive at the same point $(x_2, z = L)$ in the second zone (Fig. 9.55b). The phases of these molecules may differ by $\Delta\varphi \gg \pi$, and for all molecules starting at different x_1 the phases are randomly distributed and therefore no macroscopic polarization is observed at (x_2, L).

Note The requirement $\delta\theta < \lambda/2d$ for molecules that experience nearly the same phase in the first zone is equivalent to the condition that the residual Doppler width $\delta\omega_{\mathrm{D}}$ of the absorption profile for a laser transverse to the beam axis does not exceed the transit-time broadening $\delta\omega_{\mathrm{t}} = \pi v/d$. This can be seen immediately from the relations $v_x = v_z\delta\theta$ and $\delta\omega_{\mathrm{D}} = \omega v_x/c = \omega\delta\theta v_z/c = \delta\theta v_z 2\pi/\lambda$. For

$$\delta\theta < \lambda/2d \quad \Rightarrow \quad \delta\omega_{\mathrm{D}} < \pi v_z/d = \delta\omega_{\mathrm{t}}. \tag{9.54}$$

The phase difference $\Delta\varphi(v_x)$ of a molecular dipole starting from the point $(x_1, 0)$ in the first zone can be plotted as a function of the transverse velocity v_x, as indicated in Fig. 9.56. Although $\Delta\varphi(v_x, z_1 = d)$ shows a flat distribution at the end of the first laser beam, it exhibits a modulation with the period $\Delta v_x = \lambda/2T = \lambda v_z/2L$ in the

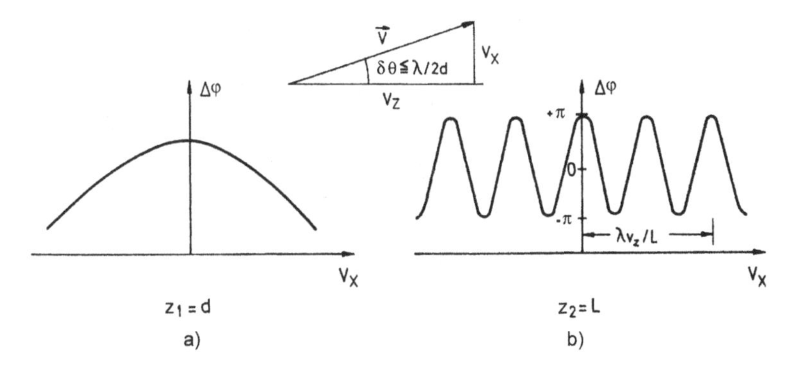

Fig. 9.56 Relative phase $\Delta\varphi(v_x)$ between the oscillating dipole molecule and the EM field: (**a**) at the end of the first zone $z = d$; (**b**) in the second zone at $z = L \gg d$

second zone. However, this modulation cannot be detected because it is washed out by integrating over all contributions from molecules with different starting points x_1 arriving at the point (x_2, L) with different transverse velocities v_x.

Fortunately, several methods have been developed that overcome these difficulties and that allow ultranarrow Ramsey resonances to be obtained. One of these methods is based on Doppler-free two-photon spectroscopy, while another technique uses saturation spectroscopy but introduces a third interaction zone at the distance $z = 2L$ downstream from the first zone to recover the Ramsey fringes [1257–1259]. We briefly discuss both methods.

9.3.2 Two-Photon Ramsey Resonance

In Sect. 2.4 we saw that the first-order Doppler effect can be exactly canceled for a two-photon transition if the two photons $\hbar\omega_1 = \hbar\omega_2$ have opposite wave vectors, i.e., $\boldsymbol{k}_1 = -\boldsymbol{k}_2$. A combination of Doppler-free two-photon absorption and the Ramsey method therefore avoids the phase dependence $\varphi(v_x)$ on the transverse velocity component. In the first interaction zone the molecular dipoles are excited with the transition amplitude a_1 and precess with their eigenfrequency $\omega_{12} = (E_2 - E_1)/\hbar$. If the two photons come from oppositely traveling waves with frequency ω, the detuning

$$\Omega = \omega + kv_x + \omega - kv_x - \omega_{12} = 2\omega - \omega_{12}, \tag{9.55}$$

of the molecular eigenfrequency ω_{12} from 2ω is independent of v_x. The phase factor $\cos(\Omega T)$, which appears after the transit time $T = L/v_z$ at the entrance of the molecular dipoles into the second field, can be composed as $\cos(\varphi_2^- + \varphi_2^+ - \varphi_1^- - \varphi_1^+)$, where each φ comes from one of the four fields (two oppositely traveling waves in each zone). If we denote the two-photon transition amplitudes in the first and second field zone by c_1 and c_2, respectively, we obtain the total transition probability as

$$\mathcal{P}_{12} = |c_1|^2 + |c_2|^2 + 2|c_1||c_2|\cos\Omega T. \tag{9.56}$$

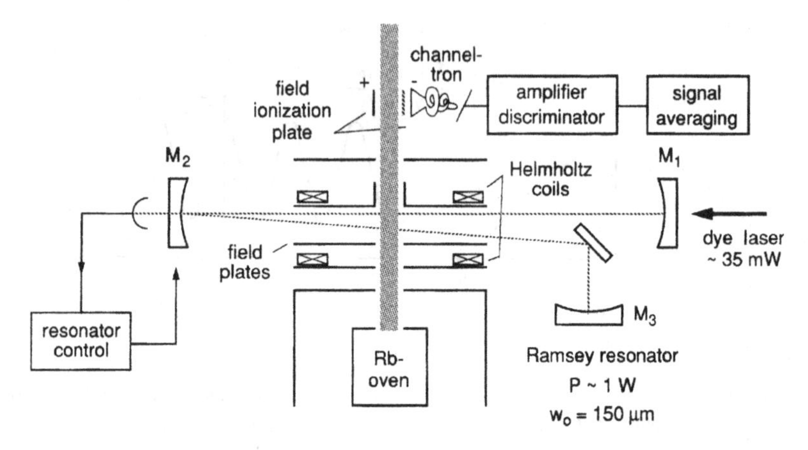

Fig. 9.57 Experimental arrangement for the observation of two-photon Ramsey fringes [1261]

The first two terms describe the conventional two-photon transitions in the first and second zones, while the third term represents the interference leading to the Ramsey resonance. Due to the *longitudinal* thermal velocity distribution $f(v_z)$, only the central maximum of the Ramsey resonance is observed with a theoretical halfwidth (for negligible natural width) of

$$\Delta\Omega = (2/3)\pi/T = 2\pi v/3L \tag{9.57}$$

$$\Rightarrow \quad \Delta\nu = \frac{1}{3T}, \quad \text{with } T = L/v.$$

The higher interference orders are washed out.

Example 9.15 With a field separation of $L = 2.5$ mm and a mean velocity $\overline{v} = 270$ m/s at 400 K, we obtain a halfwidth (FWHM) of $\Delta\nu = 1/3T = 36$ kHz for the central Ramsey fringe if the other contributions to the linewidth are negligible.

The experimental arrangement is depicted in Fig. 9.57. Rubidium atoms are collimated in an atomic beam that traverses the two laser beams. The two counterpropagating waves in both fields are generated from a single laser beam by reflection from the resonator mirrors M1, M2, M3. The radiative lifetimes of the excited atomic levels must be longer than the transit time $T = L/v$. Otherwise, the phase information obtained in the first zone would be lost at the second zone. Therefore the method is applied to long-living states, such as Rydberg states or vibrational levels of the electronic ground state. The excited Rydberg atoms are detected by field ionization. The Helmholtz coils allow the investigation of Zeeman splittings of Rydberg transitions.

The achievable spectral resolution is demonstrated by Fig. 9.58, which shows a two-photon Ramsey resonance for a hyperfine component of the two-photon Ryd-

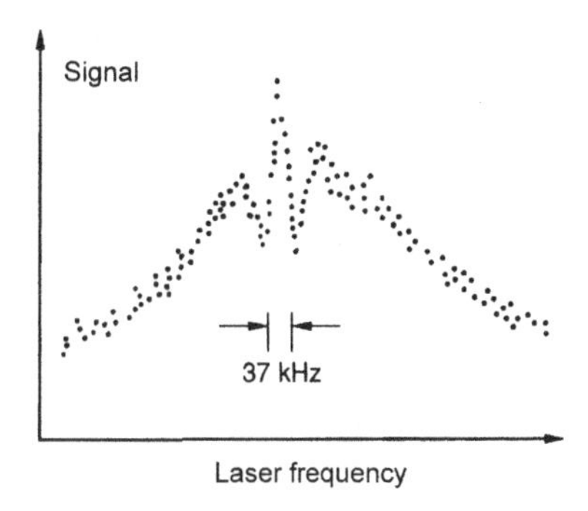

Fig. 9.58 Two-photon optical Ramsey resonance of the transition $32^2S \leftarrow 5^2S$, $F = 3$ in ^{86}Rb for a field separation of $L = 2.5$ mm [1261]

berg transition $32^2S \leftarrow 5^2S$ of rubidium atoms ^{86}Rb [1260, 1261], measured with the arrangment of Fig. 9.57. The length of the Ramsey resonator must always be kept in resonance with the laser frequency. This is achieved by piezo-tuning elements and a feedback control system (Vol. 1, Sect. 5.4.5). The halfwidth of the central Ramsey maximum was measured as $\Delta\nu = 37$ kHz for a field separation of 2.5 mm and comes close to the theoretical limit. With $L = 4.5$ mm an even narrower signal with $\Delta\nu = 18$ kHz could be achieved, whereas the two-photon resonance width in a single zone with a beam waist of $\omega_0 = 150$ μm was limited by transit-time broadening to about 600 kHz [1261].

The quantitative description of the two-photon Ramsey resonance [1262] starts from the transition amplitude per second

$$c_{if}(t) = \frac{D_{if} I}{4\hbar^2 \Delta\omega} \left(\mathrm{e}^{-\mathrm{i}\Delta\omega t} - 1\right), \tag{9.58}$$

for a two-photon transition $|i\rangle \to |f\rangle$ at the detuning $\Delta\omega = 2\omega - (\omega_{ik} + \omega_{kf}) = 2\omega - \omega_{if}$ and the laser intensity I. The two-photon transition dipole element is

$$D_{if} = \sum_k \frac{R_i k R_{kf}}{\omega - \omega_{ki}}, \tag{9.59}$$

where R_{ik}, R_{kf} are the one-photon matrix elements (Vol. 1, (2.93)).

After a molecule has passed the first interaction zone with the transit time $\tau = d/v$, the amplitude for the transition $|i\rangle \to |f\rangle$ is

$$c_{if}(1) \cdot \tau = \frac{D_{if} I_1 \tau}{4\hbar^2 \Delta\omega} \left(\mathrm{e}^{-\mathrm{i}\Delta\omega\tau} - 1\right). \tag{9.60}$$

After the transit time $T = L/v$ through the field-free region and a passage through the second interaction zone, the transition amplitude becomes

$$c_{if}(2) = c_{if}(1) + \frac{D_{if} I_2 \tau}{4\hbar^2 \Delta\omega} \left(\mathrm{e}^{-\mathrm{i}\Delta\omega(T+\tau)} - \mathrm{e}^{-\mathrm{i}\Delta\omega T}\right), \tag{9.61}$$

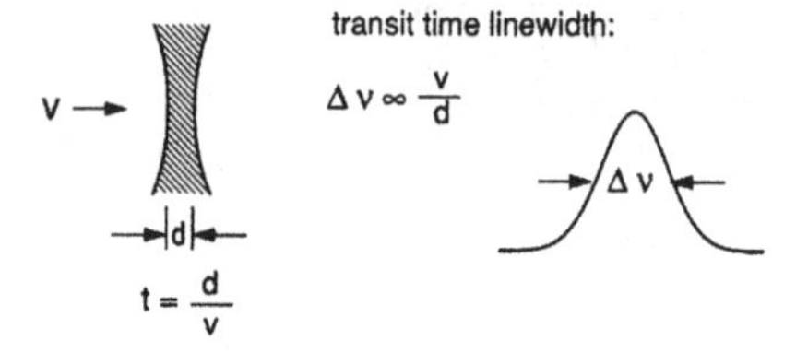

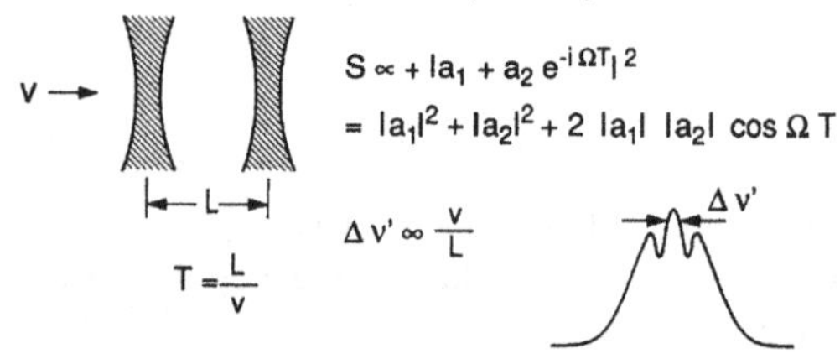

Fig. 9.59 Illustration of two-photon resonance [1261]

which yields the transition probability

$$\begin{aligned}\mathcal{P}_{if}^{(2)}(2) &= \left|c_{if}(1)\right|^2 + \left|c_{if}(2)\right|^2 + 2c_{if}(1)c_{if}(2)\cos\Delta\omega T \\ &= \frac{|D_{if}|^2\tau^2}{\hbar^2}\left[I_1^2 + I_2^2 + 2I_1I_2\cos(\Delta\omega T)\right],\end{aligned} \tag{9.62}$$

which is identical to (9.56). The signal is proportional to the power $S(\Omega)$ absorbed in the second zone (Fig. 9.59).

When the upper level $|f\rangle$ has the spontaneous lifetime $\tau_f = 1/\gamma_f$, part of the excited molecules decay before they reach the second zone, and the transition amplitude becomes

$$c_{if}(2) = c_{if}(1)\left(\mathrm{e}^{-\gamma_f T} + \mathrm{e}^{-\mathrm{i}\Delta\omega T}\right), \tag{9.63}$$

which yields with $I_1 = I_2 = I$ the smaller signal

$$S(\Delta\omega) \propto \mathcal{P}^{(2)}(2) = \frac{|D_{if}|^2 I^2\tau^2}{\hbar^2}\left[1 + \mathrm{e}^{-2\gamma_f T} + 2\mathrm{e}^{-\gamma_f T}\cos\Delta\omega T\right]. \tag{9.64}$$

9.3.3 Nonlinear Ramsey Fringes Using Three Separated Fields

Another solution to restore the Ramsey fringes, which are generally washed out in the second field, is based on the introduction of a third field at the distance $2L$ downstream from the first field. The idea of this arrangement was first pointed out by Chebotayev and coworkers [1257]. The basic idea may be understood as follows: In Sect. 2.2 it was discussed in detail that the nonlinear absorption of a molecule in

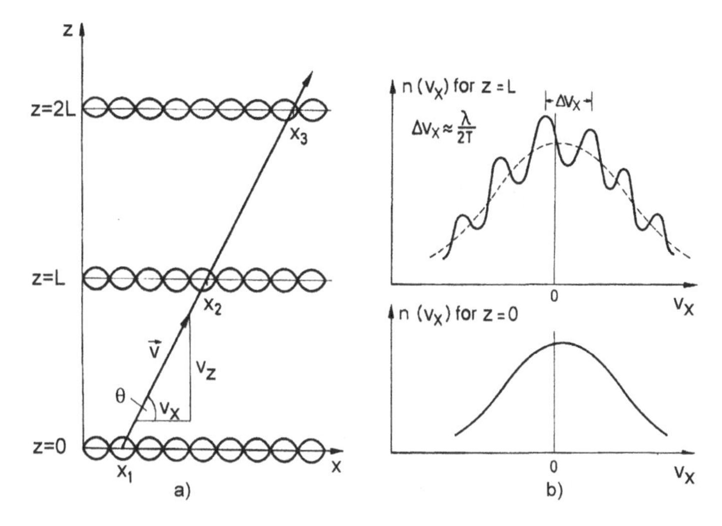

Fig. 9.60 (**a**) Straight path of a molecule through three separated fields at $z = 0$, $z = L$, and $z = 2L$. (**b**) Modulation of the population density $N(v_x)$ after the interaction with the second field [1257]

a monochromatic standing wave field leads to the formation of a narrow Lamb dip at the center ω_0 of a Doppler-broadened absorption line (Fig. 2.7). The Lamb-dip formation may be regarded as a two-step process: the depletion of a small subgroup of molecules with velocity components $v_x = 0 \pm \Delta v_x$ by the pump wave (*hole burning*), and the successive probing of this depletion by a second wave. In the standing light wave of the second zone, the nonlinear saturation depends on the relative phase between the molecular dipoles and the field. This phase is determined by the starting point $(x_1, z = 0)$ in the first zone and by the transverse velocity component v_x. Figure 9.60a depicts the collision-free straight path of a molecule with transverse velocity component v_x, starting from a point $(x_1, z = 0)$ in the first zone, traversing the second field at $x_2 = x_1 + v_x T = x_1 + v_x L/v_z$, and arriving at the third field at $x_3 = x_1 + 2v_x T$. The relative phase between the molecule and the field at the entrance (L, x_2) into the second field is

$$\Delta\varphi = \varphi_1(x_1) + \Delta\omega \cdot T - \varphi_2(x_2), \quad \text{with } \Delta\omega = \omega_{12} - \omega .$$

The macroscopic polarization at (L, x_2), which equals the vector sum of all induced atomic dipoles, averages to zero because molecules with different velocity components v_x arrive at x_2 from different points $(0, x_1)$. Note, however, that the population depletion ΔN_a in the second field depends on the relative phase $\Delta\varphi$ and therefore on v_x. If the phases $\varphi(x_1)$ and $\varphi(x_2)$ of the two fields are made equal for $x_1 = x_2$, the phase difference $\varphi(x_1) - \varphi(x_2) = \varphi(x_1 - x_2) = \varphi(v_x T)$ between the two fields at the intersection points depends only on v_x and not on x_1. *After the nonlinear interaction with the second field the number $n(v_x)$ of molecular dipoles shows a characteristic modulation* (Fig. 9.60b). This modulation cannot be detected in the second field because it appears in v_x but not in x. For the interaction of all molecules with the

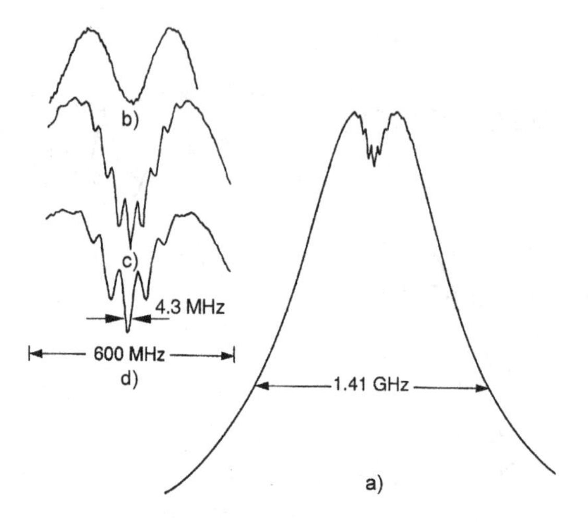

Fig. 9.61 Lamb dip of nonlinear Ramsey resonance for the neon transition $1s_5 \rightarrow 2p_2$ at $\lambda = 588.2$ nm, measured in a fast beam of metastable neon atoms: (**a**) reduced Doppler profile in the collimated beam with the Lamb dip obtained with three laser beams; (**b**)–(**d**) expanded section of the Lamb dip for three different beam geometries: the atom interacts only with two standing waves (**b**), three equally spaced interaction zones (**c**), and four zones (**d**) [1265]

probe wave at $z = z_2$, which has a spatially varying phase $\varphi(x_2)$, this modulation is completely washed out. That is, however, not true in the third field. Since the intersection points x_1, x_2, and x_3 are related to each other by the transverse velocity v_x, the modulation $N(v_x)$ in the second beam results in a nonvanishing macroscopic polarization in the third beam, given by

$$P(\Delta\omega) = 2\,\mathrm{Re}\left\{E_3 \int_{x=0}^{x_0} \int_{t=2T}^{2T+\tau} \left[P^0(z,t)\cos(kx + \varphi_3)\mathrm{e}^{\mathrm{i}\omega t}\right] \mathrm{d}x\,\mathrm{d}t\right\}. \tag{9.65}$$

The detailed calculation, using third-order perturbation theory [1257, 1263], demonstrates that the power absorbed in the third field zone leads to the signal

$$S(\Delta\omega) = \frac{\hbar\omega}{2}\left|G_1 G_2^2 G_3\right|\tau^2 \cos^2(\Delta\omega T)\cos(2\varphi_2 - \varphi_1 - \varphi_3), \tag{9.66}$$

where $G_n = \mathrm{i}D_{21}E_n/\hbar\,(n = 1, 2, 3)$ and $\varphi_1, \varphi_2, \varphi_3$ are the spatial phases of the three fields

$$E_n(x, z, t) = 2E_n(z)\cos(k_n + \varphi_n)\cos\omega t. \tag{9.67}$$

Adjusting the phases φ_n properly, such that $2\varphi_2 = \varphi_1 + \varphi_3$, allows optimization of the signal in the third zone. A detailed calculation of nonlinear Ramsey resonances, based on the density matrix formalism, has been performed by Bordé [1264].

The capability of this combination of saturation spectroscopy with optical Ramsey fringes has been demonstrated by Bergquist et al. [1265] for the example of the neon transition $1s_5 \rightarrow 2p_2$ at $\lambda = 588.2$ nm (Fig. 9.61). A linewidth of 4.3 MHz for the Lamb dip has been achieved with the distance $L = 0.5$ cm between the interaction zones. This corresponds to the natural linewidth of the neon transition.

With four interaction zones the contrast of the Ramsey resonances can still be increased [1266]. This was also demonstrated by Bordé and coworkers [1267], who used four traveling waves instead of three standing waves, which crossed a supersonic molecular beam of SF_6 perpendicularly. The Ramsey signal for vibration–rotation transitions in SF_6 around $\lambda = 10$ μm was monitored with improved contrast by an optothermal detector (Sect. 1.3.3).

For transitions with small natural linewidths the Ramsey resonances can be extremely narrow. For instance, the nonlinear Ramsey technique applied to the resolution of rotational–vibrational transitions in CH_4 in a methane cell at 2 mbar pressure yielded a resonance width of 35 kHz for the separation $L = 7$ mm of the laser beams. With the distance $L = 3.5$ cm the resonance width decreased to 2.5 kHz, and the Ramsey resonances of the well-resolved hyperfine components of the CH_4 transition at $\lambda = 3.39$ μm could be measured [1263].

It is also possible to observe the Ramsey resonances at $z = 2L$ without the third laser beam. If two standing waves at $z = 0$ and $z = L$ resonantly interact with the molecules, we have a situation similar to that for photon echoes. The molecules that are coherently excited during the transit time τ through the first field suffer a phase jump at $t = T$ in the second zone, because of their nonlinear interaction with the second laser beam, which reverses the time development of the phases of the oscillating dipoles. At $t = 2T$ the dipoles are again in phase and emit coherent radiation with increased intensity for $\omega = \omega_i k$ (*photon echo*) [1258, 1263].

9.3.4 Observation of Recoil Doublets and Suppression of One Recoil Component

The increased spectral resolution obtained with the Ramsey technique because of the increased interaction time allows the direct observation of recoil doublets in atomic or molecular transitions (Sect. 9.1.1). An example is the Ramsey spectroscopy of the intercombination line ${}^1S_0$–${}^3P_1$ in calcium at $\lambda = 657$ nm [1268], where a linewidth of 3 kHz was obtained for the central Ramsey maximum and the recoil doublet with a separation of 23 kHz could be clearly resolved (Fig. 9.62).

Although transit-time broadening is greatly reduced by the Ramsey technique, the quadratic Doppler effect is still present and may prevent the complete resolution of the recoil components. This may cause asymmetric line profiles where the central frequency cannot be determined with the desired accuracy. As was shown by Helmcke et al. [1269, 1277], one of the recoil components can be eliminated if the upper level ${}^3P_1$ of the Ca transition is depopulated by optical pumping with a second laser. In Fig. 9.63 the relevant level scheme, the experimental setup, and the measured central Ramsey maximum of the remaining recoil component are shown.

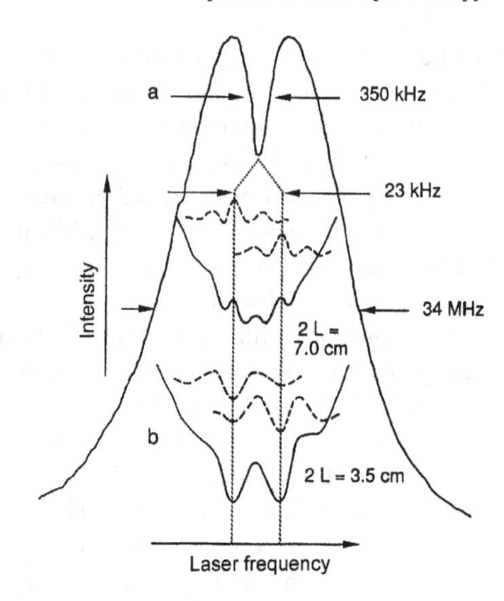

Fig. 9.62 Ramsey resonances of the calcium intercombination line at $\lambda = 657$ nm, measured in a collimated Ca atomic beam: (**a**) Doppler profile with a reduced Doppler width and a central Lamb dip, if only one interaction zone is used; (**b**) expanded section of the Lamb dip with the two recoil components observed with three interaction zones with separations $L = 3.5$ cm and $L = 1.7$ cm. The *dashed curves* show results with partial suppression of one recoil component [1268]

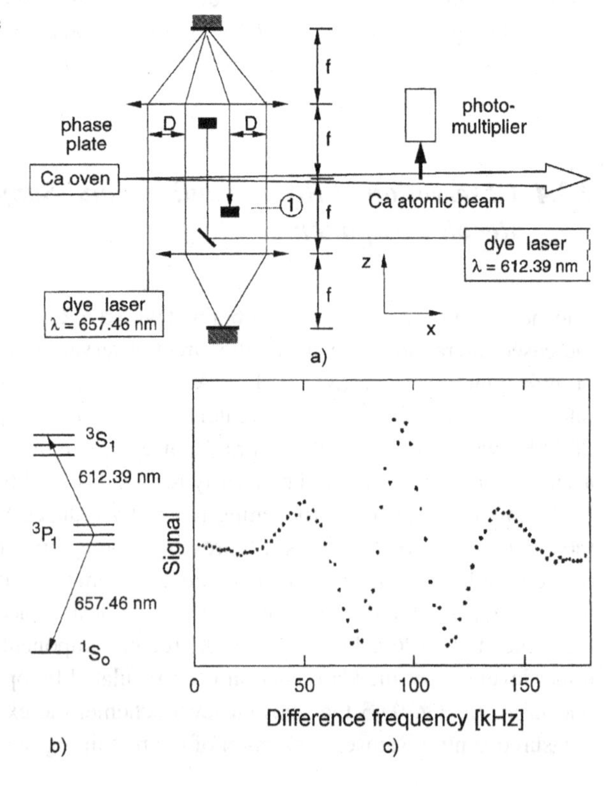

Fig. 9.63 Suppression of one recoil component by optical pumping with a second laser: (**a**) experimental setup; (**b**) level scheme; and (**c**) Ramsey resonance of the remaining recoil component [1269]

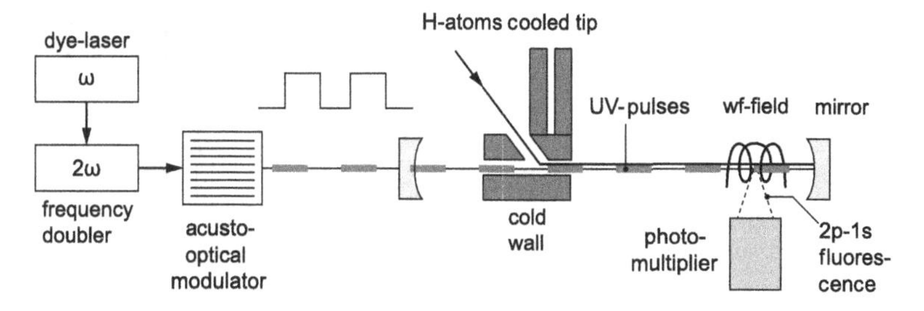

Fig. 9.64 Precise spectroscopy of the hydrogens $1S \to 2S$ transition using the Ramsey technique with UV-pulses [1270]

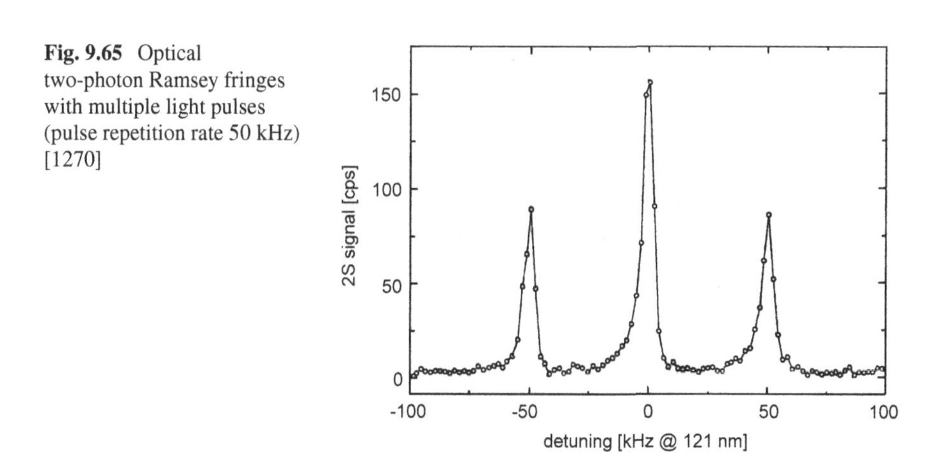

Fig. 9.65 Optical two-photon Ramsey fringes with multiple light pulses (pulse repetition rate 50 kHz) [1270]

9.3.5 *Optical Ramsey Resonances Obtained Through an Equidistant Train of Laser Pulses*

The Ramsey principle for the interaction of an atom moving in time through different electro-magnetic fields can be also realized in a completely different way. Atoms in a collimated beam are irradiated with an equidistant pulse train from a laser beam travelling collinearly with the atomic beam (Fig. 9.64). The distance Δt between successive pulses corresponds to the flight time of molecules between the separated fields in the classical Ramsey method.

This modification of the Ramsey technique was used by Hänsch and coworkers for precision measurements on the hydrogen atom. The fundamental radiation of a cw dye laser at $\lambda = 486$ nm was frequency-doubled and through an acousto-optical modulator formed into an equidistant sequence of pulses. These pulses were sent into a high finesse resonator where they formed a standing wave with a corresponding pattern of spatially separated amplitude peaks [1270]. The achieved line width of the two-photon transition at $\lambda = 121$ nm was below 5 kHz (Fig. 9.65).

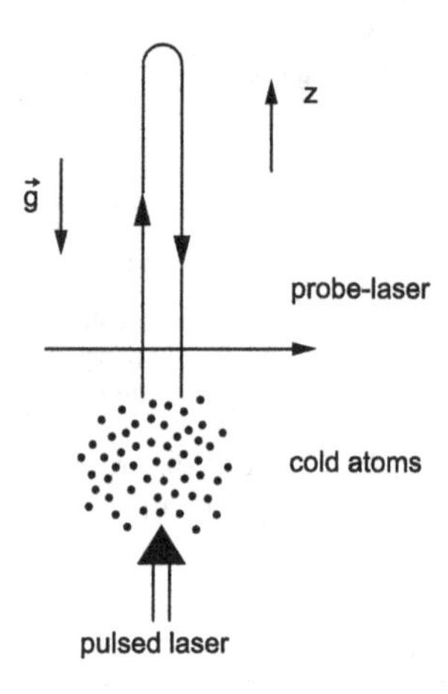

Fig. 9.66 Atom fountain. g = acceleration of Earth's gravity

9.3.6 Atomic Fountain

A very ingenious idea for obtaining narrow Ramsey fringes was the invention of the atomic fountain. Instead of a horizontal atomic beam here a vertical beam of cold atoms with initial velocity v_0 released from a trap is used which rise against gravitation from $z = 0$ to a maximum height $z_m = h = v_0^2/2g$ where they come at rest and start to fall down again along the same path (Fig. 9.66). A horizontal laser beam crosses the atomic fountain at a position z closely above $z = 0$. During their rise and their fall the atoms cross the same laser beam. At the first crossing during their rise the atoms acquire an induced oscillating dipole moment. The phase of these dipoles develops during the time until they cross the laser field again during their fall. The absorption during the second crossing depends on the difference between the phases of the atomic dipole and the laser field.

Because of the small velocity the time difference $\Delta t = 2v_0/g$ is very long. According to Eq. (9.57) the spectral width $\Delta\nu$ of the central Ramsey fringe is accordingly small.

The reservoir for the cold atom can be either a magneto-optical trap or a BEC. A short laser pulse with a wavelength tuned to the atomic absorption line travelling into the $+z$-direction pushes the atoms out of the trap. The recoil momentum $h\nu/c$ gives the atoms the initial velocity $v_0 = h \cdot \nu/(mc)$.

Example 9.16 The mean velocity of sodium atoms with $m = 23$ AMU at a temperature of $T = 10$ mK is obtained from $mv_0^2/2 = 1/2kT$ as $v_0 = (kT/m)^{1/2} = 1.9$ m/s $\rightarrow h = 0.18$ m. The rise time until the turning point is $t_h = v_0/g \approx 0.2$ s. The time between the two crossings is then $\Delta t \approx 0.4$ s. The minimum spectral width of the central Ramsey fringe is $\Delta\nu = 1/\Delta t = 2.5$ Hz. Since there are other limiting factors for the spectral width one reaches in real experiments about $\Delta\nu = 10$ Hz. The initial velocity caused

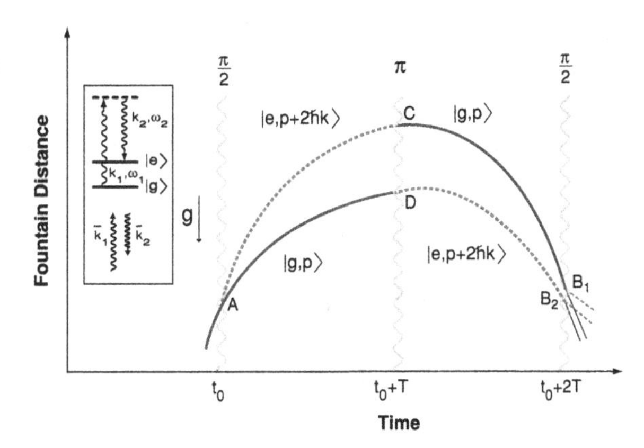

Fig. 9.67 Recoil space diagram of the atoms through the interferometer showing the separation (exaggerated) of the atomic wavepackets. The area enclosed by the two paths is proportional to the mean value of acceleration over the path, g. [1275]

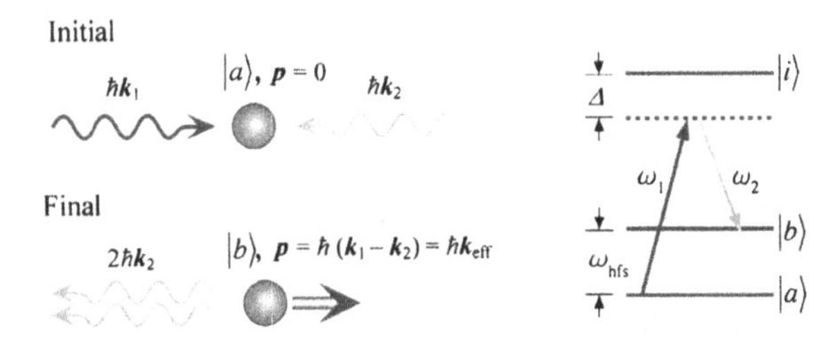

Fig. 9.68 Momentum transfer and recoil momentum in a stimulated Raman transition

by the recoil of one absorbed photon is only 0.03 m/s, which is small compared to the mean thermal velocity of 1.9 m/s. If the two contribution should be equal each atom has to absorb about 70 photons from the "pushing laser".

If such an atomic fountain is used as atomic clock [1271, 1272] one can obtain a relative frequency stability of $\Delta\nu/\nu = 5 \times 10^{-16}$!!

The atomic fountain can be used to determine the local gravitational field of the earth by measuring the acceleration of the atoms in the fountain through observing the Doppler shift of an atomic line with a weak laser beam propagating collinearly with the fountain [1273]. An accuracy of $\Delta g/g < 2 \times 10^{-8}$ could be achieved [1274] with an atomic fountain of cold Cs-atoms. The Doppler-shift of the atoms was measured using velocity selective stimulated Raman transitions. The atoms experience on their way upwards and downwards in the atomic fountain two different sequences of excitations (Fig. 9.67): At time t_0 they are excited by a $\pi/2$-pulse which creates a superposition of ground-and excited state. The atoms in the ground-state travel with the momentum p upwards. The atoms in the excited state, reached by a Raman transition, experience a recoil $2\hbar k$ and have therefore a momentum $p + 2\hbar k$ (Fig. 9.68). After a time T they reach the upper point in the fountain which is for the excited atoms slightly higher because of their additional momentum. Here they are irradiated by a π-pulse which converts the populations of $|g\rangle$ and $|e\rangle$

and transfers again a recoil momentum to the atoms. On their way downwards they reach at time $t_0 + 2T$ the lower laser beam and experience again $\pi/2$-pulse. The interference of the atoms on the two different pathes depends on the phase difference, which is affected by their acceleration on their way due to the gravitational force.

9.4 Atom Interferometry

Atomic particles moving with the momentum p can be characterized by their de Broglie wavelength $\lambda = h/p$. If beams of such particles can be split into coherent partial beams, which are recombined after traveling different path lengths, matter-wave interferometry becomes possible. This has been demonstrated extensively for electrons and neutrons, and recently also for neutral atoms [1278, 1279].

9.4.1 Mach–Zehnder Atom Interferometer

A beam of neutral atoms may be split into coherent partial beams by diffraction from two slits [1280] or from microfabricated gratings [1281], but also by photon recoil in laser beams. The atoms leave the source (Na/Ar mixture) through a nozzle and are formed into a collimated beam by two slits (Fig. 9.69). They traverse a grating where part of the atoms are diffracted which splits the beam into two parts After a flight path of more than 0.5 m they pass through a double slit where again diffraction occurs and one partial beam is bended which interferes with the non-deflected beam in the third zone (detection zone) where they are detected by a hot wire Langmuir detector. A He–Ne-laser is used to align the positions of grating and slits. Since the De Broglie-wavelength of the atoms is much smaller than the optical wavelength in the visible, the sensitivity of such an atom interferometer is much higher. Any interaction which shifts the two partial beams in a slightly different way changes the interference pattern.

While this method is analogous to the optical phenomena (Young's two-slit experiment) although with a much smaller wavelength λ, a new technique, using an optical Ramsey scheme, where the beam is split by photon recoil has no counterpart in optical interferometers. Therefore, we will briefly discuss this method, which uses a four-zone Ramsey excitation as an atomic interferometer. It was proposed by Bordé [1289], and experimentally realized by several groups. The explanation, which follows that of Helmcke and coworkers [1290], is as follows: atoms in a collimated beam pass through the four interaction zones of the Ramsey arrangement, exhibited in Fig. 9.70. Atoms in state $|a\rangle$ that absorb a laser photon in the interaction zone 1 are excited into state $|b\rangle$ and suffer a recoil momentum $\hbar k$, which deflects them from their straight path. If these excited atoms undergo an induced emission in the second zone, they return to the state $|a\rangle$ and fly parallel to the atoms that have

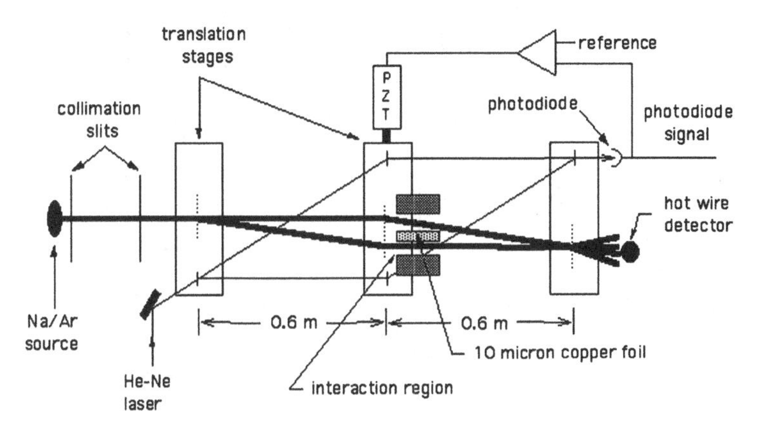

Fig. 9.69 Atomic interferometer [1284]

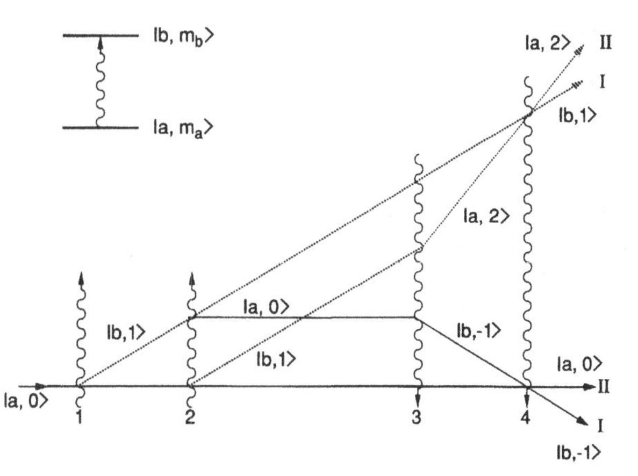

Fig. 9.70 Optical Ramsey scheme of an atomic beam passing through four traveling laser fields, interpreted as matter-wave interferometer. *Solid lines* represent the high-frequency recoil components, *dotted lines* the low-frequency components (only those traces leading to Ramsey resonances in the fourth zone are drawn) [1290]

not absorbed a photon in both zones 1 and 2. In Fig. 9.70 the atoms in the different zones are characterized by their internal state, $|a\rangle$ or $|b\rangle$, and the integer m of their transverse momentum $m\hbar k$. The figure illustrates that in the fourth zone there are two pairs of exit ports where the two components of each pair pass through the same location and can therefore interfere. The phase difference of the corresponding matter waves depends on the path difference and on the internal state of the atoms. The excited atoms in state $|b\rangle$ arriving at the fourth zone can be detected through their fluorescence. These atoms differ in the two interference ports by their transverse momentum $\pm\hbar k$, and their transition frequencies are therefore split by recoil splitting (Sect. 9.1.1). The signals in the exit ports in zone 4 depend critically on the difference $\Omega = \omega_L - \omega_0$ between the laser frequency ω_L and the atomic resonance frequency ω_0. The two trapezoidal areas in Fig. 9.70, which are each enclosed by the two interfering paths, may be regarded as two separate Mach–Zehnder interferometers (Vol. 1, Sect. 4.2.4) for the two recoil components.

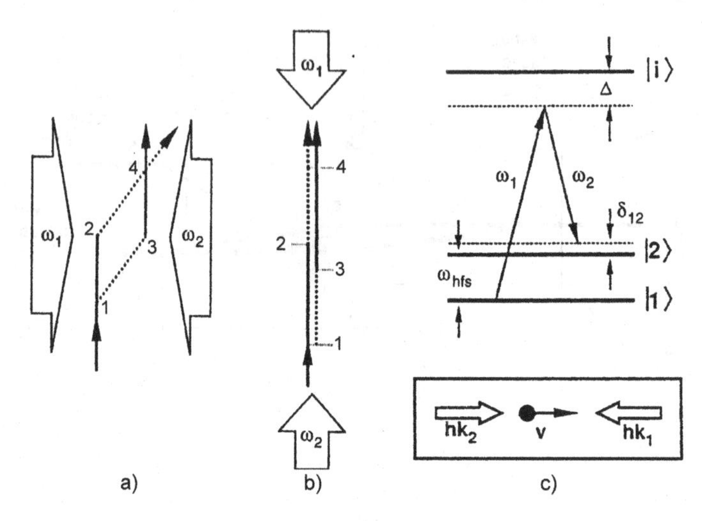

Fig. 9.71 (**a**) Paths of the atom in a Mach–Zehnder-type atomic interferometer; (**b**) momentum transfer by stimulated Raman transitions applied to rubidium atoms in an atomic fountain where the atoms move parallel to the laser beams [1291]; (**c**) level scheme for Raman transitions

Matter-wave interferometry has found wide applications for testing basic laws of physics. One advantage of interferometry with massive particles, for instance, is the possibility of studying gravitational effects. Compared to the neutron interferometer, atomic interferometry can provide atomic fluxes that are many orders of magnitude higher than thermalized neutron fluxes from reactors. Furthermore, the detection sensitivity by laser-induced fluorescence is much higher than for neutrons. The sensitivity is therefore higher and the costs are much lower.

One example of an application is the measurement of the gravitational acceleration g on earth with an accuracy of $3 \times 10^{-8} g$ with a light-pulse atomic interferometer [1291]. Laser-cooled wave packets of sodium atoms in an atomic fountain (Sect. 9.1.9) are irradiated by a sequence of three light pulses with properly chosen intensities. The first pulse is chosen as $\pi/2$-pulse, which creates a superposition of two atomic states $|1\rangle$ and $|2\rangle$ and results in a splitting of the atomic fountain beam at position 1 in Fig. 9.71 into two beams because of photon recoil. The second pulse is a π-pulse, which deflects the two partial beams into opposite directions; the third pulse finally is again a $\pi/2$-pulse, which recombines the two partial beams and causes the wave packets to interfere. This interference can, for example, be detected by the fluorescence of atoms in the upper state $|2\rangle$.

For momentum transfer stimulated Raman transitions between the two hyperfine levels $|1\rangle$ and $|2\rangle$ of the $\mathrm{Na}(3\,^2S_{1/2})$ state are used, which are induced by two laser pulses with light frequencies ω_1 and ω_2 ($\omega_1 - \omega_2 = \omega_{\mathrm{HFS}}$) traveling into opposite directions (Fig. 9.71b). Each transition transfers the momentum $\Delta p \simeq 2\hbar k$. The gravitational field causes a deceleration of the upwards moving atoms in the fountain. This changes their velocity, which can be detected as a Doppler shift of the Raman transitions. Because the Raman resonance $\omega_1 - \omega_2 = \omega_{\mathrm{HFS}}$ has an extremely narrow width (Sect. 9.6), even small Doppler shifts can be accurately measured.

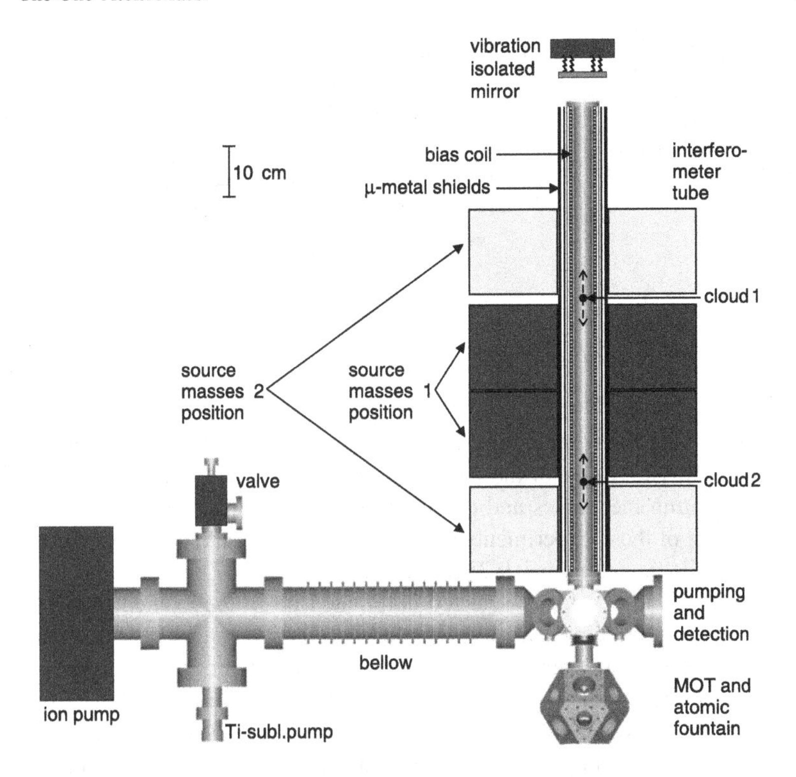

Fig. 9.72 Graphical illustration of the experimental set-up with the relevant part of the vacuum system, the atomic trajectories and the source mass positions. Not included are the laser systems, the detection units and the source mass holder. The atomic trajectories during the time of the interferometer pulse sequence are sketched (*dashed arrows*) [1276]

The gravitational constant G is from all physical constants the one with the largest experimental uncertainty. Therefore new measuring techniques have been invented to reach a higher accuracy. One of these techniques is based on atom interferometry. Its basic principle can be understood as follows: The acceleration of atoms in two clouds in an atomic fountain is measured with stimulated Raman transitions (Fig. 9.71). Two heavy masses at two different positions are added (Fig. 9.72). In the first position one mass was located above the upper atom cloud and the other below the lower cloud. Then the masses were shifted into a position where one mass was below the upper cloud and the other above the lower cloud. The change of the atom acceleration for the two positions were determined [1276].

9.5 The One-Atom Maser

In Sect. 9.3 we discussed techniques to store and observe single ions in traps. We will now present some recently performed experiments that allow investigations of

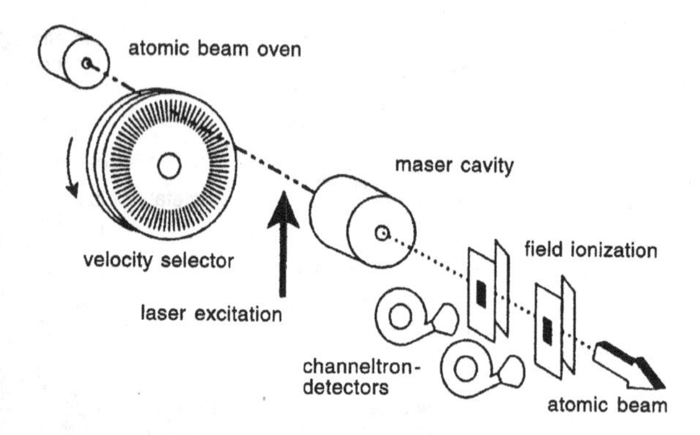

Fig. 9.73 Schematic experimental setup for the "one-atom maser" with resonance cavity and state-selective detection of the Rydberg atoms [1294]

single atoms and their interaction with weak radiation fields in a microwave resonator [1294]. The results of these experiments provide crucial tests of basic problems in quantum mechanics and quantum electrodynamics (often labeled "cavity QED"). Most of these experiments were performed with alkali atoms. The experimental setup is shown in Figs. 9.73 and 9.74.

Alkali atoms in a velocity-selected collimated beam are excited into Rydberg levels with large principal quantum numbers n either by stepwise excitation with two diode lasers, or by a single frequency-doubled dye laser. The spontaneous lifetime $\tau(n)$ increases with n^3 and is, for sufficiently large values of n, longer than the transit time of the atoms from the excitation point to the detection point. When the excited atoms pass through a microwave resonator that is tuned into resonance with a Rydberg transition frequency $\nu = (E_n - E_{n-1})/h$, a fluorescence photon emitted by a Rydberg atom during its passage through the resonator may excite a cavity mode [1295].

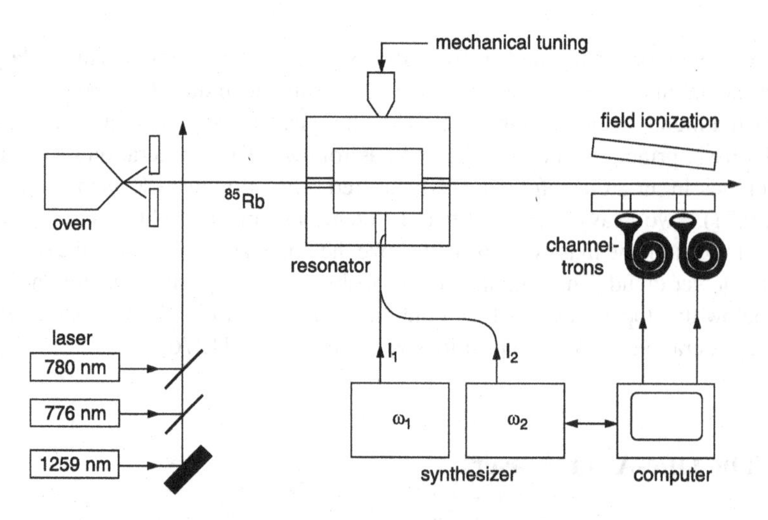

Fig. 9.74 Experimental setup for experiments with the one-atom maser with three excitation lasers, cavity resonance microwave generation and a state-selective detection system

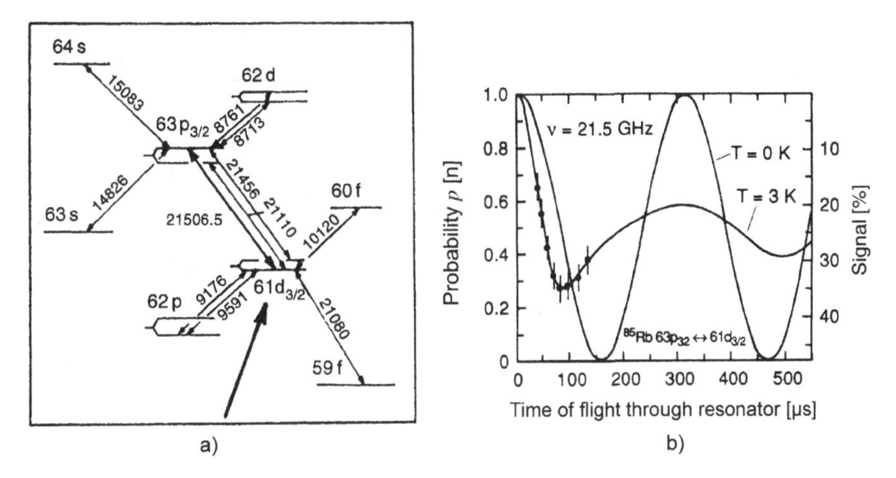

Fig. 9.75 (**a**) Level diagram of the maser transition in rubidium. The transition frequencies are given in MHz. (**b**) Measured (*points*) and calculated (*solid curve*) Rabi oscillations of an atom in the cavity field at $T = 0$ K and at $T = 3$ K, damped by the statistical influence of the thermal radiation field. The experimental points were measured with a velocity selector that allowed transit times of 30–140 µs [1297, 1299]

If the cavity resonator is cooled down to temperatures of a few Kelvin, the cavity walls become superconducting and its losses decrease drastically. Cavity Q-values of larger than 5×10^{10} can be achieved, which corresponds to a decay time of an excited mode of $T_R > 1$ s at a resonance frequency of $\nu = 21$ GHz. This decay time is much longer than the transit time of the atom through the cavity. If the density of atoms in the beam is decreased, one can reach the condition that only a single atom is present in the cavity resonator during the transit time $T = d/c \simeq 100$ µs. This allows investigations of the interaction of a single atom with the EM field in the cavity (Fig. 9.75). The resonance frequency of the cavity can be tuned continuously within certain limits by mechanical deformation of the cavity walls [1296].

Because of its large transition dipole moment $D_{n,n-1} \propto n^2$ (!) the atom that emitted a fluorescence photon exciting the cavity mode may again absorb a photon from the EM field of the cavity and return to its initial state $|n\rangle$. This can be detected when the Rydberg atom passes behind the cavity through two static electric fields (Fig. 9.73). The field strength in the first field is adjusted to field-ionize Rydberg atoms in level $|n\rangle$ but not in $|n-1\rangle$, while the second, slightly stronger field also ionizes atoms in level $|n-1\rangle$. This allows one to decide in which of the two levels the Rydberg atoms have left the microwave resonator.

It turns out that the spontaneous lifetime τ_n of the Rydberg levels is shortened if the cavity is tuned into resonance with the frequency ω_0 of the atomic transition $|n\rangle \rightarrow |n-1\rangle$. It is prolonged if no cavity mode matches ω_0 [1297]. This effect, which had been predicted by quantum electrodynamics, can intuitively be understood as follows: in the resonant case, that part of the thermal radiation field that is in the resonant cavity mode can contribute to stimulated emission in the transition $|n\rangle \rightarrow |n-1\rangle$, resulting in a shortening of the lifetime (Sect. 6.3). For the

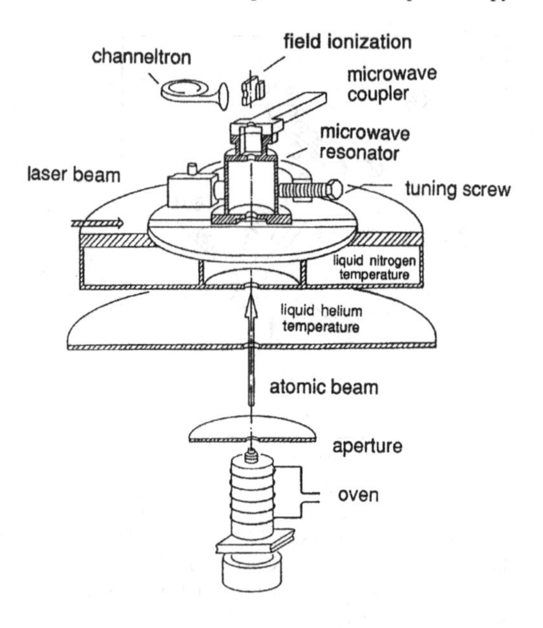

Fig. 9.76 Schematic drawing of the cooled cavity with atom source and field ionization detector

detuned cavity the fluorescence photon does not "fit" into the resonator. The boundary conditions impede the emission of the photon. If the rate $\mathrm{d}N/\mathrm{d}t$ of atoms in the level $|n\rangle$ behind the resonator is measured as a function of the cavity resonance frequency ω, a minimum rate is measured for the resonance case $\omega = \omega_0$ (Fig. 9.75).

Without a thermal radiation field in the resonantly tuned cavity, the populations $N(n,T)$ and $N(n-1,T)$ of the Rydberg levels should be a periodic function of the transit time $T = d/v$, with a period T_R that corresponds to the Rabi oscillation period. The incoherent thermal radiation field causes induced emission and absorption with statistically distributed phases. This leads to a damping of the Rabi oscillation (Fig. 9.75b). This effect can be proved experimentally if the atoms pass a velocity selector before they enter the resonator, which allows a continuous variation of the velocity and therefore of the transit time $T = d/v$. In Fig. 9.76 the schematic drawing of the experimental setup with microwave resonator, atom source and detector is shown.

The one-atom maser can be used to investigate the statistical properties of non-classical light [1298, 1299]. If the cavity resonator is cooled down to $T \leq 0.5$ K, the number of thermal photons becomes very small and can be neglected. The number of photons induced by the atomic fluorescence can be measured via the fluctuations in the number of atoms leaving the cavity in the lower level $|n-1\rangle$. It turns out that the statistical distribution does not follow Poisson statistics, as in the output of a laser with many photons per mode, but shows a sub-Poisson distribution with photon number fluctuations 70 % below the vacuum-state limit [1300]. In cavities with low losses, pure photon number states of the radiation field (Fock states) can be observed (Fig. 9.77) [1301], with photon lifetimes as high as 0.2 s! At very low

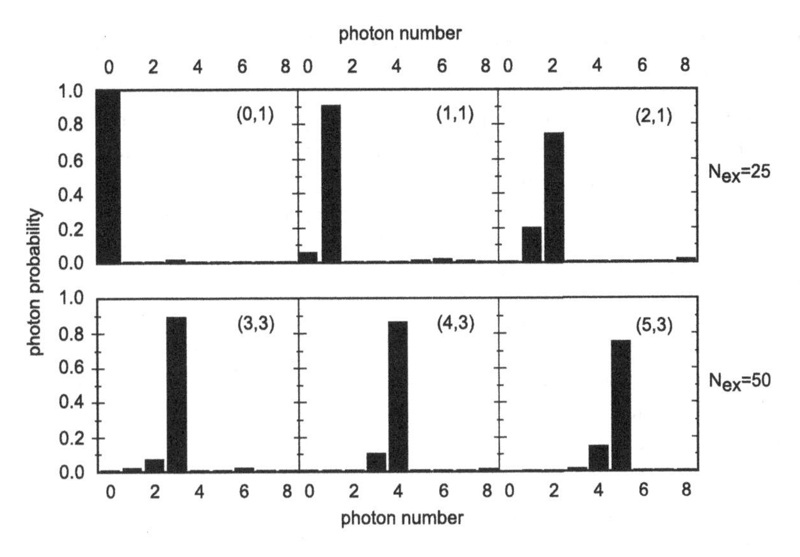

Fig. 9.77 Photon number distributions for various trapping states (nq) for two different numbers N_{ex} of photons in the resonator, where n gives the number of photons interacting with the atom and q the number of Rabi-oscillation periods [S.St. Brattke; PhD thesis, LMU Munich 2000]

temperatures of the cavity ($T < 0.2$ K) the number of thermal photons is negligibly small. Under three these conditions trapped states can be realized where an atom undergoes an integer number of Rabi oscillation periods and leaves the cavity in the same state as it has entered it.

Another interesting effect that is now accessible to measurements is the quantum electrodynamic energy shift of atomic levels when the atoms pass between two parallel metal plates (Casimir–Polder effect) [1302].

9.6 Spectral Resolution Within the Natural Linewidth

Assume that all other line-broadening effects except the natural linewidth have been eliminated by one of the methods discussed in the previous chapters. The question that arises is whether the natural linewidth represents an insurmountable natural limit to spectral resolution. At first, it might seem that Heisenberg's uncertainty relation does not allow outwit the natural linewidth (Vol. 1, Sect. 3.1). In order to demonstrate that this is not true, in this section we give some examples of techniques that do allow observation of structures *within* the natural linewidth. It is, however, not obvious that all of these methods may really increase the amount of information about the molecular structure, since the inevitable loss in intensity may outweigh the gain in resolution. We discuss under what conditions spectroscopy within the natural linewidth may be a tool that really helps to improve the quality of spectral information.

9.6.1 Time-Gated Coherent Spectroscopy

The first of these techniques is based on the selective detection of those excited atoms or molecules that have survived in the excited state for times $t \gg \tau$, which are long compared to the natural lifetime τ.

If molecules are excited into an upper level with the spontaneous lifetime $\tau = 1/\gamma$ by a light pulse ending at $t = 0$, the time-resolved fluorescence amplitude is given by

$$A(t) = A(0)\mathrm{e}^{-(\gamma/2)t} \cos \omega_0 t. \tag{9.68}$$

If the observation time extends from $t = 0$ to $t = \infty$, a Fourier transformation of the measured intensity $I(t) \propto A^2(t)$ yields, for the line profile of the fluorescence emitted by atoms at rest, the Lorentzian profile (Vol. 1, Sect. 3.1)

$$I(\omega) = \frac{I_0}{(\omega - \omega_0)^2 + (\gamma/2)^2}, \quad \text{with } I_0 = \frac{\gamma}{2\pi} \int I(\omega)\,\mathrm{d}\omega. \tag{9.69}$$

If the detection probability for $I(t)$ is not constant, but follows the time dependence $f(t)$, the detected intensity $I_\mathrm{g}(t)$ is determined by the *gate function* $f(t)$

$$I_\mathrm{g}(t) = I(t) f(t).$$

The Fourier transform of $I_\mathrm{g}(t)$ now depends on the form of $f(t)$ and may no longer be a Lorentzian. Assume the detection is gated by a step function

$$f(t) = \begin{Bmatrix} 0 & \text{for } t < T \\ 1 & \text{for } t \geq T \end{Bmatrix},$$

which may simply be realized by a shutter in front of the detector that opens only for times $t \geq T$ (Fig. 9.78).

The Fourier transform of the amplitude $A(t)$ in (9.68) is now

$$A(\omega) = \int_\mathrm{T}^\infty A(0)\mathrm{e}^{-\gamma/2t} \cos(\omega_0 t)\mathrm{e}^{-\mathrm{i}\omega t}\,\mathrm{d}t,$$

which yields in the approximation $\Omega = |\omega - \omega_0| \ll \omega_0$, with $\exp(-\mathrm{i}\omega t) = \cos(\omega t) - \mathrm{i}\sin(\omega t)$, the cosine and sine Fourier transforms:

$$\begin{aligned} A_\mathrm{c}(\omega) &= \frac{A_0}{2} \frac{\mathrm{e}^{-(\gamma/2)T}}{\Omega^2 + (\gamma/2)^2} \left[\frac{\gamma}{2} \cos(\Omega T) - \Omega \sin(\Omega T) \right], \\ A_\mathrm{s}(\omega) &= \frac{A_0}{2} \frac{\mathrm{e}^{-(\gamma/2)T}}{\Omega^2 + (\gamma/2)^2} \left[\frac{\gamma}{2} \sin(\Omega T) + \Omega \cos(\Omega T) \right]. \end{aligned} \tag{9.70}$$

If only the incoherent fluorescence intensity is observed without obtaining any information on the phases of the wave function of the excited state, we observe the

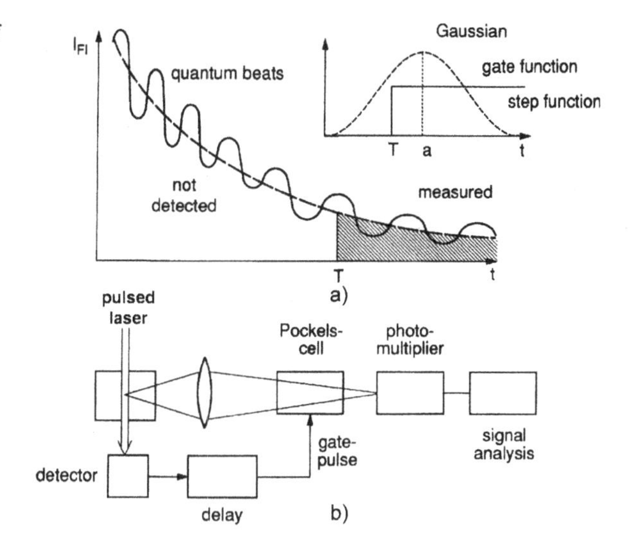

Fig. 9.78 Gated detection of exponentional fluorescence decay with a gate function $f(t)$: (**a**) schematic scheme; (**b**) experimental realization

intensity

$$I(\omega) \propto \left|A_c(\omega) - \mathrm{i}A_s(\omega)\right|^2 = A_c^2 + A_s^2 = \frac{A_0^2}{2} \frac{\mathrm{e}^{-\gamma T}}{\Omega^2 + (\gamma/2)^2}. \tag{9.71}$$

This is again a Lorentzian with a halfwidth (FWHM) of $\gamma = 1/\tau$, which is independent of the gate time T! Because of the delayed detection one loses the factor $\exp(-T/\tau)$ in intensity, since all fluorescence events occurring for times $t < T$ are missing. *This shows that with incoherent techniques no narrowing of the natural linewidth can be achieved, even if only fluorescence photons from selected long-living atoms with $t > T \gg \tau$ are selected* [1303].

The situation changes, however, if instead of the intensity (9.71), the amplitude (9.68) or an intensity representing a coherent superposition of amplitudes can be measured, where the phase information and its development in time is preserved. Such measurements are possible with one of the coherent techniques discussed in Chap. 7.

One example is the quantum-beat technique. According to (7.27a) the fluorescence signal at the time t after the coherent excitation at $t = 0$ of two closely spaced levels separated by $\Delta\omega$, with equal decay times $\tau = 1/\gamma$, is

$$I(t) = I(0)\mathrm{e}^{-\gamma t}(1 + a\cos\Delta\omega t). \tag{9.72}$$

The term $\cos(\Delta\omega \cdot t)$ contains the wanted information on the phase difference

$$\Delta\varphi(t) = \Delta\omega \cdot t = (E_i - E_k)t/\hbar,$$

between the wave functions $\psi_n(t) = \psi_n(0)\mathrm{e}^{-\mathrm{i}E_n t/\hbar}$ $(n = i, k)$ of the two levels $|i\rangle$ and $|k\rangle$.

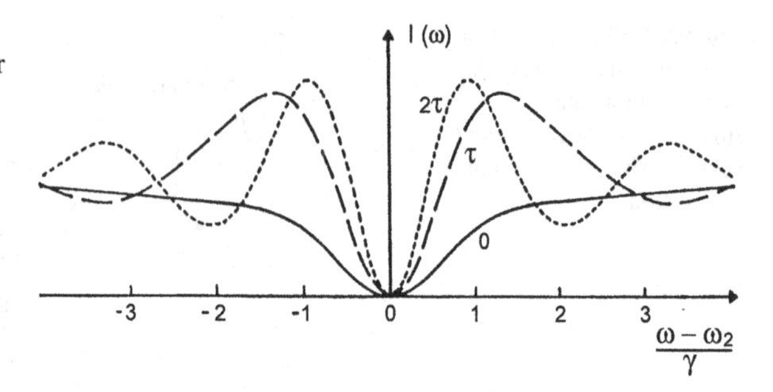

Fig. 9.79 Oscillatory structure of the cosine Fourier transform with narrowed central maximum for increasing values of the gate delay time $T = 0$, τ, and 2τ. The peak intensity has been normalized

If the detector is gated to receive fluorescence only for $t > T$, the Fourier transform of (9.72) becomes

$$I(\omega) = \int_T^\infty I(0) \cdot \mathrm{e}^{-\gamma t}(1 + a\cos\Delta\omega t)\mathrm{e}^{-\mathrm{i}\omega t}\,\mathrm{d}t, \tag{9.73}$$

where $\omega \gg \Delta\omega$ is the mean light frequency of the fluorescence. The evaluation of the integral yields the cosine Fourier transform

$$I_{\mathrm{c}}(\omega) = \frac{I_0}{2}\frac{\mathrm{e}^{-\gamma T}}{(\Delta\omega - \omega)^2 + \gamma^2}\left[\gamma\cos(\Delta\omega - \omega)T - (\Delta\omega - \omega)\sin(\Delta\omega - \omega)T\right]. \tag{9.74}$$

For $T > 0$ the intensity $I_{\mathrm{c}}(\omega)$ exhibits an oscillatory structure (Fig. 9.79) with a central maximum at $\omega \simeq \Delta\omega$ (because of the last two terms in (9.74) the center is not exactly at $\omega = \Delta\omega$) and the halfwidth

$$\Delta\omega_{12} = \frac{2\gamma}{\sqrt{1 + \gamma^2 T^2}}. \tag{9.75}$$

For $T = 0$ the width of the quantum beat signal is the sum of the level widths of the two coherently excited levels, contributing to the beat signal.

Example 9.17 For $T = 5\tau = 5/\gamma$ the *halfwidth* of the central peak has decreased from γ to 0.4γ. The peak intensity, however, has drastically decreased by the factor $\exp(-\gamma T) = \exp(-5) \simeq 10^{-2}$ to less than 1 % of its value for $T = 0$.

The decrease of the peak intensity results in a severe decrease of the signal-to-noise ratio. This may, in turn, lead to a larger uncertainty in determining the line center.

The oscillatory structure can be avoided if the gate function $f(t)$ is not a step function but a Gaussian $f(t) = \exp[-(t - T)^2/b^2]$ with $b = (2T/\gamma)^{1/2}$ [1304].

Another coherent technique that can be combined with gated detection is level-crossing spectroscopy (Sect. 7.1). If the upper atomic levels are excited by a pulsed

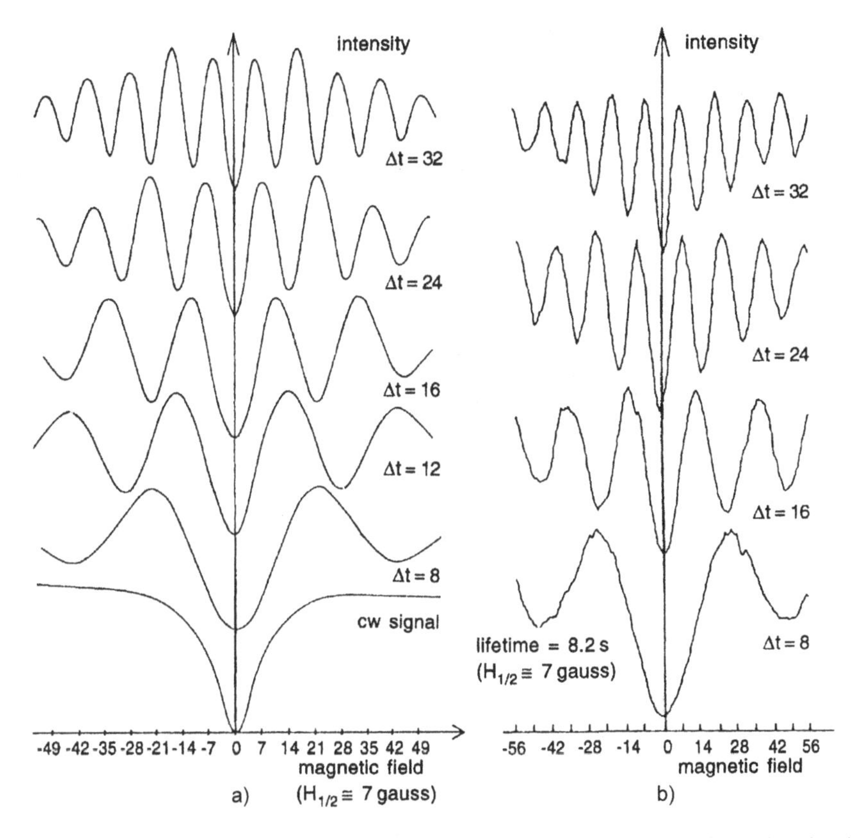

Fig. 9.80 Comparison of calculated (**a**) and measured (**b**) line profiles of level-crossing signals under observation with different gate-delay times (in ns). The *different curves* have been renormalized to equal centered peak intensities [1306]

laser and the fluorescence intensity $I_{\mathrm{Fl}}(B, t \geq \tau)$ as a function of the magnetic field is observed with increasing delay times $\Delta t = T$ for the opening of the detector gate, the central maximum of the Hanle signal becomes narrower with increasing Δt. This is illustrated by Fig. 9.80, which shows a comparison of measured and calculated line profiles for different gate-delay times Δt for Hanle measurements of the $\mathrm{Ba}(6s6p\,^1P_1)$ level [1305]. Similar measurements have been performed for the $\mathrm{Na}(3P)$ level [1306].

It should be emphasized that line narrowing is only observed if the time development of the phase of the upper-state wave function can be measured. This is the case for all methods that utilize interference effects caused by the superposition of different spectral components of the fluorescence. Therefore, an interferometer with a spectral resolution better than the natural linewidth can be used, too. However, a narrowing of the observed fluorescence linewidth with increasing gate-delay time T is only observed if the gating device is placed between the interferometer and the detector; it is *not* observed if the gate is placed between the emitting source and the interferometer [1303].

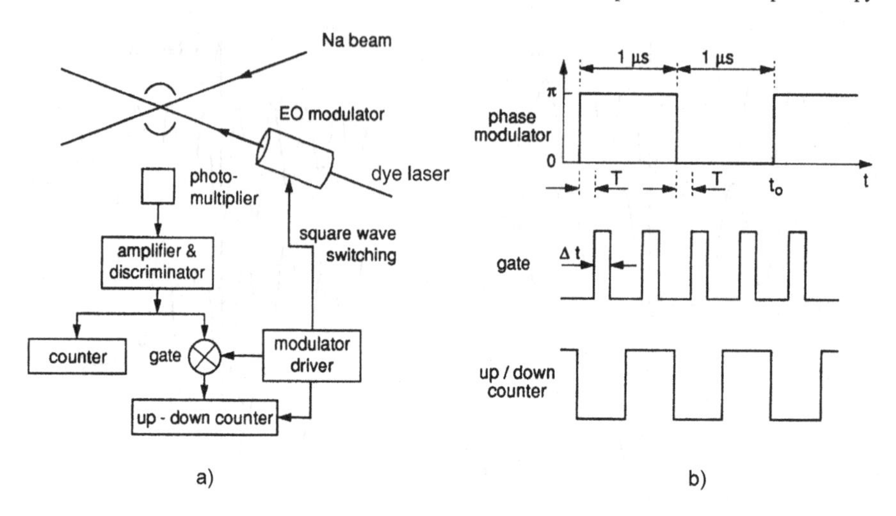

Fig. 9.81 Laser phase modulation and gating time sequence for obtaining subnatural linewidths under cw excitation [1307]

Instead of a gated switch in the fluorescence detector for pulsed excitation, one may also use excitation by a cw laser that is phase-modulated with a modulation amplitude of π (Fig. 9.81). The fluorescence generated under sub-Doppler excitation in a collimated atomic beam is observed during a short time interval Δt, which is shifted by the variable delay T against the time t_0 of the phase jump. If the fluorescence intensity $I_{\mathrm{Fl}}(\omega, T)$ is monitored as a function of the laser frequency ω, a narrowing of the line profile is found with increasing T [1307].

9.6.2 *Coherence and Transit Narrowing*

For transitions between two atomic levels $|a\rangle$ and $|b\rangle$ with the lifetimes $\tau_a \ll \tau_b$, a combination of level-crossing spectroscopy with saturation effects allows a spectral resolution that corresponds to the width $\gamma_b \ll \gamma_a$ of the longer-living level $|b\rangle$, although the natural linewidth of the transition should be $\gamma_{ab} = (\gamma_a + \gamma_b)$. The method was demonstrated by Bertucelli et al. [1308] for the example of the $\mathrm{Ca}(^3P_1{-}^3S_1)$ transition. Calcium atoms in the metastable $^3P_1$ level in a collimated atomic beam pass through the expanded laser beam in a homogeneous magnetic field $\boldsymbol{B}$. The electric vector $\boldsymbol{E}$ of the laser beam is perpendicular to $\boldsymbol{B}$. Therefore transitions with $\Delta M = \pm 1$ are induced. For $B = 0$ all M-sublevels are degenerate and the saturation of the optical transition is determined by the square of the sum $\sum D_{M_a M_b}$ of the transition matrix elements $D_{M_a M_b}$ between the allowed Zeeman components $|a, M_a\rangle$ and $|b, M_b\rangle$. If the Zeeman splitting for $B \neq 0$ becomes larger than the natural linewidth of the metastable levels in the ^3P state, the different Zeeman components saturate separately, and the saturation is proportional to

Fig. 9.82 Level scheme for transient line narrowing [1309]

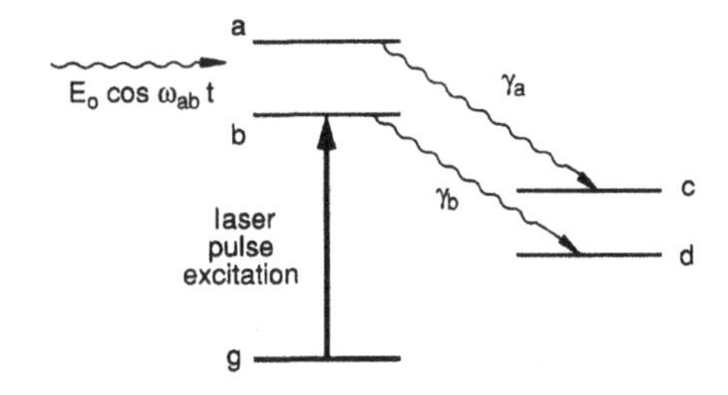

$\sum |D_{M_a M_b}|^2$ instead of $|\sum D_{M_a M_b}|^2$ for the degenerate case. Measuring the intensity of the laser-induced fluorescence $I_{\mathrm{Fl}}(B)$ as a function of the magnetic field B, one obtains a Hanle signal that depends nonlinearly on the laser intensity and that has a halfwidth $\gamma_{\mathrm{B}}(I_{\mathrm{L}}) = \gamma_{\mathrm{B}} \times \sqrt{1+S}$, where $S = I_{\mathrm{L}}/I_{\mathrm{S}}$ is the saturation parameter (Sect. 2.2).

Another technique of subnatural linewidth spectroscopy is based on transient effects during the interaction of a two-level system with the monochromatic wave of a cw laser. Assume that the system is irradiated by the cw monochromatic wave $E = E_0 \cos \omega t$, which can be tuned into resonance with the energy separation $[\omega_{ab} = (E_a - E_b)/\hbar]$ of the two levels $|a\rangle$ and $|b\rangle$ with decay constants γ_a, γ_b (Fig. 9.82).

If the level $|b\rangle$ is populated at the time $t = 0$ by a short laser pulse one can calculate the probability $P_a(\Delta, t)$ to find the system at time t in the level $|a\rangle$ as a function of the detuning $\Delta = \omega - \omega_{ab}$ of the cw laser. Using time-dependent perturbation theory one gets [1309, 1310]:

$$\mathcal{P}(\Delta, t) = \left(\frac{D_{ab} E_0}{\hbar}\right)^2 \frac{\mathrm{e}^{-\gamma_a t} + \mathrm{e}^{-\gamma_b t} - 2\cos(t \cdot \Delta)\mathrm{e}^{-\gamma_{ab} t}}{\Delta^2 + (\delta_{ab}/2)^2}, \tag{9.76}$$

where the Lorentzian factor contains the *difference* $\delta_{ab} = (\gamma_a - \gamma_b)/2$ of the level widths instead of the sum $\gamma_{ab} = (\gamma_a + \gamma_b)$.

If the fluorescence emitted from $|a\rangle$ is observed only for times $t \geq T$, one has to integrate (9.76). This yields the signal

$$\begin{aligned} S(\Delta, T) &\sim \gamma_a \int_{\mathrm{T}}^{\infty} \mathcal{P}(\Delta, t)\, \mathrm{d}t \\ &= \frac{\gamma_a (D_{ab} E_0/\hbar)^2}{\Delta^2 + (\delta_{ab}/2)^2} \\ &\quad \times \left\{ \frac{\mathrm{e}^{-\gamma_a T}}{\gamma_a} + \frac{\mathrm{e}^{-\gamma_b}}{\gamma_b} + \frac{2\mathrm{e}^{-\gamma_{ab} T}}{\Delta^2 + (\gamma_{ab}/2)^2} [\Delta \sin(\Delta \cdot T) - \gamma_{ab} \cos(\Delta \cdot T)] \right\}, \end{aligned} \tag{9.77}$$

which represents an oscillatory structure with a narrowed central peak. For $T \to 0$ this again becomes a Lorentzian

$$S(\Delta, T = 0) = \frac{\gamma_{ab}}{\gamma_b} \frac{(D_{ab} E_0/\hbar)^2}{\Delta^2 + (\gamma_{ab}/2)^2}. \tag{9.78}$$

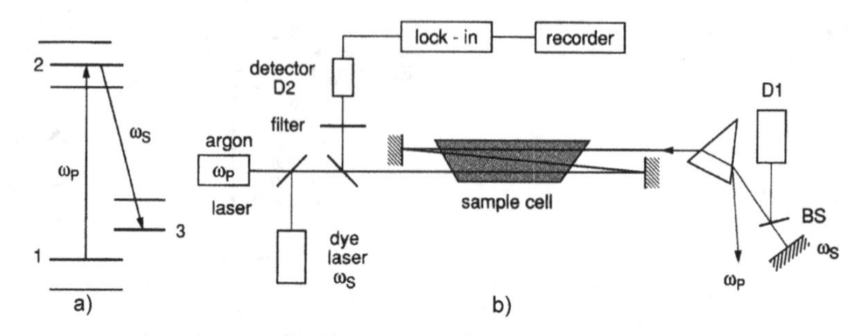

Fig. 9.83 Stimulated resonant Raman transition spectroscopy with subnatural linewidth resolution: (**a**) level scheme; (**b**) experimental arrangement

9.6.3 Raman Spectroscopy with Subnatural Linewidth

A very interesting method for achieving subnatural linewidths of optical transitions is based on induced resonant Raman spectroscopy, which may be regarded as a special case of optical–optical double resonance (Sect. 5.4). The pump laser L_1 is kept on the molecular transition $|1\rangle \to |2\rangle$ (Fig. 9.83), while the tunable probe laser L_2 induces downward transitions. A double-resonance signal is obtained if the frequency ω_s of the probe laser matches the transition $|2\rangle \to |3\rangle$. This signal can be detected by monitoring the change in absorption or polarization of the transmitted probe laser.

When both laser waves with the wave vectors $\boldsymbol{k}_p$ and $\boldsymbol{k}_s$ are sent collinearly through the sample, energy conservation for the absorption of a photon $\hbar\omega_p$ and emission of a photon $\hbar\omega_s$ by a molecule moving with the velocity $\boldsymbol{v}$ demands

$$(\omega_p - \boldsymbol{k}_p \cdot \boldsymbol{v}) - (\omega_s - \boldsymbol{k}_s \cdot \boldsymbol{v}) = (\omega_{12} - \omega_{23}) \pm (\gamma_1 + \gamma_3), \tag{9.79}$$

where γ_i is the homogeneous width of level $|i\rangle$. The quadratic Doppler effect and the photon recoil have been neglected in (9.79). Their inclusion would, however, not change the argument.

Integration over the velocity distribution $N(v_z)$ of the absorbing molecules yields with (9.79) the width γ_s of the double-resonance signal [1311]

$$\gamma_s = \gamma_3 + \gamma_1(\omega_s/\omega_p) + \gamma_2(1 \mp \omega_s/\omega_p), \tag{9.80}$$

where the minus sign holds for copropagating, the plus sign for counterpropagating laser beams. If $|1\rangle$ and $|3\rangle$ denote vibrational–rotational levels in the electronic ground state of homonuclear diatomic molecules, their radiative lifetimes are very small and γ_1, γ_2 are mainly limited at higher pressures by collision broadening and at low pressures by transit-time broadening (Vol. 1, Sect. 3.4). For $\omega_s \simeq \omega_p$ the contribution of the level width γ_2 to γ_s becomes very small for copropagating beams and the halfwidth γ_s of the signal may become much smaller than the natural linewidths of the transitions $|1\rangle \to |2\rangle$ or $|2\rangle \to |3\rangle$.

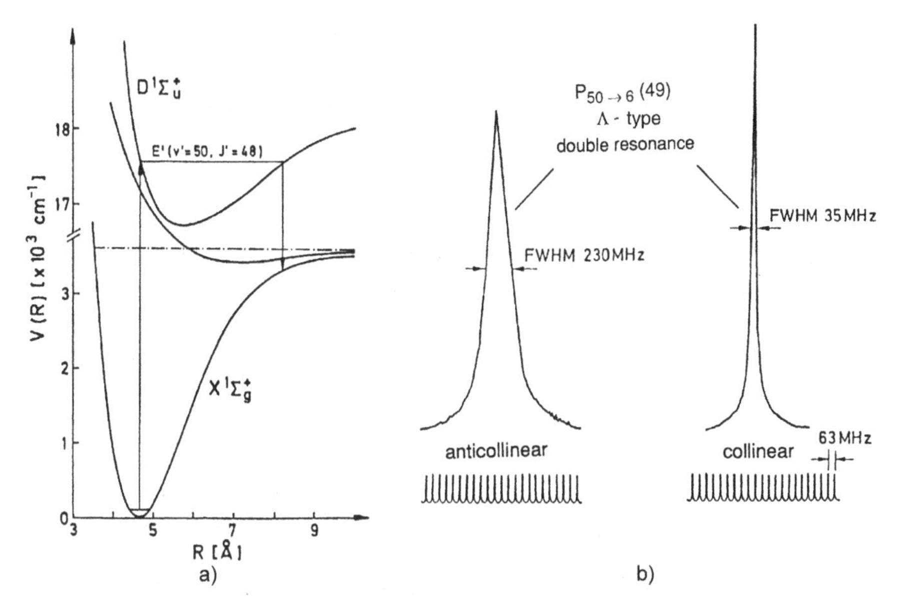

Fig. 9.84 (**a**) Potential diagram of the Cs_2 molecule with the stimulated resonance Raman transitions to high-lying vibration–rotation levels of the $X\,^1\Sigma_g^+$ ground state. (**b**) Comparison of linewidths of a Doppler-free transition $|1\rangle \to |2\rangle$ to a predissociating upper level $|2\rangle$ with an effective lifetime $\tau_{\text{eff}} \simeq 800$ ps and the much narrower linewidths of the Raman signal $S(\omega_s)$ [1313]

Of course, one cannot get accurate information on the level width γ_2 from (9.80) unless all other contributions to γ_s are sufficiently well known. The width γ_s is comparable to the width of the direct transition $|1\rangle \to |3\rangle$. However, if $|1\rangle \to |2\rangle$ and $|2\rangle \to |3\rangle$ are allowed electric dipole transitions, the direct transition $|1\rangle \to |3\rangle$ is dipole-forbidden.

Figure 9.83 exhibits the experimental arrangement used by Ezekiel et al. [1312] to measure subnatural linewidths in I_2 vapor. An intensity-modulated pump beam and a continuous dye laser probe pass the iodine cell three times. The transmitted probe beam yields the forward scattering signal (detector D1) and the reflected beam the backward scattering (detector D2). A lock-in amplifier monitors the change in transmission from the pump beam, which yields the probe signal with width γ_s. Ezekiel et al. achieved a linewidth as narrow as 80 kHz on a transition with the natural linewidth $\gamma_n = 141$ kHz. The theoretical limit of $\gamma_s = 16.5$ kHz could not be reached because of dye laser frequency jitter. This extremely high resolution can be used to measure collision broadening at low pressures caused by long-distance collisions (Sect. 8.1) and to determine the collisional relaxation rates of the levels involved.

The small signal width of the OODR signal γ_s becomes essential if the levels $|3\rangle$ are high-lying rotation–vibration levels of heavy molecules just below the dissociation limit, where the level density may become very high. As an example, Fig. 9.84 shows such an OODR signal of a rotation–vibration transition $D\,^1\Sigma_u(v' = 50, J' = 48) \to X\,^2\Sigma_g(v'' = 125, J'' = 49)$ in the Cs_2 molecule. A comparison with a Doppler-free polarization signal on a transition $|1\rangle \rightarrow |2\rangle$ to the upper predissociating level $|2\rangle = D\,^1\Sigma_u(v' = 50, J' = 48)$, which has an effective lifetime of $\tau_{\text{eff}} \simeq 800$ ps because of predissociation, shows that the OODR signal is not influenced by γ_2 [1313].

9.7 Absolute Optical Frequency Measurement and Optical Frequency Standards

In general, frequencies can be measured more accurately than wavelengths because of the deviation of light waves from the ideal plane wave caused by diffraction and local inhomogenities of the refractive index. This leads to local deviations of the phase fronts from ideal planes and therefore to uncertainties in measurements of the wavelength λ, which is defined as the distance between two phase fronts differing by 2π. While the wavelength $\lambda = c/\nu = c_0/(n\nu)$ of an EM wave depends on the refractive index n, the frequency ν is independent of n.

In order to achieve the utmost accuracy for the determination of physical quantities that are related to the wavelength λ or frequency ν of atomic transitions, it is desirable to measure the optical frequency ν instead of the wavelength λ. From the relation $\lambda_0 = c_0/\nu$ the vacuum wavelength λ_0 can then be derived, since the speed of light in vacuum

$$c_0 \underset{\text{Def}}{=} 229{,}792{,}458 \text{ m/s},$$

is now *defined* as a fixed value, taken from a weighted average of the most accurate measurements [1314, 1315]. In this section we will learn about some experimental techniques to measure frequencies of EM waves in the infrared and visible range.

9.7.1 Microwave–Optical Frequency Chains

With fast electronic counters, frequencies up to a few gigahertz can be measured directly and compared with a calibrated frequency standard, derived from the cesium clock, which is still the primary frequency standard [1316]. For higher frequencies a heterodyne technique is used, where the unknown frequency ν_x is mixed with an appropriate multiple $m\nu_R$ of the reference frequency ν_R $(m = 1, 2, 3, \ldots)$. The integer m is chosen such that the difference frequency $\Delta\nu = \nu_x - m\nu_R$ at the output of the mixer lies in a frequency range that can be directly counted.

This frequency mixing in suitable nonlinear mixing elements is the basis for building up a frequency chain from the Cs atomic beam frequency standard to the optical frequency of visible lasers. The optimum choice for the mixer depends on the spectral range covered by the mixed frequencies. When the output beams of two infrared lasers with known frequencies ν_1 and ν_2 are focused together with

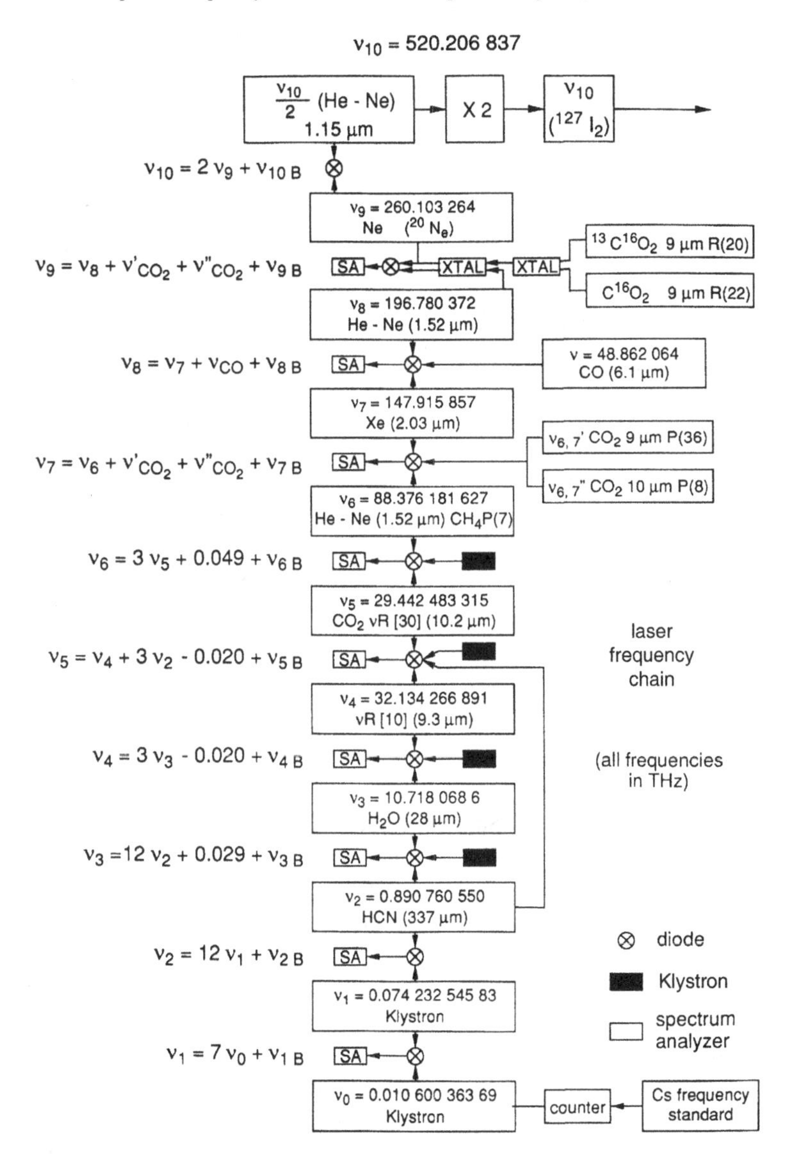

Fig. 9.85 Optical frequency chain that connects the frequency of stabilized optical lasers with the Cs frequency standard [1317]

the superimposed radiation from the unknown frequency ν_x of another laser, the frequency spectrum of the detector output contains the frequencies

$$\nu = \pm\nu_x \pm m\nu_1 \pm n\nu_2 \quad (m, n: \text{integers}), \tag{9.81}$$

which can be measured with an electronic spectrum analyser if they are within its frequency range.

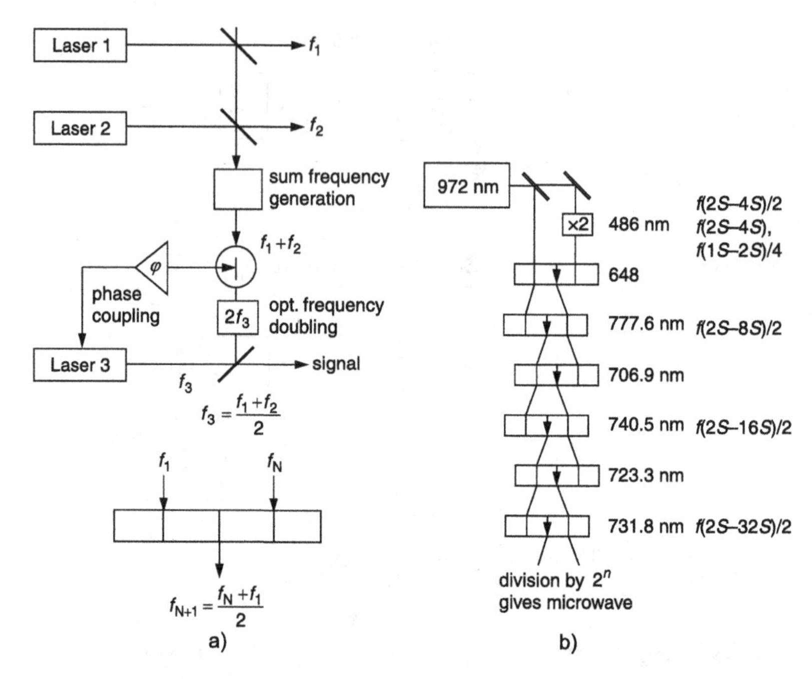

Fig. 9.86 Optical frequency division. (**a**) Schematic setup. (**b**) Coupling to transition frequencies in the H atom [1322c]

In the National Institute of Standards and Technology, NIST (the former National Bureau of Standards, NBS) the frequency chain, shown in Fig. 9.85, has been built [1317], which starts at the frequency of the hyperfine transition in cesium at $\nu = 9.1926317700$ GHz, currently accepted now as the primary frequency standard. A klystron is frequency-offset locked (Sect. 2.3.4) to the Cs clock and stabilized with the accuracy of this clock at the frequency $\nu_0 = 10.600363690$ GHz. The seventh harmonic of ν_0 is mixed with the frequency ν_1 of another klystron in a nonlinear diode, and ν_1 is again frequency-offset locked to the frequency $\nu_1 = 7\nu_0 + \nu_{1B}$, where ν_{1B} is provided by a RF generator and can be directly counted. The 12th harmonic of ν_1 is then mixed with the radiation of a HCN laser at $\nu = 890$ GHz ($\lambda = 337$ μm). The 12th overtone of this radiation is mixed with the radiation of the H_2O laser at 10.7 THz ($\lambda = 28$ μm), etc. The frequency chain goes up to the frequencies of visible HeNe lasers stabilized to HFS components of a visible transition in the I_2 molecule (Vol. 1, Sect. 5.4.5). MIM (metal–insulator–metal) mixing diodes (Vol. 1, Fig. 4.99) consisting of a thin tungsten wire with a sharp-edged tip, which contacts a nickel surface coated by a thin oxyde layer [1318], allow the mixing of frequencies above 1 THz [1319]. Also, Schottky diodes have successfully been used for mixing optical frequencies with difference frequencies up to 900 GHz [1320]. The physical processes relevant in such fast Schottky diodes have been understood only recently [1321].

In addition, other frequency chains have been developed that start from stabilized CO_2 lasers locked to the cesium frequency standard in a similar way, but then use infrared color-center lasers to bridge the gap to the I_2-stabilized HeNe laser [1322a].

A very interesting proposal for a frequency chain based on transition frequencies of the hydrogen atom has been made by Hänsch [1322b]. The basic idea is illustrated in Fig. 9.86. Two lasers are stabilized onto two transitions of the hydrogen atoms at the frequencies f_1 and f_2. Their output beams are superimposed and focused into a nonlinear crystal, forming the sum frequency $f_1 + f_2$. The second harmonics of a laser oscillating at the frequency f_3 are phase-locked to the frequency $f_1 + f_2$. This stabilizes the frequency f_3 by a servo loop to the value $(f_2 + f_1)/2$. The original frequency difference $f_1 - f_2$ is therefore halved to $f_1 - f_3 = (f_1 - f_2)/2$. By connecting several such stages in cascade until the frequency difference becomes measurable by a counter, the difference $f_1 - f_2$ can be directly connected to the cesium clock. If a laser frequency f and its second harmonic are used as the starting frequencies, the beat signal at $f/2^n$ can be observed after n stages. In order to link the $2s$–$4s$ hydrogen transition at $\lambda = 486$ nm to the 9-GHz cesium frequency standard by this method, 16 stages would be required [1322c].

9.7.2 Optical Frequency Combs

The optical frequency chains discussed in the previous section are difficult to build. Many lasers and optical harmonic generators have to be phase-locked and frequency-stabilized, and the whole setup can easily fill a large laboratory. Furthermore, each of these chains is restricted to a single optical frequency, which is linked to the cesium clock, just like the present frequency standard.

Recently, a new technique has been developed [1323] that allows the direct comparison of widely different reference frequencies and thus considerably simplifies the frequency chain from the cesium clock to optical frequencies by reducing it to a single step. Its basic principle can be understood as follows (Fig. 9.91): The frequency spectrum of a mode-locked continuous laser emitting a regular train of short pulses with repetition rate $1/\Delta T$ consists of a comb of equally spaced frequency components (the modes of the laser resonator). The spectral width $\Delta w = 2\pi/T$ of this comb spectrum depends on the temporal width $T/\Delta T$ of the laser pulses (Fourier theorem). Using femtosecond pulses from a Ti:sapphire Kerr lens mode-locked laser, the comb spectrum extends over more than 30 THz.

The electric field of the femtosecond pulses is shown in Fig. 9.87. If there were no phase shifts, each pulse would be an exact replica of the previous pulse, i.e. $E(t) = E(t - T)$. However, due to intracavity dispersion in the laser resonator, the group velocity and the phase velocity may be different. This causes a phase shift $\Delta\phi = T \cdot \omega_{CE}$ of the peak of the carrier wave $E(t)$ with respect to the peak of the pulse envelope.

The Fourier transformation of $E(t)$ gives the frequency spectrum $E(\omega)$, shown in the lower part of Fig. 9.87.

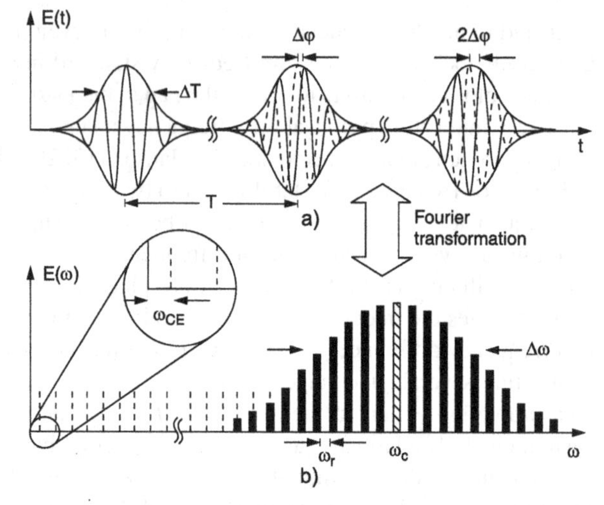

Fig. 9.87 (**a**) Electric field amplitude of regularly spaced femtosecond pulses and its phase shift against the maximum of the envelope; (**b**) Frequency spectrum $E(\omega)$ as the Fourier-transform of $E(t)$ [1325]

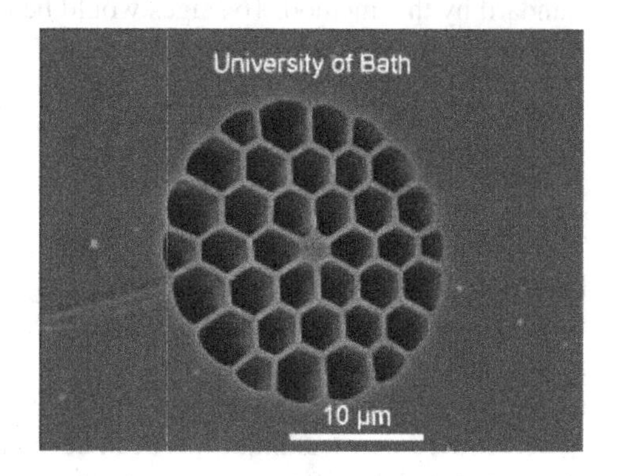

Fig. 9.88 Photonic fiber

The spectral width $\Delta\omega$ can be further increased by focusing the laser pulses into a special optical fiber, which consists of a photonic crystal (Fig. 9.88) where by self-phase modulation the spectrum is considerably broadened and extends over one decade (e.g., from 1064 nm to 532 nm) (Fig. 9.89). This corresponds to a frequency span of 300 THz [1327]! It was found by interference experiments, that the coherence properties were preserved in this broadened spectrum, i.e. the nonlinear processes in the optical fiber did not destroy the coherence of the original frequency comb.

It turns out that the spectral modes of the comb are precisely equidistant, even in the far wings of the comb [1325]. This is also true when the spectrum is broadened by self-phase modulation. These strictly regular frequency spacings are essential for optical frequency measurements [1325]. Any optical frequency ω_n can be expressed as

$$\omega_n = n \cdot \omega_r + \omega_{CE} \tag{9.82}$$

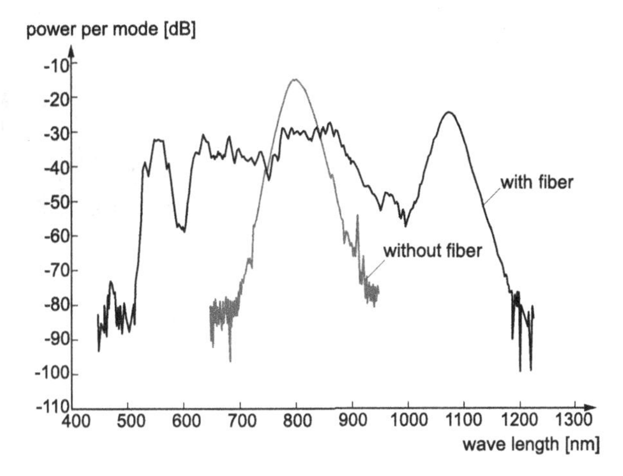

Fig. 9.89 Spectral broadening of a femtosecond laser pulse with and without the photonic fiber [1326]

where n is a large integer (e.g., $n = 10^5$), and $\omega_r = 2\pi/T$ is the mode spacing of the laser resonator, which can be exactly determined from the repetition time T of the mode-locked femtosecond laser pulses and equals the roundtrip time $T = c/L$ through the laser resonator with roundtrip length L. The offset frequency $\Delta\omega$ takes into account the fact that ω_n may not exactly coincide with one of the mode frequencies of the frequency comb. Its value lies between 0 and ω_r. Equation (9.82) links two radiofrequencies ω_r and ω_{offs} to the optical frequency ω_n to be measured.

If the frequency comb extends over more than one octave, the carrier-envelope offset frequency ω_{CE} can be determined directly. This is illustrated in Fig. 9.90. A laser is stabilized onto a comb mode with frequency $\omega_1 = n_1\omega_r + \omega_{CE}$ at the red frequency side of the comb. Its frequency ω_1 is doubled in a nonlinear crystal to $2\omega_1$ and compared with a frequency $\omega_2 = n_2\omega_r + \omega_{CE}$ at the blue end of the comb. The lowest beat frequency

$$\Delta\omega = 2\omega_1 - \omega_2 = (2n_1 - n_2)\omega_r + \omega_{CE} = \omega_{CE} \tag{9.83}$$

appears for $2n_1 = n_2$. In this case the beat frequency $\Delta\omega$ gives the offset frequency ω_{CE} directly.

A simplified picture for comparing the frequency ω_L of a laser with the optical frequency comb is shown in Fig. 9.90b. Beat frequencies

$$\Delta\omega = \omega_L - (n\omega_r + \omega_{CE})$$

between the laser frequency ω_L and the frequencies $n \cdot \omega_r + \omega_{CE}$ of the comb modes are generated when the laser beam is superimposed on the output of the frequency comb, because the laser frequency ω_L does not necessarily coincide with the frequency $n \cdot \omega_r + \omega_{CE}$ of a comb mode. The beat frequency for the mode frequency closest to the laser frequency falls into the radio frequency range and can be readily measured by a frequency counter. The correct integer n can be determined by measuring the laser frequency with a wave meter which only needs an accuracy of

Fig. 9.90 (**a**) Self-referencing of optical frequencies [1325]. (**b**) Schematic diagram for the measurement of the absolute frequency ω_L of a laser

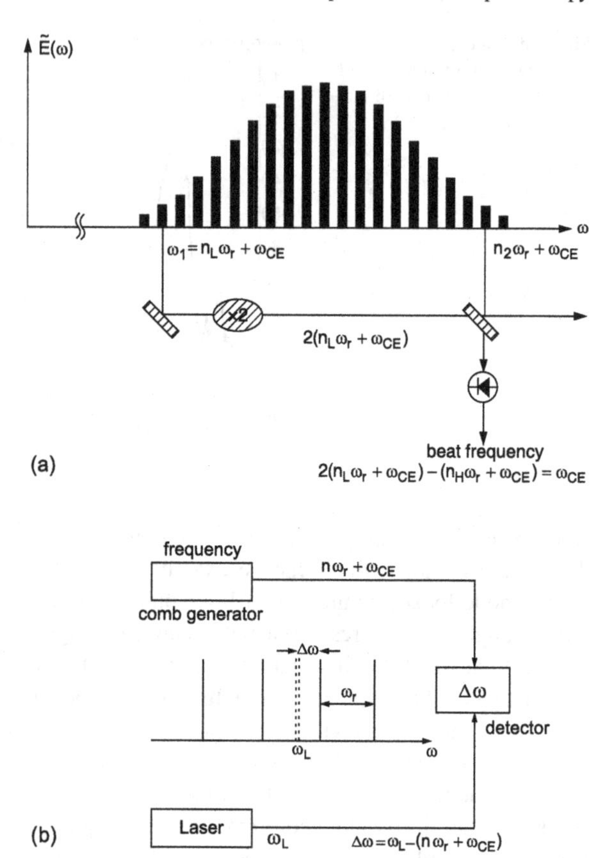

the mode spacing (about 150 MHz, corresponding at $\lambda = 600$ nm to a wavelength accuracy of $\Delta\lambda = 2 \times 10^{-3}$ Å).

The integer n can be also determined by performing repeated measurements of beat frequencies with different values of ω_r (changing the length L of the laser resonator) [1327].

Of particular interest is the determination of the absolute frequency of the $1S \to 2S$ transition in the H-atom. This is a dipole-forbidden transition and can be excited by two-photon absorption. It has a very small line-width of about 0.3 Hz where the line center can be determined with high precision. It is important for fundamental physics because its accurate measurement allows a precise comparison with the results of the quantum-electrodynamic theory of the H-atom.

The experimental realization is based on the comparison of the optical frequency comb with the Cs-atomic clock, which represents up to date the frequency standard. A dye laser at $\lambda = 486$ nm is stabilized to a frequency ν_1 which differs from the frequency ν_c of one of the teeth of the optical frequency comb by the radio-frequency f_1. The output of the dye laser is frequency doubled in a BBO-crystal and is then sent collinearly to the H-atom beam into an enhancement cavity, where it excites the two-photon transition $1S \to 2S$ of the H-atom [1326]. The radio-frequency

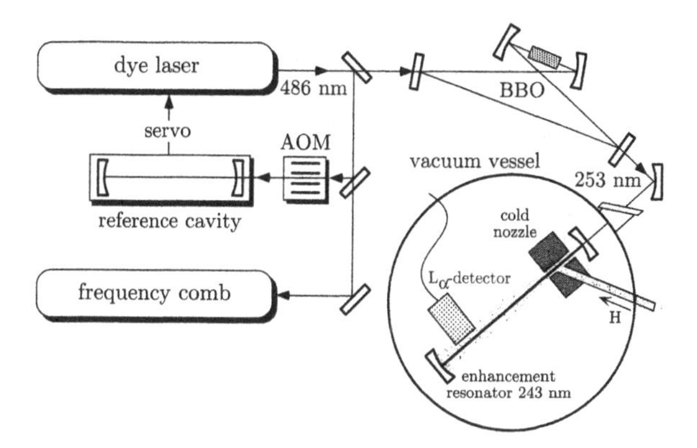

Fig. 9.91 Exciting the hydrogen $1S$–$2S$ transition with two counterpropagating photons in a standing-wave field at 243 nm. This radiation is obtained from a dye laser frequency doubled in a BBO crystal and stabilized to a reference cavity. While scanning the hydrogen resonance the frequency of this laser is measured with a frequency comb to be 2 466 061 413 187 074 (34) Hz for the hyperfine centroid [1327]

f_1 is tuned in such a way, that $4\nu_1$ coincides with the center frequency of one of the hyperfine components of the $1S \rightarrow 2S$ transition. The frequency of this transition is then

$$\nu(1S \rightarrow 2S) = 4\nu_1 = 4(\nu_c - f_1).$$

The comb-frequency can be directly compared with the cesium standard.

A possible direct link between the UV frequency of the two-photon $1S \rightarrow 2S$ transition in the H atom and the Cs clock in the microwave range with the optical frequency comb is illustrated in Fig. 9.91.

With this technique a relative uncertainty of the absolute frequency determination of about 10^{-15} is possible. This means an accuracy of the $1S$–S transition frequency at $2.4 \times 10^{15}\ \mathrm{s}^{-1}$ of better than 3 Hz!

The optical frequency comb allows a very accurate determination of the Rydberg constant from measurements of the absolute frequency of the $1S \rightarrow 2S$ transition in the hydrogen atom (see also Sect. 2.5.4).

A user friendly version of optical frequency combs has been realized with fibre lasers [1328]. Mode-locked fibre lasers doped with Er of Yb ions have only a few adjustable optical components which make them easy to handle and facilitates their maintenance. Meanwhile optical frequency combs are commercially available.

9.7.3 Spectral Extension of Frequency Combs

The spectral range of the optical frequency comb can be extended into the infrared and the ultraviolet region down into the vacuum ultraviolet (VUV) and even into the soft X-ray region (XUV). The experimental realization is shown in Fig. 9.92a.

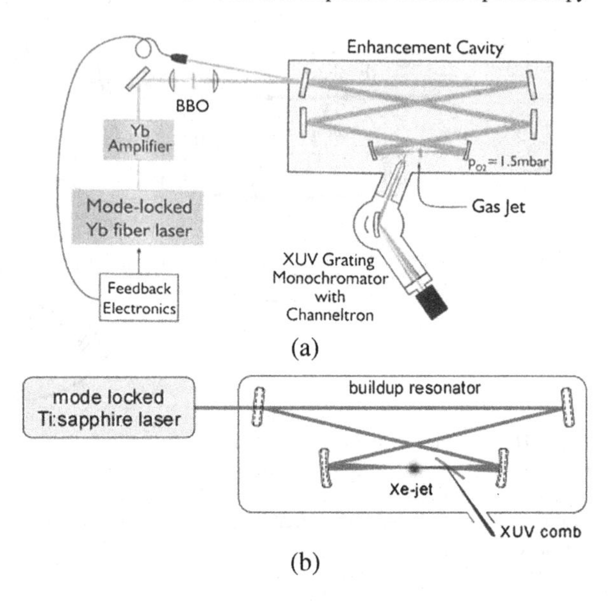

Fig. 9.92 VUV frequency comb. The high harmonics generation occurs in the gas jet. (**a**) Experimental realization [1329]. (**b**) Simplified diagram of XUV frequency comb with extraction of the VUV beam by reflection at a Brewster plate inside the enhancement resonator

The amplified output of the mode-locked femtosecond laser with optical frequency ω is sent through a double folded external enhancement ring resonator. A xenon atomic nozzle beam crosses the laser beam at the beam waist. Due to the nonlinear interaction of the intense laser beam with the Xe-atoms high harmonics are generated with frequencies $n \cdot \omega$ which have the same repetition rate as the optical pulses at ω. The width Δt of the VUV pulses $I(n\omega, t)$ is smaller than that of the optical pulses and reaches down into the attosecond range (1 as $= 10^{-18}$ s), because the intensity $I(n \cdot \omega)$ is proportional to $I(\omega)^n$ which means that the peak intensity $I(t_m)$ at the pulse maximum t_m has a higher conversion factor than the line wings. The VUV-pulses are coupled out of the resonator either through a small hole in one of the resonator mirrors (Fig. 9.92b) or by total reflection at a Brewster plate of VUV-transparent material inside the resonator which transmits the fundamental wave at ω without losses but shows total reflection of the high harmonics $n\omega$, because the refractive index $n(n\omega) < 1$ (Fig. 9.92a). The Fourier-transform of the VUV pulses gives the frequency comb in the VUV.

9.7.4 Applications of Optical Frequency Combs

The optical frequency comb (for which Th.W. Hänsch and J. Hall received the Nobel Prize) has meanwhile found numerous applications, as for instance the very accurate measurement of optical frequencies, which has been discussed in the previous section. It is much simpler than former techniques but reaches an accuracy that is two to three orders of magnitude higher.

A second example is the use of frequency combs in satellites for the global positional system GPS for navigation purposes. The accuracy of navigation depends on the frequency stability of transmitters and receivers.

An interesting application of frequency combs in astronomy allows the accurate determination of the Doppler-shift of spectral lines from stars due to their velocity relative to the earth. While former measurements had an uncertainty of $\Delta v = 10\,\mathrm{m/s}$ the comb measurements could resolve velocity changes down to 10 cm/s [1338].

In order to achieve more accurate values for the ratio of electron mass to proton mass from which the fine-structure constant can be derived, a frequency comb in the infrared region was developed which can be used to measure vibrational–rotational transitions in HD^+-ions. The frequency distance ω_r between the modes of the comb, which equals the repetition rate of the femto-second pulses, is stabilized on the resonance frequency of a cryogenic ultra-stable sapphire resonator. For the measurement of the transitions in the HD^+-ion diode lasers with external cavity and a grating for achieving single mode operation are used. Their frequency is stabilized onto the centre of the molecular transition and is then compared with the adjacent mode of the frequency comb.

The frequency comb is now not only directly combined with the Cs-clock but also with an atomic clock which is stabilized on the very narrow transition $^3P \leftrightarrow {}^1S$ of the ^{171}Yb-atom which is more stable than the Cs transition.

9.7.5 Molecular Spectroscopy with Optical Frequency Combs

The application of frequency combs to molecular spectroscopy opens the way to new and very promising techniques, which are based on a combination of laser spectroscopy and Fourier-transform spectroscopy [1342].

At first we will discuss the use of a single optical frequency comb where the probe is placed inside an optical resonator with resonance frequencies matching the modes of the comb. Therefore all absorption frequencies of the molecule under investigation contribute simultaneously to the absorption of the comb. Outside the resonator a prism or grating disperses the different frequencies and a CCD array can detect simultaneously all absorption lines inside the spectral range $\Delta\nu$ of the comb. For a measuring time Δt the signal-to-noise ratio is increased by a factor $\sqrt{\Delta t \cdot \Delta\nu}$.

The sensitivity can be further enhanced if a second frequency comb with a slightly different mode spacing is applied where the difference amounts to about 200–600 Hz. While the output of the first comb passes through the absorption cell that of the second comb does not (Fig. 9.93). The output beams of the two combs are superimposed and the resulting difference frequency spectrum is monitored with a time resolving detector.

This technique is quite similar to Fourier-transform spectroscopy, where the phase difference between the two beams in a Michelson interferometer are generated by changing the path difference as a function of time. In the dual comb spectroscopy the phase difference it generated by the different repetition frequencies of the two

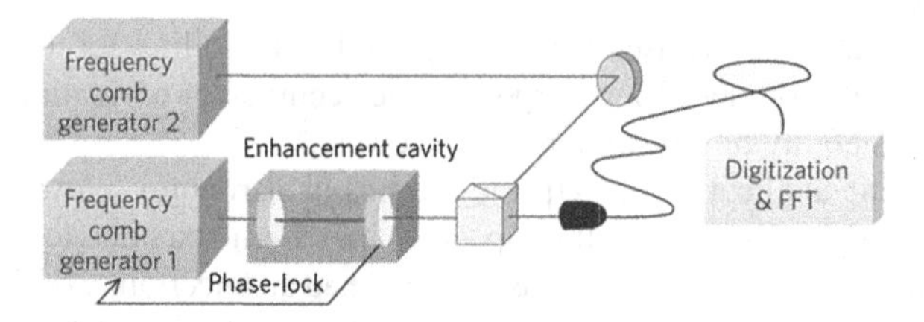

Fig. 9.93 Two FC generators (combs 1 and 2) have slightly different line spacings. Comb 1 is transmitted through the resonant high-finesse cavity, which contains the sample. The repetition frequency of comb 1 is phase-locked onto the cavity free-spectral range. The light escaping from the cavity is heterodyned against comb 2 on a single fast photodetector, yielding a down-converted radiofrequency comb containing information on the ultrasensitive absorption losses experienced by each line of comb 1. The electrical signal is digitized and is Fourier-transformed using a fast Fourier transform (FFT) algorithm [1341]

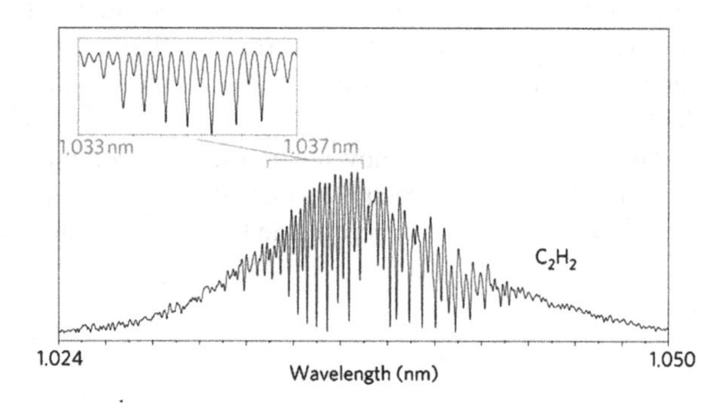

Fig. 9.94 Dual frequency comb spectrum of a vibrational overtone band of C_2H_2 [1341]

combs. This superposition transfers the detection frequencies into the rf-region and can be therefore readily monitored with time resolving detectors. The dramatically increased sensitivity was demonstrated by recording of extended spectral ranges the overtone spectrum of the $3\nu_3$ vibrational band of C_2H_2 (Fig. 9.94) over a spectral range from 1024 nm to 1050 nm ($\Delta\nu = 4.6 \times 10^{12}\ \mathrm{s}^{-1}$) within milliseconds. 18 μs at a gas pressure of 3 hPa [1341].

9.8 Squeezing

For very low light intensities the quantum structure of light becomes evident by statistical fluctuations of the number of detected photons, which lead to corresponding fluctuations of the measured photoelectron rate (Sect. 7.8). This *photon noise*, which is proportional to $\sqrt{N}$ at a measured rate of N photoelectrons per second, imposes a principal detection limit for experiments with low-level light detection [1330]. Additionally, the frequency stabilization of lasers on the millihertz scale is limited by photon noise of the detector that activates the electronic feedback loop [1331].

It is therefore desirable to further decrease the photon-noise limit. At first, this seems to be impossible because the limit is of principal nature. However, it has been shown that under certain conditions the photon-noise limit can be overcome without violating general physical laws. We will discuss this in some more detail, partly following the representation given in [1332, 1333].

Example 9.18 The shot-noise limit of an optical detector with the quantum efficiency $\eta < 1$ irradiated by N photons per second leads to a minimum relative fluctuation $\Delta S/S$ of the detector signal S, which is for a detection bandwidth Δf given by

$$\frac{\Delta S}{S} = \frac{\sqrt{N\eta\Delta f}}{N\eta} = \sqrt{\frac{\Delta f}{\eta N}}. \tag{9.84}$$

For a radiation power of 100 mW at $\lambda = 600$ μm $\rightarrow N = 3 \times 10^{17}$ s^{-1}. With a bandwidth of $\Delta f = 100$ Hz (time constant $\simeq 10$ ms) and $\eta = 0.2$, the minimum fluctuation is $\Delta S/S \geq 4 \times 10^{-8}$.

9.8.1 Amplitude and Phase Fluctuations of a Light Wave

The electric field of a single-mode laser wave can be represented by

$$\begin{aligned} E_\mathrm{L}(t) &= E_0(t)\cos\left[\omega_\mathrm{L} t + \boldsymbol{k}_\mathrm{L}\boldsymbol{r} + \phi(t)\right] \\ &= E_1(t)\cos(\omega_\mathrm{L} t + \boldsymbol{k}_\mathrm{L}\boldsymbol{r}) + E_2(t)\sin(\omega_\mathrm{L} t + \boldsymbol{k}_\mathrm{L}\boldsymbol{r}), \end{aligned} \tag{9.85}$$

with $\tan\phi = E_2/E_1$.

Even a well-stabilized laser, where all "technical noise" has been eliminated (Vol. 1, Sect. 5.4.5) still shows small fluctuations ΔE_0 of its amplitude and $\Delta\phi$ of its phase, because of quantum fluctuations. While technical fluctuations may be at least partly eliminated by difference detection (Fig. 1.9), this is not possible with classical means for photon noise caused by uncorrelated quantum fluctuations.

These fluctuations are illustrated in Fig. 9.95 in two different ways: the time-dependent electric field $\boldsymbol{E}(t)$ and its mean fluctuations of the amplitude E_0 and phase ϕ are shown in an $E(t)$ diagram and in a polar phase diagram with the axes E_1 and E_2. In the latter, amplitude fluctuations cause an uncertainty of the radius $r = |E_0|$, whereas phase fluctuations cause an uncertainty of the phase angle ϕ (Fig. 9.95b). Because of Heisenberg's uncertainty relation it is not possible that both uncertainties of amplitude and phase become simultaneously zero.

In order to gain a deeper insight into the nature of these quantum fluctuations, let us regard them from a different point of view: the EM field of a well-stabilized single-mode laser can be described by a coherent state (called a *Glauber state* [1334])

$$\langle\alpha_k| = \exp\left(-|\alpha_k|^2/2\right)\sum_{N=0}^{\infty}\frac{\alpha_k^N}{\sqrt{N!}}|N_k\rangle, \tag{9.86}$$

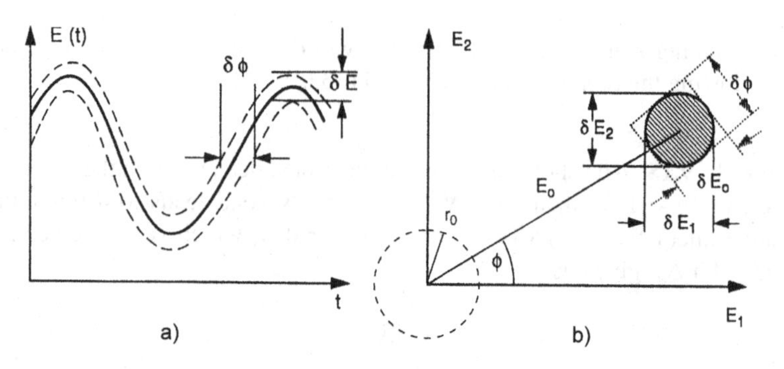

Fig. 9.95 Amplitude and phase uncertainties of a laser wave shown in an amplitude–time diagram (**a**) and in a polar phase diagram (**b**)

which is a linear combination of states with photon occupation numbers N_k. The probability of finding N_k photons in this state is given by a Poisson distribution around the mean value $N = \langle N_k \rangle = \alpha_k^2$ with the width $\langle \Delta N_k \rangle = \sqrt{\langle N_k \rangle} = |\alpha_k|$.

Although the laser field is concentrated into a single mode, all other modes of the vacuum with different frequencies ω and wave vectors $\boldsymbol{k}$ still have an average occupation number of $N = 1/2$, corresponding to the zero-point energy $\hbar\omega/2$ of a harmonic oscillator. When the laser radiation falls onto a photodetector, all these other modes are also present (*vacuum zero-field fluctuations*) and beat signals are generated at the difference frequencies by the superposition of the laser wave with these vacuum fluctuations. The magnitudes of these beat signals are proportional to the product of the amplitudes of the two interfering waves. Since the amplitude of the laser mode is proportional to $\sqrt{N}$, and that of the vacuum modes to $\sqrt{1/2}$, the magnitude of the beat signals is proportional to $\sqrt{N/2}$. Adding up all beat signals in the frequency interval Δf of the detector bandwidth gives the shot noise treated in Example 9.17. In this model the shot noise is regarded as the beat signal between the occupied laser mode and all other vacuum modes. This view will be important for the understanding of interferometric devices used for squeezing experiments.

If the field amplitude E in (9.85) is normalized in such a way that

$$\langle E^2 \rangle = \langle E_1^2 \rangle + \langle E_2^2 \rangle = \frac{\hbar\omega}{2\epsilon_0 V}, \tag{9.87}$$

where V is the mode volume, and ϵ_0 the dielectric constant, the uncertainty relation can be written as [1330, 1334]

$$\Delta E_1 \cdot \Delta E_2 \geq 1. \tag{9.88a}$$

For coherent states (9.86) of the radiation field, and also for a thermal-equilibrium radiation field one obtains the symmetric relations

$$\Delta E_1 = \Delta E_2 = 1, \tag{9.88b}$$

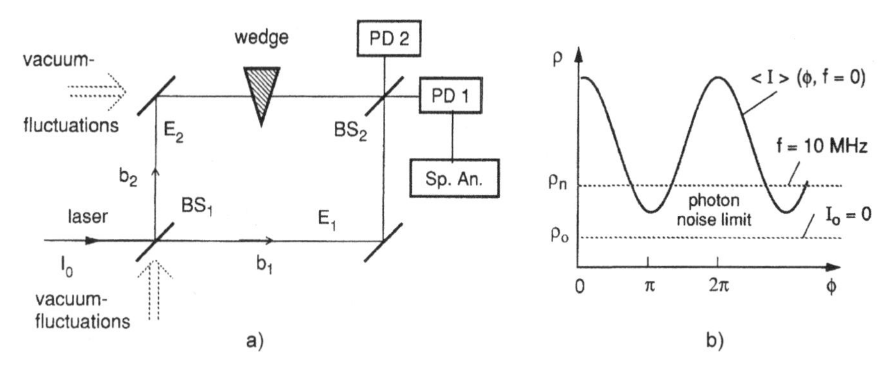

Fig. 9.96 (**a**) Mach–Zehnder interferometer with variable phase delay ϕ realized by an optical wedge. (**b**) Detected mean intensity $\langle I \rangle$ measured at $f = 0$, and phase-independent photon noise power density, measured at $f = 10$ MHz with and without input intensity I_0

and therefore the minimum possible value of the product $\Delta E_1 \cdot \Delta E_2$. In the phase diagram displayed in Fig. 9.95b the relation (9.88b) yields a circle as the *uncertainty area*.

Coherent light shows phase-independent noise. This can be demonstrated by a two-beam interferometer, such as the Mach–Zehnder interferometer shown in Fig. 9.96. The monochromatic laser beam with the mean intensity I_0 is split by the beam splitter BS_1 into two partial beams b_1 and b_2 with amplitudes E_1 and E_2, which are superimposed again by BS_2. The beam b_2 passes a movable optical wedge, causing a variable phase shift ϕ between the two partial beams. The detectors PD1 and PD2 receive the intensities

$$\langle I \rangle = \frac{1}{2} c \cdot \epsilon_0 \left[\langle E_1^2 \rangle + \langle E_2^2 \rangle \pm 2 E_1 E_2 \cos\phi \right], \tag{9.89}$$

averaged over many cycles of the optical field with frequency ω. The different signs in (9.89) for the two detectors are caused by the phase shifts of the wave after reflections by the beam splitters, where both beams undergo one reflection for PD1, while for PD2 the beam b_1 undergoes one reflection and beam b_2 three reflections. For equal amplitudes $E_1 = E_2$, the detected intensities are

$$\langle I_1 \rangle = \langle I_0 \rangle \cos^2 \phi/2, \quad \langle I_2 \rangle = \langle I_0 \rangle \cdot \sin^2 \phi/2 \quad \Rightarrow \quad \langle I_1 \rangle + \langle I_2 \rangle = \langle I_0 \rangle.$$

A variation of the phase ϕ by the wedge in one arm of the interferometer corresponds to a rotation of the vector $\boldsymbol{E}$ in the phase diagram of Fig. 9.95b. The two detectors in Fig. 9.96a measure the two projections of $\boldsymbol{E}$ onto the axes, E_1 and E_2. The arrangement of Fig. 9.96a therefore allows the separate determination of the fluctuations $\langle \delta E_1 \rangle$ and $\langle \delta E_2 \rangle$ by a proper choice of the phase ϕ.

If the frequency spectrum $I(f)$ of the detector signals is measured with a spectrum analyzer at sufficiently high frequencies f, where the technical noise is negligible, one obtains a noise power spectrum $\rho_{\mathrm{n}}(f)$, which is essentially independent

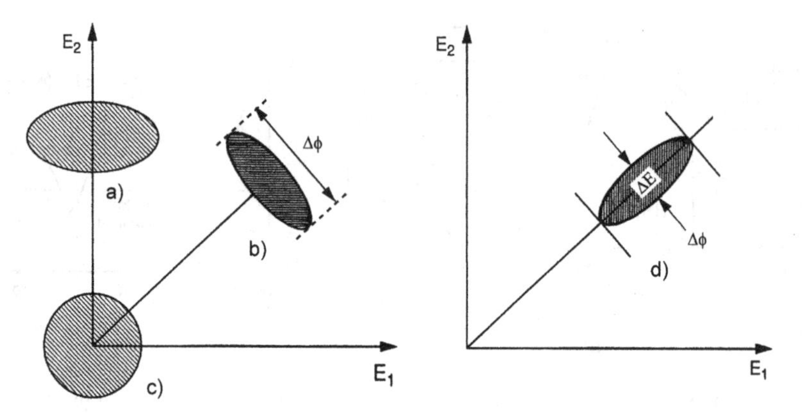

Fig. 9.97 Uncertainty areas for different squeezing conditions: (**a**) $\langle E_1 \rangle = 0$, ΔE_2 is squeezed but $\Delta\phi$ is larger than in the nonsqueezed case; (**b**) general case of squeezing of ΔE at the expense of increasing $\Delta\phi$; (**c**) uncertainty area of zero-point fluctuations with $\langle E_1 \rangle = \langle E_2 \rangle = 0$; (**d**) minimized $\Delta\phi$ at the expense of larger ΔE

of the phase ϕ (dotted line in Fig. 9.96b), but depends only on the number of photons entering the interferometer. It is proportional to $\sqrt{N}$. It is surprising that the noise power density $\rho_\mathrm{n}(f)$ of each detector is independent of the phase ϕ. This can be understood as follows: because of the statistical emission of photons the intensity fluctuations are uncorrelated in the two partial beams. Although the mean intensities $\langle I_1 \rangle$ and $\langle I_2 \rangle$ depend on ϕ, their fluctuations do not! The detected noise power $\rho_\mathrm{n} \propto \sqrt{N}$ shows the same noise level $\rho_\mathrm{n} \propto \sqrt{I_0}$ for the minimum of $I(\phi)$ as for the maximum.

If the incident laser beam in Fig. 9.96a is blocked, the mean intensity $\langle I \rangle$ becomes zero. However, the measured noise power density $\rho_\mathrm{n}(f)$ does *not* go to zero but approaches a lower limit ρ_0 that is attributed to the zero-point fluctuations of the vacuum field, which is also present in a dark room. The interferometer in Fig. 9.96a has two inputs: the coherent light field and a second field, which, for a dark input part, is the vacuum field. Because the fluctuations of these two inputs are uncorrelated, their noise powers add. Increasing the input intensity I_0 will increase the signal-to-noise-ratio

$$\frac{S}{\rho_\mathrm{n}} \propto \frac{N}{\sqrt{N} + \rho_0/h\nu}, \tag{9.90}$$

where the fundamental limitation is set by the quantum noise ρ_0.

In the phase diagram of Fig. 9.97 the radius $r = \sqrt{\rho_0}$ of the circular uncertainty area around $E_1 = E_2 = 0$ corresponds to the vacuum fluctuation noise power density ρ_0.

The preparation of *squeezed* states tries to minimize the uncertainty of one of the two quantities ΔE or $\Delta\phi$ at the expense of the uncertainty of the other. Although the *uncertainty area* in Fig. 9.97, which is squeezed into an ellipse, does increase compared to the symmetric case (9.88b), one may still gain in signal-to-noise ratio

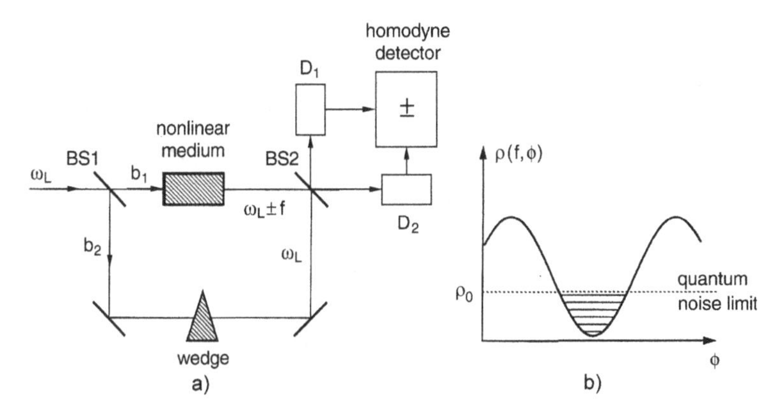

Fig. 9.98 Schematic diagram of squeezing experiments with a nonlinear medium in a Mach–Zehnder interferometer: (**a**) experimental arrangement; (**b**) noise power density $\rho(\phi)$ and quantum-noise limit ρ_0, which is independent of the phase ϕ

if the minimized quantity determines the noise level of the detected signal. This will be illustrated by some examples.

9.8.2 Experimental Realization of Squeezing

A typical layout of a squeezing experiment based on a Mach–Zehnder interferometer (Vol. 1, Sect. 4.2.4) is shown in Fig. 9.98. The output of a well-stabilized laser is split into two beams, a pump beam b_1 and a reference beam b_2. The pump beam with the frequency ω_L generates by nonlinear interaction with a medium (e.g., four-wave mixing or parametric interaction) new waves at frequencies $\omega_L \pm f$. After superposition with the reference beam, which acts as a local oscillator, the resulting beat spectrum at the frequency f is detected by the photodetectors D1 and D2 as a function of the phase difference $\Delta\phi$, which can be controlled by a wedge in one of the interferometer arms. The difference between the two detector output signals is monitored as a function of the phase difference $\Delta\phi$. Contrary to the situation in Fig. 9.96, the spectral noise power density $\rho_n(f,\phi)$ ($=P_{NEP}$ per frequency interval $\mathrm{d}f = 1\ \mathrm{s}^{-1}$) shows a periodic variation with ϕ. This is due to the nonlinear interaction of one of the beams with the nonlinear medium, which preserves phase relations. At certain values of ϕ the noise power density $\rho_n(f,\phi)$ drops below the photon noise limit $\rho_0 = (n \cdot h\nu/\eta\Delta f)^{1/2}$ that is measured without squeezing, when n photons fall onto the detector with bandwidth Δf and quantum efficiency η. This yields a signal-to-noise ratio

$$SNR_{sq} = SNR_0 \frac{\rho_0}{\rho_n}, \tag{9.91}$$

which is larger than that without squeezing SNR_0.

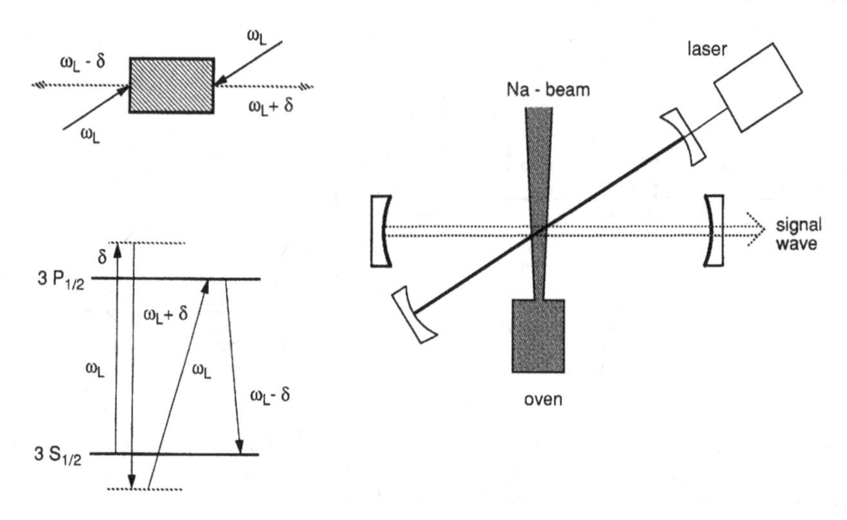

Fig. 9.99 Generation of squeezed states by four-wave mixing in an atomic Na beam. Both the pumped wave and signal and idler waves are resonantly enhanced by two optical resonators [1335]

The degree of squeezing V_{sq} is defined as

$$V_{sq} = \frac{\rho_0 - \rho_{min}(f)}{\rho_0}. \tag{9.92}$$

The first successful experimental realization of squeezing was reported by Slusher and coworkers [1335], who used the four-wave mixing in a Na-atomic beam as nonlinear process (Fig. 9.99). The Na atoms are pumped by a dye laser at the frequency $\omega_L = \omega_0 + \delta$, which is slightly detuned from the resonance frequency ω_0. In order to increase the pump power, the Na beam is placed inside an optical resonator tuned to the pump frequency ω_L. Because of parametric processes (Vol. 1, Sect. 6.7) of the two pump waves $(\omega_L, \pm \boldsymbol{k}_L)$ inside the resonator, new waves at $\omega_L \pm \delta$ are generated at this four-wave mixing process (signal and idler wave). Conservation of energy and momentum demands

$$2\omega_L \to \omega_L + \delta + \omega_L - \delta, \tag{9.93a}$$

$$\boldsymbol{k}_L + \boldsymbol{k}_i = \boldsymbol{k}_L + \boldsymbol{k}_s. \tag{9.93b}$$

A second resonator with a properly chosen length, such that the mode spacing is $\Delta\nu = \delta$, enhances the signal as well as the idler wave.

The essential point is that there are definite phase relations between pump, signal and idler waves, and this fact establishes a correlation between amplitude and phase of the signal wave. This correlation is illustrated in Fig. 9.100: The four-wave mixing generates new sidebands. If the input contains the frequencies ω_L and $\omega_L + \delta$, the output has the additional sideband $\omega_L - \delta$. Thus, the amplitude of one sideband is increased at the cost of the other sideband until both sidebands have equal amplitudes. This maximizes the amplitude modulation of the output. Since the phases

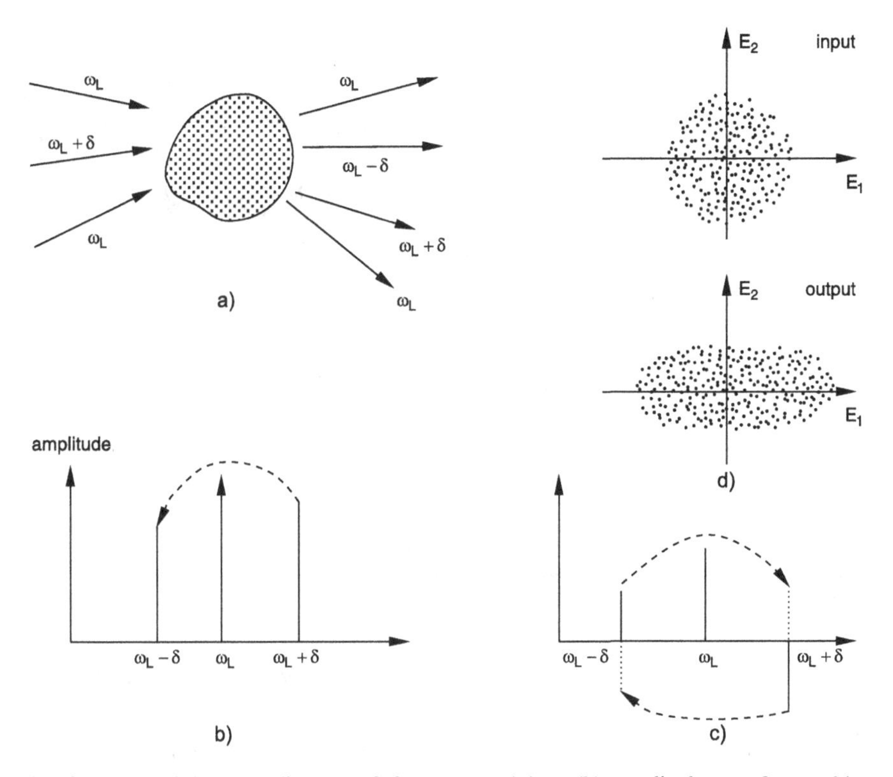

Fig. 9.100 (**a**) Schematic diagram of four-wave mixing; (**b**) amplitude transfer resulting in increased amplitude modulation; (**c**) phase transfer causing decreased phase modulation; (**d**) schematic phase diagram

of the two sidebands are opposite, this transfer minimizes the phase modulation. As was shown in detail in [1333], this correlation results in a phase-dependent noise that is, for certain phase ranges, below the quantum noise power ρ_0. In these experiments a squeezing degree of 0.1 was obtained, which means that $\rho_{\min} = 0.9\rho_0$.

The best squeezing results with a noise suppression of 60 % (≈−4 dB) below the quantum noise limit ρ_0 were obtained by Kimbel and coworkers [1336] with an optical parametric oscillator, where the parametric interaction in a $MgO:LiNbO_3$ crystal was used for squeezing.

Another example, where a beam of squeezed light was realized with 3.2 mW and 52 % noise reduction, is the second-harmonic generation with a cw Nd:YAG laser in a monolithic resonator [1337]. The experimental setup is shown in Fig. 9.101a.

The output of the cw Nd:YAG laser is frequency doubled in a $MgO:LiNbO_3$ nonlinear crystal. The endfaces of the crystal form the resonator mirrors and by a special modulation technique the resonator is kept in resonance both for the fundamental and for the second harmonic wave. A balanced homodyne interferometric detection of the fundamental wave yields the intensity noise of the light beam (I_+ detected by D1) and the shot-noise level (I_- detected by D2) as a reference (Fig. 9.101b).

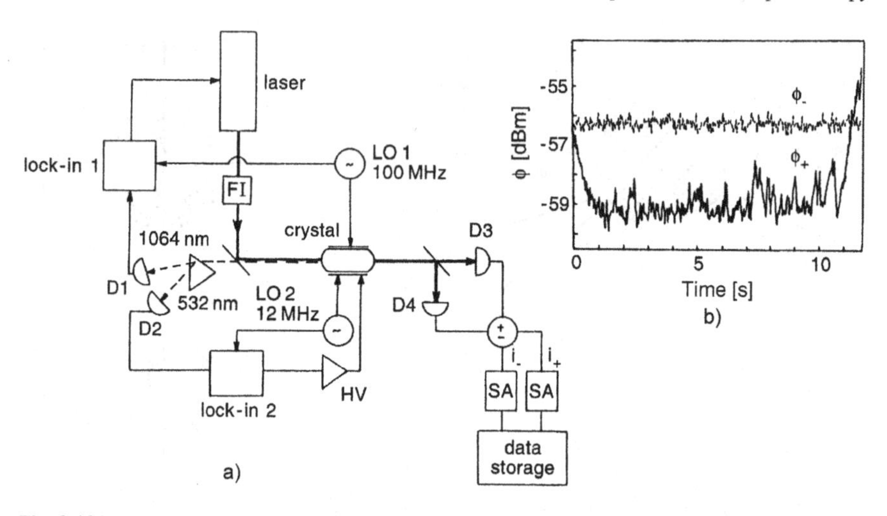

Fig. 9.101 (**a**) Experimental setup for the generation of bright squeezed light by second-harmonic generation in a monolithic resonator; (**b**) spectral power density of photocurrent fluctuations for squeezed light input to a balanced homodyne detector. The *upper curve* ϕ_- gives the shot-noise level [1337]

9.8.3 Application of Squeezing to Gravitational Wave Detectors

Nowadays, it is believed that the most sensitive method for detecting gravitational waves is based on optical interferometry, where the change in length of an interferometer arm by gravitational waves can be monitored. Although until now no gravitational waves have been detected, there are great efforts in many laboratories to increase the sensitivity of laser gravitational wave detectors in order to discover those waves, even from very distant supernovae or rotating heavy binary neutron stars [1340, 1345, 1346].

The basic part of such a detector (Fig. 9.102) is a Michelson interferometer with long arms (several kilometers) and an extremely well stabilized laser. If a gravitational wave causes a length difference ΔL between the two arms of the interferometer, a phase difference $\Delta\phi = (4\pi/\lambda)\Delta L$ appears between the two partial waves at the exit of the interferometer. Typical expected relative changes are about $\Delta L/L = 10^{-21}$. In order to give a feeling for the magnitude of ΔL, note that for $L = 2$ km for a gravitational wave emitted from a stellar supernova explosion in a neighboring galaxy, the difference ΔL will be smaller than 10^{-17} m. This is less than 1 % of the diameter of a proton! In order to increase the sensitivity, the length of the interferometer arms must be multiplied by a large factor by using multiple reflection arrangements (Fig. 9.102b) or Fabry–Perot interferometers in both arms of the Michelson interferometer. Such arrangements are expected to achieve a limit of $\Delta L \leq 10^{-11}\lambda_{\text{Laser}}$.

The sensitivity of a gravitational wave detectors depends essentially on the stability and output power of the laser. In an improved version of the advanced LIGO detector called aLIGO the following laser system is used [1357].

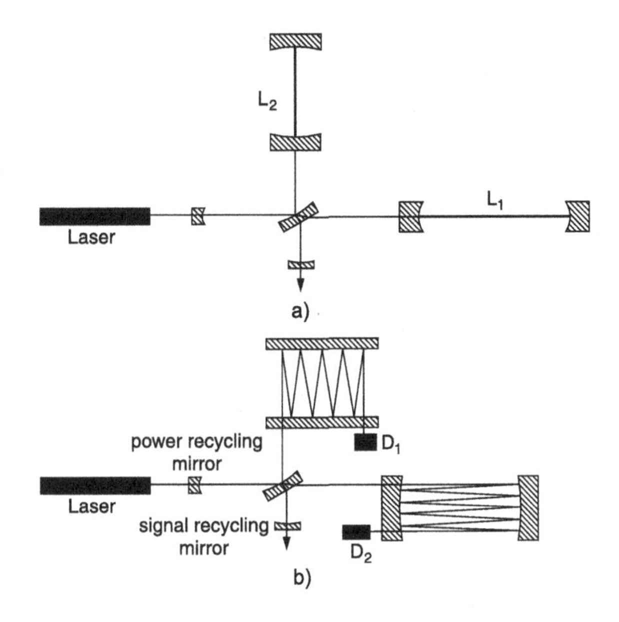

Fig. 9.102 Gravitational wave detector based on a Michelson interferometer: (**a**) basic arrangement using optical resonators; (**b**) multiple path arrangement for increasing the sensitivity

The aLIGO PSL (pre-stabilized seeded Laser system) consists of a 2 W Nd:YAG seed laser which is amplified via a Nd:YVO amplifier stage to 35 W. This 35 W laser serves as a master laser in an injection-locking scheme to seed a 200 W Nd:YAG high power oscillator stage. The output of the high power stage is spatially filtered by a newly developed bow-tie pre-mode cleaner (PMC) resonator which transmits 165 W in the fundamental Gaussian mode with only 1 % of the power in higher order modes. The PMC has two auxiliary ports to pick-off beam samples for the power and frequency stabilization. A diagnostic breadboard (DBB) is an integral part of the PSL and allows for a fully automated characterization of the 35 W and the 200 W laser with regard to temporal and spatial fluctuations and the spatial beam quality.

Since their installation the aLIGO PSLs operate stable and provide reliable light sources for the gravitational wave detectors.

In order to get an impression about the required characteristic data of a gravitational wave detector, in Table 9.1 the technical data of GEO 600 are compiled.

The minimum detectable phase change $\Delta\phi$ is limited by the phase noise $\delta\phi_{\mathrm{n}}$. If the laser delivers N photons $h\nu$ per second, the two detectors with quantum efficiencies of η measure the intensities

$$I_1(\phi) = \frac{1}{2} N h\nu\eta\big[1 + \cos(\phi + \Delta\phi)\big], \tag{9.94a}$$

$$I_2(\phi) = \frac{1}{2} N h\nu\eta\big[1 - \cos(\phi + \Delta\phi)\big]. \tag{9.94b}$$

This gives for $\phi = \pi/2$ a difference signal

$$S = \Delta I \approx N h\nu\eta\Delta\phi. \tag{9.95}$$

Table 9.1 Technical specifications of GEO600

Arm length	2×600 m
Orientation	NNW and ENE
Tube diameter	60 cm
Laser type	Diode-pumped Nd:YAG laser at 1064 nm
Laser power	30 W single mode at 1064 nm
Power recycling	factor 1000
Optics	25 cm diameter fused-silica mirrors
Signal recycling	factor 10
Frequency range	50 Hz to 6 kHz
Vacuum	pressure $<10^{-8}$ mbar
Sensitivity	depending on bandwidth $h \sim 10^{-20} \dots 10^{-21}$ (for bursts) $h \sim 10^{-26}$ (for cw signals and 1 yr integration time)

The noise power of both detectors adds quadratically. The signal-to-noise ratio for a detection bandwidth Δf

$$\mathrm{SNR} = \frac{Nh\nu\Delta\phi}{\sqrt{Nh\nu\Delta f}} = \left(\frac{Nh\nu}{\Delta f}\right)^{1/2} \Delta\phi, \tag{9.96}$$

becomes larger than unity for $\Delta\phi > \sqrt{\Delta f/(N \cdot h\nu)}$. The minimum detectable phase change is therefore limited by the maximum available photon flux N. Therefore, high-power ultrastable solid-state lasers are used.

In Sect. 9.8.1 it was shown that the noise limit without squeezing is caused by the zero-point field coupled into the second input part of the interferometer. If a beam of squeezed light is fed into this input part, the noise level can be decreased and the detectable length change ΔL can be further decreased [1356].

The combination of Ramsey spectroscopy (see Sect. 9.3) and squeezed light has brought an increased spectral resolution. While traditional Ramsey spectroscopy allows the resolution $\Delta\omega = 2\pi/3T$ where T is the flight time of the molecules between the two separated fields, using two-mode squeezed light and two-atom excitation with joint detection a spectral resolution of $\pi/3T$ could be achieved [1355].

For a more detailed representation of squeezing and the most recent experiments on the realization of squeezing, or for further applications of this interesting technique, the reader is referred to the literature [1347–1353].

9.9 Problems

9.1 What is the probability that a photon, emitted from a Ca atom at the centre of a spherical cloud with radius $R = 1$ cm on the intercombination line ${}^3P_1 \rightarrow {}^1S_0$ at $\lambda = 657$ nm with the natural line-width $\Delta\nu = 3$ kHz is reabsorbed by another

Ca atom, when the absorption cross section $\sigma(\omega_0)$ at the line centre is $\sigma(\omega_0) = 10^{-17}\ \mathrm{cm}^2$, the density of Ca atoms is $n = 10^{17}\ \mathrm{cm}^{-3}$ and the collision cross section $\sigma_{\mathrm{coll}} = 10^{-16}\ \mathrm{cm}^2$? How large is the recoil splitting?

9.2 Derive (9.12b).

9.3 Which fraction of photons of a monochromatic laser tuned to the Na-transition $F'' = 2 \rightarrow F' = 3$ in Fig. 9.5 is absorbed on the transition $F'' = 2 \rightarrow F' = 2$ when the lifetime of the upper level is $\tau = 16$ ns and the saturation parameter $S = 1$?

9.4 Na atoms in an atomic beam traversing a magnetic field are slowed down by counter propagating photons from a fixed frequency laser at $\lambda = 589$ nm. What is the field dependence $B(z)$ at the optimum deceleration?

9.5 Calculate the Doppler-limit of the minimum cooling temperature for Li-atoms ($\tau(3P_{3/2}) = 27$ ns) and for K-atoms ($\tau(5P_{3/2}) = 137$ ns).

9.6 What is the critical density for Cs atoms to achieve BEC at $T = 1\ \mu\mathrm{K}$?

9.7 Assume that the comb separation $\Delta\nu = 100$ MHz of an optical frequency comb can be determined with a relative accuracy of 10^{-16}. One of the comb frequencies coincides with the 6×10^4th multiple of the caesium clock frequency. How accurate can the frequency of a molecular transition at $\lambda = 750$ nm be measured?

9.8 Calculate the potential depth of the inhomogeneous magnetic field $B = B_0 r^2$ necessary for trapping Na atoms with a magnetic moment of $\mu = 2\mu_B$ in their $3\,^2S_{1/2}$ ground state within a volume of $1\ \mathrm{cm}^3$ at a temperature of $T = 10\ \mu\mathrm{K}$.

Chapter 10
Applications of Laser Spectroscopy

The relevance of laser spectroscopy for numerous applications in physics, chemistry, biology, and medicine, or to environmental studies and technical problems has rapidly gained enormous significance. This is manifested by an increasing number of books and reviews. This chapter can only discuss some examples that are selected in order to demonstrate how many fascinating applications already exist and how much research and development is still needed. For a detailed representation of more examples, the reader is referred to the references given in the sections that follow as well as to some monographs and reviews [1358–1364].

10.1 Applications in Chemistry

For many fields of chemistry, lasers have become indispensable tools [1365–1370]. They are employed in analytical chemistry for the ultrasensitive detection of small concentrations of pollutants, trace elements, or short-lived intermediate species in chemical reactions. Important analytical applications are represented by measurements of the internal-state distribution of reaction products with LIF (Sect. 1.3) and spectroscopic investigations of collision-induced energy-transfer processes (Sects. 8.3–8.6). These techniques allow a deeper insight into reaction paths of inelastic or reactive collisions, and their dependence on the interaction potential and the initial energy of the reactants.

Time-resolved spectroscopy with ultrashort laser pulses gives, for the first time, access to a direct view of the short time interval in which molecules are formed or fall apart during a collision. This femtosecond chemistry is discussed in Sect. 10.1.3.

The possibility of controlling chemical reactions by selective excitation of the reactants and by coherent control of the excited state wavefunction (as illustrated in Sect. 10.1.4) may open a new fascinating field of laser-induced chemistry.

W. Demtröder, *Laser Spectroscopy 2*, DOI 10.1007/978-3-662-44641-6_10

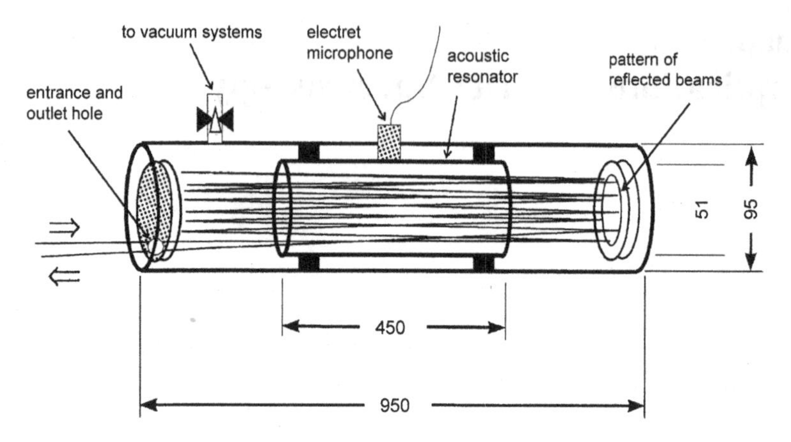

Fig. 10.1 Multipass cell of an optoacoustic spectrometer. All laser beams pass through regions in the acoustic resonator where the radial acoustic resonance has maximum amplitude. Measurements are in millimeter

10.1.1 Laser Spectroscopy in Analytical Chemistry

The first aspect of laser applications in analytical chemistry is the sensitive detection of small concentrations of impurity atoms or molecules. With the laser-spectroscopic techniques discussed in Chap. 1, detection limits down into the parts-per-billion (ppb) range can be achieved for molecules, which corresponds to a relative concentration of 10^{-9}. Atomic species and some favorable molecules can even be traced in concentrations within the parts-per-trillion (ppt) ($\hat{=}10^{-12}$) range. Recently "single molecule detection" in solids, solutions, and gases has become possible.

A very sensitive detection scheme is the photoacoustic method in combination with a multipass optical resonator, illustrated in Fig. 10.1. With this apparatus absorption coefficients down to $\alpha = 10^{-10}\ \mathrm{cm}^{-1}$ can be measured.

Example 10.1 With a diode laser spectrometer and a multipass absorption cell (Fig. 10.1) NO_2 concentrations down to the 50-ppt level in air were detected on a vibrational–rotational transition at $1900\ \mathrm{cm}^{-1}$, NO concentrations down to 300 ppt, while for SO_2 at $1335\ \mathrm{cm}^{-1}$ a sensitivity limit of 1 ppb was reached [1371].

If the spectroscopic detection of atoms can be performed on transitions that represent a true two-level system (Sect. 9.1.5), atoms with the radiative lifetime τ may undergo up to $T/2\tau$ absorption–emission cycles during their transit time T through the laser beam (photon burst). If the atoms are detected in carrier gases at higher pressures, the mean free path Λ becomes small ($\Lambda \ll d$) and T is only limited by the diffusion time. Although quenching collisions may decrease the fluorescence

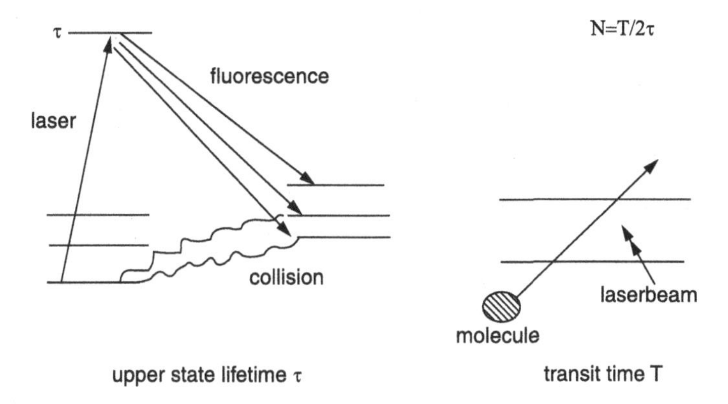

Fig. 10.2 Single-molecule detection

(Sect. 6.3), the ratio $T/2\tau$ becomes larger and the magnitude of the photon bursts may still increase inspite the decreasing quantum yield.

Example 10.2 For gases at low pressures where the mean free path Λ is larger than the diameter d of the laser beam, we obtain the typical value $T = d/\overline{v} = 10\,\mu\text{s}$ for $d = 5$ mm and $\overline{v} = 5 \times 10^2$ m/s. For an upper-state lifetime of $\tau = 10$ ns the atom emits 500 fluorescence photons (photon burst), allowing the detection of single atoms. With noble gas pressures of 1 mbar the mean-free path is $\approx$0.03 mm and the diffusion time through the laser beam may become 100 times longer. Although the lifetime is quenched to 5 ns, which means a fluorescence quantum yield of 0.5, this increases the photon burst to 5×10^4 photons.

Although molecules do not represent two-level systems, because the fluorescence from the excited upper level terminates at many vibrational–rotational levels in the electronic ground state, one can obtain many fluorescence photons for a single molecule in a liquid solution. Through collisions with the solution molecules, the terminating levels will rapidly be transferred to neighboring levels and in particular the depleted initial level will be repopulated by collisional transfer, which restores thermal equilibrium rapidly within pico- to nanoseconds (Fig. 10.2). For a transit time T of the molecules through the laser beam and an upper-state lifetime τ, the maximum number N_{phot} of fluorescence photons from a single molecule is $N_{\text{phot}} = T/2\tau$ if the laser intensity is sufficiently high to reach saturation.

Another very sensitive detection scheme is based on resonant two- or three-photon ionization of atoms and molecules in the gas phase (Sect. 1.3). With this technique even liquid or solid samples can be monitored if they can be vaporized in a furnace or on a hot wire. If, for instance, a heated wire or plate in a vacuum system is covered by the sample, the atoms or molecules are evaporated during the pulsed heating period and fly through the superimposed laser beams $L_1+L_2(+L_3)$ in front

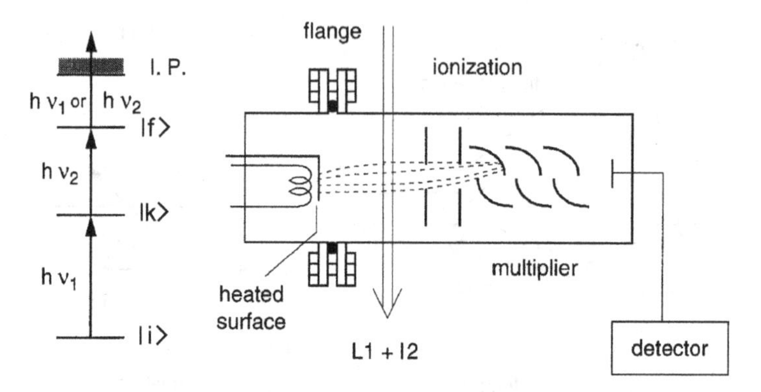

Fig. 10.3 Resonant multiphoton ionization (REMPI) as a sensitive detection technique for small quantities of atoms or molecules evaporated from a heated surface

of the heated surface (Fig. 10.3). The laser L_1 is tuned to the resonance transition $|i\rangle \rightarrow |k\rangle$ of the wanted atom or molecule while L_2 further excites the transition $|k\rangle \rightarrow |f\rangle$. Ions are formed if E_f is above the ionization potential *IP*. The ions are accelerated toward an ion multiplier. If L_2 has sufficient intensity, all excited particles in the level $|k\rangle$ can be ionized and all atoms in the level $|i\rangle$ flying through the laser beam during the laser pulse can be detected (*single-atom detection*) [1372–1374]. If $E_f < IP$ a third photon provided by L_1 and L_2 can ionize the species since there is no resonance condition for the ionizing step.

The selective detection sensitivity for spurious species with absorption spectra overlapping those of more abundant molecules or atoms can be enhanced further by a combination of mass spectrometer and resonant two-photon ionization (Fig. 10.4). This is, for instance, important for the detection of rare isotopic components in the presence of other more abundant isotopes [1374, 1375].

10.1.2 Single-Molecule Detection

The combination of illuminating the sample with a strongly focused laser beam and observing the laser-induced fluorescence or the non-resonant scattered laser light with confocal microscopy allows the detection of single molecules and their diffusion through a liquid medium. Each molecule can emit $N = T/2\tau$ photons (see Fig. 10.2) where T is the diffusion time through the laser beam and τ the upper state lifetime.

Example 10.3 $T = 1$ ms, $\tau = 10$ ns $\Rightarrow N = 5 \times 10^4$.

The experimental setup is illustrated in Fig. 10.5. The light scattered by molecules in a diluted solution with concentrations in the nanomole range, which

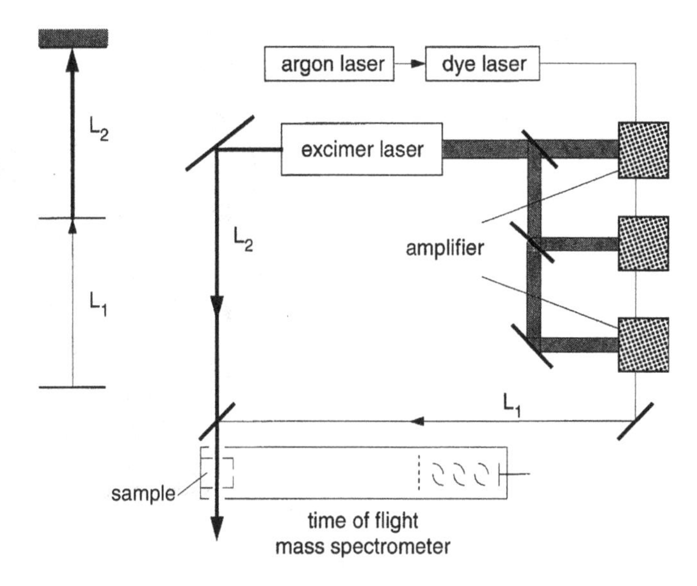

Fig. 10.4 Selective excitation of a wanted isotope by a pulsed, amplified single-mode dye laser with subsequent ionization by an excimer laser and mass-selective detection behind a time-of-flight spectrometer

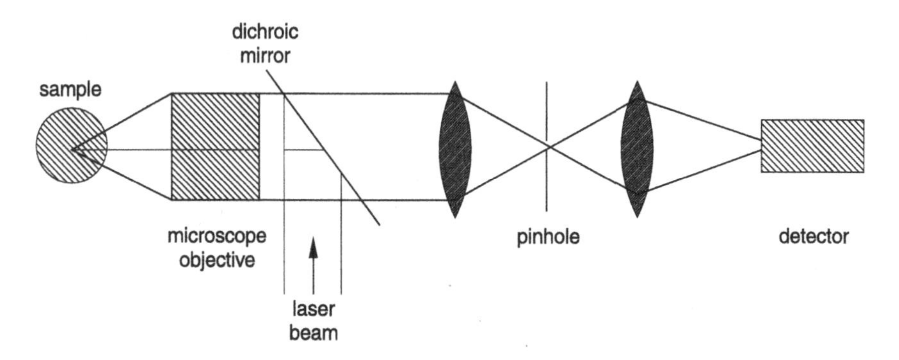

Fig. 10.5 Confocal microscopy for single-molecule detection

diffuse into the laser focus, is imaged onto a small pinhole in front of the detector. The spatial resolution in the x–y-plane perpendicular to the incident laser beam can reach 500 nm. Using two-photon excitation the spatial resolution can be even improved to about 200 nm. The resolution in the z-direction (axial resolution) depends on the Rayleigh length of the focused laser beam. Typical values are 10–30 μm, where two-photon excitation also decreases this value below 1000 nm. This gives the extremely small detection volume of less than 10^{-15} ℓ. The fluorescence is de-

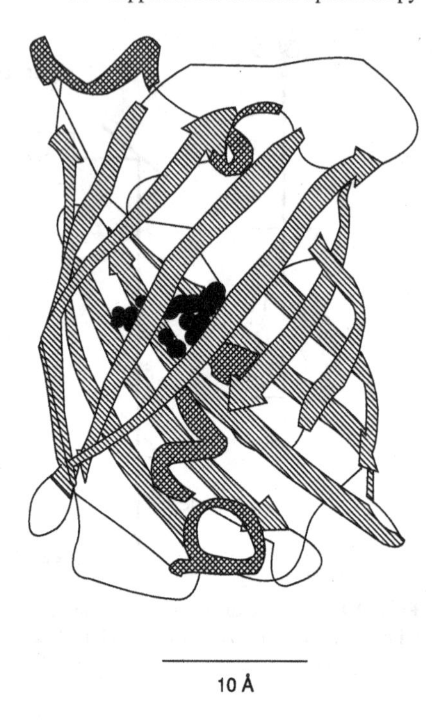

Fig. 10.6 Green fluorescence protein

tected by a high quantum efficiency avalanche photodiode [1376]. Small concentrations with extinction coefficients of 10^{-4} are still detectable. If stimulated emission pumping is used too, the radial spatial resolution can be increased still further (see Sect. 10.4.4) [1377].

With time-resolved detection the temporal behavior of a single molecule can be followed. One example is the measurement of the relaxation time constant of intersystem crossing from the excited singlet state into the triplet state of a dye molecule during the diffusion time through the detection volume [1380].

Molecular impurities in solids can now be detected down to the single-molecule level [1381]. These molecules and their interaction with their surroundings can be probed by high-resolution laser spectroscopy. The relevance of such techniques in biology is obvious [1382]. The development of single-molecule detection for the in vitro and in vivo quantification of biomolecular dynamics is essential for the understanding of biomolecular reactions and for the realization of evolutionary biotechnology [1383]. One important biological molecule is the green fluorescent protein (Fig. 10.6), which is excited by blue laser light and emits fluorescence in the green with a high quantum efficiency. It can be attached to bacteria or a specific virus, and when combined with the technique of single-molecule detection it permits real-time observations of such bacteria and allows their paths through cell membranes or inside a cell to be followed in vivo [1378, 1379].

More examples of lasers in analytical chemistry can be found in [1384–1386].

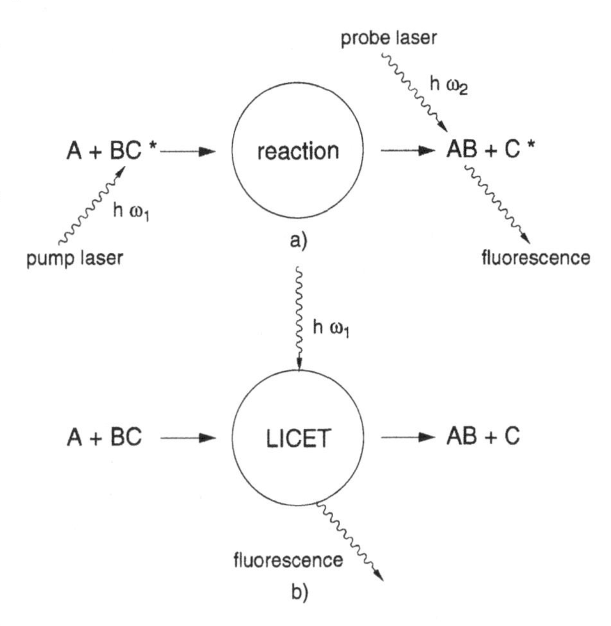

Fig. 10.7 Schematic diagram of laser-induced chemical reactions with state-selective detection of the reaction products: (**a**) by excitation of the reactants; (**b**) by excitation of the collision pair (ABC)

10.1.3 Laser-Induced Chemical Reactions

The basic principle of laser-induced chemical reactions is schematically illustrated in Fig. 10.7. The reaction is initiated by one- or multiphoton excitation of one or more of the reactants. The excitation can be performed before the reactants collide (Fig. 10.7a) or during the collision (Fig. 10.7b and Sect. 8.6).

For the selective enhancement of the wanted reaction channel by laser excitation of the reactants, the time span Δt between photon absorption and completion of the reaction is of fundamental importance. The excitation energy $n \cdot \hbar\omega$ $(n = 1, 2, \ldots)$ pumped by photon absorption into a selected excited molecular level may be redistributed into other levels by unwanted relaxation processes *before* the system ends in the wanted reaction channel. It can, for instance, be radiated by spontaneous emission, or it may be redistributed by intramolecular radiationless transitions due to vibrational or spin-orbit couplings onto many other nearly degenerate molecular levels. However, these levels may not lead to the wanted reaction channel. At higher pressures collision-induced intra- or intermolecular energy transfer may also play an important role in enhancing or suppressing a specific reaction channel.

We distinguish between three different time regimes:

(a) The excitation by the pump laser and the laser-induced reaction proceeds within a very short time Δt (femtosecond to picosecond range), which is shorter than the relaxation time for energy redistribution by fluorescence or inelastic non-reactive collisions. In this case, the above-mentioned loss mechanisms cannot play an important role, and the selectivity of driving the system into a wanted reaction channel by optical absorption is possible.

(b) Within an intermediate time range (nanosecond to microsecond range), depending on the pressure in the reaction cell, the reaction may happen before the initial excitation energy is completely redistributed by collisions. If the excited reagent is a large molecule, the intramolecular energy transfer within the many accessible degrees of freedom is, however, fast enough to redistribute the excitation energy onto many excited levels. The excited molecule may still react with a larger probability than in the ground state, but the selectivity of reaction control is partly lost.
(c) For still longer time scales (microsecond to continuous wave excitation) the excitation energy is distributed by collisions statistically over all accessible levels and is finally converted into translational, vibrational, and rotational energy at thermal equilibrium. This raises the temperature of the sample, and the effect of laser excitation with respect to the wanted reaction channel does not differ much from that for thermal heating of the sample.
(d) Recently, the new technique of coherent control of chemical reactions has been developed, where definitely shaped femtosecond pulses are used (see Sect. 6.1) to prepare a coherent wavefunction in the excited state of a reactant molecule. The phase and amplitude of this wavefunction are chosen in a way that favors the decay of the molecule into the wanted reaction channel.

For the first time range, ultrashort laser pulses generated by mode-locked lasers in the femto- to picosecond range (Chap. 6) are needed, whereas for the second class of experiments pulsed lasers in the nanosecond to microsecond range (Q-switched CO_2 lasers, excimers, or dye lasers) can be employed. Most experiments performed until now have used pulsed CO_2 lasers, chemical lasers, or excimer lasers. For femtosecond pulses Kerr-lens mode-locked Ti:sapphire lasers with subsequent amplifier stages are available (see Sect. 6.1).

Let us consider some specific examples: the first example is the laser-induced bimolecular reaction

$$\mathrm{HCl}(v=1,2)+(\mathrm{O}^3\mathrm{P})\rightarrow \mathrm{OH}+\mathrm{Cl}, \tag{10.1}$$

which proceeds after vibrational excitation of HCl with a HCl laser [1387]. The internal-state distribution of the OH radicals is monitored by LIF, induced by a frequency-doubled dye laser, which is tuned to a specific rotational line of the $(v'=0 \leftarrow v''=0)$ band in the $^2\Pi \leftarrow {}^2\Sigma$ system of OH at $\lambda = 308$ nm or the $(1 \leftarrow 1)$ band at 318 nm.

The second example is the spatially and temporally resolved observation of the explosion of an O_2/O_3 mixture in a cylindrical cell, initiated by a TEA CO_2 laser [1388]. The progress of the reaction is monitored through the decrease of the O_3 concentration, which is detected by a time-resolved measurement of the UV absorption in the Hartley continuum of O_3. If the UV probe beam is split into several spatially separated beams with separate detectors (Fig. 10.8), the spatial progression of the explosion front in time can be monitored.

The high output power of pulsed CO_2 lasers allows excitation of high vibrational levels by multiphoton absorption, which eventually may lead to the dissociation of

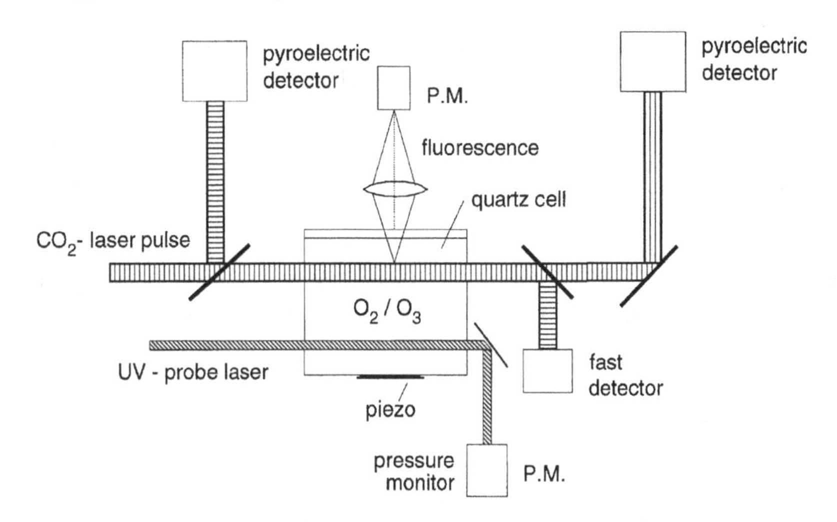

Fig. 10.8 Experimental setup for the CO_2 laser-initiated explosion of O_3. The expansion of the explosion front is monitored through the time-resolved O_3 absorption of several spatially separated UV probe laser beams. The piezo pressure detector allows a time-resolved measurement of the pressure [1388]

the excited molecule. In favorable cases the excited molecules or the dissociation fragments react selectively with other added components [1389]. Such selectively initiated chemical reactions induced by CO_2 lasers are particularly advantageous due to the large efficiency of these lasers, which makes the CO_2 photons cheap. As an example, let us consider the synthesis of SF_5NF_2 by multiphoton absorption of CO_2 photons in a mixture of S_2F_{10} and N_2F_4, which proceeds according to the following scheme:

$$\begin{aligned} S_2F_{10} + nh\nu &\to 2SF_5, \\ N_2F_4 + nh\nu &\to 2NF_2, \\ SF_5 + NF_2 &\to SF_5NF_2. \end{aligned} \tag{10.2}$$

While conventional synthesis without a laser takes about 10–20 hours at a temperature of 425 K and demands high pressures of S_2F_{10}, the laser-driven reaction proceeds much more quickly, even at the lower temperature of 350 K [1390, 1391].

Another example of CO_2 laser-initiated reactions is the gas-phase telomerization of methyl-iodide CF_3I with C_2F_4, which represents an exothermic radical chain reaction

$$\begin{aligned} CF_3I + nh\nu &\to (CF_3I)^*, \\ (CF_3I)^* + nC_2F_4 &\to CF_3(C_2F_4)_nI \quad (n = 1, 2, 3), \end{aligned} \tag{10.3}$$

producing $CF_3(C_2F_4)_nI$ with low values of n. The CO_2 laser is in near resonance with the $\nu_2 + \nu_3$ band of CF_3I. The quantum yield for the reaction (10.3) increases with increasing pressure in the irradiated reaction cell [1392].

In many cases, electronic excitation by UV lasers is necessary for laser-initiated reactions. An example is the photolysis of vinyl chloride

$$C_2H_3Cl + h\nu \rightarrow C_2H_3 + Cl \tag{10.4a}$$

$$\rightarrow C_2H_2 + HCl, \tag{10.4b}$$

induced by a XeCl excimer laser (Vol. 1, Sect. 5.7). In spite of the small absorption cross section ($\sigma = 10^{-24}\,\mathrm{cm}^2$ at $\lambda = 308\,\mathrm{nm}$) the yield ratio of the two reaction branches (10.4a, 10.4b) and its dependence on temperature could be accurately measured [1393].

As another example, let us mention the chain reaction of organic bromides with olefins induced by a KrF excimer laser [1394], which proceeds as follows:

$$\begin{aligned}
&\mathrm{RBr} + h\nu \rightarrow \mathrm{R} + \mathrm{Br} \quad \text{(initiation)},\\
&\mathrm{R}_n + \mathrm{C_2H_4} \rightarrow \mathrm{R}_{n+2} \quad \text{(propagation)},\\
&\mathrm{R}_n + \mathrm{RBr} \rightarrow \mathrm{R}_n\mathrm{Br} + \mathrm{R} \quad \text{(chain transfer)},\\
&\mathrm{Br} + \mathrm{Br} \rightarrow \mathrm{Br_2} \quad \text{(termination)},
\end{aligned} \tag{10.5}$$

where R_n denotes any radical with n carbon atoms.

Many chemical reactions can be enhanced by catalytic effects at solid surfaces. The prospect of increasing these catalytic enhancements further by laser irradiation of the surface has initiated intense research activity [1395, 1396]. The laser may either excite atoms or molecules adsorbed at the surface, or it may excite desorbed molecules just above the surface. In both cases the desorption or adsorption process is altered because excited molecules have a different interaction potential between the molecule M* and the surface than the ground-state molecules. Furthermore, the laser may evaporate surface material, which can react with the molecules.

10.1.4 Coherent Control of Chemical Reactions

Coherent control means the coherent preparation of a molecular wavefunction through the absorption of coherent radiation. The dream of chemists is the controlled selection of wanted reaction channels and the suppression of unwanted channels in a photoinduced reaction. This is illustrated by Fig. 10.9, where the excitation of the triatomic molecule ABC can induce either of the reactions AB + C or AC + B, depending on the wavefunction in the excited state of ABC, which is controlled by the form of the excitation pulse.

Controlled coherent excitation can be achieved in different ways. One method is based on interference effects, which become essential when an excited state can be populated on two ore more excitation paths. The population rates then depend on the phase relations between the optical waves that are inducing these excitations.

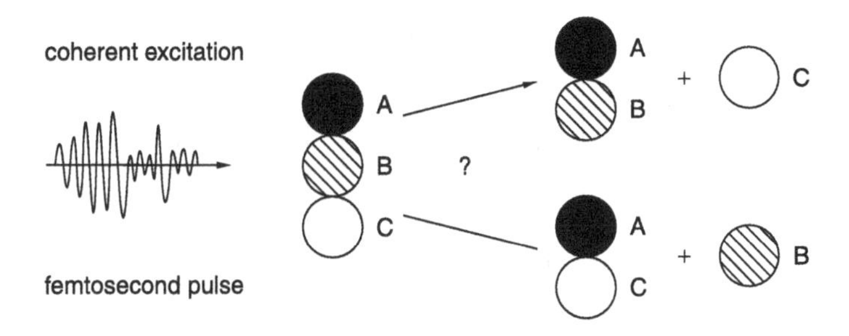

Fig. 10.9 Selection of the reaction channel by the controlled coherent excitation of a molecule

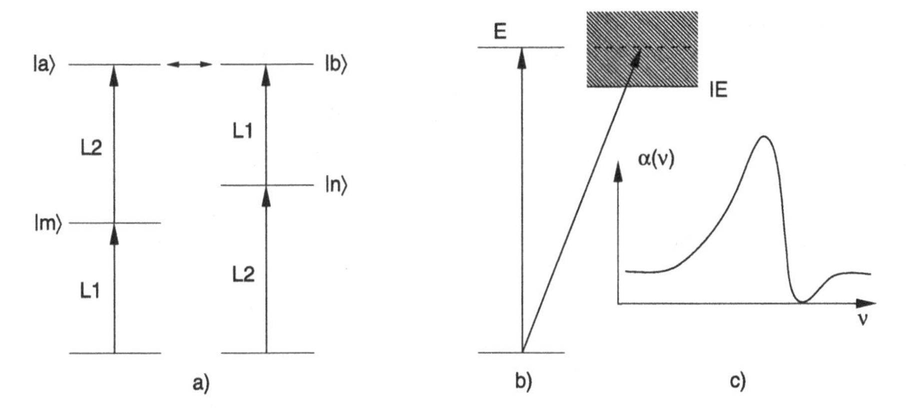

Fig. 10.10 (**a**) Coherent control of the population of the coupled levels $|a\rangle$ and $|b\rangle$ by interference between two different excitation paths; (**b**) excitation of autoionizing level; (**c**) Fano profile of absorption line

Coherent control has to be performed on a very short time scale in order to win over dephasing processes. Therefore femtosecond lasers are generally demanded [1399].

Two different schemes have been proposed [1400, 1401]. In the first scheme (Fig. 10.10), proposed by P. Brumer and M. Shapiro, two coupled levels $|a\rangle$ and $|b\rangle$ are simultaneously populated on two different excitation paths. Depending on the relative phase between the field amplitudes of the optical waves, there can be constructive interference, leading to maximum population of the mixed state $|a\rangle + |b\rangle$, or destructive interference, minimizing the population. An instructive example is the optical excitation of autoionizing states above the ionization limit, which interact with the ionization continuum (Fig. 10.10b). While the phase of the wavefunction of the discrete level changes rapidly when the optical frequency is tuned over the absorption profile, that of the continuum is only slightly affected. Therefore the phase difference of the two excitation paths depends on the optical frequency and changes in such a way that constructive interference occurs (maximum absorption) or destructive interference occurs, where the absorption is nearly zero. The resulting asymmetric absorption profile is called the Fano profile. The produced ion rate can be therefore controlled by slight changes of the excitation frequency.

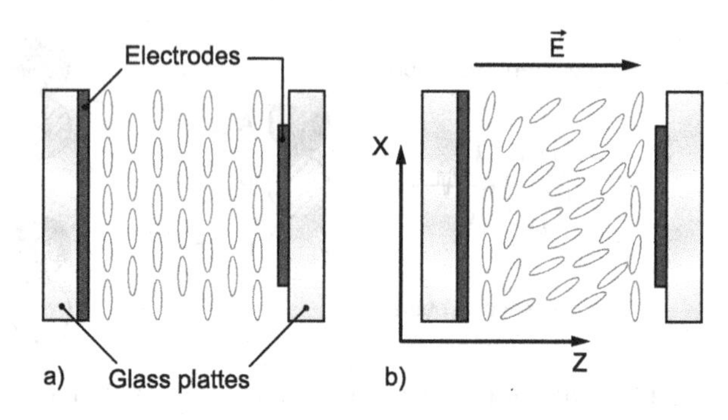

Fig. 10.11 Orientation of molecules in a liquid crystal (**a**) with, (**b**) without an external electric field

The second scheme of coherent control, proposed by D.J. Tanner and A. Rice, utilizes the time difference between two femtosecond laser pulses interacting with a molecule on two different transitions sharing a common level, which was illustrated in Sect. 6.4.4 by the example of the Na_2 molecule. Here the phase of the wave packet produced by the first pulse in the excited state develops in time, and the controlled time lag between the first and the second pulse selects a favorable phase for further excitation or deexcitation of the molecule by the second pulse.

In the first scheme it is not necessary to use two different lasers if a femtosecond pulse with a broad spectral range is used for excitation. The different spectral components in the pulse give rise to many different excitation paths. In order to achieve optimum population in the excited state, the relative phases of these different spectral components have to be optimized. This can be realized by the pulse-shaping techniques discussed in Sect. 6.1.11 (Fig. 10.11). Here a plate of many liquid crystal pixels are placed in the laser beam, which changes the phases of the lightwave by orientation of the molecules where a feedback loop with a learning algorithm is used to maximize or minimize the wanted decay channel of the excited state [1402, 1403].

One example is the dissociation of laser-excited iron pentacarbonyl $Fe(CO)_5$, where the ratio $Fe(CO)_5/Fe$ can be varied between 0.06 to 4.8 by coherent control [1404]. Another example is the photodissociation of $C_5H_5Fe(CO)_2Cl$ into selected fragments. Optimum shaped femtosecond pulses can alter the ratio $C_5H_5COCl/FeCl$ from 1 to 5 [1405], or the selected bond dissociation of acetone $(CH_3)_2CO$ or acetophenone $C_6H_5COCH_3$ [1406]. The experimental arrangement for the coherent control of chemical reactions in the gas and liquid phase is shown in Fig. 10.12 [1407]. The femtosecond pulse is shaped by the pulse shaper on the right hand side of Fig. 10.12. The shaped pulses are sent into a molecular beam or into the liquid sample. The reaction products in the gas phase are detected by a mass spectrometer, and its output signal is fed into a learning algorithm which optimizes the pulse shape with regard to the optimum signal of the wanted reaction product. In the liquid phase, the laser pulse is divided into two parts which are sent into two identical sample cells. The fluorescence from the reaction products A and B is selected by filters, and the ratio $I(\mathrm{A})/I(\mathrm{B})$ of the fluorescence intensities is monitored and used as the input for the learning algorithm. For molecules which demand UV excitation, the shaped femtosecond pulse can be frequency-doubled and the UV intensity is monitored to allow for normalization of the fluorescence signals.

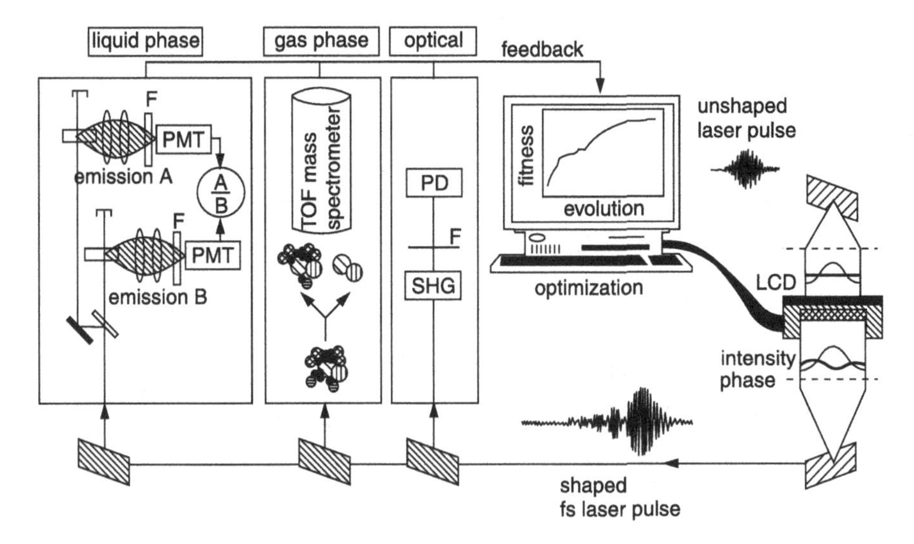

Fig. 10.12 Experimental setup for the coherent control of reactions in the gas and liquid phase [1397] (Prof. Gerber, homepage http://wep1101.physik.uni-wuerzburg.de)

The curve on the computer screen illustrates the evolution (optimization) of the wanted reaction as a function of the number of learning cycles. More examples and a detailed description of the technique can be found in [1402–1409].

10.1.5 Laser Femtosecond Chemistry

Chemical reactions are based on atomic or molecular collisions. These collisions, which may bring about chemical-bond formation or bond breaking, occur on a time scale of 10^{-11} to 10^{-13} s. In the past, the events that happened in the transition state between reagents and reaction products could not be time resolved. Only the stages "before" or "after" the reaction could be investigated [1411].

The study of chemical dynamics concerned with the ultrashort time interval when a chemical bond is formed or broken may be called *real-time femtochemistry* [1412]. It relies on ultrafast laser techniques with femtosecond time resolution [1413].

Example 10.4 Assume a dissociating molecule with a fragment velocity of 10^3 m/s. Within 0.1 ps the fragment separation changes by

$$\Delta x = \left(10^3 \times 10^{-13} = 10^{-10}\right) \text{m} = 0.1 \text{ nm}.$$

With a time resolution of 10 fs, the accuracy with which Δx can be measured is 0.01 nm = 10 pm!

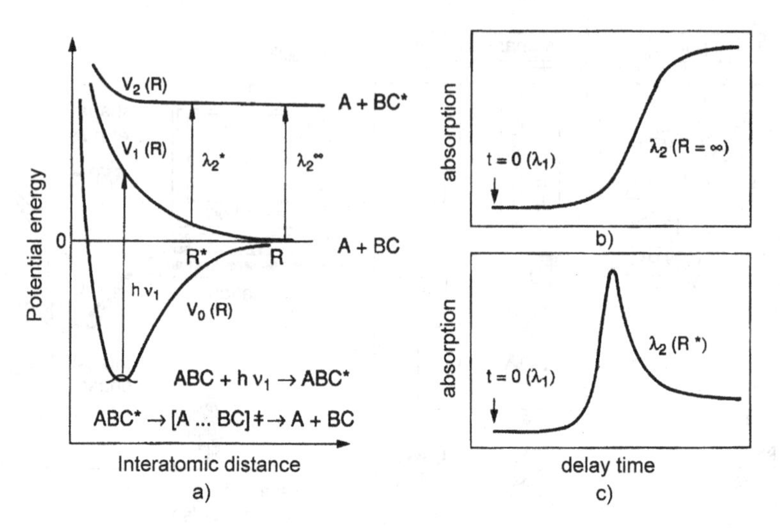

Fig. 10.13 (**a**) Potential energy curves for a bound molecule (V_0) and the first and second dissociative curves V_1, V_2; (**b**) the expected femtosecond transient signals $S(\lambda_2, t)$ versus the delay time t for $\lambda_2(R = \infty)$ and (**c**) for $\lambda_2(R^*)$ [1412]

Let us regard the photodissociation process

$$\mathrm{ABC} + h\nu \rightarrow [\mathrm{ABC}]^* \rightarrow \mathrm{A} + \mathrm{BC}, \tag{10.6}$$

which represents an unimolecular reaction. The real-time spectroscopy of bond breaking is explained by the potential diagram of Fig. 10.13.

The molecule ABC is excited by a pump photon $h\nu_1$ into the dissociating state with a potential curve $V_1(R)$. A second probe laser pulse with a tunable wavelength λ_2 is applied with the time delay Δt. If λ_2 is tuned to a value that matches the potential energy difference $h\nu = V_2(R) - V_1(R)$ at a selected distance R between A and the center of BC, the probe radiation absorption $\alpha(\lambda_2, \Delta t)$ shows a time dependence, as schematically depicted in Fig. 10.13b. When λ_2 is tuned to the transition $\mathrm{BC} \rightarrow (\mathrm{BC})^* = V_2(R = \infty) - V_1(R = \infty)$ of the completely separated fragment BC, the curve in Fig. 10.31c is expected. These signals yield the velocity $v(R)$ of the dissociation products from which the energy difference $V_2(R) - V_1(R)$ can be derived.

The experimental arrangement for such femtosecond experiments is exhibited in Fig. 10.14. The output pulses from a femtosecond pulse laser (Sect. 6.1.5) are focused by the same lens into the molecular beam. The probe pulses are sent through a variable optical-delay line and the absorption $\alpha(\Delta t)$ of the probe pulse as a function of the delay time Δt is monitored via the laser-induced fluorescence. Cutoff filters suppress scattered laser light.

Another example is the real-time observation of ultrafast ionization and fragmentation of mercury clusters $(\mathrm{Hg})_n$ $(n \leq 110)$ with femtosecond laser pulses. In pump–probe experiments short-time oscillatory modulations of the transient Hg_n^+

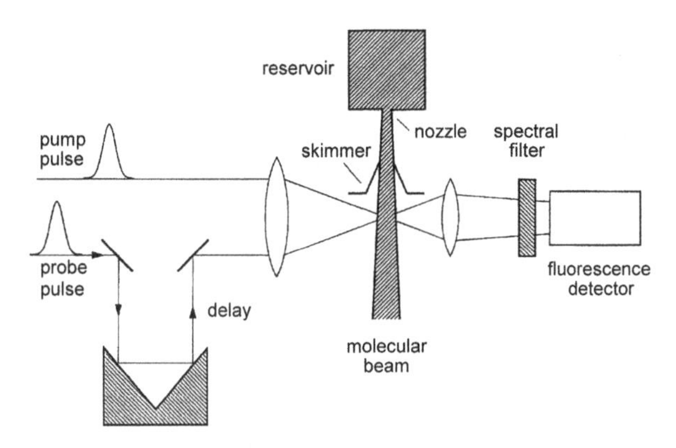

Fig. 10.14 Experimental arrangement for femtosecond spectroscopy in a molecular beam

and Hg_n^{++} signals indicate an intermediate-state dynamics common to all cluster sizes [1414].

Photoinduced reactions in the liquid phase are influenced much more by collisions than those in the gas phase. In order to study such reactions on a time scale below the mean collision time, femtosecond spectroscopy is needed. One example is the exploration of transition-state dynamics and rotational dynamics of the fragments in the photolysis of mercuric iodide HgI_2 in ethanol solution [1415, 1416].

Another example is the detailed investigation of the femtosecond dynamics of iron carbonyl $Fe(CO)_5$ [1417], where the photodissociation after excitation with 267 nm pulses was studied by transient ionization. Five consecutive processes with time constants 21, 15, 30, 47, and 3300 fs were found. The first four short-time processes represent ionization from different excited configurations, which are reached by a pathway from the initially excited Franck–Condon region to other configurations through a chain of Jahn–Teller-induced conical intersections. The experiments also showed that intersystem crossing to the triplet ground states of $Fe(CO)_4$ and $Fe(CO)_3$ takes more than 500 ps. More detailed information on experiments on laser femtosecond chemistry can be found in [1418, 1419].

10.1.6 Isotope Separation with Lasers

The classical methods of isotope separation on a large technical scale, such as the thermal diffusion or the gas-centrifuge techniques, are expensive because they require costly equipment or consume much energy [1422]. Although the largest impetus for the development of efficient new methods for isotope separation was provided by the need for uranium ^{235}U separation, increasing demands exist for the use of isotopes in medicine, biology, geology, and hydrology. Independent of the future of nuclear reactors, it is therefore worthwhile to think about new, efficient techniques of isotope separation on a medium scale. Some of the novel techniques, which are based on a combination of isotope-selective excitation by lasers with subsequent photochemical reactions, represent low-cost processes that have already been

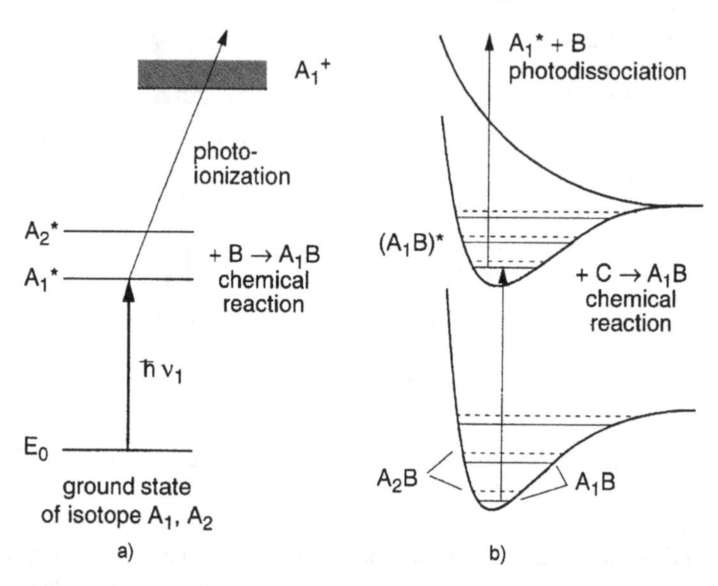

Fig. 10.15 Different possible ways of isotope separation following selective excitation of the wanted isotope: (**a**) photoionization; and (**b**) dissociation or excitation of predissociating molecular states

proved feasible in laboratory experiments. Their extension to an industrial scale, however, demands still more efforts and improvements [1423–1428].

Most methods of laser isotope separation are based on the selective excitation of the desired atomic or molecular isotope in the gas phase. Some possible ways for separating the excited species are depicted schematically in Fig. 10.15, where A and B may be atoms or molecules, such as radicals. If the selectively excited isotope A_1 is irradiated by a second photon during the lifetime of the excited state, photoionization or photodissociation may take place if

$$E_0 + h\nu_1 + h\nu_2 > E(\mathrm{A}^+), \quad \text{or} \quad > E_{\text{Diss}}. \tag{10.7}$$

The ions can be separated from the neutrals by electric fields, which collect them into a Faraday cup. This technique has been used, for example, for the separation of ^{235}U atoms in the gas phase by resonant two-photon ionization with copper-vapor laser-pumped dye lasers at high repetition frequencies [1429]. Since the line density in the visible absorption spectrum of ^{235}U is very high, the lasers are crossed perpendicularly with a cold collimated beam of uranium atoms in order to reduce the line density and the absorption linewidth.

In the case of *molecular* isotopes, the absorption of the second photon may also lead to photodissociation. The fragments R are often more reactive than the parent molecules and may react with properly added scavenger reactants S to form new compounds RS, which can be separated by chemical means.

In favorable cases no second photon is necessary if reactants can be found that react with the excited isotopes M* with a much larger probability than with the

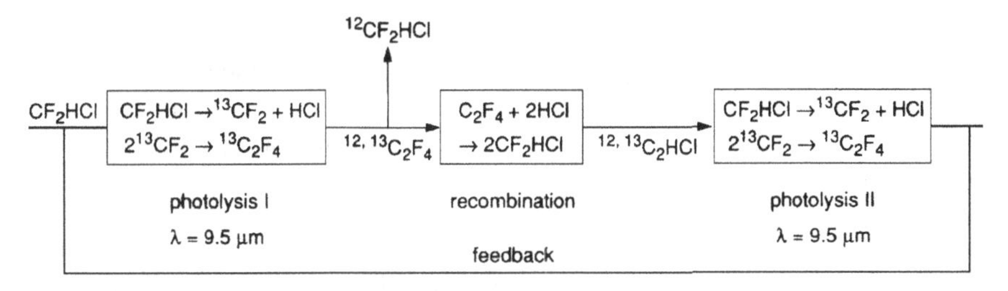

Fig. 10.16 Isotope enrichment of $^{13}CF_2$ by a cyclic process following isotope-enhanced multiphoton dissociation of Freon CF_2HCl [1433]

ground-state molecules M. An example of such a chemical separation of laser-excited isotopes is the reaction

$$\begin{aligned} I^{37}Cl + h\nu &\to \left(I^{37}Cl^*\right), \\ \left(I^{37}Cl\right)^* + C_6H_5Br &\to {}^{37}ClC_6H_5Br + I, \\ {}^{37}ClC_6H_5Br &\to C_6H_5{}^{37}Cl + Br. \end{aligned} \tag{10.8}$$

The isotope $I^{37}Cl$ can be selectively excited at $\lambda = 605$ nm by a cw dye laser. The excited molecules react in collisions with bromine benzene and form the unstable radical $^{37}ClC_6H_5Br$, which dissociates rapidly into $C_6H_5{}^{37}Cl + Br$. In a laboratory experiment, several milligrams of C_6H_5Cl were produced during a two hour exposure. Enrichment factors $K = {}^{37}Cl/{}^{35}Cl$ of $K = 6$ have been achieved [1426].

Radioactive isotopes and isotopes with nuclear spin $I \neq 0$ play an important role in medical diagnostics. The radioactive technetium isotope Tc, for example, is now used instead of iodine ^{137}I for thyroid gland diagnostics and treatment because it has a shorter decay time and therefore the diagnosis can be performed with a smaller dose. The carbon isotope ^{13}C is employed, in addition to hydrogen ^1H, in NMR tomography for monitoring the brain, or to follow up the metabolism and its possible anomalies. The isotope ^{13}C can be separated by isotope-selective excitation of formaldehyde $^{13}CH_2O$ by an UV laser into predissociating states [1430], or by multiphoton dissociation of Freon

$$CF_2HCl + n \cdot h\nu \to CF_2 + HCl, \tag{10.9}$$

with a CO_2 laser, which leads to an enrichment of $^{13}CF_2$ [1432]. The efficiency of the process (10.9) has been improved by enhancing collisions between the fragments

$$\begin{aligned} {}^{13}CF_2 + {}^{13}CF_2 &\to {}^{13}C_2F_4, \\ {}^{13}C_2F_4HCL &\to {}^{13}CF_2HCl + {}^{13}CF_2, \end{aligned}$$

which restores the parent molecule, now isotopically enriched (Fig. 10.16). This allows a continuous repetition of the enrichment cycle [1433].

Even multiphoton dissociation of larger molecules such as SF_6 by CO_2 lasers may be isotope selective [1424]. For the heavier molecule UF_6 the necessary selectivity of the first step, excited at $\lambda = 16\,\mu m$, can only be reached in a collimated cold UF_6 beam. The vibrationally excited UF_6 isotope can then be ionized by a XeCl excimer laser at $\lambda = 308$ nm [1427a]. The absolute amount of isotopes separated in this way is, however, very small [1427b].

The last examples illustrate that the most effective method of isotope separation is a combination of isotope-selective excitation with selective chemical reactions. The laser plays the role of an isotope-selective initiator of the chemical reactions [1434].

10.1.7 Summary of Laser Chemistry

The main advantages of laser applications in chemistry may be summarized as follows:

- Laser spectroscopy offers different techniques of increased sensitivity for the detection of tiny concentrations of impurities, pollutant gases, or rare isotopes down to the ultimate level of single-molecule detection.
- The combination of spectral and time resolutions allows detailed investigations of transition states and intermediate reaction transients. Femtosecond spectroscopy gives direct "real-time" information on the formation or breaking of chemical bonds during the collision process.
- Selective excitation of reaction products results (in favorable cases) in an enhancement of wanted reaction channels. It often turns out to be superior to temperature-enhanced nonselective reaction rates, particularly when coherent control techniques are used [1434].
- The preparation of reagents in selected states and the study of the internal-state distribution of the reaction products by LIF or REMPI brings complete information on the "state-to-state" molecular dynamics [1411].

More aspects of laser chemistry and many more examples of applications of laser spectroscopy in chemistry can be found in [1358–1366, 1434–1447].

10.2 Environmental Research with Lasers

A detailed understanding of our environment, such as the earth's atmosphere, the water resources, and the soil, is of fundamental importance for mankind. Since in densely populated industrial areas air and water pollution has become a serious problem, the study of pollutants and their reactions with natural components of our environment is urgently needed [1448]. Various techniques of laser spectroscopy have been successfully employed in atmospheric and environmental research: direct absorption measurements, laser-induced fluorescence techniques, photoacoustic detection, spontaneous Raman scattering and CARS (Chap. 3), resonant two-photon ionization, and many more of the sensitive detection techniques discussed in Chap. 1 can be applied to various environmental problems. This section illustrates the potential of laser spectroscopy in this field by some examples.

10.2.1 Absorption Measurements

The concentration N_i of atomic or molecular pollutants in the lower part of our atmosphere just above the ground can be determined by measurements of the direct absorption of a laser beam propagating through the atmosphere. The detector receives, at a distance L from the source, the fraction

$$P(L)/P_0 = \mathrm{e}^{-a(\omega)L}, \tag{10.10}$$

of the emitted laser power P_0. The attenuation coefficient

$$a(\omega) = \alpha(\omega) + S = N_i \sigma_i(\omega, p, T) + \sum_k N_k \sigma_k^{\text{scat}}, \tag{10.11}$$

is represented by the sum of the absorption coefficient $\alpha(\omega) = N_i \sigma_i^{\text{abs}}$ (which equals the product of density N_i of absorbing molecules in level $|i\rangle$ and the absorption cross section σ_i^{abs}) and the scattering coefficient $S = \sum N_k \sigma_k^{\text{scat}}$ (which is due to light scattering by all particles present in the atmosphere). The dominant contribution to scattering is the Mie scattering [1449] by small particles (dust, water droplets), and only a minor part is due to Rayleigh scattering by atoms and molecules.

The absorption coefficient $\alpha(\omega)$ assumes nonnegligible values only within the small spectral ranges $\Delta\omega$ of absorption lines (a few gigahertz around the center frequencies ω_0). On the other hand, the scattering cross section, which is proportional to ω^4 for Rayleigh scattering, does not vary appreciably over these small ranges $\Delta\omega$. Therefore, measurements of the laser beam attenuation at the two different frequencies ω_1 and ω_2 inside and just outside an absorption profile yields the ratio

$$\frac{P(\omega_1, L)}{P(\omega_2, L)} = \mathrm{e}^{-[a(\omega_1) - a(\omega_2)]L} \simeq \mathrm{e}^{-N_i[\sigma_i(\omega_1) - \sigma_i(\omega_2)]L}, \tag{10.12}$$

of two transmitted laser powers, from which the wanted concentration N_i of the absorbers can be derived if the absorption cross sections σ_i are known. A possible experimental realization is illustrated in Fig. 10.17. The laser beam, enlarged by a telescope, reaches a cats-eye reflector at the distance $L/2$, which reflects the beam exactly back into itself. The reflected beam is sent to the detector by the beam splitter BS. For larger distances L, beam deviations from spatially inhomogeneous fluctuations of the refractive index of the air impose a severe problem. It may be partly solved by several measures: the switching cycle from ω_1 to ω_2 is performed in a statistical sequence, the switching frequency is chosen as high as possible, and

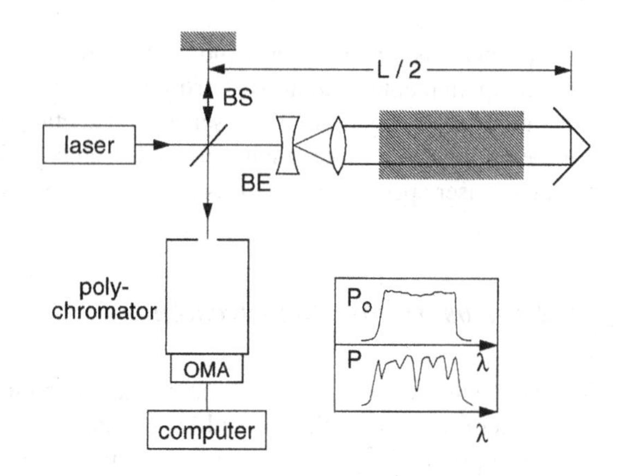

Fig. 10.17 Schematic diagram of an experimental setup for measuring the density of absorbing species integrated over the path length L. The polychromator with optical multichannel analyzer allows the simultaneous determination of several absorbing components

the areas of the retroreflector and the detector are made so large that the detector still receives the full beam in spite of small beam deviations [1450].

For such absorption measurements infrared lasers can be used which coincide with vibrational-rotational transitions of the investigated molecules (CO_2 laser, CO laser, HF or DF lasers, etc.). Particularly useful are tunable infrared lasers (diode lasers, color-center lasers, or optical parametric oscillators, Vol. 1, Sect. 5.7), which may be tuned to selected transitions. The usefulness of diode lasers has been demonstrated by many examples [1451]. For instance, with a recently-developed automated diode laser spectrometer, which is tuned by computer control through the spectral intervals of interest, up to five atmospheric trace gases can be monitored in unattended operation. Sensitivities down to 50 ppt for NO_2 and 300 ppt for NO have been reported [1371].

The infrared lasers have the advantage that the contribution of scattering losses to the total beam attenuation is much smaller than in the visible range. For measurements of very low concentrations, on the other hand, visible dye lasers may be more advantageous because of the larger absorption cross sections for electronic transitions and the higher detector sensitivity.

Often a broadband laser (for example, a pulsed dye laser without etalons), or a multiline laser (for example, a CO_2 or CO laser without grating) may simultaneously cover several absorption lines of different molecules. In such cases the reflected beam is sent to a polychromator with a diode array or an optical multichannel analyzer (OMA, Vol. 1, Sect. 4.5). If a fraction of the laser power $P_0(\omega)$ is imaged onto the upper part of the OMA detector and the transmitted power onto the lower part (insert in Fig. 10.17), electronic difference and ratio recording allows the simultaneous determination of the concentrations N_i of all absorbing species. A retroreflector arrangement is feasible for measurements at low altitudes above ground, where buildings or chimneys can support the construction. Examples are measurements of fluorine concentrations in an aluminum plant [1452], or the detection of different constituents in the chimney emission of power plants, such as NO_x and SO_x components [1453]. Often, ammonia is added to the exhaust of power stations in order to reduce the amount of NO_x emission. In such cases, the optimum concentration of NH_3 has to be controlled in situ. A recently developed detection system for these purposes has demonstrated its sensitivity and reliability [1454].

Measurements of relative isotope abundances in atmospheric gases can provide very useful information on chemical reactions in the atmosphere. For instance, the stable isotope content of atmospheric CO_2 provides information about the ecosystems carbon–water and biosphere–atmosphere and the carbon exchange between these subsystems. In many cases samples of air are collected and analysed in the lab. This is, however, a more tedious and not momentary analysis. Absorption spectroscopy with a tunable diode laser allows in situ measurements on vibrational transitions of all relevant molecules in the atmosphere and can measure the isotope ratios with high accuracy.

In many cases samples of the air with its spurious pollutant molecules are taken and measured in an absorption cell. Here tunable diode lasers, which can be tuned over the vibrational bands of the molecules, have proved to be very useful. A review on recent work in this field can be found in [1455].

For measurements over larger distances or in higher altitudes of the atmosphere, this absorption measurement with a retroreflector cannot be used. Here the LIDAR system, which is discussed in Sect. 10.2.2, has proven to be the best choice.

10.2.2 Atmospheric Measurements with LIDAR

The principle of LIDAR (light detection and ranging) is illustrated by Fig. 10.18. A short laser pulse $P_0(\lambda)$ is sent at time $t = 0$ through an expanding telescope into the atmosphere. A small fraction of $P_0(\lambda)$ is scattered back into the telescope due to Mie scattering by droplets and dust particles, and Rayleigh scattering by the atmospheric molecules. This backscattered light yields a photomultiplier signal $S(\lambda, t)$, which is measured with spectral and time resolution. The signal at time $t_1 = 2R/c$ depends on the scattering by particles at the distance R. If the detector is gated during the time interval $t_1 \pm \frac{1}{2}\Delta t$, the time-integrated measured signal

$$S(\lambda, t_1) = \int_{t_1-\frac{\Delta t}{2}}^{t_1+\frac{\Delta t}{2}} S(\lambda, t)\, \mathrm{d}t,$$

is proportional to the light power scattered by particles within a distance $R \pm \frac{1}{2}\Delta R = \frac{1}{2}c(t_1 \pm \frac{1}{2}\Delta t)$. The magnitude of the received signal $S(\lambda, t)$ depends on the attenuation of the emitted power $P_0(\lambda)$ on its way back and forth, on the solid angle $\mathrm{d}\Omega = D^2/R^2$ covered by the telescope with the diameter D, and on the concentration N and the backscattering cross section σ^{scatt} of the scattering particles:

$$S(\lambda, t) = P_0(\lambda)\mathrm{e}^{-2a(\lambda)R} N\sigma^{\text{scatt}}(\lambda) D^2/R^2. \tag{10.13}$$

The quantity of interest is the factor $\exp[-2a(\lambda)R]$, where a is the sum of absorption and scattering coefficients, which gives with (10.11) the necessary information on the concentration of absorbing species. Similar to the method described in the

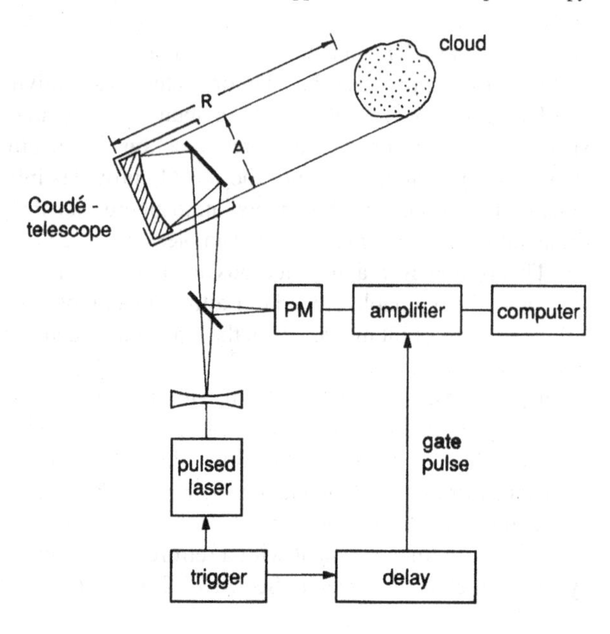

Fig. 10.18 Schematic diagram of a LIDAR system

previous section, the laser wavelength λ is tuned alternately to an absorption line at λ_1 and to a wavelength λ_2 where absorption by the wanted molecules is negligible. For sufficiently small values of $\Delta\lambda = \lambda_1 - \lambda_2$ the variation of the scattering cross section can be neglected. The ratio

$$\begin{aligned} Q(t) = \frac{S(\lambda_1, t)}{S(\lambda_2, t)} &= \exp\left\{2\int_0^R \left[\alpha(\lambda_2) - \alpha(\lambda_1)\right] \mathrm{d}R\right\} \\ &\simeq \exp\left\{2\int_0^R N_i(R)\sigma(\lambda_1)\,\mathrm{d}R\right\}, \end{aligned} \tag{10.14}$$

yields the concentration $N_i(R)$ integrated over the total absorption path length. The dependence $N_i(R)$ on the distance R can be measured with a differential technique: the sequence of the quantities $S(\lambda_1, t)$, $S(\lambda_2, t)$ $S(\lambda_1, t + \Delta t)$, and $S(\lambda_2, t + \Delta t)$ is measured alternately. The ratio

$$\frac{Q(t + \Delta t)}{Q(t)} = \mathrm{e}^{-[\alpha(\lambda_2) - \alpha(\lambda_1)]\Delta R} \simeq 1 - \left[\alpha(\lambda_2) - \alpha(\lambda_1)\right]\Delta R,$$

yields the absorption within the spatial interval $(R + \Delta R) - R = c\Delta t/2$. If the absorption coefficient under the atmospheric conditions (pressure, temperature) within the detection volume $\Delta V = \Delta R \cdot A$ (A is the area of the laser beam at the distance R) is known, the concentration $N_i = \alpha_i/\sigma_i$ of the wanted species can be determined. The lower limit $\Delta R = c\Delta t/2$ is given by the time resolution Δt of the LIDAR system and by the attainable signal-to-noise ratio. This differential-absorption LIDAR *(DIAL) method* represents a very sensitive technique for atmospheric research.

Fig. 10.19 LIDAR-system with locally separated detector telescopes

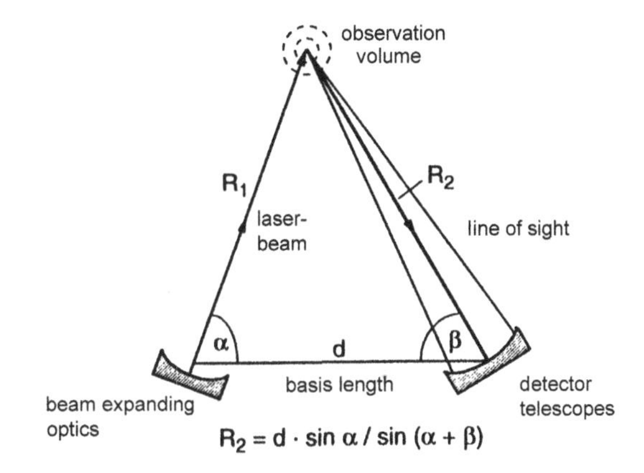

In another arrangement the emitter telescope and the receiving telescope are located at different places with a distance d (Fig. 10.19). The knowledge of d and the angles α and β allows the determination of the distances R_1 and R_2 to the inspected volume.

In this way a complete *air-pollution map* of industrial and urban areas can be recorded and pollution sources can be localized. With pulsed dye lasers NO_2 concentrations in the parts-per-million (ppm) range at distances up to 5 km can be monitored [1457]. Recent developments of improved LIDAR systems with frequency-doubled lasers have greatly increased the sensitivity as well as the spatial and spectral ranges that can be covered [1458–1460].

A further example of the application of the LIDAR technique is the measurement of the atmospheric *ozone* concentration, and its daily and annual variations as a function of altitude and latitude [1461]. In order to reach reliable and stable laser operation, with reproducible wavelength switching even under unfavorable conditions (in an aircraft or on a ship), a XeCl excimer laser at $\lambda_1 = 308$ nm was used instead of a dye laser, and the second wavelength $\lambda_2 = 353$ nm was produced by Raman shifting in a hydrogen high-pressure cell. While radiation at λ_1 is strongly absorbed by O_3 the absorption of 353 nm is negligible [1462, 1463a]. However, one has to be sure that the absorption by other gaseous components is zero for both wavelengths, or at least equal. Otherwise, serious errors can arise [1463b]. Besides the ozone layer at altitudes in the range 30–60 km, the ozone concentration just above ground is of vital interest. Here LIDAR measurements are able to provide a complete map of ozone concentrations in urban and rural areas and to find the different reactions that lead to ozone production and destruction as well as the sources of the reactants [1464].

A third example of the usefulness of differential LIDAR is given by the spectroscopic determination of daily and annual variations of the temperature profile $T(h)$ of the atmosphere as a function of the height h above ground. The everywhere-present Na atoms can be used as tracer atoms since the Doppler width of the Na-D line, which is measured with a pulsed, narrow-band tunable dye laser, is a measure of the temperature [1465, 1466].

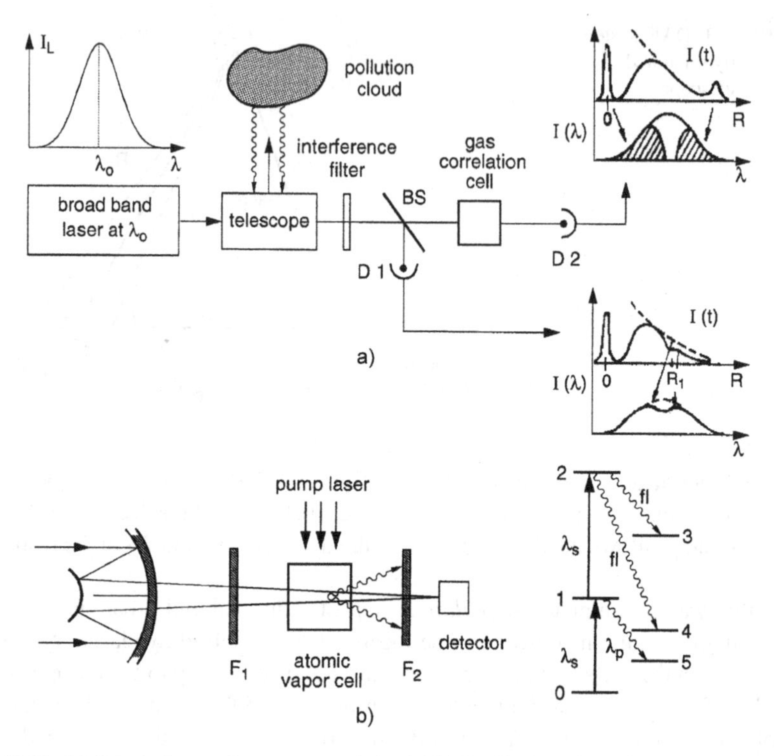

Fig. 10.20 (**a**) Principal setup for cross-correlation LIDAR with time-dependent signals and spectral distribution as detected by D1 and D2 [1470]; (**b**) optically active atomic filter [1471]

In the higher atmosphere, the aerosol concentration decreases rapidly with increasing altitudes. Mie scattering therefore becomes less effective and other techniques have to be used for the measurements of concentration profiles $N(h)$. Either the fluorescence induced by UV lasers or Raman scattering can provide the wanted signals. Fluorescence detection is sufficiently sensitive only if quenching collisions are not the dominant deactivation process for the excited levels. This means that either the radiative lifetime τ^{rad} of the excited levels must be sufficiently short or the pressure $p(h)$ should be low, that is, the altitude h high. If quenching collisions are not negligible, the effective lifetimes and the quenching cross sections have to be known in order to derive quantitative values of the concentration profiles $N_i(h)$ from measured fluorescence intensities. This difficulty may be overcome by Raman spectroscopy, which, however, has the disadvantage of smaller scattering cross sections [1467–1469].

For daytime measurements the bright, continuous background of sunlight scattered in the atmosphere limits the attainable signal-to-noise ratio. With a narrow spectral filter, which is matched to the detected fluorescence wavelength, or in the case of LIDAR to the laser wavelength, the background can be substantially suppressed. An elegant technical trick of LIDAR measurements is based on a cross-correlation method (Fig. 10.20). The back-scattered light is split into two parts: one part is reflected by the beam splitter BS onto a detector D1, while the transmitted part is sent through an absorption cell of length ℓ, which contains the investigated molecular components under comparable conditions of total pressure and temperature as in the atmosphere. Their partial density N_i is, however, chosen so high that at the center of an absorption line $\alpha(\omega_0)\ell \gg 1$. The laser bandwidth is set to be slightly larger than the linewidth of the absorbing transition. The amplification of the two amplifiers following D1 and D2 is adjusted to obtain the difference signal $S_1 - S_2 = 0$ for a wanted concentration N_i. Any deviation of $N_i(R)$ from this balanced value is monitored behind the difference amplifier as the signal $\Delta S(N_i)$, which is essentially independent of fluctuations in laser intensity and frequency, since they affect both arms of the difference detection [1470].

The cross-correlation technique represents a special case of a more general method, where the absorption lines of atomic vapors are used as narrow optical filters, matched to the special problem [1471]. This method is illustrated in Fig. 10.20b. The backscattered laser light, with wavelength λ_L, collected by the telescope is sent through a narrow spectral filter F_1 with a transmission peak at λ_L and then through an absorption cell with atomic or molecular vapor that absorbs at λ_L. The atoms or molecules excited by absorption of the photons $\hbar\omega_L$ emit fluorescence with the wavelengths $\lambda_{Fl} > \lambda_L$. This radiation is detected behind a cutoff filter, which suppresses the wavelengths $\lambda < \lambda_{Fl}$. In this way the background radiation can essentially be eliminated. These *passive atomic filters* are restricted to those wavelengths λ_L that match atomic or molecular resonance transitions starting from the thermally populated ground state. The number of possible coincidences can be greatly enlarged if transitions between *excited* states of the filter atoms or molecules can be utilized. This can be achieved with a tunable pump laser that generates a sufficiently large population density of absorbing atoms or molecules in selectively excited states (*active atomic filters*, Fig. 10.20b). A further advantage of these active filters is the fact that the fluorescence induced by the signal wave may be shifted into the shorter wavelength range. This allows the detection of infrared radiation via visible or UV fluorescence.

A new interesting approach to air-pollution measurements is based on high-power terawatt femtosecond laser pulses that are sent into the atmosphere [1473]. The experimental setup is shown in Fig. 10.21. High intensities are reached in the focal area of the parabolic sending telescope due to self-focusing, which leads to air breakdown and the production of a high-temperature plasma (Fig. 10.22). The intensity-dependent nonlinear index of refraction

$$n = n_0 + n_2 \cdot I \quad \text{with } n_2 = 3 \times 10^{-19}\ \text{cm}^2\,\text{W}^{-1} \text{ in air}$$

has its maximum value at the central axis of the Gaussian laser beam profile. This leads to a focusing of the laser beam (Fig. 10.22a). The breakdown of the air produces a plasma in the focal plane with an electron density $\rho(I)$. This plasma results in a negative change

$$\Delta n = -\rho(I)/\rho_c \quad \text{with } \rho_c = 2 \times 10^{21}\ \text{cm}^{-3}$$

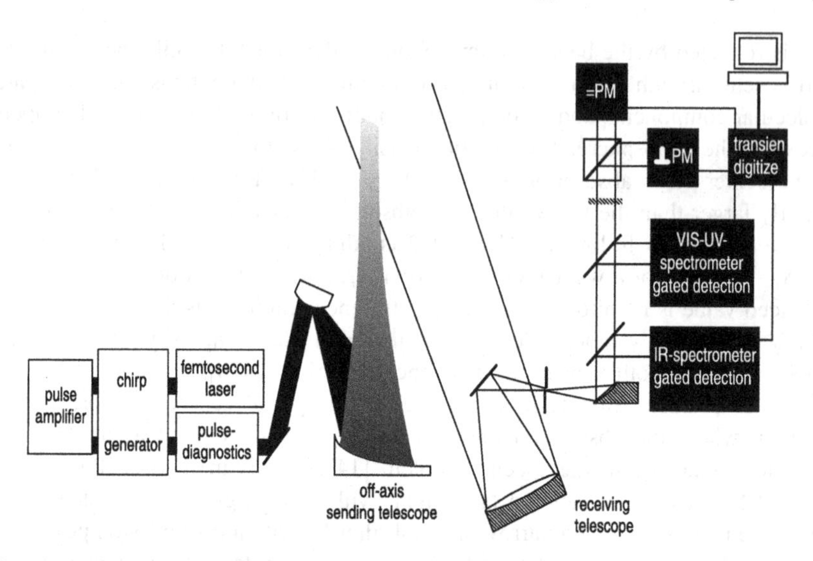

Fig. 10.21 Setup of fs-LIDAR: the laser source and sending telescope are shown on the *left hand side*, while the receiving telescope and time-resolved detection apparatus (covering a range from the UV to the IR) are depicted on the *right hand side* [1472]

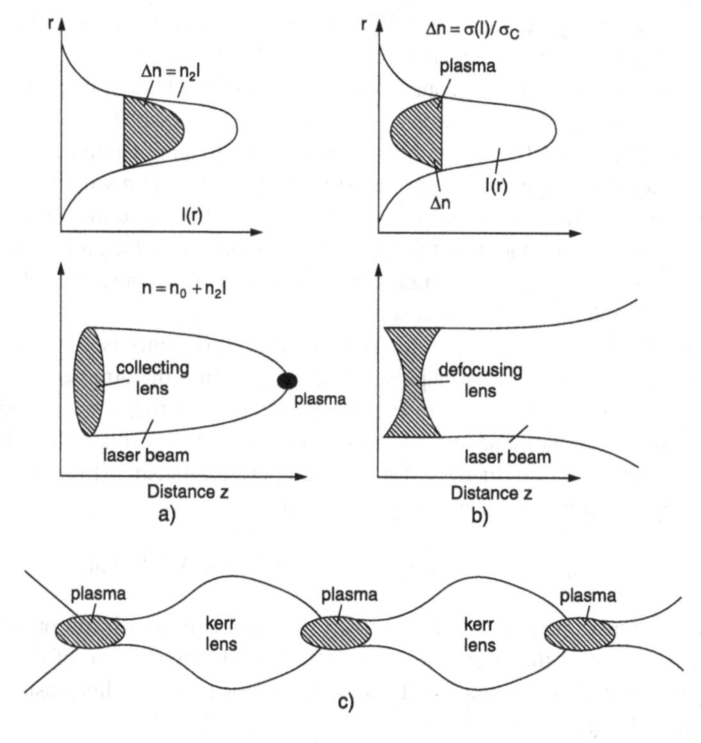

Fig. 10.22 Focusing (**a**) and defocusing (**b**) of a laser beam in the atmosphere due to the Kerr lens effect and plasma formation. The laser beam cross-section looks like a row of short sausages

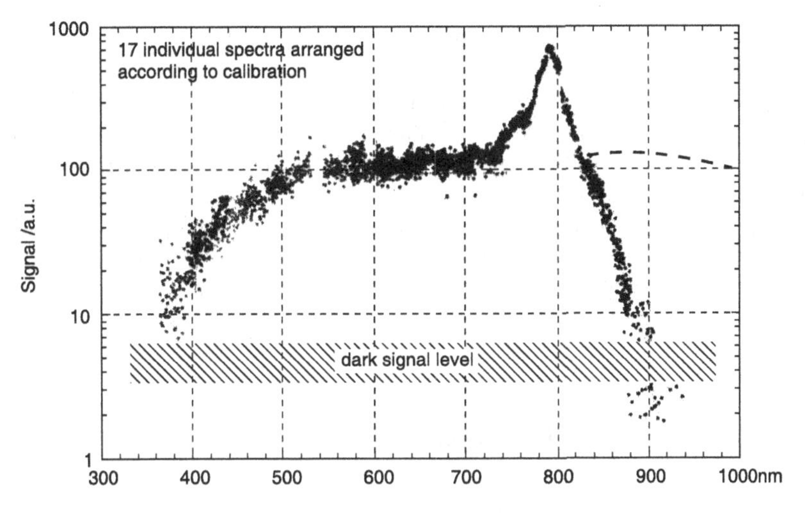

Fig. 10.23 Spectral continuum emitted by the plasma spots along the propagation path of terawatt laser pulses in the atmosphere [1472]

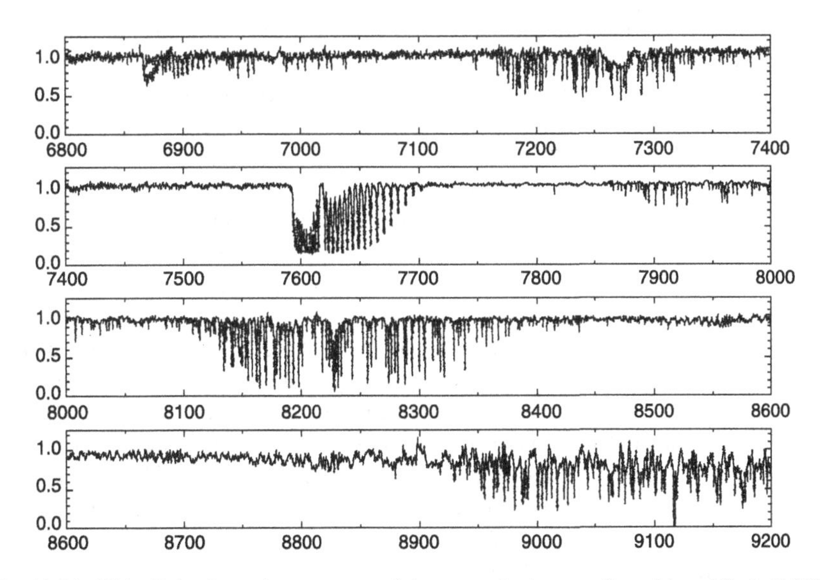

Fig. 10.24 White-light absorption spectrum of the atmospheric gases from 4 km altitude [1473]

in the refractive index which depends on the ratio of electron density ρ to the critical density ρ_c, and acts like a defocusing lens (Fig. 10.22b). Therefore, along the path of the high-power femtosecond pulses a series of plasma spots appear that look like small sausages on a straight cord (Fig. 10.22c).

These plasma spots along the laser beam can extend up to an altitude of 10 km. Because of these many bright spots, the laser beam can be seen by the naked eye kilometers away [1472]. They represent white-light sources with a continuous spectrum extending from the UV into the near infrared (Fig. 10.23). These white-light spots can be used as spectroscopic light sources, where the continuum radiation propagates through the atmosphere to the detector (Fig. 10.21), which monitors the absorption spectrum of the air constituents between sources and detector. Figure 10.24 illustrates this by showing the white-light absorption spectrum of atmospheric gases at an altitude of 4 km.

A detailed representation of the various laser-spectroscopic techniques for atmospheric research can be found in [1468, 1474]. Many examples were given in [1474–1478]. The basic physics of laser beam propagation through the atmosphere was discussed in [1479, 1480].

10.2.3 Spectroscopic Detection of Water Pollution

Unfortunately, the pollution of water by oil, gasoline, or other pollutants is increasing. Several spectroscopic techniques have been developed to measure the concentrations of specific pollutants. These techniques are not only helpful in tracing the polluting source but they can also be used to initiate measures against the pollution.

Often the absorption spectra of several pollutants overlap. It is therefore not possible to determine the specific concentrations of different pollutants from a single absorption measurement at a given wavelength λ. Either several well-selected excitation wavelengths λ_i have to be chosen (which is time consuming for in situ measurements) or time-resolved fluorescence excitation spectroscopy can be used. If the excited states of the different components have sufficiently different effective lifetimes, time-gated fluorescence detection at two or three time delays Δt_i after the excitation pulse allows a clear distinction between different components.

Often the combination of spectral and temporal resolution is helpful to simultaneously determine the individual concentrations of the different components in a mixture of pollutants, even if their absorption spectra overlap. An example is the pollution of water by different types of oil. In Fig. 10.25a the transmission curves of three different oil sorts (Diesel oil, gasoline and heavy oil) are shown and in Fig. 10.25b their emission spectra, while Fig. 10.25c shows the decay curves of the different sorts. Such measurements can help to find the polluter. In Fig. 10.26 the total fluorescence spectrum of a mixture of different aromatic hydrocarbons is shown together with the contributions of the different components, obtained by the different detection techniques discussed above [1481].

This was demonstrated by Schade [1482], who measured laser-induced fluorescence spectra of diesel oil and gasoline, excited by a nitrogen laser at $\lambda = 337.1$ nm. Time-resolved spectroscopy exhibits two different lifetimes that are determined by measuring the intensity ratio of the LIF in two different time windows (Fig. 10.25). This time-resolved spectroscopy increases the detection sensitivity for field measurements. Mineral-oil pollutions of 0.5 mg/l in water or 0.5 mg/kg in polluted soil can be detected [1482]. Using photoacoustic spectroscopy with a dye laser, concentrations down to 10^{-6}–10^{-9} mol/l of transuranium elements could be detected in

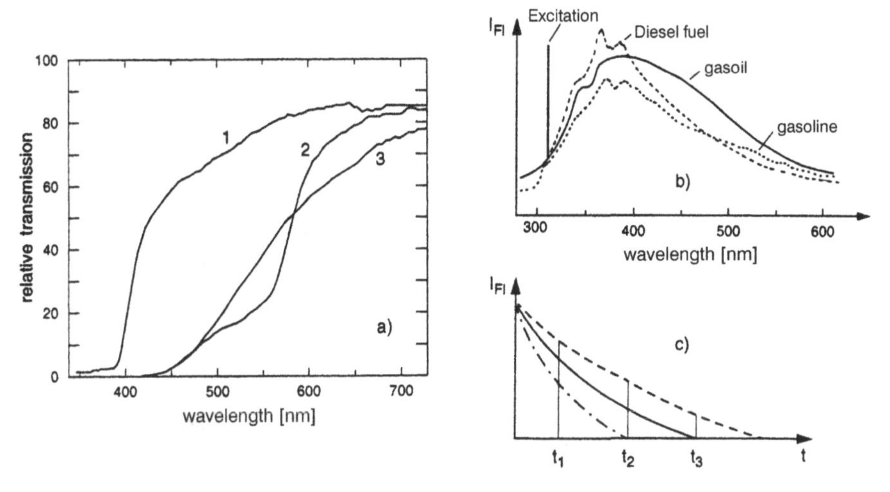

Fig. 10.25 (**a**) Transmission, (**b**) fluorescence, (**c**) time decay of fluorescence for different oil brands [1482]

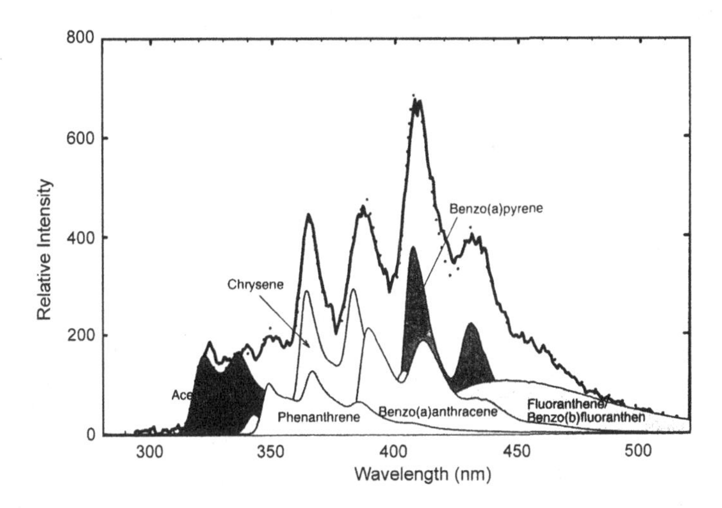

Fig. 10.26 Deconvolution of the total fluorescence into the contributions of different aromatic hydro-carbon compounds [1482]

the ground water. Such techniques are important for safety control around military test grounds and plutonium recycling factories [1483].

When a laser beam is vertically incident on the ocean surface the refraction and inclination of the reflected beam gives information on the momentary slope of the water surface at the point where the laser beam hits the surface. Rapid scanning of the laser over a certain area allows to determine the momentary amplitudes and frequencies of water waves and gives a detailled picture of the wave spectrum. Spectral resolution of the reflected light provides previously unobtainable information on the

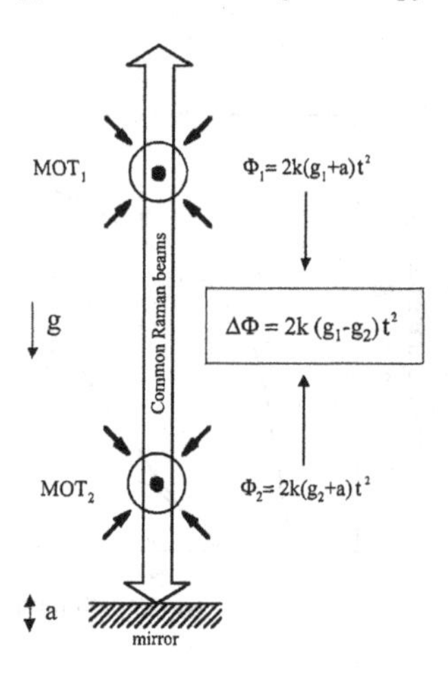

Fig. 10.27 Quantum gravity gradient gravitometer based on atomic interferometry in an atomic fountain [1487]

physico-chemical structure and processes of the ocean microlayer [1484]. Methods range from laboratory techniques to in situ ocean microlayer probes including second harmonic generation by the surface layer of the ocean.

10.2.4 Earth Science Applications

Quantum gravity gradiometers based on atomic interferometry provide high sensitivity for the high resolution mapping of mass distribution above and below the surface of the earth (see Sect. 9.1.7). They can be used in the lab as ground-based version or in a satellite as space-born system. They measure differential acceleration with high precision. Their basic principle is shown in Fig. 10.27 [1487].

Two magneto-optical traps each containing a cloud of ultracold cesium atoms are located at different heights h_1 and h_2. Detuning the frequency of the vertical laser beams which trap the atoms in the vertical directions, pushes the atoms upwards (see Sect. 9.1.7) forming an atomic fountain. Atomic interferometry on the atoms in the fountain proceeds in the following way (Fig. 10.28). A sequence of $\pi/2$–π–$\pi/2$ pulses traveling in the $\pm$ z-directions induce stimulated Raman-transitions between the two ground state hyperfine level of the Cs-atoms (Fig. 9.68). The first $\pi/2$-pulse at time t_1 creates an equal superposition of populations in the two hyperfine levels. Only atoms in the excited level can absorb the laser and experience a recoil momentum (Figs. 9.71 and 10.28) and therefore travel with slightly different velocities realizing the analogue to the beam-splitting in the Mach–Zehnder interferometer (Fig. 9.70). The subsequent π-pulse at time $t_2 = t_1 + \Delta t$ converts the populations giving a recoil momentum to atoms in the lower level while the $\pi/2$-pulse at $t_3 = t_1 + 2\Delta t$ leaves all atoms with the same momentum $p' = p + \hbar(k_1 - k_2) \approx p + 2 \cdot \hbar k_1$. The last two steps are the analogue to the redirection and recombination of the atom waves in the Mach–Zehnder interferometer

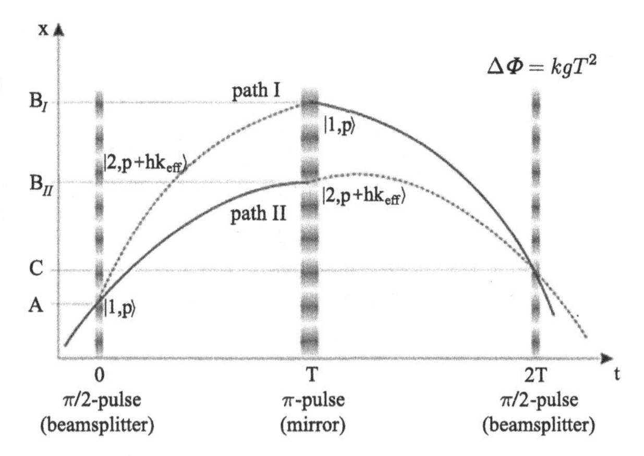

Fig. 10.28 Pulse sequence and aquired atomic momentum due to recoil. The atoms in different hyperfine levels travel different paths because only atoms in a selected hf-level suffer recoil [1487b]

(Fig. 9.70). Because of these different momentum transfers there exists a phase difference $\Delta\Phi = \Phi(t_1, z_1) - 2\Phi(t_2, z_2) + \Phi(t_3, z_3)$ between the atoms with the level population sequence $\langle 1|-\langle 2|-\langle 1|$ and those with the sequence $\langle 2|-\langle 1|-\langle 1|$. It can be shown, that $\Delta\Phi$ is related to the acceleration **a** by

$$\Delta\Phi = 2\mathbf{k}_1 \cdot \mathbf{a} \cdot \Delta t^2,$$

where the acceleration $a(z)$ in a lab experiment is equal to the gravitational acceleration $g(z)$. With an interogation time $\Delta t = 1$ s an acceleration difference $\Delta g = g(z_1) - g(z_2) = 10^{-7}$ g causes already a phase shift of 2π. If two atomic clouds in MOT_1 and MOT_2 are placed at different heights h_1 and h_2 the earth acceleration has slightly different values g_1 and g_2. The measured phase difference is then

$$\Delta\Phi = 2k(g_1 - g_2)t^2$$

This interferometric technique therefore allows the determination of changes $\Delta g = 10^{-8}$ g which are caused by changes of the mass distribution, for instance due to metal deposits in the earth crust.

10.3 Applications to Technical Problems

Although the main application range of laser spectroscopy is related to basic research in various fields of physics, chemistry, biology, and medicine, there are quite a few interesting technical problems where laser spectroscopy offers elegant solutions. Examples include investigations and optimization of flames and combustion processes in fossil power stations, in car engines, or in steel plants; analytical spectroscopy of surfaces or liquid alloys for the production of high-purity solids; or measurements of flow velocities and turbulences in aerodynamics and hydrodynamic problems.

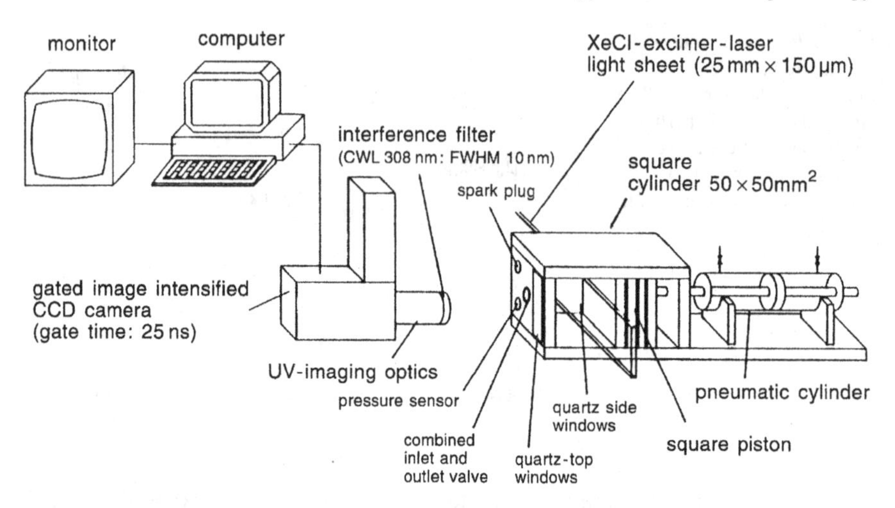

Fig. 10.29 Experimental setup for a two-dimensional analysis of combustion processes by measuring the spatial distribution of OH radicals by LIF [1488]

10.3.1 Spectroscopy of Combustion Processes

A detailed knowledge of the chemical reactions and gas-dynamical processes occurring during the combustion process is indispensible for the optimization of the thermodynamic efficiency and the minimization of pollutants. Spatially and time-resolved spectroscopy allow measurements of the concentrations of different reaction products during the combustion, which can lead to a detailed understanding of the different stages during its development and their dependence on temperature, pressure, and geometry of the combustion chamber.

The technical realization uses a one-, two-, or even three-dimensional grid of laser beams that pass through the combustion chamber. If the laser wavelength is tuned to absorption lines of atoms, molecules, or radicals, the spatial distribution of the LIF can be monitored with a video camera. With pulsed lasers and appropriately chosen gates in the electronic detection system, time-resolved spectroscopy is possible. The spatial distribution of the investigated reaction products during selected time intervals after the ignition can then be observed on a monitor. This gives direct information of the flame-front development, which can be followed on the screen in slow motion mode.

In many combustion processes OH radicals are produced as an intermediate product. These radicals can be excited by a XeCl laser at 308 nm. The UV fluorescence of OH can be discriminated against the bright background of the flame by interference filters. A possible experimental setup is depicted in Fig. 10.29. The beam of the XeCl laser is imaged into the combustion chamber in such a way that it forms a cross section of $0.15 \times 25\ \mathrm{mm}^2$. A CCD camera (Vol. 1, Sect. 4.5) with UV optics and a gate time of 25 ns monitors the OH fluorescence with spatial and time resolution [1488].

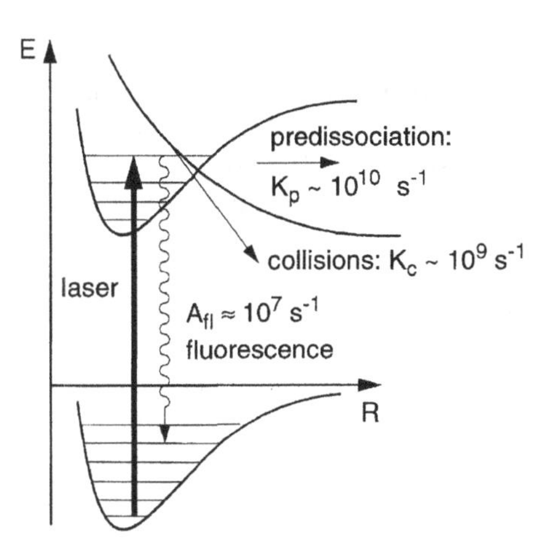

Fig. 10.30 LIF spectroscopy of predissociating levels, where the predissociating rate is fast compared to the collisional quenching rates (A_{Fl}, fluorescence rate per molecule; k_p, predissociating rate; k_c, collisional quenching rate)

For quantitative measurements of molecular concentrations by LIF the ratio A_i/R_i of radiative and radiationless deactivation probabilities of the excited level $|i\rangle$ must be known. At high pressures the collisional quenching of $|i\rangle$ becomes important, which may change the ratio A_i/R_i considerably during the combustion process. However, if the laser excites predissociating levels with very short effective lifetimes (Fig. 10.30) the predissociation rate generally far exceeds the collisional quenching rate. Although the fluorescence rate is weakened, since most molecules predissociate before they emit a fluorescence photon, the fluorescence efficiency is not much influenced by collisions [1489a]. With tunable excimer lasers such predissociating levels can be excited for most radicals that are relevant in combustion

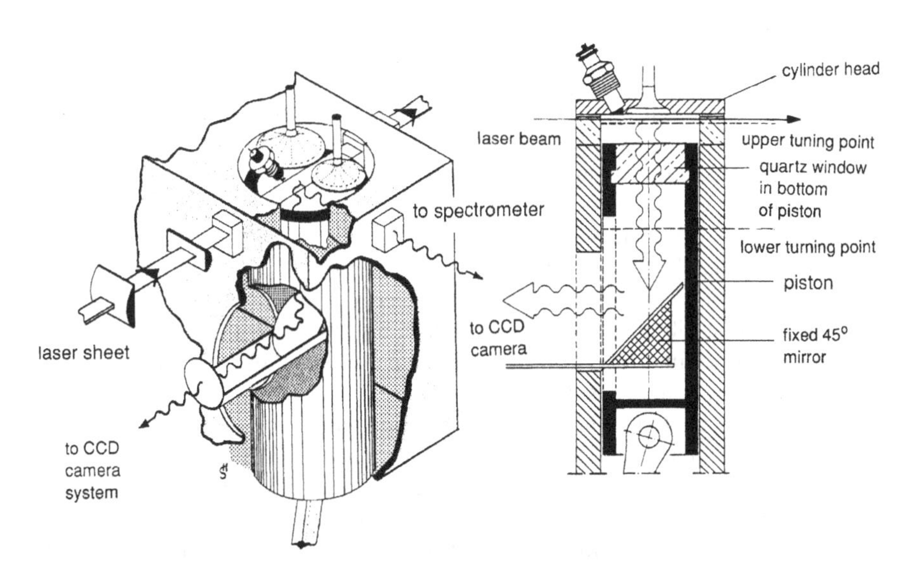

Fig. 10.31 Measurements of radical concentrations during combustion in a slightly modified car engine with windows for laser entrance and fluorescence exit [1490]

processes. The high intensity of excimer lasers has the additional advantage that the absorbing transition can be saturated. The LIF intensity is then independent of the absorption probability, but depends only on the concentration of the absorbing species [1489b]. The experimental setup for measuring the concentrations of OH, NO, CO, DH, etc. radicals during the combustion in a car engine (Otto motor) is illustrated in Fig. 10.31, where the LIF is reflected by a mirror on the moving piston through an exit window onto the CCD camera [1490].

With picosecond or femtosecond lasers the excitation efficiency and the saturation of the absorbing level is nearly independent of collisions. At a pressure of 1 bar the mean time between inelastic collisions is about 10^{-9}–10^{-10} s, which is long compared to the laser pulse width.

The spatial temperature variation in flames and combustions can also be determined by CARS (Sect. 3.3), which yields the population distribution of rotational–vibrational levels over the combustion region [1491, 1492]. A computer transforms the spatially resolved CARS signal into a colorful temperature profile on the screen.

10.3.2 Applications of Laser Spectroscopy to Materials Science

For the production of materials for electronic circuits, such as chips, the demands regarding purity of materials, their composition, and the quality of the production processes become more and more stringent. With decreasing size of the chips and with increasing complexity of the electronic circuits, measurements of the absolute concentrations of impurities and dopants become important. The following two examples illustrate how laser spectroscopy can be successfully applied to the solution of problems in this field.

Irradiating the surface of a solid with a laser, material can be ablated in a controlled way by optimizing intensity and pulse duration of the laser (*laser ablation* [1493]). Depending on the laser wavelength, the ablation is dominated by thermal evaporation (CO_2 laser) or photochemical processes (excimer laser). Laser-spectroscopic diagnostics can distinguish between the two processes. Excitation spectroscopy or resonant two-photon ionization of the sputtered atoms, molecules, or fragments allows their identification (Fig. 10.32). The velocity distribution of particles emitted from the surface can be obtained from the Doppler shifts and broadening of the absorption lines, and their internal energy distribution from the intensity ratios of different vibrational–rotational transitions [1494]. With a pulsed ablation laser the measured time delay between ablation pulses and probe laser pulses allows the determination of the velocity distribution.

Resonant two-photon ionization in combination with a time-of-flight mass spectrometer gives the mass spectrum. In many cases, one observes a broad mass range of clusters. The question is whether these clusters were emitted from the solid or whether they were formed by collisions in the evaporated cloud just after emission. Measurements of the vibrational-energy distributions can answer this question. If the mean vibrational energy is much higher than the temperature of the solid, the molecules were formed in the gas phase, where an insufficient number of collisions

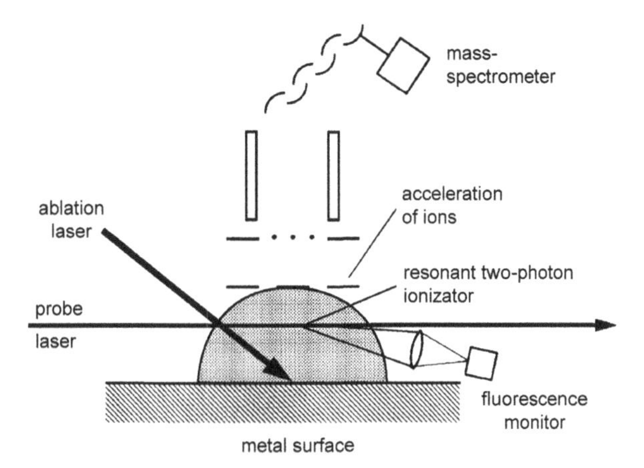

Fig. 10.32 Detection of atoms and molecules sputtered by ion bombardment or laser ablation from a surface and measurements of their energy distribution

cannot fully transfer the internal energy of molecules formed by recombination of sputtered atoms into kinetic energy [1495].

Whereas laser ablation of graphite yields thermalized C_2 molecules with a rotational–vibrational energy distribution following a Boltzmann distribution at the temperature T of the solid, ablation of electrical isolators, such as AlO, produces AlO molecules with a large kinetic energy of 1 eV, but a "rotational temperature" of only 500 K [1496].

For the production of thin amorphous silicon layers (for example, for solar photovoltaic energy converters) often the condensation of gaseous silane (SiH_4) or Si_2H_6, which is formed in a gas discharge, is utilized. During the formation of Si(H) layers, the radical SiH_2 plays an important role, which has absorption bands within the tuning range of a rhodamine 6G dye laser. With spectral- and time-resolved laser spectroscopy the efficiency of SiH_2 formation by UV laser photodissociation of stable silicon–hydrogen compounds can be investigated, as well as its reactions with H_2, SiH_4, or Si_2H_6. This gives information about the influence of SiH_2 concentration on the formation or elimination of dangling bonds in amorphous silicon [1497].

Of particular interest for in situ determinations of the composition of alloys is the technique of laser microspectral analysis [1498], where a microspot of the material surface is evaporated by a laser pulse and the fluorescence spectrum of the evaporated plume serves to monitor the composition.

Surface science is a rapidly developing field that has gained a lot from applications of laser spectroscopy [1499, 1500]. The sensitive technique of surface-enhanced Raman spectroscopy, which gives information of molecules adsorbed on surfaces, was discussed in Sect. 3.4.2.

10.3.3 Laser-Induced Breakdown Spectroscopy (LIBS)

LIBS is a sensitive technique for analyzing the chemical or atomic compositions of solid or liquid materials [1501]. Here a laser pulse is focused onto the surface

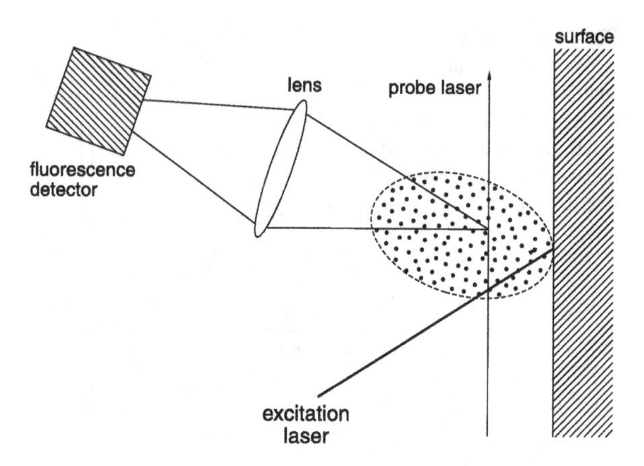

Fig. 10.33 Laser-induced breakdown spectroscopy

of a solid or liquid material. Due to the high peak intensity, fast evaporation of a small volume in the focus of the laser pulse occurs. The gaseous plume ejected from the surface contains molecules, atoms and ions contained in the focal volume. The fluorescence emitted by the excited species is collected by a lens, focused into an optical fiber and sent to the entrance slit of a spectrograph (Fig. 10.33). If the detector at the exit of the spectrograph is time-gated, the spectra can be recorded at specific times after the generation of the plume. Since the plume cools during its expansion, the ions recombine to excited states of neutral atoms or molecules and the emission from these states is a measure of the concentration of the atoms in the sample.

With moderate laser power, molecules can evaporate without fragmentation. This is particularly useful for investigations of biological samples or for inspecting tissues in vitro.

When a second weak probe laser is used, it can cross the plume at different locations, where it generates spatially resolved laser-induced fluorescence, which can be detected and analyzed. Using REMPI techniques, ionization of the neutral species can be achieved with subsequent mass-selective detection in a mass spectrometer.

An important application of LIBS is the analysis of samples from a steel converter during the melting process [1502]. A small sample is taken from the liquid steel, rapidly cooled and then inspected by LIBS in order to get information on the relative concentrations of the different components in the steel. The whole process takes only a few minutes and therefore the composition of the liquid steel can be corrected before the molten metal solidifies.

LIBS has been also successfully applied to the analysis of geochemical samples [1485]. The most accurate information can be obtained for the relative concentrations of different elements in a sample. This is important, for example, for the classification of minerals on earth or in meteroites when it is not clear whether two different samples come from the same source. Also for archeological samples the precise knowledge of elemental composition is very helpful for the exact dating and assignment. The applications of LIBS has benefitted from the use of fiber optics which allows remote sensing, where the laser and detection systems are far away

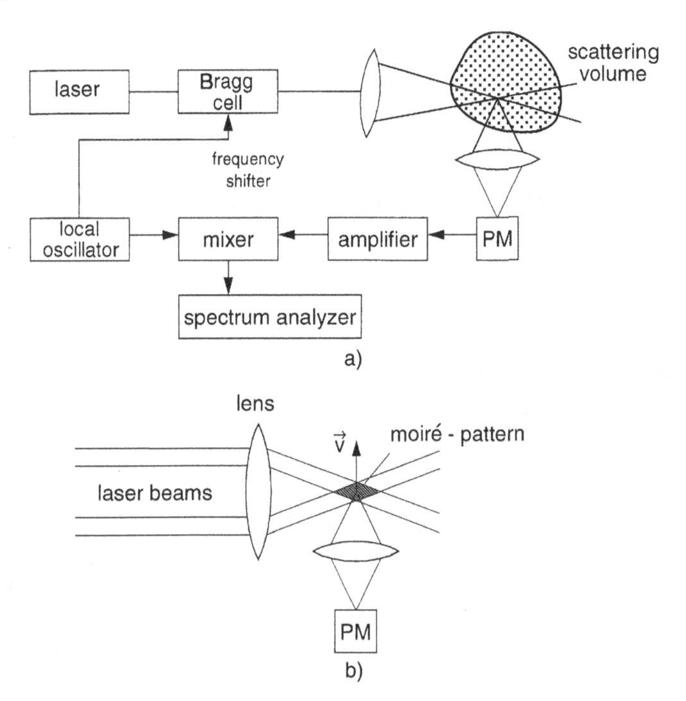

Fig. 10.34 Laser Doppler anemometry. (**a**) Schematic diagram of the whole system; (**b**) generation of the Moiré-pattern

from the place to be inspected. This is in particular important for high temperature sources such as vulcanos or steel melting furnaces and chemical reactors [1486].

An interesting application is the quantitative remote inspection of radioactive contaminations on the surface of confinement walls of nuclear reactors.

Also the contamination of soil by oil or other toxic material can be probed either by LIBS or by laser-induced fluorescence.

10.3.4 Measurements of Flow Velocities in Gases and Liquids

In many technical problems in hydrodynamics or aerodynamics, the velocity profile $\boldsymbol{v}(\boldsymbol{r}, t)$ of a flowing medium in pipes or around solid bodies is of great importance. Doppler anemometry (Sect. 7.12) is a heterodyne laser spectroscopy technique where these velocity profiles are determined from the measured Doppler shifts of the scattered light [1503–1505]. The beam of a HeNe or Ar^+ laser with wave vector $\boldsymbol{k}_\mathrm{L}$ passes through a volume element $\mathrm{d}V$ of the flowing medium. The frequency ω' of light scattered in the direction $\boldsymbol{k}_\mathrm{s}$ by particles with velocity $\boldsymbol{v}$ (Fig. 10.34) is Doppler-shifted to

$$\omega' = \omega_\mathrm{L} - (\boldsymbol{k}_\mathrm{L} - \boldsymbol{k}_\mathrm{s}) \cdot \boldsymbol{v}.$$

The scattered light is imaged onto a detector, where it is superimposed on part of the laser beam. The detector output contains the difference-frequency spectrum $\Delta\omega = \omega_L - \omega' = (\boldsymbol{k}_L + \boldsymbol{k}_s) \cdot \boldsymbol{v}$, which is electronically monitored with a heterodyne technique. One example is an airborne CO_2 laser anemometer that was developed for measuring wind velocities in the stratosphere, in order to improve long-term weather forecasts [1506]. Further examples are measurements of the velocity profiles in the exhausts of turbine engines of planes, in pipelines for gases and liquids, or even in the arteries of the human body.

10.4 Applications in Biology

Three aspects of laser spectroscopy are particularly relevant for biological applications. These are the high spectral resolution combined with high temporal resolution, and the high detection sensitivity. Laser light, focused into cells, also offers a high spatial resolution. The combination of LIF and Raman spectroscopy has proven very helpful in the determination of the structure of biological molecules, whereas time-resolved spectroscopy plays an essential role in the study of fast dynamical processes, such as the isomerization during photosynthesis or the formation of antenna molecules during the first steps of the visual process. Many of these spectroscopy techniques are based on the absorption of laser photons by a biological system, which brings the system into a nonequilibrium state. The time evolution of the relaxation processes that try to bring the system back to thermal equilibrium can be followed by laser spectroscopy [1507].

We will give several examples to illustrate the many possible applications of laser spectroscopy in the investigation of problems in molecular biology [1508].

10.4.1 Energy Transfer in DNA Complexes

DNA molecules, with their double-helix structure, provide the basis of the genetic code. The four different bases (adenine, guanine, cytosine, and thymine), which are the building blocks of DNA, absorb light in the near UV at slightly different wavelengths, but overlapping absorption ranges. By inserting dye molecules between the bases, the absorption can be increased and shifted into the visible range. The absorption spectrum and fluorescence quantum yield of a dye molecule depend on the specific place within the DNA molecule where the dye molecule has been built in. After absorption of a photon, the excitation energy of the absorbing dye molecule may be transferred to the neighboring bases, which then emit their characteristic fluorescence spectrum (Fig. 10.35). On the other hand, excitation of the DNA by UV radiation may invert the direction of the energy transfer from the DNA bases to the dye molecule.

The efficiency of the energy transfer depends on the coupling between the dye molecule and its surroundings. Measurements of this energy transfer for different base sequences allow the investigation of the coupling strength and its variation

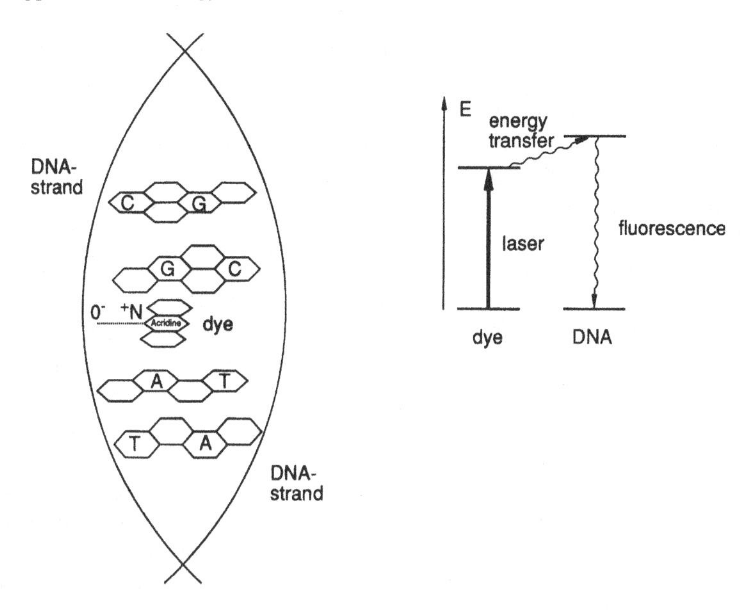

Fig. 10.35 Example of energy transfer within a DNA–dye complex with acridine dye molecules inserted between the bases adenine and guanine after laser excitation [1510]

with the base sequence [1509]. For example, the base guanine within a DNA–dye complex can be selectively excited at $\lambda = 300$ nm without affecting the other bases. The energy transfer rate is determined from the ratio of quantum efficiencies under excitation with visible light (excitation of the acridine dye molecule) and UV light (direct excitation of guanine), respectively [1510].

Since the use of dye molecules in cells plays an important role in the diagnosis and therapy of cancer (Sect. 10.5) a detailed knowledge of the relevant energy transfer processes and the photoactivity of different dyes is of vital interest.

10.4.2 Time-Resolved Measurements of Biological Processes

The detailed knowledge of the different steps of biological processes on a molecular level is one of the ambitious goals of molecular biology. The importance of this field was underlined by the award of the Nobel Prize in chemistry in 1988 to J. Deisenhofer, R. Huber, and H. Michel for the elucidation of the primary steps in photosynthesis and the visual process [1511]. This subsection illustrates the importance of time-resolved Raman spectroscopy in combination with pump-and-probe techniques (Sect. 6.4) for the investigation of fast biological processes.

Hemoglobin (Hb) is a protein that is used in the body of mammals for the transportation of O_2 and CO_2 through blood circulation. Although its structure has been uncovered by X-ray diffraction, not much is known about the structural change of Hb when it absorbs O_2 and becomes oxyhemoglobin HbO_2, or when it releases O_2

again. With laser Raman spectroscopy (Chap. 3) its vibrational structure can be studied, giving information on the force constants and the molecular dynamics. Based on high-resolution Raman spectroscopy with cw lasers, empirical rules have been obtained about the relationship between vibrational spectra and the geometrical structure of large molecules. The change in the Raman spectra of Hb before and after the attachment of O_2 therefore gives hints about the corresponding structural change. If HbO_2 is photodissociated by a short laser pulse, the Hb molecule is left in a nonstationary state. The relaxation of this nonequilibrium state back into the ground state of Hb can be followed by time-resolved Raman or LIF spectroscopy [1512].

Excitation with polarized light creates a partial orientation of selectively excited molecules. The relaxation rate of these excited molecules and its dependence on the degree of orientation can be studied either by the time-resolved absorption of polarized light by the excited molecules, or by measuring the polarization of the fluorescence and its time dependence [1513].

Of particular interest is the investigation of the primary processes of vision. The light-sensitive layer in the retina of our eye contains the protein *rhodopsin* with the photoactive molecule *retinal*. Rhodopsin is a membrane protein whose structure is not yet fully investigated. The polyene molecule retinal exists in several isomeric forms. Since the vibrational spectra of these isomers are clearly distinct, at present Raman spectroscopy provides the most precise information on the structure and dynamical changes of the different retinal configurations. In particular, it has allowed the assignment of the different retinal configurations present in the isomers rhodopsin and isorhodopsin before photoabsorption, and bathorhodopsin after photoexcitation. With pico- and femtosecond Raman spectroscopy it was proved that the isomer bathorhodopsin is formed within 1 ps after the photoabsorption. It transfers its excitation energy within 50 ns to transducin, which triggers an enzyme cascade that finally results, after several slower processes, in the signal transport by nerve conduction to the brain [1514, 1515].

Probably the most important biochemical process on earth is the photosynthesis within the chlorophyl cells in green plants. The photosynthesis process in the chlorophyll complex consists of two parts. In the primary process, light-harvesting molecules with a broad absorption spectrum in the visible absorb photons from sunlight. This leads to molecular electronic excitation. The excited molecules surrounding the central reaction center (Fig. 10.36) can transfer the excited electron in several steps to the molecules in the reaction center, where the secondary process, namely the chemical reaction

$$6H_2O + 6CO_2 \rightarrow C_6H_{12}O_6 + 6O_2$$

takes place. In this, water and carbon dioxide are synthesized to glucose and oxygen. The energy gained by this process is sufficient to synthesize a phosphor group on adenosine diphosphate ADP, resulting in the production of adenosine triphosphate ATP, which serves as an energy donor for processes in plant cells (Fig. 10.36b). Recently, it was discovered that the primary processes proceed on a time scale of 30 to 100 fs. The excitation energy is used for a transfer, which finally delivers the energy for the photosynthetic reaction [1517].

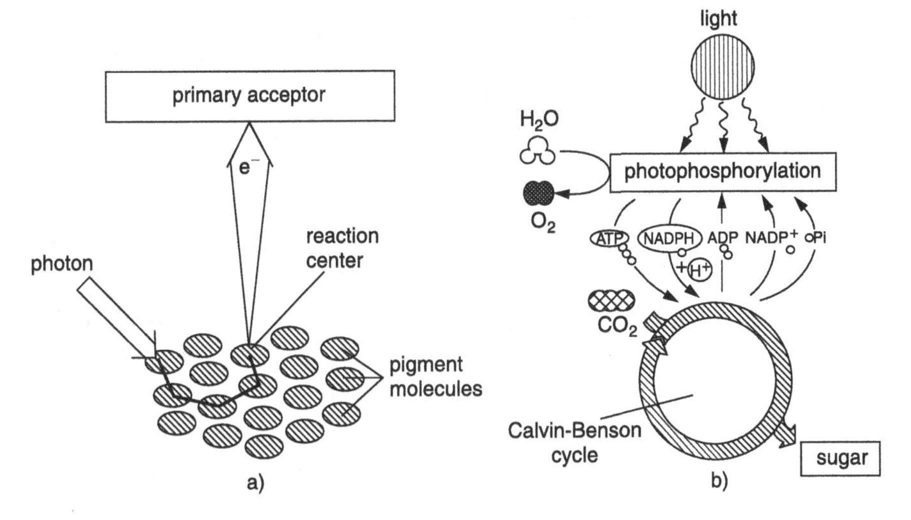

Fig. 10.36 Photosynthesis. (**a**) Primary process; (**b**) reaction cycle [1515]

Another example is the femtosecond-transient absorption and fluorescence after two-photon excitation of carotenoids. The excited β-carotene decays with a time constant of 9 ± 0.2 ps. The energy transfer process from the excited S_1 state in light-harvesting proteins can be monitored by the observed chlorophyl fluorescence [1537].

An important piece of information is the change in molecular structure during these fast processes. Here time-resolved Raman spectroscopy and X-ray diffraction with femtosecond laser-produced brilliant X-ray sources are powerful tools that are more and more applied to molecular biology.

These examples reveal that these extremely fast biochemical processes could not have been studied without time-resolved laser spectroscopy. It not only provides the necessary spectral resolution but also the sensitivity essential for those investigations. More examples and details on the spectroscopy of ultrafast biological processes can be found in [1518–1522].

10.4.3 Correlation Spectroscopy of Microbe Movements

The movement of microbes in a fluid can be observed with a microscope. They move for several seconds a straight path, but suddenly change direction. If they are killed by chemicals added to the fluid, their characteristic movement changes and can be described by a Brownian motion if there is no external perturbation. With correlation spectroscopy (Sect. 7.12), the average mean-square velocity $\langle v^2 \rangle$ and the distribution $f(v)$ of the velocities of living and dead microbes can be measured.

The sample is irradiated with a HeNe laser and the scattered light is superimposed with part of the direct laser light on the photocathode of a photomultiplier. The

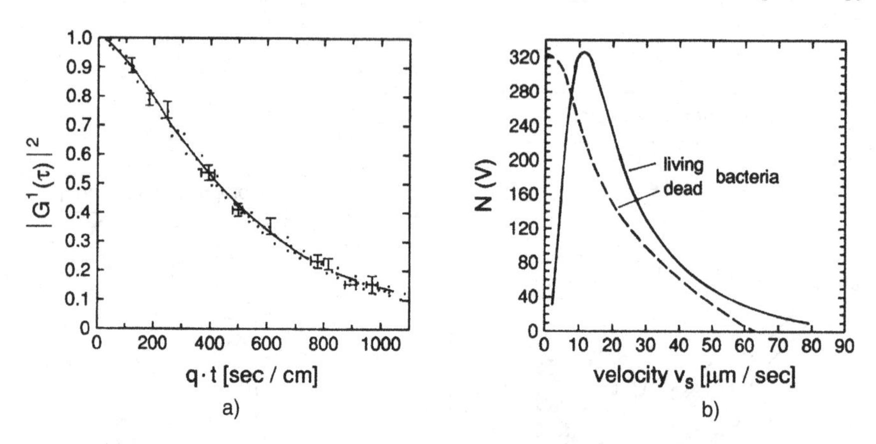

Fig. 10.37 Measured correlation function $G^1(\tau)$ (**a**) and velocity distribution (**b**) computed from $G^1(\tau)$ for living (*solid curve*) and dead (*dashed curve*) *E. coli* bacteria in solution. The *dashed curve* corresponds to the Poisson distribution of a random walk (Brownian motion) [1523]

scattered light experiences a Doppler shift $\Delta\nu = \nu(v/c) \times (\cos\vartheta_1 - \cos\vartheta_2)$, where ϑ_1 and ϑ_2 are the angles of the velocity vector $\boldsymbol{v}$ against the incident laser beam and the direction of the scattered light, respectively. The frequency distribution of this heterodyne spectrum is a measure for the velocity distribution.

Measurements of the velocity distribution of *E. coli* bacteria in a temperature-stabilized liquid gave an average velocity of 15 μm/s (Fig. 10.37), where the maximum speed ranges up to 80 μm/s. Since the size of *E. coli* is only about 1 μm, this speed corresponds to 80 body lengths per second. In comparison, the swimming champion Ian Thorpe only reached 2 m/s, which corresponds to 1 body length per second. If $CuCl_2$ is added to the solution the bacteria die, and the velocity distribution changes to that of a Brownian motion, which shows another correlation spectrum $I(K,t) \propto \exp(-D\Delta K^2 t)$, where $\Delta\boldsymbol{K} = \boldsymbol{K}_0 - \boldsymbol{K}_s$ is the difference between the wave vectors of incident and scattered light. From the correlation spectrum the diffusion coefficient $D = 5 \times 10^{-9}$ cm²/s and a Stokes diameter of 1.0 μm can be derived [1523].

Another technique uses a stationary Moiré fringe pattern produced by the superposition of two inclined beams of the same laser (Fig. 10.34b). The distance between the interference maxima is $\Delta = \lambda \sin(\frac{1}{2}\alpha)$, where α is the angle between the two wave vectors. If a particle moves with a velocity $\boldsymbol{v}$ across the maxima, the scattered light intensity $I_s(t)$ exhibits periodic maxima with a period $\Delta t = \Delta/(v\cos\beta)$, where β is the angle between $\boldsymbol{v}$ and $(\boldsymbol{k}_1 + \boldsymbol{k}_2)$.

10.4.4 Laser Microscope

A beam of a TEM_{00} laser (Vol. 1, Sect. 5.3) with a Gaussian intensity profile is focused to a diffraction-limited spot with the diameter $d \simeq 2\lambda f/D$ by an adapted lens system with the focal length f and the limiting aperture D. For example, with

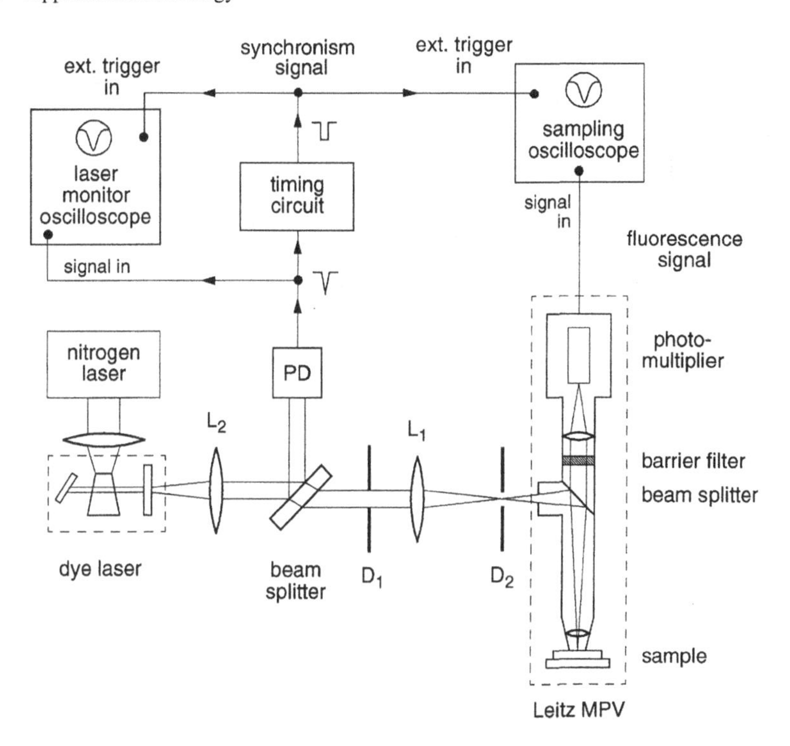

Fig. 10.38 Laser microscope [1524]

$f/D = 1$ at $\lambda = 500$ nm, a focal diameter of $d \simeq 1.0$ μm can be achieved with a corrected microscope lens system. This allows the spatial resolution of single cells and their selective excitation by the laser.

The LIF emitted by the excited cells can be collected by the same microscope and may be imaged either onto a video camera or directly observed visually. A commercial version of such a laser microscope is displayed in Fig. 10.38. For time-resolved measurements a nitrogen-laser-pumped dye laser can be used. The wavelength λ is tuned to the absorption maximum of the biomolecule under investigation. For absorption bands in the UV, the dye-laser output can be frequency doubled. Even if only a few fluorescence photons can be detected per laser pulse, the use of video-intensified detection and signal averaging over many pulses may still give a sufficiently good signal-to-noise ratio [1524].

Many of the spectroscopic techniques discussed above can be applied in combination with the laser microscope, which gives the additional advantage of spatial resolution. One example is the spectrally and spatially resolved laser-induced fluorescence excited by the laser within a certain part of a living cell. The migration of the excitation energy through the cells to their membrane within a few seconds was observed. In addition, the migration of receptor cells through cell membranes can be studied with this technique [1525]. One example is the measurement of the intracellular distribution of injected photosensitizing porphyrin and its aggregation [1526].

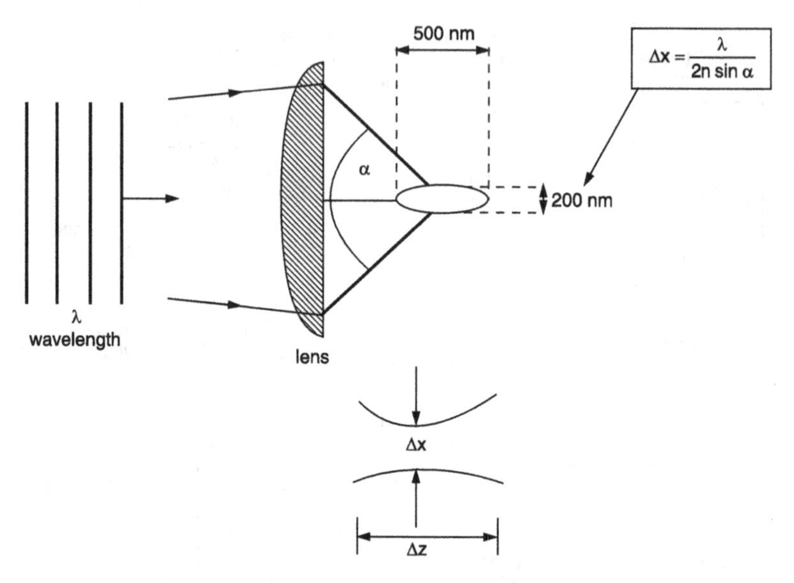

Fig. 10.39 Spatial resolution limits of the microscope: diffraction limit Δr and Rayleigh length Δz for a Gaussian beam

Porphyrin fluorescence was localized in the plasma membrane, the cytoplasm, the nuclear membrane, and the nucleoli. A redistribution of the porphyrin molecules from the plasma membrane to the nuclear membrane and adjacent intracellular sites was observed with increasing incubation time.

Damage to the respiratory chain is correlated with a decrease in ATP production and a lack of certain enzymes and cytochromes. These defects can be detected by measuring the autofluorescence of flavine molecules in intact and respiratory-deficient yeast strains with advanced microscopic techniques [1527a].

The spatial resolution of any microscope [1528] is limited in the directions x and y perpendicular to the laser by diffraction to

$$\Delta x = \Delta y = \lambda/(n \cdot \sin\alpha),$$

where λ is the wavelength of the illuminating lens, n is the refractive index on the sample side and α is the divergence angle after the collimating lens (Fig. 10.39).

In the z-direction, the spatial resolution is limited by the Raleigh length (see Vol. 1, Sect. 5.9)

$$\Delta z_R = \pi w_0^2/\lambda,$$

where w_0 is the beam waist in the focal plane. The spatial resolution Δz can be greatly improved by the "4π technique" first proposed by St. Hell [1529]. Here the sample is illuminated by a focused laser beam which is collimated after the focus, reflected by a mirror and focused again from the other side into the sample (Fig. 10.40). The coherent superposition of the two waves results in a standing wave

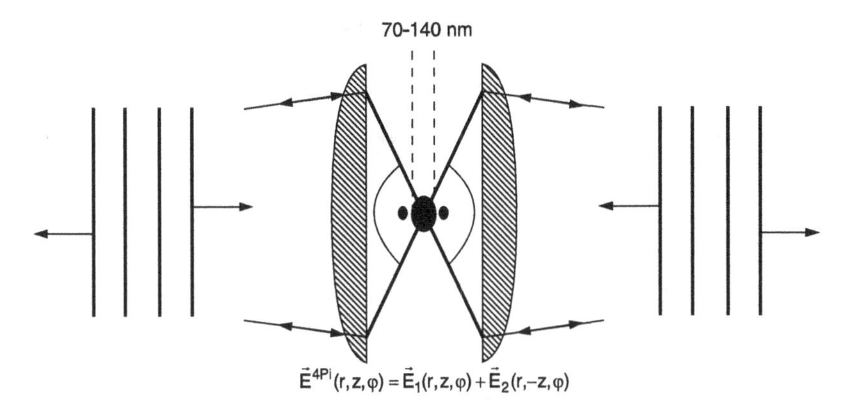

Fig. 10.40 4π microscopy [1529]

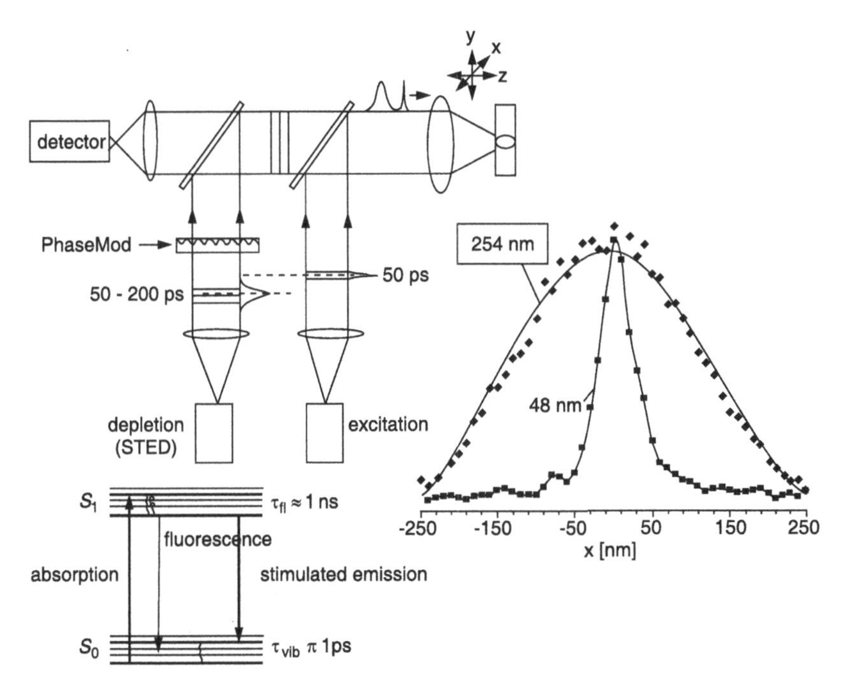

Fig. 10.41 Stimulated emission microscope [1530]

with an electric field amplitude

$$E(x, y, z) = E_1(x, y)\cos(\omega t - kz) + E_2(x, y)\cos(\omega t + kz).$$

The time-averaged total intensity of the standing wave is, for $E_1 = E_2 = E/\sqrt{2}$:

$$I(x, y, z) = E^2\cos^2(kz).$$

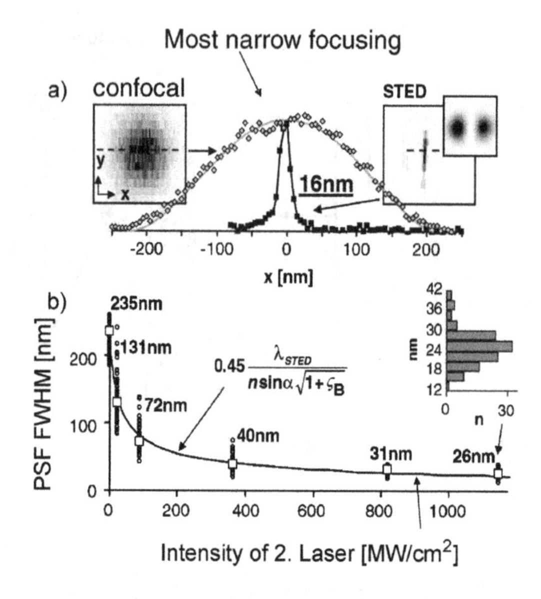

Fig. 10.42 Improvement of radial resolution by stimulated emission with denit intensity profile [1530]

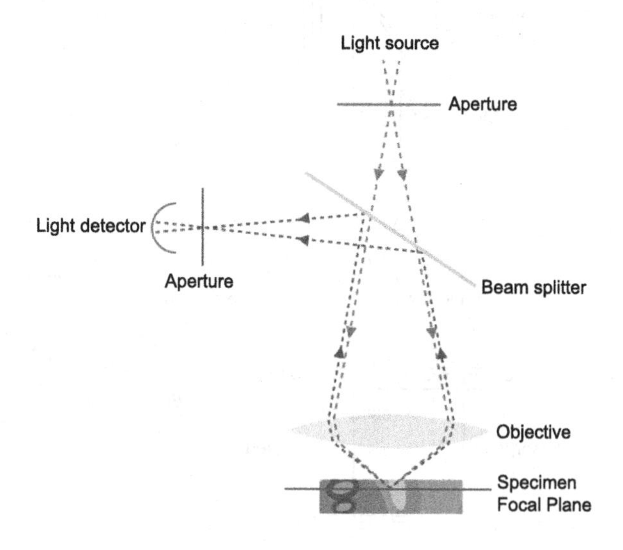

Fig. 10.43 Principle of confocal microscopy

The full halfwidth Δz of the central interference maximum is smaller than $\lambda/6$ because the field amplitude decreases rapidly away from the focal plane. Therefore, the next interference maxima, at a distance $\pm\lambda/2$ from the central maximum, are much less intense.

The 4π technique does not increase the radial resolution. This can be achieved by a stimulated emission technique (Fig. 10.41). Here, the excitation laser in the fundamental Gaussian TEM_{00} mode, which induces the sample fluorescence, is superimposed on a second laser in the donut mode TEM_{11}, which induces stimulated downward transitions from the molecular level excited by the first laser. This quenches the fluorescence around the central radial part and, depending on the intensity of the stimulating laser, only fluorescence from a narrow radial part $0 < r < r_{max}$ is observed (Fig. 10.42). This increases the radial resolution by one order of magnitude [1530].

The diffraction limit for the radial resolution can be overcome be confocal microscopy (Fig. 10.43). Here the laser beam is focussed into the sample and the scattered light is imaged onto a small aperture. Only light from the center of the focal spot is transmitted through the aperture. The spatial resolution is determined by the radius of the aperture [1534].

10.4.5 Detection of Single Biological Molecules

For many biological applications it is important to study the influence of single molecules on the behaviour of living cells. Different techniques have been developed, which allow the detection of fluorescence from single laser-excited molecules in a liquid with a microscope [1531]. The concentration of molecules has to be sufficiently low in order to insure, that within the volume seen through the microscope and within the time resolution interval Δt only a single molecule is present. After the laser excitation the molecule releases its excitation energy by emission of fluorescence within typical times τ ranging from 10^{-9} to 10^{-6} s which brings the molecules in any of the many rotational-vibrational levels in the electronic ground state (Fig. 1.57). Collisions with molecules in the liquid bring the molecule back into the depleted initial level within a few picoseconds from where the molecule can be again excited b the laser. During the time T which spends the molecule within the laser beam it can emit $N = 1/2T/\tau$ fluorescence photons, because the minimum time for an excitation-emission cycle is two times the excited state lifetime τ.

Example 10.5 $\tau = 10^{-8}$ s, $T = 10^{-2}$ s $\rightarrow N = 10^5$ photons per single molecule.

It is therefore possible to observe single molecules through a microscope with the naked eye and follow its path due to diffusion through the liquid (Fig. 10.44). With sensitive detectors, such as photomultipliers, even photon numbers of $N = 100$ can be readily detected.

In particular suited for single molecule detection is the green fluorescence molecule (Fig. 10.6), which has a large quantum efficiency and which can be attached to bacteria and viruses. When these bacteria penetrate into living cells, their path can be followed in detail due to the detection of the fluorescing molecule [1532].

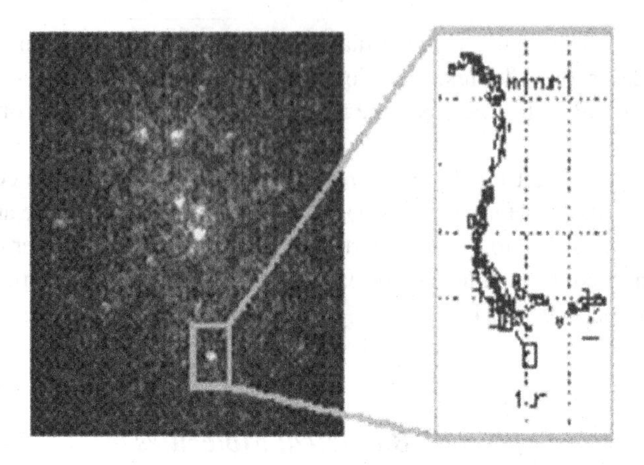

Fig. 10.44 Path of a single molecule in a liquid, obtain by sequential detection through a microscope [1532]

10.5 Medical Applications of Laser Spectroscopy

Numerous books have been published on laser applications in medical research and in hospital practice [1538–1540, 1542]. Most of these applications rely on the high laser-output power, which can be focused into a small volume. The strong dependence of the absorption coefficient of living tissue on the wavelength allows selection of the penetration depth of the laser beam by choosing the proper laser wavelength [1540].

The penetration depth of radiation into tissue is determined by absorption, scattering and reflection [1544]. Attenuation of light in deep biological tissue depends on the effective attenuation coefficient (μ_{eff}), which is defined as

$$\mu_{\mathrm{eff}} = \sqrt{3\mu_{\mathrm{a}}(\mu_{\mathrm{a}} + \mu_{\mathrm{s}}')}$$

where μ_{a} is the absorption coefficient and μ_{s} the scattering coefficient which is dependent o the size of the cells in the tissue and the anisotropy of the tissue. In Fig. 10.45 the different processes for light penetrating into tissue are schematically illustrated.

Since water is a major component in living tissue the absorption of water gives a first hint to the wavelength-dependence of the penetration depth. In Fig. 10.46 the wavelength dependence of the absorption coefficient $\alpha(\lambda)$ of water is shown on a logarithmic scale together with lasers that are used for medical applications.

For a special medical treatment the laser wavelength should be chosen such that the penetration depth $d_{\mathrm{p}} = 1/\alpha$ into the tissue reaches the tissue-layer that should be treated.

The different components of the human body have different absorption coefficients. As example the molar extinction curves of oxy-hemoglobin in the arteries and of desoxy-hemoglobin in the veins are shown in Fig. 10.47.

Fig. 10.45 Schematic illustration of the various processes which occur when light penetrates into tissue

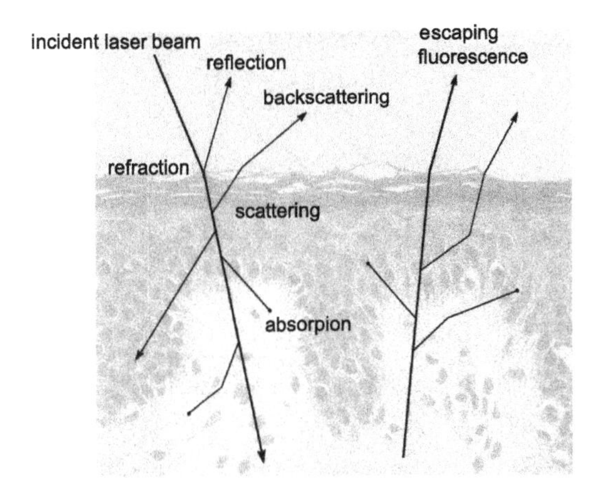

For example, skin carcinoma or port-wine marks should be treated at wavelengths for a small penetration depth in order to protect the deeper layers of the epidermia from being damaged, while cutting of bones with lasers or treatment of subcutaneous cancer must be performed at wavelengths with greater penetration depth. The most spectacular outcomes of laser applications in medicine have been achieved in laser surgery, dermatology, ophthalmology, and dentistry.

There are, however, also very promising direct applications of laser *spectroscopy* for the solution of problems in medicine. They are based on new diagnostic techniques and are discussed in this section.

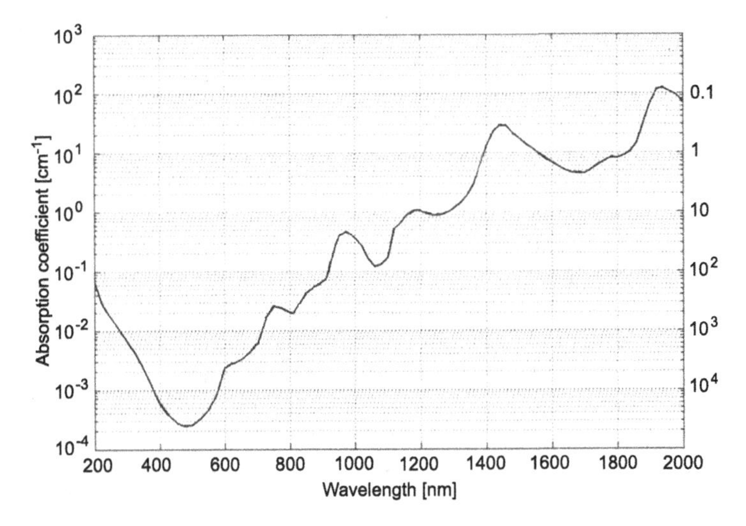

Fig. 10.46 Absorption coefficient $\alpha(\lambda)$ (*left scale*) and penetration depth $d_p/\text{mm} = 10/\alpha\ [\text{cm}^{-1}]$ (*right scale*) of water

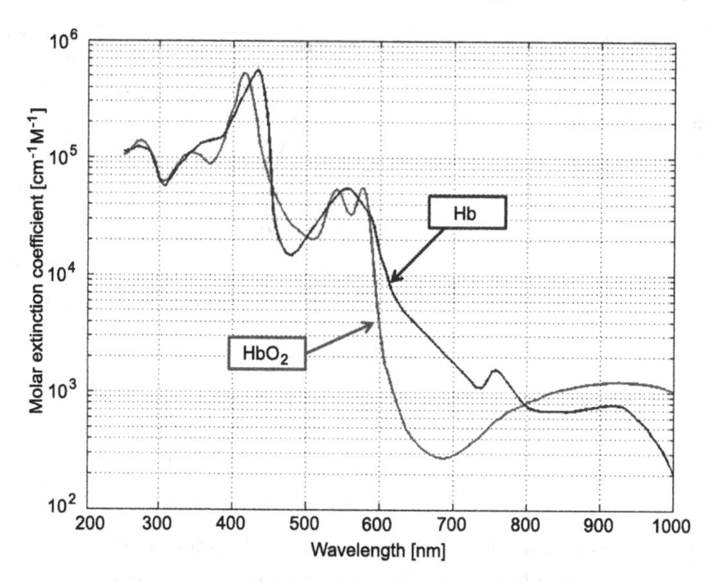

Fig. 10.47 Molar extinction coefficients for oxy-hemoglobin HbO_2 and deoxy-hemoglobin Hb [1541]

10.5.1 Infrared and Raman Spectroscopy of Respiratory Gases

During surgery on a patient, the optimum concentration and composition of narcotic gases can be indicated by the composition of the respiratory gases, that is, with the concentration ratio of N_2:O_2:CO_2. This ratio can be measured in vivo with Raman spectroscopy [1546]. The gas flows through a cell that is placed inside a multipass arrangement for an argon laser beam (Fig. 10.48). Several detectors with special spectral filters are arranged in a plane perpendicular to the beam axis. Each detector monitors a selected Raman line, which allows the simultaneous detection of all molecular components of the gas.

The sensitivity of the method is illustrated by Fig. 10.49, which depicts the time variations of the CO_2, O_2, and N_2 concentrations in the exhaled air of a human patient. Note the variation of the concentrations with changing breathing periods. The technique can be used routinely in clinical practice for anesthetic control during operations and obviously also for alcohol tests of car drivers. Instead of Raman spectroscopy, infrared absorption spectroscopy can be used.

Many bioactive molecules have absorption bands in the infrared, which are accessible by existing infrared lasers or OPOs (Fig. 10.50). When enhancement cavities and cavity ringdown or cavity leak out spectroscopy are used detection sensitivities down into the ppb or even ppt range can be achieved [1545]. This is important for sensitive measurements of breathing gases, where the composition of the exhaled gas gives information about possible diseases. One famous example is the presence of the bacteria *Helicobacter pylori* in the stomach, which can cause gastritis or even stomach cancer. Their presence and concentration can be monitored by the spec-

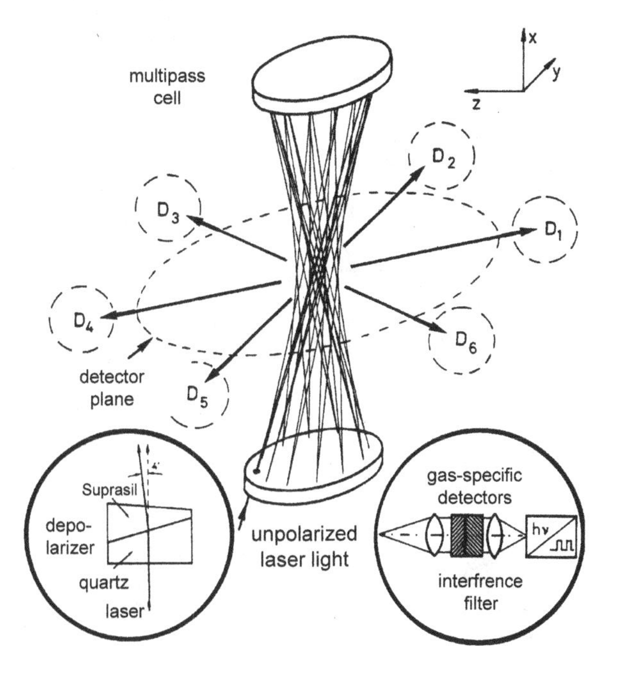

Fig. 10.48 Multipass cell and spectrally selective detector arrangement for the sensitive Raman spectroscopy and diagnostics of molecular gases [1546]

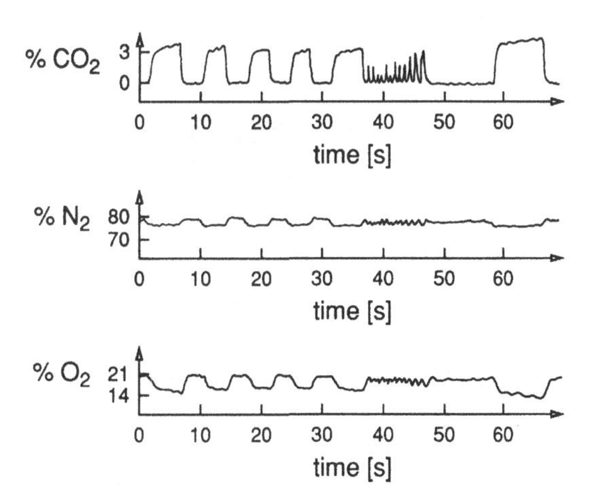

Fig. 10.49 CO_2, N_2, and O_2 concentrations of respiratory gases for various breathing periods, measured in vivo with the arrangement of Fig. 10.48 [1546]

troscopic detection of isotopes of carbon monoxide ^{13}CO or methane $^{13}CH_4$ in the gas exhaled by the patient after being given a drink containing urea $(NH_2)_2{}^{13}CO$. The bacteria decomposes this molecule and produces CO and CH_4, which can be sensitively detected by laser absorption of the fundamental bands.

In Fig. 10.51 a possible experimental arrangement for breath analysis is shown [1547, 1548]. The breathing gases are dried, mixed with a buffer gas and injected into the multipath absorption cell.

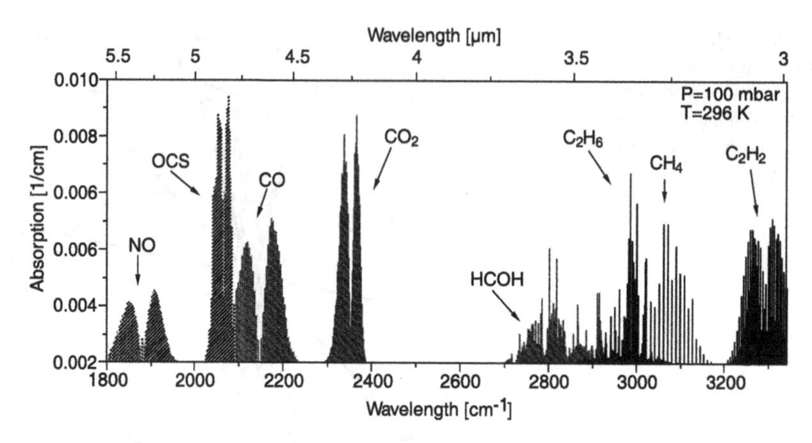

Fig. 10.50 Fundamental infrared absorption bands of some molecules relevant in biomedicine (P. Hering, Institut für Medizinische Physik, Univ. Düsseldorf)

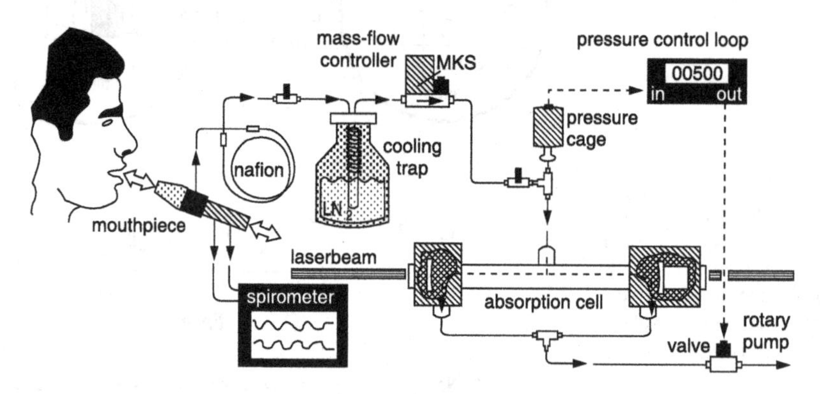

Fig. 10.51 Setup for real-time analysis of breathing gases [1547]

Such sensitive detection techniques also allow the quantitative measurement of hemoglobin in the blood of athletes, enabling checks to be performed for possible doping.

10.5.2 Lasers for Eye-Diagnostics

Lasers are not only used for eye-surgery but more and more also for the diagnosis of eye imaging defects, damages or eye diseases. One example is the simultaneous measurement of several imaging errors such as abberation, astigmatism or coma. Here a plane laser wave is imaged by the eye lens onto the retina, which reflects it. The deviations of the reflected wave phase fronts from those of a plane wave are monitored by interferometric techniques. The reflected wave is imaged by a lens system onto a CCD camera where it is superimposed with an ideal plane wave. The

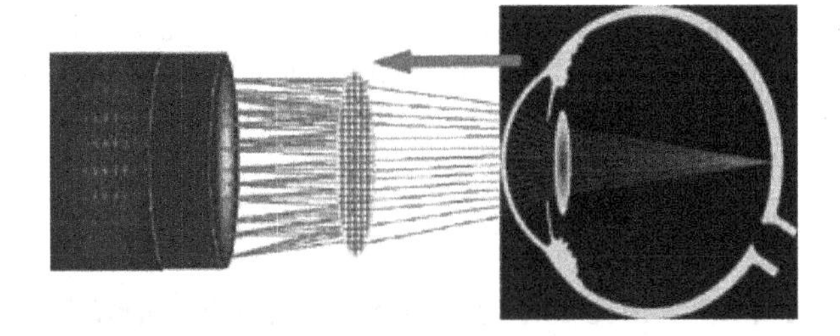

Fig. 10.52 Aerrometer for the diagnostics of imaging errors of the eye

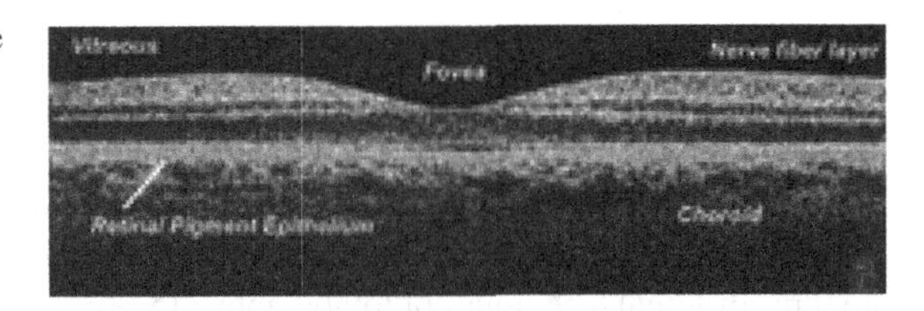

Fig. 10.53 Optical coherence tomography of the retina

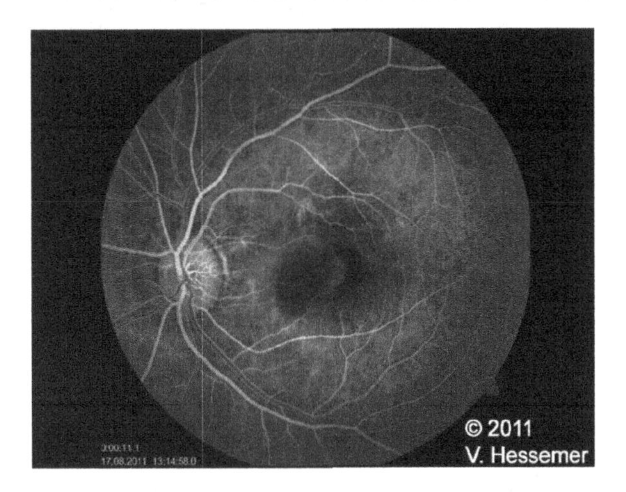

Fig. 10.54 Visualization of the blood vessels through the retina

interference pattern is analysed by a computer program. Such an optical analysing system (*Aberrometer*) computes all imaging errors including abberations of higher order and calculates the necessary optical corrections (Fig. 10.52).

Laser Biometry provides accurate measurements of the eye-length, the curvature of the cornea and the depth of the eye chamber, which are superior to the older ultrasonic technique. These measurements are important for the optimum adaption of plastic lenses implanted after cataract operations.

The high resolution *optical coherence tomography* (Sect. 7.13) is used to measure the thickness and possible wrinkles of the retina, which might lead to its detachment resulting in blindness. In Fig. 10.53 such an optical coherence tomogram of the retina is shown. Measurements at definite time intervals can show changes of the macula degeneration, a decease which is often found for elder people.

Fluorescence angiography allows a very accurate measurement of the blood vessels in the retina. A dye solutions, such as fluorescine is injected into the veins and is transported into the blood vessels of the eye within a few seconds. Here the dye solution is excited by a blue laser and the laser-induced fluorescence is detected. Its spatial distribution gives information on the blood circulation through the retina and the glass ball (Fig. 10.54).

10.5.3 Detection of Tissue Anomalies and Cancer

The most often used optical non-invasive techniques for the detection of anomalies in living tissue are optical coherence tomography (OCT) and measurements of backscattered light. Any local anomaly of the tissue causes a change of the absorption- and scattering cross sections. In OCT the different layers of the tissue are inspected and anomalies show up as a change in the interference pattern. There are three different techniques of tomography [1549]:

(a) Diffraction-limited tomography
(b) Tomography of diffuse scattered light
(c) Coherent optical tomography

Optical Tomography can inspect layers of tissue up to penetration depths of some centimeters and provides three-dimensional pictures with a spatial resolution of a few micrometers. The axial resolution depends on the spectral width $\Delta\lambda$ of the incident radiation while the transversal resolution is limited by the aperture of the optical imaging system. The axial resolution is

$$\Delta z = \lambda_0^2/(\pi \cdot \Delta\lambda)$$

where λ_0 is the central wavelength of the radiation.

Example 10.6 $\lambda = 500$ nm, $\Delta\lambda = 100$ nm $\rightarrow$ $\Delta z = 0.8$ μm.

The main benefits of OCT are:

- Live sub-surface images at near-microscopic resolution
- Instant, direct imaging of tissue morphology
- No preparation of the sample or subject
- No ionizing radiation

A practical design uses optical fibers with a fiber beam splitter (Figs. 10.55b and 10.56), which can be used for inspections of stomach cancer because the fiber can be readily inserted into the stomach and layers of the stomach wall can be detected, that are not accessible to classical gastroscopy [1551, 1552]. Examples for further medical applications are the localization of brain tumors, mamma carcinomas or the visualization of the layer structure of the retina (Fig. 10.53).

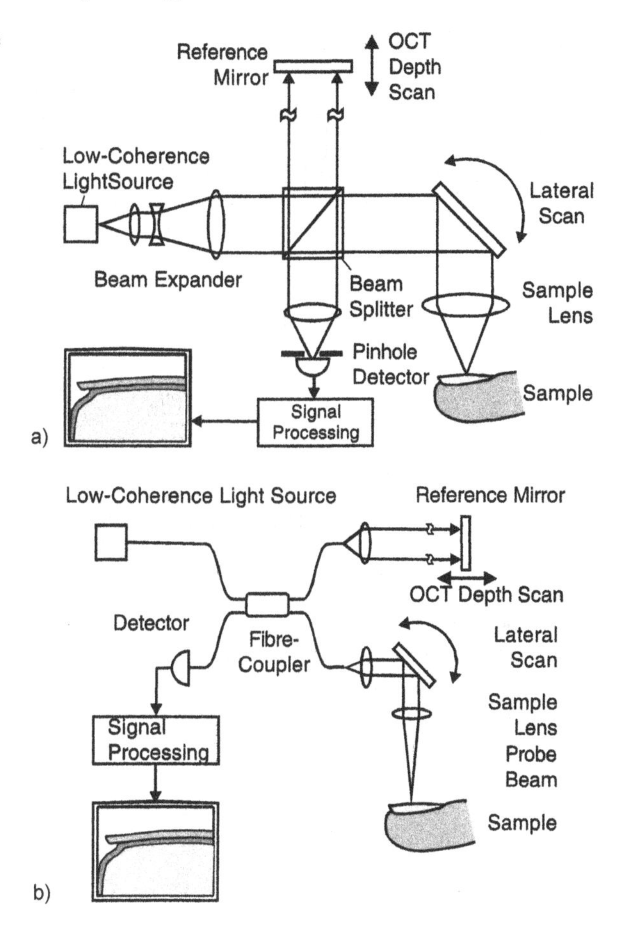

Fig. 10.55 Optical coherence tomography (**a**) design with beam expander and prism beam splitter, (**b**) design with optical fibers [1550]

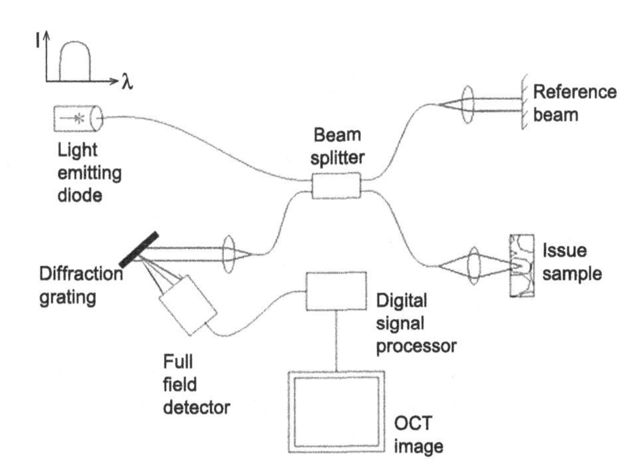

Fig. 10.56 Fiber version of OCT with spectral resolution [1550]

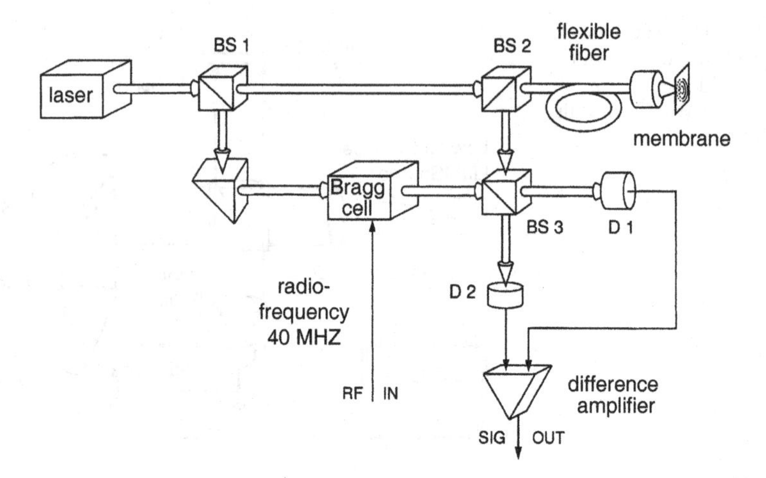

Fig. 10.57 Principle of the laser Doppler vibrometer

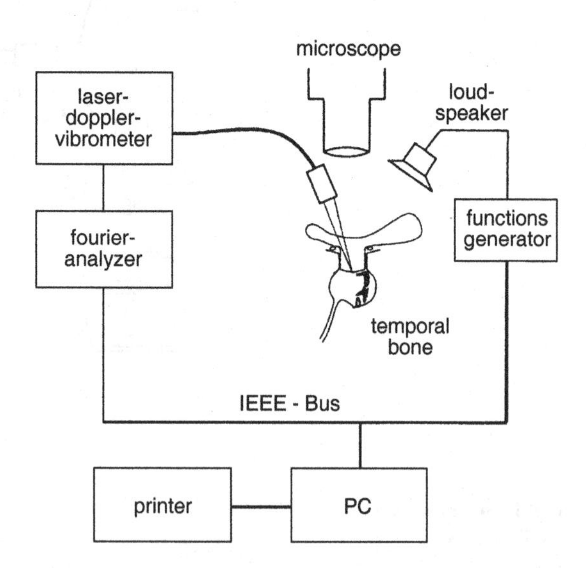

Fig. 10.58 Heterodyne measurements of frequency-dependent vibrations of the ear drum and their local variations [1554]

10.5.4 Heterodyne Measurements of Ear Drums

A large fraction of ear diseases of elderly people is due to changes in the frequency response of the ear drum. While until now investigations of such changes had to rely on the subjective response of the patient, novel laser-spectroscopic techniques allow objective studies of frequency-dependent vibrational amplitudes of the ear drum and their local variation for different locations on the drum with a laser Doppler vibrometer (Fig. 10.57). The experimental arrangement is illustrated in Fig. 10.58. The output of a diode laser is fed through an optical fiber to the ear drum. The light reflected by the drum is collected by a lens at the end of the fiber and is sent back through the fiber, where it is superimposed on a photodetector behind a beam splitter with part of the direct laser light. The ear is exposed to the sound waves of a loudspeaker with variable audio frequency f. The frequency ω of the light reflected by the vibrating ear drum is Doppler shifted. This reflected light is superimposed on part of the laser light, resulting in a heterodyne spectrum with the difference frequency. The amplitude $A(f)$ of the illuminated area of the vibrating drum can be derived from the frequency spectrum of the heterodyne signals (Sect. 7.11). In order to transfer the heterodyne spectrum in a region with less noise, the laser light is modulated at the frequency $\Omega \approx 40$ MHz by an optoacoustic modulator producing sidebands at $\omega \pm \Omega$ [1554]. The heterodyne frequencies are then around 40 MHz, where they can be sensitively detected. The intensity of the light illuminating the ear drum should always be below the threshold for ear damage, which is 160 dB for the hair cells in the inner ear.

10.5.5 Cancer Diagnostics and Photodynamic Therapy

Recently, a method for diagnostics and treatment of cancer has been developed that is based on photoexcitation of fluorescing substances, such as hematoporphyrin derivative (HPD) [1555, 1556]. The disadvantage of HPD is the long photosensitivity (4–6 weeks) where the patient has to avoid sunlight. A new drug 5-Aminolaevulinic acid (ALA) which is a precursor of protoporphyrin IX (Pp IX) in the biosynthetic pathway for haem, has a much shorter decay time and is therefore preferred. A solution of this substance is injected into the veins and is distributed in the whole body after a few hours. While HPD is released by normal cells after 2–4 days, it is kept by cancer cells for a longer time [1557]. If a tissue containing HPD is irradiated by a UV laser, it emits a characteristic fluorescence spectrum, which can be used for a diagnostic of cancer cells. Figure 10.59 shows the emission

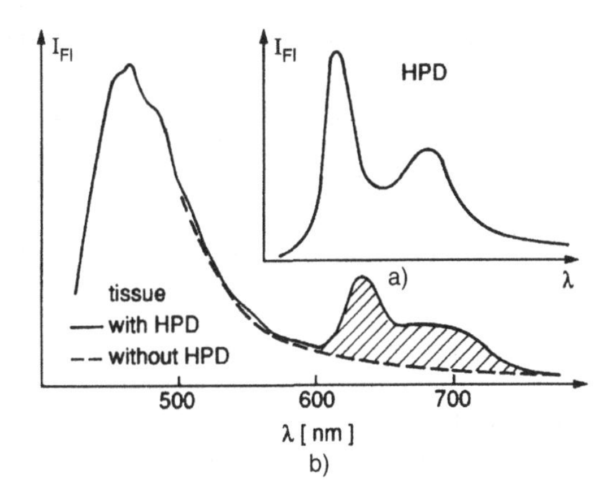

Fig. 10.59 Nitrogen laser-excited fluorescence spectrum of HPD in solution (**a**) and of tissue (**b**) without HPD (*dashed curve*), and with HPD (*solid line*) two days after injection. The *hatched* area represents the additional absorption of HPD [1558]

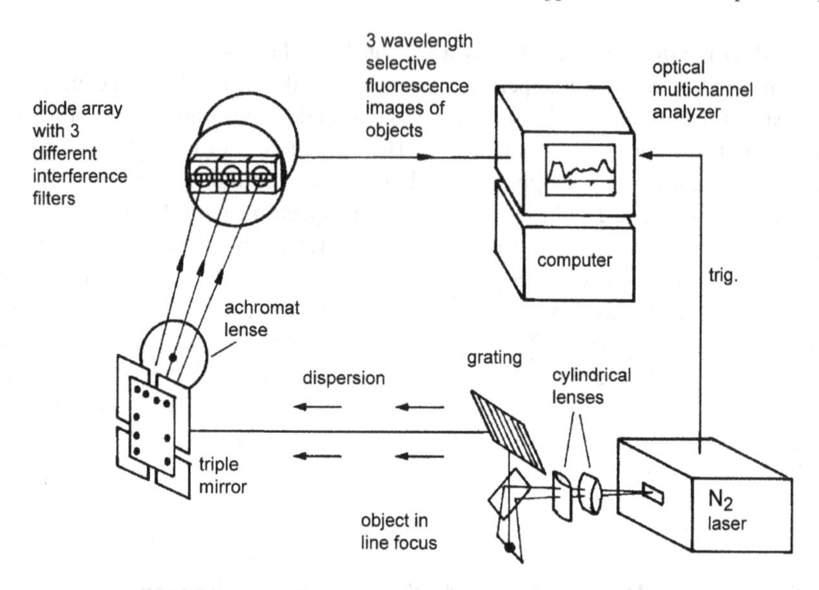

Fig. 10.60 Experimental arrangement for cancer diagnostics of rat tissue [1558]

spectrum of a tissue with and without HPD, and also the fluorescence of pure HPD in a liquid solution excited by a nitrogen laser at $\lambda = 337$ nm. The experimental arrangement for detecting cancerous tissue in rats is exhibited in Fig. 10.60 [1558]. The fluorescence is spectrally resolved by a grating and spatially separated by three slightly folded mirrors, which image a cancer region and a region of normal cells onto different parts of the diode array of an optical multichannel analyzer (Vol. 1, Sect. 4.5). A computer subtracts the fluorescence of the normal tissue from that of a cancerous tissue.

Absorption of photons in the range 500–690 nm brings HPD into an excited S_1 state, which reacts with normal oxygen in the $O_2(^3\Sigma_g^-)$ state and transfers it into the $O_2(^1\Delta)$ state (Fig. 10.61), which apparently reacts with the surrounding cells and destroys them. Although the exact mechanism of these processes is not yet completely understood, it seems that this HPD method allows a rather selective destruction of cancer cells without too much damage to the normal cells. The technique was developed in the USA, intensively applied in Japan [1559], and has since been applied successfully to patients with esophageal cancer, cervical carcinoma, and other kinds of tumors that can be reached by optical fibers without invasive surgery [1560].

For many applications the photodynamic therapy has some definite advantages over other cancer treatments:

- Cancer cells can be singled out and destroyed but most normal cells are spared.
- The damaging effect of the photosensitizing agent happens only when the drug is exposed to light.
- The side effects are fairly mild.

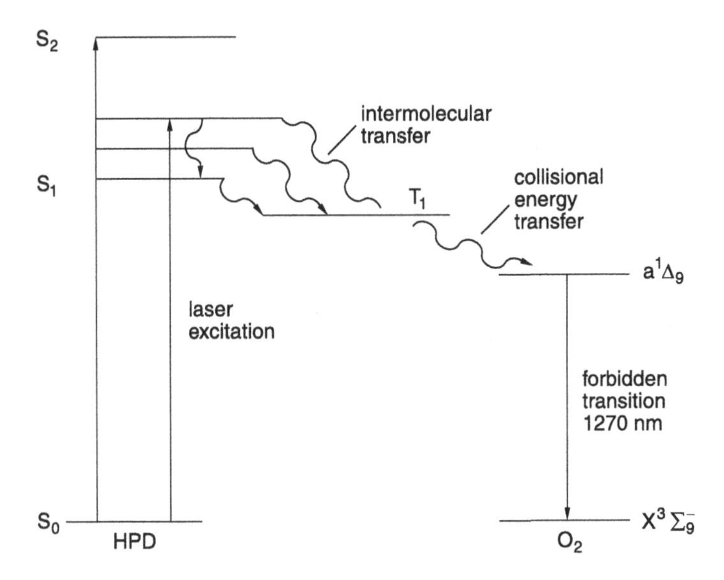

Fig. 10.61 Level scheme for laser-excitation of HPD and energy transfer to O_2 molecules

10.5.6 Laser Lithotripsy

Thanks to the development of thin, flexible optical fibers with a high damage threshold for the radiation [1562, 1563], inner organs of the human body, such as the stomach, bladder, gallbladder or kidneys, can be selectively irradiated by laser radiation. A new technique for breaking kidney stones to pieces by irradiation with pulsed lasers (laser lithotripsy) has found increasing interest because it has several advantages compared to the ultrasonic shockwave lithotripsy [1564–1566].

An optical fused quartz fiber is inserted through the urinary tract until it nearly touches the stone that is to be broken. This can be monitored by X-ray diagnosis or by endoscopy through a fiber bundle that contains, besides the fiber for guiding the laser beam, other fibers for illumination, viewing and monitoring the laser-induced fluorescence (Fig. 10.62).

If the pulse of a flashlamp-pumped dye laser is transported through the fiber and focused onto the kidney stone, the rapid evaporation of the stone material results in a shock wave in the surrounding liquid, which leads to destruction of the stone after several laser shots [1565]. The necessary laser power and the optimum wavelength depend on the chemical composition of the stone, which generally varies for different patients. It is therefore advantageous to know the stone composition before distruction in order to choose optimum laser conditions. This information can be obtained by collecting the fluorescence of the evaporated stone material at low laser powers through an optical fiber (Fig. 10.62). The fluorescence spectrum is monitored with an optical multichannel analyzer and a computer gives, within seconds, the wanted information about the stone composition [1567].

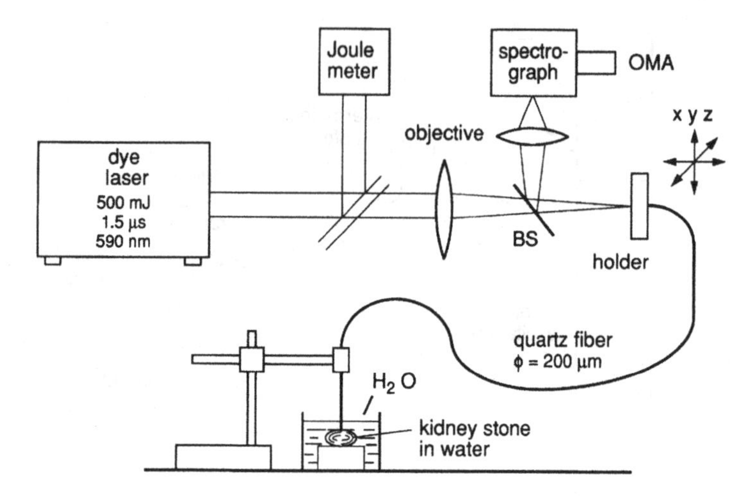

Fig. 10.62 Experimental arrangement for the spectral analysis of kidney stones for the determination of stone composition

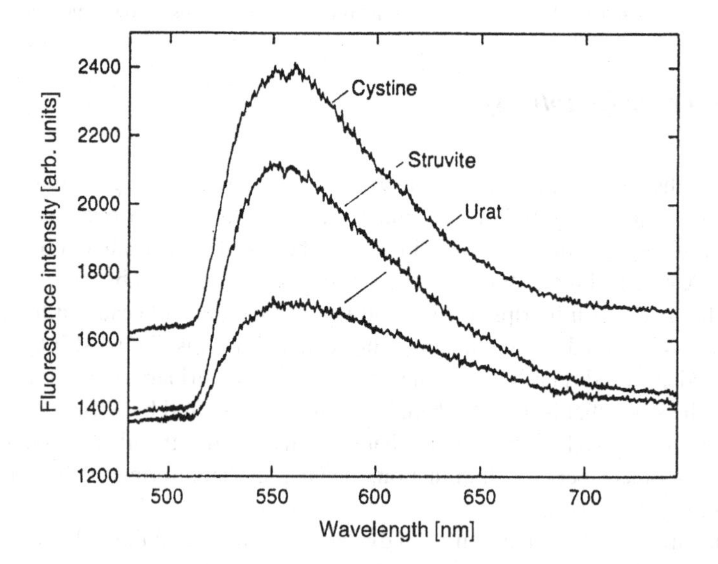

Fig. 10.63 Fluorescence of three different kidney stone materials excited with a dye laser at $\lambda = 497$ nm at low intensities to prevent plasma breakdown [1567]

First demonstrations of the capability of spectral analysis of kidney stones in vitro are illustrated in Fig. 10.63, where the fluorescence spectra of different kidney stones that had been irradiated in a water surrounding outside the body, and that were detected with the arrangement of Fig. 10.60 are shown [1567]. Further information on laser lithotripsy and spectroscopic control of this technique can be found in [1568, 1569].

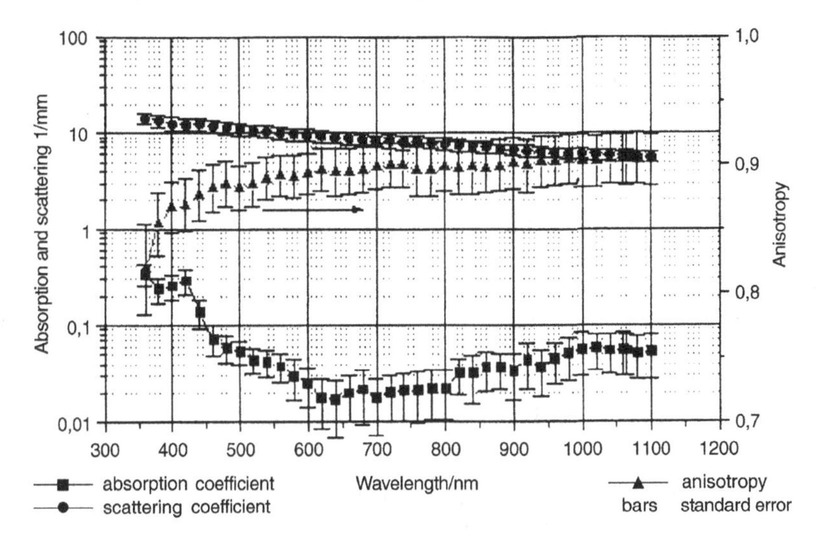

Fig. 10.64 Optical properties of human gray matter determined in vitro with an integrating-sphere setup and an inverse Monte Carlo technique [1570]

10.5.7 Laser-Induced Thermotherapy of Brain Cancer

Laser-induced interstitial thermotherapy represents a minimally invasive therapy, where the cancerous tissue is irradiated by laser light guided through an optical fiber. Planning the operation anticipates the knowledge of the absorption and scattering properties of cancerous tissue compared to healthy tissue. Several optical and computational techniques have been developed that can be used to localize the cancerous tissue and to optimize the optical radiation dose. Here the wavelength dependence of the absorption and scattering coefficients is determined in foregoing experiments. Figure 10.64 shows for illustration the spectral dependence of both coefficients and also the anisotropy of the scattering in human brain tissue determined in vitro [1570].

10.5.8 Fetal Oxygen Monitoring

Monitoring the oxygen concentration of a baby during longer birth processes is of crucial importance for the lasting health of the child. Until now the available equipment was not well adapted for clinical applications. Here a laser-based technique, using light scattering measurements seems to be very successful. The laser light is transported to the skull of the baby through an optical fiber and the scattered light is collected by a second fiber, placed some centimeters distant from the first. The collected light is then monitored as a function of the wavelength [1571]. Since the scattering cross section depends on the oxygen concentration, after calibration this method allows the determination of the wanted O_2 concentration.

More information on application of lasers and particular laser spectroscopy in medicine can be found in the Journal "Medical Laser Applications" [1574] and in reviews and books [1577].

10.6 Concluding Remarks

The preceding selection of examples of laser spectroscopic applications is somewhat arbitrary and by no means complete. The progress in this field may be measured by the increasing variety of conferences and workshops on applications of laser spectroscopy in science and technology. A good survey can be found in many conference proceedings of the Society of Photo-optical Instrumentation and Engineering (SPIE) [1572]. The reasons for this rapid expansion of applications are manifold:

- First, more types of reliable and "easy to handle" lasers in all spectral regions of interest are now commercially available.
- Second, spectroscopic equipment has greatly improved during recent years.
- Last, but not least, our understanding of many processes on a molecular level has become much more advanced. This allows a better analysis of spectral information and its transformation into reliable models of structures and processes.

In particular the development of femtosecond lasers and time-resolved detection techniques has opened the wide field of fast dynamical processes for detailed research [1574].

Solutions

Chapter 1

1. **(a)**

$$n_i = n \cdot \frac{(2J_i + 1)}{Z_{\text{rot}} \cdot Z_{\text{vib}}} \cdot e^{-E_{\text{rot}}/kT} \cdot e^{-E_{\text{vib}}/kT}$$

where $Z_{\text{rot}} = \sum_{J=0}^{\infty}(2J + 1)e^{-J(J+1)hcB/kT} \approx \frac{kT}{h \cdot c \cdot B}$ is the rotational partition function and $Z_{\text{vib}} = \sum_{v=0}^{\infty} e^{-hc\omega_e(v+\frac{1}{2})/kT}$ the vibrational partition function.

For $T = 300\ \text{K} \Rightarrow kT = 4.1 \times 10^{-21}$ J.

For $J_i'' = 20 \Rightarrow E_{\text{rot}} = hcBeJ_i''(J_i'' + 1) \approx 1.25 \times 10^{-20}$ J

$$\Rightarrow \quad \exp[-E_{\text{rot}}/kT] = \exp[-3.05] = 4.7 \times 10^{-2}$$

$$\exp[-E_{\text{vib}}/kT] = 0.617 \quad \text{for } v'' = 0$$

$$\Rightarrow \quad \exp[-E_{\text{vib}}/kT] = 0.235 \quad \text{for } v'' = 1 \quad \text{and} \quad 0.09 \quad \text{for } v'' = 2$$

$$\Rightarrow \quad Z_{\text{vib}} \approx 1.1; \quad Z_{\text{rot}} = \frac{4.1 \times 10^{-21}}{hcB} = 138$$

$$\Rightarrow \quad \frac{n_i}{n} = \frac{41}{138 \cdot 1.1} \cdot 4.7 \times 10^{-2} \cdot 0.617 = 7.8 \times 10^{-3}.$$

(b) The absorption coefficient α is

$$\alpha_{ik} = n_i \cdot \sigma_{ik}.$$

At $p = 10\ \text{mb} \Rightarrow n \approx 2.5 \times 10^{19}/\text{cm}^3$

$$\Rightarrow \quad n_i = 2 \times 10^{17}/\text{cm}^3$$

$$\Rightarrow \quad \alpha_{ik} = 2 \times 10^{17} \times 10^{-18}\ \text{cm}^{-1} = 0.2\ \text{cm}^{-1}.$$

(c)

$$P_{\text{t}} = P_i e^{-\alpha x} = 100 \cdot e^{-2}\ \text{mW} = 13.5\ \text{mW}.$$

W. Demtröder, *Laser Spectroscopy 2*, DOI 10.1007/978-3-662-44641-6

2. The molecular density in the beam is

$$n_i = N_i/v = 10^{12}/(5 \times 10^4)\ \mathrm{cm}^{-3} = 2 \times 10^7/\mathrm{cm}^3.$$

The absorption coefficient is

$$\alpha_i = n_i \sigma_i = 2 \times 10^7 \times 10^{-16}\ \mathrm{cm}^{-1} = 2 \times 10^{-9}\ \mathrm{cm}^{-1}.$$

At a pathlength of 1 mm the absorbed power is

$$P_0 - P_t = P_0\left(1 - e^{-\alpha x}\right) = P_0 \times 2 \times 10^{-10} = 2 \times 10^{-13}\ \mathrm{W}.$$

This corresponds at $\lambda = 623$ nm to

$$n_{ph} = 2 \times 10^{-13}/(h\nu) = 2 \times 10^{-13}\lambda/(hc) = 6.3 \times 10^5 \text{ absorbed photons/s}.$$

Each absorbed photon causes a fluorescence photon. With a collection efficiency of

$$\delta = \left(\frac{1}{4}\pi D^2 \Big/ L^2\right)\Big/ 4\pi = 1/64, \quad \eta = 0.2$$

$$\Rightarrow \quad n_{\mathrm{ph\,el}} = 0.2 \times \frac{1}{64} \times 6.3 \times 10^5 = 1.96 \times 10^3\ /\mathrm{s}$$

i.e. about 2000 photoelectrons/s are produced.

3. The transmitted peak power at ω_0 is

$$P_t(\omega_0) = P_0 \cdot e^{-\alpha \cdot L} = P_0 \cdot e^{-10^{-6}} \approx P_0\left(1 - 10^{-6}\right) \approx P_0 = 1\ \mathrm{mW}.$$

The dc signal is then $S_{DC} = 1$ V.

The absorption changes for $\omega \neq \omega_0$, according to the Doppler profile

$$\alpha(\omega) = \alpha(\omega_0) \cdot e^{-\left(\frac{\omega - \omega_0}{0.36\delta\omega_0}\right)^2}.$$

With $\delta\omega_0 = 2\pi \times 10^9\ \mathrm{s}^{-1}$ and $(\omega - \omega_0) = 2\pi \times 10^8\ \mathrm{s}^{-1}$ we obtain

$$\alpha(\omega) = \alpha(\omega_0) \cdot e^{-\left(\frac{0.1}{0.36}\right)^2} \approx \alpha(\omega_0) \cdot (1 - 0.08) = 0.92\alpha(\omega_0)$$

$$\Rightarrow \quad \alpha(\omega_0) - \alpha(\omega) = 0.08\alpha(\omega_0)$$

$$\Rightarrow \quad P_t(\omega) = P_0 \cdot e^{-8 \times 10^{-8}} \approx P_0\left(1 - 8 \times 10^{-8}\right).$$

The ac signal is then

$$S(\Delta\omega) = \left[P_t(\omega) - P_t(\omega_0)\right] = P_0\left(10^{-6} - 8 \times 10^{-8}\right)\ \mathrm{V}$$
$$= 0.92 \times 10^{-6} P_0 = 9.2 \times 10^{-7}\ \mathrm{V} = 0.92\ \mu\mathrm{V}.$$

4. The number of ions per s is

$$N_{\text{ion}} = N_{\text{a}} \frac{10^7}{10^7 + 10^8} = 0.091 N_{\text{a}}$$

where N_{a} is the rate of absorbing molecules.

With $\mathrm{d}N_{\text{i}}/\mathrm{d}t = N_{\text{a}} = 10^5\ \text{s}^{-1} \Rightarrow N_{\text{ion}} = 9.1 \times 10^3\ \text{s}^{-1}$.

5. **(a)** The kinetic energy of the molecules impinging per s on the bolometer is

$$\begin{aligned} E_{\text{kin}} &= n \cdot v \cdot \frac{m}{2} v^2 (0.3 \times 0.3) \\ &= 10^8 \times 4 \times 10^4 \times \frac{1}{2} \times 1.66 \times 10^{-27} \times \left(4 \times 10^4\right)^2 \times 9 \times 10^{-6} \cdot 28\ \text{W} \\ &= 1.34 \times 10^{-9}\ \text{W} = P_0. \end{aligned}$$

(b) $\Delta T = P_0 / G = 0.134$ K (dc-temperature rise).

(c) The absorbed laser power is

$$\begin{aligned} \Delta P &= P_0 - P_{\text{t}} = P_0\left(1 - \mathrm{e}^{-\alpha L}\right) = P_0\left(1 - \mathrm{e}^{-10^{-9}}\right) \approx P_0 \times 10^{-9} \\ &= 10^{-8}\ \text{mW} = 10^{-11}\ \text{W}. \end{aligned}$$

The rate of absorbed photons is

$$N_{\text{ph}} = \Delta P / h\nu = \Delta P \cdot \lambda / hc = 7.6 \times 10^7\ \text{s}^{-1}.$$

The ac-temperature rise is

$$\Delta T = \frac{\Delta P}{G} = 10^{-11}/10^{-8}\ \text{K} = 1\ \text{mK}.$$

6. The Zeeman shift for a term with magnetic quantum number M_J $(-J \le M_J \le +J)$ is

$$\Delta\nu = \mu \cdot M_J \cdot B / h$$

with the magnetic moment $\mu = 0.5 \mu_{\text{B}}$ we obtain for the frequency shift of the highest Zeeman component $M_J = +J = 2$

$$\begin{aligned} &\Delta\nu = +0.5 \mu_{\text{B}} \cdot J \cdot B / h = 7\ \text{GHz/Tesla} \\ \Rightarrow \quad &B = \frac{10^8 h}{0.5 \times 9.27 \times 10^{-24} \times 2}\ \text{Tesla} = 7.2 \times 10^{-3}\ \text{Tesla}. \end{aligned}$$

The necessary magnetic field for tuning the laser frequency into resonance with the Zeeman component $M'_J = 2 \leftarrow M''_J = 1$ is 7.2 mT.

For the linearly polarized transition $M'_J = 1 \leftarrow M''_J = 1$ the necessary field is twice as high.

For linearly polarized light one observes three Zeeman components when looking perpendicular to $\boldsymbol{B}$. For circular polarization three components appear for σ^+-light and three for σ^--light.

7. **(a)** For a uniform electric field the field strength is

$$E = V/L = 2\,\text{kV/m}.$$

(b) The force acting on the ions is $\boldsymbol{F} = q \cdot \boldsymbol{E}$.
Their acceleration is

$$a = F/m = \frac{2 \times 10^3 \times 1.6 \times 10^{-19}}{40 \times 1.66 \times 10^{-27}}\ \text{m/s}^2 = 4.8 \times 10^9\ \text{m/s}^2.$$

Their velocity before the next collision occurs is $v = a \cdot \tau$ where τ is the mean time between two collisions.

The mean free path is $\Lambda = \frac{1}{2} a\tau^2 \Rightarrow \tau = \sqrt{2\Lambda/a} = 6.4 \times 10^{-7}$ s

$$\Rightarrow \quad v_{\max} = \sqrt{2a \cdot \Lambda} = 3.1 \times 10^3\ \text{m/s} = (\Delta v_z)_{\max}.$$

The frequency of the ac-voltage is 1 kHz.
$\Rightarrow$ The period of the electric field is long compared to τ.
$\Rightarrow$ The mean velocity of the ions is $\bar{v} = \frac{1}{2} v_{\max} = 1.6 \times 10^3$ m/s.
(c) The maximum frequency modulation is

$$\Delta\nu = \nu_0 \cdot \Delta v_z / c \approx 10^9\ \text{s}^{-1}.$$

This equals the Doppler width of the absorption lines.
(d) For $\alpha(\nu_0) = 10^{-6}\ \text{cm}^{-1}$ the transmitted power at the line center ν_0 is

$$P_\text{t} = P_0 \cdot \text{e}^{-\alpha z} = P_0 \cdot \text{e}^{-10^{-4}} \approx P_0\bigl(1 - 10^{-4}\bigr).$$

At the frequency ν which is 1 GHz away from ν_0 the absorption decreases to

$$\alpha(\nu) = \alpha(\nu_0) \cdot \text{e}^{-\frac{1}{0.36}} = \alpha(\nu_0) \cdot \text{e}^{-2.78} \approx 0.06\alpha(\nu_0).$$

The ac-power modulation is

$$\Delta P_\text{t} = 10\ \text{mW} \times 0.94 \times 10^{-4} = 0.94\ \mu\text{W}.$$

$\Rightarrow$ The detected signal is

$$S = 0.94\ \text{mV}.$$

8. **(a)** According to Eq. (1.6) is

$$\frac{\Delta P}{P_0} = \frac{g_0}{g_0 - \gamma} \cdot \frac{\Delta\gamma}{\gamma + \Delta\gamma}$$

with $\Delta\gamma = 2L \times \alpha = 8 \times 5 \times 10^{-8} = 4 \times 10^{-7}$

$$g_0 = 4 \times 10^{-2}, \qquad \gamma = 2 \times 10^{-2}, \qquad P_0 = 1 \text{ mW}$$

$$\Rightarrow \quad \Delta P = 4 \times 10^{-8} \text{ W} = 40 \text{ nW}.$$

(b) The intracavity power is

$$P_{\text{int}} = P_0/T = 10^{-3}/(5 \times 10^{-3}) \text{ W} = 200 \text{ mW}.$$

The absorbed laser power is

$$P_{\text{abs}} = 2L \cdot \alpha \cdot P_{\text{int}} = 4 \times 10^{-7} \times 0.2 \text{ W} = 8 \times 10^{-8} \text{ W}.$$

The number of absorbed laser photons per s is

$$n_{\text{a}} = P_{\text{abs}}/(h \cdot \nu) = 2 \times 10^{11} \text{ s}^{-1}.$$

The number of fluorescence photons is

$$n_{\text{Fl}} = \frac{1}{2} n_{\text{a}} = 1 \times 10^{11} \text{ s}^{-1}.$$

(c) The fluorescence emitted from the line of the laser beam is best imaged by a cylindrical mirror on one side of the laser beam in the focal line of the mirror which images the line-source into a parallel light beam, directed into the opposite side of the laser beam.

The light beam can be imaged by a spherical lens onto the photo cathode. With this arrangement a collection efficiency of 20 % can be reached.

The number of fluorescence photons impinging onto the cathode is then $0.2 \times 10^{11} \text{ s}^{-1} = 2 \times 10^{10} \text{ s}^{-1}$ producing

$$n_{\text{PE}} = 0.15 \times 2 \times 10^{10} \text{ s}^{-1} = 3 \times 10^9 \text{ photoelectrons/s}.$$

(d) The statistical fluctuation of the photoelectron rate is

$$\delta n_{\text{PE}} = \sqrt{n_{\text{PE}}} \approx 5.5 \times 10^4 \text{ s}^{-1}.$$

The dark current of the PM at the anode is 10^{-9} A, which gives $10^{-9}/G = 10^{-15}$ A at the cathode.

This corresponds to $n_{\text{D}} = I_\alpha/e = \frac{10^{-15}}{1.6 \times 10^{-19}} = 6.2 \times 10^3$ electrons/s.

The shot noise of the signal current is therefore about 9 times larger than the dark current. The signal to noise ratio is $n_{PE}/\sqrt{n_{PE}} \approx 5 \times 10^4$.

If the laser power is detected by a photodiode, the signal is $S = (10\ \mathrm{V/W}) \times 10^{-3}\ \mathrm{W} = 10\ \mathrm{mV}$.

The change ΔS due to the change ΔP of the laser power is

$$\Delta S = (10\ \mathrm{V/W}) \times 4 \times 10^{-8}\ \mathrm{W} = 4 \times 10^{-7}\ \mathrm{V}.$$

If the laser power can be stabilized within $10^{-3} P_0$ the power fluctuations are $1\ \mu\mathrm{W}$ and the corresponding signal fluctuation $\delta S = 10\ \mu\mathrm{V}$, i.e. about 25 times larger than the signal ΔS. One therefore needs lock-in detection or other special noise-suppressing detection techniques.

9. The total potential energy is the superposition of the Coulomb-potential and the potential $-E_0 \cdot x$ due to the external electric field $\boldsymbol{E} = -E_0\hat{x}$

$$E_{\mathrm{pot}}^{\mathrm{eff}} = -\frac{Z_{\mathrm{eff}} \cdot \mathrm{e}^2}{4\pi\varepsilon_0 r} - \mathrm{e} \cdot E_0 \cdot x.$$

With $x = r \cdot \cos\vartheta$ we obtain for $\vartheta = 0$

$$E_{\mathrm{pot}}^{\mathrm{eff}} = -\frac{Z_{\mathrm{eff}} \cdot \mathrm{e}^2}{4\pi\varepsilon_0 r} - \mathrm{e} \cdot E_0 \cdot r.$$

This function has a maximum at $r = r_{\mathrm{m}}$ for $\mathrm{d}E_{\mathrm{pot}}^{\mathrm{eff}}/\mathrm{d}r = 0$

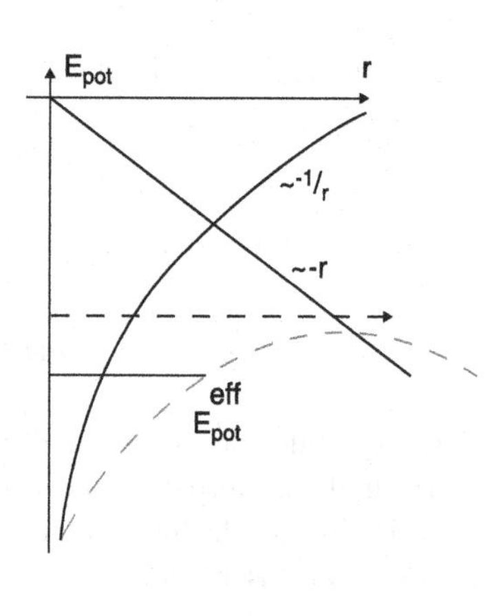

$$\Rightarrow \quad \frac{Z_{\mathrm{eff}} \cdot \mathrm{e}^2}{4\pi\varepsilon_0 r_{\mathrm{m}}^2} - \mathrm{e} \cdot E_0 = 0$$

$$\Rightarrow \quad r_{\mathrm{m}} = \left(\frac{Z_{\mathrm{eff}} \cdot \mathrm{e}}{4\pi\varepsilon_0 E_0}\right)^{1/2}$$

$$\Rightarrow \quad E_{\mathrm{pot}}^{\mathrm{eff}}(r_{\mathrm{m}}) = \sqrt{\frac{Z_{\mathrm{eff}} \cdot \mathrm{e}^3 E_0}{\pi\varepsilon_0}}.$$

The ionization potential $IP = E_1 - E_{\mathrm{pot}}(\infty)$ is lowered to

$$IP^{\mathrm{eff}} = IP - \sqrt{\frac{Z_{\mathrm{eff}} \cdot \mathrm{e}^3 E_0}{\pi\varepsilon_0}}.$$

Chapter 2

1. **(a)** With the lifetime $\tau_K = 16$ ns of the upper level with $F' = 2$ the total transition probability is

$$A_K = A_{K1} + A_{K2} = \frac{1}{\tau_K} = 6.3 \times 10^7\ \text{s}^{-1}.$$

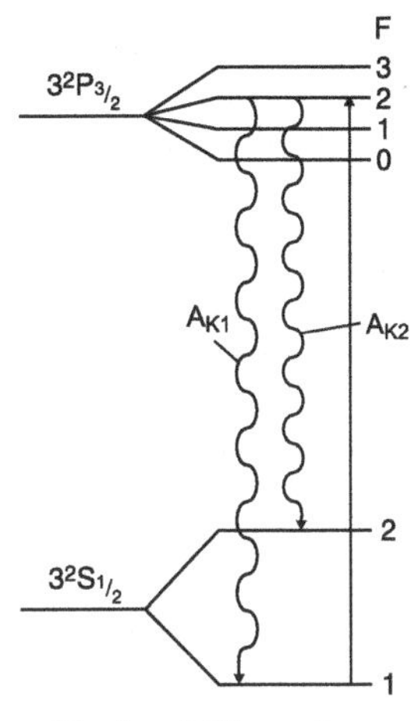

The intensity ratio of the two hf-components is

$$\frac{I(F' = 2 \to F_1'' = 1)}{I(F' = 2 \to F_2'' = 2)} = \frac{2F_1'' + 1}{2F_2'' + 1} = \frac{3}{5}.$$

This gives for the Einstein coefficients

$$A_{K1} = \frac{3}{8} A_K; \qquad A_{K2} = \frac{5}{8} A_K.$$

The transit time t_T of Na-atoms through the laser beam with $d = 0.01$ cm is

$$t_T = d/\bar{v} = \frac{10^{-2}}{5 \times 10^4}\ \text{s} = 2 \times 10^{-7}\ \text{s}.$$

The total relaxation rate into the lower level $F'' = 1$ is

$$R(F'' = 1) = N \cdot A + n_K(F' = 2) \cdot A \cdot d \cdot A_{K1}$$

where $N \cdot A = n(F'' = 1) \cdot A \cdot \bar{v}$ is the flux of atoms in level $F'' = 1$ through the focal area $A = 0.2 \times 0.01\ \text{cm}^2$, n [cm^{-3}] is the atomic density and $d = 0.01$ cm the pathlength through the focussed laser beam.

According to (2.20) the saturation intensity is then

$$\begin{aligned} I_S &= 2\sqrt{2} \cdot h\nu(A_{K1} + 1/t_T)/\lambda^2 \quad \text{with } t_T = d/\bar{v} \\ &= 2.8hc(A_{K1} + 1/t_T)/\lambda^3 \\ &= 77\ \text{W/m}^2. \end{aligned}$$

The necessary laser power is

$$P = I \cdot A = 15.4\ \mu\text{W}.$$

(b) The Doppler width is here larger than the hfs-splittings of the upper state. The pressure broadening is 250 MHz. The lower hfs-level $F'' = 1$ can be populated from the upper levels with $F' = 0, 1, 2$ and by collisional transfer from these levels and from $F'' = 2$.

The total refilling rate of level $F'' = 1$ is given by the sum of radiative transfer

$$R^{\text{rad}} = N_2' A_{21} + N_1' A_{11} + N_0' A_{01} = (5A_{21} + 3A_{11} + A_{01})N'/9 \approx A_k \cdot N_K$$

and the collisional transfer rate.

We have assumed that all hfs levels are equally populated apart from their statistical weights $(2F+1)$, because the collision broadening is larger than the natural linewidth and the hfs-splittings. The collisional refilling probability is $R_{\text{coll}} = 2\pi \times 2.5 \times 10^8\ \text{s}^{-1} \approx 1.5 \times 10^9\ \text{s}^{-1}$.

This gives for the saturation intensity

$$I_S = \frac{2.8hc}{\lambda^3}(A_K + R_{\text{coll}}) = 4.3 \times 10^3\ \text{W/m}^2$$

and a laser power

$$P = I \cdot A = 8.6 \times 10^{-4}\ \text{W} = 0.86\ \text{mW}.$$

2. The transmitted power is

$$P_t = P_0 \cdot e^{-N_i \sigma x}.$$

From $p = 10\ \text{mbar} = 10^3\ \text{Pa} \Rightarrow n = p/kT = 2.4 \times 10^{23}\ \text{m}^{-3}$.

The density of absorbing atoms is

$$n_i = 2.4 \times 10^{17}\ \text{cm}^{-3}$$

$$\Rightarrow \quad P_t = P_0 \cdot e^{-10^{-3}} \quad \text{for } x = 1\ \text{cm}$$

$$\Rightarrow \quad \frac{\Delta P}{P_0} = \frac{(P_t - P_0)}{P_0} \approx 10^{-3}.$$

The number of absorbed photons is

$$n_a = \frac{1}{3}\frac{\Delta P \cdot \Delta T}{h\nu} = 1.5 \times 10^{10}\ \text{photons/cm}^3.$$

The factor 1/3 takes into account that the laser bandwidth is three times the absorption linewidth.

$\Rightarrow$ The fraction of excited atoms is than

$$\frac{n_a}{n_i} = \frac{1.5 \times 10^{10}}{2.4 \times 10^{17}} = 6.3 \times 10^{-8}.$$

3. Because of the Kramers–Kronig-relation the change Δn of the refractive index is related to the change $\Delta\alpha$ of the absorption coefficient by

$$\Delta n = \frac{2c}{\omega_0}\Delta\alpha.$$

The phase shift between the σ^+ and σ^- component is

$$\Delta\phi = \frac{\omega \cdot L}{c}\Delta n = 2L \cdot \Delta\alpha = 2 \times 10^{-2}\alpha_0 \cdot L$$

$$\Rightarrow \quad \Delta\phi = 2 \times 10^{-2} \times 5 \times 10^{-2} = 10^{-3}.$$

The angle of the polarization plane has changed by

$$\Delta\varphi = \frac{1}{2}\Delta\phi = \frac{1}{2} \times 10^{-3}\ \text{rad} \mathrel{\hat{=}} 0.03°.$$

4. The number of detected fluorescence photons per s is

$$n_{\text{Fl}} = 0.05 \cdot 0.2 \cdot n_{\text{a}}/2$$

where n_{a} is the rate of absorbed laser photons.

The rate of incident laser photons is

$$n_{\text{ph}} = (I/h\nu) \cdot A = 10^{-1}/\left(6.6 \times 10^{-34} \times 5 \times 10^{14}\right) = 3 \times 10^{12}\ \text{s}^{-1}$$

where $A = \pi\omega_0^2$ is the cross section of the laser beam.

The absorption probability per atom is

$$P_{\text{if}} = \sigma \cdot n_{\text{ph}}/A = a \cdot I \cdot n_{\text{ph}}/A = a \cdot h\nu n_{\text{ph}}^2/A^2 = 6 \times 10^{-4}.$$

This gives the number of absorbed photons $= \frac{1}{2}$ number of excited atoms

$$n_{\text{a}} = n \cdot A \cdot L \cdot P_{\text{if}} = 10^{12} \times \pi \times 10^{-4} \times 1 \times 6 \times 10^{-4}$$
$$= 1.9 \times 10^5\ \text{s}^{-1}$$
$$\Rightarrow \quad n_{\text{Fl}} = 10^{-2} \times \frac{1}{2} \times 1.9 \times 10^5 = 9.5 \times 10^2\ \text{s}^{-1}.$$

With a quantum efficiency of 20 % of the photo cathode one obtains a counting rate of 1.9×10^2 counts/s.

5. The saturation signal is proportional to $\alpha^0 - \alpha_{\text{S}}$, where α is the absorption coefficient.

$$\left(\alpha^0 - \alpha_{\text{S}}\right) \propto S \cdot \alpha^0.$$

For the transition $(F'' = 1 \to F' = 1)$ the saturation parameter is $S_1 = 2$. The transition probability ratio R for the two transitions $(F'' = 1 \to F_1' = 1)$ and $(F'' = 1 \to F_2' = 2)$ is

$$R = \frac{2F_1' + 1}{2F_2' + 1} = \frac{3}{5}.$$

Therefore the saturation parameter $S_2 = \frac{5}{3} \cdot S_1 = 3.3$.

The saturation signal amplitudes A are

$$\left(\Delta N^0 - \Delta N\right) \cdot I \propto \Delta N_0 \cdot S \cdot I.$$

For the cross-over signal only $I/2$ is acting on each transition. Therefore the saturation parameters are $S_1/2$ and $S_2/2$

$$A_{\text{co}} = \frac{\Delta N_0}{2}(S_1 + S_2) \cdot I/2 = \frac{\Delta N_0}{4}(2 + 3.3) = \frac{5.3}{4}\Delta N_0 I$$

while the saturation signals are

$$A_1 = 2\Delta N_0 I \quad \text{and} \quad A_2 = 3.3\Delta N_0 I$$

$$A_{co} = \frac{1}{4}(A_1 + A_2).$$

6. The linewidth γ is mainly determined by the transit time t_T of the H atoms through the laser beam

$$t_T = d/\bar{v} = \frac{10^{-3}}{10^3}\,\text{s} = 10^{-6}\,\text{s} \quad \Rightarrow \quad \gamma = \frac{1}{t_T} = 10^6\,\text{s}^{-1}.$$

From $h\nu = \frac{1}{2}[E(2S_{1/2}) - E(1S_{1/2})] \Rightarrow \nu = 1.23 \times 10^{15}\,\text{s}^{-1}$

$$\Rightarrow \quad P_{if} = \sigma_0^2 \frac{I^2}{(\gamma \cdot h\nu)^2} = 1.5 \times 10^{-6}$$

$\Rightarrow$ 1.5×10^{-6} of all atoms are excited into the $2^2S_{1/2}$ level.

Chapter 3

1. The flux density of incident photons, focussed to the area A from a laser with output power P is

$$N_{ph} = \frac{P}{h\nu \cdot A} = 2.5 \times 10^{21}\,\text{photons}/(\text{s}\,\text{cm}^2).$$

The rate of scattered photons at a molecular density N_i in the volume V is

$$N_{sc} = N_{ph} \cdot N_i \cdot V \cdot \sigma$$
$$= 1.25 \times 10^{-11}\,N_i.$$

The rate of photoelectrons is

$$N_{PE} = \delta \cdot \eta \cdot N_{sc} = 0.1 \times 0.25 \times 1.25 \times 10^{-11}\,N_i/\text{s}$$
$$= 3.1 \times 10^{-13}\,N_i/\text{s}.$$

This number should be larger than 30

$$\Rightarrow \quad N_i > \frac{30}{3.1 \times 10^{-13}}\,\text{cm}^{-3} = 9.7 \times 10^{13}\,\text{cm}^{-3}.$$

2. The $3N - 6 = 3$ normal modes of H_2O are: The symmetric stretch ν_1, the bending vibration ν_2 and the asymmetric stretch ν_3

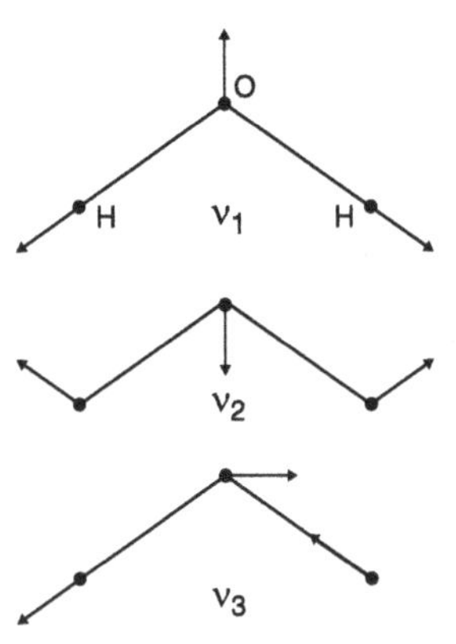

For all three normal vibrations the polarizability changes. Therefore all three are Raman-active.
The vibrations ν_1, ν_2, and ν_3 are also infrared active, because the electric dipole moment changes its magnitude for ν_1 and ν_2 and its direction for ν_3.

3. The rate of scattered photons is (see Problem 3.1)

$$N_{sc} = N_{ph} \cdot N_i \cdot V \cdot \sigma$$

$$N_{ph} = \frac{10\,\mathrm{W}}{h\nu \cdot A} = 2.4 \times 10^{21}\ \text{photons}/(\mathrm{s\,cm^2})$$

$$N_i \cdot V = 10^{21}$$

$$\Rightarrow \quad N_{sc} = 2.4 \times 10^{21} \times 10^{21} \times 10^{-29}/\mathrm{s}$$

$$= 2.4 \times 10^{13}/\mathrm{s}.$$

The energy released per second is

$$\mathrm{d}W_H/\mathrm{d}t = N_{sc} \cdot h(\nu_i - \nu_S) = N_{sc} \cdot h \cdot c \cdot (\bar{\nu}_i - \bar{\nu}_S) = N_{sc} \cdot hc \times 10^3\ \mathrm{W}$$

$$= 4.75 \times 10^{-7}\ \mathrm{W}.$$

If the laser wavelength is close to resonance with an absorption line, the additional heat is

$$\left(\frac{\mathrm{d}W_H}{\mathrm{d}t}\right)_{abs} = P_0\left(1 - \mathrm{e}^{-\alpha x}\right) \approx P_0\left(1 - \mathrm{e}^{-0.1 \cdot 0.5}\right) \approx P_0 \cdot 0.05 = 0.5\ \mathrm{W}.$$

4. The number of Raman photons is

$$N_{sc} = N_{ph} \cdot N_i \cdot V \cdot \sigma$$

$$N_{ph} = \frac{P_L}{h\nu \cdot A} = \frac{1}{1.6 \times 10^{-19} \times 2.2\pi \times 25 \times 10^{-4}}\ \mathrm{cm^{-2}}$$

$$= 10^{18}\ \mathrm{s^{-1}}.$$

$$N_i = 10^{21}\ \mathrm{cm^{-3}}; \quad V = L \cdot \pi R^2 = 0.78\ \mathrm{cm^3}; \quad \sigma = 10^{-30}\ \mathrm{cm^2}$$

$$\Rightarrow \quad N_{sc} = 7.8 \times 10^8/\mathrm{s}.$$

The intensity of the Raman radiation at $\lambda_S = 550$ nm is at the exit of the fibre

$$I = \frac{N_{sc} \cdot h\nu}{\pi R^2} = 3.2 \times 10^{-6}\ \mathrm{W/cm^2} = 3.2\ \mu\mathrm{W/cm^2}.$$

5. The diffraction-limited beam waist of the incident beams is in the focal plane

$$w_0 = \frac{f \cdot \lambda}{\pi w_S}$$

where $w_S = \frac{1}{2} \times 3$ mm is the beam waist at the lens.

With $f = 5$ cm we obtain

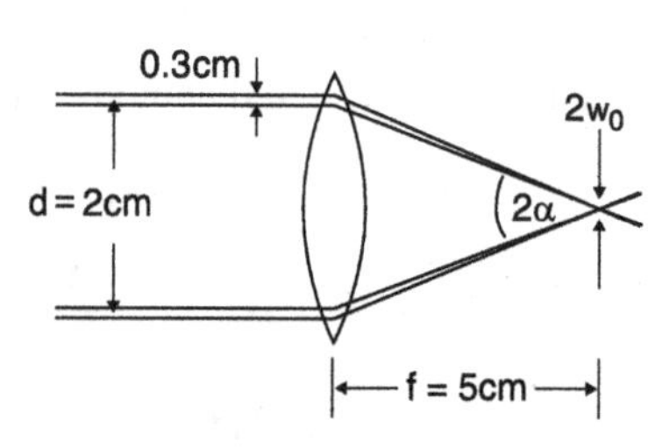

$$w_0 = 5.3 \times 10^{-4}\ \mathrm{cm} = 5.3\ \mu\mathrm{m}.$$

The angle α is determined by

$$\tan\alpha = \frac{d}{2f} = \frac{1}{10} = 0.1$$

$$\Rightarrow \quad \alpha = 6°.$$

The length of the overlap region is given by

$$l \cdot \tan\alpha \le 4w_0$$

$$\Rightarrow \quad l \le \frac{4 \times 5.3 \times 10^{-4}}{0.1}\ \mathrm{cm} = 2 \times 10^{-2}\ \mathrm{cm} = 0.2\ \mathrm{mm}.$$

Chapter 4

1. **(a)** The Doppler broadening and shift is caused only by $v_\parallel$

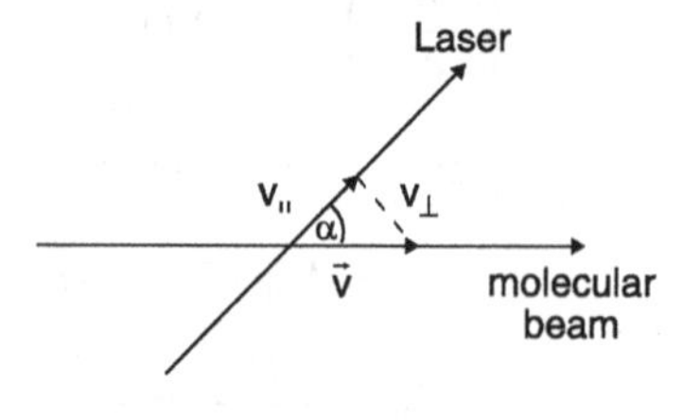

$$v_\parallel = v \cdot \cos\alpha.$$

The Doppler shift of the line center is

$$\omega' = \omega_0 + (v_\parallel/c)\omega_0 = \omega_0\left(1 + (v/c)\cos 45°\right) = \omega_0\left(1 + \frac{1}{2}\sqrt{2}v/c\right).$$

The absorption profile is

$$\alpha(\omega) = \alpha_0 \exp\left[-(\omega - \omega')/\delta\omega_D\right]^2$$

$$\delta\omega_D = 2\sqrt{\ln 2}\cdot\omega_0\cdot\cos\alpha\cdot v_p/c$$

where v_p is the most probable velocity

$$\delta\omega_D = \cos\alpha\cdot(\omega_0/c)\sqrt{8kT_{eff}\ln 2/m}.$$

(**b**) For a divergent molecular beam with a collimation angle $\varepsilon = 5°$ additional broadening occurs.

The velocity component $v_\parallel$ has now two contributions

$$v_\parallel = v\cos\alpha + v_\perp\sin\alpha.$$

$$\text{With } v_\perp = v\tan\varepsilon \quad\Rightarrow\quad v_\parallel = v(\cos\alpha + \sin\alpha\cdot\tan\varepsilon)$$
$$= v\left(\frac{1}{2}\sqrt{2}\pm\frac{1}{2}\sqrt{2}\cdot 0.09\right)$$

because ε ranges from $-5°$ to $+5°$.

The parallel component $v_\parallel$ covers a range

$$0.643v \le v_\parallel \le 0.771v.$$

The additional shift of the line center is negligible, the additional broadening is obtained as

$$\delta\omega_0 = 2\sqrt{\ln 2}\omega_0\cos\alpha(v_p/c)(1+\sqrt{2}\cdot 0.09).$$

2. Only molecules with velocity components

$$v_x < \Delta\omega_h/k = 2\pi\,\Delta\nu_h/k = \lambda\cdot\Delta\nu_h$$

can absorb the monochromatic laser radiation.

With $v_x = v\tan\varepsilon$ this gives

$$\tan\varepsilon \le \lambda\cdot\Delta\nu_h/v.$$

At a distance d from the nozzle

$$\Delta x = 2d\cdot\tan\varepsilon$$
$$\Rightarrow\quad \Delta x \le 2d\cdot\lambda\cdot\Delta\nu_h/v.$$

Inserting the numerical values $d = 10$ cm, $\lambda = 500$ nm, $\Delta\nu_h = 10$ MHz, $v = 500$ m/s gives $\Delta x \le 0.2$ cm.

3. (**a**) The velocity distribution of the thermal molecular beam is

$$n(v) = C\cdot v^2\cdot e^{-mv^2/2kT}$$

where $n(v)$ is the density, $v = v_z$ the velocity component in the direction of the beam, and C is a constant factor.

$$\text{With } \omega_a = \omega_0(1 - v_z/c) \quad \Rightarrow \quad v = v_z = (\omega_0 - \omega_a) \cdot c/\omega_0$$

we obtain ($\omega_a = \omega$ is the absorption frequency of a molecule moving with velocity v)

$$\alpha(\omega) \propto n(\omega) = C^*(\omega_0 - \omega)^2 \cdot \mathrm{e}^{-\frac{m}{2kT}[(\omega_0-\omega)\cdot c/\omega_0]^2}.$$

This differs from the Gaussian profile because of the factor $(\omega_0 - \omega)^2$. It is *not* symmetric with respect to the absorption frequency ω_0 of a molecule at rest.
(b) For complete saturation the absorption is proportional to the flux $N = n \cdot V$, not the density n of molecules into the observation region. Therefore

$$N(v) = C \cdot v^3 \mathrm{e}^{-mv^2/2kT} \quad \text{and}$$

$$\alpha(\omega) = C^*(\omega_0 - \omega)^3 \mathrm{e}^{-\frac{m}{2kT}[(\omega_0-\omega)\cdot c/\omega_0]^2}.$$

4. **(a)** The flight time between z_1 and z_2 is

$$t = d/v \quad \Rightarrow \quad \mathrm{d}t = -\left(d/v^2\right)\mathrm{d}v.$$

The velocity distribution is

$$N(v)\,\mathrm{d}v = C \cdot u^3 \cdot \mathrm{e}^{-m(v-u)^2/2kT}\,\mathrm{d}v$$

where u is the flow velocity $u = \bar{v}$.
With $t_0 = d/u$ and $t = d/v \Rightarrow \mathrm{d}t = -(d/v^2)\,\mathrm{d}v$ we obtain

$$\begin{aligned} N(t)\,\mathrm{d}t &= C\left(\frac{d^3}{t_0^3} \cdot \frac{v^2}{d}\right) \cdot \mathrm{e}^{-\frac{m}{2kT}\left(\frac{d}{t}-\frac{d}{t_0}\right)^2}\,\mathrm{d}t \\ &= C\left(\frac{d^4}{t^2 t_0^3}\right) \cdot \mathrm{e}^{-\frac{d^2 \cdot m}{2kT}\left(\frac{t_0-t}{t_0 \cdot t}\right)^2}\,\mathrm{d}t. \end{aligned}$$

Although $N(v)$ is symmetric with respect to u, $N(t)$ is not, because of the factor $1/t^2$.
(b) For the triangle approximation

$$N(v) = a\left(u - 10|u - v|\right), \quad 0.9u \le v \le 1.1u.$$

We obtain with $C = a \cdot d/t_0$

$$N(t)/\mathrm{d}t = C \cdot \frac{t^2}{d}\left(1 - 10\left|1 - \frac{t_0}{t}\right|\right)\mathrm{d}t$$

with $0.9t_0 \le t \le 1.1t_0$ where $N(t) = 0$ for $t = 0.9t_0$ and $t = 1.1t_0$.

5. For the vibrational levels of diatomic molecules is $g_i = 1$.

$$\Rightarrow \quad N(v'') = (N/Z)\mathrm{e}^{-E_{\mathrm{vib}}/kT_{\mathrm{vib}}}.$$

For $v'' = 0$:

$$N(v'' = 0) = (N/Z)\mathrm{e}^{-hc\omega_e/(2kT_{\mathrm{vib}})}.$$

The partition function is

$$Z = \sum \mathrm{e}^{-hc(n+1/2)\omega_e/kT} = \left(1 - \mathrm{e}^{-hc\omega_e/kT}\right)^{-1}$$

$$\Rightarrow \quad \frac{1}{Z} = 0.88 \quad \Rightarrow \quad N(v'' = 0) = 0.88N \cdot 0.34 = 0.3N.$$

30 % of all molecules are at $T = 100$ K in the lowest vibrational level $v'' = 0$.
For $v'' = 1 \Rightarrow \mathrm{e}^{-hc\omega_e \cdot 3/(2kT_{\mathrm{vib}})} = 0.04$

$$\Rightarrow \quad N(v'' = 1) = 0.034N.$$

Only 3.4 % of all molecules are in $v'' = 1$.
For the rotational distribution we assume that T_{rot} is independent of v''. Then

$$N(J'' = 20) = \left[(2J'' + 1)/Z_{\mathrm{rot}}\right] \cdot N_{\mathrm{rot}} \cdot \mathrm{e}^{-E_{\mathrm{rot}}/kT_{\mathrm{rot}}}$$

$$\mathrm{e}^{-E_{\mathrm{rot}}/kT_{\mathrm{rot}}} = \mathrm{e}^{-J(J+1)\cdot hcBe/kT_{\mathrm{rot}}}$$

$$= \mathrm{e}^{-20\cdot 21\cdot hc\cdot 0.15/k\cdot 10} = \mathrm{e}^{-9.04} = 1.2 \times 10^{-4}$$

$$\Rightarrow \quad N(J'' = 20) = \frac{N_{\mathrm{rot}} \cdot 41}{46} \times 1.2 \times 10^{-4} = 1.07 \times 10^{-4} N_{\mathrm{rot}}$$

$$\text{because } Z_{\mathrm{rot}} = \frac{kT_{\mathrm{rot}}}{hcB} = 46$$

$$\Rightarrow \quad N_{\mathrm{i}}(v_{\mathrm{i}}'' = 0, J_{\mathrm{i}}'' = 20) = 0.3 \times 1.07 \times 10^{-4} N = 3.2 \times 10^{-5} N.$$

The rotational level with the maximum population is obtained from $\mathrm{d}N(J)/\mathrm{d}J = 0$

$$\Rightarrow \quad J\left(N_{\mathrm{rot}}^{\max}\right) = 4$$

$$N_{\mathrm{i}}(J_{\mathrm{i}}'' = 4, v_{\mathrm{i}}'' = 0) = 0.475N.$$

Chapter 5

1. The linear polarized laser excites only the level $|2\rangle$ with $M = 0$.

$$\frac{\mathrm{d}N_2}{\mathrm{d}t} = R_1 N_1 - (R_2 + R_3) N_2 = 0$$

with $R_1 =$ excitation rate, $R_2 =$ collision induced mixing rate, $R_3 =$ spontaneous decay.

$$\Rightarrow \quad \frac{N_2}{N_1} = \frac{R_1}{R_2 + R_3} = \frac{10^7}{5 \times 10^6 + 10^8} = 9.5 \times 10^{-2}$$

$$\frac{dN_3}{dt} = R_2 N_2 - R_3 N_3 = 0$$

$$\Rightarrow \quad \frac{N_3}{N_2} = \frac{R_2}{R_3} = \frac{5 \times 10^6}{10^8} = 0.05.$$

The alignment is then

$$A = \frac{N_2 - N_3}{N_2 + N_3} = \frac{1 - N_3/N_2}{1 + N_3/N_2} = \frac{1 - 2 \times 0.05}{1 + 2 \times 0.05} = \frac{0.9}{1.1} = 0.82$$

where the factor 2 comes from the equal populations of levels with $M = +1$ and $M = -1$.

2. The optical pumping probability is

$$P_{12} \propto |\boldsymbol{D} \cdot \boldsymbol{E}|^2 = D^2 E^2 \cdot \cos^2 \alpha.$$

Only the upper level with $M = 0$ can be excited.

Without collisional mixing the alignment becomes $A = 1$ because $N_3 = 0$ (see Problem 5.1, Vol. 1).

The saturation parameter for a transition $|1\rangle \to |2\rangle$ is

$$S = \frac{2\sigma_{12} \cdot I(\omega)}{\hbar\omega \cdot A_{12}}.$$

With $\sigma_{12} = 10^{-13}$ cm^2, $I(\omega) = 1$ W/cm^2, $\hbar\omega = 4 \times 10^{-19}$ Ws and $A_{21} = 5 \times 10^6$ s^{-1} $\Rightarrow S = 0.1$.

3. **(a)** The force on the molecules is

$$\boldsymbol{F} = \frac{d\boldsymbol{B}}{dx} \cdot \mu.$$

The acceleration in the x-direction is $\boldsymbol{a} = \boldsymbol{F}/m$. The deviation Δx after a flight time t is

$$\Delta x = \frac{1}{2} a t^2 = \frac{1}{2} a \left(\frac{L}{v}\right)^2 = \frac{1}{2m}\left(\frac{d\boldsymbol{B}}{dx}\right)\mu \cdot \frac{L^2}{v^2}.$$

The deviation angle ϑ is

$$\tan \vartheta = \frac{\Delta x}{L} = \frac{1}{2m}\left(\frac{d\boldsymbol{B}}{dx}\right)\mu \cdot \frac{L^2}{v^2}.$$

Inserting the numerical values give

$$\tan\vartheta = 1.76 \times 10^{-2} \quad \Rightarrow \quad \vartheta = 1.04^\circ.$$

(b) The distance between the beginning of the magnetic field and the detector is $d = (0.1 + 0.5)\ \text{m} = 0.6\ \text{m}$.

If $N_0(x = 0)$ has a rectangular profile, than $N(x = d)$ has also a rectangular profile, which is shifted in the x-direction by the deflection angle ϑ, i.e. $N(x = d) = a \cdot N_0$ for $-d \cdot (\varepsilon - \vartheta) \le x \le d(\varepsilon + \vartheta)$ otherwise $N = 0$, if the width $\Delta x(d = 0)$ can be neglected. The width $\Delta x(d)$ of the distribution is $\Delta x = 2d \cdot \varepsilon = 100 \cdot 0.017 = 1.7$ cm.

(c) The profile for $N_1(x)$ is

$$-50 \cdot (0.0175 - 0.0176) \le x \le 50 \cdot (0.0175 + 0.0176)$$

$$= +5 \times 10^{-3} \le x \le 1.75\ \text{cm}.$$

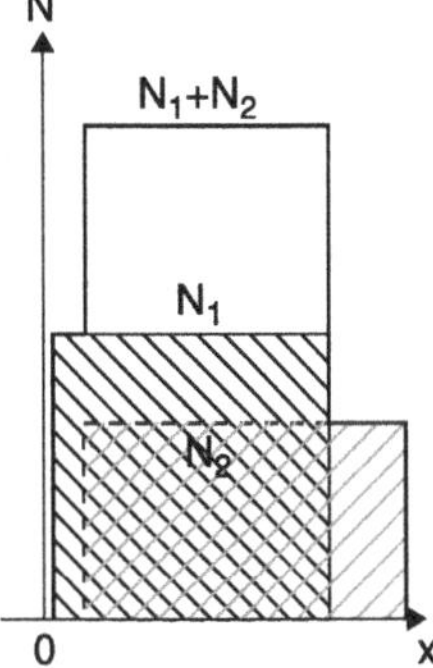

The profile for N_2 is with $\tan\vartheta = 1.98 \times 10^{-2}$

$$10^{-2} \cdot 50 \cdot 0.23 \le x \le 50 \cdot 3.73 \times 10^{-2}\ \text{cm}$$

$$= 0.115\ \text{cm} \le x \le 1.86\ \text{cm}.$$

For $N_1 = N_2$ the two rectangular profiles have the same height.

4. Under stationary conditions we have

$$\frac{\mathrm{d}N_k}{\mathrm{d}t} = R_{ik}N_i - N_k\left(\frac{1}{\tau_k} + 5 \times 10^7\ \text{s}^{-1}\right) = 0$$

$$\Rightarrow \quad \frac{N_k}{N_i} = \frac{R_{ik}}{10^8 + 5 \times 10^7}$$

$$\Rightarrow \quad \frac{N_k}{N_i} = \frac{2}{3} \times 10^{-8} R_{ik}.$$

For the lower state $|i\rangle$ we obtain

$$\frac{\mathrm{d}N_i}{\mathrm{d}t} = 0 = -R_{ik} \cdot N_i + R_{\text{coll}}(N_{i0} - N_i)$$

where N_{i0} is the population without optical pumping

$$\Rightarrow \quad \frac{N_i}{N_{i0}} = \frac{R_{\text{coll}}}{R_{ik} + R_{\text{coll}}} = 0.8 \quad \Rightarrow \quad R_{ik} = 0.25 R_{\text{coll}} = 1.25 \times 10^7\ \text{cm}^{-1}$$

$$\Rightarrow \quad \frac{N_k}{N_i} = \frac{2}{3} \times 10^{-8} \times 1.25 \times 10^7 = 8.3 \times 10^{-2} = 0.083.$$

The closely lying levels $|i\rangle$ and $|n\rangle$ in the ground state have without optical pumping nearly the same thermal population. With optical pumping the population difference is therefore $\Delta N = N_n - N_i = 0.2 N_{i0}$. In the upper state all levels are empty without optical pumping.

With optical pumping $N_k = 0.083 N_i = 0.066 N_{i0}$. The microwave signal is therefore larger in the ground state by a factor $0.2/0.066 = 3$.

5. The excitation rate is

$$R_{\text{exc}} = N_i \cdot (I/h\nu) \cdot \sigma_{ik}.$$

The refilling rate of level $|i\rangle$ is

$$R_{\text{ref}} = 10^7 (N_{i0} - N_i)$$

where N_{i0} is the population without pump laser.

Under stationary conditions we obtain

$$\frac{\mathrm{d}N_i}{\mathrm{d}t} = 0 = -R_{\text{exc}} + R_{\text{ref}}$$

$$\Rightarrow \quad N_i\left[(I/h\nu)\sigma_{ik} + 10^7\ \text{s}^{-1}\right] = 10^7\ \text{s}^{-1} \cdot N_{i0}$$

$$\Rightarrow \quad \frac{N_i}{N_{i0}} = \frac{10^7}{(I/h\nu)\sigma_{ik} + 10^7\ \text{s}^{-1}} = \frac{1}{2}$$

$$\Rightarrow \quad (I/h\nu)\sigma_{ik} = 10^7\ \text{s}^{-1}$$

$$\Rightarrow \quad I = \frac{10^7 \times 6.6 \times 10^{-34} \times 3 \times 10^8}{10^{-18} \times 5 \times 10^{-7}}\ \text{W/m}^2$$

$$= 4 \times 10^6\ \text{W/m}^2 = 400\ \text{W/cm}^2.$$

6. **(a)** Choosing the energy scale with $E(r = \infty) = 0$, the energy of the Rydberg level is

$$E_n = -\frac{Ry \cdot hc}{(n - \delta)^2}.$$

For $n = 50$ and $\delta = 2.18 \Rightarrow E_n = -9.5 \times 10^{-22}\ \text{J} \mathrel{\hat{=}} 5.9 \times 10^{-3}\ \text{eV}$.

(b) The appearance potential AP for an atom in an external electric field with amplitude E_0 is, according to (5.12)

$$AP = -IP - \sqrt{\frac{\mathrm{e}^3 E_0}{\pi \varepsilon_0}}.$$

Field ionization takes place for $AP = 0$

$$\Rightarrow \quad \sqrt{\frac{\mathrm{e}^3 E_0}{\pi \varepsilon_0}} = -IP = 9.5 \times 10^{-22}\ \text{J}$$

$$\Rightarrow \quad E_0 = \frac{(IP)^2 \pi \varepsilon_0}{e^3} = 6.1 \times 10^3 \text{ V/m}.$$

(c)

$$h \cdot \nu = Ry \cdot hc \left(\frac{1}{(n-\delta_1)^2} - \frac{1}{(n-\delta_2)^2} \right)$$

$$\Rightarrow \quad \omega_{\text{rf}} = 2\pi Ry \cdot c \left(\frac{1}{47.82^2} - \frac{1}{48.29^2} \right)$$

$$= 2\pi \cdot 28 \times 10^9 \text{ s}^{-1}$$

$$\nu_{\text{rf}} = 28 \text{ GHz}.$$

7. According to (5.18) the linewidth is

$$\Delta\Gamma = \gamma_3 + \left[(k_2/k_1)\gamma_1 + 1 \mp (k_2/k_1)\gamma_2\right]\sqrt{1+S}.$$

For $\lambda_1 = 580$ nm, $\lambda_2 = 680$ nm $\Rightarrow k_2/k_1 = \lambda_1/\lambda_2 = 0.85$.

The natural level widths are

$$\gamma_{n1} = 0, \qquad \gamma_{n3} = 0, \qquad \gamma_2 = \frac{2\pi}{\tau_2} = 6.3 \times 10^8 \text{ s}^{-1}.$$

The transit time broadening for molecules passing with velocity v through a Gaussian beam with waist v is $\delta\omega_{\text{tr}} = 2(v/w) \cdot \sqrt{2\ln 2} \approx 2.4 \cdot v/w$. The transit time is $t_{\text{tr}} = w/v$.

The level widths are

$$\gamma_1 = \frac{2\pi}{\tau_1} + \delta\omega_{\text{tr}} = 0 + 8 \times 10^7 \text{ s}^{-1}$$

$$\gamma_2 = \frac{2\pi}{\tau_2} + \delta\omega_{\text{tr}} = \left(6.3 \times 10^8 + 8 \times 10^7\right) \text{s}^{-1} = 7.1 \times 10^8 \text{ s}^{-1}$$

$\gamma_3 = 0$ because the transition $2 \to 3$ is a stimulated Raman process, where the transit broadening has no effect

$$\Rightarrow \quad \Delta\Gamma = \left[0.85 \times 8 \times 10^7 + (1 \mp 0.85) \times 7.1 \times 10^8\right] \text{s}^{-1} \quad \text{for } S = 0.$$

For collinear propagation the minus sign applies

$$\Delta\Gamma_{\text{coll}} = 1.75 \times 10^8 \text{ s}^{-1} \quad \Rightarrow \quad \Delta\nu = \Delta F_{\text{coll}}/2\pi = 2.78 \times 10^7 \text{ s}^{-1}.$$

The linewidth of the signal is 27.8 MHz. This is about four times smaller than the natural linewidth.

For anticollinear propagation the plus-sign applies

$$\Rightarrow \quad \Delta\Gamma_{\text{anticell}} = 1.38 \times 10^9 \text{ s}^{-1} \quad \Rightarrow \quad \Delta\nu_{\text{anticell}} = 220 \text{ MHz}.$$

Chapter 6

1. If the maximum transmission of the Pockels cell is achieved for $U = 0$, the transmission $T(\theta)$ is

$$T = T_0 \cos^2\theta.$$

The turning angle θ of the plane of polarization is proportional to the applied voltage.

$$\theta = a \cdot U \quad \Rightarrow \quad T = T_0 \cos^2(a \cdot U).$$

For $U = 2\,\text{kV} \Rightarrow T = 0 \rightarrow a \cdot U = \pi/2 \Rightarrow a = \frac{\pi}{2 \cdot 2 \times 10^3\,\text{V}} = 7.8 \times 10^{-4}\,\text{V}^{-1}$.

In order to prevent laser oscillation, the total losses must exceed the gain

$$\Rightarrow \quad G_a \cdot T_0 \cos^2 aU < 0.3$$

$$\Rightarrow \quad \cos aU \leq 0.178$$

$$\Rightarrow \quad a \cdot U = 1.4 \quad \Rightarrow \quad U = 1.8\,\text{kV}.$$

2. The Fourier-transform of $f(x) = \mathrm{e}^{-ax^2}$ is

$$F(t) = \frac{1}{\sqrt{2a}} \mathrm{e}^{-t^2/4a}.$$

The Doppler profile is

$$f(\nu) = C \cdot \mathrm{e}^{-\frac{(\nu-\nu_0)^2}{0.36\delta\nu_\mathrm{D}^2}} \quad \Rightarrow \quad a = \frac{1}{0.36\delta\nu_\mathrm{D}^2}, \quad x = (\nu - \nu_0)$$

$$\Rightarrow \quad F(t) = \frac{\sqrt{0.36\delta\nu_\mathrm{D}}}{\sqrt{2}} \mathrm{e}^{-0.09\delta\nu_\mathrm{D}^2 t^2}.$$

With $\delta\nu_\mathrm{D} = 8 \times 10^9\,\text{s}^{-1} \Rightarrow F(t) = C^* \cdot \mathrm{e}^{-5.76 \times 10^{18} t^2}$.

With a full halfwidth $\Delta t = 2\sqrt{\frac{\ln 2}{5.76 \times 10^{18}}} = 0.7 \times 10^{-9}\,\text{s} = 0.7\,\text{ns}$.

3. **(a)** $\Delta z = \frac{c}{n} \cdot \tau = 10^{-4}\,\text{m} = 100\,\mu\text{m}$.

(b) The spectral width $\Delta\nu$ of the pulse is $\Delta\nu = \frac{1}{\tau} = 2 \times 10^{12}\,\text{s}^{-1}$

$$\Rightarrow \quad \Delta\lambda = -\frac{c}{\nu^2}\Delta\nu = -\frac{\lambda^2}{c}\Delta\nu = 2.4 \times 10^{-9}\,\text{m}$$

$$\Delta n = \frac{\mathrm{d}n}{\mathrm{d}\lambda}\Delta\lambda = 10^3 \cdot 2.4 \times 10^{-7} = 2.4 \times 10^{-4}.$$

The broadening of the time profile due to dispersion $\Delta\tau$ can be obtained as follows:

$$\Delta\tau = z_1\left(\frac{n + \Delta n}{c} - \frac{n}{c}\right) = z_1 \cdot \frac{\Delta n}{c}$$

$$\Rightarrow \quad z_1 = c \cdot \frac{\Delta\tau}{\Delta n} = \frac{3\times10^8 \cdot 5\times10^{-13}}{2.4\times10^{-4}}\ \text{m} = 0.625\ \text{m}.$$

(c)

$$\omega = \omega_0 - A\frac{\mathrm{d}I}{\mathrm{d}t}$$

$$A = n_2\omega \cdot \frac{z_1}{c} = n_2\frac{2\pi}{\lambda}z_1 = 6.54\times10^{-14}\ \text{m}^2/\text{W}$$

$$\frac{\mathrm{d}I}{\mathrm{d}t} \approx \frac{10^{13}}{5\times10^{-13}}\ \frac{\text{W}}{\text{m}^2\,\text{s}} = 2\times10^{25}\ \text{Wm}^{-2}\,\text{s}^{-1}$$

$$\Rightarrow \quad \omega_0 - \omega = A\cdot\frac{\mathrm{d}I}{\mathrm{d}t} = 1.3\times10^{12}\ \text{s}^{-1}.$$

The frequency broadening at z_1 is $2(\omega_0-\omega) = 2.6\times10^{12}\ \text{s}^{-1} \Rightarrow \Delta\nu = \frac{2.6}{\pi}\times 10^{12} = 4.1\times10^{11}\ \text{Hz} = 410\ \text{GHz}$.

4. According to (6.21) is

$$D = \frac{1}{\lambda}\frac{\mathrm{d}S}{\mathrm{d}\lambda}d^2\left[1-(\lambda/d-\sin\alpha)^2\right]^{3/2}.$$

Inserting the numerical values $\lambda = 600$ nm, $\mathrm{d}S/\mathrm{d}\lambda = 10^5$, $d = 1\ \mu$m, $\sin 30° = \frac{1}{2} \Rightarrow D = 16.6$ cm.

5. According to (6.57),

$$\frac{1}{\tau_{\text{eff}}} = \frac{1}{\tau_{\text{rad}}} + \left(\frac{8}{\pi\mu kT}\right)^{1/2}\cdot\sigma\cdot p.$$

The lifetime τ_{eff} is reduced to half of its radiative value if

$$\left(\frac{8}{\pi\mu kT}\right)^{1/2}\cdot\sigma\cdot p = \frac{1}{\tau_{\text{rad}}} = 6.25\times10^7\ \text{s}^{-1}$$

$$\Rightarrow \quad p = 6.25\times10^7\cdot\frac{1}{\sigma}\cdot\left(\frac{\pi\mu kT}{8}\right)^{1/2} \quad \text{with } \mu = \frac{m_{\text{Na}_2}\cdot m_{\text{Ar}}}{m_{\text{Na}_2}+m_{\text{Ar}}} = 21.4\ \text{AMU}$$

$$= 613\ \text{Pa} \mathrel{\hat{=}} 6.13\ \text{mbar}.$$

Chapter 7

1. For a level with $J = 1$, $L = 1$, $S = 0$, $I = 0$ the Landé factor is $g = 1$. The Zeeman shift is:

$$\Delta E = \mu_{\text{B}}\cdot m\cdot B \quad \text{with } -J \le m \le +J.$$

The halfwidth $\Delta B_{1/2}$ of the Hanle signal is obtained when $2\Delta E = \mu_B B - (-\mu_B B)$ between the levels with $m = 1$ and $m = -1$ equals the natural linewidth $\Delta\nu_n = \frac{1}{2\pi\tau}$

$$\Rightarrow \quad 2\mu_B \Delta B_{1/2} = \frac{h}{2\pi\tau}$$

$$\Delta B_{1/2} = \frac{h}{2\mu_B \cdot 2\pi\tau} \text{ [T]}$$

$$\Delta B_{1/2} = 3.8 \times 10^{-4} \text{ Tesla.}$$

For the molecular level with $J = 20$, $L = 1$, $F = 21$, $I = 0$, $S = 0$, the Landé g-factor becomes $g = 1 \times 10^{-4}$. Therefore

$$\Delta B_{1/2} = \frac{h}{2g\mu_B \cdot 2\pi\tau} = 3.8 \text{ Tesla.}$$

2. The Landé g-factor is

$$g = g_J \cdot g_F \quad \text{with } g_F = \frac{F(F+1) + J(J+1) - I(I+1)}{2F(F+1)}.$$

For the $5\,^2S_{1/2}$ state is $g_J = 2$. For the hyperfine levels with $F = 3$ and $I = 5/2$ is $g_F = +\frac{1}{6}$, for $F = 2$ is $g_F(F = 2) = -\frac{1}{6}$.

The Zeeman splitting of $F = 2$ ($M_F = -1$ and $M_F = +1$) is

$$\Delta E = g_J \cdot g_F \cdot \Delta M \cdot B \cdot \mu_B = 2 \cdot \frac{1}{6} \cdot 2 \cdot \mu_B \cdot B = \frac{2}{3}\mu_B \cdot B.$$

The width of the Hanle signal is obtained from $\frac{2}{3}\mu_B \Delta B = \Delta\nu_{tr} \cdot h = \frac{0.4h}{T}$ (see (3.63) in Vol. 1)

$$\Rightarrow \quad \Delta B = 0.6\frac{h}{\mu_B T} = 4.27 \times 10^{-10} \text{ Tesla} = 4.27 \times 10^{-6} \text{ Gauss.}$$

This is a very narrow signal because the lifetime of the level is $\tau = \infty$ and the linewidth is solely limited by the transit time T, which can be prolonged by enlarging the laser beam or by adding a buffer gas.

3. The energy separation of the two Zeeman components with $M_J = \pm\frac{1}{2}$ is

$$\Delta E = g_J \cdot \mu_B \cdot B \quad \text{and} \quad g_J = 1 + \frac{J(J+1) + S(S+1) - L(L+1)}{2J(J+1)} = \frac{2}{3}.$$

$\Rightarrow$ The quantum beat frequency is

$$\nu = \frac{\Delta E}{h} = \frac{2}{3} \cdot \frac{\mu_B \cdot B}{h} = 9.4 \times 10^7 \text{ s}^{-1} = 94 \text{ MHz}$$

$$\Rightarrow \quad T_p = \frac{1}{\nu} = 1.06 \times 10^{-8} \text{ s.}$$

4. The oscillation frequency is $\nu = (\nu_1 + \nu_2)/2$

$$\nu = c \cdot \bar{\nu} = 3 \times 10^{10} \cdot 108.5\ \mathrm{s}^{-1} = 3.26 \times 10^{12}\ \mathrm{s}^{-1},$$

where $\bar{\nu}$ is the wavenumber

$$\Rightarrow \quad T_1 = \frac{1}{\nu} = 3 \times 10^{-13}\ \mathrm{s} = 307\ \mathrm{fs}.$$

The period of the envelope is

$$T_2 = \frac{2}{\nu_1 - \nu_2} = \frac{2}{c(\bar{\nu}_1 - \bar{\nu}_2)} = \frac{2}{3 \times 10^{10}}\ \mathrm{s} = 6.7 \times 10^{-11}\ \mathrm{s} = 67\ \mathrm{ps}.$$

5. The uncertainty of the difference frequency is

$$\delta\nu = \sqrt{\delta\nu_1^2 + \delta\nu_2^2} = \sqrt{50 \times 10^{10}}\ \mathrm{s}^{-1} = 7.1 \times 10^5\ \mathrm{s}^{-1}.$$

With a signal to noise ratio of 50 the difference frequency can be measured within an uncertainty of $\delta\nu = 1.4 \times 10^4\ \mathrm{s}^{-1} = 14\ \mathrm{kHz}$.

6. The spectrum of the heterodyne signal shows a Gaussian distribution around $\omega = 0$, because the difference frequency of the superposition of a monochromatic line profile with a Doppler-broadened profile gives a Gaussian profile around $\omega = 0$. The spectral width is

$$\delta\nu_\mathrm{D} = \frac{\nu_0}{c} \cdot \sqrt{\frac{8kT \ln 2}{m}} = \frac{\nu_0}{c}\sqrt{\langle v^2 \rangle} \cdot \sqrt{\frac{8}{3}\ln 2} = 1.36\frac{\nu_0}{c}\sqrt{\langle v^2 \rangle} = \frac{1.36}{\lambda}\sqrt{\langle v^2 \rangle}$$

$$\Rightarrow \quad \delta\nu_\mathrm{D} = 2.2 \times 10^3\ \mathrm{s}^{-1} = 2.2\ \mathrm{kHz}.$$

Chapter 8

1. The density of atoms B is

$$n = \frac{P}{kT} = 2.42 \times 10^{22}/\mathrm{m}^3 = 2.42 \times 10^{16}\ \mathrm{cm}^{-3}.$$

The pressure broadening is $2\pi\Delta\nu_p = n \cdot \sigma^{\mathrm{inel}}\bar{v}$ with $\bar{v} = \sqrt{8kT/\pi \cdot \mu} = 4.0 \times 10^2\ \mathrm{m/s}$. $\Delta\nu_p = 1.54 \times 10^5\ \mathrm{s}^{-1}$.

The natural linewidth is

$$\Delta\nu_n = \frac{1}{2\pi\tau} \approx 1.6 \times 10^7\ \mathrm{s}^{-1}$$

and therefore two orders of magnitude larger than $\Delta\nu_p$.

The saturation broadening is $\Delta\nu_S \approx \Delta\nu_h \cdot \sqrt{1+S} = \sqrt{2} \cdot \Delta\nu_h$ with the homogeneous linewidth $\Delta\nu_h = \sqrt{\Delta\nu_n^2 + \Delta\nu_p^2} \approx \Delta\nu_n$.

$$\Rightarrow \quad \Delta\nu_S = 2.26 \times 10^7\ \mathrm{s}^{-1}.$$

The resultant line profile is Lorentzian and forms the kernel of a broader background caused by velocity changing collisions.

Only collisions with $\Delta v_z = \Delta \boldsymbol{v} \cdot \boldsymbol{k} = \Delta v \cdot \lambda \cos\vartheta > \lambda \Delta \nu_S$ bring the molecules out of resonance with the laser line. Collisions with $\Delta v_z < \lambda \Delta \nu_S$ only redistribute the velocities and do not contribute essentially to the line broadening. The cross section for these collisions with $\Delta v_z > \lambda \Delta \nu_S$ is $\sigma_v(\Delta v) \le \sigma_0 \exp[-(\lambda \Delta \nu_S/\bar{v})^2]$.

With $\bar{v} = 4 \times 10^2$ m/s, $\lambda = 600$ nm, $\Delta \nu_S = 2.26 \times 10^7\ \mathrm{s}^{-1} \Rightarrow \sigma_v(\Delta v) \le 10^{-14}\ \mathrm{cm}^2$.

2.

$$\frac{1}{\tau_{\mathrm{eff}}} = \frac{1}{\tau_n} + n \cdot \sigma \cdot \bar{v} = \frac{1}{\tau_n} + p \cdot \sqrt{\frac{8}{\pi \mu k T}} \cdot \sigma$$

$$p_1 = 5\ \mathrm{mbar} = 5 \times 10^2\ \mathrm{Pa}, \qquad \tau_{1\,\mathrm{eff}} = 8 \times 10^{-9}\ \mathrm{s}$$

$$p_2 = 1\ \mathrm{mbar} = 1 \times 10^2\ \mathrm{Pa}, \qquad \tau_{2\,\mathrm{eff}} = 12 \times 10^{-9}\ \mathrm{s}.$$

Subtraction of the two equations yields

$$\frac{1}{\tau_{1\,\mathrm{eff}}} - \frac{1}{\tau_{2\,\mathrm{eff}}} = (p_1 - p_2)\sqrt{\frac{8}{\pi \mu k T}} \cdot \sigma$$

$$\Rightarrow \quad \sigma = 2.1 \times 10^{-14}\ \mathrm{cm}^2.$$

Inserting the value into the first equation gives for the natural lifetime

$$\frac{1}{\tau_n} = \frac{1}{\tau_{1\,\mathrm{eff}}} - p_1 \sigma \sqrt{\frac{8}{\pi \mu k T}} = 7.25 \times 10^7\ \mathrm{s}^{-1}$$

$$\Rightarrow \quad \tau_n = 17.2 \times 10^{-9}\ \mathrm{s}.$$

$$\Delta \nu_n = \frac{1}{2\pi \tau_n} = 9.25 \times 10^6\ \mathrm{s}^{-1}$$

$$\Delta \nu_p = \frac{1}{2\pi \tau_{\mathrm{eff}}} = 1.33 \times 10^7\ \mathrm{s}^{-1} \quad \text{for } p = 100\ \mathrm{Pa}$$

$$= 2 \times 10^7\ \mathrm{s}^{-1} \quad \text{for } p = 500\ \mathrm{Pa}.$$

3. The density of sodium atoms is

$$n = \frac{p}{kT} = 1.6 \times 10^{13}\ \mathrm{cm}^{-3}.$$

The collision rate per sodium atom is

$$R_{\mathrm{coll}} = n \cdot \sigma \cdot \bar{v} = 8 \times 10^5\ \mathrm{s}^{-1} \quad \Rightarrow \quad \Delta \nu_{\mathrm{coll}} = 1.3 \times 10^5\ \mathrm{s}^{-1}$$

and therefore negligible compared to the spontaneous decay. The saturation parameter is

$$S=\frac{B_{12}\cdot I}{c\cdot R}\quad \text{where } B_{12}=\frac{c}{h\nu}\int \sigma_{12}\,\mathrm{d}\nu\approx\frac{c}{h\nu}\sigma_{12}\Delta\nu_n$$

$$\Rightarrow\quad S=\frac{\sigma_{12}\cdot\Delta\nu_n\cdot I}{h\cdot\nu R}\quad \text{with } R=\frac{1}{\tau_n}=2\pi\Delta\nu_n=6.3\times10^{7}\ \mathrm{s}^{-1}$$

with a beam radius w the intensity is $P/\pi w^2$.

$$\Delta\nu_n\cdot I=\frac{P}{\pi w^2}=\frac{10^{-2}}{\pi\times10^{-6}}\frac{\mathrm{W}}{\mathrm{m}^2}=3.15\times10^{3}\ \mathrm{W/m}^2$$

$$\Rightarrow\quad S=0.74.$$

(Note: I [Ws/m^2] is the spectral intensity.) Saturation broadening plays only a minor role. The absorption coefficient is determined only by those atoms within the natural linewidth.

$$\alpha_S(\omega_0)=\sigma_{\mathrm{abs}}\cdot n\cdot\frac{\delta\omega_n}{\delta\omega_{\mathrm{D}}}=1.6\times10^{13}\cdot5\times10^{-11}\cdot\frac{10}{10^3}\ \mathrm{cm}^{-1}=8\ \mathrm{cm}^{-1}.$$

The density N_2 of excited atoms can be estimated from

$$\frac{\mathrm{d}N_2}{\mathrm{d}t}=R_1N_1-R_2N_2=0\quad\Rightarrow\quad\frac{N_2}{N_1}=\frac{R_1}{R_2},$$

where $R_1=B_{12}\varrho=B_{12}(I/c)\Delta\nu_n=5.3\times10^{6}\ \mathrm{s}^{-1}$ is the excitation probability and $R_2=A_2=\frac{1}{\tau_2}=6.2\times10^{7}\ \mathrm{s}^{-1}$ the spontaneous decay rate of level $|2\rangle$.

$$\Rightarrow\quad\frac{N_2}{N_1}=0.085\quad\Rightarrow\quad N_2=8.5\ \%\quad\text{of } N_1.$$

Energy-pooling collisions are collisions of N_2 atoms with N_2 atoms. Their rate is:

$$\Rightarrow\quad\dot{N}_{\mathrm{EP}}=\sigma_{\mathrm{EP}}\cdot N_2^2\cdot\bar{v}\cdot V\ \left[\mathrm{s}^{-1}\right],$$

where V is the excitation volume. $N_2=0.085\cdot1.6\times10^{13}\ \mathrm{cm}^{-3}=1.36\times10^{12}\ \mathrm{cm}^{-3}$, $\bar{v}=4\times10^{4}\ \mathrm{cm/s}$, $\sigma_{\mathrm{EP}}=3\times10^{-14}\ \mathrm{cm}^2$, $V=\pi w^2\cdot L=0.28\ \mathrm{cm}^3$ where $L=1$ cm is the observable length.

$$\Rightarrow\quad\dot{N}_{\mathrm{EP}}=3\times10^{-14}\cdot1.36^{-2}\times10^{24}\times4\times10^{4}\times0.28=6.2\times10^{14}\ \mathrm{s}^{-1}.$$

Rydberg states with $E(n)\le2E_2=2\cdot3.36\times10^{-19}\ \mathrm{J}=6.72\times10^{-19}$ J can be excited.

$E(n)=IP-\frac{Ry\cdot hc}{(n-\delta)^2}<6.72\times10^{-19}$ J where $IP=5.138\ \mathrm{eV}=8.2\times10^{-19}$ J is the ionization potential.

$\Rightarrow n - \delta < \sqrt{\frac{hc \cdot Ry}{IP - 6.72 \times 10^{-19}}} = \sqrt{12.2} = 3.49$ with a quantum defect of $\delta = 0.5 \Rightarrow n \le 4$.

4. At $p = 1$ mbar $\Rightarrow n = 1.6 \times 10^{16}$ cm^{-3}
   The refilling rate is

$$\begin{aligned} R &= n \cdot \sigma \cdot \bar{v} \\ &= 1.6 \times 10^{16} \times 10^{-14} \cdot 4 \times 10^4\ \mathrm{s}^{-1} \\ &= 6.4 \times 10^6\ \mathrm{s}^{-1}. \end{aligned}$$

The chopping period T must be long compared to $1/R \Rightarrow T \gg \frac{1}{R} = 1.5 \times 10^{-7}$ s.

In order to reach thermodynamic equilibrium between all rotational levels, which thermalize with a cross section of 3×10^{-16} cm$^2 \Rightarrow T \gg 4.5 \times 10^{-6}$ s.

Chapter 9

1. The difference between emission- and absorption frequencies is

$$\Delta\omega \approx \frac{\hbar\omega_2{}^2}{Mc^2}.$$

With $\omega_0 = \frac{2\pi c}{\lambda_0} = 2.87 \times 10^{15}$ s^{-1}, $M = 40 \times 1.66 \times 10^{-27}$ kg $= 6.6 \times 10^{-26}$ kg $\Rightarrow \Delta\omega = 3.6 \times 10^5$ s$^{-1} \Rightarrow \Delta\nu = 56$ kHz.

The collisional broadening is $\Delta\nu_{\mathrm{coll}} = \frac{1}{2\pi} \cdot (n \cdot \sigma_{\mathrm{coll}} \cdot \bar{v}) = 8 \times 10^4$ s^{-1} = 80 kHz.

The probability that another atom with the same velocity as the emitting atom absorbs the photon is

$$P = nR \cdot \sigma_{\mathrm{abs}}(\omega_0) \cdot \frac{\gamma^2}{\Delta\omega + \gamma^2} \quad \text{with } \gamma = 2\pi\Delta\nu_{\mathrm{coll}} = 5 \times 10^5\ \mathrm{s}^{-1}$$

$$\Rightarrow \quad P = 10^{17} \times 1 \times 10^{-17} \cdot \frac{\gamma^2}{3.6^2 \times 10^{10} + 25 \times 10^{10}} = 0.66.$$

Since the Doppler broadening is large compared to the recoil shift and the homogeneous linewidth, the probability that the emitted photon is absorbed by any atom is $P_{\mathrm{total}} = n \cdot R \cdot \sigma = 1$.

2. The force on an atom with mass M excerted by photon recoil is

$$\boldsymbol{F} = M \cdot \boldsymbol{a} = M \frac{\mathrm{d}\boldsymbol{v}}{\mathrm{d}t} = \frac{\mathrm{d}\boldsymbol{p}}{\mathrm{d}t}.$$

The momentum $\mathrm{d}p$ transferred by n photons is $\mathrm{d}\boldsymbol{p} = n\hbar\boldsymbol{k}$.

The maximum rate dn/dt absorbed by a two-level system equals the rate of absorption-emission cycles and is limited by the lifetime τ_K of the upper level $|k\rangle$.

The fluorescence rate is $N_k \cdot A_k = N_k/\tau_K$.

The maximum value of N_K is

$$N_k = (N_i + N_K)/2 = N/2$$

which is achieved for $S \to \infty$ because for $N_K > N_i$ the rate of stimulated emission (which compensates the recoil of absorption) becomes larger than the absorption rate. For an arbitrary saturation parameter S we obtain

$$\frac{N_k}{N} = \frac{S}{1+2S} \quad \Rightarrow \quad N_k = N \cdot \frac{S}{1+2S}.$$

The rate of absorbed photons is then $N_K/(N \cdot \tau_K) = \frac{1}{\tau_K}\frac{S}{1+2S}$ and the transferred momentum per sec

$$F = \frac{dp}{dt} = \frac{\hbar k}{\tau_K}\frac{S}{1+2S}.$$

3. The natural linewidth of the optical transition is

$$\gamma = \frac{1}{\tau_K} = 6.25 \times 10^7\ \mathrm{s}^{-1} = 2\pi\,\Delta\nu_n.$$

With a saturation parameter $S = 1 \Rightarrow \gamma_S = \gamma \cdot \sqrt{2} = 8.8 \times 10^7\ \mathrm{s}^{-1} \Rightarrow \Delta\nu_S =$ 14 MHz.

The separation of the levels $F' = 2$ and $F' = 3$ is $\Delta\nu = 60$ MHz.

The absorption probability on the transition $F'' = 2 \to F' = 2$ is for a laser frequency ν_L:

$$P_{\mathrm{abs}} \propto \frac{\sigma(2\text{--}2)}{(\nu-\nu_0)^2 + \Delta\nu_S^2} \cdot \frac{\Delta\nu_S^2}{\sigma(2\to 3)}$$

$$\frac{\sigma(2\text{--}2)}{\sigma(2\text{--}3)} = \frac{2F_2'+1}{2F_3'+1} = \frac{5}{7}.$$

For $\nu_0 = \nu(F'' = 2 \to F' = 3) \Rightarrow \nu - \nu_0 = 60$ MHz

$$\Rightarrow \quad P_{\mathrm{abs}} = \frac{5}{7}\frac{14^2 \times 10^{12}}{60^2 \times 10^{12} + 14^2 \times 10^{12}} = 3.69 \times 10^{-2}.$$

4. For the optimum deceleration the recoil force on the atoms is (see Problem 9.2)

$$F = \frac{\hbar k}{\tau_K}\frac{S}{2S+1} \quad \Rightarrow \quad \frac{\hbar k}{2\tau_K} \quad \text{for } S \to \infty.$$

The deceleration is then $a = F/m$

$$|a| = \frac{\hbar|k|}{2m\tau_K} \simeq \frac{h}{2m\lambda \cdot \tau_K} = 1.47 \times 10^7 \text{ m/s}^2$$

$$\Rightarrow \quad B = B_0\sqrt{1 - 2az/v_0^2} \quad \text{with } v_0 = 1000 \text{ m/s}$$

$$= B_0\sqrt{1 - 29.4 \cdot z \text{ [m]}}.$$

After $z = 3.4$ cm the magnetic field is zero and the velocity of the atoms has become so small that they are Doppler-shifted by less than the natural linewidth.

5. The Doppler limit $T_D = \frac{\hbar}{2\tau \cdot k_B}$ is for Li atoms ($\tau(3\,^2P_{3/2}) = 27$ ns)

$$T_D(\text{Li}) = 134\ \mu\text{K}$$

and for K atoms ($\tau(5\,^2P_{3/2}) = 137$ ns)

$$T_D(\text{K}) = 26\ \mu\text{K}.$$

6. The critical density n_c is, according to (9.44),

$$n_c = 2.612/\lambda_{DB}^3$$

with the de Broglie wavelength $\lambda_{DB} = \frac{h}{m \cdot v}$ and $v = (3k_B T/m)^{1/2}$

$$\Rightarrow \quad n_c = 13.57(m \cdot k_B T)^{3/2}/h^3$$

with $T = 10^{-6}$ K, $m = 133$ AMU $= 133 \cdot 1.66 \times 10^{-27}$ kg $= 2.2 \times 10^{-25}$ kg

$$\Rightarrow \quad n_c = 1.83 \times 10^{19} \text{ m}^{-3} = 1.83 \times 10^{13} \text{ /cm}^3.$$

If 10^7 Cs atoms can be trapped in the magnetic trap, their volume must then be compressed to $V = 5.5 \times 10^{-7}$ cm^3. At such a density, collisions would in fact destroy BEC. Therefore the temperature has to be lower than 1 μK.

7. The frequency of the cesium clock is:

$$\nu_{Cs} = 9.192631770 \times 10^9 \text{ s}^{-1}.$$

The 6×10^4th multiple is:

$$\nu_1 = 5.51558 \times 10^{14} \text{ s}^{-1}.$$

The frequency of the transition at $\lambda = 750$ nm is

$$\nu_2 = \frac{c}{\lambda} = \frac{2.99792458 \times 10^8}{750 \times 10^{-9}} \text{ s}^{-1} = 3.99723 \times 10^{14} \text{ s}^{-1}.$$

The frequency difference is

$$\Delta\nu = \nu_1 - \nu_2 = 1.51835 \times 10^{14}.$$

This corresponds to $m = \Delta\nu/10^8 = 1.51835 \times 10^6$ comb spacings. If the comb spacing can be measured with an uncertainty of $10^{-16} \Rightarrow$ the accuracy of the molecular transition frequency is: $1.518 \times 10^6 \times 10^{-16} = 1.518 \times 10^{-10}$, which gives as absolute uncertainty:

$$\delta\nu = 1.518 \times 10^{-10} \times 3.997 \times 10^{14} = 6.07 \times 10^4 = 60.7\ \text{kHz}.$$

8. We assume a magnetic field

$$B = B_0 \cdot r^2.$$

Atoms with a magnetic moment μ have the potential energy

$$E_{\text{pot}} = -\mu \cdot B = -\mu B_0 r^2.$$

The kinetic energy of the atoms at the temperature T is

$$E_{\text{kin}} = \frac{3}{2}kT.$$

Only the radial component of the velocity has to be confined. This gives:

$$E_{\text{kin}}^{\text{rad}} = \frac{1}{2}kT.$$

For a spherical volume $V = \frac{4}{3}\pi r_c^3 = 1\ \text{cm}^3 \Rightarrow r_c = 0.62$ cm.

$$\Rightarrow \quad \mu B_0 r_c^2 = \frac{1}{2}kT \quad \Rightarrow \quad B_0 = \frac{kT}{2\mu r_c^2}$$

$$\text{with } \mu = 2\mu_B \quad \Rightarrow \quad B_0 = \frac{kT}{4\mu_B r_c^2} = 9.6 \times 10^{-2}\ \text{T/m}^2$$

$$\text{for } r = 0.62\ \text{cm} \quad \Rightarrow \quad B(r_c) = 3.7 \times 10^{-6}\ \text{T} = 37\ \text{mGauss}.$$

References

Chapter 1

1. R.J. Bell, *Introductory Fourier Transform Spectroscopy* (Academic Press, New York, 1972);
   P. Griffiths, J.A. de Haset, *Fourier Transform Infrared Spectroscopy* (Wiley, New York, 1986);
   J. Kauppinen, J. Partanen, *Fourier Transforms in Spectroscopy* (Wiley, New York, 2001)
2. D.G. Cameron, D.J. Moffat, A generalized approach to derivative spectroscopy. Appl. Spectrosc. **41**, 539 (1987);
   G. Talsky, *Derivative Spectrophotometers* (VCH, Weinheim, 1994)
3. G.C. Bjorklund, Frequency-modulation spectroscopy: A new method for measuring weak absorptions and dispersions. Opt. Lett. **5**, 15 (1980)
4. M. Gehrtz, G.C. Bjorklund, E. Whittaker, Quantum-limited laser frequency-modulation spectroscopy. J. Opt. Soc. Am. B **2**, 1510 (1985)
5. G.R. Janik, C.B. Carlisle, T.F. Gallagher, Two-tone frequency-modulation spectroscopy. J. Opt. Soc. Am. B **3**, 1070 (1986)
6. F.S. Pavone, M. Inguscio, Frequency- and wavelength-modulation spectroscopy: Comparison of experimental methods, using an AlGaAs diode laser. Appl. Phys. B **56**, 118 (1993)
7. R. Grosskloss, P. Kersten, W. Demtröder, Sensitive amplitude and phase-modulated absorption spectroscopy with a continuously tunable diode laser. Appl. Phys. B **58**, 137 (1994)
8. P.C.D. Hobbs, Ultrasensitive laser measurements without tears. Appl. Opt. **36**, 903 (1997)
9. P. Wehrle, A review of recent advances in semiconductor laser gas monitors. Spectrochim. Acta, Part A **54**, 197 (1998);
   M.W. Sigrist (ed.), Tunable diode laser spectroscopy. Appl. Phys. B (2008) (Special Issue)
10. J.A. Silver, Frequency modulation spectroscopy for trace species detection. Appl. Opt. **31**, 707 (1992)
11. W. Brunner, H. Paul, On the theory of intracavity absorption. Opt. Commun. **12**, 252 (1974)
12. K. Tohama, A simple model for intracavity absorption. Opt. Commun. **15**, 17 (1975)
13. A. Campargue, F. Stoeckel, M. Chenevier, High sensitivity intracavity laser spectroscopy: applications to the study of overtone transitions in the visible range. Spectrochim. Acta Rev. **13**, 69 (1990)
14. A.A. Kaschanov, A. Charvat, F. Stoeckel, Intracavity laser spectroscopy with vibronic solid state lasers. J. Opt. Soc. Am. B **11**, 2412 (1994)
15. V.M. Baev, T. Latz, P.E. Toschek, Laser intracavity absorption spectroscopy. Appl. Phys. B **69**, 171 (1999);
    V.M. Baev, Intracavity spectroscopy with diode lasers. Appl. Phys. B **55**, 463 (1992)

W. Demtröder, *Laser Spectroscopy 2*, DOI 10.1007/978-3-662-44641-6

16. V.R. Mironenko, V.I. Yudson, Quantum noise in intracavity laser spectroscopy. Opt. Commun. **34**, 397 (1980);
V.R. Mironenko, V.I. Yudson, Quantum statistics of multimode lasing and noise in intracavity laser spectroscopy. Sov. Phys. JETP **52**, 594 (1980)
17. P.E. Toschek, V.M. Baev, One is not enough: intracavity laser spectroscopy with a multimode laser, in *Laser Spectroscopy and New Ideas*, ed. by W.M. Yen, M.D. Levenson. Springer Ser. Opt. Sci., vol. 54 (Springer, Berlin, 1987)
18. E.M. Belenov, M.V. Danileiko, V.R. Kozuborskii, A.P. Nedavnii, M.T. Shpak, Ultrahigh resolution spectroscopy based on wave competition in a ring laser. Sov. Phys. JETP **44**, 40 (1976)
19. E.A. Sviridenko, M.P. Frolov, Possible investigations of absorption line profiles by intracavity laser spectroscopy. Sov. J. Quantum Electron. **7**, 576 (1977)
20. T.W. Hänsch, A.L. Schawlow, P. Toschek, Ultrasensitive response of a CW dye laser to selective extinction. IEEE J. Quantum Electron. **8**, 802 (1972)
21. R.N. Zare, Laser separation of isotopes. Sci. Am. **236**, 86 (1977)
22. R.G. Bray, W. Henke, S.K. Liu, R.V. Reddy, M.J. Berry, Measurement of highly forbidden optical transitions by intracavity dye laser spectroscopy. Chem. Phys. Lett. **47**, 213 (1977)
23. H. Atmanspacher, B. Baldus, C.C. Harb, T.G. Spence, B. Wilke, J. Xie, J.S. Harris, R.N. Zare, Cavity-locked ring-down spectroscopy. J. Appl. Phys. **83**, 3991 (1998)
24. W. Schrepp, H. Figger, H. Walther, Intracavity spectroscopy with a color-center laser. Laser Appl **77** (1984)
25. N. Picqé, F. Gueye, G. Guelachvili, E. Sorokin, I.T. Sorokina, Time-resolved Fourier-Transform Intracavity spectroscopy with a Cr^{2+}:ZnSe-laser. Opt. Lett. **30**, 24 (2005)
26. V.M. Baev, K.J. Boller, A. Weiler, P.E. Toschek, Detection of spectrally narrow light emission by laser intracavity spectroscopy. Opt. Commun. **62**, 380 (1987);
V.M. Baev, Intracavity spectroscopy with diode lasers. Appl. Phys. B **49**, 315 (1989); **55**, 463 (1992); and **69**, 171 (1999)
27. V.M. Baev, A. Weiler, P.E. Toschek, Ultrasensitive intracavity spectroscopy with multimode lasers. J. Phys. (Paris) **48**(C7), 701 (1987)
28. T.D. Harris, Laser intracavity-enhanced spectroscopy, in *Ultrasensitive Laser Spectroscopy*, ed. by D.S. Kliger (Academic, New York, 1983)
29. E.H. Piepmeier (ed.), *Analytical Applications of Lasers* (Wiley, New York, 1986)
30. H. Atmanspacher, H. Scheingraber, C.R. Vidal, Dynamics of laser intracavity absorption. Phys. Rev. A **32**, 254 (1985);
H. Atmanspacher, H. Scheingraber, C.R. Vidal, Mode-correlation times and dynamical instabilities in a multimode CW dye laser. Phys. Rev. A **33**, 1052 (1986)
31. H. Atmanspacher, H. Scheingraber, V.M. Baev, Stimulated Brillouin scattering and dynamical instabilities in a multimode laser. Phys. Rev. A **35**, 142 (1987)
32. G. Stewart, K. Atherton, H. Yu, B. Culshaw, Cavity-Enhanced Spectroscopy in Fibre Cavities. Opt. Lett. **29**, 442 (2004)
33. A. Stark et al., Intercavity Absorption Spectroscopy with Thulium-doped Fibre Laser. Opt. Commun. **215**, 113 (2003)
34. K. Strong, T.J. Johnson, G.W. Harris, Visible intracavity laser spectroscopy with a step-scan Fourier-transform interferometer. Appl. Opt. **36**, 8533 (1997)
35. J. Cheng et al., Infrared intracavity laser absorption spectroscopy with a continuous-scan Fourier-transform interferometer. Appl. Opt. **39**(13), 2221 (2000)
36. B. Löhden, S. Kuznetsova, K. Sengstock, V.M. Baev, A. Goldman, S. Cheskis, B. Pálsdóttir, Fiber laser intracavity absorption spectroscopy for in situ multicomponent gas analysis in the atmosphere and combustion environments. Appl. Phys. B, Lasers Opt. **102**, 331–344 (2011)
37. A. Goldman, I. Rahinov, S. Cheskis, B. Löhden, S. Wexler, K. Engstock, V.M. Baev, Fiber laser intracavity absorption spectroscopy of ammonia and hydrogen cyanide in low pressure hydrocarbon flames. Chem. Phys. Lett. **423**, 147–151 (2006)

38. P. Zalicki, R.N. Zare, Cavity ringdown spectroscopy for quantitative absorption measurements. J. Chem. Phys. **102**, 2708 (1995)
39. D. Romanini, K.K. Lehmann, Ring-down cavity absorption spectroscopy of the very weak HCN overtone bands with six, seven and eight stretching quanta. J. Chem. Phys. **99**, 6287 (1993)
40. M.D. Levenson, B.A. Paldus, T.G. Spence, C.C. Harb, J.S. Harris, R.N. Zare, Optical heterodyne detection in cavity ring-down spectroscopy. Chem. Phys. Lett. **290**, 335 (1998); M.D. Levenson, B.A. Paldus, T.G. Spence, C.C. Harb, J.S. Harris, R.N. Zare, Frequency switched cavity ring down spectroscopy. Opt. Lett. **25**, 920 (2000)
41. B.A. Baldus, R.N. Zare et al., Cavity-locked ringdown spectroscopy. J. Appl. Phys. **83**, 3991 (1998)
42. J.J. Scherer, J.B. Paul, C.P. Collier, A. O'Keefe, R.J. Saykally, Cavity-ringdown laser absorption spectroscopy and time-of-flight mass spectroscopy of jet-cooled gold silicides. J. Chem. Phys. **103**, 9187 (1995);
A. O'Keefe, J.J. Scherer, J.B. Paul, R.J. Saykally, Cavity-ringdown laser spectroscopy: history, development, and applications, in *1997 ACS Symposium Series 720 on Cavity-Ringdown Spectroscopy: An Ultratrace-Absorption Measurement Technique*, ed. by K. Busch, M. Busch (1999), pp. 71–92
43. K.H. Becker, D. Haaks, T. Tartarczyk, Measurements of C_2-radicals in flames with a tunable dye lasers. Z. Naturforsch. A **29**, 829 (1974)
44. A. O'Keefe, Integrated cavity output analysis of ultraweak absorption. Chem. Phys. Lett. **293**, 331 (1998)
45. A. Popp et al., Ultrasensitive mid-infrared cavity leak-out spectroscopy using a cw optical parametric oscillator. Appl. Phys. B **75**, 751 (2003)
46. D. Halmer, G. von Basum, P. Hering, M. Mürtz, Mid-infrared cavity leak-out spectroscopy for ultrasensitive detection of carbonyl sulfide. Opt. Lett. **30**, 2314 (2005)
47. A. Deev, Cavity ringdown spectroscopy of atmospherically important radicals. PhD thesis, CALTEC, Pasadena, 2005
48. M. Mürtz, B. Frech, W. Urban, High-resolution cavity-leak-out absorption spectroscopy in the 10 μm region. Appl. Phys. B **68**, 243 (1999)
49. M. Muertz et al., Recent developments in cavity ring-down spectroscopy with tunable cw lasers in the mid-infrared. Environ. Sci. Pollut. Res. **4**, 61–67 (2002). Special Issue
50. J.J. Scherer, J.B. Paul, C.P. Collier, A. O'Keefe, R.J. Saykally, Cavity ringdown laser absorption spectroscopy history, development and application to pulsed molecular beams. Chem. Rev. **97**, 25 (1997)
51. G. Berden, R. Peèters, G. Meijer, Cavity ringdown spectroscopy: experimental schemes and applications. Int. Rev. Phys. Chem. **19**, 565 (2000)
52. K.W. Busch, M.A. Busch, *Cavity Ringdown Spectroscopy* (Oxford Univ. Press, Oxford, 1999)
53. E. Namers, D. Schramm, R. Engels, Fourier-transform phase shift cavity ringdown spectroscopy. Chem. Phys. Lett. **365**, 237 (2002)
54. G. Berden, G. Meijer, W. Ubachs, Spectroscopic Applications using ring-down cavities. Exp. Methods Phys. Sci. **40**, 49 (2002)
55. G. Berden, R. Engeln (eds.), *Cavity Ringdown Spectroscopy: Techniques and Applications* (Wiley/Blackwell, New York/Oxford, 2009)
56. R. Engeln, G. Meijer, A Fourier-transform phase-shift cavity ring down spectrometer. Rev. Sci. Instrum. **67**, 2708 (1996)
57. W.M. Fairbanks, T.W. Hänsch, A.L. Schawlow, Absolute measurement of very low sodium-vapor densities using laser resonance fluorescence. J. Opt. Soc. Am. **65**, 199 (1975)
58. H.G. Krämer, V. Beutel, K. Weyers, W. Demtröder, Sub-Doppler laser spectroscopy of silver dimers Ag_2 in a supersonic beam. Chem. Phys. Lett. **193**, 331 (1992)
59. P.J. Dagdigian, H.W. Cruse, R.N. Zare, Laser fluorescence study of AlO, formed in the reaction $Al + O_2$: product state distribution, dissociation energy and radiative lifetime. J. Chem. Phys. **62**, 1824 (1975)

60. W.E. Moerner, L. Kador, Finding a single molecule in a haystack. Anal. Chem. **61**, 1217A (1989);
W.E. Moerner, Examining nanoenvironments in solids on the scale of a single, isolated impurity molecule. Science **265**, 46 (1994)
61. K. Kneipp, S.R. Emory, S. Nie, Single-molecule Raman-spectroscopy: fact or fiction? Chimica **53**, 35 (1999)
62. T. Plakbotnik, E.A. Donley, U.P. Wild, Single molecule spectroscopy. Annu. Rev. Phys. Chem. **48**, 181 (1997)
63. H.J. Bauer, Son et lumiere or the optoacoustic effect in multilevel systems. J. Chem. Phys. **57**, 3130 (1972) (references to the historical development)
64. Y.-H. Pao (ed.), *Optoacoustic Spectroscopy and Detection* (Academic Press, New York, 1977)
65. A. Rosencwaig, *Photoacoustic Spectroscopy* (Wiley, New York, 1980)
66. K.H. Michaelian, *Photoacoustic IR Spectroscopy: Instrumental, Applications and Data Analysis* (Wiley/VCH, Weinheim, 2011)
67. V.P. Zharov, V.S. Letokhov, *Laser Optoacoustic Spectroscopy*. Springer Ser. Opt. Sci., vol. 37 (Springer, Berlin, 1986)
68. M.W. Sigrist (ed.), *Air Monitoring by Spectroscopic Techniques* (Wiley, New York, 1994);
J. Xiu, R. Stroud, *Acousto-Optic Devices: Principles, Design and Applications* (Wiley, New York, 1992)
69. P. Hess, J. Pelzl (eds.), *Photoacoustic and Photothermal Phenomena*. Springer Ser. Opt. Sci., vol. 58 (Springer, Berlin, 1988)
70. P. Hess (ed.), *Photoacoustic, Photothermal and Photochemical Processes in Gases*. Topics Curr. Phys, vol. 46 (Springer, Berlin, 1989)
71. J.C. Murphy, J.W. Maclachlan Spicer, L.C. Aamodt, B.S.H. Royce (eds.), *Photoacoustic and Photothermal Phenomena II*. Springer Ser. Opt. Sci., vol. 62 (Springer, Berlin, 1990)
72. L.B. Kreutzer, Laser optoacoustic spectroscopy: a new technique of gas analysis. Anal. Chem. **46**, 239A (1974)
73. W. Schnell, G. Fischer, Spectraphone measurements of isotopes of water vapor and nitric oxide and of phosgene at selected wavelengths in the CO- and CO_2-laser region. Opt. Lett. **2**, 67 (1978)
74. C. Hornberger, W. Demtröder, Photoacoustic overtone spectroscopy of acetylene in the visible and near infrared. Chem. Phys. Lett. **190**, 171 (1994)
75. C.K.N. Patel, Use of vibrational energy transfer for excited-state opto-acoustic spectroscopy of molecules. Phys. Rev. Lett. **40**, 535 (1978)
76. G. Stella, J. Gelfand, W.H. Smith, Photoacoustic detection spectroscopy with dye laser excitation. The 6190 Å CH_4 and the 6450 NH_3-bands. Chem. Phys. Lett. **39**, 146 (1976)
77. A.M. Angus, E.E. Marinero, M.J. Colles, Opto-acoustic spectroscopy with a visible CW dye laser. Opt. Commun. **14**, 223 (1975)
78. E.E. Marinero, M. Stuke, Quartz optoacoustic apparatus for highly corrosive gases. Rev. Sci. Instrum. **50**, 31 (1979)
79. A.C. Tam, Photoacoustic spectroscopy and other applications, in *Ultrasensitive Laser Spectroscopy*, ed. by D.S. Kliger (Academic Press, New York, 1983), pp. 1–108
80. V.Z. Gusev, A.A. Karabutov, *Laser Optoacoustics* (Springer, Berlin, 1997)
81. A.C. Tam, C.K.N. Patel, High-resolution optoacoustic spectroscopy of rare-earth oxide powders. Appl. Phys. Lett. **35**, 843 (1979)
82. J.F. NcClelland et al., Photoacoustic spectroscopy, in *Modern Techniques in Applied Molecular Spectroscopy*, ed. by F.M. Mirabella (Wiley, New York, 1998)
83. K.H. Michaelian, *Photo-acoustic Infrared Spectroscopy* (Wiley Interscience, New York, 2003)
84. T.E. Gough, G. Scoles, Optothermal infrared spectroscopy, in *Laser Spectroscopy V*, ed. by A.R.W. McKeller, T. Oka, B.P. Stoicheff. Springer Ser. Opt. Sci., vol. 30 (Springer, Berlin, 1981), p. 337;

T.E. Gough, R.E. Miller, G. Scoles, Sub-Doppler resolution infrared molecular beam spectroscopy. Faraday Disc. **71**, 6 (1981)
85. M. Zen, Cryogenicbolometers, in *Atomic and Molecular Beams Methods*, vol. 1 (Oxford Univ. Press, London, 1988)
86. R.E. Miller, Infrared laser spectroscopy, in *Atomic and Molecular Beam Methods*, ed. by G. Scoles (Oxford Univ. Press, London, 1992), pp. 192 ff.;
D. Bassi, Detection principles, in *Atomic and Molecular Beam Methods*, ed. by G. Scoles (Oxford Univ. Press, London, 1992), pp. 153 ff.
87. T.B. Platz, W. Demtröder, Sub-Doppler optothermal overtone spectroscopy of ethylene. Chem. Phys. Lett. **294**, 397 (1998)
88. K.K. Lehmann, G. Scoles, Intramolecular dynamics from Eigenstate-resolved infrared spectra. Annu. Rev. Phys. Chem. **45**, 241 (1994)
89. H. Coufal, Photothermal spectroscopy and its analytical application. Fresenius J. Anal. Chem. **337**, 835 (1990)
90. F. Träger, Surface analysis by laser-induced thermal waves. Laser Optoelektron. **18**, 216 (1986);
H. Coufal, F. Träger, T.J. Chuang, A.C. Tam, High sensitivity photothermal surface spectroscopy with polarization modulation. Surf. Sci. **145**, L504 (1984)
91. P.E. Siska, Molecular-beam studies of Penning ionization. Rev. Mod. Phys. **65**, 337 (1993)
92. Y.Y. Kuzyakov, N.B. Zorov, Atomic ionization spectrometry. Crit. Rev. Anal. Chem. **20**, 221 (1988)
93. G.S. Hurst, M.G. Payne, S.P. Kramer, J.P. Young, Resonance ionization spectroscopy and single atom detection. Rev. Mod. Phys. **51**, 767 (1979)
94. G.S. Hurst, M.P. Payne, S.P. Kramer, C.H. Cheng, Counting the atoms. Phys. Today **33**, 24 (1980)
95. M. Keil, H.G. Krämer, A. Kudell, M.A. Baig, J. Zhu, W. Demtröder, W. Meyer, Rovibrational structures of the pseudo-rotating lithium trimer Li_3. J. Chem. Phys. **113**, 7414 (2000)
96. L. Wöste, Zweiphotonen-Ionisation. Laser Optoelektron. **15**, 9 (1983)
97. G. Delacretaz, J.D. Garniere, R. Monot, L. Wöste, Photoionization and fragmentation of alkali metal clusters in supersonic molecular beams. Appl. Phys. B **29**, 55 (1982)
98. H.J. Foth, J.M. Gress, C. Hertzler, W. Demtröder, Sub-Doppler laser spectroscopy of Na_3. Z. Phys. D **18**, 257 (1991)
99. V.S. Letokhov, *Laser Photoionization Spectroscopy* (Academic, Orlando, 1987)
100. G. Hurst, M.G. Payne, *Principles and Applications of Resonance Ionization Spectroscopy* (Hilger, Bristol, 1988)
101. K.T. Flanagan et al., Collinear Resonance Ionization Spectroscopy of Neutron deficient Francium Isotopes. Phys. Rev. Lett. **111**, 212501 (2013)
102. J.C. Travis, G.C. Turk (eds.), *Laser-Enhanced Ionization Spectroscopy*. Chemical Analysis, vol. 13 (1996)
103. D.H. Parker, Laser ionization spectroscopy and mass spectrometry, in *Ultrasensitive Laser Spectroscopy*, ed. by D.S. Kliger (Academic, New York, 1983)
104. V. Beutel, G.L. Bhale, M. Kuhn, W. Demtröder, The ionization potential of Ag_2. Chem. Phys. Lett. **185**, 313 (1991)
105. H.J. Neusser, U. Boesl, R. Weinkauf, E.W. Schlag, High-resolution laser mass spectrometer. Int. J. Mass Spectrom. **60**, 147 (1984)
106. J.E. Parks, N. Omeneto (eds.), *Resonance Ionization Spectroscopy*. Inst. Phys. Conf. Ser., vol. 114 (1990);
D.M. Lübman (ed.), *Lasers and Mass Spectrometry* (Oxford Univ. Press, London, 1990)
107. P. Peuser, G. Herrmann, H. Rimke, P. Sattelberger, N. Trautmann, W. Ruster, F. Ames, J. Bonn, H.J. Kluge, V. Krönert, E.W. Otten, Trace detection of plutonium by three-step photoionization with a laser system pumped by a copper vapor laser. Appl. Phys. B **38**, 249 (1985)
108. http://mars.jpl.nasa.gov.msl

109. V. Seibert, A. Wiesner, Th. Buschmann, J. Meuer, Surface-enhanced laser desorption ionization time-of-flight mass spectrometry (SELDI TOF-MS) and ProteinChip technology in proteomics research. Pathol. Res. Practice **200**, 83 (2004)
110. M. de Vries, Laser mass spectrometry. http://web.chem.ucsb.edu/~devries/groupsite/pub/analytreview.pdf
111. D. Popescu, M.L. Pascu, C.B. Collins, B.W. Johnson, I. Popescu, Use of space charge amplification techniques in the absorption spectroscopy of Cs and Cs_2. Phys. Rev. A **8**, 1666 (1973)
112. K. Niemax, Spectroscopy using thermionic diode detectors. Appl. Phys. B **38**, 1 (1985)
113. R. Beigang, W. Makat, A. Timmermann, A thermionic ring diode for high resolution spectroscopy. Opt. Commun. **49**, 253 (1984)
114. R. Beigang, A. Timmermann, The thermionic annular diode: a sensitive detector for highly excited atoms and molecules. Laser Optoelektron. **4**, 252 (1984)
115. D.S. King, P.K. Schenck, Optogalvanic spectroscopy. Laser Focus **14**, 50 (1978)
116. J.E.M. Goldsmith, J.E. Lawler, Optogalvanic spectroscopy. Contemp. Phys. **22**, 235 (1981)
117. B. Barbieri, N. Beverini, A. Sasso, Optogalvanic spectroscopy. Rev. Mod. Phys. **62**, 603 (1990)
118. V.N. Ochkin, N.G. Preobrashensky, N.Y. Shaparev, *Opto-galvanic Effect in Ionized Gases* (CRC Press, Boca Raton, 1999)
119. M.A. Zia, B. Sulemar, M.A. Baig, Two-photon laser optogalvanic spectroscopy of the Rydberg states of Mercury by RF-discharge. J. Phys. B, At. Mol. Opt. Phys. **36**, 4631 (2003)
120. K. Narayanan, G. Ullas, S.B. Rai, A two step optical double resonance study of a Fe–Ne hollow cathode discharge using optogalvanic detection. Opt. Commun. **184**, 102 (1991)
121. C.R. Webster, C.T. Rettner, Laser optogalvanic spectroscopy of molecules. Laser Focus **19**, 41 (1983);
D. Feldmann, Optogalvanic spectroscopy of some molecules in discharges: NH_2, NO_2, A_2 and N_2. Opt. Commun. **29**, 67 (1979)
122. K. Kawakita, K. Fukada, K. Adachi, S. Maeda, C. Hirose, Doppler-free optogalvanic spectrum of $He_2(b^3\Pi_g - f^3\Delta_u)$ transitions. J. Chem. Phys. **82**, 653 (1985)
123. K. Myazaki, H. Scheingraber, C.R. Vidal, Optogalvanic double-resonance spectroscopy of atomic and molecular discharge, in *Laser Spectroscopy VI*, ed. by H.P. Weber, W Lüthy. Springer Ser. Opt. Sci., vol. 40 (Springer, Berlin, 1983), p. 93
124. J.C. Travis, Analytical optogalvanic spectroscopy in flames, in *Analytical Laser Spectroscopy*, ed. by S. Martellucci, A.N. Chester (Plenum, New York, 1985), p. 213
125. D. King, P. Schenck, K. Smyth, J. Travis, Direct calibration of laser wavelength and bandwidth using the optogalvanic effect in hollow cathode lamps. Appl. Opt. **16**, 2617 (1977)
126. V. Kaufman, B. Edlen, Reference wavelength from atomic spectra in the range 15 Å to 25,000 Å. J. Phys. Chem. Ref. Data **3**, 825 (1974)
127. A. Giacchetti, R.W. Stanley, R. Zalubas, Proposed secondary standard wavelengths in the spectrum of thorium. J. Opt. Soc. Am. **60**, 474 (1969)
128. J.E. Lawler, A.I. Ferguson, J.E.M. Goldsmith, D.J. Jackson, A.L. Schawlow, Doppler-free optogalvanic spectroscopy, in *Laser Spectroscopy IV*, ed. by H. Walther, K.W. Rothe. Springer Ser. Opt. Sci., vol. 21 (Springer, Berlin, 1979), p. 188
129. W. Bridges, Characteristics of an optogalvanic effect in cesium and other gas discharge plasmas. J. Opt. Soc. Am. **68**, 352 (1978)
130. A. Persson et al., Evaluation of intracavity optogalvanic spectroscopy for radio-carbon measurements. Anal. Chem. **85**(14), 6790–6798 (2013)
131. R.S. Stewart, J.E. Lawler (eds.), *Optogalvanic Spectroscopy* (Hilger, London, 1991)
132. R.J. Saykally, R.C. Woods, High resolution spectroscopy of molecular ions. Annu. Rev. Phys. Chem. **32**, 403 (1981)
133. C.S. Gudeman, R.J. Saykally, Velocity modulation infrared laser spectroscopy of molecular ions. Annu. Rev. Phys. Chem. **35**, 387 (1984)
134. C.E. Blom, K. Müller, R.R. Filgueira, Gas discharge modulation using fast electronic switches. Chem. Phys. Lett. **140**, 489 (1987)

135. M. Gruebele, M. Polak, R. Saykally, Velocity modulation laser spectroscopy of negative ions: the infrared spectrum of SH^-. J. Chem. Phys. **86**, 1698 (1987)
136. J.W. Farley, Theory of the resonance lineshape in velocity-modulation spectroscopy. J. Chem. Phys. **95**, 5590 (1991)
137. G. Lan, H.D. Tholl, J.W. Farley, Double-modulation spectroscopy of molecular ions: eliminating the background in velocity-modulation spectroscopy. Rev. Sci. Instrum. **62**, 944 (1991)
138. M.B. Radunsky, R.J. Saykally, Electronic absorption spectroscopy of molecular ions in plasmas by dye laser velocity modulation spectroscopy. J. Chem. Phys. **87**, 898 (1987)
139. I. Schechter, V. Palleschi, A.W. Miziolek, *Laser-Induced Breakdown Spectroscopy* (Cambridge Univ. Press, Cambridge, 2006)
140. K.J. Button (ed.), *Infrared and Submillimeter Waves* (Academic Press, New York, 1979)
141. (a) Wikipedia, List of molecules in interstellar space. http://en.wikipedia.org/wiki/List_of_molecules_in_interstellar_space;
H.S.P. Müller, F. Schlöder, J. Stutzki, G. Winnewisser, The cologne database for molecular spectroscopy. J. Mol. Struct. **742**, 215 (2005);
(b) K.M. Evenson, R.J. Saykally, D.A. Jennings, R.E. Curl, J.M. Brown, Far infrared laser magnetic resonance, in *Chemical and Biochemical Applications of Lasers*, ed. by C.B. Moore (Academic Press, New York, 1980), Chap. V
142. P.B. Davies, K.M. Evenson, Laser magnetic resonance (LMR) spectroscopy of gaseous free radicals, in *Laser Spectroscopy II*, ed. by S. Haroche, J.C. Pebay-Peyroula, T.W. Hänsch, S.E. Harris. Lect. Notes Phys., vol. 43 (Springer, Berlin, 1975)
143. W. Urban, W. Herrmann, Zeeman modulation spectroscopy with spin-flip Raman laser. Appl. Phys. **17**, 325 (1978)
144. K.M. Evenson, C.J. Howard, Laser magnetic resonance spectroscopy, in *Laser Spectroscopy*, ed. by R.G. Brewer, A. Mooradian (Plenum, New York, 1974)
145. A. Hinz, J. Pfeiffer, W. Bohle, W. Urban, Mid-infrared laser magnetic resonance using the Faraday and Voigt effects for sensitive detection. Mol. Phys. **45**, 1131 (1982)
146. Y. Ueda, K. Shimoda, Infrared laser Stark spectroscopy, in *Laser Spectroscopy II*, ed. by S. Haroche, J.C. Pebay-Peyroula, T.W. Hänsch. Lecture Notes Phys., vol. 43 (Springer, Berlin, 1975), p. 186
147. K. Uehara, T. Shimiza, K. Shimoda, High resolution Stark spectroscopy of molecules by infrared and far infrared masers. IEEE J. Quantum Electron. **4**, 728 (1968)
148. K. Uehara, K. Takagi, T. Kasuya, Stark modulation spectrometer, using a wideband Zeeman-tuned He–Xe laser. Appl. Phys. **24**, 187–195 (1981)
149. L.R. Zink, D.A. Jennings, K.M. Evenson, A. Sasso, M. Inguscio, New techniques in laser Stark spectroscopy. J. Opt. Soc. Am. B **4**, 1173 (1987)
150. K.M. Evenson, R.J. Saykally, D.A. Jennings, R.F. Curl, J.M. Brown, Far infrared laser magnetic resonance, in *Chemical and Biochemical Applications of Lasers*, vol. V, ed. by C.B. Moore (Academic Press, New York, 1980)
151. Ch.P. Slichter, *Principles of Magnetic Resonance* (Springer, Berlin, 2010)
152. M. Inguscio, Coherent atomic and molecular spectroscopy in the far infrared. Phys. Scr. **37**, 699 (1989)
153. W.H. Weber, K. Tanaka, T. Kanaka (eds.), Stark and Zeeman techniques in laser spectroscopy. J. Opt. Soc. Am. B **4**, 1141 (1987)
154. J.L. Kinsey, Laser-induced fluorescence. Annu. Rev. Phys. Chem. **28**, 349 (1977)
155. A. Delon, R. Jost, Laser-induced dispersed fluorescence spectroscopy of 107 vibronic levels of NO_2 ranging from 12,000 to 17,600 cm^{-1}. J. Chem. Phys. **114**, 331 (2001)
156. M.A. Clyne, I.S. McDermid, Laser-induced fluorescence: electronically excited states of small molecules. Adv. Chem. Phys. **50**, 1 (1982)
157. J.R. Lakowicz, *Topics in Fluorescence Spectroscopy* (Plenum, New York, 1991);
J.N. Miller, *Fluorescence Spectroscopy* (Ellis Harwood, Singapore, 1991);
O.S. Wolflich (ed.), *Fluorescence Spectroscopy* (Springer, Berlin, 1992)
158. C. Schütte, *The Theory of Molecular Spectroscopy* (North-Holland, Amsterdam, 1976)

159. G. Herzberg, *Molecular Spectra and Molecular Structure*, vol. I (Van Nostrand, New York, 1950)
160. G. Höning, M. Cjajkowski, M. Stock, W. Demtröder, High resolution laser spectroscopy of Cs_2. J. Chem. Phys. **71**, 2138 (1979)
161. C. Amiot, W. Demtröder, C.R. Vidal, High resolution Fourier-spectroscopy and laser spectroscopy of Cs_2. J. Chem. Phys. **88**, 5265 (1988)
162. C. Amiot, Laser-induced fluorescence of Rb_2. J. Chem. Phys. **93**, 8591 (1990)
163. R. Bacis, S. Chunassy, R.W. Fields, J.B. Koffend, J. Verges, High resolution and sub-Doppler Fourier transform spectroscopy. J. Chem. Phys. **72**, 34 (1980)
164. R. Rydberg, Graphische Darstellung einiger bandenspektroskopischer Ergebnisse. Z. Phys. **73**, 376 (1932)
165. O. Klein, Zur Berechnung von Potentialkurven zweiatomiger Moleküle mit Hilfe von Spekraltermen. Z. Phys. **76**, 226 (1938)
166. A.L.G. Rees, The calculation of potential-energy curves from band spectroscopic data. Proc. Phys. Soc. Lond., Sect. A **59**, 998 (1947)
167. R.N. Zare, A.L. Schmeltekopf, W.J. Harrop, D.L. Albritton, J. Mol. Spectrosc. **46**, 37 (1973)
168. G. Ennen, C. Ottinger, Laser fluorescence measurements of the ^{7}LiD($X\,^1\Sigma^+$)-potential up to high vibrational quantum numbers. Chem. Phys. Lett. **36**, 16 (1975)
169. M. Raab, H. Weickenmeier, W. Demtröder, The dissociation energy of the cesium dimer. Chem. Phys. Lett. **88**, 377 (1982)
170. C.E. Fellows, The NaLi $1\,^1\Sigma + (X)$ electronic ground state dissociation limit. J. Chem. Phys. **94**, 5855 (1991)
171. A.G. Gaydon, *Dissociation Energies and Spectra of Diatomic Molecules* (Chapman and Hall, London, 1968)
172. H. Atmanspacher, H. Scheingraber, C.R. Vidal, Laser-induced fluorescence of the MgCa molecule. J. Chem. Phys. **82**, 3491 (1985)
173. R.J. LeRoy, *Molecular Spectroscopy, Specialist Periodical Reports*, vol. 1 (Chem. Soc., Burlington Hall, London, 1973), p. 113
174. W. Demtröder, W. Stetzenbach, M. Stock, J. Witt, Lifetimes and Franck–Condon factors for the $B\,^1\Pi_u \to X\,^1\Sigma_g^+$-system of Na_2. J. Mol. Spectrosc. **61**, 382 (1976)
175. E.J. Breford, F. Engelke, Laser-induced fluorescence in supersonic nozzle beams: applications to the NaK $D\,^1\Pi \to X\,^1\Sigma$ and $D\,^1\Pi \to X\,^3\Sigma$ systems. Chem. Phys. Lett. **53**, 282 (1978);
E.J. Breford, F. Engelke, J. Chem. Phys. **71**, 1949 (1979)
176. J. Tellinghuisen, G. Pichler, W.L. Snow, M.E. Hillard, R.J. Exton, Analysis of the diffuse bands near 6100 Å in the fluorescence spectrum of Cs_2. Chem. Phys. **50**, 313 (1980)
177. H. Scheingraber, C.R. Vidal, Discrete and continuous Franck–Condon factors of the Mg_2 $A\,^1\Sigma_u - X\,^1I_s$ system and their J dependence. J. Chem. Phys. **66**, 3694 (1977)
178. C.A. Brau, J.J. Ewing, Spectroscopy, kinetics and performance of rare-gas halide lasers, in *Electronic Transition Lasers*, ed. by J.I. Steinfeld (MIT Press, Cambridge, 1976)
179. D. Eisel, D. Zevgolis, W. Demtröder, Sub-Doppler laser spectroscopy of the NaK-molecule. J. Chem. Phys. **71**, 2005 (1979)
180. E.V. Condon, Nuclear motions associated with electronic transitions in diatomic molecules. Phys. Rev. **32**, 858 (1928)
181. J. Tellinghuisen, The McLennan bands of I_2: a highly structured continuum. Chem. Phys. Lett. **29**, 359 (1974)
182. H.J. Vedder, M. Schwarz, H.J. Foth, W. Demtröder, Analysis of the perturbed NO_2 $^2B_2 \to$ $^2A_1$ system in the 591.4–592.9 nm region based on sub-Doppler laser spectroscopy. J. Mol. Spectrosc. **97**, 92 (1983)
183. A. Delon, R. Jost, Laser-induced dispersed fluorescence spectra of jet-cooled NO_2. J. Chem. Phys. **95**, 5686 (1991)
184. Th. Zimmermann, H.J. Köppel, L.S. Cederbaum, G. Persch, W. Demtröder, Confirmation of random-matrix fluctuations in molecular spectra. Phys. Rev. Lett. **61**, 3 (1988)

185. K.K. Lehmann, St.L. Coy, The optical spectrum of NO_2: is it or isn't it chaotic? Ber. Bunsenges. Phys. Chem. **92**, 306 (1988)
186. J.M. Gomez-Llorentl, H. Taylor, Spectra in the chaotic region: a classical analysis for the sodium trimer. J. Chem. Phys. **91**, 953 (1989)
187. K.L. Kompa, *Chemical Lasers*. Topics Curr. Chem., vol. 37 (Springer, Berlin, 1975)
188. R. Schnabel, M. Kock, Time-Resolved nonlinear LIF-techniques for a combined lifetime and branching fraction measurements. Phys. Rev. A **63**, 125 (2001)
189. P.J. Dagdigian, H.W. Cruse, A. Schultz, R.N. Zare, Product state analysis of BaO from the reactions $Ba + CO_2$ and $Ba + O_2$. J. Chem. Phys. **61**, 4450 (1974)
190. J.G. Pruett, R.N. Zare, State-to-state reaction rates: $Ba + HF(v = 0) \rightarrow BaF(v = 0 - 12) + H''$. J. Chem. Phys. **64**, 1774 (1976)
191. H.W. Cruse, P.J. Dagdigian, R.N. Zare, Crossed beam reactions of barium with hydrogen halides. Faraday Discuss. Chem. Soc. **55**, 277 (1973)
192. Y. Nozaki et al., Identification of Si and SiH. J. Appl. Phys. **88**, 5437 (2000)
193. V. Hefter, K. Bergmann, Spectroscopic detection methods, in *Atomic and Molecular Beam Methods*, vol. I, ed. by G. Scoles (Oxford Univ. Press, New York, 1988), p. 193
194. J.E.M. Goldsmith, Recent advances in flame diagnostics using fluorescence and ionisation techniques, in *Laser Spectroscopy VIII*, ed. by S. Svanberg, W. Persson. Springer Ser. Opt. Sci., vol. 55 (Springer, Berlin, 1987), p. 337
195. J. Wolfrum (ed.), Laser diagnostics in combustion. Appl. Phys. B **50**, 439 (1990)
196. T.P. Hughes, *Plasma and Laser Light* (Hilger, Bristol, 1975)
197. J.R. Lakowicz, *Principles of Fluorescence Spectroscopy*, 3rd edn. (Springer, Berlin, 2006)
198. L.J. Radziemski, D.A. Cremers, *Handbook of Laser-Induced Breakdown Spectroscopy* (Wiley, New York, 2006). ISBN 0-470-09299-8
199. I. Schechter, A.W. Miziolek, V. Palleschi, *Laser-Induced Breakdown Spectroscopy (LIBS): Fundamentals and Applications* (Cambridge University Press, Cambridge, 2006). ISBN 0-521-85274-9
200. R. Ahmed, M.-A. Baig, On the optimization for enhanced dual-pulse laser-induced breakdown spectroscopy. IEEE Trans. Plasma Sci. **38**(8), 2052–2055 (2010). doi:10.1109/TPS.2010.2050784. ISSN 0093-3813
201. M. Bellini, P. DeNatale, G. DiLonardo, L. Fusina, M. Inguscio, M. Prevedelli, Tunable far infrared spectroscopy of $^{16}O_3$ ozone. J. Mol. Spectrosc. **152**, 256 (1992);
J.R. Albani, *Principles and Applications of Fluorescence Spectroscopy* (Wiley-Blackwell, New York, 2007)

Chapter 2

202. W.R. Bennet Jr., Hole-burning effects in a He–Ne-optical maser. Phys. Rev. **126**, 580 (1962)
203. V.S. Letokhov, V.P. Chebotayev, *Nonlinear Laser Spectroscopy*. Springer Ser. Opt. Sci., vol. 4 (Springer, Berlin, 1977)
204. S. Mukamel, *Principles of Nonlinear Optical Spectroscopy* (Oxford Univ. Press, Oxford, 1999)
205. M.D. Levenson, *Introduction to Nonlinear Spectroscopy* (Academic, New York, 1982)
206. W.E. Lamb, Theory of an optical maser. Phys. Rev. A **134**, 1429 (1964)
207. H. Gerhardt, E. Matthias, F. Schneider, A. Timmermann, Isotope shifts and hyperfine structure of the $6s$–$7p$-transitions in the cesium isotopes 133, 135 and 137. Z. Phys. A **288**, 327 (1978)
208. S.L. Chin, *Fundamentals of Laser Optoelectronics* (World Scientific, Singapore, 1989), pp. 281 ff.
209. M.S. Sorem, A.L. Schawlow, Saturation spectroscopy in molecular iodine by intermodulated fluorescence. Opt. Commun. **5**, 148 (1972)

210. M.D. Levenson, A.L. Shawlow, Hyperfine interactions in molecular iodine. Phys. Rev. A **6**, 10 (1972)
211. H.J. Foth, Sättigungsspektroskopie an Molekülen. Diplom thesis, University of Kaiserslautern, Germany, 1976
212. R.S. Lowe, H. Gerhardt, W. Dillenschneider, R.F. Curl Jr., F.K. Tittel, Intermodulated fluorescence spectroscopy of BO_2 using a stabilized dye laser. J. Chem. Phys. **70**, 42 (1979)
213. A.S. Cheung, R.C. Hansen, A.J. Nerer, Laser spectroscopy of VO: analysis of the rotational and hyperfine structure. J. Mol. Spectrosc. **91**, 165 (1982)
214. L.A. Bloomfield, B. Couillard, Ph. Dabkiewicz, H. Gerhardt, T.W. Hänsch, Hyperfine structure of the 2^3S–5^3P transition in ^{3}He by high resolution UV laser spectroscopy. Opt. Commun. **42**, 247 (1982)
215. Ch. Hertzler, H.J. Foth, Sub-Doppler polarization spectra of He, N_2 and Ar^+ recorded in discharges. Chem. Phys. Lett. **166**, 551 (1990)
216. H.J. Foth, F. Spieweck, Hyperfine structure of the R(98), (58-1)-line of I_2 at $\lambda = 514.5$ nm. Chem. Phys. Lett. **65**, 347 (1979)
217. W.G. Schweitzer, E.G. Kessler, R.D. Deslattes, H.P. Layer, J.R. Whetstone, Description, performance and wavelength of iodine stabilised lasers. Appl. Opt. **12**, 2927 (1973)
218. R.L. Barger, J.B. West, T.C. English, Frequency stabilization of a CW dye laser. Appl. Phys. Lett. **27**, 31 (1975)
219. C. Salomon, D. Hills, J.L. Hall, Laser stabilization at the millihertz level. J. Opt. Soc. B **5**, 1576 (1988)
220. V. Bernard et al., CO_2-Laser stabilization to 0.1 Hz using external electro-optic modulation. IEEE J. Quantum Electron. **33**, 1288 (1997)
221. J.C. Hall, J.A. Magyar, High resolution saturation absorption studies of methane and some methyl-halides, in *High-Resolution Laser Spectroscopy*, ed. by K. Shimoda. Topics Appl. Phys., vol. 13 (Springer, Berlin, 1976), p. 137
222. J.L. Hall, Sub-Doppler spectroscopy, methane hyperfine spectroscopy and the ultimate resolution limit, in *Colloq. Int. due CNRS*, vol. 217 (Edit. due CNRS, Paris, 1974), p. 105
223. B. Bobin, C.J. Bordé, J. Bordé, C. Bréant, Vibration-rotation molecular constants for the ground and ($\nu_3 = 1$) states of SF_6 from saturated absorption spectroscopy. J. Mol. Spectrosc. **121**, 91 (1987)
224. M. de Labachelerie, K. Nakagawa, M. Ohtsu, Ultranarrow $^{13}C_2H_2$ saturated absorption lines at 1.5 μm. Opt. Lett. **19**, 840 (1994)
225. C. Wieman, T.W. Hänsch, Doppler-free laser polarization spectroscopy. Phys. Rev. Lett. **36**, 1170 (1976)
226. R.E. Teets, F.V. Kowalski, W.T. Hill, N. Carlson, T.W. Hänsch, Laser polarization spectroscopy, in *Advances in Laser Spectroscopy*. SPIE Proc., vol. 113 (1977), p. 80
227. R. Teets, Laser polarisation spectroscopy. PhD thesis, Physics Department, Stanford University, 1978
228. M.E. Rose, *Elementary Theory of Angular Momentum* (Wiley, New York, 1957), reprint paperback: Dover, 1995;
A.R. Edmonds, *Angular Momentum in Quantum Mechanics* (Princeton Univ. Press, Princeton, 1996)
229. R.N. Zare, *Angular Momentum: Understanding Spatial Aspects in Chemistry and Physics* (Wiley, New York, 1988)
230. V. Stert, R. Fischer, Doppler-free polarization spectroscopy using linear polarized light. Appl. Phys. **17**, 151 (1978)
231. H. Gerhardt, T. Huhle, J. Neukammer, P.J. West, High resolution polarization spectroscopy of the 557 nm transition of KrI. Opt. Commun. **26**, 58 (1978)
232. M. Raab, G. Höning, R. Castell, W. Demtröder, Doppler-free polarization spectroscopy of the Cs_2 molecule at $\lambda = 6270$ Å. Chem. Phys. Lett. **66**, 307 (1979)
233. M. Raab, G. Höning, W. Demtröder, C.R. Vidal, High resolution laser spectroscopy of Cs_2. J. Chem. Phys. **76**, 4370 (1982)

234. W. Ernst, Doppler-free polarization spectroscopy of diatomic molecules in flame reactions. Opt. Commun. **44**, 159 (1983)
235. M. Francesconi, L. Gianfrani, M. Inguscio, P. Minutolo, A. Sasso, A new approach to impedance atomic spectroscopy. Appl. Phys. B **51**, 87 (1990)
236. L. Gianfrani, A. Sasso, G.M. Tino, F. Marin, Polarization spectroscopy of atomic oxygen by dye and semiconductor diode lasers. Il Nuovo Cimento D **10**, 941 (1988)
237. M. Göppert-Mayer, Über Elementarakte mit zwei Quantensprüngen. Ann. Phys. **9**, 273 (1931)
238. W. Kaiser, C.G. Garret, Two-photon excitation in LLCA F_2: Eu^{2+}. Phys. Rev. Lett. **7**, 229 (1961)
239. J.J. Hopfield, J.M. Worlock, K. Park, Two-quantum absorption spectrum of KI. Phys. Rev. Lett. **11**, 414 (1963)
240. P. Bräunlich, Multiphoton spectroscopy, in *Progress in Atomic Spectroscopy*, ed. by W. Hanle, H. Kleinpoppen (Plenum, New York, 1978)
241. J.M. Worlock, Two-photon spectroscopy, in *Laser Handbook*, ed. by F.T. Arrecchi, E.O. Schulz-Dubois (North-Holland, Amsterdam, 1972)
242. B. Dick, G. Hohlneicher, Two-photon spectroscopy of dipole-forbidden transitions. Theor. Chim. Acta **53**, 221 (1979); and J. Chem. Phys. **70**, 5427 (1979)
243. J.B. Halpern, H. Zacharias, R. Wallenstein, Rotational line strengths in two- and three-photon transitions in diatomic molecules. J. Mol. Spectrosc. **79**, 1 (1980)
244. K.D. Bonin, T.J. McIlrath, Two-photon electric dipole selection rules. J. Opt. Soc. Am. B **1**, 52 (1984)
245. G. Grynberg, B. Cagnac, Doppler-free multiphoton spectroscopy. Rep. Prog. Phys. **40**, 791 (1977)
246. F. Biraben, B. Cagnac, G. Grynberg, Experimental evidence of two- photon transition without Doppler broadening. Phys. Rev. Lett. **32**, 643 (1974)
247. G. Grynberg, B. Cagnbac, F. Biraben, Multiphoton resonant processes in atoms, in *Coherent Nonlinear Optics*, ed. by M.S. Feld, V.S. Letokhov. Topics Curr. Phys., vol. 21 (Springer, Berlin, 1980)
248. T.W. Hänsch, K. Harvey, G. Meisel, A.L. Shawlow, Two-photon spectroscopy of Na $3s$–$4d$ without Doppler-broadening using CW dye laser. Opt. Commun. **11**, 50 (1974)
249. M.D. Levenson, N. Bloembergen, Observation of two-photon absorption without Doppler-broadening on the $3s$–$5s$ transition in sodium vapor. Phys. Rev. Lett. **32**, 645 (1974)
250. A. Timmermann, High resolution two-photon spectroscopy of the $6p^2\,{}^3P_0$–$7p\,{}^3P_0$ transition in stable lead isotopes. Z. Phys. A **286**, 93 (1980)
251. S.A. Lee, J. Helmcke, J.L. Hall, P. Stoicheff, Doppler-free two-photon transitions to Rydberg levels. Opt. Lett. **3**, 141 (1978)
252. R. Beigang, K. Lücke, A. Timmermann, Singlet–Triplet mixing in $4s$ and Rydberg states of Ca. Phys. Rev. A **27**, 587 (1983)
253. S.V. Filseth, R. Wallenstein, H. Zacharias, Two-photon excitation of CO $(A^1\Pi)$ and N_2 $(a^1\Pi_g)$. Opt. Commun. **23**, 231 (1977)
254. E. Riedle, H.J. Neusser, E.W. Schlag, Electronic spectra of polyatomic molecules with resolved individual rotational transitions: benzene. J. Chem. Phys. **75**, 4231 (1981)
255. H. Sieber, E. Riedle, J.H. Neusser, Intensity distribution in rotational line spectra I: experimental results for Doppler-free $S_1 \leftarrow S_0$ transitions in benzene. J. Chem. Phys. **89**, 4620 (1988);
E. Riedle, Doppler-freie Zweiphotonen-Spektroskopie an Benzol. Habilitation thesis, Inst. Physikalische Chemie, TU München, Germany, 1990
256. E. Riedle, H.J. Neusser, Homogeneous linewidths of single rotational lines in the "channel three" region of C_6H_6. J. Chem. Phys. **80**, 4686 (1984)
257. U. Schubert, E. Riedle, J.H. Neusser, Time evolution of individual rotational states after pulsed Doppler-free two-photon excitation. J. Chem. Phys. **84**, 5326 (1986); and 6182 (1986)

258. W. Bischel, P.J. Kelley, Ch.K. Rhodes, High-resolution Doppler-free two-photon spectroscopic studies of molecules. Phys. Rev. A **13**, 1817 (1976); and **13**, 1829 (1976)
259. R. Guccione-Gush, H.P. Gush, R. Schieder, K. Yamada, C. Winnewisser, Doppler-free two-photon absorption of NH_3 using a CO_2 and a diode laser. Phys. Rev. A **23**, 2740 (1981)
260. G.F. Bassani, M. Inguscio, T.W. Hänsch (eds.), *The Hydrogen Atom* (Springer, Berlin, 1989)
261. M. Weitz, F. Schmidt-Kaler, T.W. Hänsch, Precise optical Lamb-shift measurements in atomic hydrogen. Phys. Rev. Lett. **68**, 1120 (1992);
S.A. Lee, R. Wallenstein, T.W. Hänsch, Hydrogen $1S$–$2S$-isotope shift and $1S$ Lamb shift measured by laser spectroscopy. Phys. Rev. Lett. **35**, 1262 (1975)
262. J.R.M. Barr, J.M. Girkin, J.M. Tolchard, A.I. Ferguson, Interferometric measurement of the $1S_{1/2}$–$2S_{1/2}$ transition frequency in atomic hydrogen. Phys. Rev. Lett. **56**, 580 (1986)
263. M. Niering et al., Measurement of the hydrogen $1S$–$2S$ transition frequency by phase coherent comparison with a microwave cesium fountain clock. Phys. Rev. Lett. **84**, 5496 (2000)
264. F. Biraben, J.C. Garreau, L. Julien, Determination of the Rydberg constant by Doppler-free two-photon spectroscopy of hydrogen Rydberg states. Europhys. Lett. **2**, 925 (1986)
265. F.H.M. Faisal, R. Wallenstein, H. Zacharias, Three-photon excitation of xenon and carbon monoxide. Phys. Rev. Lett. **39**, 1138 (1977)
266. B. Cagnac, Multiphoton high resolution spectroscopy, in *Atomic Physics 5*, ed. by R. Marrus, M. Prior, H. Shugart (Plenum, New York, 1977), p. 147
267. V.I. Lengyel, M.I. Haylak, Role of autoionizing states in multiphoton ionization of complex atoms. Adv. At. Mol. Phys. **27**, 245 (1990)
268. E.M. Alonso, A.L. Peuriot, V.B. Slezak, CO_2-laser-induced multiphoton absorption of CF_2Cl_2. Appl. Phys. B **40**, 39 (1986)
269. V.S. Lethokov, Multiphoton and multistep vibrational laser spectroscopy of molecules. Comments At. Mol. Phys. **8**, 39 (1978)
270. W. Fuss, J. Hartmann, IR absorption of SF_6 excited up to the dissociation limit. J. Chem. Phys. **70**, 5468 (1979)
271. F.V. Kowalski, W.T. Hill, A.L. Schawlow, Saturated-interference spectroscopy. Opt. Lett. **2**, 112 (1978)
272. R. Schieder, Interferometric nonlinear spectroscopy. Opt. Commun. **26**, 113 (1978)
273. S. Tolanski, *An Introduction to Interferometry* (Longman, London, 1973)
274. C. Delsart, J.C. Keller, Doppler-free laser induced dichroism and birefringence, in *Laser Spectroscopy of Atoms and Molecules*, ed. by H. Walther. Topics Appl. Phys., vol. 2 (Springer, Berlin, 1976), p. 154
275. M.D. Levenson, G.L. Eesley, Polarization selective optical heterodyne detection for dramatically improved sensitivity in laser spectroscopy. Appl. Phys. **19**, 1 (1979)
276. M. Raab, A. Weber, Amplitude-modulated heterodyne polarization spectroscopy. J. Opt. Soc. Am. B **2**, 1476 (1985)
277. K. Danzmann, K. Grützmacher, B. Wende, Doppler-free two-photon polarization spectroscopy measurement of the Stark-broadened profile of the hydrogen H_α line in a dense plasma. Phys. Rev. Lett. **57**, 2151 (1986)
278. T.W. Hänsch, A.L. Schawlow, C.W. Series, The spectrum of atomic hydrogen. Sci. Am. **240**, 72 (1979)
279. R.S. Berry, How good is Niels Bohrs atomic model? Contemp. Phys. **30**, 1 (1989)
280. F. Schmidt-Kalen, D. Leibfried, M. Weitz, T.W. Hänsch, Precision measurement of the isotope shift of the $1S$–$2S$ transition of atomic hydrogen and deuterium. Phys. Rev. Lett. **70**, 2261 (1993)
281. V.S. Butylkin, A.E. Kaplan, Y.G. Khronopulo, *Resonant Nonlinear Interaction of Light with Matter* (Springer, Berlin, 1987)
282. J.J.H. Clark, R.E. Hester (eds.), *Advances in Nonlinear Spectroscopy* (Wiley, New York, 1988)
283. S.S. Kano, *Introduction to Nonlinear Laser Spectroscopy* (Academic, New York, 1988)

284. T.W. Hänsch, Nonlinear high-resolution spectroscopy of atoms and molecules, in *Nonlinear Spectroscopy, Proc. Int. School of Physics "Enrico Fermi" Course LXIV* (North-Holland, Amsterdam, 1977), p. 17
285. D.C. Hanna, M.Y. Yunatich, D. Cotter, *Nonlinear Optics of Free Atoms and Molecules*. Springer Ser. Opt. Sci., vol. 17 (Springer, Berlin, 1979)
286. St. Stenholm, *Foundations of Laser Spectroscopy* (Wiley, New York, 1984)
287. R. Altkorn, R.Z. Zare, Effects of saturation on laser-induced fluorescence measurements. Annu. Rev. Phys. Chem. **35**, 265 (1984)
288. B. Cagnac, Laser Doppler-free techniques in spectroscopy, in *Frontiers of Laser Spectroscopy of Gases*, ed. by A.C.P. Alves, J.M. Brown, J.H. Hollas. Nato ASO Series C, vol. 234 (Kluwer, Dondrost, 1988)
289. S.H. Lin (ed.), *Advances in Multiphoton Processes and Spectroscopy* (World Scientific, Singapore, 1985–1992)
290. M. Fischer, T.W. Hänsch, Precision spectroscopy of atomic hydrogen and variations of fundamental constants, in *Astrophysics, Clocks and Fundamental Constants*. Lecture Notes in Physics, vol. 648 (Springer, Heidelberg, 2004)

Chapter 3

291. A. Anderson, *The Raman Effect*, vols. 1, 2 (Dekker, New York, 1971/1973)
292. D.A. Long, *Raman Spectroscopy* (McGraw-Hill, New York, 1977);
D.A. Long, *The Raman Effect: A Unified Treatment of the Theory of Raman Scattering by Molecules* (Wiley, New York, 2001);
E. Smith, G. Dent, *Modern Raman Spectroscopy* (Wiley, New York, 2005)
293. B. Schrader, *Infrared and Raman Spectroscopy* (Wiley VCH, Weinheim, 1993);
M.J. Pelletier (ed.), *Analytical Applications of Raman Spectroscopy* (Blackwell Science, Oxford, 1999);
J.R. Ferrano, *Introductory Raman Spectroscopy*, 2nd edn. (Academic Press, New York, 2002)
294. J.R. Ferraro, K. Nakamato, *Introductory Raman Spectroscopy* (Academic Press, New York, 1994)
295. I.R. Lewis, H.G.M. Edwards (eds.), *Handbook of Raman Spectroscopy* (Dekker, New York, 2001)
296. M.C. Tobin, *Laser Raman Spectroscopy* (Wiley Interscience, New York, 1971)
297. A. Weber (ed.), *Raman Spectroscopy of Gases and Liquids*. Topics Curr. Phys, vol. 11 (Springer, Berlin, 1979)
298. G. Placzek, Rayleigh-Streuung und Raman Effekt, in *Handbuch der Radiologie*, vol. VI, ed. by E. Marx (Akademische Verlagsgesellschaft, Leipzig, 1934)
299. L.D. Barron, Laser Raman spectroscopy, in *Frontiers of Laser Spectroscopy of Gases*, ed. by A.C.P. Alves, J.M. Brown, J.M. Hollas. NATO ASI Series, vol. 234 (Kluwer, Dordrecht, 1988)
300. N.B. Colthup, L.H. Daly, S.E. Wiberley, *Introduction to Infrared and Raman Spectroscopy*, 3rd edn. (Academic Press, New York, 1990)
301. R.J.H. Clark, R.E. Hester (eds.), *Advances in Infrared and Raman Spectroscopy*, vols. 1–17 (Heyden, London, 1975–1990)
302. J. Popp, W. Kiefer, Fundamentals of Raman spectroscopy, in *Encyclopedia of Analytical Chemistry* (Wiley, New York, 2001)
303. P.P. Pashinin (ed.), *Laser-Induced Raman Spectroscopy in Crystals and Gases* (Nova Science, Commack, 1988)
304. J.R. Durig, J.F. Sullivan (eds.), *XII Int. Conf. on Raman Spectroscopy* (Wiley, Chichester, 1990)

305. H. Kuzmany, *Festkörperspektroskopie* (Springer, Berlin, 1989)
306. M. Cardona (ed.), *Light Scattering in Solids*, 2nd edn. Topics Appl. Phys., vol. 8 (Springer, Berlin, 1983);
M. Cardona, G. Güntherodt (eds.), *Light Scattering in Solids II–VI*, Topics Appl. Phys., vols. 50, 51, 54, 66, 68 (Springer, Berlin, 1982, 1984, 1989, 1991)
307. K.W. Szymanski, *Raman Spectroscopy I & II* (Plenum, New York, 1970)
308. G. Herzberg, *Molecular Spectra and Molecular Structure, Vol. II, Infrared and Raman Spectra of Polyatomic Molecules* (van Nostrand Reinhold, New York, 1945)
309. H.W. Schrötter, H.W. Klöckner, Raman scattering cross sections in gases and liquids, in *Raman Spectroscopy of Gases and Liquids*, ed. by A. Weber. Topics Curr. Phys., vol. II (Springer, Berlin, 1979), pp. 123 ff.
310. D.L. Rousseau, The resonance Raman effect, in *Raman Spectroscopy of Gases and Liquids*, ed. by A. Weber. Topics Curr. Phys., vol. II (Springer, Berlin, 1979), pp. 203 ff.
311. S.A. Acher, UV resonance Raman studies of molecular structures and dynamics. Annu. Rev. Phys. Chem. **39**, 537 (1988)
312. R.J.H. Clark, T.J. Dinev, Electronic Raman spectroscopy, in *Advances in Infrared and Raman Spectroscopy*, vol. 9, ed. by R.J.H. Clark, R.E. Hester (Heyden, London, 1982), p. 282
313. J.A. Koningstein, *Introduction to the Theory of the Raman Effect* (Reidel, Dordrecht, 1972)
314. D.J. Gardner, *Practical Raman Spectroscopy* (Springer, Berlin, 1989)
315. A. Weber, High-resolution rotational Raman spectra of gases, in *Advances in Infrared and Raman Spectroscopy*, vol. 9, ed. by R.J.H. Clark, R.E. Hester (Heyden, London, 1982), Chap. 3
316. E.B. Brown, *Modern Optics* (Krieger, New York, 1974), p. 251
317. J.R. Downey, G.J. Janz, Digital methods in Raman spectroscopy, in *XII Int. Conf. on Raman Spectroscopy*, ed. by J.R. Durig, J.F. Sullivan (Wiley, Chichester, 1990), pp. 1–34
318. W. Knippers, K. van Helvoort, S. Stolte, Vibrational overtones of the homonuclear diatomics N_2, O_2, D_2. Chem. Phys. Lett. **121**, 279 (1985)
319. K. van Helvoort, R. Fantoni, W.L. Meerts, J. Reuss, Internal rotation in CH_3CD_3: Raman spectroscopy of torsional overtones. Chem. Phys. Lett. **128**, 494 (1986); and Chem. Phys. **110**, 1 (1986)
320. W. Kiefer, Recent techniques in Raman spectroscopy, in *Adv. Infrared and Raman Spectroscopy*, vol. 3, ed. by R.J.H. Clark, R.E. Hester (Heyden, London, 1977)
321. G.W. Walrafen, J. Stone, Intensification of spontaneous Raman spectra by use of liquid core optical fibers. Appl. Spectrosc. **26**, 585 (1972)
322. H.W. Schrötter, J. Bofilias, On the assignment of the second-order lines in the Raman spectrum of benzene. J. Mol. Struct. **3**, 242 (1969)
323. D.B. Chase, J.E. Rabolt, *Fourier-Transform Raman Spectroscopy* (Academic Press, New York, 1994)
324. D.A. Long, The polarisability and hyperpolarizability tensors, in *Nonlinear Raman Spectroscopy and Its Chemical Applications*, ed. by W. Kiefer, D.A. Longer (Reidel, Dordrecht, 1982)
325. L. Beardmore, H.G.M. Edwards, D.A. Long, T.K. Tan, Raman spectroscopic measurements of temperature in a natural gas laser flame, in *Lasers in Chemistry*, ed. by M.A. West (Elsevier, Amsterdam, 1977)
326. A. Leipert, Laser Raman-Spectroskopie in der Wärme- und Strömungstechnik. Phys. Unserer Zeit **12**, 107 (1981)
327. K. van Helvoort, W. Knippers, R. Fantoni, S. Stolte, The Raman spectrum of ethane from 600 to 6500 cm^{-1} Stokes shifts. Chem. Phys. **111**, 445 (1987)
328. J. Lascombe, P.V. Huong (eds.), *Raman Spectroscopy: Linear and Nonlinear* (Wiley, New York, 1982)
329. E.J. Woodbury, W.K. Ny, Ruby laser operation in the near IR. IRE Proc. **50**, 2367 (1962)
330. G. Eckardt, Selection of Raman laser materials. IEEE J. Quantum Electron. **2**, 1 (1966)

331. A. Yariv, *Quantum Electronics*, 3rd edn. (Wiley, New York, 1989)
332. W. Kaiser, M. Maier, Stimulated Rayleigh, Brillouin and Raman spectroscopy, in *Laser Handbook*, ed. by F.T. Arrecchi, E.O. Schulz-Dubois (North-Holland, Amsterdam, 1972), pp. 1077 ff.
333. E. Esherik, A. Owyoung, High resolution stimulated Raman spectroscopy, in *Adv. Infrared and Raman Spectroscopy*, vol. 9 (Heyden, London, 1982)
334. H.W. Schrötter, H. Frunder, H. Berger, J.P. Boquillon, B. Lavorel, G. Millet, High resolution CARS and inverse Raman spectroscopy, in *Adv. Nonlinear Spectroscopy*, vol. 3 (Wiley, New York, 1987), p. 97
335. R.S. McDowell, C.W. Patterson, A. Owyoung, Quasi-CW inverse Raman spectroscopy of the ω_1 fundamental of $^{13}CH_4$. J. Chem. Phys. **72**, 1071 (1980)
336. E.K. Gustafson, J.C. McDaniel, R.L. Byer, CARS measurement of velocity in a supersonic jet. IEEE J. Quantum Electron. **17**, 2258 (1981)
337. A. Owyoung, High resolution CARS of gases, in *Laser Spectroscopy IV*, ed. by H. Walther, K.W. Roth. Ser. Opt. Sci., vol. 21 (Springer, Berlin, 1979), p. 175
338. N. Bloembergen, *Nonlinear Optics*, 3rd edn. (Benjamin, New York, 1977); D.L. Mills, *Nonlinear Optics* (Springer, Berlin, 1991)
339. C.S. Wang, The stimulated Raman process, in *Quantum Electronics: A Treatise*, vol. 1, ed. by H. Rabin, C.L. Tang (Academic Press, New York, 1975), Chap. 7
340. M. Mayer, Applications of stimulated Raman scattering. Appl. Phys. **11**, 209 (1976)
341. G. Marowski, V.V. Smirnov (eds.), *Coherent Raman Spectroscopy*. Springer Proc. Phys., vol. 63 (Springer, Berlin, 1992)
342. W. Kiefer, Nonlinear Raman spectroscopy: applications, in *Encyclopedia of Spectroscopy and Spectrometry* (Academic Press, New York, 2000), p. 1609
343. J.W. Nibler, G.V. Knighten, Coherent anti-Stokes Raman spectroscopy, in *Raman Spectroscopy of Gases and Liquids*, ed. by A. Weber. Topics Curr. Phys., vol. II (Springer, Berlin, 1979), Chap. 7
344. J.W. Nibler, Coherent Raman spectroscopy: techniques and recent applications, in *Applied Laser Spectroscopy*, ed. by W. Demtröder, M. Inguscio. NATO ASI, vol. 241 (Plenum, London, 1990), p. 313
345. S.A.J. Druet, J.P.E. Taran, CARS spectroscopy. Prog. Quantum Electron. **7**, 1 (1981)
346. I.P.E. Taran, CARS spectroscopy and applications, in *Appl. Laser Spectroscopy*, ed. by W. Demtröder, M. Inguscio (Plenum, London, 1990), pp. 313–328
347. W. Kiefer, D.A. Long (eds.), *Nonlinear Raman Spectroscopy and Its Chemical Applications* (Reidel, Dordrecht, 1982)
348. F. Moya, S.A.J. Druet, J.P.E. Taran, Rotation-vibration spectroscopy of gases by CARS, in *Laser Spectroscopy II*, ed. by S. Haroche, J.C. Pebay-Peyroula, T.W. Hänsch, S.E. Harris. Springer Notes Phys., vol. 34 (Springer, Berlin, 1975), p. 66
349. S.A. Akhmanov, A.F. Bunkin, S.G. Ivanov, N.I. Koroteev, A.I. Kourigin, I.L. Shumay, Development of CARS for measurement of molecular parameters, in *Tunable Lasers and Applications*, ed. by A. Mooradian, T. Jaeger, P. Stokseth. Springer Ser. Opt. Sci., vol. 3 (Springer, Berlin, 1976)
350. J.P.E. Taran, Coherent anti-Stokes spectroscopy, in *Tunable Lasers and Applications*, ed. by A. Mooradian, T. Jaeger, P. Stokseth. Springer Ser. Opt. Sci., vol. 3 (Springer, Berlin, 1976), p. 315
351. Q.H.F. Vremen, A.J. Breiner, Spectral properties of a pulsed dye laser with monochromatic injection. Opt. Commun. **4**, 416 (1972)
352. T.J. Vickers, Quantitative resonance Raman spectroscopy. Appl. Spectrosc. Rev. **26**, 341 (1991)
353. B. Attal Debarré, K. Müller-Dethlets, J.P.E. Taran, Resonant coherent anti-Stokes Raman spectroscopy of C_2. Appl. Phys. B **28**, 221 (1982)
354. A.C. Eckbreth, BOX CARS: crossed-beam phase matched CARS generation. Appl. Phys. Lett. **32**, 421 (1978)

355. Y. Prior, Three-dimensional phase matching in four-wave mixing. Appl. Opt. **19**, 1741 (1980)
356. S.J. Cyvin, J.E. Rauch, J.C. Decius, Theory of hyper-Raman effects. J. Chem. Phys. **43**, 4083 (1965)
357. P.D. Maker, Nonlinear light scattering in methane, in *Physics of Quantum Electronics*, ed. by P.L. Kelley, B. Lax, P.E. Tannenwaldt (McGraw-Hill, New York, 1960), p. 60
358. K. Altmann, G. Strey, Enhancement of the scattering intensity for the hyper-Raman effect. Z. Naturforsch. A **32**, 307 (1977)
359. S. Nie, L.A. Lipscomb, N.T. Yu, Surface-enhanced hyper-Raman spectroscopy. Appl. Spectrosc. Rev. **26**, 203 (1991)
360. J. Reif, H. Walther, Generation of Tunable 16 μm radiation by stimulated hyper-Raman effect in strontium vapour. Appl. Phys. **15**, 361 (1978)
361. M.D. Levenson, J.J. Song, Raman-induced Kerr effect with elliptical polarization. J. Opt. Soc. Am. **66**, 641 (1976)
362. S.A. Akhmanov, A.F. Bunkin, S.G. Ivanov, N.I. Koroteev, Polarization active Raman spectroscopy and coherent Raman ellipsometry. Sov. Phys. JETP **47**, 667 (1978)
363. J.W. Nibler, J.J. Young, Nonlinear Raman spectroscopy of gases. Annu. Rev. Phys. Chem. **38**, 349 (1987)
364. Z.Q. Tian, B. Ren (eds.), *Progress in Surface Raman Spectroscopy* (Xiaman Univ. Press, Xiaman, 2000)
365. H. Wackerbarth et al., Detection of explosives based on surface enhanced Raman spectroscopy. Appl. Opt. **49**, 4362 (2010)
366. B. Eckert, H.D. Albert, H.J. Jodl, Raman studies of sulphur at high pressures and low temperatures. J. Phys. Chem. **100**, 8212 (1996)
367. P. Dhamelincourt, Laser molecular microprobe, in *Lasers in Chemistry*, ed. by M.A. West (Elsevier, Amsterdam, 1977), p. 48
368. G. Mariotto, F. Ziglio, F.L. Freire Jr., Light-emitting porous silicon: a structural investigation by high spatial resolution Raman spectroscopy. J. Non-Cryst. Solids **192**, 253 (1995)
369. L. Quin, Z.X. Shen, S.H. Tang, M.H. Kuck, The modification of a spex spectrometer into a micro-Raman spectrometer. Asian J. Spectrosc. **1**, 121 (1997)
370. W. Kiefer, Femtosecond coherent Raman spectroscopy. J. Raman Spectrosc. **31**, 3 (2000)
371. M. Danfus, G. Roberts, Femtosecond transition state spectroscopy and chemical reaction dynamics. Comments At. Mol. Phys. **26**, 131 (1991)
372. L. Beardmore, H.G.M. Edwards, D.A. Long, T.K. Tan, Raman spectroscopic measurements of temperature in a natural gas/air-flame, in *Lasers in Chemistry*, ed. by M.A. West (Elsevier, Amsterdam, 1977), p. 79
373. M.A. Lapp, C.M. Penney, Raman measurements on flames, in *Advances in Infrared and Raman Spectroscopy*, vol. 3, ed. by R.S.H. Clark, R.E. Hester (Heyden, London, 1977), p. 204
374. X. He et al., ZnO–Ag hybrids for ultrasensitive detection of trinitrotoluene by surface-enhanced Raman spectroscopy. Phys. Chem. Chem. Phys. **16**, 14706 (2014)
375. E.L. Izake, Forensic and homeland security applications of modern portable Raman spectroscopy. Forensic Sci. Int. **202**, 1–8 (2010)
376. F. Burkhart, Optical sensors come of age. Spie Prof., April 2014
377. J.P. Taran, CARS: techniques and applications, in *Tunable Lasers and Applications*, ed. by A. Mooradian, P. Jaeger, T. Stokseth. Springer Ser. Opt. Sci., vol. 3 (Springer, Berlin, 1976), p. 378
378. T. Dreier, B. Lange, J. Wolfrum, M. Zahn, Determination of temperature and concentration of molecular nitrogen, oxygen and methane with CARS. Appl. Phys. B **45**, 183 (1988)
379. H.D. Barth, C. Jackschath, T. Persch, F. Huisken, CARS spectroscopy of molecules and clusters in supersonic jets. Appl. Phys. B **45**, 205 (1988)
380. F. Adar, J.E. Griffith (eds.), Raman and luminescent spectroscopy in technology. SPIE Proc. **1336** (1990)

381. A.C. Eckbreth, Laser diagnostics for combustion temperature and species, in *Energy and Engineering Science*, ed. by A.K. Gupta, D.G. Lilley (Abacus Press, Cambridge, 1988)
382. A.S. Haka et al., Diagnosing breast cancer by Raman spectroscopy. Proc. Natl. Acad. Sci. USA **102**(25), PMC 1194905n (2005)
383. R. McCreery, *Raman Spectroscopy for Chemical Analysis* (Wiley Interscience, New York, 2007)
384. I. Lewis, G. Howell, M. Edwards, *Handbook of Raman Spectroscopy* (CRC Press, Boca Raton, 2001)
385. G. Marowsky, V.V. Smirnov (eds.), *Coherent Raman Spectroscopy*. Springer Proc. Phys., vol. 63 (Springer, Berlin, 1992)
386. M.D. Fayer, *Ultrafast Infrared and Raman Spectroscopy* (CRC Press, Boca Raton, 2001)

Chapter 4

387. R. Campargue, *Atomic and Molecular Beams; the State of the Art* (Springer, Berlin, 2001); S.Y.T. van de Meerakker, H.L. Bethlem, N. Vanhaecke, G. Meijer, Manipulation and control of molecular beams. Chem. Rev. **112**, 4828 (2012);
S.Y.T. van de Meerakker, H.L. Bethlem, N. Vanhaecke, G. Meijer, Taming molecular beams. Nat. Phys. **4**, 595 (2008);
M. Farnik, Molecular beams. http://www.jh-inst.cas.cz/~farnik/lectures/MF-1-MolBeams_ICONIC-TrSchool-2010.pdf
388. R. Abjean, M. Leriche, On the shapes of absorption lines in a divergent atomic beam. Opt. Commun. **15**, 121 (1975)
389. R.W. Stanley, Gaseous atomic beam light source. J. Opt. Soc. Am. **56**, 350 (1966)
390. J.B. Atkinson, J. Becker, W. Demtröder, Hyperfine structure of the 625 nm band in the a $^3\Pi_u \leftarrow X\,^1\Sigma_g$ transition for Na_2. Chem. Phys. Lett. **87**, 128 (1982); and **87**, 92 (1982)
391. R. Kullmer, W. Demtröder, Sub-Doppler laser spectroscopy of SO_2 in a supersonic beam. J. Chem. Phys. **81**, 2919 (1984)
392. W. Demtröder, F. Paech, R. Schmiedle, Hyperfine-structure in the visible spectrum of NO_2. Chem. Phys. Lett. **26**, 381 (1974)
393. R. Schmiedel, I.R. Bonilla, F. Paech, W. Demtröder, Laser spectroscopy of NO_2 under very high resolution. J. Mol. Spectrosc. **8**, 236 (1977)
394. U. Diemer, Dissertation, Universität Kaiserslautern, Germany, 1990;
U. Diemer, H.M. Greß, W. Demtröder, The $2\,^3\Pi_g \leftarrow X\,^3\Sigma_u$-triplet system of Cs_2. Chem. Phys. Lett. **178**, 330 (1991);
H. Bovensmann, H. Knöchel, E. Tiemann, Hyperfine structural investigations of the excited AO^+ state of Tl I. Mol. Phys. **73**, 813 (1991)
395. C. Duke, H. Fischer, H.J. Kluge, H. Kremling, T. Kühl, E.W. Otten, Determination of the isotope shift of ^{190}Hg by on line laser spectroscopy. Phys. Lett. A **60**, 303 (1977)
396. P. Jacquinot, Atomic beam spectroscopy, in *High-Resolution Laser Spectroscopy*, ed. by K. Shimoda. Topics Appl. Phys., vol. 13 (Springer, Berlin, 1976), p. 51
397. G. Nowicki, K. Bekk, J. Göring, A. Hansen, H. Rebel, G. Schatz, Nuclear charge radii and nuclear moments of neutrons deficient Ba-isotopes from high resolution laser spectroscopy. Phys. Rev. C **18**, 2369 (1978)
398. G. Ewald et al., Nuclear charge radii of ^{8,9}Li determinated by laser spectroscopy. Phys. Rev. Lett. **93**, 113002 (2004)
399. R. Sanchez et al., The nuclear charge radii of the radioactive lithium isotopes. GSI report, 2004
400. L.A. Hackel, K.H. Casleton, S.G. Kukolich, S. Ezekiel, Observation of magnetic octople and scalar spin-spin interaction in I_2 using laser spectroscopy. Phys. Rev. Lett. **35**, 568 (1975); and J. Opt. Soc. Am. **64**, 1387 (1974)

401. W. Lange, J. Luther, A. Steudel, Dye lasers in atomic spectroscopy, in *Adv. Atomic and Molecular Phys*, vol. 10 (Academic Press, New York, 1974)
402. G. Scoles (ed.), *Atomic and Molecular Beam Methods*, vols. I/II (Oxford Univ. Press, New York, 1988/1992);
C. Whitehead, Molecular beam spectroscopy. Europ. Spectrosc. News **57**, 10 (1984)
403. J.P. Bekooij, High resolution molecular beam spectroscopy at microwave and optical frequencies. Dissertation, University of Nijmwegen, The Netherlands, 1983
404. W. Demtröder, Visible and ultraviolet spectroscopy, in *Atomic and Molecular Beam Methods II*, ed. by G. Scoles (Oxford Univ. Press, New York, 1992);
W. Demtröder, H.J. Foth, Molekülspektroskopie in kalten Düsenstrahlen. Phys. Bl. **43**, 7 (1987)
405. S.A. Abmad et al. (eds.), *Atomic, Molecular and Cluster Physics* (Narosa Publ. House, New Delhi, 1997)
406. R. Campargue (ed.), *Atomic and Molecular Beams—The State of the Art 2000* (Springer, Berlin, 2001)
407. P.W. Wegner (ed.), *Molecular Beams and Low Density Gas Dynamics* (Dekker, New York, 1974)
408. K. Bergmann, W. Demtröder, P. Hering, Laser diagnostics in molecular beams. Appl. Phys. **8**, 65 (1975)
409. H.-J. Foth, Hochauflösende Methoden der Laserspektroskopie zur Interpretation des NO_2-Moleküls. Dissertation, F.B. Physik, Universität Kaiserslautern, Germany, 1981
410. K. Bergmann, U. Hefter, P. Hering, Molecular beam diagnostics with internal state selection. Chem. Phys. **32**, 329 (1978);
K. Bergmann, U. Hefter, P. Hering, Molecular beam diagnostics with internal state selection. J. Chem. Phys. **65**, 488 (1976)
411. G. Herzberg, *Molecular Spectra and Molecular Structure* (van Nostrand, New York, 1950)
412. N. Ochi, H. Watanabe, S. Tsuchiya, Rotationally resolved laser-induced fluorescence and Zeeman quantum beat spectroscopy of the V^1B state of jet-cooled CS_2. Chem. Phys. **113**, 271 (1987)
413. D.H. Levy, L. Wharton, R.E. Smalley, Laser spectroscopy in supersonic jets, in *Chemical and Biochemical Applications of Lasers*, vol. II, ed. by C.B. Moore (Academic Press, New York, 1977)
414. H.J. Foth, H.J. Vedder, W. Demtröder, Sub-Doppler laser spectroscopy of NO_2 in the $\lambda =$ 592–595 nm region. J. Mol. Spectrosc. **88**, 109 (1981)
415. D.H. Levy, The spectroscopy of supercooled gases. Sci. Am. **251**, 86 (1984)
416. E. Pebay-Peyroula, R. Jost, S_1–S_0 laser excitation spectra of glyoxal in a supersonic jet. J. Mol. Spectrosc. **121**, 167 (1987);
B. Soep, R. Campargue, Laser spectroscopy of biacetyl in a supersonic jet and beam, in *Rarefied Gas Dynamics*, vol. II, ed. by R. Campargue (Commissariat A L'Energie Atomique, Paris, 1979)
417. M. Ito, Electronic spectra in a supersonic jet, in *Vibrational Spectra and Structure*, vol. 15, ed. by J.R. Durig (Elsevier, Amsterdam, 1986);
M. Ito, T. Ebata, N. Mikami, Laser spectroscopy of large polyatomic molecules in supersonic jets. Ann. Rev. Phys. Chem. **39**, 123 (1988)
418. W.R. Gentry, Low-energy pulsed beam sources, in *Atomic and Molecular Beam Methods I*, ed. by G. Scoles (Oxford Univ. Press, New York, 1988), p. 54
419. S.B. Ryali, J.B. Fenn, Clustering in free jets. Ber. Bunsenges. Phys. Chem. **88**, 245 (1984)
420. P. Jena, B.K. Rao, S.N. Khanna (eds.), *Physics and Chemistry of Small Clusters* (Plenum, New York, 1987)
421. P.J. Sarre, Large gas phase clusters. Faraday Trans. **13**, 2343 (1990)
422. G. Benedek, T.P. Martin, G. Paccioni (eds.), *Elemental and Molecular Clusters*. Springer Ser. Mater. Sci., vol. 6 (Springer, Berlin, 1987);
U. Kreibig, M. Vollmer, *Optical Properties of Metal Clusters*. Springer Ser. Mater. Sci., vol. 25 (Springer, Berlin, 1995)

423. M. Kappes, S. Leutwyler, Molecular beams of clusters, in *Atomic and Molecular Beam Methods*, vol. 1, ed. by G. Scoles (Oxford Univ. Press, New York, 1988), p. 380
424. H.J. Foth, J.M. Greß, C. Hertzler, W. Demtröder, Sub-Doppler spectroscopy of Na_3. Z. Phys. D **18**, 257 (1991)
425. M.M. Kappes, M. Schär, U. Röthlisberger, C. Yeretzian, E. Schumacher, Sodium cluster ionization potentials revisited. Chem. Phys. Lett. **143**, 251 (1988)
426. C. Brechnignac, P. Cahuzac, J.P. Roux, D. Davolini, F. Spiegelmann, Adiabatic decomposition of mass-selected alkali clusters. J. Chem. Phys. **87**, 3694 (1987)
427. J.M. Gomes Llorente, H.S. Tylor, Spectra in the chaotic region: a classical analysis for the sodium trimer. J. Chem. Phys. **91**, 953 (1989)
428. M.M. Kappes, Experimental studies of gas-phase main-group metal clusters. Chem. Rev. **88**, 369 (1988)
429. M. Broyer, G. Delecretaz, P. Labastie, R.L. Whetten, J.P. Wolf, L. Wöste, Spectroscopy of Na_3. Z. Phys. D **3**, 131 (1986)
430. C. Brechignac, P. Cahuzac, F. Carlier, M. de Frutos, J. Leygnier, Alkali-metal clusters as prototype of metal clusters. J. Chem. Soc. Faraday Trans. **86**, 2525 (1990)
431. J. Blanc, V. Boncic-Koutecky, M. Broyer, J. Chevaleyre, P. Dugourd, J. Koutecki, C. Scheuch, J.P. Wolf, L. Wöste, Evolution of the electronic structure of lithium clusters between four and eight atoms. J. Chem. Phys. **96**, 1793 (1992)
432. W.D. Knight, W.A. deHeer, W.A. Saunders, K. Clemenger, M.Y. Chou, M.L. Cohen, Alkali metal clusters and the jellium model. Chem. Phys. Lett. **134**, 1 (1987)
433. V. Bonacic-Koutecky, P. Fantucci, J. Koutecky, Systemic ab-initio configuration–interaction studies of alkali-metal clusters. Phys. Rev. B **37**, 4369 (1988)
434. A. Kiermeier, B. Ernstberger, H.J. Neusser, E.W. Schlag, Benzene clusters in a supersonic beam. Z. Phys. D **10**, 311 (1988)
435. K.H. Fung, W.E. Henke, T.R. Hays, H.L. Selzle, E.W. Schlag, Ionization potential of the benzene–argon complex in a jet. J. Phys. Chem. **85**, 3560 (1981)
436. C.M. Lovejoy, M.D. Schuder, D.J. Nesbitt, Direct IR laser absorption spectroscopy of jet-cooled CO_2HF complexes. J. Chem. Phys. **86**, 5337 (1987)
437. L. Zhu, P. Johnson, Mass analyzed threshold ionization spectroscopy. J. Chem. Phys. **94**, 5769 (1991)
438. E.L. Knuth, Dimer-formation rate coefficients from measurements of terminal dimer concentrations in free-jet expansions. J. Chem. Phys. **66**, 3515 (1977)
439. J.M. Philippos, J.M. Hellweger, H. van den Bergh, Infrared vibrational predissociation of van der Waals clusters. J. Phys. Chem. **88**, 3936 (1984)
440. J.B. Hopkins, P.R. Langridge-Smith, M.D. Morse, R.E. Smalley, Supersonic metal cluster beams of refractory metals: spectral investigations of ultracold Mo_2. J. Chem. Phys. **78**, 1627 (1983);
J.M. Hutson, Intermolecular forces and the spectroscopy of van der Waals molecules. Ann. Rev. Phys. Chem. **41**, 123 (1990)
441. H.W. Kroto, J.R. Heath, S.C. O'Brian, R.F. Curl, R.E. Smalley, C_{60}: Buckminsterfullerene. Nature **318**, 162 (1985)
442. S. Grebenev, M. Hartmann, M. Havenith, B. Sartakov, J.P. Toennies, A.F. Vilesov, The rotational spectrum of single OCS molecules in liquid ^{4}He droplets. J. Chem. Phys. **112**, 4485 (2000);
J.P. Toennies et al., Superfluid helium droplets. Phys. Today **54**(2), 31 (2001); and Annu. Rev. Phys. Chem. **49**, 1 (1998)
443. S. Grebenev et al., Spectroscopy of molecules in helium droplets. Physica B **280**, 65 (2000)
444. S. Grebenev et al., *Spectroscopy of OCS-Hydrogen Clusters in* He-*Droplets*. Proc. Nobel Symposium, vol. 117 (World Scientific, Singapore, 2001), pp. 123 ff.
445. S. Grebenev et al., The structure of OCS-H_2 van der Waals complexes embedded in ^{4}He/^{3}He-droplets. J. Chem. Phys. **114**, 617 (2001)
446. F. Madeja, M. Havenith et al., Polar isomer of formic acid dimer formed in helium droplets. J. Chem. Phys. **120**, 10554 (2004)

447. K. von Haeften, A. Metzelthin, S. Rudolph, V. Staemmler, M. Havenith, High resolution spectroscopy of NO in helium droplets. Phys. Rev. Lett. **95**, 215301 (2005)
448. C.P. Schulz, P. Claus, D. Schumacher, F. Stienkemeier, Formation and Stability of High Spin Alkali Clusters. Phys. Rev. Lett. **92**, 013401 (2004)
449. F. Stienkemeyer, K. Lehmann, Spectroscopy and Dynamics in He-nano-droplets. J. Phys. B **39**, R127 (2006)
450. F. Stienkemeyer, W.E. Ernst, J. Higgins, G. Scoles, On the use of liquid He-cluster beams for the preparation and spectroscopy of alkali dimers and often weakly bound complexes. J. Chem. Phys. **102**, 615 (1995)
451. J. Higgins et al., Photo-induced chemical dynamics of high spin alkali trimers. Science **273**, 629 (1996)
452. R. Michalak, D. Zimmermann, Laser-spectroscopic investigation of higher excited electronic states of the KAr molecules. J. Mol. Spectrosc. **193**, 260 (1999)
453. F. Bylicki, G. Persch, E. Mehdizadeh, W. Demtröder, Saturation spectroscopy and OODR of NO_2 in a collimated molecular beam. Chem. Phys. **135**, 255 (1989)
454. T. Kröckertskothen, H. Knöckel, E. Tiemann, Molecular beam spectroscopy on FeO. Chem. Phys. **103**, 335 (1986)
455. G. Meijer, B. Janswen, J.J. ter Meulen, A. Dynamus, High resolution Lamb-dip spectroscopy on OD and SiCl in a molecular beam. Chem. Phys. Lett. **136**, 519 (1987)
456. G. Meijer, Structure and dynamics of small molecules studied by UV laser spectroscopy. Dissertation, Katholicke Universiteit te Nijmegen, Holland, 1988
457. H.D. Barth, C. Jackschatz, T. Pertsch, F. Huisken, CARS spectroscopy of molecules and clusters in supersonic jets. Appl. Phys. B **45**, 205 (1988)
458. E.K. Gustavson, R.L. Byer, High resolution CW CARS spectroscopy in a supersonic expansion, in *Laser Spectroscopy VI*, ed. by H.P. Weber, W. Lüthy. Springer Ser. Opt. Sci., vol. 40 (Springer, Berlin, 1983), p. 326
459. J.W. Nibler, J. Yang, Nonlinear Raman spectroscopy of gases. Annu. Rev. Phys. Chem. **38**, 349 (1987)
460. J.W. Nibler, Coherent Raman spectroscopy: techniques and recent applications, in *Applied Laser Spectroscopy*, ed. by W. Demtröder, M. Inguscio. NATO ASI Series B, vol. 241 (Plenum, New York, 1991), p. 313
461. J.W. Nibler, G.A. Puhanz, *Adv. Nonlinear Spectroscopy*, vol. 15 (Wiley, New York, 1988), p. 1
462. S.L. Kaufman, High resolution laser spectroscopy in fast beams. Opt. Commun. **17**, 309 (1976)
463. W.H. Wing, G.A. Ruff, W.E. Lamb, J.J. Spezeski, Observation of the infrared spectrum of the hydrogen molecular ion HD^+. Phys. Rev. Lett. **36**, 1488 (1976)
464. M. Kristensen, N. Bjerre, Fine structure of the lowest triplet states in He_2. J. Chem. Phys. **93**, 983 (1990)
465. D.C. Lorents, S. Keiding, N. Bjerre, Barrier tunneling in the He_2 $c^3\Sigma_g^+$ state. J. Chem. Phys. **90**, 3096 (1989)
466. H.J. Kluge, Nuclear ground state properties from laser and mass spectroscopy, in *Applied Laser Spectroscopy*, ed. by W. Demtröder, M. Inguscio. NATO ASI Series B, vol. 241 (Plenum, New York, 1991)
467. E.W. Otten, Nuclei far from stability, in *Treatise on Heavy Ion Science*, vol. 8 (Plenum, New York, 1989), p. 515
468. R. Jacquinot, R. Klapisch, Hyperfine spectroscopy of radioactive atoms. Rep. Prog. Phys. **42**, 773 (1979)
469. J. Eberz et al., Collinear laser spectroscopy of ^{108g108m}In using an ion source with bunched beam release. Z. Phys. A **328**, 119 (1986)
470. B.A. Huber, T.M. Miller, P.C. Cosby, H.D. Zeman, R.L. Leon, J.T. Moseley, J.R. Peterson, Laser-ion coaxial beam spectroscopy. Rev. Sci. Instrum. **48**, 1306 (1977)

471. M. Dufay, M.L. Gaillard, High-resolution studies in fast ion beams, in *Laser Spectroscopy III*, ed. by J.L. Hall, J.L. Carlsten. Springer Ser. Opt. Sci., vol. 7 (Springer, Berlin, 1977), p. 231
472. S. Abed, M. Broyer, M. Carré, M.L. Gaillard, M. Larzilliere, High resolution spectroscopy of N_2O^+ in the near ultraviolet, using FIBLAS (Fast-Ion-Beam Laser Spectroscopy). Chem. Phys. **74**, 97 (1983)
473. D. Zajfman, Z. Vager, R. Naaman et al., The structure of C_2H^+ and $C_2H_2^+$ as measured by Coulomb explosion. J. Chem. Phys. **94**, 6379 (1991)
474. L. Andric, H. Bissantz, E. Solarte, F. Linder, Photofragment spectroscopy of molecular ions: design and performance of a new apparatus using coaxial beams. Z. Phys. D **8**, 371 (1988)
475. J. Lermé, S. Abed, R.A. Hold, M. Larzilliere, M. Carré, Measurement of the fragment kinetic energy distribution in laser photopredissociation of N_2O^+. Chem. Phys. Lett. **96**, 403 (1983)
476. H. Stein, M. Erben, K.L. Kompa, Infrared photodissociation of sulfur dioxide ions in a fast ion beam. J. Chem. Phys. **78**, 3774 (1983)
477. N.J. Bjerre, S.R. Keiding, Long-range ion-atom interactions studied by field dissociation spectroscopy of molecular ions. Phys. Rev. Lett. **56**, 1458 (1986)
478. D. Neumark, High resolution photodetachment studies of molecular negative ions, in *Ion and Cluster Ion Spectroscopy and Structure*, ed. by J.P. Maier (Elsevier, Amsterdam, 1989), pp. 155 ff.
479. R.D. Mead, V. Hefter, P.A. Schulz, W.C. Lineberger, Ultrahigh resolution spectroscopy of C_2^-. J. Chem. Phys. **82**, 1723 (1985)
480. O. Poulsen, Resonant fast-beam interactions: saturated absorption and two-photon absorption, in *Atomic Physics 8*, ed. by I. Lindgren, S. Svanberg, A. Rosén (Plenum, New York, 1983), p. 485
481. D. Klapstein, S. Leutwyler, J.P. Maier, C. Cossart-Magos, D. Cossart, S. Leach, The $B\,^2A_2'' \to \tilde{X}\,^2F''$ transition of 1,3,5-$C_6F_3H_3^+$ and 1,3,5-$C_6F_3D_3^+$ in discharge and supersonic free jet emission sources. Mol. Phys. **51**, 413 (1984)
482. S.C. Foster, R.A. Kennedy, T.A. Miller, Laser spectroscopy of chemical intermediates in supersonic free jet expansions, in *Frontiers of Laser Spectroscopy*, ed. by A.C.P. Alves, J.M. Brown, J.M. Hollas. NATO ASI Series C, vol. 234 (Kluwer, Dordrecht, 1988)
483. P. Erman, O. Gustafssosn, P. Lindblom, A simple supersonic jet discharge source for sub-Doppler spectroscopy. Phys. Scr. **38**, 789 (1988)
484. D. Pflüger, W.E. Sinclair, A. Linnartz, J.P. Maier, Rotationally resolved electronic absorption spectra of triacethylen cation in a supersonic jet. Chem. Phys. Lett. **313**, 171–178 (1999)
485. M.A. Johnson, R.N. Zare, J. Rostas, L. Leach, Resolution of the $\tilde{A}$ photoionization branching ratio paradox for the $^{12}CO_2$ state. J. Chem. Phys. **80**, 2407 (1984)
486. A. Kiermeyer, H. Kühlewind, H.J. Neusser, E.W. Schlag, Production and unimolecular decay of rotationally selected polyatomic molecular ions. J. Chem. Phys. **88**, 6182 (1988)
487. U. Boesl, Multiphoton excitation and mass-selective ion detection for neutral and ion spectroscopy. J. Phys. Chem. **95**, 2949 (1991)
488. K. Walter, R. Weinkauf, U. Boesl, E.W. Schlag, Molecular ion spectroscopy: mass-selected resonant two-photon dissociation spectra of CH^3I^+ and CD_3I^+. J. Chem. Phys. **89**, 1914 (1988)
489. J.P. Maier, Mass spectrometry and spectroscopy of ions and radicals, in *Encyclopedia of Spectroscopy and Spectrometry*, ed. by J.C. Lindon, G.E. Trauter, J.L. Holm (Academic Press, New York, 1999), p. 2181
490. C.-Y. Ng, *Photoionization and Photodetachment*. Advances in Physical Chemistry, vol. 10 (2000);
C.Y. Ng, Spectroscopy and dynamics of neutrals and ions by high-resolution infrared-vacuum ultraviolet photoionization and photoelectron methods, in *Frontiers of Molecu-*

lar Spectroscopy, ed. by J. Laane (Elsevier Science and Technology, Amsterdam, 2009), Chap. 19, pp. 659–691
491. C.W. Walter, N.D. Gibson, P. Andersson, C.M. Janczak, K.A. Starr, A.P. Snedden, R.L. Field III, Infrared photodetachment of Ce: threshold spectroscopy and resonance structure. Phys. Rev. A **76**, 052702 (2007)
492. P. Anderson, PhD thesis, Gothenburg University, 2009. https://gupea.ub.gu.se/bitstream/2077/20071/1/gupea_2077_20071_1.pdf
493. R. Otto, A.W. Ray, J.S. Daluz, R.E. Continetti, Direct IR excitation in a fast ion beam: application to NO^- photodetachment cross sections. EPJ Techniques Instrum. **1**, 3 (2014)
494. V. Esanlov (ed.), *Negative Ions* (Cambridge University Press, Cambridge, 1996)
495. E. Schlag (ed.), *Time of Flight Mass Spectrometry and Its Applications* (Elsevier, New York, 1994), 420 pp.
496. V. Beutel, H.G. Krämer, G.L. Bahle, M. Kuhn, K. Weyers, W. Demtröder, High resolution isotope selective laser spectroscopy of Ag_2 molecules. J. Chem. Phys. **98**, 2699 (1997)
497. H.G. Krämer, Laser Spektroskopie an kleinen Alkali Clustern. PhD thesis, TU, Kaiserslautern, 1998
498. D. Bing, Untersuchungen zur Mehrstufen-Laser Spektroskopie an gespeicherten H_3^+-Ionen. PhD thesis, Heidelberg, 2010
499. D.M. Lubman (ed.), *Lasers and Mass Spectrometry* (Oxford University Press, Oxford, 1990)
500. E.J. Bieske, M.W. Rainbird, A.E.W. Knight, Suppression of fragment contribution to mass-selected resonance enhanced multiphoton ionization spectra of van der Waals clusters. J. Chem. Phys. **90**, 2086 (1989); and **94**, 7019 (1991);
U. Boesl, Laser mass spectrometry for environmental and industrial chemical trace analysis. J. Mass Spectrom. **35**, 289 (2000)
501. U. Boesl, J. Grotemeyer, K. Walter, E.W. Schlag, Resonance ionization and time-of-flight mass spectroscopy. Anal. Instrum. **16**, 151 (1987)
502. C.W.S. Conover, Y.J. Twu, Y.A. Yang, L.A. Blomfield, A time-of-flight mass spectrometer for large molecular clusters produced in supersonic expansions. Rev. Sci. Instrum. **60**, 1065 (1989)
503. C. Brechignac, P. Cahuzac, R. Pflaum, J.P. Roux, Photodissociation of size-selected K_n^* clusters. J. Chem. Phys. **88**, 3022 (1988);
C. Brechignac, P. Cahuzac, R. Pflaum, J.P. Roux, Adiabatic decomposition of mass-selected alkali clusters. J. Chem. Phys. **87**, 5694 (1987)
504. J.A. Syage, J.E. Wessel, Molecular multiphoton ionization and ion fragmentation spectroscopy, in *Appl. Spectrosc. Rev.*, vol. 24 (Dekker, New York, 1988), p. 1

Chapter 5

505. R.A. Bernheim, *Optical Pumping, an Introduction* (Benjamin, New York, 1965)
506. B. Budick, Optical pumping methods in atomic spectroscopy, in *Adv. At. Mol. Phys.*, vol. 3 (Academic Press, New York, 1967), p. 73
507. R.N. Zare, Optical pumping of molecules, in *Int'l Colloquium on Doppler-Free Spectroscopic Methods for Simple Molecular Systems* (CNRS, Paris, 1974), p. 29
508. M. Broyer, G. Gouedard, J.C. Lehmann, J. Vigue, Optical pumping of molecules, in *Adv. At. Mol. Phys.*, vol. 12 (Academic Press, New York, 1976), p. 164
509. G. zu Putlitz, Determination of nuclear moments with optical double resonance, in *Springer Tracts Mod. Phys.*, vol. 37 (Springer, Berlin, 1965), p. 105
510. C. Cohen-Tannoudji, Optical pumping with lasers, in *Atomic Physics IV*, ed. by G. zu Putlitz, E.W. Weber, A. Winnacker (Plenum, New York, 1975), p. 589
511. R.N. Zare, *Angular Momentum* (Wiley, New York, 1988)

512. R.E. Drullinger, R.N. Zare, Optical pumping of molecules. J. Chem. Phys. **51**, 5532 (1969)
513. K. Bergmann, State selection via optical methods, in *Atomic and Molecular Beam Methods*, ed. by G. Scoles (Oxford Univ. Press, Oxford, 1988), p. 293
514. H.G. Weber, P. Brucat, W. Demtröder, R.N. Zare, Measurement of $NO_2{}^2B_2$ state g-values by optical radio frequency double-resonance. J. Mol. Spectrosc. **75**, 58 (1979)
515. W. Happer, Optical pumping. Rev. Mod. Phys. **44**, 168 (1972);
R.J. Knize, Z. Wu, W. Happer, Optical pumping and spin exchange in gas cells. Adv. At. Mol. Phys. **24**, 223 (1987)
516. B. Budick, Optical pumping methods in atomic spectroscopy. Adv. At. Mol. Phys. **3**, 73 (1967);
M. Broyer et al., Optical pumping of molecules. Adv. At. Mol. Phys. **12**, 165 (1976)
517. B. Decomps, M. Dumont, M. Ducloy, Linear and nonlinear phenomena in laser optical pumping, in *Laser Spectroscopy of Atoms and Molecules*, ed. by H. Walther. Topics Appl. Phys., vol. 2 (Springer, Berlin, 1976), p. 284
518. G.W. Series, Thirty years of optical pumping. Contemp. Phys. **22**, 487 (1981)
519. P.R. Hemmer, M.K. Kim, M.S. Shahriar, Observation of sub-kilohertz resonances in RF-optical double resonance experiment in rare earth ions in solids. J. Mod. Opt. **47**, 1713 (2000)
520. I.I. Rabi, Zur Methode der Ablenkung von Molekularstrahlen. Z. Phys. **54**, 190 (1929)
521. H. Kopfermann, *Kernmomente* (Akad. Verlagsanstalt, Frankfurt, 1956)
522. N.F. Ramsay, in *Molecular Beams*, 2nd edn. (Clarendon, Oxford, 1989)
523. J.C. Zorn, T.C. English, Molecular beam electric resonance spectroscopy, in *Adv. At. Mol. Phys.*, vol. 9 (Academic, New York, 1973), p. 243
524. D.D. Nelson, G.T. Fraser, K.I. Peterson, K. Zhao, W. Klemperer, The microwave spectrum of K = O states of $Ar–NH_3$. J. Chem. Phys. **85**, 5512 (1986)
525. A.E. DeMarchi (ed.), *Frequency Standards and Metrology* (Springer, Berlin, 1989), pp. 46 ff.
526. W.J. Childs, Use of atomic beam laser RF double resonance for interpretation of complex spectra. J. Opt. Soc. Am. B **9**, 191 (1992)
527. S.D. Rosner, R.A. Holt, T.D. Gaily, Measurement of the zero-field hyperfine structure of a single vibration-rotation level of Na_2 by a laser-fluorescence molecular-beam resonance. Phys. Rev. Lett. **35**, 785 (1975)
528. A.G. Adam, Laser-fluorescence molecular-beam-resonance studies of Na_2 lineshape due to HFS. PhD thesis, Univ. of Western Ontario, London, Ontario, 1981;
A.G. Adam, S.D. Rosner, T.D. Gaily, R.A. Holt, Coherence effects in laser-fluorescence molecular beam magnetic resonance. Phys. Rev. A **26**, 315 (1982)
529. W. Ertmer, B. Hofer, Zerofield hyperfine structure measurements of the metastable states $3d^2 4s^4 F_{3/2} 9/2$ of *SC using laser-fluorescence-atomic beam magnetic resonance technique. Z. Phys. A **276**, 9 (1976)
530. G.D. Domenico et al., Sensitivity of double resonance alignment magnetometers. arXiv: 0706.0104v1 [physics.atom-phy], 1 Juni 2007
531. J. Pembczynski, W. Ertmer, V. Johann, S. Penselin, P. Stinner, Measurement of the hyperfine structure of metastable atomic states of ^{55}Mm, using the ABMR-LIRF-method. Z. Phys. A **291**, 207 (1979); and **294**, 313 (1980)
532. N. Dimarca, V. Giordano, G. Theobald, P. Cérez, Comparison of pumping a cesium beam tube with D_1 and D_2 lines. J. Appl. Phys. **69**, 1159 (1991)
533. G.W. Chantry (ed.), *Modern Aspects of Microwave Spectroscopy* (Academic Press, London, 1979)
534. K. Shimoda, Double resonance spectroscopy by means of a laser, in *Laser Spectroscopy of Atoms and Molecules*, ed. by H. Walther. Topics Appl. Phys., vol. 2 (Springer, Berlin, 1976), p. 197
535. K. Shimoda, Infrared-microwave double resonance, in *Laser Spectroscopy III*, ed. by J.L. Hall, H.L. Carlsten. Springer Ser. Opt. Sci., vol. 7 (Springer, Berlin, 1975), p. 279

536. H. Jones, Laser microwave-double-resonance and two-photon spectroscopy. Comments At. Mol. Phys. **8**, 51 (1978)
537. F. Tang, A. Olafson, J.O. Henningsen, A study of the methanol laser with a 500 MHz tunable CO_2 laser. Appl. Phys. B **47**, 47 (1988)
538. R. Neumann, F. Träger, G. zu Putlitz, Laser microwave spectroscopy, in *Progress in Atomic Spectroscopy*, ed. by H.J. Byer, H. Kleinpoppen (Plenum, New York, 1987)
539. J.C. Petersen, T. Amano, D.A. Ramsay, Microwave-optical double resonance of DND in the $A\,^1A''(000)$ state. J. Chem. Phys. **81**, 5449 (1984)
540. R.W. Field, A.D. English, T. Tanaka, D.O. Harris, P.A. Jennings, Microwave-optical double resonance with a CW dye laser, BaO $X\,^1\Sigma$ and $A\,^1\Sigma$. J. Chem. Phys. **59**, 2191 (1973)
541. R.A. Gottscho, J. Brooke-Koffend, R.W. Field, J.R. Lombardi, OODR spectroscopy of BaO. J. Chem. Phys. **68**, 4110 (1978); and J. Mol. Spectrosc. **82**, 283 (1980)
542. J.M. Cook, G.W. Hills, R.F. Curl, Microwave-optical double resonance spectrum of NH_2. J. Chem. Phys. **67**, 1450 (1977)
543. W.E. Ernst, S. Kindt, A molecular beam laser-microwave double resonance spectrometer for precise measurements of high temperature molecules. Appl. Phys. B **31**, 79 (1983)
544. W.J. Childs, The hyperfine structure of alkaline-earth monohalide radicals: new methods and new results 1980–1982. Comments At. Mol. Phys. **13**, 37 (1983)
545. W.E. Ernst, S. Kindt, T. Törring, Precise Stark-effect measurements in the $^2\sigma$-ground state of CaCl. Phys. Rev. Phys. Lett. **51**, 979 (1983); and Phys. Rev. A **29**, 1158 (1984)
546. M. Schäfer, M. Andrist, H. Schmutz, F. Lewen, G. Winnewisser, F. Merkt, A 240–380 GHz millimeter wave source for very high resolution spectroscopy of high Rydberg states. J. Phys. B, At. Mol. Opt. Phys. **39**, 831 (2006);
A. Osterwalder, A. Wuest, F. Merkt, C. Jungen, High resolution millimeter wave spectroscopy and MQDT of the hyperfine structure in high Rydberg states of molecular hydrogen H_2. J. Chem. Phys. **121**, 11810 (2004)
547. W. Demtröder, D. Eisel, H.J. Foth, G. Höning, M. Raab, H.J. Vedder, D. Zevgolis, Sub-Doppler laser spectroscopy of small molecules. J. Mol. Struct. **59**, 291 (1980)
548. F. Bylicki, G. Persch, E. Mehdizadeh, W. Demtröder, Saturation spectroscopy and OODR of NO_2 in a collimated molecular beam. Chem. Phys. **135**, 255 (1989)
549. M.A. Johnson, C.R. Webster, R.N. Zare, Rotational analysis of congested spectra: application of population labelling to the BaI C–X system. J. Chem. Phys. **75**, 5575 (1981)
550. M.A. Kaminsky, R.T. Hawkins, F.V. Kowalski, A.L. Schawlow, Identification of absorption lines by modulated lower-level population: spectrum of Na_2. Phys. Rev. Lett. **36**, 671 (1976)
551. A.L. Schawlow, Simplifying spectra by laser labelling. Phys. Scr. **25**, 333 (1982)
552. D.P. O'Brien, S. Swain, Theory of bandwidth induced asymmetry in optical double resonances. J. Phys. B **16**, 2499 (1983)
553. S.A. Edelstein, T.F. Gallagher, Rydberg atoms, in *Adv. At. Mol. Phys.*, vol. 14 (Academic Press, New York, 1978), p. 365
554. I.I. Sobelman, *Atomic Spectra and Radiative Transitions*, 2nd edn. Springer Ser. Atoms and Plasmas, vol. 12 (Springer, Berlin, 1992)
555. R.F. Stebbings, F.B. Dunnings (eds.), *Rydberg States of Atoms and Molecules* (Cambridge Univ. Press, Cambridge, 1983)
556. T. Gallagher, *Rydberg Atoms* (Cambridge Univ. Press, Cambridge, 1994)
557. H. Figger, Experimente an Rydberg-Atomen und Molekülen. Phys. Unserer Zeit **15**, 2 (1984)
558. J.A.C. Gallas, H. Walther, E. Werner, Simple formula for the ionization rate of Rydberg states in static electric fields. Phys. Rev. Lett. **49**, 867 (1982)
559. C.E. Theodosiou, Lifetimes of alkali-metal-atom Rydberg states. Phys. Rev. A **30**, 2881 (1984)
560. J. Neukammer, H. Rinneberg, K. Vietzke, A. König, H. Hyronymus, M. Kohl, H.J. Grabka, Spectroscopy of Rydberg atoms at $n = 500$. Phys. Rev. Lett. **59**, 2847 (1987)

561. K.H. Weber, K. Niemax, Impact broadening of very high Rb Rydberg levels by Xe. Z. Phys. A **312**, 339 (1983)
562. K. Heber, P.J. West, E. Matthias, Pressure shift and broadening of SnI Rydberg states in noble gases. Phys. Rev. A **37**, 1438 (1988)
563. R. Beigang, W. Makat, A. Timmermann, P.J. West, Hyperfine-induced n-mixing in high Rydberg states of ^{87}Sr. Phys. Rev. Lett. **51**, 771 (1983)
564. T.F. Gallagher, W.E. Cooke, Interaction of blackbody radiation with atoms. Phys. Rev. Lett. **42**, 835 (1979)
565. L. Holberg, J.L. Hall, Measurements of the shift of Rydberg energy levels induced by blackbody radiation. Phys. Rev. Lett. **53**, 230 (1984)
566. H. Figger, G. Leuchs, R. Strauchinger, H. Walther, A photon detector for submillimeter wavelengths using Rydberg atoms. Opt. Commun. **33**, 37 (1980)
567. D. Wintgen, H. Friedrich, Classical and quantum mechanical transition between regularity and irregularity. Phys. Rev. A **35**, 1464 (1987)
568. G. Raithel, M. Fauth, H. Walther, Quasi-Landau resonances in the spectra of rubidium Rydberg atoms in crossed electric and magnetic fields. Phys. Rev. A **44**, 1898 (1991)
569. G. Wunner, Gibt es Chaos in der Quantenmechanik? Phys. Bl. **45**, 139 (1989); M. Gutzwiller, *Chaos in Classical and Quantum Mechanics* (Springer, Berlin, 1990)
570. A. Holle, J. Main, G. Wiebusch, H. Rottke, K.H. Welge, Laser spectroscopy of the diamagnetic hydrogen atom in the chaotic region, in *Atomic Spectra and Collisions in External Fields*, ed. by K.T. Taylor, M.H. Nayfeh, C.W. Clark (Plenum, New York, 1988)
571. P. Meystre, M. Sargent III, *Elements of Quantum Optics*, 2nd edn. (Springer, Berlin, 1991)
572. H. Held, J. Schlichter, H. Walther, Quantum chaos in Rydberg atoms. Lect. Notes Phys. **503**, 1 (1998)
573. A. Holle, G. Wiebusch, J. Main, K.H. Welge, G. Zeller, G. Wunner, T. Ertl, H. Ruder, Hydrogenic Rydberg atoms in strong magnetic fields. Z. Phys. D **5**, 271 (1987)
574. H. Rottke, K.H. Welge, Photoionization of the hydrogen atom near the ionization limit in strong electric field. Phys. Rev. A **33**, 301 (1986)
575. R. Seiler, T. Paul, M. Andrist, F. Merkt, Generation of programmable near Fourier-limited pulses of narrow band laser radiation from the near infrared to the vacuum ultraviolet. Rev. Sci. Instrum. **76**, 103103 (2005)
576. C. Fahre, S. Haroche, Spectroscopy of one- and two-electron Rydberg atoms, in *Rydberg States of Atoms and Molecules*, ed. by R.F. Stebbings, F.B. Dunnings (Cambridge Univ. Press, Cambridge, 1983)
577. J. Boulmer, P. Camus, P. Pillet, Autoionizing double Rydberg states in barium, in *Electronic and Atomic Collisions*, ed. by H.B. Gilbody, W.R. Newell, F.H. Read A.C. Smith (Elsevier, Amsterdam, 1988)
578. J. Boulmer, P. Camus, P. Pillet, Double Rydberg spectroscopy of the barium atom. J. Opt. Soc. Am. B **4**, 805 (1987)
579. I.C. Percival, Planetary atoms. Proc. R. Soc. Lond. A **353**, 289 (1977)
580. R.S. Freund, High Rydberg molecules, in *Rydberg States of Atoms and Molecules*, ed. by R.F. Stebbing, F.B. Dunning (Cambridge Univ. Press, Cambridge, 1983); G. Herzberg, Rydberg molecules. Annu. Rev. Phys. Chem. **38**, 27 (1987)
581. R.A. Bernheim, L.P. Gold, T. Tipton, Rydberg states of $^7Li_2$ by pulsed optical–optical double resonance spectroscopy. J. Chem. Phys. **78**, 3635 (1983); D. Eisel, W. Demtröder, W. Müller, P. Botschwina, Autoionization spectra of Li_2 and the $X\,^2\Sigma_g^+$ ground state of Li_2^+. Chem. Phys. **80**, 329 (1983)
582. M. Schwarz, R. Duchowicz, W. Demtröder, C. Jungen, Autoionizing Rydberg states of Li_2: analysis of electronic-rotational interactions. J. Chem. Phys. **89**, 5460 (1988)
583. C.H. Greene, C. Jungen, Molecular applications of quantum defect theory, in *Adv. At. Mol. Phys*, vol. 21 (Academic Press, New York, 1985), p. 51
584. F. Merkt, Molecules in high Rydberg states. Annu. Rev. Phys. Chem. **48**, 675 (1997)
585. A. Osterwalder, F. Merkt, High resolution spectroscopy of high Rydberg states. Chimica **54**, 89 (2000)

586. S. Fredin, D. Gauyacq, M. Horani, C. Jungen, G. Lefevre, F. Masnou-Seeuws, *S* and *d* Rydberg series of NO probed by double resonance multiphoton ionization. Mol. Phys. **60**, 825 (1987);
R. Zhao, I.M. Konen, R.N. Zare, Optical–optical double resonance photoionization spectroscopy of nf Rydberg states of nitric oxide. J. Chem. Phys. **121**, 9938 (2004)
587. U. Aigner, L.Y. Baranov, H.L. Selzle, E.W. Schlag, Lifetime enhancement of ZEKE-states in molecular clusters and cluster fragmentation. J. Electron Spectrosc. Relat. Phenom. **112**, 175 (2000)
588. M. Sander, L.A. Chewter, K. Müller-Dethlefs, E.W. Schlag, High-resolution zero-kinetic-energy photoelectron spectroscopy of NO. Phys. Rev. A **36**, 4543 (1987)
589. K. Müller-Dethlefs, E.W. Schlag, High-resolution ZEKE photoelectron spectroscopy of molecular systems. Annu. Rev. Phys. Chem. **42**, 109 (1991);
E.R. Grant, M.G. White, ZEKE threshold photoelectron spectroscopy. Nature **354**, 249 (1991);
M.S. Ford, R. Lindner, K. Müller-Dethlefs, Fully rotationally resolved ZEKE photoelectron spectroscopy of C_6H_6 and C_6D_6. Mol. Phys. **101**, 705 (2003)
590. C.E.H. Descent, K. Müller-Dethlefs, Hydrogen-bonding and van der Waals Complexes Studies by ZEKE and REMP Spectroscopy. Chem. Rev. **100**, 3999 (2000)
591. R. Signorelli, U. Hollenstein, F. Merkt, PFI–ZEKE photo electron spectroscopy study of the first electronic states of Kr_2^+. J. Chem. Phys. **114**, 9840 (2001);
S. Willitsch, F. Innocenti, J.M. Dyke, F. Merkt, High resolution pulse-field-ionization ZEKE photoelectron spectroscopic study of the two lowest electronic states of the ozone cation O_3^+. J. Chem. Phys. **122**, 024311 (2005)
592. P. Goy, M. Bordas, M. Broyer, P. Labastie, B. Tribellet, Microwave transitions between molecular Rydberg states. Chem. Phys. Lett. **120**, 1 (1985)
593. P. Filipovicz, P. Meystere, G. Rempe, H. Walther, Rydberg atoms, a testing ground for quantum electrodynamics. Opt. Acta **32**, 1105 (1985)
594. C.J. Latimer, Recent experiments involving highly excited atoms. Contemp. Phys. **20**, 631 (1979)
595. J.C. Gallas, G. Leuchs, H. Walther, H. Figger, Rydberg atoms: high resolution spectroscopy, in *Adv. At. Mol. Phys.*, vol. 20 (Academic Press, New York, 1985), p. 414
596. G. Alber, P. Zoller, Laser-induced excitation of electronic Rydberg wave packets. Contemp. Phys. **32**, 185 (1991)
597. K. Harth, M. Raab, H. Hotop, Odd Rydberg spectrum of ^{20}Ne: high resolution laser spectroscopy and MQDT analysis. Z. Phys. D **7**, 219 (1987)
598. V.S. Letokhov, V.P. Chebotayev, *Nonlinear Laser Spectroscopy*. Springer Ser. Opt. Sci., vol. 4 (Springer, Berlin, 1977), Chap. 5
599. T. Hänsch, P. Toschek, Theory of a three-level gas laser amplifier. Z. Phys. **236**, 213 (1970)
600. C. Kitrell, E. Abramson, J.L. Kimsey, S.A. McDonald, D.E. Reisner, R.W. Field, D.H. Katayama, Selective vibrational excitation by stimulated emission pumping. J. Chem. Phys. **75**, 2056 (1981)
601. H.-L. Da (guest ed.), Molecular spectroscopy and dynamics by stimulated-emission pumpings. J. Opt. Soc. Am. B **7**, 1802 (1990)
602. G. Zhong He, A. Kuhn, S. Schiemann, K. Bergmann, Population transfer by stimulated Raman scattering with delayed pulses and by the stimulated-emission pumping method: a comparative study. J. Opt. Soc. Am. B **7**, 1960 (1990)
603. K. Yamanouchi, H. Yamada, S. Tsuciya, Vibrational levels structure of highly excited SO_2 in the electronic ground state as studied by stimulated emission pumping spectroscopy. J. Chem. Phys. **88**, 4664 (1988)
604. U. Brinkmann, Higher sensitivity and extended frequency range via stimulated emission pumping SEP. Lambda Phys. Highlights, 1 (1990)
605. H. Weickenmeier, U. Diemer, M. Wahl, M. Raab, W. Demtröder, W. Müller, Accurate ground state potential of Cs_2 up to the dissociation limit. J. Chem. Phys. **82**, 5354 (1985)

606. H. Weickemeier, U. Diemer, W. Demtröder, M. Broyer, Hyperfine interaction between the singlet and triplet ground states of Cs_2. Chem. Phys. Lett. **124**, 470 (1986)
607. M. Kabir, S. Kasabara, W. Demtröder, A. Doi, H. Kato, Doppler-free laser polarization spectroscopy and optical–optical double resonance polarization spectroscopy of a large molecule: naphthalene. J. Chem. Phys. **119**, 3691 (2003)
608. R. Teets, R. Feinberg, T.W. Hänsch, A.L. Schawlow, Simplification of spectra by polarization labelling. Phys. Rev. Lett. **37**, 683 (1976)
609. N.W. Carlson, A.J. Taylor, K.M. Jones, A.L. Schawlow, Two step polarization-labelling spectroscopy of excited states of Na_2. Phys. Rev. A **24**, 822 (1981)
610. B. Hemmerling, R. Bombach, W. Demtröder, N. Spies, Polarization labelling spectroscopy of molecular Li_2 Rydberg states. Z. Phys. D **5**, 165 (1987)
611. W.E. Ernst, Microwave optical polarization spectroscopy of the $X\,^2S$ state of SrF. Appl. Phys. B **30**, 2378 (1983)
612. W.E. Ernst, T. Törring, Hyperfine Structure in the $X\,^2S$ state of CaCl, measured with microwave optical polarization spectroscopy. Phys. Rev. A **27**, 875 (1983)
613. W.E. Ernst, O. Golonska, Microwave transitions in the Na_3 cluster. Phys. Rev. Lett. (2002, submitted)
614. Th. Weber, E. Riedle, H.J. Neusser, Rotationally resolved fluorescence dip and ion-dip spectra of single rovibronic states of benzene. J. Opt. Soc. Am. B **7**, 1875 (1990)
615. M. Takayanagi, I. Hanazaki, Fluorescence dip and stimulated emission-pumping laser-induced-fluorescence spectra of van der Waals molecules. J. Opt. Soc. Am. B **7**, 1878 (1990)
616. H.S. Schweda, G.K. Chawla, R.W. Field, Highly excited, normally inaccessible vibrational levels by sub-Doppler modulated gain spectroscopy. Opt. Commun. **42**, 165 (1982)
617. M. Elbs, H. Knöckel, T. Laue, C. Samuelis, E. Tiemann, Observation of the last bound levels near the Na_2 ground state asymptote. Phys. Rev. A **59**, 3665 (1999)
618. A. Crubellier, O. Dulieu, F. Masnou-Seeuws, M. Elbs, H. Knöckel, E. Tiemann, Simple determination of Na_2 scattering lengths using observed bound levels of the ground state asymptote. Eur. Phys. J. D **6**, 211 (1999)
619. J. Léonard et al., Giant helium dimers produced by photoassociation of ultracold metastable atoms. Phys. Rev. Lett. **91**, 073203 (2003)
620. M. Semczuk et al., High resolution photo-association spectroscopy of the $^6Li_2\,^3\Sigma_g$ state. Phys. Rev. A **87**, 052505 (2013)
621. E.C. Stanca-Kaposta, J.P. Simon, High-resolution infrared–ultraviolet (IR–UV) Double-resonance spectroscopy of biological molecules, in *Handbook of High Resolution Spectroscopy*, ed. by M. Quack, F. Merkt (Wiley, New York, 2011), p. 1911
622. W.C. Stwalley, H. Wang, Photoassociation of ultracold atoms: a new spectroscopic technique. J. Mol. Spectrosc. **194**, 228 (1999)
623. K.M. Jones, E. Tiesinger, P.D. Lett, P.J. Julienne, Ultracold photoassociation spectroscopy: long range molecules and atomic scattering. Rev. Mod. Phys. **78**, 1041 (2006)
624. N. Vanhaecke et al., Photoassociation spectroscopy of ultracold long-range molecules. C. R. Phys. **5**, 161 (2004)

Chapter 6

625. J. Herrmann, B. Wilhelmi, *Lasers for Ultrashort Light Impulses* (North Holland, Amsterdam, 1987)
626. J.C. Diels, W. Rudolph, *Ultrashort Laser Pulse Phenomena*, 2nd edn. (Academic Press, San Diego, 2006);
C. Rulliere (ed.), *Femtosecond Laser Pulses* (Springer, Berlin, 1998)

627. S.A. Akhmanov, V.A. Vysloukhy, A.S. Chirikin, *Optics of Femtosecond Laser Pulses* (AIP, New York, 1992)
628. V. Brückner, K.H. Felle, V.W. Grummt, *Application of Time-Resolved Optical Spectroscopy* (Elsevier, Amsterdam, 1990)
629. J.G. Fujimoto (ed.), Special issue on ultrafast phenomena. IEEE J. Quantum Electron. **25**, 2415 (1989)
630. G.R. Fleming, Sub-picosecond spectroscopy. Annu. Rev. Phys. Chem. **37**, 81 (1986)
631. W.H. Lowdermilk, Technology of bandwidth-limited ultrashort pulse generation, in *Laser Handbook*, vol. 3, ed. by M.L. Stitch (North Holland, Amsterdam, 1979), pp. 361–420, Chap. B1
632. L.P. Christov, Generation and propagation of ultrashort optical pulses, in *Progress in Optics*, vol. 24 (North Holland, Amsterdam, 1991), p. 201
633. W. Kaiser (ed.), *Ultrashort Laser Pulses*, 2nd edn. Topics Appl. Phys., vol. 60 (Springer, Berlin, 1993);
S.L. Shapiro (ed.), *Ultrashort Light Pulses*. Topics Appl. Phys., vol. 18 (Springer, Berlin, 1977)
634. *Picosecond/Ultrashort Phenomena I–IX*, Proc. Int'l Confs. 1978–1994:
Picosecond Phenomena I, ed. by K.V. Shank, E.P. Ippen, S.L. Shapiro. Springer Ser. Chem. Phys., vol. 4 (Springer, Berlin, 1978);
Picosecond Phenomena II, ed. by R.M. Hochstrasser, W. Kaiser, C.V. Shank. Springer Ser. Chem. Phys., vol. 14 (Springer, Berlin, 1980);
Picosecond Phenomena III, ed. by K.B. Eisenthal, R.M. Hochstrasser, W. Kaiser, A. Laubereau. Springer Ser. Chem. Phys., vol. 38 (Springer, Berlin, 1982);
Ultrashort Phenomena IV, ed. by D.H. Auston, K.B. Eisenthal. Springer Ser. Chem. Phys., vol. 38 (Springer, Berlin, 1984);
Ultrashort Phenomena V, ed. by G.R. Fleming, A.E. Siegman. Springer Ser. Chem. Phys., vol. 46 (Springer, Berlin, 1986);
Ultrashort Phenomena VI, ed. by T. Yajima, K. Yoshihara, C.B. Harris, S. Shionoya. Springer Ser. Chem. Phys., vol. 48 (Springer, Berlin, 1988);
Ultrashort Phenomena VII, ed. by E. Ippen, C.B. Harris, A. Zewail. Springer Ser. Chem. Phys., vol. 53 (Springer, Berlin, 1990);
Ultrafast Phenomena VIII, ed. by J.-L. Martin, A. Migus, G.A. Mourou, A.H. Zewail. Springer Ser. Chem. Phys., vol. 55 (Springer, Berlin, 1993);
Ultrafast Phenomena IX, ed. by P.F. Barbara, W.H. Knox, G.A. Mourou, A.H. Zewail. Springer Ser. Chem. Phys., vol. 60 (Springer, Berlin, 1994);
Ultrafast Phenomena X, ed. by P.F. Barbard, J.G. Fujimoto. Springer Ser. Chem. Phys. (Springer, Berlin, 1996);
Ultrafast Phenomena XI, ed. by T. Elsaesser, J.G. Fujimoto, D.A. Wiersma, W. Zinth. Springer Ser. Chem. Phys. (Springer, Berlin, 1998);
Ultrafast Phenomena XII, ed. by T. Elsaesser, S. Mukamel, M.M. Murnane. Springer Ser. Chem. Phys. (Springer, Berlin, 2000)
Ultrafast Phenomena XIII, ed. by D.R. Miller et al. Springer Ser. Chem. Phys., vol. 71 (Springer, Heidelberg, 2003)
Ultrafast Phenomena XIV, ed. by T. Kobayashi et al. Springer Ser. Chem. Phys., vol. 79 (Springer, Berlin, 2005)
Ultrafast Phenomena XV, ed. by P. Corkum, D. Jonas et al. Springer Ser. Chem. Phys., vol. 88 (Springer, Berlin, 2007)
635. T.R. Gosnel, A.J. Taylor (eds.), *Ultrafast Laser Technology*. SPIE Proc., vol. 44 (1991)
636. E. Niemann, M. Klenert, A fast high-intensity-pulse light source for flash-photolysis. Appl. Opt. **7**, 295 (1968)
637. L.S. Marshak, *Pulsed Light Sources* (Consultants Bureau, New York, 1984)
638. P. Richter, J.D. Kimel, G.C. Moulton, Pulsed nitrogen laser: dynamical UV behaviour. Appl. Opt. **15**, 756 (1976)
639. D. Röss, *Lasers, Light Amplifiers and Oscillators* (Academic Press, London, 1969)

640. A.E. Siegman, *Lasers* (University Science Books, Mill Valley, 1986)
641. F.P. Schäfer (ed.), *Dye Lasers*, 3rd edn. Topics Appl. Phys., vol. 1 (Springer, Berlin, 1990); F.J. Duarte (ed.), *High Power Dye Lasers*. Springer Ser. Opt. Sci., vol. 65 (Springer, Berlin, 1991)
642. F.J. McClung, R.W. Hellwarth, Characteristics of giant optical pulsation from ruby. IEEE Proc. **51**, 46 (1963)
643. R.B. Kay, G.S. Waldman, Complete solutions to the rate equations describing Q-spoiled and PTM laser operation. J. Appl. Phys. **36**, 1319 (1965)
644. O. Kafri, S. Speiser, S. Kimel, Doppler effect mechanism for laser Q-switching with a rotating mirror. IEEE J. Quantum Electron. **7**, 122 (1971)
645. G.H.C. New, The generation of ultrashort light pulses. Rep. Prog. Phys. **46**, 877 (1983)
646. E. Hartfield, B.J. Thompson, Optical modulators, in *Handbook of Optics*, ed. by W. Driscal, W. Vaughan (McGraw Hill, New York, 1974)
647. W.E. Schmidt, Pulse stretching in a Q-switched Nd:YAG laser. IEEE J. Quantum Electron. **16**, 790 (1980)
648. Spectra Physics, Instruction manual on model 344S cavity dumper
649. A. Yariv, *Quantum Electronics* (Wiley, New York, 1975)
650. P.W. Smith, M.A. Duguay, E.P. Ippen, Mode-locking of lasers, in *Progr. Quantum Electron*, vol. 3 (Pergamon, Oxford, 1974)
651. M.S. Demokan, *Mode-Locking in Solid State and Semiconductor-Lasers* (Wiley, New York, 1982)
652. W. Koechner, *Solid-State Laser Engineering*, 4th edn. Springer Ser. Opt. Sci., vol. 1 (Springer, Berlin, 1996)
653. C.V. Shank, E.P. Ippen, Mode-locking of dye lasers, in *Dye Lasers*, ed. by F.P. Schäfer, 3rd edn. (Springer, Berlin, 1990), Chap. 4
654. W. Rudolf, Die zeitliche Entwicklung von Mode-Locking-Pulsen aus dem Rauschen. Dissertation, Fachbereich Physik, Universität Kaiserslautern, 1980
655. P. Heinz, M. Fickenscher, A. Lauberau, Electro-optic gain control and cavity dumping of a Nd:glass laser with active passive mode-locking. Opt. Commun. **62**, 343 (1987)
656. W. Demtröder, W. Stetzenbach, M. Stock, J. Witt, Lifetimes and Franck–Condon-factors for the BÔX system of Na_2. J. Mol. Spectrosc. **61**, 382 (1976)
657. H.A. Haus, *Waves and Fields in Optoelectronics* (Prentice Hall, New York, 1982)
658. R. Wilbrandt, H. Weber, Fluctuations in mode-locking threshold due to statistics of spontaneous emission. IEEE J. Quantum Electron. **11**, 186 (1975)
659. B. Kopnarsky, W. Kaiser, K.H. Drexhage, New ultrafast saturable absorbers for Nd:lasers. Opt. Commun. **32**, 451 (1980)
660. E.P. Ippen, C.V. Shank, A. Dienes, Passive mode-locking of the cw dye laser. Appl. Phys. Lett. **21**, 348 (1972)
661. G.R. Flemming, G.S. Beddard, CW mode-locked dye lasers for ultrashort spectroscopic studies. Opt. Laser Technol. **10**, 257 (1978)
662. D.J. Bradley, Methods of generations, in *Ultrashort Light Pulses*, ed. by S.L. Shapiro. Topics Appl. Phys., vol. 18 (Springer, Berlin, 1977), Chap. 2
663. P.W. Smith, Mode-locking of lasers. Proc. IEEE **58**, 1342 (1970)
664. L. Allen, D.G.C. Jones, Mode-locking of gas lasers, in *Progress in Optics*, vol. 9 (North-Holland, Amsterdam, 1971), p. 179
665. C.K. Chan, Synchronously pumped dye lasers, in *Laser Tech. Bull. Spectra Phys.*, vol. 8 (1978)
666. J. Kühl, H. Klingenberg, D. von der Linde, Picosecond and subpicosecond pulse generation in synchroneously pumped mode-locked CW dye lasers. Appl. Phys. **18**, 279 (1979)
667. G.W. Fehrenbach, K.J. Gruntz, R.G. Ulbrich, Subpicosecond light pulses from synchronously pumped mode-locked dye lasers with composite gain and absorber medium. Appl. Phys. Lett. **33**, 159 (1978)
668. D. Kühlke, V. Herpers, D. von der Linde, Characteristics of a hybridly mode-locked CW dye lasers. Appl. Phys. B **38**, 159 (1978)

669. R.H. Johnson, Characteristics of acousto-optic cavity dumping in a mode-locked laser. IEEE J. Quantum Electron. **9**, 255 (1973)
670. B. Couillaud, V. Fossati-Bellani, Mode locked lasers and ultrashort pulses I and II. Laser Appl. **4**, 79 (1985); and 91 (1985)
671. http://www.unikassel.de/fb10/fileadmin/datas/fb10/physik/femtosek/downloads/Sonstiges/Skript-Version30.pdf
672. C. Rulliere, *Femtosecond Laser Pulses*, 2nd edn. (Springer, Heidelberg, 2004); Appl. Phys. B **65** (1997), Special issue on "Ultrashort pulse generation"
673. R.L. Fork, O.E. Martinez, J.P. Gordon, Negative dispersion using pairs of prisms. Opt. Lett. **9**, 150 (1984);
D. Kühlke, Calculation of the colliding pulse mode locking in CW dye ring lasers. IEEE J. Quantum Electron. **19**, 526 (1983)
674. S. DeSilvestri, P. Laporta, V. Magni, Generation and applications of femtosecond laser-pulses. Europhys. News **17**, 105 (1986)
675. R.L. Fork, B.T. Greene, V.C. Shank, Generation of optical pulses shorter than 0.1 ps by colliding pulse mode locking. Appl. Phys. Lett. **38**, 671 (1981)
676. K. Naganuma, K. Mogi, 50 fs pulse generation directly from a colliding-pulse mode-locked Ti:sapphire laser using an antiresonant ring mirror. Opt. Lett. **16**, 738 (1991)
677. M.C. Nuss, R. Leonhardt, W. Zinth, Stable operation of a synchronously pumped colliding pulse mode-locking ring dye laser. Opt. Lett. **10**, 16 (1985)
678. P.K. Benicewicz, J.P. Roberts, A.J. Taylor, Generation of 39 fs pulses and 815 nm with a synchronously pumped mode-locked dye laser. Opt. Lett. **16**, 925 (1991)
679. L. Xu, G. Tempea, A. Poppe, M. Lenzner, C. Spielmann, F. Krausz, A. Stingl, K. Ferencz, High-power sub-10-fs Ti:Sapphire oscillators. Appl. Phys. B **65**, 151 (1997);
A. Poppe, A. Führbach, C. Spielmann, F. Krausz, Electronics on the time scale of the light oscillation period, in *OSA Trends in Optics and Photonics*, vol. 28 (Opt. Soc. Amer., Washington, 1999)
680. A.M. Kovalevicz Jr., T.R. Schihli, F.X. Kärtner, J.G. Fujiimoto, Ultralow-threshold Kerr-lens mode-locked Ti:Al_2O_3 laser. Opt. Lett. **27**, 2037 (2002)
681. L.E. Nelson, D.J. Jones, K. Tamura, H.A. Haus, E.P. Ippen, Ultrashort-pulse fiber ring lasers. Appl. Phys. B **65**, 277 (1997)
682. G.P. Agrawal, *Nonlinear Fiber Optics* (Academic Press, London, 1989)
683. S.A. Akhmanov, A.P. Sukhonukov, A.S. Chirkin, Nonstationary nonlinear optical effects and ultrashort light pulse formation. IEEE J. Quantum Electron. **4**, 578 (1968);
W.J. Tomlinson, R.H. Stollen, C.V. Shank, Compression of optical pulses chirped by self-phase modulation in fibers. J. Opt. Soc. Am. B **1**, 139 (1984)
684. D. Marcuse, Pulse duration in single-mode fibers. Appl. Opt. **19**, 1653 (1980)
685. E.B. Treacy, Optical pulse compression with diffraction gratings. IEEE J. Quantum Electron. **5**, 454 (1969)
686. C.V. Shank, R.L. Fork, R. Yen, R.H. Stolen, W.J. Tomlinson, Compression of femtosecond optical pulses. Appl. Phys. Lett. **40**, 761 (1982)
687. E. Treacy, Optical pulse compression with diffraction gratings. IEEE J. Quantum Electron. **5**, 454 (2003)
688. J.G. Fujiimoto, A.M. Weiners, E.P. Ippen, Generation and measurement of optical pulses as short as 16 fs. Appl. Phys. Lett. **44**, 832 (1984)
689. R.L. Fork, C.H. BritoCruz, P.C. Becker, C.V. Shank, Compression of optical pulses to six femtoseconds by using cubic phase compensation. Opt. Lett. **12**, 483 (1987)
690. R. Ell, U. Morgner, F.X. Kärtner, J.G. Fujimoto, E.P. Ippen, V. Scheuer, G. Angelow, T. Tschudi, Generation of 5 fs pulses and octave-spanning spectra directly from a Ti:sapphire laser. Opt. Lett. **26**(6), 373–375 (2001)
691. S. DeSilvestri et al., Few-cycle pulses by external compression, in *Few-Cycle Laser Pulses Generation and Its Application*, ed. by F.X. Kärtner. Topics Appl. Phys., vol. 95 (Springer, Berlin, 2004)

692. R. Szipöcs, K. Ferencz, C. Spielmann, F. Krausz, Chirped multilayer coatings for broadband dispersion control in femtosecond lasers. Opt. Lett. **19**, 201 (1994)
693. (a) R. Szipöcz, A. Köbázi-Kis, Theory and designs of chirped dielectric laser mirrors. Appl. Phys. B **65**, 115 (1997);
(b) R. Paschotta, *Encyclopedia for Photonics and Laser Technology* (Photonics Consulting GmbH, www.rp-photonics.com/encyclopedia.html)
(c) N. Matuschek, F.X. Kärtner, U. Keller, Theory of double-chirped mirrors. IEEE J. Sel. Top. Quantum Electron. **4**(2), 197 (1998)
694. I.D. Jung, F.X. Kärtner, N. Matuschek, D.H. Sutter, F. Morier-Genoud, Z. Shi, V. Scheuer, M. Milsch, T. Tschudi, U. Keller, Semiconductor saturable absorber mirrors supporting sub 10 fs pulses. Appl. Phys. B **65**, 137 (1997)
695. F.X. Kärtner et al., Ultra broadband double-chirped mirror pairs for generation of octave spectrum. J. Opt. Soc. Am. B **19**, 302 (2001); and Topics Appl. Phys. **95**, 73 (2004);
E.R. Morgner et al., Octave spanning spectra directly from a two-foci Ti:sapphire laser with enhanced self-phase modulation, in *Laser and Electro-optics (CLEO 2001)*, vol. 2001, p. 26
696. J.E. Midwinter, *Optical Fibers for Transmission* (Wiley, New York, 1979)
697. E.G. Neumann, *Single-Mode Fibers*. Springer Ser. Opt. Sci., vol. 57 (Springer, Berlin, 1988)
698. E.P. Ippen, D.J. Jones, L.E. Nelson, H.A. Haus, Ultrafast fiber lasers, in *Ultrafast Phenomena XI*, ed. by T. Elsässer et al. (Springer, Berlin, 1998)
699. M.J.F. Digonnet, *Rare Earth—Doped Fiber Lasers and Amplifiers*, 2nd edn. (CRC Press, Boca Raton, 2001)
700. M. Kimura, *Fiber Lasers: Research, Technology and Applications* (Nova Pub., New York, 2009)
701. Z. Cocagne, Fiber lasers: basic theory manufacturing and applications. http://mechanical.illinois.edu/media/uploads/course_websites/cocagne_fiber_lasers.20100420.4bce8145cec420.36594601.pdf
702. V.E. Zakharov, A.B. Shabat, Exact theory of two-dimensional self-focussing and one-dimensional self-modulation of waves in nonlinear media. Sov. Phys. JETP **37**, 823 (1973)
703. A. Hasegawa, *Optical Solitons in Fibers*, 2nd edn. (Springer, Berlin, 1990)
704. J.R. Taylor, *Optical Solitons—Theory and Experiment* (Cambridge Univ. Press, Cambridge, 1992)
705. G.P. Agrawal, *Nonlinear Fiber Optics* (Academic Press, San Diego, 1989)
706. L.F. Mollenauer, R.H. Stolen, The soliton laser. Opt. Lett. **9**, 13 (1984)
707. F.M. Mitschke, L.F. Mollenauer, Stabilizing the soliton laser. IEEE J. Quantum Electron. **22**, 2242 (1986)
708. E. Dusuvire, *Erbium-Doped Fiber Amplifiers* (Wiley, New York, 1994)
709. M.E. Fermann, A. Galvanauskas, G. Sucha, D. Harter, Fiber-lasers for ultrafast optics. Appl. Phys. B **65**, 259 (1997);
U. Keller, Ultrafast all solid-state laser technology. Appl. Phys. B **58**, 349 (1994)
710. K. Tamura, H.A. Haus, E.P. Ippen, Self-starting additive pulse mode-locked erbium fiber ring laser. Electron. Lett. **28**, 2226 (1992)
711. E.P. Ippen, D.J. Jones, L.E. Nelson, H.A. Haus, Ultrafast fiber lasers, in *Ultrafast Phenomena XI*, ed. by T. Elsaesser et al. (Springer, Berlin, 1998), p. 30
712. F.M. Mitschke, L.F. Mollenauer, Ultrashort pulses from the soliton laser. Opt. Lett. **12**, 407 (1987)
713. F.M. Mitschke, Solitonen in Glasfasern. Laser Optoelektron. **4**, 393 (1987);
I.N. Duling III, All-fiber ring soliton laser mode locked with a nonlinear mirror. Opt. Lett. **16**(8), 539 (1991);
F.M. Mitschke, L.F. Mollenauer, Ultrashort pulses from the soliton laser. Opt. Lett. **12**(6), 407 (1987);
N. Nichizawa, T. Goto, Novel supercontinuum fiber lasers and wavelength tunable soliton pulses. spie.org/documents/newsroom/imported/519/2006120519.pdf

714. B. Wilhelmi, W. Rudolph (eds.), *Light Pulse Compression* (Harwood Academic, Chur, 1989)
715. M.E. Fermann et al., Ultrawide tunable Er soliton fibre laser amplifier in Yb-doped fibre. Opt. Lett. **24**, 1428 (1999)
716. R. Huber, H. Satzger, W. Zinth, J. Wachtveitl, Noncollinear optical parametric amplifiers with output parameters improved by the application of a white light continuum generated in CaF_2. Opt. Commun. **194**, 443 (2001);
M.A. Arbore et al., Frequency doubling of femtosecond erbium fiber soliton lasers in periodically poled lithium niobate. Opt. Lett. **22**, 13 (1997)
717. E. Riedle, www.bmo.physik.uni-muenchen.de
718. T. Wilhelm, E. Riedle, 20 femtosecond visible pulses go tunable by noncollinear parametric amplification. Opt. Photonics News **8**, 50 (1997);
E. Riedle et al., Generation of 10 to 50 fs pulses tunable through all of the visible and the NIR. Appl. Phys. B **71**, 457 (2000)
719. T. Wilhelm, J. Piel, E. Riedle, Sub-20-fs pulses tunable across the visible from a blue-pumped single-pass noncollinear parametric converter. Opt. Lett. **22**, 1414 (1997);
J. Piel, M. Beutter, E. Riedle, 20–50 fs pulses tunable across the near infrared from a blue-pumped noncollinear parametric amplifier. Opt. Lett. **25**, 180 (2000)
720. www.physik.uni-freiburg.de/terahertz/graphics/nopa-setup.gif
721. P. Baum, S. Lochbrunner, E. Riedle, Generation of tunable 7-fs ultraviolet pulses. Appl. Phys. B **79**, 1027 (2004)
722. P. Matousek, A.W. Parker, P.F. Taday, W.T. Toner, T. Towrie, Two independently tunable and synchronized femtosecond pulses generated in the visible at the repetition rate 40 kHz using optical parametric amplifiers. Opt. Commun. **127**, 307 (1996)
723. P. Baum, E. Riedle, M. Greve, H.R. Telle, Phase-locked ultrashort pulse trains at separate and independently tunable wavelengths. Opt. Lett. **30**, 2028 (2005)
724. E. Riedle, www.bmo.physik.uni-muenchen.de/~wwwriedle/projects/NOPA_overview (homepage);
I.Z. Kozma, P. Baum, S. Lochbrenner, E. Riedle, Widely tunable sub-30-fs ultraviolet pulses by chirped sum-frequency mixing. Opt. Express **11**, 3110 (2003)
725. T. Brixner, M. Strehle, G. Gerber, Feedback-controlled optimization of amplified femtosecond laser pulses. Appl. Phys. B **68**, 281 (1999)
726. T. Hornung, R. Meier, M. Motzkus, Optimal control of molecular states in a learning loop with a parametrization in frequency and time domain. Chem. Phys. Lett. **326**, 445 (2000)
727. T. Baumert, T. Brixner, V. Seyfried, M. Strehle, G. Gerber, Femtosecond pulse shaping by an evolutionary algorithm with feedback. Appl. Phys. B **65**, 779 (1997)
728. A. Pierce, M.A. Dahleh, H. Rubitz, Optimal control of quantum-mechanical systems. Phys. Rev. A **37**, 4950 (1988)
729. R.W. Schoenlein, J.Y. Gigot, M.T. Portella, C.V. Shank, Generation of blue-green 10 fs pulses using an excimer pumped dye amplifier. Appl. Phys. Lett. **58**, 801 (1991)
730. C.V. Shank, E.P. Ippen, Subpicosecond kilowatt pulses from a mode-locked CW dye laser. Sov. Phys. JETP **34**, 62 (1972)
731. R.L. Fork, C.V. Shank, R.T. Yen, Amplification of 70-fs optical pulses to gigawatt powers. Appl. Phys. Lett. **41**, 233 (1982)
732. S.R. Rotman, C. Roxlo, D. Bebelaar, T.K. Yee, M.M. Salour, Generation, stabilization and amplification of subpicosecond pulses. Appl. Phys. B **28**, 319 (1982)
733. A. Rundquist et al., Ultrafast laser and amplifier sources. Appl. Phys. B **65**, 161 (1997)
734. E. Salin, J. Squier, G. Mourov, G. Vaillancourt, Multi-kilohertz Ti:Al_2O_3 amplifier for high power femtosecond pulses. Opt. Lett. **16**, 1964 (1991)
735. G. Sucha, D.S. Chenla, Kilohertz-rate continuum generation by amplification of femtosecond pulses near 1.5 μm. Opt. Lett. **16**, 1177 (1991)
736. A. Sullivan, H. Hamster, H.C. Kapteyn, S. Gordon, W. White, H. Nathel, R.J. Blair, R.W. Falcow, Multiterawatt, 100 fs laser. Opt. Lett. **16**, 1406 (1991)

737. T. Elsässer, M.C. Nuss, Femtosecond pulses in the mid-infrared generated by downconversion of a travelling-wave dye laser. Opt. Lett. **16**, 411 (1991)
738. J. Heling, J. Kuhl, Generation of femtosecond pulses by travelling-wave amplified spontaneous emission. Opt. Lett. **14**, 278 (1991)
739. Y. Stepanenko, C. Radzewicz, Multipass noncollinear optical parametric amplifier for femtosecond pulses. Opt. Express **14**, 779 (2006)
740. U. Keller, Ultrafast solid-state laser technology. Appl. Phys. B **58**, 347 (1994)
741. N. Ishii, L. Tuni, F. Krause et al., Multi-millijoule chirped parametric amplification of few-cycle pulses. Opt. Lett. **30**, 562 (2005)
742. M. Drescher et al., Time-resolved atomic innershell spectroscopy. Nature **419**, 803 (2001)
743. M. Wickenhauser, J. Burgdörfer, F. Krausz, M. Drescher, Time Resolved Fano Resonance. Phys. Rev. Lett. **94**, 023002 (2005)
744. E. Goulielmakis et al., Direct Measurement of Light Waves. Science **305**, 1267 (2004)
745. G.A. Mourou, C.P.J. Barty, M.D. Pery, Ultrahigh-intensity lasers: physics of the extreme on a tabletop. Phys. Today, 22 (1998)
746. E. Seres, I. Seres, F. Krausz, C. Spielmann, Generation of soft X-ray radiation. Phys. Rev. Lett. **92**, 163002 (2004)
747. J. Seres et al., Source of kiloelectron X-rays. Nature **433**, 596 (2005)
748. F. Krausz, Progress report MPQ ,Garching, 2005/2006, p. 195 (2007);
R. Kienberger et al., Single sub-fs soft-X-ray pulses: generation and measurement with the atomic transient recorder. J. Mod. Opt. **52**, 261 (2005)
749. M. Drescher, F. Krausz, Attosecond Physics: Facing the wave-particle duality. J. Phys. B **38**, 727 (2005);
F. Krausz, Report of MPQ, Garching, 2005
750. D.B. Milosevic, G.G. Paulus, W. Becker, Ionization by few-cycle pulses. Phys. Rev. A **71**, 061404(R) (2005);
G.G. Paulus et al., Measurement of the phase of few-cycle laser pulses. Phys. Rev. Lett. **91**, 253004 (2003)
751. W.H. Knox, R.S. Knox, J.F. Hoose, R.N. Zare, Observation of the O-fs pulse. Opt. Photonics News **1**, 44 (1990)
752. M. Wollenhaupt, A. Assion, Th. Baumert, Short and ultrashort laser pulses, in *Springer Handbook of Lasers and Optics*, 2nd edn., ed. by F. Träger (Springer, Berlin, 2007)
753. H. Niikura, D.M. Villeneuve, P.B. Corkum, Mapping Attosecond Electron Wave Packet Motion. Phys. Rev. Lett. **94**, 083003 (2005);
P. Corkum, Attosecond imaging, in *Max Born, A Celebration* (Max Born Institut Berlin, Berlin, 2004);
J. Levesque, P.B. Corkum, Attosecond science and technology. Can. J. Phys. **84**, 1 (2006);
D.M. Villeneuve, Attoseconds: at a glance. Nature **449**, 997 (2007)
754. M.F. Kling et al., Control of Electron Localization in Molecular Dissociation. Science **312**, 246 (2006);
P.B. Corkum, F. Krausz, Attosecond science. Nature Phys. **3**, 381 (2007);
H. Niikura, P.B. Corkum, Attosecond and Angström science. Adv. At. Mol. Opt. Phys. **54**, 511 (2007)
755. T. Fuji et al., Attosecond control of optical waveforms. New J. Phys. **7**, 116 (2005);
S. Chelkowski, G.L. Kudin, A.D. Bandrauk, Observing electron motions in molecules. J. Phys. B **39**, S409 (2006);
J. Levesque et al., Probing the electronic structure of molecules with high harmonics. J. Mod. Opt. **53**, 182 (2006)
756. E. Sorokin, Solid-state materials for few-cycle pulse generation and amplification, in *Few Cycle Laser Pulses Generation and Its Applications*, ed. by F.X. Kärtner. Topics Appl. Phys., vol. 95 (Springer, Berlin, 2004), pp. 3–72
757. S. Svanberg et al., Applications of terawatt lasers, in *Laser Spectroscopy XI*, ed. by L. Bloomfield, T. Gallagher, D. Lanson (AIP, New York, 1993)

758. R.R. Alfano (ed.), *The Supercontinuum Laser Source* (Springer, New York, 1989); J.D. Kmetec, J.I. MacKlin, J.F. Young, 0.5 TW, 125 fs Ti:sapphire laser. Opt. Lett. **16**, 1001 (1991)
759. L. Xu et al., High-power sub-10-fs Ti:sapphire oscillators. Appl. Phys. B **65**, 151 (1997); T. Brabec, F. Krausz, Intense few cycle laser fields: frontiers of nonlinear optics. Rev. Mod. Phys. **77**, 545 (2000)
760. C.H. Lee, *Picosecond Optoelectronic Devices* (Academic Press, New York, 1984)
761. Hamamatsu, FESCA-200 (Femtosecond Streak camera C6138, information sheet, August 1988) and actual information under http://usa.hamamatsu.com/sys-streak/guide.htm
762. http://www.hamamatsu.com/resources/pdf/sys/e_streakh.pdf
763. F.J. Leonberger, C.H. Lee, F. Capasso, H. Morkoc (eds.), *Picosecond Electronics and Optoelectronics II*. Springer Ser. Electron. Photon., vol. 28 (Springer, Berlin, 1987)
764. D.J. Bradley, Methods of generation, in *Ultrashort Light Pulses*, ed. by S.L. Shapiro. Topics Appl. Phys., vol. 18 (Springer, Berlin, 1977), Chap. 2
765. D.J. Bowley, Measuring ultrafast pulses. Laser Optoelectron. **6**, 81 (1987)
766. H.E. Rowe, T. Li, Theory of two-photon measurement of laser output. IEEE J. Quantum Electron. **6**, 49 (1970)
767. H.P. Weber, Method for pulsewidth measurement of ultrashort light pulses, using nonlinear optics. J. Appl. Phys. **38**, 2231 (1967)
768. J.A. Giordmaine, P.M. Rentze, S.L. Shapiro, K.W. Wecht, Two-photon Excitation of fluorescence by picosecond light pulses. Appl. Phys. Lett. **11**, 216 (1967); see also [11.26]
769. I.Z. Kozma, P. Baum, U. Schmidhammer, S. Lochbrunnen, E. Riedle, Compact autocorrelator for the online measurement of tunable 10 femtosecond pulses. Rev. Sci. Instrum. **75**, 2323 (2004)
770. W.H. Glenn, Theory of the two-photon absorption-fluorescence method of pulsewidth measurement. IEEE J. Quantum Electron. **6**, 510 (1970)
771. E. Riedle, Lectures, Winterschool on Laser Spectroscopy, Trieste, 2001
772. C. Iaconis, I.A. Walmsley, Spectral phase interferometry for direct electric-field reconstruction of ultrashort optical pulses. Opt. Lett. **23**, 792 (1998)
773. E.P. Ippen, C.V. Shank, Techniques for measurement, in *Ultrashort Light Pulses*, ed. by S.L. Shapiro. Topics Appl. Phys., vol. 18 (Springer, Berlin, 1977), Chap. 3
774. D.H. Auston, Higher order intensity correlation of optical pulses. IEEE J. Quantum Electron. **7**, 465 (1971)
775. J.C. Diels, W. Rudolph, *Ultrashort Laser Pulse Phenomena*, 2nd edn. (Academic Press, New York, 2006)
776. R. Trebino, D.J. Kane, Using phase retrieval to measure the intensity and phase of ultrashort pulses: frequency resolved optical gating. J. Opt. Soc. Am. A **11**, 2429 (1993)
777. D.J. Kane, R. Trebino, Single-shot measurement of the intensity and phase of an arbitrary ultrashort pulse using frequency-resolved optical gating. Opt. Lett. **18**, 823 (1993)
778. F. Trebino, A. Baltuska, M.S. Pshenichnikov, D.A. Wiersma, Measuring ultrashort pulses in the single-cycle regime: frequency-resolved optical gating, in *Few-Cycle Laser Pulse Generation and Its Application*, ed. by F.X. Kärtner (Springer, Heidelberg, 2004), p. 231
779. T.C. Wong, R. Trebino, Recent developments in experimental techniques for measuring two pulses simultaneously. Appl. Sci. **3**, 299–313 (2013)
780. Th. Schultz, M. Vrakking (eds.), *Attosecond and XUV Spectroscopy: Ultrafast Dynamics and Spectroscopy* (Wiley/VCH, Weinheim, 2014)
781. D. Austin, High-resolution interferometric diagnostics for ultrashort pulses. PhD thesis, Oxford University, 2010
782. J. Ratner, G. Steinmeyer, R. Bartels, R. Trebino, The coherent artifact in modern pulse measurements. Opt. Lett. **37**(14), 2874–2876 (2012)
783. R. Horstmeyer, Reconstructing ultrashort pulses. http://citeseerx.ist.psu.edu/viewdoc/summary?doi=10.1.1.310.3842
784. S. Akturk, M. Kimmel, P. O'Shea, R. Trebino, Extremely simple device for measuring 20 fs pulses. Opt. Lett. **19**, 1025 (2004)

785. S. Cormier, I.A. Walmsley, E.M. Kosik, A.S. Wyatt, L. Corner, Spectral phase interferometry for complete reconstruction of attosecond pulses. Laser Phys. **15**, 909 (2005)
786. M.E. Anderson, A. Monmayrant, S.-P. Gorza, P. Wasylczyk, I.A. Walmsley, SPIDER: A decade of measuring ultrashort pulses. Laser Phys. Lett. **5**(4), 259–266 (2008)
787. P. O'Shea, S. Akturk, M. Kimmel, R. Trebino, Practical issues in ultra-short-pulse measurements with 'GRENOUILLE'. Appl. Phys. B **79**, 683 (2004)
788. I.A. Walmsley, Characterization of ultrashort optical pulses in the few-cycle regime using spectral phase interferometry for direct electric field reconstruction, in *Few Cycle Laser Pulse Generation and Its Applications*, ed. by F.X. Kärtner (Springer, Heidelberg, 2004), p. 265
789. P. Baum, E. Riedle, Design and calibration of zero-additional-phase SPIDER. J. Opt. Soc. Am. B **22**, 1875 (2005)
790. I. Walmsley, Attosecond pulse measurement techniques. http://www.attosecond.org/theproject/metrology/xuvspider.asp
791. Y. Mairesse, F. Quere, Frequency-resolved optical gating for complete Reconstruction of attosecond bursts. Phys. Rev. A **71**, 011401 (2005)
792. S. Birger, S.A. Heinrich, A method for unique phase retrieval of ultrafast optical fields. Meas. Sci. Technol. **20**, 015303 (2009)
793. A. Unsöld, B. Baschek, *The New Cosmos*, 5th edn. (Springer, Berlin, 1991)
794. R.E. Imhoff, F.H. Read, Measurements of lifetimes of atoms, molecules. Rep. Prog. Phys. **40**, 1 (1977)
795. M.C.E. Huber, R.J. Sandeman, The measurement of oscillator strengths. Rep. Prog. Phys. **49**, 397 (1986)
796. J.R. Lakowvicz, B.P. Malivatt, Construction and performance of a variable-frequency phase-modulation fluorometer. Biophys. Chem. **19**, 13 (1984); and Biophys. J. **46**, 397 (1986)
797. J. Carlson, Accurate time resolved laser spectroscopy on sodium and bismuth atoms. Z. Phys. D **9**, 147 (1988)
798. D.V. O'Connor, D. Phillips, *Time Correlated Single Photon Counting* (Academic Press, New York, 1984)
799. W. Wien, Über Messungen der Leuchtdauer der Atome und der Dämpfung der Spektrallinien. Ann. Phys. **60**, 597 (1919)
800. P. Hartmetz, H. Schmoranzer, Lifetime and absolute transition probabilities of the $^2P_{10}$ ($^3S_1$) level of NeI by beam-gas-dye laser spectroscopy. Z. Phys. A **317**, 1 (1984)
801. D. Schulze-Hagenest, H. Harde, W. Brandt, W. Demtröder, Fast beam-spectroscopy by combined gas-cell laser excitation for cascade free measurements of highly excited states. Z. Phys. A **282**, 149 (1977)
802. L. Ward, O. Vogel, A. Arnesen, R. Hallin, A. Wännström, Accurate experimental lifetimes of excited levels in NaII, SdII. Phys. Scr. **31**, 149 (1985)
803. H. Schmoranzer, P. Hartmetz, D. Marger, J. Dudda, Lifetime measurement of the $B\,^2\Sigma_u^+$ ($v = 0$) state of $^{14}N_2^+$ by the beam-dye-laser method. J. Phys. B **22**, 1761 (1989)
804. A. Arnesen, A. Wännström, R. Hallin, C. Nordling, O. Vogel, Lifetime in KII with the beam-laser method. J. Opt. Soc. Am. B **5**, 2204 (1988)
805. Z.G. Zhang, S. Svanberg, P. Quinet, P. Palmeri, E. Biemont, Time-resolved laser spectroscopy of multiple ionized atoms. Phys. Rev. Lett. **87**, 27 (2001);
E. Biemont, P. Palmeri, P. Quinet, Z. Zhang, S. Svanberg, Doubly ionized thorium: laser lifetime measurements and transition probability determination of intersection cosmochronology. Astrophys. J. **567**, 54602 (2002)
806. A. Lauberau, W. Kaiser, Picosecond investigations of dynamic processes in polyatomic molecules and liquids, in *Chemical and Biochemical Applications of Lasers II*, ed. by C.B. Moore (Academic Press, New York, 1977)
807. W. Zinth, M.C. Nuss, W. Kaiser, A picosecond Raman technique with resolution four times better than obtained by spontaneous Raman spectroscopy, in *Picosecond Phenomena III*,

ed. by K.B. Eisenthal, R.M. Hochstrasser, W. Kaiser, A. Laubereau. Springer Ser. Chem. Phys., vol. 38 (Springer, Berlin, 1982), p. 279

808. A. Seilmeier, W. Kaiser, Ultrashort intramolecular and intermolecular vibrational energy transfer of polyatomic molecules in liquids, in *Picosecond Phenomena III*, ed. by K.B. Eisenthal, R.M. Hochstrasser, W. Kaiser, A. Laubereau. Springer Ser. Chem. Phys., vol. 38 (Springer, Berlin, 1982), p. 279
809. M. Nisoli et al., Highly efficient parametric conversion of femtosecond Ti:sapphire laser pulses at 1 kHz. Opt. Lett. **19**, 1973 (1994)
810. E. Riedle, M. Beutter, S. Lochbrunner, J. Piel, S. Schenkl, S. Spörlein, W. Zinth, Generation of 10 to 50 fs pulses, tunable through all of the visible and the NIR. Appl. Phys. B **71**, 457 (2000)
811. W. Shuicai, H. Junfang, X. Dong, Z. Changjun, H. Xun, A three-wavelength Ti:sapphire femtosecond laser for use with the multi-excited photosystem II. Appl. Phys. B **72**, 819 (2001)
812. L.-S. Ma et al., Synchronization and phase-locking of two independent femtosecond lasers, in *Conference Proceedings of ICOLS 2001, Snowbird, Utah*, 2001, pp. 2–26
813. W. Zinth, W. Holzapfel, R. Leonhardt, Femtosecond dephasing processes of molecular vibrations, in *Ultrafast Phenomena VI* (1988), p. 401
814. G. Angel, R. Gagel, A. Lauberau, Vibrational dynamics in the S_1 and S_0 states of dye molecules studied separately by femtosecond polarization spectroscopy, in *Ultrafast Phenomena VI* (1988), p. 467
815. F.J. Duarte (ed.), *High-Power Dye Lasers*. Springer Ser. Opt. Sci., vol. 65 (Springer, Berlin, 1991)
816. W. Kütt, K. Seibert, H. Kurz, High density femtosecond excitation of hot carrier distributions in InP and InGaAs, in *Ultrafast Phenomena VI* (1988), p. 233
817. W.Z. Lin, R.W. Schoenlein, M.J. LaGasse, B. Zysset, E.P. Ippen, J.G. Fujimoto, Ultrafast scattering and energy relaxation of optically excited carriers in GaAs and AlGaAs, in *Ultrafast Phenomena VI* (1988), p. 210
818. M. Chinchetti, M. Aschlimann et al., Spin-flop processes and ultrafast magnetization dynamics in Co: unifying the microscopic and macroscopic view of femtosecond magnetism. Phys. Rev. Lett. **97**, 177201 (2006)
819. L.R. Khundkar, A.H. Zewail, Ultrafast molecular reaction dynamics in real-time. Annu. Rev. Phys. Chem. **41**, 15 (1990);
A.H. Zewail (ed.), *Femtochemistry: Ultrafast Dynamics of the Chemical Bond, I and II* (World Scientific, Singapore, 1994)
820. A.H. Zewail, Femtosecond transition-state dynamics. Faraday Discuss. Chem. Soc. **91**, 207 (1991)
821. T. Baumert, M. Grosser, R. Thalweiser, G. Gerber, Femtosecond time-resolved molecular multiphoton ionisation: the Na_2 system. Phys. Rev. Lett. **67**, 3753 (1991)
822. T. Baumert, B. Bühler, M. Grosser, R. Thalweiser, V. Weiss, E. Wiedemann, R. Gerber, Femtosecond time-resolved wave packet motion in molecular multiphoton ionization and fragmentation. J. Phys. Chem. **95**, 8103 (1991)
823. E. Schreiber, *Femtosecond Real Time Spectroscopy of Small Molecules and Clusters* (Springer, Berlin, 1998)
824. O. Svelto, S. DeSilvestry, G. Denardo (eds.), *Ultrafast Processes in Spectroscopy* (Plenum, New York, 1997);
T. Brixner, N.H. Damrauer, G. Gerber, Femtosecond quantum control. Adv. At. Mol. Phys. **46**, 1 (2001)
825. R. Kienberger, F. Krausz, *Sub-femtosecond XUV Pulses: Attosecond Metrology and Spectroscopy*. Topics Appl. Phys., vol. 95 (Springer, Berlin, 2004), p. 343
826. M. Drescher et al., Time-resolved atomic inner shell spectroscopy. Nature **419**, 803 (2001)
827. H.J. Eichler, P. Günther, D.W. Pohl, *Laser-Induced Dynamic Gratings*. Springer Series in Optical Sciences, vol. 50 (Springer, Berlin, 1986)

828. F.X. Kärtner (ed.), *Few Cycle Laser Pulse Generation and Its Application*. Topics Appl. Phys., vol. 95 (Springer, Berlin, 2004)
829. F. Krausz, G. Korn, P. Corkum, I.A. Walmsley (eds.), *Ultrafast Optics IV*. Springer Ser. Opt. Sci., vol. 95 (Springer, Berlin, 2004)
830. S. Watanabe, K. Midorikawa (eds.), *Ultrafast Optics V*. Springer Ser. Opt. Sci., vol. 132 (Springer, Berlin, 2007)

Chapter 7

831. W. Hanle, Über magnetische Beeinflussung der Polarisation der Resonanzfluoreszenz. Z. Phys. **30**, 93 (1924)
832. M. Norton, A. Gallagher, Measurements of lowest-S-state lifetimes of gallium, indium and thallium. Phys. Rev. A **3**, 915 (1971)
833. F. Bylicki, H.G. Weber, H. Zscheeg, M. Arnold, On NO_2 excited state lifetime and g-factors in the 593 nm band. J. Chem. Phys. **80**, 1791 (1984)
834. H.H. Stroke, G. Fulop, S. Klepner, Level crossing signal line shapes and ordering of energy levels. Phys. Rev. Lett. **21**, 61 (1968)
835. M. McClintock, W. Demtröder, R.N. Zare, Level crossing studies of Na_2, using laser-induced fluorescence. J. Chem. Phys. **51**, 5509 (1969)
836. R.N. Zare, Molecular level crossing spectroscopy. J. Chem. Phys. **45**, 4510 (1966)
837. P. Franken, Interference effects in the resonance fluorescence of "crossed" excited states. Phys. Rev. **121**, 508 (1961)
838. G. Breit, Quantum theory of dispersion. Rev. Mod. Phys. **5**, 91 (1933)
839. R.N. Zare, Interference effects in molecular fluorescence. Acc. Chem. Res. **4**, 361 (1971)
840. J. Alnis et al., The Hanle effect and level crossing spectroscopy in Rb-vapour under strong laser excitation. J. Phys. B, At. Mol. Opt. Phys. **36**, 1161 (2003)
841. B. Cahn et al., Zeeman-tuned rotational level crossing spectroscopy in a diatomic free radical. arXiv:1310.6450 [physics]
842. G.W. Series, Coherence effects in the interaction of radiation with atoms, in *Physics of the One- and Two-Electron Atoms*, ed. by F. Bopp, H. Kleinpoppen (North-Holland, Amsterdam, 1969), p. 268
843. W. Happer, Optical pumping. Rev. Mod. Phys. **44**, 168 (1972)
844. J.N. Dodd, R.D. Kaul, D.M. Warrington, The modulation of resonance fluorescence excited by pulsed light. Proc. Phys. Soc. **84**, 176 (1964)
845. M.S. Feld, A. Sanchez, A. Javan, Theory of stimulated level crossing, in *Int'l Colloq. on Doppler-Free Spectroscopic Methods for Single Molecular Systems*, ed. by J.C. Lehmann, J.C. Pebay-Peyroula. (Ed. du Centre National Res. Scient., Paris, 1974), p. 87
846. H. Walther (ed.), *Laser Spectroscopy of Atoms and Molecules*. Topics Appl. Phys., vol. 2 (Springer, Berlin, 1976)
847. G. Moruzzi, F. Strumia (eds.), *The Hanle Effect and Level Crossing Spectroscopy* (Plenum, New York, 1992);
M. Auzinsky et al., Level-Crossing Spectroscopy of the 7, 9 and $10D_{5/2}$ states of ^{133}Cs. Phys. Rev. A **75**, 022502 (2007)
848. P. Hannaford, Oriented atoms in weak magnetic fields. Phys. Scr. T **70**, 117 (1997)
849. J.C. Lehmann, Probing small molecules with lasers, in *Frontiers of Laser Spectroscopy*, vol. 1, ed. by R. Balian, S. Haroche, S. Liberman (North-Holland, Amsterdam, 1977)
850. M. Broyer, J.C. Lehmann, J. Vigue, G-factors and lifetimes in the B-state of molecular iodine. J. Phys. **36**, 235 (1975)
851. H. Figger, D.L. Monts, R.N. Zare, Anomalous magnetic depolarization of fluorescence from the NO_2 $^2B_2$-state. J. Mol. Spectrosc. **68**, 388 (1977)

852. J.R. Bonilla, W. Demtröder, Level crossing spectroscopy of NO_2 using Doppler-reduced laser excitation in molecular beams. Chem. Phys. Lett. **53**, 223 (1978)
853. H.G. Weber, F. Bylicki, NO_2 lifetimes by Hanle effect measurements. Chem. Phys. **116**, 133 (1987)
854. F. Bylicki, H.G. Weber, G. Persch, W. Demtröder, On g factors and hyperfine structure in electronically excited states of NO_2. J. Chem. Phys. **88**, 3532 (1988)
855. H.J. Beyer, H. Kleinpoppen, Anticrossing spectroscopy, in *Progr. Atomic Spectroscopy*, ed. by W. Hanle, H. Kleinpoppen (Plenum, New York, 1978), p. 607
856. P. Cacciani, S. Liberman, E. Luc-Koenig, J. Pinard, C. Thomas, Anticrossing effects in Rydberg states of lithium in the presence of parallel magnetic and electric fields. Phys. Rev. A **40**, 3026 (1989)
857. G. Raithel, M. Fauth, H. Walther, Quasi-Landau resonances in the spectra of rubidium Rydberg atoms in crossed electric and magnetic fields. Phys. Rev. A **44**, 1898 (1991)
858. J. Bengtsson, J. Larsson, S. Svanberg, C.G. Wahlström, Hyperfine-structure study of the $3d^{10}p^2P_{3/2}$ level of neutral copper using pulsed level crossing spectroscopy at short laser wavelengths. Phys. Rev. A **41**, 233 (1990)
859. G. Hermann, G. Lasnitschka, J. Richter, A. Scharmann, Determination of lifetimes and hyperfine splittings of Tl states $nP_{3/2}$ by level crossing spectroscopy with two-photon excitation. Z. Phys. D **10**, 27 (1988)
860. G. von Oppen, measurements of state multipoles using level crossing techniques. Comments At. Mol. Phys. **15**, 87 (1984)
861. A.C. Luntz, R.G. Brewer, Zeeman-tuned level crossing in $^1\Sigma$ CH_4. J. Chem. Phys. **53**, 3380 (1970)
862. A.C. Luntz, R.G. Brewer, K.L. Foster, J.D. Swalen, Level crossing in CH_4 observed by nonlinear absorption. Phys. Rev. Lett. **23**, 951 (1969)
863. J.S. Levine, P. Boncyk, A. Javan, Observation of hyperfine level crossing in stimulated emission. Phys. Rev. Lett. **22**, 267 (1969)
864. G. Hermann, A. Scharmann, Untersuchungen zur Zeeman-Spektroskopie mit Hilfe nichtlinearer Resonanzen eines Multimoden Lasers. Z. Phys. **254**, 46 (1972)
865. W. Jastrzebski, M. Kolwas, Two-photon Hanle effect. J. Phys. B **17**, L855 (1984)
866. L. Allen, D.G. Jones, The helium-neon laser. Adv. Phys. **14**, 479 (1965)
867. C. Cohen-Tannoudji, Level-crossing resonances in atomic ground states. Comments At. Mol. Phys. **1**, 150 (1970)
868. S. Haroche, Quantum beats and time resolved spectroscopy, in *High Resolution Laser Spectroscopy*, ed. by K. Shimoda. Topics Appl. Phys., vol. 13 (Springer, Berlin, 1976), p. 253
869. H.J. Andrä, Quantum beats and laser excitation in fast beam spectroscopy, in *Atomic Physics 4*, ed. by G. zu Putlitz, E.W. Weber, A. Winnacker (Plenum, New York, 1975), p. 635
870. H.J. Andrä, Fine structure, hyperfine structure and Lamb-shift measurements by the beam foil technique. Phys. Scr. **9**, 257 (1974)
871. R.M. Lowe, P. Hannaford, Observation of quantum beats in sputtered metal vapours, in *19th EGAS Conference, Dublin* (1987)
872. W. Lange, J. Mlynek, Quantum beats in transmission by time resolved polarization spectroscopy. Phys. Rev. Lett. **40**, 1373 (1978)
873. J. Mlynek, W. Lange, A simple method of observing coherent ground-state transients. Opt. Commun. **30**, 337 (1979)
874. H. Harde, H. Burggraf, J. Mlynek, W. Lange, Quantum beats in forward scattering: subnanosecond studies with a mode-locked dye laser. Opt. Lett. **6**, 290 (1981)
875. J. Mlynek, K.H. Drake, W. Lange, Observation of transient and stationary Zeeman coherence by polarization spectroscopy, in *Laser Spectroscopy IV*, ed. by A.R.W. McKellar, T. Oka, B.P Stoicheff. Springer Ser. Opt. Sci., vol. 30 (Springer, Berlin, 1981), p. 616
876. G. Leuchs, S.J. Smith, E. Khawaja, H. Walther, Quantum beats observed in photoionization. Opt. Commun. **31**, 313 (1979)

877. M. Dubs, J. Mühlbach, H. Bitto, P. Schmidt, J.R. Huber, Hyperfine quantum beats and Zeeman spectroscopy in the polyatomic molecule propynol CHOCCHO. J. Chem. Phys. **83**, 3755 (1985)
878. H. Bitto, J.R. Huber, Molecular quantum beat spectroscopy. Opt. Commun. **80**, 184 (1990); R.T. Carter, R. Huber, Quantum beat spectroscopy in chemistry. Chem. Soc. Rev. **29**, 305 (2000)
879. H. Ring, R.T. Carter, R. Huber, Creation and phase control of molecular coherences using pulsed magnetic fields. Laser Phys. **9**, 253 (1999)
880. W. Scharfin, M. Ivanco, St. Wallace, Quantum beat phenomena in the fluorescence decay of the $C(^1B_2)$ state of SO_2. J. Chem. Phys. **76**, 2095 (1982)
881. P.J. Brucat, R.N. Zare, NO_2 $A\,^2B_2$ state properties from Zeeman quantum beats. J. Chem. Phys. **78**, 100 (1983); and **81**, 2562 (1984)
882. N. Ochi, H. Watanabe, S. Tsuchiya, S. Koda, Rotationally resolved laser-induced fluorescence and Zeeman quantum beat spectroscopy of the $V\,^1B_2$ state of jet cooled CS_2. Chem. Phys. **113**, 271 (1987)
883. P. Schmidt, H. Bitto, J.R. Huber, Excited state dipole moments in a polyatomic molecule determined by Stark quantum beat spectroscopy. J. Chem. Phys. **88**, 696 (1988)
884. J.N. Dodd, G.W. Series, Time-resolved fluorescence spectroscopy, in *Progress in Atomic Spectroscopy*, ed. by W. Hanle, H. Kleinpoppen (Plenum, New York, 1978)
885. J. Mlynek, Neue optische Methoden der hochauflösenden Kohärenzspektroskopie an Atomen. Phys. Bl. **43**, 196 (1987)
886. A. Corney, *Atomic and Laser Spectroscopy* (Oxford Univ. Press, London, 1977)
887. B.J. Dalton, Cascade Zeeman quantum beats produced by stepwise excitation using broad-line laser pulses. J. Phys. B **20**, 251, 267 (1987)
888. N.V. Vitanov, T. Halfmann, B.W. Shore, K. Bergmann, Laser-induced population transfer adiabatic passage technique. Annu. Rev. Phys. Chem. **52**, 763 (2001)
889. R. Garcia-Fernandez, A. Ekers, L.P. Yatsenko, N.V. Vitanov, K. Bergmann, Control of Population flow in coherently driven quantum ladders. Phys. Rev. Lett. **95**(1–4), 043001 (2005)
890. N.V. Vitanov, M. Fleischhauer, B.W. Shore, K. Bergmann, Coherent manipulation of atoms and molecules by sequential pulses, in *Adv. At. Mol. Opt. Phys.*, vol. 46 (Academic Press, New York, 2001), pp. 55–190
891. G. Alber, P. Zoller, Laser-induced excitation of electronic Rydberg wave packets. Contemp. Phys. **32**, 185 (1991)
892. A. Wolde, I.D. Noordam, H.G. Müller, A. Lagendijk, H.B. van Linden, Observation of radially localized atomic electron wave packets. Phys. Rev. Lett. **61**, 2099 (1988)
893. T. Baumert, V. Engel, C. Röttgermann, W.T. Strunz, G. Gerber, Femtosecond pump-probe study of the spreading and recurrance of a vibrational wave packet in Na_2. Chem. Phys. Lett. **191**, 639 (1992)
894. M. Gruebele, A.H. Zewail, Ultrashort reaction dynamics. Phys. Today **43**, 24 (1990)
895. M. Gruebele, G. Roberts, M. Dautus, R.M. Bowman, A.H. Zewail, Femtosecond temporal spectroscopy and direct inversion to the potentials: application to iodine. Chem. Phys. Lett. **166**, 459 (1990)
896. F.C. deSchryver, S.E. Fyter, G. Schweitzer, *Femtochemistry* (Wiley, New York, 2001)
897. E. Schreiber, *Femtosecond Real-Time Spectroscopy of Small Molecules and Clusters*. Springer Tracts in Modern Physics, vol. 143 (Springer, Berlin, 1999)
898. B.D. Bruner, H. Suchowski, N.V. Vitanov, Y. Silberberg, Spatiotemporal coherent control of three level systems. Phys. Rev. A **81**, 063410 (2010);
P.W. Brumer, M. Shapiro, *Principles of the Quantum Control of Molecular Processes* (Wiley, New York, 2003)
899. H. Rabitz, R. de Vivie-Riedle, M. Motzkus, K. Kompa, Chemistry—whither the future of controlling quantum phenomena? Science **288**, 824–828 (2000)
900. P. Brumer, M. Shapiro, Control of unimolecular reactions using coherent-light. Chem. Phys. Lett. **126**, 541–546 (1986)

901. J. Mlyneck, W. Lange, H. Harde, H. Burggraf, High resolution coherence spectroscopy using pulse trains. Phys. Rev. A **24**, 1099 (1989)
902. H. Lehmitz, W. Kattav, H. Harde, Modulated pumping in Cs with picosecond pulse trains, in *Methods of Laser Spectroscopy*, ed. by Y. Prior, A. Ben-Reuven, M. Rosenbluth (Plenum, New York, 1986), p. 97
903. R.H. Dicke, Coherence in spontaneous radiation processes. Phys. Rev. **93**, 99 (1954)
904. E.L. Hahn, Spin echoes. Phys. Rev. **80**, 580 (1950);
C.P. Slichter, *Principles of Magnetic Resonance*, 3rd edn. Springer Ser. Solid-State Sci., vol. 1 (Springer, Berlin, 1990)
905. I.D. Abella, Echoes at optical frequencies, in *Progress in Optics*, vol. 7 (North-Holland, Amsterdam, 1969), p. 140
906. S.R. Hartmann, Photon echoes, in *Lasers and Light, Readings from Scientific American* (Freeman, San Francisco, 1969), p. 303
907. C.K.N. Patel, R.E. Slusher, Photon echoes in gases. Phys. Rev. Lett. **20**, 1087 (1968)
908. R.G. Brewer, Coherent optical transients. Phys. Today **30**, 50 (1977)
909. R.G. Brewer, A.Z. Genack, Optical coherent transients by laser frequency switching. Phys. Rev. Lett. **36**, 959 (1976)
910. R.G. Brewer, Coherent optical spectroscopy, in *Frontiers in Laser Spectroscopy*, ed. by R. Balian, S. Haroche, S. Lieberman (North-Holland, Amsterdam, 1977)
911. L.S. Vasilenko, N.Y. Rubtsova, Coherent spectroscopy of gaseous media: ways of increasing spectral resolution. Bull. Acad. Sci. USSR Phys. Ser. **53**(12), 54 (1989)
912. R.G. Brewer, R.L. Shoemaker, Photon echo and optical nutation in molecules. Phys. Rev. Lett. **27**, 631 (1971)
913. P.R. Berman, J.M. Levy, R.G. Brewer, Coherent optical transient study of molecular collisions. Phys. Rev. A **11**, 1668 (1975)
914. Y.Y. Lion, I.-C. Chen, R.-K. Lee, Few Cycle self-induced transparency solitons. Phys. Rev. A **83**, 043828 (2011)
915. G. He, S.H. Liu, *Physics of Nonlinear Optics* (World Scientific, Singapore, 2003)
916. S.L. McCall, E.L. Hahn, Self-induced transparency. Phys. Rev. **183**, 457 (1969)
917. E. Arimondo, Coherent population trapping in laser spectroscopy. Prog. Opt. **35**, 257 (1996)
918. A. Nagel et al., Experimental realization of a coherent dark state magnetometer. Europhys. Lett. **44**, 31–36 (1998);
C. Affolderbach, M. Stähler, S. Knappe, R. Wynabds, An all optical high sensitivity magnetic gradiometer. Appl. Phys. B **75**, 605–612 (2002)
919. C. Cohen-Tannoudji, J. Dupont-Roc, G. Grynberg, *Atoms-Photon Interactions: Basic Processes and Applications* (Wiley, New York, 1992);
C. Cohen-Tannoudji, Dark resonances: from optical pumping to cold atoms and molecules. http://www.phys.ens.fr/~cct/articles/Kosmos/Kosmos.pdf
920. C. Freed, D.C. Spears, R.G. O'Donnell, Precision heterodyne calibration, in *Laser Spectroscopy*, ed. by R.G. Brewer, A. Mooradian (Plenum, New York, 1974), p. 17
921. F.R. Petersen, D.G. McDonald, F.D. Cupp, B.L. Danielson, Rotational constants of $^{12}C^{16}O_2$ from beats between Lamb-dip stabilized laser lines. Phys. Rev. Lett. **31**, 573 (1973), also in *Laser Spectroscopy*, ed. by R.G. Brewer, A. Mooradian (Plenum, New York, 1974), p. 555
922. T.J. Bridge, T.K. Chang, Accurate rotational constants of CO_2 from measurements of CW beats in bulk GaAs between CO_2 vibrational-rotational laser lines. Phys. Rev. Lett. **22**, 811 (1969)
923. L.A. Hackel, K.H. Casleton, S.G. Kukolich, S. Ezekiel, Observation of magnetic octupole and scalar spin–spin interaction in I_2 using laser spectroscopy. Phys. Rev. Lett. **35**, 568 (1975)
924. W.A. Kreiner, G. Magerl, E. Bonek, W. Schupita, L. Weber, Spectroscopy with a tunable sideband laser. Phys. Scr. **25**, 360 (1982)
925. J.L. Hall, L. Hollberg, T. Baer, H.G. Robinson, Optical heterodyne saturation spectroscopy. Appl. Phys. Lett. **39**, 680 (1981)

926. P. Verhoeve, J.J. terMeulen, W.L. Meerts, A. Dynamus, Sub-millimeter laser-sideband spectroscopy of H_3O^+. Chem. Phys. Lett. **143**, 501 (1988)
927. E.A. Whittaker, H.R. Wendt, H. Hunziker, G.C. Bjorklund, Laser FM spectroscopy with photochemical modulation: a sensitive high resolution technique for chemical intermediates. Appl. Phys. B **35**, 105 (1984)
928. F.T. Arecchi, A. Berné, P. Bulamacchi, High-order fluctuations in a single mode laser field. Phys. Rev. Lett. **88**, 32 (1966)
929. H.Z. Cummins, H.L. Swinney, *Light beating spectroscopy*. Progress in Optics, vol. 8 (North-Holland, Amsterdam, 1970), p. 134
930. E.O. DuBois (ed.), *Photon Correlation Techniques*. Springer Ser. Opt. Sci., vol. 38 (Springer, Berlin, 1983)
931. F. Yang, The molecular structure of green fluorescent protein. PhD thesis, Rice University, Houston, Texas, April 1999
932. R. Rigler, E.S. Elson (eds.), *Fluorescence Correlation Spectroscopy: Theory and Applications*. Springer Ser. Chem. Phys., vol. 65 (Springer, Berlin, 2001)
933. R. Rigler, S. Wennmalm, L. Edman, Fluorescence correlation spectroscopy, in *Single Molecule Spectroscopy*. Springer Series in Chemical Physics, vol. 65 (Springer, Berlin, 2001), p. 459
934. P. Schwille, E. Haustein, Fluorescence correlation spectroscopy. https://www.biophysics.org/Portals/1/PDFs/Education/schwille.pdf
935. N. Wiener, Generalized harmonic analysis. Acta Math. **55**, 117 (1930)
936. C.L. Mehta, *Theory of photoelectron counting*. Progress in Optics, vol. 7 (North Holland, Amsterdam, 1970), p. 373
937. B. Saleh, *Photoelectron Statistics*. Springer Ser. Opt. Sci., vol. 6 (Springer, Berlin, 1978)
938. L. Mandel, *Fluctuation of Light Beams*. Progress in Optics, vol. 2 (North-Holland, Amsterdam, 1963), p. 181
939. A.J. Siegert, MIT Rad. Lab. Rpt. No. 465, 1943
940. E.O. Schulz-DuBois, High-resolution intensity interferometry by photon correlation, in [930], p. 6
941. P.P.L. Regtien (ed.), *Modern Electronic Measuring Systems* (Delft Univ. Press, Delft, 1978); P. Horrowitz, W. Hill, *The Art of Electronics* (Cambridge Univ. Press, Cambridge, 1980)
942. H.Z. Cummins, E.R. Pike (eds.), *Photon Correlation and Light Spectroscopy* (Plenum, New York, 1974)
943. E. Stelzer, H. Ruf, E. Grell, Analysis and resolution of polydispersive systems, in [930], p. 329
944. N.C. Ford, G.B. Bennedek, Observation of the spectrum of light scattered from a pure fluid near its critical point. Phys. Rev. Lett. **15**, 649 (1965)
945. R. Hanbury Brown, *The Intensity Interferometer* (Taylor and Francis, London, 1974)
946. F.T. Arecchi, A. Berné, P. Bulamacchi, High-order fluctuations in a single mode laser field. Phys. Rev. Lett. **88**, 32 (1966)
947. M. Adam, A. Hamelin, P. Bergé, Mise au point et étude d'une technique de spectrographie par battements de photons hétérodyne. Opt. Acta **16**, 337 (1969)
948. E. Haustein, P. Schwille, Fluorescence correlation spectroscopy: novel variations of an established technique. Annu. Rev. Biophys. Biomol. Struct. **36**, 151 (2007)
949. http://www.chm.bris.ac.uk/motm/GFP/GFPh.htm
950. B. Indge, M. Baker, M. Rowland, *A New Introduction to Human Biology (AQA Specification A)* (Hodder Education, Murray, 2000)
951. W. Drexler, J.G. Fujimoto (eds.), *Optical Coherence Tomography: Technology and Applications* (Springer, Berlin, 2008)
952. R. Rigler, S. Wennmalm, L. Edman, *Fluorescence Correlation Spectroscopy in Single Molecule Analysis*. Springer Ser. Chem. Phys., vol. 65 (Springer, Berlin, 2001), p. 459
953. A.F. Harvey, *Coherent Light* (Wiley, London, 1970)
954. J.I. Steinfeld (ed.), *Laser and Coherence Spectroscopy* (Plenum, New York, 1978)
955. B.W. Shore, *The Theory of Coherent Atomic Excitation*, vols. 1, 2 (Wiley, New York, 1990)

956. P. Schwille, E. Haustein, Fluorescence correlation spectroscopy. http://www.biophysics.org/education/schwille.pdf
957. L. Mandel, E. Wolf, *Coherence and Quantum Optics I–IX, Proc. Rochester Conferences* (Plenum, New York, 1961/1967/1973/1978/1984/1990/1996/2002/2008)

Chapter 8

958. R.B. Bernstein, *Chemical Dynamics via Molecular Beam and Laser Techniques* (Clarendon, Oxford, 1982)
959. J.T. Yardle, *Molecular Energy Transfer* (Academic Press, New York, 1980)
960. W.H. Miller (ed.), *Dynamics of Molecular Collisions* (Plenum, New York, 1976)
961. P.R. Berman, Studies of collisions by laser spectroscopy. Adv. At. Mol. Phys. **13**, 57 (1977)
962. G.W. Flynn, E. Weitz, Vibrational energy flow in the ground electronic states of polyatomic molecules. Adv. Chem. Phys. **47**, 185 (1981)
963. V.E. Bondeby, Relaxation and vibrational energy redistribution processes in polyatomic molecules. Annu. Rev. Phys. Chem. **35**, 591 (1984)
964. S.A. Rice, Collision-induced intramolecular energy transfer in electronically excited polyatomic molecules. Adv. Chem. Phys. **47**, 237 (1981)
965. P.J. Dagdigian, State-resolved collision-induced electronic transitions. Annu. Rev. Phys. Chem. **48**, 95 (1997)
966. J.J. Valentini, State-to-State chemical reaction dynamics in polyatomic systems. Annu. Rev. Phys. Chem. **52**, 15 (2001)
967. C.A. Taatjes, J.F. Herschberger, Recent progress in infrared absorption techniques for elementary gas phase reaction kinetics. Annu. Rev. Phys. Chem. **52**, 41 (2001)
968. *Proc. Int'l Conf. on the Physics of Electronic and Atomic Collisions, ICPEAC I–XVII* (North-Holland, Amsterdam)
969. *Proc. Int'l Conf. on Spectral Line Shapes*, vols. I–XI. Library of Congress Catalog
970. J. Hinze (ed.), *Energy Storage and Redistribution in Molecules* (Plenum, New York, 1983)
971. F. Aumeyer, H. Winter (eds.), *Photonic, Electronic and Atomic Collisions* (World Scientific, Singapore, 1998)
972. G. Peach, *Collisional Broadening of Spectral Lines*. Springer Handbook of Atomic, Molecular and Optical Physics (2006), pp. 875–888
973. J. Ward, J. Cooper, Correlation effects in the theory of combined Doppler and pressure broadening. J. Quant. Spectrosc. Radiat. Transf. **14**, 555 (1974)
974. J.O. Hirschfelder, Ch.F. Curtiss, R.B. Bird, *Molecular Theory of Gases and Liquids* (Wiley, New York, 1954)
975. Th.W. Hänsch, P. Toschek, On pressure broadening in a He–Ne laser. IEEE J. Quantum Electron. **6**, 61 (1969)
976. J.L. Hall, Saturated absorption spectroscopy, in *Atomic Physics*, vol. 3, ed. by S.J. Smith, G.K. Walthers (Plenum, New York, 1973), pp. 615 ff.
977. J.L. Hall, *The Line Shape Problem in Laser-Saturated Molecular Absorption* (Gordon & Breach, New York, 1969)
978. S.N. Bagayev, Spectroscopic studies into elastic scattering of excited particles, in *Laser Spectroscopy IV*, ed. by H. Walther, K.W. Rothe. Springer Ser. Opt. Sci., vol. 21 (Springer, Berlin, 1979), p. 222
979. K.E. Gibble, A. Gallagher, Measurements of velocity-changing collision kernels. Phys. Rev. A **43**, 1366 (1991)
980. C.R. Vidal, F.B. Haller, Heat pipe oven applications: production of metal vapor-gas mixtures. Rev. Sci. Instrum. **42**, 1779 (1971)
981. R. Bombach, B. Hemmerling, W. Demtröder, Measurement of broadening rates, shifts and effective lifetimes of Li_2 Rydberg levels by optical double-resonance spectroscopy. Chem. Phys. **121**, 439 (1988)

982. K.D. Heber, P.J. West, E. Matthias, Collisions between Sr-Rydberg atoms and intermediate and high principal quantum number and noble gases. J. Phys. D **21**, 563 (1988)
983. R.B. Kurzel, J.I. Steinfeld, D.A. Hazenbuhler, G.E. LeRoi, Energy transfer processes in monochromatically excited iodine molecules. J. Chem. Phys. **55**, 4822 (1971)
984. J.I. Steinfeld, Energy transfer processes, in *Chemical Kinetics Phys. Chem. Ser. One*, vol. 9, ed. by J.C. Polany (Butterworth, London, 1972)
985. G. Ennen, C. Ottinger, Rotation-vibration-translation energy transfer in laser excited Li_2 $(B\,^1\Pi_u)$. Chem. Phys. **3**, 404 (1974)
986. Ch. Ottinger, M. Schröder, Collision-induced rotational transitions of dye laser excited Li_2 molecules. J. Phys. B **13**, 4163 (1980)
987. K. Bergmann, W. Demtröder, Inelastic cross sections of excited molecules. J. Phys. B **5**(7), 1386 (1972); and **5**(7), 2098 (1972)
988. D. Zevgolis, Untersuchung inelastischer Stoßprozesse in Alkalidämpfen mit Hilfe spektral- und zeitaufgelöster Laserspektroskopie. Dissertation, Faculty of Physics, Kaiserslautern, 1980
989. T.A. Brunner, R.D. Driver, N. Smith, D.E. Pritchard, Rotational energy transfer in Na_2–Xe collisions. J. Chem. Phys. **70**, 4155 (1979);
T.A. Brunner, et al., Simple scaling law for rotational energy transfer in Na_2^*–Xe collisions. Phys. Rev. Lett. **41**, 856 (1978)
990. R. Schinke, *Theory of Rotational Transitions in Molecules, Int'l Conf. Phys. El. At. Collisions XIII* (North-Holland, Amsterdam, 1984), p. 429
991. M. Faubel, Vibrational and rotational excitation in molecular collisions. Adv. At. Mol. Phys. **19**, 345 (1983)
992. K. Bergmann, H. Klar, W. Schlecht, Asymmetries in collision-induced rotational transitions. Chem. Phys. Lett. **12**, 522 (1974)
993. H. Klar, Theory of collision-induced rotational energy transfer in the 1/4-state of diatomic molecule. J. Phys. B **6**, 2139 (1973)
994. A.J. McCaffery, M.J. Proctor, B.J. Whitaker, Rotational energy transfer: polarization and scaling. Annu. Rev. Phys. Chem. **37**, 223 (1986)
995. G. Sha, P. Proch, K.L. Kompa, Rotational transitions of N_2 $(a\,^1\Pi_g)$ induced by collisions with Ar/He studied by laser REMPI spectroscopy. J. Chem. Phys. **87**, 5251 (1987)
996. C.R. Vidal, Collisional depolarization and rotational energy transfer of the $Li_2(B\,^1\Pi_u)$–$Li(^2S_{1/2})$ system from laser-induced fluorescence. Chem. Phys. **35**, 215 (1978)
997. W.B. Gao, Y.Q. Shen, H. Häger, W. Krieger, Vibrational relaxation of ethylene oxide and ethylene oxide-rare-gas mixtures. Chem. Phys. **84**, 369 (1984)
998. M.H. Alexander, A. Benning, Theoretical studies of collision-induced energy transfer in electronically excited states. Ber. Bunsenges. Phys. Chem. **14**, 1253 (1990)
999. E. Nikitin, L. Zulicke, *Theorie Chemischer Elementarprozesse* (Vieweg, Braunschweig, 1985)
1000. E.K. Kraulinya, E.K. Kopeikana, M.L. Janson, Excitation energy transfer in atom-molecule interactions of sodium and potassium vapors. Chem. Phys. Lett. **39**, 565 (1976); and Opt. Spectrosc. **41**, 217 (1976)
1001. W. Kamke, B. Kamke, I. Hertel, A. Gallagher, Fluorescence of the $Na^* + N_2$ collision complex. J. Chem. Phys. **80**, 4879 (1984)
1002. L.K. Lam, T. Fujimoto, A.C. Gallagher, M. Hessel, Collisional excitation transfer between Na and Na_2. J. Chem. Phys. **68**, 3553 (1978)
1003. H. Hulsman, P. Willems, Transfer of electronic excitation in sodium vapour. Chem. Phys. **119**, 377 (1988)
1004. G. Ennen, Ch. Ottinger, Collision-induced dissociation of laser excited Li_2 $B\,^1\Pi_\mu$. J. Chem. Phys. **40**, 127 (1979); and **41**, 415 (1979)
1005. J.E. Smedley, H.K. Haugen, St.R. Leone, Collision-induced dissociation of laser-excited Br_2 $[B\,^3\Pi(O_u^+);\, v',\, J']$. J. Chem. Phys. **86**, 6801 (1987)
1006. E.W. Rothe, U. Krause, R. Dünen, Photodissociation of Na_2 and Rb_2: analysis of atomic fine structure of ^2P products. J. Chem. Phys. **72**, 5145 (1980)

1007. G. Brederlow, R. Brodmann, M. Nippus, R. Petsch, S. Witkowski, R. Volk, K.J. Witte, Performance of the Asterix IV high power iodine laser. IEEE J. Quantum Electron. **16**, 122 (1980)
1008. V. Diemer, W. Demtröder, Infrared atomic Cs laser based on optical pumping of Cs_2 molecules. Chem. Phys. Lett. **176**, 135 (1991)
1009. St. Lemont, G.W. Flynn, Vibrational state analysis of electronic-to-vibrational energy transfer processes. Annu. Rev. Phys. Chem. **28**, 261 (1977)
1010. A. Tramer, A. Nitzan, Collisional effects in electronic relaxation. Adv. Chem. Phys. **47**(2), 337 (1981)
1011. S.A. Rice, Collision-induced intramolecular energy transfer in electronically excited polyatomic molecules. Adv. Chem. Phys. **47**(2), 237 (1981)
1012. K.B. Eisenthal, Ultrafast chemical reactions in the liquid state, in *Ultrashort Laser Pulses*, 2nd edn., ed. by W. Kaiser. Topics Appl. Phys., vol. 60 (Springer, Berlin, 1993)
1013. P. Hering, S.L. Cunba, K.L. Kompa, Coherent anti-Stokes Raman spectroscopy study of the energy paritioning in the $Na(3P)$–H_2 collision pair with red wing excitation. J. Phys. Chem. **91**, 5459 (1987)
1014. H.G.C. Werij, J.F.M. Haverkort, J.P. Woerdman, Study of the optical piston. Phys. Rev. A **33**, 3270 (1986)
1015. A.D. Streater, J. Mooibroek, J.P. Woerdman, Light-induced drift in rubidium: spectral dependence and isotope separation. Opt. Commun. **64**, 1 (1987)
1016. M. Allegrini, P. Bicchi, L. Moi, Cross-section measurements for the energy transfer collisions $Na(3P)+Na(3P)\rightarrow Na(5S,4D)+Na(3S)$. Phys. Rev. A **28**, 1338 (1983)
1017. J. Huenneckens, A. Gallagher, Radiation diffusion and saturation in optically thick Na vapor. Phys. Rev. A **28**, 238 (1983)
1018. J. Huenneckens, A. Gallagher, Associative ionization in collisions between two $Na(3P)$ atoms. Phys. Rev. A **28**, 1276 (1983)
1019. S.A. Abdullah, M. Allegrini, S. Gozzini, L. Moi, Three-body collisions between laser excited Na- and K-atoms in the presence of buffer gas. Nuovo Cimento D **9**, 1467 (1987)
1020. H.G.C. Werij, M. Harris, J. Cooper, A. Gallagher, Collisional energy transfer between excited Sr atoms. Phys. Rev. A **43**, 2237 (1991)
1021. S.G. Leslie, J.T. Verdeyen, W.S. Millar, Excitation of highly excited states by collisions between two excited cesium atoms. J. Appl. Phys. **48**, 4444 (1977)
1022. A. Ermers, T. Woschnik, W. Behmenburg, Depolarization and fine structure effects in half-collisions of sodium-noble gas systems. Z. Phys. D **5**, 113 (1987)
1023. P.W. Arcuni, M.L. Troyen, A. Gallagher, Differential cross section for Na fine structure transfer induced by Na and K collisions. Phys. Rev. A **41**, 2398 (1990)
1024. G.C. Schatz, L.J. Kowalenko, S.R. Leone, A coupled channel quantum scattering study of alignment effects in $Na(^2P_{3/2})+He\rightarrow Na(^2P_{1/2})+He$ collisions. J. Chem. Phys. **91**, 6961 (1989)
1025. T.R. Mallory, W. Kedzierski, J.B. Atkinson, L. Krause, $9\,^2D$ fine structure mixing in rubidium by collisions with ground-state Rb and noble-gas atoms. Phys. Rev. A **38**, 5917 (1988)
1026. A. Sasso, W. Demtröder, T. Colbert, C. Wang, W. Ehrlacher, J. Huennekens, Radiative lifetimes, collisional mixing and quenching of the cesium 5 Dy levels. Phys. Rev. A **45**, 1670 (1992)
1027. D.L. Feldman, R.N. Zare, Evidence for predissociation of Rb_2^* $(C\,^1\Pi_u)$ into Rb^* $(^2P_{3/2})$ and Rb $(^2S_{1/2})$. Chem. Phys. **15**, 415 (1976);
E.J. Breford, F. Engelke, Laser induced fluorescence in supersonic nozzle beams: predissociation in the Rb_2 $C\,^1\Pi_u$ and $D\,^1\Pi_u$ states. Chem. Phys. Lett. **75**, 132 (1980)
1028. K.F. Freed, Collision-induced intersystem crossing. Adv. Chem. Phys. **47**, 211 (1981)
1029. J.P. Webb, W.C. McColgin, O.G. Peterson, D. Stockman, J.H. Eberly, Intersystem crossing rate and triplet state lifetime for a lasing dye. J. Chem. Phys. **53**, 4227 (1970)
1030. X.L. Han, G.W. Schinn, A. Gallagher, Spin-exchange cross sections for electron excitation of Na $3S$–$3P$ determined by a novel spectroscopic technique. Phys. Rev. A **38**, 535 (1988)

1031. T.A. Cool, Transfer chemical laser, in *Handbook of Chemical Lasers*, ed. by R.W.F. Gross, J.F. Bott (Wiley, New York, 1976)
1032. S.A. Ahmed, J.S. Gergely, D. Infaute, Energy transfer organic dye mixture lasers. J. Chem. Phys. **61**, 1584 (1974)
1033. B. Wellegehausen, Optically pumped CW dimer lasers. IEEE J. Quantum Electron. **15**, 1108 (1979)
1034. W.J. Witteman, *The CO_2 Laser*. Springer Ser. Opt. Sci., vol. 53 (Springer, Berlin, 1987)
1035. W.H. Green, J.K. Hancock, Laser excited vibrational energy exchange studies of HF, CO and NO. IEEE J. Quantum Electron. **9**, 50 (1973)
1036. W.B. Gao, Y.Q. Shen, J. Häger, W. Krieger, Vibrational relaxation of ethylene oxide and ethylene oxide-rare gas mixtures. Chem. Phys. **84**, 369 (1984)
1037. S.R. Leone, State-resolved molecular reaction dynamics. Annu. Rev. Phys. Chem. **35**, 109 (1984)
1038. J.O. Hirschfelder, R.E. Wyatt, R.D. Coalson (eds.), *Lasers, Molecules and Methods* (Wiley, New York, 1989)
1039. E. Hirota, K. Kawaguchi, High resolution infrared studies of molecular dynamics. Annu. Rev. Phys. Chem. **36**, 53 (1985)
1040. R. Feinberg, R.E. Teets, J. Rubbmark, A.L. Schawlow, Ground-state relaxation measurements by laser-induced depopulation. J. Chem. Phys. **66**, 4330 (1977)
1041. J.G. Haub, B.J. Orr, Coriolis-assisted vibrational energy transfer in D_2CO/D_2CO and HDCO/HDCO-collisions. J. Chem. Phys. **86**, 3380 (1987)
1042. A. Lauberau, W. Kaiser, Vibrational dynamics of liquids and solids investigated by picosecond light pulses. Rev. Mod. Phys. **50**, 607 (1978)
1043. T. Elsaesser, W. Kaiser, Vibrational and vibronic relaxation of large polyatomic molecules in liquids. Annu. Rev. Phys. Chem. **43**, 83 (1991)
1044. L.R. Khunkar, A.H. Zewail, Ultrafast molecular reaction dynamics in real-time. Annu. Rev. Phys. Chem. **41**, 15 (1990)
1045. R.T. Bailey, F.R. Cruickshank, Spectroscopic studies of vibrational energy transfer. Adv. Infrared Raman Spectrosc. **8**, 52 (1981)
1046. Ch.E. Hamilton, J.L. Kinsey, R.W. Field, Stimulated emission pumping: new methods in spectroscopy and molecular dynamics. Annu. Rev. Phys. Chem. **37** (1986)
1047. A. Geers, J. Kappert, F. Temps, J.W. Wiebrecht, Preparation of single rotation-vibration states of $CH_3O(C(^2E))$ above the H–CH_2O dissociation threshold by stimulated emission pumping. Ber. Bunsenges. Phys. Chem. **94**, 1219 (1990)
1048. M. Becker, U. Gaubatz, K. Bergmann, P.L. Jones, Efficient and selective population of high vibrational levels by stimulated near resonance Raman scattering. J. Chem. Phys. **87**, 5064 (1987)
1049. N.V. Vitanov, T. Halfmann, B.W. Shore, K. Bergmann, Laser-induced population transfer by adiabatic passage techniques. Annu. Rev. Phys. Chem. **52**, 763 (2001)
1050. G.W. Coulston, K. Bergmann, Population transfer by stimulated Raman scattering with delayed pulses. J. Chem. Phys. **96**, 3467 (1992)
1051. W.R. Gentry, State-to-state energy transfer in collisions of neutral molecules, in *ICPEAC XIV, 1985* (North-Holland, Amsterdam, 1986), p. 13
1052. H.L. Dai (guest ed.), Molecular spectroscopy and dynamics by stimulated emission pumping. J. Opt. Soc. Am. B **7**, 1802 (1990) (Special issue)
1053. P.J. Dagdigian, Inelastic scattering: optical methods, in *Atomic and Molecular Methods*, ed. by G. Scoles (Oxford Univ. Press, Oxford, 1988), p. 569
1054. J.C. Whitehead, Laser studies of reactive collisions. J. Phys. B, At. Mol. Opt. Phys. **23**, 3443 (1996)
1055. J.B. Pruett, R.N. Zare, State-to-state reaction rates: $Ba + HF(v = 0.1) \to BaF(v = 0-12) + H$. J. Chem. Phys. **64**, 1774 (1976)
1056. P.J. Dagdigian, H.W. Cruse, A. Schultz, R.N. Zare, Product state analysis of BaO from the reactions $Ba + CO_2$ and $Ba + O_2$. J. Chem. Phys. **61**, 4450 (1974)

1057. K. Kleinermanns, J. Wolfrum, Laser stimulation and observation of elementary chemical reactions in the gas phase. Laser Chem. **2**, 339 (1983)
1058. K.D. Rinnen, D.A.V. Kliner, R.N. Zare, The $H + D_2$ reaction: prompt HD distribution at high collision energies. J. Chem. Phys. **91**, 7514 (1989)
1059. D.P. Gerrity, J.J. Valentini, Experimental determination of product quantum state distributions in the $H + D_2 \rightarrow HD + D$ reaction. J. Chem. Phys. **79**, 5202 (1983)
1060. H. Buchenau, J.P. Toennies, J. Arnold, J. Wolfrum, $H + H_2$: the current status. Ber. Bunsenges. Phys. Chem. **94**, 1231 (1990)
1061. W. Chapman, B.W. Blackman, D.J. Nesbitt, State-to-state reactive scattering of $F + H_2$ in supersonic jets. J. Chem. Phys. **107**, 8193 (1997);
B.W. Blackman, PhD thesis, JILA, Boulder, CO, 1996
1062. V.M. Bierbaum, S.R. Leone, Optical studies of product state distribution in thermal energy ion-molecule reactions, in *Structure, Reactivity and Thermochemistry of Ions*, ed. by P. Ausloos, S.G. Lias (Reidel, New York, 1987), p. 23
1063. M.N.R. Ashfold, J.E. Baggott (eds.), *Molecular Photodissociation Dynamics, Adv. in Gas-Phase Photochemistry* (Royal Chem. Soc., London, 1987)
1064. E. Hasselbrink, J.R. Waldeck, R.N. Zare, Orientation of the CN $X\,^2\Sigma^+$ fragment, following photolysis by circularly polarized light. Chem. Phys. **126**, 191 (1988)
1065. F.J. Comes, Molecular reaction dynamics: sub-Doppler and polarization spectroscopy. Ber. Bunsenges. Phys. Chem. **94**, 1268 (1990)
1066. M.H. Kim, L. Shen, H. Tao, T.J. Martinez, A.G. Suits, Conformationally controlled chemistry. Science **315**, 1561 (2007)
1067. J.F. Black, J.R. Waldeck, R.N. Zare, Evidence for three interacting potential energy surfaces in the photodissociation of ICN at 249 nm. J. Chem. Phys. **92**, 3519 (1990)
1068. P. Andresen, G.S. Ondrey, B. Titze, E.W. Rothe, Nuclear and electronic dynamics in the photodissociation of water. J. Chem. Phys. **80**, 2548 (1984)
1069. St.R. Leone, Infrared fluorescence: a versatile probe of state selected chemical dynamics. Acc. Chem. Res. **16**, 88 (1983)
1070. Many contributions in Laser Chem. Int. J. (Harwood, Chur)
1071. G.E. Hall, P.L. Houston, Vector correlations in photodissociation dynamics. Annu. Rev. Phys. Chem. **40**, 375 (1989)
1072. R.N. Dixon, G.G. Balint-Kurti, M.S. Child, R. Donovun, J.P. Simmons (guest eds.), Dynamics of molecular photofragmentation. Faraday Discuss. Chem. Soc. **82** (1986) (Special issue)
1073. A. Gonzales Urena, Influence of translational energy upon reactive scattering cross sections of neutral–neutral collisions. Adv. Chem. Phys. **66**, 213 (1987)
1074. M.A.D. Fluendy, K.P. Lawley, *Chemical Applications of Molecular Beam Scattering* (Chapman and Hall, London, 1973)
1075. K. Liu, Crossed-beam studies of neutral reactions: state specific differential cross sections. Annu. Rev. Phys. Chem. **52**, 139 (2001)
1076. V. Borkenhagen, M. Halthau, J.P. Toennies, Molecular beam measurements of inelastic cross sections for transition between defined rotational states of CsF. J. Chem. Phys. **71**, 1722 (1979)
1077. K. Bergmann, R. Engelhardt, U. Hefter, J. Witt, State-resolved differential cross sections for rotational transition in $Na_2 + Ne$ collisions. Phys. Rev. Lett. **40**, 1446 (1978)
1078. K. Bergmann, State selection via optical methods, in *Atomic and Molecular Beam Methods*, vol. 12, ed. by G. Coles (Oxford Univ. Press, Oxford, 1989)
1079. K. Bergmann, U. Hefter, J. Witt, State-to-state differential cross sections for rotational transitions in $Na_2 +$ He-collisions. J. Chem. Phys. **71**, 2726 (1979); and **72**, 4777 (1980)
1080. H.G. Rubahn, K. Bergmann, The effect of laser-induced vibrational band stretching in atom-molecule collisions. Annu. Rev. Phys. Chem. **41**, 735 (1990)
1081. V. Gaubatz, H. Bissantz, V. Hefter, I. Colomb de Daunant, K. Bergmann, Optically pumped supersonic beam lasers. J. Opt. Soc. Am. B **6**, 1386 (1989)

1082. R. Düren, H. Tischer, Experimental determination of the K($4\,^2P_{3/3}$–Ar) potential. Chem. Phys. Lett. **79**, 481 (1981)
1083. I.V. Hertel, H. Hofmann, K.A. Rost, Electronic to vibrational-rotational energy transfer in collisions of Na($3\,^2P$) with simple molecules. Chem. Phys. Lett. **47**, 163 (1977)
1084. P. Botschwina, W. Meyer, I.V. Hertel, W. Reiland, Collisions of excited Na atoms with H_2-molecules: ab initio potential energy surfaces. J. Chem. Phys. **75**, 5438 (1981)
1085. E.E.B. Campbell, H. Schmidt, I.V. Hertel, Symmetry and angular momentum in collisions with laser excited polarized atoms. Adv. Chem. Phys. **72**, 37 (1988)
1086. R. Düren, V. Lackschewitz, S. Milosevic, H. Panknin, N. Schirawski, Differential cross sections for reactive and nonreactive scattering of electronically excited Na from HF molecules. Chem. Phys. Lett. **143**, 45 (1988)
1087. R.T. Skodje et al., Observation of transition state resonance in the integral cross section of the F + HD reaction. J. Chem. Phys. **112**, 4536 (2000)
1088. J.J. Harkin, R.D. Jarris, D.J. Smith, R. Grice, Reactive scattering of a supersonic fluorine-atom beam F + C_2H_5I; C_3H_7I; $(CH_3)_2CHI$. Mol. Phys. **71**, 323 (1990)
1089. J. Grosser, O. Hoffmann, S. Klose, F. Rebentrost, Optical excitation of collision pairs in crossed beams: determination of the NaKr $B\,^2\Sigma$-potential. Europhys. Lett. **39**, 147 (1997)
1090. C. Cohen-Tannoudji, S. Reynaud, Dressed-atom description of absorption spectra of a multilevel atom in an intense laser beam. J. Phys. B **10**, 345 (1977)
1091. S. Keynaud, C. Cohen-Tannoudji, Collisional effects in resonance fluorescence, in *Laser Spectroscopy V*, ed. by A.R.W. McKellar, T. Oka, B.P. Stoicheff. Springer Ser. Opt. Sci., vol. 30 (Springer, Berlin, 1981), p. 166
1092. S.E. Harris, R.W. Falcone, W.R. Green, P.B. Lidow, J.C. White, J.F. Young, Laser-induced collisions, in *Tunable Lasers and Applications*, ed. by A. Mooradian, T. Jaeger, P. Stockseth. Springer Ser. Opt. Sci., vol. 3 (Springer, Berlin, 1976), p. 193
1093. S.E. Harris, J.F. Young, W.R. Green, R.W. Falcone, J. Lukasik, J.C. White, J.R. Willison, M.D. Wright, G.A. Zdasiuk, Laser-induced collisional and radiative energy transfer, in *Laser Spectroscopy IV*, ed. by H. Walther, K.W. Rothe. Springer Ser. Opt. Sci., vol. 21 (Springer, Berlin, 1979), p. 349
1094. E. Giacobino, P.R. Berman, Cooling of vapors using collisionally aided radiative excitation. NBD special publication No. 653, US Dept. Commerce, Washington, DC, 1983
1095. A. Gallagher, T. Holstein, Collision-induced absorption in atomic electronic transitions. Phys. Rev. A **16**, 2413 (1977)
1096. A. Birnbaum, L. Frommhold, G.C. Tabisz, Collision-induced spectroscopy: absorption and light scattering, in *Spectral Line Shapes 5*, ed. by J. Szudy (Ossolineum, Wroclaw, 1989), p. 623;
A. Bonysow, L. Frommhold, Collision-induced light scattering. Adv. Chem. Phys. **35**, 439 (1989)
1097. F. Dorsch, S. Geltman, P.E. Toschek, Laser-induced collisional and energy in thermal collisions of lithium and strontium. Phys. Rev. A **37**, 2441 (1988)
1098. C. Figl, J. Grosser, O. Hoffmann, F. Rebentrost, Repulsive K–Ar potentials from differential optical collisions. J. Phys. B, At. Mol. Opt. Phys. **37**, 3369 (2004)
1099. F. Rebentrost, C. Figl, R. Goldmann, O. Hoffmann, D. Spelsberg, J. Grosser, Nonadiabatic electron dynamics in the exit channel of Na-molecule optical collisions. J. Chem. Phys. **128**, 224307 (2008)
1100. K. Burnett, Spectroscopy of collision complexes, in *Electronic and Atomic Collisions*, ed. by J. Eichler, I.V. Hertel, N. Stoltenfoth (Elsevier, Amsterdam, 1984), p. 649
1101. N.K. Rahman, C. Guidotti (eds.), *Photon-Associated Collisions and Related Topics* (Harwood, Chur, 1982)
1102. J. Szudy, W.E. Baylis, Profiles of linewings and rainbow satellites associated with optical and radiative collisions. Phys. Rep. **266**, 127 (1996)

1103. E.W. McDaniel, E.J. Mansky, Guide to bibliographies, books, reviews and compendia of data on atomic collisions. Adv. At. Mol. Opt. Phys. **33**, 389 (1994)
1104. G. Boato, G.G. Volpi, Experiments on the dynamics of molecular processes: a chronicle of fifty years. Annu. Rev. Phys. Chem. **50**, 23 (1999)

Chapter 9

1105. J.L. Hall, Some remarks on the interaction between precision physical measurements and fundamental physical theories, in *Quantum Optics, Experimental Gravity and Measurement Theory*, ed. by P. Meystre, M.V. Scully (Plenum, New York, 1983)
1106. A.I. Miller, *Albert Einstein's Special Theory of Relativity* (Addison-Wesley, Reading, 1981);
J.L. Heilbron, *Max Planck* (Hirzel, Stuttgart, 1988)
1107. H.G. Kuhn, *Atomic Spectra*, 2nd edn. (Longman, London, 1971);
I.I. Sobelman, *Atomic Spectra and Radiative Transitions*, 2nd edn. Springer Ser. Atoms Plasmas, vol. 12 (Springer, Berlin, 1992)
1108. W.E. Lamb Jr., R.C. Retherford, Fine-structure of the hydrogen atom by a microwave method. Phys. Rev. **72**, 241 (1947); and **79**, 549 (1959)
1109. C. Salomon, J. Dalibard, W.D. Phillips, A. Clairon, S. Guellati, Laser cooling of cesium atoms below 3 μK. Europhys. Lett. **12**, 683 (1990)
1110. H.J. Metcalf, P. van der Straaten, *Laser Cooling and Trapping* (Springer, Berlin, 1999)
1111. K. Sengstock, W. Ertmer, Laser manipulation of atoms. Adv. At. Mol. Opt. Phys. **35**, 1 (1995)
1112. H. Frauenfelder, *The Mössbauer Effect* (Benjamin, New York, 1963);
U. Gonser (ed.), *Mössbauer Spectroscopy*. Topics Appl. Phys., vol. 5 (Springer, Berlin, 1975)
1113. J.L. Hall, Sub-Doppler spectroscopy: methane hyperfine spectroscopy and the ultimate resolution limit, in *Laser Spectroscopy II*, ed. by S. Haroche, J.C. Pebay-Peyroula, T.W. Hänsch, S.E. Harris. Lecture Notes Phys., vol. 43 (Springer, Berlin, 1975), p. 105
1114. C.H. Bordé, Progress in understanding sub-Doppler-line shapes, in *Laser Spectroscopy III*, ed. by J.L. Hall, J.L. Carlsten. Springer Ser. Opt. Sci., vol. 7 (Springer, Berlin, 1977), p. 121
1115. S.N. Bagayev, A.E. Baklanov, V.P. Chebotayev, A.S. Dychkov, P.V. Pokuson, Superhigh resolution laser spectroscopy with cold particles, in *Laser Spectroscopy VIII*, ed. by W. Pearson, S. Svanberg. Springer Ser. Opt. Sci., vol. 55 (Springer, Berlin, 1987), p. 95
1116. J.C. Berquist, R.L. Barger, D.L. Glaze, High resolution spectroscopy of calcium atoms, in *Laser Spectroscopy IV*, ed. by H. Walther, K.W. Rothe. Springer Ser. Opt. Sci., vol. 21 (Springer, Berlin, 1979), p. 120; and Appl. Phys. Lett. **34**, 850 (1979)
1117. U. Sterr, K. Sengstock, J.H. Müller, D. Bettermann, W. Ertmer, The magnesium Ramsey interferometer: applications and prospects. Appl. Phys. B **54**, 341 (1992)
1118. B. Bobin, C. Bordé, C. Breaut, Vibration-rotation molecular constants for the ground state of SF_6 from saturated absorption spectroscopy. J. Mol. Spectrosc. **121**, 91 (1987)
1119. E.A. Curtis, C.W. Oates, L. Hollberg, Observation of Large Atomic Recoil-Induced Asymmetrics in Cold Atomic Spectroscopy. J. Opt. Soc. Am. B **20**, 977 (2003)
1120. T.W. Hänsch, A.L. Schawlow, Cooling of gases by laser radiation. Opt. Commun. **13**, 68 (1975)
1121. W. Ertmer, R. Blatt, J.L. Hall, Some candidate atoms and ions for frequency standards research using laser radiative cooling techniques. Prog. Quantum Electron. **8**, 249 (1984)
1122. W. Ertmer, R. Blatt, J.L. Hall, M. Zhu, Laser manipulation of atomic beam velocities: demonstration of stopped atoms and velocity reversal. Phys. Rev. Lett. **54**, 996 (1985)
1123. R. Blatt, W. Ertmer, J.L. Hall, Cooling of an atomic beam with frequency-sweep techniques. Prog. Quantum Electron. **8**, 237 (1984)

1124. W.O. Phillips, J.V. Prodan, H.J. Metcalf, Neutral atomic beam cooling, experiments at NBS, in *NBS Special Publication*, No. 653 (US Dept. of Commerce, Washington, 1983); and Phys. Lett. **49**, 1149 (1982)
1125. H. Metcalf, Laser cooling and magnetic trapping of neutral atoms, in *Methods of Laser Spectroscopy*, ed. by Y. Prior, A. Ben-Reuven, M. Rosenbluth (Plenum, New York, 1986), p. 33
1126. J.V. Prodan, W.O. Phillips, Chirping the light-fantastic? in *Laser Cooled and Trapped Atoms*. NBS Special Publication, No. 653 (US Dept. Commerce, Washington, 1983)
1127. D. Sesko, C.G. Fam, C.E. Wieman, Production of a cold atomic vapor using diode-laser cooling. J. Opt. Soc. Am. B **5**, 1225 (1988)
1128. R.N. Watts, C.E. Wieman, Manipulating atomic velocities using diode lasers. Opt. Lett. **11**, 291 (1986)
1129. B. Sheeby, S.Q. Shang, R. Watts, S. Hatamian, H. Metcalf, Diode laser deceleration and collimation of a rubidium beam. J. Opt. Soc. Am. B **6**, 2165 (1989)
1130. H. Metcalf, Magneto-optical trapping and its application to helium metastables. J. Opt. Soc. Am. B **6**, 2206 (1989)
1131. I.C.M. Littler, St. Balle, K. Bergmann, The CW modeless laser: spectral control, performance data and build-up dynamics. Opt. Commun. **88**, 514 (1992)
1132. J. Hoffnagle, Proposal for continuous white-light cooling of an atomic beam. Opt. Lett. **13**, 307 (1991)
1133. I.C.M. Littler, H.M. Keller, U. Gaubatz, K. Bergmann, Velocity control and cooling of an atomic beam using a modeless laser. Z. Phys. D **18**, 307 (1991)
1134. R. Schieder, H. Walther, L. Wöste, Atomic beam deflection by the light of a tunable dye laser. Opt. Commun. **5**, 337 (1972)
1135. I. Nebenzahl, A. Szöke, Deflection of atomic beams by resonance radiation using stimulated emission. Appl. Phys. Lett. **25**, 327 (1974)
1136. J. Nellesen, J.M. Müller, K. Sengstock, W. Ertmer, Large-angle beam deflection of a laser cooled sodium beam. J. Opt. Soc. Am. B **6**, 2149 (1989)
1137. S. Villani (ed.), *Uranium Enrichment*. Topics Appl. Phys., vol. 35 (Springer, Berlin, 1979)
1138. C.E. Tanner, B.P. Masterson, C.E. Wieman, Atomic beam collimation using a laser diode with a self-locking power buildup-cavity. Opt. Lett. **13**, 357 (1988)
1139. J. Dalibard, C. Salomon, A. Aspect, H. Metcalf, A. Heidmann, C. Cohen-Tannoudji, Atomic motion in a standing wave, in *Laser Spectroscopy VIII*, ed. by S. Svanberg, W. Persson. Springer Ser. Opt. Sci., vol. 55 (Springer, Berlin, 1987), p. 81
1140. St. Chu, J.E. Bjorkholm, A. Ashkin, L. Holberg, A. Cable, Cooling and trapping of atoms with laser light, in *Methods of Laser Spectroscopy*, ed. by Y. Prior, A. Ben-Reuven, M. Rosenbluth (Plenum, New York, 1986), p. 41
1141. T. Baba, I. Waki, Cooling and mass analysis of molecules using laser-cooled atoms. Jpn. J. Appl. Phys. **35**, 1134 (1996)
1142. J.T. Bahns, P.L. Gould, W.C. Stwalley, Formation of Cold ($T < 1$ K) Molecules. Adv. At. Mol. Opt. Phys. **42**, 171 (2000)
1143. W.C. Stwalley, Making molecules at microKelvin, in *Atomic and Molecular Beams*, ed. by R. Campargue (Springer, Berlin, 2001), p. 105
1144. P. Pillet, F. Masnou-Seeuws, A. Crubelier, Molecular photoassociation and ultracold molecules, in *Atomic and Molecular Beams*, ed. by R. Campargue (Springer, Berlin, 2001), p. 113
1145. J.M. Doyle, B. Friedrich, J. Kim, D. Patterson, Buffer-gas loading of atoms and molecules into a magnetic trap. Phys. Rev. A **52**, R2515 (1995)
1146. E. Lusovoj, J.P. Toennies, S. Grebenev et al., Spectroscopy of molecules and unique clusters in superfluid He-droplets, in *Atomic and Molecular Beams*, ed. by R. Campargue (Springer, Berlin, 2001), p. 775
1147. S. Grebenev, M. Hartmann, M. Havenith, B. Sartakov, J.P. Toennies, A.F. Vilesov, The rotational spectrum of single OCS molecules in liquid ^{4}He droplets. J. Chem. Phys. **112**, 4485 (2000)

1148. K. von Haeften, S. Rudolph, I. Simanowski, M. Havenith, R.E. Zillich, K.B. Whaley, Probing phonon-rotation coupling in He-nano droplets: Infrared spectra of CO. Phys. Rev. B **73**, 054502 (2006)
1149. S. Rudolph, G. Wollny, K. von Haeften, M. Havenith, Probing collective excitations in He-nano-droplets observation of phonon wings in the infrared spectrum of methane. J. Chem. Phys. **126**, 124318 (2007)
1150. V.S. Letokhov, V.G. Minogin, B.D. Pavlik, Cooling and capture of atoms and molecules by a resonant light field. Sov. Phys. JETP **45**, 698 (1977); and Opt. Commun. **19**, 72 (1976)
1151. V.S. Letokhov, B.D. Pavlik, Spectral line narrowing in a gas by atoms trapped in a standing light wave. Appl. Phys. **9**, 229 (1976)
1152. A. Ashkin, J.P. Gordon, Cooling and trapping of atoms by resonance radiation pressure. Opt. Lett. **4**, 161 (1979)
1153. J.P. Gordon, Radiation forces and momenta in dielectric media. Phys. Rev. A **8**, 14 (1973)
1154. M.H. Mittelman, *Introduction to the Theory of Laser-Atom Interaction* (Plenum, New York, 1982)
1155. J.E. Bjorkholm, R.R. Freeman, A. Ashkin, D.B. Pearson, Transverse resonance radiation pressure on atomic beams and the influence of fluctuations, in *Laser Spectroscopy IV*, ed. by H. Walther, K.W. Rothe. Springer Ser. Opt. Sci., vol. 21 (Springer, Berlin, 1979), p. 49
1156. R. Grimm, M. Weidemüller, Y.B. Ovchinnikov, Optical dipole traps for neutral atoms. Adv. At. Mol. Opt. Phys. **42**, 95 (2000)
1157. D.E. Pritchard, E.L. Raab, V. Bagnato, C.E. Wieman, R.N. Watts, Light traps using spontaneous forces. Phys. Rev. Lett. **57**, 310 (1986);
H. Metcalf, Magneto-optical trapping and its application to helium metastables. J. Opt. Soc. Am. B **6**, 2206 (1989)
1158. (a) J. Nellessen, J. Werner, W. Ertmer, Magneto-optical compression of a monoenergetic sodium atomic beam. Opt. Commun. **78**, 300 (1990);
(b) C. Monroe, W. Swann, H. Robinson, C. Wieman, Very cold trapped atoms in a vapor cell. Phys. Rev. Lett. **65**, 1571 (1990)
1159. W. Phillips, Nobel Lecture, Stockholm, 1995
1160. A.M. Steane, M. Chowdhury, C.J. Foot, Radiation force in the magneto-optical trap. J. Opt. Soc. Am. B **9**, 2142 (1992)
1161. J. Söding et al., Gravitational laser trap for atoms with evanescent wave cooling. Opt. Commun. **119**, 652 (1995);
Ochinnikov et al., Surface trap for Cs-atoms based on evanescent wave cooling. Phys. Rev. Lett. **79**, 2225 (1997)
1162. A. Shevchenko, Atom traps on an evanescent wave mirror. PhD thesis, Helsinki University of Technology, 2004
1163. M.R. de Saint Vincent, J.P. Brantut, Ch. Borde, A. Aspect, T. Bourdel, P. Bouye, A quantum trampoline for ultracold atoms. Europhys. Lett. **89**, 10002 (2010)
1164. W. Hensel, PhD thesis, Faculty of Physics, LMU, München, 2000
1165. W.D. Phillips, Laser cooling and trapping of neutral atoms. Rev. Mod. Phys. **70**, 721 (1998)
1166. *Feynman Lectures on Physics I* (Addison-Wesley, Reading, 1965)
1167. S. Stenholm, The semiclassical theory of laser cooling. Rev. Mod. Phys. **58**, 699 (1986)
1168. A. Aspect, E. Arimondo, R. Kaiser, N. Vansteenkiste, C. Cohen-Tannoudji, Laser cooling below the one-photon recoil energy by velocity-selective coherent population trapping. J. Opt. Soc. Am. B **6**, 2112 (1989)
1169. J. Dalibard, C. Cohen-Tannoudji, Laser cooling below the Doppler limit by polarization gradients: simple theoretical model. J. Opt. Soc. Am. B **6**, 2023 (1989);
C. Cohen-Tannoudji, New laser cooling mechanism, in *Laser Manipulation of Atoms and Ions*, ed. by A. Arimondo, W.D. Phillips, F. Strumia (North-Holland, Amsterdam, 1992), p. 99
1170. D.S. Weiss, E. Riis, Y. Shery, P.J. Ungar, St. Chu, Optical molasses and multilevel atoms: experiment. J. Opt. Soc. Am. B **6**, 2072 (1989)

1171. P.J. Ungar, D.S. Weiss, E. Riis, St. Chu, Optical molasses and multilevel atoms: theory. J. Opt. Soc. Am. B **6**, 2058 (1989)
1172. C. Cohen-Tannoudji, W.D. Phillips, New mechanisms for laser cooling. Phys. Today **43**, 33 (1990)
1173. S. Chu, C. Wieman (eds.), Laser Cooling and Trapping. J. Opt. Soc. Am. B **6** (1989)
1174. E. Arimondo, W.D. Phillips, F. Strumia (eds.), *Laser Manipulation of Atoms and Ions, Varenna Summerschool, 1991* (North-Holland, Amsterdam, 1992)
1175. W. Phillips (ed.), *Laser-cooled and Trapped Ions*. Spec. Publ., No. 653 (National Bureau of Standards, Washington, 1984);
E. Arimondo, W. Phillips, F. Strumia (eds.), *Laser Cooling and Trapping of Neutral Atoms*. Proc. Int. School of Physics Enrico Fermi, Course CXVIII (North Holland, Amsterdam, 1992)
1176. S. Martelucci (ed.), *Bose–Einstein Condensates and Atom Laser* (Kluwer Academic, New York, 2000);
A. Griffin, D.W. Snoke, S. Stringari (eds.), *Bose–Einstein Condensation* (Cambridge Univ. Press, Cambridge, 1995)
1177. W. Ketterle, N.J. van Druten, Evaporative Cooling. Adv. At. Mol. Opt. Phys. **37**, 181 (1996)
1178. W. Ketterle, N.J. van Druten, Evaporative cooling of trapped atoms. Adv. At. Mol. Opt. Phys. **37**, 181 (1996)
1179. I. Bloch, T.W. Hänsch, T. Esslinger, Atom Laser with a cw output coupler. Phys. Rev. Lett. **82**, 3008 (1999);
S. Martelucci, A.N. Chester, A. Aspect, M. Inguscio (eds), *Bose–Einstein Condensates and Atom Lasers* (Springer, Berlin, 2000)
1180. P. Julienne, Cold binary collisions in a Light Field. J. Res. Natl. Inst. Stand. Technol. **101**, 487 (1996)
1181. E. Tiesinga et al., A spectroscopic method for determination of scattering length of sodium atom collisions. J. Res. Natl. Inst. Stand. Technol. **101**, 505 (1996)
1182. S.E. Pollack et al., Collective Excitation of a Bose–Einstein condensate by modulation of the atomic scattering length. Phys. Rev. A **81**, 053627 (2010)
1183. W. Ketterle, The atom laser. http://cua.mit.edu/ketterle_group/projects_1997/atomlaser
1184. V.S. Letokhov, B.D. Pavlik, Spectral line narrowing in a gas by in a standing light wave. Appl. Phys. **9**, 229 (1976)
1185. E.A. Burt et al., Coherence, correlations, and collisions: what one learns about Bose-Einstein condensates from their decay. Phys. Rev. A **79**, 337 (1997)
1186. F. Schreck et al., Bose–Einstein condensation of ^{86}Sr. Phys. Rev. A **82**, 011608(R) (2010)
1187. I. Bloch, Th.W. Hänsch, T. Esslinger, Atom laser with cw output coupler. Phys. Rev. Lett. **82**, 3008 (1999)
1188. J. Klärs, J. Schmitt, T. Damm, F. Vewinger, M. Weitz, Bose–Einstein condensation of paraxial light. Proc. SPIE **8600**, 86000L (2013)
1189. W. Ketterle, Workshop on ultracold Fermi-gases, Levico, 2005
1190. W. Ketterle, J.-i. Shin, Fermi gases go with the superfluid flow. Phys. World **June**, 38 (2007)
1191. C.A. Regal, M. Greiner, D.S. Jin, Observation of resonance condensation of fermionic atom pairs. Phys. Rev. Lett. **92**, 040403 (2004)
1192. I. Bloch, Ultracold quantum gases in optical lattices. Nat. Phys. **1**, 23–30 (2005)
1193. I. Bloch, M. Greiner, Exploring quantum matter with ultracold atoms in optical lattices. Adv. At. Mol. Phys. **52**, 1–47 (2005)
1194. A. Crubellier, O. Dulieu, F. Masnou-Seeuws, H. Knöckel, E. Tiemann, Simple determination of scattering length using observed bound levels at the ground state asymptote. Europhys. J. D **6**, 211 (1999)
1195. S. Stellmer, M.K. Tey, R. Grimm, F. Schreck, Observation of Bose–Einstein condensation of molecules. Phys. Rev. A **82**, 041602 (2010)
1196. F. Lang, K. Winkler, C. Strauss, R. Grimm, J. Hecker Denschlag, Ultracold triplet molecules in the rovibrational ground state. Phys. Rev. Lett. **101**, 133005 (2008). arXiv:0809.0061

1197. K. Winkler, F. Lang, G. Thalhammer, P.v.d. Straten, R. Grimm, J. Hecker Denschlag, Coherent optical transfer of Feshbach molecules to a lower vibrational state. Phys. Rev. Lett. **98**, 043201 (2007). cond-mat/0611222
1198. S. Stellmer, M.K. Tey, R. Grimm, F. Schreck, Bose–Einstein condensation of ^{86}Sr. Phys. Rev. A **82**, 041602(R) (2010). arXiv:1009.1701
1199. S. Jochim et al., Bose–Einstein condensation of molecules. Science **301**, 1510 (2003); M. Bartenstein et al., Crossover from a molecular Bose–Einstein condensate to degenerate Fermi gas. Phys. Rev. Lett. **92**, 120401 (2004)
1200. M. Mark, T. Kraemer, J. Harbig, C. Chin, H.C. Nägerl, R. Grimm, Efficient creation of molecules from a cesium Bose–Einstein condensate. Europhys. Lett. **69**, 706 (2005)
1201. C. Chim et al., Observation of Feshbach-like resonances in collisions between ultra cold molecules. Phys. Rev. Lett. **94**, 123201 (2005)
1202. (a) M. Mark et al., Spectroscopy of ultra cold trapped Cs-Feshbach molecules. Phys. Rev. A **76**, 033610 (2007);
(b) J. Weiner, Advances in ultracold collisions. Adv. At. Mol. Opt. Phys. **35**, 332 (1995);
(c) T. Walker, P. Feng, Measurements of collisions between laser-cooled atoms. Adv. At. Mol. Opt. Phys. **34**, 125 (1994)
1203. M. Endres, M. Cheneau, T. Fukuhara, C. Weitenberg, P. Schauß, C. Gross, L. Mazza, M.C. Banuls, L. Pollet, I. Bloch, S. Kuhr, Single-site- and single-atom-resolved measurement of correlation functions. Appl. Phys. B (2013). doi:10.1007/s00340-013-5552-9
1204. M. Greiner, O. Mandel, T. Esslinger, T.W. Hänsch, I. Bloch, Quantum phase transition from a superfluid to a Mott insulator in a gas of ultracold atoms. Nature **415**(6867), 39–44 (2002). doi:10.1038/415039a
1205. M. Atala, M. Aidelsburger, M. Lohse, J.T. Barreiro, B. Paredes, I. Bloch, Observation of the Meissner effect with ultracold atoms in bosonic ladders. arXiv:1402.0819
1206. H. Weickenmeier, U. Diemer, W. Demtröder, M. Broyer, Hyperfine-interaction between the singlet and triplet ground states and Cs_2. Chem. Phys. Lett. **124**, 470 (1986)
1207. K. Rubin, M.S. Lubell, A proposed study of photon statistics in fluorescence through high resolution measurements of the transverse deflection of an atomic beam, in *Laser Cooled and Trapped Atoms*. NBS Special Publ., No. 653 (1983), p. 119
1208. Y.Z. Wang, W.G. Huang, Y.D. Cheng, L. Liu, Test of photon statistics by atomic beam deflection, in *Laser Spectroscopy VII*, ed. by T.W. Hänsch, Y.R. Shen. Springer Ser. Opt. Sci., vol. 49 (Springer, Berlin, 1985), p. 238
1209. V.M. Akulin, F.L. Kien, W.P. Schleich, Deflection of atoms by a quantum field. Phys. Rev. A **44**, R1462 (1991)
1210. W. Ertmer, S. Penselin, Cooled atomic beams for frequency standards. Metrologia **22**, 195 (1986);
C. Salomon, Laser cooling of atoms and ion trapping for frequency standards, in *Metrology at the Frontiers of Physics and Technology*, ed. by L. Crovini, T.J. Quinn (North-Holland, Amsterdam, 1992), p. 405
1211. J.L. Hall, M. Zhu, P. Buch, Prospects for using laser prepared atomic fountains for optical frequency standards applications. J. Opt. Soc. Am. B **6**, 2194 (1989)
1212. E.D. Commins, Electric dipole moments of leptons. Adv. At. Mol. Opt. Phys. **40**, 1 (1999)
1213. F.M.H. Crompfoets, H.L. Bethlem, R.T. Jongma, G. Meyer, A prototype storage ring for neutral molecules. Nature **411**, 174 (2001)
1214. B. Friedrich, Slowing of supersonically cooled atoms and molecules by time-varying nonresonant dipole forces. Phys. Rev. A **61**, 025403 (2000)
1215. W. Paul, M. Raether, Das elektrische Massenfilter. Z. Phys. **140**, 262 (1955);
W. Paul, Elektromagnetische Käfige für geladene und neutrale Teilchen. Phys. Bl. **46**, 227 (1990)
1216. E. Fischer, Die dreidimensionale Stabilisierung von Ladungsträgern in einem Vierpolfeld. Z. Phys. **156**, 1 (1959)
1217. G.H. Dehmelt, Radiofrequency spectroscopy of stored ions. Adv. At. Mol. Phys. **3**, 53 (1967); and **5**, 109 (1969)

1218. G. Werth, Storage of particles in a Paul trap. http://www.physik.uni-mainz.de/werth/nlinres/pe_eng.html;
K. Blaum, High accuracy mass spectroscopy with stored ions. Phys. Rep. **425**, 1 (2006);
F.G. Major et al., *Physics and Techniques of Charged Particle Traps*. Springer Ser. Atom. Optical and Plasma Physics, vol. 37 (Springer, Berlin, 2005);
G. Werth, H. Häffner, W. Quint, Continuous Stern–Gerlach effect on atomic ions. Adv. At. Mol. Opt. Phys. **48**, 191 (2002);
F. Galve, G. Werth, Motional frequencies in a planar Penning trap. Hyperf. Interact. **174**, 41 (2007)
1219. J.F. Todd, R.E. March, *Quadrupole Ion Trap*, 2nd edn. (Wiley, New York, 2005)
1220. R.E. Drullinger, D.J. Wineland, Laser cooling of ions bound to a penning trap. Phys. Rev. Lett. **40**, 1639 (1978)
1221. E.T. Whittacker, S.N. Watson, *A Course of Modern Analysis* (Cambridge Univ. Press, Cambridge, 1963);
J. Meixner, F.W. Schaefke, *Mathieusche Funktionen und Sphäroidfunktionen* (Springer, Berlin, 1954)
1222. P.E. Toschek, W. Neuhauser, Spectroscopy on localized and cooled ions, in *Atomic Physics*, vol. 7, ed. by D. Kleppner, F.M. Pipkin (Plenum, New York, 1981)
1223. W. Neuhauser, M. Hohenstatt, P.E. Toschek, H.G. Dehmelt, Visual observation and optical cooling of electrodynamically contained ions. Appl. Phys. **17**, 123 (1978)
1224. Y. Stalgies, I. Siemens, B. Appasamy, T. Altevogt, P.E. Tuschek, The spectrum of single-atom resonances fluorescence. Europhys. Lett. **35**, 259 (1996)
1225. P.E. Toschek, W. Neuhauser, Einzelne Ionen für die Doppler-freie Spektroskopie. Phys. Bl. **36**, 1798 (1980)
1226. T. Sauter, H. Gilhaus, W. Neuhauser, R. Blatt, P.E. Toschek, Kinetics of a single trapped ion. Europhys. Lett. **7**, 317 (1988)
1227. R.E. Drullinger, D.J. Wineland, Laser cooling of ions bound to a Penning trap, in *Laser Spectroscopy IV*, ed. by H. Walther, K.W. Rother. Springer Ser. Opt. Sci., vol. 21 (Springer, Berlin, 1979), p. 66; and Phys. Rev. Lett. **40**, 1639 (1978)
1228. D.J. Wineland, W.M. Itano, Laser cooling of atoms. Phys. Rev. A **20**, 1521 (1979)
1229. W. Neuhauser, M. Hohenstatt, P.E. Toschek, H. Dehmelt, Optical sideband cooling of visible atom cloud confined in a parabolic well. Phys. Rev. Lett. **41**, 233 (1978)
1230. H.G. Dehmelt, Proposed $10^{14}\Delta\nu < \nu$ laser fluorescence spectroscopy on a $\mathrm{T1}^+$ mono-ion oscillator. Bull. Am. Phys. **20**, 60 (1975)
1231. P.E. Toschek, Absorption by the numbers: recent experiments with single trapped and cooled ions. Phys. Scr. T **23**, 170 (1988)
1232. T. Sauter, R. Blatt, W. Neuhauser, P.E. Toschek, Quantum jumps in a single ion. Phys. Scr. **22**, 128 (1988); and Opt. Commun. **60**, 287 (1986)
1233. W.M. Itano, J.C. Bergquist, R.G. Hulet, D.J. Wineland, The observation of quantum jumps in Hg^+, in *Laser Spectroscopy VIII*, ed. by S. Svanberg, W. Persson. Springer Ser. Opt. Sci., vol. 55 (Springer, Berlin, 1987), p. 117
1234. F. Diedrich, H. Walther, Nonclassical radiation of a single stored ion. Phys. Rev. Lett. **58**, 203 (1987)
1235. R. Blümel, J.M. Chen, E. Peik, W. Quint, W. Schleich, Y.R. Chen, H. Walther, Phase transitions of stored laser-cooled ions. Nature **334**, 309 (1988)
1236. F. Diedrich, E. Peik, J.M. Chen, W. Quint, H. Walther, Ionenkristalle und Phasenübergänge in einer Ionenfalle. Phys. Bl. **44**, 12 (1988)
1237. F. Diedrich, E. Peik, J.M. Chen, W. Quint, H. Walther, Observation of a phase transition of stored laser-cooled ions. Phys. Rev. Lett. **59**, 2931 (1987)
1238. R. Blümel, C. Kappler, W. Quint, H. Walther, Chaos and order of laser-cooled ions in a Paul trap. Phys. Rev. A **40**, 808 (1989)
1239. J. Javamainen, Laser cooling of trapped ion-clusters. J. Opt. Soc. Am. B **5**, 73 (1988)
1240. D.J. Wineland, J.C. Bergquist, W.M. Itano, J.J. Bollinger, C.H. Manney, Atomic-ion Coulomb clusters in an ion trap. Phys. Rev. Lett. **59**, 2935 (1987);

W. Quint, Chaos und Ordnung von lasergekühlten Ionen in einer Paulfalle. Dissertation, MPQ-Berichte 150, MPQ für Quantenoptik, Garching, 1990
1241. J.N. Tan, J.J. Bollinger, B. Jelenkovic, D.J. Wineland, Long-Range Order in Laser-Cooled, Atomic-Ion Wigner Crystals Observed by Bragg-Scattering. Phys. Rev. Lett. **75**, 4198 (1995)
1242. Th.V. Kühl, Storage ring laser spectroscopy. Adv. At. Mol. Opt. Phys. **40**, 113 (1999)
1243. H. Poth, Applications of electron cooling in atomic, nuclear and high energy physics. Nature **345**, 399 (1990)
1244. (a) J.P. Schiffer, Layered structure in condensed cold one-component plasma confined in external fields. Phys. Rev. Lett. **61**, 1843 (1988);
(b) I. Waki, S. Kassner, G. Birkl, H. Walther, Observation of ordered structures of laser-cooled ions in a quadrupole storage ring. Phys. Rev. Lett. **68**, 2007 (1992)
1245. J.S. Hangst, M. Kristensen, J.S. Nielsen, O. Poulsen, J.P. Schiffer, P. Shi, Laser cooling of a stored ion beam to 1 mK. Phys. Rev. Lett. **67**, 1238 (1991)
1246. U. Schramm et al., Observation of laser-induced recombination in merged electron and proton beams. Phys. Rev. Lett. **67**, 22 (1991)
1247. P. Schindler, D. Nigg, T. Monz, J.T. Barreiro, E. Martinez, S.X. Wang, S. Quint, M.F. Brandl, V. Nebendahl, C.F. Roos, M. Chwalla, M. Hennrich, R. Blatt, A quantum information processor with trapped ions. New J. Phys. **15**, 123012 (2013). arXiv:1308.3096;
R. Blatt, Ionen in Reih und Glied. Quantencomputer mit gespeicherten Ionen. Phys. J. **4**(11), 37 (2005)
1248. J.I. Cirac, P. Zoller, Qubits, Gatter and register. Phys. J. **4**(11), 30 (2005)
1249. B. Lanyon et al., Universal digital quantum simulations with trapped ions. Science **334**, 57 (2011)
1250. J. Stolze, D. Suter, *Quantum Computing: A Short Course from Theory to Experiment* (Wiley/VCH, Weinheim, 2008)
1251. B.P. Lanyon, P. Jurcevic, M. Zwerger, C. Hempel, E.A. Martinez, W. Dür, H.J. Briegel, R. Blatt, C.F. Roos, Measurement-based quantum computation with trapped ions. Phys. Rev. Lett. **111**, 210501 (2013). arXiv:1308.5102
1252. J.S. Pedernales, R. Di Candia, P. Schindler, T. Monz, M. Hennrich, J. Casanova, E. Solano, Entanglement measures in ion-trap quantum simulators without full tomography. arXiv:1402.4409
1253. T.C. English, J.C. Zorn, Molecular beam spectroscopy, in *Methods of Experimental Physics*, vol. 3, ed. by D. Williams (Academic Press, New York, 1974)
1254. I.I. Rabi, Zur Methode der Ablenkung von Molekularstrahlen. Z. Phys. **54**, 190 (1929)
1255. N.F. Ramsey, *Molecular Beams*, 2nd edn. (Clarendon, Oxford, 1989)
1256. J.C. Bergquist, S.A. Lee, J.L. Hall, Ramsey fringes in saturation spectroscopy, in *Laser Spectroscopy III*, ed. by J.L. Hall, J.L. Carlsten (Springer, Berlin, 1977)
1257. Y.V. Baklanov, B.Y. Dubetsky, V.P. Chebotayev, Nonlinear Ramsey resonance in the optical region. Appl. Phys. **9**, 171 (1976)
1258. V.P. Chebotayev, The method of separated optical fields for two level atoms. Appl. Phys. **15**, 219 (1978)
1259. C. Bordé, Sur les franges de Ramsey en spectroscopie sans élargissement Doppler. C.R. Acad. Sci., Sér. B **282**, 101 (1977)
1260. S.A. Lee, J. Helmcke, J.L. Hall, P. Stoicheff, Doppler-free two-photon transitions to Rydberg levels. Opt. Lett. **3**, 141 (1978)
1261. S.A. Lee, J. Helmcke, J.L. Hall, High-resolution two-photon spectroscopy of Rb Rydberg levels, in *Laser Spectroscopy IV*, ed. by H. Walther, K.W. Rothe. Springer Ser. Opt. Sci., vol. 21 (Springer, Berlin, 1979), p. 130
1262. Y.V. Baklanov, V.P. Chebotayev, B.Y. Dubetsky, The resonance of two-photon absorption in separated optical fields. Appl. Phys. **11**, 201 (1976)
1263. S.N. Bagayev, V.P. Chebotayev, A.S. Dychkov, Continuous coherent radiation in methane at $\lambda = 3.39$ μm in spatially separated fields. Appl. Phys. **15**, 209 (1978)

1264. C.J. Bordé, Density matrix equations and diagrams for high resolution nonlinear laser spectroscopy: application to Ramsey fringes in the optical domain, in *Advances in Laser Spectroscopy*, ed. by F.T. Arrecchi, F. Strumia, H. Walther (Plenum, New York, 1983), p. 1
1265. J.C. Bergquist, S.A. Lee, J.L. Hall, Saturated absorption with spatially separated laser fields. Phys. Rev. Lett. **38**, 159 (1977)
1266. J. Helmcke, D. Zevgolis, B.U. Yen, Observation of high contrast ultra narrow optical Ramsey fringes in saturated absorption utilizing four interaction zones of travelling waves. Appl. Phys. B **28**, 83 (1982)
1267. C.J. Bordé, C. Salomon, S.A. Avrillier, A. Van Lerberghe, C. Breant, D. Bassi, G. Scoles, Optical Ramsey fringes with travelling waves. Phys. Rev. A **30**, 1836 (1984)
1268. J.C. Bergquist, R.L. Barger, P.J. Glaze, High resolution spectroscopy of calcium atoms, in *Laser Spectroscopy IV*, ed. by H. Walther, K.W. Rothe. Springer Ser. Opt. Sci., vol. 21 (Springer, Berlin, 1979), p. 120
1269. J. Helmcke, J. Ishikawa, F. Riehle, High contrast high resolution single component Ramsey fringes in Ca, in *Frequency Standards and Metrology*, ed. by A. De Marchi (Springer, Berlin, 1989), p. 270
1270. A. Huber, B. Gross, M. Weitz, Th.W. Hänsch, Two-photon optical Ramsey spectroscopy of the $1S$–$2S$ transition in atomic hydrogen. Phys. Rev. A **58**, R2631 (1998);
B. Gross, A. Huber, M. Niering, M. Weitz, T.W. Hänsch, Optical Ramsey spectroscopy of atomic hydrogen. Europhys. Lett. **44**, 186
1271. M. Inguscio, L. Fallini, *Atomic Physics: Precise Measurements and Ultracold Matter* (Oxford University Press, Oxford, 2013)
1272. F. Riehle, *Frequency standards: Basics and Applications* (Wiley/VCH, Weinheim, 2005)
1273. M.A. Kasevich, *Atomic Interferometry in an Atomic Fountain* (Stanford Dep. of Appl. Phys., Stanford, 1992)
1274. A. Peters, K.Y. Chunb, S. Chu, High Precision gravity measurements using an atomic interferometer. Metrologia **38**, 25 (2001)
1275. J.B. Fixler, G.T. Foster, J.M. McGuirk, M.A. Kasevich, Atom Interferometer Measurement of the Newtonian Constant of Gravity. Science **315**, 74 (2007)
1276. J. Stuhler, M. Fattori, T. Petelski, E.M. Tino, MAGIA-using atom interferometry to determine the Newtonian gravitational constant. J. Opt. B, Quantum Semiclass. Opt. **5**, 75 (2003)
1277. F. Riehle, J. Ishikawa, J. Helmcke, Suppression of recoil component in nonlinear Doppler-free spectroscopy. Phys. Rev. Lett. **61**, 2092 (1988)
1278. J. Mlynek, V. Balykin, P. Meystere (guest eds.), Atom interferometry. Appl. Phys. B **54**, 319–368 (1992);
C.S. Adams, M. Siegel, J. Mlynek, Atom optics. Phys. Rep. **240**, 144 (1994)
1279. P. Bermann (ed.), *Atom Interferometry* (Academic Press, San Diego, 1997);
J. Arlt, G. Birkl, F.M. Rasel, W. Ertmer, Atom optics, guided atoms, and atom interferometry. Adv. At. Mol. Opt. Phys. **50**, 55–89 (2005)
1280. O. Carnal, J. Mlynek, Young's double slit experiment with atoms: a simple atom interferometer. Phys. Rev. Lett. **66**, 2689 (1991)
1281. D.W. Keith, C.R. Ekstrom, Q.A. Turchette, D.E. Pritchard, An interferometer for atoms. Phys. Rev. Lett. **66**, 2693 (1991)
1282. D.E. Pritchard, A.D. Cronin, S. Gupta, D.A. Kokorowski, Atom optics: old ideas, current technology, and new results. Ann. Phys. **10**, 35–54 (2001)
1283. J. Schmiedmayer, M.S. Chapman, C.R. Ekstrom, T.D. Hammond, D.W. Keith, A. Lenef, R.A. Rubenstein, E.T. Smith, D.E. Pritchard, Optics and interferometry with atoms and molecules. Adv. At. Mol. Phys. **37**(Suppl. 3), 407 (1997)
1284. http://www.physics.arizona.edu/~cronin/Research/Introduction/IFMdescription.html
1285. J. Arlt, G. Birkl, E. Rasel, W. Ertmer, Atom optics, guided atoms, and atom interferometry. Adv. At. Mol. Opt. Phys. **50**, 55–89 (2005)
1286. A.D. Cronin, J. Schmiedmayer, D.E. Pritchard, Optics and interferometry with atoms and molecules. Rev. Mod. Phys. **81**(3), 1051–1129 (2009). doi:10.1103/RevModPhys.81.1051

1287. A. Louchet-Chauvet et al., Comparison of 3 absolute gravimeters based on different methods for the e-MASS project. IEEE Trans. Instrum. Meas. **60**, 2527 (2011)
1288. H. Müntinga et al., Interferometry with Bose–Einstein condensates in microgravity. Phys. Rev. Lett. **110**, 093602 (2013)
1289. C.J. Bordé, Atomic interferometry with internal state labelling. Phys. Lett. A **140**, 10 (1989)
1290. F. Riehle, A. Witte, T. Kisters, J. Helmcke, Interferometry with Ca atoms. Appl. Phys. B **54**, 333 (1992)
1291. M. Kasevich, S. Chu, Measurement of the gravitational acceleration of an atom with a light-pulse atom interferometer. Appl. Phys. B **54**, 321 (1992)
1292. M.R. Andrews, C.G. Townsend, H.J. Miesner, D.S. Durfee, D.M. Kurn, W. Ketterle, Observation of interference between two Bose condensates. Science **275**, 637 (1997)
1293. S. Martellucci, A.N. Chester, A. Aspect, M. Inguscio (eds.), *Bose–Einstein Condensates and Atom Lasers* (Kluwer/Plenum, New York, 2000)
1294. F. Diedrich, J. Krause, G. Rempe, M.O. Scully, H. Walther, Laser experiments on single atoms and the test of basic physics. Physica B **151**, 247 (1988); and IEEE J. Quantum Electron. **24**, 1314 (1988)
1295. S. Haroche, J.M. Raimond, Radiative properties of Rydberg states in resonant cavities. Adv. At. Mol. Phys. **20**, 347 (1985)
1296. G. Rempe, H. Walther, The one-atom maser and cavity quantum electrodynamics, in *Methods of Laser Spectroscopy*, ed. by Y. Prior, A. Ben-Reuven, M. Rosenbluth (Plenum, New York, 1986)
1297. H. Walther, Single-atom oscillators. Europhys. News **19**, 105 (1988);
H. Walther, One atom maser and other experiments on cavity quantum electrodynamics. Phys. Usp. **39**, 727 (1996)
1298. G. Rempe, M.O. Scully, H. Walther, The one-atom maser and the generation of nonclassical light, in *Proc. ICAP 12, Ann Arbor* (1990)
1299. P. Meystre, G. Rempe, H. Walther, Very low temperature behaviour of a micromaser. Opt. Lett. **13**, 1078 (1988)
1300. G. Rempe, H. Walther, Sub-Poissonian atomic statistics in a micromaser. Phys. Rev. A **42**, 1650 (1990)
1301. B.T. Varcoe, S. Brattke, M. Weidinger, H. Walther, Preparing pure photon number states of the radiation field. Nature **403**, 743 (2000)
1302. M. Marrocco, M. Weidinger, R.T. Sang, H. Walther, Quantum electrodynamic shifts of Rydberg energy levels between two parallel plates. Phys. Rev. Lett. **81**, 5784 (1998)
1303. H. Metcalf, W. Phillips, Time resolved subnatural width spectroscopy. Opt. Lett. **5**, 540 (1980)
1304. J.N. Dodd, G.W. Series, Time-resolved fluorescence spectroscopy, in *Progr. Atomic Spectroscopy A*, ed. by W. Hanle, H. Kleinpoppen (Plenum, New York, 1978)
1305. S. Schenk, R.C. Hilburn, H. Metcalf, Time resolved fluorescence from Ba and Ca, excited by a pulsed tunable dye laser. Phys. Rev. Lett. **31**, 189 (1973)
1306. H. Figger, H. Walther, Optical resolution beyond the natural linewidth: a level crossing experiment on the $3\,^2P_{3/2}$ level of sodium using a tunable dye laser. Z. Phys. **267**, 1 (1974)
1307. F. Shimizu, K. Umezu, H. Takuma, Observation of subnatural linewidth in Na D_2-lines. Phys. Rev. Lett. **47**, 825 (1981)
1308. G. Bertuccelli, N. Beverini, M. Galli, M. Inguscio, F. Strumia, Subnatural coherence effects in saturation spectroscopy using a single travelling wave. Opt. Lett. **10**, 270 (1985)
1309. P. Meystre, M.O. Scully, H. Walther, Transient line narrowing: a laser spectroscopic technique yielding resolution beyond the natural linewidth. Opt. Commun. **33**, 153 (1980)
1310. A. Guzman, P. Meystre, M.O. Scully, Subnatural spectroscopy, in *Adv. Laser Spectroscopy*, ed. by F.T. Arecchi, F. Strumia, H. Walther (Plenum, New York, 1983), p. 465;
D.P. O'Brien, P. Meystre, H. Walther, Subnatural linewidth in atomic spectroscopy. Adv. At. Mol. Phys. **21**, 1 (1985)
1311. V.S. Letokhov, V.P. Chebotayev, *Nonlinear Laser Spectroscopy*. Springer Ser. Opt. Sci., vol. 4 (Springer, Berlin, 1977)

1312. R.P. Hackel, S. Ezekiel, Observation of subnatural linewidths by two-step resonant scattering in I_2-vapor. Phys. Rev. Lett. **42**, 1736 (1979)
1313. H. Weickenmeier, U. Diemer, W. Demtröder, M. Broyer, Hyperfine interaction between the singlet and triplet ground states of Cs_2. Chem. Phys. Lett. **124**, 470 (1986)
1314. E.R. Cohen, B.N. Taylor, The 1986 CODATA recommended values of the fundamental physical constants. J. Phys. Chem. Ref. Data **17**, 1795 (1988)
1315. F. Bayer-Helms, Neudefinition der Basiseinheit Meter im Jahr 1983. Phys. Bl. **39**, 307 (1983);
F. Bayer-Helms, Documents concerning the new definition of the metre. Metrologia **19**, 163 (1984)
1316. K.M. Baird, Frequency measurements of optical radiation. Phys. Today **36**, 1 (1983)
1317. K.M. Evenson, D.A. Jennings, F.R. Peterson, J.S. Wells, Laser frequency measurements: a review, limitations, extension to 197 THz (1.5 μm), in *Laser Spectroscopy III*, ed. by J.L. Hall, J.L. Carlsten. Springer Ser. Opt. Sci., vol. 7 (Springer, Berlin, 1977);
D.A. Jennings, F.R. Peterson, K.M. Evenson, Direct frequency measurement of the 260 THz (1.15 μm) ^{20}Ne laser: and beyond, in *Laser Spectroscopy IV*, ed. by H. Walther, K.W. Rothe. Springer Ser. Opt. Sci., vol. 21 (Springer, Berlin, 1979), p. 39
1318. K.M. Evenson, M. Inguscio, D.A. Jennings, Point contact diode at laser frequencies. J. Appl. Phys. **57**, 956 (1985)
1319. L.R. Zink, M. Prevedelli, K.M. Evenson, M. Inguscio, High resolution far infrared spectroscopy, in *Applied Laser Spectroscopy*, ed. by M. Inguscio, W. Demtröder (Plenum, New York, 1991), p. 141
1320. H.V. Daniel, B. Maurer, M. Steiner, A broadband Schottky point contact mixer for visible laser light and microwave harmonics. J. Appl. Phys. B **30**, 189 (1983)
1321. H.P. Roeser, R.V. Titz, G.W. Schwaab, M.F. Kimmit, Current-frequency characteristics of submicron GaAs Schottky barrier diodes with femtofarad capacitances. J. Appl. Phys. **72**, 3194 (1992)
1322. (a) B.G. Whitford, Phase-locked frequency chains to 130 THz at NRC, in *Frequency Standards and Metrology*, ed. by A. De Marchi (Springer, Berlin, 1989);
(b) T.W. Hänsch, High resolution spectroscopy of hydrogen, in *The Hydrogen Atom*, ed. by G.F. Bussani, M. Inguscio, T.W. Hänsch (Springer, Berlin, 1989);
(c) S.G. Karshenboim, F.S. Pavone, G.F. Bussani, M. Inguscio, T.W. Hänsch (eds.), *The Hydrogen Atom* (Springer, Berlin, 2001)
1323. J. Reichert, M. Niering, R. Holzwarth, M. Weitz, T. Udem, T.W. Hänsch, Phase coherent vacuum ultraviolet to radiofrequency comparison with a mode-locked laser. Phys. Rev. Lett. **84**, 3232 (2000)
1324. M. Stowe et al., Direct frequency comb spectroscopy. Adv. At. Mol. Opt. Phys. **55**, 1 (2008)
1325. R. Holzwarth et al., Optical frequency synthesizer for precision spectroscopy. Phys. Rev. Lett. **85**, 2264 (2000);
L.S. Ma et al., Optical frequency synthesis and comparison with uncertainty at the 10^{-19} level. Science **303**, 1843 (2004);
T.W. Hänsch, Frequency Comb project. http://www.mpq.de/~haensch/comb/research/combs.html
1326. J.L. Hall, Defining and measuring optical frequencies. Nobel lecture, December 2005. Available at http://nobelprize.org/physics/laureates/2005/hall-lecture.html;
M. Fischer et al., New limits on the drift of fundamental constants from laboratory measurements. Phys. Rev. Lett. **92**, 230802 (2004)
1327. S.A. Diddams, T.W. Hänsch et al., Direct link between microwave and optical frequencies with a 300 THz femtosecond pulse. Phys. Rev. Lett. **84**, 5102 (2000);
M.C. Stove et al., Direct frequency comb spectroscopy. Adv. At. Mol. Opt. Phys. **55**, 1 (2008)
1328. T. Udem, R. Holzwarth, T.W. Hänsch, Optical frequency metrology. Nature **416**, 233–237 (2002)

1329. B. Bernhardt et al., Vacuum ultraviolet frequency comb generated by a femtosecond enhancement cavity in the visible. Opt. Lett. **37**, 503 (2012)
1330. R. Loudon, *The Quantum Theory of Light* (Clarendon, Oxford, 1973)
1331. H. Gerhardt, H. Welling, A. Güttner, Measurements of laser linewidth due to quantum phase and quantum amplitude noise above and below threshold. Z. Phys. **253**, 113 (1972);
M. Zhu, J.L. Hall, Stabilization of optical phase/frequency of a laser system. J. Opt. Soc. Am. B **10**, 802 (1993)
1332. H.A. Bachor, P.J. Manson, Practical implications of quantum noise. J. Mod. Opt. **37**, 1727 (1990);
H.A. Bachor, P.T. Fisk, Quantum noise—a limit in photodetection. Appl. Phys. B **49**, 291 (1989)
1333. H.A. Bachor, *A Guide to Experiments in Quantum Optics* (Wiley/VCH, Weinheim, 1998)
1334. R.J. Glauber, Optical coherence and photon statistics, in *Quantum Optics and Electronics*, ed. by C. DeWitt, A. Blandia, C. Cohen-Tannoudji (Gordon & Breach, New York, 1965), p. 65;
J.D. Cresser, Theory of the spectrum of the quantized light field. Phys. Rep. **94**, 48 (1983);
H. Paul, Squeezed states—nichtklassische Zustände des Strahlungsfeldes. Laser Optoelektron. **19**, 45 (1987)
1335. R.E. Slusher, L.W. Holberg, B. Yorke, J.C. Mertz, J.F. Valley, Observation of squeezed states generated by four wave mixing in an optical cavity. Phys. Rev. Lett. **55**, 2409 (1985)
1336. M. Xiao, L.A. Wi, H.J. Kimble, Precision measurements beyond the shot noise limit. Phys. Rev. Lett. **59**, 278 (1987)
1337. H.J. Kimble, D.F. Walls (guest eds.), Feature issue on squeezed states of the electromagnetic field. J. Opt. Soc. Am. B **4**, 1449 (1987);
P. Kurz, R. Paschotta, K. Fiedler, J. Mlynek, Bright squeezed light by second harmonic generation and monolytic resonator. Europhys. Lett. **24**, 449 (1993)
1338. T. Steinmetz et al., Laser frequency combs for astronomical observations. Science **321**, 1335 (2008)
1339. P.R. Saulson, *Fundamentals of Gravitational Wave Detectors* (World Scientific, Singapore, 1995)
1340. T.M. Niebaum, A. Rüdiger, R. Schilling, L. Schnupp, W. Winkler, K. Danzmann, Pulsar search using data compression with the Garching gravitational wave detector. Phys. Rev. D **47**, 3106 (1993)
1341. B. Bernhardt et al., Cavity-enhanced dual comb spectroscopy. Nat. Photonics **4**, 55 (2010); and Appl. Phys. B **100**, 3 (2010);
B. Bernhardt, PhD thesis, LMU, München, 2007;
C.H. Gohle et al., Frequency comb vernier spectroscopy for broadband high resolution high sensitivity absorption and dispersion spectra. Phys. Rev Lett. **99**, 263902 (2007)
1342. T.W. Hänsch, N. Picqué, *Laser Spectroscopy and Frequency Combs ICOLS 2013*. J. Physics: Conference Series, vol. 467 (2013)
1343. M. Stowe et al., Direct frequency comb spectroscopy. Adv. At. Mol. Opt. Phys. **55**, 1 (2008)
1344. J. Jones, R. Kevin, D. Moll, M.J. Thorpe, J. Ye, Phase-coherent frequency combs in the vacuum ultraviolet via high-harmonic generation inside a femtosecond enhancement cavity. Phys. Rev. Lett. **94**, 193201 (2005)
1345. P.G. Blair (ed.), *The Detection of Gravitational Waves* (Cambridge Univ. Press, Cambridge, 1991)
1346. P.S. Saulson, *Fundamentals of Interferometric Gravitational Wave Detectors* (World Scientific, Singapore, 1994)
1347. K. Zaheen, M.S. Zubairy, Squeezed states of the radiation field. Adv. At. Mol. Phys. **28**, 143 (1991)
1348. H.J. Kimble, Squeezed states of light. Adv. Chem. Phys. **38**, 859 (1989)
1349. P. Tombesi, E.R. Pikes (eds.), *Squeezed and Nonclassical Light* (Plenum, New York, 1989)
1350. E. Giacobino, C. Fabry (guest eds.), Quantum noise reduction in optical systems. Appl. Phys. B **55**, 187–297 (1992)

1351. D.F. Walls, G.J. Milburn, *Quantum Optics*, study edn. (Springer, Berlin, 1995)
1352. H.A. Haus, *Electromagnetic Noise and Quantum Optical Measurements* (Springer, Berlin, 2000)
1353. H.J. Carmichael, R.J. Glauber, M.O. Scully (eds.), *Directions in Quantum Optics* (Springer, Berlin, 2001)
1354. P.R. Saulson, *Fundamentals of Interrefractive Gravitational Wave Detectors* (World Scientific, Singapore, 1994)
1355. K. Qu, G.S. Agaral, Ramsey spectroscopy using squeezed light. Opt. Lett. **38**, 2563 (2013)
1356. H. Vahlbruch, A. Khalaidovski, N. Lastzka, C. Gräf, K. Danzmann, R. Schnabel, The GEO600 squeezed light source. Class. Quantum Gravity **27**, 084027 (2010)
1357. http://www.geo600.org/1265074/Advanced_LIGO

Chapter 10

1358. A. Mooradian, T. Jaeger, P. Stokseth (eds.), *Tunable Lasers and Applications*. Springer Ser. Opt. Sci., vol. 3 (Springer, Berlin, 1976)
1359. C.T. Lin, A. Mooradian (eds.), *Lasers and Applications*. Springer Ser. Opt. Sci., vol. 26 (Springer, Berlin, 1981)
1360. J.F. Ready, R.K. Erf (eds.), *Lasers and Applications*, vols. 1–5 (Academic Press, New York, 1974–1984)
1361. S. Svanberg, *Atomic and Molecular Spectroscopy*, 2nd edn. Springer Ser. Atoms Plasmas, vol. 6 (Springer, Berlin, 1991);
W. Rettig, B. Strehmel, S. Schrader, H. Seifert (eds.), *Applied Fluorescence in Chemistry, Biology and Medicine* (Springer, Berlin, 2002)
1362. A.Y. Spasov (ed.), *Lasers: Physics and Applications*. Proc. 5th Int. School on Quantum Electronics, Sunny Beach, Bulgaria, 1988 (World Scientific, Singapore, 1989)
1363. H.D. Bist, *Advanced Laser Spectroscopy and Applications* (Allied Publishers, New Delhi, 1996)
1364. D. Andrews, *Applied Laser Spectroscopy* (Wiley, New York, 1992)
1365. C.B. Moore, *Chemical and Biochemical Applications of Lasers*, vols. 1–5 (Academic Press, New York, 1974–1984)
1366. D.K. Evans, *Laser Applications in Physical Chemistry* (Dekker, New York, 1989);
D.L. Andrews, *Lasers in Chemistry* (Springer, Berlin, 1986);
A.H. Zewail (ed.), *Advances in Laser Chemistry*. Springer Ser. Chem. Phys., vol. 3 (Springer, Berlin, 1978)
1367. G.R. van Hecke, K.K. Karukstis, *A Guide to Lasers in Chemistry* (Jones & Bartlett, Boston, 1997)
1368. R.T. Rizzo, A.B. Myers, *Laser Techniques in Chemistry* (Wiley, New York, 1995)
1369. H. Telle, A. Gonzales Ureña, R.J. Donovan, *Laser Chemistry: Spectroscopy, Dynamics and Applications* (Wiley, New York, 2007)
1370. J.W. Hepburn, R.E. Contunetti, M.A. Johnson, *Laser Techniques for State-Selected and State-to-State Chemistry*. SPIE Proc., vol. 3271 (1998)
1371. G. Schmidtke, W. Kohn, U. Klocke, M. Knothe, W.J. Riedel, H. Wolf, Diode laser spectrometer for monitoring up to five atmospheric trace gases in unattended operation. Appl. Opt. **28**, 3665 (1989)
1372. G.S. Hurst, M.P. Payne, S.P. Kramer, C.H. Cheng, Counting the atoms. Phys. Today **33**, 24 (1980)
1373. V.S. Letokhov, *Laser Photoionization Spectroscopy* (Academic Press, Orlando, 1987)
1374. P. Peuser, G. Herrmann, H. Rimke, P. Sattelberger, N. Trautmann, Trace detection of plutonium by three-step photoionization with a laser system pumped by a copper vapor laser. Appl. Phys. B **38**, 249 (1985)

1375. T. Whitaker, Isotopically selective laser measurements. Lasers Appl. **5**, 67 (1986)
1376. H. Kano, H.T.M. van der Voort, M. Schrader, G.M.P. van Kampen, S.W. Hell, Avalanche photodiode detection with object scanning and image restoration provides 2–4 fold resolution increase in two-photon fluorescence microscopy. Bioimaging **4**, 187 (1996)
1377. S.W. Hell, J. Wichmann, Breaking the diffraction resolution limit by stimulated emission. Opt. Lett. **19**, 780 (1994)
1378. H. Gugel et al., Cooperative 4Pi excitation yields sevenfold sharper optical sections in live-cell microscopy. Biophys. J. **87**, 4146 (2004)
1379. C. Bräuchle, G. Seisenberger, T. Endreß, M.V. Ried, H. Büning, M. Hallek, Single virus tracing: visualization of the infection pathway of a virus into a living cell. Chem. Phys. Chem. **3**, 229 (2002)
1380. J. Widengren, Ü. Mets, R. Rigler, Fluorescence correlation spectroscopy of triplet states in solution. J. Chem. Phys. **99**, 13368 (1995)
1381. W.E. Moerner, R.M. Dickson, D.J. Norris, Single-molecule spectroscopy and quantum optics in solids. Adv. At. Mol. Opt. Phys. **38**, 193 ff. (1997)
1382. G. Jung, J. Wiehler, B. Steipe, C. Bräuchle, A. Zumbusch, Single-molecule microscopy of the green fluorescent protein using two-color excitation. Chem. Phys. Chem. **2**, 392 (2001)
1383. P. Schwille, U. Haupts, S. Maiti, W.W. Web, Molecular dynamics in living cells observed by fluorescence correlation spectroscopy. Biophys. J. **77**, 2251 (1999)
1384. E.H. Piepmeier (ed.), *Analytical Applicability of Lasers* (Wiley, New York, 1986);
J. Sneddon, T.L. Thiem, Y. Lee (eds.), *Lasers in Analytical Atomic Spectroscopy* (Wiley/VCH, Weinheim, 1997)
1385. I. Osad'ko, *Selective Spectroscopy of Single Molecules*. Springer Ser. Chem. Phys., vol. 69 (Springer, Berlin, 2003);
R. Rigler, M. Orrit, T. Basche (eds.), *Single Molecule Spectroscopy* (Springer, Berlin, 2002);
Chr. Gell, D. Brockwell, *Handbook of Single Molecule Fluorescence Spectroscopy* (Oxford Univ. Press, Oxford 2006)
1386. K. Niemax, *Analytical Aspects of Atomic Laser Spectrochemistry* (Harwood Acad. Publ., Philadelphia, 1989)
1387. A. Baronarski, J.W. Butler, J.W. Hudgens, M.C. Lin, J.R. McDonald, M.E. Umstead, Chemical applications of lasers, in *Advances in Laser Chemistry*, ed. by A.H. Zewail. Springer Ser. Chem. Phys., vol. 3 (Springer, Berlin, 1986), p. 62
1388. B. Raffel, J. Wolfrum, Spatial and time resolved observation of CO_2-laser induced explosions of O_2–O_3-mixtures in a cylindrical cell. Z. Phys. Chem. **161**, 43 (1989)
1389. R.L. Woodin, A. Kaldor, Enhancement of chemical reactions by infrared lasers. Adv. Chem. Phys. **47**, 3 (1981);
M. Quack, Infrared laser chemistry and the dynamics of molecular multiphoton excitation. Infrared Phys. **29**, 441 (1989)
1390. C.D. Cantrell (ed.), *Multiple-Photon Excitation and Dissociation of Polyatomic Molecules*. Springer Topics. Curr. Phys., vol. 35 (Springer, Berlin, 1986);
V.N. Bagratashvili, V.S. Letokhov, A.D. Makarov, E.A. Ryabov, *Multiple Photon Infrared Laser Photophysics and Photochemistry* (Harwood, Chur, 1985)
1391. J.H. Clark, K.M. Leary, T.R. Loree, L.B. Harding, Laser synthesis chemistry and laser photogeneration of catalysis, in *Laser Applications in Physical Chemistry*, ed. by D.K. Evans (Dekker, New York, 1989), p. 74
1392. G. Mengxiong, W. Fuss, K.L. Kompa, CO_2 laser induced chain reaction of $C_2F_4 + CF_3I$. J. Phys. Chem. **94**, 6332 (1990)
1393. M. Schneider, J. Wolfrum, Mechanisms of by-product formation in the dehydrochlorination of dichloromethane. Ber. Bunsenges. Phys. Chem. **90**, 1058 (1986)
1394. Z. Linyang, W. Fuss, K.L. Kompa, KrF laser induced telomerization of bromides with olefins. Ber. Bunsenges. Phys. Chem. **94**, 867 (1990)
1395. K.L. Kompa, Laser photochemistry at surfaces. Angew. Chem. **27**, 1314 (1988)

1396. M.S. Djidjoev, R.V. Khokhlov, A.V. Kieselev, V.I. Lygin, V.A. Namiot, A.I. Osipov, V.I. Panchenko, Y.B.I. Provottorov, Laser chemistry at surfaces, in *Laser Applications in Physical Chemistry*, ed. by D.K. Evans (Dekker, New York, 1989), p. 7
1397. H.-L. Dai, W. Ho (eds.), *Laser Spectroscopy and Photochemistry on Metal Surfaces*. Advanced Series in Physical Chemistry, vol. 5 (World Scientific, Singapore, 1995)
1398. D. Bäuerle, *Laser Processing and Chemistry*, 2nd edn. (Springer, Berlin, 1996)
1399. de Vivie-Riedle, H. Rabitz, K.L. Kompa (eds.), Laser control of quantum dynamics. Chemical Physics **267** (2001) (Special Issue)
1400. P. Brumer, M. Shapiro, Control of unimolecular reactions using coherent light. Chem. Phys. Lett. **126**, 541 (1986);
M. Shapiro, P. Brumer, Coherent control of atomic, molecular and electronic processes. Adv. At. Mol. Opt. Phys. **42**, 287 (2000)
1401. D.J. Tannor, R. Kosloff, S.A. Rice, Coherent pulse sequence induced control of selectivity reactions. J. Chem. Phys. **85**, 5805 (1986)
1402. M. Shapiro, Association, dissociation and the acceleration and suppression of reactions by laser pulses. Adv. Chem. Phys. **114**, 123–192 (1999);
A. Assion, T. Baumert, M. Bergt, T. Brixner, B. Kiefer, V. Seyfried, M. Strehle, G. Gerber, Control of chemical reactions by feedback-optimized phase-shaped femtosecond laser pulses. Science **282**, 919 (1998)
1403. D. Zeidler, S. Frey, K.L. Kompa, M. Motzkus, Evolutionary algorithms and their applications to optimal control studies. Phys. Rev. A **64**, O23420 (2001)
1404. M. Bergt, T. Brixner, B. Kiefer, M. Strehle, G. Gerber, Controlling the femtochemistry of $Fe(CO)_5$. J. Phys. Chem. **103**, 10381 (1999)
1405. T. Brixner, B. Kiefer, G. Gerber, Problem complexity in femtosecond quantum control. Chem. Phys. **267**, 241 (2001)
1406. R.J. Levis, G.M. Menkir, H. Rabitz, Selective bond dissociation and rearrangement with optimally tailored, strong field laser pulses. Science **292**, 709 (2001)
1407. T. Brixner, G. Gerber, Quantum control of gas-phase and liquid-phase femtochemistry. Chem. Phys. Chem. **4**, 418 (2003)
1408. A. Rice, M. Zhao, *Optical Control of Molecular Dynamics* (Wiley, New York, 2000)
1409. P. Gaspard, I. Burghardt (eds.), *Chemical Reactions and Their Control on the Femtosecond Time Scale*. Adv. Chem. Phys., vol. 101 (1997)
1410. J.L. Herek, W. Wohlleben, R.J. Cogdell, D. Zeidler, M. Motzkus, Quantum control of the energy flow in light harvesting. Nature **417**, 533 (2002)
1411. R.N. Zare, R.B. Bernstein, State to state reaction dynamics. Phys. Today **3**, 43 (1980)
1412. A.H. Zewail, Laser femtochemistry. Science **242**, 1645 (1988);
A.H. Zewail, The birth of molecules. Sci. Am. **262**, 76 (1990);
J. Manz, L. Wöste (eds.), *Femtosecond Chemistry*, vols. I and II (VCH, Weinheim, 1995)
1413. A.H. Zewail, *Femtochemistry* (World Scientific, Singapore, 1994)
1414. B. Bescós, B. Lang, J. Weiner, V. Weiss, E. Wiedemann, G. Gerber, Real-time observation of ultrafast ionization and fragmentation of mercury clusters. Eur. Phys. J. D **9**, 399 (1999)
1415. H. Bürsing, P. Vöhringer, Transition state probing and fragment rotational dynamics of HgI_2. Phys. Chem. Chem. Phys. **2**, 73 (2000)
1416. St. Hess, H. Bürsing, P. Vöhringer, Dynamics of fragment recoil in the femtosecond photodissociation of triiodide ions. J. Chem. Phys. **111**, 5461 (1999)
1417. S.A. Trushin, W. Fuss, K.L. Kompa, W.E. Schmid, Femtosecond dynamics of $Fe(CO)_5$ photodissociation at 267 nm studied by transient ionization. J. Phys. Chem. A **104**, 1997 (2000)
1418. A.H. Zewail, Femtosecond transition-state dynamics. Faraday Discuss. Chem. Soc. **91**, 1 (1991)
1419. M. Kimble, W. Castleman Jr., *Femtochemistry VII: Fundamental Fast Processes in Chemistry, Physics, and Biology* (Elsevier Science, Amsterdam, 2006)
1420. M.M. Martin, J.T. Hynes, *Femtochemistry and Femtobiology* (Elsevier Science, Amsterdam, 2004)

1421. P. Hannaford, *Femtosecond Laser Spectroscopy* (Springer, Berlin, 2004)
1422. S. Villani, *Isotope Separation* (Am. Nucl. Soc., Hinsdale, 1976);
S. Villani, *Uranium Enrichment*. Topics Appl. Phys., vol. 35 (Springer, Berlin, 1979);
W. Ehrfeld, *Elements of Flow and Diffusion Processes in Separation Nozzles*. Springer Tracts Mod. Phys., vol. 97 (Springer, Berlin, 1983)
1423. C.D. Cantrell, S.M. Freund, J.L. Lyman, Laser induced chemical reactions and isotope separation, in *Laser Handbook*, vol. 3, ed. by M.L. Stitch (North-Holland, Amsterdam, 1979);
R.N. Zare, Laser separation of isotopes. Sci. Am. **236**, 86 (1977);
F.S. Becker, K.L. Kompa, Laser isotope separation. Europhys. News **12**, 2 (1981);
R.D. Alpine, D.K. Evans, Laser isotope separation by the selective multiphoton decomposition process. Adv. Chem. Phys. **60**, 31 (1985)
1424. J.P. Aldridge, J.H. Birley, C.D. Cantrell, D.C. Cartwright, Experimental and studies of laser isotope separation, in *Laser Photochemistry, Tunable Lasers*, ed. by S.E. Jacobs, S.M. Sargent, M.O. Scully, C.T. Walker (Addison-Wesley, Reading, 1976)
1425. J.I. Davies, J.Z. Holtz, M.L. Spaeth, Status and prospects for lasers in isotope separation. Laser Focus **18**, 49 (1982)
1426. M. Stuke, Isotopentrennung mit Laserlicht. Spektrum Wiss. **4**, 76 (1982)
1427. (a) F.S. Becker, K.L. Kompa, The practical and physical aspects of uranium isotope separation with lasers. Nuc. Technol. **58**, 329 (1982);
(b) C.P. Robinson, R.J. Jensen, Laser methods of uranium isotope separation, in *Uranium Enrichment*, ed. by S. Villani. Topics Appl. Phys., vol. 35 (Springer, Berlin, 1979), p. 269
1428. L. Mannik, S.K. Brown, Laser enrichment of carbon 14. Appl. Phys. B **37**, 79 (1985)
1429. A. von Allmen, *Laser-Beam Interaction with Materials*, 2nd edn. Springer Ser. Mater. Sci., vol. 2 (Springer, Berlin, 1995);
P.N. Bajaj, K.G. Manohar, B.M. Suri, K. Dasgupta, R. Talukdar, P.K. Chakraborti, P.R.K. Rao, Two colour multiphoton ionization spectroscopy of uranium from a metastable state. Appl. Phys. B **47**, 55 (1988)
1430. A. Outhouse, P. Lawrence, M. Gauthier, P.A. Hacker, Laboratory scale-up of two stage laser chemistry separation of ^{13}C from CF_2HCL. Appl. Phys. B **36**, 63 (1985);
I. Deac, V. Cosma, D. Silipas, L. Muresan, V. Tosa, Parametric study of the IRMPD of CF_2HCl molecules with the $9P22$ CO_2 laser time. Appl. Phys. B **51**, 211 (1990)
1431. M. Lackner (ed.), *Lasers in Chemistry* (Wiley/VCH, Weinheim, 2008)
1432. C. D'Ambrosio, W. Fuss, K.L. Kompa, W.E. Schmid, S. Trushin, ^{13}C separation by a continuous discharge CO_2 laser Q-switched at 10 kHz. Infrared Phys. **29**, 479 (1989); and Appl. Phys. B **47**, 19 (1988)
1433. K. Kleinermanns, J. Wolfrum, Laser in der Chemie – Wo stehen wir heute? Angew. Chem. **99**, 38 (1987);
J. Wolfrum, Laser spectroscopy for studying chemical processes. Appl. Phys. B **46**, 221 (1988)
1434. D.J. Neshitt, St.R. Leone, Laser-initiated chemical chain reactions. J. Chem. Phys. **72**, 1722 (1980)
1435. D. Bäuerle, *Chemical Processing with Lasers*. Springer Ser. Mater. Sci., vol. 1 (Springer, Berlin, 1986)
1436. V.S. Letokhov (ed.), *Laser Analytical Spectrochemistry* (Hilger, Bristol, 1985)
1437. K. Peters, Picosecond organic photochemistry. Annu. Rev. Phys. Chem. **38**, 253 (1987)
1438. M. Gruehele, A.H. Zewail, Ultrafast reaction dynamics. Phys. Today **13**, 24 (1990)
1439. J. Wolfrum, Laser spectroscopy for studying chemical processes. Appl. Phys. B **46**, 221 (1988);
J. Wolfrum, Laser stimulation and observation of simple gas phase radical reactions. Laser Chem. **9**, 171 (1988)
1440. E. Hirota, From high resolution spectroscopy to chemical reactions. Annu. Rev. Phys. Chem. **42**, 1 (1991)
1441. J.I. Steinfeld, *Laser-Induced Chemical Processes* (Plenum, New York, 1981)

1442. L.J. Kovalenko, S.L. Leone, Innovative laser techniques in chemical kinetics. J. Chem. Educ. **65**, 681 (1988)
1443. A. Ben-Shaul, Y. Haas, K.L. Kompa, R.D. Levine, *Lasers and Chemical Change*. Springer Ser. Chem. Phys., vol. 10 (Springer, Berlin, 1981)
1444. K.L. Kompa, S.D. Smith (eds.), *Laser-Induced Processes in Molecules*. Springer Ser. Chem. Phys., vol. 6 (Springer, Berlin, 1979)
1445. K.L. Kompa, J. Warner, *Laser Applications in Chemistry* (Plenum, New York, 1984)
1446. A.H. Zewail (ed.), *The Chemical Bond: Structure and Dynamics* (Academic Press, Boston, 1992)
1447. L.R. Khundar, A.H. Zewail, Ultrafast reaction dynamics in real times. Annu. Rev. Phys. Chem. **41**, 15 (1990)
1448. J. Steinfeld, *Air Pollution* (Wiley, New York, 1986)
1449. C.F. Bohren, D.R. Huffman, *Absorption and Scattering of Light by Small Particles* (Wiley, New York, 1983)
1450. R. Zellner, J. Hägele, A double-beam UV-laser differential absorption method for monitoring tropospheric trace gases. Opt. Laser Technol. **17**, 79 (1985)
1451. E.D. Hinkley (ed.), *Laser Monitoring of the Atmosphere*. Topics Appl. Phys., vol. 14 (Springer, Berlin, 1976);
B. Stumpf, D. Göring, R. Haseloff, K. Herrmann, Detection of carbon monoxide, carbon dioxide with pulsed tunable $Pb_{1-x}Se_x$ diode lasers. Collect. Czechoslov. Chem. Commun. **54**, 284 (1989)
1452. A. Tönnissen, J. Wanner, K.W. Rothe, H. Walther, Application of a CW chemical laser for remote pollution monitoring and process control. Appl. Phys. **18**, 297 (1979)
1453. W. Meinburg, H. Neckel, J. Wolfrum, Lasermeßtechnik und mathematische Simulation von Sekundärmaßnahmen zur NO_x-Minderung in Kraftwerken. Appl. Phys. B **51**, 94 (1990);
A. Arnold, H. Becker, W. Ketterle, J. Wolfrum, Combustion diagnostics by two dimensional laser-induced fluorescence using tunable excimer lasers. SPIE Proc. **1602**, 70 (1991)
1454. W. Meienburg, H. Neckel, J. Wolfrum, In situ measurement of ammonia with a $^{13}CO_2$-waveguide laser system. Appl. Phys. B **51**, 94 (1990)
1455. P. Wehrle, A review of recent advances in semiconductor laser based gas monitors. Spectrochim. Acta, Part A **54**, 197 (1998)
1456. R. Kormann, H. Fischer, C. Gurk, F. Helleis, T. Klüpfel, K. Kowalski, R. Königstedt, U. Parchatka, V. Wagner, Application of a multi-laser tunable diode laser absorption spectrometer for atmospheric trace gas measurements at sub-ppbv levels. Spectrochim. Acta, Part A **58**, 2489 (2002)
1457. K.W. Rothe, U. Brinkmann, H. Walther, Remote measurement of NO_2-emission from a chemical factory by the differential absorption technique. Appl. Phys. **4**, 181 (1974)
1458. H.J. Kölsch, P. Rairoux, J.P. Wolf, L. Wöste, Simultaneous NO and NO_2 DIAL measurements using BBO crystals. Appl. Opt. **28**, 2052 (1989)
1459. J.P. Wolf, H.J. Kölsch, P. Rairoux, L. Wöste, Remote detection of atmospheric pollutants using differential absorption LIDAR techniques, in *Applied Laser Spectroscopy*, ed. by W. Demtröder, M. Inguscio (Plenum, New York, 1991), p. 435
1460. A.L. Egeback, K.A. Fredrikson, H.M. Hertz, DIAL techniques for the control of sulfur dioxide emissions. Appl. Opt. **23**, 722 (1984)
1461. J. Werner, K.W. Rothe, H. Walther, Monitoring of the stratospheric ozone layer by laser radar. Appl. Phys. B **32**, 113 (1983)
1462. W. Steinbrecht, K.W. Rothe, H. Walther, Lidar setup for daytime and nighttime probing of stratospheric ozone and measurements in polar and equatorial regimes. Appl. Opt. **28**, 3616 (1988)
1463. (a) J. Shibuta, T. Fukuda, T. Narikiyo, M. Maeda, Evaluation of the solarblind effect in ultraviolet ozone lidar with Raman lasers. Appl. Opt. **26**, 2604 (1984);
(b) C. Weitkamp, O. Thomsen, P. Bisling, Signal and reference wavelengths for the elimination of SO_2 cross sensitivity in remote measurements of tropospheric ozone with lidar. Laser Optoelectron. **24**, 246 (1992)

1464. A. Asmann, R. Neuber, P. Rairoux (eds.), *Advances in Atmospheric Remote Sensing with LIDAR* (Springer, Berlin, 1997)
1465. U.v. Zahn, P. von der Gathen, G. Hansen, Forced release of sodium from upper atmospheric dust particles. Geophys. Res. Lett. **14**, 76 (1987)
1466. F.J. Lehmann, S.A. Lee, C.Y. She, Laboratory measurements of atmospheric temperature and backscatter ratio using a high-spectral-resolution lidar technique. Opt. Lett. **11**, 563 (1986)
1467. M.M. Sokolski (ed.), *Laser Applications in Meteorology and Earth- and Atmospheric Remote Sensing*. SPIE Proc., vol. 1062 (1989)
1468. R.M. Measure, *Laser Remote Sensing: Fundamentals and Applications* (Wiley, Toronto, 1984)
1469. J. Looney, K. Petri, A. Salik, Measurements of high resolution atmospheric water vapor profiles by use of a solarblind Raman lidar. Appl. Opt. **24**, 104 (1985)
1470. H. Edner, S. Svanberg, L. Uneus, W. Wendt, Gas-correlation LIDAR. Opt. Lett. **9**, 493 (1984)
1471. J.A. Gelbwachs, Atomic resonance filters. IEEE J. Quantum Electron. **24**, 1266 (1988)
1472. P. Rairoux, H. Schillinger, S. Niedermeier, M. Rodriguez, F. Ronneberger, R. Sauerbrey, B. Stein, D. Waite, C. Wedekind, H. Wille, L. Wöste, Remote sensing of the atmosphere, using ultrashort laser pulses. Appl. Phys. B **71**, 573 (2000)
1473. L. Wöste, S. Frey, J.P. Wolf, LIDAR—monitoring of the air with femtosecond plasma channels. Adv. At. Mol. Opt. Phys. **53**, 413 (2006)
1474. S. Svanberg, Fundamentals of atmospheric spectroscopy, in *Surveillance of Environmental Pollution and Resources by El. Mag. Waves*, ed. by I. Lund (Reidel, Dordrecht, 1978); Ph.N. Slater, *Remote Sensing* (Addison-Wesley, London, 1980)
1475. R.M. Measures, *Laser Remote Chemical Analysis* (Wiley, New York, 1988)
1476. D.K. Killinger, A. Mooradian (eds.), *Optical and Laser Remote Sensing*. Springer Ser. Opt. Sci., vol. 39 (Springer, Berlin, 1983)
1477. R.N. Dubinsky, Lidar moves towards the 21st century. Laser Optron. **7**, 93 (1988); S. Svanberg, Environmental monitoring using optical techniques, in *Applied Laser Spectroscopy*, ed. by W. Demtröder, M. Inguscio (Plenum, New York, 1991), p. 417
1478. H. Walther, Laser investigations in the atmosphere, in *Festkörperprobleme*, vol. 20 (Vieweg, Braunschweig, 1980), p. 327
1479. E.J. McCartney, *Optics of the Atmosphere* (Wiley, New York, 1976)
1480. J.W. Strohbehn (ed.), *Laser Beam Propagation in the Atmosphere*. Topics Appl. Phys., vol. 25 (Springer, Berlin, 1978)
1481. J. Bublitz, W. Schade, M. Dickenhausen, Quantitative analysis of PAH-molecules by time-resolved LIF-spectroscopy in water and in the ground, in *Laser in Remote Sensing*, ed. by C. Werner, W. Waidelich (Springer, Berlin, 1994), p. 118
1482. W. Schade, Experimentelle Untersuchungen zur zeitaufgelösten Fluoreszenzspektroskopie mit kurzen Laserpulsen. Habilitation-Thesis, Math.-Naturw. Fakultät, Univ. Kiel, Germany, 1992
1483. J. Ilkin, R. Stumpe, R. Klenze, Laser-induced photoacoustic spectroscopy for the speciation of transuranium elements in natural aquatic systems. Top. Curr. Chem. **157**, 129 (1990)
1484. G.M. Korenowski, Applications of laser technology and laser spectroscopy in studies of the ocean microlayer. in *The Sea Surface and Global Change*, ed. by P.S. Liss et al. (1997), pp. 445–470
1485. F. Anabitarte, A. Cobo, J.M. Lopez-Higuera, Laser-Induced Breakdown Spectroscopy: Fundamentals, Applications, and Challenges. Int. Sch. Res. Not. **2012**, 2852 (2012)
1486. D.A. Cremers, L.J. Radziemski, J. Wiley, *Handbook of Laser-Induced Breakdown Spectroscopy* (Wiley, New York, 2006); W. Miziolek, V. Palleschi, I. Schechter, *Laser-Induced Breakdown Spectroscopy (LIBS): Fundamentals and Applications* (Cambridge University Press, Cambridge, 2006)
1487. (a) N. Yu, J.M. Kohel, L. Romans, L. Maleki, Quantum gravity gradiometer sensor for earth science applications, in *ESTC 2002, Pasadena, CA* (2002);

(b) N. Yu, J.M. Kohel, J.R. Kellogg, L. Maleki, Development of an atom-interferometer gravity gradiometer for gravity measurement from space. Appl. Phys. B **84**, 647 (2006);
(c) M.J. Snaden et al., Measurement of the earth's gravity gradient wit an atom interferometry based gravity gradiometer. Phys. Rev. Lett. **81**, 971 (1998);
(d) J.H. Olbertz, A mobile high-precision gravimeter based on atom interferometry. PhD Thesis, Humboldt University, Berlin, 2011

1488. R. Suntz, H. Becker, P. Monkhouse, J. Wolfrum, Two-dimensional visualization of the flame front in an internal combustion engine by laser-induced fluorescence of OH radicals. Appl. Phys. B **47**, 287 (1988)
1489. (a) A.M. Wodtke, L. Hüwel, H. Schlüter, H. Voges, G. Meijer, P. Andresen, High sensitivity detection of NO in a flame using a tunable Ar–F-laser. Opt. Lett. **13**, 910 (1988);
(b) M. Schäfer, W. Ketterle, J. Wolfrum, Saturated 2D-LIF of OH and 2D determination of effective collisional lifetimes in atmospheric pressure flames. Appl. Phys. B **52**, 341 (1991)
1490. P. Andresen, G. Meijer, H. Schlüter, H. Voges, A. Koch, W. Hentschel, W. Oppermann, Zweidimensionale Konzentrationsmessungen im Brennraum des Transparentmotors mit Hilfe von Laser-Fluoreszenzverfahren. Bericht 11/1989, MPI für Strömungsforschung Göttingen, 1989;
Combustion optimization pushed forward by excimer LIF-methods. Lambda-Physic Highlights No. 14, December 1988
1491. M. Alden, K. Fredrikson, S. Wallin, Application of a two-colour dye laser in CARS experiments for fast determination of temperatures. Appl. Opt. **23**, 2053 (1984)
1492. J.P. Taran, CARS spectroscopy and applications, in *Applied Laser Spectroscopy*, ed. by W. Demtröder, M. Inguscio (Plenum, New York, 1991), p. 365;
A. D'Allescio, A. Cavaliere, Laser spectroscopy applied to combustion, in *Applied Laser Spectroscopy*, ed. by W. Demtröder, M. Inguscio (Plenum, New York, 1991), p. 393
1493. R.W. Dreyfus, Useful macroscopic phenomena due to laser ablation, in *Desorption Induced by Electronic Transitions DIET IV*. Springer Ser. Surf. Sci., vol. 19 (Springer, Berlin, 1990), p. 348;
J.C. Miller, R.F. Haglund (eds.), *Laser Ablation: Mechanisms and Applications*. Lecture Notes Phys., vol. 389 (Springer, Berlin, 1991);
J.C. Miller (ed.), *Laser Ablation*. Springer Ser. Mater. Sci., vol. 28 (Springer, Berlin, 1994)
1494. R. DeJonge, Internal energy of sputtered molecules. Comments At. Mol. Phys. **22**, 1 (1988)
1495. H.L. Bay, Laser induced fluorescence as a technique for investigations of sputtering phenomena. Nucl. Instrum. Methods B **18**, 430 (1987)
1496. R.W. Dreyfus, J.M. Jasinski, R.E. Walkup, G. Selwyn, Laser spectroscopy in electronic materials processing research. Laser Focus **22**, 62 (1986);
R.W. Dreyfus, R.W. Walkup, R. Kelly, Laser-induced fluorescence studies of excimer laser ablation of Al_2O_3. J. Appl. Phys. **49**, 1478 (1986)
1497. J.M. Jasinski, E.A. Whittaker, G.C. Bjorklund, R.W. Dreyfus, R.D. Estes, R.E. Walkup, Detection of SiH_2 in silane and disilane glow discharge by frequency modulated absorption spectroscopy. Appl. Phys. Lett. **44**, 1155 (1984)
1498. H. Moenke, L. Moenke-Blankenburg, *Einführung in die Laser Mikrospektralanalyse* (Geest und Portig, Leipzig, 1968)
1499. D. Bäuerle, *Laser Processing and Chemistry*, 3rd edn. (Springer, Berlin, 2000)
1500. H.L. Dai, W. Ho (eds.), *Laser Spectroscopy and Photochemistry on Metal Surfaces*. Adv. Ser. Phys. Chem., vol. 5 (World Scientific, Singapore, 1995)
1501. A.W. Miziolek, V. Paleschi, I. Schechter, *Laser-Induced Breakdown Spectroscopy* (Cambridge Univ. Press, Cambridge, 2006);
D.A Cremers, L.J. Radziemski, *Handbook of Laser-Induced Breakdown Spectroscopy* (Wiley, New York, 2006)
1502. C.M. Li et al., Quantitative analysis of phosphorus in steel using laser-induced breakdown spectroscopy in air atmosphere. J. Anal. At. Spectrom. **29**, 1432–1437 (2014) (Advance article)

1503. F. Durst, A. Melling, J.H. Whitelaw, *Principles and Practice of Laser-Doppler Anemometry*, 2nd edn. (Academic Press, New York, 1981)
1504. T.S. Durrani, C.A. Greated, *Laser Systems in Flow Measurement* (Plenum, New York, 1977)
1505. L.E. Drain, *The Laser Doppler Technique* (Wiley, New York, 1980)
1506. F. Durst, G. Richter, Laser Doppler measurements of wind velocities using visible radiation, in *Photon Correlation Techniques in Fluid Mechanics*, ed. by E.O. Schulz-Dubois. Springer Ser. Opt. Sci., vol. 38 (Springer, Berlin, 1983), p. 136
1507. R.M. Hochstrasser, C.K. Johnson, Lasers in biology. Laser Focus **21**, 100 (1985)
1508. B. Valeur, J.C. Brochon (eds.), *New Trends in Fluorescence Spectroscopy: Application to Chemistry and Life Sciences* (Springer, Berlin, 2001);
M. Hof, R. Hütterer, V. Fiedler, *Fluorescence Spectroscopy in Biology*. Springer Ser. Fluoresc. (Springer, Berlin, 2005)
1509. A. Anders, Dye-laser spectroscopy of bio-molecules. Laser Focus **13**, 38 (1977);
A. Anders, Selective laser excitation of bases in nucleic acids. Appl. Phys. **20**, 257 (1979)
1510. A. Anders, Models of DNA-dye-complexes: energy transfer and molecular structure. Appl. Phys. **18**, 373 (1979);
M.E. Michel-Beyerle (ed.), *Antennas and Reaction Centers of Photosynthetic Bacteria*. Springer Ser. Chem. Phys., vol. 42 (Springer, Berlin, 1983)
1511. R.R. Birge, B.M. Pierce, The nature of the primary photochemical events in bacteriorhodopsin and rhodopsin, in *Photochemistry and Photobiology*, ed. by A.H. Zewail (Harwood, Chur, 1983), p. 841
1512. P. Cornelius, R.M. Hochstrasser, Picosecond processes involving CO, O_2 and NO derivatives of hemoproteins, in *Picosecond Phenomena III*, ed. by K.B. Eisenthal, R.M. Hochstrasser, W. Kaiser, A. Lauberau. Springer Ser. Chem. Phys., vol. 23 (Springer, Berlin, 1982)
1513. D.P. Millar, R.J. Robbins, A.H. Zewail, Torsion and bending of nucleic acids, studied by subnanosecond time resolved depolarization of intercalated dyes. J. Chem. Phys. **76**, 2080 (1982)
1514. L. Stryer, The molecules of visual excitation. Sci. Am. **157**, 32 (1987)
1515. R.A. Mathies, S.W. Lin, J.B. Ames, W.T. Pollard, From femtoseconds to biology: mechanisms of bacterion rhodopsin's light driven proton pump. Annu. Rev. Biophys. Biophys. Chem. **20**, 1000 (1991)
1516. W.H. Freeman, D. Savadan et al., *Life: The Science of Biology*, 4th edn. (Sinauer Associates, Sunderland, 2004)
1517. D.C. Youvan, B.L. Marrs, Molecular mechanisms of photosynthesis. Sci. Am. **256**, 42 (1987)
1518. A.H. Zewail (ed.), *Photochemistry and Photobiology* (Harwood, London, 1983);
V.S. Letokhov, *Laser Picosecond Spectroscopy and Photochemistry of Biomolecules* (Hilger, London, 1987);
R.R. Alfano (ed.), *Biological Events Probed by Ultrafast Laser Spectroscopy* (Academic Press, New York, 1982);
D. Purves, R.B. Lotto, *Why We See What We Do: An Empirical Theory of Vision* (Sinauer Associates, Sunderland, 2003);
R.E. Goldmann, *Live Cell Imaging* (Cold Spring Harbor Laboratory Press, 2004);
C. Gell, D. Brockwell, A. Smith, *Handbook of Single-Molecule Fluorescence* (Oxford Univ. Press, Oxford, 2006)
1519. W. Kaiser (ed.), *Ultrashort Laser Pulses*, 2nd edn. Topics Appl. Phys., vol. 60 (Springer, Berlin, 1993)
1520. E. Klose, B. Wilhelmi (eds.), *Ultrafast Phenomena in Spectroscopy*. Springer Proc. Phys., vol. 49 (Springer, Berlin, 1990)
1521. J.R. Lakowicz (ed.), *Time-Resolved Laser Spectroscopy in Biochemistry*. SPIE Proc., vol. 909 (1988)
1522. R.R. Birge, L.A. Nufie (eds.), *Biomolecular Spectroscopy*. SPIE Proc., vol. 1432 (1991)

1523. R. Nossal, S.H. Chen, Light scattering from mobile bacteria. J. Phys. Suppl. **33**, C1-171 (1972)
1524. A. Andreoni, A. Longoni, C.A. Sacchi, O. Svelto, Laser-induced fluorescence of biological molecules, in *Tunable Lasers and Applications*, ed. by A. Mooradian, T. Jaeger, P. Stokseth. Springer Ser. Opt. Sci., vol. 3 (Springer, Berlin, 1976), p. 303
1525. G.N. McGregor, H.G. Kaputza, K.A. Jacobsen, Laser-based fluorescence microscopy of living cells. Laser Focus **20**, 85 (1984)
1526. H. Schneckenburger, A. Rück, B. Baros, R. Steiner, Intracellular distribution of photosensitizing porphyrins measured by video-enhanced fluorescence microscopy. J. Photochem. Photobiol. B **2**, 355 (1988)
1527. (a) H. Schneckenburger, A. Rück, O. Haferkamp, Energy transfer microscopy for probing mitochondrial deficiencies. Anal. Chim. Acta **227**, 227 (1988);
(b) P. Fischer, Time-resolved methods in laser scanning microscopy. Laser Opt. Elektron. **24**, 36 (1992)
1528. E. Abbé, Arch. Mikrosk. Anat. **9**, 413 (1873);
B.R. Masters, Ernst Abbé and the foundations of scientific microscopes. Opt. Photonics. News **18**, 18 (2007)
1529. S.W. Hell, E. Stelzer, Fundamental improvement of resolution with a 4 Pi-confocal fluorescence microscope using two-photon excitation. Opt. Commun. **93**, 277 (1992);
J. Bewersdorf, R. Schmidt, S.W. Hell, Comparison of I^5M and 4 Pi-microscopy. J. Microsc. **222**, 105 (2006)
1530. V. Westphal, S.W. Hell, Nanoscale resolution in the focal plane of an optical microscope. Phys. Rev. Lett. **94**, 143903 (2005);
K. Willig, J. Keller, M. Bossi, S.W. Hell, STED-microscopy resolves nanoparticle assemblies. New J. Phys. **8**, 106 (2006)
1531. T. Lebold, J. Michaelis, T. Blin, C. Bräuchle, *Single Molecule Spectroscopy* (Wiley/VCVH, Weinheim, 2012)
1532. C. Bräuchle, G. Leisenberger, T. Endress, M.C. Ried, Single virus tracing. Vizualisation of the infection pathway of a virus into a living cell. Chem. Phys. Chem. **3**, 299 (2002)
1533. P.P. Mondal, A. Diaspro, *Fundamentals of Fluorescence Microscopy* (Springer, Berlin, 2014)
1534. http://en.wikipedia.org/wiki/Confocal_microscopy; http://www.physics.emory.edu/faculty/weeks//confocal/
1535. H. Scheer, Chemistry and spectroscopy of chlorophylls, in *CRC Handbook of Organic Photochemistry and Photobiology*, ed. by W.M. Horspool, P.S. Song (CRC, New York, 1995), p. 1402;
P. Mathis, Photosynthetic reaction centers, in *CRC Handbook of Organic Photochemistry and Photobiology*, ed. by W.M. Horspool, P.S. Song (CRC, New York, 1995), p. 1412
1536. I. Lutz, W. Zinth et al., Primary reactions of sensory rhodopsins, in *Ultrafast Phenomena XII*, ed. by T. Elsäser et al. Springer Series in Chem. Phys., vol. 66 (Springer, Berlin, 2000), pp. 677, 680;
W. Zinth, et al., Femtosecond spectroscopy and model calculations for an understanding of the primary reactions in bacterio-rhodopsin, in *Ultrafast Phenomena XII*, ed. by T. Elsäser et al. Springer Series in Chem. Phys., vol. 66 (Springer, Berlin, 2000), p. 680
1537. P.J. Walla, P.A. Linden, G.R. Fleming, Fs-transient absorption and fluorescence upconversion after two-photon excitation of carotenoids in solution and in LHC II, in *Ultrafast Phenomena XII*, ed. by T. Elsäser et al. Springer Series in Chem. Phys., vol. 66 (Springer, Berlin, 2000), p. 671
1538. L. Goldstein (ed.), *Laser Non-surgical Medicine. New Challenges for an Old Application*, (Lancaster, Basel, 1991)
1539. G. Biamino, G. Müller (eds.), *Advances in Laser Medicine I* (Ecomed. Verlagsgesell., Berlin, 1988)
1540. S.L. Jacques (ed.), *Proc. Laser Tissue Interaction II*. SPIE Proc., vol. 1425 (1991);
A. Anders, I. Lamprecht, H. Schacter, H. Zacharias, The use of dye lasers for spectroscopic

investigations and photodynamics therapy of human skin. Arch. Dermatol. Res. **255**, 211 (1976)

1541. http://en.wikipedia.org/wiki/Optical_window_in_biological_tissue
1542. H.P. Berlien, G. Müller (eds.), *Angewandte Lasermedizin* (Ecomed, Landsberg, 1989)
1543. G. Ackermann, PhD thesis, University Regensburg, 2001
1544. A.M.K. Enejder, Light scattering and absorption in tissue—models and measurements. Dissertation thesis, Lund Institute of Technology, Lund, Sweden, 1997
1545. A. Popp et al., Ultrasensitive mid-infrared cavity-leak out spectroscopy using a cw optical parametric oscillator. Appl. Phys. B **75**, 751 (2002)
1546. H. Albrecht, G. Müller, M. Schaldach, Entwicklung eines Raman- spektroskopisches Gasanalysesystems. Biomed. Tech. **22**, 361 (1977); and *Proc. VII Int'l Summer School on Quantum Optics*, Wiezyca, Poland (1979)
1547. M. Mürtz, D. Halmer, M. Horstian, S. Thelen, P. Hering, Ultrasensitive trace gas detection for biomedical applications. Spectrochim. Acta A **63**, 963 (2006)
1548. M. Mürtz, T. Kayser, D. Kleine, S. Stry, P. Hering, W. Urban, Recent developments on cavity ringdown spectroscopy with tunable cw lasers in the mid-infrared. Proc. SPIE **3758**, 7 (1999)
1549. W. Drexler, *Optical Coherence Tomography* (Springer, Berlin, 2008)
1550. A.F. Fercher, W. Drexler, C.K. Hitzenberger, T. Lasser, Optical coherence tomography: principles and applications. Rep. Prog. Phys. **66**, 239 (2003)
1551. K. König, Laser in der Medizin. www.uni-jena.de/clm
1552. A. Katzir, *Fiber-Optics in Medicine*. Proc. SPIE, vol. 906 (1988)
1553. http://en.wikipedia.org/wiki/Optical_coherence_tomography
1554. H.J. Foth, N. Stasche, K. Hörmann, Measuring the motion of the human tympanic membrane by laser Doppler vibrometry. SPIE Proc. **2083**, 250 (1994)
1555. T.J. Dougherty, J.E. Kaulmann, A. Goldfarbe, K.R. Weishaupt, D. Boyle, A. Mittleman, Photoradiation therapy for the treatment of malignant tumors. Cancer Res. **38**, 2628 (1978);
B.W. Henderson, T.J. Dougherty (eds.), *Photodynamic Therapy: Basic Principles and Applications* (Dekker, New York, 1992);
D. Kessel, Components of hematoporphyrin derivatives and their tumor-localizing capacity. Cancer Res. **42**, 1703 (1982)
1556. G. Jori, Photodynamic therapy: basic and preclinical aspects, in *CRC Handbook of Organic Photochemistry and Photobiology*, ed. by W.M. Horspool, P.S. Song (CRC, New York, 1995), p. 1379;
T.J. Dougherty, Clinical applications of photodynamic therapy, in *CRC Handbook of Organic Photochemistry and Photobiology*, ed. by W.M. Horspool, P.S. Song (CRC, New York, 1995), p. 1384
1557. P.J. Bugelski, C.W. Porter, T.J. Dougherty, Autoradiographic distribution of HPD in normal and tumor tissue in the mouse. Cancer Res. **41**, 4606 (1981)
1558. A.S. Svanberg, Laser spectroscopy applied to energy, environmental and medical research. Phys. Scr. **23**, 281 (1988)
1559. Y. Hayata, H. Kato, Ch. Konaka, J. Ono, N. Takizawa, Hematoporphyrin derivative and laser photoradiation in the treatment of lung cancer. Chest **81**, 269 (1982)
1560. A. Katzir, *Optical Fibers in Medicine IV*. SPIE Proc., vol. 1067 (1989); and vol. 906 (1988)
1561. P. Agostinis, K. Berg, K.A. Cengel et al., Photodynamic therapy of cancer: an update. CA Cancer J. Clin. **61**(4), 250–280 (2011)
1562. L. Prause, P. Hering, Lichtleiter für gepulste Laser: Transmissionsverhalten, Dämpfung und Zerstörungsschwellen. Laser Optoelektron. **19**, 25 (1987); **20**, 48 (1988)
1563. A. Katzir, Optical fibers in medicine. Sci. Am. **260**, 86 (1989)
1564. H. Schmidt-Kloiber, E. Reichel, Laser lithotripsy, in *Angewandte Lasermedizin*, ed. by H.P. Berlien, G. Müller (Ecomed, Landsberg, 1989), VI, Sect. 2.12.1
1565. R. Steiner (ed.), *Laser Lithotripsy* (Springer, Berlin, 1988);
R. Pratesi, C.A. Sacchi (eds.), *Lasers in Photomedicine and Photobiology*. Springer Ser.

Opt. Sci., vol. 31 (Springer, Berlin, 1982);
L. Goldmann (ed.), *The Biomedical Laser* (Springer, Berlin, 1981)
1566. W. Simon, P. Hering, Laser-induzierte Stoßwellenlithotripsie an Nieren- und Gallensteinen. Laser Optoelektron. **19**, 33 (1987)
1567. D. Beaucamp, R. Engelhardt, P. Hering, W. Meyer, Stone identification during laser-induced shockwave lithotripsy, in *Proc. 9th Congress Laser 89*, ed. by W. Waidelich (Springer, Berlin, 1990)
1568. R. Engelhardt, W. Meyer, S. Thomas, P. Oehlert, Laser-induzierte Schockwellen-Lithotripsie mit Mikrosekunden Laserpulsen. Laser Optoelektron. **20**, 36 (1988)
1569. S.P. Dretler, Techniques of laser lithotripsy. J. Endourol. **2**, 123 (1988);
B.C. Ihler, Laser lithotripsy: system and fragmentation processes closely examined. Laser Optoelektron. **24**, 76 (1992)
1570. S. Willmann, A. Terenji, I.V. Yaroslavsky, T. Kahn, P. Hering, Determination of the optical properties of a human brain tumor using a new microspectrophotometric technique. Proc. SPIE **3598**, 233 (1999)
1571. A.N. Yaroslavsky, I.V. Yaroslavsky, T. Goldbach, H.J. Schwarzmaier, Influence of the scattering phase function approximation on the optical properties of blood. J. Biomed. Opt. **4**, 47 (1999)
1572. C.-Y. Ng (ed.), *Optical Methods for Time- and State Resolved Chemistry*. SPIE Proc., vol. 1638 (1992);
B.L. Feary (ed.), *Optical Methods for Ultrasensitive Dilution and Analysis*. SPIE Proc., vol. 1435 (1991);
J.L. McElroy, R.J. McNeal, *Remote Sensing of the Atmosphere*. SPIE Proc., vol. 1491 (1991);
S.A. Akhmanov, M. Poroshina (eds.), *Laser Applications in Life Sciences*. SPIE Proc., vol. 1403 (1991);
L.O. Jvassand (ed.), *Future Trends in Biomedical Applications of Lasers*. SPIE Proc., vol. 1535 (1991);
J.R. Lakowicz (ed.), *Time-Resolved Laser Spectroscopy in Biochemistry*. SPIE Proc., vol. 1204 (1991);
R.R. Birge, L.A. Nafie, *Biomolecular Spectroscopy*. SPIE Proc., vol. 1432 (1991)
1573. Medical Laser Application. Int. J. Laser Treat. Res. Continued as Medical Photonics (Elsevier, Amsterdam). Special Issues: Basic investigations for diagnostic purposes **26**(3). Published August 2011
1574. F. Dausinger, F. Lichtner, H. Lubatschowski (eds.), *Femtosecond Technology for Technical and Medical Applications*. Topics in Appl. Phys., vol. 96 (Springer, Berlin, 2004)
1575. S. Harris, Lasers Med. (2011) (SPIE Professional magazine)
1576. M.L. Wolbarsht (ed.), *Laser Applications in Medicine and Biology*, vols. 1–5 (Springer, Berlin, 1971–1991)
1577. S. Svanberg, Medical applications of laser spectroscopy. Phys. Scr. **90** (1989). doi:10.1088/0031-8949/1989/T26/012
1578. P.V. Gupta, D.D. Bhawalkar, Laser applications in medicine. Curr. Sci. **77**(7), 925–933 (1999)
1579. P. Hering, Laser spectroscopy: biological and medical applications, in *Int. Conf. on Laser Spectroscopy* (2010). http://icole.mutes.ac.ir/downloads/Sci_Sec/W04/Laser spectroscopy_Applications_L2.pdf

Index

W. Demtröder, *Laser Spectroscopy 2*, DOI 10.1007/978-3-662-44641-6

Printed in the United States
By Bookmasters